S0-AET-449

FRANCO

ALSO BY PAUL PRESTON

Editor, *Spain in Crisis: Evolution and Decline of the Franco Regime* (Harvester Press, 1976).

The Coming of the Spanish Civil War: Reform Reaction and Revolution in the Second Spanish Republic 1931–1936 (Macmillan, 1978).

(with Denis Smyth) *Spain, the EEC and NATO* (Royal Institute of International Affairs & Routledge and Kegan Paul, 1984).

Editor, *Revolution and War in Spain 1931–1939* (Methuen, 1984).

Las derechas españolas en el siglo veinte: autoritarismo, fascismo, golpismo (Editorial Sistema, Madrid, 1986).

The Triumph of Democracy in Spain (Methuen, 1986).

The Spanish Civil War 1936–1939 (Weidenfeld & Nicolson, 1986).

Editor (with Helen Graham), *The Popular Front in Europe* (Macmillan, 1987).

Salvador de Madariaga and the Quest for Liberty in Spain (Clarendon Press, 1987).

The Politics of Revenge: Fascism and the Military in 20th Century Spain (Unwin Hyman, 1990)

Editor (with Frances Lannon), *Elites and Power in Twentieth-Century Spain: Essays in Honour of Sir Raymond Carr* (Clarendon Press, 1990)

PAUL PRESTON

FRANCO

A Biography

BasicBooks
A Division of HarperCollins*Publishers*

The author and publishers are grateful to the following
agencies, archives, and private collectors for their assistance
in the location of the photographs used in this book: *La
Actualidad Española*, A. G. E. Fotostock (1989), Archiv für
Kunst und Geschichte Berlin, Associated Press/Topham,
the British Newspaper Library, Centre for Contemporary
Spanish Studies, Manuel Fraga Iribarne, the Hemeroteca
Municipal de Madrid, the Instituto de España, Laureano
López Rodó, the Ministerio de Cultura, the Ministerio de
Información, Enrique Moradiellos García, Popperfoto,
María Salorio, Romón Serrano Suñer, *La Voz de Asturias*.

Copyright © 1994 by BasicBooks,
A Division of HarperCollins*Publishers*, Inc.

First published in the United Kingdom in 1993 by
HarperCollins*Publishers*.

All rights reserved. Printed in the United States of America.
No part of this book may be reproduced in any manner
whatsoever without written permission except in the case of
brief quotations embodied in critical articles and reviews.
For information, address BasicBooks, 10 East 53rd Street,
New York, NY 10022-5299.

LIBRARY OF CONGRESS CATALOGING-IN-PUBLICATION DATA
Preston, Paul, 1946–
   Franco : a biography / Paul Preston.
    p. cm.
   Includes index.
   ISBN 0–465–02515–3 (cloth)
   ISBN 0–465–02516–1 (paper)
    1. Franco, Francisco, 1892–1975. 2. Heads of state—
Spain—Biography. I. Title.
DP264.F7P74 1994
946.082'092—dc20
[B]   94–28636
                                   CIP

96 97 98 99 ◆/HC 9 8 7 6 5 4 3 2 1

For James and Christopher

ACKNOWLEDGEMENTS

I have been working on this book for many years. Inevitably, in that time I have incurred a burden of debt to many people who, in one way or another – sharing memories and ideas, helping me to obtain rare material and, latterly, reading the various drafts – have made the final product better than it might otherwise have been. I would like to take this opportunity to thank them now.

The assistance of the staffs of the following libraries and archives are gratefully acknowledged: the Library of Queen Mary and Westfield College and, in particular, Miss Susan Richards of the Inter-Library Loans section; the Institute of Historical Research and, in particular, Miss Bridget Taylor; the British Library of Political and Economic Sciences, the Public Record Office, the Archive of the Ministerio de Asuntos Exteriores, Madrid, the British Library, Bloomsbury and Colindale, the Cambridge University Library, the Allison Peers Collection at the University Library of Liverpool. I would like to thank the Controller of HM Stationery Office by whose permission quotations from the papers in the Public Record Office appear. I should also like to thank Mrs Carol Toms of Queen Mary & Westfield College and Mrs Pat Christopher of the London School of Economics for their unstinting support and for the good-humoured way in which they helped me fulfil the administrative duties that keep professors from research and teaching.

For talking and/or writing to me about their own experiences of Franco and his regime, I am immensely grateful to the following: the late Ignacio Arenillas de Chaves, the late Rafael Calvo Serer, Joaquín Calvo Sotelo, Cirilo Cánovas García, Fabián Estapé Rodríguez, the late Joseba Elosegi, General Hernando Espinosa de los Monteros, Ignacio Espinosa de los Monteros, Manuel Fraga Irribarne, Ramón Garriga Alemany, the late

José María Gil Robles, the late Ernesto Giménez Caballero, Folke von Knobloch, Juan Cristóbal von Knobloch, Laureano López Rodó, the late Adolfo Muñoz Alonso, José Joaquín Puig de la Bellacasa, General Ramón Salas Larrazabal, María Salorio, Ramón Serrano Suñer, Fernando Serrano-Suñer Polo and the late Eugenio Vegas Latapie.

For their invaluable aid in locating important documentary material, I am indebted to the following friends and colleagues: Angelines Alonso, John Costello, Lesley Denny, Chris Ealham, Agustín Gervás, Antonio Gómez Mendoza, Ian Gibson, Joe Harrison, Santos Juliá, Qasim bin Ahmed, and Francisco Villacorta Baños. Ricardo Figueiras Iglesias provided invaluable assistance with material from Galicia. For crucial help in tracing information for the Asturian part of Franco's life, I owe a particular debt to Carmen Benito del Pozo and Victoria Hidalgo Nieto. For advice on the paintings of Franco and Carrero Blanco, I am indebted to Nigel Glendinning. On the musical background to Bernhardt's trip to Bayreuth, I would like to thank Norman Cooper and Barry Millington. On aeronautical aspects of the Civil War, I have learned much from Gerald Howson, and on other military aspects of German intervention from Williamson Murray. On the psychology of Franco, I was fortunate to be able to call on the expertise of Nina Farhi. On various medical matters discussed in the last chapter, I have benefited from the advice of Dr Roy MacGregor and Anthony Ashford-Hodges FRCS.

I am grateful to Michael Alpert, Brian Bond, George Hills and David Wingeate Pike for advice on particular points of detail regarding Franco's role in the Civil War and the Second World War. I have derived much stimulus and encouragement from discussions about Franco over many years and around many tables with Alicia Alted Vigil, Joan Ashford-Hodges, José María Coll Comín, Elías Díaz, Musa Farhi, Jerónimo Gonzalo, Juan Antonio Masoliver, Florentino Portero, Denis Smyth, Javier Tusell and Manuel Vázquez Montalbán. A number of friends have been even more implicated in the book's gestation, discussing Franco with me, providing me with inaccessible material and reading and commenting on early drafts of the text. My debt in this regard to Nicolás Belmonte, Sheelagh Ellwood, Enrique Moradiellos, Ismael Saz, Herbert R. Southworth and Angel Viñas is enormous.

The book's faults remain my own. That they are not even greater is due to the help that I have received from the friends already mentioned and from two others. Mía Rodríguez Salgado gave generously of her time to provide close and perceptive readings of successive drafts of the manuscript which together with the implacable but always creative criticisms to which the text was subjected by Jonathan Gathorne-Hardy immeasurably improved it. Indeed, one of the greatest incidental pleasures of work on the book came from discussions with both about the problems of biography and of narrative. Philip Gwyn Jones of Harper-Collins guided the book through its production stages with a sure and sensitive hand.

For many years, my wife Gabrielle put up with the presence in our home of an uncongenial uninvited guest in the person of Francisco Franco. Without her tolerance and support, life with the Caudillo would have reached breaking point long before the book was completed. Moreover, many of the judgements within the text have benefited greatly from her acute critical sense. Finally, I would like to thank my sons James and Christopher to whom this book is dedicated. Without them, it would have been completed much sooner and it, and I, would have been the poorer.

CONTENTS

MAPS

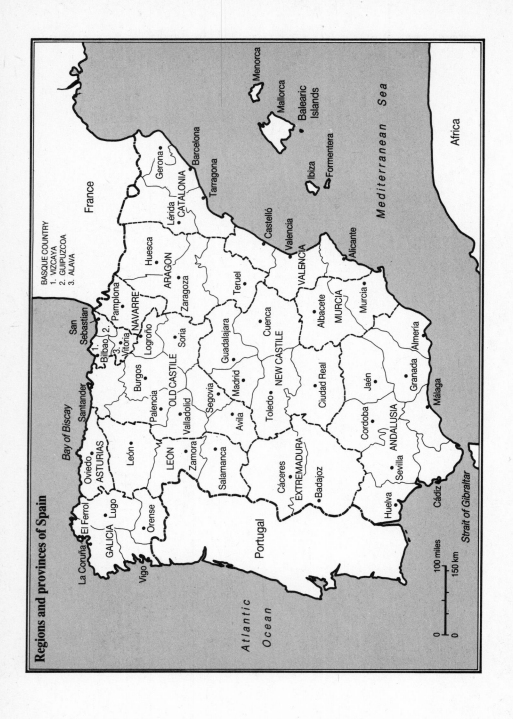

Regions and provinces of Spain

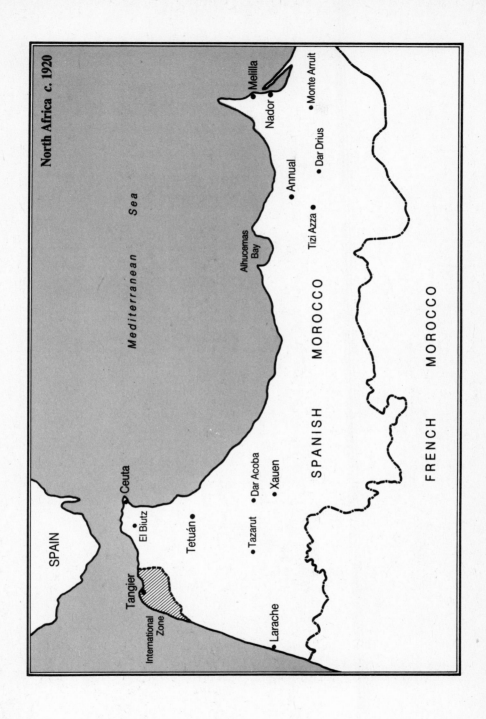

North Africa c. 1920

PROLOGUE

The Enigma of General Franco

DESPITE fifty years of public prominence and a life lived well into the television age, Francisco Franco remains the least known of the great dictators of the twentieth century. That is partly because of the smoke screen created by hagiographers and propagandists. In his lifetime, he was compared with the Archangel Gabriel, Alexander the Great, Julius Caesar, Charlemagne, El Cid, Charles V, Philip II, Napoleon and a host of other real and imaginary heroes.[1] After a lunch with Franco, Salvador Dalí said 'I have reached the conclusion that he is a saint'.[2] For others, he was much more. A children's textbook explained that 'a Caudillo is a gift that God makes to the nations that deserve it and the nation accepts him as an envoy who has arisen through God's plan to ensure the nation's salvation', in other words, the messiah of the chosen people.[3] His closest collaborator and *eminence grise*, Luis Carrero Blanco, declared in 1957 in the Francoist Cortes: 'God granted us the immense mercy of an exceptional Caudillo whom we can judge only as one of those gifts which, for some really great purpose, Providence makes to nations every three or four centuries'.[4]

Such adulation may be dismissed as typical of the propaganda machine of a despotic regime. Nonetheless, there were many who spontaneously accepted these comparisons and many others, who by dint of their relentless repetition, failed to question them. This is not an obstacle to knowing Franco. What does render him more enigmatic is the fact that Franco saw himself in the inflated terms of his own propaganda. His inclination to compare himself to the great warrior heroes and empire-builders of Spain's past, particularly El Cid, Charles V, Philip II, came to be second nature, and only partly as a consequence of reading his own press or listening to the speeches of his supporters. That Franco revelled in the wild exaggerations of his own propaganda seems at odds with the many eyewitness accounts of a man who was shy in private and inhibited and ill-at-ease on public occasions. Similarly, his cruelly repressive politics may seem to be contradicted by the personal timidity which led many who met him to comment just how little he coincided with their image of a dictator. In fact, the hunger for adulation, the icy cruelty and the tongue-tied shyness were all manifestations of a deep sense of inadequacy.[5]

The inflated judgements of the Caudillo and his propagandists are at the other extreme from the left-wing view of Franco as a vicious and unintelligent tyrant, who gained power only through the help of Hitler and Mussolini, and survived for forty years through a combination of savage repression, the strategic necessities of the great powers and luck. This view is nearer the truth than the wild panegyrics of the Falangist press, but it explains equally little. Franco may not have been El Cid but was neither so untalented nor so lucky as his enemies suggest.

How did Franco get to be the youngest general in Europe since Napoleon? How did he win the Spanish Civil War? How did he survive the Second World War? Does he deserve credit for the great Spanish economic growth of the 1960s? These are important questions with a crucial bearing on Spanish and European history in the twentieth century and they can be answered only by close observation of the man. He was a brave and outstandingly able soldier between 1912 and 1926, a calculating careerist between 1927 and 1936, a competent war leader between 1936 and 1939 and a brutal and effective dictator who survived a further thirty-six years in power. Even close observation, however, has to grapple with mysteries such as the contrast between the skills and qualities required to achieve his successes and a startling intellectual mediocrity which led him to believe in the most banal ideas.

The difficulties of explanation are compounded by Franco's own efforts at obfuscation. In maturity, he cultivated an impenetrability which ensured that his intentions were indecipherable. His chaplain for forty years, Father José María Bulart, made the ingenuously contradictory comment that 'perhaps he was cold as some have said, but he never showed it. In fact, he never showed anything'.[6] The key to Franco's art was an ability to avoid concrete definition. One of the ways in which he did that was by constantly keeping his distance, both politically and physically. Always reserved, at innumerable moments of crisis throughout his years in power, Franco was simply absent, usually uncontactable while hunting in some remote sierra.

The greatest obstacle of all to knowing Franco is that, throughout his life, he regularly rewrote his own life story. In late 1940, when his propagandists would have us believe that he was keeping a lonely and watchful vigil to prevent Hitler pulling Spain into the World War, he found the time and emotional energy to write a novel-cum-filmscript. *Raza* (Race) was transparently autobiographical. In it, and through its heroic central character, he put right all of the frustrations of his own life.[7]

Raza was merely the most extreme, and self-indulgent, manifestation of Franco's tireless efforts to create a perfect past. Like his war diary of 1922, it provides invaluable insight into his psychology. In his scattered writings and thousands of pages of speeches, in his fragments of unfinished memoirs and in innumerable interviews, he endlessly polished his role and remarks in certain incidents, consistently putting himself in the best light and providing the raw material to ensure that any biography would be hagiography. The persistence of many favourable myths is a testimony to his success.

The need to tamper with reality which is revealed by Franco's tinkering with his own past was indicative of considerable insecurity. He dealt with this not just in his writings but also in his life by creating for himself successive public personae. The security provided by these shields permitted Franco almost always to seem contained and composed. Everyone who came into contact with him remarked on his affably courteous, but always distant, manner. Behind the public display, Franco remained intensely private. He was abundantly imbued with the inscrutable pragmatism or *retranca* of the *gallego* peasant. Whether that was because of his origins as a native of Galicia, or the fruit of his Moroccan experiences is impossible to say. Whatever its roots in Franco, *retranca* may be defined as an evasion of commitment and a taste for the imprecise. It is said that if you meet a *gallego* on a staircase, it is impossible to deduce if he is going up or down. Franco perhaps embodied that characteristic more than most *gallegos*. When those close to him tried to get hints about forthcoming ministerial changes, they were rebuffed with skill: 'People are saying that in the next reshuffle of civil governors so-and-so will go to Province X', tries the friend; 'Really?' replies the sinuous Franco, 'I've heard nothing'. 'It's being said that Y and Z are going to be ministers', ventures his sister. 'Well', replies her brother, 'I haven't met either of them'.[8]

The monarchist aviator Juan Antonio Ansaldo wrote of him 'Franco is a man who says things and unsays them, who draws near and slips away, he vanishes and trickles away; always vague and never clear or

categoric'.[9] John Whitaker met him during the Civil War: 'He was effusively flattering, but he did not give a frank answer to any question I put
to him. A less straightforward man I never met.'[10]

Mussolini's Ambassador Roberto Cantalupo met him some months
later and found Franco to be 'icy, feminine and elusive [*sfuggente*]'.[11] The
day after first meeting Franco in 1930, the poet and noted wit José María
Pemán was introduced by a friend as 'the man who speaks best in all
Spain' and remarked 'I think I've just met the man who keeps quiet best
in all Spain' ('*Tengo la sospecha de haber conocido al hombre que mejor se calla
en España*').[12]

In his detailed chronicles of their almost daily contact during more
than seventy years of friendship, his devoted cousin and aide-de-camp,
Francisco Franco Salgado-Araujo, 'Pacón', presents a Franco who issued
instructions, recounted his version of events or explained how the world
was threatened by freemasonry and Communism. Pacón never saw a
Franco open to fruitful dialogue or to creative self-doubt. Another lifelong friend, Admiral Pedro Nieto Antúnez, presented a similar picture.
Born, like Franco, in El Ferrol, 'Pedrolo' was to be successively ADC to
the Caudillo in 1946, Assistant Head of the *Casa Civil* in 1950, and
Minister for the Navy in 1962. He was one of Franco's constant companions on the frequent and lengthy fishing trips on his yacht, the *Azor*.
When asked what they talked about during the long days together,
'Pedrolo' said 'I have never had a dialogue with the General. I have heard
very long monologues from him, but he wasn't speaking to me but to
himself'.[13]

The Caudillo remains an enigma. Because of the distance that Franco
so assiduously built around himself through deliberate obfuscations and
silences, we can be sure only of his actions, and, provided they are judici-
ously evaluated, of the opinions and accounts of those who worked with
him. This book is an attempt to observe him more accurately and in
more detail than ever before. Unlike many books on Franco, it is not a
history of twentieth-century Spain nor an analysis of every aspect of the
dictatorship, but rather a close study of the man. Through memoirs and
interviews, his collaborators have provided ample material and there are
copious despatches by foreign diplomats who dealt with him face-to-face
and reported on his activities. Franco's own writings, his speeches – in
which he often held a kind of dialogue with himself – and his recently
published papers also constitute a rich, if not easy, source for the biographer. They are the instrument of his own obfuscations but they also
provide remarkable insight into his own self-perception.

By use of these sources, it is possible to follow Franco closely as he became successively a conspirator, Generalísimo of the military rebels of 1936 and Caudillo of the victorious Nationalists. Several myths do not survive a comprehensive investigation of his survival of the Second World War and the Cold War and of his devious dealings with Hitler, Mussolini, Churchill, Roosevelt, Truman and Eisenhower. Equally striking is the picture which emerges of his passage from the active dictator of the 1950s to the somnolent figurehead of his last days. By following him step by step and day by day, a more accurate and convincing picture can emerge than has hitherto been current. Indeed, only by such an exhaustive examination can the enigma of the elusive Franco begin to be resolved.

I

THE MAKING OF A HERO

1892–1922

FRANCISCO FRANCO BAHAMONDE was born at 12.30 a.m. on 4 December 1892 in the calle Frutos Saavedra 108, known locally as the calle María, in El Ferrol in the remote north-western region of Galicia. He was christened Francisco Paulino Hermenegildo Teódulo on 17 December in the nearby military parish church of San Francisco.*

At the time, El Ferrol, an inward-looking and still walled town, was a small naval base with a population of twenty thousand. The Franco family had lived there since the early eighteenth century and had a tradition of work in the *intendencia naval* (pay corps/administration).[†1] Franco's grandfather, Francisco Franco Vietti, was Intendente Ordenador de la Marina (naval paymaster) with a rank equivalent to brigadier general in the Army. He had married Hermenegilda Salgado-Araujo, with whom he had two children. The first, Nicolás Franco Salgado-Araujo, the father of the future Caudillo, was born on 22 November 1855, his sister Hermenegilda on 1 December 1856.

Nicolás followed his father into the administrative branch of the Spanish navy in which, after fifty years service, he rose to be Intendente-General, a rank also equivalent to brigadier general. As a young man, stationed first in Cuba then in the Philippines, Nicolás acquired a reputation for fast living.‡

* Francisco in memory of his paternal grandfather, Hermenegilda in memory of his paternal grandmother and in honour of his godmother, Paulino in honour of his godfather and Teódulo because he was christened on the feast of Saint Teódulo.
† There has been much idle speculation that his family was Jewish, on the basis of his appearance and because both Franco and Bahamonde are common Jewish surnames in Spain.
‡ Indeed, after Franco's death, there were press revelations concerning Nicolás's relationship in Manila with a fourteen-year-old girl, Concepción Puey, with whom he was said to have had an illegitimate son, Eugenio Franco Puey, who made himself known to Francisco Franco in 1950 – *Opinión*, 28 February 1977; *Interviu*, No. 383, 14–20 September 1983.

On 24 May 1890, when he was nearly thirty-five, Nicolás Franco Salgado-Araujo married the twenty-four year-old María del Pilar Bahamonde y Pardo de Andrade in the Church of San Francisco in El Ferrol. She was the pious daughter of Ladislao Bahamonde Ortega, commissar of naval equipment at the port. The union of this free-thinking bon viveur with the conservative, moralistic Pilar was not a success. Nevertheless, they had five children, of whom Nicolás was the first, Francisco the second, followed by Paz, Pilar and Ramón.*[2]

Franco's family had been concerned for over a century with the administration of the naval base in El Ferrol. When Franco was born, the town was remote and isolated, separated from La Coruña by a twelve-mile steamer journey to the south across the bay or by forty miles of poor, and in bad weather, often impassable, road. La Coruña was in turn 375 miles, or two days by bone-shaking railway, from Madrid. El Ferrol was hardly a cosmopolitan place. It was a town of rigid social hierarchies in which the privileged caste consisted of naval officers and their families. Naval administrators or merchant navy officers were considered to be of a lower category. Social barriers cut the lower middle-class Franco family off from 'proper' naval officers since the administration corps was regarded as inferior to the sea-going Navy, or *Cuerpo General*. The idea of a heroic family naval tradition, so carefully nurtured by Franco himself in later life, was an aspiration rather than a reality. That can be perceived in Nicolás Franco Salgado-Araujo's determination that his sons become 'real' naval officers.

Partly because a naval commission was a common ambition among the Ferrolano middle class and because of his father's job, Francisco developed an interest in things of the sea. As a child he played pirates in the harbour with the gangplanks of the ferries and rowed in the tranquil waters of the virtually enclosed *ría* (firth or fjord) of El Ferrol.[3] As an adolescent, he tried to join the Navy. His two primary schools, the Colegio del Sagrado Corazón and the Colegio de la Marina, both specialised in preparing children for the Navy entrance examinations.[4] Nicolás Franco Bahamonde did manage to fulfil his father's expectations, but Francisco's naval ambitions were to be thwarted. His failure to enter the navy would weigh heavily on him. In Salamanca during the Civil War, it was common knowledge that to please him or deflect his anger it was always worth trying to change the subject to naval matters.[5] As Caudillo, he spent as much time as he could aboard his yacht *Azor*, wore an admiral's

* Nicolás was born 1 July 1891, Pilar 27 February 1895 and Ramón 2 February 1896.

uniform at every opportunity and, when visiting coastal cities, liked to arrive from the sea on board a warship.

His childhood was dominated by the efforts of his mother to cope with the overbearing severity and later the constant absences of his father, the shadow of whose infidelities hung over the home. He was brought up by Doña Pilar in an atmosphere of piety and stifling provincial lower middle class gentility. Marriage had only briefly diminished the number and length of Nicolás Franco Salgado-Araujo's card games and drinking sessions at the officers' club. After the birth of his daughter Paz, in 1898, Nicolás had returned to his bachelor habits. The distress that this caused his wife was compounded by the death of Paz in 1903, after an undiagnosed illness lasting four months. Pilar Bahamonde was devastated.[6] Nicolás Franco was, at home, a bad-tempered authoritarian who easily lost control of himself if contradicted. His daughter Pilar described him as running the house like a general, although she also claimed that he beat his sons no more than was the norm at the time, a double-edged claim which leaves it difficult to evaluate the scale and intensity of his violence. The young Nicolás bore the brunt of his anger and Ramón also carried a deep resentment of his father and his uncontrolled violence all through his life. Until Nicolás Franco left home in 1907, his children and his wife were often the victims of his frequent rages.

Francisco was too well-behaved, too much of a 'little old man' (*niño mayor*), in his sister's phrase, to arouse his father's anger with any frequency. Nevertheless, Pilar recounts the deep sulk that came over him whenever he was cuffed unjustly by his father.[7] Unable to win his father's acceptance and affection, Francisco seems to have turned in on himself. He was a lonely child, withdrawn to the point of icy detachment. A story is told that when he was aged about eight, Pilar heated a long needle until the tip was red-hot and pressed it onto his wrist. Allegedly, gritting his teeth as his flesh burnt, he said only 'how shocking the way burnt flesh smells'.[8] Within the family, Francisco was long overshadowed by his two brothers, Nicolás and Ramón, who were extroverts and took after their father. Nicolás, who became a naval engineer, was the father's favourite. Interviewed in the press in 1926, Franco *père* dismissed as unremarkable the achievements of his two younger sons, Francisco as commander of the Foreign Legion and Ramón who had become the first man to fly the south Atlantic.[9] Even in later life, when Francisco was Head of State, his father, when asked about 'his son', would perversely talk about Nicolás or sometimes Ramón. Only when pressed would Don Nicolás talk about the person he called 'my other son'.

In total contrast to her despotic husband, Pilar Bahamonde was a gentle, kindly and serene woman. She responded to the humiliations suffered at the hands of the gambling and philandering Nicolás by presenting to the world a facade of quiet dignity and religious piety that hid her shame and the economic difficulties she had to face. That is not to say that the family suffered privations, since she received financial help from her father, Ladislao Bahamonde Ortega, who lived with her after the death of his wife, and also from her husband. Nevertheless, once her husband established residence in Madrid from 1907, what Pilar Bahamonde received from him must necessarily have been limited. There was always a maid in the house, but some sacrifices were required to keep up appearances. Sending all four children to private schools put a strain on the family economy. It has been suggested, although strenuously denied by the family, that she had to take in lodgers.[10] Despite these difficulties, her kindness extended to her relations and she helped to bring up the seven younger children of her brother-in-law Hermenegildo Franco.[11]

Pilar Bahamonde tried to imbue her children with a determination to get on in life and to escape from their situation by study and hard work, a philosophy which seems to have taken root principally with her second son and her daughter Pilar. Nevertheless, all four of her surviving children were to be fearless and powerfully ambitious in one way or another. Nicolás Franco Salgado-Araujo was a liberal, sympathetic to freemasonry and critical of the Catholic Church. In contrast, Pilar Bahamonde was politically conservative and a deeply pious Catholic. Given the circumstances of his childhood, and the nature and ideas of his father, it is hardly surprising that an enduring and unsubtle Catholicism, sexual prurience and a hatred for liberalism and freemasonry should be part of the legacy which the young Franco was left by his mother.[12] What is more intriguing is the fact that his brothers followed in the footsteps of Don Nicolás rather than those of Doña Pilar. After her husband left, Doña Pilar always wore black. It seems too that, as Francisco witnessed her introspective piety becoming an effective shield against her misfortunes, he suppressed his own emotional vulnerability at the cost of developing a cold inner emptiness.

Doña Pilar's unhappiness and stoical attempts to put a good face on her plight made it difficult for her to compensate her children for the behaviour of her husband. Each responded differently: Francisco identified with his mother, denying the need for his father's approval which he longed for and never achieved. His hedonistic elder brother Nicolás grew up to be as pleasure-loving as his father, free with money and with women. His wild younger brother Ramón would be an irresponsible adventurer, famous for his exploits as an air-ace and notorious for his decadent private life in the 1920s and for a superficial involvement with both anarchism and freemasonry. Francisco was much more deeply attached to his mother than were either of his brothers. He regularly accompanied her to communion and was a pious child. He cried when he made his first communion. When on leave in El Ferrol, the adult Francisco would never fail to fulfil any religious duty for fear of upsetting his mother.*[13]

It is impossible to say with any precision what effect the separation of his parents and the departure of his father had on Francisco, although there is surely some significance in the fact that one of the few remarks that he ever made on the subject of children was: 'small children should never be separated from their parents. It is not good to let that happen. The child needs to have the security provided by the support of his parents and they should not forget that their children are their personal responsibility.'[14] As Caudillo, Franco denied vehemently that there was anything abnormal in Don Nicolás's relationship with his wife or his children. On one occasion, however, when given irrefutable evidence of his father's pecadillos, his reaction was revealing. He snapped 'Alright but they never diminished his paternal authority'.[15] The difficulties of Franco's relationship with his father were later reflected in various efforts to reconstruct it in an idealised way. In his diary of his first year in the Spanish Foreign Legion, he told a clearly apocryphal story in which can be discerned his own longings. A young officer in Morocco is crossing the street when a grizzled veteran soldier salutes him. The officer goes to return the salute, their eyes meet, they look at each other and embrace in tears. It is the officer's long-lost father.[16] It was a trial run for his autobiographical novel, *Raza*, in which he created the father he would rather have had as a naval hero of total moral rectitude. When his father died, he had the body seized and implicitly reinvented the second part of his life by having him buried with a pomp which, while in accord with military regulations, was hardly appropriate to the bohemianism of Don Nicolás's final years. Franco's own lifelong avoidance of drink, gambling and women bore testimony to a determination to create an existence which was the antithesis of his father's life.

Franco would implacably reject all the things he associated with his

* Often he would join her in the difficult trek up the Pico Douro to the east of El Ferrol to pray to the Virgen de Chamorro in fulfilment of promises she had made in her prayers for his safe return.

father, from the pleasures of the flesh to the ideas of the Left. Franco's
repudiation of his father was matched by a deep identification with his
mother, something which might perhaps be seen in many aspects of his
personal style, a gentle manner, a soft voice, a propensity to weep, an
enduring sense of deprivation. A tone of self-pitying resentment runs
through his speeches as Caudillo, a continual echo of the hard-done-by
little boy that he must have been, and was one of the motivating forces
of his drive to greatness.

Two great political events of Franco's early youth were to dominate
his later development – the loss of Cuba in 1898 and the involvement of
Spain in a costly colonial war in Morocco. Imperial disaster provoked
civilian distrust of an incompetent Army and intensified military resentment of the political establishment and of civilian hostility to conscription. Throughout his life, Franco would remark on the profound effect
that the 1898 'disaster' had on him. In 1941, when he was near to declaring war on the Axis side, he declared 'when we began our life, . . . we
saw our childhood dominated by the contemptible incompetence of those
men who abandoned half of the fatherland's territory to foreigners'. He
would see his greatest achievement as wiping out the shame of 1898.[17]

Francisco was five and a half when the great naval defeat at the hands
of the United States occurred in Santiago de Cuba on 3 July 1898. Spain
lost the remnants of her empire – Cuba, Puerto Rico and the Philippines.
Although it is highly unlikely that, at such an age, he was aware of what
was happening, a disaster of that dimension could not but have a profound
effect on a small naval garrison town like El Ferrol. Many of his schoolfriends lost relatives and wore mourning. Mutilated men were seen
around the town for many years. More importantly, when he became a
cadet in the Army, he went directly into an atmosphere which had festered
since 1898. Defeat was attributed to the treachery of politicians who had
sent naval and military forces into battle with inadequate resources. That
it took the massively superior US forces three months to defeat the
ramshackle Spanish fleet left Franco convinced that bravery was worth
hundreds of tons of superior equipment.[18]

The defeat of 1898 had an immediate impact on Franco because of
the consequent budget cuts. The *Escuela de Administración Naval*, the
usual channel for boys of the Franco family into the Navy, was closed in
1901. It was decided that Nicolás and Francisco would prepare instead
for the entrance examinations of the Cuerpo General de la Armada. They
went to the local middle-class school, the *Escuela del Sagrado Corazón*. At
this time, before his father abandoned the family home, Francisco was,

according to contemporaries who saw him outside the family, a meticulous plodder, 'good at drawing but otherwise quite average, quite ordinary. He was a nice lad, of a happy disposition, thoughtful; he took his
time in answering questions but he was a playful lad.'[19] He was of sickly
appearance and so thin that his playmates nicknamed him *cerillito* (little
match-stick). Within the family, his sister was struck by the extent to
which Francisco emulated his mother's quiet seriousness. He was an
obedient, well-behaved and affectionate child, although timid, rather sad
and uncommunicative. Then, as later, he had little spontaneity. He was
very particular about his appearance, a trait that would follow him
throughout his life. Even then he seemed older than his years and his
obstinacy, astuteness and caution were evident. Among his closest childhood friends was his cousin Ricardo de la Puente Bahamonde, who would
be executed in Morocco in 1936, with Franco's acquiescence.[20] As an
adolescent, Francisco showed a normal interest in girls, favouring slim
brunettes, mainly from among his sister's schoolfriends. He wrote them
poems and was mortified when they were shown to his sister.[21]

The loss of Cuba was to have serious domestic consequences. It
hastened the rise of a regionalist movement in Catalonia and imbued
Army officers with a determination to wipe away the ignominy of defeat
through a colonial enterprise in Morocco. Catalan regionalism and the
Moroccan adventure were to interract in an explosive manner. The demonstration in 1898 of Spain's international impotence shook the faith of
the Catalan élites in the central government. The Catalan economy had
depended on the Cuban market and now the previously latent sense that
Madrid was an incompetent and parasitical obstacle to Catalan dynamism
found ever more vocal expression, above all in the appearance in early
1901 of the Catalanist party, the Lliga Regionalista.[22] In the context of
insecurity and humiliation provoked by the loss of Cuba, military anger
at what was seen as as political betrayal during the war with the USA
was compounded by the emergence of militant Catalanism, which soldiers
perceived as an aggressive separatist threat to the unity of the *Patria*.[23]

In November 1905, the Barcelona offices of both the Catalan satirical
magazine, *Cu-Cut!* and of the Lliga Regionalista's newspaper, *La Veu de
Catalunya*, were ransacked by three hundred fiercely centralist junior
officers to the applause of the officer corps throughout Spain. Given
widespread military approval of what was happening, the government
was unable to impose discipline or to resist military demands for measures
to punish offences against the honour of the Army. In 1906, politicians
bowed to military readiness to interfere in politics by introducing the *Ley*

de Jurisdicciones which gave the Army jurisdiction over perceived offences against the *Patria*, the King and the Army itself.[24] It was a considerable boost to the Army's sense of superiority over civilian society.

On reaching the age of twelve, first Nicolás and then Francisco, together with their fourteen year-old cousin, Francisco Franco Salgado Araujo, entered the Naval Preparatory School run by Lieutenant-Commander Saturnino Suanzes. There they became friendly with Camilo Alonso Vega, who was to remain a lifelong comrade. Nicolás, and a friend of the two brothers, Juan Antonio Suanzes, were successful in their efforts to join the Cuerpo General de la Armada. Nicolás chose to go to the Naval Engineering School. Franco and his lanky cousin Pacón* nurtured hopes of going to the *Escuela Naval Flotante*, the naval cadet ship. Then a decree was published restricting entry which closed the way to them. There was never any question of seeking a career other than a military one and so the now fourteen year-old Franco was sent to the Academia Militar de Infantería in Toledo. Pacón failed the entrance examination for 1907 but was successful the following year.[25]

When he accepted a post in Madrid in 1907, Nicolás Franco Salgado-Araujo went alone and gradually severed his links with his family. Members of his family have suggested that he was obliged to take the post, but having been able to spend nearly twenty years in El Ferrol without threat of being moved away, it seems more likely that he deliberately sought the posting to the capital in order to escape an unhappy marriage.[26] Although there was no divorce from Pilar, he later 'married' his lover Agustina Aldana in an informal non-religious ceremony in Madrid and lived with her in the calle Fuencarral in Madrid until his death in 1942. A child who lived with them and to whom they were devoted has been variously described as their illegitimate daughter or Agustina's niece whom they had informally adopted. The scandalized family referred to Agustina as his 'housekeeper' (*ama de llaves*).[27]

Accordingly, it was an embittered home in El Ferrol which the young Francisco left in July 1907 to take the entrance examinations for the military academy. He was accompanied on the long journey from La Coruña to Toledo by his father. Despite the fascination of the new landscapes through which he passed, the tension between him and his father made it a less than pleasant experience. Don Nicolás was unbending and rigid in the course of a journey during which his son needed encourage-

* 'Pacón' means 'big Frank' which he was always called to distinguish him from Franco, who was known in the family as 'Paquito' or 'little Frank'.

ment and affection.[28] Despite these inauspicious beginnings, Franco passed his examinations and entered the Academy on 29 August 1907 along with 381 other aspirants, including many future comrades-in-arms such as Juan Yagüe and Emilio Esteban Infantes. The Academy occupied the Alcázar built by Carlos V which dominated the hill around which the town was built. Far from the misty green valleys of Galicia and the placid *ría* in which he used to sail, dusty Toledo in the arid Castilian plain must have constituted a brutal shock. Although there is no evidence of his being sensitive to the wealth of religious art with which Toledo abounded, it appears that he responded to the sense of the past which pulsates in its streets. In his novel *Raza*, the character representing Franco (the cadet José Churruca) 'got more from the stones [of Toledo] than from his books'.[29] A growing obsession with the greatness of imperial Spain made him receptive to Toledo as a symbol of that greatness. His later identification with the figure of El Cid may also have had its origins in his adolescent ramblings around the historic streets of the town.

Life as an Army cadet would itself have strengthened his interest in Spanish history. Even by his own restrained account in later life, it is clear that he suffered some considerable agonies. Away from the loving care of his mother for the first time, young Franco had to grit his teeth and find inner reserves of determination to get on. In the austere conditions of the Alcázar, he would also have to deal with the problems arising from his anything but imposing physique (1.64 metres/5'4" tall, and painfully thin). Already vulnerable because of the desertion of his father, the separation from his mother, his central refuge, must inevitably have forced him to cope with acute insecurity. He seems to have dealt with it in two related ways. First, he threw himself into Army life, fulfilling his tasks with the most thorough sense of duty and making a fetish of heroism, bravery and the military virtues. The rigid structures of military hierarchy and the certainties of orders gave him a framework to which he could relate. At the same time, he began to create another identity. The insecure teenager from Galicia would become the tough desert hero and eventually, as Caudillo, the El Cid-like 'saviour of Spain'.[30]

On account of his size and high-pitched voice, he was soon called *Franquito* (little Franco) by his companions and, during his three years in the Academy subjected to various minor humiliations. He was forced to drill with a rifle which had had fifteen centimetres sawn off the barrel. He worked hard, with a particular interest in topography and the uncritical and idealized military history of Spain served up to the cadets. Having no interest in sexual or alcoholic safaris into the more disreputable parts of the town, he became a target for the cruel initiation ceremonies (*novat-adas*) of his fellow students, against which he reacted with some violence. In his own muted version, recollected nearly seventy years later, he spoke of the 'sad welcome offered to those of us who came full of illusions to join the great military family' and described the *novatadas* as a 'heavy cross to bear' (*un duro calvario*).[31] Other accounts, seeking traces of the later hero in the young cadet, recount his virile reactions. One oft-repeated story tells how his books were hidden and he was punished for not having them in the correct place. They were hidden again. The cadet officer was about to punish him again when Franco threw a candlestick at his head. When taken before the C.O., he refused to name those who had picked on him.[32] Such behaviour helped him to make some friends, including Camilo Alonso Vega, Juan Yagüe and Emilio Esteban Infantes, although he was never to be close to any of them.

In Britain and America, the Army cadet at the turn of the century began his military studies only after completing his civilian education. In Toledo, young, relatively uneducated boys began to absorb Army discipline and the conventions of the military view of the world when they were that much more ignorant and impressionable.[33] In professional terms, Franco can have learned little beyond the practical skills of horsemanship, shooting and fencing. The basic text-book was the *Reglamento provisional para la instrucción teórica de las tropas de Infantería* which was based on the lessons of the Franco-Prussian war and ignored the sweeping changes which had taken place in German military thinking since 1870. The increasing prominence given in both the German and British armies to the artillery and engineers was not replicated in Spain where the infantry remained dominant. The recent experience in Cuba was not used to draw any military conclusions, although they would have been immensely useful for the colonial adventures in North Africa. The stress was rather on discipline, military history and moral virtues – bravery in the face of the enemy, unquestioning faith in military regulations, absolute obedience and loyalty to superior officers.[34] Cadets were also imbued with an acute sense of the Army's moral responsibilities as guardian of the essence of the nation. No slight or insult to the Army, to the flag, to the monarch, to the nation could ever be tolerated. By extension, when a government brought the nation into disrepute by permitting disorder then it was the duty of the patriotic Army officer to rise up against the government in defence of the nation.

The method of training was usually the rote learning of masses of facts, in particular of the details of the great battles of the Spanish past.

However, these battles were examined as exemplars of bravery and resistance to the last rather than analysed for their tactical or strategic lessons. Franco's own central memory of his time at the Academy was of a major on the teaching staff who had been decorated for heroism with the Cruz Laureada de San Fernando (the Spanish equivalent of the Victoria Cross). He had been given the medal for a hand-to-hand knife fight in Morocco from which, Franco recalled with pleasure, 'he still had the glorious scars on his head'. The impact on Franco's way of thinking – and, indeed, on his own methods when Director of the Spanish General Military Academy at Zaragoza twenty years later – was revealed in his remark that 'this alone taught us more than all the other disciplines'.[35] When the cadets eventually went into the field, they had to improvise since they had been taught very little of practical application.

While Francisco was studying in Toledo, the events known as the *semana trágica* broke out in Barcelona in late July 1909. To military eyes, these disturbances were triply disturbing, with their connotations of anti-militarism, anti-clericalism and Catalan separatism. The government of Antonio Maura was under pressure from both Army officers close to Alfonso XIII and Spanish investors in Moroccan mines. Moreover, attacks by tribesmen on the railway leading to the port of Melilla had given rise to French threats to export their ore through Algeria. Maura also feared that France might use the apparent Spanish inability to keep order in her protectorate as an excuse to absorb it. Accordingly, he took advantage of an attack by tribesmen on the railway at Melilla on 9 July to send an expeditionary force to expand Spanish territory as far as the mineral deposits of the nearby mountains. The Minister of War decided to send a brigade of light infantry garrisoned in Barcelona. The brigade's reservists, mainly married men with children, were called up and, without adequate preparations, embarked from the port of Barcelona over the next few days. Over the next week, there were anti-war protests in Aragón, Valencia and Catalonia in the home towns of the reservists. In Barcelona, on Sunday 18 July 1909, a spontaneous demonstration broke out against the war. On that same day, Rif tribesmen launched an attack on Spanish supply lines in Morocco. On the following day, news began to reach Spain of new military disasters in Melilla. Untrained, ill-equipped and devoid of basic maps, the appallingly ramshackle state of the Spanish Army was revealed again. Throughout the week, the scale of the defeat and of the casualties was inflamed by rumours. There were anti-war demonstrations in Madrid, Barcelona and cities with railway stations from which conscripts were departing for the war.

During the following weekend, anarchists and socialists in Barcelona agreed to call a general strike. On Monday 26 July, the strike spread quickly, although it was not directed against the employers, some of whom supported its anti-war purpose. The Captain-General of the region, Luis de Santiago, decided to treat it as an insurrection, overruling the civil governor, Angel Ossorio y Gallardo, and declared martial law. Barricades were set up in the streets of outlying working class districts and anti-conscription protests debouched into anti-clerical disturbances and church-burnings. General de Santiago could do no more than defend the principal points of the city because he feared that his conscripts would fraternize with the rioters. Reinforcements were delayed by the fact that the attention of the military high command and of the government was distracted by the battle of Barranco del Lobo in Morocco. By 29 July, however, units had arrived and the movement was put down over the next two days with the use of artillery. There were numerous prisoners taken and 1,725 people were subsequently tried, of whom five were sentenced to death. Among them was Francisco Ferrer Guardia, the free-thinking founder of the libertarian school, the *Escuela Moderna*.[36]

Particularly spine-chilling accounts of what was happening were given to the cadets in Toledo by their instructors. There was outrage that pacifists and revolutionaries should be on the loose while part of the Army was fighting for survival in Morocco. The many international demonstrations on behalf of Francisco Ferrer were seen by the young Franco as the work of international freemasonry. The circle of cadets in which Franco moved regarded the events in Barcelona, and the defeat at Barranco del Lobo, as evidence that the political establishment was weak and incompetent.[37]

The gulf between the military and civil society was widening dramatically at this time. It is impossible to comprehend Franco either personally or politically without understanding the extent to which he first assumed and then expressed the attitudes of the typical Army officer of his day. The milestones along the road to the civil-military divorce – the 'disaster' of 1898, the *Cu-Cut!* incident of 1905, the 'tragic week' of 1909 – were reached either shortly before Franco joined the Army or during his early, formative, years in the service. These events and their professional and political implications were inevitably the talk of military academies and officers' messes. For someone as single-mindedly, not to say obsessively, committed to the military career as the young Franco, it was impossible for the resentments arising from these events not to be burnt deep into his consciousness.

Franco completed his studies at the Academy in June 1910. His ambition, like that of most of those who graduated at that time, was to go and fight in Morocco, where rapid promotion was possible and where he could help wipe out the shame of Cuba. On 13 July 1910, Franco was formally incorporated into the officer corps of the Army as a second lieutenant with the mediocre position of no. 251 of the 312 cadets of his year (of the original 381) who survived to graduate. Despite this mediocre start, Franco would be the first of his class to become a general.

It has been claimed that the young Franco applied immediately for a posting in Morocco, and was refused on the grounds of age, tough competition and his low place in the seniority list.[38] In fact, there would have been no point making a formal application for a posting in Morocco since, at the time, only first lieutenants and above could be posted to Africa.[39] He was posted to the *Regimiento de Zamora no. 8*, which was stationed in his home town of El Ferrol. There, from 22 August 1910 until February 1912, he was able to be near his mother and to show off his uniform to his contemporaries. He also had to face the crushing boredom of garrison duty in a small provincial town. Mornings were given over to parades and drills, afternoons to riding. Then there were guard duties. He was often able to eat at home. During this time, the continuing influence of his mother was reflected in the fact that he joined the religious confraternity *Adoración Nocturna* on 11 June 1911.[40] He also consolidated his friendship with Camilo Alonso Vega and with his cousin Pacón. At the end of 1911, the order prohibiting second lieutenants from being posted to Morocco was lifted and all three began to make frequent transfer requests.

Perhaps suffocated by the gloomy domestic situation, probably driven by patriotism, certainly aware of a second lieutenant's poor pay and that opportunities for promotion would come easier in Morocco than in a Peninsular garrison, Franco was anxious to be on his way and to overcome his 251st placing. While he was harkening to the siren calls of Africa, the Left was campaigning vigorously against the colonial war in general and against conscription in particular. Like many young soldiers, Franco developed what would be a lifelong contempt for left-wing pacifism. With the situation of the Spanish Army deteriorating in Morocco, the transfer requests of the three young officers were finally accepted on 6 February 1912. They were posted as reserves to Melilla. Franco and his two companions immediately set off on the long and difficult journey. With the road to the nearest railway station flooded by rain storms, and the port for the normal ferry service to La Coruña closed, they decided to go to the Naval Headquarters in El Ferrol in search of a ride. They were allowed to travel on board the merchant ship *Paulina*, which involved a hair-raising storm-tossed six hour journey standing in a gangway. From La Coruña, they carried on by rail to Málaga where they arrived after two days travel. They reached Morocco on 17 February 1912.[41]

The thin, boy soldier with round staring eyes who arrived in Melilla found a filthy, run-down colonial town.[42] The nineteen year-old Franco reported for duty at the Fort of Tifasor which was part of the outer defences of Melilla. Tifasor was under the command of Colonel José Villalba Riquelme, who had been Director of the Academia de Infantería when Franco was a cadet. Villalba Riquelme's first order to him was to cover his sword scabbard in mat leather to stop it glinting and providing a target for snipers. Indeed, in the shortest time possible, Franco had to learn this and all the other practicalities of life in combat that he had not been taught in the Academy in Toledo nor learned on garrison duty in El Ferrol. Like most young officers, he can have had little expectation of the difficulties that faced the Spanish Army in the field.

The most obvious problem was the warlike local population's bitter hatred of the occupying troops. Given the poor technological level of the Spanish armed forces, the Moroccan adventure would be no push-over. The Army was inefficient, weighed down by bureaucracy and inadequately supplied with obsolete equipment: it had more generals and fewer artillery pieces per thousand men than the armies of such countries as Montenegro, Romania and Portugal. Its eighty thousand men were commanded by more than tweny-four thousand officers of whom 471 were generals.[43] In the eyes of Army officers, the most damaging source of difficulty was the inability of the Spanish political establishment to provide either the resources or the decisive policy necessary to give the professional soldiers any chance of success. Indeed, the political élite's awareness of the growing pacificism of much of public opinion merely confirmed many Army officers in their belief that Spain could not be properly ruled by civilians. Moreover, there was Spain's subordinate position to France in the area. Spain was burdened with indefensible frontiers in Morocco which simply ignored the realities of tribal boundaries. French dominance also inhibited Madrid's policy-making.

How this came to be so is almost inextricably complicated. Morocco was ruled by a Sultan who had to impose by terror his authority and his tax-collection system on the other tribal leaders. In the early years of the century, tribal leaders rebelled against the dissolute Sultan Abd el Aziz. In the general upheaval, two major revolts took place. The first was that of Bu Hamara in the lands between Fez and the Algerian border. The more important was that of El Raisuni, a vicious cattle rustler and tribal leader, in the Jibala mountains of the north-west. In the context of the still incomplete scramble for Africa, it was a situation that attracted the great powers.

For many years, Britain had maintained influence in Morocco to guarantee safe passage through the Straits of Gibraltar. However, since the humiliating debâcle of the Fashoda incident in 1898 which had blocked their Egyptian ambitions, the French had been seeking to consolidate their empire to the west. They were anxious to find a way to take over the Moroccan Sultanate which was the obvious gap in an imperial chain from Equatorial Africa to Tunisia. By 1903, Britain, weakened by the Boer War, was apprehensive of the rise of Germany and open to a French Alliance. Unable in any case to prevent a French take-over, the British wanted above all to safeguard Gibraltar. In April 1904, in the Anglo-French Agreement, Britain consented to French ambitions in Morocco provided that the area opposite Gibraltar be in weaker, Spanish, hands.[44]

It was left to the French to square things with the Spanish. In October 1904, the French granted northern Morocco to Spain. Tangier was given international status. Using the pretext of tribal disorders, the French then took over Morocco by instalments. By 1912, a formal French Protectorate was established. In November 1912, France signed an entente with Spain giving her a similar protectorate in the north. Subsequent political arrangements meant that the Sultan maintained nominal political control of all of Morocco under French tutelage. However, in the Spanish zone, local authority was vested in the Sultan's representative, known as the Khalifa, who was selected by the Sultan from a short-list of two names drawn up by Madrid.

It was a situation fraught with difficulties. The Moroccans never accepted the arrangement, which they found deeply humiliating, and they fought it until they regained their independence in 1956. Spain's long-standing military enclaves, Ceuta and Melilla, had to communicate by sea. The recently acquired Protectorate of the interior was a roadless, infertile mountain wilderness. Moreover, because it ignored crucial tribal boundaries, the French gift to Spain was almost impossible to police. Thus, the Spaniards were to be involved in a ruinously expensive and virtually pointless war.[45] They did not enjoy the technological and logistical superiority which characterised other imperial adventures of the time. Curiously, Spain's officers in general, and Franco in particular, nurtured two myths. The first was that Moroccans loved them; the second that the French had stood in the way of a Spanish Moroccan empire.

At the time of Franco's arrival on African soil, the initiative in Spain's Moroccan war lay with the Berber tribesmen who inhabited the two barren mountain regions of the Jibala to the west and the Rif. Battle-hardened, ruthless in the defence of their lands, familiar with the terrain, they were the opposite of the poorly trained and totally unmotivated Spanish conscripts who faced them. Franco claimed years later that he spent his first night in the field sleepless, with a pistol in his hand, out of distrust of his own men.[46] The recently arrived Franco was a small part of a series of military operations aimed at building a defensive chain of blockhouses and forts between the larger towns. That this was the Spanish tactic showed that nothing had been learned from the Cuban War where similar procedures had been adopted. Officers felt considerable resentment at the contradictory orders to advance or retreat emanating from the Madrid government.

After the insecurities of his childhood, the great formative experience of Franco's life was his time as a colonial officer in Africa. The Army provided him with a framework of certainties based on hierarchy and command. He revelled in the discipline and happily lost himself in a military machine built on obedience and a shared rhetoric of patriotism and honour. Having arrived in Morocco in 1912, he spent ten and a half of the next fourteen years there. As he told the journalist Manuel Aznar in 1938, 'My years in Africa live within me with indescribable force. There was born the possibility of rescuing a great Spain. There was founded the idea which today redeems us. Without Africa, I can scarcely explain myself to myself, nor can I explain myself properly to my comrades in arms.'[47] In Africa, he acquired the central beliefs of his political life: the Army's role as the arbiter of Spain's political destiny and, most importantly of all, his own right to command. He was always to see political authority in terms of military hierarchy, obedience and discipline, referring to it always as *el mando*.

As a young second lieutenant, Franco immediately threw himself into his duties, soon demonstrating the cold-blooded bravery born of his ambition. On 13 June 1912 he was confirmed as first lieutenant. It was his first and only promotion solely for reasons of seniority. On 28 August, Franco was sent to command the position of Uixan, which protected the mines of Banu Ifrur. The Moroccan war was intensifying but Franco was paying assiduous court to Sofía Subirán, the beautiful niece of the High Commissioner, General Luis Aizpuru. Bored by his elaborate formality and inability to dance, she successfully resisted a determined postal assault which lasted for nearly a year.[48] In the spring of 1913, stoical about his disappointment in love, he applied for a transfer to the recently formed native police, the *Regulares Indígenas*, aware that they were always in the vanguard of attacks and presented endless opportunities for displays of courage and rapid promotion. On 15 April 1913, Franco's posting to the *Regulares* came through. At this time, El Raisuni began a major mobilization of his men. The Spanish base of Ceuta was reinforced by, among others, Franco and the *Regulares*. On 21 June 1913, he arrived at the camp of Laucien and was then posted to the garrison of Tetuán. Between 14 August and 27 September, he took part in several operations and began to make a name for himself. On 22 September, with his fierce Moorish mercenaries, he gained a small local victory for which, 12 October 1913, he was rewarded with the Military Merit Cross first class. In their relatively short existence, the *Regulares* had developed a tradition of exaggerated *machismo* scorning protection when under enemy fire. When Franco eventually reached the point at which he had the right to lead his men on horseback, he favoured a white horse, out of a mixture of romanticism and bravado.

For a brief period, the situation was stabilized in the Spanish Protectorate: the towns of Ceuta, Larache and Alcazarquivir were under control but communications in the harsh territory in between were threatened by El Raisuni's guerrillas and snipers. Attempting to hold this area was ruinously expensive in men and money. The lines of communication were dotted with wooden blockhouses, six metres long by four metres wide, protected up to a height of one and a half metres by sandbags and surrounded by barbed wire. Building them under Moorish sniper fire was immensely dangerous. They were garrisoned by platoons of twenty-one men who lived in the most appallingly isolated conditions and had to be provisioned every few days with water, food and firewood. Provisioning required escorts who were vulnerable to sniper fire. Very occasionally, the chains of blockhouses communicated by heliograph and signal lamps.[49]

For his bravery in a battle at Beni Salem on the outskirts of Tetuán on 1 February 1914, the twenty-one year-old Franco was promoted to captain '*por méritos de guerra*', with effect from that date although it was not announced until 15 April 1915. He was building a reputation as a meticulous and well-prepared field officer, concerned about logistics, provisioning his units, map-making, camp security. Twenty years later, Franco told a journalist that to stave off boredom in Morocco, he had devoured military treatises, memoirs of generals and descriptions of battles.[50] By 1954, he had inflated this to the point of telling the English journalist S.F.A. Coles rather implausibly that, in his off-duty hours in Morocco, he had studied history, the lives of the great military commanders, the ancient Stoics and philosophers and works of political science.[51] This later reconstruction by Franco contrasted curiously with the assertion of his friend and first biographer that he spent every available moment either at the parapet watching for the enemy through his binoculars or else surveying the terrain on horseback in order to improve his unit's maps.[52]

Whatever Franco did in his spare time, it was during this period that anecdotes began to be told about his apparent imperturbability under fire. He was said to be cold and serene in his risk-taking rather than recklessly brave. He was already making good his low position in the pass list of his year at the Academy (*promoción*). This came near to costing him his life during a large-scale clean-up operation against guerrilla tribesmen who were massing in the hills around Ceuta in June 1916. The guerrillas had their main support point about six miles to the west of the town, in the mountain top village of El Biutz, which dominated the road from Ceuta to Tetuán and was protected by a line of trenches manned by machine-gunners and riflemen. Rigidly constrained by their own field regulations, the Spaniards could be expected to make a frontal assault up the slope. As they were advancing, being decimated by fire from the trenches above, other tribesman planned to pour down the back of the hill, sweep around below the Spaniards and trap them in a cross-fire.

In the early hours of the morning of 29 June 1916, with high losses being recorded, Franco was part of the leading company of the Segundo Tabor (second battalion) of *Regulares* which was heading the advance. When the company commander was badly wounded, Franco assumed command. With men dropping all around him, he broke through the enemy encirclement and played a significant role in the fall of El Biutz. However, he was shot in the stomach. Normally, in Africa, abdominal wounds were fatal. That night's report referred to Captain Franco's 'incomparable bravery, gift for command and energy deployed in combat'. The tone of the report implied that his death was inevitable. He was carried to a first aid post at a place called Cudia Federico. The medical officer staunched the bleeding and refused for two weeks to send him the six miles by stretcher to the casualty clearing station outside Ceuta. He believed that for the wounded man to be moved would kill him and the delay saved Franco's life. By 15 July, Franco had recovered sufficiently to be transferred to the military hospital in Ceuta. There an X-ray showed that the bullet had not hit any vital organ. A fraction of an inch in any direction and he would have died.[53]

In a war which, during his time in Africa, claimed the lives of nearly one thousand officers and sixteen thousand soldiers, it was to be Franco's only serious wound. His luck gave rise to many later anecdotes about his daring. It also led his Moorish troops to believe that he was blessed with *baraka*, the mystical quality of divine protection which kept him invulnerable. Their belief seems to have infected him with his lifelong conviction that he had enjoyed the benevolent glance of providence. He later said somewhat portentously 'I have seen death walk by my side many times, but fortunately, she did not know me'.[54] The location of the wound was also the basis of speculation about Franco's apparent lack of interest in sexual matters. What little medical evidence is available does not support any such interpretation. Moreover, long before receiving the wound, Franco had refrained from participating in the sexual adventures of his comrades in his time as a cadet in the Academy and in subsequent postings in both mainland Spain and in Africa.[55] His distaste for his father's behaviour is sufficient to account for the extreme propriety of his sex life.

The High Commissioner in Morocco, General Francisco Gómez Jordana, father of the future foreign minister, recommended Franco for promotion to major again '*por méritos de guerra*' and the procedure also began for him to be awarded Spain's highest award for bravery, the Gran Cruz Laureada de San Fernando. Both proposals were opposed by the Ministry of War. The military advisers of the Ministry cited the twenty-three year-old Franco's age for denying the promotion. Franco reacted fiercely and appealed against the decision, seeking the support of the High Commissioner for a petition (*recurso reglamentario*) to the Commander-in-Chief of the Army, King Alfonso XIII. In the face of such determination, the King granted the appeal and on 28 February 1917, Franco was promoted to Major with effect from 29 June 1916. He had taken exactly six years to rise from second lieutenant to major. Along the way, he had gained the reputation at the palace of being the officer who with the greatest cheek asked for help or made complaints about his career.[56] The nomination for the Laureada was turned down on 15 June 1918. It is a reasonable assumption that, having gained his promotion by going above the heads of the Ministerial advisers, Franco's case was not reviewed with any great sympathy.[57]

There can be little doubt that, at the time, Franco preferred the promotion to the medal.* The contrast between the natural timidity of the young second lieutenant who had arrived in Africa five years earlier and the determined drive to gain promotion holds an important clue to his psychology. Franco's petition to Alfonso XIII revealed unfettered ambition. His bravery under fire was a means to the same end. The courage of the young soldier, like the cold authoritarianism of the dictator later on, may be interpreted as different manifestations of public personae which both protected him from any sense of inadequacy and provided ways of fulfilling his ambition. Franco left ample written evidence that he was not satisfied with the reality of his own life, most notably in his novel *Raza*. It is difficult not to suspect that Franco invented his own persona as the hero of the desert almost as deliberately as he did that of his hero José Churruca in *Raza*.

Promoted to major, Franco was obliged to return to mainland Spain since there were no vacant positions for officers of that rank in Morocco. He was posted instead to Oviedo in the spring of 1917 in command of a battalion of the *Regimiento de Infantería del Príncipe*. In Oviedo, he lived in the Hotel París where he became friends with a university student, Joaquín Arrarás, who would be his first biographer twenty years later. A year later, he was joined by his two companions Pacón and Camilo Alonso Vega. Despite his rank, his reputation for bravery and his brutal experiences in the Moroccan inferno, Franco's adolescent appearance and his diminutive size led to him being known locally as '*el comandantín*' (the little major).[58] Always reserved and never gregarious, he can hardly have enjoyed the routine of garrison life in Oviedo. The rainy climate and green hills of Asturias may have reminded him of his native Galicia but now the call of Africa was more powerful than that of home. As Arrarás put it, he had 'the poison of Africa in his veins'.[59]

In the daily colonial skirmishes, Franco had come to be admired and successful yet few of his comrades knew him. He was never to allow himself to become close to anyone, perhaps for fear of revealing his

---

* In retrospect, he nurtured considerable resentment about his failure to receive the Gran Cruz for what happened at El Biutz. Forty-five years later, when he reconstructed the episode, he said that the wound had been to the liver rather than the lower abdomen, which might suggest some sensitivity about its alleged consequences for his masculinity. He claimed that, despite the gravity of the wound, he had heroically continued directing operations from his stretcher. In this fanciful recollection, he had missed the medal only because the doctor who attended him had reported later that he had been on the verge of collapse, in the mistaken belief that this would strengthen his case for the award. As it was, according to Franco, this led the adjudicators to conclude that his state of health would not have permitted him to continue in command. Ramón Soriano, *La mano izquierda de Franco* (Barcelona, 1981) pp. 141–2.

essential insecurity. Nevertheless, he had forged professional and even personal links which would remain a central part of his life. He had become an *Africanista*, one of those officers who believed that, in their commitment to fighting to conquer Morocco, they alone were concerned with the fate of the *Patria*. The *esprit de corps* consequent on shared hardship and daily risk developed into a shared contempt both for professional politicians and the pacifist left-wing masses whom the *Africanistas* regarded as obstacles to the successful execution of their patriotic duty. Life in a mainland posting also signified a drastic slowing down of the promotion process. Moreover, his high rank relative to his age must have made him the target of some resentment. In Morocco, for all his youth and his lack of social skills, he was recognized as a brave and competent soldier to be trusted under fire. In Oviedo, among officers who were twice his age but still only majors or captains, or generals who saw in him only a dangerous climber, he was not popular and was driven in on himself.[60]

He was put in charge of the instruction of *oficiales de complemento* (auxiliary officers) which permitted him to establish relations with some important local families in the closed society of Oviedo. In the late summer of 1917, at a village fair (*romería*), he met an attractive local girl, María del Carmen Polo y Martínez Valdés, the daughter of a rich local family, albeit not as illustrious as it once had been. At the time, the slender dark-eyed Carmen was a fifteen year-old school-girl at the convent of Las Salesas. Franco wanted them 'to walk out together' but she refused on the grounds that, being a soldier, he could disappear as quickly as he had appeared. She also thought fifteen was too young for a steady relationship. Nevertheless, when she returned to the convent in the autumn of 1917, he wrote to her, although his letters were intercepted by the nuns and handed over to her family. With the imperturbable optimism and determination which characterized his professional behaviour, he began a dogged siege. Carmen, her school friends, and even the nuns, were thrilled to note that the famous Major now began to be a daily attender at 7 a.m. mass. He could catch a glimpse of her through a wrought-iron grill.[61] The willowy and elegant Carmen Polo carried herself with a certain aristocratic hauteur. The deeply conservative Franco felt a near reverence for the aristocracy and admired his fiancée's family and their way of life.[62]

The incipient romance with the young Army officer of modest family, even more modest prospects and a dangerous occupation met with the initial opposition of the bride's widowed father, Felipe Polo. He declared that to let his daughter marry Franco would be tantamount to letting her marry a bullfighter, a comment which carried with it considerable snobbery as well as a recognition of the risks of service in Africa.[63] Even more determined was the opposition of Carmen's aunt Isabel, Felipe Polo's sister, who, since the death of his wife had taken responsibility for the upringing of his four children. Like her brother, Isabel Polo hoped for a better match than a soldier for her niece.[64] However, despite this parental opposition, Franco pursued Carmen Polo tenaciously. He would pass messages to her in the hat-band of a mutual friend or else place them in the pockets of her coat while it hung in a café. They would meet clandestinely.[65] Ultimately, Carmen's own determination would overcome the resistance of her family. Thereafter, that determination would be put at the service of her future husband's career.

The relationship developed in a socially divided city. The inflation and shortages which resulted from the First World War were intensified by the militancy of the local working class. The Socialist Party took the lead in agitation against the deteriorating living standards along with attacks on the 'criminal war in Morocco' which deeply offended and infuriated Franco and other soldiers. Outrage that such attacks should be permitted was part of a general disgust with a political system which was blamed for the many disasters faced by the Army. Military discontent now came to the boil because of a simultaneous internal squabble between those who had volunteered to fight in Africa and those who had remained in the Peninsula, *Africanistas* and *peninsulares*. For those who had fought in Africa, the risks were enormous but the prizes, in terms of adventure and rapid promotion, high. The mainland signified a more comfortable, but boring, existence and promotion only by strict seniority. When salaries began to be hit, like those of civilians, by inflation, there was resentment among the *peninsulares* for those like Franco who had gained quick promotion. Some arms, such as the Artillery, had managed to impose a system of totally rigid seniority with an agreement by all members of its officer corps to refuse any promotion by merit. So-called *Juntas de Defensa*, rather like trade unions, were founded in many garrisons to protect the seniority system and to seek better pay.

What might have been an internal military issue was to contribute to a catastrophic upheaval in national politics. The coming of the First World War had already aroused political passions by giving rise to a bitter debate involving senior generals about whether Spain should intervene. Given the country's near bankruptcy and the parlous state of the Army, neutrality was inevitable, much to the chagrin of many officers.

Massive social upheaval came as a consequence of Spain's position as a non-belligerent. Her economically privileged position of being able to supply both the Entente and the Central Powers with agricultural and industrial products saw coalmine-owners from Asturias, Basque steel barons and shipbuilders, and Catalan textile magnates experience a spiralling boom which constituted the first dramatic take-off for Spanish industry. The balance of power within the economic elite shifted. Agrarian interests remained pre-eminent but industrialists were no longer prepared to tolerate their subordinate political position. Their dissatisfaction came to a head in June 1916 when the Liberal Minister of Finance, Santiago Alba, attempted to impose a tax on the notorious war profits of northern industry without a corresponding measure to deal with those made by the agrarians. Although the move was blocked, it so underlined the arrogance of the landed elite that it precipitated a bid by the industrial bourgeoisie to carry through political modernisation.

In the kaleidoscopic confusion of rapid economic growth, social dislocation, regionalist agitations and a bourgeois reform movement, the military was to play an active and contradictory role. The discontent of the Basque and Catalan industrialists had already caused them to challenge the Spanish establishment by sponsoring regionalist movements which infuriated the profoundly centralist military mentality. Now the self-interested reforming zeal of industrialists determined to hold on to their war profits coincided with the more desperate bid for change from a proletariat impoverished by the war. Boom industries attracted rural labour to towns where the worst conditions of early capitalism prevailed. This was especially true of Asturias and the Basque Country. At the same time, massive exports created shortages, rocketing inflation and plummeting living standards. The Socialist trade union, the *Unión General de Trabajadores* (General Union of Workers) and the anarcho-syndicalist *Confederacion Nacional del Trabajo* (National Confederation of Labour) were drawn together in the hope that a joint general strike might bring about free elections and then reform.[66] While industrialists and workers pushed for change, middle-rank Army officers were protesting at low wages, antiquated promotion structures and political corruption. A bizarre and short-lived alliance was forged in part because of a misunderstanding about the political stance of the Army.

Military complaints were couched in the language of reform which had become fashionable after Spain's loss of empire in 1898. Known as 'Regenerationism', it associated the defeat of 1898 with political corruption. Ultimately, 'Regenerationism' was open to exploitation by either the Right or the Left since among its advocates there were those who sought to sweep away the degenerate political system based on the power of local bosses or *caciques* by democratic reform and those who planned simply to destroy *caciquismo* by the authoritarian solution of 'an iron surgeon'. However, in 1917 the officers who mouthed 'Regenerationist' cliches were acclaimed as the figureheads of a great national reform movement. For a brief moment, workers, capitalists and the military were united in the name of cleansing Spanish politics of the corruption of *caciquismo*. As things turned out, the great crisis of 1917 was not resolved by the successful establishment of a political system capable of permitting social adjustment but instead consolidated the power of the entrenched landed oligarchy.

Despite a rhetorical coincidence in their calls for reform, the ultimate interests of workers, industrialists and officers were contradictory and the existing system survived by skilfully exploiting these differences. The Prime Minister, the Conservative Eduardo Dato, conceded the officers' economic demands. He then provoked a strike of Socialist railway workers in Valencia, forcing the UGT to act before the anarcho-syndicalist CNT was ready. Now at peace with the system, the Army was happy to defend it by crushing with excessive harshness the strike which broke out on 10 August 1917. In Asturias, where the strike was pacific, the military governor General Ricardo Burguete y Lana declared martial law on 13 August. He accused the strike organizers of being the paid agents of foreign powers. Announcing that he would hunt down the strikers 'like wild beasts', he sent columns of regular troops and Civil Guards into the mining valleys to cow the population. A curfew was imposed by a campaign of terror. The severity of Burguete's response, with eighty dead, one hundred and fifty wounded and two thousand arrested of whom many were severely beaten and tortured, guaranteed the failure of the strike.[67]

One of the columns was under the command of the young Major Franco. Consisting of a company of the *Regimiento del Rey*, a machine-gun section from the *Regimiento del Príncipe* and a detachment of Civil Guards, he played a significant role in re-establishing order after the strike. Indeed, the official historian of the Civil Guard referred to him as 'the man responsible for restoring order'.[68] Despite several allegations that his actions at this time established his reliability in the eyes of the local bourgeoisie, Franco himself claimed years later, before a huge audience of Asturian miners, that his column had seen no action.[69] That seems unlikely but it is impossible to reconstruct now the exact role that he played in the repression. Certainly, his job was to protect the mines from sabotage and, within the terms of martial law, to pass judgement on cases of fighting between individual strikers and Civil Guards since the strike had been declared. Implausibly, in 1963, he told George Hills, then head of the BBC Spanish services, that the appalling conditions which he saw led him to start a huge programme of reading in sociology and economics.[70] In contrast to Franco's paternalist recollections, Manuel Llaneza, the moderate leader of the Asturian mineworkers union wrote at the time of the '*odio africano*' (African hatred) that had been unleashed against the mining villages, in an orgy of rape, looting, beatings and torture.[71]

The growing hostility of many Army officers to the existing political system was intensified in the years following 1917 by the major campaign carried out by the *Partido Socialista Obrero Español* (the Spanish Socialist Workers Party) against the Moroccan war and by the indecision shown by successive governments. Army officers simply wanted to be given the resources and the liberty to elaborate policy without political hindrance. Successive governments, inhibited by ever greater popular hostility to the loss of life in Morocco, reduced material support and imposed an essentially defensive strategy upon the Army. In the eyes of the military high command, the hypocritical politicians were playing a double game, demanding of the soldiers cheap victories while remaining determined not to be seen sinking resources into a colonial war.[72] Accordingly, instead of proceeding to the full-scale occupation of the Rif which the military regarded as the only proper solution, the Army was obliged to keep to the limited strategy of guarding important towns and the communications between them. Inevitably, the tribal guerrillas were able to attack the supply convoys, involving the military in a seemingly interminable war of attrition which they blamed on the civilian politicians. An effort to change the trend of events was made in August 1919 when, on the death of General Gómez Jordana, the prime minister, the Conde de Romanones, named the forty-six year-old General Dámaso Berenguer as High Commissioner for the Moroccan Protectorate. A brilliant officer with an outstanding record, Berenguer had risen to be Minister of War in November 1918.[73]

One of the difficulties faced by Berenguer was the ambition and jealousy of the military commander of Ceuta, General Manuel Fernández Silvestre. Although they liked and respected each other, and were both favourites of Alfonso XIII, their working relationship was complicated by the fact that Silvestre was two years older than Berenguer, had once been his commanding officer and outranked him, albeit by only one number, in the seniority list. That seniority, together with Silvestre's personal friendship with the King, fuelled his tendency towards insubordination. There were major policy differences between them, Silvestre wanting an all-out showdown with the Moroccan tribes; Berenguer inclining towards a peaceful domination of the tribes by the skilful use of indigenous forces.[74] Berenguer drew up a three year plan for the pacification of the zone. It aimed at the eventual linking of Ceuta and Melilla by land. The first part envisaged the conquest of the tribal territory to the east of Ceuta, known as Anyera, including the town of Alcazarseguir. This was to be followed by the domination of the Jibala with its two major towns, Tazarut and Xauen. With government approval, the plan was initiated with the occupation of Alcazarseguir on 21 March 1919. This led El Raisuni to retaliate with a campaign of attacks on Spanish supply convoys.

At this time, Franco was sufficiently removed from events in Morocco to have joined the *Juntas de Defensa* despite the fact that they advocated promotion by rigid seniority. It may be supposed that he did so without conviction and in response to the jealousy of junior officers, much older than himself, who had not served in Africa. After all, the *Juntas'* policy, if generally applied, would remove the major incentive for officers to volunteer to serve in Morocco. Before Franco could get too involved in the concerns of the Peninsular Army, seeds of dramatic changes in his existence and in his future prospects had been sown on 28 September 1918, when he travelled from his unit in Oviedo to Valdemoro near Madrid. He remained there until 16 November taking part in an obligatory marksmanship course for majors. There he met Major José Millán Astray, a man thirteen years older than himself and about to be promoted to Lieutenant-Colonel. Renowned for his manic bravery and consequent serious injuries, Millán explained to Franco his ideas for creating special units of volunteers for Africa along the lines of the French Foreign Legion. Franco was excited by their discussions and impressed Millán Astray as a possible future collaborator.[75]

Franco returned to garrison duty in Oviedo where he remained throughout 1919 and for most of 1920. During that time, Millán Astray had presented his ideas to the then Minister of War, General Tovar. In his turn, Tovar had passed them on to the General Staff and Millán was sent to Algeria to observe the structure and tactics of the French Foreign Legion. After he returned, a royal order was published approving the principle of a foreign volunteer unit. Tovar was then replaced by General Villalba Riquelme who shelved the idea pending the more thorough-going reorganization of the African Army then being contemplated. In May 1920, Villalba was in turn replaced by the Vizconde de Eza who happened to hear Millán Astray lecture on the subject of the new unit at the *Círculo Militar* in Madrid. Eza was sufficiently convinced to authorize its recruitment.

In June 1920, Millán met Franco again in Madrid to offer him the job of second-in-command of the Spanish Legion. At first, given his now flourishing relationship with Carmen and the fact that Morocco seemed, for the moment at least, to be as quiet as mainland Spain, he was not particularly excited by the offer.[76] However, after a brief hesitation, and faced with the prospect of kicking his heels interminably in Oviedo, he accepted. It was to be the beginning of a difficult period for Carmen Polo which was to show that she could match her husband in patience and determination. Speaking about the experience eight years later, she said 'I had always dreamed that love would be an existence lit up by joy and laughter; but it brought me nothing but sadness and tears. The first tears that I shed as a woman were for him. When we were engaged, he had to leave me to go to Africa to organize the first *bandera* of the Legion. You can imagine my constant anxiety and unease, terribly intensified on the days that the newspapers talked about operations in Morocco or when his letters were delayed more than usual.'[77]

The Legion was formally established on 31 August 1920 under the name *Tercio de Extranjeros* (*Tercio*, or third, was the name used in the sixteenth century for regiments in the Army of Flanders which had been composed of three groups, pikemen, crossbowmen and arquebusiers). At its inception, it also had three *banderas*, ('colours' or 'flags') or battalions. Millán Astray disliked the name *Tercio* and always insisted on calling the new force 'the Legion', a name Franco also favoured. In the immediate aftermath of the First World War, there had been no problem recruiting volunteers. On 27 September 1920, Franco was named commander of its *primera bandera* (first battalion). Putting aside his plans for a life with Carmen Polo, he set off on the Algeciras ferry on 10 October 1920, accompanied by the first two hundred mercenaries, a motley band of desperados, misfits and outcasts, some tough and ruthless, others simply pathetic. They were hard cases, ranging from common criminals, via foreign First World War veterans who had been unable to adjust to peacetime, to the gunmen (*pistoleros*) who fought in the social war then tearing Barcelona apart. This short, slight, pallid twenty-eight year-old major, with his high-pitched voice, seemed poorly fitted to be able to command such a crew.

Millán Astray was obsessed with death and offered his new recruits little more than the chance to fight and die. The romantic notion that the Legion would offer its outcast recruits redemption through sacrifice, discipline, hardship, violence and death was held dear by both Millán and Franco throughout their lives. It underlies Franco's diary of its first two years, *Diario de una bandera*, a curious mixture of sentimentalised *Beau Geste*-style adventure-story romanticism and cold insensitivity in the face of human bestiality. In his speech of welcome to the first recruits, a hysterical Millán told them that, as thieves and murderers, their lives had been at an end before joining the Legion. Inspired by a frenzied and contagious fervour, he offered them a new life but the price to be paid would be their deaths. He called them '*los novios de la muerte*' (the bridegrooms of death).[78] They gave the Legion a mentality of brutal ruthlessness which Franco was to share to the full even though he remained outwardly reserved. Discipline was savage. Men could be shot for desertion and for even minor infractions of discipline.[79] Throughout the time that he was second-in-command to Millán Astray, Franco never wavered in his obedience, discipline and loyalty, although the temptation to contradict his manic commander must have been considerable.[80]

On the night of their arrival in Ceuta, the legionaries terrorised the town. A prostitute and a corporal of the guard were murdered. In the course of chasing the culprits, there were two more deaths.[81] Franco was obliged to take the *primera bandera* to Dar Riffien, where an old arch was rebuilt with the inscription '*Legionarios a luchar; legionarios a morir*' (Legionaries onward to fight; Legionaries onward to die'). They had arrived in Africa at a difficult moment. Berenguer had proceeded to the second stage of his grand plan for the occupation of the Spanish zone. On 14 October 1920, El Raisuni's headquarters, the picturesque mountain town of Xauen, had been occupied by Spanish troops. To the Moors, Xauen was 'the Sacred City' or 'the mysterious'. Tucked into a deep gorge, the historic fortified redoubt of Xauen was theoretically unconquerable. Its capture was an almost bloodless triumph thanks to the military Arabist, Colonel Alberto Castro Girona, who had entered the city disguised as a Moorish charcoal burner and, by a mixture of threats and bribes, persuaded the notables to surrender.[82] However, since the marauding tribes between Xauen and Tetuán were not subdued, an expensive policing operation had now to be undertaken. Within a week of arriving, Franco's legionaries were sent to Uad Lau to guard the road to Xauen.

Franco would soon be joined by his eternal cronies, his cousin Pacón, and Camilo Alonso Vega. He charged Alonso Vega with creating a battalion farm to provide funds to permit decent provisioning and the building of better barracks. The farm was a great success, not only providing fresh meat and vegetables for the troops but also making a profit. Similarly, Franco made the arrangements for a permanent fresh water supply from the nearby mountains to Dar-Riffien.[83] It was typical of his methodical and thoughtful approach to the practicalities of both camp life and hostilities against the Moors. His concerns were narrowly military. Encased in the shell of his public persona, he apparently shared few of the feelings and appetites of his comrades, becoming known as the man without fear, women or masses, (*sin miedo, sin mujeres, y sin misa*). With no interests or vices other than his career, his study of terrain, map work and general preparations for action made the units at his command stand out in an Army notorious for indiscipline, inefficiency and low morale.

In addition, in the Legion, Franco was to show a merciless readiness to impose his power over men physically bigger and harder than himself, compensating for his size with an unnerving coldness. Despite fierce discipline in other matters, no limits were put by Millán Astray or by Franco on the atrocities which were committed against the Moorish villages which they attacked. The decapitation of prisoners and the exhibition of severed heads as trophies was not uncommon. The Duquesa de la Victoria, a philanthropist who organized a team of volunteer nurses, would receive in 1922 a tribute from the Legion. She was given a basket of roses in the centre of which lay two severed Moorish heads.[84] When the Dictator General Primo de Rivera visited Morocco in 1926, he was appalled to find one battalion of the Legion awaiting inspection with heads stuck on their bayonets.[85] Indeed, Franco and other officers came to feel a fierce pride in the brutal violence of their men, revelling in their grim reputation. That notoriety was itself a useful weapon in keeping down the colonial population and its efficacy taught Franco much about the exemplary function of terror. In his *Diario de una bandera*, he adopted a tone of benevolent paternalism about the savage antics of his men.[86] In Africa, as later in the Peninsula during the Civil War, he condoned the killing and mutilation of prisoners. There can be little doubt that the years of early manhood spent amidst the inhuman savagery of the Legion contributed to the dehumanizing of Franco. It is impossible to say whether he arrived in Africa already so cut off from normal emotional responses as to be untouched by the pitiless brutality which surrounded him. When Franco had been in the *Regulares*, a somewhat older officer, Gonzalo Queipo de Llano, was struck with the imperturbability and

satisfaction with which he presided over the cruel beatings to which
Moorish troops were subjected in punishment for minor misdemean-
ours.[87] The ease with which he now became accustomed to the bestiality
of his troops certainly suggests a lack of sensitivity bordering on inner
emptiness. That would account for the unflinching, indeed insouciant,
way he was able to use terror in the Civil War and the subsequent years
of repression.

To survive and prosper in the Legion, the officers had to be as hard
and ruthless as their men. At one point, preoccupied by a rash of indiscipline and desertions, Franco wrote to Millán Astray requesting permission
to resort to the death penalty. Millán consulted with higher authorities
and then told Franco that death sentences could be passed only within
the strict rules laid down by the code of military justice. A few days later,
a legionaire refused to eat his food and then threw it at an officer. Franco
quietly ordered the battalion to form ranks, picked a firing squad, had
the offending soldier shot, and then made the entire battalion file past
the corpse. He informed Millán that he took full responsibility for an
action which he regarded as a necessary and exemplary punishment to
re-establish discipline.[88] On another occasion, Franco was informed that
two legionaires who had committed a robbery and then deserted had
been captured. 'Shoot them', he ordered. In reply to a protest from
Vicente Guarner, his one-time contemporary at the Toledo military academy who happened to be visiting the unit, Franco snapped 'Shut up. You
don't realize what kind of people they are. If I didn't act with an iron
hand, this would soon be chaos.'[89] According to one sergeant of the
Legion, both men and officers were frightened of him and of the eery
coldness which enabled him to have men shot without batting an eyelid.
'You can be certain of getting everything that you have a right to, you
can be sure that he knows where he's taking you but as for how he treats
you . . . God help you if there is anything missing from your equipment,
or if your rifle is dirty or you are a loafer'.[90]

At the beginning of 1921, General Berenguer's long-term scheme of
slow occupation, fanning out from Ceuta, was prospering. At the same
time, General Manuel Fernández Silvestre was engaged in a more
ambitious, indeed reckless, campaign to advance from Melilla westwards
to the bay of Alhucemas. On 17 February 1921, Silvestre had occupied
Monte Arruit and was making plans to cross the Amekran River. Advancing into inaccessible and hostile territory, Silvestre's success was more
apparent than real. Abd-el-Krim, the aggressive new leader who had
begun to impose his authority on the Berber tribes of the Rif, warned

Silvestre that, if he crossed the Amekran, the tribes would resist in force. Silvestre just laughed.[91] However, Berenguer was satisfied that Silvestre had the situation under control and had decided to squeeze El Raisuni's territory by capturing the Gomara mountains. The Legion was ordered to join the column of one of the outstanding officers in the Spanish Moroccan Army, Colonel Castro Girona. Their task was to help in the establishment of a continuous defensive line of blockhouses between Xauen and Uad Lau. When that line met the other which joined Xauen to Alcazarquivir, El Raisuni was surrounded. On 29 June 1921, the legionaries were in the vanguard of the force sent to assault El Raisuni's headquarters.

However, before the attack was mounted, on 22 July 1921, one of the *banderas* of the Legion was ordered to proceed to Fondak without being given any reason. Lots were drawn and Franco's *bandera* was selected. After an exhausting forced march, they arrived to be ordered to carry on to Tetuán and then to Ceuta. When they reached Tetuán, they heard rumours of a military disaster near Melilla. On arrival at Ceuta, the rumours were confirmed and they were put aboard the troop transport *Ciudad de Cádiz* and sent to Melilla.[92] What they did not know was the scale of the disaster. General Fernández Silvestre had over-extended his lines across the Amekran towards the Bay of Alhucemas and suffered a monumental defeat at the hands of Abd-el-Krim. Known by the name of the village Annual, where it began, the defeat was in fact a rout which took place over a period of three weeks and rolled back the Spanish occupation to Melilla itself. As the Spanish troops fled, enthusiastic tribesmen joined the revolt. Garrison after garrison was slaughtered. The fragility and artificiality of the Spanish protectorate was brutally exposed. All of the gains of the last decade, five thousand square kilometres of barren scrub, won at the cost of huge sums of money and thousands of lives, disappeared in a matter of hours. There would be horrific massacres at outposts near Melilla, Dar Drius, Monte Arruit and Nador. Within a few weeks, nine thousand Spanish soldiers died. The tribesmen were on the outskirts of a panic-stricken Melilla yet, too preoccupied with looting, they failed to capture it, unaware that the town was virtually undefended.[93]

At that point, reinforcements arrived, among them Franco and his men who reached Melilla on 23 July 1921 and were given orders to defend the town at all costs.[94] The Legion was used first to mount an immediate holding operation, then to consolidate the outer defences of Melilla to the south. From their defensive position in the hills outside the town, Franco could observe the siege of the last remnants of the garrison at the village of Nador but his request for permission to take a detachment of volunteers to relieve them was denied. Defeat followed defeat, Nador falling on 2 August and Monte Arruit on 9 August.[95] The Legion was sent out piecemeal to strengthen other units in the area, to escort supply columns, to hold the most exposed blockhouses. It was an exhausting task, with officers and men on duty round the clock.[96] Through the press and his published diary, the role played by Franco in the defence of Melilla contributed to his conversion into a national hero. In particular, he enhanced his reputation in the relief of the advanced position at Casabona, by unexpectedly using his escort column to attack the besieging Moroccans.[97] He had learned from fighting the Moorish tribesmen how, contrary to peninsular field regulations, effective use could be made of ground cover.[98]

By 17 September 1921, Berenguer was able to order a counter-attack to recoup some of the territory lost. The Legion was once more in the vanguard. On the first day of the offensive, near Nador, Millán Astray was seriously wounded in the chest. He fell to the ground shouting 'they've killed me, they've killed me' then sat up to shout '¡Viva el Rey! ¡Viva España! ¡Viva la Legión!'. As stretcher-bearers came to carry him away, he handed over command to Franco.

When the young major and his men entered Nador, they found heaps of the unburied, rotting corpses of their comrades killed six weeks earlier. Franco wrote later that Nador, with the bodies lying in the midst of the scattered booty of the attackers, was 'an enormous cemetery'.[99] In the following weeks, he and his men were used in many similar operations, taking part in the recapture of Monte Arruit on 23 October. He saw no contradiction in the fact that, although he approved of the atrocities committed by his own men, he was appalled by the mutilation of the hundreds of corpses of Spanish soldiers found at Monte Arruit. He and his men left Monte Arruit 'feeling in our hearts a desire for revenge, for the most exemplary punishment ever seen down the generations'.[100] Franco himself recounted that, on one occasion during the campaign, a captain ordered his men to cease firing because their targets were women. One old Legionarie muttered 'but they are factories for baby Moors'. 'We all laughed', wrote Franco in his diary, 'and we remembered that during the disaster [at Melilla], the women were the most cruel, finishing off the wounded and stripping them of their clothes, in this way paying back the welfare that civilization brought them.'[101]

On 8 January 1922, Dar Drius fell to Berenguer's column and much of what had been lost at Annual had been recaptured. Franco was indignant about the fate of Spanish soldiers massacred by the Moors at Dar Drius in 1921 and outraged that the Legion was not permitted to enter the village and take its revenge.[102] However, they had their chance a few days later. An incident took place which led the press in Galicia to praise 'the *sang froid*, the fearlessness and the contempt for life' of the 'beloved Paco Franco'. A blockhouse near Dar Drius was attacked by tribesmen and the defending *legionarios* were forced to appeal for help. The Commander of the Spanish forces in the village ordered the entire detachment of the Legion there to go to their aid. Franco said that twelve would be enough and asked for volunteers. When the entire unit stepped forward, he chose twelve and they set off. The attack on the blockhouse was driven off. The next morning Franco and his twelve volunteers returned carrying 'as trophies the bloody heads of twelve *harqueños* (tribesmen)'.[103]

When occasional leave permitted, Franco would visit Carmen Polo in Asturias. On these trips to Oviedo, as an ever more celebrated military hero, he was a welcome guest at the dinner parties of the local aristocracy. His presence was entirely compatible with a reverence for the nobility which would remain constant throughout his life.[104] Here, as he socialised, he began to make contacts which would be useful in later life and he also began to make an investment in his public image which suggests the scale of his ambition. The press began to seek him out. In interviews, speeches made at banquets given in his honour and in his publications, he began consciously to project the image of the selfless hero. Shortly after he had taken over command of the Legion from Millán Astray, Franco had received a telegram of congratulations from the *Alcalde* (mayor) of El Ferrol. In the heat of battle, he found time to make a self-deprecatory reply: 'The Legion is honoured by your greeting. I merely fulfil my duties as a soldier. An affectionate greeting to the town from the *legionarios*'.[105] It was typical of Franco's perception of himself at the time as the brave but self-effacing officer who is interested only in his duty. It was an image in which he believed implicitly and also one which he made some effort to project publicly. On leaving an audience with the King in early 1922, he told reporters that the King had embraced him and congratulated him on his success commanding the Tercio during Millán Astray's absence: 'What he has been said about me is a bit exaggerated. I merely fulfil my duty. The rank-and-file soldiers are truly valiant. You could go anywhere with them'.[106] It would be wrong to say that when Franco spoke in such terms he was merely being cynical. There is little doubt that the young major sincerely saw himself in the *Beau Geste* terms of his own diary. Nonetheless, his behaviour in interviews and the fact that he published the diary in late 1922, freely giving away copies of it, suggest an awareness of the value of a public presence in the longed-for transition from hero to general.

II

THE MAKING OF A GENERAL

1922–1931

FRANCO WAS beginning to evince signs of cultivating his public image, but he was genuinely popular with his men because of his methodical thoroughness and his insistence on always leading assaults himself. He was a keen advocate of the use of bayonet charges in order to demoralize the enemy. With his exploits well reported in the national press, he was being converted into a national hero, 'the ace of the Legion'. The rotund and plain-speaking General José Sanjurjo, himself one of the heroes of the African campaign and Franco's superior officer, said to him 'you won't be going to hospital as a result of shot fired by a Moor but because I'm going to knock you down with a stone the next time I see you on horseback in action'.[1]

In June 1922, Sanjurjo recommended Franco for promotion to Lieutenant-Colonel for his role in the recapture of Nador. Because enquiries were still being held into the disaster of Annual, the request was turned down. Nevertheless, Millán Astray was promoted to full colonel and Sanjurjo himself to Major-General. Franco merely received the military medal and remained a Major. Outraged by civilian criticisms of the Army and by indications that the government was contemplating withdrawal from Morocco, Millán Astray made a number of injudicious speeches and was removed from command of the Legion on 13 November 1922. To his chagrin, Franco was not invited to take his place since, still a major, he was too junior. Command was given instead to Lieutenant-Colonel Rafael de Valenzuela of the *Regulares*. Having been passed over for command, Franco then left the Legion. For the man who had built it up from scratch with Millán, the prospect of being second-in-command to a newcomer must have seemed unacceptable.[2] He requested a mainland posting and was eventually sent back to the *Regimiento del Príncipe* in Oviedo.

To the dismay of most Army officers, the collapse at Annual reinforced the pacifism of the Left and diminished the public standing of both the Army and the King. Alfonso XIII was widely suspected of having encouraged Silvestre to make his rash advance.[3] In August 1921, General José Picasso had been appointed to head an investigation into the defeat. The Picasso report led to the indictment of thirty-nine officers including Berenguer, who was obliged to resign as High Commissioner on 10 July 1922. Throughout the autumn of 1922, the Picasso report was the object of hostile scrutiny by a committe of the Cortes, known as the 'Responsibilities Commission', set up to examine political responsibilities for the disaster. The brilliant Socialist orator Indalecio Prieto denounced the corruption which had weakened the colonial Army and so ensured that Silvestre's temerity would turn into overwhelming defeat. The Socialist deputy called for the closure of the military academies, the dissolution of the quartermasters corps and the expulsion from the Army of the senior officers in Africa. His speech was printed as a pamphlet and one hundred thousand copies were distributed free of charge.[4]

Berenguer had been replaced by General Ricardo Burguete, under whom Franco had served in Oviedo in 1917. Burguete as High Commissioner followed government orders to attempt to pacify the rebels by bribery rather than by military action. On 22 September 1922, he made a deal with the now obese and burnt-out El Raisuni whereby, in return for controlling the Jibala on behalf of Spain, he was given a free hand and a large sum of money. Since he was already under siege in his headquarters at Tazarut in the Jibala, El Raisuni's power might have been squashed definitively had the Spaniards had the imagination and daring to occupy the centre of the Jibala. The policy of accommodation was a major error. Spanish troops were withdrawn from the territory of a man on the verge of defeat. He was enriched and his reputation and power inflated.

Burguete's aim was to pacify the tribes in the west in order to have more freedom in his efforts to crush the altogether more dangerous Abd-el-Krim in the east. After first pursuing negotiations with him for the ransom of Spanish prisoners of war, Burguete passed to the offensive in the autumn. Burguete intended to use, as his forward base, the hill-top fortified position of Tizi Azza, to the south of Annual. However, before his attack could get under way, the Rif tribes struck at the beginning of November 1922. Safely ensconced in the slopes above the town, they fired down on the garrison causing two thousand casualties and obliging the Spaniards to dig in for the winter.[5]

The worsening situation in Morocco and the compromises pursued by Burguete may have convinced Franco that he was right to have left the Legion, whatever his reasons might have been. He was showered with honours as he passed through Madrid en route to Asturias. The King bestowed on him the Military Medal on 12 January 1923 and the honour of being named *gentilhombre de cámara*, one of an élite group of military courtiers.[6] Franco was also the guest of honour at a dinner given by his admirers.

He was also the subject of an immensely flattering and revealing profile written by the Catalan novelist and journalist Juan Ferragut. It constitutes a portrait of Franco at the point when, with marriage around the corner, heroism was giving way to a more calculated ambition.* In Ferragut's profile, there can still be heard the tone of the eager man of action which would soon disappear from Franco's repertoire. Nevertheless, the clichéd patriotism and romanticised heroism of many of his remarks suggest that the persona of the intrepid desert hero was not entirely natural and spontaneous. When asked why he had left Morocco, Franco replied 'because we aren't doing anything there anymore. There's no shooting. The war has become a job like any other, except that it's more exhausting. Now all we do is vegetate.' There was a contrived element about Franco's answers which suggested an intense consciousness of his public image. When Ferragut asked him if he liked action, the thirty year-old Major replied 'yes ... at least up to now. I believe that a soldier has two periods, one of war and one of study. I've done the first and now I want to study. War used to be more simple; all you needed was heart. But today it is more complicated; it is, perhaps, the most difficult science of them all'. Ferragut described him as boyish: 'his sunburnt face, his black, brilliant eyes, his curly hair, a certain timidity in his speech and gestures and his quick and open smile make him seem like a child. When he is praised, Franco blushes like a girl who has been flattered.' He brushes aside the praise, as befits a hero, 'but I've done nothing really! The dangers are less than people think. It's all a question of endurance'.

Asked about his most emotional memory of the war, he replied 'I remember the day at Casabona, perhaps the hardest day of the war. That day we saw what the Legion was made of. The Moors were pressing strongly and we were fighting at twenty paces. We had a company and a half and we suffered one hundred casualties. Handfuls of men were

* Ferragut had written the fictionalised *Memorias de un legionario* and had been rumoured to have ghost-written Franco's *Diario de una bandera*, although the article made a great point of the interview being their first meeting.

falling, almost all wounded in the head or the stomach, yet our strength never wavered for a second. Even the wounded, dragging themselves along covered in blood, cried '¡Viva la Legión! Seeing them, so manly, so brave, I felt an emotion which choked me.' Asked if he had ever felt fear, he smiled as if puzzled, and shyly replied 'I don't know. No one knows what courage and fear are. In a soldier, all this is summed up in something else: the concept of duty, of patriotism.' The romantic note continued with references to the anxious vigils of his mother and bride-to-be. Ferragut asked him directly, 'Are you in love, Franco?' to which the affable interviewee replied '¡Hombre!, what do you think? I'm just off to Oviedo to get married.'[7]

On 21 March 1923, Franco arrived in Oviedo where his exploits ensured that he would be feted. At the beginning of June, local society turned out in force for a banquet at which he was presented with a gold key, the symbol of his recently acquired status as Gentilhombre de cámara, purchased for him after a local subscription. The King had still not granted the reglamentary permission for his wedding. Since this was a mere formality, the ceremony was being planned for June. However, while Carmen and Francisco were waiting to hear from the Palace, their plans suffered another reversal. Franco had gone to El Ferrol where he spent most of May with his family. In early June, Abd el Krim launched another attack on Tizi Azza, the key to the outreaches of the Melilla defence lines. If Tizi Azza fell, it would have been relatively easy for other Spanish positions to collapse in a domino effect. On 5 June 1923, the new commander of the Legion, Lieutenant-Colonel Valenzuela, was killed in a successful action aimed at breaking the siege.[8]

An emergency cabinet meeting three days later, on 8 June 1923, decided that the most suitable replacement for Valenzuela was Franco. The Minister of War, General Aizpuru, telegrammed to inform him that he had been promoted to Lieutenant-Colonel with retrospective effect from 31 January 1922 and that the King had bestowed command of the Legion upon him. His marriage would have to be postponed again. The ambitious Carmen may have found consolation for the loss of her bride-groom in his promotion, the signs of royal patronage and the enormous local prestige that he thereby enjoyed, although, interviewed in 1928, she talked of her anxieties while he was away and of his principal defect being his love for Africa.[9]

Prior to leaving Spain, Franco was the guest of honour at banquets in both the Automobile Club in Oviedo and at the Hotel Palace in Madrid.

One of the principal Asturian newspapers dedicated an entire front page to his promotion and his prowess, complete with extravagant tributes from General Antonio Losada, the military governor of Oviedo, from the Marqués de la Vega de Anzó and other local dignitaries.[10] Interviewed on arrival at the Automobile Club banquet on the evening of Saturday 9 June, Franco showed himself to be the public's ideal young hero, dashing, gallant and, above all, modest. He dismissed talk of special bravery and showed himself perplexed by all the fuss that was being made. Clearly conscious of the dash he was cutting, he interrupted the journalist's attempted eulogy by saying 'I just did what all the Legionaires did, we fought with a desire to win and we did win'. A discreet reminder of what he was leaving behind brought a delicate glimpse of emotion which Franco quickly put behind him. The journalist remarked sycophantically 'how the brave Legionaires will rejoice at your appointment!'. Franco replied 'Rejoice? Why? I'm an officer just like . . .', only to be interrupted by a passing ex-Legionaire who said 'say yes, that they will all rejoice, of course they will rejoice'. Like a hero of romantic fiction, Franco replied with a modest laugh, saying 'Don't go overboard. Yes, you're right, the lads care for me a lot.'

The interview ended with Franco being asked about his plans, at which he gave another hint of the sacrifice he was having to make. He replied with a curious mixture of virile enthusiasm and self-regarding pomposity: 'Plans? What happens will decide that. I repeat that I am a simple soldier who obeys orders. I will go to Morocco. I will see how things are. We will work hard and as soon as I can get some leave I will come back to Oviedo to . . . to do what I thought was virtually done, which for the moment duty prevents, taking precedence over any feelings, even those with roots deep in the soul. When the *Patria* calls, we have only rapid and concise response: *¡Presente!*'[11] There is no doubt that this, and other interviews from this period, show an altogether more attractive figure than the one that Franco was later to become, in large measure as a consequence of the corrupting influence of constant adulation. The Minister of War and future President of the Second Republic, Niceto Alcalá Zamora, thought Franco's near contemporary and rival, Manuel Goded, a more promising officer than Franco. However, he liked Franco's air of modesty, 'the loss of which when he became a general damaged him significantly'.[12]

Within a week of passing through Madrid, Franco had taken up his new command in Ceuta and was soon in the thick of the action. Shortly after Franco's arrival in Africa, Abd el Krim followed up his attack on

Tizi Azza with another on Tifaruin, a Spanish outpost near the River
Kert to the west of Melilla. Nearly nine thousand men besieged Tifaruin
and they were dislodged on 22 August by two *banderas* of the Legion
under Franco's command.[13]

Such was the accumulated military discontent about what was perceived as civilian betrayal of the Army in Morocco that since early in
1923 two groups of senior generals, one in Madrid and the other in
Barcelona under Miguel Primo de Rivera, had been toying with the idea
of a military coup.[14] The incident which triggered it off took place on
23 August. There were a number of public disturbances in Málaga involving conscripts being embarked for Africa. Civilians were jostled and Army
officers assaulted. Some of the recruits were merely drunk, others were
Catalan and Basque nationalists making political protests. Order was
finally restored by the Civil Guard. An NCO in the Engineering Corps
(*suboficial de ingenieros*), José Ardoz, was killed and the crime was attributed
to a *gallego*, Corporal Sánchez Barroso. Sánchez Barroso was immediately
tried and sentenced to death. Since Annual, there had been a widespread
public revulsion against the Moroccan enterprise and, in consequence,
there was a huge outcry about the death sentence. On 28 August, Sánchez
Barroso was given a royal pardon, at the request of the cabinet. The
officer corps was outraged by the humiliation of the Málaga incidents,
by the subsequent public rejection of their cause in Morocco and by what
they saw as the slight involved in the pardon.[15]

On 13 September, the Falstaffian eccentric General Miguel Primo de
Rivera launched his military coup supported by the garrisons of his own
military region of Catalonia and by that of Aragón, under the control of
his intimate friend General Sanjurjo. There is considerable debate about
the King's complicity in the coup. What is certain is that he acquiesced
in the military demolition of the constitutional monarchy and happily
embarked on a course of authoritarian rule. After six years of sporadic
bloodshed and instability since 1917, and the fashionable 'regenerationist'
calls for an 'iron surgeon', the Military Directory set up by Primo de
Rivera met with only token resistance and, given widespread disillusion
with the *caciquista* system, much benevolent expectation.[16] Despite the
mutual admiration which, at this stage of their careers, united Franco
and Sanjurjo, neither Franco nor most of the officers of the Legion
were particularly enthusiastic about the coup. They regarded most of the
officers who supported Primo as primarily members of the *Juntas de
Defensa* and therefore enemies of promotion by merit. In addition, they
were fully aware of the belief of Primo himself that Spain should abandon

her Moroccan protectorate.[17] It is clear, however, that Franco had no objections in principle to the military taking political power, particularly as royal approval was quickly forthcoming. In any case, his mind was on other things – his new command and his impending marriage, for which royal permission had finally been granted on 2 July.[18]

The thirty year-old Francisco Franco was married to the twenty-one year-old María del Carmen Polo in the Church of San Juan el Real in Oviedo at midday on Monday 22 October 1923. His fame and popularity as a hero of the African war ensured that substantial crowds of well-wishers and casual onlookers would gather round the church and on the pavements of the streets traversed by the wedding party. By 10.30 a.m., the Church was full and the crowd had spilled out and packed the surrounding streets. The police had difficulty maintaining the flow of local traffic. As befitted his position as a *gentilhombre de cámara*, Franco's *padrino* (best man) was Alfonso XIII, by the proxy of the military governor of Oviedo. General Losada took Carmen's arm and they entered the Church under the royal canopy (*palio*). That honour, combined with Franco's growing reputation, was reflected in the fact that his marriage was reported in the society pages not only of local newspapers but also of the national press. A bemedalled Franco wore the field uniform of the Legion. The ceremony was carried out by a military chaplain while the organ played Franco's choice of military marches. On leaving the church, the couple were greeted by wild cheering and applause. The crowd followed the cars back to the Polo house and continued to cheer.[19] The marriage constituted a major social occasion in Oviedo, the centre-piece of which was a spectacular wedding banquet.[20] Franco's father, Nicolás Franco Salgado-Araujo, was not present. As might have been expected, it was to be a solidly enduring, if not a passionate, marriage.* Five years later, Carmen would recall her wedding day, 'I thought I was dreaming or reading a beautiful novel . . . about me'.[21] Among the mountains of telegrams was a collective greeting from the married men of the Legion and another from a Legion battalion which welcomed Carmen as their new mother.[22]

The social position of both bride and groom was reflected in the fact that those who signed the marriage certificate as witnesses included two local aristocrats, the Marqués de la Rodriga and the Marqués de la Vega de Anzó. The unctuous tone of local reporting not only gives an indi-

---

* In later life, particularly after Franco gained power, the relationship seemed more formal than spontaneously affectionate. Pacón commented that Franco seemed morose and inhibited in the company of Doña Carmen.

cation of the prestige that Franco already enjoyed but it also reflects the kind of adulation with which he was bombarded. 'Yesterday, Oviedo enjoyed moments of intimate and longed-for satisfaction and of jubilant delight. It was the wedding of Franco, the brave and popular head of the Legion. If the desire of the couple to see their love blessed before the altar was great, the interest of the public was no less immense on seeing them happy with their dream of love come true. In this pure love, all of us who know Franco and Carmina have given something of their own hearts and have suffered with them their worries, their anguish, their justified impatience. From the King down to the last of the hero's admirers there was a unanimous desire that this love, so beset by ill-luck, should have the divine sanction which would lead them to the supreme happiness.'[23] 'The pause in the struggle of the brave Spanish warrior has had its triumphant apotheosis. Those polite and gallant phrases whispered by the noble soldier in the ear of his beautiful beloved have had the divine epilogue of their consecration.'[24] One journal in Madrid headed its commentary with the headline 'The Wedding of an heroic Caudillo' (warrior-leader).[25] This was one of the first ever public uses of the term Caudillo with respect to Franco. It can easily be imagined how such adulatory prose moulded Franco's perception of his own importance.

By tradition, on marrying, a senior officer was required to 'kiss the hands' of the King. After a few days honeymoon spent at the Polo summer house, La Piniella near San Cucao de Llanera outside Oviedo, and prior to setting up home in Ceuta, the newly weds travelled to Madrid and called at the royal palace in late October. In 1963, the Queen recalled lunch with a silent and timid young officer.[26]

In later years, Franco himself twice recounted the interview with the King to his cousin and also to George Hills. Franco alleged in these accounts that the King was anxious to know how the Army in Africa felt about the recent coup and the military situation in Morocco. Franco claimed to have told the King that the Army was doubtful about Primo because of his belief in the need to abandon the protectorate. When the King demonstrated an equally pessimistic inclination to pull out, Franco boldly replied with his opinion that the 'rebels' (the local inhabitants) could be defeated and the Spanish protectorate consolidated. He allegedly pointed out that, so far, Spanish operations had been piecemeal, pushing back the Moors from one small piece of ground after another, attempting to hold it, and to retake it after it had been recaptured. Rather than this endless drain on men and materials, Franco suggested an idea long favoured by *Africanistas*, a major attack on the headquarters of Abd el

Krim in the region of the Beni Urriaquel tribe. The most direct route was by sea to the Bay of Alhucemas.

The King arranged for Franco to dine with General Primo de Rivera and tell him of his plan.[27] Primo was hardly likely to be sympathetic given both his long-standing conviction that Spain should withdraw from Morocco and his determination, as Dictator, to reduce military expenditure.[28] When he met Franco, Primo would almost certainly not have been surprised to hear that the young Lieutenant-Colonel shared the commitment of the *Africanistas* to remaining in Morocco. Franco had long since published his variant of the view that Spain's Moroccan problem would be solved at Alhucemas, 'the heart of anti-Spanish rebellion, the road to Fez, the short exit to the Mediterranean, and there is to be found the key to much propaganda which will end the day that we set foot on that coast.'[29] The idea of a landing at Alhucemas had been in the air for some years and the general staff had prepared detailed contingency plans in the event of the politicians giving the go-ahead. According to his own account, by the time Franco managed to put his case for a landing to the Dictator, it was in the early hours of the morning. The anything but abstemious Primo was somewhat merry, and Franco was convinced that he would never remember their conversation. Nevertheless, Primo suggested that he submit his scheme in written form.

In this subsequent version of events, Franco's narrative is tailored to show that the plan for the Alhucemas landing was his own. That he should remember it as his own brainchild was entirely understandable after years of being told so by sycophants and given the fact that he did play a prominent role in putting the case against withdrawal from Morocco.[30] At the beginning of 1924, he had been one of the founders, along with General Gonzalo Queipo de Llano, of a journal called *Revista de Tropas Coloniales* which advocated that Spain maintain its colonial presence in Africa. At the start of 1925, he would become head of its editorial board. Franco was to write more than forty articles for the journal. In one published in April 1924, entitled 'passivity and inaction', he argued that the weakness of Spanish policy, 'the parody of a protectorate', was encouraging rebellion among the indigenous tribes.[31] It made a considerable impact.

Shortly after visiting the King, the newly wed Franco and his bride took up residence in Ceuta. The situation in Morocco seemed ominously quiet. In fact, by the spring of 1924, Abd el Krim's power had grown enormously and he no longer recognized the authority of the Sultan. He was presenting himself as the figurehead of a vaguely nationalistic Berber

movement and talking in terms of establishing an independent socialist
state. Numerous tribes accepted his leadership and, under his self-
bestowed title of 'Emir of the Rif', in 1924, he formally requested
membership of the League of Nations.[32] After the defeat of Annual, the
Spanish counter-offensive had recaptured an area around Melilla. Apart
from that, the Spanish foothold consisted only of the towns of Ceuta,
Tetuán, Larache and Xauen. The local garrisons were confident that they
could hold the territory but were seriously disturbed by rumours that
they were about to receive orders to withdraw. Anticipating difficulties,
the military commander of Ceuta, General Montero, during the fesival
of the Pascua Militar on 5 January called upon the officers under his
command to give their word that they would obey orders no matter what
they were. Franco took the lead in pointing out that they could not be
asked to obey orders that were contrary to military regulations.[33]

Possibly alerted by these objections, Primo de Rivera finally decided
to inspect the situation personally. In the meantime, Sanjurjo was sent
to take over as commander of Melilla. Abd el Krim greeted him with an
offensive on Sidi Mesaud only to be driven back by the Legion commanded by Franco. When the Dictator arrived in June 1924, he quickly
grasped the essential absurdity of the Spanish military predicament. His
inclination was to abandon the Protectorate on the grounds that to pacify
it fully would be too expensive and to go on holding it on the basis of
strings of waterless, indefensible blockhouses was ludicrous. For part of
his tour, the Dictator insisted on being accompanied by Franco. At the
time, the young Lieutenant-Colonel was deeply concerned about
rumours that Primo had come to arrange a Spanish withdrawal. He had
just tried to convince the High Commissioner, General Aizpuru, that the
publication of orders to abandon the inland towns would provoke a major
offensive by the forces of Abd el Krim. Franco had agreed with Lieutenant-Colonel Luis Pareja of the Regulares that, in the event of a withdrawal
from Xauen, they would both apply for transfers to the mainland. In a
letter to Pareja in July 1924, Franco declared that when the time came
the bulk of his officers would do the same.[34]

At one notorious dinner, in Ben Tieb on 19 July 1924, there was an
incident involving the Legion and the Dictator which has become the
basis of subsequent myth. This was the dinner at which, legend in the
Legion would have it, Franco had arranged for the Dictator to be served
a menu consisting entirely of eggs.[35] *Huevos* (eggs) being the Spanish
slang for testicles, the machista symbolism was obvious: the visitor needed
huevos and the Legion had plenty to spare. However, given Franco's

fanatical respect for discipline and his ambitious concern for his career, it is difficult to believe that he would so blatantly insult a senior officer and head of the government. In 1972, Franco denied that such a menu had been served.

At the dinner, Franco made a harsh but careful speech against *aban-donismo*. What he said revealed his lifelong commitment to Spanish Morocco: 'where we tread is Spanish soil, because it has been bought at the highest price and with the most precious coin: the Spanish blood shed here. We reject the idea of pulling back because we are convinced that Spain is in a position to dominate her zone.' Primo responded with an equally strong speech explaining the logic behind plans for a with-drawal and a call for blind obedience. When a Colonel of Primo's staff said '*muy bien*' (hear, hear), the irascible and diminutive Major José Enrique Varela, unable to contain himself, shouted '*muy mal*'. Primo's speech was interrupted by hissing and hostile remarks. Sanjurjo, who accompanied him, later told José Calvo Sotelo, the Dictator's Minister of Finance, that he had kept his hand on the butt of his pistol throughout the speeches, fearing a tragic incident. When the Dictator finished he was greeted with total silence. Franco, ever careful, hastened to visit Primo immediately after the dinner to clarify his position. He said that if what had happened required punishment, he was prepared to resign. Primo made light of Franco's part in the affair and permitted him to return later and again put his point of view about a landing in Alhu-cemas.[36] In his own 1972 version, he claimed, implausibly, to have given Primo de Rivera a dressing down. As a consequence, he said, Primo de Rivera promised to do nothing without consulting the 'key officers'.[37]

Shortly after the Ben Tieb dinner, the Dictator prepared an operation to fold up 400 positions and block-houses. As Franco and others had warned, the talk of withdrawal encouraged Abd el Krim and stimulated the desertion of large numbers of Moroccan troops from the Spanish ranks. Lieutenant-Colonel Pareja understood that this meant that the conditions agreed with Franco for their joint resignations had arrived. He presented his transfer request and was disgusted to discover that Franco had not kept his word. Franco, always cautious, particularly after his confrontation with Primo de Rivera, remained in his post.[38] Shortly after the return of Primo to Madrid, Abd el Krim attacked in force, cutting the Tangier-Tetuan road and threatening Tetuan. A communiqué was issued on 10 September 1924 announcing the evacuation of the zone. Anxiety about the consequences of the proposed withdrawal led a number of officers in Africa to toy with the idea of a coup against Primo. The

ring-leader was Queipo de Llano, who claimed in 1930 that Franco had
visited him on 21 September 1924 to ask him to lead a coup against the
Dictator. In 1972, Franco did not deny that the conversation had taken
place. However, as had happened in the case of his pact with Lieutenant-
Colonel Pareja, nothing came of an uncharacteristically frank expression
of discontent. Where military discipline was concerned, habitual caution
always prevailed.[39]

Franco and the Legion were thrown into service at the head of a
column led by General Castro Girona which set off from Tetuan on 23
September in order to relieve the besieged garrison at Xauen, 'the sacred
city', in the mountains. It took them until 2 October to fight the forty
miles there. Over the next month, units from isolated positions drifted
in until at the beginning of November there were ten thousand men in
Xauen, many of them wounded, most of them exhausted. An evacuation
was then undertaken. Primo won over much of the Army of Africa by
assuming complete responsibility for whatever might happen, naming
himself High Commissioner on 16 October. He returned to Morocco
and set up his general staff in Tetuán. The evacuation of the Spanish,
Jewish and friendly Arab inhabitants of Xauen was an awesome task.
Children, women, and other civilians, the old and the sick, were packed
into trucks. The immensely long and vulnerable column set off on
15 November. Moving slowly at night, their rear was covered by the
Legion under Franco. Constantly harrassed by raiding tribesmen, and
severely slowed down by rain storms which turned the tracks into impassable mud, it took four weeks to return to Tetuan where the survivors
arrived on 13 December. It was a remarkable feat of dogged determination though nothing approaching the 'magisterial military lesson' perceived by Franco's hagiographers.[40]

Franco was deeply disappointed to be a party to the abandonment of
any fragment of the territory in defence of which much of his life had
been spent. He published an article on the tragedy of the withdrawal,
based on his diary. Vividly and passionately written, it reflects the resignation and sadness of the day before the retreat.[41] However, he was
consoled by being awarded another Military Medal and by being promoted to full colonel on 7 February 1925 with effect from twelve months
earlier, 31 January 1924. He was also allowed to keep the command of
the Legion, although that post should have been held by a Lieutenant-
Colonel. He was further consoled when Primo de Rivera in late 1924
changed his mind about abandoning Morocco. The Dictator decided
sometime in late November or early December to pursue the Alhucemas

landing and ordered that detailed plans be drawn up. In early 1925, Franco experimented with amphibious landing craft. It was during one of these exercises, on 30 March 1925, on board the Spanish coastal patrol vessel *Arcila*, that he was offered a plate of breakfast by a young naval lieutenant called Luis Carrero Blanco who would, from 1942 to 1973, be his closest collaborator. Franco refused the offer on the grounds that, since being wounded in El Biutz, he always went into action on an empty stomach.[42]

In March 1925, on a visit to Morocco, General Primo de Rivera presented Franco with a letter from the King and a gold religious medal. The letter was fulsome: 'Dear Franco, On visiting the [Virgin of the] Pilar in Zaragoza and hearing a prayer for the dead before the tomb of the leader of the Tercio, Rafael Valenzuela, gloriously killed at the head of his *banderas*, my prayers and my thoughts were for you all. The beautiful history that you are writing with your lives and your blood is a constant example of what can be done by men who reckon everything in terms of the fulfilment of their duty ... you know how much you are loved and appreciated by your most affectionate friend who embraces you – Alfonso XIII.'[43]

After entering Xauen, the triumphant Abd el Krim had celebrated his hegemony by capturing El Raisuni. He then made a colossal mistake. At precisely the moment that the French were moving into the no-man's-land between the two protectorates, his long-term ambition of creating a more or less socialist republic led him to try to overthrow the Sultan, who was the instrument of French colonial rule. Taking on the French, initially he defeated them. His advance skirmishers came within twenty miles of Fez. This led to an agreement in June 1925 between Primo de Rivera and the French commander in Africa, Philippe Pétain, for a combined operation. The plan was for a substantial French force of one hundred and sixty thousand colonial troops to attack from the south while seventy-five thousand Spanish soldiers moved down from the north. The Spanish contingent was to land at Alhucemas under the overall command of General Sanjurjo. Franco was in command of the first party of troops to go ashore and had responsibility for establishing a bridgehead.

There was no effort at secrecy either in the planning or on the night of 7 September, when Spanish ships arrived in the bay with lights ablaze and the troops singing. As a result of poor reconnaissance, the landing took place on a beach where the landing craft hit shoals and sand-banks too far out for tanks to be disembarked. Moreover, the water was at a depth of over one and a half metres and many of the Legionaires could not swim. Their attack was awaited by rows of entrenched Moors who immediately began to fire. The naval officer in charge of the landing craft radioed the fleet where the High Command awaited news. In view of his signal, the vessels were ordered to withdraw. Franco decided that a retreat at that point would shatter the morale of his men and boost that of the Moorish defenders. Accordingly, he countermanded the order and told his bugler to sound the attack. His Legionaires jumped overboard, waded to the shore and succeeded in establishing the bridgehead. Franco was later called before his superiors to explain himself which he did by reference to military regulations which granted officers a degree of initiative under fire.[44]

The entire operation was a condemnation of the appalling organization of the Spanish Army and poor planning by Sanjurjo. After the bridgehead was established there was insufficient food and ammunition to permit an advance. There was extremely poor ship-to-shore communication and very limited artillery support. Two weeks passed before the order was given to move beyond the bridgehead. Then the advance was subject to the mortar batteries placed by Abd el Krim. In part because of the tenacity of Franco himself, the Spanish attack continued. However, with the French moving up from the south, it was only a matter of time before Abd el Krim surrendered. On 26 May 1926, he gave himself up to the French authorities.[45] The resistance of the Rif and Jibala tribes collapsed.

Franco produced a vividly, if somewhat romantically, written diary of his participation in the landing, entitled *Diario de Alhucemas*. It was published over four months from September to December 1925 in the *Revista de Tropas Coloniales* and again in 1970 in a version which he himself censored.[46] Referring to an attack on a hill which took place in the first hours after the landing, he wrote in 1925 'those defenders who are too tenacious are put to the knife' changing it in 1970 to 'those defenders who are too tenacious fell beneath our fire'. Even after editing the text in 1970, Franco left in phrases reminiscent of the adventure stories of his youth. Men were not shot but 'scythed down by enemy lead'. 'Fate has snatched away from us the flower of our officers. Our time has come. Tomorrow we will avenge them!'[47] Years later, he told his doctor that, during the Alhucemas campaign, a deserter from the Legion was brought in and, with no hesitation other than the time taken to confirm his identity, he ordered a firing squad to be formed and the man shot.[48]

On 3 February 1926, Franco was promoted to Brigadier General, which made the front page of the newspapers in Galicia.[49] At the age of 33, he was the youngest general in Europe, and was finally obliged by his seniority to leave the Legion. On being promoted, Franco's service record had the following added: 'He is a positive national asset and surely the country and the Army will derive great benefit from making use of his remarkable aptitudes in higher positions'.[50] He was given command of the most important brigade in the Army, the First Brigade of the First Division in Madrid, composed of two aristocratic regiments, the *Regimiento del Rey* and the *Regimiento de León**.

On returning to Spain, Franco brought with him a political baggage acquired in Africa which he would carry through the rest of his life. In Morocco, Franco had come to associate government and administration with the endless intimidation of the ruled. There was an element too of the patronizing superiority which underlay much colonial government, the idea that the colonised were like children who needed a firm paternal hand. He would effortlessly transfer his colonial attitudes to domestic politics. Since the Spanish Left was pacifist and hostile to the great adventure in Morocco, associated in his mind with social disorder and regional separatism, he considered leftists to be as dire an enemy as rebellious tribesmen.[51] He regarded the poisonous ideas of the Left as acts of mutiny to be eradicated by iron discipline which, when it came to governing an entire population, meant repression and terror. The paternal element would later be central to his own perception of his rule over Spain as a strong and benevolent father figure.

In Africa, Franco had also learned many of the stratagems and devices which were to be his political hallmark after 1936. He had observed that political success came from a cunning game of divide and rule among the tribal chiefs. That is what the Sultan did; it was what the better Spanish High Commissioners aspired to do. At a lower level, local garrison commanders had to do something similar. Astute, greedy, envious and resentful chieftains were played off against one another in a shifting game of alliances, betrayals and lightning strikes. His assimilation of such skills would permit him to run rings around his political enemies, rivals and collaborators inside Spain from 1936 until well into the 1960s. Although he acquired such skills, he had never developed any serious interest in the Moroccans. Like most colonial officers, Franco did not

* At the time, each military region of Spain had two divisions, each composed of two brigades. However, given the shortage of recruits, in practice only the first brigade of each Captaincy General was at operational strength. (Suárez Fernández, *Franco*, I, pp. 187, 191.)

learn more than a smattering of the language of those he fought and ruled. Later in life, he would also fail in his attempts to learn English. Absorbed in military matters, he could never muster much interest in other cultures and languages.[52]

On the day on which his promotion to general was announced, Franco's success had been somewhat overshadowed by the spectacular national newspaper coverage given to his brother Ramón. Major Ramón Franco was crossing the South Atlantic with Captain Julio Ruiz de Alda, one of the future founders of the Falange, in the *Plus Ultra*, a Dornier DoJ Wal flying boat.[53] The regime and the press was treating Ramón as a modern Christopher Columbus. A committee was set up in El Ferrol to organize various tributes to the two brothers, including the unveiling of a plaque on the wall of the house in which they had been born. It read 'In this house were born the brothers Francisco and Ramón Franco Baamonde, valiant soldiers who, at the head of the Tercio of Africa and crossing the Atlantic in the seaplane 'Plus Ultra', carried out heroic deeds which constitute glorious pages of the nation's history. The town of El Ferrol is honoured by such brilliant sons to whom it dedicates this tribute of admiration and affection.'[54]

Franco took up his important post in Madrid in time to admire the achievements of the Primo de Rivera dictatorship. What the officer corps perceived as regional separatism had been suppressed and labour unrest dramatically diminished. Anarchist and Communist unions had been suppressed while the Socialist union, the Unión General de Trabajadores, was given control of a newly created state arbitration machinery. The UGT became the semi-official trade union organization of the regime. A massive programme of infrastructural investment in roads and railways created a high degree of prosperity and near full employment. For an Army officer, particularly after the disorders of the period 1917 to 1923, it was a good time to be on active service. The constant criticism of the Army which officers associated with the parliamentary monarchy had been silenced. The triumph of Alhucemas had revived military popularity. It is little wonder that, like many Army officers and civilian rightists, Franco would come to look back on the six years of the Primo de Rivera dictatorship as a golden age. He often commented during the 1930s that they were the only period of good government that Spain had enjoyed in modern times. In his view, Primo's error was to have announced that he would hold power only for a short time until he had solved Spain's problems. Franco said reprovingly to his Oviedo acquaintance, the monarchist, Pedro Sainz Rodríguez 'that was a mistake; if you accept a command you have to take it as if it was going to be for the rest of your life'.[55]

The Dictatorship was also a period in which Franco experienced further inflation of his ego. On the evening of 3 February 1926, his fellow cadets of the fourteenth intake (*promoción no.14*) at the Academia de Infantería de Toledo met to pay homage to the first of their number to become a general. They presented him with a dress sword and a parchment with the following inscription: 'When the passage through the world of the present generation is no more than a brief comment in the book of History, there will endure the memory of the sublime epic written by the Spanish Army in the development of the nation. And the glorious names of the most important *caudillos* will be raised on high, and above them all will be lifted triumphantly that of General Francisco Franco Bahamonde to reach the sublime heights achieved by other illustrious men of war, Leiva, Mondragón, Valdivia and Hernán Cortés. His comrades pay this tribute of admiration and affection to him in recognition of his patriotism, his intelligence and his bravery'.[56]

In the course of the next few days, Franco would receive many telegrams from the local authorities of El Ferrol recounting the acts of homage mounted for his mother. On Sunday 7 February, bands played, firework displays were organized and ships in the bay sounded their horns. The town turned out to acclaim the historic flight by Ramón, who was still in Argentina, although Franco was not forgotten in the endless tributes made to Doña Pilar Bahamonde y Pardo de Andrade. 12 February was declared a holiday in El Ferrol in honour of both brothers. The streets of the town were illuminated and a Te Deum was sung in the Church of San Julián to celebrate their achievements. The plaque was unveiled in the calle María. Messages of congratulation to Doña Pilar for both her sons arrived from the *Alcaldes* (mayors) of El Ferrol, the four provincial capitals of Galicia and from many towns across Spain.[57] On 10 February, a massive crowd turned out in the Plaza de Colón (Columbus) in Madrid to acclaim Ramón's achievement. In part, the media coverage and public enthusiasm were orchestrated by the Primo de Rivera dictatorship in order to profit in propaganda terms from the flight of the *Plus Ultra*.

The adulation was largely directed at Ramón, but there is no reason to believe that Franco was resentful at seeing the black sheep of the family suddenly converted into a national hero. The adulation of his brother as a twentieth century Christopher Columbus may, however, have inspired Franco's later efforts to present himself as a modern-day

El Cid. Franco always had an intense loyalty to his family and, over the years, was to use his own position to help and protect Ramón from the consequences of his wilder actions. In any case, his own triumphs and popularity were sufficient in frequency and intensity for him not to need to feel envy. At Easter 1926, during the Corpus Christi procession at Madrid's San Jerónimo Church, he commanded the troops which lined the streets and escorted the host. As the legendary hero of Africa, he was the object of the admiring attention of the upper class *Madrileños* who made up the congregation.[58] In the late summer of 1927, Franco accompanied the King and Queen on an official visit to Africa during which new colours were given to the Legion at their headquarters in Dar Riffien.[59]

On 14 September 1926, Franco's first and only child María del Carmen was born in Oviedo where Carmen had gone to be with her dying father.[60] The new arrival was to become the focus of his emotional life. Years later, he was to say 'when Carmen was born, I thought that I would go mad with joy. I would have liked to have had more children but it was not to be'.[61] There have been insistent rumours that Carmen was not really Francisco's daughter but was adopted, and that the father may have been his promiscuous brother Ramón. There is no evidence to support this theory, which seems to have arisen entirely from the fact that there are no known photographs of Carmen Polo noticeably pregnant and from Ramón's notorious sexual adventurism.[62] Franco's sister, Pilar, went out of her way in her memoirs to make a point of saying that she saw Carmen Polo pregnant although her dates are wrong by two years.[63]

The posting to Madrid began a period in which Franco had plenty of spare time. Rather than make the lives of his colonels a misery by frequent surprise inspections, he left them to get on with running their own barracks, a pattern that he would later follow with his ministers. He rented an apartment on the elegant Castellana avenue and enjoyed a busy social life. He regularly met military friends from Africa and the Toledo Academy at the regular social gatherings or *tertulias* of the upper-class club, *La Gran Peña*, and in the cafés of Alcalá and the Gran Vía. He was relatively close to Millán Astray, Emilio Mola, Luis Orgaz, José Enrique Varela and Juan Yagüe.[64] While living in Madrid, he acquired a passion for cinema and became a member of the *tertulia* of the politician and writer Natalio Rivas, a member of the Liberal Party.[65] At Rivas's invitation, he appeared along with Millán Astray in a film entitled *La Malcasada* made by the director Gómez Hidalgo in Rivas's house. Franco's small part was as an Army officer recently returned from the African wars.[66]

At this stage of his life, as later, Franco had little interest in day-to-day politics. None the less, he began to think that ultimately he might play a political role of some kind. The popular acclaim which he had received after Alhucemas, the rapidity of his promotions, and the company which he now kept in Madrid, all pushed him to take for granted his own importance as a national figure. As he put it in retrospect 'I was, as a result of my age and my prestige, called to render the highest services to the nation'. The Army's apparent political success under Primo de Rivera may also have increased his tendency to dream of higher things for himself. He claimed later that, in preparation for his transcendental tasks, and taking advantage of the fact that his command in Madrid left him with very little to do, he began to read books on contemporary Spanish history and political economy.[67] How much reading he did is impossible to say; his books were lost in Madrid in 1936 when his flat was ransacked by anarchists. Certainly, neither his speeches nor his own writings indicate any significant insight into history or economics.

Given his propensity to chat, he probably talked rather than read about economics. As he claimed later, he started at this time 'with some frequency to visit the manager of the Banco de Bilbao, where Carmen had a few savings (*unos ahorrillos*)'. The banker in question was affable and intelligent and stimulated in Franco an interest in economics. Franco also discussed contemporary political issues with his immediate circle of friends and acquaintances. It is likely that such café conversations with friends, the bulk of whom were *Africanistas* like himself, can only have cemented his prejudices. Nevertheless, in later life he was to place enormous value on these conversations.[68]

His reading and his *tertulias* boosted Franco's confidence in his own opinions to an inordinate degree. While on holiday in Gijón in 1929, he bumped into General Primo de Rivera on the beach. The ministers of Primo's government were spending a few days together away from Madrid and the Dictator invited Franco to lunch with them – a mark of considerable favour by Primo towards the young general. His self-esteem duly inflated, Franco found himself seated next to José Calvo Sotelo, the brilliant Finance Minister, who was in the midst of trying to defend the value of the peseta against the consequences of a massive balance of payments deficit, a bad harvest and the first signs of the great Depression. Franco assured an intensely irritated Calvo Sotelo that there was no point in using Spain's gold and foreign currency reserves to support the value of the peseta and that the money so used would be better spent on industrial investment. The reasoning by which Franco reached the interpretation he put before the Minister revealed a simple cunning: he based his argument on the belief that there need be no link between the exchange rate of a currency and the nation's gold and foreign reserves provided their value were kept secret.[69]

The economic difficulties discussed at this lunch were not the only problems besetting the Dictatorship. The military was deeply divided and some sections of the Army were turning against the regime. Franco was paradoxically to be the beneficiary of one of the most serious errors made by the Dictator in this regard. Primo de Rivera was anxious to reform the antiquated structures of the Spanish Army and in particular to slim down the inflated officer corps. His ideal was a small professional Army but, as a result of the reversal of his original policy of *abandonismo* in Morocco, it had grown significantly in size and cost in the mid-1920s. By 1930, the officer corps would be reduced in size by only about 10 per cent and the Army as a whole by more than 25 per cent, at an inordinately high price in terms of internal military discontent. Large sums were spent on efforts at modernization although the final increase in the number of mechanized units was immensely disappointing.[70]

The relative failure of Primo's technical reforms was overshadowed by the legacy of one bitterly divisive issue. Most publicity was generated, and most damage caused in terms of morale, by the Dictator's efforts to eradicate the divisions between the artillery and the infantry over promotions. To a large extent, this was the question which had given birth to the *Juntas de Defensa* in 1917. Divisions between the infantry, and particularly the *Africanistas*, on the one hand, and the artillery and the engineers on the other arose from the fact that it was much more difficult for an engineer or the commander of an artillery battery to gain promotion by merit than for an infantry officer leading charges against the Moors. To underline their discontent with a promotion system which favoured the colonial infantry, the Artillery corps had sworn in 1901 to accept no promotions which were not granted on grounds of strict seniority and to seek instead other rewards or decorations.

Although on coming to power Primo de Rivera had been thought within the Army to be more sympathetic to the artillery position, he seems to have changed his mind as a result of his contacts with the infantry officer corps in Morocco before and during the Alhucemas operation.[71] By decrees of 21 October 1925 and 30 January 1926, he introduced greater flexibility into the promotion system. This gave him the freedom to promote brave or capable officers but it was also perceived as opening a Pandora's box of favouritism. There was already tension when, in a typically precipitate manner, on 9 June 1926, the Dictator

issued a decree specifically obliging the artillery to accept the principle
of promotions by merit. Those who had accepted medals instead of promotions were now deemed retrospectively to have been promoted. Hostility within the mainland officer corps to a whole range of tactless
encroachments on military sensibilities by the Dictator was already leading to contacts between some officers and the liberal opposition to the
regime. It came to a head in a feeble attempt at a coup known as the
Sanjuanada on 24 June 1926.[72] In August, the imposition of promotions
upon the artillery provoked a near mutiny by artillery officers who confined themselves to their barracks. In Pamplona, shots were fired by
infantrymen sent to put an end to one such 'strike' of artillerymen. The
Director of the Artillery Academy of Segovia was sentenced to death, a
sentence later commuted to life imprisonment, for refusing to hand over
the Academy.[73] Throughout the issue, Franco was careful not to get
involved. He, more than anyone in the entire armed forces, had reason
to be grateful to the system of promotions by merit.

Primo de Rivera won, but at the cost of dividing the Army and of
undermining its loyalty to the King. His policy on promotions was to
provide much of the cause for the grievances which lay behind some
officers moving in the direction of the Republican movement. Thus,
when the time came, some sectors of the Army would be ready to stand
aside and permit first Primo's own demise and then the coming of the
Second Republic in April 1931.[74] Broadly speaking, the *Africanistas*
remained committed to the Dictatorship and thereafter were to be bitterly
hostile to the democratic Republic which followed it in 1931.[75] Indeed,
the fault lines of the divisions created in the 1920s would run right
through to the Civil War in 1936. Many of those who moved into opposition against Primo would be favoured by the subsequent Republican
regime. In contrast, the *Africanistas*, including Franco, would see their
previously privileged position dismantled.

The artillery/infantry, *juntero/Africanista*, issue had an immediate and
direct impact on Franco's life. In 1926, the Dictator was convinced that
part of the promotions problem derived from the fact that there were
separate academies for the officers of the four major corps, the infantry
in Toledo, the artillery in Segovia, the cavalry in Valladolid and the
engineers in Guadalajara. He concluded that Spain needed a single
General Military Academy and decided to revive the Academia General
Militar which had existed briefly during its so-called 'first epoch' between
1882 and 1893.[76] By this time, and particularly after Alhucemas, Primo
had developed a great liking for Franco. He told Calvo Sotelo that Franco

was 'a formidable chap, and he has an enormous future not only because
of his purely military abilities but also because of his intellectual ones'.[77]
The Dictator was clearly grooming Franco for an important post. He
sent him to the École Militaire de St Cyr, then directed by Philippe
Pétain, in order to examine its structure. On 20 February 1927, Alfonso
XIII approved a plan for a similar Spanish academy, and on 14 March
1927 Franco was made a member of a commission to prepare the way
for it. By Royal Decree of 4 January 1928, he was appointed its first
director. He expressed a preference for it to be sited at El Escorial but
the Dictator insisted that it be in Zaragoza. Years later, Franco was
alleged to have said that, if the Academy had been located at El Escorial
instead of 350 kilometres from the capital, the fall of the monarchy in
1931 could have been avoided.[78]

In moving to the Academia General Militar, Franco was leaving behind
him the kind of soldiering in which he made his reputation. Never again
would he lead units of assault troops in the field. It was a major change,
which taken with his marriage in 1923 and the birth of his daughter in
1926, would affect him profoundly. Until 1926, Franco was an heroic
field soldier, an outstanding column commander, fearless if not reckless.
Henceforth, as befitted his changing sense of his public *persona*, he would
take ever fewer risks. In Morocco, he had been a ruthless disciplinarian,
an abstemious and isolated individual with few friends.[79] After his return
to the Peninsula, he seems to have relaxed slightly, although he was
always to remain obsessed with the primacy of unquestioning obedience
and discipline. He became readier to turn a blind eye to laziness or
incompetence in his subordinates, getting the best out of willing collaborators by manipulation and rewards. He became a relatively convivial
frequenter of clubs and cafés where he would take an aperitif and give
rein to his inclination to chat, recounting anecdotes and reminiscences
among a group of military friends.[80]

Until the late 1920s, he showed few signs of being the archetypal
gallego, slow, cunning and opaque, of his later years. He was a man of
action, obsessed with his military career and little else. His early military
writings are relatively straightforward and decently written, with some
sensitivity to people and places. He was, of course, reserved, and predis-
posed by his military experience, and particularly by Africa, to certain
political ideas, hostile to the Left and to regional autonomy movements.
If he did read about politics, economics and recent history, it was probably
more to confirm his prejudices than in search of enlightenment. From
this time, a convoluted style and a pomposity of tone begins to be discern-

ible in his speeches. In part, family responsibilities account for a greater caution but the more potent motive for his self-regard was a perception of his potential political importance. He was the object of public adulation in certain circles and had had plenty of indications that he was the general with the most brilliant prospects.[81] He was showered with promotions, honours and plum postings. The talk of his being the youngest general in Europe cannot have failed to have affected him, as must the idea of providence watching over him, an idea particularly dear to his wife. To her influence in this respect must be added that of his near inseparable cousin, Pacón, now a major, who had become his ADC in the late summer of 1926.[82]

At the end of May 1929 there appeared in the magazine *Estampa*, in the section called 'The woman in the home of famous men', a rare interview with Carmen Polo and her husband. Conducted by Luis Franco de Espés, the Barón de Mora, a fervent admirer of Franco, the interview was as much concerned with 'the famous man' as with 'the woman in the home'. Asked if he was satisfied to be what he was, Franco replied sententiously 'I am satisfied to have served my fatherland to the full'. The Barón asked him what he would have liked to be if not a soldier to which he replied 'architect or naval officer. However, aged fourteen I entered the Infantry Academy in Toledo against the will of my father.' This was the first time that Franco had indicated any paternal opposition to his joining the military academy. There is no reason why his father should have opposed the move and, if he had done, there can be little doubt that he would have imposed his will. Apparently, Franco was trying to put distance between his beloved military career and his hated father.

'All this', he said, 'is only with regard to my profession because my real inclination has always been towards painting'. On lamenting that he had no time to practice any particular genre, Carmen interrupted to point out that he painted rag dolls for their daughter, 'Nenuca'. Then, the interview turned to the 'the beautiful companion of the general, hiding the supreme delicacy of her figure behind a subtle dress of black crêpe'. Blushing, she recounted how she and her husband had fallen in love at a *romería* (country fair) and how he had pursued her doggedly thereafter. Playing the role of the faithful hand-maiden to the great man, she revealed her husband's major defects to be that 'he likes Africa too much and he studies books which I don't understand'. Turning back to Franco, the Barón de Mora asked him about the three greatest moments of his life to which he responded with 'the day that the Spanish Army landed at Alhucemas, the moment of reading that Ramón had reached Pernambuco

and the day we got married'. The fact that the birth of his daughter
Carmen did not figure in the list suggests that he was more anxious to
project an image of patriotism untrammelled by 'unmanly' emotions. He
was then asked about his greatest ambition which he revealed as being
'that Spain should become as great again as she was once before.' Asked
if he was political, Franco replied firmly 'I am a soldier' and declared
that his most fervent desire was 'to pass unnoticed. I am very grateful for
certain demonstrations of popularity but you can imagine how annoying it
is to feel that you're often being looked at and talked about'. Carmen
listed her greatest love as music and her greatest dislike as 'the Moors'.
She had few happy memories of her time as an Army wife in Morocco
spent consoling widows.[83]

Franco had arrived in Zaragoza on 1 December 1927 to supervise
the building and installation of the new institution. The first entrance
examinations were held in June 1928. On 5 October of that year, with
the new buildings still unfinished, the Academy opened for its first intake
in a nearby barracks. The new Director's speech on opening the Academy
reflected the philosophy that he had learned from his mother. Its theme
was 'he who suffers overcomes'.[84] He also instructed the cadets to follow
the 'ten commandments' or 'decálogo' which he had compiled on the
basis of a similar 'decálogo' elaborated for the Legion by Millán Astray.
Expressed in the most sententious terms, the commandments were: 1)
Make great love for the Fatherland and fidelity to the King manifest
in every act of your life; 2) Let a great military spirit be reflected in your
vocation and your discipline; 3) Link to your pure chivalry a constant
jealous concern for your reputation; 4) Be faithful in the fulfilment of your
duties, being scrupulous in everything that you do; 5) Never grumble, nor
tolerate others doing so; 6) Make yourself loved by those of lower rank
and highly regarded by your superiors; 7) Volunteer for every sacrifice
at times of greatest risk and difficulty; 8) Feel a noble comradeship,
sacrificing yourself for your comrades and taking delight in their successes, prizes and progress; 9) Love responsibility and be decisive; 10) Be
brave and self-denying.[85]

The generation educated under Franco's close supervision at the Academia General Militar de Zaragoza, in its so-called second epoch between
1928 and 1931, was to receive significantly more practical training than
had hitherto been the practice in the Toledo infantry academy. Franco
insisted that no textbooks be used and that all classes be based on the
practical experiences of the instructors.[86] Skill in the use and care of
weapons was insisted upon. The horsemanship of the graduates was of a

high standard. Franco himself would direct from horseback the toughest manoeuvres. However, the central stress, derived from the *decálogo*, was on 'moral' values: patriotism, loyalty to the King, military discipline, sacrifice, bravery.[87] The idea that 'moral' values could triumph over superior numbers or technology was one of the constant refrains of Franco's military thought. Reflecting the Director's own experiences in the primitive Moroccan war, the level of tactical and technological education at Zaragoza was not highly advanced and considerable effort went into denouncing democratic politics.

During the Civil War, officers who had trained at the Academy under Franco remembered him as a martinet who had laid traps for unwary cadets. In the streets of Zaragoza, he would pretend to be looking in shop windows to catch those who tried to get past without saluting their Director. As they went on, they would be called back by Franco's soft, high, feared, voice. Remembering the nightly activities of his own contemporaries at Toledo, he insisted that all cadets carry at least one condom while walking in the city. Occasionally, he would stop them in the street and demand to see their protective equipment. There were strict penalties for those unable to produce it.[88] In his farewell speech to the Academy in 1931, he listed among the great patriotic achievements of his time in the post the elimination of venereal disease among the cadets through 'vigilance and prophylaxis'.[89] His pride in that achievement was reflected when, in 1936, he boasted to his English teacher that he had 'put down vice ruthlessly' among the cadets at Zaragoza.*[90]

Franco's period at the Academy was viewed in retrospect as a triumph by *Africanistas* and other right-wing Army officers and a disaster by liberal and left-wing officers. His brother Ramón wrote to him to complain of the 'troglodytic education' imparted there. For the distinguished *Africanista*, General Emilio Mola, in contrast, it was the peak of excellence.[91] The Academy's regulations demanded that the teaching staff be chosen on the basis of *méritos de guerra*, irrespective of the subject being taught. Accordingly, the teaching staff was dominated by *Africanista* friends of Franco, most of whom had been brutalized by their experiences in a pitiless colonial war and were noted more for their ideological rigidity than for their intellectual attainments. Of 79 teachers, 34 were infantrymen and 11 from the Legion. The assistant director of the Academy was Colonel Miguel Campins, a good friend and comrade in arms from

---

* It would be an abiding obsession. On a visit to the Zaragoza Military Academy in 1942, he told one of the staff that an additional bed should be put in rooms that had two 'to avoid marriages' – Baón, *La cara humana*, p. 117.

Africa who had been with him at the battle of Alhucemas. A highly competent professional, Campins elaborated the training programme at the Academy.[92] The other senior members of staff included Emilio Esteban-Infantes, later to be involved in the attempted Sanjurjo coup of 1932; Bartolomé Barba-Hernández, who was to be, on the eve of the Civil War, leader of the conspiratorial organization Unión Militar Española; and Franco's lifelong close friend Camilo Alonso Vega, later to be a dour Minister of the Interior. Virtually without exception, the Academy's teachers were to play prominent roles in the military uprising of 1936. With such men on the staff, the Academy concentrated on inculcating the ruthless arrogance of the Foreign Legion, the idea that the Army was the supreme arbiter of the nation's political destiny and a sense of discipline and blind obedience. A high proportion of the officers who passed through the Academy were later to be involved in the Falange. An even higher proportion fought on the Nationalist side during the Civil War.[93]

During his period at the head of the Military Academy, Franco developed the *dejar hacer* (turning a blind eye) style of delegation which was to be taken to extremes when he was Head of State. Those of the teaching staff who did not pull their weight were not punished but nor were they favoured. Those who had an enthusiasm or a speciality were allowed full initiative in that area – the instructor who liked football delegated to coach the team, the one who liked gardening given control of the Academy gardens, the amateur photographer put in charge of the dark room. Of the lazy or incompetent, Franco would simply comment '*A Fulano, no le veo la gracia*' (I don't see what So-and-So has going for him) but would never reprimand those who did not pull their weight (*arrimar el hombro* – a favourite phrase of Franco's).

Franco's arrival in Zaragoza provoked considerable popular attention. The Academy, the Director and his senior staff became a major focus of local social life and Franco indulged his penchant for socializing and for interminable late-night after-dinner *tertulias* with military friends and minor aristocrats. Encouraged by Doña Carmen, he began to mix with the dominant families of the local establishment. It perhaps reflected Franco's own small-town and lower middle class origins that he always preferred provincial social life, in Oviedo, Ceuta or Zaragoza, to that of Madrid.[94] Even so, contemporary photographs of Franco in evening dress or lounge suit show him significantly less at ease than when in uniform. He was happier hunting. Far from his African exertions, he turned increasingly to hunting for exercise, pleasure and, it may be supposed, as an outlet for his aggression.

It was during his period in Zaragoza that Franco began to intensify his anti-Communist and authoritarian ideas. Shortly before leaving Madrid for Zaragoza, he had been given, along with several other young officers, a subscription to a journal of anti-Comintern affairs from Geneva, the *Bulletin de L'Entente Internationale contre la Troisième Internationale*. The Entente, founded by the Swiss rightist Théodore Aubert and the White Russian emigré Georges Lodygensky, was vehemently anti-Bolshevik and praised the achievements of fascism and military dictatorships as bulwarks against Communism. An emissary from the Entente, Colonel Odier, visited Madrid and arranged with General Primo de Rivera for several subscriptions to be purchased by the Ministry of War and to be distributed to a few key officers.[95] It clinched what was to be a lifelong obsession with anti-Communism. It also played its part in the transition of Franco from the adventurous soldier of the 1920s to the suspicious and conservative general of the 1930s. Receiving the bulletin uninterruptedly until 1936, he came to see the Communist threat everywhere and to believe that the entire Spanish Left was wittingly or unwittingly working in the interests of the Comintern. In 1965, Franco revealed to both Brian Crozier and George Hills the influence that the Entente had had over him. He told Hills that the Entente had alerted him to the need to be ready for the flank attack from the invisible (Communist) enemy. Indeed, he left Crozier with the impression that his acquaintance with its work was an event in his life equal in importance in its impact on him to the birth of Nenuca.[96]

Another influence in Franco's life was initiated as a result of an invitation in the spring of 1929 to the German Army's General Infantry Academy in Dresden. He was thrilled by the organization and discipline of the German Army. On his return, he made it clear to his cousin Pacón that he had been especially impressed by the Academy's cult of reverence for the regiments which had achieved the great German military triumphs of the recent past. He was particularly sympathetic to German efforts to break free of the shackles of the Versailles Treaty.[97] It was the beginning of a love affair which would intensify during the Civil War, reach its peak in 1940, and not begin to die until 1945.

The Dictatorship fell on 30 January 1930. The bluff Primo de Rivera had ruled by a form of personal improvisation which had ensured that he would bear the blame for the regime's failures. By 1930, there was barely a section of Spanish society which he had not estranged. He had offended Catalan industrialists both by his anti-Catalanism and because of the rise in raw material prices in the wake of the collapse in value of the peseta. He had outraged landowners by trying to introduce paternalist labour legislation for land-workers. The Socialist Unión General Traba-jadores had supported him as long as public works projects had kept up levels of employment. With the coming of the slump, many Socialists had allied with the banned anarcho-syndicalist union, the Confederación Nacional del Trabajo, in opposition. Most damagingly, the divisions in the Army provoked by Primo's promotions policy were instrumental in the Captains-General and the King withdrawing support for the regime. Unlike most twentieth century dictators, Primo withdrew quietly once he had recognised that his support had disappeared. He went into exile in Paris where he died on 16 March 1930. A return to the pre-1923 constitutional system was impossible, not least because the King could no longer count on the loyalty of the old monarchist political élite which he had so irresponsibly abandoned in favour of Primo. Alfonso XIII was forced to seek another general. His choice of General Dámaso Berenguer, irrevocably associated with the disaster of Annual, infuriated the Left. For nearly a year, Berenguer's mild dictatorship, the so-called *Dictablanda*, would flounder along in search of formula for a return to the consti-tutional monarchy. A combination of working class agitation fuelled by the economic depression, military sedition provoked by Primo's policies, and republican conspiracy ensured Berenguer's eventual failure.

The fall of the Dictator disappointed Franco but little more: he was oblivious to the implicit threat to the monarchy itself. Among Franco's staff, the artillerymen and engineers were understandably pleased by Primo's demise. However, Franco ensured that the demise of Primo would provoke no public clashes in the Academy between *junteros* and *Africanistas* by imposing an iron ban on speaking about politics.[98] By withdrawing his confidence from Primo, the King also lost the loyalty of General Sanjurjo, now Director-General of the Civil Guard. Franco did not blame the King for the fall of the Dictatorship. In any case, he was the object of special attention, not to say flattery, from Alfonso XIII. On 4 June 1929, in a solemn ceremony in the Madrid Retiro, the King had personally presented him with the Medalla Militar which he had won in 1925.[99] On 5 June 1930, Alfonso XIII visited the Academy and three days later Franco took the entire body of cadets to the capital to take part in the swearing of the flag by the Madrid garrison. Led by Franco on a prancing horse, they headed the parade, to the wild applause of those present. On the following day, the cadets took the guard at the Royal

Palace and Franco appeared on the balcony with the King. The crowd on that day included several hundred members of the *Juventud Monárquica* (monarchist youth), who would soon form the élite of the conservative extreme right during the Republic.[100]

Accordingly, it was a cause of the greatest embarrassment to Franco that his brother Ramón had moved into the orbit of the republican opposition to the regime. From the later part of 1929, their relations became very strained. Franco had been annoyed and embarrassed in July 1924 when Ramón had married Carmen Díaz Guisasola without seeking the King's permission.[101] The breach between his brother and the King had been forgotten in the wake of his Atlantic crossing in 1926. However, Ramón's ever more frantic efforts to repeat that success had ended in disgrace. The reasons for his fall from grace were complex. In the summer of 1929, to boost the domestic aircraft industry, the Spanish government agreed to sponsor an attempt by Ramón to cross the North Atlantic in a Dornier Super Wal flying boat built under licence in Spain. Because of doubts about the reliability of the Spanish aeroplane, Ramón used a German-built one bought in Italy, fraudulently switching the registration markings. The flight was a disaster: the aircraft was blown off course near the Azores, and it and the crew were lost for days and only found at the end of June after a massive and immensely costly search involving the Spanish, British and Italian navies.[102] When he was found, there was widespread rejoicing and a tearful General Franco was publicly embraced by an equally lacrimose General Primo de Rivera.[103] Franco led a massive demonstration to the British Embassy in Madrid to express thanks for the role of the Royal Navy.[104] It then emerged that the planes had been switched and rumours began to circulate that Ramón had been promised a fabulous sum of money if he broke the world seaplane distance record flying a German aircraft. Colonel Alfredo Kindelán, the head of Military Aviation, was furious and had Ramón expelled from the Air Force on 31 July 1929. Thereafter, he moved rapidly to the left, became a free-mason and got involved in anarcho-syndicalist conspiracies aimed at bringing down the monarchy.[105]

After this disgrace, Ramón's relations with his brother were virtually non-existent and were reduced to letters; patronizing, sententious, though ultimately kindly ones from Franco, mischievously disrespectful ones from Ramón. On 8 April 1930, Franco wrote a long letter to Ramón revealing of his loyalty both to his family and to the established order. In an effort to head off his brother's demise, Franco warned him that his activities within the Army, inciting garrisons and officers to rebel, were known to the authorities. Regarding the Berenguer regime as entirely legal, Franco was worried that his brother was risking the loss of his prestige and his good name. He appealed to him to think of 'the great sorrow that such things cause Mamá, a sorrow which the rest of us share' and ended fondly, 'Your brother loves and embraces you, Paco'.[106]

Its tone of tolerant restraint is remarkable given that, in Francisco's eyes, Ramón's behaviour would not only bring dishonour on the family but also possibly impede his own chances of advancement. There is also a typical readiness to attribute the lowest motives to Ramón's revolutionary friends while assuming that Ramón himself is free of such baseness. The letter also revealed a political naïvety in Franco's suggestion that the dictatorship of General Berenguer was more legal than that of Primo de Rivera. Ramón was not slow to comment on that in his reply on 12 April. Ramón was shocked by what he called his brother's 'healthy advice' and 'vain bourgeois counsels' and invited him to step down from his 'little general's throne'. He also took the opportunity to comment that the education being given the cadets in Zaragoza would ensure that they would be bad citizens.[107]

Engrossed in his work at the Zaragoza military academy, Franco paid little attention to the rising tide of political agitation in 1930 except in so far as it involved his brother. The anti-monarchical movement was growing with labour unrest intensifying by the day. A broad front of Socialists, middle class Republicans, Basque and Catalan regionalists and renegade monarchists who, repelled by the mistakes of the King, had become conservative republicans, joined together in mid-August 1930. United by the so-called Pact of San Sebastián, they established a provisional government-in-waiting which began to plot the downfall of the monarchy.[108] Ramón Franco was an important element in the republican conspiracies. In late 1930, watched by agents of the Dirección General de Seguridad, he was travelling around Spain liaising with other conspirators, trying to buy arms and organizing the making of bombs.[109] General Emilio Mola, now Director-General de Seguridad, had taken the decision to arrest him but, as an admirer of his heroic exploits and as a friend of Franco, he decided to give Ramón a last chance to avoid the consequences of his activities. Mola asked Franco to try to persuade his brother to desist. Although he agreed to try, Franco showed no optimism that he might succeed but he was immensely faithful to the family and still felt a protective loyalty towards his madcap brother. He visited Madrid and they dined together on 10 October but Ramón remained committed to the planned republican rising. Mola then had Ramón brought in for questioning on the evening of 11 October and detained in military prison on the following morning. Mola again called Franco in and informed him of the charges against his brother which included bomb-making, gun-smuggling and involvement in the attempted murder of a monarchist aviator, the Duque de Esmera. Franco and Mola hoped to use these charges to frighten Ramón into abandoning his revolutionary activities: Franco visited his brother in his cell and recited them to him. This merely provoked him into escaping from prison on 25 November. Thereafter, he took part, with General Queipo de Llano, in the revolutionary movement of mid-December 1930. Both Ramón's escape and his participation in the events of December would cause Franco intense chagrin both as an officer and as a monarchist.[110]

Having failed in his efforts to make his brother see sense, Francisco returned hastily to Zaragoza where he had to receive the visit of a French delegation led by André Maginot. On 19 October, Maginot presented Franco with the Légion d'Honneur for his part in the Alhucemas landing. On his return to France, he declared that the Zaragoza Academy was the most modern of its kind in the world.[111] Maginot's ideas of modernity had yet to be put to the test by the armies of the Third Reich.

In November, Franco was approached by an emissary from the most prominent figure of the San Sebastián coalition, the grand old man of Spanish republicanism, the wily and cynical Alejandro Lerroux. He was invited to join in the Republican conspiracies along with so many other officers including his brother. According to Lerroux, Franco refused point blank but then insinuated, at a later meeting, that he *would* rebel against the constituted power but only if the *Patria* were in danger of being overwhelmed by anarchy.[112] Despite warnings from his cousin Pacón and the attitude of his brother, Franco was so far distanced from day-to-day politics that he was convinced that the monarchy was in no danger.[113]

The revolutionary plot in which Ramón was implicated aimed to bring the San Sebastián provisional government to power. One of its ramifications was to be a rebellion by the garrison of the tiny Pyrenean mountain town of Jaca in the province of Huesca. Anticipating what was supposed to be a nationally co-ordinated action, the Jaca rebellion was precipitated on 12 December. Its leaders, Captains Fermín Galán, Angel García Hernández and Salvador Sediles, hoped to march south from Jaca and spark off a pro-Republican movement in the garrisons of Huesca, Zaragoza and Lérida.[114] Along the road to Huesca, Galán's column was challenged by a small group of soldiers led by the military governor of Huesca, General Manuel Lasheras, who was wounded in the clash. When the news of the actions of the Jaca rebels reached Madrid in the early hours of the morning of 13 December, the government declared martial law in the entire Aragonese military region. A sporadic general strike broke out in Zaragoza. Franco put the Academy in a state of readiness and armed the cadets. The Captain-General of the Aragonese military region, General Fernández de Heredia, put together a large column and sent them to Huesca, half way between Zaragoza and Jaca. In case the rebels should have left Huesca already and headed south, he ordered Franco to use his cadets to hold the Huesca-Zaragoza road. In the event, it was not necessary. Galán's cold, wet and hungry column was stopped at Cillas, three kilometres from Huesca, and the Jaca revolt was put down.[115]

Galán and García Hernández were seen as being the two ringleaders and were shot after summary courts martial on 14 December.[116] As far as Franco was concerned, their punishment was entirely appropriate since they were mutineers. He was perhaps fortunate that he did not have to make similar considerations about his brother, who was heavily involved in the central action of the plot in the capital. On 15 December, Ramón had flown over the royal Palacio de Oriente in Madrid, planning to bomb it but, in the event, seeing civilians strolling in the gardens, had merely dropped leaflets calling for a general strike. He had then fled to Portugal and then on to Paris.[117] Franco did not vacillate in his condemnation of the revolutionary events of mid-December, but his sense of family solidarity prevented him applying the same standards to his brother. Hours after Ramón's flight over the Palacio Real, another aircraft flew over Madrid and dropped leaflets directed at the city's inhabitants denouncing Ramón as a 'bastard apparently drunk on your blood'. Franco was so incensed on behalf of his mother (if not his brother) that he left Zaragoza for Madrid where he demanded explanations from Berenguer, the Head of the Government, General Federico Berenguer, the Captain-General of Madrid and Mola, the Director-General of Security, all of whom assured him that the flight and the pamphlets had no official status.[118]

On 21 December, Franco sent another letter to Ramón. Not surprisingly, in the light of the scandal that Ramón's activities had occasioned, the distress of their mother and the fact that he was in danger of being shot, the letter is deeply sorrowful. Despite the gulf between their political views, Francisco showed compassionate concern for 'My beloved and unfortunate brother' and enclosed two thousand pesetas. He ended sanctimoniously 'May you break away from the vice-ridden ambience in which you have lived for the last two years, in which the hatred and the passion of the people who surround you deceive you in your chimeras. May your forced exile from our *Patria* calm your spirit and lift you above all passions and egoisms. May you rebuild your life far from these sterile struggles which fill Spain with misfortunes. And may you find well-being and peace in your path. These are the wishes of your brother who embraces you.' The money which accompanied the letter was a substantial sum at the time. Grateful as Ramón was for his brother's help, he was repelled by his reactionary notions and surprised by his lack of awareness of the tide of popular feeling.[119]

If Franco had any doubts about the legitimacy of the executions of Galán and García Hernández, they would have been resolved on 26 December when General Lasheras died from an infection and uraemia which may have been related to the wound that he had received when trying to stop Galán. Franco attended his funeral.[120] The public outcry about the execution of Galán and García Hernández damaged the monarchy in a way that the Jaca revolt itself had failed to do. As the two executed rebels were being turned into martyrs, to the outrage of many senior military figures including Franco, the Liberals in the government withdrew their support and General Berenguer was obliged to resign on 14 February.[121] After an abortive attempt by the Conservative politician José Sánchez Guerra to form a government with the support of the imprisoned Republican leaders, Berenguer was finally replaced as prime minister on 17 February by Admiral Juan Bautista Aznar. He did, however, continue in the cabinet as Minister for the Army.[122]

Since the Jaca rebellion of Galán and García Hernández had taken place within the military region of Aragón, Franco was appointed a member of the tribunal which was to court martial Captain Salvador Sediles and other officers and men who had been involved. It took place between 13 and 16 March when the campaign for the municipal elections of 12 April had already begun. There was no more potent subject during that campaign than that of the executions of Galán and García Hernández. Admiral Aznar declared in advance of the verdicts in the supplementary court martial that he was of a mind to ask the King for clemency whatever the sentences. Franco, however, declared: 'it is necessary that military crimes committed by soldiers be judged by soldiers who are accustomed to command', within which category he clearly included a readiness to punish indiscipline by death. In the event, there was one more death sentence, for Captain Sediles, five life sentences and other lesser sentences, all of which were commuted.[123]

In the municipal elections of 12 April 1931, Franco voted for the monarchist candidacy in Zaragoza.[124] The results would go against Alfonso XIII, provoke his withdrawal from Spain and open the way to the establishment of the Second Republic. For Franco, the deeply conservative monarchist and royal favourite, it would be a severe shock. To the ambitious young general, it would seem to be the end of a meteoric rise. That fact, taken with Franco's prominence in the military uprising of 1936, has led the Caudillo's hagiographers to portray him as working towards that glorious denouement from the very first. This was far from being the case. Franco had still to undergo many experiences before he became an implacable enemy of the Republic.

Ironically, in early 1931, there was an event in Franco's personal life which was to reveal its full significance only in 1936. In 1929, the Director of the Military Academy had met a brilliant lawyer, Ramón Serrano Suñer, who was working in Zaragoza as a member of the élite legal corps of *Abogados del Estado* (State lawyers) and they had become friends. Serrano Suñer often lunched or dined with the Franco family.[125] As a result, Serrano Suñer came to know Doña Carmen's beautiful younger sister, Zita. In February 1931, Serrano Suñer married her, then aged nineteen, in Oviedo. The groom's witness was José Antonio Primo de Rivera, son of the Dictator and future founder of the Falange, the bride's Francisco Franco.[126] The marriage clinched the close relationship between Serrano Suñer and Franco out of which would be forged the Caudillo's National-Syndicalist State. The wedding ceremony also provided the occasion for a historic first meeting for the eventual dictator and fascist leader whose names were to be tied together for forty years after 1936. At the time, none of the three could have had any idea of the imminent political cataclysm which would link their fates.

III

IN THE COLD

Franco and the Second Republic, 1931–1933

THE MUNICIPAL elections of 12 April 1931 were intended by the government to be the first stage of a controlled return to constitutional normality after the collapse of the Primo de Rivera dictatorship. However, on the evening of polling day, as the results began to be known, people started to drift onto the streets of the cities of Spain and, as the crowds grew, Republican slogans were shouted with increasing excitement. In the countryside, the power of the local bosses or *caciques* was unbroken but in the towns, where the vote was much freer, monarchist candidates had suffered a disaster. With the artillerymen on his staff at the Academy openly rejoicing at the Republican triumph, Franco was deeply worried about the situation.[1] While he mused in his office in Zaragoza, his one-time commanding officer and a man whom he admired, General Sanjurjo, was clinching the fate of the King. Sanjurjo now Director-General of the para-military Civil Guard, the monarchy's most powerful instrument of repression, had informed several cabinet ministers that he could not guarantee the loyalty of the men under his command in the event of mass demonstrations against the monarchy.[2] In fact, there was little reason to suspect the loyalty of the Civil Guard, a brutal and conservative force. Sanjurjo's fear was rather that the defence of the monarchy could be attempted only at the cost of copious bloodshed, given the scale of the popular hostility to the King.

That Sanjurjo was not prepared to risk a bloodbath on behalf of Alfonso XIII reflected the fact that he had personal reasons for feeling resentment towards the King. He felt that he had been snubbed by the King for marrying beneath his rank and he had not forgiven Alfonso XIII for failing to stand by Primo de Rivera in January 1930.[3] Sanjurjo's reluctance to defend his King may also have reflected two conversations that he had with Alejandro Lerroux in February and April 1931, during which the Republican leader had tried to persuade him to ensure the benevolent neutrality of the Civil Guard during a change of regime. Sanjurjo informed the Director-General of Security, General Mola, of the first of these meetings and assured him that he had not agreed to Lerroux's request.[4] His subsequent actions during the crisis of 12, 13 and 14 April, together with the favourable treatment which he received afterwards from the new regime, were to lead Franco to suspect that perhaps Sanjurjo had been bought by Lerroux and betrayed the monarchy.

Franco was unaware of what Sanjurjo was saying to the cabinet ministers on 12 April, but he was in telephone contact with Millán Astray and other generals. He considered marching on Madrid with the cadets from the Academia but refrained from doing so after a telephone conversation with Millán Astray at 11.00 a.m. on the morning of 13 April.[5] Millán Astray asked him if he thought that the King should fight to keep his throne. Franco replied that everything depended on the attitude of the Civil Guard. For the next five and a half years, the stance of the Civil Guard would be Franco's first concern in thinking about any kind of military intervention in politics. Most of the Spanish Army, apart from its Moroccan contingent, was made up of untried conscripts. Franco was always to be intensely aware of the problems of using them against the hardened professionals of the Civil Guard. Now, Millán Astray told Franco that Sanjurjo had confided in him that the Civil Guard could not be relied upon and that Alfonso XIII therefore had no choice but to leave Spain. Franco commented that, in view of what Sanjurjo said, he too thought that the King should go.[6]

Franco had also been greatly influenced by the telegram that Berenguer sent in the early hours of 13 April to the Captains-General of Spain. The Captains-General in command of the eight military regions into which the country was divided were effectively viceroys. In the telegram, Berenguer instructed them to keep calm, maintain the discipline of the men under their command and ensure that no acts of violence impede 'the logical course that the supreme national will imposes on the destinies of the Fatherland'.[7] Berenguer's attitude derived from his own pessimism about Army morale. He believed that some Army officers were simply blasé about the danger to the monarchy. More seriously, he suspected that many others were indifferent and even hostile to its fate in the wake of the divisions created in the 1920s. Nevertheless, despite his telegram and his own inner misgivings, on the morning of 14 April, out of loyalty to the monarchy, Berenguer told the King that the Army was ready to overturn the result of the elections. Alfonso XIII refused.[8] Shortly after Berenguer's interview with the King, Millán Astray told Berenguer about his conversation with the Director of the Zaragoza Academy on the previous day repeating, as 'an opinion which has to be taken into account', Franco's view that the King had no choice but to leave.[9]

The King decided to leave Spain but not to abdicate, in the hope that his followers might be able to engineer a situation in which he would be begged to return. Power was assumed on 14 April 1931 by the Provisional Government whose membership had been agreed in August 1930 by the Republicans and Socialists who had made the Pact of San Sebastián. Although led by Niceto Alcalá Zamora, a conservative Catholic landowner from Córdoba who had once been a Minister under the King, the Provisional Government was dominated by Socialists and centre and left Republicans committed to sweeping reform.

In a number of ways in the first week of the Republic, Franco displayed unmistakably, if guardedly, a repugnance for the new regime and a lingering loyalty to the old. There was nothing unusual in his feeling such loyalty – a majority of Army officers were monarchists and would have been unlikely to change their convictions overnight. Franco was ambitious but took discipline and hierarchy very seriously. On 15 April, he issued an order to the cadets, in which he announced the establishment of the Republic and insisted on rigid discipline: 'If discipline and total obedience to orders have been the invariable practice in this Centre, they are even more necessary today when the Army is obliged, with serenity and unity, to sacrifice its thoughts and its ideology for the good of the nation and the tranquility of the *Patria*.'[10] It was not difficult to decipher the hidden meaning: Army officers must grit their teeth and overcome their natural repugnance towards the new regime.

For a week, the red and gold monarchist flag continued to fly over the Academia. The Captain-General of Aragón, Enrique Fernández de Heredia, had been instructed by the Provisional Government to raise the Republican tricolour throughout the region. With the military headquarters in Zaragoza surrounded by hostile crowds demanding that *Cacahuete* (peanut), as the vegetarian Fernández de Heredia was known, fly the Republican flag, he refused. At midnight on 14 April, the new Minister of War, Manuel Azaña, ordered him to hand over command of the region to the military governor of Zaragoza, Agustín Gómez Morato, who was considered loyal to the Republican cause and who, indeed, was to be imprisoned by the Nationalists in July 1936 for opposing the military rebellion in Morocco. Gómez Morato undertook the substitution and

telephoned all units in Aragón to order them to do the same. At the
Military Academy, Franco informed his superior that changes of insignia
could be ordered only in writing. It was not until after 20 April when
the new Captain-General of the region, General Leopoldo Ruiz Trillo,
had signed an order to the effect that the Republican flag should be
flown, that Franco ordered the monarchist ensign struck.[11]

   In 1962, Franco wrote a partisan and confused interpretation of the fall
of the monarchy in his draft memoirs in which he blamed the guardians of
the monarchist fortress for opening the gates to the enemy. The enemy
consisted of a group of 'historic republicans, freemasons, separatists and
socialists'. The freemasons were 'atheistic traitors in exile, delinquents,
swindlers, men who betrayed their wives'.[12] The narrowness of his
interpretation is striking in several ways. Franco's admiration for the
dictatorship is understandable. His assumption that the King had not
contravened the constitution in acquiescing in a military coup d'état in
1923 and that the situation in April 1931 was therefore one of constitutional legality was clearly the view of a soldier who never questioned
the Army's right to rule. The clear implication is that the monarchy
should, and but for Sanjurjo and the Civil Guard could, have been
defended by force in April 1931, which was certainly not his view at
the time. Franco conveniently forgot his own ruthless pragmatism. The
mistake having been made by others, he had made the best of a bad job
and got on with his career.

   Nonetheless, the flag incident suggested that Franco was sufficiently
affected by the fall of the monarchy to want to establish some distance
between himself and the Republic. It was not a question of outright
indiscipline nor is it plausible that he was trying well in advance to build
up credit with conservative political circles. In keeping the monarchist
flag flying, Franco was advertising the fact that, unlike some officers who
had been part of, or at least in touch with, the Republican opposition,
he could not be considered as in any way tainted by disloyalty to the
monarchy. Perhaps even more than from the pro-Republican officers
whom he despised anyway, he was marking distance between himself and
his brother Ramón who had been one of the most notorious military
traitors to the King. Francisco clearly saw his own position as altogether
more praiseworthy than that of General Sanjurjo whom he later came to
regard, with Berenguer, as responsible for the fall of the monarchy.[13]
However, he would not permit his regret at the fall of the monarchy to
stand in the way of his career. As military monarchism went, Franco's
pragmatic stance was a long way from, for instance, that of the founder

of the Spanish Air Force, General Kindelán, who went into voluntary exile on 17 April rather than live under the Republic.[14] Nonetheless, Franco felt great repugnance for those officers who had opposed the monarchy and were rewarded by being given important posts under the Republic. On 17 April, General Gonzalo Queipo de Llano became Captain-General of Madrid, General Eduardo López Ochoa of Barcelona and General Miguel Cabanellas of Seville. All three would play crucial roles in Franco's later career and he never trusted any of them.

It was perhaps with these promotions in mind that, on 18 April, Franco wrote a letter to the Director of *ABC*, the Marqués de Luca de Tena. The monarchist *ABC* was the most influential newspaper on the Right in Spain. The issue of that morning had published his photograph alongside a news item that he was about to go to Morocco as High Commissioner, the most coveted post in the Army and one which was, at the time, the peak of Franco's ambition. The basis of the item was a suggestion by Miguel Maura, the Minister of the Interior, to Manuel Azaña, the Minister of War, that Franco be appointed to the post. It would have been a sensible way of buying his loyalty. In fact, the plum Moroccan job was given to General Sanjurjo, who held it briefly in conjunction with the headship of the Civil Guard – such preferment no doubt feeding Franco's suspicions that Sanjurjo was being paid off for his treachery. The ostensible objective of Franco's letter was to request that the newspaper publish a correction but it was another gesture aimed at establishing his distance from Spain's new rulers. In convoluted and ambiguous language, he denied that he had been offered any appointment and asserted that 'I could not accept any such post unless I was ordered to do so. To accept such a post might be interpreted in some circles as suggesting that there had been some prior understanding on my part with the regime which has just been installed or else apathy or indifference in the fulfilment of my duties'.[15] That Franco believed that he needed to make his position clear in the leading conservative daily reflects both his ambition and his sense of himself as a public figure. Having clarified his loyalty to the monarchy, he then went on to mend his fences with the Republican authorities by proclaiming his respect for the 'national sovereignty', a reflection of his cautious pragmatism and of the flexibility of his ambitions.

The limits of military loyalty were to be severely tried under the Republic. The new Minister of War, Azaña, had studied military politics and was determined to remedy the technical deficiencies of the Spanish

Army and to curtail its readiness to intervene in politics. Azaña was
an austere and brilliantly penetrating intellectual who, despite laudable
intentions, was impatient of Army sensibilities and set about his task
without feeling the need to massage the collective military ego. The
Army which he found on taking up his post was under-resourced and
over-manned, with a grossly disproportionate officer corps. Equipment
was obsolete and inadequate and there was neither ammunition nor fuel
enough for exercises and manoeuvres. Azaña wished to reduce the Army
to a size commensurate with the nation's economic possibilities to
increase its efficiency and to eradicate the threat of militarism from
Spanish politics. Even those officers who approved of these aims were
uneasy about a decimation of the officer corps. Nevertheless, implemented with discretion, Azaña's objectives might have found some support within the Army. However, conflict was almost inevitable. Azaña
and the government in which he served were determined to eliminate
where possible the irregularities of the Dictatorship of Primo de Rivera.
There were those, Franco foremost among them, who admired the Dictatorship and had been promoted by it. They could not view with equanimity any assault on its works. Secondly, Azaña was inclined to be
influenced by, and to reward the efforts of, those sections of the Army
which were most loyal to the Republic. That necessarily meant military
opponents of the Dictatorship, who were *junteros* and largely artillerymen.
That in turn infuriated the *Africanistas* who had opposed the *junteros* since
1917.[16]

The many measures which Azaña promulgated in the first months of
the Republic divided the Army and were seized upon by the rightist press
in order to generate the idea that the military, along with the Church,
was being singled out for persecution by the new regime. That was a
distortion of Azaña's intentions. By a decree of 22 April 1931, Army
officers were required to take an oath of loyalty (*promesa de fidelidad*) to
the Republic just as previously they had to the monarchy. It did not
matter what an officer's inner convictions were and no mechanism was
set up to purge or investigate those who were monarchists. According to
the decree, to stay in the ranks, an officer simply had to make the promise
'to serve the Republic well and faithfully, obey its laws and defend it by
arms'. In the case of those who refused to give the promise, it was to be
assumed that they wished to leave the service. Most officers had no difficulty about making the promise. For many, it was probably a routine
formula without special significance and was made by many whose real
convictions were anti-Republican.[17] After all, few had felt bound by their

oath of loyalty to the monarchy to spring to its defence on 14 April. On the other hand, although a reasonable demand on the part of the new Minister and the new regime, the oath could easily be perceived by the more partisan officers as an outrageous imposition. Adept at manipulating the military mentality, the right-wing press generated the impression that those whose convictions prevented them swearing the oath were being hounded penniless out of the Army.[18] In fact, those who opted not to swear were considered members of the reserve and were to receive their pay accordingly.

A prominent right-wing general, Joaquín Fanjul, retrospectively summed up the feelings of many officers: 'When the Republic came into being, it placed many officers in a dilemma: respect it and undertake formally to defend it or else leave the service. The formula was rather humiliating, offspring as it was of the person who conceived it. I thought about it for four days, and finally I offered up my humiliation to my *Patria* and I signed as did most of my comrades.'[19] In so far as Franco was forced to decide between his profession and his convictions in April 1931, he opted, understandably and without any apparent difficulty, for his profession. Franco was a more sinuous and pragmatic individual than Fanjul as was shown by a conversation which he had in 1931 with an artilleryman of his acquaintance, General Reguera, who had retired under the terms of the Azaña law. 'I believe that you have committed a mistake,' said Franco. 'The Army cannot lose its senior officers just for the sake of it at times as difficult as these.' When Reguera explained the disgust he felt at 'serving those people and their dishcloth of a flag', Franco replied 'It's a pity that you and others like you are leaving the service precisely when you could be of most use to Spain and are leaving the way clear to those whom we all know who would do anything to climb a few rungs of the ladder. Those of us who have stayed on will have a bad time, but I believe that by staying we can do much more to avoid what neither you nor I want to happen than if we had just packed up and gone home'.[20]

On 25 April, the announcement was made of the decree which came to be known as the *Ley Azaña*. It offered voluntary retirement on full pay to all members of the officer corps, a generous and expensive way of trying to reduce its size. However, the decree stated that after thirty days, any officer who was surplus to requirements but had not opted for the scheme would lose his commission without compensation. This caused massive resentment and further encouragement of the belief, again fomented by the rightist press, that the Army was being persecuted by the Republic. Since the threat was never carried out, its announcement was a gratuitously damaging error on the part of Azaña or his ministerial advisers.

As soon as the decree was made public, the most alarmist rumours were spread about unemployment and even exile for officers who were not enthusiastic Republicans.[21] A large number accepted, rather more than one third of the total, and as many as two thirds among those colonels who had no hope of ever being promoted to general.[22] Franco of course did not. He was visited by a group of officers from the Academy who asked his advice on how to respond to the new law. His reply gave a revealing insight into his notion that the Army was the ultimate arbiter of Spain's political destinies. He said that a soldier served Spain and not a particular regime and that, now more than ever, Spain needed the Army to have officers who were real patriots.[23] At the very least, Franco was keeping his options open.

Like many officers, Franco found his relationship with the new regime subject to constant frictions. Before April was out, he became embroiled in the so-called 'responsibilities' issue. General Berenguer had been arrested on 17 April, for alleged offences committed in Africa, as Prime Minister and later as Minister of War during the summary trial and execution of Galán and García Hernández.[24] General Mola was arrested on 21 April for his work as Director-General of Security under Berenguer.[25] These arrests were part of a symbolic purge of significant figures of the monarchy which did the nascent Republic far more harm than good. The issue of 'responsibilities' harked back to the Annual disaster and the role played in it by royal interference, military incompetence and the deference of politicians towards the Army. It was popularly believed that the military coup of 1923 had been carried out in order to protect the King from the findings of the 'Responsibilities Commission' set up in 1921. Accordingly, the issue was still festering. To the 'responsibilities' contracted by Army officers and monarchist politicians before 1923 the Republican movement had added the acts of political and fiscal abuse and corruption carried out during the Dictatorship and after. The greatest of these was considered to be the execution of Galán and García Hernández. With the Dictator dead and the King in exile, it was inevitable that Berenguer would be an early target of Republican wrath.

The campaign 'for responsibilities' helped keep popular Republican fervour at boiling point in the early months of the Regime but at a high price in the long term. In fact, relatively few individuals were imprisoned or fled into exile but the 'responsibilities' issue created a myth of a vindictive and implacable Republic, and increased the fears and resentments of powerful figures of the old regime, inducing them to see the threat posed by the Republic as greater than it really was.[26] In the eyes of officers like Franco, Berenguer was being tried unjustly for his part in a war to which they had devoted their lives, and for following military regulations in court-martialling Galán and García Hernández. Far from being heroes and martyrs, they were simply mutineers. Mola was a hero of the African war who, as Director-General of Security, had merely been doing his job of controlling subversion. What enraged Franco and many other *Africanistas* was that officers whom they considered courageous and competent were being persecuted while those who had plotted against the Dictator were being rewarded with the favour of the new regime. The 'responsibilities' trials were to provide the *Africanistas* with a further excuse for their instinctive hostility to the Republic. Franco would move more circumspectly along this road than many others like Luis Orgaz, Manuel Goded, Fanjul and Mola, but he would make the journey all the same. Like them, he came to see the officers who received the preferment of the Republic as lackeys of freemasonry and Communism, weaklings who pandered to the mob.

In this context, Franco had an ambiguous attitude towards Berenguer. Although he approved of his actions in connection with the Jaca rising, he would soon come to question his failure to fight for the monarchy in April 1931. Moreover, he harboured considerable personal resentment towards Berenguer. Having informed Franco in 1930 that he was going to promote him to *General de División* (Major-General), Berenguer had then realised that his friend General León was about to reach the age at which he should have passed into the reserve. To avoid this, and on the grounds that Franco had plenty of time before him, Berenguer gave the promotion instead to León.[27] It is thus slightly surprising that, at the end of April, Franco agreed to act as defender in Berenguer's court martial. Along with Pacón Franco Salgado-Araujo, his ADC, he visited Madrid on 1 May and interviewed Berenguer in his cell on the following day. On 3 May, Franco was informed that the Minister of War refused authorization for him to act on behalf of Berenguer on the grounds that he was resident outside the military region in which the trial was taking place.[28] It was the beginning of the mutual distrust which would characterize the momentous relationship between Franco and Azaña. It was during the trip to Madrid that Franco's attitude to Sanjurjo began to sour. His friend Natalio Rivas told him about Sanjurjo's interview with Lerroux on 13 April. Franco concluded that some offer of future preferment had been made which accounted for Sanjurjo's failure to mobilize the Civil Guard in defence of the King.[29]

Franco's latent hostility to the Republic was brought nearer to the surface with Azaña's military reforms. In particular, he was appalled by the abolition of the eight historic military regions which were no longer to be called *Capitanías Generales* but were converted into 'organic divisions' under the command of a Major-General who would have no legal powers over civilians. The viceregal jurisdictional powers held by the old Captains-General were eliminated and the rank of Lieutenant-General was deemed unnecessary and was also suppressed.[30] These measures were a break with historic tradition: they removed the Army's jurisdiction over public order. They also wiped out the possibility for Franco of reaching the pinnacles of the rank of Lieutenant-General and the post of Captain-General. He would reverse both measures in 1939. However, he was hardly less taken aback by Azaña's decree of 3 June 1931 for the so-called *revisión de ascensos* (review of promotions) whereby some of the promotions on merit given during the Moroccan wars were to be re-examined. It reflected the government's determination to wipe away the legacy of the Dictatorship – in this case to reverse some of the arbitrary promotions made by Primo de Rivera. The announcement raised the spectre that, if all of those promoted during the *Dictadura* were to be affected, Goded, Orgaz and Franco would go back to being colonels, and many other senior *Africanistas* would be demoted. Since the commission carrying out the revision would not report for more than eighteen months, it was to be at best an irritation, at worst a gnawing anxiety for those affected. Nearly one thousand officers expected to be involved, although in the event only half that number had their cases examined.[31]

The right-wing press and specialist military newspapers mounted a ferocious campaign alleging that Azaña's declared intention was to '*triturar el Ejército*' (crush the Army).[32] Azaña never made any such remark, although it has become a commonplace that he did. He made a speech in Valencia on 7 June in which he praised the Army warmly and declared his determination to *triturar* the power of the corrupt bosses who dominated local politics, the *caciques* in the same way as he had dismantled 'other lesser threats to the Republic'. This was twisted into the notorious phrase.[33] To the fury of the *Africanistas*, it was rumoured that Azaña was being advised by a group of Republican officers known among his rightist opponents as the 'black cabinet'. The abolition of promotion by merit reflected the commitment of the artillery to promotion only by strict seniority. Azaña's informal military advisers included artillery officers, such as Majors Juan Hernández Saravia and Arturo Menéndez López, and consisted largely of *junteros* who had taken part in the movement against the Dictatorship and the Monarchy. Franco regarded these officers as contemptible. There was ill feeling elsewhere in the officer corps that, instead of using the most senior Major-Generals, Azaña should listen to such relatively junior men.[34]

However, Hernández Saravia complained to a comrade that Azaña was too proud to listen to advice from anyone. Moreover, far from setting out to persecute monarchist officers, Azaña seems rather to have cultivated many of them, such as Sanjurjo or the monarchist General Enrique Ruiz Fornells whom he kept on as his under-secretary. Indeed, there were even some leftist officers who took retirement out of frustration at what they saw as Azaña's complaisance with the old guard and the offensive and threatening language which Azaña was accused of using against the Army is difficult to find. Azaña, although firm in his dealings with officers, spoke of the Army in public in controlled and respectful terms.[35]

Franco was well known for his repugnance for day-to-day politics. His daily routine at the Military Academy was a full and absorbing one. Nevertheless, he was soon obliged to think about the changes that had taken place. The conservative newspapers which he read, *ABC*, *La Época*, *La Correspondencia Militar*, presented the Republic as responsible for Spain's economic problems, mob violence, disrespect for the Army and anticlericalism. The press, and the material which he received and devoured from the *Entente Internationale contre la Troisième Internationale*, portrayed the regime as a Trojan Horse for Communists and freemasons determined to unleash the Godless hordes of Moscow against Spain and all its great traditions.[36] The challenges to military certainties constituted by Azaña's reforms cannot have failed to provoke, at the very least, nostalgia for the monarchy. Similarly, news of the rash of church burnings which took place in Madrid, Málaga, Seville, Cadiz and Alicante on 11 May did not pass him by. The attacks were carried out largely by anarchists, provoked by the belief that the Church was at the heart of the most reactionary activities in Spain. Franco was probably unaware of accusations that the first fires were started with aviation spirit secured from Cuatro Vientos aerodrome by his brother Ramón. He cannot, however, have failed to learn of his brother's published statement that 'I contemplated with joy those magnificent flames as the expression of a people which wanted to free itself from clerical obscurantism'.[37] In notes made for his projected memoirs, jotted down nearly thirty years after the event, Franco described the church burnings as the event which defined the Republic.[38] That reflects not only his underlying Catholicism, but also the extent to which the Church and the Army were increasingly flung together as the self-perceived victims of Republican persecution.

However, more than for anything else that had happened since 14 April, Franco was to bear Azaña the deepest grudge of all for his order of 30 June 1931 closing the Academia General Militar de Zaragoza. The first news of it reached him while on manoeuvres in the Pyrenees. His initial reaction was disbelief. When it sank in, he was devastated. He had loved his work there and he would never forgive Azaña and the so-called 'black cabinet' for snatching it from him. He and other *Africanistas* believed that the Academy had been condemned to death merely because it was one of Primo de Rivera's successes. He was also convinced that the 'black cabinet' wanted to bring him down because of their envy of his spectacular military career. In fact, Azaña's decision was based on doubts about the efficacy of the kind of training imparted in the Academy and also on a belief that its cost was disproportionate at a time when he was trying to reduce military expenses. Franco controlled his distress with difficulty.[39] He wrote to Sanjurjo hoping that he might be able to intercede with Azaña. Sanjurjo replied that he must resign himself to the closure. A few weeks later, Sanjurjo commented to Azaña that Franco was 'like a child who has had a toy taken away from him'.[40]

Franco's anger glimmered through the formalised rhetoric of his farewell speech which he made on the parade-ground at the Academy on 14 July 1931. He opened by commenting with regret that there would be no *jura de bandera* (swearing on the flag) since the laic Republic had abolished the oath. He then surveyed the achievements of the Academy under his direction, including the elimination of vice. He made much of the loyalty and duty that the cadets owed to the *Patria* and to the Army. He commented on discipline, saying that it 'acquires its full value when thought counsels the contrary of what is being ordered, when the heart struggles to rise in inward rebellion against the orders received, when one knows that higher authority is in error and acting out of hand'. He made a rambling and convoluted, but nonetheless manifestly bitter, allusion to those who had been rewarded by the Republic for their disloyalty to the monarchy. He made an oblique reference to the Republican officers who held the key posts in Azaña's Ministry of War as 'a pernicious example within the Army of immorality and injustice'. His speech ended with the cry '*¡Viva España!*'.[41] He was to comment proudly more than thirty years later 'I never once shouted '*¡Viva la República!*'.[42]

After his speech, Franco returned to his office only to be called out several times to appear on the balcony to receive the frenetic applause of those present. When he said farewell to Pacón, who had worked with him as an instructor in tactics and weaponry and as his ADC, the future Caudillo was crying. He packed his things and travelled to his wife's country house, La Piniella, at Llanera near Oviedo.[43]

The speech was published as Franco's order of the day and reached Azaña. Azaña wrote in his diary two days later, 'Speech by General Franco to the cadets of the Academia General on the occasion of the end of the course. Completely opposed to the Government, guarded attacks against his superiors; a case for immediate dismissal, if it were not the case that today he ceased to hold that command.' As it was, Azaña limited himself to a formal reprimand (reprensión) in Franco's service record for the speech to the cadets.[44]

Acutely jealous of his spotless military record, Franco's resentment on being informed of this reprimand on 23 July may be imagined. Nevertheless, his concern for his career led him to swallow his pride and to write on the next day an ardent, if less than convincing, self-defence, in the form of a letter to the Chief of the General Staff of the V Military Division within whose jurisdiction the Academy lay. It requested him to pass on to the Minister of War, 'my respectful complaint and my regret for the erroneous interpretation given to the ideas contained in the speech ... which I endeavoured to limit to the purest military principles and essences which have been the norm of my entire military career; and equally my regret at his apparent assumption that there is something lukewarm or reserved about the loyal commitment that I have always given, without officious ostentation which is against my character, to the regime which the country has proclaimed, whose ensign hoisted in the central parade ground of the Academy flew over the military solemnities and whose national anthem closed the proceedings.'[45]

Azaña did not regard the obligatory flying of the Republican flag and the playing of the new national anthem as special merits and was not convinced. He seems to have believed that the once favourite soldier of the monarchy needed bringing down a peg or two. His contacts with Franco, in this letter and at a meeting in August, convinced him that he was sufficiently ambitious and time-serving to be easily bent to his purposes. In his basic assessment, Azaña was probably correct, but he seriously misjudged how easy it would be to act on it. If Azaña had given Franco the degree of preferment to which he had become accustomed under the monarchy, it was entirely possible that he might have become the darling of the Republic. As it was, Azaña's policy towards Franco was to be altogether more restrained although, from the point of view of the Republican Minister of War, it was indeed generous. After losing the Academy, Franco was kept without a posting for nearly eight months which gave him time to devote to his reading of anti-Communist and anti-masonic literature but left him with only 80 per cent of his salary. Without a personal fortune, living in his wife's house, his career apparently curtailed, Franco harboured considerable rancour for the Republican regime. Doña Carmen encouraged his bitterness.[46]

Throughout the summer of 1931, Army officers fumed at both the military reforms and at what they saw as the anarchy and disorder constituted by a number of strikes involving the anarchosyndicalist Confederación Nacional del Trabajo in Seville and Barcelona.[47] Given the discontent occasioned by Azaña's reforms and the monarchist quest for praetorian champions to overthrow the Republic, there were well-founded rumours of possible military conspiracy. The names of Generals Emilio Barrera and Luis Orgaz were the most often cited and they were both briefly put under house arrest in mid-June. Eventually, in September, after evidence of further monarchist plots, Azaña would have Orgaz exiled to the Canary Islands. Azaña was convinced by reports reaching the Ministry that Franco was conspiring with Orgaz and regarded him as the more fearsome of the two (*'el más temible'*).[48] As the summer wore on, Azaña continued to believe that he was on the fringe of some kind of plot. In reports on contacts between Franco's friend, the militantly right-wing Colonel José Enrique Varela, and the powerful hard-line monarchist boss of Cádiz, Ramón de Carranza, the names of Franco and Orgaz had been mentioned. The Minister wrote in his diary 'Franco is the only one to be feared', a tribute to his reputation for seriousness and efficiency. Azaña gave instructions that Franco's activities be monitored. In consequence, when he visited Madrid in mid-August, the Director-General of Security, Angel Galarza, had him under the surveillance of three policemen.[49]

On 20 August, during his stay in Madrid, Franco visited the Ministry of War and spoke with the under-secretary who reminded him that he was obliged to call on the Minister. He returned on the following day. Azaña criticized his farewell speech to the Academy in Zaragoza. Franco had to swallow the criticism but Azaña was not fooled, writing later in his diary 'he tries to seem frank but all rather hypocritically'. Azaña warned him, somewhat patronizingly, not to be carried away by his friends and admirers. Franco made protests of his loyalty, although he admitted that monarchist enemies of the Republic had been seeking him out, and seized the opportunity to inform the Minister that the closure of the Academy had been a grave error. When Azaña hinted that he would like to make use of Franco's services, the young general commented with an ironic smile 'and to use my services, they have me followed everywhere by a police car! They will have seen that I don't go anywhere.' An embarrassed Azaña had the surveillance lifted.[50]

The hypocritical Franco of Azaña's account is entirely consistent with the document which he had submitted in defence of his speech at the closure of the Academy.* Azaña was rather condescending towards Franco, confident that he could bring him to heel.[51] It is likely that his miscalculations about Franco derived in part from an assumption that he was as manipulable as his brother Ramón for whom Azaña, who knew him well, felt only impatience and contempt.

At the beginning of May, Franco had been refused permission to act as defender of Berenguer. In fact, the Consejo Supremo del Ejército had annulled the warrant against Berenguer soon afterwards and the Tribunal Supremo ordered the release of Mola on 3 July. However, the issue of 'responsibilities' remained deeply divisive, with moderate members of the government, including Azaña, keen to play it down. After a venomous debate, on 26 August, the Cortes empowered the 'Responsibilities Commission' to investigate political and adminstrative offences in Morocco, the repression in Catalonia between 1919 and 1923, Primo de Rivera's 1923 coup, the Dictatorships of Primo and Berenguer and the Jaca court martial.[52] To the fury of Azaña, who rightly believed that the Commission was dangerously damaging to the Republic, a number of aged generals who had participated in Primo's Military Directory were arrested at the beginning of September.[53]

The hostility of some officers and the doubts of the many about the direction the Republic was taking were intensified by the bitter debate over the proposed new constitution which took place between mid-August and the end of the year. Its laic clauses, particularly those which aimed to break the clerical stranglehold on education, provoked hysterical press reaction on the Right. The determination of the Republican and Socialist majority in the Cortes to push these clauses through provoked

---

* This differs from the version given by Franco to his friend and biographer, Joaquín Arrarás. According to this version, Azaña said 'I have re-read your extraordinary order to the cadets and I would like to believe that you did not think through what you wrote', to which Franco claims to have replied, 'Señor Ministro, I never write anything that I haven't thought through beforehand'. Azaña's version, written on the day, is altogether more plausible than that recounted by Franco six years later in the heat of the civil war. Joaquín Arrarás, *Franco* (Valladolid, 1937) p. 166.

the resignation of the two most prominent deeply Catholic members of the government, the conservative prime minister Niceto Alcalá Zamora and his Minister of the Interior, Miguel Maura Gamazo. Azaña became prime minister. The right-wing press screamed that 'the very existence of Spain is threatened'.[54]

Apocalyptic accounts in the right-wing press of anarchy and the implications of the constitutional proposals, together with the continuing determination of the Republican Left to press ahead with the 'responsibilities' issue, intensified the fears of Army officers. In the eyes of most of them, some senior generals were being accused of rebellion when all they had done was to put a stop to anarchy in 1923 while others, Berenguer and Fernández de Heredia, were being tried for dealing with the mutiny of Jaca. As the then Captain-General of Aragón, Fernández de Heredia was the man who had signed the death sentences. Posters, books and even a play by Rafael Alberti, *Fermín Galán*, glorified 'the martyrs of the Repúblic'. Ramón Franco dedicated his book *Madrid bajo las bombas* (Madrid beneath the bombs) to 'the martyrs for freedom, Captains Galán and García Hernández, assassinated on Sunday 14 December 1930 by Spanish reaction incarnated in the monarchy of Alfonso XIII and his government, presided by General Dámaso Berenguer'. The beatification of Galán and García Hernández was something which infuriated all but committed Republicans in the officer corps. Franco was especially outraged that the Republic appeared to be applying double standards in trying to eradicate unsound promotions granted during the 1920s at the same time as pursuing favouritism towards those who had collaborated in its establishment. Ironically, Ramón Franco had been appointed Director-General de Aeronáutica. Franco's brother abused his position to participate in anarchist conspiracies against the Republic, lost his post and was only saved from a prison sentence by his election as a parliamentary deputy for Barcelona and by the solidarity of his masonic colleagues.[55]

When the Responsibilities Commission began to gather evidence for the forthcoming trial of those involved in the executions after the Jaca uprising, Franco appeared as a witness. In the course of his cross-examination on 17 December 1931, Franco's answers were dry and to the point. He reminded the court that the code of military justice permitted summary executions to take place without the prior approval of the civilian authorities. However, when asked if he wished to add anything to his statement, he revealingly went on to defend military justice as 'a juridical and a military necessity, by which military offences, of a purely military nature, and committed by soldiers, are judged by persons militarily prepared for the task'. Accordingly, he declared that, since the members of the Commission had no military experience, they were not competent to judge what had happened at the Jaca court martial.

When proceedings recommenced on the following day, Franco effectively lined himself against one of the cherished myths of the Republic by stating that Galán and García Hernández had committed a military offence, dismissing the central premiss of the Commission that they had carried out a political rebellion against an illegitimate regime. Franco declared 'receiving in sacred trust the arms of the nation and the lives of its citizens, it would be criminal in any age and in any situation for those who wear a uniform to use those arms against the nation or against the state which gave us them. The discipline of the Army, its very existence and the health of the state demand of us soldiers the bitter disappointments of having to apply a rigid law'.[56] Although carefully ringed around by declarations of respect for parliamentary sovereignty, it was implicitly a statement that he regarded the defence of the monarchy by the Army in December 1930 to have been legitimate, a view contrary to those held by many in authority in the Republic. His views on the canonization of the Jaca rebels could also easily be deduced from the statement. However, in its implications about a disciplined acceptance of the Republic, his statement was entirely consistent with both his order of the day on 15 April and his farewell speech at the Academy. It may therefore be taken as further evidence that, unlike hotheads such as Orgaz, he was still far from turning his discontent into active rebellion. After a protracted ordeal, both Berenguer and Fernández de Heredia were found innocent by the Tribunal Supremo in 1935.[57]

Franco's obscure declarations of disciplined loyalty were some distance from the enthusiastic commitment which might have gained him official favour. After the loss of the Academy, the questioning of his promotions, and the working class unrest highlighted by the right-wing press, Franco's attitude to the Republic could hardly be other than one of suspicion and hostility. It is not surprising that he had to wait some considerable time before he got a posting, but it was an indication both of his professional merits and of Azaña's recognition of them that, on 5 February 1932, he was posted to La Coruña as Commander of the *XV Brigada de Infantería de Galicia*, where he arrived at the end of the month. The local press greeted his arrival with the headline 'A Caudillo of the Tercio' and praised not only his bravery and military skill but also 'his noble gifts as a correct and dignified gentleman'. He again took Pacón with him as his ADC.

He was delighted to be in La Coruña, near to his mother, whom he visited every weekend.[58]

That Azaña believed that he was treating Franco well may be deduced from the fact that the posting saved the young general from the consequences of a decree published in March 1932 establishing the obligatory retirement of those who had spent more than six months without a posting. The appointment came only a few days before the end of the period after which Franco would have had to go into the reserve and he must have suffered considerable anxiety during the months of waiting. Azaña had deliberately kept him in a state of limbo as a punishment for the farewell speech to the Military Academy and to tame the arrogance of the soldier seen as the golden boy of the monarchy.[59] In fact, by the point at which he posted Franco to La Coruña, Azaña seems to have decided that he had learned his lesson and might now be recruited to the new regime. Knowing Ramón Franco well, Azaña seemed again to be judging his older brother in the same terms. If that was so, it reflected an under-estimate of Franco's capacity for resentment. Rather than reacting with gratitude and loyalty as Azaña had hoped, Franco harboured a grudge against him for the rest of his life.

Before their next meeting seven months later, a major crisis in civilian-military relations had occurred, and been resolved. It took the form of a military uprising in August 1932, the origins of which went back to the end of 1931. At that time, in the course of an otherwise peaceful general strike of landworkers in the province of Badajoz in Extremadura, there was bloodshed involving the Civil Guard in Castilblanco, a remote village in the heart of the arid zone known as the *Siberia extremeña*. Like most of the area, Castilblanco suffered high unemployment. On 30 and 31 December, the workers of the village held peaceful demonstrations. As they were dispersing to their homes, the *alcalde* (mayor) panicked and instructed the local four-man Civil Guard unit to intervene to break up the crowd. After some scuffling, a Civil Guard opened fire killing one man and wounding two others. In response, the villagers set upon the four guards, beating them to death with stones and knives.[60] There was an outcry in the right-wing press and the Republican-Socialist government headed by Azaña was accused of inciting the landless labourers against the Civil Guard. Sanjurjo visited Castilblanco, in his capacity as Director-General of the Civil Guard, and blamed the outrage on the extreme leftist Socialist deputy for Badajoz, Margarita Nelken. In a revealing association of the working class and the Moors, he declared that during the collapse of Melilla, even at Monte Arruit, he had not seen similar atrocities. He also demanded justice for the Civil Guard.[61] It was part of a process whereby the military was being convinced that the Republic signified disorder and anarchy. No issue was more indicative of the social abyss which divided Spain. For the Right, the Civil Guard was the beloved *benemérita*, the guardian of the social order; for the Left, it was a brutal and irresponsible Army of occupation at the service of the rich.

While the country was still reeling from the horror of Castilblanco, there occurred another tragedy. In the village of Arnedo in the province of Logroño in northern Castile, some of the employees of the local shoe factory had been sacked for belonging to the socialist trade union, the Unión General de Trabajadores. During a protest meeting, the Civil Guard, with no apparent provocation, opened fire killing four women, a child and a worker as well as wounding thirty other by-standers, some of whom died in the course of the next few days. In the light of the remarks made by General Sanjurjo after Castilblanco, it was difficult for the incident not to be seen as an act of revenge.[62] Azaña reluctantly bowed to pressure in the left-wing press and by left-wing deputies in the Cortes to remove Sanjurjo from the command of the Civil Guard and transfer him to the less important post of head of the *Carabineros*, the frontier and customs police.[63] On 5 February 1932, in the batch of postings which sent Franco to Galicia, Sanjurjo was replaced as Director of the Civil Guard by General Miguel Cabanellas.[64]

Under any circumstances, Sanjurjo would have objected to losing the post of Director-General of the Civil Guard. In the context of the leftist campaign against him, his removal was interpreted by the right-wing press, and by himself, as an outrage and a further blow in favour of anarchy. Many on the Right began to see Sanjurjo as a possible saviour and encouraged him to think about overthrowing the Republic. The Castilblanco and Arnedo incidents had wiped away Sanjurjo's original sin in the eyes of the extreme Right, his failure to act on behalf of the monarchy in April 1931. Now he was seen as the most likely guarantor of law and order, something which was transmuted in rightist propaganda into the defence of 'the eternal essences of Spain'. Throughout 1932, as the agrarian reform statute and the Catalan autonomy statute painfully passed through the Cortes, the Right would grow ever more furious at what it perceived as assaults on property rights and national unity. Across Spain, petitions in favour of Sanjurjo were signed by many Army officers, although not by Franco. Several efforts were made to push Sanjurjo towards a coup d'état and he began to plot against the Republic.

General Emilio Barrera informed the Italian Ambassador Ercole

Durini di Monzo in February that a movement to 'oppose bolshevism and restore order' could count on widespread military support including that of Generals Goded and Sanjurjo.[65] Lerroux, who was determined to see Azaña's Left Republican-Socialist coalition evicted from power, was in contact with Sanjurjo. They were united in resenting the presence of the Socialists in the government and talked about a possible coup. [66] Any military conspiracy would have benefited enormously from the participation of Franco. However, he kept his distance out of innate caution when faced with an ill-prepared and highly questionable coup attempt. He distrusted Sanjurjo and had no reason to risk everything when he could continue to exercise his chosen profession within the Republic.

Franco was anxious not to jeopardize his new found comforts. Despite his proven capacity to put up with physical discomfort and to work hard in the most difficult conditions, Franco always enjoyed physical comfort when it was available. In the interval between leaving Morocco and taking on the task of building up the Zaragoza Academy, he had enjoyed a light work load and a full social life. Now, in La Coruña, he was effectively military governor, and had a splendid life-style, with a large house and white-gloved servants. La Coruña was then a beautiful and peaceful seaport and not the bustling and anonymous town that it was to become during the later years of his dictatorship. Franco's minimal duties as military commander permitted him to be a frequent visitor to the yacht club (*Club Náutico*) where he was able to indulge, on a small scale, his love of sailing. It was there that he made the acquaintance of Máximo Rodríguez Borrell, who after the war would become his regular fishing and hunting companion. Max Borrell was to be one of his very few close civilian friends and to remain so until his final illness.[67]

The fact that Franco was not prepared to take risks for Sanjurjo does not mean that he was enthusiastic about the political situation. However, he was altogether more cautious than many of his peers and he carefully distanced himself from the coup attempt of 10 August 1932. Nonetheless, as might have been expected given his long African association with Sanjurjo, he knew about its preparation. On 13 July, Sanjurjo visited La Coruña to inspect the local carabineros and had dinner with Franco, discussing with him the forthcoming uprising. According to his cousin, Franco told Sanjurjo at this meeting that he was not prepared to take part in any kind of coup. [68] The monarchist plotter Pedro Sainz Rodríguez organized a further, and elaborately clandestine, meeting in a restaurant on the outskirts of Madrid. Franco expressed considerable doubts about the outcome of the coup and said he was still undecided about what his own position would be when the moment arrived, promising Sanjurjo that, whatever he decided, he would not take part in any action launched by the government against him.[69]

Franco was sufficiently vague for Sanjurjo to assume that he would support the rising. According to Major Juan Antonio Ansaldo, an impetuous monarchist aviator, conspirator and devoted follower of Sanjurjo, Franco's 'participation in the 10 August coup was considered certain', but 'shortly before it took place, he freed himself of any undertaking and advised several officers to follow his example'.[70] It is probably going too far to suggest that Franco first supported Sanjurjo's plot and then changed his mind. However, given Franco's labyrinthine ambiguity, it would have been easy for Sanjurjo and his fellow-plotters to allow themselves to take his participation for granted. His hesitations and vagueness while he waited for the outcome to become clear would have permitted such an assumption. It is certainly the case that Franco did nothing to report what was going on to his superiors.

Franco's final refusal to become part of the conspiracy was based largely on his view that it was inadequately prepared, as he indicated to the right-wing politician, José María Gil Robles, at a dinner in the home of their mutual friend, the Marqués de la Vega de Anzó.[71] He was afraid that a failed coup would 'open the doors to Communism'.[72] He was, however, also highly suspicious of the links between Sanjurjo and Lerroux whose involvement in what was being prepared could be perceived in a speech which he made in Zaragoza on 10 July. Aligning himself with the cause of the plotters, Lerroux was trying to push the government to adopt a more conservative line, tacitly threatening the military intervention which would follow if it did not. As ever the outrageous cynic and flatterer of the military, Lerroux declared that, when he came to power, he would reopen the Academia General Militar and reinstall Franco as Director.[73]

Franco himself visited Madrid at the end of July in order 'to choose a horse'.[74] It was rumoured, to his annoyance, that he had come to join the plot. When asked by other officers, as he was repeatedly, if he were part of the conspiracy, he replied that he did not believe that the time had yet come for a rising but that he respected those who thought that it had. He was outraged to discover that some senior officers were openly stating that he was involved. He told them that, if they continued to 'spread these calumnies', he would 'take energetic measures'. By chance, he met Sanjurjo, Goded, Varela and Millán Astray at the Ministry of War. Varela told him that Sanjurjo wanted to sound him out about the forthcoming coup. Sanjurjo at first denied this but agreed to meet Franco and Varela together. Over lunch, Franco told them categorically that they should not count on his participation in any kind of military uprising. In a barely veiled rebuke to Sanjurjo for his behaviour in April 1931, Franco justified his refusal to join the plot on the grounds that, since the Republic had come about because of the military defection from the cause of the monarchy, the Army should not now try to change things.[75] This meeting could account for the caustic remark made by Sanjurjo in the summer of 1933 during his imprisonment after the coup's failure: '*Franquito es un cuquito que va a lo suyito*' ('little Franco is a crafty so-and-so who looks after himself').[76]

The Sanjurjo coup was poorly organized and, in Madrid, easily dismantled. It was briefly successful in Seville but, with a column of troops loyal to the government marching on the city, Sanjurjo fled.[77] The humiliation of part of the Army and the reawakening of the mood of popular fiesta which had initially greeted the establishment of the Republic occasioned by Sanjurjo's defeat cannot have failed to convince Franco of the wisdom of his prognostications about the rising.[78] The fact that the armed urban police, the *Guardias de Asalto* and the Civil Guard had played no part in the rising had underlined their importance. Franco was more convinced than ever that any attempted *coup d'état* needed to count on their support.

Azaña had long been worried that Franco might be involved in a plot against the regime and in the course of the Sanjurjada had feared that he might be part of the coup. However, when he telephoned La Coruña on 10 August, he was relieved to find that Franco was at his post. Curiously, he very nearly was not. Franco had requested permission for a brief spell of leave in order to take his wife and daughter on a trip around the beautiful fjord-like bays of Galicia, the *rías bajas*, but it had been refused since his immediate superior, Major-General Félix de Vera, had also been about to go away. Accordingly, when the coup took place, Franco had been in acting command of military forces in Galicia.[79]

The conspiratorial Right, both civilian and military, reached the more general conclusion which Franco had drawn in advance – that they must never again make the mistake of inadequate preparation. A monarchist 'conspiratorial committee' was set up by members of the extreme rightist group Acción Española and Captain Jorge Vigón of the General Staff in late September 1932 to begin preparations for a future military rising. The theological, moral and political legitimacy of a rising against the Republic was argued in the group's journal *Acción Española*, of which Franco had been a subscriber since its first number in December 1931.[80]

The group operated from Ansaldo's house in Biarritz. Substantial sums of money were collected from rightist sympathizers to buy arms and to finance political destabilization. One of the earliest operations was to set up subversive cells within the Army itself, and the responsibility for this task was given to Lieutenant-Colonel Valentín Galarza of the General Staff.[81] Galarza had been involved in the Sanjurjada but nothing could be proved against him. Azaña wrote in his diary, 'I have left without a posting another Lieutenant-Colonel of the General Staff, Galarza, an intimate of Sanjurjo and Goded, who before the Republic was one of the great *mangoneadores* (meddlers) of the Ministry. Galarza is intelligent, capable and obliging, slippery and obedient. But he is definitely on the other side. There is nothing against him in the prosecution case. Nevertheless, he is one of the most dangerous'.[82] All that Azaña could do was to leave Galarza without an active service posting. Galarza aimed to recruit key generals and Franco, already a friend, was one of his prime targets.[83]

Azaña seems to have assumed that Franco's presence at his post during the *Sanjurjada* meant that they were now totally reconciled. When the Prime Minister visited La Coruña from 17 to 22 September 1932, however, Franco made slight efforts to disabuse him of the idea. Franco, according to his own account, was no more than stiffly polite to the Prime Minister. In the course of a stay in Galicia during which he was received enthusiastically, Azaña made an effort to be friendly but Franco did not respond with any warmth.[84] If indeed Franco set out to put distance between himself and the Prime Minister, Azaña seems not to have noticed.*

Franco's account probably reflects his desire to wipe away the disagreeable memory of the time when he was Azaña's subordinate. In fact, at this time, Franco was immensely careful.[85] When Sanjurjo requested that he appear as his defender in his trial, he refused. His glacial coldness was

---

* He later claimed that he had gone to great lengths not to be photographed with the Prime Minister, pointing out that his superior, Major General Vera, took priority. Franco also said that, by using the pretext that Doña Carmen was unwell, he had avoided being present at a morning reception given on Sunday 19 September by the La Coruña Sporting Club for Azaña and his friend and host, Santiago Casares Quiroga, the Minister of the Interior, and a prominent *gallego*. There exist photographs of them together during the visit to the city, next to each other and certainly with Franco nearer to Azaña than was General Vera. Similarly, the local press of the time reported Franco's presence at Azaña's table at a much more lavish occasion than the morning function, a dinner given that same evening at the Hotel Atlántida, in La Coruña and again at another lunch on Wednesday 21 September. See the photograph in Xosé Ramón Barreiro Fernández, *Historia contemporánea de Galicia* 4 vols (La Coruña, 1982) II, p. 241.

revealed when he said to his one-time commander, 'I could, in fact, defend you, but without hope of success. I think in justice that by rebelling and failing, you have earned the right to die'.[86] Nor did he join the conspiratorial efforts which led eventually to the creation of the Unión Militar Española, the clandestine organization of monarchist officers founded by Lieutenant-Colonel Emilio Rodríguez Tarduchy, a close friend of Sanjurjo, and Captain Bartolomé Barba Hernández, like Galarza an officer of the general staff. The UME emerged finally in late 1933 and was linked, through Galarza, to the activities of Ansaldo and Vigón.[87]

On 28 January 1933, the results of the *revisión de ascensos* were announced. Franco's promotion to colonel was impugned, that to general validated. Goded's promotions to brigadier and major-general were both annulled. However, they were not demoted but rather frozen in their present position in the seniority scale until a combination of vacancies arising and seniority permitted them to catch up with their accelerated promotions. So Franco kept his rank with effect from the date of his promotion in 1926. He nevertheless dropped from number one in the *escalafón* (list) of brigadier generals to 24, out of 36. Like most of his comrades, Franco smouldered with resentment at what was perceived as a gratuitous humiliation and nearly two years of unnecessary anxiety.[88] Years later, he still wrote of promotions being 'pillaged' (*despojo de ascensos*) and of the injustice of the entire process.[89]

In February 1933, Azaña had him posted to the Balearic Islands as comandante general, 'where he will be far from any temptations'.[90] It was a post which would normally have gone to a Major-General and may well have formed part of Azaña's efforts to attract Franco into the Republican orbit, rewarding him for his passivity during the *Sanjurjada*. After the preferments with which he had been showered by the King and Primo de Rivera, Franco did not perceive command of the Balearic Islands as a reward. In his draft memoirs, he wrote that it was less than his seniority merited (*postergación*).[91] More than two weeks after the appointment, he had still not made the reglamentary visit to the Ministry of War to report on his impending move. The Socialist leader, Francisco Largo Caballero, told Azaña that Franco had been heard to boast that he would not go.[92] Finally on 1 March, having been in Madrid for two days, he came to say his farewells to Azaña, in his capacity as Minister of War. The delay was a carefully calculated act of disrespect. Azaña perceived that Franco was still furious about the annulment of promotions but the subject did not arise, and they spoke merely of the situation in the Balearic Islands.[93] The new military commander arrived at Palma de Mallorca on

16 March 1933, and with Mussolini's ambitions heightening tension in the Mediterranean, dedicated himself to the job of improving the defences of the islands.

Throughout 1933, the fortunes of the Azaña government declined. By the beginning of September, the Republican-Socialist coalition was in tatters. Right-wing success in blocking reform had undermined the faith of the Socialists in Azaña's Left Republicans. On 10 September, the increasingly conservative and power-hungry Lerroux began to put together an all-Republican cabinet. It was reported in *ABC* that he had offered Franco the job of Minister or undersecretary of War. Although he came from the Balearic Islands to Madrid for discussions with the Radical leader, Franco finally declined the offer.[94] The post was one of those to which he aspired, but the Lerroux cabinet of 12 September was expected to last for no more than a couple of months since it could not command a parliamentary majority. Convinced that the only way to implement reform was to form a government on their own, the Socialists refused to rejoin a coalition with Azaña and it was widely assumed that President Alcalá Zamora would soon be forced to call general elections. In such conditions, taking over a ministry would have given Franco no opportunity to introduce the changes which he regarded as essential.

During the campaign for the November 1933 elections, with the possibility that the Socialists might win and establish a government bent on sweeping reform, Franco, although busy and fulfilled in the Balearics, was pessimistic about the prospects for the armed forces. He talked to friends of leaving the Army and going into politics. According to Arrarás, rumours to this effect reached rightist circles in Madrid and he was visited in Palma by a messenger from the increasingly powerful Catholic authoritarian party, the Confederación Española de Derechas Autónomas (the Spanish Confederation of Autonomous Right-Wing Groups). The envoy allegedly offered Franco inclusion as a candidate in both the CEDA's Madrid list and in another provincial list in order to guarantee his election. He refused outright.[95] He did, however, vote for the CEDA in the elections.[96] With the Left divided and the anarchists abstaining, a series of local alliances between the Radicals and the CEDA ensured their victory. The Radicals got 104 deputies and the CEDA 115 to the Socialists' 58 and the Left Republicans' 38. The subsequent period of government by a coalition of the ever-more corrupt Radicals and the CEDA would see Franco come in from the cold, as he perceived his comfortable exile in the Balearics, and much nearer to the centre of political preferment.

IV

IN COMMAND

Franco and the Second Republic, 1934–1936

AFTER THE vexations of the previous two years, the period of Centre-Right government, which came to be known by the Spanish Left as the *bienio negro* (two black years), moved Franco back into the sunlight. After what he perceived as the harsh persecution to which he and like-minded officers had been subjected by Azaña, the forty-two year-old general found himself lionized by politicians as he had not been since the Dictatorship. The reasons were obvious. He was the Army's most celebrated young general of rightist views, and was untainted by collaboration with the Republic. His renewed celebrity and favour coincided with, and indeed to an extent fed upon, the bitter polarization of Spanish politics in this period.

The Right saw its success in the November 1933 elections as an opportunity to put the clock back on the attempted reforms of the previous nineteen months of Republican-Socialist coalition government. In a context of deepening economic crisis, with one in eight of the workforce unemployed nationally and one in five in the south, a series of governments bent on reversing reform could provoke only desperation and violence among the urban and rural working classes. Employers and landowners celebrated victory by slashing wages, cutting their work forces, in particular sacking union members, evicting tenants and raising rents. The labour legislation of the previous governments was simply ignored.

Within the Socialist movement, rank-and-file bitterness at losing the elections and outrage at the vicious offensive of the employers soon pushed the leadership into a tactic of revolutionary rhetoric in the vain hope of frightening the Right into restraining its aggression and pressuring the President of the Republic, Niceto Alcalá Zamora, into calling new elections. In the long term, this tactic was to contribute to the feeling on the Right, and particularly within the high command of the Army, that strong authoritarian solutions were required to meet the threat from the Left.

Alcalá Zamora had not invited the sleek and pudgy CEDA leader, José María Gil Robles, to form a government despite the fact that the Catholic CEDA was the biggest party in the Cortes. The President suspected the immensely clever and energetic Gil Robles of planning to establish an authoritarian, corporative state and so turned instead to the cynical and corrupt Alejandro Lerroux, leader of the increasingly conservative Radicals, the second largest party. But Lerroux's power-hungry Radicals were dependent on CEDA votes and became the puppets of Gil Robles. In return for introducing the harsh social policies sought by the CEDA's wealthy backers, the Radicals were allowed to enjoy the spoils of office. The Socialists were angered by the corruption of the Radicals but the first working class protest came from the anarchists. With irresponsible naïvety, a violent uprising was called for 8 December 1933. However, the government had been forewarned of the anarcho-syndicalists' plans and quickly declared a state of emergency (*Estado de alarma*). Leaders of the CNT and the FAI were arrested, press censorship was imposed, and union buildings were closed down.

In traditionally anarchist areas – Aragón, the Rioja, Catalonia, the Levante, parts of Andalusia and Galicia – there were sporadic strikes, some trains were derailed and Civil Guard posts were attacked. After desultory skirmishes with the Civil Guard and the Assault Guards, the revolutionary movement was soon suppressed in Madrid, Barcelona and the provincial capitals of Andalusia, Alicante and Valencia. Throughout Aragón and in the regional capital, Zaragoza, however, the rising enjoyed a degree of success. Anarchist workers raised barricades, attacked public buildings, and engaged in armed combat with the forces of order. The government sent in several companies of the Army which, with the aid of tanks, took four days to crush the insurrection.[1] The movement reinforced the conviction of many of the more right-wing officers that, even with a conservative government in power, the Republic had to be overthrown.[2]

The difficulties experienced in the suppression of the revolt led, on 23 January 1933 to the resignation of the Minister of the Interior, Manuel Rico Avello, who was packed off to Morocco as High Commissioner. He was replaced by Diego Martínez Barrio, the Minister of War, who was replaced in turn by the conservative Radical deputy for Badajoz and crony of Lerroux, Diego Hidalgo who knew more about the agrarian problem than about military questions.* However, with engaging humility, he admitted his lack of military knowledge and his need for professional advice.³ He also set out to cultivate military sympathies for his party by softening the impact of some of the measures introduced by Azaña and reversing others.⁴ When the new Minister of War had been in post barely a week, at the beginning of February, Franco made his acquaintance in Madrid. Clearly impressed by the young general, at the end of March 1934, Hidalgo successfully placed before the cabinet a proposal for his promotion from Brigadier to Major-General (*General de División*), in which rank he was again the youngest in Spain.⁵ Hidalgo, expecting an effusive response, was dismayed by the cold and impersonal telegram which Franco sent him on receiving the news of his promotion. Reflecting on it later, Hidalgo commented, 'I never ever saw him either joyful or depressed'.⁶

The relationship between Franco and Hidalgo was consolidated in June during a four-day visit made by the Minister to the Balearic Islands where Franco was Comandante General. Hidalgo was much taken by the general's considerable capacity for work, his obsession with detail, his cool deliberation in resolving problems. One incident stuck in his mind. It was the Minister's custom on visiting garrisons to request that the commanding officer celebrate his visit by releasing any soldier currently under arrest. Although there was only one prisoner, a captain, in Menorca, Franco refused, saying 'if the Minister orders me I will do it; if he merely makes a request, no.' When Hidalgo asked what crime could be so heinous, Franco replied that it was the worst that any officer could commit: he had slapped a soldier. It was a surprising remark from the officer who had had a soldier shot for refusing to eat his rations. Both incidents in fact showed his obsession with military discipline. Hidalgo was so impressed by Franco that, before leaving Palma de Mallorca, and contrary to military protocol, he invited him to join him as an adviser that September during military manoeuvres in the hills (*montes*) of León.⁷

As 1934 progressed, Franco became the favourite general of the Radicals just as, when the political atmosphere grew more conflictive after October, he was to become the general of the more aggressively right-wing CEDA. The favour of Hidalgo contrasted strongly with the treatment Franco perceived himself to have suffered at the hands of Azaña.

---

* Family responsibilities had obliged him to avoid military service in 1907 by the device of buying himself out. This, together with the fact that he was the author of a book on the Russian revolution, ensured that his appointment was greeted with trepidation on the Right.

Moreover, with the Radical government, backed in the Cortes by the CEDA, pursuing socially conservative policies and breaking the power of one union after another, the Republic began to seem altogether more acceptable to Franco. For many conservatives, 'catastrophist' solutions to Spain's problems seemed for the moment less urgent. The extreme Right, however, remained unconvinced and so continued to prepare for violence. The most militant group on the ultra Right were the Carlists of the Traditionalist Communion, break-away royalists who had rejected the liberal heresy of the constitutional monarchists and advocated an earthly theocracy under the guidance of warrior priests. The Carlists were collecting arms and drilling in the north and the spring of 1934 saw Fal Conde, the movement's secretary, recruiting volunteers in Andalusia. The Carlists, together with the fascist Falange Española, and the influential and wealthy 'Alfonsists', the conventional supporters of Alfonso XIII and General Primo de Rivera, constituted the self-styled 'catastrophist' Right. They were so-called because of their determination to destroy the Republic by means of a cataclysm rather than by the more gradual legalist tactic favoured by the CEDA. Their plans for an uprising would eventually come to fruition in the summer of 1936.

On 31 March 1934, two Carlist representatives accompanied by the leader of the Alfonsist monarchist party, Renovación Española, Antonio Goicoechea, and General Barrera saw Mussolini in Rome. They signed a pact which promised money and arms for a rising.[8] In May 1934, the monarchists' most dynamic and charismatic leader, José Calvo Sotelo, was granted amnesty and returned to Spain after the three years' exile suffered as he fled the 'responsibilities' campaign. Henceforth, the extreme rightist press, in addition to criticizing Gil Robles for alleged weakness, began to talk of the need to 'conquer the State' – a euphemism for the violent seizure of its apparatus, as the only certain way to guarantee a permanent authoritarian, corporative regime.

Although Franco was careful to distance himself from the generals who were part of monarchist conspiracies, he certainly shared some of their preoccupations. His ideas on political, social and economic issues were still influenced by the regular bulletins which had been receiving since 1928 from the *Entente Internationale contre la Troisième Internationale* of Geneva. In the spring of 1934, he took out a new subscription at his own expense, writing to Geneva on 16 May expressing his admiration for 'the great work which you carry out for the defence of nations from Communism' and his 'wish to co-operate, in our country, in your great effort'.[9] An ultra-right-wing organization which now had contacts with

Dr Goebbels' Antikomintern, the Entente skilfully targeted and linked up influential people convinced of the need to prepare for the struggle against Communism, and supplied subscribers with reports which purported to expose plans for forthcoming Communist offensives. The many strikes which took place during 1934, when seen through the prism of the Entente's publications, helped convince Franco that a major Communist assault on Spain was under way.[10]

If Franco was circumspect with regard to extreme Right monarchist conspirators, he had even less to do with the nascent fascist groups which were beginning to appear on the scene. Gil Robles' youth movement, the *Juventud de Acción Popular* (JAP) held great fascist-style rallies were held at which Gil Robles was hailed with the cry '*¡Jefe! ¡Jefe! ¡Jefe!*' (the Spanish equivalent of *Duce*) in the hope that he might start a 'March on Madrid' to seize power. However, the JAP was not taken seriously by the 'catastrophist' Right. Monarchist hopes focused rather more on the openly fascist group of José Antonio Primo de Rivera, the Falange, as a potential source of shock troops against the Left. As a southern land-owner, an aristocrat and eligible socialite, and above all as the son of the late dictator, José Antonio Primo de Rivera was a guarantee to the upper classes that Spanish fascism would not get out of their control in the way of its German and Italian equivalents. The Falange remained insignificant until 1936, important until then only for the role played by its political vandalism in screwing up the tension which would eventually erupt into the Civil War. José Antonio was a close friend of Ramón Serrano Suñer, Franco's brother-in-law, but despite Serrano's efforts to bring them together, the cautious, hard-working general and the flamboyant playboy would never hit it off.

Indeed, during the first half of 1934, Franco's interest in politics was minimal. In late February, his mother Pilar Bahamonde de Franco had decided to go on pilgrimage to Rome. Franco travelled to Madrid in order to escort her to Valencia to catch a boat to Italy. While in the capital, staying at the home of her daughter Pilar, she caught pneumonia. After an illness lasting about ten days, she died on 28 February, aged sixty-six. It is the unanimous affirmation of those close to him that the loss affected Francisco profoundly despite the fact that he had not lived with his mother for twenty-seven years. He had adored her.[11] Outside the family, he showed no signs of his bereavement. After her death, Franco rented a large apartment in Madrid where he and his wife regularly received the visits of other generals, prominent right-wing politicians, aristocrats and the elite of Oviedo when they passed through the capital. The most frequent recreations of Francisco and Carmen were visits to the cinema and to the flea-market (*Rastro*) in search of antiques, often accompanied by their favourite niece Pilar Jaraiz Franco.[12]

While Franco concerned himself with family and professional matters, the political temperature was rising throughout Spain. The Left was deeply sensitive to the development of fascism and was determined to avoid the fate of their Italian, German and Austrian counterparts. Encouraged by Gil Robles, the Radical Minister of the Interior Rafael Salazar Alonso was pursuing a policy of breaking the power of the Socialists in local administration and provoking the unions into suicidal strikes. The gradual demolition of the meagre Republican-Socialist achievements of 1931–1933 reached its culmination on 23 April with the amnesty of those accused of responsibilities for the crimes of the Dictatorship, like Calvo Sotelo, and those implicated in the coup of 10 August 1932, most notably Sanjurjo himself. Lerroux resigned in protest after Alcalá Zamora had hesitated before signing the amnesty bill. While Lerroux ran the government from the wings, one of his lieutenants, Ricardo Samper, took over as prime minister. Socialists and Republicans alike felt that the entire operation was a signal from the Radicals to the Army that officers could rise whenever they disliked the political situation.[13] The Left was already suspicious of the government's dependence on CEDA votes, because the monarchist Gil Robles refused to affirm his loyalty to the Republic.

Political tension grew throughout 1934. Successive Radical cabinets were incapable of allaying the suspicion that they were merely Gil Robles' Trojan Horse. By repeatedly threatening to withdraw his support, Gil Robles provoked a series of cabinet crises as a result of which the Radical government took on an ever more rightist colouring. On each occasion, some of the remaining liberal elements of Lerroux's party would be pushed into leaving it and its rump became progressively more dependent on CEDA whims. With Salazar Alonso provoking strikes throughout the spring and summer of 1934 and thereby picking off the most powerful unions one by one, the government widened its attacks on the Republic's most loyal supporters and also began to mount an assault against the Basques and, even more so, the Catalans.

In Catalonia, the regional government or *Generalitat* was governed by a left Republican party, the *Esquerra*, under Luis Companys. In April, Companys had passed an agrarian reform, the *Ley de Cultivos*, to protect tenants from eviction by landowners. Although Madrid declared the reform unconstitutional, Companys went ahead and ratified it. Meanwhile, the government began to infringe the Basques' tax privileges and, in an attempt to silence protest, forbade their municipal elections. Such high-handed centralism could only confirm the Left's fears of the Republic's rapid drift to the right. That anxiety was intensified by Salazar Alonso's provocation and crushing defeat of a major national strike by the Socialist landworkers' union during the summer. There were hundreds of arrests of trade union leaders and thousands of internal deportations, with peasants herded onto trucks and driven hundreds of miles from their homes to be left to make their way back without food or money. In the meantime, Army conscripts brought in the harvest. Workers' societies were closed down and leftists on town councils forcibly replaced by government nominees. In the Spanish countryside, the clock was being put back to the 1920s.[14]

The vengeful policies pursued by the Radical governments and encouraged by the CEDA divided Spain. The Left saw fascism in every action of the Right; the Right, and many Army officers, smelt Communist-inspired revolution in every demonstration or strike. In the streets, there was sporadic shooting by Socialist and Falangist youths. The Government's attacks on regional autonomy and the increasingly threatening attitude of the CEDA were driving sections of the Socialist movement to place their hopes in a revolutionary rising to forestall the inexorable destruction of the Republic. On the Right, there was a belief that, if the Socialists could be provoked into an insurrection, an excuse would be provided to crush them definitively. Gil Robles' youth movement, the JAP, held a rally on 9 September at Covadonga in Asturias, the site of a battle in 732 considered to be the starting point for the long reconquest of Spain from the Moors. The symbolic association of the right-wing cause with the values of traditional Spain and the identification of the working class with the Moorish invaders was a skilful device that would help secure military sympathy. It foreshadowed the Francoist choreography of the *Reconquista* developed after 1936 with Franco himself cast as the medieval warrior king.

At the rally, Gil Robles spoke belligerently of the need to crush the 'separatist rebellion' of the Catalans and the Basque Nationalists.[15] The wily Gil Robles – the politician on the Right with the greatest strategic vision – knew that the Left considered him a fascist and was determined to prevent the CEDA coming to power. He therefore pushed for the CEDA to join the government precisely in order to provoke a Socialist reaction. This is in fact what happened. CEDA ministers entered the cabinet; there was an uprising in Asturias and it was smashed by the Army.[16] Gil Robles said later: 'I asked myself this question: "I can give

Spain three months of tranquillity if I do not enter the government. If we enter, will the revolution break out? Better that it do so before it is well prepared, before it defeats us." This is what we did, we precipitated the movement, met it and implacably smashed it from within the government'.[17]

In September, Franco left the Balearics and travelled to the mainland to take up Diego Hidalgo's invitation to join him as his personal technical adviser during the Army manoeuvres taking place in León at the end of the month under the direction of General Eduardo López Ochoa. Since López Ochoa had been part of the opposition against Primo de Rivera and was implicated in the December 1930 military rebellion, Franco regarded him with some hostility. It is possible that the large-scale military manoeuvres, planned in the late spring, were part of a wider project by Salazar Alonso, Hidalgo and Gil Robles to crush the Left. The manoeuvres were held in an area contiguous, and of nearly identical terrain, to Asturias where the final left-wing bid to block the CEDA's passage to power was likely to come.[18] In retrospect, it seems more than a coincidence that the Minister of War should have arranged for Franco to accompany him as his personal adviser on those manoeuvres and should then put him in charge of the repression of the revolutionary strike.

It is not clear why the Minister needed a 'personal technical adviser' when López Ochoa and other senior officers, including the Chief of the General Staff, were there under his orders. On the other hand, if the central concern was the ability of the Army to crush a left-wing action, Franco was more likely to give firm advice than López Ochoa or General Carlos Masquelet, the Chief of Staff. Franco's first biographer, Joaquín Arrarás, claimed that when Hidalgo invited Franco to leave the Balearics and come to the mainland, 'his real intention was to ensure that the general would be in Madrid at the Minister's side during the hazardous days which were expected'.[19] There can be no doubt that Hidalgo was aware of a possible left-wing insurrection. At the end of August, he had named General Fanjul to head an investigation into the loss of weapons from the state small-arms factories.[20] Then, in early September, when some members of the cabinet had been in favour of cancelling the manoeuvres, Hidalgo insisted that they go ahead precisely because of imminent left-wing threats. Three days before the manoeuvres began, Hidalgo ordered the Regiment no.3 from Oviedo which was to have taken part not to leave the Asturian capital again because he expected a revolutionary outbreak.[21] Moreover, the astonishing speed with which

Franco was later able to get the Spanish Legion from Africa to Asturias suggests some prior consideration of the problem.

On the Right, the readiness of the Army to deal with a likely leftist initiative was an issue of frequent discussion. Salazar Alonso raised it at cabinet meetings and in press interviews. At this time, secret contacts between the CEDA and senior military figures had provided assurances that the Army was confident of being able to crush any leftist uprising provoked by CEDA entry into the cabinet.[22] Curiously, during the manoeuvres, José Antonio Primo de Rivera made an effort to cultivate a relationship with Franco. On the fringe of events, but clearly impressed by indications of Franco's likely influence on what was about to happen, the Falange leader wrote him a frantic letter* claiming that Socialist victory was imminent and equivalent to 'a foreign invasion' since France would seize the opportunity to annex Catalonia. It is indicative of Franco's confidence in Diego Hidalgo at this time that he read José Antonio's letter without interest and did not bother to reply.[23]

Nevertheless, the political crisis was soon to absorb Franco totally. On 26 September, Gil Robles made his move and announced that the CEDA could no longer support a minority government. In dutiful response, Lerroux formed a new cabinet including three CEDA ministers. There was outrage among even conservative Republicans. The UGT called a general strike. In most parts of Spain, the prompt action of the government in declaring martial law and arresting the hesitant Socialist leaders guaranteed its failure.[24] In Barcelona events were more dramatic. Pushed by extreme Catalan nationalists, and alarmed by developments in Madrid, Companys proclaimed an independent state of Catalonia 'within the Federal Republic of Spain' in protest against what was seen as the betrayal of the Republic. It was a largely rhetorical gesture since the rebellion of the *Generalitat* was doomed when Companys refused to arm the workers. The futile defence of the short-lived Catalan Republic was undertaken by a small group of officers from the local security services. They were soon overwhelmed.[25] The only place where the Left's protest was not easily brushed aside was in Asturias. There, the emergence of spontaneous rank-and-file revolutionary committees impelled the local Socialist leaders to go along with a movement organized jointly by the UGT, the CNT and, belatedly, the Communists, united in the *Alianza Obrera* (workers' alliance).[26]

---

* Once more Ramón Serrano Suñer served as the intermediary between them, entrusting delivery of the letter to his brother José.

During the September manoeuvres, Franco had asked the Minister for permission to visit Oviedo on family business before returning to the Balearics – Franco planned to sell some land belonging to his wife. However, before he could set off from Madrid, the Asturian revolutionary strike broke out. Diego Hidalgo decided that Franco should stay on at the Ministry as his personal adviser.[27] The situation worsened and, on 5 October, the Civil Governor of Asturias handed over control of the region to the military commander of Oviedo, Colonel Alfredo Navarro, who immediately declared martial law. At a tense cabinet meeting on 6 October, chaired by the President of the Republic Alcalá Zamora, it was decided to name General López Ochoa to command the troops sent to fight the revolutionary miners. The choice of López Ochoa for this difficult task reflected both his position as Inspector General del Ejército in the region and his reputation as a loyal republican and a freemason. López Ochoa later confided to the Socialist lawyer Juan-Simeón Vidarte that Alcalá Zamora had asked him to undertake the task precisely because he thereby hoped to keep bloodshed to a minimum. This created serious friction with Hidalgo, Salazar Alonso and the three new CEDA ministers who, urged on by Gil Robles, had been in favour of sending General Franco. They then tried unsuccessfully to have Franco named Chief of the General Staff instead of the more liberal incumbent, Masquelet, a friend of Azaña.[28]

Although the proposal to put Franco formally in command of troops in Asturias was rejected by Alcalá Zamora, Diego Hidalgo informally put him in overall charge of operations. Franco thus received an intoxicating taste of unprecedented politico-military power. The Minister used his 'adviser' as an unofficial Chief of the General Staff, marginalising his own staff and slavishly signing the orders which Franco drew up.[29] In fact, the powers informally exercised by Franco went even further than might have been apparent at the time. The declaration by decree of martial law (*estado de guerra*) effectively transferred to the Ministry of War the responsibilities for law and order normally under the jurisdiction of the Ministry of the Interior. Diego Hidalgo's total reliance on Franco effectively gave him control of the functions of both Ministries.[30] The Minister's desire to have Franco by his side in Madrid is comprehensible. He admired him and Franco had specific knowledge of Asturias, its geography, communications and military organization. He had been stationed there, had taken part in the suppression of the general strike of 1917 and had been a regular visitor since marrying Carmen. Nevertheless, the particularly harsh manner in which Franco directed the repression from

Madrid gave a stamp to the events in Asturias which they might not have had if control had been left to the permanent staff of the Ministry.

The idea that a soldier should exercise such responsibilities came naturally to Franco. It harked back to the central ideas on the role of the military in politics which he had absorbed during his years as a cadet in the Toledo Academy. It was a step back in the direction of the golden years of the Primo de Rivera dictatorship. He took for granted the implicit recognition of his personal capacity and standing. All in all, it was to be a profoundly formative experience for him, deepening his messianic conviction that he was born to rule. He would try unsuccessfully to repeat it after the Popular Front election victory in February 1936 before doing so definitively in the course of the Civil War.

Hidalgo's decision to use Franco derived also from his distrust, fuelled by Gil Robles, of both General Masquelet and other liberal officers in the Ministry of War who had been close to Azaña.[31] At the time, the unusual appointment provoked criticisms from the under-secretary of the Ministry of War, General Luis Castelló.[32] Franco's approach to the events of Asturias was coloured by his conviction, fed by the material he received from the Entente Anticomuniste of Geneva, that the workers' uprising had been 'deliberately prepared by the agents of Moscow' and that the Socialists 'with technical instructions from the Communists, thought they were going to be able to install a dictatorship'.[33] That belief no doubt made it easier to use troops against Spanish civilians as if they were a foreign enemy.

In the telegraph room of the Ministry of War, Franco set up a small command unit consisting of himself, his cousin Pacón and two naval officers, Captain Francisco Moreno Fernández and Lieutenant-Commander Pablo Ruiz Marset. Having no official status, they worked in civilian clothes. For two weeks, they controlled the movement of the troops, ships and trains to be used in the operation of crushing the revolution. Franco even directed the naval artillery bombardments of the coast, using his telephone in Madrid as a link between the cruiser *Libertad* and the land forces in Gijón.[34] Uninhibited by the humanitarian considerations which made some of the more liberal senior officers hesitate to use the full weight of the armed forces against civilians, Franco regarded the problem before him with icy ruthlessness.

The rightist values to which he was devoted had as their central symbol the reconquest of Spain from the Moors. Yet, doubting the readiness of working class conscripts to fire on Spanish workers, and anxious not to encourage the spread of revolution by weakening garrisons elsewhere in the mainland, Franco had no qualms about shipping Moorish mercenaries to fight in Asturias, the only part of Spain where the crescent had never flown. There was no contradiction for him in using the Moors in the simple sense that he regarded left-wing workers with the same racialist contempt with which he had the tribesmen of the Rif. 'This is a frontier war', he commented to a journalist, 'against socialism, Communism and whatever attacks civilization in order to replace it with barbarism'.[35] Two *banderas* of the Legion and two *tabores* of Regulares were sent to Asturias with unusual speed and efficiency.

When it became known that one of the officers in charge of the troops coming from Africa, Lieutenant-Colonel López Bravo, had expressed doubts as to whether they would fire on civilians, Franco recommended his immediate replacement. He placed his Academy contemporary and close friend Colonel Juan Yagüe in overall charge of the African troops. He also ordered the removal of the commander of the León Air Force base, his cousin and childhood friend, Major Ricardo de la Puente Bahamonde, because he suspected that he sympathized with the miners and was ordering his pilots not to fire on the strikers in Oviedo. Almost immediately, Franco ordered the bombing and shelling of the working class districts of the mining towns. Some of the more liberal generals regarded such orders as excessively brutal.[36]

The losses among women and children, along with the atrocities committed by Yagüe's Moroccan units, contributed to the demoralization of the virtually unarmed revolutionaries. Yagüe sent an emissary to Madrid to complain to both Franco and Gil Robles about the humanitarian treatment given by López Ochoa to the miners. López Ochoa's pact with the miners' leader Belarmino Tomás permitted an orderly and bloodless surrender and so provoked Franco's suspicions.[37] In contrast, Franco showed total confidence in Yagüe during the active hostilities, in the course of which a savage repression was carried out by the African troops. When Gijón and Oviedo were recaptured by government troops, summary executions of workers were carried out.[38]

Thereafter, Franco also put his stamp on the political mopping up. After the miners surrendered, Hidalgo and Franco regarded their task as unfinished until all those involved had been arrested and punished. After Hidalgo 'took advice', presumably Franco's, the police operations were entrusted to the notoriously violent Civil Guard Major Lisardo Doval who was appointed on 1 November 'delegate of the Ministry of War for public order in the provinces of Asturias and León'. Doval was widely considered an expert on left-wing subversion in Asturias. His fame as a crusader against the Left had made him immensely popular among the upper and middle classes of the region. He was given special powers to by-pass any judicial control or other legal obstacles to his activities. As Franco knew he would, Doval carried out his task with a relish for brutality which provoked horror in the international press. It has been suggested that Franco was unaware of either Doval's methods or his reputation as a torturer.[39] This is unlikely given that they had coincided as boys in El Ferrol, in the Infantry Academy at Toledo and in Asturias in 1917.

The right-wing press presented Franco, rather than López Ochoa, as the real victor over the revolutionaries and as the mastermind behind such a rapid success. Diego Hidalgo was unstinting in his praise for Franco's value, military expertise and loyalty to the Republic and the rightist press began to refer to him as the 'Saviour of the Republic'.[40] In fact, Franco's handling of the crisis had been decisive and efficient but hardly brilliant. His tactics, however, were interesting in that they prefigured his methods during the Civil War. They had consisted essentially of building up local superiority to suffocate the enemy and, as the use of Yagüe and Doval indicated, sowing terror within the enemy ranks.[41]

After the victory over the Asturian rebels, Lerroux and Gil Robles agonized over the issue of death penalties for the revolutionaries in Asturias and the officers who had defended the short-lived Catalan Republic. The trials which would make most impact on Franco were those involving charges of military rebellion. On 12 October 1934, the officers who had supported the rebellion in Catalonia had been tried and sentenced to death. Sergeant Diego Vázquez, who had deserted to join the strikers in Asturias, was tried and sentenced to death on 3 January 1935.[42] The bulk of the Right howled for vengeance but Alcalá Zamora favoured clemency and Lerroux was inclined to agree. Many on the Right wanted Gil Robles to withdraw CEDA support for the government if the death sentences were not carried out. He refused for fear of Alcalá Zamora giving power to a more liberal cabinet.

Franco, always rigidly in favour of the severest penalties for mutiny and of the strictest application of military justice, believed that Gil Robles was totally mistaken. He told the Italian *Chargé d'Affaires*, Geisser Celesia, 'The victory is ours and not to apply exemplary punishments to the rebels, not to castigate energetically those who have encouraged the revolution and have caused so many casualties among the troops, would signify trampling on the just rights of the military class and encourage an early extremist response.'[43] The fact that pardons were eventually granted would contribute in 1936 to Franco's decision to take part in the military uprising which opened the Civil War.

In 1934, however, Franco was hostile to any military intervention in politics. His part in suppressing the Asturian insurrection had left him satisfied that a conservative Republic ready to use his services could keep the Left at bay. Not all his comrades-in-arms shared his complacency. Fanjul and Goded were discussing with senior CEDA figures the possibility of a military coup to forestall the commutation of the death sentences. Gil Robles told them through an intermediary that the CEDA would not oppose a coup. It was agreed that they would consult other generals and the commanders of key garrisons to see if it might be possible 'to put Alcalá Zamora over the frontier'. After checking with Franco and others, they concluded that they did not have the support necessary for a coup.[44]

Franco exercised a similarly restraining influence over other would-be rebels. In late October, Jorge Vigón and Colonel Valentín Galarza believed that the moment had come to launch the military rising which they had been preparing since the autumn of 1932. Their plan was for the monarchist aviator, Juan Antonio Ansaldo, to fly to Portugal, pick up Sanjurjo and take him to the outskirts of Oviedo where he would link up with Yagüe. It was assumed that together Sanjurjo and Yagüe would easily persuade the bulk of the Army to join them in rebellion against the Republic. While the conspirators waited in the home of Pedro Saínz Rodríguez for the order to proceed, the journalist Juan Pujol arrived to say that he had spoken with Franco at the Ministry of War and Franco believed that it was not the right moment.[45] Enjoying considerable power and confident of his ability to use it decisively against the Left, he had no reason to want to risk his career in an ill-prepared coup. The fact that other prominent officers now deferred to his views, as they had not in 1932, was a measure of the dramatic increase in prestige bestowed upon him by the events in Asturias.

Although delighted with the repression of the Asturian rising, Gil Robles sought to strengthen his own political position and so he joined Calvo Sotelo in deriding the Radical government for weakness. Diego Hidalgo was one of the sacrificial victims.[46] Accordingly, from 16 November 1934 to 3 April 1935, the Prime Minister, Alejandro Lerroux, himself took over the Ministry of War. He awarded Franco the Gran Cruz de Mérito Militar and kept him in his extraordinary post of ministerial adviser until February 1935. Lerroux had intended to reward Franco by making him High Commissioner in Morocco but was prevented from doing so by the opposition of Alcalá Zamora.[47] Instead, he kept on the existing civilian High Commissioner, the conservative Republican Manuel Rico Avello, and made Franco Commander-in-Chief of the Spanish Armed Forces in Morocco.

Despite any disappointment that he might have felt at not being made High Commissioner, being an *Africanista*, Franco perceived the post of head of the African Army as a substantial reward for his work in repressing the revolution. As he put it himself, 'the Moroccan Army constituted the most important military command'.[48] On arrival, he hastened to inform the *Entente Internationale contre la Troisième Internationale* of his change of address.[49] Although he was to be there barely three months, it was a period which he enjoyed immensely. As Commander-in-Chief, he consolidated his existing influence within the armed forces in Morocco and established new and important contacts which were to facilitate his intervention at the beginning of the Civil War. His relationship with Rico Avello was similar in many respects to that which he had enjoyed with Diego Hidalgo. The High Commissioner, recognizing his own ignorance of Moroccan affairs, relied on Franco for advice of all kinds. Franco also established an excellent working relationship with the Chief of the General Staff of the Spanish forces in Morocco, Colonel Francisco Martín Moreno. This was to be crucial in 1936.[50]

On the road to civil war, there could be no going back from the events of October 1934. The Asturian rising had frightened the middle and upper classes. Equally, the vengeful repression urged by the Right and carried out by the Radical-CEDA coalition convinced many on the Left that electoral disunity must never be risked again. The publicity given to Franco's role in the military repression of the uprising ensured that thereafter he would be regarded as a potential saviour by the Right and as an enemy by the Left. Franco himself was to draw certain conclusions from the Asturian uprising. Convinced by the material received from Geneva that a Communist assault on Spain was being planned, he saw the events of October 1934 in those terms. He was determined that the Left should never be allowed to enjoy power even if won democratically.[51]

Nothing was done by successive conservative governments in the fifteen months after October 1934 to eliminate the hatreds aroused by the revolution itself or by its brutal repression. The CEDA claimed that it would remove the need for revolution by a programme of moderate land and tax reforms. Even if this claim was sincere in the mouths of the party's few convinced social Catholics, the limited reforms proposed were blocked by right-wing intransigence from the majority. Thousands of political prisoners remained in jail; the Catalan autonomy statute was suspended and a vicious smear campaign was waged against Azaña in a vain effort to prove him guilty of preparing the Catalan revolution. Azaña was thereby converted into a symbol for all those who suffered from the repression.[52]

The CEDA made a significant advance towards its goal of the legal introduction of an authoritarian corporate state on 6 May 1935 when five *Cedistas*, including the *Jefe* himself as Minister of War, entered a new cabinet under Lerroux. Gil Robles appointed known opponents of the regime to high positions – Franco was recalled from Morocco to become Chief of the General Staff; Goded became Inspector General and Director of the Air Force, and Fanjul became Under-Secretary of War. The President, Alcalá Zamora, was hostile to the appointment of Franco, regularly remarking that 'young generals aspire to be fascist caudillos'. Eventually, threats of resignation from both Lerroux and Gil Robles overcame the President's opposition.[53] There was a fierce rivalry and mutual dislike between Franco and Goded. Goded had wanted the job of Chief of the General Staff and was heard to comment bitterly that he awaited the failure of Franco.[54]

Franco in mid-1935 was still some way from thinking in terms of military intervention against the Republic. Indeed, it would be wrong to assume that he spent much time thinking about overthrowing the Republic. As long as he had a posting which he considered to be appropriate to his merits, he was usually content to get on with his job in a professional manner. He had been extremely happy during his three months in Morocco and, while sad to leave an interesting job, he was thrilled by this even more important posting. In his new post, able to carry on the job which he had done in October, he can have felt little urge or need to rebel at this time. In any case, he remained deeply influenced by the failure of Sanjurjo's coup of 10 August 1932. Moreover, given the ease of his relationship with Gil Robles, his day-to-day work gave him enormous satisfaction.[55]

As Chief of Staff, Franco worked long hours to fulfil his central task which he saw as being to 'correct the reforms of Azaña and return to the components of the armed forces the internal satisfaction which had been lost with the coming of the Republic'. He neglected his family, obsessively working until late at night, at weekends and on holidays.[56] Azaña's revisions of promotions by merit were set aside. Many loyal Republican officers were purged and removed from their posts, because of their 'undesirable ideology'. Others, of known hostility to the Republic, were reinstated and promoted. Emilio Mola was made General in command of Melilla and shortly afterwards head of military forces in Morocco. José Enrique Varela was promoted to general. Medals and promotions were distributed to those who had excelled in the repression of the October uprising.[57] Gil Robles and Franco had secretly brought Mola to Madrid to prepare detailed plans for the use of the colonial Army in mainland Spain in the event of further left-wing unrest.[58]

Alcalá Zamora remained deeply suspicious of Gil Robles' political motives in fostering the careers of anti-Republican officers and in trying to transfer control of the Civil Guard and the police from the Ministry of the Interior to the Ministry of War. In some ways – regimental reorganization, motorization, equipment procurement – Gil Robles continued the reforms of Azaña.[59] The CEDA-Radical government was anxious for the Army to re-equip to ensure its efficacy in the event of having to face another left-wing rising. As Chief of Staff, Franco was involved in establishing contacts with arms manufacturers in Germany as part of the projected rearmament.[60] There can be little doubt that he enjoyed his new job as much as he had liked being Director of the Military Academy in Zaragoza. Despite the later deterioration of their relationship after 1936, he and Gil Robles worked well together in a spirit of co-operation and mutual admiration. Like Diego Hidalgo and Manuel Rico Avello, Gil Robles recognized his own ignorance in military affairs and was happy to leave Franco to get on with things. Franco looked back on his period as Chief of the General Staff with great satisfaction because his achievements facilitated the later Nationalist war effort.[61]

After earlier doubts, in the late summer of 1935, Franco made contact, through Colonel Valentín Galarza, with the *Unión Militar Española*, the extreme rightist conspiratorial organization run by his one-time subordinate Captain Bartolomé Barba Hernández. Galarza, who organized UME liaison between the various garrisons across the country, kept Franco informed about the morale and readiness of the organization's members. In retrospect, Franco saw his approach to the UME as being to prevent it 'organizing a premature coup along the lines of a nineteenth century *pronunciamiento*'.[62] It is entirely in character that he would want any military action in which he might be involved to be fully prepared.

On 12 October 1935, Don Juan de Borbón, the son of Alfonso XIII, married in Rome. It was to be an excuse for monarchists, among them the plotters of Acción Española, such as José Calvo Sotelo, Jorge Vigón, Eugenio Vegas Latapie, Juan Antonio Ansaldo, to travel en masse to Italy. Franco was not among their number. Nevertheless, he did contribute to the wedding present given by the officers who had once been *gentilhombres* of Alfonso XIII.[63]

Franco's readiness to make contact with the UME reflected his concern at the fact that, despite the strength of the repression, the organized Left was growing in strength, unity and belligerence. The economic misery of large numbers of peasants and workers, the savage persecution of the October rebels and the attacks on Manuel Azaña combined to produce an atmosphere of solidarity among all sections of the Left. A series of gigantic mass meetings were addressed by Azaña in the second half of 1935 and the enthusiasm for unity shown by the hundreds of thousands who attended them helped clinch mass enthusiasm for what became the Popular Front.

The tiny Spanish Communist Party joined the Popular Front, an electoral coalition which, contrary to rightist propaganda and the material sent to Franco by the *Entente contre la Troisième Internationale* of Geneva, was not a Comintern creation but the revival of the 1931 Republican-Socialist coalition. The Left and centre Left joined together on the basis of a programme of amnesty for prisoners, of basic social and educational reform and trade union freedom. However, Comintern approval of the Popular Front strategy, ratified at its VII Congress on 2 August 1935, was used by the *Entente* to convince its subscribers, including Franco, that Moscow planned a revolution in Spain.[64]

Gil Robles' tactic of gradually breaking up successive Radical cabinets was overtaken in the autumn by the revelation of two massive financial scandals involving followers of Lerroux. In mid-September, Alcalá Zamora invited the dour conservative Republican, Joaquín Chapaprieta, to form a government. With the Radical Party on the verge of disintegration, Gil Robles provoked the resignation of Chapaprieta on 9 December in the belief that he would be asked to form a government. Alcalá Zamora, however, had no faith in Gil Robles's commitment to the Republic. Instead, when he spoke with the President on 11 December, Gil Robles learned with rage that he was not being asked to be prime minister. Alcalá Zamora pointed out that the degree of government instability demonstrated the need for new elections. Gil Robles could hardly argue that it would now stop since he had provoked that instability in order to pave the way to firm government by himself. He had overplayed his hand. The President was so suspicious of Gil Robles that, throughout the subsequent political crisis, he had the Ministry of War surrounded by Civil Guards and the principal garrisons and airports placed under special vigilance.[65]

The only choice now open to Gil Robles was to patch together some compromise which would enable the CEDA to avoid elections and thus carry on in the government or else arrange a coup d'état. He tried both options simultaneously. On the same evening a messenger was sent to Cambó, head of the Catalan Lliga, to ask him to join the CEDA and the Radicals in a coalition government. Cambó refused. Meanwhile, in the Ministry of War, Gil Robles was discussing the situation with Fanjul. Fanjul claimed enthusiastically that he and General Varela were prepared to bring the troops of the Madrid garrison onto the streets that very night to prevent the President from going through with his plans to dissolve the Cortes. There were plenty of officers only too willing to join them, especially if a coup had the blessing of the Minister of War and could therefore be seen as an order. However, Gil Robles was worried that such an action might fail, since it would certainly face the resistance of the Socialist and anarchist masses. Nevertheless, he told Fanjul that, if the Army felt that its duty lay in a coup, he would not stand in its way and, indeed, would do all that he could to maintain the continuity of government while it took place. Only practical doubts held him back and so he suggested that Fanjul check the opinion of Franco and other generals before making a definite decision. He then passed a sleepless night while Fanjul, Varela, Goded and Franco weighed up the chances of success. All were aware of the problem presented by the fact that there was every likelihood that the Civil Guard and the police would oppose a coup.[66]

Calvo Sotelo, confined to bed with a fierce attack of sciatica, also sent Juan Antonio Ansaldo to see Franco, Goded and Fanjul to urge them to make a coup against the plans of Alcalá Zamora. Franco, however, convinced his comrades that, in the light of the strength of working class resistance during the Asturian events, the Army was not yet ready for a coup. [67] When the young monarchist plotter, the Conde de los Andes, telephoned Madrid from Biarritz to hear the details of the expected coup, Ansaldo replied 'The usual generals, and especially the *gallego*, say that they cannot answer for their people and that the moment has not yet arrived'.[68] The government of Joaquín Chapaprieta was replaced by the interim cabinet of Manuel Portela Valladares. Thus, on 12 December, Gil Robles was obliged to abandon the Ministry of War with 'infinite bitterness'. When the staff of the Ministry said goodbye to Gil Robles on 14 December, a tearful Franco made a short speech in which he declared 'the Army has never felt itself better led than in this period.[69]

In response to the move towards a more liberal cabinet, José Antonio Primo de Rivera sent his lieutenant Raimundo Fernández Cuesta to Toledo on 27 December with a wild proposal to Colonel José Moscardó, military governor and Director of the *Escuela Central de Gimnasia* (Central School of Physical Education) there. The suggestion was that several hundred Falangist militants would join the cadets in the Alcázar of Toledo to launch a coup. Common sense should have told Moscardó that it was a ridiculous idea. However, he felt that he could not make a decision without discussing it first with Franco. Leaving Fernández Cuesta waiting in Toledo, he drove to Madrid and consulted with the Chief of the General Staff who, as could have been foreseen, told him that the scheme was impracticable and badly timed.[70]

Franco made it clear that he resented these initiatives from civilians as attempts to take advantage of the 'most distinguished officers' for their own partisan purposes. Moscardó was one of a number of officers, to whom he referred as 'simplistic comrades', who brought such proposals to him. He told them all that to precipitate matters was to guarantee failure. The job of the Army was to maintain its unity and discipline to be ready to intervene if and when the Republic proved itself totally unviable. What the Army could not do was to try to destroy the Republic before the population was ready.[71] After Gil Robles was replaced as Minister of War by General Nicolás Molero, Franco was left as Chief of the General Staff. Like his predecessor, Molero was happy for Franco to get on with a job which he did well. Franco wrote to a friend on 14 January 1936, 'I am still here in my post and I don't think they'll move me'. His contentment, along with his natural caution, may well have contributed to his inclination against conspiratorial adventures.[72]

The elections were scheduled for 16 February 1936. Throughout January, rumours of a military coup involving Franco were so insistent that, late one night, the interim prime minister Manuel Portela Valladares sent the Director-General de Seguridad, Vicente Santiago, to the Ministry of War to see Franco and clarify the situation. The Chief of the General Staff was clearly still in the same cautious mood in which he had greeted Moscardó a few days earlier. Nevertheless, there was a double-edge in his reply. 'The rumours are completely false; I am not conspiring and I will not conspire as long as there is no danger of Communism in Spain; and to put your mind at rest even more, I give you my word of honour, with all the guarantees that this carries between comrades in arms. While you are in the Dirección General de Seguridad, I have complete confidence that law and order, which is of such importance to all Spaniards and above all to the Army, will not be overthrown. Our job is to co-operate.' The Director-General de Seguridad then said something which was uncannily prophetic: 'If you and your comrades at any time feel that the circumstances which you mention come about and you are pushed to a rising, I dare say that if you don't win in forty-eight hours there will follow misfortunes the like of which were never seen in Spain or in any revolution.' Franco replied 'We will not make the same mistake as Primo de Rivera in putting the Army in charge of the government'.[73] That Franco should discount the possibility of military government after a coup reflected his recent discussions with Goded and Fanjul about the plan to put Gil Robles in power, a plan rejected as unsafe.

Inevitably, the election campaign was fought in an atmosphere of violent struggle. In propaganda terms, the Right enjoyed an enormous advantage. Rightist electoral funds dramatically exceeded those of the poverty-stricken Left, although Franco was to remain convinced that the reverse was the case. He believed that the Left was awash with gold sent from Moscow and money stolen by the revolutionaries in October 1934.[74] Ten thousand posters and 50 million leaflets were printed for the CEDA. They presented the elections in terms of a life-or-death struggle between good and evil, survival and destruction. The Popular Front based its campaign on the threat of fascism and the need for an amnesty for the prisoners of October.

In fact, Franco was absent from Spain during part of the election campaign, attending the funeral of George V in London. He was chosen to attend because he was Chief of Staff and because he had once served in the Eighth Infantry Regiment of which the King of England was Honorary Colonel. He attended the funeral service at Westminster Abbey on Wednesday 28 January and, along with other foreign dignitaries, accompanied the coffin to its final resting place in St George's Chapel, Windsor.[75] On the return journey by cross-channel ferry, Franco made some significant remarks to Major Antonio Barroso, the Spanish military attaché in Paris, who had accompanied him on the trip. He told Barroso that the Popular Front was the direct creation of the Comintern and was intended as a Trojan Horse to introduce Communism into Spain. He said that Mola and Goded were equally worried and everything now hinged on what the Popular Front did if it won the elections. The Army had to be ready to intervene if necessary.[76]

The Chief of the General Staff returned to Madrid on 5 February. Franco's instinctive caution was to the fore during a meeting that he held with José Antonio Primo de Rivera, at the home of Ramón Serrano

Suñer's father and brothers, just before the elections in mid-February. The leader of the Falange was obsessed with the need for a military intervention of surgical precision as a prelude to the creation of a national government to stop the slide into revolution. In fact, despite a seductive charm which made him the darling of Spanish high society, the young fascist leader had never attracted or impressed Franco, who, at this meeting, was evasive, rambling and cautious. Almost certainly, at the back of his mind was the madcap scheme which José Antonio Primo de Rivera had recently put to Colonel Moscardó. Franco was not about to become the accomplice in conspiracy of a young Falangist leader whom he did not respect and who had little popular support. Rather than get to the point of the meeting, he chatted aimlessly. José Antonio was deeply disillusioned and irritated, saying 'my father for all his defects, for all his political disorientation, was something else altogether. He had humanity, decisiveness and nobility. But these people . . .'.[77]

The elections held on 16 February resulted in a narrow victory for the Popular Front in terms of votes, but a massive triumph in terms of seats in the Cortes.[78] In the early hours of the morning of 17 February, as the first results were coming in, the popular enthusiasm of the masses was sending panic through right-wing circles. Franco and Gil Robles, in a co-ordinated fashion, worked tirelessly to hold back the decision of the ballot boxes. The main target of their efforts was the Prime Minister (who was also Minister of the Interior). Gil Robles and Franco both saw clearly that it was crucial to persuade him to stay on in order to ensure that the Civil Guard and the crack police units (the *Guardias de Asalto*) would not oppose the Army's measures to reimpose 'order'.

At about 3.15 a.m. on 17 February, Gil Robles presented himself at the Ministerio de la Gobernación and asked to see Portela. The CEDA leader was outraged to discover that Portela had gone to his rooms at the Hotel Palace. Portela was woken to be told that Gil Robles was waiting to see him. Three quarters of an hour later, the Prime Minister arrived. Gil Robles, claiming to speak in the name of all the forces of the right, told him that the Popular Front successes meant violence and anarchy and asked him to declare martial law. Portela replied that his job had been to preside over the elections and no more. He was, nevertheless, sufficiently convinced by Gil Robles to agree to declare a State of Alert (a stage prior to martial law) and to telephone Alcalá Zamora and ask him to authorize decrees suspending constitutional guarantees and imposing martial law.[79]

At the same time, Gil Robles sent his private secretary, the Conde de

Peña Castillo, to instruct his one-time aide Major Manuel Carrasco Verde to contact Franco. Carrasco was to inform Franco of what was happening and urge him to add his weight to Gil Robles' pleas urging Portela not to resign and to bring in the Army. Carrasco woke the Chief of the General Staff at home with the message. Franco leapt to the unjustified conclusion that the election results were the first victory of the Comintern plan to take over Spain. Accordingly, he sent Carrasco to warn Colonel Galarza and instruct him to have key UME officers alerted in provincial garrisons. Franco then telephoned General Pozas, Director-General of the Civil Guard, an old *Africanista* who was nonetheless loyal to the Republic. He told Pozas that the results meant disorder and revolution. Franco proposed, in terms so guarded as to be almost incomprehensible, that Pozas join in an action to impose order. Pozas dismissed his fears and told him calmly that the crowds in the street were merely 'the legitimate expression of republican joy'.

Disappointed by Pozas's cool reception, Franco was driven by further news of crowds in the streets and sightings of clenched fist salutes to put pressure on the Minister of War, General Nicolás Molero. He visited him in his rooms and tried unsuccessfully to get him to seize the initiative and declare martial law. Finally convinced by Franco's arguments about the Communist danger, Molero agreed to force Portela to call a cabinet meeting to discuss the declaration of martial law. Primed by Franco as to what to say, Molero rang Portela and a cabinet meeting was arranged for later that morning. Franco was convinced that the session was called because of his pressure on Molero although it is likely that a meeting would have been held anyway.[80]

Franco decided that it was essential to get Portela to use his authority and order Pozas to use the Civil Guard against the populace. He approached their mutual friend, Natalio Rivas, to see if he could arrange a meeting. By mid-morning, Franco had managed to get an appointment to see Portela, but not until 7 p.m. In the meanwhile, at mid-day, the cabinet met, under the chairmanship of Alcalá Zamora, and declared, as Portela had promised Gil Robles, a State of Alert for eight days. It also approved, and the President signed, a decree of martial law to be kept in reserve and used as and when Portela judged necessary.[81] Franco had gone to his office and been further alarmed by reports of minor incidents of disorder which arrived in the course of the morning. So he sent an emissary to General Pozas, asking him, rather more directly than some hours earlier, to use his men 'to hold back the forces of the revolution'. Pozas again refused. General Molero was totally ineffective and Franco was virtually running the Ministry. He spoke to Generals Goded and Rodríguez del Barrio to see if the units under their command could be relied upon if necessary. Shortly after the cabinet meeting ended, Franco took it upon himself to try to put into action the blank decree of martial law, which Portela had been granted by the cabinet. Franco had learned of the existence of the decree from Molero who had been at the cabinet meeting.[82]

Within minutes of being telephoned by Molero, Franco used the existence of the decree as a threadbare cloak of legality behind which to try to get local commanders to declare martial law. Franco was effectively trying to revert to the role that he had played during the Asturian crisis, assuming the de facto powers of both Minister of War and Minister of the Interior. In fact, the particular circumstances of October 1934 – a workers' uprising, the formal declaration of martial law and the total confidence placed in him by the then Minister of War, Diego Hidalgo, – did not now exist. The Chief of the General Staff had no business usurping the job of the Head of the Civil Guard. However, Franco followed his instincts and, in response to orders emanating from his office in the Ministry of War, martial law (*estado de guerra*) was actually declared in Zaragoza, Valencia, Oviedo and Alicante. Similar declarations were about to made in Huesca, Córdoba and Granada.[83] Too few local commanders responded, the majority replying that their officers would not support a movement if it had to be against the Civil Guard and the Assault Guards. When local Civil Guard commanders rang Madrid to check if it were true that martial law had been declared, Pozas assured them that it had not.[84] Franco's initiative came to naught.

So, when Franco finally saw the Prime Minister in the evening, he was careful to play it both ways. In the most courteous terms, Franco told Portela that, in view of the dangers constituted by a possible Popular Front government, he offered him his support and that of the Army if he would stay in power. He made it clear that Portela's agreement would remove the obstacle to an Army take-over most feared by the officer corps, the opposition of the police and the Civil Guard to military action. 'The Army does not have the moral unity at this moment to undertake the task of saving Spain. Your intervention is necessary because you have authority over Pozas and can draw on the unlimited resources of the State, with the police at your orders.' However, Franco spent much of the short interview shoring up his own personal position by trying to convince the Prime Minister that he personally was not involved in any kind of conspiracy. Franco told Portela's political secretary, his nephew

José Martí de Veses, that he was completely indifferent to politics and was concerned only with his military duties.[85]

Despite Portela's outright refusal to take up the offers of support from both Gil Robles and Franco, efforts to organise military intervention continued. The key issue remained the attitude of the Civil Guard. In the evening of 17 February, in an attempt to build on Franco's efforts earlier in the day, General Goded tried to bring out the troops of the Montaña barracks in Madrid. However, the officers of that and other garrisons refused to rebel without a guarantee that the Civil Guard would not oppose them. It was believed in government circles that Franco was deeply involved in Goded's initiative. Pozas, backed up by General Miguel Núñez de Prado, head of the police, was convinced that Franco was conspiring. However, they reassured Portela on the 18th with the words 'the Civil Guard will oppose any coup attempt (*militarada*)', and Pozas surrounded all suspect garrisons with detachments of the Civil Guard.[86] Just before midnight on the 18th, José Calvo Sotelo and the militant Carlist Joaquín Bau went to see Portela in the Hotel Palace and urged him to call on Franco, the officers of the Madrid military garrison and the Civil Guard to impose order.[87] All this activity around Portela and the failure of Goded justified Franco's instinctive suspicions that the Army was not yet ready for a coup.

A last despairing effort was made by Gil Robles who secretly met Portela under some pine trees at the side of the road from Chamartín to Alcobendas on the outskirts of the capital at 8.30 a.m. on the morning of 19 February.[88] It was to no avail and the efforts of Gil Robles, Calvo Sotelo and Franco did not divert Portela and the rest of the cabinet from their determination to resign and, in all probability, frightened them into doing so with greater alacrity. At 10.30 a.m. on the morning of 19 February, they agreed to hand over power to Azaña immediately, instead of waiting for the opening of the Cortes. Before Portela could inform Alcalá Zamora of this decision, he was told that General Franco had been waiting for him for an hour since 2.30 p.m. at the Ministerio de la Gobernación. During that hour, Franco told Portela's secretary that he was apolitical but that the threats to public order meant that the decree of martial law which Portela had in his pocket should be put into effect. Martí de Veses said that this would divide the Army. Franco replied confidently that the use of the Legion and the Regulares would hold the Army together. That remark confirmed again not only his readiness to use the colonial Army on mainland Spain, but also his conviction that it was essential to do so if the Left was to be decisively defeated. When he was admitted to the Prime Minister's office, Franco did a repeat performance of his double game of the previous evening. He insisted on his own innocence of conspiracy but, aware of his failure with Pozas, again begged Portela not to resign. Portela could not be swayed from his decision which he communicated shortly afterwards to Alcalá Zamora.[89]

To the chagrin of the Right and, indeed, to his own annoyance, Azaña was forced to accept power prematurely, in the late afternoon of 19 February. Franco may have covered his back effectively, but there can be little doubt that he had come nearer during the crisis of 17–19 February to engaging in a military coup than ever before. In the last resort, he had been prevented only by the determined attitude of Generals Pozas and Núñez de Prado. It was scarcely surprising under those circumstances that, when Azaña became prime minister again, Franco should be removed from his position at the head of the general staff. It was to be a major step in turning Franco's latent resentments into outright aggression against the Republic.

V

THE MAKING OF A CONSPIRATOR

Franco and the Popular Front, 1936

THE IMPACT on Franco of the left-wing election victory was almost immediate. On 21 February, the new Minister of War, General Carlos Masquelet, put a number of proposed postings before the cabinet. Amongst them was that of Franco to be Comandante General of the Canary Islands, of Goded to be Comandante General of the Balearic Islands and of Mola to be military governor of Pamplona. Franco was not remotely pleased with what was, in absolute terms, an important post. He sincerely believed that, as Chief of the General Staff, he could play a crucial role in holding back the threat of the Left. As his activities in the wake of the elections showed, his experience in October 1934 had given him a taste for power. That was one reason why the new government wanted him far from the capital.

The Military Region of the Canary Islands, like that of the Balearics, was not traditionally, even prior to Azaña's abolition of the post, a Captaincy-General. Nevertheless, in importance, both jobs counted only marginally below the eight peninsular Military Regions and were held by a Major-General. After all, Franco was only number 23 in the list of 24 Major-Generals on active service. General Mola, four points lower at number three on the list of Brigadier Generals, was made military commander of Pamplona and so subordinate to the regional commander in Zaragoza.[1] Franco was fortunate to get such a senior posting from the new Minister of War but he perceived it as a demotion and another slight at the hands of Azaña. Years later, he spoke of the posting as a 'banishment' (*destierro*). Above all, he was worried that his work in removing liberal officers would be reversed.[2]

Before leaving Madrid, Franco made the obligatory visits to the new Prime Minister Azaña and to the President of the Republic, Alcalá

Zamora. The only accounts of these two meetings derive from Franco's own testimony to his cousin Pacón and to his biographer Joaquín Arrarás. Even from his partial accounts, it is clear that his motives were complex. Ostensibly, he was trying to convince them to do something about the danger of Communism. It is clear that he thought their best course would have been to keep him on as Chief of the General Staff. In large part, as with his efforts in 1931 to hold onto the Military Academy, this was because he wanted to keep a post in which he felt fulfilled and for which he thought that he was the best man. It is impossible to discern whether he also hoped by staying in Madrid to be able to take part in military conspiracy.

In Franco's jaundiced eyes, Alcalá Zamora was dangerously sanguine about the situation. Franco told him that there were insufficient means available to oppose the revolution. The President replied that the revolution had been defeated in Asturias. Franco said 'Remember, Mr President, what it cost to hold back the revolution in Asturias. If the assault is repeated right across the country, it will be really difficult to contain it. The Army lacks the basic means to do so and there are generals who have been put back into key positions who do not want the revolution to be defeated.' Alcalá did not take the hint and merely shook his head. When Franco rose to leave, the President said 'You can leave without worrying, general. There will be no Communism in Spain', to which Franco claimed, with hindsight, to have replied 'Of one thing I am certain, and I can guarantee, that, whatever circumstances may arise, wherever I am, there will be no Communism'.

Again by his own account, Franco appears to have got short shrift from Azaña. His gloomy predictions that the replacement of 'capable' officers by Republicans would open the gates to anarchy were greeted with a sardonic smile. Franco said 'you are making a mistake in sending me away because in Madrid I could be more useful to the Army and for the tranquillity of Spain'. Azaña ignored the offer: 'I don't fear uprisings. I knew about Sanjurjo's plot and I could have avoided it but I preferred to see it defeated'.[3] Neither Azaña's diaries nor Alcalá Zamora's memoirs contain references to these interviews. However, even if Franco's versions of the conversations are apocryphal, they reflect a vivid recollection of his embittered state of mind at the time and of his disgust at what he saw as Azaña's frivolous and malicious insouciance in the face of the Communist menace.

Removed once more from a job he loved, Franco was more than ever a general to be feared. He was not the only one. The narrowness of the left-wing electoral victory reflected the polarization of Spanish society. The savage repression of the previous period ensured that there would be little spirit of conciliation on either side of the political divide. After the failure of the various efforts by Gil Robles and Franco to persuade Portela Valladares to stay in power with Army backing, the Right abandoned all pretence of legalism. The hour of the 'catastrophists' had struck. Gil Robles's efforts to use democracy against itself had failed. Henceforth, the Right would be concerned only with destroying the Republic rather than with taking it over. Military plotting began in earnest.

While waiting to leave for the Canary Islands, Franco spent time talking about the situation with General José Enrique Varela, Colonel Antonio Aranda and other like-minded officers. Everywhere he went, he was followed by agents of the Dirección General de Seguridad.[4] On 8 March, the day before setting out for Cádiz on the first stage of his journey, Franco met a number of dissident officers at the home of José Delgado, a prominent stockbroker and crony of Gil Robles. Among those present were Mola, Varela, Fanjul and Orgaz, as well as Colonel Valentín Galarza. They discussed the need for a coup. They were all agreed that the exiled General Sanjurjo should head the rising.

The impetuous Varela favoured an audacious coup in Madrid; the more thoughtful Mola proposed a co-ordinated civilian/military uprising in the provinces. Mola believed that the movement should not be overtly monarchist. Franco said little other than to suggest shrewdly that any rising should have no specific party label. He made no firm commitments. They departed, having agreed to begin preparations with Mola as overall director and Galarza, as liaison chief. They undertook to act if the Popular Front dismantled the Civil Guard or reduced the size of the officer corps, if revolution broke out or if Largo Caballero was asked to form a government.[5]

After leaving the meeting, Franco collected his family and the inevitable Pacón and headed for the Atocha station to catch the train to Cádiz where they would embark for Las Palmas. At Atocha, a group of generals, including Fanjul and Goded came to wish him farewell. On arrival at Cádiz, Franco was shocked by the scale of disorder which greeted his party, churches having been attacked by anarchists. When the military governor of Cádiz informed him that 'Communists' had set fire to a convent near his barracks, Franco was furious: 'Is it possible that the troops of a barracks saw a sacrilegious crime being committed and that you just stood by with your arms folded?' The colonel replied that he had been ordered by the civilian authorities not to intervene. Franco barked 'Such orders, since they are unworthy, should never be obeyed by an officer of our Army' and he refused to shake hands with the colonel.

Franco's anger reflected his own deep-seated attachment to Catholicism inherited from his mother. It was inextricably entangled with his military-hierarchical view of society. From revulsion at the Left's disrespect for God and the Church it was but a short step to thinking that the use of military force to defend the social order was both necessary and justified. He was even more dismayed when a crowd on the quay which had arrived complete with a band to see off the new civil governor of Las Palmas sang the *Internationale* with their fists raised in the Communist salute. The constant reminders of popular enthusiasm for the Republic led Franco to comment to his cousin that his comrades were wrong to imagine that a swift coup was possible. 'It's going to be difficult, bloody and it'll last a long time – yet there seems to be no other way, if we're going to be one step ahead of the Communists'.[6]

The boat, *Dómine*, reached the Canary Islands at 7 p.m. on the evening 11 March 1936. On arriving at Las Palmas, Franco was greeted by the military governor of the island, General Amado Balmes. After a short tour, he set off again with his family in the *Dómine* for Tenerife where they docked on 12 March at 11.00 a.m. On the dockside, they were awaited by a mass of Popular Front supporters. The local Left had decreed a one-day strike for workers to go to the port in order to boo and whistle the man who had put down the miners' rising in Asturias. Ignoring the banners which denounced 'the butcher of Asturias', Franco remained calm, said goodbye to the ship's captain, descended the steps and inspected the company of troops which awaited him. According to his cousin, his display of cool indifference impressed the crowd whose derision turned to applause.[7]

Franco immediately set to work on a defence plan for the islands and especially on the measures to be taken to put down political disturbances. He also took advantage of the opportunities offered by the Canary Islands and began to learn golf and English. According to his English teacher, Dora Lennard, he took lessons three times a week from 9.30 to 10.30 and was an assiduous student. He wrote two exercises for homework three times a week and only once failed to do so because of pressure of work. Five out of six of his exercises were about golf for which he had quickly become an obsessive enthusiast. He acquired a reading knowledge but could not follow spoken English. His favourite subjects in their conversation classes were the Popular Front's enslavement to the agents of Moscow and his love for his time at the Academia General Militar in

Zaragoza.[8] Franco's own later efforts to wipe away his hesitations during the spring of 1936 led him to imply, in numerous interviews, that he had been anxiously overseeing the conspiracy. As so often in his life, he remoulded reality. It is a telling comment on this particular case of remembered glory that, in fact, in early July 1936, he was planning a golfing holiday in Scotland to improve his game.[9]

Golf and English lessons aside, Franco and Carmen led a full social life. Their guides to the society of the Canaries were Major Lorenzo Martínez Fuset and his wife. Martínez Fuset, a military lawyer, and an amiable and accommodating character, became Franco's local confidant.[10] Otherwise, Franco's activities were slightly inhibited by the scale of surveillance to which he was subject. His correspondence was tampered with, his telephone tapped, and he was being watched both by the police and by members of the Popular Front parties. This reflected the fear that he inspired in both the central government and in the local Left in the Canary Islands. There were rumours inside his headquarters that an assassination attempt was likely. Pacón and Colonel Teódulo González Peral, the head of the divisional general staff, organized the officers under Franco's command into a round-the-clock bodyguard. Franco was reported to have declared proudly 'Moscow sentenced me to death two years ago'.[11] If indeed he made the remark, it reflected the heady propaganda that he was receiving from the *Entente* in Geneva rather than any interest in his activities on the part of the Kremlin.

Despite the air of clandestinity which seems to have surrounded Franco's activities in the Canary Islands, he was openly being talked about as the leader of a forthcoming coup.[12] Pro-fascist and anti-Republican remarks made by him, some in public, suggest that he was not as totally cautious as is usually assumed. On the occasion of the military parade to celebrate the fifth anniversary of the foundation of the Second Republic, Franco spoke with the Italian consul in the Canary Islands and loudly (*ad alta voce*) expressed to him his enthusiasm for Mussolini's Italy. He was particularly fulsome in his congratulations for Italy's role in the Abyssinian war and said how anxiously he awaited news of the fall of Addis Ababa. He appears to have made a point of ensuring that he was overheard by the British Consul. On the next day, the Italian Consul visited Franco to thank him and was delighted when the general's anti-British sentiments led him to speak of his sympathy for Italy as a 'new, young, strong power which is imposing itself on the Mediterranean which has hitherto been kept as a lake under British control'. Franco also talked of his belief that Gibraltar could easily be dominated by modern artillery placed in Spanish territory and talked enticingly, for his listener, of the ease with which a fleet anchored in Gibraltar harbour could be destroyed by air attack.[13]

On 27 April, Ramón Serrano Suñer made a journey to the Canary Islands with the difficult task of persuading his brother-in-law to withdraw his candidacy for the re-run elections about to take place in Cuenca. In the wake of the so-called Popular Front elections of 16 February 1936, the parliamentary committee entrusted with examining the validity of the outcome, the *comisión de actas*, had declared the results null and void in certain provinces. One of these was Cuenca, where there had been falsification of votes. Moreover, once the defective votes were discounted, no list of candidates reached the 40 per cent of votes necessary to win the majority block of seats.[14] In the re-run elections scheduled for the beginning of May 1936, the right-wing slate included both José Antonio Primo de Rivera and General Franco. The Falange leader was included in the hope of securing for him the parliamentary immunity which would ensure his release from jail where he had been since 17 March.[15]

Serrano Suñer was behind Franco's late inclusion in the right-wing list announced on 23 April.[16] On 20 April, a letter from Franco to the secretary of the CEDA expressed his interest in being a candidate in one of the forthcoming re-run elections, preferably Cuenca. Gil Robles discussed the matter with Serrano Suñer. When he approved Franco's candidacy, Serrano Suñer set off immediately for the Canary Islands to inform his brother-in-law. The monarchist leader Antonio Goicoechea offered to give up his place in the right-wing list but Gil Robles simply instructed the CEDA provincial chief in Cuenca, Manuel Casanova, to stand down. The support for Franco manifested by the CEDA and Renovación Española was not replicated by the third political party involved in Cuenca, the Falange. When the revised list of right-wing candidates was published, Gil Robles received a visit from Miguel Primo de Rivera who came to inform him that his brother was firmly opposed to the list, regarding the inclusion of Franco as a 'crass error'.

Since Varela was also standing in the simultaneous rerun at Granada, José Antonio Primo de Rivera shrewdly wished to avoid his chances of election being diminished if the rightist eagerness for military candidates were too transparent. He also, in the wake of his unfortunate meeting with Franco before the February elections, regarded the general as likely to be a disaster in the Cortes. He threatened to withdraw from the Cuenca list if Franco's name was not removed, something which Gil Robles felt unable to do. Efforts by various right-wing leaders including Serrano

Suñer failed to persuade the Falange leader to withdraw his opposition
to Franco. José Antonio said to Serrano Suñer: 'This is not what he's
good at and, given that what is brewing is something more conclusive
than a parliamentary offensive, let him stay in his territory and leave me
where I have already proved myself'. Serrano was then obliged to inform
Franco. He managed to persuade his brother-in-law that he would not
take well to the cut-and-thrust of parliamentary debate. The argument
that Franco would be risking public humiliation did the trick. On 27 April,
Franco withdrew and Manuel Casanova returned to the list.[17] Franco was
aware of the Falangist leader's hostility to his candidacy and subsequent
events would show that he neither forgave nor forgot.

The Left, and Prieto in particular, were concerned that Franco planned
to use his parliamentary seat as a base from which to engage in military
plotting. This was a reasonable interpretation and was indeed adopted
by Francoist propaganda once the Civil War was under way. However,
it is not clear whether Franco's quest for a parliamentary seat was motivated by the need to effect his transfer from the Canary Islands to the
mainland in order to play a key role in the conspiracy or by more selfish
motives. Gil Robles suggested that the desire to go into politics reflected
Franco's doubts about the success of a military rising. As yet undeclared
vis-à-vis the conspiracy, he wanted a safe position in civilian life from
which to await events.[18] Fanjul confided a similar opinion to Basilio
Alvarez, who had been a Radical deputy for Orense in 1931 and 1933:
'perhaps Franco wants to protect himself from any governmental or disciplinary inconvenience by means of parliamentary immunity.'[19]

Certainly, the versions of the Cuenca episode produced by Franco and
his propagandists make it clear that it was to be an abiding source of
embarrassment. Within a year, Franco was to be found rewriting it,
through his official biographer Joaquín Arrarás. In his 1937 version, the
parties of the Right offered Franco a place in the list for Cuenca, because
he was a persecuted man and to allow him the freedom 'to organize the
defence of Spain'. Franco 'publicly rejected' the offer because he neither
believed in the honesty of the election process nor expected anything
from the Republican parliament.[20] This ludicrously inaccurate version of
the events surrounding the Cuenca elections implied that, if the electoral
system had been honest, Franco would have stood. Subsequently in 1940,
Arrarás eliminated this inadvertent proclamation of faith in democracy
and claimed that Franco had withdrawn his candidacy because of 'the
twisted interpretations' to which it was subject.[21] A decade after the
events, Franco himself claimed in a speech to the Falangist Youth in

Cuenca that his desire to be a parliamentary deputy was occasioned by 'dangers for the *Patria*'.[22]

By the early 1960s, Franco was eschewing any hint that he might have been seeking a bolt-hole. Writing in the third person, he claimed rather that 'General Franco was looking for a way of legally leaving the archipelago which would permit him to establish a more direct contact with the garrisons in order to have a more direct link with those places where there was a danger of the Movement being a failure'. There is an outrageous re-casting of history in this account. Franco attributes to himself the credit for securing a place for José Antonio Primo de Rivera in the right-wing candidacy, which is simply untrue. With equal inaccuracy, he claims that General Fanjul had stood down as a candidate to make way for Franco himself when he had done so for José Antonio. He then fudges the reasons for the eventual withdrawal of his own candidacy with the vague and incorrect statement that, on the morning that candidates were to be announced, he received a telegram from those concerned (*los afectados*) to the effect that 'it was impossible to maintain his candidacy because his name had been 'burned' (*quemado*).[23]

That Franco should omit to mention the rift with the leader of the Falange was entirely understandable. After all, after 1937, the Nationalist propaganda machine would work frenetically to convert Franco into the heir to José Antonio in the eyes of the Falangist masses. Similarly, in writing that his intention was to be able to oversee the preparations for a coup, Franco inadvertently revealed his desire to diminish Mola's posthumous glory as the sole director of the rising. In his third and most plausible attempt to rewrite the Cuenca episode, Arrarás wrote that Franco withdrew 'because he preferred to attend to his military duties, by which means he believed he could better serve the national interest'. The suggestion of any friction between Franco and José Antonio Primo de Rivera remained taboo.[24]

Left-wing suspicions of Franco's motives were expressed by Indalecio Prieto shortly after Franco's candidacy was dropped, in a celebrated speech in Cuenca. He commented that 'General Franco, with his youth, with his gifts, with his network of friends in the Army, is a man who could at a given moment be the caudillo of a movement with the maximum chances of success'. Accordingly, without attributing such intentions to Franco, Prieto claimed that other right-wing plotters were seeking to get parliamentary immunity for him in order to facilitate his conversion into 'the caudillo of a military subversion'.[25] In any case, the Cuenca election was declared at the last minute to be technically a re-run.

Since the electoral law required that candidates in a re-run should have secured 8 per cent of the vote in the first round, new candidates could not be admitted by the provincial Junta del Censo. Accordingly, although José Antonio Primo de Rivera gained sufficient votes to win a seat, his election was not recognized.[26]

Helpless before the rising numbers of strikes and deaf to the background hum of military conspiracy stood the minority government. Only Republicans sat in the Cabinet, because Largo Caballero refused to let Socialists join a coalition. He pinned his hopes on two naïve scenarios: either the Republicans would quickly find themselves incapable of implementing their own reform programme and have to make way for an exclusively Socialist cabinet or else there would be a fascist coup which would be crushed by popular revolution. In May, Largo used his immense influence inside the Socialist leadership to prevent the formation of a government by the more realistic Prieto. As long as Azaña was prime minister, authority could be maintained. However, in order to put together an even stronger team, Azaña and Prieto plotted to remove the more conservative Alcalá Zamora from the presidency. Azaña would become president and Prieto take over as prime minister. The first part of the plan worked but not the second as a result of Largo Caballero's opposition and Prieto's failure to fight it. The consequences were catastrophic. The last chance of avoiding civil war was missed. Spain lost a shrewd and strong prime minister, and, to make matters worse, on assuming the presidency, Azaña increasingly withdrew from active politics. The new Prime Minister, Santiago Casares Quiroga, suffering from tuberculosis, was incapable of generating the determination and energy required in the circumstances.

Unemployment was rocketing and the election results had dramatically raised the expectations of workers in both town and countryside. To the outrage of employers, trade unionists sacked in the aftermath of the Asturian events were forcibly reinstated. There were sporadic land seizures as frustrated peasants took into their own hands the implementation of the new government's commitment to rapid reform. What most alarmed the landlords was that labourers whom they expected to be servile were assertively determined not to be cheated out of reform as they had between 1931 and 1933. Many landowners withdrew to Seville or Madrid, or even to Biarritz or Paris, where they enthusiastically joined, financed, or merely awaited news of, ultra-rightist plots against the Republic.

Under the energetic leadership of General Mola, the plot was developing fast. It was more thoroughly prepared than any previous effort, taking full account of the lesson of the *Sanjurjada* of 10 August 1932 that casual *pronunciamientos* could not work where the Civil Guard was in opposition and where the proletariat was ready to use the weapon of the general strike. The tall bespectacled Mola, as '*El Director*', having learnt plenty of police procedure during his time as Director-General of Security in 1930–1931, took to conspiracy with gusto. Brave and of adventurous spirit, he enjoyed the danger.[27] Pamplona was an excellent place from which to direct the conspiracy, being the headquarters of the most militant group of the ultra-Right, the Carlists.[28] Mola had plenty of willing and competent assistants. Through Valentín Galarza, known among the plotters as 'the technician' (*el técnico*), the right-wing conspiratorial organization, Unión Militar Española, was at his disposal. He drew up his first directive in April, 'The objective, the methods and the itineraries'. In it, aware of the deficiencies of the preparations of *Sanjurjada*, he specified in detail the need for a complex civilian support network and above all for political terror: 'the action must be violent in the extreme in order to crush the strong and well-organized enemy as soon as possible. All leaders of political parties, societies or unions not committed to the Movement will be imprisoned and exemplary punishments administered to such individuals in order to strangle movements of rebellion or strikes'.[29]

In the middle of May, Mola was visited secretly by a Lieutenant-Colonel Seguí of the general staff of the African Army, who informed him that the garrisons of Morocco were ready to rise. Among the *Africanista* officers, Mola relied on Yagüe as the most tireless in the preparation of the rising in Morocco. In May too, Mola was in contact with a group of generals who would each play a crucial role in the Civil War: the brutal Gonzalo Queipo de Llano, head of the *Carabineros* (the Spanish frontier guards), the austere monarchist Alfredo Kindelán, the key link with conspirators in the Air Force and the easy-going Miguel Cabanellas, head of the Zaragoza military division.[30] Franco was fully informed through Galarza. As part of the post-1939 propaganda effort to wipe away the memory of Franco's minimal participation in the preparations, it was claimed that he carried on a twice-weekly correspondence with Galarza. These thirty coded letters have never been traced.[31] In fact, Franco was anything but enthusiastic, commenting to the optimistically headstrong Orgaz, who had been banished to the Canary Islands in the early spring, 'You are really mistaken. It's going to be immensely difficult and very bloody. We haven't got much of an army, the intervention of the Civil Guard is looking doubtful and many officers will side with the constituted power, some because it's easier, others because of their convictions. Nobody should forget that the soldier who rebels against the constituted power can never turn back, never surrender, for he will be shot without a second thought'.[32] At the end of May, Gil Robles complained to the American journalist H. Edward Knoblaugh that Franco had refused to head the coup, allegedly saying 'not all the water in the Manzanares could wash out the stain of such a move'. Discounting the choice of a less than torrential river, this and other remarks suggest that the experience of the *Sanjurjada* of 1932 was on his mind.[33] Not to be able to turn around or change his mind must have been Franco's idea of hell.

With the conspiracy developing rapidly, Franco's caution was stoking up the impatience of his *Africanista* friends. On 25 May, Mola had drawn up his second directive to the plotters, a broad strategic plan of regional risings to be followed by concerted attacks on Madrid from the provinces.[34] Clearly, it would be an enormous advantage to have Franco as part of the team. Captain Bartolomé Barba was sent by Goded to the Canary Islands on 30 May to tell Franco to make his mind up and abandon 'so much prudence'. Colonel Yagüe told Serrano Suñer that he was in despair at Franco's mean-minded carefulness and his refusal to take risks.[35] Serrano Suñer himself was baffled when Franco told him that what he really would have liked was to tranfer his residence to the south of France and direct the conspiracy from there. Given Mola's position, there was no question of Franco organizing the rising. The clear implication was that he was more concerned with covering his personal retreat in the event of failure.[36] This inevitably suggests that selfless commitment to the rising had not been his main reason for trying to stand for election in Cuenca.

The rationale for the conspiracy was the fear of the middle and upper classes that an inexorable wave of Godless, Communist-inspired violence was about to inundate society and the Church. Their panic was generated assiduously by the rightist press and by the widely reported parliamentary speeches of the insidious Gil Robles and the belligerent monarchist leader José Calvo Sotelo. Their denunciations of disorder found a spurious justification in the street violence provoked by the Falange's terror squads. In their turn, the activities of Falangist gangs were financed by the same monarchists who were behind the military coup. The startling rise of the Falange was a measure of the changing political climate. Cashing in on middle class disillusionment with the CEDA's legalism, the Falange expanded rapidly. Moreover, attracted by its code of violence, the bulk of the CEDA's youth movement, the JAP, went over *en masse*. The rise

of the Falange was matched by the ascendancy within the Socialist movement of Largo Caballero. Intoxicated by Communist flattery – *Pravda*
had called him 'the Spanish Lenin'– he undermined Prieto's efforts at
a peaceful solution. Largo toured Spain, prophesying the triumph of the
coming revolution to crowds of cheering workers. The May Day marches,
the clenched fist salutes, the revolutionary rhetoric and the violent attacks
on Prieto were used by the rightist press to generate an atmosphere of
terror among the middle classes and to convince them that only a military
coup could save Spain from chaos.

Certain factors made the conspirators' task much easier than it might
otherwise have been. The government failed to act decisively on the
repeated warnings that it received of the plot. At the beginning of June,
Casares Quiroga, as Minister of War, set out to decapitate the conspiracy
in Morocco by removing the officers in charge of the two Legions into
which the *Tercio* was now organized. On 2 June, he sent for Yagüe who
was head of the so-called *Segunda Legión*. On the following day, he
removed Yagüe's fellow-conspirator Lieutenant-Colonel Heli Rolando
de Tella from command of the *Primera Legión*. When Yagüe was received
by the Minister on 12 June, Casares Quiroga offered him a transfer either
to a desirable post on the Spanish mainland or to a plum position as a
military attaché abroad. Yagüe told Casares that he would burn his uniform rather than not be able to serve with the Legion. After giving him
forty-eight hours to reconsider, Casares weakly acquiesced in Yagüe's
vehemently expressed desire to return to Morocco. It was a major political
error given Yagüe's key role in the conspiracy.[37] A comparable stroke of
luck protected the overall director of the plot. The Director-General of
Security, Alonso Mallol, pointed the finger at Mola. On 3 June, Mallol
made an unannounced visit to Pamplona with a dozen police-filled trucks
and undertook searches allegedly aimed at arms smuggling across the
French frontier. Having been warned of the visit by Galarza who in turn
had been informed by a rightist police superintendent, Santiago Martín
Báguenas, Mola was able to ensure that no evidence of the conspiracy
would be found.[38]

The ineffective efforts of the Republican authorities to root out the
conspirators helps explain one of the mysteries of the period, a curious
warning to Casares Quiroga from the pen of General Franco. He wrote
to the Prime Minister on 23 June 1936 a letter of labyrinthine ambiguity,
both insinuating that the Army was hostile to the Republic and suggesting
that it would be loyal if treated properly. The letter focused on two issues.
The first was the recently announced reintegration into the Army of the

officers tried and sentenced to death in October 1934 for their part in the defence of the Generalitat. The rehabilitation of these officers went directly against one of Franco's greatest obsessions, military discipline.[39] The second cause of Franco's outrage was that senior officers were being posted for political reasons. The removal of Heli Rolando de Tella from the Legion and the near loss of Yagüe must have been on his mind. He informed the Minister that these postings of brilliant officers and their replacement by second-rate sycophants were arbitrary, breached the rules of seniority and had caused immense distress within the ranks of the Army. No doubt he regarded his own transfer from the general staff to the Canary Islands as the most flagrant case.

He then wrote something which, although absolutely untrue, was probably written with sincerity. In Franco's value system, the movement being organized by Mola, and about which he was fully informed, merely constituted legitimate defensive precautions by soldiers who had the right to protect their vision of the nation above and beyond particular political regimes. 'Those who tell you that the Army is disloyal to the Republic are not telling you the truth. Those who make up plots in terms of their own dark passions are deceiving you. Those who disguise the anxiety, dignity and patriotism of the officer corps as symbols of conspiracy and disloyalty do a poor service to the *Patria*.' The anxieties which he shared with his brother officers about the law and order problem led Franco to urge Casares to seek the advice 'of those generals and officers who, free of political passions, live in contact with their subordinates and are concerned with their problems and morale'. He did not mention himself by name but the hint was unmistakeable.[40]

The letter was a masterpiece of ambiguity. The clear implication was that, if only Casares would put Franco in charge, the plots could be dismantled. At that stage, Franco would certainly have preferred to reimpose order, as he saw it, with the legal sanction of the government rather than risk everything in a coup. In later years, his apologists were to spill many gallons of ink trying to explain away this letter either as a skilful effort by Franco the conspirator to put Casares off the scent and make him halt his efforts to replace subversives with loyal Republicans or else as a prudent warning by Franco the loyal officer which was stupidly ignored by the Minister of War.[41] In fact, the letter had exactly the same purpose as Franco's appeals to Portela in mid-February. Franco was ready to deal with revolutionary disorder as he had done in Asturias in 1934 and was now, in guarded terms, offering his services. If Casares had accepted his offer, there would have been no need for an uprising.

That was certainly Franco's retrospective view.[42] The government of the Popular Front did not share his commitment to suppressing the aspirations of the masses. In any case, Casares took no notice of him. If he had, the eventual outcome would certainly have been very different. If Franco was within his rights to send such a letter, Casares should have acknowledged his concern. If he believed that Franco had abused his position then Casares should have taken disciplinary measures against him. The Prime Minister's failure to reply can only have helped to incline Franco towards rebellion.

Franco's letter was a typical example of his ineffable self-regard, his conviction that he was entitled to speak for the entire army. At the same time, its convoluted prose reflected his *retranca*, the impenetrable cunning associated with the peasants of Galicia. At the time of writing, Franco was still distancing himself from the conspirators. His determination to be on the winning side without taking any substantial risks hardly set him apart as a likely charismatic leader although it did prefigure his behaviour towards the Axis in the Second World War. At the same time as he wrote to Casares, Franco also wrote to two Army colleagues. The first letter was to Colonel Miguel Campins, his assistant in the Zaragoza Academy, currently in command of a light infantry battalion in Catalonia. The other was to Colonel Francisco Martín Moreno, chief of the general staff of Spanish forces in Morocco with whom Franco had worked in early 1935 when he had been Commander-in-Chief there. The letters suggest clearly that Franco was not yet a committed conspirator, expressing merely his anxiety that the political situation might worsen to the point at which the Army would have to intervene. He asked if they would collaborate with him if such an occasion were to arise. Martín Moreno wrote back to say that, if Franco appeared in Tetuán, he would place himself at his orders, 'but at no one else's'. Campins, in contrast, replied that he was loyal to the government and to the Republic and that he did not favour any intervention by the Army. He had signed his own death warrant.[43]

A few days after Franco wrote his letter to Casares, the division of duties among the conspirators was settled. Franco was expected to be in command of the rising in Morocco. Cabanellas would be in charge in Zaragoza, Mola in Navarre and Burgos, Saliquet in Valladolid, Villegas in Madrid, González Carrasco in Burgos, Goded in Valencia. Goded insisted on exchanging cities with González Carrasco.[44] For several reasons, Mola and the other conspirators were loath to proceed without Franco. His influence within the officer corps was enormous, having been both Director of the Military Academy and Chief of the General Staff. He also enjoyed the unquestioning loyalty of the Spanish Moroccan Army. The coup had little chance of succeeding without the Moroccan Army and Franco was the obvious man to lead it. Yet, in the early summer of 1936, Franco still preferred to wait in the wings. Calvo Sotelo frequently cornered Serrano Suñer in the corridors of the Cortes to badger him impatiently 'what is your brother-in-law thinking about? What is he doing? Doesn't he realize what is on the cards?'[45]

His coy hesitations saw his exasperated comrades bestow upon him the ironic nickname of 'Miss Canary Islands 1936'. Sanjurjo, still bitter about Franco's failure to join him in 1932, commented that 'Franco will do nothing to commit himself; he will always be in the shadows, because he is crafty' (*cuco*). He was also heard to say that the rising would go ahead 'with or without Franquito'.[46] There were plenty of other good generals who were in on the conspiracy and many more who were not. Why Franco's hesitations infuriated Mola and Sanjurjo was not just because of the danger and inconvenience involved in having to plan around a doubtful element. They were anxious to have him aboard because they rightly sensed that his decision would clinch the involvement of many others. He was 'the traffic light of military politics', in the words of José María Pemán.*[47]

When Franco did eventually commit himself, his role was of the first importance without being the crucial one. The Head of State after the coup triumphed was to be Sanjurjo. As technical master-mind of the plot, Mola was then expected to have a decisive role in the politics of the victorious regime. Then came a number of generals each of whom was assigned a region, among them Franco with Morocco. Several of them were of equal prominence to Franco, especially Fanjul in Madrid and Goded in Barcelona. Moreover, leaving aside the roles allotted to Sanjurjo and Mola, Franco's future in the post-coup polity could only lie in the shadow of the two charismatic politicians of the extreme Right, José Calvo Sotelo and José Antonio Primo de Rivera. In fact, given his essential caution, Franco seems not to have nurtured high-flying ambitions in the spring and early summer of 1936. When Sanjurjo asked what prizes his fellow-conspirators aspired to, Franco had opted for the job of High Commissioner in Morocco.[48] As the situation changed, Franco would adjust his ambitions with remarkable agility and uninhibited by any self-

---

* Pemán, a sardonically witty poet and playwright, was member of the extreme right-wing monarchist group, *Acción Española*.

doubts. The hierarchy of the plotters would in fact soon be altered with astonishing rapidity.

The arrangements for Franco's part in the coup were first mooted in Mola's Directive for Morocco. Colonel Yagüe was to head the rebel forces in Morocco until the arrival of 'a prestigious general'. To ensure that this would be Franco, Yagüe wrote urging him to join in the rising. He also planned with the CEDA deputy Francisco Herrera to present Franco with a *fait accompli* by sending an aircraft to take him on the 1,200 kilometre journey from the Canary Islands to Morocco. Francisco Herrera, a close friend of Gil Robles, was the liaison between the conspirators in Spain and those in Morocco. Yagüe, for his part, was devoted to Franco. As a consequence of his clashes with General López Ochoa during the Asturian campaign, he had been transferred to the First Infantry Regiment in Madrid. A personal intervention by Franco had got him back to Ceuta.[49] After meeting Yagüe on 29 June in Ceuta, Herrera undertook the lengthy journey to Pamplona where he arrived somewhat the worse for wear on 1 July to make arrangements for an aircraft for Franco. Apart from the financial and technical difficulties of getting an aircraft at short notice, Mola still had grave doubts about whether Franco would join the rising.

However, after consulting with Kindelán, he gave the go-ahead for this plan on 3 July. Herrera proposed going to Biarritz to see if the exiled Spanish monarchists at the resort could resolve the money problem. On 4 July, he spoke to the millionaire businessman Juan March who had got to know Franco in the Balearic Islands in 1933. He agreed to put up the cash. Herrera then got in touch with the Marqués de Luca de Tena, owner of the newspaper *ABC*, to get his assistance. March gave Luca de Tena a blank cheque and he set off for Paris to make the arrangements. Once there on 5 July, Luca de Tena rang Luis Bolín, the *ABC* correspondent in England, and instructed him to charter a seaplane capable of flying direct from the Canary Islands to Morocco or else the best possible conventional aircraft. Bolín in turn rang the Spanish aeronautical inventor and rightist, Juan de la Cierva who lived in London. La Cierva flew to Paris and told Luca de Tena that there was no suitable seaplane and recommended instead a De Havilland Dragon Rapide. Knowing the English private aviation world well, La Cierva recommended using Olley Air Services of Croydon. Bolín went to Croydon on 6 July and hired a Dragon Rapide.*[50]

* The necessary funds to hire Dragon Rapide G-ACYR – £2000 – were supplied by Juan March through the Fenchurch Street branch of Kleinwort's Bank.

La Cierva and Bolín arranged for a set of apparently holidaying passengers to mask the aeroplane's real purpose. On 8 July, Bolín went to Midhurst in Sussex to speak to Hugh Pollard, a retired army officer and adventurer, and make the arrangements. Pollard, his nineteen year-old daughter Diana and her friend Dorothy Watson would travel as tourists to provide Bolín with a cover for his flight. Leaving Croydon in the early hours of the morning of 11 July, the plane was piloted by Captain William Henry Bebb, ex-RAF. Despite poor weather, it reached Bordeaux at 10.30 a.m. where Luca de Tena and other monarchist plotters awaited Bolín with last-minute instructions. They arrived in Casablanca, via Espinho in Northern Portugal and Lisbon, on the following day, 12 July.[51]

Although the date for his journey to Morocco was now imminent, Franco was having ever more serious doubts, obsessed as usual with the experience of 10 August 1932. On 8 July, Alfredo Kindelán managed to speak briefly with Franco by telephone and was appalled to learn that he was still not ready to join. Mola was informed two days later.[52] On the same day that the Dragon Rapide reached Casablanca, 12 July, Franco sent a coded message to Kindelán in Madrid for onward transmission to Mola. It read *'geografía poco extensa'* and meant that he was refusing to join in the rising on the grounds that he thought that the circumstances were insufficiently favourable. Kindelán received the message on 13 July. On the following day, he sent it on to Mola in Pamplona in the hands of a beautiful socialite, Elena Medina Garvey, who acted as messenger for the conspirators. Mola flew into a rage, furiously hurling the paper to the ground. When he had cooled down, he ordered that the pilot Juan Antonio Ansaldo be found and instructed to take Sanjurjo to Morocco to do the job expected of Franco. The conspirators in Madrid were informed by Mola that Franco was not to be counted on. However, two days later, a further message arrived to say that Franco was with them again.[53]

The reason for Franco's sudden change of mind were dramatic events in Madrid. On the afternoon of 12 July, Falangist gunmen had shot and killed a leftist officer of the Republican Assault Guards, Lieutenant José del Castillo. Castillo was number two on a black list of pro-Republican officers allegedly drawn up by the ultra-rightist Unión Militar Española, an association of conspiratorial officers linked to Renovación Española. The first man on the black list, Captain Carlos Faraudo, had already been murdered. Enraged comrades of Castillo responded with an irresponsible reprisal. In the early hours of the following day, they set out to avenge his death by murdering a prominent Right-wing politician. Failing to find Gil Robles who was holidaying in Biarritz, they kidnapped and shot

Calvo Sotelo. On the evening of the 13th, Indalecio Prieto led a delegation of Socialists and Communists to demand that Casares distribute arms to the workers before the military rose. The Prime Minister refused, but he could hardly ignore the fact that there was now virtually open war.

The political outrage which followed the discovery of Calvo Sotelo's body played neatly into the hands of the military plotters. They cited the murder as graphic proof that Spain needed military intervention to save her from disaster. It clinched the commitment of many ditherers, including Franco. When he received the news in the late morning of 13 July, he exclaimed to its bearer, Colonel González Peral, 'The *Patria* has another martyr. We can wait no longer. This is the signal!'.[54] Fuming with indignation, he told his cousin that further delay was out of the question since he had lost all hope of the government controlling the situation. Shortly afterwards, Franco sent a telegram to Mola. Later in the afternoon, he also ordered Pacón to buy two tickets for his wife and daughter on the German ship *Waldi* which was due to leave Las Palmas on 19 July bound for Le Havre and Hamburg.[55] His foresight did not extend to warning other members of his family. His sister-in-law Zita Polo underwent enormous dangers in escaping from Madrid with her children. Pilar Jaraiz, his niece, was imprisoned with her new-born son.[56]

Franco's English teacher wrote later that 'the morning after the news of Calvo Sotelo's murder had reached us, I had found him a changed man, when he came for his lessons. He looked ten years older, and had obviously not slept all night. For the first time, he came near to something like losing his iron self-control and unalterable serenity ... It was with visible effort that he attended to his lesson.'[57] The heady decisiveness with which Franco responded to the news is not incompatible with Dora Lennard's comment on his sleepless night.* The decision was of sufficient enormity to provoke agonizing doubts, as his precautions for the safety of his wife and daughter demonstrated.

Later, the assassination of Calvo Sotelo was used to obscure the fact that the coup of 17–18 July had been long in the making. It also deprived the conspirators of a powerful and charismatic leader. As a cosmopolitan rightist of wide political experience, Calvo Sotelo would have been the senior civilian after the coup and unlike many of the ciphers that were

---

* On other occasions, Franco would show a similar determination to move on, apparently indifferent to the tragedy just recounted to him. The demise of Alfonso XIII in 1931, the death of Mola in April 1937 and Mussolini's fall from power in 1943 all produced nearly identical responses.

to be used by Franco. It is difficult not to imagine that he would have imposed his personality on the post-war state. His death, even if no one could have judged it in such terms at the time, removed an important political rival to Franco.

In the short term, Calvo Sotelo's assassination gave a new urgency to plans for the uprising. The Dragón Rapide had left Bolín in Casablanca and was still en route for the Canary Islands. It arrived at 14.40 on 14 July at the airport of Gando near Las Palmas on the island of Gran Canaria. Hugh Pollard and the two girls took a ferry to Tenerife where he was to make known his arrival by presenting himself at the Clínica Costa with the password 'Galicia saluda a Francia'. Bebb was left with the aircraft on Gran Canaria to await instructions from an unknown emissary who would make himself known with the password 'Mutt and Jeff'. Mean- while, at 2 a.m. on the morning of 15 July, the sleek diplomat José Antonio Sangróniz appeared at Pacón's hotel room in Santa Cruz de Tenerife with news of the latest developments and the date scheduled for the rising. At 7.30 a.m. on the same morning, Pollard went to the clinic where he contacted Doctor Luis Gabarda, a major of the military medical service, who was acting on behalf of Franco. He was told to return to his hotel and await an emissary from Franco with his instructions.[58]

Franco had acute immediate problems which took precedence over any long-term ambitions. As military commander of the Canary Islands, his headquarters were in Santa Cruz de Tenerife. The Dragon Rapide from Croydon had been instructed to land at the airport of Gando on Gran Canaria in part because it was nearer to mainland Africa, also because it was known that Franco was being watched by the police but, above all, because of the low cloud and thick fog which afflicts Tenerife. In order to travel from Santa Cruz to Gran Canaria, Franco needed the authorization of the Ministry of War. His request for permission to make an inspection tour of Gran Canaria was likely to be turned down, not least because it was barely a fortnight since his last one. The rising was scheduled to start on 18 July, so Franco would have to leave for Morocco on that day at the latest. In the event he did so, yet none of his biographers seem to regard it as odd that the Dragon Rapide should have been directed to Gran Canaria with confidence in Franco's ability to get there too. That he got there at all was the result of either a remarkable coinci- dence or foul play.

On the morning of 16 July, Franco failed to appear for his scheduled English lesson.[59] On the same morning, General Amado Balmes, military commander in Gran Canaria, and an excellent marksman, was shot in the stomach while trying out various pistols in a shooting range. Francoist historiography has played down the incident as a tragic, but fortunately timed, accident. Allegedly, a pistol blocked and in trying to free it, holding it against his stomach, it went off.[60] To counter suggestions that Balmes was assassinated, Franco's official biographers have claimed that Balmes was himself an important figure in the plot. His cousin has portrayed Balmes as an intimate friend of Franco. Balmes was allegedly to organize the coup in Las Palmas and thus had to be replaced by Orgaz who was conveniently exiled there.[61] Strangely, however, Balmes never figured in the subsequent Pantheon of heroes of the 'Crusade'. Moreover, it is extraordinary that, despite the fact that Madrid did indeed refuse permission for Franco to travel to Gran Canaria to make an inspection, he and his immediate circle never doubted that they would find a way of getting to Las Palmas. Other sources suggest that Balmes was a loyal Republican officer and member of the Unión Militar Republicana Antifascista who had withstood intense pressure to join the rising.[62] If that was true, he had, like many other Republican officers, put his life in mortal danger. It is virtually impossible now to say if his death was accidental, suicide or murder.

What is certain is that he died at the exact moment urgently needed by Franco. The duty of presiding at the funeral gave Franco the perfect excuse to travel to Gran Canaria on the overnight boat. Franco was determined to go without seeking permission for fear that it might be denied. His cousin persuaded him that it would be altogether less suspicious for him to ring the Ministry and inform the under-secretary, General De la Cruz Boullosa. Franco agreed with what turned out to be good advice. The under-secretary expressed surprise that Franco had not been in touch earlier to report on the death of Balmes. He gave the excuse that he had been seeking fuller information on what had happened and was granted permission to preside over the burial. Franco left Tenerife for Las Palmas in the mail-boat *Viera y Clavijo* shortly after midnight on 16 July. He was accompanied by his wife and daughter, Lieutenant-Colonel Franco Salgado-Araujo, Major Lorenzo Martínez Fuset and an escort consisting of five other officers. They arrived at Las Palmas at 8.30 a.m. on Friday 17 July. Pollard had returned to Las Palmas on the same ferry. Before leaving Tenerife, Franco had collected Sangróniz's diplomatic passport and gave Colonel González Peral the proclamation of the military rebellion to be used on the following morning. Bebb and Pollard made the final arrangements with General Orgaz. The funeral ceremony for Balmes occupied most of the morning. Franco then took his wife and daughter for a drive around the town. Later, they dined with Pacón and Orgaz.[63]

Coordinated risings were planned to take place all over Spain on the following morning. However, indications that the conspirators in Morocco were about to be arrested led to the action being brought forward there to the early evening of 17 July. The garrisons rose in Melilla, Tetuan and Ceuta in Morocco. At 4 a.m. in the morning of 18 July, Franco was woken in his hotel room to be given the news. Colonel Luis Solans, Lieutenant-Colonel Seguí and Colonel Darío Gazapo had seized Melilla 'in Franco's name' and arrested the overall military commander in Morocco, the Republican General Gómez Morato. Yagüe had taken charge in Ceuta and Colonels Eduardo Saénz de Buruaga, Juan Beigbeder and Carlos Asensio Cabanillas had taken Tetuán. Franco was to have reason to be grateful for the role of Beigbeder, an accomplished Arabist, in taking over the Spanish High Commission and subsequently securing Moroccan acquiescence in the rising.[64]

On hearing of their successes, Franco set out for military headquarters in Las Palmas accompanied by his cousin and Major Martínez Fuset and sent for Orgaz to join them there. Franco then sent a telegram to the eight divisional headquarters and the other main military centres of the peninsula. The news that Franco and the Army of Africa were on the side of the rebels was meant as a rallying cry to the conspirators in other areas: 'Glory to the Army of Africa. Spain above all. Receive the enthusiastic greeting of these garrisons which join you and other comrades in the peninsula in these historic moments. Blind faith in our triumph. Long live Spain with honour. General Franco.' The sending of such a telegram was an unequivocal indication that Franco attributed to himself a central national role in the rising. At 5.00 a.m. on 18 July, he signed a declaration of martial law. It was to be announced in Las Palmas by an infantry company complete with bugles and drums. At about the same time, a desperate telephone call for Franco came from the undersecretary of the Ministry of War in Madrid, General De la Cruz Boullosa. Martínez Fuset answered and claimed that Franco was out inspecting barracks.[65]

At 5.15 a.m. in the morning of 18 July, Inter-Radio of Las Palmas began to broadcast Franco's manifesto. The rather confused text was later attributed to Lorenzo Martínez Fuset.[66] The typed copy sent to the radio station had a post-script in Franco's handwriting, 'accursed be those who, instead of doing their duty, betray Spain. General Franco'. It avoided commitment to either the Republic or the Monarchy justifying the rising entirely in terms of defending the *Patria* by putting an end to anarchy. The text also claimed that Franco's action was necessary because of a power vacuum in Madrid. Some of it was entirely fanciful: the Constitution, it alleged, was in tatters; the government was blamed for failing to defend Spain's frontiers 'when in the heart of Spain, foreign radio stations can be heard calling for the destruction and division of our soil'. It threatened 'war without quarter against the exploiters of politics' and 'energy in the maintenance of order in proportion to the magnitude of the demands that arise' which was an obscure way of saying all resistance would be crushed.[67]

Franco himself made contact with trusted officers on the island and, on his orders, they seized the post office, the telegraph and telephone centres, the radio stations, power generators, and water reservoirs. He had rather more difficulty persuading the head of the local Civil Guard, Colonel Baraibar, to join the rising.[68] While Baraibar wavered, Franco, his family and his group of fellow rebels were in serious danger. Crowds were gathering outside the Gobierno Civil and groups of workers from the port were heading into Las Palmas. Pacón managed to keep the two groups from uniting by use of small artillery pieces and before 7 a.m. had dispersed the crowds. The beleagured group was then joined by retired officers, Falangists and right-wingers who were given arms. The situation remained tense and Franco was anxious to be on his way to Africa. Accordingly, he handed over command to Orgaz. Carmen Polo and Carmencita Franco were taken by Franco's escort to the port and hidden on board the naval vessel *Uad Arcila* until the arrival of the German liner *Waldi* which was to take them to Le Havre.*[69]

With fighting still going on, Franco himself set off at 11 a.m. on a naval tugboat for Gando airport where Bebb's Dragon Rapide awaited him. It would have been virtually impossible to reach Gando by a road journey through villages controlled by the Popular Front. The tug went in as near to shore as possible and Franco and his party were then carried to the beach by sailors.[70] At 14.05 hours on 18 July, the aircraft took off for Morocco. It has been suggested that, for fear of his plane being intercepted, Franco carried a letter to the Prime Minister announcing his decision to go to Madrid to fight for the Republic.[71] This seems to be contradicted by the fact that, armed with Sangróniz's passport, Franco

---

* There they were met by Franco's friend, the Spanish military attaché in Paris, Major Antonio Barroso who escorted them to Bayonne. They were to remain for the first three months of the Civil War in the home of the Polo family's old governess Madame Claverie, under the protection of Lorenzo Martínez Fuset.

was passing himself off as a Spanish diplomat. He thus changed from his uniform into a dark grey suit, Pacón into a white one and both threw their military identification papers out of the aircraft.[72] Franco put on a pair of glasses and, at some point on the journey, shaved off his moustache.

There is considerable dispute about the details of the journey. Arrarás and Bolín have a dark grey suit for Franco, Franco Salgado-Araujo white summer suits for both. All three are more plausible than Hills who claims that Franco changed into Arab dress and Crozier who adds, bizarrely, a turban. Arab dress would have been an odd choice of disguise for someone travelling on Sangróniz's Spanish diplomatic passport. Franco Salgado-Araujo claims that they put their uniforms in a suitcase and threw it out of the aircraft. Given the difficulty of throwing a suitcase out of an aircraft in flight and the fact that they emerged from the aircraft in uniform at the end of their journey, it appears that Pacón's memory failed him. There is also contention about the when and where of the demise of the moustache. The issue is whether he shaved on board the aircraft or later, during the stop-over at Casablanca. Pacón and Arrarás place the event on the aircraft but it is unlikely that Franco had a dry shave in a bumpy aircraft in the early stages of his journey. Luis Bolín, who shared a hotel room with Franco in Casablanca, claims that he shaved there. The emergency pilot also claimed the credit for removing the moustache.[73] Whenever the momentous shave took place, it gave rise to Queipo de Llano's later jibe that the only thing that Franco ever sacrificed for Spain was his moustache.[74]

They made a stop at Agadir in the late afternoon where they had some difficulty in getting petrol. The Dragon Rapide then flew onto Casablanca, where, arriving late at night, they were surprised by the sudden disappearance of the landing lights. With fuel running out, there were moments of intense anxiety. The airport was officially closed but Bolín had bribed an official to open up. The light fault was only a blown fuse. When they had landed safely and were eating a sandwich, they decided on the advice of Bebb not to continue the journey north until morning. They then spent a few hours in a hotel. At first light, on 19 July, the aircraft took off for Tetuán. Franco, who had barely slept for three days, was full of vitality at 5.00 a.m. On crossing the frontier into Spanish Morocco, Franco and Pacón changed back into uniform. Unsure as to the situation that awaited them, they circled the aerodrome at Tetuán until they saw Lieutenant-Colonel Eduardo Saenz de Buruaga, an old *Africanista* crony of Franco. Totally reassured, Franco cried '*podemos aterrizar, he visto al rubito*' ('we can land, I've just seen blondy'), and they landed to receive the enthusiastic welcome of the waiting insurgents.[75]

Quickly made aware of the dramatic shortage of aircraft available to the rebels, Franco decided that Bolín should accompany Bebb in the Dragon Rapide as far as Lisbon to report to Sanjurjo and then go on to Rome to seek help. Two hours after depositing its passengers, the Dragon Rapide set off for Lisbon at 9.00 a.m. carrying Bolín with a piece of paper from General Franco which read 'I authorize Don Luis Antonio Bolín to negotiate urgently in England, Germany or Italy the purchase of aircraft and supplies for the Spanish non-Marxist Army'. When Bolín asked for more details, Franco scribbled in pencil on the bottom of the paper '12 bombers, 3 fighters with bombs (and bombing equipment) of from 50 to 100 kilos. One thousand 50-kilo bombs and 100 more weighing about 500 kilos.' In Lisbon, Bolín was to get the further authorization of Sanjurjo for his mission. On 20 July, the aircraft went from Lisbon to Biarritz. On 21 July, Bebb* took Bolín to Marseille whence he travelled on to Rome in order to seek military assistance from Mussolini.[76]

The fact that Franco should so quickly have decided to do something about the rebels' need for foreign help is immensely revealing both of his self-confidence and his ambition. Sanjurjo was convinced that Franco aspired to nothing more than to be Alto Comisario in Morocco. However, his experience during the repression of the Asturian rising had given Franco a rather greater sense of his abilities and a significantly higher aspiration. How far-reaching those ambitions were to be was as yet something even Franco did not know. The situation would change rapidly as rivals were suddenly eliminated, as relationships were forged with the Germans and Italians and as the politics of the rebel zone fluctuated. Ever flexible, Franco would adjust his ambitions as, in the dramatic events ahead, more enticing possibilities arose.

---

* After the civil war, Bebb and Pollard were decorated with the Falangist decoration the Knight's Cross of the Imperial Order of the Yoke and the Arrows. Dorothy Watson and Diana Pollard were given the medal of the same order.

VI

THE MAKING OF A GENERALÍSIMO

July–August 1936

THERE CAN be no doubt that the unlikely figure of Franco, short and with a premature paunch, had a remarkable power to lift the morale of those around him. It was a quality which would play a crucial role in the Nationalist victory and would single him out as leader of the rebel war effort. Having finally shaken himself out of his spring-time hesitations, he once again temporarily resumed the adventurous persona which had served him so well in his rise to the rank of general. It could not have been better suited to the early days of the rising and would see him victoriously through the first months of the Civil War and take him to the doors of absolute power. At that point, caution would reassert itself.

When he drove into Tetuán from the aerodrome at 7.30 a.m. on the morning of Sunday 19 July, the streets were already lined with people shouting '¡Viva España!' and '¡Viva Franco!'. He was greeted at the offices of the Spanish High Commission by military bands and gushingly enthusiastic officers. One of his first acts in his new headquarters was to draw up an address to his fellow military rebels throughout Morocco and in Spain. The text throbbed with self-confidence. Declaring that 'Spain is saved', it ended with words which summed up Franco's unquestioning confidence, 'Blind faith, no doubts, firm energy without vacillations, because the *Patria* demands it. The *Movimiento* sweeps all before it and there is no human force that can stop it'. Broadcast repeatedly by local radio stations, it had the instant effect of raising rebel spirits. When he reached Ceuta in the early afternoon, the scenes which he encountered were more consistent with the beginning of a great adventure than of a bloody civil war. Later in the day, he drove to the headquarters of the Legion in Dar Riffien. Nearly sixteen years earlier, he had arrived there for the first time to become second-in-command of the newly created force. His sense of destiny cannot fail to have been excited by the fact that now he was met by wildly euphoric soldiers chanting 'Franco! Franco! Franco!'. Yagüe made a short and emotional speech: 'Here they are, just as you left them . . . Magnificent and ready for anything. You, Franco, who so many times led them to victory, lead them again for the honour of Spain'. The newly arrived leader, on the verge of tears, embraced Yagüe and spoke to the *Legionarios*. He recognized that they were hungry for combat and raised their pay, already double that of the regular Army, by one peseta per day.[1]

That practical gesture was evidence that, behind the rhetoric, he was aware of the need to consolidate the support of those on whom he would have to rely in the next crucial weeks. Immediately on arriving at the High Commission, he had spent time in conclave with Colonels Saenz de Buruaga, Beigbeder and Martín Moreno discussing ways of recruiting Moorish volunteers.[2] Now, on his return to Tetuán from Dar-Riffien, he took a further measure to secure Moroccan goodwill. He awarded the Gran Visir Sidi Ahmed el Gamnia Spain's highest medal for bravery, the *Gran Cruz Laureada de San Fernando*, for his efforts in containing single-handed an anti-Spanish riot in Tetuán.[3] It was a gesture which was to facilitate the subsequent recruiting of Moroccan mercenaries to fight in peninsular Spain.[4]

The readiness of Franco to use Moroccan troops in Spain had already been demonstrated in October 1934. The gruesome practices of the Legion and the *Regulares* were to be repeated with terrible efficacy during the bloodthirsty advance of the Army of Africa towards Madrid in 1936. At a conscious level, it was no doubt for him a simple military decision. The Legion and the *Regulares* were the most effective soldiers in the Spanish armed forces and it was natural that he would use them without agonizing over the moral implications. The central epic of Spanish history, deeply embedded in the national culture and especially so in right-wing culture, was the struggle against the Moors from 711 to 1492. In more recent times, the conquest of the Moroccan protectorate had cost tens of thousands of Spanish lives. Accordingly, the use of Moorish mercenaries against Spanish civilians was fraught with significance. It showed just how partial and partisan in class terms was the Nationalists' interpretation of patriotism and their determination to win whatever the price in blood.

Franco believed that he was rebelling to save the *Patria*, or rather his version of it, from Communist infiltration, and any means to do so were licit. He did not view liberal and working class voters for the Popular

Front as part of the *Patria*. In that sense, as the Asturian campaign of 1934 had suggested, Franco would regard the working class militiamen who were about to oppose his advance on Madrid in the same way as he had regarded the Moorish tribesmen whom it had been his job to pacify between 1912 and 1925. He would conduct the early stages of his war effort as if it were a colonial war against a racially contemptible enemy. The Moors would spread terror wherever they went, loot the villages they captured, rape the women they found, kill their prisoners and sexually mutilate the corpses.[5] Franco knew that such would be the case and had written a book in which his approval of such methods was clear.[6] If he had any qualms, they were no doubt dispelled by an awareness of the enormity of the task facing himself and his fellow rebels. Franco knew that, if they failed, they would be shot. In such a context, the Army of Africa was a priceless asset, a force of shock troops capable of absorbing losses without there being political repercussions.[7] The use of terror, both immediate and as a long-term investment, was something which Franco understood instinctively. During, and long after the Civil War, those of his enemies not physically eliminated would be broken by fear, terrorised out of opposition and forced to seek survival in apathy.

Because of his cool resolve and his infectious optimism, the decision of Franco to join the rising and to take over the Spanish forces in Morocco was a considerable boost to the morale of the rebels everywhere. Described as 'brother of the well-known airman' and 'a turncoat general' by *The Times*, he was stripped of his rank by the Republic on 19 July.[8] He was one of only four of the twenty-one Major-Generals on active service to declare against the government, the others being Goded, Queipo and Cabanellas.[9] There were officers whose decision to join the rising was clinched by hearing about Franco.[10] More than one rebel officer in mainland Spain reacted to the news with a spontaneous shout of '*¡Franquito está con nosotros! ¡Hemos ganado!*' (Franco's with us. We've won).[11] They were wrong in the sense that the plotters, with the partial exception of Franco, who expected the struggle to last a couple of months, had not foreseen that the attempted coup would turn into a long civil war. Their plans had been for a rapid *alzamiento*, or rising, to be followed by a military directory like that established by Primo de Rivera in 1923, and they had not counted on the strength of working class resistance.

Nevertheless, the plotters were fortunate that their two most able generals, Franco and Mola, had been successful in the early hours of the coup. While Franco to the far south of Spanish territory could rely on the brutal military forces of the Moroccan protectorate, Mola, in the north enjoyed the almost uniformly committed support of the local civilian Carlists of Navarre. In Pamplona, the Carlist population had turned the coup into a popular festival, thronging the streets and shouting *¡Viva Cristo Rey!* (long live Christ the King). These two successes permitted the implementation of the rebel plan of simultaneous marches on Madrid.

On 18 July, that broad strategy was still in the future. The rising had been successful only in the north and north-west of Spain, and in isolated pockets of the south. With a few exceptions, rebel triumphs followed the electoral geography of the Republic. In Galicia and the deeply Catholic rural regions of Old Castile and León, where the Right had enjoyed mass support, the coup met little opposition. The conservative ecclesiastical market towns – Burgos, Salamanca, Zamora, Segovia and Avila, fell almost without struggle. In contrast, in Valladolid, after Generals Andrés Saliquet and Miguel Ponte had arrested the head of the VII Military Region, General Nicolás Molero, it took their men, aided by local Falangist militia, nearly twenty-four hours to crush the Socialist railway workers of Valladolid.[12] Elsewhere, in most of the Andalusian countryside, where the landless labourers formed the mass of the population, the left took power. In the southern cities, it was a different story. A general strike in Cádiz seemed to have won the town for the workers but after the arrival of reinforcements from Morocco, the rebels under Generals José López Pinto and José Enrique Varela, gained control. Córdoba, Huelva, Seville and Granada all fell after the savage liquidation of working class resistance. Seville, the Andalusian capital and the most revolutionary southern city, fell to the lanky eccentric Queipo de Llano and a handful of fellow-conspirators who seized the divisional military headquarters by bluff and bravado. Related to Alcalá Zamora by marriage, Queipo had been considered a republican until the demise of the President inspired a seething hatred of the regime. Perhaps in expiation of his republican past, he would soon be notorious for the implacable ferocity first demonstrated by the bloody repression of working class districts during his take-over of Seville.[13]

In most major urban and industrial centres – Madrid, Barcelona, Valencia, Bilbao – the popular forces by-passed the dithering Republican government and seized power, defeating the military rebels in the process. In Madrid, the general in charge of the rising, Rafael Villegas, was in hiding and sent his second-in-command, General Fanjul, to take command of the one post they held, the Montaña barracks. Besieged by local working class forces, Fanjul was captured and subsequently tried and executed.[14] After defeating the rebels at the Montaña barracks, left-wing militiamen from the capital headed south to reverse the success of the rising in Toledo. With loyal regular troops, they captured the town. However, the rebels under Colonel José Moscardó, the town's military commander, retreated into the Alcázar, the impregnable fortress which dominates both Toledo and the river Tagus which curls around it on the southern, eastern and western sides.

The defeat of the rising in Barcelona deprived the conspirators of one of their most able generals, Manuel Goded, a potential rival to Franco both militarily and politically. In Barcelona, Companys refused to issue arms but depots were seized by the CNT. In the early hours of 19 July, rebel troops began to march on the city centre. They were met by anarchists and the local Civil Guard which, decisively, had stayed loyal. The CNT stormed the Atarazanas barracks, where the rebels had set up headquarters. When Goded arrived by seaplane from the Balearic Islands to join them, the rising was already defeated. Captured, he was forced to broadcast an appeal to his followers to lay down their arms. The defeat of the rebellion in Barcelona was vital for the government, since it ensured that all of Catalonia would remain loyal.[15]

In the Basque Country, divided between its Catholic peasantry and its urban Socialists, the Republic's support for local national regionalist aspirations tipped the balance against the rebels. As Franco had foreseen, the role of the Civil Guard and the Assault Guards was to be crucial. Where the two police forces remained loyal to the government, as they did in most large cities, the conspirators were defeated. In Zaragoza, the stronghold of the CNT, where they did not, the decisive united action of the police and the military garrison had taken over the city before the anarcho-syndicalist masses could react. In Oviedo, the audacious military commander, Colonel Antonio Aranda, seized power by trickery and bravery. He persuaded both Madrid and the local Asturian left-wing forces that he was true to the Republic. Several thousand miners confidently left the city to assist in the defence of Madrid only for many of them to be massacred in a Civil Guard ambush in Ponferrada. Aranda, after speaking with Mola on the telephone, declared for the rebels. By the following day, Oviedo was under siege from enraged miners.[16] The insurgent triumphs in Oviedo, Zaragoza and the provincial capitals of Andalusia had faced sufficient popular hostility to suggest that a full-scale war of conquest would have to be fought before the rebels would control of all of Spain.

After three days, the conspirators held about one third of Spain in a huge block including Galicia, León, Old Castile, Aragón and part of

Extremadura, together with isolated enclaves like Oviedo, Seville and Córdoba. Galicia was crucial for its ports, agricultural products and as a base for attacks on Asturias. The rebels also had the great wheat-growing areas, but the main centres of both heavy and light industry in Spain remained in Republican hands. They faced the legitimate government and much of the Army, although its loyalty was sufficiently questionable for the Republican authorities to make less than full use of it. The government was unstable and indecisive. Indeed, the rebels received a promising indication of the real balance of power when Casares Quiroga resigned to be replaced briefly by a cabinet bent on some form of compromise with the rebels. When Casares withdrew, President Azaña held consultations with the moderate Republican, Diego Martínez Barrio, with the Socialists Largo Caballero and Prieto and with his friend, the conservative Republican, Felipe Sánchez Román. As the basis of a compromise, Sánchez Román suggested a package of measures including the prohibition of strikes and a total crack-down on left-wing militias. The outcome was a cabinet of the centre under Martínez Barrio. Convinced that this was a cabinet ready to capitulate to military demands, the rebels were in no mood for compromise.[17]

It was now too late. Neither Mola nor the Republican forces to the left of Martínez Barrio were prepared to accept any deal. When Martínez Barrio made his fateful telephone call to Mola at 2 a.m. on 19 July, the conversation was polite but sterile. Offered a post in the government, Mola refused on the grounds that it was too late and an accommodation would mean the betrayal of the rank-and-file of both sides.[18] On the following day, Martínez Barrio was replaced by José Giral, a follower of Azaña. After his Minister of War, General José Miaja, also tried unsuccessfully to negotiate Mola's surrender, Giral quickly grasped the nature of the situation and took the crucial step of authorizing the arming of the workers. Thereafter, the defence of the Republic fell to the left-wing militias. In consequence, the revolution which Franco believed himself to have been forestalling was itself precipitated by the military rebellion. In taking up arms to fight the rebels, the Left picked up the power abandoned by the bourgeois political establishment which had crumbled. The middle class Republican Left, the moderate Socialists and the Communist Party then combined to play down the revolution and restore power to the bourgeois Republic. By May 1937, they would be successful, suffocating the revolutionary élan of the working class along the way.

In the interim, a beleagured state, under attack from part of its Army and unable to trust most of those who declared themselves loyal, with its judiciary and police force at best divided, saw much day-to-day power pass to *ad hoc* revolutionary bodies. Under such circumstances, the Republican authorities were unable, in the early weeks and months, to prevent extremist elements committing atrocities against rightists in the Republican zone. This gave a retrospective justification for a military rising which had no prior agreed objectives. The fact that it would be the Communists who eventually took the lead in the restoration of order and the crushing of the revolution was simply ignored by officers like Franco who believed that they had risen to defeat the Communist menace. That generalised objective was the nearest that the conspirators had to a political plan. Franco's own bizarre declaration in the Canary Islands before setting off for Africa ended 'Fraternity, Liberty and Equality'. Many of the declarations by other officers ended with the cry '*¡Viva la República!*'. At most, they knew that they planned to set up a military dictatorship, in the specific form of a military directory.[19]

Equally vague were the military prognostications. There were those, like General Orgaz, who believed that the rising would have achieved its objectives within a matter of hours or at most days.[20] Mola, realizing the crucial importance of Madrid, and anticipating a possible failure in the capital, expected that a dual advance from Navarre and the south would be necessary and therefore require a short civil war lasting two or three weeks. The reverses of the first few days sowed doubts in the minds of the early optimists. Almost alone among the conspirators, Franco, with his obsessions about the importance of the Civil Guard, had taken a more realistic view. Not even he had anticipated a war which would have gone on much beyond mid-September. However, he took the disappointments of the first few days phlegmatically, resourcefully seeking new solutions and insisting to all around him that they must have 'blind faith' in victory. There can be little doubt that his 'blind faith' was sincere. It reflected both his temperament and his long-held conviction that superior morale won battles, something learned in Africa. From his first days with the Legion, he retained the belief that morale had to be backed up by iron discipline. The categorical optimism of his first radio broadcasts in Tetuán was complemented with dire warnings about what would happen to those who opposed the rebels. On 21 July, he promised that the disorders ('*hechos vandálicos*') of the Popular Front would receive 'exemplary punishment'. On 22 July, he said 'for those who persist in opposing us or hope to surrender at the last minute, there will be no pardon'.[21]

Unaware as yet of the fate of the rising on the mainland, Franco had set up headquarters in the officers of the Spanish High Commission in

Tetuán. One of the first issues with which he had to deal provided an opportunity to demonstrate precisely the kind of iron discipline from which he believed the will to win would grow. On arrival at Tetuán, he was informed that his first cousin Major Ricardo de la Puente Bahamonde had been arrested and was about to undergo a summary court martial for having tried to hold the Sania Ramel airport of Tetuán for the Republic and then, when that was no longer possible, disabling the aircraft there. According to Franco's niece, he and Ricardo de la Puente were more like brothers than cousins. As adults, their ideological differences became acute. Franco had had him removed from his post during the Asturian rising. In one of their many arguments, Franco once exclaimed 'one day I'm going to have you shot'. De la Puente was now condemned to death and Franco did nothing to save him. Franco believed that a pardon would have been taken as a sign of weakness, something he was not prepared to risk. Rather than have to decide between approving the death sentence or ordering a pardon, he briefly handed over command to Orgaz and left the final decision to him.[22]

While Franco consolidated his hold on Morocco, things were not going well for the Nationalists on the other side of the Straits. The losses of Fanjul in Madrid and Goded in Barcelona were substantial blows.[23] Now, as Mola and other successful conspirators awaited Sanjurjo's arrival from his Portuguese exile to lead a triumphal march on Madrid, at dawn on 21 July, they received more bad news.[24] Sanjurjo had been killed in bizarre circumstances. On 19 July, Mola's envoy, Juan Antonio Ansaldo, the monarchist air-ace and playboy who had once organized Falangist terror squads, had arrived in Estoril at the summer house where General Sanjurjo was staying.[25] His tiny Puss Moth bi-plane seemed an odd choice for the mission the more so as the far more suitable Dragon Rapide used by Franco had just landed in Lisbon almost certainly with a view to picking up Sanjurjo. The journey could also have been made by road. However, when Ansaldo arrived, he announced dramatically to an enthusiastic group of Sanjurjo's hangers-on that he was placing himself at the orders of the Spanish Chief of State. Overcome with emotion at this theatrical display of public respect, Sanjurjo agreed to travel with him.[26]

To add to the problems posed by the minuscule scale of Ansaldo's aeroplane, the Portuguese authorities now intervened. Although Sanjurjo was legally in the country as a tourist, the Portuguese government did not want trouble with Madrid. Accordingly, Ansaldo was obliged to clear customs and depart alone from the airport of Santa Cruz. He was then

to return towards Estoril and collect Sanjurjo on 20 July at a disused
race-track called La Marinha at Boca do Inferno (the mouth of hell) near
Cascaes. In addition to his own rather portly self, Sanjurjo had, according
to Ansaldo, a large suitcase containing uniforms and medals for his ceremonial entry into Madrid. The wind forced Ansaldo to take off in the
direction of some trees. The overweight aircraft had insufficient lift to
prevent the propeller clipping the tree tops. It crashed and burst into
flames. Sanjurjo died although his pilot survived.[27] Contrary to Ansaldo's
version, it was later claimed in Portugal that the crash was the result of
an anarchist bomb.[28]

Whatever the cause, the death of Sanjurjo was to have a profound
impact on the course of the war and on the career of General Franco.
He was the conspirator's unanimous choice as leader. Now, with Fanjul
and Goded eliminated, his death left Mola as the only general to be a
future challenger to Franco. Mola's position as 'Director' of the rising
was in any case more than matched by Franco's control of the Moroccan
Army which would soon emerge as the cornerstone of Nationalist success.
When war broke out, the military forces in the Peninsula, approximately
one hundred and thirty thousand men in the Army and thirty-three
thousand Civil Guards, were divided almost equally between rebels and
loyalists. However, that broad stalemate was dramatically altered by the
fact that the entire Army of Africa was with the rebels. Against the battle-
hardened colonial Army, the improvised militiamen and raw conscripts,
with neither logistical support nor overall commanders, had little
chance.[29] Apart from Mola, the only other potential challenger to
Franco's pre-eminence was the Falangist leader, José Antonio Primo de
Rivera, but he was in a Republican prison in Alicante.

In these early days of the rising, it is unlikely that even the quietly
ambitious Franco would have been thinking of anything but winning the
war. The death of Sanjurjo was a harsh demonstration to the conspirators
that the *alzamiento* was far from the instant success for which they had
hoped. The collapse of the revolt in Madrid, Barcelona, Valencia, Málaga
and Bilbao obliged the insurgents to evolve a plan of attack to conquer
the rest of Spain. Since Madrid was seen as the hub of Republican resistance, their strategy was to take the form of drives on the Spanish capital by
Mola's northern Army and Franco's African forces. The rebels, however,
confronted unexpected problems. Mola's efforts were to be dissipated by
the need to send troops to San Sebastián and to Aragón. Moreover, the
mixed columns of soldiers, Carlist *Requetés* and Falangists sent by Mola
against Madrid were surprisingly halted at the Somosierra pass in the

Sierra to the north and at the Alto del León to the north-west by the improvised workers' militias from the capital. Threatened from the Republican-held provinces of Santander, Asturias and the Basque Country, the northern Army was also impeded by lack of arms and ammunition.

Franco's Army was paralysed by the problem of transport to the mainland. The conspirators had taken for granted that the fleet would be with them but their hopes had been dashed by a below-decks mutiny. In facing the daunting problem of being blockaded in Morocco, Franco displayed a glacial *sang froid*. His apparent lack of nerves prevented his being dismayed by the numerous reverses that the rebels had encountered in the first forty-eight hours. Even the worst news never disturbed his sleep.[30] Franco's optimism and his determination to win was the dominant theme of an interview which he gave to the American reporter Jay Allen in Tetuán on 27 July. When Allen asked him how long the killing would continue now that the coup had failed, Franco replied 'there can be no compromise, no truce. I shall go on preparing my advance to Madrid. I shall advance. I shall take the capital. I shall save Spain from Marxism at whatever cost.' Denying that there was a stalemate, Franco declared 'I have had setbacks, the defection of the Fleet was a blow, but I shall continue to advance. Shortly, very shortly, my troops will have pacified the country and all of this will soon seem like a nightmare.' Allen responded 'that means that you will have to shoot half Spain?', at which a smiling Franco said 'I repeat, at whatever cost.'[31]

Before Franco had arrived in Tetuán, on 18 July, the destroyer *Churruca* and two merchant steamers, the *Cabo Espartel* and the *Lázaro* and a ferry boat had managed to get 220 men to Cádiz. However, within a matter of hours the crew of the *Churruca*, like those of many other Spanish naval vessels, mutinied against their rebel officers. On 19 July, the gunboat *Dato* and another ferry got a further 170 to Algeciras. In the following days, only a few more troops were able to cross in Moroccan lateen-rigged feluccas (*faluchos*).[32] These men were to have a crucial impact on the success of the rising in Cádiz, Algeciras and La Línea. Within hours of arriving in Tetuán, Franco had discussed with his cousin Pacón and Colonel Yagüe the urgent problem of getting the Legion across the Straits of Gibraltar. The Moroccan Army was effectively immobilized. However, Franco did have two major strokes of luck in this regard. The first was the sympathy for his cause of the authorities on the Rock who refused facilities for the Republican fleet. The second was that the tall, incorruptible General Alfredo Kindelán, the founder of the Spanish Air

Force and a prominent monarchist conspirator, happened to be in Cádiz as Mola's liaison with senior naval officers. In the confusion, and with his contact with Mola broken, Kindelán linked up with the troops recently arrived from Morocco. From Algeciras, he spoke by telephone with Franco who made him head of his Air Force.[33] Kindelán was to be a useful asset in organizing the crossing of the Straits.

Cut off by sea from mainland Spain, Franco, advised by Kindelán, began to toy with the then revolutionary idea of getting his Army across the Straits by air and to seek a way of breaking through the blockade by sea.[34] The few aircraft available at Tetuán had been damaged by the sabotage efforts of Major de la Puente Bahamonde. Those units and others at Seville were soon repaired and in service. A few *Legionarios* able to cross the Straits by air landed at Tablada Aerodrome at Seville and helped consolidate Queipo de Llano's hold on the city.[35] Thereafter, from dawn to late in the evening each day, a constant shuttle was maintained by three Fokker F.VIIb3m trimotor transports and one Dornier DoJ Wal flying boat. Each aircraft did four trips per day; the Fokkers carrying sixteen to twenty soldiers and equipment every time, the Dornier able to carry only twelve and having to land in Algeciras Bay. From 25 July, the original four aircraft were joined by a Douglas DC-2 capable of carrying twenty-five men and, from the end of the month, by another Dornier DoJ Wal flying boat.[36]

The airlift was as yet far too slow. Ironically, the main worry of Franco and his cousin was that Mola might get to Madrid before them. At one point, Franco commented 'in September, I'll be back in the Canary Islands, happy and contented, after obtaining a rapid triumph over Communism'.[37] Even before German and Italian assistance arrived, Franco was fortunate that Kindelán, the energetic Major Julio García de Cáceres and the Air Force pilots who had joined the uprising worked miracles, both repairing the flying boats which had been out of action and putting eight aged Breguet XIX biplane light bombers and two Nieuport 52 fighters at his disposal. These would provide the escorts whose harassment of the Republican navy would sow panic among the inexperienced left-wing crews when Franco decided to risk sea crossings.[38] Franco recognized the importance of the contribution that was being made by Kindelán, by naming him on 18 August, *General Jefe del Aire*.[39]

Even before the early limited airlift was properly under way, Franco was seeking a way of breaking through the sea blockade. On the evening of 20 July, he called a meeting of his staff, attended by Yagüe, Beigbeder, Saenz de Buruaga and Kindelán, as well as naval and Air Force officers.

Assured by Kindelán that the aircraft available could deal with any hostile vessels, Franco decided to send a troop convoy by sea from Ceuta at the earliest opportunity. He overruled strong expressions of doubt, particularly from Yagüe and the naval officers present, who were concerned at the threat posed by the Republican navy. Franco, however, convinced as always of the importance of moral factors in deciding battles, believed that the Republican crews, without trained officers to navigate, oversee the engine rooms or direct the guns, would present little danger. He acknowledged the validity of the objections, but simply brushed them aside. 'I have to get across and I will get across'. It would be one of the few times that Franco the cautious and meticulous planner would take an audacious risk. He decided against a night crossing because his one major advantage, the Republican naval crews' fear of air attack would be neutralized. The precise date of the convoy would be left until the Nationalists had better air cover and more intelligence of Republican fleet movements.[40] It would eventually take place on 5 August.

Ultimately, the conversion of the rising into a long drawn-out war of attrition was to favour Franco's political position and the establishment of a personal dictatorship. At first, however, Franco's isolation in Africa left the political leadership of the coup in the hands of Mola. Nevertheless, although Franco's every thought may have been on winning the war, he still took for granted that he was the leading rebel once Sanjurjo was dead, informing both the Germans and the Italians of this. His ambitions were, however, pre-empted by events in the north.

On 19 July, having made his declaration of martial law in Pamplona, Mola had sketched out an amplified version of his earlier document on the military directory and its corporative policies.[41] On 23 July, he set up a seven-man Junta de Defensa Nacional in Burgos under the nominal presidency of General Cabanellas, the most senior Major-General in the Nationalist camp after the death of Sanjurjo. It consisted of Generals Mola, Miguel Ponte, Fidel Dávila and Andrés Saliquet and two colonels from the general staff, Federico Montaner and Fernando Moreno Calderón. Mola also sought some civilian input from the Renovación Española group. [42] Having been a deputy for Jaén in Lerroux's Radical Party between 1933 and 1935, Cabanellas was regarded by his fellow members as dangerously liberal. His elevation to preside the Junta reflected not simply his seniority but Mola's anxiety to get him away from active command in Zaragoza. Mola himself had visited Zaragoza on 21 July and had been appalled to find Cabanellas exercising restraint in crushing opposition to the rising and contemplating using ex-members of the Radical Party to create a municipal government.[43] On 24 July, the Junta named Franco head of its forces on the southern front. On 1 August, Captain Francisco Moreno Fernández, was named Admiral in command of the section of the navy which had not remained loyal to the Republic, and was added to the Junta.[44]

Only on 3 August, after his first units had crossed the Straits would Franco be added to the Junta de Burgos along with Queipo de Llano and Orgaz. The functions of the Junta were extremely vague. Indeed, the powers of Cabanellas were no more than symbolic. Queipo quickly established *de facto* a kind of vice-royal *fief* in Seville from which he would eventually govern most of the south.[45] There was potential friction between Queipo and Franco. Queipo loathed Franco personally and Franco distrusted Queipo as one of the generals who had betrayed the monarchy in 1931. In addition, there was a more immediate source of tension. Queipo wanted to use the troops being sent from Africa for a major campaign to spread out from the Seville-Huelva-Cádiz triangle which he controlled. He was eager to conquer all of Andalusia, the central and eastern hinterland of which was experiencing a process of revolutionary collectivisation.[46] Franco simply ignored Queipo's aspirations.

In order to resolve the immediate difficulties over transporting the Moroccan Army across the Straits, Franco had turned to fellow rightists abroad for help. On 19 July, the Dragon Rapide had set off for Lisbon and then Marseille, en route back to London. Aboard the aircraft, Luis Bolín carried the paper scribbled by Franco authorizing him to negotiate the purchase of aircraft and other supplies. Bolín left the Dragon Rapide at Marseille and continued on to Rome by train.[47] Franco's early efforts to gain foreign assistance were ultimately successful but they involved several days of frantic effort and frustration. Moreover, it was to be his own efforts, rather than those of Bolín or the monarchist emissaries sent by Mola, which would secure Italian aid since Mussolini was highly suspicious of Spanish rightists eternally announcing that their revolution was about to start.[48]

While Bolín was still travelling, Franco spoke on 20 July to the Italian military attaché in Tangier, Major Giuseppe Luccardi and asked for his help in obtaining transport aircraft. Luccardi telegraphed military intelligence in Rome, where there was grave doubt about the wisdom of helping the Spanish rebels, doubts shared to the full by Mussolini.[49] On 21 July, Franco spoke again to Major Luccardi, stressing the desperate difficulties that he faced in getting his troops across the Straits. Luccardi was sufficiently impressed to put Franco in touch with the Italian Minister

Plenipotentiary in Tangier, Pier Filippo de Rossi del Lion Nero. Franco convinced him on 22 July to send a telegram to Rome requesting twelve bombers or civilian transport aircraft. Mussolini simply scribbled 'NO' in blue pencil at the bottom of the telegram. On a desperate follow-up telegram, the *Duce* wrote only 'FILE'.[50] Meanwhile, Bolín had arrived in Rome on 21 July. At first, he and the Marqués de Viana, armed with a letter of presentation from the exiled Alfonso XIII, were received enthusiastically by the new Italian foreign minister, Count Galeazzo Ciano. Fresh from his long conversation with Franco in Casablanca in the early hours of 19 July, Bolín assured Ciano that, with Sanjurjo dead, Franco would be undisputed leader of the rising. Despite Ciano's initial sympathy, after consulting Mussolini, he turned Bolín away.[51] However, Ciano had been sufficiently intrigued by De Rossi's telegram to request further assessments from Tangier of the seriousness of Franco's bid for power.[52]

While he was still evaluating the information coming in from Tangier, Ciano received on 25 July a more prestigious delegation sent by General Mola. Unaware of Franco's efforts to secure Italian assistance, Mola had called a meeting in Burgos on 22 July with six important monarchists.* Mola outlined the need for foreign help and it was decided that José Ignacio Escobar, the aristocratic owner of *La Época*, would go to Berlin and Antonio Goicoechea, who had signed a pact with Mussolini in March 1934, would lead a delegation to Rome. When Goicoechea's group spoke to Ciano they revealed that Mola was more concerned with rifle cartridges than with aeroplanes.[53] Mola's plea for ammunition seemed small-scale in comparison with Franco's ambitious appeal. Mussolini was by this time beginning to get interested in the Spanish situation as a consequence of the news that the French were about to aid the Republic.[54] Accordingly, in response more to Franco's personal efforts with the Italian authorities in Tangier than to the efforts of monarchists in Rome, Ciano finally responded to Franco's request for aircraft on 28 July with twelve Savoia-Marchetti S.81 Pipistrello bombers.[55]

The bombers were despatched from the Sardinian capital Cagliari in the early hours of the morning of 30 July. As a result of unexpectedly strong headwinds, three ran out of fuel, one crashing into the sea, one crashing while attempting an emergency landing at Oujda near the Algerian border and a third landing safely in the French zone of Morocco

* Antonio Goicoechea, the head of Renovación Española, the intellectual Pedro Sáinz Rodríguez, the Conde de Vallellano, José Ignacio Escobar, owner of the monarchist newspaper *La Época*, the lawyer José María de Yanguas y Messía and Luis María Zunzunegui.

where it was impounded.[56] On 30 July, Franco was informed that the remaining nine had landed at the aerodrome of Nador. However, they were grounded for the next five days until a tanker of high-octane fuel for their Alfa Romeo engines was sent from Cagliari. Since there were insufficient Spaniards able to fly them, the Italian pilots enrolled in the Spanish Foreign Legion.[57] German aircraft also soon began to arrive and the operation for getting the troops of the Moroccan Army across the Straits intensified.

The history of the negotiations for Italian aid shows Franco seizing the initiative and pursuing it with dogged determination. It also shows that Mussolini and Ciano unequivocally placed their bets on Franco rather than on Mola. The exchange of telegrams between Ciano and De Rossi refers to the 'Francoist' rebellion and to 'Franco's movement'.[58] In Germany too, Franco's contacts prospered more. In fact, Mola had substantial prior connections but his various emissaries got entangled in the web of low level bureaucracy in Berlin. In contrast, Franco had the good fortune to secure the backing of energetic Nazis resident in Morocco who had good party contacts through the *Auslandorganisation*. Moreover, as it had with the Italians, his command of the most powerful section of the Spanish Army weighed heavily with the Germans.[59]

Franco's first efforts to get German help were unambitious. Among his staff in Tetuán, the person with the best German contacts was Beigbeder. Accordingly, on 22 July, Franco and Beigbeder asked the German consulate at Tetuán to send a telegram to General Erich Kühlental, the German military attaché to both France and Spain, an admirer of Franco who was based in Paris. The telegram requested that he arrange for ten troop-transport planes with German crews to be sent to Spanish Morocco and ended 'The contract will be signed afterwards. Very urgent! On the word of General Franco and Spain'. This modest telegram was incapable of instigating the sort of official help that Franco needed. It received a cool reception when it reached Berlin in the early hours of the morning of 23 July.[60] However, almost immediately after its despatch, Franco had decided to make a direct appeal to Hitler.

On 21 July, the day before sending the telegram to Kühlental, Franco had been approached by a German businessman resident in Morocco, Johannes Eberhard Franz Bernhardt, who was an active Nazi Party member and friend of Mola, Yagüe, Beigbeder and other *Africanistas*. Bernhardt was to be the key to decisive German assistance. Uneasy about the telegram to Kühlental, Franco decided later in the day on 22 July to use Bernhardt to make a formal approach to the Third Reich for transport aircraft. Bernhardt informed the *Ortsgruppenleiter* of the Nazi Party in Morocco, another resident Nazi businessman, Adolf Langenheim.[61] Langenheim reluctantly agreed to go to Germany with Bernhardt, and Captain Francisco Arranz, staff chief of Franco's minuscule Air Forces.[62] The plan was facilitated by the arrival in Tetuán on 23 July of a Lufthansa Junkers Ju-52/3m mail plane which, on Franco's orders, Orgaz had requisitioned in Las Palmas on 20 July. The Bernhardt mission was a bold initiative by Franco which would make him the beneficiary of German assistance and constitute a giant step on his path to absolute power.

When the party arrived in Germany on 24 July, Hitler was staying at Villa Wahnfried, the Wagner residence, while attending the annual Wagnerian festival in Bayreuth. The delegation was rebuffed by Foreign Ministry officials in Berlin fearful of the international repercussions of granting aid to the Spanish military rebels. However, they were welcomed by Ernst Wilhelm Bohle, the head of the *Auslandorganisation* who enabled them to travel on to Bavaria and provided a link with Rudolf Hess which in turn gained them access to the Führer.[63] Hitler received Franco's emissaries on the evening of 25 July on his return from a performance of *Siegfried* conducted by Wilhelm Fürtwängler. They brought a terse letter from Franco requesting rifles, fighter and transport planes and anti-aircraft guns. Hitler's initial reaction to the letter was doubtful but in the course of a two hour monologue he worked himself into a frenzy of enthusiasm, although noting the Spanish insurgents' lack of funds, he exclaimed, 'That's no way to start a war'. However, after an interminable harangue about the Bolshevik threat, he made his decision. He immediately called his Ministers of War and Aviation, Werner von Blomberg and Hermann Göring, and informed them of his readiness to launch what was to be called *Unternehmen Feuerzauber* (Operation Magic Fire) and to give Franco twenty aircraft rather than the ten requested. The choice of name for the operation suggests that the Führer was still under the influence of the 'Magic Fire' music which accompanies Siegfried's heroic passage through the flames to liberate Brünnhilde. Göring, after initially expressing doubts about the risks, became an enthusiastic supporter of the idea.[64]

Ribbentrop's immediate thought was that the Reich should keep out of Spanish affairs for fear of complications with Britain. Hitler, however, stuck to his decision because of his opposition to Communism.[65] The Führer was determined that the operation would remain totally secret and suggested that a private company be set up to organize the aid and the subsequent Spanish payments. This was to be implemented in the

form of a barter system based on two companies, HISMA and ROWAK.*
Although not the motivating factor, the contribution of Spanish minerals
to Germany's rearmament programme was soon a crucial element in
relations between Franco and Germany.[66]

It has been suggested that Hitler also consulted Admiral Canaris, the
enigmatic head of the *Abwehr*, German Military Intelligence. The dapper
Canaris knew Spain well, having spent time there as a secret agent during
the First World War, and spoke fluent Spanish. It is unlikely that he was
at Bayreuth during the Bernhardt visit, but it is certainly true that once
Hitler decided to aid Franco, Canaris would be the link between them,
much to the irritation of Göring. He was regularly sent to Spain to
resolve problems and in the process established a relationship with
Franco.[67] Canaris quickly began to oversee German aid to Spain, from
4 August liaising with the recently promoted General Mario Roatta, the
flamboyant head of Italian military intelligence. They agreed at the end
of the month that Italian and German assistance would be channelled
exclusively to Franco.[68]

Despite Mola's endeavours, Franco had emerged as the man with international backing.[69] The differences between their approaches to the
Germans were significant. Franco's emissaries had direct links with the
Nazi Party, arrived with credible documentation and relatively ambitious
requests. Mola's envoy, José Ignacio Escobar, had neither papers nor
specific demands other than for rifle cartridges. He had to seek out old
contacts within the conservative German diplomatic corps which was
hostile to any adventurism in Spain. On the basis of the information
before the German authorities, Franco was clearly the leading rebel general, confident and ambitious, while Mola seemed unprofessional and
lacking vision.[70] Franco's own aspirations glimmered through his mendacious statement to Langenheim that he presided over a directorate
consisting of himself, Mola and Queipo de Llano.[71]

Hitler's decision to send twenty bombers to Franco helped turn a *coup
d'état* going wrong into a bloody and prolonged civil war, although it is
clear that Franco would eventually have got his men across the Straits
without German aid. Ten of the Junkers Ju-52/3m, together with the
armaments and military fittings of all twenty, embarked by sea from
Hamburg for Cádiz on 31 July and arrived on 11 August. The other ten,

* German equipment would be imported to Spain by the Compañía Hispano-Marroquí
de Transportes (HISMA) set up on 31 July by Franco and Berhardt and Spanish raw
materials imported into Germany by the Rohstoffe-und-Waren-Einkaufsgesellschaft
(ROWAK) created on 7 October 1936 at the initiative of Marshal Göring.

disguised as civilian transport aircraft, flew directly to Spanish Morocco between 29 July and 9 August. All were accompanied by spare parts and technicians.[72] On 29 July, a delighted Franco telegrammed Mola 'today the first transport aircraft arrives. They will go on arriving at the rate of two per day until we have twenty. I am also expecting six fighters and twenty machine guns.' The telegram ended on a triumphant note, 'We have the upper hand (*Somos los amos*). ¡Viva España!'. All arrived but one, which blew off course and landed in Republican territory.[73]

Despite the consequent intensification of the Nationalist air-lift, there was considerable exaggeration in Hitler's much-quoted remark of 1942 that 'Franco ought to erect a monument to the glory of the Junkers Ju-52. It is this aircraft that the Spanish revolution has to thank for its victory.'[74] The Ju-52 was only one part, albeit a crucial one, of the airlift. What is equally remarkable at this stage of the military rebellion is Franco's unquenchable optimism which not only kept up morale among his own men but also consolidated his authority with his fellow rebels elsewhere in Spain. In Burgos, Mola was in despair at the delay in getting the Army of Africa to the mainland. He telegrammed Franco on 25 July that he was contemplating a retreat behind the line of the river Duero after his initial attack on Madrid had been repulsed. With characteristic firmness and optimism, Franco replied: 'Stand firm, victory certain'.[75]

On 1 August, Franco again telegrammed Mola: 'we will ensure the successful passage of the convoy, crucial to the advance'.[76] On 2 August, accompanied by Pacón, Franco flew to Seville to galvanize the preparations being made by Colonel Martín Moreno for the march on Madrid which was to begin that day.[77] He could see that, even with the Italian and German transport aircraft, the airlift was far too slow. His plan for a convoy to break the blockade had been scheduled for 2 and then 3 August but cancelled. So, on returning to Morocco on 3 August, Franco held a meeting of his staff to fix a new date for the flotilla to make its dash across the Straits. Franco insisted that the troop convoy go by sea from Ceuta at dawn on 5 August despite concerns about the risks expressed by Yagüe and the naval officers. Convinced that the Republican crews were ineffective, Franco side-stepped the objections.[78] He knew too that the Republican navy would be inhibited by the presence of German warships which were patrolling the Moroccan coasts.[79] Accordingly, he sent another reassuring telegram to Mola on 4 August.[80]

On the morning of 5 August, air attacks were launched on the Republican ships in the Straits and the convoy set out but was forced back by thick fog. Meanwhile, Franco telephoned Kindelán in Algeciras and asked him to request the British authorities at Gibraltar to refuse access to the port to the Republican destroyer, *Lepanto*. This request was met and the Republican ship was allowed only to let off its dead and wounded before being obliged to leave Gibraltar. The convoy of ferry boats and naval vessels with three thousand men again set forth in the late afternoon, watched by Franco from the nearby hill of El Hacho. Air cover was provided by the two Dornier flying boats, the Savoia-81 bombers and the six Breguet fighters. The Republican vessels in the vicinity, incapable of manoeuvring to avoid air attack, made little effort to impede their passage. The success of the so-called 'victory convoy' brought the number of soldiers transported across the Straits to eight thousand together with large quantities of equipment and ammunition.[81]

The convoy's success was a devastating propaganda blow to the Republic. The news that the ruthless Army of Africa was on the way depressed Republican spirits as much as it boosted those in the Nationalist zone. By 6 August, there were troop-ships regularly crossing the Straits under Italian air cover. The Germans also sent six Heinkel He-51 fighters and ninety-five volunteer pilots and mechanics from the Luftwaffe. Within a week, the rebels were receiving regular supplies of ammunition and armaments from both Hitler and Mussolini. The airlift was the first such operation of its kind on such a scale and constituted a strategic innovation which redounded to the prestige of General Franco. Between July and October 1936, 868 flights were to carry nearly fourteen thousand men, 44 artillery pieces and 500 tons of equipment.[82]

At this time, Mola made a significant error in the internal power stakes. On 1 August, heir to the Spanish throne, the tall and good-natured Don Juan de Borbón, the third son of Alfonso XIII, arrived in Burgos in a chauffeur-driven Bentley.* Anxious to fight on the Nationalist side, he had left his home in Cannes on 31 July, despite the fact that on that day his wife Doña María de Mercedes was giving birth to a daughter. Mola ordered the Civil Guard to ensure that he left Spain immediately. The fact that he did so abruptly and without consultation with his fellow generals revealed both Mola's lack of subtlety and his anti-monarchist

---

* Alfonso XIII's eldest son, Alfonso, was afflicted by haemophilia and had formally accepted the loss of his right to the throne in June 1933 when he contracted a morganatic marriage with Edelmira Sampedro, the daughter of a rich Cuban landowner. The King's second son, Jaime, immediately renounced his own rights on the grounds of a disablement (he was deaf and dumb). Jaime would, in any case, have lost his rights when, in 1935, he also married morganatically an Italian, Emmanuela Dampierre Ruspoli, who although an aristocrat was not of royal blood. Alfonso died in September 1938 after a car crash in Miami.

sentiments. The incident contributed to deeply monarchist officers transferring their long-term political loyalty to Franco.[83] In contrast, when
Franco later took a similar step, preventing Don Juan volunteering to
serve on the battleship *Baleares,* he was careful to pass off his action as
an effort to guarantee that the heir to the throne should be 'King of all
Spaniards' and not be compromised by having fought on one side in the
war.[84]

Two days after the successful 'victory convoy', Franco flew to Seville
and established his headquarters in the magnificent palace of the
Marquesa de Yanduri.[85] Marking a clear distinction with Queipo's more
modest premises, the palace's grandeur revealed more about Franco's
political ambitions than his military necessities. He began to use a Douglas DC-2 to visit the front or travel to meet Mola for consultations.[86] In
Seville, he began to gather around him the basis of a general staff. Apart
from two ADCs, Pacón and an artillery Major Carlos Díaz Varela, there
were Colonel Martín Moreno, General Kindelán and, a recent arrival,
General Millán Astray.[87] This reflected the fact that finally he had an
army on the move.

Even before the 'victory convoy', Franco had already, on 1 August,
ordered a column under the command of the tough Lieutenant-Colonel
Carlos Asensio Cabanillas to occupy Mérida and deliver seven million
cartridges to the forces of General Mola. The column had set out on
Sunday 2 August in trucks provided by Queipo de Llano and advanced
eighty kilometres in the first two days. Facing fierce resistance from
untrained and poorly armed Republican militiamen, they took another
four days to reach Almendralejo in the province of Badajoz. Asensio's
column had been followed on 3 August by another column led by Major
Antonio Castejón which had advanced somewhat to the east and on
7 August by a third under Lieutenant-Colonel Heli Rolando de Tella.
Franco telegrammed Mola on 3 August to make it clear that the ultimate
goal of these columns was Madrid. After the frenetic efforts of the previous two weeks to secure international support and get his troops across
the Straits, Franco's mood was euphoric.

Franco placed Yagüe in overall field command of the three columns.
He ordered them to make a three-pronged attack on Mérida, an old
Roman town near Cáceres, and an important communications centre
between Seville and Portugal. The columns advanced with the Legionaires on the roads and the Moorish *Regulares* fanning out on either side
to outflank any Republican opposition. With the advantage of local air
superiority provided by Savoia-81 flown by Italian Air Force pilots and

Junkers Ju-52 flown by Luftwaffe pilots, they easily took villages and towns in the provinces of Seville and Badajoz, El Real de la Jara, Monesterio, Llerena, Zafra, Los Santos de Maimona, annihilating any leftists or supposed Popular Front sympathisers found and leaving a horrific trail of slaughter in their wake. The execution of captured peasant militiamen was jokingly referred to as 'giving them agrarian reform'. After the capture of Almendralejo, one thousand prisoners were shot including one hundred women. Mérida fell on 10 August. In a little over a week, Franco's forces had advanced 200 kilometres. Shortly afterwards, initial contact was made with the forces of General Mola.[88] Thus, the two halves of rebel Spain were joined into what came to be called the Nationalist zone.

The terror which surrounded the advance of the Moors and the Legionaries was one of the Nationalists' greatest weapons in the drive on Madrid. After each town or village was taken by the African columns, there would be a massacre of prisoners and women would be raped.[89] The accumulated terror generated after each minor victory, together with the skill of the African Army in open scrub, explains why Franco's troops were initially so much more successful than those of Mola. The scratch Republican militia would fight desperately as long as they enjoyed the cover of buildings or trees. However, they were not trained in elementary ground movements nor even in the care and reloading of their weapons. Thus, even the rumoured threat of being outflanked by the Moors would send them fleeing, abandoning their equipment as they ran.[90] Franco was fully aware of the Nationalists' superiority over untrained and poorly armed militias and he and his Chief of Staff, Colonel Francisco Martín Moreno, planned their operations accordingly. Intimidation and the use of terror, euphemistically described as *castigo* (punishment), were specified in written orders.[91]

Given the iron discipline with which Franco ran military operations, there is little possibility that the use of terror was merely a spontaneous or inadvertent side effect. There was little that was spontaneous in Franco's way of running a war. On being informed of the bravery of a group of Falangist militiamen in capturing some Republican fortifications, Franco ordered them to be shot if they ever again contravened the day's orders, 'even though I have to go and place the highest decorations on their coffins'.[92] In late August, Franco boasted to a German emissary of the measures taken by his men 'to suppress any Communist movement'.[93] The massacres were useful from several points of view. They indulged the blood-lust of the African columns, eliminated large numbers of potential opponents – anarchists, Socialists and Communists whom Franco despised as rabble – and, above all, they generated a paralysing terror.

He wrote Mola on 11 August an extraordinarily significant letter, revealing his expectations of a quick end to the war, his strategic vision and the colonial mentality behind his views on the conquest of territory. He agreed that the priority should be the occupation of Madrid but stressed the need to annihilate all resistance in the 'occupied zones', especially in Andalusia. Franco mistakenly assumed that the early capture of Madrid would precede attacks on the Levante, Aragón, the north and Catalonia. He suggested that Madrid be squeezed into submission by 'tightening a circle, depriving it of water supplies and aerodromes, cutting off communications'. Crucially, in the light of his later remarkable diversion of troops away from Madrid, he ended with the words: 'I did not know that [the Alcázar of] Toledo was still being defended. The advance of our troops will take the pressure off and relieve Toledo without diverting forces which might be needed'.[94]

At the time that Franco's letter was being written, Mola was complaining about the difficulties of liaison.[95] Telephone contact between Seville and Burgos was established immediately after the capture of Mérida. The two generals spoke on 11 August. Apparently oblivious to any eventual political implications, Mola agreed with Franco that there was no point duplicating his successful international contacts and therefore ceded to him the control of supplies. Mola's political allies were appalled at his naïvety. José Ignacio Escobar asked him if he had therefore agreed on the telephone that the head of the movement be Franco. Mola replied guilelessly, 'It is an issue which will be resolved when the time comes. Between Franco and I there are neither conflicts nor personal ambitions. We see entirely eye-to-eye and to leave in his hands this business of the procurement of arms abroad is just a way of avoiding a harmful duplication of effort.' When Escobar insisted that this made Mola the second-in-command as far as the Germans were concerned, he brushed aside his remarks. The control of arms supplies guaranteed that Franco and not Mola, with all the attendant political implications, would dominate the assault on the capital.[96]

After the occupation of Mérida, Yagüe's troops turned back south-west towards Portugal to capture Badajoz, the principal town of Extremadura, on the banks of the River Guadiana near the Portuguese frontier. Although encircled, the walled city was still in the hands of numerous but ill-armed left-wing militiamen who had flocked there before the advancing Nationalist columns. Many were armed only with scythes and hunting shotguns. Most of the regular troops garrisoned there had been called away to reinforce the Madrid front.[97] If Yagüe had pressed on to Madrid, the Badajoz garrison could not seriously have threatened his column from the rear. It has been suggested that Franco's decision to turn back to Badajoz was a strategic error, contributing to the delay which allowed the government to organize its defences. Accordingly, Nationalist historians have blamed Yagüe but the decision smacks of Franco's caution rather than Yagüe's frenetic impetuousness. Franco made all the major daily decisions merely leaving their implementation to Yagüe. He had personally supervised the operation against Mérida and, on the evening of 10 August, received Yagüe in his headquarters to discuss the capture of Badajoz and the next objectives.[98] He wanted to knock out Badajoz to clinch the unification of the two sections of the Nationalist zone and to cover completely the left flank of the advancing columns.

On 14 August, after heavy artillery and bombing attacks, the walls of Badajoz were breached by suicidal attacks from Yagüe's *Legionarios*. Then a savage and indiscriminate slaughter began during which nearly two thousand people were shot, including many innocent civilians who were not political militants. According to Yagüe's biographer, in 'the paroxysm of war', it was impossible to distinguish pacific citizens from leftist militiamen, the implication being that it was perfectly acceptable to shoot prisoners.[99] The *Legionarios* and *Regulares* unleashed an orgy of looting and the carnage left streets strewn with corpses, a scene of what one eyewitness called 'desolation and dread'. After the heat of battle had cooled, two thousand prisoners were rounded up and herded to the bull-ring, and any with the bruise of a rifle recoil on their shoulders were shot. The shootings went on for weeks thereafter. Yagüe told the American journalist John T. Whitaker, who accompanied him for most of the march on Madrid, 'Of course, we shot them. What do you expect? Was I supposed to take four thousand Reds with me as my column advanced racing against time? Was I expected to turn them loose in my rear and let them make Badajoz Red again?'.[100] In fact, the savagery unleashed on Badajoz reflected both the traditions of the Spanish Moroccan Army and the outrage of the African columns at encountering a solid resistance and, for the first time, suffering serious casualties. In retrospect, it can be seen that the events of Badajoz might have been taken to anticipate what would happen when the columns reached Madrid. The clear lesson was that the easy victories of the *Legionarios* and *Regulares* in open country were not replicated in built-up cities. This was not widely perceived in the Nationalist camp but the stiffening of Republican resistance does seem to have dented Franco's earlier optimism.

The distant cloud of potential difficulties at Madrid could hardly dim Franco's appreciation of the benefits won at Badajoz. Now, crucially, there was unrestricted access to the frontier of Portugal, the Nationalists' first international ally. From the beginning, Oliveira Salazar had permitted the rebels to use Portuguese territory to link their northern and southern territories.[101] It was access to Portuguese help which, as much as any other factor, had decided Franco to swing his columns westwards through the province of Badajoz rather than the more direct route along the main road from Seville to Madrid, across the Sierra Morena via Córdoba.*[102]

On 14 August, General Miguel Campins, Franco's one-time friend and second-in-command at the Academia General Militar de Zaragoza, was tried in Seville for the crime of 'rebellion'. The court martial was presided over by General José López Pinto. Campins was sentenced to death and shot on 16 August.[103] His crime was to have refused to obey Queipo's demand on 18 July that he declare martial law in Granada and to have delayed two days before joining the rising. Franco was unable to overcome the determination of Queipo de Llano to have Campins shot. According to Franco's cousin, despite refusing Queipo's order, Campins had in fact telegraphed Franco putting himself under his orders. Franco wrote a number of letters to Queipo requesting that mercy be shown to Campins. Queipo simply tore them up but Franco did not push the matter further for fear of undermining the unity of the Nationalist camp.[104] According to his sister Pilar, Franco was upset by the death of his friend.[105] Queipo's determination to execute Campins despite pleas for mercy reflected both his brutal character and his long-standing loathing of Franco. Franco took his revenge in 1937 by ignoring Queipo's own pleas for mercy for his friend General Domingo Batet, who was condemned to death for opposing the rising in Burgos.[106]

While Campins was being tried and shot, Franco made a cunning move which boosted his stock in the eyes of Spanish rightists at the expense of his rivals in the Junta. In Seville on 15 August, flanked by Queipo, he announced the decision to adopt the monarchist red-yellow-red flag. Queipo acquiesced cynically, reluctant to draw attention to his own republicanism. Mola, who barely two weeks before had expelled the

* Assuming that Franco would attack through Cordoba, and believing the Yagüe columns to be engaged only in local operations, the Republican General Miaja had concentrated his exiguous defensive forces on the Córdoba–Madrid line.

heir to the throne, was not consulted. Only with acute misgivings did
General Cabanellas sign a decree of the Junta de Defensa Nacional two
weeks later ratifying the use of the flag.[107] Franco had managed to present
himself to conservatives and monarchists as the one certain element
among the leading rebel generals. It was a clear indication that while the
others thought largely of eventual victory, Franco kept a sharp eye on
his own long-term political advantage.

In fact, Mola and Franco were worlds apart in both political preferences and in temperament. In the words of Mola's secretary José María
Iribarren, Mola 'was neither cold, imperturbable nor hermetic. He was
a man whose face transmitted the impressions of each moment, whose
stretched nerves reflected disappointments'.[108] Mola himself seemed
totally oblivious to security, strolling around Burgos alone and in civilian
clothes. His headquarters were chaotic with visitors wandering in at all
times.[109] Queipo de Llano was equally casual about visitors. In contrast,
Franco had a bodyguard and the tightest security arrangements at his
headquarters. Visitors were searched thoroughly and during interviews
with Franco, the door was kept ajar and one of the guards kept watch
via a strategically placed mirror.[110]

Those who did get in to see him did not find a daunting war lord.
Many aspects of Franco's demeanour, his eyes, his soft voice, the apparent
outer calm struck many commentators as somehow feminine. John
Whitaker, the distinguished American journalist, described him thus: 'A
small man, his hand is like a woman's and always damp with perspiration.
Excessively shy, as he fences to understand a caller, his voice is shrill and
pitched on a high note which is slightly disconcerting since he speaks
very softly – almost in a whisper.'[111] The femininity of Franco's appearance was frequently, and inadvertently, underlined by his admirers. 'His
eyes are the most remarkable part of his physiognomy. They are typically
Spanish, large and luminous with long lashes. Usually they are smiling
and somewhat reflective, but I have seen them flash with decision and,
though I have never witnessed it, I am told that when roused to anger
they can become as cold and hard and steel.'[112]

Franco certainly had heated arguments in Seville with Queipo de Llano
who had difficulty concealing his contempt for the man who was below
him in the seniority scale. In contrast, Mola remained on good terms
with Franco.[113] A German agent reported to Admiral Canaris in mid-
August on the view from Franco's headquarters. The report showed the
wily *gallego* subtly consolidating his position and confirming the fears of
Mola's supporters that he had sold the pass to Franco on 11 August.

The agent's report stated that German aid must be channelled through Franco.[114] Mola continued to recognize Franco's superior position in terms of foreign supplies and battle-hardened troops. Their correspondence in August shows Franco as the distributor of largesse in terms of financial backing and military hardware. Franco could boast of the fact that foreign suppliers made few if any demands upon him in terms of early payment. He could offer to send Mola aircraft.[115]

On 16 August, Franco, accompanied by Kindelán, flew to Burgos where Mola could not have failed to notice the manic fervour with which his comrade was received by the local population. A solemn high mass was said in the Cathedral by the Archbishop.[116] At dinner that night, Franco's optimism about the progress of the war was as unshakeable as ever. The only glimmer of anxiety came in a comment to Mola that he was worried that he had had no news of his wife Carmen and his daughter Nenuca.[117] After dinner, Franco and Mola spent several hours locked in secret conclave. Although no decision was taken, it was obvious to both of them that the efficient prosecution of the war required a single overall military command.[118] It was obvious too that some kind of centralised diplomatic and political apparatus was necessary. Franco and his small staff were working ceaselessly to maintain foreign logistical support. The Junta de Burgos which used to meet late at night was also finding itself overwhelmed with work.[119] Given Franco's near monopoly of contacts with the Germans and Italians and the apparently unstoppable progress of his African columns, Mola must have realized that the choice of Franco to assume the necessary authority would be virtually inevitable. Franco's staff had already loaded the dice by convincing German Military Intelligence that the victory in Extremadura had indisputably established him as 'Commander-in-Chief'. Portuguese newspapers and other sections of the international press described him as 'Commander-in-Chief' presumably on the basis of information supplied by his headquarters. The Portuguese consul in Seville referred to him as 'the supreme commander of the Spanish Army' as early as mid-August.[120]

Mola was gradually being forced towards the same view. On 20 August, he sent a message to Franco pointing out his own troops were having difficulties on the Madrid front and asking to be informed of Franco's plans for his advance on the capital. In the event of Franco's advance being delayed Mola would make arrangements to concentrate his activities on another front.[121] The text of his telegram suggested less a deferential subordination to Franco's greater authority than a rational desire to co-ordinate their efforts in the interests of the war effort. Mola was not thinking in terms of a power struggle but three days later he was brutally made aware of the extent to which Franco was consolidating his own position. On 21 August, Mola received a visit from Johannes Bernhardt in Valladolid. Bernhardt came with the good news that an anxiously awaited German shipment of machine-guns and ammunition was on its way by train from Lisbon. Mola's delight was severely diminished when Bernhardt said to him 'I have received orders to tell you that you are receiving all these arms not from Germany but from the hands of General Franco'. Mola went white but quickly accepted the inevitable. It had already been agreed with General Helmuth Wilberg, head of the inter-service commission sent by Hitler to co-ordinate *Unternehmen Feuerzauber*, that German supplies would be sent only on Franco's request and to the ports indicated by him.[122]

After the capture of Badajoz, Yagüe's three columns had begun to advance rapidly up the roads to the north-east in the direction of the capital. Tella's column had moved to Trujillo on the road towards Madrid while Castejón's column had raced towards Guadalupe on Tella's southern flank. By 17 August, Tella had reached the bridge across the Tagus at Almaraz and shortly afterwards arrived at Navalmoral de la Mata on the borders of the province of Toledo. Castejón's column would capture Guadalupe on 21 August. Castejón, Tella and Asensio would join together on 27 August before the last town of importance on the way to Madrid, Talavera de la Reina. In two weeks, they had advanced three hundred kilometres.[123]

Despite these heady successes, Franco's telegram in reply to Mola suggested that his unflappable optimism was beginning to be eroded by Republican resistance. He made it clear that, on the advance to Talavera de la Reina, he feared strong Republican flank attacks at Villanueva de la Serena and Oropesa. 'A well-defended town can hold up the advance. I'm down to six thousand men and have to guard long lines of communication. Flank attacks limit my capacity for movement.' He outlined to Mola the next stages of the push, on to the important road junction at Maqueda in Toledo, then from Maqueda diagonally north-east to Navalcarnero on the road to Madrid.*[124] Within a month, the bold and direct strategy outlined to Mola would be abandoned in the interests of ensuring that Franco would be the undisputed Generalísimo.

* The Francoist military historian, Colonel José Manuel Martínez Bande, has seen this message as the first sign of Franco's decision to relieve the Alcázar de Toledo. His view is based entirely on the presence in the message of the words: 'Maqueda-Toledo', which he arbitrarily takes to mean 'relief of the Alcázar'. However, the rest of Franco's text shows rather that after Maqueda the column would make a continued thrust to Madrid in a direct line to Navalcarnero rather than make any diversion to Toledo.

VII

THE MAKING OF A CAUDILLO

August–November 1936

THE SUCCESSES of the African columns and the imminent attack on Talavera led, on 26 August, to Franco transferring his headquarters from Seville to the elegant sixteenth century Palacio de los Golfines de Arriba in Cáceres. He was anxious to move on from Seville in order to establish his total autonomy, free from the interference or disdain of Queipo de Llano in whose presence he always felt uncomfortable.[1] Like his earlier choice of the Palacio de Yanduri in Seville, it indicated a jealous concern for his public status. Franco was beginning to build a political apparatus capable of daily dealings with the Germans and Italians. Already he had a diplomatic office, headed by José Antonio de Sangróniz. Lieutenant-Colonel Lorenzo Martínez Fuset acted as legal adviser and political secretary. Franco was also accompanied from time to time by his brother Nicolás, who travelled between Cáceres and Lisbón where he was working for the Nationalist cause. Nicolás would soon be acting as a kind of political factotum. Millán Astray was in charge of propaganda. Even at this early stage, the tone of Franco's entourage was sycophantic.[2]

The sheer volume of work facing Franco, effectively co-ordinating Nationalist 'foreign policy' and logistical organization, as well as maintaining close overall supervision of the advance of the African columns, obliged him to work immensely long hours. His resistance to discomfort and the powers of endurance which he had displayed as a young officer in Africa were undiminished but he began to age noticeably. The manic Millán Astray boasted to Ciano that 'our Caudillo spends fourteen hours at his desk and doesn't get up even to piss'.[3] When his wife and daughter returned to Spain after their two-month exile in France – on 23 September – he responded to the announcement of their arrival by sending them a message that he had important visitors waiting. They were obliged to wait for more than an hour. He had little time for family life.[4] Such concentration and strain perhaps contributed to the quenching of his early optimism but the re-emergence of a cautious Franco after the brief reincarnation of the impetuous African hero denoted both the prospect of power and the growing strength of Republican resistance.

The difficulties that were now slowing down the advance of the African columns impelled Franco's Italian and German allies to step up their assistance. On 27 August, accompanied by Lieutenant-Colonel Walter Warlimont of the War Ministry staff, Canaris met Roatta in Rome to co-ordinate their views on the scale and nature of future assistance from Italy and Germany to the Nationalists. At a further meeting on the following day, they were joined by Ciano. Canaris again insisted that assistance be provided 'only to General Franco, because he holds the supreme command of operations'. Joint Italo-German planning required a recognizable overall Nationalist commander with whom to communicate.[5]

Talavera was encircled by the three columns. The propaganda value for the Nationalists of the massacre at Badajoz was revealed when large numbers of militiamen fled in buses 'like a crowd after a football match'. The town fell on 3 September. Another savage and systematic massacre ensued.[6] While Franco's forces had been moving through Extremadura and into New Castile, Mola had begun an attack on the Basque province of Guipúzcoa to cut the province off from France. Irún and San Sebastián were attacked daily by Italian bombers and bombarded by the Nationalist fleet. Irún's poorly armed and untrained militia defenders fought bravely but were overwhelmed on 3 September. San Sebastián fell on 12 September. It was a key victory for the Nationalists. Guipúzcoa was a rich agrarian province which also contained important heavy industries. The Nationalist zone was now united in a single block from the Pyrenees through Castille and western Spain to the far south. The Republican provinces of Vizcaya, Santander and Asturias were isolated, able to communicate with the rest of the Republic only by sea or air.[7]

The losses of Talavera and Irún provoked the fall of the government of José Giral. A cabinet which more clearly reflected the working class bases of the Republic was introduced under the leadership of Francisco Largo Caballero. The clearer definition of the Republic and its move towards a stronger central authority was the corollary of the ever fiercer resistance being mounted against Franco's advancing columns. The reduction of political indecision on the Republican side intensified the feeling among the senior Nationalist commanders that a unified command was an urgent necessity. Franco's ambitions could be deduced from a statement to the Germans in Morocco that he wished 'to be looked upon not only as the saviour of Spain but also as the saviour of Europe from the spread of Communism'.[8] Now, the issue of a single command opened an opportunity for him. Mola flew to Cáceres on 29 August and discussed the matter with him.[9]

In the meanwhile, the bulk of Nationalist success was being chalked up by Franco's Army of Africa. Protected to the south by the Tagus, Yagüe's troops secured their northern flank by linking up with Mola's forces. With the road to Madrid now open, for the next two weeks desperate Republican counter-attacks sought to recapture Talavera, but Franco showed a dogged resolve not to give up an inch of captured ground. Stiffening resistance and Franco's determination to purge terri-tory of leftists as it was captured account for the slowing down of his advance. In fact, he was on the verge of slowing it down even further by a momentous decision.

Among the issues crowding in on him, Franco gave some thought to the besieged garrisons of Toledo and Santa María de la Cabeza in Jaén. He regularly released his own Douglas DC-2 aircraft and his pilot Captain Haya for missions to both fortresses. On 22 August, he had sent a message to the Alcázar de Toledo promising to bring relief.[10] The fortress was still unsuccessfully besieged by Republican militiamen who had wasted time, energy and ammunition in trying to capture this strategically un-important stronghold. The one thousand Civil Guards and Falangists who had retreated into the Alcázar in the early days of the rising, had taken with them as hostages many women and children, the families of known leftists.[11] However, the resistance of the Alcázar was being turned into the great symbol of Nationalist heroism. Subsequently, the reality of the siege would be embroidered beyond recognition, in particular through the famous, and almost certainly apocryphal, story that Moscardó was telephoned and told that, unless he surrendered, his son would be shot.* Naturally, the existence, and subsequent fate, of the hostages was entirely forgotten.[12]

Franco's troops took more than two weeks to cover the ground from Talavera to the town of Santa Olalla in the province of Toledo on the road to Madrid.[13] On 20 September, Yagüe's forces captured Santa Olalla and imposed another 'exemplary punishment' on the militiamen they captured.[14] Maqueda, at the cross-roads where the road divided to go

---

* Before the myth-makers began to work, *ABC*, Seville, 3 October 1936 claimed that 'communications with the outside were totally cut throughout the siege'.

either north to Madrid or east to Toledo, also fell to Yagüe on
21 September. At this point, that is to say *after* the fall of Maqueda,
Franco had to make the decision whether to let the African columns race
onto Madrid or else turn eastwards to relieve Toledo. It was a complex
decision with political as well as military implications. While Yagüe was
capturing Santa Olalla and Maqueda, Franco had been engaged in meetings with the other generals of the Junta de Defensa Nacional to discuss
the need for a single Commander-in-Chief for the Nationalist forces. It
is immensely difficult to reconstruct in precise detail the where, when,
why and how of Franco's decision but a key is to be found in the role of
Yagüe.

On the day after Maqueda fell, an 'officially' sick and exhausted Yagüe
handed over command to Asensio.[15] It has been suggested that Franco's
decision to relieve Yagüe of his command was influenced by Mola's
intense hostility to him.[16] It is possible, but highly unlikely, that Franco
would have relieved the highly successful Yagüe at the insistence of
Mola.* It has also been suggested that Yagüe's replacement had less to
do with his illness than with his opposition to Franco's decision to interrupt the march on Madrid to relieve the Alcázar de Toledo.[17] Either of
these possibilities would make sense if, in replacing Yagüe, Franco was
punishing him for indiscipline. However, it seems unlikely that Yagüe
was in disgrace of any kind since his withdrawal from the front was
accompanied by promotion to full colonel and his immediate incorporation into Franco's close entourage.[18] By 22 September, Yagüe was
already installed in the Palacio de los Golfines de Arriba, a curious resting
place for a man in disgrace.[19]

There is, however, a third and altogether more likely possibility which
fits the facts of Yagüe's health, his promotion and his activities over the
next few weeks. Yagüe's substitution was made necessary because he had
a weak heart consequent on problems with his aorta: he was genuinely
exhausted and not really fit for further uninterrupted campaigning.
Recognizing Yagüe's priceless contribution at the head of the African
columns, Franco was happy to give him a respite, promote him and use

* At some point on either 20 or 21 September, Yagüe and Mola met to discuss the
co-ordination of operations between their forces which had recently made contact over a
long front. Their disagreements became increasingly heated. Mola told Yagüe that his
behaviour constituted mutiny for which he could have him shot. Turning to his column
commanders, Asensio, Castejón and Tella, Yagüe said 'We don't think so' (*¡Verdad que
no!*) at which Mola was forced to make a joke of his original remark and back down.
(Letter to the author from General Ramón Salas Larrazabal, 9 May 1991, recounting the
testimony of one of the column chiefs present at the meeting, probably Asensio Cabanillas.)

his immense prestige within the Legion for another task, as part of the orchestration of his bid to become Generalísimo. The ever faithful Yagüe, despite his obvious need for rest, threw himself into the job with a gusto which makes it difficult to imagine that there was serious friction between him and Franco.

Franco was fully aware of the possible military consequences of diverting his troops to Toledo. He would lose an unrepeatable chance to sweep onto the Spanish capital before its defences were ready. Both Kindelán and his Chief of Operations, Lieutenant-Colonel Antonio Barroso, warned him that opting to go to Toledo might cost him Madrid. Yagüe's opposition seems to have been the most outspoken. He reiterated the point made by Franco to Mola in his message of 11 August that the mere proximity of his columns to Madrid would have sent the besieging militiamen racing back to the capital. However, as had happened with Yagüe's doubts over the crossing of the Straits in early August, his unquestioning faith in Franco brought him round. Franco disagreed with his staff that the delay of a week would undermine his chances of capturing Madrid. Nevertheless, he openly stated that, even if he knew for certain that going to Toledo would lose him the capital, he would still fulfil his promise to liberate the besieged garrison.[20] He was more interested in the political benefits of the relief of the Alcázar and to maximise those benefits he needed Yagüe at his side rather than in the field.

As a result of Franco's decision, there would be a delay from 21 September to 6 October before the march on Madrid could continue. The two weeks were lost by Franco while he took Toledo and was involved in the process of his own political elevation. That delay would constitute the difference between an excellent chance to pluck Madrid easily and having to engage in a lengthy siege as a result of the reorganization of the capital's defences and the arrival of foreign aid. At precisely this time, the Germans began to voice their impatience with 'extraordinary' and 'incomprehensible' delays which were permitting the Republican government to receive help from abroad.[21] Given that Franco never ceased to complain to his allies about Soviet assistance to the Republic, it is ironic that he should so dramatically have underestimated its impact on the defence of Madrid. In moving his forces to Toledo, Franco gave a higher priority to the inflation of his own political position by means of an emotional victory and a great propagandistic coup than to the early defeat of the Republic. After all, had he moved onto Madrid immediately, he would have done so before his own political position had been irrevocably consolidated. The entire process of choosing a Caudillo would have been delayed. Then the triumph, and therefore the future, would have had to be shared with the other generals of the Junta.

Convinced of Franco's monarchist good faith, Kindelán had long been urging Franco to raise the question of the need for a single command. Ostensibly at least, Franco showed little interest.[22] Since his arrival in Tetuán on 19 July, Franco had been swamped every day by pressing problems. However, in the course of solving them, his self-confidence and ambitions had grown. In addition to organizing a combat Army without the normal logistical and financial support of the State to feed, arm and pay his troops, he had extended his activities into the international arena, acquiring a monopoly of arms and ammunitions deliveries. However, it was only in September as co-ordination with Mola's forces for the final push on Madrid became likely that a formally recognized Commander-in-Chief became an urgent necessity.

There is no reason to doubt that Franco's faith in his own abilities had already convinced him that, if there was to be a single command, then he should exercise it. He had long since presented himself to the agents of Berlin and Rome as the effective leader of the Nationalist cause. In early September, the Italian military mission under General Mario Roatta presented its credentials to Franco and thereby conveyed Mussolini's *de facto* recognition of his leadership.[23] Any scruples which he expressed to Kindelán and Pacón reflected slow-moving prudence rather than modesty. Instinctive caution inclined him to avoid possible failure and humiliation by taking care not to be seen to have sought the post of Commander-in-Chief. A show of hesitation would disarm the jealousy of his rivals.

From the earliest moments of the uprising, Franco had been concerned about political unity within the Nationalist zone. Shocked by the Aladdin's cave of uniforms and militias which he had encountered on arrival at Seville, he had commented to José María Pemán in mid-August 1936, 'everyone will have to sacrifice things in the interests of a rigid discipline which should not lend itself to divisions or splinter groups'.[24] His interest in establishing overall authority over both the military and political spheres, however, quickened as a result of pressures from the Third Reich.

Herr Messerschmidt, the representative in Spain of the German War Matériel Export Cartel met Franco at the end of August. Messerschmidt's report concluded 'It goes without saying that everything must remain concentrated in Franco's hands so that there may be a leader who can hold everything together'.[25] In mid-September, Johannes Bernhardt informed Franco that Berlin was anxious to see him installed as Chief of State.

Franco replied cautiously that he had no desire to get mixed up in politics. Bernhardt made it clear that further arms shipments were in doubt unless Berlin had a sovereign chief with whom to negotiate and who could take responsibility for future commitments. Characteristically, Franco did not respond and left Bernhardt to fill the ensuing silence. Bernhardt informed him that he would shortly be travelling to Berlin with Lieutenant-Colonel Walter Warlimont, the head of Hitler's unofficial military mission, in order to report to the Führer and Göring about the progress of the war. One of the issues that Warlimont would be discussing was the political leadership of Nationalist Spain. The clear implication was that Franco's favoured position as the exclusive channel for German aid could be endangered unless he could show that his grip on power was unshakeable. Disappointed by the general's non-committal response, Bernhardt approached Nicolás Franco who undertook to work on his brother. Since Franco was not easily manipulable, Nicolás's efforts may be supposed to have been confined to underlining that now was an ideal moment to make a bid for power.[26]

In the meanwhile, Kindelán, Nicolás Franco, Orgaz, Yagüe and Millán Astray formed a kind of political campaign staff committed to ensuring that Franco became first Commander-in-Chief and then Chief of State. It is clear from Kindelán's own account that this was done with Franco's knowledge and approval. Not surprisingly, Franco maintained sufficient reserve to enable him to disown their efforts should they have proved unsuccessful. It thus appeared that they were taking the lead although Franco was anything but a passive shuttlecock in their game. Kindelán suggested that a gathering of the Junta de Defensa Nacional together with other senior Nationalist generals be called to resolve the issue. The meeting was convoked at Franco's request, an initiative which clearly indicated his interest in the single command and his availability as a candidate. The choice of additional generals who were invited was also deeply significant. They were Orgaz, Gil Yuste and Kindelán, all totally committed to Franco and all monarchists. In the wake of Mola's expulsion of Don Juan, they looked to Franco to hold the fort until victory over the Republic permitted the restoration of the monarchy.

The historic gathering was held on 21 September at the same time as the African columns were taking Maqueda. The meeting took place in a wooden cabin (barracón) at a recently improvised airfield near Salamanca. General Cabanellas was in the chair and the others present were the members of the Junta, Franco, Mola, Queipo de Llano, Dávila, Saliquet, and Colonels Montaner and Moreno Calderón and the three additional generals. During the morning session, three and a half hours went by without Kindelán and Orgaz managing to get a discussion started on the question of a Commander-in-Chief, despite three attempts. There exist no minutes of the meeting, and the only record is constituted by Kindelán's notes. In those notes, there is no indication that there was any discussion of the decision to interrupt the attack on Madrid in order to relieve the Alcázar at Toledo. At lunch on the estate of Antonio Pérez Tabernero, a bull-breeder, Kindelán and Orgaz decided to overcome the reluctance of their comrades and insisted that the subject be discussed in the afternoon session. Mola surprisingly supported them, saying 'I believe the single command to be of such interest that if we haven't named a Generalísimo within a week, I am not going on'. When the discussion was resumed, all showed themselves to be in favour, except Cabanellas, who advocated leadership by a junta or directory.[27]

The choice was effectively limited to the *cuatro generales* of the Republican song. The most senior, Cabanellas was not possible. He had rebelled against the Primo de Rivera dictatorship, had been Radical parliamentary deputy for Jaén between 1933 and 1935, and was thought to be a freemason. His role in the 18 July rising was unclear and he had no special standing as a combat general. The next in seniority, Queipo de Llano, had betrayed Alfonso XIII in 1930 and, for that reason and because of his family links with Alcalá Zamora, was considered to have been the beneficiary of favouritism under the Republic. He was also privately despised for the obscene radio broadcasts which he delivered nightly from Seville against the Republic. Mola, the most junior, was somewhat discredited by the initial failures of the rising and by the difficulties faced by his northern forces relative to the spectacular successes of Franco's Army of Africa. He also knew that he could not match Franco's contacts with the Germans and Italians.[28]

When it came to the vote on who should be Generalísimo, the two colonels abstained because of their inferior rank. Kindelán voted first, proposing that the single command be entrusted to Franco. He was followed by Mola, then Orgaz and the others, except Cabanellas who said that he could not take part in an election for a post which he considered unnecessary.[29] Although he cannot but have reflected wryly on Franco's hesitations about joining the rising in June and the first half of July, Mola took his rival's elevation with good grace. On leaving the meeting, Mola told his adjutants that it had been decided to create the job of Generalísimo. When they asked him if he had been nominated, he replied 'Me? Why? Franco.' Mola later told his adjutants that he had proposed the

name of Franco as Generalísimo, 'he is younger than me, has higher
rank, is immensely well-liked and is famous abroad'.[30] Shortly afterwards,
Mola told the monarchist politician Pedro Sainz Rodríguez that he had
supported Franco in the power stakes because of his military abilities and
the fact that he was likely to get the most votes. However, he made it
clear that he regarded Franco's leadership as transitory and was assuming
that he himself would play a major role in moulding the political future
after the war.[31] Many years later, Queipo de Llano, on criticizing Franco,
was asked by the monarchist Eugenio Vegas Latapié why he had voted
for him. 'And who else could we appoint?', he replied. 'It couldn't be
Cabanellas. He was a convinced Republican and everyone knew that he
was a freemason. Nor could we name Mola because we would have lost
the war. And my prestige was seriously impaired.'[32] Nonetheless, Queipo
made no secret of his dissatisfaction with the decision that had been
taken.[33]

The half-heartedness shown by some of Franco's peers about his elevation was to have an immediate impact on his conduct of the war. It is
impossible to say with total certainty when exactly Franco took the
decision to direct his troops towards Toledo. The timing is crucial to
any assessment of his motives. His official biographer has claimed, without any proof, that it was before the airfield meeting at which he was
elected as Generalísimo. Such a timing would conveniently diminish any
suspicion of self-serving about the decision.[34] However, the decision
became a matter of urgency only after the capture of Maqueda and that
did not take place until the early evening of 21 September. The Salamanca
meeting started in the morning and Franco and his staff had to make an
early start to travel there from Cáceres. In fact, there is little doubt that
the decision was taken sometime after the fall of Maqueda and therefore
after the meeting of the generals at the airfield.[35] Whether taken in the
evening of 21 September or later, it was after Franco had been elected
Generalísimo. He did not draw up specific orders until three days later.[36]
Whenever Franco made his decision, which Mola's secretary described
as 'completely personal', he did so in a context of knowledge of the events
of 21 September.[37]

The meeting on that day had left him with gnawing doubts about
his election as Generalísimo. Behind the near unanimous vote and the
expressions of support for Franco could be discerned coolness and hesitations on the part of the other generals. The simple election to the status
of *primus inter pares* was merely a step on the road to absolute power and
there was still some distance to go. At the time, it was assumed, even by

those involved in his election, that what they were doing was merely guaranteeing the unity of command necessary for victory and putting it temporarily in the hands of the most successful general amongst them.[38] The agreement to keep the decision secret until it was formally approved and published by the Junta de Burgos reflected their doubts. It would have been entirely characteristic of Franco to seek to tip the balance by the propaganda coup of the relief of the Alcázar. If that is so, the soundness of his judgement that further efforts were required was confirmed when several days went by and nothing happened about his election being announced formally.

The silence was rightly interpreted by Kindelán as a symptom of the lack of conviction of some of the generals at the meeting. Cabanellas was procrastinating precisely because he feared the implications of dictatorial powers being granted to Franco. In the meanwhile, Nicolás Franco, who had recently arrived in Cáceres from Lisbon, brought the news that the German and Italian envoys to Portugal had told him that their governments wanted to see a single command and preferably in the hands of Franco. Nicolás also used his own recent encounter with Johannes Bernhardt to overcome his brother's apparent qualms about taking on political responsibilities. The lure of being Head of State, the interlocutor of Hitler and Mussolini, must have been seductive, as Nicolás seems to have perceived. However, even more than with the single command, it could be dangerous to be seen to be bidding for such power. With his customary caution, Franco preferred to let others make the running and wait for the new honour to be thrust upon him.

Accordingly, Kindelán, Nicolás Franco, Yagüe and Millán Astray proposed a further meeting at which the powers of the new Generalísimo would be clearly laid out and a proposal made that the post carried with it the Headship of State. Worried about his brother's hesitations, Nicolás asked Yagüe to put pressure on him. On 27 September, Yagüe told Franco that if he refused to seek the single command, the Legion would seek another candidate, a prospect which decisively guaranteed that he would seek full powers for himself.[39] By the time that such a meeting could take place, Franco would have chalked up the great propaganda victory of the relief of the Alcázar at Toledo.

It has been suggested that Franco's attitude to the garrison at Toledo was affected by bitter memories of his own inability to help the soldiers trapped at Nador in July 1921 after the disaster of Annual.[40] The fact that he had been a cadet at Toledo may also have influenced him but would scarcely have justified the decision to make a strategically secondary

objective into the first priority. There is little doubt that the relief of the
siege would have appealed to the romantic side of a soldier deeply imbued
with the ethos of *Beau Geste*, all the more so as it could be made into a
tale which might have come straight out of the legends of El Cid. However, when so much was at stake, the ruthlessly pragmatic Franco would
not have let himself be swayed by such considerations unless there were
other advantages to be gained.

In December 1936, he revealed more of the truth than perhaps he
intended when he told a Portuguese journalist that 'we committed a
military error and we committed it deliberately. Taking Toledo required
diverting our forces from Madrid. For the Spanish Nationalists, Toledo
represented a political issue that had to be resolved'.[41] Whatever Franco's
motives, his decision did his personal ambitions no harm although it was
to have serious consequences for the Nationalist cause. By permitting
Madrid to organize its defences, the diversion was to swing the advantage
back to the Republic almost as starkly as the crossing of the Straits had
given it to the military rebels.

In fact, the pace of the Army of Africa had already been slowed considerably. It took as long to get the 80 km from Talavera to Toledo as
it had to travel the nearly 400 km from Seville to Talavera, a reflection
of the fact that the Republic was gradually beginning to get some trained
men into the field.[42] This was reason enough to hasten the attack on the
capital. Nevertheless, on 25 September, three columns of the Moroccan
Army, since 24 September under the overall command of the African
veteran and Carlist sympathizer, General Varela, swept to the north of
Toledo. Under the individual commands of Colonel Asensio, Major
Castejón and Colonel Fernando Barrón, they cut off the road to Madrid
and then moved south against the city on the following day. After fierce
fighting, the militia began to retreat. On 27 September, the world's war
correspondents, 'who previously had been permitted to "participate" in
the bloodiest battles of the war', were prevented from accompanying the
attacking Legionaires and *Regulares* as they unleashed another massacre.
No prisoners were taken. The streets were strewn with corpses and liter-
ally ran with rivulets of blood which gathered in puddles. The American
journalist Webb Miller told the US Ambassador that he had seen the
beheaded corpses of militiamen. Hand grenades were tossed in among
the helpless wounded Republicans in the San Juan Bautista hospital. On
the next day, 28 September, General Varela entered the Alcázar to be
greeted with Moscardó's laconic report '*Sin novedad en el Alcázar, mi
general*' (all quiet in the Alcázar, general).[43]

On the evening of Sunday 27 September, in the flush of the victory at Toledo, Franco, Yagüe and Millán Astray addressed a frenetically cheering crowd from the balcony of the Palacio de los Golfines in Cáceres. Franco spoke hesitantly, his fluting voice anything but inspirational. Yagüe, recalling the threatening conversation which he had had with Franco earlier in the day, was carried away with enthusiasm. He declared vehemently 'tomorrow we will have in him our Generalísimo, the Head of State'. Millán Astray said 'Our people, our Army, guided by Franco, are on the way to victory'. There were parades by the Falange and the Legion while the band played the anthem of the Legion *Los Novios de la Muerte* (bridegrooms of death) and the Falangist song *Cara al sol* (face to the sun). The crowd chanted 'Franco! Franco! Franco!'. The scenes of popular acclamation for Franco were described lavishly in the press of the entire Nationalist zone.[44]

As the crowd melted away, Nicolás Franco and Kindelán were drawing up a draft project to be put to the following day's meeting of the Junta that was to decide the powers of the new Generalísimo. Yagüe had already played a key role by announcing in his speech that the Legion wanted Franco as single commander. Nicolás Franco and Kindelán continued to play their part, arranging that, on arrival at the airfield at Salamanca for the proposed meeting, Franco would be met by a guard of honour, consisting not only of a number of airmen, but also of a detachment of Carlist *Requetés* and another of Falangists. Thus, the somewhat intimidating symbolism of his political, as well as his military, leadership would be established before the meeting.[45] On the morning of Monday 28 September, Franco, Orgaz, Kindelán and Yagüe flew to Salamanca, 'determined', in Kindelán's words, 'to achieve their patriotic purpose whatever the cost'.*

At the morning session of the meeting, the other generals showed some disinclination to discuss the question of the powers to be exercised by the single commander and some were in favour of putting off the decision for some weeks. After all, a week previously when, with more or less goodwill, they had agreed to make Franco military Commander-in-Chief, there had been no hint that he might also have political powers. With the fall of Madrid and the end of the war assumed to be imminent,

* The myth propagated by Franco's hagiographers (Luis Galinsoga & Francisco Franco-Salgado, *Centinela de occidente* [Barcelona, 1956] p. 21) that he did not attend the meeting has no basis other than a determination to give the impression that the Generalísimo had power thrust upon him. Brian Crozier, *Franco: A Biographical History* (London, 1967) p. 212, mistakenly places the meeting on 29 September and so assumes Franco's absence on the grounds that, on that day, he was in Toledo congratulating Moscardó.

the generals were reluctant to bestow wide-ranging authority on Franco since they suspected how difficult it would be to persuade him to relinquish it. However, Kindelán insisted and read out the draft decree. In article 1, it proposed the subordination of the Army, Navy and Air Force to a single command, in article 2 that the single commander be called Generalísimo, and in article 3 that the rank of Generalísimo carry with it the function of Chief of State, 'as long as the war lasts', a phrase which guaranteed Franco the support of the monarchist generals. The proposal, which implied the demise of the Junta de Defensa Nacional, was received with hostility, particularly by Mola. He recognized that Franco was the superior general but that did not mean that he wanted to give him absolute political power. Even Orgaz wavered in his support for Kindelán.

Over lunch, Kindelán and Yagüe worked on their comrades, describing the scenes of popular rejoicing in Cáceres on the previous evening. No doubt Yagüe stressed the will of the Legion and Nicolás Franco emphasized the German pressures to which he had been subjected. Before the afternoon session began, Queipo and Mola returned to their respective headquarters. On the basis of Kindelán's proposal, a reluctant agreement was reached to the effect that Franco would be head of the government as well as Generalísimo. Cabanellas' undertook to put it into practice within two days.[46] On leaving the meeting, an exultant Franco said to his host, Antonio Pérez Tabernero, 'this is the most important moment of my life'.[47] In fact, Cabanellas still harboured doubts and decided to sign the decree only late in the night of 28 September after lengthy telephone consultations with Mola and Queipo. According to Cabanellas's son, Queipo said 'Franco is a swine.* I have never liked him and never will. However, we've got to go along with his game until we can block it'. A more cautious Mola made it clear that he saw no alternative to the reluctant acceptance of Franco's nomination.[48]

Cabanellas entrusted to a professor of international law, José Yanguas Messía, the wording of the Junta's decree formally recording the decision. Its first article stated that 'in fulfilment of the agreement made by the Junta de Defensa Nacional, the Head of the Government of the Spanish State will be Excelentísimo Sr. General Don Francisco Bahamonde, who will assume all the powers of the new State'. There have been claims that, before being printed, the decree was tampered with either by Franco or his brother. Ramón Garriga, who was later to be part of Franco's press

---

* What Queipo called Franco is deemed by Cabanellas to be 'unprintable' and so 'swine' is merely a guess.

service in Burgos, alleged that the reference in the draft to Franco being head of government of the Spanish State only provisionally 'while the war lasted' was read by Franco and crossed out before it was submitted to Cabanellas for signature. Tampering was not necessary. Made Head of the Government of the Spanish State, Franco simply referred to himself as, and arrogated to himself the full powers of, Head of State. The hopes of monarchists like Kindelán, Orgaz and Yanguas were totally misplaced. Having reached the peak of his power, Franco had no intention of handing over in his lifetime to a King, although he would always skilfully keep alive the hopes of the monarchists.[49] The bulk of the Nationalist press announced that Franco had been named *Jefe del Estado Español* (Head of the Spanish State). Only the Carlist *Diario de Navarra* committed the sin of referring to Franco as *Jefe del Gobierno del Estado Español* (Head of the Government of the Spanish State).[50]

Cabanellas commented 'You don't know what you've just done, because you don't know him like I do since I had him under my command in the African Army as officer in charge of one of the units in my column. If, as you wish, you give him Spain, he is going to believe that it is his and he won't let anyone replace him either during the war or after until he is dead.'[51] Cabanellas's comment was uncannily similar to one made some years later by Colonel Segismundo Casado, also a one-time *Africanista*, 'Franco incarnates the mentality of a Captain of the Tercio. That is all there is to it. We are told, "Take so many men, occupy such-and-such a position and do not move from there until you get further orders". The position occupied by Franco is the nation and since he has no superior officer, he will not move from there.'[52]

Franco derived incalculable political capital from his decision to divert his forces from Madrid. The liberation of the Alcázar was re-staged two days later and cinema audiences across the world saw Franco touring the rubble with a haggard Moscardó. In front of reporters, Moscardó repeated his famous phrase, *sin novedad* (all quiet), to Franco.[53] Overnight Generalísimo Franco became an international name, a name which symbolized the Nationalist war effort. In Nationalist Spain, he became the saviour of the besieged heroes. Not the least of his pleasure must have derived from emulating the great warrior heroes of medieval Spain.

The analogy was given the sanction of the Church on 30 September by the long pastoral letter, entitled 'The Two Cities', issued by the Bishop of Salamanca Dr Enrique Plá y Deniel. The Church had long since come out in favour of the military rebels but not hitherto as explicitly as Plá y Deniel. His pastoral built on the blessing given by Pius XI to exiled

Spaniards at Castelgandolfo on 14 September in which the Pope had distinguished between the Christian heroism of the Nationalists and the savage barbarism of the Republic. Plá y Deniel's text quoted St Augustine to distinguish between the earthly city (the Republican zone) where hatred, anarchy and Communism prevailed, and the celestial city (the Nationalist zone) where the love of God, heroism and martyrdom were the rule. For the first time, the word 'crusade' was used to describe the Civil War.[54]

The text was submitted to Franco before being published. He not only approved it but adjusted his own rhetoric subsequently to derive from it the maximum political advantage. By latching onto the idea of a religious crusade, Franco could project himself not just as the defender of his Spain but also as the defender of the universal faith. Leaving aside the gratifying boost to his own ego, such a propaganda ploy could bring only massive benefit in terms of international support for the rebel cause.[55] Many British Conservative MPs, for instance, intensified their support for Franco after he began to stress Christian rather than fascist credentials. Sir Henry Page Croft (Bournemouth) declared him to be 'a gallant Christian gentleman' and Captain A.H.M. Ramsay (Peebles) believed Franco to be 'fighting the cause of Christianity against anti-Christ'. They and many others used their influence with banks and government to incline British policy towards the Nationalists' interests.[56]

On 1 October 1936, the investiture of the new Chief of State took place. The pomp and the ceremony that were mounted were a long way from the improvisation of Franco's first days as a military rebel barely ten weeks ago. A large guard of honour consisting of soldiers as well as Falangist and Carlist militias awaited his arrival in front of the Capitanía General of Burgos. An enormous and delirious crowd erupted into applause and cheers when his motor car entered the square in front of military headquarters. In the throne room, in the presence of the diplomats of Italy, Germany and Portugal, Cabanellas formally handed over the powers of the Junta de Defensa to a visibly delighted Franco. An anything but impressive figure, short, balding and now with an incipient double chin and paunch, Franco stood apart on a raised dais. Cabanellas said 'Head of the Government of the Spanish State: in the name of the Junta de Defensa Nacional, I hand over to you the absolute powers of the State.'

Franco's reply was shot through with hauteur, regal self-confidence and easily assumed authority: 'General, Generals and Officers of the Junta, You can be proud, you received a broken Spain and you now

deliver up to me a Spain united in a unanimous and grandiose ideal.
Victory is on our side. You give me Spain and I assure you that the
steadiness of my hand will not waver and will always be firm.' After the
ceremony, he appeared on the balcony and made a speech to the sea of
arms raised in the fascist salute. The grandiloquent tone of his words in
the throne room was replaced by a rhetorical commitment to social
reform which can only have reflected a desire to be in tune with his Nazi
and Fascist sponsors. Its cynical promises were to remain long unfulfilled:
'Our work requires sacrifices from everyone, principally from those who
have more in the interests of those who have nothing. We will ensure
that there is no home without light or a Spaniard without bread.'
Altogether more credible was his declaration that night on Radio Castilla
to the effect that he planned a totalitarian State for Spain.[57]

Thereafter, from his very first decree, Franco simply referred to himself as *Jefe del Estado*. At that stage, of course, there was not much in the
way of a State for Franco to be Head of. The task of constructing it
began immediately, although with little immediate success. The Junta de
Burgos was dissolved and replaced by a Junta Técnica del Estado, presided
over by General Fidel Dávila.* General Orgaz was made High Commissioner in Morocco with the job of maintaining the flow of Moorish
mercenaries. The Junta Técnica remained in Burgos while Franco set up
his headquarters in Salamanca, near the Madrid battle front without being
too near and merely one hour's drive from Portugal should things turn
out badly. Mola was given command of the Army of the North, newly
formed by merging his troops with the Army of Africa. Queipo de Llano
was given command of the Army of the South, consisting of the scattered
forces operating in Andalusia, Badajoz and Morocco. Cabanellas was mar-
ginalised in punishment for his lukewarm response to Franco's elevation,
being given the purely symbolic title of Inspector of the Army. Franco
could rarely find time to receive him in Salamanca. No doubt he resented
the fact that Cabanellas had once been his superior and usually referred
to him, like Sanjurjo had done, as '*Franquito*' (little Franco).[58] He was
equally unforgiving with other one-time superiors, like Gil Robles, who
found himself cold-shouldered.†

One of the first things that Franco did after being elected as Nationalist

* It had a Secretaría General del Jefe del Estado, a Secretaría General of Foreign Relations
and a Gobierno General. There were also seven ministerial departments or 'commissions',
Finance; Justice; Industry, Commerce and Supply; Agriculture; Labour; Culture and Education; Public Works and Communications.
† Gil Robles told the author in Madrid in 1970 of his belief that Franco could not tolerate
having around anyone who had been his superior.

leader was to send fulsome telegrams to Hitler and Rudolf Hess. Hitler responded with a verbal, rather than a written, message via the aristocratic German diplomat, the Count Du Moulin-Eckart, who was received by Franco on 6 October. Hitler claimed that he could better help Franco by not appearing to have recognized the Nationalist Government until after the capture of Madrid. On the eve of renewing the assault on Madrid, Franco responded in terms of with 'heartfelt thanks for the Führer's gesture and complete admiration for him and the new Germany.' Du Moulin was impressed by the conviction of his enthusiasm for Nazi Germany, reporting that 'the cordiality with which Franco expressed his veneration for the Führer and Chancellor and his sympathy for Germany, and the decided friendliness of my reception, permitted not even a moment of doubt as to the sincerity of his attitude toward us'.[59]

In tune with the warmth of such sentiments, there began a massive propaganda campaign in fascist style to elevate Franco into a national figure. An equivalent title to *Führer* and *Duce* was adopted in the form of *Caudillo* – a term linking Franco to the warrior leaders of Spain's medieval past. Franco considered himself, like them, to be a warrior of God against the infidels who would destroy the nation's faith and culture.*[60] All newspapers in the Nationalist zone had to carry under their masthead the slogan '*Una Patria, Un Estado, Un Caudillo*' (a deliberate echo of Hitler's *Ein Volk, ein Reich, ein Führer*). The ritual chants of 'Franco! Franco! Franco!' were heard with insistent frequency. The sayings and speeches of Franco were reproduced everywhere.

Almost immediately, Nicolás Franco made tentative plans for the creation of a Francoist political party along the lines of General Primo de Rivera's Unión Patriótica. It would have consisted of conservative elements, largely from the CEDA, and therefore encountered the hostility of the Falange. Realizing how ill-advised it was to work against the ever larger Falange, the brothers dropped the idea.[61] There was an element of irony about what was happening. The new powers that had been granted to Franco were given in the belief that a single command would hasten an already imminent victory. In fact, the Nationalist triumph was soon to become a distant long-term prospect. In part that was for reasons

* The seed had been first planted in Franco's mind in the late 1920s. At that period, he spent time at a small Asturian estate owned by his wife known as La Piniella, situated near San Cucao de Llanera, thirteen kilometres from Oviedo. A particularly sycophantic local priest who fancied himself as the chaplain to the house was constantly telling both Doña Carmen and Franco himself that he would repeat the epic achievements of El Cid and the great medieval Caudillo Kings of Asturias. Franco's wife had often reminded him of the priest's comments.

beyond the Caudillo's control, such as the arrival of the International Brigades and Russian tanks and aircraft, and the creation of the Popular Army. However, that such things were able to have the effect that they did was largely Franco's responsibility, attributable to the delay of nearly two weeks in the march on Madrid as a result of the diversion to Toledo and then of the time devoted to the orchestration of his elevation to supreme power. Increasingly thereafter, it would begin to seem that Franco had an interest in the prolongation of the war in order to have time both to annihilate his political enemies on the Left and his rivals on the Right and to consolidate the mechanisms of his power.

Once established as Head of State, and with the eyes of Nationalist Spain now upon him, Franco's propagandists built him up as a great Catholic crusader and his public religiosity intensified. From 4 October 1936 until his death, he had a personal chaplain, Father José María Bulart.[62] He now began each day by hearing mass, a reflection of both political necessity and the influence of Doña Carmen. In order to please his wife, when he was available he would join in her regular evening rosary, although, at this stage of his career at least, without any great piety.[63] No one can say with total certainty what part Carmen Polo played in encouraging her husband's ambition nor how much he had been affected by Bishop Plá y Deniel's declaration of a crusade. Doña Carmen believed in his divine mission and such fulsome ecclesiastical support made it easier for her to convince him of it.[64]

As Franco came to believe in his own special relationship with divine providence, and as he became more isolated and weighed down with power and responsibility, his religiosity became more pronounced.* Apart from any spiritual consolation it may have given him, his new found religiosity also reflected a realistic awareness of the immeasurable assistance which the endorsement of the Catholic Church could give him in terms of clinching foreign and domestic support. In the Generalísimo's elevated concept of his own importance, the official approbation and blessing of the Church was essential. It was not just a question of broad Catholic support for the Nationalist cause but rather of specific recognition by the universal Church of his personal status as its champion. The speed with which Franco sought such recognition mirrored the speed with which he began to manifest monarchical pretensions. Religious ritual had traditionally played a crucial part in elevating the figure of the King

---

* It was said that religious ceremonial bored Franco almost more than anything else and, in power, he suffered agonies when he had to receive religious delegations, commenting 'we're doing saints today' ('hoy estamos de santos').

in the great age of early modern Spain. Believing that he represented continuity with the glories of the Golden Age, he took it for granted that the Church would validate his rule. Accordingly, he arrogated the royal prerogative of entering and leaving churches under a canopy (*bajo palio*).

On 1 October, the Primate of Spain Cardinal Isidro Gomá y Tomás sent a telegram congratulating him on the relief of the Alcázar and on his elevation to the Headship of State. Franco replied on 2 October with one of his grandiloquent messages, beginning 'on assuming the powers of the Headship of the Spanish State with all their responsibilities I could receive no better help than the blessing of Your Eminence.'[65] It was the beginning of a close relationship with Gomá.

Franco's fellow generals were somewhat taken aback by the ease with which the new Generalísimo adopted a distant and elevated style. He set up his headquarters in the Episcopal Palace in Salamanca which was graciously ceded to him by Bishop Plá y Deniel. Within two weeks of his investiture, visitors to the Palace, often known as the *cuartel general*, were being required to attend audiences in morning suit.[66] He was already surrounded by the Guardia Mora, the Moorish Guard, which would accompany him everywhere until the late 1950s. In resplendent uniforms, they stood like statues throughout the palace, a graphic indication of the Asiatic despotism in the making. German specialists arrived and built a special air-raid shelter.[67] Franco's picture appeared everywhere, on cinema screens, on the walls of shops, offices and schools. Along with his portrait, slogans were stencilled on walls, 'the Caesars were undefeated generals. Franco!' An entire propaganda apparatus was erected and then devoted to the inflation of the myth of the all-seeing political and military genius Franco. The scale of adulation to which he was subjected inevitably took its toll on his personality.[68]

In the process of moving from the improvised bureaucracy appropriate to a military campaign to the erection of a State apparatus, Franco made several errors in his choice of collaborators until the entire enterprise was taken over by his brother-in-law Ramón Serrano Suñer. His brother Nicolás may have been an excellent kingmaker but he was less successful as a chancellor. By dint of his relationship with the new Generalísimo, and operating out of an office next to that of his brother, he quickly accumulated enormous power. Nicolás, who resembled his father in his tastes and appetites much more than he did his brother, was an amusing and popular *bon viveur* whose bohemian and chaotic life-style was the despair of all who had to deal with him. He would rise at 1 p.m. and receive visitors until 3 p.m. when he would disappear for lunch until

7 p.m. followed by an evening's socializing. Reappearing around mid-night, he would then work until 4 or 5 a.m., often keeping those who had come to see him waiting for seven or eight hours at a time. Given his relationship to the Generalísimo, few complained, although his practices especially infuriated the Germans.[69] Yet despite the power and the favour that he enjoyed, Nicolás did little or nothing to begin the task of creating a State infrastructure.

However, the most disastrous of Franco's appointments was that of Millán Astray as Head of Press and Propaganda. It is possible that Franco enjoyed Millán's adulation but most of his activities were counter-productive. Within days of Franco's elevation, Millán was proclaiming that Franco was 'the man sent by God to lead Spain to liberation and greatness', 'the man who saved the situation during the Jaca rising' and the 'greatest strategist of the century'.[70] He ran the Nationalist press office like a barracks, summoning the journalists in his team with a whistle and then haranguing them much as he had the Legion prior to an action. Franco seems to have seen him as a kind of mascot, but his antics ended up bringing the Nationalist cause into disrepute.[71] Millán's own choice of collaborators was especially unfortunate. Because of the link established between Franco and Luis Bolín during the flight of the Dragon Rapide, Millán named Bolín chief of press in the south and gave him the honorific title of Captain in the Legion.[72] Bolín started to use the uniform and throw his weight about accordingly, attempting to control the flow of news about Nationalist Spain by intimidating foreign journalists. Millán Astray encouraged his subordinates to threaten foreign journalists with execution. Bolín followed the order with gusto, most notoriously in the case of Arthur Koestler, the mistreatment of whom provoked an inter-national scandal which led to his release from prison. As a result of the subsequent publication of Koestler's book *Spanish Testament*, Bolín fell into disgrace.[73]

Press liaison in the north was put in the hands of the notorious Captain Gonzalo de Aguilera, Conde de Alba y Yeltes, a polo-playing ex-cavalryman, mainly on the grounds of his manic bigotry and the fact that he could speak excellent English, German and French. Captain Aguilera did more harm than good by outrageous and eminently quotable remarks to journalists. Much of what he said merely reflected the common beliefs of many officers on the Nationalist side. On the grounds that the Spanish masses were 'like animals', he told the foreign newspapermen that 'We've got to kill and kill and kill'. He boasted to them of shooting six of his labourers on the day the Civil War broke out *'Pour encourager les autres'*.

He regularly explained to any who would listen that the fundamental cause of the Civil War was 'the introduction of modern drainage: prior to this, the riff-raff had been killed by various useful diseases; now they survived and, of course, were above themselves.' 'Had we no sewers in Madrid, Barcelona, and Bilbao, all these Red leaders would have died in their infancy instead of exciting the rabble and causing good Spanish blood to flow. When the war is over, we should destroy the sewers. The perfect birth control for Spain is the birth control God intended us to have. Sewers are a luxury to be reserved for those who deserve them, the leaders of Spain, not the slave stock.'[74] He believed that husbands had the right to shoot their unfaithful wives. When accompanying the influential journalist Virginia Cowles, Aguilera maintained a constant flow of sexist remarks which he occasionally interrupted to say things like 'Nice chaps, the Germans, but a bit too serious; they never seem to have any women around, but I suppose they didn't come for that. If they kill enough Reds, we can forgive them anything'.[75]

That Millán was hardly the best man to present the cause of Franco's New State to the outside world was made starkly clear on 12 October 1936, during the celebrations in Salamanca of the Day of the Race, the anniversary of Christopher Columbus's 'discovery' of America. The magnificent and regal choreography stressed the permanence of the New State. A tribune was erected in the Cathedral for the distinguished guests. Franco was not present but was represented by General Varela and by Doña Carmen. A sermon by the Dominican priest Father Fraile praised Franco's recuperation of the 'the spirit of a united, great and imperial Spain'. The political, military and ecclesiastical dignitaries then transferred to the University for a further ceremony under the presidency of the *Rector Perpétuo*, the seventy-two year-old philosopher and novelist Miguel de Unamuno. He announced that he was taking the chair in place of General Franco who could not attend because of his many pressing commitments.

A series of speeches stressed the importance of Spain's imperialist past and future. One in particular, by Francisco Maldonado de Guevara, who described the Civil War in terms of the struggle of Spain, traditional values and eternal values against the anti-Spain of the reds and the Basques and Catalans, seems to have outraged Unamuno, who was already devastated by the 'logic of terror' and the arrest and assassination of friends and acquaintances. (A week earlier Unamuno had visited Franco in the Bishop's Palace to plead vainly on behalf of several imprisoned friends.)[76] The vehemence of Maldonado's speech stimulated a

Legionaire to shout '*¡Viva la muerte!*' (long live death), the battle cry of the Legion. Millán Astray then intervened to begin the triple Nationalist chant of '*¡España!*' and back came the three ritual replies of '*¡Una!*', '*¡Grande!*' and '*¡Libre!*' (United! Great! Free!). When Unamuno spoke, it was to counter the frenzied glorification of the war and the repression. He said that the civil war was an uncivil war, that to win was not the same as to convince (*vencer no es convencer*), that the Catalans and Basques were no more anti-Spanish than those present. 'I am a Basque and I have spent my life teaching you the Spanish language which you do not know'. At this point he was interrupted by a near apoplectic Millán Astray who stood up to justify the military uprising. As Millán worked himself into a homicidal delirium, Unamuno stood his ground pointing out the necrophiliac inanity of the slogan 'Long live death'. Millán shouted 'Death to intellectuals' to which Unamuno replied that they were in the temple of intelligence and that such words were a profanity.

With shouting and booing rising to a crescendo and Unamuno being threatened by Millan Astray's armed bodyguards, Doña Carmen intervened. With great presence of mind and no little courage, she took the venerable philosopher by the arm, led him out and took him home in her official car. It has been suggested by two eyewitnesses that Millán Astray himself ordered Unamuno to take the arm of the wife of the Head of State and leave.[77] Such was the ambience of fear in Salamanca at the time that Unamuno was shunned by his acquaintances and removed at the behest of his colleagues from his position in the University.[78] Under virtual house arrest, Unamuno died at the end of December 1936 appalled at the repression, the 'collective madness' and 'the moral suicide of Spain'.[79] Nevertheless, he was hailed at his funeral as a Falangist hero.[80] Nearly thirty years later, Franco commented to his cousin on what he saw as Unamuno's 'annoying attitude, unjustifiable in a patriotic ceremony, on such an important day and in a Nationalist Spain which was fighting a battle with a ferocious enemy and encountering the greatest difficulties in achieving victory'. In retrospect, he regarded Millán Astray's intervention as an entirely justified response to a provocation. Nevertheless, at the time, it was thought prudent to have Millán Astray replaced.[81]

The incident with Unamuno was a minor embarrassment in the process of consolidation of Franco as undisputed leader. In political terms, everything was going his way. In the course of the attack on Madrid, Franco was fortunate to see, indeed to an extent to facilitate, the removal from the scene of one of his last remaining potential rivals. The panic provoked by the advance on the capital and the broadcast of boasts by Mola about

the imminent capture of Madrid by his 'Fifth Column' of secret Nationalist sympathisers had seen violent reprisals taken among rightists, either
against individual saboteurs who were caught or against the large groups
of prisoners taken from Madrid jails and massacred at Paracuellos de
Jarama.[82] The conservatives and other middle class victims of atrocities
in Madrid were not the only Nationalist civilians to lose their lives. The
most celebrated was José Antonio Primo de Rivera. Although the Falangist leader had been in a Republican jail in Alicante since his arrest on
14 March 1936, an escape bid or a prisoner exchange was not inconceivable.* Obviously, given the pre-eminence of José Antonio Primo de
Rivera, his release or escape would not be easy. In the event, however,
lack of co-operation by Franco ensured that it would not happen.

This was entirely understandable. Franco needed the Falange both as
a mechanism for the political mobilization of the civilian population and
as a way of creating an identification with the ideals of his German and
Italian allies. However, if the charismatic José Antonio Primo de Rivera
were to have turned up at Salamanca, Franco could never have dominated
and manipulated the Falange as he was later to do. After all, since before
the war, José Antonio had been wary about too great a co-operation with
the Army for fear that the Falange would simply be used as cannon fodder
and fashionable ideological decoration for the defence of the old order.
In his last ever interview, with Jay Allen, on 3 October, published in the
Chicago Daily Tribune on 9 October and in the *News Chronicle* on
24 October 1936, the Falangist leader had expressed his dismay that the
defence of traditional interests was being given precedence over his party's
rhetorical ambitions for sweeping social change.[83] Even taking into
account the possibility that José Antonio was exaggerating his revolutionary aims to curry favour with his jailers, the implied clash with the political
plans of Franco was clear. In fact, Allen told the American Ambassador,
Claude G. Bowers, that José Antonio's attitude was defiant and contemptuous rather than conciliatory and that he had been obliged to cut
short the interview 'because of the astounding indiscretions of Primo'.[84]

Franco, as something of a social climber, might have been expected
to admire the dashing and charismatic socialite José Antonio who was
after all son of the dictator General Primo de Rivera. However, despite

---

* Several prominent Nationalists crossed the lines in these ways. The exchanges (*canjes*)
included important Falangists like Raimundo Fernández Cuesta who was officially
exchanged for a minor Republican figure, Justino de Azcárate, and Miguel Primo de Rivera
who was exchanged for the son of General Miaja. Among the more significant escapees
was Ramón Serrano Suñer.

the efforts of Ramón Serrano Suñer over the previous six years, their relationship had never prospered. José Antonio had come to regard Franco as pompous, self-obsessed and possessed of a caution verging on cowardice. Their relationship had definitively foundered in the spring of 1936, during the re-run elections in Cuenca when José Antonio had vehemently opposed the general's inclusion in the right-wing list of candidates. Franco had never forgiven him.

For some time before his elevation to the overall leadership of the Nationalist side, Franco had been considering plans to subordinate the various political strands of the Nationalist coalition to a single authority. In late August, he had told Messerschmidt that the CEDA would have to disappear. In his conversation on 6 October with Count Du Moulin-Eckart, the new Head of State had informed his first diplomatic visitor that his main preoccupation was the 'unification of ideas' and the establishment of a 'common ideology' among the Army, the Falange, the monarchists and the CEDA. He confided in his visitor his cautious belief that 'it would be necessary to proceed with kid gloves'. Given his own essential conservatism and the links of the elite of the Nationalist coalition with the old order, such delicacy would indeed be required. Unification could only be carried out at the cost of the political disarmament of the ever more numerous and vociferous Falange. Such an operation would be easier to perform if the Falangist leader were not present.

Early attempts to liberate José Antonio were initially approved by Franco. His grudging consent was given for the obvious reason that to withhold it would be to risk losing the goodwill of the Falange which was providing useful para-military and political assistance throughout the rebel zone. The first rescue attempt had been the work of isolated groups of Falangists in Alicante. Then in early September, when the Germans had come to see the Falange as the Spanish component of a future world political order, more serious efforts were made. German aid came from the highest levels on the understanding that the operation was approved by General Franco something for which there were precedents.

Franco had already intervened personally with the Germans to get help for the rescue of the family of Isabel Pascual de Pobil, the wife of his brother Nicolás. Thanks to the efforts of Hans Joachim von Knobloch, the German consul in Alicante, eighteen members of the Pasqual de Pobil family were disguised as German sailors and taken aboard a ship of the German Navy. The efforts to free the Falangist leader hinged largely on the co-operation of German naval vessels anchored at Alicante and of von Knobloch. Knobloch co-operated with the rash and excitable Falangist

Agustín Aznar in an ill-advised scheme to get Primo de Rivera out by bribery which fell through when Aznar was caught and only narrowly escaped. An attempt was made on von Knobloch's life and shortly after he was expelled from Alicante by the Republic on 4 October.[85]

On arriving at Seville on 6 October, von Knobloch and Aznar renewed their efforts to liberate José Antonio. Von Knobloch elaborated a scheme to bribe the Republican Civil Governor of Alicante while Aznar prepared a violent prison break-out. They were received in Salamanca by Franco who, after thanking von Knobloch for securing the escape from Alicante of the family of his brother Nicolás, gave his permission for them to continue their efforts. However, that verbal permission obscured the fact that his backing was less than enthusiastic. While von Knobloch returned to Alicante to implement his scheme, Franco informed the German authorities that he insisted on a number of conditions for the continuation of the operation. These were that efforts be made to rescue José Antonio without handing over any money, that if it was necessary to give money then the amount should be haggled over, and that von Knobloch should not take part in the operation. These strange conditions considerably diminished the chances of success but the Germans in Alicante decided to go ahead. Franco then issued even more curious instructions. In the event of the operation being a success, total secrecy was to be maintained about José Antonio being liberated. He was to be kept apart from von Knobloch, who was the main link with the Falangist leadership. He was to be interrogated by someone sent by Franco. He was not to be landed in the Nationalist zone without the permission of Franco. He informed the Germans that there existed doubts about the mental health of Primo de Rivera. The operation was aborted.[86]

A further possibility for Primo de Rivera's release arose from a suggestion by Ramon Cazañas, Falangist *Jefe* (chief) in Morocco. He proposed that an exchange be arranged for General Miaja's wife and daughters who were imprisoned in Melilla. Franco apparently refused safe-conducts for the negotiators although he later agreed to the family of General Miaja being exchanged for the family of the Carlist, Joaquín Bau. The Caudillo also refused permission for another Falangist, Maximiano García Venero, to drum up an international campaign to save José Antonio's life.[87] Similarly, Franco sabotaged the efforts of José Finat, Conde de Mayalde, a friend of José Antonio. Mayalde was married to a granddaughter of the Conde de Romanones and he persuaded the venerable politician to use his excellent contacts in the French government to get Blum to intercede with Madrid on behalf of Primo de Rivera. Franco delayed

permission for Romanones to go to France until after the death sentence
was announced.[88]

José Antonio Primo de Rivera was shot in Alicante prison on
20 November 1936. Franco made full use of the propaganda opportunities
thereby provided, happy to exploit the eternal absence of the hero while
privately rejoicing that he now could not be inconveniently present. The
news of the execution reached Franco's headquarters shortly after it took
place.[89] It was in any case published in the Republican and the French
press on 21 November. Until 16 November 1938, Franco chose publicly
to refuse to believe that José Antonio was dead. The Falangist leader
was more use 'alive' while Franco made his political arrangements. An
announcement of his death would have opened a process whereby the
Falange leadership could have been settled at a time when Franco's own
position was only just in the process of being consolidated. The provisional leader of the Falange, the violent but unsophisticated Manuel
Hedilla, made the tactical error of acquiescing in Franco's manoeuvre.
The first news of the execution coincided with the Third *Consejo Nacional*
of the *Falange Española y de las JONS* in Salamanca on 21 November but
Hedilla failed to make an announcement, out of a vain hope, built on a
hundred rumours, that by some subterfuge or other, his leader had survived. Thereafter, Franco would have to deal only with a decapitated
Falange.[90]

Franco's attitude to José Antonio Primo de Rivera's 'absence' was
enormously revealing of his peculiarly repressed way of thinking. 'Probably', he told Serrano Suñer in 1937, 'they've handed him over to the
Russians and it is possible that they've castrated him'.[91] Franco used the
cult of *el ausente* (the absent one) to take over the Falange. All its external
symbols and paraphernalia were used to mask its real ideological disarmament. Some of Primo de Rivera's writings were suppressed and his designated successor, Hedilla, would be imprisoned under sentence of death
in April 1937. While the public cult was manipulated to build up Franco
as the heir to José Antonio, the Caudillo in private expressed his contempt
for the Falangist leader. Serrano Suñer was always aware that praise for
José Antonio was guaranteed to irritate Franco. On one occasion, the
Generalísimo exploded '*Lo ves, siempre a vueltas con la figura de ese muchacho
como cosa extraordinaria*' ('see, always going on about that lad as if he was
something out of the ordinary'). On another, Franco claimed delightedly
to have proof that Primo de Rivera had died a coward's death.[92]

It is possible that José Antonio might have worked to bring an early
end to the carnage although whether, in the hysterical atmosphere of the

times, he would have had any success is entirely a different matter. He was certainly open to the idea of national reconciliation in a way never approached by Franco either during the war or in the thirty-five years that followed. In his last days in prison, José Antonio was sketching out the possible membership and policies of a government of 'national concord' whose first act was to have been a general amnesty. His attitude to Franco was revealed clearly in his comments on the implications of a military victory which he feared would merely consolidate the past. He saw such a victory as the triumph of 'a group of generals of depressing political mediocrity, committed to a series of political clichés, supported by old-style intransigent Carlism, the lazy and short-sighted conservative classes with their vested interests and agrarian and finance capitalism'.

The papers in which he put these thoughts down were sent to Prieto by the military commander of Alicante, Colonel Sicardo. Eventually, the Socialist leader forwarded copies to his two executors, Ramón Serrano Suñer and Raimundo Fernández Cuesta, in the hope of provoking dissent among the Falangist purists. This was a political error on Prieto's part. With José Antonio dead, the validation of Serrano Suñer and Fernández Cuesta as his executors gave them his authority to carry out Franco's policy.[93] Had José Antonio Primo de Rivera reached Salamanca, he would have been a certain, and influential, critic of Franco. Franco's exploitation of the Falange as a ready-made political base would have been made significantly more difficult.[94] However, to assume that Franco would not have seen off Primo de Rivera in the same way as he disposed of so many rivals is to take too much for granted.

In contrast to the ruthlessness with which Franco disposed of his rivals was the alacrity with which he bent rules in the interests of his family. The examples of this during the Civil War presaged the protection under which the so-called 'Franco clan' would prosper in the post-war years. His intervention on behalf of Nicolás's in-laws was an example of his readiness to do things for his family. Even more striking was the rehabilitation of his left-wing extremist brother Ramón despite the vehement opposition of many important military figures. In September 1936, Ramón Franco who was in Washington as Spanish air attaché, wrote to a friend in Barcelona to ascertain how he would be received in the Republican zone. Azaña allegedly said to the mutual friend 'he shouldn't come, he'd have a really hard time'. In the wildly precipitate way that had always characterized his behaviour, Ramón decided to go instead to the Nationalist zone shortly after hearing of his brother's elevation to the Headship of State.[95]

Despite his past as an anarchist agitator and as a freemason and his involvement in various revolutionary activities, all 'crimes' for which others paid with their lives, Ramón was welcomed by his brother. In Seville, Queipo de Llano had already executed Blas Infante, the Andalusian Nationalist lawyer who had stood with Ramón in the revolutionary candidacy in the 1931 elections. The exquisite care for appearances which had allegedly prevented Franco opposing the execution of his cousin Ricardo de la Puente Bahamonde at the beginning of the military uprising did not apply in the case of his brother. Ramón was sent to Mallorca to take over as head of the Nationalist forces there and given the acting rank of Lieutenant-Colonel. This caused very considerable ill feeling within the Nationalist Air Force and planted the seeds of a rift between Franco and his kingmaker, Alfredo Kindelán. On 26 November, Kindelán wrote the Generalísimo a fierce protest against his high-handed action. Couched in formally respectful terms, it accepted Franco's right to command as he felt best but spoke of the 'personal mortification' felt by Kindelán at not even having been consulted and of the ill feeling which had been provoked among Nationalist airmen whose reaction ranged 'from those who accept that he be allowed to work in aeronautical matters outside Spain to those who demand that he be shot'.[96] Franco simply ignored the letter and took his revenge against Kindelán by dropping him at the end of the war. Franco had taken to the prerogatives of his power with the skill and arbitrariness of a Borgia: they were attributes he was to need and to use to the full in the months ahead.

VIII

FRANCO AND THE SIEGE OF MADRID

October 1936–February 1937

IRONICALLY, Franco had hoped, by the day on which the disagreeable incident between Millán and Unamuno had taken place, to have been celebrating the capture of Madrid. There had been a significant slowing down of the rhythm of operations during the two weeks in which he was otherwise occupied clinching his elevation to power. The war could not be delayed indefinitely and, on 6 October, Franco announced to journalists that his offensive against the capital was about to begin. Under the overall direction of Mola, the Nationalist forces began a co-ordinated push against Madrid on the following day. An extremely tired Army of Africa resumed its northward march under the command of General Varela, assisted by Colonel Yagüe as his second-in-command.[1] The ten thousand-strong force was organized in five columns under Asensio, Barrón, Castejón, Colonel Francisco Delgado Serrano and Tella. Supplies of arms had been collected and they were augmented by the arrival of substantial quantities of Italian artillery and light tanks. Italian instructors quickly trained Spaniards in their use and, on 18 October, Franco, accompanied by the Italian military mission, was able to inspect the first Italo-Spanish motorised armoured units.[2]

After frequent consultations with Franco, Mola developed a two-part final strategy to take Madrid which was already surrounded on the west from due north to due south. The idea was first for the Nationalist forces to march on Madrid, simultaneously reducing the length of the front and tightening their grip on the capital, and then for Varela's Army of Africa to make a frontal assault through the northern suburbs. The push which began on 7 October saw an advance from Navalperal in the north, near El Escorial, Cebreros to the west and Toledo in the south. The forward defences of the city were demoralised by Nationalist bombing and then brushed aside by motorised columns armed with fast Italian whippet tanks. Desperate counter-attacks from the capital were easily repelled, thereby intensifying the optimism of the attacking forces.[3]

However, a different kind of war was about to begin. From 18 July until 7 October, the brunt of the Nationalist effort had been borne by the Army of Africa, on a forced march, frontally attacking towns and villages and opposed only by untrained amateur militiamen. It was little different from the kind of colonial war in which Franco and the other *Africanistas* had received their early military experiences. In this type of warfare, the advantage was entirely with the Legion and the *Regulares*. Henceforth, there was to be a move towards a war of fronts. Paradoxically, as the Germans, Italians and Russians poured in material assistance in the form of the latest weaponry, in part at least by way of experiment for the *next* war, Franco would remain fixed in the strategic world of the Great War.

More than with the attack on Madrid, the Generalísimo was occupied with the operation to break the siege of Oviedo and the city's liberation on 17 October gave him enormous pleasure. He seems to have taken less direct interest in the campaign for Madrid. It was not until 20 October, considerably after the diversion of the Army of Africa to Toledo, that he seemed to wake up to the extent to which the capital was being strengthened and issued the order to 'concentrate maximum attention and available combat forces on the fronts around Madrid'.[4] Indeed, his absence from the operations to take Madrid, and from the subsequent Nationalist chronicles thereof, was quite remarkable. Perhaps Franco suspected that there was little easy glory to be won and thus slyly left Mola to take responsibility.

Mola himself was happy to seize the opportunity to make good his failure to capture Madrid at the beginning of the war.[5] His optimism was widely shared: a Nationalist *alcalde* (mayor) and city councillors had already been named.[6] Nationalist radio stations broadcast the news that Mola was preparing to enter the Puerta del Sol in the centre of Madrid on a white horse. He even offered to meet the *Daily Express* correspondent there for a coffee and Republican wags set up a table to await him.[7] Nationalist aircraft showered Madrid with leaflets containing an ultimatum for the evacuation of the civilian population and total surrender. The situation was deteriorating so rapidly that there seemed little hope.[8] Then on 15 October, the first arms and equipment from the Soviet Union began to be unloaded at Cartagena. Once the fifty tanks, twenty armoured cars and 108 fighter aircraft were assembled and transported to the

Madrid front, giving the Republic a brief parity of force, there would be no quick victory for the Nationalists.[9]

By the end of the month, Mola's forces had taken a ring of small towns and villages near the capital, including Brunete, Móstoles, Fuenlabrada, Villaviciosa de Odón, Alcorcón and Getafe. Madrid was inundated with refugees from the surrounding villages along with their sheep and other farm animals.[10] There were major problems of food and water distribution. Harassed by Nationalist aircraft, the militia columns were also falling back along the roads to Madrid in considerable disarray. On 31 October, with twenty-five thousand Nationalist troops under Varela about to reach the western and southern suburbs of Madrid, Mola issued a warning about the dangers of further delay.[11]

However, from 1 to 6 November, there was a serious slowing-down of the advance, usually attributed to the Nationalists' need to rest their troops and their confidence that they had time to do so. However, it has been alleged that the hesitation was in part caused by Franco making long consultations with his German and Italian advisers.[12] It would also appear that between 4 and 6 November, an acrimonious debate took place within the Nationalist camp as to how to go about seizing the capital. Yagüe and Varela proposed daring blitzkrieg attacks through the suburbs, while Mola called for a broad frontal assault in the belief that Madrid would offer no more resistance than Toledo.* A cautious Franco rejected the plans of Yagüe and Varela for fear of losing the crack African columns.[13]

Franco thus left Mola free to push his own over-optimistic strategy of a full-scale assault from the west across the River Manzanares and through the University City and the Casa del Campo, the old royal hunting ground of sparsely wooded scrub. By 7 November, the Nationalists were ready to begin what they assumed would be their final frontal assault.[14] On 28 October, the Falange and the Carlists drew up lists of the buildings, hotels, cinemas, theatres, radio stations and newspapers that they planned to occupy after the victory.[15] Civilian rightists who followed in the wake of the Army of Africa had packed their suitcases in anticipation of an early return to their homes in Madrid's better neighbourhoods. It was believed in the Francoist camp that, within hours, *Legionarios* would be in the Puerta del Sol.[16]

However, the news of the arrival of Russian weaponry and technicians

---

* Yagüe wanted to penetrate along a line through the poorly defended north-eastern suburbs of Puerta de Hierro, Dehesa de la Villa and Cuatro Caminos while Varela favoured a similar thrust through the south-eastern suburbs of Vallecas and Vicálvaro.

along with the first 1,900 men of the International Brigades diminished the optimism at the Generalísimo's headquarters. Heavy Russian tanks were put into action from the end of October to blunt the advance of the fast-moving Nationalist columns, although the lack of skilled drivers and gunners dramatically diminished their efficacy. Soviet I-15 and I-16 fighter aircraft piloted by Russian airmen went into action for the first time on 4 November and would, for about six months at least, reverse the easy air superiority enjoyed by the Nationalists during the drive on Madrid.[17] Without knowing fully the scale of the Russian aid to the Republic, the Germans were already becoming frustrated with the slowness of Franco's progress towards Madrid.

The German Foreign Minister Constantin von Neurath complained to Ciano on 21 October about Franco's inactivity on the Madrid front.[18] Shared concern about the fate of the Nationalist cause was one of the many factors pushing Italy and Germany together. Indeed, Mussolini was soon to start talking of the Rome-Berlin Axis. Both Ciano and von Neurath expected Madrid to fall by the end of the month or in the first week of November at which point they planned to extend formal recognition to Franco.[19] At the end of October, however, the German Minister of War, General von Blomberg, sent Admiral Canaris and General Hugo Sperrle to Salamanca to investigate the reasons for Franco's failure to take Madrid. Von Blomberg had instructed both Canaris and Sperrle to inform Franco 'most emphatically' that the German government did not consider his ground and air combat tactics 'promising of success' and that 'continued adherence to this hesitant and routine procedure (failure to exploit the present favourable ground and air situation, scattered employment of the Air Force) is even endangering what has been gained so far.'

Canaris and Sperrle were to inform Franco of the conditions under which he would receive future reinforcements. The German units would be under the command of a German officer, who would be Franco's sole adviser on their use and responsible only to him. Franco's command would be maintained only 'outwardly'. The consolidation of German forces was conditional on the 'more systematic and active conduct of the war' and the Generalísimo's acceptance of these demands 'without reservation'.[20] Once the Generalísimo had agreed, a complete battle group under General Sperrle, known as the Condor Legion, was assembled and despatched with astonishing speed. Within a matter of days, a force of specialised units, equipped with the latest developments in German bomber and fighter aircraft and tanks and other motorised weapons was en route to Seville. Five thousand Germans landed in Cádiz on 16 November and a further seven thousand on 26 November along with artillery, aircraft and armoured transport.[21]

So sure was the Republican government that Madrid would fall that, after acrimonious discussions, it left for Valencia on 6 November. With Nationalist artillery shells falling on the suburbs, it seemed to be the beginning of the end.[22] The organization of the city's defence was placed in the hands of a Defence Junta presided over by the recently appointed Captain-General of New Castile, José Miaja.[23] The portly, balding fifty-eight year-old Miaja was despised by Franco as incompetent and scruffy and regarded by Queipo de Llano as inept, stupid and cowardly.[24] Known largely for the abortive counter-attacks which had failed to stop Franco's advance through Extremadura, Miaja was assumed by many, including himself, to have been chosen as the scapegoat to take the blame for the fall of the capital.[25]

The bluff and good-humoured Miaja quickly surrounded himself with a staff of highly competent assistants, of whom the most outstanding was to be his Chief of Staff, Lieutenant-Colonel Vicente Rojo. While Rojo planned the defence, Miaja worked on raising the morale of the defenders. Unaware that Miaja was anything more than a sacrificial victim, Franco announced on 7 November that he would attend mass in Madrid on the next day. On the morning of 8 November, congratulatory telegrams to Generalísimo Franco from the governments of Austria and Guatemala were delivered at the Ministry of War in the capital.[26] Lisbon Radio also jumped the gun by describing in detail the frenetic welcome that he received from the people of Madrid. The American Hearst Press's sensationalist correspondent, H.R. Knickerbocker, wrote a detailed description of the victory parade, 'from the steps of the Telefónica', which even included the customary barking dog following behind.[27] The British journalist Henry Buckley was told by a news editor in London that his story of fighting in the outskirts must be wrong because it was known that Franco's forces were in the centre of the city.[28]

Miaja and Rojo faced a frightening situation. They had little or no idea of the scale, disposition or readiness of the forces at their disposal. There was a shortage of rifles and ammunition, no anti-aircraft cover and little or no radio liaison between the random collection of arbitrarily armed irregulars whose only asset was their determination to defend the city to the death. Miaja and Rojo were fully aware of the skill and aggression of the *Legionarios* and *Regulares* about to hit them. They also knew of the numerous and well-organized fifth column of Nationalist supporters carrying out sabotage and ready to rise in the city.[29]

Varela, understandably confident that Madrid would fall easily in the light of the government's desertion, delayed in launching the attack in order to allow his troops to rest. He had faced virtually no resistance on 5 November. Had he attacked on 6 November when demoralization still gripped the population, he might have had an easy victory.[30] As it was, Rojo and Miaja were able to spend the night of 6 November and the entire day and night of the seventh organizing the disparate forces at their disposal. Rojo was blessed even more by the fact that on the night of 7 November Varela's detailed battle plan was found in a captured Nationalist tank.[31] Curiously, the departure of the indecisive government of Largo Caballero seemed to take with it the blanket of pessimism and the proximity of Franco's forces wiped away internecine political squabbles.[32]

In the silent streets of the capital on the night of 7 November, the defenders were united by tormenting thoughts of what had happened after the Army of Africa had entered Badajoz and Toledo. Nevertheless, there was a popular determination to fight to the last.[33] Along with the Communist Party's Fifth Regiment, the most highly organised and disciplined force in the central zone, the 1,900 men of the Eleventh International Brigade helped Miaja to lead the entire population of Madrid in a desperate and remarkable defence. Inspired by Miaja's jocular bluster and guided by Rojo's brilliant use of Varela's battle plan, the ordinary citizens of Madrid, with aged rifles and insufficient cartridges, dressed only in their civilian clothes, halted the Nationalist forces.[34] In the course of the attack – launched in brilliant autumn sunshine on 8 November – the Army of Africa suffered casualties on a scale hitherto unknown as it battled to cross the Manzanares, which is dominated from above by the terrace-like avenue known as the Paseo de Rosales. Major Antonio Castejón, the most fiercely energetic of Franco's column commanders, was seriously wounded. With his hip shattered, Castejón, depressed by the high casualties among his Moors, told the American journalist John Whitaker, 'We made this revolt and now we are beaten.'[35]

Varela's attack through the Casa de Campo had faltered by 10 November at the cost of the lives of one third of the men of the International Brigades. When the Manzanares was finally crossed on 15 November, there was hand-to-hand fighting between them and the Moors in the University buildings.[36] Defending their city, with their backs to its walls, the working-class militia were much more of a match for the Moors than they had been in open scrub land. However, after the arrival on 12 November of the Condor Legion, working-class districts were shelled and bombed more systematically than before, although the Generalísimo was careful to try to spare the plush Barrio de Salamanca, the residential district where many of his fifth columnists lived and other important rightists with his forces had their homes. The Germans were anxious to experiment with terror bombing. The damage was massive, the military impact negligible.[37] In deciding to try to terrorize Madrid into submission, and permitting the incendiary bombing of a city bulging with Spain's art treasures, Franco had cast aside the pretence that he was not prepared to damage the capital. He had told Portuguese journalists that he would destroy Madrid rather than leave it to the Marxists.[38] The American Ambassador wrote to Washington: 'it is currently reported that the former King, Alfonso, has protested against this policy to Franco. If he is responsible it can only come from the fact that in his humiliation over his failure to take Madrid in a few days, he has permitted his resentment to get the better of his judgement.'[39]

By 22 November, the Nationalist attack was repulsed.[40] On the following day, Franco and his Chief of Staff, Colonel Martín Moreno, travelled from Salamanca to Leganés on the outskirts of Madrid. The Generalísimo addressed a meeting of Mola, Saliquet, Varela and their respective general staffs. Without massive reinforcements which he simply did not have, there was no choice but to abandon the attack. The Generalísimo ordered an end to frontal assaults on the grounds of the weakness of his forces, the foreign assistance received by the Republic and the difficult tactical situation of the Nationalist Army, given its reliance on long exposed lines of supply and communication.[41] Orgaz would take over the forces on the Madrid front, Mola those in the north. Franco's forces had suffered their first major reverse.[42] However, instead of taking the militarily sensible decision of withdrawing to easily defended lines four or five kilometres from the city, Franco revealed his obstinate determination never to give up an inch of conquered ground. He thus ordered Asensio to fortify the positions taken in the University City in order, as he perceived it, to maintain a psychological and moral advantage, irrespective of the cost which, in the next three months, would be considerable.[43]

Franco was immensely fortunate that the Republican forces in Madrid were too depleted to mount a serious counter-offensive. If they had, the tide might well have turned decisively in their favour. Totally disconcerted by the losses suffered by their men, Varela and Yagüe had told Captain Roland von Strunk, a German military observer in Spain, in the presence of John Whitaker, 'We are finished. We cannot stand at any point if the Reds are capable of undertaking counter-attacks.' Captain von Strunk was in total agreement, convinced that only German reinforcements could save Franco from defeat. He commented bitterly to the US Consul in Seville that 'Franco could have captured Madrid on the first day' and added that he had informed Franco that he must accept German direction of the campaign or else Germany would withdraw its material and Franco had accepted.[44] In Paris, in Rome, in Morocco, as well as in the Nationalist tents around Madrid, it was believed that if Franco did not get more help from Germany and Italy, his movement would collapse.[45]

Before the Republic could test the new confidence forged in the flames of Madrid, Franco's battered columns would receive massive reinforcements from Fascist Italy. It is ironic that only four days before Franco's tacit acknowledgement – in his change of strategy – that he had been defeated, he had secured the co-ordinated recognition of Germany and Italy. In near-identical terms, Berlin and Rome justified their action on the grounds that Franco controlled 'the greater part of Spanish territory'.[46] On 18 November in Salamanca, a visibly emotional Franco appeared before crowds wildly cheering for Hitler and Mussolini. He told them that Nazi Germany and Fascist Italy were 'the bulwarks of culture, civilization and Christianity in Europe'.[47] On the same day, Hitler instructed the new German *Chargé d'Affaires* in Spain about his duties. The man selected was the retired General Wilhelm Faupel, one-time organizer of the Freikorps, adviser to the Argentinian and Peruvian Armies, and Director of the Ibero-Amerikanisches Institut. A staunch Nazi, he was told not to interfere in military affairs.[48] Faupel presented his credentials to Franco on 30 November.[49]

Franco's delight with the signs of co-ordinated fascist help would no doubt have been tarnished had he known of the contempt with which the Italians viewed his military achievements. On 25 November, Mussolini told the German Ambassador to Rome, Ulrich von Hassell, that the Nationalists were lacking in offensive spirit and personal bravery. After negotiating with Franco the Italo-Spanish agreement on military and economic co-operation, Filippo Anfuso, Ciano's representative, reported on 3 December that the Nationalists acted as if they were taking part in a colonial war, concerned with tiny tactical actions rather than with striking great strategic blows. He concluded that Franco needed Italian generals, an Italian column under the orders of Roatta and a sense of urgency.[50] It was only because Mussolini wanted a fascist Spain to put pressure on France and was hopeful that Franco could be coached in the ways of fascism that the Duce contemplated sending further aid to the Caudillo. But, like the Germans, he insisted on certain conditions. The most important was an undertaking 'to conduct future Spanish policy in the Mediterranean in harmony with that of Italy'.[51]

That Franco, conventionally considered to be fiercely proud, should have been happy to accept German and Italian aid on humiliating conditions was not at all puzzling. In the first place, he was desperate. Moreover, he still felt a certain deference towards both Hitler and Mussolini. It was to be his good fortune that, as the American Ambassador in Berlin, William E. Dodd, observed, 'having recognized Franco as conqueror when this has yet to be proved, Mussolini and Hitler must see to it that he is successful or be associated with a failure'.[52] Italy was already racing down the slippery slope to total commitment. In a matter of four months, Mussolini had gone almost imperceptibly from his initial reluctant decision to supply twelve transport aircraft, via the shipping of substantial quantities of aircraft and armoured vehicles in August, September and October, to formal recognition. That gesture would soon involve Mussolini in an irrevocable commitment to Franco's cause which was now facing possible defeat and needed massive assistance.

Faupel telegrammed the Wilhelmstrasse on 5 December with the stark message 'We are now faced with the decision either to leave Spain to herself or to throw in additional forces.' In the German Foreign Office, State-Secretary Weizsäcker feared that to comply would require sending a sea convoy which would attract the hostile attention of England. He believed that Italy should bear the brunt of helping Franco.[53] Immediately after signing his secret agreement with Franco on 28 November, Mussolini called a staff conference to examine the possibility of stepping up Italian military aid to Franco and asked Hitler to send a representative. On 6 December, the Duce, Ciano and Roatta met a pessimistic Admiral Canaris at the Palazzo Venezia. Mussolini suggested that Germany and Italy each prepare a division for Spain, that German and Italian instructors be sent to train Franco's troops and that a joint Italo-German general staff direct and co-ordinate operations alongside Franco's staff. Canaris agreed to co-ordination of the continued delivery of military aircraft and naval and submarine support for Franco in the Mediterranean but repeated the views of Hitler, of von Blomberg, of other senior Wehrmacht officers and of State-Secretary Weizsäcker that Germany could not be seen to send large numbers of troops to Franco without risking international repercussions which might undermine her rearmament plans. Nevertheless, Mussolini decided to go ahead with Italy's commitment of substantial ground forces. It was also agreed that a joint Italo-German general staff be set up to galvanize Franco's operations despite the fears of Canaris that Franco would narrow-mindedly resist.[54]

It is clear from the minutes of this meeting on 6 December that Mussolini, in a spirit of disdain towards Franco, had decided to take the outcome of the Spanish Civil War into his own hands. Although, for obvious reasons, Franco was not informed about what had been said at the meeting, he could in general terms be confident that the Italians could now withdraw their support for him only with the greatest difficulty. On the following day, Mussolini wrote to General Roatta giving him command of all Italian land and air forces already in Spain and soon to be sent. The Duce instructed Roatta to liaise with Franco and the newly arrived German *Chargé d'Affaires*, General Faupel, over the creation of a joint headquarters staff. Two days after the 6 December conference, Mussolini set up a special office, the *Ufficio Spagna*, to co-ordinate the various ministerial contributions to Italian aid for Franco.*[55]

The clinching of external assistance was paralleled inside Spain by the consolidation of the Generalísimo's undisputed authority. Franco had already sabotaged what limited chances there had been of rescuing José Antonio Primo de Rivera. Now, in December 1936, the Generalísimo provided another stark illustration of the speed and skill with which he could act when he felt himself threatened. As the numbers of casualties suffered by the Moroccan Army grew, Franco had to reconcile himself to relying more and more on the recruitment of militia whose first loyalty was to a political group. Inevitably, that increased the political weight of the two parties which made the most substantial contribution, the Falange and the Carlist *Comunión Tradicionalista*. There was no immediate difficulty or doubt about their commitment to the Nationalist cause but, in the long run, their political ambitions differed considerably. Having gone to some trouble to start building his own absolute power, Franco was sensitive to potential threats both to the efficacy of the Nationalist war effort and to his own hegemony. The absence of José Antonio left the Falange disorientated. The veil of secrecy about his death maintained that situation. The Carlists were then, in the short term, more of a threat to Franco's hegemony within the Nationalist zone. The President of their National War Junta, Manuel Fal Conde, had been asserting the autonomy of Carlism since late October.[56] The Carlists saw a chance to make a

---

* Formally directed by Conte Luca Pietromarchi, the *Ufficio Spagna* was under the authority of Ciano, and enjoyed virtual autonomy in military decisions.

more overt bid to consolidate their independence within the Nationalist camp when a decision was announced giving regular army rank to militia officers, and creating short-term training courses to turn them into *alféreces provisionales* (provisional second lieutenants).

On 8 December, with the permission of Mola, they set up a separate *Real Academia Militar de Requetés* for the technical and ideological training of Carlist officers. They claimed that their purpose was no more than to ensure the replacement of casualties and those *Requeté* officers who had gone into the regular forces. The Falangists had two such academies, but had taken the precaution of securing Franco's approval. The Generalísimo was quietly furious and took the opportunity to flex his muscles. After carefully consulting, cultivating and neutralizing Fal Conde's more malleable rival, the languid Conde de Rodezno, Franco moved. Fal Conde was informed through General Dávila, the administrative head of the *Junta Técnica del Estado*, that Franco considered the establishment of a Carlist Academy to be tantamount to a *coup d'état*. Fal Conde was given forty-eight hours either to leave the Nationalist zone or else to face a court martial. Franco gave serious thought to executing the Carlist leader. As it was, since he was loath to risk undermining the morale of the *Requetés* fighting at the front, the Caudillo contented himself with his exile to Portugal.[57] To clinch his control over the autonomous militias, Franco issued a decree militarizing all three militia groups, those of the Falange, of the Carlists and of the CEDA, and placing them under the command of Colonel Monasterio.

By a curious coincidence, just as Franco was dealing with the threat to his authority posed by the Carlists, another hazard placed itself uninvited on his agenda. Don Juan de Borbón, the heir to the throne of Alfonso XIII, remained anxious to take part in the Nationalist war effort. He wrote to the Generalísimo on 7 December 1936, reminded him that he had served in the Royal Navy on HMS *Enterprise* and HMS *Iron Duke* and respectfully requested permission to join the crew of the battlecruiser *Baleares* which was then nearing completion. Although the young prince promised to remain inconspicuous, not go ashore at any Spanish port and to abstain from any political contacts, Franco was quick to perceive the dangers both immediate and distant.[58] If Don Juan were to fight on the Nationalist side, intentionally or otherwise, he would soon become a figurehead for the large numbers of Alfonsine monarchists, especially in the Army, who, for the moment, were content to leave Franco in charge while waiting for victory and an eventual restoration. There was the danger that the Alfonsists would become a distinct group alongside the Falangists and the Carlists, adding their voice to the political diversity which was beginning to come to the surface in the Nationalist zone. Having just been liberated from the problem of José Antonio Primo de Rivera and in the process of cutting down Fal Conde, Franco was hardly likely to welcome Don Juan de Borbón with open arms.

His response was a masterpiece of duplicity. He delayed some weeks before replying to Don Juan. 'It would have given me great pleasure to accede to your request, so Spanish and so legitimate, to fight in our navy for the cause of Spain. However, the need to keep you safe would not permit you to live as a simple officer since the enthusiasm of some and the officiousness of others would stand in the way of such noble intentions. Moreover, we have to take into account the fact that the place which you occupy in the dynastic order and the obligations which arise from that impose upon us all, and demand of you, the sacrifice of desires which are as patriotic as they are noble and deeply felt, in the interests of the *Patria*. . . . It is not possible for me to follow the dictates of my soldier's heart and to accept your offer.'[59] Not only did he thus gracefully refuse a dangerous offer, and so dissipate the threat, but he also squeezed considerable political capital out of so doing. He let it be known 'secretly' among Falangists that he had prevented the heir to the throne from entering Spain because of his own commitment to the future Falangist revolution. He also gave publicity to what he had done and gave reasons which consolidated his own position among the monarchists. 'My responsibilities are great and among them is the duty not to put his life in danger, since one day it may be precious to us . . . If one day a King returns to rule over the State, he will have to come as a peace-maker and should not be found among the victors.'[60] The cynicism of such sentiments could only be appreciated after nearly four decades had elapsed during which Franco had dedicated his efforts to institutionalizing the division of Spain into victors and vanquished and omitting to restore the monarchy.

For the moment, however, Don Juan was a minor problem compared with the military task facing the Generalísimo. At the end of November, Varela had launched an operation to relieve the Nationalist troops tied down to the north-west of Madrid in the Casa de Campo and the Ciudad Universitaria. Little was achieved and the casualties were enormous on both sides. A further effort was made on 15 and 16 December, also at the cost of heavy losses.[61] Both sides had dug in to regroup, and for more than three weeks, the Madrid front saw only partial, albeit bitterly contested, actions. The daring and decisiveness with which Franco had confronted the problems of crossing the Straits and the first precipitate dash northwards of the African columns were now consigned to the past.

General Faupel was shocked when Franco boasted to him in early December 'I will take Madrid; then all of Spain, including Catalonia, will fall into my hands more or less without a fight'. Faupel regarded this as a frivolous assessment since Franco was now faced with a complex war of manoeuvre. The retired German general concluded that Franco's 'military training and experience do not fit him for the direction of operations on their present scale'. In fact, despite the bravado of his words, Franco faced the task with a plodding, indeed hesitant, prudence. He also accepted with deference the overbearing advice of Faupel who, despite Hitler's admonition to keep out of military affairs, was profligate with his opinions. The Generalísimo, who regarded himself as the most meticulous officer in the Spanish army, exercised iron self-control and swallowed Faupel's peremptory and patronizing instruction to issue 'sharp orders for the better care of equipment, rifles and machines guns in particular.' He was playing for higher stakes and on 9 December asked Faupel 'that one German and one Italian division be placed at his disposal as soon as possible'.[62]

Subsequently the Caudillo claimed that he had requested German and Italian arms not troops.[63] However, that became true only much later in 1937 after a massive conscription and recruiting operation. In December 1936, with his armies exhausted and decimated at Madrid, he was desperate for reinforcements.* The Generalísimo was immensely lucky that, within two weeks of the offensive against Madrid breaking down because of his own shortage of reliable troops, the Duce should have decided to send massive aid. On 9 December 1936, Franco received the formal offer of Italian help in the form of officers, NCOs, specialist tank crews, radio operators, artillerymen and engineers, to be incorporated into mixed brigades of Spanish and Italian troops. Rome offered uniforms, armaments and equipment for these brigades and asked Franco how many brigades could be organized. Franco was delighted and arrangements for the creation of two such mixed brigades were made in mid-December.

---

* The only explicit evidence of a request by Franco is Faupel's telegram to the Baron von Neurath, which was reported in the French press at the time and not denied. Moreover, the alleged request closely coincides with the decision by Mussolini on 6 December to send substantial reinforcements. Mussolini's appreciation of Franco's needs was made on the basis of reports from various agents in Spain including Anfuso and General Roatta. Given the close contact between Franco and Roatta since September, it is improbable that Roatta would have made recommendations likely to be disowned by the Generalísimo.

The necessary regular Italian army officers, specialists and ordinary ground troops would begin to arrive in mid-January.[64]

In the meanwhile, Hitler held a conference in the German Chancellery on 21 December with Göring, von Blomberg, Faupel, Warlimont, Friedrich Hossbach, the Wehrmacht liaison officer to the Führer and Werner von Fritsch, the Commander-in-Chief of the German Army. They discussed further assistance to Franco. Faupel asked for three divisions to be sent to Spain but was vehemently opposed by the others for fear of prematurely risking a general war. The Führer therefore decided not to send large numbers of German troops because his wider diplomatic game would derive more benefit from a prolongation of the Spanish Civil War than from a quick victory for Franco. It had been thought in Berlin since late November that the longer the war went on, the more likely Italy was to be drawn into the German orbit. Nevertheless, it was decided that Germany would send sufficient help in the form of aircraft, arms and equipment to ensure that Franco was not defeated.[65] The Generalísimo was thus immensely fortunate to be able to count on support from Hitler and Mussolini which would be greater and more consistent than anything that the Republic could hope for from the Soviet Union.

In addition to the specialist regular troops necessary for the creation of the mixed Spanish-Italian brigades, Mussolini decided 'in view of the unsatisfactory situation' to send, two contingents of three thousand Black Shirts each, in self-contained units with their own officers, artillery and transport. On 14 December Roatta's assistant, Lieutenant-Colonel Emilio Faldella, gave Franco a note to the effect that the Italian government wished the volunteers to be organized in autonomous Italian companies with Italian officers. It was made clear that these contingents would be additional to the proposed mixed brigades.[66] Franco wanted troops but not in autonomous units under Italian command. His annoyance was revealed when he asked Faldella 'Who requested them?' and snapped 'When one sends troops to a friendly country, one at least asks permission'.[67]

It is clear that Franco was glad to have the Black Shirts but had hoped simply to incorporate them into his own units as foreign legionaries. His suspicions of the efficacy of Falangist militias were not replicated with regard to the Italian Fascist volunteers since he had been told that they had been battle-hardened in Abyssinia. He was, of course, deeply irritated by the lack of consideration of his position implicit in the blunt and unexpected terms in which their arrival was announced. The strength of the Italian contingents that arrived in late December and early January was, according to a report by Faupel, based presumably on information from Roatta, 'determined not by previous agreement with Franco but according to independent Italian estimates'.[68] Nevertheless, he hastened to use them as soon as they disembarked and, on 12 January, he would request another nine thousand Black Shirts.[69]

Such external assistance was necessary to enable Franco to go forward from the deadlock in Madrid. On 28 November, General Saliquet had written to the Generalísimo with a proposal for an encircling operation, against the Madrid-La Coruña road to the north-west and a dual thrust from the south-west of Madrid and from Soria in the north-east towards Alcalá de Henares.[70] Franco mused over this proposal for three weeks and it was not until 19 December that he issued orders which would break the stalemate prevailing since he had called off the frontal assault on Madrid at the Leganés meeting on 23 November. They envisaged a refinement of Saliquet's plan, implementing it closer to Madrid by three thrusts outwards from the exposed wedge which the Nationalists had driven into the capital's defences.[71]

In heavy rain and fog, across muddy terrain, costly and sterile battles were fought for villages like Boadilla del Monte which was virtually destroyed. Varela was wounded on Christmas Day and field command was assumed by Orgaz. After crippling losses in the fighting, the attack was briefly called off. Roatta telegrammed the *Ufficio Spagna* on 27 December complaining of apathy at Franco's headquarters and reporting that the Generalísimo's staff was incapable of mounting an operation appropriate to a large-scale war.[72] On 3 January, the assault was renewed with increased ferocity and reached the important crossroads at Las Rozas on the road to El Escorial and La Coruña. On 7 January, Pozuelo and Húmera fell. In six days, scarcely ten kilometres of road had been taken by the Nationalists. They had eased the pressure on their troops in the Casa de Campo and the Ciudad Universitaria but at enormous cost. When the fronts had stabilized by 15 January, each side had lost in the region of fifteen thousand men.[73] The various efforts to take Madrid had severely depleted Franco's forces. The Republicans were now solidly dug in and Franco was fortunate that they were unable to seize the unique opportunity to launch a counter-attack to break through his severely overstretched lines.

In the midst of the reverses around Madrid, Franco was relieved to discover that his cultivation of the Church was bearing fruit. On 22 December, Cardinal Gomá returned from Rome where he had been frantically working for Vatican recognition of Franco. The cautious Curia held back but, in order to demonstrate the Church's sympathy for Franco's cause, Gomá was appointed the Vatican's confidential *Chargé d'Affaires* in Nationalist Spain. It was the crucial first step towards full diplomatic recognition.[74] Gomá and the Generalísimo met on 29 December and agreed on a joint statement to the Vatican, in which it was made clear that, in the interests of eventual recognition, Franco was ready to do everything possible to favour the Church's position in Spain.[75]

The clinching of relations with the Vatican was of immense long-term political importance to Franco. In immediate terms, even more welcome was the military help promised by Mussolini. With the attacks around Madrid stalling, Franco had been relieved by the fact that in mid-December, the Duce had begun sending the first of what, by mid-February 1937, would be nearly fifty thousand fascist militiamen and regular troops masquerading as volunteers.[76] Whatever gloss Franco would put on it later, the arrival of Italian reinforcements was of crucial importance to his military survival. Inevitably, once the Duce had committed his own prestige to a Nationalist victory in Spain, the stalemate around Madrid quickly intensified his impatience with Franco. At the end of the year, he requested Hitler to send to a meeting in Rome in mid-January someone 'with full powers' to discuss Italo-German co-operation to bring about 'a real decision in Spain'.[77] In fact, it was becoming ever more apparent that the Italians were going to be left by Hitler to make the decisive contribution to Franco's success. Roatta reported to Rome on 12 January that Canaris had told him that Sperrle was pessimistic about both the initial efficacy of the Condor Legion and the state of the Nationalist forces. Sperrle, in turn, told Roatta that the real problem was German fear of provoking a premature war with France.[78]

At the meeting held at the Palazzo Venezia on the evening of 14 January 1937, Hitler's representative was Hermann Göring.* Mussolini was irritated that Italo-German aid, rather than spurring Franco on to greater efforts, merely permitted him to indulge his natural inclination to wear down the Republic by a slow campaign of attrition. Göring agreed that, if Franco had known how to use it properly, the Italo-German material and technical assistance was enough to have permitted him to

---

* Göring's visit to Rome was a symbolic affirmation of the growing warmth between the Nazi and Fascist regimes. During a packed programme, he visited the Fencing Academy at the Forum where he challenged Mussolini to a sabre duel. To the delight of the senior Nazis and Fascists present, they slugged it out for twenty minutes, showing remarkable agility given their respective sizes – with Mussolini the eventual victor (Ramón Garriga, *Guadalajara y sus consecuencias* [Madrid, 1974] pp. 42–3).

win already. The Air Minister declared bitterly that the recognition of Franco before the capture of Madrid had been a major error to remedy which it was agreed that he would have to be subjected to 'energetic pressure' to accelerate his operations and make full use of the lavish means put at his disposal.

Despite his expressions of solidarity with Mussolini, fear of inter-national complications impelled Göring to say that Germany could not send a division to Spain. This left the immediate task of preventing Franco being defeated to the Duce who was disappointed but not unhappy to be the senior partner in Spain. Declaring that Franco must win, he said that there were no longer any restraints on his actions in Spain. To ensure that Franco adopt a more energetic policy, it was decided to oblige him to accept the joint Italo-German general staff. Mussolini and Göring agreed that to ensure Franco's victory before, as they wrongly imagined would happen, the British erected an effective blockade to stop foreign intervention,* substantial additional aid would have to be sent to Spain by the end of January. Mussolini suggested telling Franco that thereafter there would be no more help.[79]

On the day after the meeting in the Palazzo Venezia, the chiefs of staff of the Italian military ministries met at the Palazzo Chigi with the staff of the *Ufficio Spagna* and Ciano's representative Anfuso to discuss the minimum programme of aid to Franco. Partly out of contempt for Franco's generalship and partly out of a desire to monopolize the antici-pated triumph for Fascism, it was agreed that the Italian contingent must be used as an independent force under an Italian general only nominally responsible to Franco's overall command. Three possibilities were out-lined for the decisive action by which Italian forces would win the war for Franco. Mussolini favoured a massive assault from Teruel to Valencia to cut off Catalonia from the rest of Spain. This was to be preceded by the terror bombing of Valencia. However, it was acknowledged that such an operation required the full co-operation of Franco. A second option was a march from Sigüenza to Guadalajara to tighten irrevocably the Nationalist grip on Madrid. The third more limited possibility was the

---

* The Anglo-French policy of Non-Intervention, adopted in August 1936, was a farce which favoured the Nationalists at the expense of the Republic and appeased the fascist dictators. It was described by a Foreign Office official as 'an extremely useful piece of humbug'. It is clear that a more resolute attitude by London would have inhibited the Germans and Italians in their assistance to Franco. (Enrique Moradiellos, *Neutralidad benévola: el Gobierno británico y la insurrección militar española de 1936* [Oviedo,1990] pp. 117–88; Douglas Little, *Malevolent Neutrality: The United States, Great Britain, and the Origins of the Spanish Civil War* [Ithaca, 1985] pp. 221–65.)

capture of Málaga to provide a seaport nearer to Italy and a launching pad for an attack on Valencia from the south-west.[80]

After his failures around Madrid, Franco had little choice but to grit his teeth and acquiesce in the demeaning Italo-German suggestions which were communicated to him by Anfuso on 23 January. The document presented by Anfuso made it clear that international circumstances prevented aid being continued indefinitely.[81] At first, the Generalísimo seemed perplexed.[82] However, on the following day, he gave Anfuso a note expressing his thanks for Italo-German help and a desperate plea for it to continue for at least another three months.[83] The prospect of the British imposing an effective blockade galvanized him into giving serious consideration to the three strategic proposals made by the Italians. In effusively thanking Mussolini for his assistance, Franco told Anfuso that he would now accelerate the end of the war by undertaking a great decisive action. On 26 January, he accepted Roatta's suggestion that, henceforth, the regular high-level advice of Faupel and Roatta on major strategic issues would be implemented by Franco's own staff, in which were to be included ten senior German and Italian officers.[84] Mussolini considered that he could send instructions to Franco as to a subordinate.[85]

Sensitive to any slur or slight, Franco cannot fail to have resented the clear insinuation of German and Italian disdain for his military prowess. Nevertheless, he showed no sign of it and accepted, along with the imposition of foreign staff officers, Mussolini's strategic suggestions. According to Kindelán, anxious to play down Franco's deference to the Duce, the Generalísimo was unsure of the military value of the new arrivals, despite the fact that they were well-equipped by comparison with his own troops and many had had experience in the Abyssinian war. He thus decided to test them in a relatively easy campaign in the south.[86]

It is indeed the case that, to offset the failure in Madrid, the Generalísimo had already accepted a proposal from Queipo for a piecemeal advance towards Málaga. A sporadic campaign to mop up the rest of Andalusia, as savage and bloodthirsty as the march on Madrid, had been intensified in mid-December with considerable success.[87] However, after the arrival of Italian troops, the nature of the campaign changed dramatically. Rather than Franco skilfully blooding them in a campaign of his choice, they were engaged in an operation chosen by Mussolini. As the Black Shirts were setting out, Mussolini had reminded Roatta on 18 December 1936 of his own long-held conviction that a major attack should be launched against Málaga. Roatta immediately informed Franco of the Duce's preference and found him grudgingly amenable (*sufficientemente propenso*) to it. Thereafter, the Duce followed the progress of the attack with an enthusiasm commensurate with it having been his own brainchild.[88]

Franco wanted to incorporate the newly arrived Italians into mixed units on the Madrid front but had to acquiesce in Mussolini's desire to see them operate autonomously in Andalusia.[89] In the light of the thin and scattered defences of Málaga, Roatta wanted a *guerra celere* (rapid strike) attack by his own motorised columns whereas Franco favoured Queipo's original proposal for a gradual but thorough conquest of Republican territory. Franco was not much interested in a lightning victory for which Mussolini could take the credit and which might end the war before his leadership was consolidated. On 27 December, Roatta effectively overruled the Generalísimo's preference for a slow advance backed up by political purges. They reached a compromise in which both types of assault would take place simultaneously. Franco had to bite his tongue when his request for two Italian motorised companies for the Madrid front was rejected by Roatta on the grounds of his own greater needs in preparing the attack on Málaga. On 9 January 1937, an optimistic Roatta and a sceptical Queipo agreed a division of responsibilities which reflected Franco's concessions.[90] Under the direction of Queipo de Llano who was installed on the battlecruiser *Canarias*, and of Roatta on land, two columns began to advance in mid-January. By the end of the month after the capture of Alhama on the Málaga-Granada road, they were ready for the final push.

Colonel Wolfram von Richthofen, Chief of Staff of the Condor Legion, wrote in his diary on 3 February 'nothing is known about the Italians, their whereabouts and their intentions. Franco knows nothing either. He really ought to go to Seville to put himself in the picture and hope for a share of the Málaga victory laurels.'[91] To make good his ignorance and to give the impression of overall control of events, Franco was already travelling from Salamanca to Seville on 3 February, the same day on which, in torrential rain, the Italo-Spanish forces moved on Málaga. The advance took the form of troops distributed in a large concentric circle, the Spanish units moving eastwards from Marbella and the Italian motorized columns racing south west from Alhama without concern for their flanks.[92] The Generalísimo visited the front and on 5 February at Antequera discussed the progress of the campaign with Queipo and Roatta. Convinced that the operation was going to be successful, he did not wait for the fall of Málaga but returned to Seville on 6 February and to Salamanca on the following day to oversee a new push on the Madrid front.[93]

On 7 February, after a rapid march, Nationalists and Italians reached Málaga. Its military command had been changed with alarming frequency in the preceding days, morale was abysmally low, and after bombing raids by Italian aircraft and bombardment by Nationalist warships, the city collapsed easily. Italian troops were first to enter Málaga and briefly ruled the city before ostentatiously handing it over to the Spaniards. Roatta claimed the victory for Mussolini and sent a triumphant, and implicitly wounding, telegram to Franco: 'Troops under my command have the honour to hand over the city of Málaga to Your Excellency'.[94] In fact, given the massive numerical and logistical superiority of the attackers, the triumph was less of an achievement than it seemed at the time. Neglected by the Valencia government, the defending forces were in more or less the same state of readiness as the improvised militiamen who had faced Franco's Army of Africa six months earlier.[95] Neither the Nationalists nor the Italians showed much mercy. The international outcry was less than that provoked by the massacre of Badajoz, because Franco had ordered all war correspondents to be kept out of Málaga.[96] After the battle, Queipo and Roatta sent a motorised column to pursue refugees escaping along the coast road. Within the city itself nearly four thousand Republicans were shot in the first week alone and the killings continued on a large scale for months. The refugees who blocked the road out of Málaga were shelled from the sea and bombed and machine-gunned from the air.[97]

When Roatta's news of the victory at Málaga reached Salamanca, Franco unsurprisingly showed little interest. His humiliating subordination to Mussolini had been starkly underlined. Millán Astray, who came to congratulate the Generalísimo and found him absorbed gazing at a huge wall map, exclaimed: 'I expected to find you celebrating the victory in Málaga not here on your own looking at a map. ' Franco diminished the Italian achievement by pointing at the map and saying 'Just look what remains to be conquered! I can't afford the luxury of taking time off.'[98] This gloomy and contrived effect of unceasing military dedication was out of tune with Franco's normally irrepressible faith in victory. He was certainly preoccupied by the progress of the battle in the Jarama valley which he had launched just as Málaga was about to fall but he could hardly have been immune to the fact that the loss of Málaga was a fierce blow to the Republic in terms of captured territory, prisoners and weaponry. He had gained the food-producing province of Málaga and most of Granada, deprived his enemies of a strategically crucial sea port with a population of one hundred and fifty thousand people and shortened the southern front. The feigned lack of interest revealed his resentment of the disdainful Roatta and the fact that he could take no pleasure in a triumph attributed by the world's press to Mussolini.[99]

The fall of Málaga provoked a major internecine crisis within the Republic. The Communists began to reveal their impatience with Largo Caballero and obliged him to accept the resignation of General Asensio, his under-secretary of war.[100] Ironically, the one negative consequence for Franco of such an easy victory was the totally erroneous notion that both he and Mussolini derived of the efficacy of the Italian contingent.[101] Mussolini was so delighted that he promoted Roatta to Major-General. The Duce and his Chief of Staff at the Ministry of the Army, Alberto Pariani, immediately produced ambitious plans for the Italian troops to sweep on to Almería and then through Murcia and Alicante to Valencia.[102] However, Roatta's reports to Rome on the eve of the attack on Málaga had presented a bleak picture of Italian disorganization, indiscipline and lack of technical preparation. Now he had to restrain Mussolini's enthusiasm and persuade him that a long haul along the south coast exposed to constant flank attack would be less decisive than operations envisaged by Franco in the centre.[103]

Franco was happy to get Italian help on the Madrid front and quick to deflate the euphoric Queipo who was anxious to use the triumph at Málaga as the basis for a triumphal march through Eastern Andalusia towards Almería. Franco remained obsessed with Madrid and had no reason to want to give away triumphs to Queipo de Llano. Accordingly, he prohibited further advance in Andalusia, to the bitter chagrin of Queipo.[104] It was, however, with some trepidation that Franco viewed the prospect of what seemed at the time like a fearsome Italian army, directed from Rome, allowing Mussolini graciously to hand him victories on a plate. It was a perception which would have disastrous consequences during the battle of Guadalajara.

At this time the nationalist press began to circulate a story which linked Franco's destiny with the intercession of the saints. Allegedly, in the chaos of defeat, the military commander of Málaga, Colonel José Villalba Rubio, left various items of luggage behind him when he fled. In a suitcase left in his hotel was found the holy relic of the hand of St Teresa of Avila which had been stolen from the Carmelite Convent at Ronda.[105] In fact, the relic was found in police custody. It was sent to Franco who kept it with him for the rest of his life. The recovery of the relic was the excuse for the exaltation of St Teresa as 'the Saint of the Race', the champion of Spain and her religion in the *Reconquista*, during the conquest of America and in the battles of the Counter-Reformation. Catholic and

political propagandists alike stressed the Saint's association with the Caudillo in similar exaltation of his providential role.[106] Franco himself seems
to have believed in his special relationship with St Teresa. Cardinal Gomá
reported Franco's reluctance to part with the arm as proof of his intense
Catholic faith and his belief that he was leading a religious crusade. The
Bishop of Málaga granted permission for the relic to remain in Franco's
possession and never left his side on any trip which obliged him to sleep
away from home.*[107]

Encouraged by the easy success which he anticipated in the south and
by the availability of the Condor Legion, Franco had simultaneously
renewed his efforts to take Madrid. On 6 February 1937, an army of
nearly sixty thousand well-equipped men, under the direction of General
Orgaz, had launched a huge attack through the Jarama valley towards
the Madrid-Valencia highway to the east of the capital. Still convinced
that he could capture the capital, Franco took a special interest in the
campaign.[108] Two days later, his determination to win would be intensified by a desire for a victory to overshadow the Italian triumph at Málaga.

Almost simultaneously, Mussolini had sent a new Ambassador to
Nationalist Spain, the emollient Roberto Cantalupo, who arrived shortly
after the battle for Málaga.[109] It was a reflection of Franco's seething
resentment at the behaviour of Roatta and Mussolini over the conquest
of Málaga that he kept Cantalupo waiting for days before receiving him.
Cantalupo got a sense that, although everyone knew that Málaga had
been captured by the Italians, no one said so. 'Here', he reported to
Ciano on 17 February, 'the coin of gratitude circulates hardly at all.'
When he finally met the Caudillo for an informal meeting, Cantalupo
got the impression that Franco believed in ultimate victory but was no
longer certain that it was anything other than a long way off. If anything,
the Caudillo seemed to prefer the prospect of a long war although he
put off explaining why for a future meeting. He did make it clear that
he would not contemplate a negotiated peace.[110]

The implicit conflict between Mussolini's urge for the rapid and spectacular defeat of the Republic and Franco's gradual approach quickly
came into the open. Four days after the fall of Málaga, Roatta being
wounded, he sent his Chief of Staff, Colonel Emilio Faldella, to visit the

---

* An aide was appointed specifically to to carry it and to guard it against loss or theft.
Occasionally, over the years, the nuns wrote to Franco requesting that he return the hand
if only for a period of loan of a month, three weeks or a fortnight. Franco, fearful that he
would not get it back, never complied, arranging instead for his faithful cousin Pacón to
send a charitable donation to pacify them.

Generalísimo in Salamanca and discuss the next operation in which the *Corpo di Truppe Volontarie* (CTV), as the Italian forces now came to be known, might be used. On the afternoon of 12 February, Faldella found Franco's staff jubilant about their forces' early thrust over the Jarama river and what they assumed to be an imminent and decisive victory. Faldella was told by Franco's chief of operations, Colonel Antonio Barroso, that Alcalá de Henares would be occupied within five days and Madrid cut off from Valencia. Faldella told Barroso that he was going to propose that the next operation for the CTV should be an offensive against both Sagunto, to the north of Valencia, and Valencia itself, one of the options favoured by Mussolini since mid-January and communicated to the Generalísimo by Anfuso on 22 January. Barroso advised him against even mentioning it on the grounds that Franco would never allow the Italians to carry out an autonomous assault on a politically sensitive target like the Republican capital, given his central concern with his own prestige. Accordingly, after consulting Roatta by telephone, Faldella altered the note which he had brought for Franco to suggest instead the remaining option of those contemplated by Mussolini after the meeting with Göring, a major push from Sigüenza to Guadalajara to close the circle around Madrid.[111]

When Faldella was received by Franco at 8 p.m. on 13 February, the usually polite Generalísimo ostentatiously failed to thank him for the Italian action at Málaga and said 'the note has surprised me, because it is a real imposition'. The expected success in the Jarama gave the Caudillo the confidence to speak in stronger terms than previously to Faldella, who was after all the acting military representative of Mussolini. 'When all is said and done', Franco told Faldella, 'Italian troops have been sent here without requesting my authorization. First I was told that companies of volunteers were coming to be incorporated into Spanish battalions. Then I was asked for them to be formed into independent battalions on their own and I agreed. Next senior officers and generals arrived to command them, and finally already-formed units began to arrive. Now you want to oblige me to allow these troops to fight together under General Roatta's orders, when my plans were altogether different.' Faldella replied that the reasoning behind all this was simply that Mussolini was trying to make good the failure of the Germans to supply troops to which Franco responded: 'This is a war of a special kind, that has to be fought with exceptional methods so that such a numerous mass cannot be used all at once, but spread out over several fronts it would be more useful.'[112] These remarks revealed not just Franco's resentments about

Italian aid, but also the limitations of his strategic vision. His preference for piecemeal actions over a wide area reflected both his own practical military experiences in a small-scale colonial war and his desire to conquer Spain slowly and so consolidate his political supremacy.[113]

Faldella tried to make him see the opportunity for a decisive victory offered by the determined use of the Italian CTV. Franco would not be shaken from his preference for the gradual and systematic occupation of Republican territory: 'In a civil war, a systematic occupation of territory accompanied by the necessary purge (*limpieza*) is preferable to a rapid rout of the enemy armies which leaves the country still infested with enemies.' Faldella pointed out that a rapid defeat of the Republic at Valencia would make it easier for him to root out the Left in Spain. At this point, Barroso interrupted and, as his master's voice, said 'you must take into account that the Generalísimo's prestige is the most important thing in this war, and that it is absolutely unacceptable that Valencia, the seat of the Republican government, should be occupied by foreign troops.'[114]

On the following day, Franco sent a written reply to Faldella, in which he grudgingly accepted his offer of an attack from Sigüenza to Guadalajara. He claimed that he had never wanted Italian troops used *en masse* for fear of international complications and because it was damaging for 'decisive actions against objectives of the highest political importance to be carried out other than by the joint action of Spanish and Italian units'.[115] Cantalupo believed that the Caudillo had been had brought around by an Italian promise to ensure that Spanish troops entered Madrid as the victors.[116] In fact, he was responding to sticks as well as carrots. The potential conflict between Franco and the commanders of the CTV was such that Roatta flew to Rome to discuss the problem with Mussolini. The Duce reacted firmly in support of Roatta, threatening to withdraw his forces if Franco continued to respond as he had to Faldella. To show that he meant business, twenty fighter aircraft promised to Franco were redirected to the Italian command in Spain which was given control over the Air Force units which had previously flown under the Generalísimo's orders.[117]

Mussolini's threat drew additional effect because it came as the Nationalist attack in the Jarama ground to a halt. The Jarama valley was defended fiercely by Republican troops reinforced by the International Brigades and the battle saw the most vicious fighting of the entire Civil War. As in the battle for the La Coruña road, the Nationalist front advanced a few miles, but no major strategic gain was made. Once again Madrid was saved, albeit at a high cost in blood. The Republicans lost more than ten thousand including some of the best British and American members of the Brigades, and the Nationalists about seven thousand.[118]

Franco's earlier defiance turned to desperation. Now, only six days after his churlish treatment of Faldella on 13 February, he sent Barroso to beg him to begin the offensive as soon as possible. Faldella refused, on the grounds that his planned initiative could not be rushed and so, on the following day, Millán Astray asked Faldella to see him. They dined together at CTV headquarters on 21 February and Millán spoke in 'pathetic terms' about the Nationalists' difficulties around Madrid and begged for a rapid Italian intervention. Faldella was convinced that Millán Astray had come at Franco's behest. In the event, Franco had to wait until Faldella and Roatta were ready. After all, moving the Italian Army from Málaga to central Spain was no easy task.

The Generalísimo's desire to use the Italians as reinforcements within his Jarama campaign was coldly brushed aside. A seething Franco was having to bend to what the Italians wanted. The general plan of operations which he sent to Mola on 23 February exactly followed the strategy outlined in Faldella's note of 13 February. One week later, the Italians were still not ready and, on 1 March, Barroso again pleaded with Faldella to persuade Roatta to begin an immediate action.[119] Although Orgaz and Varela had managed to hold the line at the Jarama, the Generalísimo was desperate for a diversion to relieve his exhausted forces. For Franco, an Italian attack on Guadalajara, forty miles north-east of Madrid, would be an ideal distraction. That was not what the Italians had in mind at all. A major disaster was in the making.

IX

THE AXIS CONNECTION

Guadalajara & Guernica, March–April 1937

ALTHOUGH THINGS were taking a turn for the worse militarily, Franco dismissed out of hand any suggestions of a compromise peace with the Republicans or even with the profoundly Catholic Basques. Proposals to this end made by the Vatican were discussed by the Generalísimo and Cardinal Gomá in mid-February. Although respectful with the Primate, Franco had rejected anything less than outright surrender, refusing to negotiate with, and therefore recognize the authority of, those whom he held responsible for the present situation in the Basque Country. Gomá reported to Rome that Franco saw any mediation as merely putting off the necessary solution of a political and historical problem, by which he meant the eradication of Basque nationalism. Negotiations meant concessions and concessions meant 'rewarding rebellion' and would raise the expectations of other regions.[1] Franco's negative attitude to mediation of any kind reflected his perception of the war as an all-or-nothing, life-or-death struggle which had to end with the total annihilation of the Republic and its supporters.

This was certainly the impression given to the Italians. When Cantalupo's credentials arrived from Rome, he was received officially on 1 March with a scale of splendour which not only underlined the value that Franco placed on Italian assistance but also reflected his own taste for pomp. Any hopes harboured by his fellow generals that Franco considered his headship of the State to be at all provisional must by now have started to wither. The imposing ostentation and grandeur with which the Caudillo surrounded his public appearances resounded with permanence. Cantalupo was treated to eight military bands. The colourful ranks of Falangist, Carlist and other militias, Spanish, Italian and Moorish troops formed up in a solemn procession through Salamanca's enormous but elegantly proportioned Plaza Mayor to the Palacio del Ayuntamiento. The Generalísimo arrived in the square escorted by his Moorish Guard, resplendent in their blue cloaks and shining breastplates. It recalled the entry of Alfonso XIII into Melilla in 1927, an occasion on which he was accompanied by Franco, who was increasingly indulging his own taste for royal ceremony. His arrival was greeted with the chant of 'Franco! Franco! Franco!'. He received Cantalupo in a salon magnificently adorned for the occasion with sixteenth-century Spanish tapestries and seventeenth-century porcelain. During the ceremony, Franco was accompanied by Mola, Kindelán, Cabanellas, Dávila and Queipo de Llano as well as a veritable court of other army officers and functionaries in full dress uniform. Yet, Franco himself did not match the regal show and an unimpressed Cantalupo wrote to Rome 'He stepped out with me on the balcony that offered an incredible spectacle of the immense square but was incapable of saying anything to the people that applauded and waited to be harangued; he had become cold, glassy and feminine again'.[2]

Away from the pomp of Salamanca, Roatta, Faldella and other senior Italian officers were shocked by the relentless repression behind the lines.[3] Cantalupo requested instructions from Rome and on 2 March Ciano told him to inform Franco of the Italian Government's view that some moderation in the reprisals would be prudent because unrestrained brutality could only increase the duration of the war. When Cantalupo saw Franco on 3 March, the Caudillo was fully prepared for the meeting. Cantalupo appealed to him to slow down the mass executions in Málaga in order to limit the international outcry. Denying all personal responsibility and lamenting the difficulties of controlling the situation at a distance, Franco claimed that the massacres were over 'except for those carried out by uncontrollable elements'. In fact, the slaughter hardly diminished but its judicial basis was changed. Random killings were now replaced by summary executions under the responsibility of the local military authorities. Franco claimed to have sent instructions for greater clemency to be shown to the rabble (*masse incolte*) and continued severity against 'leaders and criminals' as a result of which only one in every five of those tried was now being shot.

Nevertheless, Rome continued to receive horrifying accounts from the Italian Consul in Málaga, Bianchi.* On 7 March, Cantalupo was

* Among those responsible for the repression, Carlos Arias Navarro, a young military prosecutor, who came to be known as 'the butcher of Málaga', later became a close friend of the Franco family and in 1973 was to replace Carrero Blanco as prime minister (Rafael Borràs Betriu *et al.*, *El día en que mataron a Carrero Blanco* [Barcelona, 1974] p. 252).

instructed to go to Málaga but Franco persuaded him that the situation was too dangerous for a visit. Nevertheless, the Generalísimo did undertake to have two military judges removed.[4] Franco's proclaimed difficulties about curtailing the killings in Málaga contrasted starkly with his response to a complaint by Cardinal Gomá about the shooting of Basque Nationalist priests in late October 1936. Valuing the good opinion of the Church more than that of the Italians, he replied instantaneously: 'Your Eminence can rest assured that this stops immediately'. Shortly thereafter, Sangróniz confirmed to Gomá that 'energetic measures had been taken'.[5]

At this time, Franco himself was sufficiently concerned by the unfavourable publicity provoked by the blanket repression to give a brilliantly ambiguous interview on the subject to Randolph Churchill. It was clear that in describing his policy as one of 'humane and equitable clemency', Franco's meaning differed considerably from the way in which his words were understood by Churchill and his readers. Franco declared that 'ringleaders and those guilty of murder' would receive the death penalty, 'just retribution', but claimed mendaciously that all would be given fair trials, with defence counsel and 'the fullest opportunity to state his case and call witnesses'. He omitted to mention that the defence counsel would be named by the court and would often outdo the prosecutors in demanding fierce sentences. Similarly, when Franco said that 'when we have won, we shall have to consolidate our victory, pacify the discontented elements and unite the country', Churchill could have no idea of the scale of the blood that would be shed or of the terror which would be deployed to realize those ends.[6]

For most of the Civil War, those Republican prisoners not summarily executed as they were captured or murdered behind the lines by Falangist terror squads were subjected to cursory courts martial. Often large numbers of defendants would be tried together, accused of generalised crimes and given little opportunity to defend themselves. The death sentences passed merely needed the signature (*enterado*) of the general commanding the province. As a result of the Italian protests, from March 1937 death sentences had to be sent to the Generalísimo's headquarters for confirmation or pardon. The last word on death sentences lay with Franco, not as Head of State, but as Commander-in-Chief of the Armed Forces. In this area, his close confidant was Lieutenant-Colonel Lorenzo Martínez Fuset of the military juridical corps, who was *auditor del Cuartel General del Generalísimo* (legal adviser to headquarters). Franco insisted on seeing the death sentences personally, although he spent little time on reaching a decision. Martínez Fuset would bring folders of death sentences to Franco. Despite the regime myth of a tireless and merciful Caudillo agonizing late into the night over death sentences, the reality was much starker. In fact, in Salamanca or in Burgos, after lunch or over coffee, or even in a car speeding to the battle front, the Caudillo would flick through and then sign sheafs of them, often without reading the details but nonetheless specifying the most savage form of execution, strangulation by *garrote*. Occasionally, he would make a point of decreeing *garrote y prensa* (garrote reported in the press).[7]

Specifying press coverage was not just a way of intensifying the pain of the families of the condemned men but also had the wider objective of demoralizing the enemy with evidence of inexorable might and implacable terror. That was one of the lessons of war learnt by Franco in Morocco. At one lunch in the winter of 1936–37, the case of four captured Republican militiawomen was discussed. Johannes Bernhardt who was present was taken aback by the casual way Franco, in the same tone that he would use to discuss the weather, passed judgement, 'There is nothing else to be done. Shoot them.'[8] He could be gratuitously vindictive. On one occasion, having discovered that General Miaja's son had been tried and absolved by a Nationalist tribunal in Seville, Franco intervened personally to have him rearrested and retried in Burgos. There was some doubt as to whether Captain Miaja had voluntarily come over to the Nationalists or been captured. Accordingly, the Burgos court issued a light sentence so Franco had the unfortunate young Miaja tried again in Valladolid. In Valladolid, the military tribunal found him not guilty and set him free. At this point, Franco intervened again and quite arbitrarily had him sent to a concentration camp at Miranda del Ebro where he remained until he was freed in a prisoner exchange for Miguel Primo de Rivera.[9]

Throughout 1937 and 1938, his brother-in-law and close political adviser, Ramón Serrano Suñer often tried to persuade him to adopt more juridically sound procedures and Franco consistently refused, saying 'keep out of this. Soldiers don't like civilians intervening in affairs connected with the application of their code of justice.'[10] At one point, Serrano Suñer tried to arrange a reprieve for a Republican army officer. After first telling him that it was none of his business, Franco finally yielded to his brother-in-law's pressure and undertook to do something. If Franco had wanted to help, he could have done so. As it was, four days later, he told Serrano Suñer that 'the Army won't put up with it, because this man was head of Azaña's guard.'[11] Serrano Suñer and Dionisio Ridruejo both alleged that the Caudillo arranged for reprieves for death sentences to arrive only after the execution had already been carried out.[12]

Like Hitler, Franco had plenty of collaborators willing to undertake the detailed work of repression and, also like the Führer, he was able to distance himself from the process. Nonetheless, since he was the supreme authority within the system of military justice, there is no dispute as to where ultimate responsibility lay. Franco was aware that some of his subordinates enjoyed the bloodthirsty work of the repression. His Director-General of Prisons, Joaquín del Moral, was notorious for the prurient delight he derived from executions. General Cabanellas protested to Franco about the distasteful dawn excursions organized in Burgos by Del Moral in order to enjoy the day's shootings. Franco did nothing. He was fully conscious of the extent to which the repression not only terrified the enemy but also inextricably tied those involved in its implementation to his own survival. Their complicity ensured that they would cling to him as the only bulwark against the possible revenge of their victims.[13]

In early March, to the chagrin of Cantalupo, Mussolini sent Roberto Farinacci, the powerful Fascist boss of Cremona, as his personal envoy to inform Franco of his 'ideas about the future' which involved placing a Prince of Savoy on the throne of Spain. That idea was politely but firmly rejected by Franco. However, the Caudillo was more amenable when Farinacci tried to convince him to create a fascist-style 'Spanish National Party' in order to control every aspect of political life. Delighted to be discussing 'his' future State, and clearly unencumbered by any inhibitions about the provisional nature of his mandate, Franco said that he was not planning to rely on either the Falangist or the Carlists in his post-war reconstruction. In rejecting the idea of an Italian prince, he made it clear that the restoration of the monarchy was anything but an immediate prospect, saying 'First, I have to create the nation: then we will decide whether it is a good idea to name a king.' It encapsulated the political philosophy which was to keep him in power until his death in 1975. Farinacci was not impressed with Franco, describing him in a letter to Mussolini as 'a rather timid man whose face is certainly not that of a *condottiere*'. He was overheard by agents of the Spanish secret police declaring that Mussolini would have to take over Spain and appoint him as pro-consul. In particular, he thought, like Himmler later, that the slaughter of prisoners taking place behind the Nationalist lines was politically senseless and he protested in vain to Franco. He also made contact with the Falangist leader Manuel Hedilla as well as with Nicolás Franco in the hope of accelerating the fusion of Falangists and Carlists.[14]

The creation of a single party was clearly on Franco's agenda but he was for the moment totally absorbed by events at the Madrid front. With his forces depleted in the Jarama and in desperate need of a diversion, Franco was anxious for Faldella to implement the proposal made on 13 February for an attack on Guadalajara. Negotiations between the two sides revealed differences over the scope of the enterprise. Roatta and his staff quickly came to suspect that Franco did not want the Italian troops to secure a decisive victory but only to alleviate the pressure on Orgaz's forces after the bloody stalemate over the Jarama. The Italians regarded the Corpo di Truppe Volontarie as a force of elite shock troops and were determined not to see it worn down in the kind of piecemeal attrition favoured by Franco.[15] Anxious to get the Italians into action, on 1 March, Franco effectively agreed to the Italian plan to close the circle around Madrid, with a joint attack south-west from Sigüenza towards Guadalajara backed up by a north-eastern push by Orgaz towards Alcalá de Henares. He assured Roatta that his forces in the Jarama would operate at the same time as the Italian assault provided that they could be reinforced by one of the newly formed Italo-Spanish mixed brigades. Aware of the weakness of Orgaz's depleted troops, and fearing that they might not be ready for some days, on 4 March Roatta sent the second mixed brigade to strengthen them.[16]

On 5 March, Roatta wrote to Franco, confirming what had been agreed four days earlier and informing him that the Italian forces would start their advance on 8 March. On the same day, Roatta received a reply from Franco couched in guarded and ambiguous terms which revealed a lack of optimism about the Italian hopes of a decisive break-through. Although accepting that Orgaz's forces would move to link up with the CTV at Pozuelo del Rey to the south-east of Alcalá de Henares, the Generalísimo implied that the extent of their advance would depend entirely on how much resistance they might meet along the way. Since Franco's letter made no mention of the date of the attack, Roatta took this to signify that he had accepted 8 March.[17] This seemed to be confirmed when, on 6 March, one of Orgaz's commanders, General Saliquet, ordered an advance in the Jarama towards Pozuelo del Rey for 8 March. On 7 March, the eve of the battle, Roatta telegrammed Rome to say that he was still expecting the supporting action promised by the Spanish forces.[18]

Despite different immediate expectations of what would come of the attack, both sides certainly went into the operation talking in similar terms of closing the circle around Madrid.[19] Deceived by the ease of his triumph at Málaga, Roatta was convinced that he could reach Guadalajara before the Republicans could mount any serious counter-attack. Nearly forty-five thousand troops were gathered in three groups for the main attack. 31,218 Italians in three divisions were to be flanked by two smaller Spanish brigades consisting of Legionnaires, Moors and *Requetés*, jointly under the command of General Moscardó, the hero of the Alcázar. Amply equipped with tanks, pieces of heavy artillery, planes and trucks, it was the most heavily armed motorised force yet to go into action in the war.[20] However, its advantages were diminished by technical deficiencies in the equipment and inadequate preparation of the troops. Mussolini wanted the three Italian divisions to act as a unit because he hoped that they would score up another victory which, like Málaga, would be attributed by the world to Fascism. The mood in the Nationalist headquarters was notably more pessimistic than that of Roatta and his staff. There was considerable resentment among the Nationalist officer corps of sarcastic remarks made by the Italians about why it had taken so long to capture a defenceless city like Madrid.[21]

On 8 March, the Black Flames division under the Italian General Amerigo Coppi broke through the thin Republican defences using the *guerra celere* tactics that had brought Roatta such success at Málaga. However, the Republic was better organized around Madrid than at Málaga. Moreover, as it became apparent by the evening of that first day that the Jarama front was quiet, the Republicans were able to strip that area unhindered and concentrate their forces against the Italians. As Coppi moved rapidly towards Madrid, dangerously exposing his left flank and over-extending his lines of communication, Republican reinforcements moved up unmolested by Orgaz's troops. The Italian position was further endangered by the slowness of the Spanish columns on their right.

In general, the Black Shirts were surprised by the strength of Republican resistance and by the weather. Inadequately clothed, many dressed in colonial uniforms, they were caught in heavy snow and sleet. Their aeroplanes stranded on muddy improvised airfields, they made excellent targets for the Republican Air Force flying from permanent runways. Light Italian tanks with fixed machine guns were shown to be vulnerable to the Republic's Russian T-26 with their revolving turret-mounted cannon.[22] Now desperate for the Spanish supporting attack from the south, Roatta sent violent protests to Franco who feigned powerlessness, informing him that he had had to exert all his authority to oblige Orgaz to make a token action on 9 March which would be followed by a full-scale attack on the following day. It was extremely implausible that Orgaz would oppose an order from Franco. Moreover, the attack which began on 9 March was on the tiniest scale and it was not followed up on either 10 or 11 March. On 11 March, Orgaz was replaced as overall commander of the armies around Madrid by Saliquet. On 12 March, Varela was replaced by General Fernando Barrón. On the same day, Roatta sent a message to Franco to say that, without the guarantee of some diversionary activity in the Jarama, he could not move since his advance was being blocked by Republican units taken from the Jarama front.[23]

The Italians later discovered that, until well into the battle, Franco had refused to give the order for Orgaz and Varela to advance in the Jarama, despite the fact that Barroso pleaded with him to do so. Franco tried to obscure this by having Roatta and Cantalupo informed that he had relieved Orgaz and Varela of their immediate commands specifically as a reprimand for the inaction of the Nationalist troops on the Jarama front. A slightly placated Mussolini telegrammed Roatta 'I hope that Saliquet will not imitate the immorality of his predecessor'.[24] However, there was no question of Orgaz and Varela being in disgrace. Varela was promoted to Major-General on 15 March and posted to take command of the Avila division and Orgaz was given the crucial job of creating the new mass army which Franco needed.[25] The fact that Franco felt able to move them suggests that he did not view the promised attack from the Jarama as a major priority.* Having removed Varela and Orgaz, Franco and Saliquet promised Roatta an attack in the Jarama valley for 12 March. This also failed to materialize. On that day, Republican troops counter-attacked and the Italian advance was halted with heavy losses just south-east of the village of Brihuega. Finally, there were attacks on 13, 14 and 15 March but on a very small scale.[26]

With the lines more or less stabilised, a much chastened Roatta accepted that the advance would get no further. Aware that his troops were at their best moving forward but easily demoralised when under attack, he was anxious to avoid a total debacle. Franco, however, avoided Roatta's frantic requests for a meeting in Salamanca. Finally, during the afternoon of 15 March, the Italian general caught up with Franco, Mola and Kindelán at Arcos de Medinaceli near the front. Roatta requested permission to withdraw his troops from the attack. His hope was that the small advance made could now be defended by Spanish troops. He

---

* Roatta learned later from Italians attached to Franco's staff that Orgaz had not received any orders for an attack in the Jarama valley until the evening of 7 March. That delay effectively meant that Orgaz could not obey them for some time since, up to the moment of receiving them, he had been instructed to advance only after the Italians had occupied Guadalajara.

recognized the poor defensive qualities of his own men and suggested that perhaps they could continue to advance further outside the capital, from north to south. The Generalísimo refused outright.

Franco was either culpably deficient in hard information or else maliciously determined to use the Italians as pawns in his preferred tactic of attrition. Contrary to all the evidence, he insisted that the Republic was 'militarily and politically on the verge of defeat' and that 'the complete solution be sought in the region of Madrid, with the continuation pure and simple of the operations in course'. Roatta argued that further operations on the immediate Madrid front were doomed to failure given the apparent paralysis of the Nationalist forces in the Jarama, the sheer scale of Republican resistance and the exhaustion of the CTV. Franco simply refused to budge. He had had to accept the imposition of a joint general staff, the deployment of autonomous Italian units, the humiliating insinuation that Mussolini could run his war better than he could and the possibility that the victory would be won by the Italians to the detriment of his own political ambitions. His reluctance to help Roatta, either by fulfilling his promise for the attack from the Jarama or by relieving his troops in the line, smacked of revenge. He was rubbing the Italians' noses in their earlier arrogant confidence that they could take Madrid alone and that the advance on Guadalajara would be a walk-over. He certainly seemed to be determined not to make any sacrifices of his own troops and happy to let the Italians exhaust themselves in a bloodbath with the Republicans.

At loggerheads, Franco and Roatta reached an uncomfortable and ambiguous compromise by which the Generalísimo agreed to the Italians resting until 19 March but not deciding firmly what would happen thereafter. On returning to his headquarters, a still seriously concerned Roatta wrote to Franco that to persist with the original plan would simply consume their best troops to little avail. He proposed instead the abandonment of the present operations and a regrouping for a future decisive operation. Franco began a series of consultations with his own generals.[27] During the lull, the Republicans counter-attacked again in force on 18 March. Unaware that disaster was imminent, Roatta again visited the Generalísimo in Salamanca. They rehearsed the arguments of three days earlier, with Roatta insisting that the Italian contingent should be replaced while Franco, ever obstinate in terms of giving up territory or admitting any kind of reverse, remained adamant that the Italians should renew the attack on Guadalajara.

While Roatta banged the table and complained violently about the missing offensive in the Jarama, Franco continued to maintain, either misguidedly or malevolently, that the Italians were massively superior in men and materials to the Republicans. As Franco was explaining why the assault on Guadalajara must be continued in some form or other, news arrived of a massive Republican assault.[28] The Italians had not used the lull to strengthen their defences which was a culpable negligence on the part of Roatta. Nevertheless, the ease with which they were overrun proves Roatta to have been correct in his contention to Franco about the relative weakness of his troops. The Republicans recaptured Brihuega and routed the Italians. Roatta returned to see Franco again on 19 March requesting that his 'shock troops' not be kept in a defensive function but be allowed to regroup and be used elsewhere. The Generalísimo refused. After further attacks, a personal appeal from Cantalupo finally persuaded Franco to substitute the CTV with Spanish units.[29]

Mussolini was outraged, declaring to Ulrich von Hassell, the German Ambassador in Rome, that he had informed the Italian command in Spain that no one could return alive until a victory over the Republic had wiped out the shame of this defeat. On the basis of Roatta's reports, he also blamed the Spaniards for failing to fire a shot to back up his forces and, in a telegram to Ciano, denounced the deplorable passivity of Franco's forces.[30] The reaction of Franco and his staff was a mixture of disappointment at the defeat and *Schadenfreude* at the Italians' humiliation. Italian fascist songs were sung in the Nationalist trenches with their words changed to ridicule the retreat. Nationalist officers at the headquarters of General Monasterio's cavalry in Valdemoro, including Monasterio himself and Franco's friend, the artillery officer Luis Alarcón de la Lastra, had toasted 'Spanish heroism of whatever colour it might be'. Yagüe made no secret of the fact that he was delighted to see the arrogant Italians brought down a peg or two.[31] Cantalupo advised Farinacci, who was still in Spain, that he ought not to risk returning to Salamanca.[32]

Roatta maintained thereafter that the ultimate defeat was fundamentally the consequence of Franco's failure to keep his word.[33] That view underestimates the ferocity of Republican resistance, the role played by the weather, the poor fitness, discipline, training and morale of the Italian troops and his own mistakes. Nonetheless, if the promised attack had materialised, the Republic would have been hard pressed to mount a defence and the outcome might have been very different. Significantly, Franco was anything but abashed by the defeat. On 23 March, talking to Colonel Fernando Gelich Conte, one of the Italian staff officers attached to his headquarters, he brushed it off as militarily irrelevant.[34]

In fact, there is every reason to suppose that he was not displeased by the huge cost to the Republic of its victory in such a crippling confrontation in which the corresponding cost to the Nationalists had been borne by the Italians.

It has been suggested that Franco connived at the humiliation of the Italians.[35] That is an over-simplification since he was too cautious to risk a defeat whose consequences could not be foreseen. It is more likely that, in his desire to let the CTV confront and wear down the Republican forces around Madrid, he miscalculated the risks of not throwing his promised forces into battle. He had little desire to see the Italians win a sweeping and rapid victory when his own plans focused on a war sufficiently slow to permit thorough-going political purges.[36] It is significant that, a month before the defeat, Cantalupo reported to Rome that Mola and Queipo had insinuated to Franco that his prestige diminished in inverse proportion to the success of Italian arms.[37]

Franco clearly felt that he was obliged to justify himself to the Duce. Accordingly, he wrote to Mussolini on 19 March a letter of self-exoneration containing a number of feeble and contradictory arguments. These ranged from alleging confusion over the dates for the launching of the Guadalajara offensive to an effort to diminish the gravity of the missing Jarama push by claiming that the Republican forces which had faced the CTV were dramatically smaller than, in reality, they had been.[38] He also sent a messenger to Cantalupo with an equally mendacious claim that, in fulfilment of the agreement reached with Roatta, he had ordered advances by Orgaz on 25 February and 1 March. According to this emissary, by the time of the 8 March advance on Guadalara, Orgaz had allegedly lost more than one third of his men and was unable to attack further. That had indeed been true two weeks earlier which is why Franco had importuned Faldella on 21 February to begin the Guadalajara offensive prematurely. If Orgaz's troops were so depleted, it would imply at best an irresponsible lack of co-ordination between Franco and Roatta and at worst culpable military incompetence on Franco's part in permitting the Guadalajara advance to take place in such circumstances. To make matters worse, in an interview with Cantalupo on 23 March, in an even more crass exercise of self-justification, Franco blamed everything on Orgaz for not speaking up about the weakness of his forces. But it was precisely because Franco had told Roatta about that weakness that the Italian commander had sent the second mixed brigade to reinforce Orgaz's troops on 4 March.[39]

The inescapable conclusion is that Franco sought to let the Italians bear the brunt of the fighting at Guadalajara while Orgaz's forces regrouped after the battering they had received during the battle of Jarama. The only possible mitigation is that he did so in the post-Málaga misapprehension that the Black Shirts were near-invincible. Whatever Franco's thoughts, Mussolini could see that he had been used but he had little choice but to continue supporting Franco. Guadalajara had smashed the myth of fascist invincibility and Mussolini found himself committed to Franco until the myth was rebuilt. Equally, however galling, it was now clear that it made more sense to work with Franco for a Nationalist victory than independently.[40] Shortly after his letter of exculpation, Franco had requested help for a huge assault on Bilbao. Ignoring remarks made by Roatta about the miraculous appearance of the necessary forces for Bilbao which had never materialized during the battle of Guadalajara, Mussolini ordered his commander henceforth to obey the instructions and directives of Franco. Italian forces would henceforth be distributed in Spanish units and subject to the command of Franco's generals. When Cantalupo informed him of this on 28 March, Franco was delighted. The Italian Ambassador found him as if 'freed of a nightmare'. Franco asked him to inform the Duce of his 'joy at being understood and appreciated'.*[41]

Guadalajara finally broke Franco's determination to win the war at Madrid and imposed upon him a momentous strategic *volte-face*. For the Republic, Guadalajara was only a defensive victory which nevertheless significantly delayed ultimate defeat. It was a huge Republican triumph in terms of morale. Much valuable equipment and thousands of prisoners were captured. Documents were found which proved that many of the Italians were regular soldiers and thereby destroyed the Nationalist lie that they were all volunteers.[42] However, the Non-Intervention Committee refused to accept this damning evidence of official Italian intervention because it was not presented by a country represented on the Committee. The hypocrisy of the Committee was brutally exposed when the Italian representative, Grandi, announced on 23 March that no Italian 'volunteers' would be withdrawn until Franco's victory was complete and final.[43]

Cantalupo and Faupel were in despair, seeing Guadalajara as an irresponsible squandering of the material superiority put in Franco's hands by Italian and German assistance.[44] Indeed, Cantalupo's depressingly

---

* Later the same evening, in an episode starkly revealing of his deviousness, Franco assured Roatta him that he had interceded with the Duce to prevent his being removed as Italian commander in Spain. The arrogant but intelligent Italian general was not fooled by this transparent attempt at manipulation.

realistic assessment that there was no prospect of an early victory led to his recall to Rome in early April after barely two months in Spain.*[45] Franco, however, was jubilant that the consequences of his military miscalculations and his readiness to deceive the Italians were entirely positive. Interviewed on 22 April by H.R. Knickerbocker, he issued his considered judgement on Guadalajara and its military implications. He denied that there had been any defeat. When asked about his strategic conclusions, he replied 'Wars will not be won or lost in the air, although aircraft will play an ever more important role in future wars. Tanks are relatively useful and have, of course, a role in battle but it is only a limited one.' Further revealing his distance from the great leaps in military thinking which were taking place during the second half of the 1930s, he announced a military credo that would have been appropriate in the Middle Ages, 'success, after all, is found where there is the intelligent expertise of the commander, the bravery of the troops and faith'.[46]

The lesson to be drawn from the contrasting results of the near-contemporaneous battles for Málaga, the Jarama and Guadalajara was clear. The Republic was concentrating its best-trained and equipped troops in the centre of Spain and leaving other fronts relatively neglected. Against the Republican Army of the Centre, the Nationalists were achieving only small gains at the cost of massive bloodshed while against the militias of the periphery, substantial triumphs could come relatively easy. Accordingly, there was a case for desisting from the obsessive concentration on Madrid and destroying the Republic by instalments elsewhere. Colonel Juan Vigón Suerodiaz, chief of Mola's general staff, had written to Kindelán on 1 March 1937 to persuade him of the importance of ending the war in the north quickly. His main argument was the immense value to the Nationalist cause that could be derived from the seizure of the coal, iron and steel reserves and the armaments factories of the Basque provinces. He called on Kindelán to plead with the Generalísimo to make a priority of operations in the north.[47]

Despite the strength of such arguments, Franco was initially deaf to Kindelán's advocacy of Vigón's plans and remained obsessed with Madrid. The commander of the Condor Legion, General Hugo Sperrle, put forward similar arguments with greater insistence and held out the prospect of co-ordinated ground/air operations. It took the news of the defeat at Guadalajara to change Franco's thinking.[48] However, even after he had finally succumbed to the pressure from Sperrle and Vigón and accepted

---

* He was later replaced by the Conte Guido Viola di Campalto.

that the defeat of the Republic must be sought somewhere other than the outskirts of Madrid, Franco never entirely broke free of his obsession with the capital. He refused consistently to accept the sound advice of his general staff to make a number of small tactical retreats around Madrid to more defensible positions in order to release large numbers of troops for other fronts.[49]

The defeats around Madrid were interpreted by Franco's German advisers as pointing to the necessity for a large-scale modern army. Finally and reluctantly persuaded of the need to resort to mass conscription, Franco gradually began the process which would see one million mobilized under his command by the end of the war. The job of recruitment was given to General Orgaz, who had so successfully organized the supply of Moorish mercenaries.[50] The immense expansion of the armed forces gave an added urgency to the campaign in the north, where the prize was the heavy industry and the armaments production needed to equip the new recruits.

Accordingly, on 20 March 1937, after hearing about the scale of the Italian retreat, Franco finally responded to the pressure which had been building up from Mola and Sperrle for a major assault on the Basque Country. He believed that resistance in the north would be slight especially after the assurances made by Sperrle about the difference that would be made by concerted airborne ground attacks by the Condor Legion. On 22 March, the Generalísimo presented Kindelán with a sketchy outline of his immediate plans. They called for a portion of the Army besieging the capital to dig in near Sigüenza where the Italian advance had been blocked, and for a huge new force to be massed to attack and take Bilbao. On 23 March, he summoned Mola to Salamanca and gave him specific orders for the assault on Bilbao which derived from Vigón's suggestions and Sperrle's proposals.[51]

The operational details were hammered out at meetings held on 24 and 26 March involving General Kindelán, as head of the Nationalist Air Force, General José Solchaga and General López Pinto as field commanders, Vigón as Mola's Chief of Staff and Colonel Wolfram von Richthofen, the Condor Legion's Chief of Staff.* Richthofen held out to his Spanish counterparts the prospect of a novel strategy of 'close air support', using aircraft for sustained ground attack to smash the morale of opposing troops.[52] Accordingly, arrangements were made at these meetings for

---

* Richthofen was a cousin of the First World War air ace, Manfred von Richthofen, the 'Red Baron'.

continuous and rapid liaison between the headquarters of the Spanish ground forces and the Condor Legion. Two hours before any attack, the Air Force commanders would inform the ground headquarters in order for the necessary co-ordination to take place. It was also agreed at these meetings that attacks would proceed 'without taking into account the civilian population'.[53]

Mola gathered a large army consisting of African Army units, of *Requetés* now fully militarised as the Navarrese Brigades and of mixed Spanish-Italian brigades. It was backed by the air support of the small but well equipped Condor Legion and of units of the Italian Aviazione Legionaria under Richthofen's command.[54] Progress towards the full integration of the Italian troops was facilitated by the recall of Roatta and Faldella and their replacement by General Bastico and Colonel Gastone Gambara. However, given the scale of the reorganization undertaken by Bastico, the CTV was not ready for large-scale action until the end of April.[55] After Guadalajara, the Germans were keen to show their superiority over the Italians and to practice and develop their techniques of ground attack from the air. In this context, the relationships between Mola and Sperrle and between their chiefs of staff, Vigón and von Richthofen, were constant and close if not exactly cordial. The Condor Legion was theoretically responsible directly to Franco.[56] Sperrle was more meticulous than Roatta had ever been about observing that hierarchy and enjoyed generally good relations with Franco in consequence.[57]

In practice, however, the need to integrate joint air/ground operations on an hour-by-hour basis rendered liaison with Salamanca impracticable. So, content with Sperrle's deferential manner, Franco allowed him a free hand to liaise directly with Mola and Vigón, except on major issues. Franco was delighted to be able to consider the crack Condor Legion as part of *his* forces and to sit back and take the credit for its achievements. In the field, Mola and Vigón were also happy to accept the help and advice of Sperrle and Richthofen and the consequence was that, with Franco's conscious acquiescence, the Germans had the decisive voice in the campaign.* Sperrle wrote in 1939, 'All suggestions made by the Condor Legion for the conduct of the war were accepted gratefully and followed.' While the advance was being planned, von Richthofen wrote

---

* Lieutenant-General Attilio Teruzzi, recently arrived in Spain to take command of a division of Black Shirts, wrote to Ciano to complain that 'Franco does not command his generals: he requests and they sometimes do what he asks and often refuse or else do part of what he wants and then badly' (Teruzzi to Ciano, 6 April 1937, ASMAE, Spagna Politica Fascista, b.84).

in his diary on 24 March, 'we are practically in charge of the entire business without any of the responsibility' and, on 28 March, 'I am an omnipotent and effective commander (*Feldherr*) . . . and I have established effective ground/air command.'[58]

Confident of German backing, Mola opened the campaign on 31 March by deploying the weapon of mass fear which had been so effective for Franco in the rapid advance on Madrid. He issued a proclamation which was broadcast and also printed in a leaflet dropped on the main towns. It contained the following threat: 'If submission is not immediate, I will raze Vizcaya to the ground, beginning with the industries of war. I have the means to do so'.[59] This was followed by a massive artillery and aircraft bombardment in which the picturesque country town of Durango was destroyed. 127 civilians died during the bombing and a further 131 died shortly after as a consequence of their wounds. As was later to be the case with the more notorious bombing of Guernica, Salamanca denied that the raid on Durango had taken place and attributed the damage to the Basques themselves.[60]

Nevertheless, progress on the first three days was so slow that Sperrle sent a report to Kindelán in which he complained that 'if the troops do not advance faster, we will not enter Bilbao'. Sperrle believed that Franco had retained too much artillery and infantry on the Madrid front.[61] On 2 April, Sperrle and Richthofen complained to Mola. Equally anxious to speed things up, Mola suggested to him that the industries of Bilbao be destroyed. When the German commander asked why it made any sense to destroy industries which it was hoped to capture shortly after, Mola replied 'Spain is totally dominated by the industrial centres of Bilbao and Barcelona. Under such a domination, Spain can never be cleaned up. Spain has got too many industries which only produce discontent', adding that 'if half of Spain's factories were destroyed by German bombers, the subsequent reconstruction of Spain would be greatly facilitated'. In response to the notion that Spain's health required the elimination of the industrial proletariat,* Sperrle pointed out that the German Air Forces in Spain would attack factories only when Franco gave them specific orders to do so. According to Richthofen, Mola told Vigón to issue the order. Richthofen said that it had to come from a higher authority. Mola then signed orders himself for attacks on Basque industrial targets. Richthofen agreed to bomb the explosives factory at Galdácano on the 'next free

---

* A notion reminiscent of the deranged ideas of Millán Astray's bizarre press officer, Captain Aguilera.

day'. Sperrle and Richthofen, however, informed Franco and awaited his permission to carry out Mola's orders.[62] Sperrle even offered to put an aircraft at Franco's disposal for him to come to Vitoria to discuss the situation.[63]

In expecting the entire north of Spain to fall in under three weeks, Franco and Mola had underestimated the determination of the Basques. They were both disconcerted by the slowness of the first stage of their advance towards Bilbao's unfinished 'iron ring' of fortifications.* By 8 April, the Nationalist forces had completed only the first part of their planned offensive. After intense bombing on 4 April, they occupied the village of Ochandiano where the Basques had temporarily established their field headquarters and the heights to the north which they had intended to do on the first day. Steep, wooded hills, poor roads and heavy rain and fog had held up the advance of General Solchaga's troops. Franco visited the front, ostensibly to witness the triumph, but in fact to resolve the differences between Mola and Sperrle.[64] While he was in the north, Mola announced that it would be necessary 'to destroy systematically the war industries of the province of Vizcaya. To this effect, on 9 April we will begin the complete destruction of the power station at Burceña, the steelworks of Euskalduna and the explosives factory of Galdácano.' It seems that Franco had given permission for the partial implementation of the order signed by Mola on 2 April.[65] The dogged Basque retreat continued to exact a high price from the attacking forces but the terror provoked by artillery and aerial bombardment and political divisions within the Republican ranks ensured the gradual collapse of Basque resistance.[66]

In the early days of the Basque offensive, in the evening of 4 April, Franco received the Italian Ambassador Cantalupo for what would be their last meeting. He laid out with surprising candour the philosophy of his war effort. In a tone of regal condescension which was coming to dominate his manner away from the battle front, he spoke of himself in the third person: 'Ambassador, Franco does not make war on Spain but is merely carrying out the liberation of Spain. . . . I must not exterminate an enemy nor destroy cities, nor fields, nor industries nor production. That is why I cannot hurry.'[67] Franco had no doubts that the 'liberation' of his 'Spain' signified, as his actions showed, the thorough-going repression of all liberal and leftist elements. However, his remarks suggest

---

* The ring had been under construction since October 1936. The military engineer responsible for building the ring, Captain Alejandro Goicoechea, had deserted to the Nationalists on 27 February 1937.

that he was dwelling on the wisdom of going along with Mola's manic determination to annihilate Basque industry on which Sperrle had consulted him. The differences between Franco and Mola over the appropriate targets in the northern campaign do not indicate humanitarian preoccupations on the part of the Generalísimo. For Franco, 'Spain' had an entirely partisan meaning. He was reluctant to damage the material interests of his 'Spain'. After all, one of the central reasons for the war against the Basques was to secure a substantial industrial base, arms factories and mineral wealth, a point he had stressed at some length to Cantalupo a week earlier.[68]

There were implied rebukes for Roatta and Mussolini when Franco said, switching from Spanish to French, 'If I were in a hurry, I would be a bad Spaniard, and I would be comporting myself like a foreigner.' The rebuke was more political than moral. A week earlier, Cantalupo had reported to Mussolini that Franco resented Roatta's patronizing attitude not least because the Italian general did not understand 'the theory of the war in Spain'. In other words, the rapid victories sought by Roatta and Mussolini would not serve Franco's conception of the needs of his 'Spain'. The Caudillo's most revealing statement in this regard followed immediately, 'in sum, I must not conquer but liberate and liberating also means redeeming'.[69]

It was a statement in which messianic arrogance jostled with an icy readiness to put thousands of his countrymen to the sword. He made a reference to the destruction of the small Basque town of Durango four days previously by aircraft of the Condor Legion flying at his orders. 'Others might think that when my aircraft bomb red cities I am making a war like any other, but that is not so. My generals and I are Spaniards and we suffer in fulfilling the duty which the *Patria* has assigned to us but we must go on fulfilling it.' Referring to 'the cities and in the countryside which I have already occupied but which are still not redeemed', he declared ominously that 'we must carry out the necessarily slow task of redemption and pacification, without which the military occupation will be largely useless. The moral redemption of the occupied zones will be long and difficult because in Spain the roots of anarchism are old and deep.'[70] The kind of moral redemption which he had in mind, already seen in Badajoz and Málaga, more than explained the need for slowness. It would guarantee that there would never be any turning back, not only through the physical elimination of thousands of liberals and leftists but also in the long-term terrorizing of others into political support or apathy.

As the Generalísimo made blatantly clear, military decisions were now entirely subordinate to these wider political considerations. Accordingly, 'I limit myself to partial offensives with certain success. I will occupy Spain town by town, village by village, railway by railway . . . Nothing will make me abandon this gradual programme. It will bring me less glory but greater internal peace. That being the case, this civil war could still last another year, two, perhaps three. Dear ambassador, I can assure you that I am not interested in territory but in inhabitants. The reconquest of the territory is the means, the redemption of the inhabitants the end.' With a tone of helpless regret, he went on, 'I cannot shorten the war by even one day . . . It could even be dangerous for me to reach Madrid with a stylish military operation. I will take the capital not an hour before it is necessary: first I must have the certainty of being able to found a regime.'[71]

Despite these opinions, Franco was perplexed by the progress of the campaign in the north. Sperrle and Richthofen were also frustrated with the slowness of the advance. Since the beginning of the campaign, Richthofen had experimented with terror bombing to break the morale of the civilian population and to destroy road communications where they passed through population centres. This tactic had begun with the destruction of Durango on 31 March and been followed up by the attack on Ochandiano. In stating to Cantalupo that, when his aircraft bombed Republican towns, he and his generals were merely fulfilling their patriotic duty, Franco was admitting that he approved of such terror bombing. How fully Franco understood the German strategy is another matter. Just as his clashes with Roatta had derived from his desire to split up Italian forces and use them piecemeal on several different fronts, now there were similar strategic differences with the Germans. On 12 April, Franco disconcerted Sperrle by requesting that he send him all the aircraft which he was not using in the north to be used around Madrid. Under orders from Berlin not to split his forces, Sperrle offered to leave the Basque campaign and transfer the entire Condor Legion to the centre. Only after Colonel von Funck, the German military attaché in Salamanca, had laboriously explained to him the strategic thinking behind the German operation did Franco refuse the offer and order Sperrle to remain in the north.[72] The episode reveals not only the limitations of the Generalísimo's strategic vision, but also that Sperrle was still directly responsible to him.

On 20 April, the Nationalists began the second phase of their offensive and German air support was to play an even more crucial role. Sperrle, Richthofen, Mola and Vigón were sufficiently frustrated by the slowness of the advance to talk again of reducing Bilbao to 'debris and ash'.[73] In the event, the great morale-destroying blow would fall not on the Basque capital but on another smaller, more manageable but equally significant target. By 24 April, after merciless air bombardment and artillery pounding, the Basque forces were falling back in some disarray.[74] In the course of 25 April, as indeed throughout the entire campaign, Richthofen and Vigón were in constant contact by telephone co-ordinating aircraft, artillery and infantry. They agreed on the need to try to bottle up the retreating Basques around Guernica and Marquina. In the evening of 25 April, Richthofen again telephoned Vigón and arranged to see him at 7.00 a.m. the following morning. He wrote in his diary 'units ready for tomorrow'.[75]

They talked again at 6.00 a.m. on the morning of Monday 26 April and then met, as arranged, at 7.00 a.m. After close consulation with Vigón, Richthofen organized a series of bombing attacks aimed at impeding the retreat of the Basque forces. He seems to have decided to combine the tactical objective of blocking the retreat south of Guernica near Marquina with the broader strategic coup of the devastating blow to which Mola's broadcast had referred. Richthofen wrote in his diary of an implacable attack on the roads, bridge and suburbs of Guernica. 'There things must be closed up, it is necessary to secure finally a triumph over enemy personnel and material.'[76] Franco had made sufficient comments since 18 July 1936 about his belief that the Civil War would be won on morale for him to have few objections. If he had disapproved of what happened at Durango, Ochandiano and other villages, he had had ample time to put a stop to Richthofen's programme. His remarks to Cantalupo, in fact, make it clear that he approved and indeed took pride in what was happening.

From 4.40 to 7.45 in the late afternoon of 26 April, which was market day in the small town of Guernica, a blow consistent with Mola's threat was struck. Between the normal population, refugees and peasants coming in for the market, there were at least ten thousand people in Guernica on that day. The military authorities had tried to suspend the market because of the war but many peasants from surrounding hamlets had arrived as usual. The town had no anti-aircraft defences. It was annihilated in three hours of sustained bomb attacks by aircraft of the Condor Legion and the Italian Aviazione Legionaria under the overall command of Richthofen. The raid was carried out by twenty-three Junkers Ju 52s, four new He 111s, ten Heinkel He 51s, three Savoia-Marchetti S81 Pipistrelli and one Dornier Do 17 escorted by twelve Fiat CR32s and possibly six of the first ever Messerschmitt Bf109s.[77] It was an operation of a scale which could hardly have been organized by the Germans behind the backs of the Spanish staff with whom there was, in any case, constant liaison. Terrified civilians who fled into the surrounding fields were strafed by the machine-guns of Heinkel He 51s. The number of victims will never be known for certain because of the chaos and the fact that the Nationalists had captured the town before the debris was cleared. The Basque government estimated that 1,645 people died and a further 889 were injured in the bombing. The truth seems to lie nearer to that figure than to the much lower numbers suggested by Franco's propagandists. If the smaller scale bombing of the smaller town of Durango had left 258 dead then the number of victims at Guernica must have been considerably higher.[78]

Guernica was the ancient capital of the Basque country and of deep symbolic importance to the Basque people, a fact of which Mola and Vigón were fully aware. Richthofen's concern was less with political symbols and more with the dissemination of terror and chaos in the Republican rearguard. On 27 and 30 April, he expressed his disgust with Mola's forces for failing to move up and take advantage of the opportunity which his fliers had created, 'the town was completely blocked for at least twenty-four hours, it was the ideal precondition for a great success, if only the troops had been thrown in. As it was, just a complete technical success of our 250 kgs (explosive) and ECB1 (incendiary) bombs'. The incendiary effects of the attack cannot have been an side effect. The weight of bombs dropped on Guernica was the equivalent of half of the tonnage dropped by the entire Condor Legion on the crucial opening day of the campaign when it was necessary to make an early break-through. Moreover, Richthofen personally selected the unusual bomb load of explosive 'splinter' and incendiary bombs in line with his belief that while bomb craters in roads could be filled, massive destruction of buildings was a more effective obstacle to retreating troops.[79]

As the first such destruction by bombing of an 'open' town, Guernica was burned into the European conscience as Franco's great crime. The scale of the outrage felt by opponents of fascism was compounded by subsequent efforts on the part of the Nationalists to deny any responsibilty. However, the widespread and enduring outrage over Guernica might not have damaged the Generalísimo's cause in the way that it did had it not been for the fortuitous presence of three British and one Belgian reporter in the vicinity and the arrival of the articulate Basque priest Father Alberto Onaindía during the bombing. It was the subsequent efforts of Franco's propagandists to deny the destruction of the town attested by these and many other eyewitnesses which turned Guernica into a propaganda disaster for Franco. George Steer, correspondent of *The Times*, was one of those first four journalists to arrive at the scene. His report, published two days later in both *The Times* and *The New York Times*, provoked a world-wide storm of concern.[80]

Franco's foreign press service, under the direction of Luis Bolín, immediately denied that the bombing had taken place. Radio Nacional broadcasting from Salamanca claimed that there were no German or other foreign aircraft in Nationalist Spain. Although the Nationalists knew that Guernica had been destroyed on 26 April, they issued a statement to the effect that bad weather had prevented their Air Force flying on 27 April and that therefore they could not have bombed Guernica. When it quickly became obvious that outright denial was no longer tenable, the Nationalists claimed that Guernica had been dynamited by the Basques themselves. That story was maintained by some even up to the 1990s.[81]

The Generalísimo was himself credited at the time with penning the first denial. It has been claimed that Franco was appalled to discover later that he had been lied to by both Bolín and the Germans and that he shouted 'I will not have war made on my own people'.[82] If he made the remark, which is extremely unlikely, it can only have been in a spirit of duplicity. Not only did it imply a total *volte-face* relative to his activities since 17 July 1936 but also ignored the close liaison between the Condor Legion and his and Mola's headquarters. Franco's attitude to the moral necessity of wiping out the enemy was stated publicly often enough.[83]

It is not plausible that the Caudillo entertained qualms about the bombings against the 'red separatists' of the Basque country. He wrote a letter of thanks and congratulation to Sperrle and Richthofen for their help during the campaign.[84] In fact, all the available facts suggest that if either Franco or Mola was appalled it was because of the controversy and the damaging publicity which it generated. The only difference about Guernica was the thoroughness of the destruction and the presence of war correspondents and Father Onaindía. When the Nationalists reached Guernica on 29 April, the Carlist Jaime del Burgo asked a lieutenant-colonel of Mola's staff 'was it necessary to do this?'. With extraordinary violence, the officer barked 'this is what has to be done with all of Vizcaya and with all of Catalonia'.[85] When Salamanca began the cover-up, the pilots of the Condor Legion were ordered to deny the attack on Guernica.[86] Franco himself, having once denied the events, stuck to his guns.

Once the international furore began, he was not prepared to admit either that he had knowingly or unknowingly given a free hand to the Germans to carry out such an atrocity.

On 7 May General Sperrle, using his pseudonym Sander, telegrammed Franco to ask if his own enquiries had produced results which might allow the German government to accept British proposals for an international investigation into the events of Guernica. The Condor Legion had already sent in a team to remove bomb fins, unexploded bombs and other signs of the bombardment. Sperrle was effectively asking Franco to back him up *vis-à-vis* an embarrassed Ribbentrop. The Generalísimo replied immediately with a telegram which stressed the existence of a small-arms factory in Guernica and stated that 'Units of our front line requested the Air Force to bomb the crossroads, a request fulfilled by German and Italian aircraft, and because of poor visibility caused by smoke and clouds of dust, bombs hit the town. Therefore it is not possible to permit an investigation. The reds took advantage of the bombing to set fire to town. An investigation constitutes a propaganda manoeuvre to undermine Nationalist Spain and friendly nations.'[87] What was most striking, apart from the admission that the raid was requested by the Spaniards, was the lack of any suggestion that the German local command took independent initiatives or of any reproach for the scale of the bombing.

If Franco was as furious with Sperrle as has been suggested, it is extremely odd that he did not use this opportunity to hasten his recall to Germany. Franco's telegram suggests an anxiety to exonerate the Condor Legion of any suggestion of insubordination, lest the international repercussions induce Hitler to withdraw his forces from Spain. The fact that Franco effectively told Sperrle to lie to his superiors about the bombing and its consequences suggests that attack was planned with the approval of Salamanca and without the knowledge of Berlin. That Franco and Sperrle should take part in this conspiracy of silence suggests at the very least a high degree of complicity between them.[88]

Many of the inconsistent points of Franco's telegram were to be repeated for many years by his propagandists and in his own speeches.[89] Coming from the Generalísimo, they point only to mendaciousness or, at best, culpable ignorance. It is not credible that the Nationalists would want to destroy an arms factory that they were on the point of capturing. That the target was the crossroads or the Rentería bridge over the River Mundaca is contradicted by the weight of bombs dropped and the fact that a high proportion were incendiaries, ineffective on stone but appropriate for terrorizing the residential sector of the town which was largely of wooden construction. It is hardly surprising that there was smoke if incendiaries were dropped on a town constructed largely of wood.[90] The Basque army was indeed retreating along the roads towards Guernica, but it had not yet reached the town. Franco appears not to have asked himself why the Basque Republicans would dynamite the town and do precisely what Richthofen hoped to do – cut off the retreat by placing a massive human catastrophe in their path.[91] If the aim of the alleged saboteurs was to deny the town's advantages to the Nationalists, then neither at the time nor later did Franco seem to reflect on the curious fact that the small arms factory and the crucial bridge remained intact when Mola's forces arrived three days later on 29 April. These were remarkable omissions by dynamiters carrying out a scorched earch policy.*

The controversy made it a central symbol of the war, immortalised in the painting by Pablo Picasso. That Guernica was destroyed by explosive and incendiary bombs dropped from aircraft of the Condor Legion piloted by Germans is no longer open to any dispute. Moreover, there can no longer be any doubt that the atrocity was carried out at the behest of the Nationalist high command, and not on the initiative of the Germans.[92] Even if the bombing was not undertaken at the specific request of, but merely tolerated by, the Nationalist high command in order to destroy Basque morale and undermine the defence of Bilbao, it would make little difference to Franco's overall responsibility. The Condor Legion was in Spain at his request and Sperrle was directly subordinate to him.[93]

The only issue still debatable is the level of Franco's detailed liaison with Sperrle. The request was not unique but formed part of the consistent policy of air/ground co-operation elaborated since late March by Vigón and Richthofen. Its broad purpose throughout the campaign and, in the specific case of the devastation of Guernica, was to shatter Basque morale. That longer term purpose was confirmed two days later when Mola publicly linked the fate of Guernica with that of Bilbao. It was reported that he had declared that 'we shall raze Bilbao to the ground and its bare desolate site will remove the British desire to support Basque Bolsheviks against our will'. A shiver of fear ran through Bilbao.[94]

---

* Franco was visited a couple of days later by the Marqués del Moral, Frederick Ramón Bertodano y Wilson, an Anglo-Spanish enthusiast of his cause. Moral, who believed the story about Basque dynamiters, was distressed by the damage being done to the Nationalists by reports about the bombing. He went to Salamanca to beg Franco to consent to an enquiry to allow the 'truth' to come out. Naturally, the Generalísimo refused and promised only to renew previous statements in other forms, *Dez anos de política externa (1936–1947) a nação portuguesa e a segunda guerra mundial* I (Lisbon, 1961) pp. 333–4.

X

THE MAKING OF A DICTATOR

Franco & the Unificación, April 1937

THROUGHOUT the early spring of 1937, the Caudillo was becoming more aware than ever of political divisions within the Nationalist zone. In part, that was an immediate consequence of the decision to create a new mass army of which the most numerous elements would be militant Falangists and Carlists. However, he was concerned not just with the impact of those divisions on the war effort but also with the problems and opportunities that such frictions offered for his own political ambitions. In fact, by comparison with his Republican enemies, Franco faced only minor problems of internal political rivalry. Both Queipo and Mola bitterly resented his growing power but Franco had realized that their capacity to challenge him diminished the longer he remained in power.[1]

Nonetheless, Queipo had begun, in the aftermath of the victory at Málaga and the conquest of large areas of Andalusia, to build his own autonomous power base for a future bid against Franco. The Generalísimo was sufficiently worried to send Nicolás Franco to Seville in an unsuccessful attempt to cut Queipo's links with the local oligarchy. Queipo remained a problem if not a threat. Mola too was beginning to move politically, making speeches about the future political organization of Spain devoid of any references to Franco's leadership.[2] Cardinal Gomá informed the Vatican that Mola's energy and political astuteness would ensure him a significant role in the political future of Nationalist Spain.[3]

Given such rivalries, control over the political groups which supplied the bulk of the militia was an issue which mattered to Franco. Superficially at least, all other questions were relegated while the soldiers got on with the job of winning the war. As the Nationalist position improved, despite the reverses around Madrid, a jockeying for post-war power could be discerned. Franco firmly believed in the need for unified military and political command, especially in wartime. Moreover, he was ambitiously determined that such command should remain indefinitely in his hands. In any case, perceiving political competition as akin to mutiny, the Generalísimo was determined to impose iron discipline upon his political subordinates and rivals. It was also understandable that, having become Head of State with German and Italian encouragement, he should seek to align himself even closer to his allies by mimicking their single-party systems. After all, as adulation intensified within the Nationalist zone, the Caudillo was coming to see himself as a providential national saviour in the manner of the Führer and the Duce.

The idea of uniting the various political forces had been in the air for some time in the Nationalist zone.[4] However, in early January 1937, it seems to have taken root in Franco's mind as a result of an Italian suggestion which came from Guglielmo Danzi, officially Italian press attaché in Salamanca but in reality an important representative of the *Partito Fascista Italiano* with access to the Generalísimo. Danzi telegrammed the Ufficio Spagna on 9 January 1937: 'Accepting my suggestion, General Franco has decided to found a political association of which he will be the official head . . . he will endeavour to unite the parties into a political body along the lines of the Fascist Party.'[5]

That task was rendered easier by two factors. To begin with, there had been a high level of political co-operation between the components of his coalition during the Second Republic. The Falange, Renovación Española and the CEDA all owed a large element of their ideologies to Carlism. Despite tactical differences, they shared a broad strategic aim of constructing an authoritarian corporatist State, with the working class regimented within a State-sponsored syndical organization. Falangist terror squads had been financed by the monarchists of Renovación Española. The reprisals provoked by their activities and the publicity thereby generated was used by the CEDA to denounce the Republic as a regime of anarchy. They had collaborated in the military conspiracy, in the rising and the war effort, intensely aware that their future survival depended on the success of the enterprise. Nonetheless, each group in the Nationalist coalition nurtured ambitions of putting its own special stamp on the future authoritarian regime which was the common goal. The monarchists wanted a restoration of a military monarchy along the lines of the Primo de Rivera dictatorship; the Carlists, a virtual theocracy under their own pretender; the Falange, a Spanish equivalent of the Third Reich.

The second advantage enjoyed by Franco was that each of the groups had been decapitated, in one way or another, by the removal of its undis-

puted leader. In the case of Renovación Española, the assassination of
Calvo Sotelo had deprived the monarchists of the one national figure
remotely capable of making good their lack of popular support. Thereafter, its leaders had been obliged to throw their full weight behind
Franco in the hope of maintaining some influence. In the case of the
Falange, Franco had merely had to give fate the slightest nudge by not
backing to the full attempts to save the life of José Antonio Primo de
Rivera. The massive inflation in its numbers rendered the Falange both
a desirable political asset and a potential threat. However, the lack of
political skill and sophistication on the part of the internal aspirants to
the party's leadership made them relatively easy adversaries for Franco.
In the case of the Carlists, Franco had already eliminated Fal Conde,
showing that he could lash out ruthlessly and decisively when he judged
the moment propitious.

That left the CEDA and Franco was thought to be politically closest
to Gil Robles's party. Certainly, whether or not it was true, Franco had
let it be known that he had been approached by the party to stand in the
1933 elections. He had been on the point of standing on a CEDA ticket
in the re-run elections in Cuenca in 1936. Much had been made of his
collaboration with Gil Robles in the Ministry of War in 1935 and Franco
had been unable to contain his tears when the CEDA leader was replaced.
However, Franco now made a point of distancing himself from Gil
Robles. This was partly because, in the frenzied atmosphere of the wartime Nationalist zone, the gradualist road to the authoritarian State
espoused by Gil Robles had come to seem like treacherous weakness, a
sentiment reflected in the massive migration of CEDA militants to the
Falange and the Carlists. Franco was not about to let himself be out of
step with large numbers of his supporters. At the same time, his incipient
hostility towards Gil Robles indicated that he now saw in the CEDA
leader a potential rival who should be eclipsed as soon as possible. Gil
Robles would thus be destroyed by a whispering campaign branding him
as the political wimp who had failed to smash the Left when he had the
chance.

Franco was happy to use Gil Robles's services abroad but he made it
clear that he was not welcome in the Nationalist zone. In the early months
of the war, Gil Robles had been in Lisbon, helping Nicolás Franco to
establish a Nationalist unofficial embassy or 'Agency of the Burgos Junta'.
Composed of various aristocrats, diplomats and rightist politicians, the
Agency organised the purchase of arms and other supplies, propaganda
and financial assistance for the rebel cause.[6] Gil Robles was particularly

successful in raising foreign currency for the Nationalists.[7] He visited rebel Spain on several occasions, between late July 1936 and May 1937, but met an increasingly hostile reception. In Pamplona on 28 July, where he had gone to collect his wife and son, he was insulted in his hotel by several aristocratic ladies and accused of being responsible for what was happening in Spain, an accusation which was to be made with increasing frequency.[8] After passing through Salamanca, he arrived in Burgos on 2 September 1936 and, within a matter of hours, a group of Falangists had tried to beat him up and arrest him. General Fidel Dávila, then Civil Governor of Burgos, requested instructions from General Cabanellas who ordered that the CEDA chief be given protection.[9]

Queipo de Llano predicted to Arthur Koestler in late August 1936 that Gil Robles would play no role in the future government of Spain.[10] Once Franco was made Chief of State on 1 October 1936, there was little political future for Gil Robles and it gradually became clear to him that the gap between them was unbridgeable. Nevertheless, he wrote to their mutual friend, the Marqués de la Vega de Anzó, on 26 October 1936, asking him to pass on to the Generalísimo his belief that 'the present moment demands the disappearance of all, and let it be understood well, of *all* parties' and accepted in advance the disappearance of the militia of the *Juventud de Acción Popular*, the CEDA Youth Movement. He then wrote to Franco directly on 2 November to announce that, in the interests of the necessary national movement, he was suspending all political activity on the part of the CEDA.[11]

It gradually became obvious that these efforts and sacrifices made Gil Robles no more welcome at Franco's headquarters, the so-called *Cuartel General*. In later years, Gil Robles came to believe that Franco could not tolerate having around someone who had been his superior.[12] As Franco worked to build up his own political power, the presence of a strong personality of enormous talent such as Gil Robles would necessarily have been unwelcome. His hostility to Gil Robles can be deduced from the paranoid and clearly apocryphal story that he told a Mexican journalist. The Generalísimo alleged that some young CEDA militants had asked Gil Robles for advice at the beginning of the war only to be told that they should stand aside and let the reds and the military rebels tear each other apart in order for the CEDA to step in and take over.[13]

For some time, Gil Robles continued loyally serving Franco's cause partly out of ideological commitment, partly in the hope of playing the kind of role assumed first by Nicolás Franco and later by Ramón Serrano Suñer, and, ultimately, because there was nothing else that he could do.[14]

On 10 February 1937, he was interviewed by the newspaper *Arriba España* and declared that 'the Movement which began on 17 July marks a new direction for the *Patria*. Once victory is achieved, political parties should disappear and be integrated into a single, broad national movement. When this happy moment arrives, Acción Popular, far from being an obstacle or a stumbling block, will be proud to facilitate the process.'[15] Effectively a broken reed within the politics of the Nationalist zone, Gil Robles would 'graciously' accept Franco's forced unification of the right-ist parties in April 1937. After the Civil War, he would become a central figure in the monarchist opposition to Franco.

By the beginning of 1937, then, Franco could face the problem of the seething rivalries within Nationalist ranks from an advantageous position. Having become Generalísimo and Head of State, he was too busy with the war to give much thought to the creation of a single party along Fascist lines. However, after Danzi made his suggestion, Franco skilfully began to prepare the ground. Fal Conde and Gil Robles were eliminated and, simultaneously, in conversations with the provisional Falangist leader, Manuel Hedilla, with the moderate Carlist leader, the Conde de Rodezno, and with numerous monarchists, Franco allowed them each to believe that only by supporting him would their particular interests be safeguarded. At the battlefronts, there was little friction between Falangist and Carlist militias. Moreover, influential elements in both parties real-ized that some kind of union was inevitable and preferred to take the lead rather than risk an imposed settlement. Equally, the sharper Alfonsist monarchists and those Cedistas who had not already migrated to the Falange were not averse to some kind of single party of which they hoped to become the general staff.

More ideologically extreme or less cynical elements tried to cling to their pre-war identities and thereby attracted the hostile attention of Franco and his security services. There were some clashes in the rear-guard. However, the removal of Fal Conde, by leaving the Carlists in the hands of the pragmatic Conde de Rodezno, cleared the way for negotiations with the Falange. Talks in February 1937 went well. How-ever, when moved to Portugal, they broke down on the intransigence of Fal Conde. That was hardly surprising since the Falangists seemed to be proposing to absorb the Carlists. Minutes of these meetings circulated openly in the Nationalist zone in mid-April 1937, made available, it was alleged, by Franco's headquarters so that he might step in with a solution when the two parties had failed.[16]

In the early months of the war, a mixture of improvisation, emergency and euphoria kept political differences in the background. Franco became both Generalísimo and Head of State before the first significant reverses were suffered by the Nationalists. However, after the failure to capture Madrid and especially after the defeat at Guadalajara, it was recognised that a long struggle was on the cards. There was widespread agreement that some form of political structure would have to be produced to unite the Nationalist zone. Once the process started, the political ferment intensified as certain elements realised that it was not just a stable political context for military victory which was at stake but the long-term political future.

The brains behind the creation of a new political movement, and, indeed, chief architect of the Francoist State, was Ramón Serrano Suñer, the general's brother-in-law, who reached Salamanca on 20 February 1937. Despite his role as liaison between Franco and the military consirators during the spring of 1936, Serrano Suñer had not been warned of the date of the uprising and he and his family underwent terrifying experiences in consequence. He had witnessed the murder of friends in jail in the Republican zone and himself only just evaded the *sacas* (the forcible removal of prisoners for illegal execution) which were to claim the lives of his two brothers, José and Fernando. Serrano Suñer managed to get out of the prison in Madrid, the *Cárcel Modelo*, by telling the Republican Minister of Justice, Manuel de Irujo, that he had had nothing to do with the Falange nor any political relationship with his brother-in-law, Franco.[17] His experiences had made him an impassioned and totally committed opponent of democracy.[18] During his brother-in-law's ordeal in the *Cárcel Modelo*, it had not occurred to Franco to do anything to initiate a prisoner exchange in order to help him escape.[19] Nevertheless, on hearing of his arrival at Hendaye on 20 February 1937, Franco sent a car to bring Serrano Suñer and his family to Salamanca. The Generalísimo invited them to move into an attic in the Palacio Episcopal where he had his headquarters.[20]

The slightly built and dashingly handsome Serrano Suñer, as elegant in his speech as in his appearance, had the talent and political credentials necessary to create the political machinery lacking in the Nationalist zone. He was one of the outstanding legal minds of his generation. A prominent figure in the *Juventud de Acción Popular*, Serrano Suñer had been instrumental in bringing over much of its rank-and-file to the Falange in the spring of 1936. However, apart from his acute intelligence and political experience, one of Serrano Suñer's major attractions to Franco as a potential instrument to tame the Falange was his lack of an

independent power base. Moreover, Doña Carmen, who regarded
Nicolás Franco with some distaste, was not unhappy to see him supplanted by her brother-in-law. She disliked Nicolás for his dissolute
bohemianism and for his eccentric working practices which were so different from those of her more methodical husband. She was also jealous of
the wife of Nicolás Franco, the vivacious Isabel Pascual de Pobil, who
cut much more of a dash in Salamanca society. It has been suggested that
Doña Carmen was outraged because gifts which she believed to be meant
for her, addressed to 'Señora de Franco', were delivered in error to this
other Señora de Franco.[21] Nicolás sensed danger in the arrival of Serrano
Suñer.[22] Doña Carmen admired Serrano Suñer as a learned academic,
lawyer and parliamentary deputy. Often in conversations in the small
family group, when the loquacious Franco interrupted his brother-in-law,
Doña Carmen would say 'Shut up Paco and listen to what Ramón is
saying', which Serrano Suñer was convinced planted the first seeds of
later resentments.[23]

From the moment that he was installed in the palace, Serrano Suñer
dedicated his entire energies and his powerful intellect to the cause for
which his brothers had died. For better or for worse, he decided that
he could best do that through Franco.[24] Serrano Suñer's considerable
ambition, however, was not so much personal as abstract, a commitment
to an idea. As he put it himself, after his experiences in Madrid, he was
'traumatized, depersonalized'.[25] Franco, by nature mistrustful, but aware
of his own political inexperience, was prepared to place his confidence
in Serrano Suñer. He was astute, as so often, in doing so. Serrano Suñer's
arrival in Salamanca constituted a significant elevation of the intellectual
level within the Nationalist leadership. The combination of his friendship
with José Antonio Primo de Rivera and his relationship with the Generalísimo made him someone to reckon with. That, together with his
introspective air of commitment to implementing the ideas of José
Antonio enabled Serrano Suñer to build a bridge between Franco and
many of the best and brightest of the Falange.[26] However, if he managed
to make Francoists of the Falange's leaders, he failed utterly to make a
Falangist of Franco who remained less interested in social programmes
than in maintaining his own power.[27]

When Serrano Suñer reached the Nationalist zone, its political life
was virtually non-existent beyond personal squabbles. It was what he
called 'un Estado campamental' (a field State). The south remained the
independent fief of General Queipo de Llano. In the north, the central
political authority was the Junta Técnica del Estado which had been created

on 1 October 1936. Its offices were spread throughout Burgos, Valladolid and Salamanca. The president was General Francisco Gómez-Jordana whose offices were in Burgos but the real power lay in the hands of Nicolás Franco at the head of the *Secretaría General del Estado* which operated in Salamanca alongside the Generalísimo's headquarters. Nicolás's office consisted only of himself, two under-secretaries, José Carrión and Manuel Saco, and the diplomat José Antonio Sangróniz. A *bon viveur* like Nicolás Franco, the rotund and shrewd ex-Cedista Sangróniz dealt with foreign matters.[28] Since the autumn of 1936, Nicolás Franco had been aware of the lack of a State structure. However, he had neither the energy, the technical juridical knowledge nor the will to set about building a State. Nicolás was in any case in no hurry to create anything which might diminish the *de facto* power of his brother which, after all, had no more juridical basis than his nomination by a small group of generals. Both Franco and his brother had an instinctive feeling that the passage of time and military victory would consolidate the Generalísimo's power and had assumed that a formal government structure could wait until the capture of Madrid. As that prospect receded, the shambolic nature of Nicolás's administration became more unacceptable.*

The way forward to underpinning Franco's personal power with a formal State structure complete with popular political support now opened up with the arrival of Serrano Suñer. Until his appearance, the Nationalist administration had been concerned primarily with military matters. Little or no time or thought had been given to the question of mass political, as opposed to military, mobilization. Throughout March 1937 Serrano Suñer had discussed the problem first with Franco and then with Mola, the Carlist Conde de Rodezno, the monarchist intellectual Pedro Sainz Rodríguez and the Primate, Cardinal Gomá. Amongst those contacted by Serrano Suñer was the Falangist Manuel Hedilla, the provincial *Jefe* from Santander, who had been elected the *Jefe Nacional* of the Falange's provisional *Junta de Mando* (command council) on 2 September 1936. Their relations were anything but friendly.[29] Hedilla was a fascist thug who, if not as illiterate as his enemies maintained, was easily outmanoeuvred by Nicolás Franco and Serrano Suñer.[30]

Franco had long been wondering how to reduce the various political

* In early April 1937, Cardinal Gomá reported to the Vatican 'it is the conviction of every intelligent observer that the political aspect of the government is some way from offering the same guarantee of competence and good sense as the military' (María Luisa Rodríguez Aisa, *El Cardenal Gomá y la guerra de España: aspectos de la gestión pública del Primado 1936–1939* [Madrid, 1981] p. 153).

elements of the nationalist zone to a common denominator under his own leadership. The difference between the Generalísimo and his brother-in-law was that Franco, engrossed in the daily problems of how to win the war, saw a unification as merely a device to consolidate his political power. Serrano Suñer obliged him to think ahead and consider what kind of State should be constructed after the victory.[31] They often discussed these problems for several hours, in long strolls after lunch around the gardens of the Episcopal Palace. While the battle of Guadalajara and the campaign for the Basque Country were being waged, Franco's political future was being debated on these strolls. So engrossed did the Generalísimo become that his cousin and aide-de-camp, Pacón, worried that he was neglecting the military direction of the war.[32] Since Franco had few confidants, it was assumed that the *cuñado* (brother-in-law) Serrano Suñer was the *eminence grise* behind the Generalísimo. Serrano Suñer quickly acquired the matching nickname of *cuñadísimo* (supreme brother-in-law) and political visitors to headquarters tended to seek him out. He gave the instinctively cunning, but politically unlettered, Franco a kind of political education. The freedom and directness with which Serrano Suñer spoke to him, born of their long friendship and family links, could never be matched by the sycophants who soon surrounded the Generalísimo.[33]

There was considerable jealousy and hostility towards the proud and solitary Serrano Suñer, some political, some personal. The political simplicity of Lorenzo Martínez Fuset in particular was cruelly exposed. Others in the higher ranks of the military, especially General Alfredo Kindelán were anxious about the influence over Franco of someone of such apparently radical fascist views. Monarchist intellectuals and politicians, like Eugenio Vegas Latapie, Pedro Sainz Rodríguez and Antonio Goicoechea, were dismayed to see that the Falange, which they had become accustomed to run rings around, now had a serious champion.[34] Ironically, the group most threatened by the arrival of Serrano Suñer – the existing provisional leadership of the Falange – was the least worried.

As it was, the leadership of the Falange was a hotbed of personal rivalry and the man most feared was the naïve Manuel Hedilla. The construction of a personality cult around him was taken by other aspirants to the mantle of José Antonio Primo de Rivera to be a signal that Hedilla was harbouring ambitions to be in politics what Franco was in military terms. There is little doubt that the Generalísimo reached the same conclusion. The publication in January 1937 of an interview with Hedilla by the pro-Nazi journalist Victor de la Serna entitled 'Hedilla a 120 por hora' (Hedilla going at seventy-five miles an hour) clearly implied that Hedilla was on the point of overtaking all his rivals. Franco can hardly have been pleased by Hedilla's declaration in the course of the interview that 'I would rather have repentant Marxists than cunning rightists corrupted by politics and *caciquismo* [clientelism]'. An issue of a Falangist pictorial weekly, *Fotos*, was dedicated almost totally to Hedilla, something which can hardly have pleased Franco.[35]

Hedilla was altogether too radical for Franco's taste, making frequent declarations about the need to limit the excesses of capitalism and implying that, after the war, an overwhelmingly powerful Falange would do so.[36] In fact, even if Serrano Suñer had not arrived, there is little doubt that Franco would have manipulated Hedilla to his own satisfaction. Hedilla should have been suspicious when, at the end of February 1937, Franco said to him 'you know something, Hedilla? I have ordered a blue shirt'.[37] The Caudillo began to develop a patronizing relationship with the guileless Hedilla whom he regarded as agreeably malleable by comparison with his rivals in the Falange *Junta de Mando*. All connected in some way with the Primo de Rivera family, the so-called *legitimista* clique consisted of the loudly aggressive and unsubtle Agustín Aznar, head of the Falange militias and fiancé of José Antonio's cousin Lola, his dour sister Pilar, his ambitious cousin Sancho Dávila, the head of the Seville Falange, and his one-time law-clerk (*pasante*), the sinuous Rafael Garcerán, now provincial chief in Salamanca.[38]

Despite his preliminary talks with the various elements in the Nationalist camp, there was no effort at negotiation by Franco. That would have meant revealing his hand. Instead, he waited and watched while all the interested parties took up positions. There were rumours that Mola aspired to be head of the government, leaving the running of the war to Franco. José Ignacio Escobar met Mola at this time and found him deeply irritated with Franco and with himself for having so lightly ceded the position of Generalísimo.[39] When asked by Hedilla about the rumours, Mola was circumspect, since he suspected him of being Franco's man. Mola said, on the assumption that it would be carried back to Franco, no more than that 'my ambition consists in not taking on more things than I know how to do well. Perhaps I can run the campaign in the north with success. After all, war is my profession. But I am absolutely sure that in the present circumstances I would be a failure presiding over a government. You can tell whoever talks to you about such a possibility that General Mola never goes looking for certain failure.' Nevertheless,

Hedilla came away with the impression that Mola was cautiously hiding his real intentions.[40] If indeed Mola harboured ambitions, he was waiting until after he had clinched military success in the north before declaring himself. By the time he was ready, it would be too late.

In any potential rivalry between Mola or Queipo and Franco, control of the Falange would be crucial. Conversely, the rivalries between the various factions within the Falange were conditioned by the looming threat of a take-over by the Generalísimo. Decapitated by the execution of its founder, the Falange was in the throes of an increasingly bitter power struggle between Hedilla, as his designated successor, and the '*legitimista*' group of José Antonio's close friends led by Agustín Aznar and Sancho Dávila.* They, like Hedilla, were jockeying for position prior to Franco's feared move to unify the parties. Social snobs, they regarded Hedilla as too radical and too proletarian.[41] Sancho Dávila and Agustín Aznar were fiercely hostile to the idea of a unification of Falangists and Carlists under the auspices of Franco. Even José Antonio Primo de Rivera's sister, Pilar, told Hedilla, 'Be careful. The Falange must not be handed over to Franco. Don't give it away!'[42]

At this time, Hedilla was confident that he could leave behind Aznar, Dávila and company by means of a deal with Franco which would make him effective leader of the united party. He was led to believe that he had an arrangement with Franco. The Generalísimo took no part in this understanding, being careful to deal through several layers of intermediaries. Nicolás Franco, Ramón Serrano Suñer, José Antonio Sangróniz, Lorenzo Martínez Fuset and Lieutenant-Colonel Antonio Barroso, the Generalísimo's chief of operations, made up the political general staff operating from his headquarters or *Cuartel General*. The *Cuartel General* liaised with Hedilla through two eminently disownable figures, the mysterious Captain Ladislao López Bassa and the even more shadowy military doctor, Captain Vicente Sergio Orbaneja, a distant cousin of José Antonio Primo de Rivera.† By this deal, Hedilla thought that he would guarantee his own continued pre-eminence in return for promising not to oppose the unification and accepting a less radical social programme for the unified Falange. The Caudillo would be the figurehead as *Jefe Nacional* and Hedilla, although theoretically second-in-command, would be *de facto* head as secretary-general of the *Junta Política*, the executive committee.[43]

---

* They were backed by Falangists from Valladolid led by José Antonio Girón de Velasco and Luis González Vicén.
† The two had first made the acquaintance of Franco during his time as military commander in the Baleares.

Happy with such a deal, Hedilla at no time contemplated the possibility that he was being manipulated by a superior opponent.

To strengthen his position, in the first months of 1937, Hedilla not only maintained contact with Franco and his various emissaries but also curried favour with the Germans and Italians. His social radicalism attracted the representatives of the Nazi and Fascist Parties as much as it worried Franco. Hedilla had already had minor clashes with various generals over the scale of the repression, including one violent row with Mola. Both Nazis and Fascists saw the blanket Nationalist repression of the Left as short-sighted, believing that it made more sense to recruit a working class base for the regime. They were thus interested in fostering the career of Hedilla.[44] On his visit to Spain in March 1937, Farinacci invited Hedilla to visit Italy. Foreign patronage boosted Hedilla's self-confidence. At the beginning of February, his secretary, José Antonio Serrallach,* told the German Ambassador Faupel that the Falange cultivated Franco as the Nazis had manipulated Fieldmarshal Hindenburg prior to their take-over.[45]

Similarly, Hedilla told a sympathetic Cantalupo in early March that he and his followers were republicans and men of the Left. He described Franco contemptuously as a reactionary and said that, once Madrid fell, his Falange would come out in its true leftist colours.[46] Such indiscretions quickly reached Franco's ears. Through an intermediary, his close friend from Santander, Tito Menéndez Rubio, Hedilla made overtures to Rome, in search of Italian support in a possible power struggle. Menéndez Rubio visited Danzi, who acted as political liaison between Mussolini and Franco, and painted a picture of Hedilla as a charismatic leader of working-class origins who regarded Franco as incapable of leading the fascist Spain of the future.[47] Hedilla himself told Cantalupo that the Falange would tolerate Franco as Head of State while the war continued but not as head of a united party. In fact, Cantalupo regarded Hedilla as 'a modest man of little culture, mediocre spirit, with little impact'.[48] Although Danzi and Farinacci were working to accelerate the creation of a single party, Cantalupo was circumspect and did nothing that might provoke Franco.[49]

The Germans also cultivated Hedilla. Hitler sent him a signed copy of a luxury limited edition of *Mein Kampf*.[50] According to Cantalupo, Faupel despised Franco and, immediately after the battle of Guadalajara, told him on two separate occasions that it was necessary to eliminate Franco in order to give power to the Falange which, for the Germans,

* A Catalan Falangist educated in Germany and possibly a German agent.

meant Hedilla.[51] It is highly unlikely that Franco did not also get wind of all this particularly as he had recently put the security services of his headquarters on a much more offensive footing. He placed the organization of domestic espionage in the hands of Major Lisardo Doval, notorious for his activities during the repression in Asturias in 1934. Doval was a manic enemy of the Falange and believed the bulk of its members to be reds and freemasons in disguise. He had been given wide discretionary powers by Franco to build up files on the major figures of the Falange and to create a network of agents and informers to spy on them.[52] The Generalísimo left nothing to chance.

Under the overall strategic direction of Serrano Suñer, it was the job of the *Cuartel General* to handle the civil war in the Falange in such a way as to ensure that Franco would be the victor. Their strongest card was that, precisely because he was confident of the co-operation of Franco, the unsophisticated Hedilla felt able to concentrate on dealing with the threat from the Primo de Rivera family clique.[53] The greatest threat to his ambitions came from Franco yet Hedilla so severely underestimated his adversary as to let his own moves be directed by the Generalísimo's staff. A rumour was spread, probably by Nicolás Franco and Doval, to the effect that Aznar's group intended to assassinate Hedilla. On 12 April, Hedilla himself went to several northern cities to secure armed support. In San Sebastián, he attempted to recruit Colonel Antonio Sagardía, head of a column of militarized Falangists, to smash Aznar's militias. He also seems to have tried unsuccessfully to secure the co-operation of Yagüe.[54]

Also on 12 April, while Hedilla had been rustling up support in the north, Franco and his *Cuartel General* were busily making their own moves. Nicolás Franco complained to Danzi about the efforts of intriguers (*mestatori*) to upset the political balance in Nationalist Spain. In what was a clear message to the Italians not to foster divisions, Nicolás talked of Falangist machinations which did not help the Generalísimo's plans for unity. He spoke of 'a network of hidden, divisive dissidence of a sinister and antimilitarist character' and made it clear that Franco was about to act against those whom he held responsible. Nicolás reassured Danzi that his brother planned to unify all forces by decree into a fascist party called the Falange 'of which he would elect himself leader'.[55]

On the same day, 12 April, the Generalísimo received moderate Carlists, including the Conde de Rodezno, and informed them that he was preparing a decree instituting a single party.[56] There is reason to suppose, although no documentary proof, that Franco also had Hedilla informed of this intention. Hedilla was in regular contact with members of the inner cabinet at the *Cuartel General* throughout the tumultuous week from 12 to 19 April. His subsequent actions make sense if he had been led to believe that Franco planned to amalgamate the two parties, reserving for himself a symbolic leadership (*jefatura*) while leaving Hedilla as *de facto* chief. He would no doubt have been shocked had he learned that on 14 April, Franco told the German Ambassador that, given the poor leadership qualities of Hedilla, he was going to fuse the various parties and assume the leadership himself.[57] Cantalupo, now in Rome, informed Mussolini that Hedilla was a 'worthy adversary for Franco, like him a poor thing, a simple soul, of a seamless lack of culture, boneheaded and without political sense'. Cantalupo believed that Franco and his brother were plotting with the other faction of the Falange to eliminate Hedilla.[58]

Franco's machinations were certainly made easier by the internal power struggle inside the Falange. On 14 April, while Franco was talking to Faupel about Hedilla's incompetence, Hedilla himself met Sangróniz. The curtain was about to rise on the *Cuartel General*'s scenario. Sangróniz insinuated to Hedilla that, if he co-operated in the unification, he would end up as effective head of the united party while Franco concentrated on military affairs. The only condition was that he suppress the so-called '*legitimista* rebellion' of the Aznar-Primo de Rivera clique. This was what Hedilla was planning to do anyway and why he had been drumming up support in the provincial capitals of the north during the previous days. He told Nicolás Franco and Doval about his plans.[59] Hedilla believed that the best way to defeat Aznar and Dávila was to hold an extraordinary meeting of the Falange's supreme body, the *Consejo Nacional*, and have himself named *Jefe Nacional*. Encouraged by Faupel, and his own political secretary, José Antonio Serrallach, Hedilla also thought that, by ending the provisional status of his leadership position before the unification, he would improve his own bargaining position with Franco.[60]

Hedilla intended that the extraordinary meeting of the *Consejo Nacional* to elect a permanent party leader should take place on 25 April.[61] He was confident that such a meeting could be packed to ensure his election. Aware of what Hedilla was planning, possibly even informed of it by the *Cuartel General*, the anti-Hedilla group was quick to produce a counter-plan and began gathering its forces in Salamanca. Their strategy was to use the much smaller provisional executive committee, the *Junta de Mando* (command council), to pre-empt the *Consejo Nacional* meeting and replace Hedilla as provisional *Jefe Nacional* by a triumvirate. Both groups went onto a war footing. Just as Hedilla had brought in support from the north, Aznar's militiamen were placed in readiness and were backed up

by three car-loads from Seville armed with rifles, machine-guns and grenades. Nicolás Franco and Doval knew this was happening but they
refrained from blocking the entrances to the city precisely because any
disorder provoked by the squabbling Falangists would provide the perfect
justification for a military intervention and a take-over by the Generalísimo. Indeed, they kept Hedilla informed and possibly even exaggerated
the scale of Aznar's forces in order to push him to retaliatory action
which would escalate the conflict.[62]

On the morning of 16 April, the atmosphere in Salamanca was highly
charged. There were many armed Falangists in the city and it was assumed
that the clash so long rumoured was about to take place. Doval had some
of his own men dressed as Falangists. Aznar, Dávila and Garcerán were
going to expel Hedilla from the *Junta de Mando* on the grounds that he
had bypassed its collective authority and had betrayed the Falange by
agreeing to a unification on Franco's terms. It was known at Franco's
headquarters that this was about to happen – not just because of the
espionage activities of Doval but because Aznar and Garcerán deliberately
informed Doval, who was a friend of Aznar's father.[63] On Franco's desk,
military maps were replaced with minute-by-minute reports on the activities of the various factions. Hedilla was told by Barroso that Franco
expected him to deal with any indiscipline or disorder mounted by Aznar's
group. The anti-Hedillistas entered Falange headquarters shortly after
11 a.m. on the morning of 16 April for a meeting of the *Junta de Mando*.
They informed Hedilla of the charges against him and announced that
he was to be replaced by a triumvirate of Aznar, Dávila and José Moreno
in order to forestall what they imagined would be his sell-out to Franco.[64]

Hedilla could easily have had them overcome by force but had been
instructed by Franco's headquarters to let the plot thicken as the conspirators had planned. Accordingly, he left them in his office and immediately
went to Franco's headquarters and recounted the events of the morning
to Lieutenant-Colonel Barroso. He may also have briefly seen Franco
himself. Barroso offered him 'refuge and asylum' in the *Cuartel General*
which indicated clearly that they both expected further violence. Whether
the anti-Hedillistas then fought or backed down, Franco would appear
as the man who had saved the Falange and its leader from their sordid
personalist ambitions. Hedilla, confident that he was not in danger, and
perhaps suspecting that to accept would be to announce his political
impotence, declined the offer. He did, however, assure Barroso of his
total loyalty to the Generalísimo.[65] In their turn, the triumvirate sought
an audience with the Generalísimo later in the day and gave him their

explanation of the removal of Hedilla. They also expressed their total loyalty and desire to serve his war effort.[66] The triumvirate delivered to Radio Nacional a statement about the removal of Hedilla and their manifesto. However, Nicolás Franco prevented the broadcast of either.[67]

The Generalísimo still held all the cards and Hedilla was now in a weaker position than ever. There could be little doubt in Franco's mind that the recently deposed *Jefe* would be only too glad to accept the post of second-in-command to the Generalísimo in a united party. Hedilla, however, was determined to strengthen his position, if only as a negotiating base for the future. Although it is possible that he harboured thoughts of running the Falange on his own terms, it is more likely that he wanted merely to be able to use his position to safeguard as much as possible of its radical fascist agenda in its unified form. He now used his house as operational headquarters for his fight-back against Aznar, Dávila and Garcerán. It was decided to bring forward the proposed meeting of the *Consejo Nacional* in order to put before it the issue of the 'rebellion' of the *legitimistas*.[68] In a confused atmosphere of exaltation, machismo and boasting, Hedilla and his followers were determined, as they thought had been agreed all along with the *Cuartel General*, to finish off that very night the job of crushing the plotters. Both sides were being carefully watched by Doval's network of spies.[69]

Hedilla and Serrallach planned for Falange headquarters to be seized by José María Alonso Goya, the head of Hedilla's bodyguard, who had been brought from Burgos after the *Jefe Provisional*'s recent trip there. Goya was to be backed up by loyal cadets from the Falangist Militia School at Pedro Llen, a village outside Salamanca. The cadets were commanded by their instructor, a Finnish Nazi, Carl Magnus Gunnar Emil von Haartman. However, when given instructions on the evening of 16 April by Serrallach and Angel Alcázar de Velasco, Haartman refused to comply without a signed order from Hedilla, claiming that Doval and Nicolás Franco were looking for any excuse to act against the School. In fact, it is possible that Haartman was acting in accord with Barroso. There were now two armed camps in Salamanca, the party headquarters in the calle de Toro and Hedilla's house in calle Maizales, each with a confused mob of Falangists milling around inside and outside. Serrallach returned to the Militia School and gave Haartman the written order from Hedilla which had the desired effect. Haartman and a group of armed cadets arrived at Maizales with José María Alonso Goya at which point Hedilla ordered them to go to party HQ and arrest Aznar, Dávila and Garcerán.[70]

Haartman's men seized Falangist headquarters at around 1.30 a.m. on the morning of 17 April.[71] Dávila, Garcerán and Aznar had long since gone home. Meanwhile, José María Alonso Goya and Daniel López Puertas, together with other members of Hedilla's escort went to the *pensión* in calle Pérez Pujol 3 where Sancho Dávila was staying. They were carrying loaded pistols and hand-grenades. Dávila refused to believe that they were merely taking him to see Hedilla and was convinced that they were going to 'take him for a ride' (*pasear*). In the ensuing brawl, Goya was shot in the back of the neck by one of Dávila's bodyguards, Manuel Peral. Peral was then killed by López Puertas. López Puertas later claimed that the entire operation had been planned several days earlier, even down to the inclusion in the party sent to detain Dávila of two doctors in anticipation of bloodshed.[72] When Hedilla's bodyguards, now minus the distraught López Puertas, went to seize Garcerán, he held them at bay in the street by wildly shooting at them with a machine-gun from his window until the Civil Guard arrived.[73] Hedilla subsequently went to great lengths to deny any responsibility for what had happened.*[74]

Franco, who had just gone to bed, was immediately informed of what had happened. Dávila and Garcerán were arrested by the military authorities, as was Aznar just as he was summoning his militia to recapture Falange headquarters. Aznar was charged with provoking disorder in the rearguard. The continuing complicity between the *Cuartel General* and Hedilla was clear in the arrest of those who had been attacked and the continued freedom, for the moment, of their attackers.[75] However, Franco now seized the opportunity to clip the wings of the Falange. There was a total news blackout about the events in the calle Pérez Pujol broken only by Hedilla's review *Fotos* on 24 April.[76] The military authorities in each province immediately informed the local leadership of the Falange that they would be held responsible for any disorders.[77] The fact that Hedilla had a conversation with Serrano Suñer on 17 April suggests that he remained *persona grata* at this stage and could still have had a role to play in Franco's orchestrated take-over of the Falange.[78]

Oblivious to the subordinate role envisaged for him by Franco, Hedilla

---

* He reconstructed the events both in a series of letters from himself and his close collaborators sent to Ramón Serrano Suñer in 1948 and in a book commissioned from his wartime press chief Maximiano García Venero. In these accounts, at 11 p.m. on the night of 16 April, Goya and Daniel López Puertas arrived at Hedilla's house and suggested that they go and settle things amicably with Dávila and Aznar. In this highly unlikely version, Goya and López Puertas were instructed by Hedilla not to use violence, the bloodshed was the consequence of an unfortunate misunderstanding and the squad fired upon by Garcerán from his window consisted of nothing more than 'three or four peaceful Falangists out for a late night stroll' (*pacíficos noctámbulos*).

tried to clinch his victory over his rivals by bringing forward the proposed *Consejo Nacional* meeting to 18 April. He was encouraged in this by the fact that, while Doval's men who controlled roads into Salamanca were preventing the arrival of Aznar supporters, safe-conducts issued by Hedilla were sufficient to permit his men to attend the *Consejo Nacional*. Moreover, Dávila was under arrest on a charge of murder and would be unable to attend. These external signs of good relations with the *Cuartel General* blinded the naïve Hedilla to the fact that the deaths in Pérez Pujol, the shooting at Garcerán's house and the fighting at Falange headquarters between the two factions provided the perfect excuse for Franco to strike. The way was prepared by Doval's spy network which assiduously spread rumours to the effect that there had been an assassination attempt against Franco, that Falangists were leaving the front in order to impose their views on Franco and that some top Falangists were in touch with the Republic in search of a negotiated peace.[79]

At the meeting of the *Consejo Nacional* on the evening of 18 April, with Goya's body lying at the door, Hedilla gave an account of the events of the previous few days which made it quite clear that he had been operating in close accord with the *Cuartel General*. He was duly elected *Jefe Nacional*.[80] The careful orchestration of events had now reached the point at which Franco and his kitchen cabinet could openly take control. The hand of Mussolini might again be seen in the fact that in the course of 18 April, the Caudillo consulted with Guglielmo Danzi, who later informed Rome that he and Franco had together elaborated the fusion of the parties.[81]

Hedilla blithely hastened to announce to the Generalísimo the news of his own election. On arrival at the Episcopal Palace, Hedilla was surprised to see that Franco's office was set up with microphones as for a radio broadcast. What the proud new *Jefe Nacional* did not realize was just how minor was the role that he had been playing in the theatrical enterprise mounted by the *Cuartel General*. Franco, the star of the show, was about to take the stage. Doval had been busy organizing a 'spontaneous' demonstration of popular support for him. Barroso welcomed Hedilla and told him that the Generalísimo placed great importance on his presence, as well he might. Hedilla's participation in the evening's proceedings was essential to diminish any hint of military interference in the internal affairs of the Falange. The Generalísimo himself then greeted the new *Jefe Nacional* cordially and, on being informed of his election, remarked 'Very good; it's just what I expected.' Faced with the 'insistent demands' of the crowd, a 'reluctant' Franco then appeared on the balcony of the Episcopal Palace and made a short speech announcing the immediate fusion of the Falange and the Carlists.[82]

Coming back into the building, the Caudillo then read a fuller and pompously messianic speech over the radio at 10 p.m. and, having ended his broadcast with the proclamation that God had entrusted the life of the *Patria* into his hands, the Caudillo was informed that a large crowd was gathering outside the palace.[83] Visibly jubilant, Franco gently pushed Hedilla out onto the balcony to join him in receiving the applause. He embraced Hedilla and it appeared as if the recently elected *Jefe Nacional* was formally handing over his powers to the leader of the newly forged party, as Cabanellas had handed over the powers of the Burgos Junta to the Generalísimo on 1 October 1936. The Nationalist press underlined the symbolism when it reported the election of Hedilla as *Jefe Nacional* with full powers and his immediate visit to Franco's headquarters 'where he listened to the speech of His Excellency the Head of State, and after congratulating him, put the Falange unconditionally at his service'.[84] At the same time, units of the Army obliged Aznar's men to set off for the battlefront. Queipo de Llano intervened with Franco to secure the release of a much-chastened Sancho Dávila.[85]

On 19 April, Franco's political triumph was enshrined in the formal decree of unification, a unilateral initiative emanating from the Caudillo and Serrano Suñer, with the agreement of Generals Queipo de Llano and Mola. By its terms, the Falange was forcibly unified with the Carlists to form a single party, the *Falange Española Tradicionalista y de las Juntas de Ofensiva Nacional Sindicalista* (FET y de las JONS), remarkable more for the length of its name than for its ideological content.[86] The text had been prepared by Ernesto Giménez Caballero and Ramón Serrano Suñer without discussion of the details with either Hedilla or the Carlist leadership.[87] Franco declared that the new organization would be formed 'under my leadership' (*bajo mi jefatura*). The decree gave him total power within the new single party and the right to nominate half of the *Consejo Nacional*'s members. Those nominees would then elect the other half. With one blow, he had cut off the efforts of others to put their own stamp on the unification process.

On 20 April, the day following the decree of unification, Franco was visited by Hedilla and several members of the recently elected *Junta de Mando*, the executive of the now defunct *Falange Española*. The Caudillo greeted them with great warmth. When they asked him about the division of powers and prebends in the new organization, he gave them the impression that all their aspirations would be met.[88] It is quite clear that at this stage Franco took it for granted that, having announced that the unification would take place '*bajo mi jefatura*', Hedilla would become merely his subordinate. Hedilla, on the other hand, seems to have presumed that, under the vague overall authority of Franco, his recent election as *Jefe Nacional* of *Falange Española* made him *de facto* leader of *FET y de las JONS*. However, at the battlefronts, Carlist and Falangist cadres found themselves brusquely treated as subordinates by army officers who assumed that both organizations had now passed into military control. They considered, as did Franco himself, that *FET y de las JONS* was subject to the power and military authority of the Generalísimo.[89] Hedilla might have been able to discern that Franco intended the Falange to be reduced to an entirely subordinate position had he known that on 22 April the *Cuartel General* had given Cardinal Gomá an entirely mendacious account of recent events. The Primate was informed that the unification had been accelerated to put a stop to the clashes which had broken out on 16 April. The omission of any mention of the way in which the violence in Salamanca had been orchestrated from the *Cuartel General* suggested that Hedilla's days were numbered.[90]

Instructions from Franco's headquarters prohibited partisan propaganda on behalf of either the Falange or the Carlists. The local sections of both parties were ordered 'to orient their current propaganda towards the integration of the *Movimiento* and the exaltation of the Caudillo'.[91] There were rumours that outraged Falangist and Carlist extremists planned to withdraw volunteers from the front. Franco was determined both to undermine Hedilla's plans to be *de facto* leader and to prevent opposition to the forced merger crystallizing around him. Accordingly, he failed to fulfil his promise to give Hedilla the post of secretary-general of the *Junta Política* of *FET y de las JONS*. Instead, he cunningly set out to neutralize him by the device of publicly offering him on 22 April a position as merely another member of the new *Junta Política*.

There was no discussion since Franco did not offer Hedilla the post face-to-face. Hedilla learned of the composition of the unified *Junta Política* in the press of 23 April. The remaining members were the most docile elements of the Falange and Carlism, all more loyal to Franco than to any political ideology. Even the rather dim Hedilla must have been alerted to what was going on by the inclusion of the nonentity López Bassa, as a reward for his services to Franco during the run-up to the unification.[92] It was a neat trap. If Hedilla accepted inclusion in such a group, he was condemning himself to impotence as a decorative part of Franco's entourage. He was also ensured of the hostility of other important Falangists such as Pilar Primo de Rivera, Aznar and the influential intellectual, Dionisio Ridruejo, who were all urging him not to accept. If, on the other hand, he refused, he was guaranteed the Caudillo's enmity.[93]

Having read in the newspapers of his nomination, Hedilla did nothing and he then received visits from Barroso, Millán Astray, López Bassa and other emissaries of the *Cuartel General* urging him to accept. When Hedilla refused to join the *Junta Política*, he effectively signalled his opposition to the unification on Franco's terms. The *Cuartel General* reacted accordingly. There is considerable doubt as to exactly what happened next. In so far as it is possible to disentangle the totally partisan versions of both sides, it seems to be the case that, outraged by Franco's failure to keep his promises and confident of his own newly confirmed leadership of the Falange, Hedilla tried foolishly to mobilize his dwindling forces against the Caudillo through emissaries sent to the provinces. One of his supporters wrote later that 'we gambled blindly and lost'.[94] Franco assured Faupel that Hedilla had telegrammed the provincial chiefs (*Jefes*) instructing them to obey orders from the *Cuartel General* only if they came through him. Faupel was soon convinced by Franco that 'Hedilla was heavily compromised'.[95] Whether Hedilla's actions really constituted a serious threat to Franco is impossible to ascertain.*[96]

The fact is that Franco chose to see Hedilla's refusal to join the new *Junta Política* and his subsequent measures as acts of military indiscipline. At 7.00 p.m. on 25 April 1937, Hedilla was arrested by Franco's faithful bloodhound, the vicious Major Doval, who arrived with two lorry-loads of Civil Guards. His trial would take place on 29 May 1937. In the meanwhile, an investigation was entrusted to Doval whose interrogators now accused Hedilla of responsibility for the events of 16 April. The extent to which he had been acting on the advice of the *Cuartel General*

* Hedilla claimed later that he rejected Franco's offers believing that the united party could not remain faithful to the ideals of José Antonio Primo de Rivera. Accordingly, his refusal to join the *Junta Política* was a symbolic gesture of defiance in defeat. He thus loyally opted not to oppose Franco but refused to be his ideological adornment. Hedilla also denied the allegedly subversive message of the telegrams claiming that he and his followers had aimed at limiting internecine squabbling in the rearguard between the recently merged forces. With Falangists and Carlists each trying to take over the others' headquarters in certain places, José Sainz Nothnagel of the *Junta de Mando* sent a telegram which stated 'To prevent erroneous interpretations of the decree of unification, obey no orders other than those received through proper hierarchical channels'. Hedilla claimed that the phrase 'through proper hierarchical channels' proved that they had accepted Franco's authority. It was, however, an equally reasonable assumption that, in Falangist jargon, the phrase meant that only orders passed down through Hedilla should be obeyed.

was neatly forgotten. He was also accused of having contacts with the Republican zone and with planning to assassinate Franco. The *Cuartel General*'s rumour factory ensured that these accusations were spread around. The fact that Pilar Primo de Rivera and Aznar had denounced Hedilla as a Francoist was also forgotten, as was the embrace on the balcony of the Episcopal Palace on 19 April. At his trial, charges were put forward which in their contradictions were reminiscent of the Stalinist purges.[97] Franco's instincts as a commander made it inevitable that he would not tolerate the kind of internal squabbling which was fatally to weaken his Republican opponents. For the crime of real or imagined military rebellion, Hedilla was sentenced to death, along with Daniel López Puertas, who had killed Peral, and two others. Sancho Dávila and Rafael Garcerán were absolved.[98]

With regard to both the arrest and the sentencing of Hedilla, Franco kept his usual distance. When Dionisio Ridruejo protested to Franco about Hedilla's original detention, the Caudillo feigned surprise: 'But have they arrested Hedilla? I have still not been informed. I ordered the intelligence services to investigate the events of the last few days and to act accordingly. Doubtless, they found something incriminating against him'. Ridruejo expressed his outrage that the leader of the Falange should simply be arrested by the man who was succeeding him. To Ridruejo's surprise, Franco listened patiently, indeed rather nervously, biting his lips and avoiding direct eye contact. Only later did Ridruejo discover that during the interview during which he thought that they were alone, Franco had placed armed men behind a tapestry hanging on one of the walls of his office. Apparently, this was standard practice.[99]

When Pilar Primo de Rivera visited Doña Carmen to ask her to intercede with Franco, she replied 'there's no need, with Ramón [Serrano Suñer] here, the Falangists have a sound defender'.[100] When Serrano Suñer remonstrated with Franco about the death sentences, he replied 'I simply cannot allow agitation in the rearguard.' For Franco, it was a question of military discipline. He told Faupel that 'he was determined, since he was fighting a war, to nip in the bud any action directed against him and his Government by shooting the guilty parties'.[101] Faupel told Franco that 'the shooting of Hedilla, the only real representative of the workers, will make a very bad impression'.[102] Franco was unmoved when visited by the tearful mother of Hedilla bearing a grovelling letter. He received equally coldly a plea from Cardinal Gomá.[103] Eventually Serrano Suñer persuaded him that the executions of Hedilla and his fellow rebels would be damaging to the regime. Franco gave way with ill grace, saying 'one day, these weaknesses will come back to haunt us'.[104] Nicolás Franco told an Italian official, probably Danzi, that Hedilla had not been shot 'so as not to make a martyr out of a nonentity'.[105]

After the sentence was commuted by Franco on 19 July, Hedilla spent another four years in Francoist jails in harsh conditions, albeit not as harsh as those suffered by some of his followers. He would never again play a political role in Franco's Spain although it has been alleged that he lived well in the lower reaches of the regime's corruption.[106] When Hedilla's mother came again to plead with Franco to release him, he told her that her son was 'an innocent victim', although he did not pardon him until May 1947.[107] The death sentences on Hedilla and three of his comrades, together with lengthy prison sentences for others, effectively ended the feeble resistance to Franco's ambition to become absolute ruler of the Nationalist zone. The bulk of the Carlists were furious but, in the interests of the war effort, they silenced their outrage. Most of the Falangists who had initially opposed the unification, more malleable than Hedilla, were easily bought off. Pilar Primo de Rivera, a great prize from Franco's point of view, joined the Francoist camp as head of the *Sección Feminina* of the *FET y de las JONS* and Agustín Aznar, another prize as a *'legitimista'*, became assessor to the militias.[108] Garcerán also joined the claque of Franco adulators.[109]

The rank-and-file of the Nationalist zone welcomed the unification as a way of putting a stop to the friction between the various groups. Franco told Faupel that he had received sixty thousand telegrams in support of his action.[110] However, since the new party was now the only political formation permitted, the independence of the Spanish fascist movement was at an end. The Falange had been castrated.[111] Potential Falangist leaders learned their lesson. That was shown by the docility of Raimundo Fernández Cuesta who was soon to arrive in Salamanca on a prisoner exchange masterminded by Prieto in the vain hope that he would be a thorn in Franco's side. It was even clearer in the case of José Luis de Arrese who was arrested as an Hedillista, then rescued and given preferment by Serrano Suñer, only to turn into the most fertile sycophant that Franco would ever have. The choice was clear: loyalty to Franco and access to the privileges of power or opposition to Franco and unemployment, prison and maybe even execution. The new party was popularly known as the Falange rather than by its cumbersome full title but Falangists were just one, albeit the most dominant, of its component groups. That was reflected in the fact that the new single party also came to be called the *Movimiento* (Movement). Forced to accept Franco as their new leader, the Falangists saw their ideological role usurped by the Church, their party turned into a machine for the distribution of patronage and their 'revolution' indefinitely postponed.

Although Franco had clinched a political power to match his military power, Serrano Suñer assumed the day-to-day tasks leaving his brother-in-law free to concentrate on the war. Many of the early decrees and the choice of ministers reflected Serrano's influence. The personal relationship between the two families, Serrano Suñer's part in the preparation of the rising and his fanatical and ascetic commitment to the nationalist cause contributed to the confidence which Franco showed in him.[112] There was also an element of cunning in a process whereby Franco left Serrano Suñer to be the lightning conductor through which the ideological conflicts of the Nationalist forces could flow. During what remained of the Civil War, he was left to domesticate the Falange. After the war, Serrano Suñer bore the brunt of the internal power struggle between the Army and the Falange. In both tasks, Franco was to be the winner and Serrano Suñer the loser.

The unification enshrined Franco's determination to eliminate any political rivals. It was not a difficult task. Calvo Sotelo was dead. His deputy, the effete Antonio Goicoechea and other leaders of Renovación Española dutifully accepted the decree of unification and dissolved their party.[113] For Gil Robles, derided as having delayed the inevitable war against corrupt democracy, it served for nothing that he too had accepted the unification in a long and unctuous letter to Franco dated 22 April 1937.[114] If he expected a call to Spain, he was to be disappointed.[115]

Fal Conde had already been dealt with and the unification was greeted by the Carlist movement with demonstrations of apparent rejoicing, ecstatic newspaper articles and sacks of letters and telegrams of congratulations to Franco. There were those who expressed regret and if they did so with the necessary degree of respect, their qualms were eased when they were given important posts in the new organization. Fal Conde was gradually eliminated from the higher reaches of the united party.[116] Franco's private distaste for José Antonio Primo de Rivera did not stand in the way of their public identification. The creation of the myth of Franco the natural successor to José Antonio was only one of many myths created around him. It was a crucial one in clinching the loyalty of hundreds of thousands of Falangists and of facilitating the continued solidarity of the Axis powers.

Paradoxically, the clinching of Franco's position as the undisputed leader of a fascist single party slightly endangered his relationship with

the Catholic Church. The Vatican's relationship with both Fascism and
Nazism was an uneasy one. For that reason, Franco began to play up his
anti-clericalism to the Nazis and play down his fascist sympathies to
churchmen. The myth of Franco the Catholic Crusader was inflated
alongside the myth of Franco the Falangist *Jefe*. In both domestic and
international terms, the legitimization of the Francoist cause provided by
the Catholic Church was too valuable ever to be put at risk. However,
there were sufficiently large areas of coincidence between Franco and the
Church – hostility to rationalism, freemasonry, liberalism, socialism and
Communism – to ensure that the Church willingly accepted much of the
political rhetoric of the Nationalist zone.

The difficulties of the relationship between fascism and the Church
had already been illustrated. On 14 March 1937, just two weeks after the
launching of the Basque offensive, the Vatican had published the encyclical *Mit brennender Sorge* criticizing Hitlerian racism. It was followed five
days later by *Divini Redemptoris* condemning Communism. With German
units playing a crucial role in the war against the most Catholic part of
Spain, Franco was anxious that no publicity be given to the first of the
two encyclicals. A blanket of silence about *Mit brennender Sorge* descended
in Nationalist Spain. Asked by the Vatican about the Spanish response,
Gomá replied that there had been no references in the press and that he
feared that Nationalist politics might be taking a Hitlerian direction.
Gomá requested copies of the German original, had it translated it into
Spanish and distributed to the dioceses of Nationalist Spain. However,
after learning of the hostility of the *Cuartel General* to the encyclical, he
thought it prudent to inform bishops not to give publicity to the text.
Franco told Faupel on 23 May that he had instructed the Cardinal Archbishop of Toledo that it should be silenced. In fact, Gomá had decided,
after consultation with Plá y Deniel, not to risk a clash with Franco.
When Gomá protested to him that Radio Salamanca and the Nationalist
press had published German attacks on the encyclical, Franco simply
denied all knowledge of the matter.[117]

After the destruction of Guernica, when many Catholics began to
question the sanctity of the Francoist cause, Gomá rendered the Caudillo
another inestimable service. At Franco's request, they met in Burgos on
10 May. In the course of a two-hour interview, Franco revealed an anxiety
to reassure the Primate that his assumption of the headship of a fascist
party did not mean that Nazi ideas would in any way diminish his commitment to the inculcation of Catholic values in the new Spain. This was
the necessary prelude to two requests from the Caudillo. He asked Gomá

to persuade the other bishops to go to Rome and use their influence to hasten the Vatican's recognition of his regime. Gomá pointed out the impracticality of this notion and assured Franco that the Vatican was already well supplied with information on the Spanish situation. The second request was for the Primate to write an episcopal letter to 'dispel false information abroad', by which he meant the news about the bombing of Guernica. With the Basque issue dividing Catholic opinion around the world, and even, he claimed, having 'negative consequences in his relations with some European chancelleries', Franco wanted foreign Catholic opinion united in the democracies against any possible aid to the Republic.

In fact, Cardinal Pacelli, the Vatican Secretary of State, had already written to Gomá on 10 February 1937 suggesting some public statement about the co-operation of Basque Catholics with Communists. After consulting with the other members of the hierarchy who were enthusiastic about the idea, by early May, Gomá had begun to synthesize the various drafts which he had requested from the bishops of Spain. Now, in response to Franco's request, and after further consultation with the other bishops, Gomá drafted the collective letter 'To the Bishops of the Whole World' which was published on 1 July 1937.[118] The text legitimized the military rebellion and defended the Nationalist State from accusations of fascist statism.[119] It was signed by two Cardinals, six Archbishops, thirty-five bishops and five vicars-general. It was not signed by Cardinal Francesc Vidal i Barraquer, Archbishop of Tarragona in Catalonia, nor by Monsignor Mateo Múgica, Bishop of Vitoria in the Basque Country both of whom were deeply concerned by the possible consequences for Catholics in the Republican zone.[120] The letter did Franco's cause incalculable good. In contrast, the hesitations of the Vatican in conceding full recognition of Nationalist Spain provoked his irritation.[121] Paradoxically, as relations with the Vatican were normalized, after the sending of Monsignor Ildebrando Antoniutti as Papal *Chargé d'Affaires* in October 1937, Franco came to resent the authority of the Vatican in naming bishops as a challenge to his own. He was annoyed that the Vatican did not accord him the same privileges in this regard as it had the kings of Spain.[122]

That he could think in such terms is a measure of the way in which, now fuelled by constant adulation, his self-image was ever more grandiose. He was Generalísimo of all the armed forces, Head of State, Head of Government and *Jefe Nacional* of the Falange. Even if his fellow generals still thought of his position as provisional, he did not. There were already signs that he considered himself as Regent for life. The unification was an important landmark in his changing view of his own position. A machinery of popular adoration made it easier to think in terms of staying on after the war. At the back of his mind, he may have intended eventually to restore the monarchy but it was a distant prospect. As his speeches would soon indicate, he believed that the Spanish monarchy had started to decline after Philip II. Seeing himself as a great hero like the saintly warrior-kings of the past, he believed that he could restore the monarchy to its sixteenth-century greatness but only once he had assiduously eradicated the poison of three misspent centuries. Such a view was a natural extension of the Crusade rhetoric and the support of the Church. It was ironic that his own monarchical pretensions should bring him into potential conflict with the Church hierarchy. For the moment, however, with central control now firmly established over his own zone, the main obstacle to his ambitions remained the Republic. With the unification complete, he now turned back to the task of its annihilation.

XI

FRANCO'S WAR OF ANNIHILATION

May 1937–January 1938

FRANCO STILL had battles to win against the Republic but he had now won the political war within the Nationalist zone. 18 July was decreed to be a national holiday. A Francoist calendar was instituted from 18 July 1937, from which date ran the so-called 'Second Triumphal Year'. At the suggestion of the manic Navarrese Falangist priest, Fermín Yzurdiaga, one of the most extreme of the adulators who increasingly surrounded Franco, 1 October was also denominated a national holiday, the 'Day of the Caudillo'.[1] His accumulation of military and political powers effectively gave him control over every aspect of political life. Only the areas of jurisdiction of the Church eluded him and he would endeavour, within a context of outwardly reverential religiosity, to submit the Church to his political will. Otherwise, his powers were comparable to those of Hitler and greater than those of Mussolini.

For nearly forty years he would use them with consummate skill, striking decisively at his outright enemies but maintaining the loyalty of those within the Nationalist coalition with cunning and a perceptive insight into human weakness worthy of a man who had learnt his politics among the tribes of Morocco. The ability to calibrate almost instantly the weakness and/or the price of a man enabled Franco to know unerringly when a would-be opponent could be turned into a collaborator by some preferment, or even the promise of it – a ministry, an embassy, a prestigious military posting, a job in a State enterprise, a decoration, an import licence or just a box of cigars.[2] Within weeks of the unification decree, a huge and well-paid party bureaucracy proliferated in ironic contradiction to the continuing rhetoric of an austere crusade. Many one-time followers of Gil Robles and Lerroux saw their opportunity and embedded themselves in the system, wiping away their original sin of participation in the Republic by means of a loudly proclaimed loyalty to the Caudillo.[3]

In military terms, Franco had every confidence in his ultimate victory though with a calendar marked in years rather than months. His allies found it difficult to comprehend his long term view of the political benefits of a war of attrition. One consequence of international dismay at the duration of the Spanish war was that suggestions for a negotiated settlement began to be heard. To the intense annoyance of Franco, several proposals for a compromise peace between the Basques and the National-ists were made in the spring and early summer of 1937, including one each from the Vatican and Fascist Italy. However, given the enormous industrial wealth of the Basque Country and its deeply Catholic popu-lation, Franco, anxious to seem reasonable to world opinion in the wake of Guernica, considered the proposals as ways to hasten the surrender of an intact city. On 7 May, in response to Italian suggestions, Franco agreed with Mola that the Basques be offered the preservation of their city, strict controls on the behaviour of occupying troops, the evacuation of political leaders, no reprisals, and even special fiscal status for the Basque Country, in return for immediate surrender. In fact, in the short period during which Franco was prepared to offer such conditions, it proved impossible for the mediators to put them to the Basques. Once his troops again began to move on Bilbao, Franco refused to be held to the proposed terms.[4] After Guernica, the Basque forces had been reorganized and the eight kilometres from Guernica to Bilbao were defended every centimetre of the way. Not until late May did Mola's troops have Bilbao surrounded.

On 7 May, as he was setting off on his official visit to London to represent the Republic at the coronation of George VI, the Socialist Julián Besteiro was requested by Manuel Azaña to appeal to Anthony Eden for his mediation. Eden received Besteiro on the evening of 11 May.[5] As a consequence of these private representations, the British ambassadors to Italy, Germany, Portugal, France and the Soviet Union sought international co-operation to secure the withdrawal of foreign volunteers from Spain.[6] In the event, nothing came of the initiative, given the determination of Italy and Germany that it should not prosper and the lack of energetic Anglo-French efforts in favour of an imposed peace. In any case, the reaction of Franco was violently hostile. As he had made clear to Cantalupo barely six weeks earlier, victory for him meant the annihilation of large numbers of Republicans and the total humiliation and terrorization of the surviving population. Nothing had changed the Generalísimo's view that any mediation at this stage would benefit the Republic. A compromise peace followed by elections would leave a substantial population capable of making its wishes felt. He spoke to Faupel on 22 May and rejected Eden's proposal, declaring that he 'and all Nationalist Spaniards would rather die than place the fate of Spain once more in the hands of a Red or a democratic government'.[7]

The Vatican also tried to mediate. On 21 May 1937, Cardinal Gomá went to Lourdes to meet Monsignor Giuseppe Pizzardo, secretary of the Holy Congregation of Special Ecclesiastical Affairs. Monsignor Pizzardo wanted him to sound out Franco on the possibility of a peace initiative in the Civil War. Gomá reacted much as Franco had already done, replying that such an initiative would merely help the Republic. Franco told Gomá that the distance between the two sides precluded any negotiated peace. He would accept only the unconditional surrender of the Republicans. When Gomá put to him Vatican doubts about the ferocity of the Nationalist repression and the suggestion that the war had arisen out of social inequality, Franco rejected both ideas. He said that he would change nothing with regard to the repression since no one was condemned other than under military justice and he denied that the war had anything to do with the unjust distribution of wealth.[8]

By dismissing the prospect of a compromise and embracing a war to the death, Franco was ruthlessly assuming an awesome responsibility. It was clear that, by this stage, he regarded his own authority as untrammelled. In so far as there was anyone capable of challenging his authority, it was Mola. Although Mola had loyally accepted Franco's rise to be Generalísimo, Franco could never entirely forgive him for his preeminence in the early days of the rising. In the tirelessly suspicious mind of the Caudillo, Mola would always have been a threat. Even before 1936, when, as old *Africanista* comrades-in-arms, they were ostensibly friends, Franco spoke of him in private with contempt, as indeed he did of almost all of his contemporaries, *'es un majadero'* (he's a fool).[9] Like his Chief of Staff, Vigón, Mola had been quick to see the impossibility of capturing Madrid and keen to pursue the war on other fronts. As the war developed, the decisive, not to say pyromaniac, Mola was anxious to terminate the war in the north by the unrestrained use of terror bombing and the destruction of Basque industry. He was driven to distraction by Franco's obsession with Madrid and by his commitment to a plodding war of attrition.

Moreover, partly as a result of his own lack of interest in a swift end to the war, partly to emphasize Mola's subordination, Franco put many petty obstacles in the way of the Basque campaign. Above all, he had interfered with Mola's air forces and had taken troops away from the northern front to reinforce positions around Madrid. However, Mola's main differences with Franco were political. Just before the war had begun, Mola had produced a document which suggested a commitment to maintaining the Republican regime, eventual restoration of parliamentary government, religious freedom and protection of workers' rights.[10] The austere Mola was known to disapprove of the corruption permitted by Franco as a method of controlling his subordinates. On a visit to the *Cuartel General* in the early summer of 1937, he told Franco that while he was happy for him to be Head of State, Generalísimo and head of 'that party of yours', something would have to be done about the rearguard and the creation of a satisfactory government. The clear implication was that he was seeking some kind of central political role for himself. Since before the unification, Franco had been annoyed by rumours that he would concentrate on the military leadership and leave Mola to form a government. Now, it was assumed by Franco and Serrano Suñer that Mola's next visit in early June would bring a full-scale ultimatum.[11]

On 3 June 1937, Mola set out from Pamplona to Vitoria and thence on to Burgos. In the province of Burgos, between the villages of Castil de Peones and Alcocero,* his plane crashed and everyone on board was killed.[12] Rumours abounded that the crash had not been an accident but rather the consequence of sabotage. It is also possible that the aircraft was mistakenly shot down by Nationalist fighters since Mola was flying in an Airspeed Envoy with English markings similar to planes used to fly supplies into the Republic from France. It had originally been flown to Pamplona from Madrid by a defecting Republican pilot and requisitioned by Mola. Equally, as the official version claimed, the aircraft may simply have hit a hillside in thick fog.[13]

When the news reached Franco's headquarters, there was consternation among his subordinates and great trepidation about giving him the news. Finally, it was decided to entrust the job to Admiral Cervera, Chief of the Naval General Staff. Himself overcome with emotion, Cervera wandered around the point to the irritation of an increasingly impatient Franco. 'Out with it', ordered the Generalísimo and the Admiral gave the news. The Caudillo brushed it off lightly, 'At last, so that's all it is. I thought you were going to tell me that they'd sunk the cruiser *Canarias*.'[14] The coolness with which Franco received the news

---

* According to the Burgos judge Antonio Ruiz Vilaplana, the aircraft hit the Monte de La Brújula, the hill where those killed each night in the Falangist repression in Burgos were taken for mass burial, Ruiz Vilaplana, *Doy fe . . . un año de actuación en la España nacionalista* (Paris, n.d. [1938]) pp. 85–8.

was reported by the German ambassador, Wilhelm Faupel: 'the Generalísimo undoubtedly feels relieved by the death of General Mola. He told me recently: "Mola was a stubborn fellow, and when I gave him directives which differed from his own proposals he often asked me: 'Don't you trust my leadership any more?'".'[15] Sangróniz made it clear that Franco and his immediate staff did not regard Mola's passing as in any way a loss. He said to Vegas Latapie, 'when all is said and done, there's no reason for so much fuss ... A general who dies at the front ... It's virtually normal.' Sangróniz took his cue from the Generalísimo himself.[16] Cardinal Gomá, however, lamented Mola's passing because he felt that he would have been more resistant than Franco to Nazi and Fascist influences.[17]

Franco attended Mola's funeral and showed not the slightest trace of emotion. As the body was brought down the steps of divisional military headquarters, the Generalísimo flung his right arm out energetically in the fascist salute. Having put weight on in the previous months, his uniform split open at the arm-pit to the suppressed hilarity of some of the onlookers.[18] Hitler commented years later: 'The real tragedy for Spain was the death of Mola; there was the real brain, the real leader. Franco came to the top like Pontius in the Creed.'[19] Mola's private papers were confiscated and his role was dramatically played down by the nationalist propaganda apparatus and by subsequent Francoist historiography. On 3 June 1939, Franco inaugurated a monument to the memory of Mola in the hills around Alcocero. Thereafter, he was forgotten, the path to the monument soon overgrown.[20] On 17 July 1948, Franco would grant Mola the posthumous title of Duke, an easy gesture. He would tolerate the cult of the memory of José Antonio Primo de Rivera, in order to secure the loyalty of the Falangist masses but to cultivate the memory of Mola carried no such prize and would have involved the humiliating recollection that he had been the inspiration and planner behind the rising. One consequence of Mola's death was that, to the relief of his staff, Franco stopped travelling by air and began to visit the front by car.[21]

On 11 June, now under the command of General Fidel Dávila, the Army of the North renewed its march towards Bilbao. The diminutive Dávila was blindly faithful to Franco and took no decisions of any moment without consulting the Generalísimo. Dávila was replaced as head of the Junta Técnica del Estado in Burgos by the monarchist General Francisco Gómez Jordana, conde de Jordana. With Mola gone, and now freer of some of the political and diplomatic occupations of previous months,

Franco began to take a closer interest in the progress at the front which he visited with greater frequency than before. His general staff continued to be headed by Francisco Martín Moreno, now a General, who dealt with all routine matters, but his immediate tactical discussions were with Colonel Barroso and General Kindelán.[22]

Preceded by fierce artillery bombardment and bombing by both the Condor Legion and the Italian Aviazione Legionaria, the Nationalist forces made rapid progress, quickly closing their grip around Bilbao. Armed with the plans of the city's iron-ring fortification, which had been betrayed by the deserter Captain Goicoechea, Colonel Juan Bautista Sánchez's forces broke through at its weakest point on 12 June. Nationalist fears of a repeat of the siege of Madrid did not materialize largely thanks to the bombings of Durango and Guernica and the much-publicized threats of Mola. Believing themselves to have been abandoned to their fate by the central government, the morale of the poorly equipped Basques plummeted.[23] The Basque army withdrew and the Nationalists were allowed to enter Bilbao on 19 June almost unopposed because the authorities did not want to risk the same fate as Guernica. The armaments and explosives factories, steel-making plants, shipyards and heavy engineering works were left intact.[24] Franco ordered that only small numbers of troops should enter the city, in order to prevent the damaging publicity associated with massacres such as those of Badajoz and Málaga.[25] However, in the following month, even if the atrocities of the south were not repeated, nearly one thousand Basques were executed and a further sixteen thousand imprisoned in punishment for their nationalist ambitions. Six months later, large numbers of executions were still being carried out.[26] German economic experts were delighted with the capture of the mineral-rich north.[27] Faupel was nevertheless scathing about Franco's military leadership, pointing out to Berlin that the campaign had taken nearly three months to cover forty kilometres.[28]

Franco's satisfaction with the way things had gone in the north was clearly revealed by his manifestations of gratitude to the Condor Legion. He telegrammed Hitler 'In the moment in which the Nationalist troops march victorious into Bilbao, I send you enthusiastic greetings from myself and my army in reply to the confidence which the great German people and its Führer have shown us.'[29] Even more significant was a letter which he wrote to General Sperrle after the fall of Bilbao: 'on the completion of the part of the operation which was crowned by the capture of Bilbao and the occupation of almost the entire province of Vizcaya, during which the air forces under your command took part in such an

effective and splendid manner, I would like at this time to thank and give
my congratulations to your Excellency and to ask you particularly to
express my extreme thanks for such splendid work to Lieutenant-Colonel
von Richthofen, who worked with so much skill.'[30]

After the fall of Bilbao, the Nationalists' northern campaign met few
obstacles. Nevertheless, Kindelán fretted at Franco's failure to seize the
opportunity for a rapid sweep through the north.* Faupel also complained
that the subsequent regrouping of forces was done with agonizing slowness.[31] Three weeks went by as the next stage of the Nationalist advance
through Vizcaya and into Santander was prepared. It was to be launched
on 9 July.[32] In anticipation of a great victory, Franco moved his headquarters from the Episcopal Palace in Salamanca to the aristocratic
Palacio Muguiro in Burgos. He was to remain there for a further two
years until some months after his final victory over the Republic.

Franco's delay enabled the Republic to try to halt the seemingly inexorable process by which their territory was being whittled away. At dawn
on 6 July, a Republican offensive was launched at the village of Brunete,
in arid scrubland fifteen miles west of Madrid. It had been well planned
and thoroughly prepared by General Vicente Rojo, the Republican Chief
of Staff. On the morning of 6 July, the Generalísimo and his ADC, Pacón
Franco Salgado-Araujo, had just got into their car to go to Vizcaya to
oversee the new push in the north when they were stopped by Major
Carmelo Medrano with news of the assault on Brunete.[33] Rojo's attack
achieved initial surprise and came near to cutting off the Nationalist
besiegers of the capital.

At the weakest point in the lines of the Nationalist Army of the Centre,
between Yagüe's army corps and that of Varela, an army of more than
eighty thousand troops poured through and advanced twelve kilometres.
Franco's immediate response was to say to Barroso, 'they've smashed
down the Madrid front'. Briefly, he lost his unflappable serenity. Years
later, Barroso claimed it as the moment at which he saw Franco most
upset in the entire war.[34] However, as more units broke through in conditions of extreme heat, the deficiencies of Republican junior officers

---

* His memoirs were later censored on precisely this point. He wrote in 1941: 'the enemy
was defeated but was not pursued; the success was not exploited, the withdrawal was not
turned into a disaster. *This was due to the fact that while the tactical conception of the operation
was masterly, as was its execution, the strategic conception on the other hand was much more
modest.*' Franco's censorship held up the publication of Kindelán's memoirs until 1945 and
then suppressed the underlined passage along with many others, (cf. Alfredo Kindelán,
Mis cuadernos de guerra 1936–1939 [Madrid, n.d. (1945)] p. 86 and Kindelán, [1982 ed.],
pp. 9, 127).

were exposed. Some confusion overtook the Republican attack which broke down as the Nationalists under Colonels Barrón, Asensio and Sáenz de Buruaga held the line and rallied, helped by reinforcements rushed to them by Yagüe. As Rojo had hoped, Franco suspended the attack on the north and sent two Navarrese brigades plus the Condor Legion and the Italian Aviazione Legionaria to Madrid. He put Varela in overall command. Despite the relative insignificance of Brunete, Franco was as determined as always not to give up an inch of territory once captured. The political implications meant more to him than the military consequences of a delay in the northern campaign. At appalling human cost, he would again hammer home to Republican Spain the message of his invincibility.

Having decided to throw substantial forces into the Madrid front, Franco saw the possibility of a great success open up as the Republican advance was held. The Nationalists were able to seize air superiority helped by the introduction at Brunete of the new German fighter, the Bf 109.[35] Reinforced by the Navarrese Brigades from the north under Colonels Camilo Alonso Vega and Juan Bautista Sánchez, Varela was able to counter-attack on 18 July. As the battle swung the Nationalists' way, thanks above all to German air support, Franco faced the decision of whether to press ahead in Vizcaya as planned or keep the forces from the north for another major attempt to take Madrid. According to Kindelán, the Generalísimo vacillated for at least a week. Vigón wrote a despairing letter to Kindelán pointing out that to respond to the Republican attack would only prolong the war. Kindelán's reply to Vigón on 13 July made it clear that Franco believed that he could destroy large numbers of Republican troops on the Madrid front if he committed sufficient force. Kindelán's letter showed that the final decision was exclusively Franco's, asserting that the Generalísimo 'invents his own operational plans without permitting his General Staff or other subordinates to influence his decisions'.[36] Franco himself took a closer day-to-day interest in the fighting than he had in the Basque campaign which had been left to Speerle, Richthofen, Mola and Vigón. He set up a temporary headquarters at Dehesa del Rincón in Villa del Prado to the south of the road between Madrid and Avila. He visited Varela's headquarters every day and discussed the day's operations with him. Franco's perpetually boundless confidence again boosted the morale of his men.[37]

In conditions of sweltering heat and considerable chaos, with both sides mistakenly dropping shells on their own troops, Varela's counter-attack backed by devastating air and artillery attacks drove the Republican forces back to their starting point.[38] It was one of the bloodiest slogging combats of the war. Rojo realized that Franco was committed to the destruction of as many Republican troops as possible in a war of attrition.[39] At the debilitating cost of more than twenty thousand of its best troops and much valuable equipment, the Republic had done little at Brunete except create a breathing space during which inadequate efforts were made to reorganize its northern forces.[40] In the event, the eventual collapse of Santander was delayed by five weeks and that of the rest of the north by two months. The battle ended on 25 July, the feast day of Santiago (St James), patron saint of Spain. This permitted Franco to declare on returning to Salamanca, 'the Apostle has granted me victory on his feast-day'.[41] Varela was in favour of pursuing the Republicans back to Madrid but was restrained by Franco who, to the relief of Vigón, preferred that the Nationalist advance in the north should continue into Santander and Asturias. He told Varela plausibly that it was necessary to finish the war in the north before the coming of the fog, rain and snow of the winter. Varela's ADC, Juan Ignacio Luca de Tena, believed that Franco was loath to see Varela cover himself with glory.[42] A victory in Madrid would probably have rendered the Republican rump in the north militarily irrelevant. Again, it was as if the Generalísimo was anxious that the war should not end in a sudden Nationalist victory before each area of Spain had been cleansed of leftists and liberals.

In the midst of the fighting at Brunete, Franco had returned to Salamanca for the celebration of the first year of his '*Movimiento*'. His broadcast speech was, according to his cousin, entirely written by himself. It gave the measure of the extent to which Franco had come to see himself as a providential figure, the very embodiment of the spirit of traditional Spain. He made it clear that his achievement was to have saved 'Imperial Spain which fathered nations and gave laws to the world'.[43] On the same day there was published an interview which the Caudillo had given to the director of the monarchist newspaper *ABC*. In it, he announced the imminent formation of his first government. Asked if his references to the historic greatness of Spain implied a monarchical restoration, he replied truthfully but with masterly ambiguity that, 'on this subject, my preferences are long since known, but now we can think only of winning the war, then it will be necessary to liquidate it, then construct the State on firm bases. While all this is happening, I cannot be an interim power.'

At the end of the interview, Franco launched into an astonishing and unprompted paean to the Spanish aristocracy: 'I believe that one can count on one's fingers the families of that social class which do not have several members fighting at the front ... Some simple people who criticize a lady of noble lineage because they see her sitting in an elegant bar do not think that perhaps she has arrived there from the hospital where she is looking after the wounded.' The rich *señorito*, once rightly denigrated, 'when he is seen in the bar, often has his face tanned by the air of the battlefield and his hands rough from bearing his rifle if he is not on crutches or with his arm in a sling. And he should be regarded with respect.'[44] There is no reason to believe that this was anything other than a sincere admission of Franco's admiration for the Spanish aristocracy. If there was a political motive, it can only have been to reassure the conservative readers of *ABC* that his recent elevation to the leadership of a fascist party did not mean that he swallowed its egalitarian rhetoric.

The Navarrese brigades were now transported back to the north. After three weeks of preparation, General Dávila was ready by 14 August to begin a great operation to encircle Santander. He led an army of sixty thousand troops, amply supplied with Italian arms and equipment, backed by the Condor Legion and the Corpo di Truppe Volontarie under General Ettore Bastico. In brilliant summer weather, Dávila's forces enjoyed a virtual walk-over. They had massive air and artillery support as well as numerical superiority. The depleted and disorganized Republican forces in their way were easily brushed aside. Within seven days, the port of Santander was at the mercy of the attacking Nationalists. From his headquarters at the Palacio Muguiro in Burgos, Franco had visited the front on most days accompanied by his immediate general staff.* Wherever they happened to set up temporary headquarters would be known as *Términus*.[45]

The Italians had continued to press for a negotiated surrender of the Basque army which had now withdrawn into the province of Santander. No doubt with the atrocities of Málaga in his mind, Mussolini had written to Franco on 6 July, urging him to show moderation, refrain from reprisals against the civilian population and allow prisoners of war to be held in Italian custody. The Duce argued that a Basque surrender would undermine the entire Republican position in the north and be a great propaganda victory for Franco in the Catholic world. The Caudillo replied two days later and grudgingly agreed to accept the Duce's requests. His reservations were that he did not believe that the Basques would surrender nor, if they did, that it would in any way weaken the

---

* It consisted of Francisco Franco Salgado-Araujo, Antonio Barroso and Carmelo Medrano, with, since the death of Mola, Juan Vigón as effective Chief of Staff.

Asturian resistance.[46] Negotiations then dragged on for nearly two months as the Basques played for time. To the astonishment of the Italians, on 23 July, Nicolás Franco repeated his brother's earlier offer of no reprisals and facilities for the evacuation of political and military leaders if the Basques surrendered.[47]

With the Nationalist forces inexorably marching westwards, the Basques finally agreed to surrender to the Italians at Santoña to the east of Santander on 26 August.[48] In accordance with the agreement made, Basque political personalities embarked on two British ships, the SS *Seven Seas Spray* and the SS *Bobie*, under Italian protection. On 27 August, with Nationalist warships blockading the port, on Franco's orders, Dávila told the Italians to disembark the refugees, which they refused to do, although they advised the Basques to go ashore. The prisoners were held by the Italians for four days but, on 31 August, Franco ordered Bastico to hand them over. He hesitated and only after assurances from Barroso that the surrender conditions would be respected did he relinquish the captives on 4 September. Summary trials began at once and hundreds of death sentences were passed. The Italians were appalled by Franco's duplicity and cruelty.

Bastico sent Roatta to Salamanca to plead with Franco to stop the executions and allow the Basque leaders to leave the country. Roatta reminded the Caudillo that the Basques had surrendered after being offered such terms and pointed out that Italian honour was at stake. The Generalísimo simply ignored his arguments. As always, he found it easy to distance himself from what was happening. He conveniently forgot his brother's initiative which was merely a ploy to hasten the Basque surrender and ensure that they surrender to Franco and not the Italians. The Generalísimo believed himself entirely free of any moral obligations since he had carefully left the Italians to make the running in any agreements with the Basques and taken no public responsibility for Nicolás's initiative.[49] In any case, since the Basques had not surrendered until the last minute despite taking little active part in the defence of Santander, Franco believed that he owed them nothing.[50]

In the short term, Franco's duplicity brought his war effort enormous benefit. The Basques surrendered at little cost to his troops, allowed their industrial wealth to pass into Nationalist hands intact and were humiliated. In the longer term, Franco handled the Basque situation clumsily. Relations between the Basque ruling party, the Basque Nationalist Party (*Partido Nacionalista Vasco*) and the Republican government in Valencia had been sufficiently tense to raise the possibility of a truce between the conservative, Catholic Basques and the Nationalists.[51] In the view of one Basque Nationalist priest, 'if Franco had been clever, he would have told our troops that they had fought bravely, cleanly and surrendered honourably. Having said that, he would then have called for volunteers to join his army to take Madrid. I'm convinced that eighty per cent of our troops would have responded to such a call on the spot.'[52] As it was, Franco's continuing obsession with humiliating and annihilating his enemies ensured that the Basques would be his fiercest and most effective enemies in the later years of his dictatorship.

The Italians entered the elegant coastal resort of Santander on 26 August. They claimed this as a great triumph, and their troops paraded through Santander holding aloft giant portraits of Mussolini. In Italy, the press gloried in this revenge for Guadalajara even though in reality the Italian troops had faced virtually no resistance. Large numbers of prisoners were herded into the local bullring. Franco regarded Bastico's efforts to shield the Basque prisoners from his 'justice' as intolerable and, in late September, he wrote Mussolini a 'nasty letter' requesting his replacement. Bastico was replaced by his second-in-command, General Mario Berti.[53]

A month earlier, General Faupel was replaced as German ambassador by Baron Eberhard von Stohrer, a tall career diplomat whose figure was the more imposing because of his tendency to wear a long military cloak. Franco bitterly resented Faupel's patronizing arrogance in general and his meddling with Hedilla and the Falange in particular. There was in any case incipient friction between Berlin and Burgos during the summer of 1937. The Germans were determined to secure profit for themselves from Franco's capture of Basque mineral wealth. The scale of German armaments deliveries to Spain could not be paid for in its entirety by Spanish exports of pyrites and other minerals; accordingly, since February 1937, Göring had pursued a policy whereby Germany's favourable trade balance with Nationalist Spain was used to buy mines and mining rights. This policy, known as the *Montana Project*, was pursued aggressively by Johannes Bernhardt and German economic experts sent to Spain for that purpose.[54] After Faupel revealed to the Generalísimo Germany's price for armaments in terms of mineral concessions in Spain, Franco told Serrano Suñer over lunch, 'I would prefer to lose everything before giving up or mortgaging a particle of our national wealth.'[55] Nevertheless, despite his patriotic rhetoric, in mid-July Franco accepted a series of trade agreements with Germany that obliged him to inform the Reich of any economic dealings with third countries, to grant Germany most-favoured-nation status and to give mutual help in the form of the inter-change of raw materials, food and manufactures.[56]

Faupel visited the Caudillo at his new headquarters in Burgos on 20 August 1937 to say his farewells. Franco assured him, with character-istic hypocrisy, that his departure was most unwelcome.[57] In fact, the aggressive economic imperialism of the *Montana Project* had taken some of the rosy illusion out of Franco's attitude to the Third Reich. He derived invaluable assistance from the Axis powers and clearly enjoyed being projected as a fascist leader on a level with Hitler and Mussolini. However, Franco was never inhibited by gratitude. Moreover, just as, as a young man, he had adopted totally the persona of the courageous hero of the Legion, now he believed himself to be the warrior hero restoring Spain's greatness. His own self-esteem inflated by a chorus of sycophants, he found it easy to believe that Germany and Italy really owed him a debt since he had fought a battle in the interests of the Axis. Therefore, with little ado, in early October 1937, Franco announced, to the outrage of the Germans, that all foreign titles to mines and mining rights were null and void.

At first, implausible efforts were made to assure the Germans that the measure was not aimed at them. Eventually, Nicolás gave Bernhardt the excuse that 'the Montana affair could not be settled by the Generalísimo alone, since he could not bear the responsibility for the mortgaging of Spanish property'. Nicolás told the Germans that all would be resolved after the Generalísimo had formed a government and proper investi-gations had been carried out. The Caudillo himself assured von Stohrer that no harm would come to German interests and told him on 20 December that any delays in resolving the problems caused by his decree were due to unavoidable legal difficulties. Göring was furious about the fate of what he called his 'war booty'. When, in early November 1937, Franco ordered the controlled Nationalist press to tone down its attacks on Britain and its corrupt democracy, alarm bells rang in both Berlin and Rome. The Germans suspected that the Caudillo was hedging his bets internationally. Their suspicions were proved right on 16 November when the *de facto* recognition of Franco by London was announced in the form of the appointment of Sir Robert Hodgson as British agent to Nationalist Spain. Franco coolly ignored German insin-uations of withdrawal from Spain convinced that the wider interests of the Third Reich prevented the abandonment of his cause by Hitler. During a meeting on 20 December with Stohrer and Bernhardt, Franco treated them with unusual aloofness and expressed his dismay at the clandestine German acquisition of mining rights. The entire episode is illustrative of Franco's growing self-confidence. After the Montana conflict, Franco's admiration for the Third Reich would be tinged with a distrust of German acquisitiveness which in complex circumstances would help save him from disaster in the Second World War.[58]

On 24 August 1937, as Santander was on the verge of falling, the Republicans launched another offensive along a broad front westwards from Catalonia aimed at encircling Zaragoza. Initiated by Rojo with the specific aim of providing another diversion to gain time for the defence of Asturias, the offensive was facilitated by the presence in the area of troops sent to break the power of the anarchist Council of Aragón. The Aragón front was long and had been a forgotten area of the conflict since anarchists from Barcelona had been held up there in the early days of the war. So peaceful was the front that the opposing sides had occasionally held football matches. That now changed with a ferocious Republican assault concentrated particularly on the small fortified town of Belchite to the south-east of Zaragoza.

Surprisingly, Franco did not take the bait as he had at Brunete. He decided against delaying the assault on Asturias which suggested that he was at last listening to the advice of Vigón. Although reinforcements under Barrón and Sáenz de Buruaga were sent from the Madrid front along with substantial air support, Franco conceded ground that was of little strategic value and limited his efforts to helping General Ponte to hold the city of Zaragoza. As so often, the Republican advance was initially successful. But, after the exhausting effort of Brunete, it was insufficiently supported by reserves and petered out in the appalling heat against fierce Nationalist defence. By 6 September, after two weeks, Belchite fell but Franco was able to see that the Republic's broad strategic attack on Zaragoza had failed. The Generalísimo frequently travelled to the front near Alfaro and discussed with his staff whether to launch a counter-offensive to recover Belchite. He decided that Nationalist morale and élan which were his eternal objective would be better secured by the continued assault on Asturias and he let the front stabilize.[59]

Barely diverted by the assault on Zaragoza, the Nationalists proceeded to mop up the remainder of the north during September and October. Franco planned a great three-pronged assault on a now encircled Asturias which began on 2 September. Under the overall command of General Dávila and led in the field by Generals Antonio Aranda and José Solchaga, troops quickly moved through the rain-swept mountains. Anxious to finish the campaign before the winter, Franco imbued his staff with a greater urgency than was normally the case. The Nationalists' efforts were greatly facilitated by the fact that the Republicans had virtually no air cover. Although Asturias was geographically a strong defensive redoubt, it was tightly blockaded by sea and remorselessly bombarded from the air. The defenders' morale was shattered as the Germans perfected their ground-attack techniques with forays along the mountain valleys, using a combination of incendiary bombs and gasoline to create an early form of napalm.[60]

Gijón and Avilés fell to the Nationalists on 21 October. The balance of power had now shifted dramatically Franco's way. The Republic had lost the coal industry and its northern armies. Franco had gained one hundred thousand prisoners who could be used for forced labour and a large population from which to draw conscripts. The coal mines of Asturias could now be linked with the iron-ore production of the Basque Country. Already better off in terms of tanks and aeroplanes, the Nationalists were now able to consolidate their military superiority through control of the production of iron ore. A powerful and well-equipped army was now free for use in the centre and the east. All the ports of northern Spain were in Franco's hands. The Nationalist fleet, hitherto occupied in the blockade of the northern ports, was free to concentrate in the Mediterranean – now the Republic's only maritime supply route for food and arms imports.[61]

Confident now of his massive military and geographical advantages, Franco turned again to the political consolidation of his regime. To the delight of the Caudillo, images of the *Reconquista* of Spain from the Moors were used to exalt and reinforce the notion that he was the heroic leader of a 'Crusade' to liberate Spain from the godless hordes of Moscow.[62] *Imperio* (empire) became an ideological watchword. However, the imperial verbiage and the references to Ferdinand and Isabella were balanced by more modern borrowings from Fascism and Nazism. The Falangist symbol of the yoke and the arrows, like the swastika and the *fasces*, married the ancient and the modern. Theorists of the regime attempted to elaborate its own Führer principle, the so-called *teoría del caudillaje*, which borrowed from the doctrines of German National Socialism. Parliamentary democracy and the rule of law were dismissed as obnoxious symptoms of the liberal age.

The scripts for the identification of Franco with the great heroes of Spain's past were the work of many hands, including Fermín Yzurdiaga, Ernest Giménez Caballero and even Dionisio Ridruejo. Overall, the legend of Franco the providential Caudillo was master-minded by Serrano

Suñer through the Nationalist press and propaganda machinery which he controlled. The *cuñadísimo* admired many aspects of Nazism and the Third Reich, although, as a fervent Catholic, he was uncomfortable with Nazi atheism. Having spent much time in Italy as a student, he was a more convinced Italophile. He visited the Nuremberg rally in 1937 along with Nicolás Franco and other Nationalist dignatories and, with characteristic sensitivity, felt that he had not been treated with sufficient deference.[63] His interest in contemporary rightist political movements was reflected in his contributions to the statutes of the Falange which, signed by Franco on 4 August 1937, gave the Caudillo absolute power. According to Article 47 of those statutes 'The *Jefe* is responsible before God and before History.' Franco insisted on removing passages which considered the theoretical reasons for which the *Jefe* could be removed.[64]

The pseudo-medieval choreography of the regime made its debut as a consequence of the creation on 19 October 1937 of the first *Consejo Nacional* of the unified *FET y de las JONS*. Modelled on Mussolini's Fascist Grand Council, the *Consejo Nacional* was the supreme body of the single party, the great council through which the aspirations of the various component groups of the Nationalist coalition would be filtered. In theory, it had a Falangist majority. In fact, more than half of the fifty members named were thinly disguised or recently converted monarchists, although almost all were loyal Francoists. In fact, as Franco no doubt intended, when it acquired its full fifty members, they were too numerous and ideologically disparate to make it more than an innocuous talking shop. [65] It was decided that the members of the *Junta Política*, the executive organ of the *Consejo Nacional*, would be designated by the Caudillo himself.[66]

In tune with the historical emphasis of the regime's rhetoric, Franco declared to a French journalist on 16 November 1937 'our war is not a civil war . . . but a Crusade . . . Yes, our war is a religious war. We who fight, whether Christians or Muslims, are soldiers of God and we are not fighting against men but against atheism and materialism.'[67] History was even more to the forefront at the swearing-in ceremony of the *Consejo Nacional*. It had been carefully prepared by the Propaganda Services, under their Director-General Dionisio Ridruejo, and unmistakably revealed the hand of Fermín Yzurdiaga in a choreography of pseudo Golden Age pomp. It took place on 2 December 1937 at the Monasterio de Santa María la Real de las Huelgas to the west of Burgos. Preceded by drummers and trumpeters in seventeenth century period dress, the members of the *Consejo* filed through the cloisters. They swore loyalty to Franco before a gaunt marble Christ figure and the battle standard of the historic battle of the Navas de Tolosa (*el pendón*). When Queipo de Llano tried to protest at the arbitrary nomination by Franco of the members, he was silenced by the Caudillo who snapped 'This is not a parliament and we do not come here to make politics nor to raise trivial points'. He was right – the *Consejo Nacional* served no purpose other than to provide a ceremonial framework for the adulation of Franco and well-paid sinecures for its members.[68]

The confidence in himself and his office that was generated by such ceremonial was revealed in a harsh and dismissive letter which the Caudillo wrote to Alfonso XIII on 4 December 1937. The King, who had recently donated one million pesetas to the Nationalist cause, had written to Franco expressing concern that the restoration of the monarchy seemed not to be a priority. Franco replied coldly, insinuating that the problems which caused the Civil War were of the King's making and outlining both the achievements of the Nationalists and the tasks remaining to be carried out after the war. Expanding on his interview published on 18 July 1937 in *ABC*, the Caudillo made it clear that Alfonso XIII could expect to play no part in that future: 'the new Spain which we are forging has so little in common with the liberal and constitutional Spain over which you ruled that your training and old-fashioned political practices necessarily provoke the anxieties and resentments of Spaniards.' The letter ended with a request that the King look to the preparation of his heir, 'whose goal we can sense but which is so distant that we cannot make it out yet'.[69] It was the clearest indication yet that Franco had no intention of ever relinquishing power.

That Franco was able to give time to thinking about his political future was a sign of the way the military balance was pointing to his ultimate victory. That could be deduced from the transfer of the Republican government from Valencia to Barcelona. The evacuation was justified as facilitating the mobilization of the resources of Catalonia for the war effort, but there was an element of defeatism in the government's removal to a place nearer the French frontier prior to the feared Nationalist push against Valencia.[70]

Franco was ever more assertive and less deferential to both the Germans and the Italians. At the same time as he made a stand on the Germans' *Montana* scheme, being deeply irritated by Italian attempts to hog the glory of the victory at Santander, he let it be known to Ciano's representative Filippo Anfuso that he would be prepared to see Italian troops withdrawn, although he still had need of their artillery and air power.[71] There was more than a hint of bravado in that statement. After the mutual exhaustion caused by the Republican offensive in Aragón, there had been nearly two months of military inactivity during which the Nationalist forces were reorganized into six army corps. Since mid-September, Franco had been thinking in terms of his next great push taking place in Aragón. This would permit him to recapture Belchite and leave his forces poised for one of two great operations – either to attack Valencia and split the Republican zone or else to sweep through Catalonia and cut off the Republic from the French frontier. Many of his own generals, as well as his Axis advisers, counselled such a move through Aragón. However, in the last week of November 1937, his mind turned again to his obsession, Madrid and, in early December, he decided to launch his next attack there.[72]

His plan was to do on the Guadalajara front what the Italians had failed to do in March 1937, that is to say complete the encirclement of Madrid with a push towards Alcalá de Henares. The Generalísimo now had more than six hundred thousand men under his command.[73] With his immediate staff, he set up temporary headquarters at the *parador* of Medinaceli as an army of more than one hundred thousand men was gathered for the attack near Guadalajara.[74] Aware of what was being prepared, Vicente Rojo was urged by Indalecio Prieto to launch another diversionary offensive on 15 December in the hope of turning Franco away from Madrid. It was directed against Teruel, capital of the bleakest of the Aragonese provinces. The Nationalist lines there were weakly held and the city was already virtually surrounded by Republican forces.[75]

The strategy was skilfully elaborated by Rojo in a mere six days and, once again, complete surprise was achieved. The Nationalists, caught unawares, found their aeroplanes grounded by the weather. This allowed the Republican forces to press home their initial advantage and, in the first week, to close a pocket of one thousand square kilometres and, for the first time, to enter an enemy-held provincial capital.[76] Franco was only days away from initiating his own Guadalajara offensive when the first news arrived. The sound advice of the senior German and Italian officers in Spain was to abandon Teruel and go ahead with the planned operation to cut off Madrid. His own staff, including Generals Yagüe, Varela and Aranda, also believed that he should not let himself be diverted from his original plans. However, his determination to bring the Republic to total humiliating annihilation did not admit of allowing the enemy such successes. The capture of Madrid would have hastened the end of the war and possibly, with Rojo having thrown everything into the Teruel offensive, at little cost. In contrast, to snuff out the move against Teruel had little strategic significance and might, and indeed did, take a bloody toll. However, to Franco, its attraction was that it provided the opportunity to destroy a large body of the Republic's best forces.[77]

Franco threw troops into Aragón without entirely renouncing his Madrid offensive. Nevertheless, to the delight of Rojo, he pulled forces away from the capital and towards Teruel.[78] Ciano commented 'Our generals are restless, quite rightly. Franco has no idea of synthesis in war. His operations are those of a magnificent battalion commander. His objective is always ground, never the enemy. And he doesn't realize that it is by the destruction of the enemy that you win a war.'[79] What Ciano failed to perceive was that Franco's obsession with 'ground' was a conscious search for great battles of attrition which could, and did, destroy vast numbers of the enemy's troops.

On 20 December, Franco decided to throw an entire army corps under Varela into the battle for Teruel. On the next day, to the chagrin of many of his own officers, he definitively abandoned his projected assault on Madrid.[80] His headquarters (*Términus*) was established in a train which moved up and down the Jiloca valley to the north-west of Teruel, taking Franco each day as near as possible to the front where he received the daily reports of Dávila.[81] Franco's forces were unable to relieve Colonel Domingo Rey d'Harcourt, the besieged military governor of Teruel. Determined not to permit the surrender of any position, he telegraphed Rey d'Harcourt urging him to defend the city street by street using petrol and hand-grenades until reinforcements could get through. Despite further encouraging telegrams to the garrison from Franco, the Republicans ground down its resistance. The campaign took place in the midst of one of the cruellest winters Spain had ever suffered, the bitter cold intensified by the rocky terrain around Teruel.[82] Franco's handling of the campaign merited only the scorn of the Italian commander General Berti who reported in Rome that there was 'lack of unity in command, inadequate co-ordination, no bite, and no anxiety to finish the campaign'.[83]

Although the snow stopped on 29 December, the temperature dropped to $-20°$ C. The counter-attack ordered by Franco, headed by Varela and Aranda, was held up by the appalling weather. The trucks carrying reinforcements could not pass the snow and ice-bound roads. With the temperature dropping to the lowest recorded levels of the century, the Nationalist relief force reached the outskirts of Teruel on 30 December. Here it was held. The Republicans were subjected to a heavy battering by artillery and bombers. In the freezing conditions, morale was sapped on both sides with soldiers dying from exposure and others having frost-bitten limbs amputated. However, after bloody house to house fighting, Rey d'Harcourt and the exhausted Nationalist garrison succumbed on 8 January. Franco was bitterly upset by the loss of Teruel and got angry with his commanders in a way that surprised a staff unused to seeing him lose his equanimity. Despite his heroic resistance, Rey d'Harcourt was denounced as a vile traitor in the Nationalist zone, the scapegoat for the defeat.[84] Although Mussolini regarded Teruel as no more than a light local success for the Republic, he was worried by the postponement of the Madrid offensive. Ciano was more pessimistic, regarding the Italian position as untenable. He was talking of either seeking a decorous way of disengaging Italian forces or else obliging Franco to mount a decisive campaign against the Republic.[85]

Within ten days of the city falling into the hands of the Republic, the advancing forces of Aranda, Varela, together with a third army corps, that of Morocco, under Yagüe, became the besiegers. On 29 January, Franco told the Italian Ambassador, Count Viola, and the Commander of the CTV, General Berti, of his delight that the Republic was destroying its reserves by throwing them into 'the witches' cauldron of Teruel'.[86] At massive cost to both sides, the battle swayed back and forth until finally, on 7 February 1938, the Nationalists broke through and the Republic lost a huge swathe of territory and several thousand prisoners as well as tons of valuable equipment. It was the beginning of an inexorable advance which in two weeks led to the recapture of Teruel on 22 February, the capture of nearly fifteen thousand prisoners and the loss of more equipment.[87] The Nationalists were now poised to sweep through Aragón at their leisure. After another costly defence of a small advance, the Republicans had to retreat.

What the successive breakdown of the three Republican offensives at Brunete, Belchite and Teruel demonstrated was that the sheer material superiority of the Nationalists could always prevail over the courage of the loyalist troops. Each time, the Republicans had been unable to follow up their initial advantage. In part, this reflected political conflicts within the Republican zone. However, it was also a consequence of the fact that by early 1938 Franco had a twenty per cent advantage in terms of men and an overwhelming one in terms of aircraft, artillery and other equipment.[88] His exploitation of his logistical superiority in regaining Teruel made it the military turning point of the Civil War. Interviewed on 3 March, Franco boasted both of the material superiority bestowed by the victories in the north in 1937 and of the fact that the battle of Teruel had seen the physical annihilation of the best units of the Republican Army.[89] With the Republican army shattered, and unable to rebuild as long as the French frontier remained closed, a splendid opportunity opened up to Franco.

An indication of the strength of Franco's position could be perceived in his treatment of Mussolini during the Teruel campaign. The Duce had written to him on 2 February 1938 threatening to withdraw his assistance unless the war effort was intensified. In view of the break-through at Teruel, Berti, after consultation with Viola, did not deliver the letter for a few days. When they visited him together to put Mussolini's demand for a more energetic conduct of the war, Franco received them amiably without volunteering any information about his strategic plans.[90] Franco left the letter unanswered for two weeks. Mussolini hoped to see Italian troops used in some great strategic victory and fretted accordingly. On 23 February, a frustrated Duce sent a telegram to Franco urging him to fight, using the CTV, or else accede to an Italian withdrawal. By 26 February, still awaiting a reply to his letter of 2 February, the Duce ordered the CTV in Spain and the Italian air forces in the Balearics to abstain from all further operations. This finally galvanized Franco to reply with a letter backdated to 16 February.

The Caudillo's letter was a typical concoction of emollient waffle and imprecision. To Mussolini's advocacy of a crushing victory, the Caudillo expressed total agreement. He described the possible withdrawal of Italian forces as something which would be interpreted throughout the world as cowardice. Perceptively taking for granted that Mussolini's vanity would prevent him risking such opprobrium, Franco went on to request even more supplies. Entangled in the logic of Franco's rhetoric, Mussolini capitulated, merely requesting in a friendly letter dated 3 March, and delivered to him by Viola on 8 March, that the CTV be allowed to participate in 'one good decisive battle'.[91] The balance of power between Franco and Mussolini had changed decisively since the eve of the battle for Málaga.

Even as the battle of Teruel raged, the institutionalization of the Caudillo's rule was confirmed on 30 January 1938 with the long-promised formation of his first regular cabinet. The rule of the *Junta Técnica del Estado* of Burgos was formally brought to an end and the *Secretaría General del Estado* dissolved. Serrano Suñer became Minister of the Interior and the dominant figure. He would be a useful lightning conductor for Franco. The *cuñadísimo* was widely assumed to be responsible for the savage repression when, in fact, the trial and execution of prisoners remained the personal concern of the Generalísimo. Serrano Suñer would have considerable power, controlling the apparatus of press and propaganda. Other posts went to a carefully balanced selection of soldiers, monarchists, Carlists and tamed Falangists. The dominant tone was unadventurous, conservative and, above all, military, with links to the Primo de Rivera dictatorship.

The soldiers chosen were all too old to be serious rivals. Queipo de Llano was simply ignored. The President of the *Junta Técnica del Estado*, the sixty-one year-old General Gómez Jordana, much liked by foreign diplomats for his honest patrician style, became Minister of Foreign Affairs and vice-president of the cabinet. Dourly loyal to Franco, Jordana had served as a member of General Primo de Rivera's military directory with responsibility for Moroccan affairs. The fifty-nine year-old – and extremely short – General Dávila became Minister of Defence. Equally loyal to Franco, Dávila had little personal ambition. The seventy-five year-old General Severiano Martínez Anido became Minister of Public Order. Notorious for his savagery as Civil Governor of Barcelona in the early 1920s when the infamous *ley de fugas* (the shooting of 'escaping' prisoners) had become the norm, he had won Franco's admiration for his imposition of law and order under General Primo de Rivera. Martínez Anido intensified the purge of leftists in the territory captured by the Nationalists. He was to die shortly before the end of 1938 and his functions would then revert to the Ministry of the Interior under Serrano Suñer.

Like Jordana and Martínez Anido, the Minister of Finance, Andrés Amado, a member of Acción Española, had also served the Dictatorship as Director-General of Stamp Duty under his friend Calvo Sotelo. The young monarchist intellectual Pedro Sáinz Rodríguez became Minister of Education. The engineer Alfonso Peña Boeuf was Minister of Public Works. The moderate Carlist, Tomás Domínguez de Arévalo, the Conde de Rodezno, was rewarded for his collaboration in the unification by being made Minister of Justice. A conservative *camisa nueva* (a 'new shirt' or recent convert to the Falange), Pedro González Bueno was also rewarded for his part in the unification by being made head of Syndical Organization and Action, a kind of paternalist ministry of labour.[92]

The first government constituted an early anticipation of the shrewd balancing acts of the next twenty years in which Franco would try to satisfy and neutralize all the forces in the nationalist camp.[93] The cabinet reflected the extent to which the Caudillo would pick ministers for reasons other than competence in the field of their ministry. He wanted to make his brother Nicolás Minister of Industry and Commerce, which Serrano Suñer opposed on the grounds that it was 'too much family'. Franco desisted only when Serrano Suñer threatened to resolve the problem by resigning himself. At the *cuñadísimo's* suggestion, Franco's childhood friend, the marine engineer Juan Antonio Suanzes, was appointed. Suánzes was to be the father of some of the more excessive plans for the disastrous policy of economic autarky (self-sufficiency) later pursued under Franco. Nicolás Franco was sent to Lisbon, where he eventually became a useful intermediary between the Caudillo and Don Juan de Borbón.* Sangróniz was also despatched to a gilded exile, as Ambassador in Venezuela, partly because both the Germans and the Italians regarded him as excessively Anglophile.[94]

There had been talk of Serrano Suñer being named Secretary-General of the *FET y de las JONS* in the cabinet. In fact, when the veteran Falangist Raimundo Fernández Cuesta arrived in the Nationalist zone as a result of a prisoner exchange, he had been given the job. This reflected Franco's cunning. He wanted to make a gesture to Pilar Primo de Rivera and the *camisas viejas* ('old shirts') and also to tame a man who might have been considered heir to José Antonio Primo de Rivera. In his belief that anyone could be bought, Franco was rarely disabused. In his first cabinet, the Caudillo made Fernández Cuesta Minister of Agriculture, a truly disastrous appointment. By means of the rigid imposition of a corrupt syndical administration of the countryside, Fernández Cuesta helped to turn the great wartime agricultural surpluses of the Francoist zone into the famine of the 1940s.[95]

The Caudillo declared that the new government would organize Spain along totalitarian lines, eliminating the class struggle, political parties and the electoral practices of liberal democracy.[96] Franco himself virtually guaranteed the post-war economic difficulties of his regime by opting for autarky in slavish emulation of his Axis allies, thereby renouncing any efforts to gain credits from Britain, France or the United States. Franco's economic naïvety was striking. Within less than a year, and contrary to the evidence of the country's total prostration, he would announce that

* Nicolás was later involved in questionable business deals which exploited his influence with the Caudillo and benefited from his protection when they ended in scandal and accusations of fraud. His activities ranged from the sale of letters of introduction to ministers to profitable participation in companies with government links. Three in particular ended in disaster from the consequences of which he was saved by Franco's benevolence (Ramón Garriga, *Nicolás Franco, el hermano brujo* [Barcelona, 1980] pp. 171–84, 306–20, 269–91).

Spain was on the verge of self-sufficiency in armaments and of resolving her housing, education and health problems. He also declared his certainty that an autarkic Spain could achieve full-scale economic well-being.[97]

All of Franco's ministers coincide in the recollection that the Generalísimo left them entirely free to pursue their own departmental policies. Their only obligation was to be bound to the general direction of policy agreed at cabinet meetings. The sessions would last for hours with acrimonious debates between the Falangists and the more conservative and, usually monarchist, military ministers. Franco said very little and confined himself to listening. That was in large part because, until the late 1950s, crucial overall policy decisions would often be made by him personally outside the council of ministers. Franco's conduct of the cabinet meetings was symbolic of his technique of *dejar hacer*, ruling over the Nationalist zone as supreme arbiter. He could, and did, remove and name ministers at will. However, he would change his ministers rarely, giving them considerable leeway unless their political ambitions began to threaten him.

On the day after the announcement of the government, Franco received a number of foreign diplomatic representatives in Salamanca, including the British agent, Sir Robert Hodgson, who found the now greying Caudillo to be an attractive figure: 'He has a soft voice and speaks gently and rapidly. His charm lies in his eyes, which are of a yellowy brown, intelligent, vivacious, and have a marked kindliness of expression.' Their encounter was cordial. Hodgson assured Franco that the British attitude towards Nationalist Spain was entirely disinterested and that London hoped to maintain friendly relations with his government. Hodgson accepted that British efforts to be neutral might not have seemed sufficiently friendly to Franco. The Generalísimo replied ingratiatingly that the English lessons so brusquely interrupted in the Canary Islands in July 1936 were indicative of his affectionate feelings towards Britain. He claimed with bare-faced hypocrisy that the first legislative plans of his new government would 'harmonize with English ideas'.[98]

Two days later, the new cabinet issued a manifesto in which, as well as expressing faith in imminent victory, Franco revealed his determination for Spain to be taken seriously in foreign affairs. Spain's 'sense of honour is too great to allow her to forget those who were her friends in the days of trial during the Communist menace'.[99] That clearly implied a commitment to the Axis. Accordingly, one of the first political milestones of Franco's new government was the *Fuero del Trabajo*, a pseudo-

constitution based on the Italian *Carta del lavoro* which was approved on
9 March 1938, having been elaborated with startling haste. The *Consejo
Nacional* of *FET y de las JONS* had sponsored a draft drawn up by advocates of the Falangist 'revolution', including Ridruejo. When their radical
text was put to the cabinet on 1 March, the conservative ministers were
appalled. Franco curtailed debate by insisting on a programmatic declaration within forty-eight hours since the law had been announced in the
press as forthcoming. It was left to Ridruejo and Eduardo Aunós, of
Acción Española, to elaborate a compromise. Claiming to represent a
middle way between 'liberal capitalism and Marxist materialism', the
Fuero set out to implement the Falange's unfinished revolution (*revolución
pendiente*), granting Spaniards '*Patria*, bread and justice in a military and
in a gravely religious fashion'. The cloudy rhetoric was vaguely progressive, but reflected the influence of the more conservative elements of the
FET y de las JONS, since it dropped two of the twenty-seven points
of the original Falange – the nationalization of the banks and agrarian
reform.[100]

Since the *Consejo Nacional* was a decorative sham, the tasks of the
Falange were assumed by its executive committee, the twelve-man *Junta
Política* which met a few times under the chairmanship of Franco. Even
this came to an end after Franco was driven into a rage in the spring of
1938 by what he saw as efforts to impose a more Falangist stamp on his
regime. The background was a plot by a number of Falangist hotheads,
led by Agustín Aznar and Fernando González Vélez, to push Franco in
a more radical direction. This was paralleled by an open Falangist attempt
to change the statutes of the party to bring it more into line with the
Nazi and Fascist Parties. At the same time as the Falangists pressed for
a more totalitarian structure, the monarchists were constantly trying to
break the power of the Falange. Franco stood in between, irritated that
either group should attempt to impose their will upon him yet usually
dealing with them both with infinite patience and cunning. On this
occasion, Dionisio Ridruejo had been appointed to head a sub-committee
of the Junta Política, to examine the statutes of *FET y de las JONS*.

Ridruejo read out the more or less totalitarian proposal drawn up by
his sub-committee, to a meeting of the *Junta Política* chaired by Franco.
The monarchist Pedro Sáinz Rodríguez protested. He pointed out that
the proposal gave excessive power to the Party and thereby implied a
lack of confidence in the government. With the plot on his mind, Franco
exploded. In a totally uncharacteristic loss of temper, he shouted 'not in
the government, but in me, in the Caudillo, disloyalty to him'. Banging

the table with his fist, he shouted 'I should have shot Hedilla! Who are these Ridruejo, Aznar, González Vélez and company to define the regime?' Ridruejo did not understand the reference to Aznar and González Vélez. Nevertheless, he rose, stated calmly that he had been put on the sub-committee precisely to initiate such discussions and said that he could not see any challenge to the government in seeking greater powers for a party whose absolute leader was also head of the government. Finally he turned to Serrano Suñer and said 'Goodbye Ramón, I thought we came here to reason.' Franco, on one of the very few occasions on which he had publicly lost his temper, was annoyed with himself. He suddenly calmed down and asked Ridruejo to resume his place and to forget the incident. Two days later, the Generalísimo had González Vélez and Aznar imprisoned.[101]

This episode showed a Franco who was acquiring the political skills of a Machiavellian prince or, perhaps, rediscovering practices he had observed in Morocco between 1912 and 1925. He had travelled a long road since leaving the Canary Islands at the start of the Civil War. The adventurous and ebullient soldier of the first months of the war was coming to resemble an oriental despot, calculating and duplicitous. Having adopted the persona of semi-monarchical crusader, he was determined to stay in power. His reversion to a watchful *gallego* caution and the cunning cruelty of a Moroccan tribal chief was, in that context, a natural process. At the end of one cabinet meeting, the Minister of Industry, Suanzes, reminded Pacón Franco Salgado-Araujo of the days in El Ferrol when all three were pupils at his father's cramming institution. He contrasted the diminutive Franco of those days, always being knocked about by the other boys, with the Caudillo who now imposed respect to the extent that 'we dare not even give him a pat on the shoulder'. The ever-present Pacón observed that for all his pleasant manner, Franco always maintained a distance even with his friends.[102] With victory now apparently imminent, and surrounded by sycophants, the tendency to an icy regal hauteur would increase dramatically.

XII

TOTAL VICTORY

February 1938–April 1939

THE VICTORY at Teruel opened up the possibility of striking a series of crippling blows against the Republic. Franco lost little time in taking advantage of his now overwhelming superiority in men, aircraft, artillery and equipment relative to the nearly exhausted Republicans. Nevertheless, he was to make serious military mistakes in the spring and summer of 1938, although they were perhaps not as crass as has been made out. His concern with the physical annihilation of the enemy dominated his thinking and left no room for stylish strategic operations. That being the case, he should be seen as more than the petty-minded battalion commander so often derided by Hitler, Mussolini and Ciano. He was now to show some skill in handling a large army of several hundreds of thousands of men across a huge front.

The detailed plans for a great march to the east were drawn up by General Juan Vigón. On 24 February 1938, the Generalísimo outlined these plans to a meeting of senior commanders held in Zaragoza. Two hundred thousand men would advance across a 260-kilometre wide front following the direction of the Ebro valley with major operations to the north towards the Pyrenees and to the south towards Valencia. Franco was now throwing his massive material superiority into the game, confident that it would stretch the Republican armies to breaking point.

The offensive was entrusted to the overall command of General Dávila. The job of punching through enemy lines fell to Yagüe who was instructed to use his German and captured Russian tanks to cover the infantry but in the event was to attempt the nearest thing to a *Blitzkrieg* attack ever permitted by Franco.[1] The Generalísimo was seriously depressed by the sinking of the cruiser *Baleares* on 6 March 1938.[2] However, his optimism soon reasserted itself and he quickly pulled himself together for the new operation. The massive push was begun through Aragón on 9 March on the same day as he presided at the cabinet meeting which approved the *Fuero del Trabajo*. The objective of the new operation to destroy more Republican forces and to reach the point where the River Segre, which ran north to south through western Catalonia, met the Ebro, running west to east, near Lérida. It was because it went so unexpectedly well that, within eight days, Franco could contemplate pushing on to the sea and cutting off Catalonia from Valencia and the central Republican zone. That a Nationalist victory at Teruel might be the prelude to such an offensive was the great fear of both Prieto and Rojo, both of whom realized that, once Franco had cut the Republican zone in two, the end would be in sight.[3] Indeed, failure at Teruel was the prelude to the fall of Prieto, accused of a by now not unreasonable defeatism.

A huge Nationalist force, consisting of Yagüe's Moroccan Army Corps, Aranda's Galician Army Corps, Varela's Castilian Army Corps, Solchaga's Navarrese Army Corps, Moscardó's Aragonese Army Corps and the CTV under General Berti, advanced at speed. They were backed by the Condor Legion and, in an attempt to destroy civilian morale, by the indiscriminate Italian bombing of Barcelona in which more than one thousand people lost their lives. Although it is clear that Franco previously and subsequently permitted the bombing of industrial and military targets in Barcelona, as well as of other Republican cities, in this case he was outraged. He was indignant that Mussolini had not consulted him and, in his desire to see 'the Italians horrifying the world by their aggressiveness instead of charming it by their skill at playing the guitar', had ordered the bombing of residential areas. Franco was infuriated by what he considered to be a blunder which had merely strengthened the Catalan will to resist, the more so as he normally tried to avoid damaging the homes of his own supporters.

The Vatican representative in Spain, Monsignor Antoniutti, appealed to Franco to do something to stop the slaughter. Lord Perth, the British Ambassador in Italy, also protested about the raids. Perth was told by Ciano that the operations were initiated by Franco and that Italy could do no more than try to influence him to stop them. However, Ciano later noted in his diary that Franco knew nothing about them.* It was

---

* In 1967, discussing the entry in Ciano's diary with his cousin Francisco Franco Salgado-Araujo, Franco said that 'all bombings were by special decision of the Spanish high command'. That was generally true but, with regard to this specific case, the then seventy-five year-old Franco's memory failed him. (Franco Salgado-Araujo, *Mis conversaciones*, p. 494.)

known in Franco's *Cuartel General* that the order had emanated from Mussolini. The Caudillo requested the Duce to refrain from issuing direct orders to Italian aviators based in Mallorca.[4]

Irrespective of the bombing attacks, the new advance went spectacularly well. One hundred thousand troops, covered by two hundred tanks and nearly one thousand German and Italian aircraft, took part. Colonel Wilhelm von Thoma, in command of the Condor Legion's fast tank units, wanted to use swift *Blitzkrieg* tactics but Franco, in the style of First World War generals, planned to use tanks only as infantry support. Von Thoma made his point to the Caudillo, much to the delight of Yagüe. It hardly mattered given the Nationalists' massive material superiority.[5] Franco took a great personal interest in the campaign, setting up his headquarters from 9 March in the magnificent Palace of the Duque de Vistahermosa in the small town of Pedrola, thirty kilometres north-west of Zaragoza on the road to Logroño. He was to spend much of his time there in the last year of the war, accompanied by a small staff including Pacón, Martínez Fuset, Barroso and his chaplain Father José María Bulart.[6]

After an opening artillery and aerial bombardment, the Nationalists found their Republican opponents exhausted, short of guns and ammunition and generally unprepared. Demoralization after the defeat of Teruel was compounded by organizational confusion and the devastating air superiority enjoyed by the Nationalists. Prieto, Miaja and Rojo had underestimated the scale of Franco's advance. Assuming reasonably that the Nationalist leader could not see beyond his own obsession with Madrid, they had been reluctant to move forces from the capital to strengthen the Republican army in Aragón.[7] On 10 March, the ruins of Belchite were recaptured by Yagüe, who was embraced by an emotional Franco on the next day. It was a symbolic loss for the Republic which also entailed the destruction of large numbers of the Popular Army.[8] On 15 March, Franco issued the historic order in which he stated that the disorganization and demoralization of the enemy was now such as to permit a bid to reach the coast.[9] With the Republican forces in total disarray, on 23 March Yagüe crossed the River Ebro near Quinto, where he set up his headquarters.[10] Solchaga had reached the Pyrenees, while the Italians, and the forces of Aranda and the brilliant young Rafael García Valiño were sweeping into the rugged Maestrazgo in the south of Aragón.

By early April, with the Republican Army of the East crumbling before them, the Nationalists had reached Lérida. The city fell to Yagüe on

4 April after a brave defence by the division commanded by the Communist 'El Campesino'.[11] Yagüe argued that the rapid occupation of an isolated and badly defended Catalonia, thus sealing the French frontier, was the best way to stop deliveries of arms to the Republic. Many of Franco's own immediate staff, including Vigón and Kindelán, believed that the Generalísimo should use his now overwhelming superiority to do this and finish off the Republic. Kindelán made the point in frequent visits to Pedrola and in an avalanche of letters.[12] General Hellmuth Volkmann, who had replaced Sperrle as commander of the Condor Legion on 1 November 1937, was instructed by the German War Ministry to urge Franco to continue the drive against Catalonia until the entire region was conquered and not to call a halt there in order to go on the offensive on other fronts.[13] Had the Generalísimo followed all this advice, he could probably have brought the war to a speedier conclusion. There were no significant Republican forces between Lérida and Barcelona and the Republican armies of the south and centre were unable to help Catalonia. Victory beckoned. In Catalonia lay the Republic's remaining war industry and the seat of the Republican government. The loss of the region would therefore be a devastating blow to Republican morale.[14] In the event of Franco not choosing to advance through Catalonia, it was assumed that the next best target would be Madrid.

However, it seems that the Generalísimo did not favour an attack on Catalonia because a sudden Republican *débâcle* occasioned by the loss of Barcelona would still have left a substantial number of armed Republicans in central and southern Spain. Similarly, a swift collapse as a result of victory at Madrid would have left numerous Republican forces in Catalonia and in the south-east and Franco's aim remained the total annihilation of the Republic and its supporters.[15] His decision not to go for a swift kill may also have been motivated in part by the fear that, following the German *Anschluss* with Austria on 11 March 1938, the French were sufficiently worried by fascist triumphs around them to contemplate intervening on the side of the Republic in Catalonia. On 16 March, Franco's staff received information from a sympathizer in the French General Staff that Blum had proposed to the Permanent Committee of National Defence an ultimatum to Franco demanding that he renounce support from foreign forces. There was talk in the French press that three or possibly five French divisions were about to be sent to the Catalan front.[16] In the event, Blum did no more than permit the passage of supplies to the Republic. The re-opening of the French frontier on 17 March brought renewed hope to the defenders of the Republic.

However, Franco's anxiety about the activities of the French cannot entirely account for his cautious turn away from Catalonia.[17]

After considerable hesitation, and to the chagrin of Kindelán, Vigón and Yagüe, and to the astonishment of Azaña, Franco diverted his troops to the south for a major attack on Valencia.[18] Yagüe was ordered to dig in along the Segre and many of his best units were transferred to the south although he and Colonel Heli Rolando de Tella were keen to continue their advance to Barcelona. Yagüe's success had exceeded all expectations but Franco was reluctant to let him move on without having detailed contingency plans. He said years later 'I've never played a card without seeing what came next and at that moment I couldn't see the next card'. The international situation certainly made the next cards difficult to predict but it is hard to avoid the conclusion that Franco was motivated not just by his natural caution but also by a reluctance to move to the definitive victory before further destruction and demoralization of the Republic's human resources. Vicente Rojo was astonished when Franco turned to the Maestrazgo, writing later that, had Franco continued the attack against Catalonia, 'with less effort and in less time, he would have had in May 1938 the triumph of February 1939'.[19]

Yagüe's consequent frustration exploded at a Falangist banquet held in Burgos on 19 April 1938 to commemorate the anniversary of the Unification of *FET y de las JONS*. His irritation at Franco's military hesitations stirred him to express his even greater dislike of the conservative direction of Nationalist politics. He praised the bravery of Republican soldiers. Using language previously heard only from the lips of Hedilla, he spoke of the need to heal wounds, to establish social justice, to make overtures to thousands of Republicans in Francoist jails and to have the magnanimity to forgive. He spoke even more warmly of the need to pardon Hedilla and his imprisoned followers. Yagüe was briefly relieved of his command as part of Franco's carrot-and-stick policy. It was as successful as always. Yagüe was restored to command of the Moroccan Army Corps after a few weeks and was to be found writing affectionately to Franco within a matter of months.[20]

Franco's forces moved down the Ebro valley cutting off Catalonia from the rest of the Republic. By 15 April, Good Friday, they had reached the sea at the fishing village of Vinaroz. On the beach at Benicasim, joyful Navarrese soldiers under the command of Franco's friend Camilo Alonso Vega cavorted in the waves. Alonso Vega crossed himself with water from the Mediterranean.[21] The Nationalist press declared joyfully that the end of the war was imminent. 'The Sword of Franco has divided in two the

Spain still held by the Reds.'[22] Franco was sufficiently optimistic to talk
to the Germans and Italians about a possible withdrawal of volunteers,
albeit with the Condor Legion's equipment to remain in Spain.[23] On
23 April, a major advance across a wide front was launched under Generals
Varela, Aranda and García Valiño down through the difficult terrain of
the Maestrazgo with the aim of widening the drive to the Mediterranean.
However, the arms that the Republic had been able to secure as a result
of the re-opening of the border with France led to the Nationalist advance
being reduced to a painful crawl.[24] By the beginning of May, the Nationalist advance was grinding to a standstill on all fronts and the Generalísimo
was again forced to rely on the Condor Legion and the CTV remaining
in Spain until victory was assured. To ensure that the Condor Legion
would not be withdrawn, the Caudillo made greater concessions than
ever before to the Germans on the scale of their participation in Spanish
mining.[25]

However, the respite for the Republic was brief. Blum's second administration, which in any case had been hamstrung by the lack of a clear-cut
majority, lasted just over six weeks before Daladier took over in late April.
He closed the border with Spain again on 13 June. In the meanwhile,
after the resignation of the anti-Fascist Foreign Secretary Sir Anthony
Eden on 20 February 1938, Britain was going ever further down Chamberlain's road of appeasement at any price.[26] An Anglo-Italian treaty had
been signed in April, whereby the British tacitly condoned the Duce's
intervention in Spain.

Appalled by Franco's continued slowness, Mussolini sent nearly six
thousand new troops and large numbers of aircraft to Spain in June and
July.[27] The injection of Italian equipment revived the Nationalist advance
and blanket bombing and heavy artillery fire gradually pushed back the
Republicans. Even so, Franco's offensive against Valencia did not go as
planned. His generals, Varela, Aranda and Garcia Valiño, found progress
towards the coast slow and exhausting as the Republicans demonstrated
their usual skill and determination in defence. Through the use of well-
planned trenches and properly protected communications lines, they were
able to inflict heavy casualties on the Nationalists while suffering relatively few themselves. Nonetheless, the progress of the Nationalists was
inexorable, if painfully slow. On 26 May, Kindelán had written to Franco
pleading with him to abandon an operation which was incurring high
casualties but the Generalísimo was unperturbed.[28] A note in Franco's
handwriting dated 18 May suggests that he had contemplated regularizing
the battle lines in the Maestrazgo and launching an attack on Catalonia.

However, having committed himself to the costly drive on Valencia, he refused to be diverted.[29] On 15 June, Aranda took Castellón.[30]

As a consequence of the Nationalist success in reaching the Mediterranean coast, sea communications between the two halves of the divided Republican zone assumed a crucial importance. Franco was determined to eliminate Republican maritime commerce and authorized the Condor Legion and the Aviazione Legionaria to undertake indiscriminate bombing of the Republic's poorly defended coastal towns and of merchant shipping in the area. Since only British-registered ships enjoyed effective protection, they constituted a disproportionately high percentage of Republican shipping and therefore of the tonnage sunk by bomb attacks. Between mid-April and mid-June, twenty-two British-registered ships were attacked off the Spanish coast and eleven of them were either sunk or seriously damaged. At the same time, in an effort to smash Republican morale, savage attacks were made on open cities such as Valencia, Alicante and Barcelona. Granollers, a small town thirty kilometres to the north of Barcelona, was bombed on 2 June and several hundred women and children killed. This, together with the attacks on British ships, finally provoked protests in London where Churchill led calls for firm action.[31]

Lord Halifax and Neville Chamberlain were significantly embarrassed by what was happening although neither wanted to do anything to undermine Franco's position. However, the stock of the anti-Francoist Eden began to revive and there were fears that Chamberlain might fall. Chamberlain wrote in his diary 'I have been through every possible form of retaliation, and it is absolutely clear that none of them can be effective unless we are prepared to go to war with Franco . . . of course, it may come to that, if Franco were foolish enough'.[32] After the humiliating sinking of a British ship in sight of a Royal Navy warship, Sir Robert Hodgson, the British diplomatic agent, was recalled for consultations. Efforts were made to get Franco to restrain his attacks through representations made confidentially to both the Germans and the Italians, which only increased the contempt felt for Chamberlain in both Rome and Berlin. After a gentle request from Lord Perth that Italy intercede against Franco's bombing policy, Ciano replied that the attacks were directed by Franco's military advisers and that the Italians were not responsible for them. He then told the German Ambassador Hans Georg von Mackensen that 'we have of course done nothing and have no intention of doing anything either'.[33] Ribbentrop similarly dismissed the request of the British Ambassador to the Third Reich, Sir Nevile Henderson, by telling him 'the question of the air raids was the concern of General Franco, to whom, moreover, we could give no advice regarding his conduct of the war'.[34] Halifax told the German Ambassador in London, Dr Herbert von Dirksen, of the concern caused in Britain by the bombings, although he made it clear that 'he wished *at all events* to avoid creating any ill feeling in Germany'.[35]

Franco's initial response to British protests was to brush them off. He regretted the 'incidental' loss of British lives but refused to guarantee that no British ships would be bombed in the future on the grounds that they might be carrying materials of use to his enemies, in which he included coal and food.[36] However, it was feared in Burgos, Berlin and Rome that Chamberlain's position might be endangered by the bombings and there was agreement that nothing should be done to intensify any risk of his demise and replacement by Eden.[37] The Germans were also annoyed that Franco had permitted reports to circulate that the Condor Legion was carrying out these bombing reports on its own initiative when in fact they were carried out by both Germans and Italians operating on his orders.[38] The German Foreign Office instructed Ambassador Stohrer to insinuate to Franco that, if he continued to allow the opprobrium for these bombing raids to be borne by the Condor Legion, German forces might have to be withdrawn from Spain. In fact, Berlin had already taken the decision to reinforce the Condor Legion in recognition of Franco's decision to open the way to greater German participation in Spanish mining.[39]

Pressure from Berlin and advice from the Italians to refrain from bombing British ships in Republican harbours inclined the Caudillo to call off the raids. Despite a life-time of anti-British rhetoric, and of accusations that London supported the Republic, Franco knew well enough the value to him of Chamberlain's policies and feared the return of Eden.[40] At the beginning of July, an unofficial envoy from London, Lord Phillimore, an enthusiastic militant of various pro-Nationalist organizations, arrived in Spain. Phillimore visited Franco with the knowledge and approval of the Prime Minister. He was Chairman of the 'The Friends of National Spain' and a prominent member of the 'United Christian Front' which was devoted to proving that Franco was fighting for Christianity against anti-Christ. On receiving Phillimore, Franco, anxious to maintain the covert British support which was so useful to him, sent a grovelling message of thanks to Neville Chamberlain. A somewhat embarrassed Halifax passed on the message to an equally embarrassed Chamberlain who agreed that no publicity at all should be given to the note.[41] More importantly, Franco moderated attacks on British ships, if not on undefended Spanish towns.

On 18 July 1938, the second anniversary of the military uprising, the Nationalist government resolved to elevate to the 'dignity of Captain-General of the Army and the Navy, the Chief of State, Generalísimo of the Armies of Land, Sea and Air, and *Jefe Nacional* of the *Falange Española Tradicionalista y de las JONS*, Excelentísimo Señor don Francisco Franco Bahamonde'. The decree stated that the government wished thereby to 'pay just tribute to the man who, by divine plan, and assuming the greatest responsibility before his people and before History, had the inspiration and the wisdom and the courage to lift up the authentic Spain against the *antipatria*; and then, as the inimitable architect of our entire Movement, personally and in unequalled fashion directs one of the most difficult campaigns known to History'. When Franco was ceremonially presented with the sash and baton of his new rank, he compared himself with the great captains of history who attained triumph by being the instrument of destiny.[42] Captain-General was a rank previously reserved for the Kings of Spain. The Caudillo could now fulfil a lifelong dream. He began to indulge the caprice of appearing at some public functions in the uniform of admiral of the fleet.[43]

In his speech, the Caudillo stressed his close relationship with José Antonio Primo de Rivera, citing the letter written to him by the Falangist leader shortly before the October 1934 uprising, although omitting to mention that he had failed to answer it. Among other sleights of hand, he denounced the Republic's use of foreign volunteers as a 'foreign invasion' and declared, without irony, that the destroyers of Guernica, meaning the imaginary Basque dynamiters of his propaganda, had lost their right to call themselves Spaniards. For the rest, the speech was an imprecise account of what could be expected from his egalitarian 'National Revolution'.[44] The vaguely fascist rhetoric was matched by the choreography of the event which recalled the style of public occasions in both Hitler's Germany and Mussolini's Italy. On the previous evening, a torchlight procession of Falangists had marched through Burgos before pausing in front of the Palacio del Generalísimo where they sang '*Cara al Sol*'. On 18 July, the buildings and streets of Burgos were decorated for imperial pageantry which was an uneasy mixture of the fascist and the medieval. The streets were sanded for the military parade and, along the route, huge portraits of the Caudillo adorned the walls of public buildings and large obelisks were erected, crowned with the Falangist yoke and arrows and the Spanish imperial eagle. After the ceremony,

Franco left the palace surrounded by his exotic Moorish Guard, another symbol of his imperial status.[45]

Such theatrical ceremonial reflected the fact that the end of the war, after all, seemed quite near. By 23 July 1938, Valencia was under direct threat, with the Nationalists less than forty kilometres away. In an attempt to restore contact between Catalonia and the rest of the Republican zone, a desperate diversionary assault across the River Ebro was conceived and planned by General Rojo. A special Army of the Ebro was formed for the purpose and placed under the command of the Communist General Juan Modesto. Franco had put the onus of the Nationalist defence to the south of the Ebro on the impetuous Yagüe. There were rumours of an imminent Republican thrust across the river and specific reports were carried to the Nationalists by deserters from the Popular Army. Yagüe, however, was unable to ascertain where, when and how. Given the sheer length of the Ebro front, the forces at his disposal were insufficient to guard the river in equal depth along its course. In addition, the fact that Franco's main concern was the stalemate in the Maestrazgo ensured that Yagüe's constant requests for reinforcements were ignored by the Generalísimo's *Cuartel General*.[46]

A huge concentration of men, numbering some eighty thousand, was secretly transported to the river banks. On the night of 24–5 July, using boats, the first units of Juan Modesto's army crossed at the bend in the river near Gandesa. The remainder crossed on pontoon bridges on the following day. They surprised the thinly held Nationalist lines.[47] A combination of over-confidence and poor intelligence meant that the scope and scale of the Republican advance had been misjudged. Yagüe himself informed Franco on 22 July that he was optimistic about being able to deal with any attack. Franco was in Burgos when it finally occurred. Woken by Pacón, he immediately telephoned Kindelán to order a massive air bombardment of the Republican bridgeheads.[48] The Popular Army was able to inflict serious casualties on Yagüe's troops, although the 14th International Brigade sustained heavy losses and was forced to withdraw. Further upstream, however, the Republican forces succeeded in establishing a major bridgehead within a broad bend in the river. By 1 August, they had reached the outskirts of Gandesa forty kilometres from their starting point. Gandesa was the centre of an important network of roads which would have been the ideal launching pad for further Republican advances.

It has been claimed that, while his staff was dismayed by the crossing, the visionary Franco was delighted by the attack, immediately seeing the opportunity to tempt the Republicans into a trap, allowing them to pour across the river in order to encircle and smash them. It is certainly the case that Franco's staff were initially demoralized by the Republicans' strategic success.[49] The reaction of Franco himself, however, was icily phlegmatic but hardly visionary. As so often before, his response was simply to set out to regain the lost territory. Having initially ordered the Republican forces to be pounded from the air, he then rushed in reinforcements to contain the advance. On 2 August, he visited the front for the first time. He established headquarters again in the Palace at Pedrola. Shortly afterwards, a mobile headquarters, in the form of a heavily camouflaged, and well-guarded, convoy of lorries, was set up near Alcañiz, disguised as a radio station.[50] Within a few days, Franco began to see the possibility of turning against the Republican bridgeheads in force and unleashing a relentless battle of attrition in order, at no little cost to himself, to smash the enemy army. The push against Valencia was abandoned. A desperate and strategically meaningless battle began for the territory which had been taken involving a bloodbath worse even than those of Brunete, Belchite and Teruel. But Franco was oblivious to losses once he saw the opportunity to crush the Republican army.[51]

To the despair of some of his own staff and of both Germans and Italians, the battle at the Ebro was to last for nearly four months. Had he chosen merely to contain the Republicans with their backs to the river and proceeded to a rapid attack in the direction of a now virtually undefended Barcelona, he might have hastened the end of the war by six months. When Kindelán made this point to him, Franco merely shrugged his assent but said nothing.[52] With nearly one million men now under arms, he could afford to be careless of their lives. His background in the African wars did not incline him to behave otherwise. He preferred to turn Gandesa into the cemetery of the Republican army rather than secure a swift and imaginative victory.[53]

The dismay felt throughout the Nationalist high command took its toll on morale and was manifested in a rash of criticism of the Generalísimo's judgement. Kindelán questioned the cautious use of tanks.[54] Pacón claimed that the Ebro was the only battle in the Civil War in which disagreements between Franco and his generals became public.[55] Baron von Stohrer reported violent scenes as Franco accused his generals of not carrying out orders correctly.[56] The Italian General Mario Berti informed Franco, on Mussolini's instructions, that he must intensify his efforts to end the war and that further matériel could be sent only if the Italians had more say in the strategic conduct of the war effort. Count Viola told von Stohrer that this was a bluff but that it reflected Rome's anxiety about Franco's dilatory style.[57] The Duce was becoming ever more pessimistic about Franco's 'flabby conduct of the war'. He told Ciano, 'Put on record in your diary that today, 29 August, I prophesy the defeat of Franco. Either the man doesn't know how to make war or he doesn't want to. The reds are fighters, Franco is not.'[58] In fact, weighed down by the atmosphere of stalemate and war weariness, the determined Republican premier Juan Negrín was looking for a compromise peace but Franco was set on exacting nothing less than unconditional surrender.

The irrevocable sentence of death for the Republic came with the British reaction to the Czechoslovakian crisis during late September. Negrín had pinned his hopes on an escalation which would facilitate his alignment with the western democracies. To that end, he made the gesture of announcing that he would be withdrawing all volunteers on the Republican side. Franco agreed with Negrín that the outbreak of a general European war would jeopardize Nationalist victory. He assumed that the Republic would align itself with France and Russia against Germany and that, while supplies would flood into the Republicans, Nationalist Spain would be virtually cut off from the Axis powers and threatened by the French army. The Caudillo was dismayed that Hitler should have timed the Sudeten crisis without consideration for the problems of the Spanish Nationalists and he awaited the outcome of the Munich meeting in painful suspense. The cabinet remained in permanent session. Even Mussolini believed that the Nationalists had lost their chance of victory and that Franco must now seek a compromise peace.[59] The stalemate at the Ebro was deeply depressing and possible French incursions into Catalonia, the Basque Country and Spanish Morocco were a daunting prospect for Franco's overstretched forces.[60] Franco was perplexed and hurt because he received no information from Berlin until after the Munich agreement was signed. It is indicative of the stress that he was suffering that, for the first time in years, he was unwell, confined to his headquarters in Aragón.[61]

Despite feeling betrayed, Franco took action to decouple Nationalist Spain from any international conflict over Czechoslovakia only after he received enquiries from London and Paris about his stance. The British and French governments wanted to know what he planned to do in the event of a general European war. Jordana reflected Franco's anxieties when he told the British diplomatic agent in Burgos, Sir Robert Hodgson, of the Caudillo's 'warmest feelings of sympathy for England' and his concern for the success of Chamberlain's peace initiatives. Jordana denied that Franco would line up with the Axis and claimed that he intended to maintain strict neutrality.[62] The same message was repeated in London by his representative, the Anglophile Duke of Alba, along with a request that the Foreign Office persuade the French to respect his neutrality in the event of European war.[63]

Franco's primordial worry was that nothing should stand in the way of his own war effort inside Spain. Accordingly, he was simultaneously determined to ensure that he was not thereby alienating his Axis allies. The Germans and the Italians were thus informed that, despite Franco's deep sympathy for their cause, he regretted that Nationalist Spain was not yet strong enough to line up on their side.[64] Ciano wrote in his diary 'Disgusting! Enough to make our dead in Spain turn in their graves.' Mussolini also reacted violently and talked of pulling all his troops out of Spain although he quickly calmed down.[65] Hitler was appalled at what he saw as Franco's ingratitude but later commented to Göring 'it's a filthy trick but what else can the poor things do?'[66]

In the event, rather than risk war with Hitler, Chamberlain effectively surrendered Czechoslovakia with the Munich Agreement of 29 September.[67] A deeply relieved Franco immediately sent his 'warmest congratulations' to Chamberlain for 'his magnificent efforts for the preservation of peace in Europe'.[68] He perhaps perceived, as did Winston Churchill, that Britain had 'sustained a defeat without a war'.[69] Certainly, he hastened to send his congratulations to Hitler, not on securing peace, but on the favourable settlement of the 'Sudeten German question'. At the very least, he was anxious to counter the feeling in both Rome and Berlin that his declaration of neutrality had been made in unseemly haste. Munich delivered a devastating blow against the Republic and it was hardly surprising that Franco should also express personally to von Stohrer his enthusiasm for the Führer's triumph.[70] Shortly thereafter, Franco authorized a concerted policy of denigration of President Roosevelt and Secretary of State Cordell Hull. The American Ambassador concluded that 'the Franco regime is hostile to the United States, its leaders and its principles and policies'.[71]

By opening the dams on the Pyrenean tributaries of the Ebro, the Nationalists managed to cut off the Republican forces which were trapped in hilly country with little cover and short of supplies. Under orders not to retreat, the Republicans doggedly clung on despite fierce air and artillery bombardment. Five hundred cannon fired over 13,500 rounds at them every day for nearly four months. In sweltering heat, with little or no water, shelled from dawn to dusk, they held out. Now in the autumn,

according to von Stohrer, the Nationalists were being 'bled white'. Enter-
taining serious doubts about a Nationalist victory, he believed,
erroneously, that the Generalísimo might countenance a compromise
peace. In fact, Franco was adamant that under no circumstances would
he contemplate any kind of mediation.[72]

Conscious that Munich had shattered the last hopes of the Republican
government for salvation in a European war, and ever more determined
to smash the Republican army definitively, Franco was gathering over
thirty thousand fresh troops and seeking substantial deliveries of new
German equipment with which to arm them. From mid-October, von
Stohrer seized the opportunity of Franco's need to push him again on
the mining concessions sought by the Third Reich. The Germans stepped
up deliveries of equipment after Franco had agreed to grant more of the
mining rights pursued in the Montana project and also to meet the costs
of the Condor Legion.[73] The concession of increased German participation in Spanish mainland and Spanish Moroccan mining enterprises
was a considerable price to pay and belied Franco's claims about his
commitment to Spanish sovereignty. It was clear evidence of 'the firm
intention of Nationalist Spain to continue to orient itself towards
Germany politically and economically after the end of the war'.[74] The
Caudillo's reasons were entirely understandable. With the French frontier closed and the Republic no longer receiving help from the Soviet
Union, deliveries of German equipment would once again give him the
crucial advantage for the final push.

The decisive Nationalist counter-offensive of the Battle of the Ebro
was launched on 30 October 1938. Franco followed the battle closely
from his mobile field headquarters. He relied on the tactic of concentrating air and artillery attacks on selected small areas, and then following
these up with infantry attacks. These heavy poundings had the attraction
to him of physically smashing the Republican forces.[75] By mid-
November, at horrendous cost in casualties, the Francoists had pushed
the Republicans out of the territory captured in July. As they retreated
back across the Ebro, the Republicans left behind them many dead and
much precious material. Both sides had suffered heavily although there
remains controversy as to the number of casualties. The Nationalists lost
more than six thousand five hundred troops dead and, if von Stohrer is
to be believed, nearly thirty thousand wounded. The Republicans lost a
similar number of wounded and nearer fifteen thousand dead. They were
the heaviest casualties of the war. It had taken Franco four months to
recover the territory gained by the Republic in one week in July. With

a more adventurous strategy, pinning down the Republicans near Gandesa, and launching an attack on Barcelona from Lérida, he could probably have ended the war in the summer of 1938. However, by his preference for attrition, Franco had ensured that the Republic lost its army. The last despairing effort of the Ebro had given Franco the kind of decisive victory that meant more to him than swift strategic manoeuvres. He had achieved the physical annihilation of his enemy. There would be no negotiated truces, no conditions, no peace with honour.

A commission to oversee an exchange of prisoners was headed by the British Field-Marshal, Sir Philip Chetwode. It was successful in securing the exchange of one hundred British prisoners held by the Nationalists in return for one hundred Italians held by the Republic.[76] That success reflected Franco's desire to ingratiate himself with both Italians and British. However, Field-Marshal Chetwode was able to do little for Republican prisoners to whom Franco was merciless.[77] Chetwode wrote to Lord Halifax in mid-November 'I can hardly describe the horror that I have conceived of Spain since my interview with Franco three days ago. He is worse than the Reds and I could not stop him executing his unfortunate prisoners. And when I managed to get 140 out of the Cuban embassy in Madrid across the lines the other day, having got them across, Franco frankly refused to give anyone for them in spite of his promise. And when he did send people down nearly half of them were not the people he had promised to release but criminals who had been in jail, many of them, since before the war started.'[78]

The Generalísimo's attitude to the enemy cannot have been softened by the death of his younger brother. On 28 October 1938, Ramón Franco had been killed while flying a mission from Pollensa in Mallorca to bomb the docks in Valencia, just as he had previously participated in raids on Barcelona. Rumours have circulated to the effect that he was the victim of sabotage but they have little foundation. When the news was brought into Franco's office in Burgos, he showed not the slightest flicker of emotion. Easily moved to tears when it suited him, he rarely revealed his real feeelings when directly affected. He sent a telegram to the Nationalist Air Force: 'It is nothing to give a life joyfully for the *Patria* and I am proud that the blood of my brother, the aviator Franco, should be united to that of many aviators who have fallen.' Ramón was buried with considerable pomp in Mallorca with Nicolás representing the Caudillo.[79]

The end for the Republic was imminent. Just as the defeat at Teruel opened the way to the Nationalist sweep to the coast, the post-Ebro exhaustion invited another offensive. Indeed, the Republican will to resist was kept alive only by the fear born of Franco's much-publicized determination to eradicate liberals, socialists and Communists from Spain. Baron von Stohrer wrote to the Wilhelmstrasse on 19 November 1938: 'the main factors which still separate the belligerent parties are mistrust, fear and hatred'.[80] Franco told James Miller, Vice-president of the United Press, that a negotiated peace was out of the question 'because the criminals and their victims cannot live side-by-side'. Committed to a post-war policy of institutionalized revenge, he rejected the idea of a general amnesty and declared that the Nationalists had a list of two million reds who were to be punished for their 'crimes'.[81] The political files and documentation captured as each town had fallen to the Nationalists were gathered in Salamanca. Carefully sifted, they provided the basis for an immense card index of members of political parties, trade unions and masonic lodges. The Republican zone was kept on a war footing by terror of Nationalist reprisals.

In mid-November, despite his colossal material superiority, the Generalísimo was afflicted by a renewed bout of indecision. He was still toying with the idea of another assault on Madrid or else a renewed attack on Valencia in order to threaten the capital. Italian military advisers were hostile to any operation against Madrid since their intelligence reports suggested that the Republic had strengthened its position in central Spain. There was also opposition within Franco's high command.[82] After the Ebro, the Republic's greatest weakness was now in Catalonia and pressure from his own generals finally inclined the Generalísimo to opt for an attack there. On 26 November 1938, he issued a general order in which he underlined that the Ebro victory opened the way to the total annihilation of the remaining Republican forces. A massive army was gathered along a line surrounding Catalonia from the Mediterranean to the Ebro and to the Pyrenees. Originally planned for 10 December, the offensive was put off until 15 December. Frustrated by continual delays occasioned by a period of torrential rain, the new Italian commander, General Gastone Gambara, pushed Franco for a decision on a date for its launch. Finally, 23 December was picked.[83]

While hesitating over the next move, the Caudillo took time off to visit his native province of La Coruña to receive a 'gift' from its people. Julio Muñoz Aguilar, Civil Governor of La Coruña, and Pedro Barrié de la Maza, a local banker, had had the idea of organizing a subscription by which the people of the province could show their gratitude for their salvation at the hands of the Caudillo. A splendid country house, known as the Pazo de Meiras, which had belonged to the Galician novelist Emilia Pardo Bazán, had become available and was purchased in March 1938 by means of the sums raised through the subscription. It is possible that many gave freely but, since the sums to be donated were fixed by the authorities, it may be supposed that many others contributed for fear of being thought disloyal. It was another notch on the ratchet of adulation. It was also the beginning of the move away from the relative personal austerity which had characterized the Francos' lifestyle during the war.* The house was restored and Franco received the keys and the deeds of the donation on 5 December 1938. Muñoz Aguilar was rewarded by being given the lucrative posts of Head of Franco's Household (*Jefe de la Casa Civil del Generalísimo*) and administrator of the *Patrimonio Nacional*, consisting of the properties and art treasures of the royal family. Barrié de la Maza was later ennobled by Franco, being given the title of Conde de Fenosa in 1955.[84]

During the lull before the final push on Catalonia, there were rumblings of impatience with the Generalísimo's apparent inability to bring the war to an end.[85] In the midst of this, Franco's niece Pilar Jaraiz Franco arrived in Burgos having been released from prison in Valencia on a prisoner exchange (*canje*). Her reception by the Generalísimo and his wife was starkly revealing of the atmosphere in which they lived. She had previously been very close to them both, and had been a page at their wedding. Before the war, when they spent time in Madrid, prior to having their own apartment, they had always stayed with Pilar Jaraiz's mother Pilar Franco. Their niece had often accompanied them on expeditions to the cinema or in search of antiques at the *Rastro* (flea-market). The Generalísimo had given her away when she was married in 1935 and Doña Carmen had helped her choose her trousseau (*ajuar*) and

---

* Franco and his wife had already begun to acquire property. In November 1937, José María de Palacio y Abarzuza, Conde de las Almenas, died childless. He expressed his gratitude to Franco for 'reconquering Spain' by leaving him in his will an estate in the Sierra de Guadarrama near El Escorial, known as Canto del Pico. Consisting of 820,000 square metres, it was dominated by a large mansion called the Casa del Viento. After the Civil War, Franco bought a large estate near Móstoles on the outskirts of Madrid known as Valdefuentes. Doña Carmen acquired an entire apartment building in Madrid and, in 1962, the magnificent Palacio de Cornide in La Coruña. The family accumulated a further fifteen properties. In addition, it has been calculated that Franco received four thousand million pesetas' (approximately £4,000,000/$7,500,00) worth of gifts during his rule – Mariano Sánchez Soler, *Villaverde: fortuna y caída de la casa Franco* (Barcelona, 1990) pp. 39–51, 92–4, 122–4, 127, 131–9. That calculation probably does not include the value of the hundreds of commemorative gold medals given to Franco by towns and organizations all over Spain which Doña Carmen had melted down into ingots – Peñafiel, *El General*, p. 149.

furnish her apartment. Pilar Jaraiz arrived in Burgos after spending two appalling years in a Republican prison with her baby, who had almost died of meningitis, a period of suffering directly attributable to her relationship with Franco. Yet, he was distant and greeted her coldly, reducing her to an embarrassed silence and making her feel like 'a beetle' (*escarabajo*). His wife also greeted her with harsh indifference and left her totally perplexed by asking her 'whose side are you on?' It was an indication of the way in which, accustomed to constant adulation, Franco and his wife were responding to the recent mutterings of criticism. They already regarded themselves as a breed apart from ordinary mortals.[86]

On 23 December 1938, from his field headquarters, *Términus*, Franco oversaw his final offensive against Catalonia. He had new German equipment in abundance and sufficient Spanish and Italian reserves to be able to relieve his troops every two days. The attacking force consisted of five Spanish Army Corps, under Generals Agustín Muñoz Grandes, García Valiño, Moscardó, Solchaga and Yagüe, together with the four divisions of the CTV under Gambara. A massive artillery barrage preceded the attack. The shattered Republicans could put up only token resistance.[87] Gambara opted for a *guerra celere* tactic which soon saw the Italians thirty kilometres ahead of the more cautious Spaniards. The success of the Italian advance provoked the French government into opening the frontier to allow equipment into Spain. Mussolini was in despair, sending both Gambara and Ambassador Viola to see Franco to urge more speed.[88] Franco confessed to Viola his perpetual anxiety about a French intervention. In response, Ciano informed London and Berlin that, in the event of any such French action, Italy would make war on France on Spanish soil. The threat had the effect of blocking the possibility of substantial French aid to Catalonia.

With such crucial assistance, it was hardly surprising that Franco should dream of Nationalist Spain rearranging the world as an Axis partner. His hopes gleamed through an interview he gave to Manuel Aznar, one of his most skilled and enthusiastic panegyrists, on 31 December 1938, on the eve of his final victory. The text, widely reproduced, signposted clearly the direction that Franco planned to take in both domestic and foreign policy over the next few years. Draconian repression at home and aggressive ambition abroad were to be the order of the day. The Caudillo made it starkly clear that there could be no thoughts of amnesty or reconciliation for the defeated Republicans, to whom he referred as 'criminals' (*delincuentes*). Only punishment and repentance would open the way to their 'redemption'. Prisons and labour camps were the necessary purgatory for those with minor 'crimes'. Others could expect no better fate than death or exile.

Identification with the apparently invincible fascist dictators who had helped him to victory was reflected in Franco's belligerent declaration that Spain should henceforth be a 'nation in arms'. He boasted about an 'enormous industrial base' being created to sustain Spain's new military ambitions, capable of producing a forceful navy and a strong air force, thereby revealing the limits of his understanding of economic problems. With astonishing, and entirely misplaced, complacency, he expressed his confidence about Spain's immediate economic future. Victory in the Civil War was merely the first stage to a full-scale rebirth of Spanish imperial greatness. In a clear challenge to British and French hegemony in the Mediterranean, he claimed that Spain held the entrance to the sea and that new weaponry altered the relation of forces in the area in Spain's interests. In the language of a fascist dictator lost in dreams of empire, he boasted that 'efforts to reduce Spain to slavery in the Mediterranean' would impel him to go to war and asserted that the area's affairs could not be discussed without Spanish participation.[89]

Tarragona fell on 15 January 1939. The road to a virtually defenceless Barcelona lay open. In the three weeks of the advance, three thousand square kilometres had fallen to the Nationalists.[90] With the Republican troops in disarray, Franco issued an order on 16 January to accelerate the pace with no quarter to be given to the enemy.[91] The Republican government fled northwards to Gerona on 25 January 1939 as Yagüe crossed the Llobregat river to the south of Barcelona. The following day, the rebels entered the deserted streets of the starving Catalan capital. A savage purge began in which thousands were shot. By 10 February, all of Catalonia had fallen.[92] The Italians believed that it was Gambara's adoption of *guerra celere* tactics which had brought about the victory.[93]

The rump of the Republican Cortes held its last meeting at Figueras near the French border. On Sunday, 6 February, after Negrín had tried to persuade him to return to Madrid, the President of the Republic, Manuel Azaña, went into exile. He was followed three days later by Negrín and General Rojo. Miaja was left in authority over the remaining Republican forces. At the end of February, Azaña resigned, and his consti-tutionally designated successor, Diego Martinez Barrio, refused to return to Spain. With Britain and France having announced their recognition of the Franco government, the Republic was left in a constitutional shambles. The legal validity of Negrín's government was unclear. Never-theless, a huge area of about thirty per cent of Spanish territory still remained to the Republic. The overall command of this central zone lay with General Miaja, although he spent most of his time in Valencia. Negrín still nurtured the vain hope of hanging on until a European war started and the democracies at last realized that the Republic had been fighting their fight.

Even if further military resistance was impossible, the Communists were determined to hold on to the bitter end in order to be able to derive political capital out of the 'desertion' of their rivals. Non-Communist elements, however, wanted to make peace on the best possible terms. Such hopes seemed vain after the publication on 13 February 1939 of Franco's Law of Responsibilities, by which supporters of the Republic were declared guilty of the crime of supporting the 'illegitimate' Republic. The law was retroactive to October 1934. A Commission was set up in December 1938 to prove the illegitimacy of the Republic and another a year later to document the persecution of rightists in the Republican zone. The Law of Responsibilities declared membership of left-wing political parties or a masonic lodge to be crimes. The all-embracing clauses of the law included 'serious passivity'.[94] It was the first step in the full-scale institutionalization of a repression which had already been implacably, if informally, applied in the territory captured by the Nationalists. Now the end of the war would see a massive wave of political arrests, trials, executions and imprisonment.

On 4 March, the ascetic Colonel Segismundo Casado, commander of the Republican Army of the Centre and Miaja's effective substitute, decided to put a stop to what was increasingly senseless slaughter. Together with disillusioned anarchist leaders and the distinguished Socialist logic professor, Julián Besteiro, Casado formed an anti-Negrín National Defence Junta, in the hope that his contacts in Burgos would facilitate negotiation with Franco. He may also have hoped that by inspiring a military uprising 'to save Spain from Communism', he would somehow endear himself to Franco. Personally unambitious and a capable soldier, Casado was outraged that Negrín and the Communists talked of resistance to the bitter end while simultaneously arranging to get funds out of Spain and organizing aircraft for their flight into exile. The Casado revolt against the Republican government sparked off what was effectively a second civil war within the Republican zone.

What was happening in Madrid had echoes elsewhere. At the Cartagena naval base, a bizarre set of events was set in train when Negrín sent the Communist Major Francisco Galán to take over command. A number of artillery officers, with views similar to those of Casado, rose against

Galán on 5 March. They were embarrassed to find their action seconded
by secret Nationalist sympathizers, retired rightists and local Falangists.
The Falangists seized the local radio station. Sporadic fighting broke out
between Galán, the anti-Communist Republican artillery officers and the
Nationalists. This gave rise to one of Franco's few acts of rashness in his
conduct of the war effort. The Nationalists sent telegrams to Franco's
headquarters begging for assistance. Franco decided to send two divisions,
one from Castellón and the other from Málaga. By the morning of
6 March, the hastily assembled expedition of unescorted transport ships
was in sight of Cartagena. However, in the meanwhile, loyal Republicans
had re-established control of the port. Coastal batteries fired on the
improvised fleet, sinking one transport with the loss of more than a
thousand Nationalist troops.[95]

Meanwhile in Madrid, arrests of Communists had begun on 6 March.
General Miaja reluctantly agreed to join the Casado Junta and took over
its presidency. Most of the Communist leadership had already left Spain.
From France, they denounced the Junta in the most virulent terms. On
7 March, Major Luis Barceló, pro-Communist commander of the I Corps
of the Army of the Centre, decided to take more direct action. His troops
surrounded Madrid, and for several days there was fierce fighting in the
Spanish capital. The IV Corps, commanded by the anarchist, Cipriano
Mera, managed to gain the upper hand, and a ceasefire was arranged on
10 March. Barceló, together with some other Communist officers, was
arrested and executed. This marked the end of the dominance of the
Communist Party in the central zone. In the meantime, Casado was
attempting to negotiate terms with Franco, whose basic condition for
permitting the escape of a small number of Republicans was the surrender
of the Republican Air Force. When, for technical reasons, this had not
happened by his first deadline, Franco brusquely broke off negotiations.
Not surprisingly, he remained interested only in unconditional surrender.
He refused to give any undertakings to the British and American Governments concerning reprisals, declaring his patriotism, high-mindedness
and generosity to be an adequate guarantee.[96] He was not the only one
on the Nationalist side. Most shared the view of Serrano Suñer that, after
so much bloodshed, a compromise peace was unacceptable.[97]

With the bankruptcy of Casado's plans brutally exposed, troops all
along the line were surrendering or just going home, although some took
to the hills from where they kept up a guerrilla resistance until 1951. On
26 March, a gigantic and virtually unopposed advance was launched along
a wide front. Franco's forces simply occupied deserted positions. The

Nationalists entered an eerily silent Madrid on 27 March. A delighted
Ciano wrote in his diary 'Madrid has fallen and with the capital all the
other cities of Red Spain. The war is over. It is a new, formidable victory
for Fascism, perhaps the greatest one so far'.[98] On 30 March, Franco
failed to appear for his daily routine. For the only day in the war, he was
incapacitated by illness, a bout of influenza with a high fever. It suggests
that, at last able to relax, he had succumbed to the intense accumulated
stress of nearly three years of running the war effort. Because of his
illness, he was unable to receive a visit from Admiral Canaris. From his
sick-bed, he followed the bloodless fall of city after city, Alicante, Jaén,
Cartagena, Cuenca, Guadalajara, Ciudad Real and so on.[99] By 31 March,
all of Spain was in Nationalist hands. A final bulletin was issued by
Franco's headquarters on 1 April 1939. Hand-written by Franco himself,
it ran 'Today, with the Red Army captive and disarmed, our victorious
troops have achieved their final military objectives. The war is over.'
Franco had the gratification of a telegram from the Pope thanking him
for the immense joy which Spain's 'Catholic victory' had brought him.
It was a victory which had cost well over half a million lives. It was to
cost many more.

XIII

BASKING IN GLORY

The Axis Partnership, April–September 1939

WITH THE end of the Civil War, Franco's euphoria knew few bounds. Two closely cherished illusions had come together in the triumph. Victory gave substance to his carefully constructed self-image as the medieval warrior-crusader, defender of the faith and restorer of Spanish national greatness, with his relationship to the Church as an important plank in the theatrical panoply.[1] On 19 March, Gomá wrote to Franco that the newly elected Pontiff Pius XII (Eugenio Pacelli) had sent him his blessing. On 3 April, Gomá again wrote to him in terms which can only have inflated his notion of his God-given mission: 'God has found in Your Excellency the worthy instrument of his providential plans.'[2] The identification between the Church and the Caudillo was emphasized on 16 April in a broadcast in Spanish made by Pius XII on Vatican Radio. 'With immense joy', the Pope gave his apostolic blessing to the victors reserving special praise for 'the most noble and Christian sentiments' of the Chief of State. The text had been prepared by Gomá.[3]

Given his openly declared hatred of liberal democracy and Bolshevism, there was little doubt where Franco's sympathies would lie when Hitler unleashed his wars to exterminate both. The Caudillo's loathing of Communism was matched only by his obsession with freemasonry. He attributed the loss of empire in general and the 1898 disaster in particular to the collaboration of Spanish and North American freemasons. He regarded the Republican side in the Civil War as being controlled by a conspiracy of freemasons, Bolsheviks and Jews. In January 1937, he had ordered that all freemasons be expelled from his Army. In that year, one of the extreme right's most manic opponents of freemasonry, Father Juan Tusquets, had come to work in the Nationalist Press Service in Burgos, with the task of sniffing out masonic influence. One of Tusquets' closest cronies was Father José María Bulart, Franco's personal chaplain.[4] Over the next two years, partly at Tusquets' behest, and with the active encouragement of Franco, the *Cuartel General* built up a massive index on eighty thousand individuals suspected of being masons – despite the fact that there had been no more than ten thousand freemasons in Spain in 1936 and that fewer than one thousand remained after 1939. These files would facilitate the purges carried out in the 1940s under the infamous Law for the Repression of Freemasonry and Communism which would be introduced in February 1940. Franco himself had obsessively begun to collect masonic artefacts and publications and created his own masonic grotto.[5]

In part, the phobia about freemasonry lay behind his hostility to the democratic powers, but there were other closely connected reasons. As an *Africanista*, he had a long-standing resentment of Britain and France whom he held responsible for Spain's international subservience. Although the British policy of non-intervention had significantly favoured his victory, Franco could never forgive London for not embracing his cause more openly. He regarded France with a mixture of contempt and resentment for its wartime policy of indecisive support for the Republic. Accordingly, in the flush of victory and inflated by an incessant chorus of adulation, he saw himself as the natural partner of Hitler and Mussolini, one of the new leaders who would reorganize the world on a more equitable basis. He made no secret of his colonial ambitions. As a third fascist dictator, as his interview with Manuel Aznar had indicated, he looked forward to diminishing the power of Britain and France in the Mediterranean and also to creating a new colonial empire in North Africa as the imperial heir to Charles V and Philip II. As early as 2 February 1938, the Caudillo had adopted the imperial crown and shield of Charles V as the arms of the Spanish state, explicitly retaining the columns and the motto *plus ultra* as symbols of overseas expansion.[6] Despite his megalomanic pleasure in comparisons with the great Spanish kings of the past, Franco realized that fulfilment of his ambitions required the goodwill of the Axis powers.

Nevertheless, Franco knew that he had to bide his time. Aware that he needed to secure British and French recognition of his regime, which he had rightly assumed would be a fatal blow to Republican morale, Franco had hesitated before agreeing to join the Anti-Comintern Pact until the cabinet meeting of 20 February 1939.[7] Franco told both Baron von Stohrer and Count Viola that the decision was a foregone conclusion and 'came from the heart', blaming the delays exclusively on Jordana. He informed the Italians that it was to be kept secret until after his Civil War victory was clinched.[8] Mussolini saw in Franco's success an instrument of additional pressure on France. The Fascist press rejoiced that 'the victory of Spain is a Fascist victory'. Ciano believed that cordial relations with Franco meant that the importance of Gibraltar had been reduced, that Italy would gain access to the Atlantic and that France's overland route to Africa was cut.[9]

Mussolini shared Franco's vision of a trio of fascist dictators dismantling Anglo-French hegemony, albeit attributing a more junior role to the Caudillo. The Duce had no hesitation in patronizingly advising Franco how to run his affairs. However, when he cautioned against the return of the monarchy, the normally proud Franco was happy to listen for the simple reason that the advice coincided with his own intentions. Throughout the war, after each victory, the Generalísimo had sent a telegram to Alfonso XIII but, after the capture of Madrid, he had not done so. The outraged Alfonso XIII rightly took this to mean that Franco had no intention of restoring the monarchy.[10] The exiled King allegedly said shortly before he died, 'I picked Franco out when he was a nobody. He has double-crossed and deceived me at every turn.'[11]

On 10 March, Franco told General Gambara, the head of the Italian Military Mission, that he would be forced to remain neutral in a general war unless he received substantial military aid from the Axis.[12] He also frequently complained to the Germans and the Italians about pressure from the French in the hope of Axis assistance to enable him to become a more active player on the international stage. For their part, the Italian and German military authorities were anxious both to help and, in the process, to make Spanish forces dependent on their equipment.[13]

The Anti-Comintern document was signed in Burgos* on 27 March and his act of solidarity with the Axis was made public on 6 April.[14] His brother Nicolás, now Spanish Ambassador in Lisbon, told Oliveira Salazar that Spanish accession to the Anti-Comintern Pact 'had come to represent a political confession of faith and a clear statement of future policy.'[15] As the beginning of a double game, the diminutive Conde de Jordana was left to play down the importance of the Pact to the new British Ambassador, Sir Maurice Drummond Peterson, who had arrived in Madrid at the end of March, implying that it was no more than a gesture of ideological solidarity.[16] It was also with an eye to disarming

---

* The signatories were General Gómez Jordana, the Italian Ambassador Count Viola, the Japanese Minister Makoto Yano and the German Ambassador von Stohrer.

British suspicions that Franco told the Portuguese Ambassador that the Anti-Comintern Pact was merely 'rose water' and that Spain was not unconditionally tied to the Axis.[17] Those assurances were belied by the signing in Burgos on 31 March of an Hispano-German Treaty of Friendship. In the event of war, the Treaty committed each to avoid 'anything in the political, military and economic fields that might be disadvantageous to its treaty partner or of advantage to its opponent'.[18]

The link with Germany fostered the Caudillo's fantasies of glory. He had always been ambitious. His early military career and his giddy progress from captain to general showed that. Nevertheless, until 1936, his ambitions had been confined to the highest posts available to a soldier, Head of a Military Region, Chief of the General Staff, High Commissioner in Morocco, and so on. The military rising had opened entirely new vistas of ambition and no sooner did he begin to cherish an ambition than it was fulfilled. The military rebel who thought he would soon be back in the Canary Islands became Commander-in-Chief. As Generalísimo, he was attracted by the Headship of State. No sooner was that acquired than he began to toy with the idea of a single party like those of his Axis allies. Having tamed the various political forces of the Nationalist coalition and become *Jefe Nacional* of the *FET y de las JONS*, he began to envy the clout on the international scene enjoyed by Hitler and Mussolini. In this respect, the humiliation of the western democracies at Munich was particularly striking to him, all the more so in the light of his own impotence as he nervously awaited the outcome of the Czech crisis.

Accordingly, the ceremonial and choreography of his regime would henceforth proclaim that the Caudillo was both a worthy contemporary of the Duce and the Führer and a fitting heir to the great warrior-kings of Spain's glorious imperial past. He would watch the developing crisis in Europe with a sense that here was an opportunity that he, who had regularly adjusted his ambitions upwards in the last three years, could seize. As Head of State, Commander-in-Chief of the Armed Forces, Head of Government and National Chief of the Single Party, he had a combination of powers unknown in Spain even in the time of Philip II. Having reached the pinnacle of power and prestige within Spain, to fulfil further ambitions and to redress what he saw as the historical injustices perpetrated against Spain by Britain and France meant wielding power on the international stage. In that respect, 1939–40 would be the apogee of the ambition of the Caudillo. He cherished hopes of empire on the cheap, on the coat-tails of Hitler. But, it would slowly become clear to him in the course of the Second World War that the sky was not the limit for Francisco Franco. Then, with his infinite flexibility, unabashed, he would draw in his horns and get on with the task of keeping hold of the power which he held within Spain.

In April 1939, however, his ambition was almost limitless, tempered less than it should have been by an awareness of the political tasks awaiting him and the economic and military weakness left by the Civil War. In one area, his self-confidence was justified. In dealing with the hostility of the defeated Republican population and with the ambitions and rivalries of the various forces of the Nationalist coalition, he demonstrated skill and ruthlessness in equal measure. His instincts, developed in Africa, inclined him to solve both problems by behaving in Spain as he would in Morocco if he were High Commissioner. In other words, his rule would be that of an all-powerful military colonial ruler. The enemy, the defeated Republicans, would be savagely crushed. The 'families' of the Nationalist coalition would be manipulated like friendly tribes, bribed, enmeshed in competition among themselves, involved in corruption and repression in such a way as to make them suspicious of one another but unable to do without the supreme arbiter.

However, if Franco could feel confident in political terms, Spain's military and economic capacity hardly justified adventurism. The Army was barely in any condition to defend Spain in the major war which was about to break out, let alone to embark on any adventures of conquest. Although some modern equipment brought by the Germans and Italians was left behind after April 1939, its extensive use in combat situations had seen it depreciate very significantly. Difficulties in obtaining, let alone paying for, spare parts diminished its utility even further. Franco had done virtually nothing by way of acquiring either an air force or mechanized armoured units. After demobilizing about half of the Nationalist Army in the summer of 1939, he still commanded over half a million abysmally equipped men and 22,100 officers.[19] That he should maintain an army on such a scale reflected both the Caudillo's cautious desire for a powerful repressive force and a misplaced sense of his own military importance. Egged on by sycophants and Falangist zealots anxious to see Spain tied to the Axis, he was less aware than he should have been that the forces at his disposal hardly made him a major player in the great game that was about to begin.

In the early summer of 1939, however, Franco was ready to flex his muscles on the international scene. Significant troop concentrations were ordered on both the French border and near Gibraltar. The French Government was already sufficiently nervous about Franco's hostile intentions to have sent the eighty-four year-old Marshal Philippe Pétain to Madrid as Ambassador. Such a gesture reflected the scale of French fears. Pétain's prestige was to be used to flatter Franco in the vain hope of diminishing his hostility to France. Notwithstanding this acquaintance, they were some distance from being the close friends depicted by Francoist propagandists and by Franco himself.* For Léon Blum 'it rated the apprentice dictator altogether too high'. On being informed of the French government's intention, Franco is alleged to have wept with emotion.[20] Pétain arrived in a spirit of goodwill and collaboration.[21] However, the Caudillo's tears dried quickly enough. His self-esteem inflated by the appointment, Franco kept the venerable *Maréchal* waiting longer than usual before permitting him to present his credentials. Then, despite a formally friendly speech in response to flattering words from Pétain, the Caudillo and his ministers treated him with a surly disdain.[22] Pétain was furious at this treatment and never forgave Franco. Concerned by threatening Spanish troop movements and Franco's adherence to the Anti-Comintern Pact, the French Government quickly recalled Pétain to Paris for consultations although he was dismissive of Franco's hostile intent.[23]

Despite his initiatives near the French border, the Caudillo was careful to play down the new orientation of his foreign policy in his speech on 11 April on receiving the credentials of the new British Ambassador. The quick-witted and rather arrogant Peterson did not take to Franco. Having been told that the Caudillo spoke French, he tried to conduct the interview in that language only to find Franco oblivious to his attempts. Accordingly, he called on his extremely tall military attaché, Major Mahony of the Irish Guards, complete with bearskin, to interpret, much to the chagrin of the diminutive Franco.[24]

Constantly seeking ways in which he might emulate, and indeed identify himself with, his Axis allies, on 8 May, Franco pulled Spain out of the League of Nations. On the other hand, a proposed visit to Spain by Field Marshal Göring on 10 May was so crassly mismanaged as to guarantee thereafter the *Reichsmarschall*'s intense dislike for the Caudillo. This contrasted dramatically with the visit to Spain of Ciano later in the summer.[25] When Hitler and Mussolini signed the Pact of Steel at the end of May 1939, Franco, in a gesture of virile anti-British bellicosity, sent troops to the Gibraltar area. At the time, he was occupied in the kind of

---

* They had met before in February 1926 when Pétain visited Madrid and again in 1927 when Franco visited the École Militaire de St Cyr, then directed by Pétain.

choreographed public glorification so beloved of both Hitler and Mussolini. A series of spectacular victory parades were held between mid-April
and mid-May in the major provincial capitals of Andalusia and in Valencia
with the participation of Axis, as well as Spanish, troops. The conservative
Catalan politician Francesc Cambó noted in his diary: 'As if he did not
feel or understand the miserable, desperate situation in which Spain finds
itself or think of nothing but his own victory, he indulges the need to do
a lap of honour around the country just as a bullfighter, after a good
faena, struts around the ring collecting the applause, cigars, berets and
the odd jacket thrown in.'[26] The celebrations culminated on 19 May in
a display which managed at the same time to identify Franco with Hitler
and Mussolini, to associate him with the great medieval warrior figures
of Spanish history and to humiliate the defeated Republican population.
On 18 May, the Caudillo made his state entry into the capital whose
principal streets were draped in the red and yellow Nationalist colours.
Madrid's main avenue, the Avenida de la Castellana, was renamed the
Avenida del Generalísimo Franco. According to the press release from
his Burgos office, 'General Franco's entry into Madrid will follow the
ritual observed when Alfonso VI, accompanied by the Cid, captured
Toledo in the Middle Ages'.[27] Bonfires were lit on the highest mountains
of every province. On the following day, two hundred thousand troops
marched before Franco in a sixteen-mile victory parade.

In khaki military uniform, but wearing the blue shirt of the Falange
and the red beret of the Carlists, Franco presided. Starting at 9.00 a.m.,
the procession was led by General Andrés Saliquet. Behind the band of
the Carabinieri, a battalion of Italian black-shirted *Arditi* marched with
their daggers raised in a Roman salute. This provoked the delight of the
crowd as did the agile whippet tanks and other mechanized and cavalry
units of the regular Italian army. Thereafter, for five hours, Falangists,
Carlist *Requetés* carrying huge crucifixes, regular Spanish troops, Foreign
Legionaries and Moorish mercenaries filed through the rain-swept streets
bearing the bullet-riddled flags of the Civil War. A special item in the
procession was the cavalry militia of Andalusian señoritos mounted on
their priceless Arab steeds and polo ponies. To permit them to gallop
past the Caudillo, special arrangements in the timing of the procession
had to be made, thus continuing an unusual deference to this unit whose
military efficacy was of less importance than its use as a symbol of class
domination in Queipo's campaign to capture Andalusia. In another kind
of symbol, the rear was brought up by the Portuguese volunteers who
had fought for Franco and, led by General von Richtofen, Hitler's Condor

Legion. Overhead, a large formation of biplanes spelled out the letters 'VIVA FRANCO'. Another aeroplane wrote the name of Franco in smoke. Spain's most important decoration for bravery, the Cruz Laureada de San Fernando, was bestowed on Franco by General Varela.[28]

The parade clearly projected Franco as a full-scale military partner of the Axis. His speech on the occasion was entirely in tune with that image. He warned 'certain nations', clearly England and France, not to try to use economic pressures to control Spanish policy. He expressed his determination to stamp out the political forces which had been defeated and to remain alert against 'the Jewish spirit which permitted the alliance of big capital with Marxism'.[29] After the parade, Franco hosted a banquet at the royal palace, the Palacio de Oriente, for the senior officers of the units which had participated in the parade. In an even more theatrical production, on the next day, a carefully realized medieval symbolism underlined the association between Franco's war effort and the crusade against the Moors. Guns thundered as the Caudillo arrived to attend the solemn *Te Deum* service held at the royal basilica of Santa Barbara to give thanks for his victory. The choir from the monastery of Santo Domingo de Silos greeted him with a tenth-century Mozarabic chant written for the reception of princes. Surrounded by the glorious military relics of Spain's crusading past, including the battle flag of Las Navas de Tolosa, the great victory over the Moors in 1212, the standard used by Don Juan de Austria at the Battle of Lepanto in 1571, and the Señera of Valencia, Franco presented his 'sword of victory' to Cardinal Gomá, Archbishop of Toledo and Primate of all Spain, who solemnly blessed him. The sword was then laid on the High Altar before the great crucifix of the Christ of Lepanto which had been especially brought from Barcelona.[30]

No one could miss the allusion. The dumpy Franco was the contemporary heir to the great crusading warriors of the past. He was the El Cid and the Don Juan de Austria of his time. The Church was delighted to go along with the reinvention of an idealized relationship between the medieval Church and the great Catholic warrior-heroes of the past, something made clear in an exchange of letters between Franco and the Primate. Cardinal Gomá ordered the sword to be displayed in the Cathedral of Toledo along with other great historic relics.[31]

In 1939, Franco saw his links with contemporary fascism as the necessary prerequisite to the revival of a glorious imperial tradition. On 23 May, he bid formal farewell to von Richthofen and the Condor Legion. Ending an enthusiastic speech, the Caudillo said 'I have felt the greatest pride in having had under my orders German leaders, officers and men' asking them to take back to Germany 'the imperishable gratitude of Spain'.[32] Relations with Italy were even more cordial, with both the Caudillo and his Minister of the Interior and brother-in-law anxious to intensify links with Mussolini.[33] On 1 June 1939, Serrano Suñer left Cádiz en route for Naples accompanying the remaining Italian troops on their return home. An Italian naval convoy took him to Naples along with twelve generals, an Admiral and several other senior naval officers and numerous top Falangists. They were accompanied by three thousand Spanish troops who paraded through the streets of both Naples and Rome. The party was greeted with a spectacular degree of pomp and ceremony which greatly affected Serrano Suñer and was in stark contrast to the brusque treatment he was later to receive at the hands of the Germans.

King Vittorio Emanuele III, three royal princes, the Cardinal Archbishop of Naples, Ciano and Serrano Suñer presided over a marchpast of Italian and Spanish troops. In Rome, Serrano took the salute along with Mussolini at another parade. An emotional Serrano told both Mussolini and Ciano that Spain needed two or preferably three years in order to complete her military preparations. However, when war broke out, 'Spain will be at the side of the Axis because she will be guided by feeling and by reason. A neutral Spain would be destined to a future of poverty and humiliation.' Spain would never be free or sovereign until she had regained Gibraltar and captured French Morocco. Ciano was immensely impressed by Serrano Suñer despite his appearance as 'a slender and sickly man'. The Italian rated him highly for his intelligence and his impetuosity. Both publicly and privately, Serrano Suñer expressed profuse and sincere gratitude for Italian help during the Civil War as well as intense hostility to Britain and France. Serrano admired the Duce unreservedly, referring to him in a newspaper interview as 'one of the rare men of genius whom history throws up only once every two or three thousand years'.[34]

Ciano told the German Ambassador in Rome, Hans Georg von Mackensen, that Serrano Suñer 'really was an extremist for the Axis'. Ciano and Mussolini both felt that Serrano was 'undoubtedly the strongest Axis prop in the Franco regime'. Their view was no doubt strengthened by the undiplomatic and unrestrained manner in which Serrano Suñer criticized as Anglophile monarchists both the Spanish Foreign Minister General Jordana and the Ambassador in Rome, Pedro García Conde, whom Ciano regarded as 'a great fool'.[35] Friction between Serrano Suñer and Jordana was symptomatic both of military-Falangist rivalries in gen-

eral and of Serrano Suñer's ambitions. He left the Italians impatient for
Franco to reshuffle his government to reflect Serrano Suñer's enthusiasms
and bring Spain into the orbit of Rome.[36]

There is little doubt about Serrano Suñer's attitude to Fascist Italy. It
was widely supposed in Spanish military circles and among the diplomatic
community in Madrid that he was equally committed to Nazi Germany.
However, even before the end of the Civil War, the German Ambassador
von Stohrer, who was later to become a close friend of Serrano Suñer,
had expressed doubts about the *cuñadísimo*'s attitude to the Third Reich.
Stohrer thought him too Jesuitical and Vaticanist to be a loyal friend to
Germany.[37] The Germans eventually came to regard him as an enemy
and he himself spent considerable energy portraying himself as the man
who worked skilfully to keep Spain out of the war. What is absolutely
certain is that he bitterly hated the British and the French, partly because
he abhorred liberal democracy and more particularly because of his belief
that their Embassies in Republican Madrid had refused sanctuary to his
brothers who were shortly after to die in jail.[38] For this reason too he
opposed the release of Republican prisoners whom he held responsible
for the loss of the Falange's 'best comrades'.[39]

With the Civil War over, Spain was anxious to be part of the Axis
club in its own right. Accordingly, while in Rome, Serrano Suñer laid
the ground for Franco to make an official visit to Italy. He also confided
in Ciano his own anxiety to receive an official invitation to Germany,
something which, on his return to Madrid, he reiterated to the tall,
imposing von Stohrer.[40] Serrano Suñer's vaulting ambition was revealed
when he told Ciano of his desire to be Foreign Minister and suggested
that a hint from Mussolini to Franco might be enough to do the trick.
After discussing the matter with the Duce, who had been sufficiently
impressed with Serrano Suñer's fascist credentials to be happy to see him
promoted, Ciano undertook to write to Franco.[41] The wider question
of whether Franco should confine himself to being Head of State and
Commander-in-Chief of the Army, leaving Serrano to be head of government, was clearly on the Duce's mind.[42] Such a change would have
brought Spain more closely under Fascist influence, given Serrano
Suñer's admiration for Mussolini. However, Serrano Suñer claimed later
that he hastened to disabuse Mussolini of the idea. The motive for his
self-sacrificing alacrity was the entirely plausible one that his position
would be seriously endangered if Franco even heard of such a topic being
discussed.[43]

During the early summer of 1939, there was considerable speculation

about Spain's likely alignment in the event of a general war. A possible
clue was given in Franco's speech in Burgos to the *Consejo Nacional* of
the Falange on 5 June. Boasting of the *hábil prudencia* (astute caution)
which had allegedly characterized his policy during the Civil War, the
Generalísimo stated that his victory had been won against the wishes
of the 'false democracies', freemasonry and international Communism.
Having lumped the latter together, he then referred to 'a secret offensive'
against Spain which he implied was the work of both France and England.[44] There was very considerable friction with France and talk of
seizing French property in Spain. Serrano Suñer's press was whipping
up anti-French sentiment.[45] In July, the French consul in Madrid was
badly beaten by Spanish Army officers.[46] On the instructions of Madrid,
José Félix de Lequerica, the deeply cynical, Francophobe and anti-Semitic
Spanish Ambassador in Paris, adopted a bullying attitude when he
demanded a speedy implementation of the Jordana-Bérard Pact of February for the return to Spain of gold reserves, war matériel and refugees.[47]

The icy disdain with which the Caudillo had treated Marshal Pétain
at the credentials ceremony partly accounts for the mocking tone of the
French Ambassador's attitude to Franco. At one of Franco's elaborate
ceremonies, held in the Escorial, Pétain was heard to comment contemptuously, indicating a notice on the wall, 'I see that it is forbidden
here, not merely to smoke, but to spit'.[48] He wrote in early June 1939
that 'next to the Don Quixote of his brother-in-law, the Generalísimo
often appears to be Sancho Panza'. Sir Samuel Hoare similarly noted
the difference between 'the slow thinking and slow moving' Franco and
Serrano 'quick as a knife in word and deed'. Pétain was by now concerned
about Franco's ambitions.[49]

So too, for different reasons, were the Germans. Von Stohrer told
Jordana that 'it would not be expedient either for Spain or for us if the
Spanish Government were to show their cards in advance over the attitude
they would adopt in a possible war ... we must attach the greatest
importance to Spain's attitude in a future war remaining a completely
unknown quantity in France and Britain'. The Germans hoped that a
Spanish commitment to the Axis side would tie down French troops
on the Pyrenean frontier and inhibit both Britain and France from an
intervention 'in problems which were no concern of theirs' (a reference
to the planned attack on Poland).[50]

Franco told Count Viola on 5 July with evident regret that Spain
needed 'a period of tranquillity to devote herself to internal reconstruction
and the achievement of the economic autonomy indispensable for the

military power to which she aspired . . . in her present conditions, Spain could not face a European war'. However, he was at pains to make it clear that his idea of neutrality in the event of war would be decisively favourable to Spain's friends and 'implicitly admitted the difficulties for Spain of remaining aloof from the conflict'. As he warmed to his theme of Spain's assistance to the Axis, he asserted that he had suspended the post-Civil War demobilization in order to keep a large army available to counter the 'impositions' of the British and the French. France, he repeated with emphasis, 'would never be able to feel easy with regard to Spain' and would be obliged to tie down a large portion of her army in the south. A force of six hundred thousand men would be divided between the Pyrenees and Gibraltar and would 'permit him to make Spain's weight be felt in the unfolding of events and possibly to take advantage of circumstances'. All of this was regarded by Viola as evidence of a 'vigilant neutrality'.[51] The French Air Force attaché reported to Paris that Franco's maintenance of an army of more than half a million men was scarcely compatible with his talk of neutrality.[52]

Franco's references to the need for tranquillity was no doubt provoked by the continuing resistance of Republican stragglers all over Spain and ongoing squabbles between Falangists and Carlists. There were embarrassing signs of both as the return state visit of Count Ciano was awaited.[53] Two thousand Asturian miners were still fighting substantial units of the Spanish Army and in early July a *Requeté* lieutenant was shot dead by Falangists in Irún.[54] Pétain reported to Paris on the trials of an 'incalculable number of unfortunate so-called reds' as the guerrilla war in Asturias intensified.[55] Six weeks after Serrano Suñer's stay in Rome, the Italian Foreign Minister arrived in Barcelona on 10 July for his reciprocal visit with a flotilla consisting of four battlecruisers and several torpedo boats. After a tour of the battlefields of the Civil War, the Count was escorted into Madrid by one hundred thousand Falangists. He visited Franco in San Sebastián where they discussed the international situation. Ciano noted Franco's unease in political matters as opposed to his greater confidence in things military. It was hardly surprising that he still tended to lean on his brother-in-law.

In a way which there is no reason to believe was other than sincere, Franco told Ciano he planned to follow the line of the Rome–Berlin Axis although he suggested that Spain needed five years of peace for economic and military preparation before she could identify completely with the totalitarian states. In the event of war, he would prefer neutrality but would be on the Axis side because he did not believe that his regime could survive the victory of the democracies in a general war. Accordingly, with apparent lack of concern about Spain's bankruptcy, he talked about a major rearmament programme for both the Navy and the Air Force. The Caudillo made a show of gushing admiration for Mussolini. In a letter sent with Ciano, Mussolini had made two essential points, 'I consider the re-establishment of the monarchy to be highly dangerous for the regime gloriously founded by you through the sacrifice of so much blood' and 'Expect nothing from France and England; they are by definition the irreconcilable enemies of YOUR Spain'. On both counts, this was music to the Caudillo's ears. Franco broke free of his normal reserve and responded in kind, telling Ciano that he expected from the Duce 'instruction and directives'. Franco gave Ciano the impression of being 'completely dominated by the personality of Mussolini and feels that to face the peace he needs the Duce just as he needed him to win the war.'[56]

Dreaming of empire, flattered by his close contact with the Axis powers, Franco was less than coldly realistic at this time. He talked patronizingly of Portugal, remarking of her help in the Civil War that she had merely 'saved her own skin since she knew that she could expect nothing from the reds'. The Portuguese Ambassador found him moonstruck by the splendour of power.[57] However, if his head was in the clouds with regard to his own importance in international affairs, his feet remained firmly on the ground with regard to his domestic concerns. His thoughts were briefly distracted from foreign affairs in the third week of July in order to deal with an outburst of rebellion from a long-standing rival, General Gonzalo Queipo de Llano, who since the earliest days of the Civil War continued to rule over a virtually independent fief in Andalusia. Queipo had never bothered to conceal his low opinion of Franco both as a military leader and as a human being, regarding him as '*egoista y mezquino*' (selfish and mean). He made indiscreet references to the irregularities surrounding Franco's election as Generalísimo and coined the nickname '*Paca la culona*' (fatty Francine). There was no shortage of willing confidants to pass his comments on to Franco.[58]

Queipo was also scathing about Serrano Suñer and was regarded as a possible leader of military opposition to the Falange. In May 1939, Franco had been informed by Beigbeder from Tetuán that Queipo was making soundings for the creation of a military directory to replace him and to neutralize the power of the Falange.[59] When the German Condor Legion had returned to Germany, Queipo, without Franco's permission and to his intense annoyance, had flown ahead in order to be in Berlin to greet them. The Caudillo did not act immediately. He bided his time until

Queipo went too far with an act of public disrespect. In a speech to 104 Andalusian *alcaldes* (mayors) on 18 July, he expressed his outrage that Franco had granted the military decoration of the *Cruz Laureada de San Fernando* to the city of Valladolid but not to Seville, his own power-base. His annoyance was understandable given the role played by Seville in the 1936 *alzamiento* (rising) and the fact that Valladolid was a centre of Falangist strength. He dismissed the honour given to Valladolid as the work of Serrano Suñer. His reference to the frail Serrano was unmistakable – 'If things go on as they have been doing, idiots as fragile as clay toys will be turned into heroes' (*tontos frágiles como juguetes de barro se conviertan en héroes*).

The personal insults and the public criticism of his reliance on Serrano Suñer stung Franco into taking action. Franco now lured Queipo away from Seville on the pretext of calling him to Burgos for 'consultations' on 27 July and sent General Solchaga to take over as Captain-General of the Seville military region. Franco was taking revenge for what he perceived as a long list of humiliations suffered at the hands of the sneering Queipo. In a bitter meeting, the Generalísimo brandished a thick file of copies of letters sent by Queipo replete with insulting remarks about him. The deposed viceroy of Andalusia was confined in a Burgos hotel until he could be sent to Italy as head of a military mission.[60] The ease of the victory over Queipo was surprising even for Franco who was a master at controlling his domestic rivals. Foreign affairs provided more of a challenge.

Shortly after Ciano's visit, and while Franco was occupied with taming Queipo, the head of German Military Intelligence, the *Abwehr*, Admiral Canaris, had made a visit to Spain where he found a depressing picture of misery and continuing Republican resistance. The Caudillo confided in him his worries that an immediate outbreak of war involving Germany and Italy might provoke France into invading Spanish Morocco. Spain, he said, was in no position to sustain a war either now or in the near future and a direct assault on Gibraltar was unthinkable. However, he did suggest that a midget submarine base could be established at Tarifa in order to theaten the Straits and agreed to the setting up of logistical support points at Santander, Vigo, Cádiz and possibly Barcelona for the forthcoming German war effort in the Atlantic.[61] These supply depots would permit refuelling and crew replacements and play an important role in extending the range and operational efficacy of German submarines and other warships during 1940 and after. Hitler was suitably impressed. The concession of naval facilities reflected Franco's anxiety that, if the Axis won the coming war before he was ready, the world would be reconstructed without regard for his ambitions.

Blown along by the prevailing winds of world politics, he permitted the Falange a growing ascendancy. In a decree signed on 31 July, he reasserted its position as the only party, while keeping absolute control in his own hands. Its ruling body, the *Consejo Nacional*, was large, unwieldy and largely decorative. The small permanent executive committee, the *Junta Política*, was to serve as the link between the party and the government. It was to be headed by Serrano Suñer, who thereby gained enormous power. Members of all the armed services were automatically to be members of the Falange and required to use the fascist salute on political occasions.[62]

Franco's own position was strengthened even further by the *Ley de la Jefatura del Estado* on 8 August 1939 which gave him legislative power to make laws and decrees without consulting the cabinet. It gave him 'the supreme power to issue laws of a general nature', and to issue specific decrees and laws without discussing them first with the cabinet 'when reasons of urgency so advise'. The controlled press was lavish in its praise for the readiness of the 'supreme chief' to assume the powers necessary to allow him to fulfil his historic destiny of national reconstruction.[63] It was power of a kind previously enjoyed only by the Kings of medieval Spain and emphasized the personal nature of his dictatorship. It also stressed the extent to which he regarded ministerial and government functions as administrative rather than political. He always referred to political power as 'command' (*el mando*) and treated the machinery of government as if it were the Army.

The Caudillo had been warned by Ciano in August that war was likely between Germany and Poland. The news seems to have come as a surprise to him since he had been confident that Britain and France would force Poland to surrender to German demands.[64] He responded with troop movements and the building of fortifications near the French border and on the frontier between Spanish and French Morocco. He remained circumspect because, until the collapse of France in 1940, he believed that the French Army constituted a fearsome fighting force. When Pétain made outraged protests about the fortifications, Franco claimed that they were merely defensive. Pétain pointed out that it was a short distance from defence works to offensive support bases. The Marshal was shocked when the Caudillo pointedly failed to accompany him to the door, something even Alfonso XIII had always done.[65]

Franco had also set up a new Gibraltar command of one division.

These anti-French and anti-British measures, he informed both Italian and German Ambassadors, were by way of helping the Axis. In the wake of his interview with Pétain, he boasted to the Italian *Chargé d'Affaires* that Spanish military activities were causing grave anxieties to the French. He noted to Stohrer that France would be unable to withdraw any men from Morocco since Spain had eighty-seven thousand men there which exceeded the peacetime strength of the forces in the French zone. Through his brother in Lisbon, Franco also put pressure on Portugal to ignore its commitments to Britain and to maintain neutrality.[66]

Franco's awareness that war was imminent was also reflected in the sweeping cabinet changes of 9 August 1939.[67] The Caudillo altered his cabinet in line with the changing world situation and, within the very narrow possibilities open to him, he began to rearm by seeking financial and technical help from Italy for the rebuilding of the Spanish navy and Air Forces.[68] On the very day that his cabinet changes were announced, Franco made a remarkable boast to General Gastone Gambara, the Head of the Italian Military Mission in Spain. Delighted by effusive praise from the Duce for the recent reorganization of the Falange, the Caudillo clearly felt the need to show that he was worthy of his mentor. Accordingly, he told Gambara, with his 'usual imperturbable serenity', that he intended to close the Straits and destroy the British installations in Gibraltar with heavy artillery.[69] Plans for a state visit by Franco to Rome in September 1939 and to Berlin later in the autumn were postponed only because of the outbreak of the Second World War.[70]

The message of the cabinet changes was clear, particularly in the transfer of the Anglophile monarchist Jordana away from the Foreign Ministry to the symbolic post of President of the *Consejo del Estado* (Council of State) and in the nomination of the enthusiastic Falangist Yagüe as Minister for the Air Force. The dour and straightforward Jordana was replaced by the volatile and indiscreet Colonel Beigbeder, an early adherent of the Falange and, at the time of his appointment, fervently pro-Axis.[71] Franco told Serrano Suñer that Beigbeder was mad: 'when we were in Africa, Beigbeder was always disappearing . . . sometimes he had been in a brothel, others on retreat in a Franciscan monastery'.[72] Serrano Suñer claimed later that Beigbeder's appointment was his idea. There had long been a tension between the *cuñadísimo* and the conservative Jordana who was hostile to the Falange and to the Italians.[73] Curiously, Beigbeder shared some of Jordana's suspicions of Italian ambitions in the same parts of North Africa as were coveted by Franco.[74] Something of an Arabist, he shared Franco's Moroccan dreams and was an entirely appropriate person to pursue Spain's imperial aspirations. On the other hand, as one-time Spanish military attaché in Berlin, he was suspicious of the ruthlessness and rapaciousness of the Nazis. Eventually, doubts about his fervour for the Axis would intensify Serrano Suñer's ambitions to assume full control of foreign affairs himself.[75]

Although not formally prime minister, since the Caudillo did not share power, Serrano Suñer was the minister wielding most influence and the one with greatest freedom of action.[76] Like Beigbeder, Yagüe was also a Falangist veteran (*camisa vieja*). He had ended the war as head of the Spanish Moroccan Army. Given his talent and his popularity in both Falangist and military circles, Franco saw him as a possible rival. The Caudillo had not forgiven him for his criticisms of the Nationalist failure to make a social revolution and of the treatment of political prisoners. During the summer of 1939, Yagüe had formed part of the Spanish military mission which accompanied the Condor Legion on its return to Germany. In the course of two months in the Third Reich, he developed an unrestrained admiration for Nazi social policies, for the German Army and even more for the Luftwaffe. In consequence, Yagüe became the object of cultivation by Marshal Göring.

It was thus with characteristic cunning that Franco accepted a suggestion from Serrano Suñer and tried to neutralize Yagüe by making him Ministro del Aire. When Serrano Suñer informed him of his appointment, Yagüe said 'you are completely mistaken and involved in an impossible undertaking because with that man [Franco] we are going nowhere. He is disloyal, distrustful and a sneak (*alparcero*).' The appointment got Yagüe away from the most powerful operational command in the Spanish Army where he might have been a focus of powerful opposition.[77] It was typical of Franco's habit of choosing ministers not for any special competence in the area of the Ministry concerned but as pawns on the political chessboard. Yagüe would find himself isolated at the Air Ministry since, as an infantryman, he was significantly less competent than the obvious choice, General Alfredo Kindelán, who had been head of the Nationalist Air Force during the civil war. An additional advantage to Franco in the promotion of Yagüe was precisely that it kept Kindelán out of the cabinet. These private reasons aside, with war imminent, the appointment of an Axis enthusiast such as Yagüe was a useful public gesture in the direction of Berlin. In 1956, Franco claimed to have been obliged to pass over Kindelán because he was too pro-Allied.[78] As Minister, Yagüe worked in vain for the rebuilding of the Spanish Air Force with German help.

Disgust with Falangist malpractice in local and central government and its alleged corruption was commonplace among the military monarchist critics of Franco and particularly motivated the highly conservative Kindelán. And Kindelán was the most forthright of all the senior generals who felt that they were perfectly entitled by their seniority to treat Franco as no more than their elected leader. Since it was he who more than anyone else had been Franco's kingmaker in September 1936, and since he could count upon immense respect among senior generals, the Generalísimo was obliged to tread warily. However, there can be little doubt that he wanted to rid himself of an awkward critic.[79] The problem was Kindelán's belief that, while the appointment of Franco as Generalísimo in 1936 had been necessary at the time, it had been only for the duration of the civil war.[80] In the atmosphere of sycophancy in which Franco increasingly enveloped himself, Kindelán's dignified independence must have seemed to him impertinence. Moreover, Kindelán had made little secret of his disappointment that Franco had not fulfilled the purpose for which, in his view, the Civil War had been fought, the restoration of the monarchy. Accordingly, in the cabinet changes of 9 August 1939, Franco had taken the opportunity to begin the process of bringing down Kindelán.[81] Bitterly humiliated, he was sent to be military commander of the Balearic Islands. He remained as committed as ever to a monarchist restoration and was to become an increasingly vocal critic of the Caudillo.

The new council of ministers reflected Franco's tireless concern to strengthen his own position. However, the cabinet was no more than a talking shop. Franco kept the overall direction of policy firmly in his own hands and discussed it only with his most intimate advisers. At this stage, that meant Serrano Suñer and would later mean Carrero Blanco. This was particularly true of foreign affairs in which Franco took a special interest. The relative insignificance of the cabinet had already been made clear by the issue, on the day before the new government was announced, of the law laying down Franco's powers as Head of State.

Such gestures by Franco as Yagüe's appointment and the concession of naval supply facilities at Spanish ports had not been lost on Hitler. At a conference at the Obersalzburg on 22 August, at which he discussed the impending attack on Poland with the commanding generals of the three armed services, the Führer declared that, along with Mussolini, Germany's only certain ally was Franco.[82] Nevertheless, there was some popular Spanish mistrust towards the Third Reich expressed after the publication of the Nazi-Soviet Pact. Angry crowds demonstrated outside the German Embassy's summer retreat in San Sebastián on 22 August.

The Duke of Alba told the Portuguese Ambassador in London that there was great indignation among Spanish generals and Kindelán expressed his chagrin to the French Air Attaché.[83] Franco was taken aback by news of the German initiative but he was sufficiently alive to the strategic benefit accruing to the Third Reich to comment to Serrano Suñer 'it's odd that now we're allies of the Russians'.[84] The Falangist press praised the Third Reich for gaining such a powerful ally.

At the cabinet meeting held on 25 August, awareness of Spanish military and economic weakness rather than pique at German duplicity was behind Franco's determination to remain neutral. Franco told the Portuguese Ambassador Pedro Theotonio Pereira on the same day that Poland would submit to Germany and that there would be no war. Pereira was struck by the extent to which Franco spoke patronizingly about the most unexpected and complex subjects (*cada vez gosta mais de falar com tom doutoral sobre os assuntos mais complexos e inesperados*). It was as if he regarded the impending triumphs of the Third Reich as somehow his own. Pereira wrote to Salazar that he was increasingly worried about the ideas of the Generalísimo. 'I find him besotted with state power and with personal power. Of everyone in the Spanish government, he is the one who says the strangest things to me and who speaks in language closest to the Axis'. Unlike Beigbeder and the Minister for the Navy, Admiral Moreno, who were both outraged by the Molotov-Ribbentrop Pact, Franco told Pereira that he found nothing scandalous about the German understanding with Soviet Russia.[85] Serrano Suñer saw Pereira in the evening of 30 August and also launched into a full-scale justification of the German move. A dismayed Pereira came away convinced that Franco and Serrano Suñer were toying with the idea of Spanish participation in the war.[86]

Franco understood the cynical calculations involved in the Molotov-Ribbentrop agreement and shared the viewpoint of his brother-in-law.[87] When the French Foreign Minister Georges Bonnet asked Franco to mediate over Poland, the Caudillo, after checking with Mussolini, refused.[88] In fact, still believing in the might of the French army, Franco anticipated a long war in western Europe which he assumed would play calamitously into the hands of world Communism. On 1 September 1939, the German Ambassador called on Spain's new Foreign Minister to discuss the imminent war. He told Beigbeder that Spain could not remain really neutral 'since her future and the fulfilment of her national hopes depended on our victory'. The initially pro-German sympathies of the tall and swarthy Beigbeder would eventually swing towards Britain in the second half of 1940.[89] Nevertheless, he agreed with von Stohrer and assured him that Spain was willing to help Germany as far as possible both for reasons of self-interest and gratitude.

A visit by von Stohrer to Serrano Suñer produced an undertaking to influence the attitude of the Spanish press completely in favour of the German cause.[90] This he did so effectively that it became an unequivocal Axis propaganda weapon in Spain.* The press secretary of the German Embassy, the sinister Hans Lazar, supplied the willing Falangist press apparatus with Nazi propaganda material which was then relayed as news. Pro-Allied material virtually never appeared except in response to specific diplomatic protests.[91] In fact, German influence over the press was just one of the many ways in which Spain was becoming an informal German colony. The police apparatus was strongly influenced by the Gestapo. Embassy and Ministry telephones were tapped by Germans with official acquiescence, secured either by bribery or ideological affinity.[92]

Franco's ever-closer relations with Nazi Germany were to cause some friction with the Church. On 8 August, Gomá had issued his long pastoral letter 'Lecciones de la guerra y deberes de la paz' (lessons of the war and duties of the peace). In it, he criticized the exaltation of the power of the State. He called for social justice and political reforms to preclude the possibility of another civil war. Franco was outraged. Gomá's suggestion of forgiveness for the defeated flew in the face of his efforts to maintain a spirit of vengeful triumphalism. The censorship apparatus prohibited publication of the pastoral, beyond the bulletin of the archdiocese of Toledo where it had originally appeared.[93] At the time of the prohibition, Franco had had much more on his mind than Church-State relations. The Second World War broke out on 3 September. Franco's dreams of imperial greatness and personal glory were about to be put to the test.

---

* For instance, *Arriba*, 22 September, incited its readers to assault anyone heard criticizing Germany. The issue for 27 September 1939 praised the German submarine campaign.

XIV

THE MAN WHO WOULD BE EMPEROR

The Defeat of France, 1940

WHEN WAR was declared, Franco, like Mussolini, lamented the fact that it had happened too soon. The best that either could do was to proffer surreptitious help and take advantage where possible. Officially, Franco announced that 'the most strict neutrality' would be required of Spanish subjects.[1] The Caudillo was deeply gratified when Mussolini wrote to him in fulsome terms to express his approval. The Duce also informed him that the course of events might lead him to revise his decision to take no military initiative.[2] Given Franco's readiness to emulate Mussolini, the letter can only have intensified his determination to play as opportunistic game as possible. Since the two Mediterranean dictators had similar North African ambitions, the Caudillo must have felt a twinge of anxiety that Italy might steal a march on him in the new world order. For the moment, on the international stage, Franco cynically presented himself as a peacemaker, albeit with signal lack of success. He issued a call to the great powers to localize the conflict. The Caudillo's regrets for the demise of Catholic Poland at the hands of Hitler and Stalin did not run deep and his peace-making was intended to assist the Axis by making it more difficult for other powers to intervene on Poland's behalf.[3]

In private, his attitude was even less neutral. Both he and Serrano Suñer believed that Spain had been kept in humiliating subjugation by the arrogance of Britain and France.[4] When Marshal Pétain and Sir Maurice Peterson called on the Foreign Minister to deliver formal notes from their governments undertaking to respect Spain's neutrality, Franco refused to receive them.[5] The Caudillo's attitude was soon reflected in the tightly controlled press in the form of unrestrained anti-British and anti-French sentiments together with eager reports of German sympathy for Spain's imperial ambitions in Africa.[6] In mid-September, Franco told von Stohrer that, if England and France won the war, his revolution would suffer and that is why his attitude was one of benevolent neutrality towards Germany.[7]

His desk strewn with admiring accounts of the invincibility of the German military machine sent by the Spanish Military Attaché in Berlin, Colonel José Luis Roca de Togores, Franco enviously enthused to Stohrer about 'Germany's brilliant military victories'.[8] The press justified the Nazi-Soviet pact on the bizarre grounds that Communism was dead in Russia. The dismemberment of Catholic Poland was blamed exclusively on the Allies for rejecting Germany's claims.[9] Indeed, as the months drew on, Franco's enthusiasm for the Axis cause became less restrained. On 26 September, Franco delivered a speech in Burgos to the Second *Consejo Nacional* of the Falange in which, while affirming his Catholicism, he made no reference whatsoever to Communism. He spoke of his readiness to take 'heroic decisions if the circumstances demand it'. Resentment of England and France, and delight that they were about to get their just deserts, overcame all other emotions.[10] The Caudillo's confidence reflected his conviction that Britain would soon sue for peace and might even appeal to him to act as mediator with Hitler.[11] At the same time, however, evidence of Spain's worsening economic situation began to mount, provoking public admissions from Serrano Suñer of the deficiencies of the food supply and the distribution networks. Blame was laid on the reds.[12]

Franco was confident that economic problems would disappear as a result of the adoption of fascist-style policies of autarky and of the enhanced status of Spain as a military power. He denounced the principles of free trade as the evil sham behind which Spain had been colonized. On 8 October, he completed his own intensely simplistic ten-year plan for reviving Spain's economic fortunes. Entitled 'Foundations and Directives of a Plan for the Reorganization of our Economy in Harmony with our National Reconstruction', it was distributed to members of the cabinet. In detail, the plan probably owed much to the assistance of the Caudillo's lifelong friend and Minister of Industry in his first cabinet, Juan Antonio Suanzes. It was built on an entirely misplaced optimism in the capacity of Spain to substitute imports, to increase exports, to rely on its own raw materials and to do all this without foreign investment, despite the economic disruption of the Civil War and the fact that Spain had negligible fuel supplies.[13] Accordingly, at a time of appalling shortages of food, clothing and building materials, Franco took the personal, and entirely avoidable, decision to cut imports and not seek credits from the democracies. The advantages of neutrality which brought economic growth to Spain during the First World War were deliberately eschewed because of ideological considerations and a mistaken appreciation of economic reality. The later official line that autarky was forced on Spain by external circumstances is rendered untenable by the discovery in the mid-1980s of Franco's plan. The shortages provoked by autarky were exacerbated by the equally disastrous decision to maintain the peseta at an dramatically overvalued rate. Rationing led to black-marketeering and corruption on a spectacular scale. The suffering which the Spanish people had to undergo throughout the years of hunger in the 1940s, in large part as a result of the economic decisions taken by the Caudillo, is incalculable.[14] Shortages of essential goods, especially clothing and shoes, starvation, a massive increase in prostitution and epidemics of diseases, including some not seen in the Mediterranean since biblical times, became the daily reality of the so-called *años de hambre* (hunger years).[15]

On Sunday 1 October 1939, the regime and its press and radio networks celebrated the third anniversary of Franco's elevation to the Headship of State. A spirit of semi-religious rejoicing in 'the sacred unity of the *Patria* in Franco' inaugurated what would henceforth be an annual holiday, the '*Día del Caudillo*'.[16] On 18 October, Franco transferred his headquarters from Burgos to the capital. In the carefully scripted farewell ceremony, the identification of the Caudillo with El Cid was again to the fore. The Alcalde's (mayor's) intensely sycophantic speech included the words 'the city says with all its heart, as it did to the Caballero de Vivar [El Cid was Rodrigo de Vivar], "Caudillo, here is Burgos: glory to God on high and all praise to you, saviour of Spain".'[17]

On moving to the capital, the Caudillo had planned to take up residence in the royal palace, the Palacio de Oriente. Serrano Suñer hastened to persuade him that such a move would be taken as announcing his unlimited ambitions. The *cuñadísimo* managed to convince him that it was not in his interests to be seen to be afflicted with *folie de grandeur* nor to risk his relations with the monarchists among his supporters. As a compromise, he accepted the idea of the substantial but secluded Palace of El Pardo on the La Coruña road just outside Madrid. It had been built as a hunting lodge by Carlos I and converted into a more extensive residence by Carlos III. It was decorated by tapestries by Goya and other painters of the period in the reign of Carlos IV and extended further by Ferdinand VII. The attractions of El Pardo for both Franco and his wife were its royal past, its security and the fact that the hilly estate attached to it was ideal for hunting.

While El Pardo was being restored, Franco and Doña Carmen moved to the Castle of Viñuelas, which belonged to the Duque del Infantado, eighteen kilometres outside Madrid. During their stay at the Castle of Viñuelas, Franco showed considerable interest in establishing his salary as Head of State. After considering what Alfonso XIII and the two Republican presidents, Alcalá Zamora and Azaña, had received, and taking into account Spain's parlous economic state, his initial salary was set at the not inconsiderable sum of 700,000 pesetas.* The Caudillo and his wife remained at Viñuelas until March 1940. Apart from modern conveniences, the renovation work at El Pardo stressed the eighteenth-century aspects of the building's decor and thus reflected the Francos' identification of themselves with the royal leaders of the past. Pétain remarked on the way in which Franco was 'assuming more and more the position of king'.[18] Once installed in El Pardo, Franco insisted that his wife be accorded the aristocratic treatment of being called 'La Señora' and antagonized monarchists with a decree that the Royal March should be played whenever his wife arrived at a state function, as it had for the queen before 1931.[19] Surrounded by a court of sycophants, isolated from the real world, they would remain ensconced there for thirty-five years apart from short official visits to the provinces, three lightning foreign trips to meet Hitler, Mussolini and Salazar, and the long holidays which Franco observed with enthusiasm.

The passions of the civil war and the sense of solidarity with the Axis were fanned into a blaze in November in yet another elaborate spectacle. On the third anniversary of the execution of José Antonio Primo de Rivera by the Republicans, 20 November 1939, his body was exhumed in Alicante and began the first stages of journey to be reburied with full military honours at El Escorial, the resting place of the Kings and Queens of Spain. In a massively choreographed nation-wide operation, for ten days and ten nights, a torchlit procession escorted José Antonio's mortal remains in a five-hundred kilometre journey. Like other ceremonial exhumations, it served to keep alive the hatreds of the civil war.†

* In the spring of 1940, the peseta was officially valued at 39 to the pound sterling which would make his salary £17,950. In 1992 terms, his salary would have been £511,500. This did not include other emoluments such as his salary as a Captain-General and Generalísimo of the Armed Forces, or as *Jefe Nacional* of the Falange. His sister wrote 'Naturally, he did not pay rent for living in El Pardo, and his living expenses were included in the civil list. What I can state categorically is that he never let the State pay for his clothes. He paid personally for his own underwear.' (Franco, *Nosotros*, p. 101).
† Such as the transfer of the bodies of General Sanjurjo from Estoril in October 1939 for reburial in Pamplona, of the ultra-rightist Dr Albiñana from Madrid in April 1940 to Valencia and of General Goded from Barcelona for reburial in Madrid in July 1940.

The Falangist Youth Front, the Sección Femenina, the syndicates, and even regular army units took part. Great bonfires and Church services punctuated the journey. Falangists from every province took their turns as pall-bearers. As they were relieved, artillery salutes and bell-ringing broke out in all the towns and villages of Spain. All school classes and university lectures were interrupted for teachers and professors to raise their arms in the fascist salute and shout 'José Antonio ¡Presente!'. When the cortège arrived in Madrid, it was received by the high commands of the armed services and representatives from Nazi Germany and Fascist Italy. At the Escorial Palace of San Lorenzo, there were monumental wreathes from both Hitler and Mussolini.[20] Some members of the high command were enraged that Franco should accord higher honours to the Falange leader than those paid to Sanjurjo or Mola, let alone to the great monarchs of the past.[21] Considerable violence was engendered by these orchestrations of civil war hatreds, including beatings and murders of Republican prisoners when Alicante prison was stormed by enraged Falangists, which led to serious rifts within the regime hierarchy.

Perhaps inevitably in such a heightened atmosphere, Franco's sympathies for the Axis burned ever more fiercely. On receiving Lord Lloyd in mid-November, he revealed his belief that the best ships of the Royal Navy had been sunk, that England was on the point of starvation and that India was in the grip of revolution.[22] He attacked England and France in his anti-Semitic and imperialistic New Year radio broadcast transmitted on 31 December 1939. 'Now you will understand', he declared, 'the reasons which have led other countries to persecute and isolate those races marked by the stigma of their greed and self-interest.' In an astonishing act of deference to the Nazi-Soviet Pact, the Caudillo criticized the democracies for their 'persecution and extermination' of the Communist Party.[23] A gratified Goebbels noted in his diary 'Something, at least, for our money, our aircraft and our blood'.[24]

Much of the speech was given over to justification of the country's economic difficulties, a reflection of the increased scale of popular muttering against food and fuel shortages, rationing and the regime in general. That had already been seen in a press campaign denouncing jokes about Franco as a crime against the regime.[25] A story went the rounds that the police had captured the man who made up the subversive jokes. After being tortured, he admitted to making them all up except the one in which Franco had promised 'not a home without light nor a Spaniard without bread'. In the shadow of the regime's difficulties, Franco made some bizarre decisions. His commitment to autarky was backed up by a naïve faith in miraculous solutions to extraordinarily complex economic problems. He was easily convinced in late 1939 by geologists eager to please him that Spain possessed enormous gold deposits. Accordingly, he authorized and even went to direct in person gold-mining operations in Extremadura. He could not then resist announcing, in his New Year's eve broadcast on 31 December 1939, just as the country was descending into a period of appalling privation, that the massive gold deposits in Spain presaged a wonderful economic future.[26] The gold was never found.

The desperate nature of Spain's economic difficulties seems to have inflated the credulity of a man with little formal grounding in economics or basic science. Shortly after the announcement of the apocryphal gold discoveries, in early 1940, Franco announced that Spain would soon be self-sufficient in energy and a rich petroleum-exporting country. The basis of the claim was a bogus synthetic petrol allegedly invented by an Austrian, Albert Elder von Filek, who had persuaded Franco that by mixing water with plant extracts and other secret ingredients, the distilled product would be a fuel superior to natural gasoline. Filek had insinuated his way into Franco's confidence by presenting himself as a convinced follower of the Nationalist cause who had been imprisoned by the Republicans in Madrid during the Civil War. He claimed to have had spectacular offers to buy his invention from the world's great oil companies. However, as a gleeful Franco explained to Lequerica, von Filek's admiration for the Caudillo was such as to make him cede his invention to him at a loss. Von Filek was granted the waters of the River Jarama and land on its banks to build a factory. Franco was assured that the trucks which brought fish to Madrid from the seaports of the north had been using the fuel. The Caudillo's chauffeur was part of the sting and convinced him that his own car had been running on the fuel. Vast subterranean tanks were to be built to hold the petrol which would save Spain an annual 150 million pesetas in foreign exchange. Eventually, the fraud came to light and von Filek was imprisoned along with the chauffeur.[27]

Confident of his gold and petrol, Franco snubbed the British Ambassador at the New Year dinner given to the government and the diplomatic corps.[28] Hitler's satisfaction was reflected in the despatch of several New Year gifts to the Caudillo including a six-wheeled Mercedes identical to his own. He had reason to be grateful. The submarine support points were now in operation. In early 1940, German U-boats were using Spanish territorial waters to recharge batteries, rest crews and restock supplies.[29] The Spanish Foreign Ministry regularly provided the German Embassy with information received from Spain's diplomatic missions abroad. This gave the Germans invaluable sources of information from countries where they had no diplomatic relations. Reports from France were to be especially useful during the Franco-German hostilities in June 1940. The Spanish Foreign Ministry also regularly obtained for the Germans reports on the effect in Britain of Luftwaffe bombing raids.[30] Later in 1940, at the orders of General Vigón, Colonel Ansaldo, who had just returned from a Spanish Air Force mission to Britain, was debriefed in the Ministerio del Aire by Canaris and other German officers.[31]

On 23 April 1940, Franco revealed to the Portuguese Ambassador his conviction that the Luftwaffe was on the verge of wiping out the Royal Navy.[32] Within the highest echelons of the Spanish Army, Franco's enthusiasm for the Axis cause was shared unequivocally by Yagüe and Vigón but others harboured doubts. That was not the only cause of serious friction between the Army and the pro-Axis elements of the Falange. The attack on Alicante prison, together with *sacas* (the illegal removal of prisoners and their murder) by Falangists in Valencia had led the Captain-General of the Valencian region, Antonio Aranda, to order the summary execution of those responsible. This led to a major outburst of internecine hostility within the regime.

Many senior military figures, including Kindelán, were especially anxious lest Falangist imperialist ambitions drag Spain into war as an ally of the Axis. Sharing Kindelán's apprehension, the Minister for the Army, General Varela, had begun at the beginning of the year to collect information from the *Capitanías Generales* on the real capabilities of the Spanish Army. In March, Kindelán sent Varela a damning report which showed that Spain was totally unprepared should war break out and that her 'frontiers were still undefended'. Varela read the assessment to a meeting of the high command, gathered together as the *Consejo Superior del Ejército*, a body whose collective views carried considerable weight with Franco. Kindelán's report was endorsed and passed on to the Caudillo. In May 1940, he received from the Chief of his General Staff, General Carlos Martínez Campos, an equally depressing account of the armed force's lack of preparedness which underlined the lack of aircraft and mechanized units.[33]

As impotent, if eager, spectators of the phoney war, Franco and Mussolini were drawn together even more. The warmth of their relations was underlined by the settlement of Spain's Civil War debts in a lengthy negotiation during the summer of 1939 and the spring of 1940. It was linked to the possibility of Spanish rearmament being based on Italian weaponry.[34] Franco assured Gambara that he intended to keep in the closest contact with Italy as far as foreign policy was concerned.[35] Eventually, the Duce, ever restless and unwilling, as he put it, to sit on the side-lines while history was being written, decided to enter the war despite the fact that, after her exhausting enterprises in Abyssinia, in Spain and in Albania, Italy was barely in better shape than Spain for a military escapade. He had given Franco two months notice of his plans, on 8 April 1940.[36] The Caudillo was highly appreciative of the gesture. He told Gambara on 13 April that, despite Spain's acute shortages of grain, petrol, arms and ammunition, if she were dragged into the war, she would do her duty. Earlier in the day, the Caudillo had ordered reinforcements to the Pyrenees, Gibraltar, the Balearics and Spanish Morocco.[37]

The enthusiasm of both Franco and Serrano Suñer was inflamed by the rapidity of German successes in Norway and Denmark.[38] Serrano Suñer had already told Stohrer in the first half of April that Spain was on Germany's side and that Italy's imminent entry into the war would precipitate a Spanish entry. However, even Serrano Suñer was pessimistic about Spain's chances of waging war given the parlous state of her reserves of fuel and grain. The bread ration of five hundred grammes every second day was reduced by half at the beginning of May.[39] Nevertheless, Franco and Serrano Suñer were sorely tempted by the prospect of Spanish belligerence leading to the acquisition of Gibraltar and Tangier.[40] They, like all pro-Axis elements in Spain, must have been impressed by reports of German and Italian support for Spanish aspirations.[41] The press began to talk of the opening up of imperial opportunities for Spain.[42]

In the midst of so much adulation, it was hardly surprising that Franco should resent any signs of independence. When these came from a Cardinal, they were all the more infuriating since they exposed the brittleness of his apparently deep understanding with the Catholic Church. Franco had planned a tour of Andalusia in mid-March 1940 to end in Seville at the end of Holy Week. In the course of a religious procession in which he was hailed by the Falange and the local military, the Cardinal-Archbishop, Pedro Segura, was not seen at his side. Segura was protesting about the regime's pro-Nazism and attempts by Falangists to paint their yoke and arrows symbol on the walls of Seville Cathedral. Infuriated by the discourtesy, Franco sent an escort of Falangists to bring Cardinal Segura to pay him court. When he refused, Franco gave permission for the local Falange to begin a campaign of harassment of the Cardinal. Plans were made for a Falangist parade on the 'Day of Victory', 1 April, to end at the Cathedral as a public humiliation for Segura. The irascible Cardinal simply threatened to excommunicate all those involved. His Cathedral was the only one in Spain not to be adorned with the names of Nationalist war dead and Falangist graffiti. In his next pastoral, he denounced the silencing of Gomá's earlier call for reconciliation and the regime's closeness to the Third Reich. He sent a copy to Franco. Thereafter, he was ostentatiously followed wherever he went by armed Falangists. Franco endeavoured unsuccessfully through his Ambassador to the Holy See, José Yanguas y Messía, to have Segura removed.[43]

Hostility between the Church and the Falange echoed that between the Army and the Falange. Despite these internal tensions there is little to suggest that the Caudillo had any significant worries for the future. On the day of the victory parade to celebrate the first anniversary of his triumph over the Republic, Franco announced his personal decision to raise a colossal monument to those who had fallen in his cause during the Civil War. It was indicative of his self-regard that, like the Pharoahs, he could think in terms of a monument on a scale that would defy posterity. After a victory lunch at the Madrid Capitanía General, at which Doña Carmen was seated between the German and Italian Ambassadors, the Caudillo led a cavalcade of cars to Cuelgamuros in the Guadarrama valley near El Escorial. When members of his cabinet, Falangist leaders, senior generals and members of the diplomatic corps were assembled, Colonel Valentín Galarza, Franco's under-secretary of the Presidencia del Gobierno, read a decree announcing the construction of the monument, to be known as the *Valle de los Caídos* (valley of the fallen). After setting off the first charge of dynamite, Franco addressed the company on the magnitude of what he planned.

The decree announcing the foundation of the monument, dated 1 April 1940, vividly revealed Franco's megalomaniac thoughts about his own place in history: 'The dimension of our Crusade, the heroic sacrifices involved in the victory and the far-reaching significance which this epic has had for the future of Spain cannot be commemorated by the simple monuments by which the outstanding events of our history and the glorious deeds of Spain's sons are normally remembered in towns and villages. The stones to be erected must have the grandeur of the monuments of old, which defy time and forgetfulness . . .' The imposing valley of Cuelgamuros, in the Sierra de Guadarrama to the north-east of Madrid, with its gigantic granite outcrops, was found by Franco himself only after a careful search for exactly what he wanted in terms of natural grandeur.[44]

The basic architectural notion was Franco's and in the course of the monument's construction, he would sketch out ideas for the architect,

Pedro Muguruza. Millán Astray suggested that architecture was Franco's
secret vocation, having designed various buildings for the Legion.[45] Muguruza's task was to produce a monument that would link Franco's era to
that of the Catholic Kings, to Charles V and to Philip II. It was originally
envisaged that the job would take twelve months. In the event, it was to
take two decades and become, after hunting, Franco's greatest personal
obsession. It was said that the Valle de los Caídos became the nearest
thing in Franco's life to 'another woman'. The gigantic work fell to
captured Republicans who had escaped the executioner.

Franco's belief that the 'crimes' of the Republicans could be 'redeemed
by work' was behind the creation in the 1940s, of 'penal detachments'
and 'labour battalions' of captive Republicans used as forced labour in
the construction of dams, bridges, and irrigation canals. In the course of
the construction of the monument, twenty thousand were employed, and
fourteen died, along with many who lost limbs in accidents or were
afflicted with silicosis. It took nearly twenty years to dig the 850 foot
long basilica, to construct the monastery, carved into the hillside of the
Valle de Cuelgamuros and to erect the immense cross which towered
five hundred feet above it. The arms of the cross were the width of two
saloon cars. It cost Spain almost as much as had Philip II's Escorial in a
more prosperous era.[46]

It is a striking reflection of Franco's self-confidence, not to say complacency, that he could find time at the beginning of 1940 to make
excursions into the country in search of a site for his monument. At this
time too he took up painting. Subjected to frequent posing sessions for
portraits, he entertained himself by having a mirror placed behind the
painters so that he could watch what they were doing. One day, when
one of them, Enrique Segura y Sotomayor, forgot to take his paints away
with him, the Caudillo tried for himself.[47] No doubt both painting and
the Valle de los Caídos were the more attractive as hobbies because of
Hitler's known artistic pretensions. With Spain on the verge of economic
collapse, thousands dying of starvation and Europe ravaged by war, his
apparent lack of concern is remarkable. Even more astonishing is that by
the end of the year, claiming in retrospect to have been tortured by worry
about pressure from the Third Reich, he would find time to write a film
script.

Franco's equanimity continued to be disturbed throughout April and
May 1940 by Cardinal Segura's provocative criticisms. The Civil Governor of Seville sent the Generalísimo notes taken in the Cathedral during
one of the Cardinal's sermons. In the sermon, he had proclaimed that,

in classical literature, *caudillos* were 'captains of thieves' and that, in the writings of St Ignatius of Loyola, *caudillo* was a synonym of 'devil'. Franco's rage was such that he ordered the Cardinal expelled from Spain. The outspoken Segura had been expelled by the Republic in 1931 and for Franco to do the same would have inflicted immense damage on the regime's image both inside and outside Spain. That he could contemplate an act of such enormity is a symptom of just how much victory in the Civil War, proximity to the Axis dictatorships and constant adulation had undermined his instinctive caution. It also demonstrated the superficiality of Franco's much-vaunted commitment to the Church. Only the intervention of Serrano Suñer prevented him from committing a grave political error which might have led to a rupture of relations with the Vatican. Franco had to content himself with the existing diplomatic efforts to persuade the Vatican to withdraw Segura.[48]

Despite this, Franco was still, as always, inclined to caution in foreign affairs. On 30 April 1940, he sent what Ciano called a 'colourless message' to Mussolini which was taken in Rome as confirming 'the absolute and unavoidable neutrality of a Spain preparing to bind up her wounds'. The Caudillo's letter suggested that, at this time still an admirer of the French Army, he thought the war might be long and difficult. Accordingly, he praised Mussolini's good sense in having delayed his entry into the hostilities. Referring to Spain's economic prostration, he wrote 'You will understand how upsetting it is for me and my people that the bad timing of this struggle should catch us so far behind.'[49] He must have had on his mind the recent reports from his high command. On the one hand, he knew that an economically and militarily exhausted Spain could not sustain a long war effort but, on the other, he could not bear the thought that France and Britain might be annihilated by Hitler's Wehrmacht and Spain still not get any of the spoils. He hoped therefore to make a last-minute entry into the war to earn a seat at the table at which the booty would be distributed.

Already on 23 April, by way of doing a service to the Axis and for the consumption of the British, Franco had told the Portuguese Ambassador Pereira the outright lie that he was absolutely convinced that Italy would not go to war.[50] Beigbeder continued the game when he received the American Ambassador, the elegant Virginian Alexander Wilbourne Weddell on 4 May.[51] Beigbeder told him that Spain would maintain its neutrality by force of arms. On 14 May, in conversation with the President of ITT, Sosthenes Behn, Serrano Suñer repeated that Spain was ready to defend her neutrality but implied strongly that it would be against Britain or France.[52] All this pointed to Franco's closeness to Mussolini. However, among the factors which distinguished the Duce and the Caudillo at this time, apart from the fact that by temperament Franco was less given to irresponsible rashness, was the existence of a battle-hardened General Staff in Spain, both more pessimistic and less sycophantic than its Italian equivalent. Moreover, as a highly experienced soldier himself, Franco had a realistic notion of his country's capabilities. That did not, however, mean that he was immune to imperialist temptations. Moreover, the factor most inclining him to caution, his admiration for the French Army, would soon be removed. Franco was about to take Spain to the edge of the precipice.

When the Germans invaded Belgium and Holland on 10 May, Franco's press applauded their 'defensive action' and the justice of their success. The Caudillo reacted with appreciative enthusiasm, remarking to Beigbeder, 'The Germans have a good eye. They always pick the right place and time.'[53] On 16 May, the French Ambassador, Marshal Pétain was recalled by Prime Minister Paul Reynaud to become Vice-President in his government. Before leaving Madrid, he was instructed to visit Franco with assurances that the activities of Spanish Republican exiles were being repressed. It was a feeble device to squeeze a promise of neutrality from the Caudillo who was beginning to lose his respect for French military might.[54] When he went for his last audience with Franco, Pétain said, 'my fatherland has been defeated, I have been called to make peace and sign the armistice.' Franco's reply revealed his own ruthlessly egoistic view of politics. 'You are the victor of Verdun,' he said, 'the greatest living glory of France. You are the symbol of powerful and victorious France. Don't go. Don't give your name to what others have lost.'[55] It has also been suggested that, dependent on the defeat of France for the fulfilment of his imperial dreams, Franco was reluctant to risk Pétain returning to revive French military fortunes.[56]

Significantly, in a panic-stricken Paris ten days later, when Marshal Pétain was asked about the likelihood of Spanish intervention in the war, he was confident of Franco's neutrality. He believed that Spain's military weakness deprived her of any alternative.[57] Pétain's assessment of Spanish resources was sound but his confidence in Franco was seriously misplaced, based on a smokescreen deliberately generated by the Caudillo. In addition to the war industries of Toulouse, Angoulême and Bergerac, many factories from northern and eastern France which had been evacuated to the south on the outbreak of war were vulnerable to Spanish attack. Franco wrote to his military attaché in Paris, Colonel Barroso, at the end of May authorizing him to assure the French general staff that the southern frontier could be left unguarded.[58] It is difficult not to suspect that, just three days before he offered his services to Hitler, he was trying to remove obstacles to any Spanish military operations against France.

After earlier doubts, Franco was now certain of early German victory.[59] Washington was aware of the Caudillo's confidence but cognizant too of Spain's horrendous economic problems. There was little sympathy for Franco in the State Department and even less in the liberal press of the United States. Nonetheless, the State Department was prepared to listen sympathetically to Spanish requests for aid rather than let him drift into the arms of the Axis.[60] The British were sufficiently worried about Franco's intentions to replace their Ambassador in Madrid, Sir Maurice Peterson, on 24 May 1940 by Sir Samuel Hoare. In part, it was a question of finding a suitable post for Hoare who had just lost the Air Ministry in the war cabinet reshuffle which had seen the departure of Chamberlain and the arrival of Churchill. More substantially, it was an indication of London's fears about Franco's intentions that such a senior figure – and one thought of as an appeaser – was chosen for this 'special mission'. With France about to fall, it was crucial to prevent Franco throwing in his lot with Hitler and Mussolini. If he did, the loss of Gibraltar and the Spanish Atlantic ports to the Axis would have been a devastating blow to Britain. As Hoare put it thirteen months later 'I had come on what was really a purchasing mission for the purpose of buying time – local time for the fortification of Gibraltar and world time for British recovery after the French collapse'.[61]

The excuse given for the replacement of the Ambassador to Spain was that there had been complaints about the intelligent but extremely prickly Peterson. Although sympathetic to Franco's politics, Peterson was deeply scornful of the Caudillo in personal terms, claiming to order him about more or less at will. Peterson's caustic comments about Franco in particular and the Spaniards in general led one colleague to liken him to a director of an art gallery who had all the qualifications for the job except that he hated pictures.[62] With what at the time seemed arrogant complacency, Peterson was confident that Franco would not go to war against the western allies. He told Sir Robert Bruce Lockhart that Spain could not afford to go to war with a starving population and that 'Franco is a small man and frightened' and was refusing to see both Pétain and himself. In fact, Franco rarely saw any foreign diplomats other than the German Ambassador.[63]

Hoare's appointment was not welcomed in the Diplomatic Corps. Sir Alexander Cadogan, head of the Foreign Office, told Lady Halifax 'there is one bright spot – there are lots of Germans and Italians in Madrid and therefore a good chance of S.H. being murdered'. With the Germans already at Ostend and the retreat at Dunkirk under way, Hoare flew to Lisbon on 29 May and on to Madrid on 1 June where he found high prices, food shortages, German domination of communications, the press and aviation and his Embassy virtually besieged.[64] The job facing the dapper and precise Hoare was to keep Franco out of the war largely by persuading him that Axis defeat was, in the long term, inevitable. Despite being afraid for his life, he carried out his task with enormous skill and bravado.* Whether in formal confrontation with Franco and his Foreign Ministers or just dropping subversive hints which undermined the pro-Axis Serrano Suñer's position in the cabinet or in clandestine contacts with Franco's military and monarchist enemies, his finesse and energy were unstinting and more often than not efficacious.[65]

Franco's perception of his own strength was dramatically inflated by the collapse of the French whose Army he had once feared. The demise of France and Britain opened up imperial vistas which might also simplify his domestic problems, particularly the rivalry between the Falange and the military. The lowest moments for Britain and France seemed to be Spain's opportunity and Franco decided to take a small risk. While crowds of Falangists chanted 'Gibraltar español' outside the British Embassy, and the British Expeditionary Force retreated at Dunkirk, Franco watched with excitement. Along with the Chief of the General Staff, General Juan Vigón and the majority of his generals, the Caudillo had unquestioning faith in the Wehrmacht. They were impatient to seize the opportunity provided by German successes to take Gibraltar and French Morocco.[66]

Accordingly, Franco sent Vigón to Hitler with an effusive letter of congratulation:

Dear Führer: At the moment when the German armies, under your leadership, are bringing the greatest battle in history to a victorious close, I would like to express to you my admiration and enthusiasm

---

* The Argentine Ambassador in Madrid was much taken by Hoare's skill as a dancer of tangos and the Portuguese Ambassador in London admired his competence in acrobatics and ice-dancing, Adrián C. Escobar, *Diálogo íntimo con España: memorias de un embajador durante la tempestad europea* (Buenos Aires, 1950) p. 50; Monteiro to Salazar, 2 June 1940, *DAPE*, VII (Lisbon, 1971) p. 97.

and that of my people, who are watching with deep emotion the glorious course of a struggle which they regard as their own.

He went on at some length to explain how economic difficulties consequent upon the Civil War and fears of British naval strength in the Mediterranean obliged him to hide his support for Germany behind an official neutrality,

I do not need to assure you how great is my desire not to remain aloof from your cares and how great is my satisfaction in rendering to you at all times services which you regard as most valuable.

The letter was dated 3 June, although Vigón did not leave until 10 June.[67] The purpose of Vigón's visit and the letter was to ensure the minimal Spanish participation in the war necessary for a seat at the peace conference table.[68] It would have no effect because Hitler had no intention of paying a high price for services which he believed would not be needed since he expected the British to surrender at any moment.

It was hardly surprising that, behind a self-righteous rhetoric of commitment to peace and mediation, Franco was determined to profit from the chagrin of the French and to pander to his Axis friends. Indeed, in a stealthy and sinuous way, he contributed to the demise of France. In mid-May, for instance, when the French government wanted to send the right-wing French Basque deputy and future Vichy Minister of Youth, Jean Ybarnegaray, to Madrid to seek Spanish mediation with the Italians to head off the feared declaration of war, Franco, on the specific request of Mussolini, personally refused on the specious grounds that it might damage Spain's neutral status. He then sent a full report of his actions to Mussolini together with a request for the Duce's advice as whether he should also inform Hitler. Mussolini was delighted by such deference.[69] Franco also worked stealthily to the detriment of the French in relation to Germany. He had after all signed a treaty with the Third Reich in March 1939 which committed him to consultations with Berlin in the event of an international crisis. Accordingly, throughout June 1940, on instructions from Madrid, Franco's Ambassador in France worked to further the German cause. Lequerica cultivated Pétain, along with other figures of the French right, particularly Laval. His reports to Madrid on their conversations and on Pétain's pessimism were immediately handed on to the Germans. They were an invaluable source of information about French intentions at the highest level.[70]

Inevitably, in 1940, the strategic importance of Spain to the Axis cause made Franco the object of courtship by both sides, the Germans to bring him into the war and the British to keep him out. Despite some internal dispute as to the wisdom of such a policy, the British opted to use the carrot and stick made available to them by their ability to blockade Spanish trade and to give desperately needed credit. Since November 1939, a British delegation led by David Eccles had been in Madrid racing against time to negotiate a war trade agreement with Spain in order to give Franco a popular reason to remain neutral.[71] In contrast, the Germans, not initially interested in Spanish participation, showed little interest in wooing Franco.

The uphill struggle facing Hoare and Eccles was indicated in early June when the British and French Embassies in Madrid and the consulates in Barcelona and Málaga were stormed by Falangists and the Francoist press gleefully reported German and Italian sympathy for the return of Gibraltar.[72] On 9 June, Mussolini wrote to Franco: 'When you read this letter, Italy will have entered the war on the side of Germany. I request of you, within the broad lines of your policy, moral and economic solidarity with Italy. In the new reorganization of the Mediterranean which will result from the war, Gibraltar will be returned to Spain.'[73] Yagüe urged the Caudillo to join the Duce. With Franco's approval, Yagüe had already granted a request by Ciano for Italian bombers to refuel secretly on Spanish airfields.[74] The war fever being generated by the Falangist press provoked several letters of protest to the Army Minister, General Varela, from senior colleagues including Kindelán, Ponte and Orgaz. This pressure, in some small measure, helped temper Franco's excitement about the fate of his northern neighbour.[75]

A degree of caution prevailed to the extent of efforts to mask Spain's growing commitment to the Axis. Beigbeder told the US Ambassador Weddell, that the Italian action was 'madness'.[76] Franco in contrast wrote a highly charged reply to Mussolini's letter of 9 June, pledging his moral solidarity and as much economic help as Spain could afford. However, before sending it, he received a letter from Ciano urging him to follow Mussolini's earlier example and change Spanish neutrality to non-belligerence. This Franco did in an amended draft of the letter sent to the Duce on 10 June. Franco sought the rubber stamp of the cabinet for his decision at a meeting on 12 June.[77] Mussolini sent an effusive message of gratitude.[78] The Caudillo told the Italian *Chargé d'Affaires*, with a tone of regret, that 'the present state of the Spanish armed forces prevents the adoption of a more resolute attitude but that nonetheless he was proceeding to accelerate to the full the preparation of the Army for any eventuality'. The Caudillo also spoke of his resentment of the United States.[79] This news was eagerly related to the Germans.[80] The public announcement of non-belligerence was made on the following day. The Falangist press declared that Spain must stand by the countries which had helped her in the Civil War.[81]

In Britain and Portugal, it was assumed that non-belligerence meant, as it had for Mussolini, a prelude to a declaration of war.[82] Franco consistently tried to use the Portuguese to deceive the British. For months, he had been assuring the Portuguese Ambassador, Pereira, of his commitment to neutrality and of his lack of acquisitive plans. He did so again on 10 June, the same day that he wrote to Mussolini to offer non-belligerence.[83] On the day that non-belligerence was announced, he sent his brother Nicolás to assure the Portuguese Foreign Ministry that it constituted no divergence from Spain's existing neutral line.[84] The Caudillo saw Lisbon as a useful conduit to the Foreign Office, to be exploited, while the Axis was winning, to mask his own position. In 1943, when the outcome of the war seemed more doubtful, he would use Lisbon to endorse his neutral credentials in the eyes of the Allies. In the summer of 1940, however, he harboured predatory thoughts about Portugal.

There were many in both the Falange and the officer corps who were tempted by reports that Gibraltar was poorly defended and the French army in Morocco demoralized. Beigbeder was against any declaration of war although he told the Italian Chargé that Gibraltar 'will fall like a ripe fruit when the moment comes'.[85] Franco was sorely tempted but reluctant to do anything without explicit German support.[86] Indicative of Franco's deference to the Third Reich was the fact that he gave virtually free access to the German Ambassador. In marked contrast, he declined to move beyond platitudes and enter into serious discussions with Hoare, any more than he had with Peterson before him. When, in mid-June, Hoare presented his credentials, he was anything but impressed by Franco in the flesh. The Caudillo's 'small, rather corpulent, bourgeois figure seemed insignificant. His voice was very different from the uncontrolled shrieks of Hitler or the theatrically modulated bass of Mussolini. It was indeed the voice of a doctor with a good bedside manner, and of a doctor with a big family practice and an assured income.'[87] Franco clearly believed, as the Madrid gossip had it, that the Allies were already defeated and that Hoare had come merely to offer Gibraltar. Equally off-hand with Washington's Ambassador, the Caudillo kept Weddell at arm's length, consistently breaking appointments to receive him.

Officially, Franco did not deal with specific departmental affairs and this was used as an excuse for him to decline to receive Ambassadors, other, of course, than von Stohrer. He lived in heavily guarded isolation in his El Pardo palace where the atmosphere of grim seclusion seemed to Hoare to be more fitting for an oriental despot. The Caudillo held cabinet meetings there and received his ministers. He left only for State occasions and hunting trips. In fact, Franco's unavailability was compounded by the fact that he was increasingly taken with hunting. When audiences were granted, it was, as Hoare put it, 'difficult to penetrate the cotton-wool entanglements of his amazing complacency'. Franco would sit by a writing table decorated with signed photographs of Hitler and Mussolini and assiduously avoid any serious debate. He seemed to Hoare blithely unaware of the economic and naval strength of the British Empire or of her friend, the United States.[88]

Nevertheless, it is clear that Franco took a particular interest in the making of foreign policy. Beigbeder, Jordana and Lequerica, all Spanish Ministers of Foreign Affairs at different points during the Second World War, claimed that the Caudillo made policy while they simply dealt with issues of detail and implemented his instructions.[89] From 1945 onwards, Franco's propagandists worked hard to present Serrano Suñer as the exclusive architect of pro-German policy. That is nonsense. It is inconceivable that Franco passively let his brother-in-law make foreign policy. Serrano Suñer shared Franco's enthusiasm for German triumphs and was keener than ever to take over the reins of Spanish foreign policy. At a reception in the Brazilian Embassy, Serrano Suñer invited those present to a cocktail party in defeated London two weeks later.[90] He was already intriguing against Foreign Minister Beigbeder and establishing a close direct relationship with Stohrer. Serrano told the German Ambassador that, although there was no need for Spain automatically to follow Italy into the war, 'Spain would, however, vigilantly follow developments in order to intervene at the right moment'.[91] However, to acknowledge Serrano Suñer's ambitions is not to diminish the extent of the Caudillo's own pro-Axis fervour.

Indeed, the Spanish assistance most valuable for the Axis could not have been mounted without Franco's knowledge and explicit consent. German submarines were being provisioned and repaired in Spanish ports and relief submarine crews were permitted to travel across Spain. U-boats were thus enabled to remain longer away from their home bases. By June 1940, using Spanish facilities, German submarines could reach the north coast of Brazil and extend their operational radius further south and so threaten British supply lines. They attacked convoys in mid-Atlantic in the confidence that fuel would be available for the trip home. In case of damage by enemy action, U-boat commanders could depend upon carrying out repairs in Spanish ports and getting medical care for wounded crew. The supply system set up in late 1939 and the first half of 1940 required considerable planning and a complex infrastructure. It was set up with the approval of Franco after some early hesitation born of his fear of British naval strength. The Caudillo then passed operational responsibility to Beigbeder. Franco also permitted German reconnaissance aircraft to fly with Spanish markings and a radio station at La Coruña was at the service of the Luftwaffe. In the autumn of 1940, requests would be made for secret night-time refuelling of German destroyers in bays on Spain's northern coast.[92]

According to a post-war United Nations Security Council investigation, during the Second World War German planes operated from Spanish airfields against Allied shipping. German aircraft forced down on Spanish territory were repaired by Spaniards and the Germans were permitted to carry out detailed inspections of British and American planes forced down. German espionage and sabotage against Allied targets in Spain was facilitated by the Spanish authorities. Similarly, German observation posts on the Mediterranean coast made it possible for the German command to have exact information about the number, type and course of British and American ships entering the Mediterranean and to attack them accordingly.[93]

The claim that, with a powerful Wehrmacht on his frontiers, the Caudillo had to treat the Third Reich with caution and even benevolent neutrality has been used insistently by his apologists.[94] This is an entirely spurious argument. There was no question of hostile German action against Spain. With planning for an attack on Russia already beginning in the summer of 1940, the Wehrmacht had little spare capacity for an assault on Spain. And, given the level of valuable co-operation from Franco, Hitler had no need to contemplate one.[95]

Despite Franco's much-vaunted friendship with Pétain, at 2.30 a.m. on 14 June, as the Germans poured into Paris, Spain occupied Tangier. Lequerica merely informed the French at 6.30 p.m. on the previous evening that the action was necessary to guarantee the city's security.[96] Beigbeder boasted to the Italian *Chargé d'Affaires* Zoppi that this had been done 'when the Quai d'Orsay was so occupied with other grave matters as to be incapable of opposing the Spanish intentions'.[97] So much for Franco's later claims to have acted with benignly protective consideration towards defeated France in 1940. In fact, at a time of catastrophic defeat at the hands of the Third Reich, the French were affected by what they saw as evidence of an additional threat from a hostile Spain.

Franco and Serrano Suñer saw the seizure of Tangier as the first positive step towards a full-scale African empire. The decision was taken by the two of them alone, without discussion with other Ministers or the General Staff.[98] Fervent telegrams from Falangist organizations thanked the Caudillo for returning Africa to Spain.[99] Speaking from the balcony of his Embassy, Stohrer told Falangists in the street below that 'Spain's desires will be granted'.[100] On the following day, Vigón, after presenting Ribbentrop with Franco's gift of the chain of the Order of the Yoke and Arrows, told him of Spain's desire to take over all of Morocco.[101] On the next day, 16 June 1940, Hitler received Vigón at the Château Acoz in Belgium, to the south of Châtelet and told him that he was delighted that Franco 'had acted without talking'.[102] However, he did not take up the offer of belligerence contained in Franco's letter and merely acknowledged Spain's Moroccan ambitions. Vigón, however, returned to Madrid completely bewitched by what he saw on an organized tour of the western front and confirmed Franco's views about the invincibility of the Wehrmacht.[103]

Within a matter of hours, however, Franco was able to make a contribution to the demise of France. The government now headed by Pétain had fled to Bordeaux. At 12.30 a.m. in the morning of 17 June, Paul Baudouin, the Foreign Minister, called in the Spanish Ambassador, Lequerica, and his military attachés Ansaldo and Barroso. He requested that Spain act as intermediary with the Germans to request a cessation of hostilities and negotiate peace conditions. After enormous difficulties, Lequerica got the message through to Madrid. It was discussed by Franco and Jordana and passed on to the German Ambassador at 3 a.m.[104]

The official reason for the choice of Franco as intermediary was the high esteem in which Pétain supposedly held the Caudillo.[105] In fact, the implausible appeal to 'la espada más limpia de Europa'[106] (the cleanest sword in Europe) revealed, in the wake of the occupation of Tangier, a cold appreciation of Franco's vanity and was aimed at delaying or diverting a potential enemy who might be on the verge of opening a third front against France.[107] The complex French decision to seek an armistice had been influenced by the various anti-French gestures made by Franco during June 1940. The non-belligerency declaration, the move on Tangier and Spanish troop deployments near the Pyrenees and on the border with French Morocco had been seen in Paris as evidence that Franco

RIGHT: Nicoláo Franco Salgado-
Araujo and María del Pilar
Bahamonde y Pardo de Andrade,
carrying Francisco at his baptism,
17 December 1892

BELOW: Ramón, Pilar and
Francisco, El Ferrol, c.1906

Franco as an Army officer cadet (*standing*) with his brother Nicolás (*seated*) in the uniform of a naval cadet, *c*.1908

Second Lieutenant Franco shortly after joining his regiment in El Ferrol, September 1910

Major Franco of the newly created
Tercio Extranjero (Foreign Legion),
Morocco, 1920

Franco with General Sanjurjo
(*far right*) in Morocco, 1921

ABOVE: Franco at a dinner given by Asturian admirers, Automobile Club, Oviedo, 10 June 1922

LEFT: Doña Carmen and Franco on their wedding day, Oviedo, 22 October 1923

RIGHT: Franco with his brother Ramón, Morocco, 1925

BELOW: Colonel Franco, General Miguel Primo de Rivera and General Sanjurjo, near Alhucemas, 1925

ABOVE: Ramón Franco is hailed by admirers after his return from exile, Madrid, April 1931

LEFT: General Franco, as Military Commander of the Balearic Islands, talks with President Niceto Alcalá-Zamora during naval manoeuvres, 10 June 1934

ABOVE: Franco with (*left*)
Cavalcanti and (*right, yelling*)
Mola, Burgos, August 1936

RIGHT: Franco harangues
the survivors of the siege of the
Alcázar de Toledo, 29 September
1936

ABOVE: The Francoist assault on Madrid, November 1936

LEFT: A wartime propaganda photograph of Franco, 1937, showing perhaps why foreign correspondents were so often struck by his eyes and also how he aged during the war

Propaganda postcard, produced by
Franco's supporters in Britain, c.1937

"It is not the army alone that is fighting, with the rest of the population
holding aloof or showing hostility.   The whole of the nation is in arms ;
the whole of the civil population has mobilised spontaneously, without
distinction of class, sex or age"—General Franco.

Franco, Serrano Suñer and Mola
in Burgos, May 1937 – their last
meeting before Mola's death

ABOVE: The victory at Santander claimed for
Mussolini; Franco telegrams Mussolini
praising the enthusiasm and skill of the Italian
troops. 28 August 1937

RIGHT: Franco decorating Italian soldiers
after the fall of Santander, August 1937

ABOVE: Franco and Dávila at the front near Teruel, January 1938

LEFT: Franco at his field headquarters, summer 1938, (*left to right*) Vigón (Juan), Franco, Dávila and (*with his back to the camera*) Kindelán

Generals Garcia-Valiño (*left*) and Yagüe (*right*), December 1938

Official poster celebrating the Nationalist victory in 1939. In the bottom left-hand corner is the text of Franco's final wartime daily report (*parte*) issued on 1 April 1939

Franco presides, with General Queipo de Llano and Admiral Cervera, at the victory parade in Seville, May 1939. To Franco's right are General Dávila and Serrano Suñer; behind Admiral Cervera, General Kindelán can just be seen

Franco with Cardinal Segura, Seville, 1939

LEFT: Senior Francoists examine pictures of victory monuments, mid-1939: (*left to right*) Antonio Barroso, Franco, Serrano Suñer, Pedro Muguruza (the architect of the Valle de los Caídos), Francisco Gómez-Jordana

BELOW: El Valle de los Caídos

ABOVE: Franco presides at the first meeting of his new cabinet in Burgos, 12 August 1939. *Left to right*, seated: Yagüe, Varela, Beigbeder, Franco, Serrano Suñer, Admiral Moreno, Muñoz Grandes. Standing: Alarcón de la Lastra, Gamero del Castillo, Larraz, Esteban Bilbao, Sánchez Masas, Peña Boeuf, Benjumea, Ibáñez Martín

BELOW: The Falangist High Command at El Escorial for the burial of the remains of José Antonio Primo de Rivera, 20 November 1939. *Left to right*: Raimundo Fernández-Cuesta, Miguel Primo de Rivera, Pilar Primo de Rivera, Agustín Muñoz Grandes, Rafael Sánchez Masas, Pedro Gamero de Castillo, Ramón Serrano Suñer

Hitler greets Franco on his arrival at
Hendaye, 23 October 1940

Serrano Suñer, Franco
and Mussolini, Bordighera,
12 February 1941

was either about to declare war on France or at least to facilitate a German march on French North Africa.[108]

During the days of the French collapse, Lequerica, in collaboration with his friend Pierre Laval, was tireless in pushing Pétain's Government towards an armistice and away from the idea of continuing the war from North Africa. He insinuated to Pétain that Hitler might contemplate a negotiated peace. Lequerica held out hopes of a right-wing France joining Spain, Italy and Germany in a new order to replace the evil empire of Anglo-Jewry.[109] Franco was clearly hoping that if he were established as the middle man between a defeated France and a victorious Germany, rich colonial pickings would ensue. Nothing about the armistice suggests that the French saw Franco as anything other than a subordinate of Hitler.

As always Franco was working at more than one level. He was offering his services to both the Germans and the French. Hitler having shown no interest in Spanish belligerence, Franco hoped to exploit the French catastrophe. On the very morning of the day of the armistice request, before any detailed settlements could even be broached between France and Germany, Franco instructed Lequerica to demand the tribal territories of the Beni-Zéroual in southern Morocco near Fez and the Beni-Snassen in eastern French Morocco. With an air of embarrassment at seeming to be seeking payment for handling the armistice request, Lequerica transmitted the demands on the following day, 18 June. The loss of those districts would have seriously weakened French Morocco. Beni-Snassen would have given Spain a lever against the Oran areas of Algeria. On the next day, Franco revealed the scale of his ambitions to the Italians – the union of Spanish and French Morocco under his protectorate, part of Algeria, the extension of Spanish Sahara and the expansion of Spain's territories in the Gulf of Guinea.[110]

Two arguments were used to justify the demands to Comte Renom de la Baume, the French Ambassador in Madrid. The first was the typically Francoist one that, since France was bound to lose some of her empire, it was better for Spain to profit than Germany. The second was the entirely spurious one, in the light of the shopping list which Franco was presenting at the same time to both Germany and Italy, that Spain merely wished to control possible outbreaks of disorder among the local tribes. Nevertheless, despite Franco's barely cloaked greed, in the midst of so many disasters, there were those in the French Government ready to cave in to his demands. However, the French military commander in Algeria, General Noguès, was virulently hostile to any concession and he finally imposed his view on Pétain. Franco clearly intended to back up his arguments with force but Noguès convinced General Asensio, the Spanish High Commissioner in Morocco, that any incursion into French territory would be sharply repulsed.[111]

On the same day that he put pressure on Pétain, Franco presented the Germans with a formal offer to go to war in return for the fulfilment of his colonial aspirations. It seems likely that he was emboldened to this step, merely three days after his first offer through Vigón had been ignored, not only by French misfortunes but also by the war fever which, in Madrid at least, the Falange managed to generate. In the event of England continuing hostilities after the surrender of France, the Caudillo offered to enter the war on the Axis side in return for 'war materials, heavy artillery, aircraft for the attack on Gibraltar, and perhaps the co-operation of German submarines in the defence of the Canary Islands. Also supplies of some foodstuffs, ammunition, motor fuel and equipment, which will certainly be available from the French war stocks.'[112]

Franco's confidence both in German success and in his own standing as an Axis partner was reflected in his treatment of both Weddell and Hoare. He passed from an offhand coldness to a tactless boasting. He arrogantly revealed to Weddell, when he finally deigned to see him on 22 June, how much he was relishing the imminent division of the French and British empires. He brushed aside an American offer of economic co-operation conditional on Spanish neutrality.[113] He firmly told the British Ambassador later on the same day that an Allied victory was entirely impossible. 'Why', he asked to the consternation of Hoare, 'do you not end the war now? You can never win it. All that will happen if the war is allowed to continue is the destruction of European civilization.' Clearly confident of the outcome of his proposal to the Germans, Franco also told Hoare that Spain needed nothing from the British Empire.[114] Rumours were spread around Madrid that Hoare had offered Franco Gibraltar in return for a Spanish undertaking not to join the war on the Axis side.[115]

American Embassy staff in Madrid saw the Caudillo's indifference to offers of economic help as evidence of his unshakeable confidence in German victory and subsequent benevolence. British and American suspicions were also growing that a recent surge in Spanish oil imports suggested either preparations for war or surreptitious measures to help out Italy.[116] The suspicion was compounded by the fact that the shipments were being arranged by the pro-Axis Thorkild Rieber, the President of the Texas Oil Company, who had met Franco's oil needs during the Civil

War. At British request, US oil exports to Spain were now restricted so as to prevent stockpiling in Axis interests. It was a shrewd policy which neither gave Franco the confidence to go to war nor threw him entirely on the mercy of the Third Reich.[117]

After keeping Franco waiting for nearly a week, the Germans replied coolly to his second offer of belligerence. Convinced of Britain's imminent collapse, Hitler had little interest in Spanish participation on Franco's terms. Some of Franco's aspirations collided with the Führer's own plans to create a German empire in Africa. Spanish belligerence required grain and fuel supplies on a scale which the Third Reich could not afford. The formal response to the Spanish offer came from State Secretary, Baron Ernst von Weizsäcker: the German Government took cognizance of Spain's territorial desires in North Africa, warmly welcomed the Spanish offer to enter the war and undertook to consider requests for military equipment 'at the proper time'.[118] Hitler was not about to prejudice the armistice negotiations with Pétain in order to give gratuitous satisfaction to Franco. The Caudillo's disappointment was reflected in the fact that he immediately instructed Serrano Suñer to request permission to visit the Reich to iron out the differences over territorial ambitions and German supplies.[119]

It was at precisely this time that Franco dismissed General Yagüe as Minister for the Air Force. Frustrated by Franco's dilatoriness, Yagüe had become more explicit in his criticisms and, opposing Franco's policy of blanket vengefulness, he was rehabilitating Republican air-force officers, some of whom had been freemasons. Ever more extreme in his radical Falangism, he became involved, as did General Agustín Muñoz Grandes rather more circumspectly, in a plot to remove Franco. Exposed by the intelligence services, Yagüe had a tense and emotional meeting with Franco on 27 June 1940 after which he was sacked from his ministerial post and exiled to the village of his birth, San Leonardo in Soria. The feeble official pretext used was the fact that he had told Hoare that England was defeated and deserved to be. Yagüe's remarks were inopportune but they hardly differed from those with which Franco had affronted Hoare on 22 June.[120]

The Yagüe incident aside, Franco, Serrano Suñer and Beigbeder were all, in their different ways, obsequious towards the Third Reich and the press was virulently anti-British.[121] There was a permissive attitude to incursions across the frontier of uniformed German soldiers in tanks and armoured cars, some of whom even took part in small semi-official victory parades. General José López Pinto, Captain-General of the VI Military

Region, Burgos, accompanied by his staff, a military band, the German Ambassador and officials of the San Sebastián branch of the Nazi Party,* greeted the commander of the German troops who had reached the Spanish border on 27 June. López Pinto hosted a formal reception in their honour toasting them with a cry of *¡Viva Hitler!*. Only after repeated protests by Hoare was action taken and López Pinto removed from his post.[122]

The dark, mysterious Beigbeder was later considered to be an Anglophile, but in the summer of 1940, he was essentially his master's voice.[123] Alternately an ascetic and a womanizer, Beigbeder was reputed to be entranced by the statuesque young Baroness von Stohrer. As the summer of 1940 wore on, however, it became increasingly apparent that the conduct of foreign policy was a matter between Franco and Serrano Suñer. The issue of Beigbeder's sexual activities was used by Serrano Suñer to plant doubts in Franco's mind about his reliability.[124] In response, Beigbeder intensified his own pro-German sentiment as a desperate attempt to keep a grip on his post. For instance, on 23 June, Beigbeder offered to detain the Duke and Duchess of Windsor – who were passing through Madrid from the south of France to Lisbon – in case the Germans wanted to make contact with them.[125]

Perhaps because he regarded Beigbeder as insufficiently influential, Stohrer pursued the question of the Duke of Windsor through Serrano Suñer who, in his turn, consulted with the Caudillo. A Spanish diplomat, Javier 'Tiger' Bermejillo, was assigned to accompany the Duke and his personal reports to Franco led the Caudillo to believe that the ex-King was keen to act as a peacemaker. Throughout the summer of 1940, Serrano Suñer and Franco were willing collaborators in German machinations to prevent the Duke of Windsor taking up the post of Governor of the Bahamas in order that he might be used against 'the Churchill clique' in peace negotiations with England. Nicolás Franco, the Ambassador in Lisbon, was mobilized on numerous occasions and Miguel Primo de Rivera, head of the Falange in Madrid and a friend of the Duke, was sent to Portugal to intercede with him not to go to the Bahamas. In the hope of persuading him to be a kind of English Rudolf Hess, the Duke was told by another emissary, Serrano Suñer's close collaborator Angel Alcázar de Velasco, that the British secret service had plans to assassinate him.[126] Their efforts were in vain.

---

* The Nazi Party, through its *Auslandorganisation*, had branches in many non-German cities. Most of the members were German businessmen resident abroad.

Franco's swaggering behaviour to Hoare continued into the summer. He no doubt derived enormous satisfaction in late June from being able to inform him of Hitler's peace terms.[127] The Caudillo also instructed Beigbeder to offer Hoare the good offices of Spain for the transmission to Berlin of a British request for armistice.[128] Hoare believed that a policy of building up anti-war feeling in Spain and showing a readiness to help alleviate the near famine conditions afflicting the country – what his enemies saw as appeasement of Franco – was the only way to keep Spain out of the war. Efforts to overthrow Franco he regarded as sheer temerity since any success by the Left would be used by the Germans as an excuse to invade. On the other hand, as a way of restraining Franco, Hoare was enthusiastic about cultivating, and indeed bribing on a massive scale, senior Spanish Army officers.[129]

Hoare was sufficiently impressed by the anti-British demonstrations which had greeted his arrival to seek guidance from London on 17 June as to how he should react if the Gibraltar question were raised by the Spaniards. The War Cabinet met on 18 June and decided that he should stonewall by saying that he had to consult London. If pressed, he should say that, for obvious reasons, the question of Gibraltar could not be discussed during the war but that London would be prepared to discuss with Spain 'this or any other question of common interest after the conclusion of hostilities [without, however, referring specifically to Gibraltar]'.[130] In fact, R.A. Butler, Parliamentary Under-Secretary for Foreign Affairs, had already told Franco's Ambassador, the Duque de Alba, on 8 June that 'England is disposed to consider later on all the problems and aspirations of Spain including Gibraltar.'[131] Both Hoare and Halifax continued to press the benefits of appeasing Franco by offering concessions over the sovereignty of Gibraltar. Churchill put a temporary stop to it with a memorandum to Halifax on 26 June: 'I am sure that we shall gain nothing by offering to "discuss" Gibraltar at the end of the war. Spaniards will know that, if we win, discussions would not be fruitful; and if we lose they would not be necessary.' However, Halifax and Hoare continued to insist that a categoric refusal ever to discuss Gibraltar would favour the pro-Axis camp in Madrid. Finally Churchill yielded to their arguments. In September, Hoare did intimate to Beigbeder that Britain would be prepared to talk about Gibraltar after the war.[132]

Such vague offers of future talks were not enticing by comparison with what Franco could hope to gain from his Axis ties. It was hardly surprising then that Serrano Suñer continued to press for an official invitation to visit Germany.[133] In contrast to the Spanish efforts at ingratiation, the Germans were dismissive about the Spaniards. Hitler told the Italians that he did not want the Spaniards in French Morocco lest it provoke a British landing there. The Führer wanted air bases in Morocco and was already beginning to covet one of the Canary Islands, aspirations in no way consistent with Franco's view of his own importance in the new world order.[134] Berlin requested von Stohrer to ensure that Spain was no longer exporting strategic goods to France and Britain. It was assumed that essential Spanish raw materials would be exported to the Third Reich.[135]

Not dismayed by the Führer's offhand response to his offers, as the efforts to arrange a visit to Berlin by Serrano Suñer showed, Franco remained anxious to negotiate Spanish entry into the war. He was heartened by the arrival, at the end of June, of Admiral Canaris. Since the Civil War, Canaris was on good personal terms with Franco. Indeed, a large photograph of the Caudillo complete with a long dedication was one of the few fripperies in his austere Berlin office.[136] On this trip, he spoke to Beigbeder, Vigón and Franco. Canaris made it quite clear that, for the moment, Germany had no interest in Spanish belligerence. However, he did have a specific request for Spanish co-operation. He pressed Franco on 6 July to grant permission for German troops to cross Spanish soil in the event of a British invasion of Portugal or of Portugal joining the war on Britain's side and suggested that such troops could proceed to the recovery of Gibraltar. Franco was cautious. He was fully aware that to allow German troops to enter the peninsula would consolidate the puppet status of both Portugal and Spain. On the other hand, the Caudillo was not averse to using Spanish troops to force Portugal into dependence on Spain. Accordingly, Franco suggested that, for action in Portugal or against Gibraltar, Spanish forces would be perfectly adequate as long as they were provided with artillery and aircraft.[137]

Fresh from this meeting, when the Generalísimo met the Portuguese Ambassador, Pereira, later in the day, his tone was patronizing. Franco advised renunciation as soon as possible of Portugal's friendship with Britain and spoke of Hitler as 'an extraordinary man, moderate, sensitive, full of the spirit of humanity and with great ideas'. Franco's earlier admiration for the French army was forgotten. He claimed to have seen its defeat coming all along because the French were decadent and did not want to fight. He dismissed Dunkirk as a disgrace and remarked that 'Germany has the war won. The most Britain can do is drag it out a little longer in the hope of squeezing better peace terms than France.'[138]

Pereira feared that Franco hoped to use his relationship with the Third

Reich to clinch his dominance over Portugal in the same way as he hoped to get French colonies on the cheap. In the first week of July, Spanish troops had been deployed near the Portuguese frontier.[139] There had been calls from hard-line Falangists for the outright annexation of Portugal since the Civil War and now they were being heard again.[140] Before Canaris's request, Franco and Serrano Suñer had already used the threat of hostile German action against Portugal through Spain as a lever to break the Anglo-Portuguese alliance and to force Portugal into becoming a dependent ally of Spain. On 26 June, Serrano Suñer had taken Pereira to El Escorial and spent the entire afternoon trying to persuade him that Portugal must break free of 'the dead weight of her English alliance'.[141]

At the meeting with Franco on 6 July, Pereira passed on a suggestion from Salazar that Spain and Portugal consolidate their 1939 Treaty of Friendship. The Caudillo agreed to talks beginning on an extension of the Treaty, but he was anxious that any new Hispano-Portuguese agreement remain secret, presumably so as not to give the Germans the impression that his commitment to the Axis was wavering. The Portuguese attitude was that the whole point of tightening the Iberian front was to deter the Germans. For Franco, it was rather the opposite. His brother Nicolás conveyed an urgent message from him to Oliveira Salazar on 13 July offering all Spain's forces to assist Portugal in repelling 'any demand or abuse by the English'.[142]

The Portuguese reaction, in the confidence that their neutrality would not be violated by Britain, was to try to tie Spain down to a mutual undertaking to defend each other's neutrality. Hoare, therefore, regarded the amendment to the 1939 Hispano-Portuguese Treaty of Friendship and Non-Aggression signed on 29 July 1940 as a triumph for Beigbeder's policy of moderation. If anything, in fact, it opened the way for both Franco and Salazar to have some protection, however flimsy, against possible British or German incursions into Iberia. Salazar saw in Franco a predator who needed to be neutralized and also a possible interlocutor with the Axis, should it be victorious. Franco harboured ambitions of taking over Portugal with Axis help, but was happy to go along with the treaty both to allay suspicions of his designs and also to provide a channel to the British in the unlikely event that the war went their way.[143]

Franco's confidence in Axis triumph remained ebullient through the summer. In the course of the fourth anniversary celebrations of the Nationalist uprising on 17 July 1940, he spoke to the *Consejo Nacional* of *FET y de las JONS*. His tone was once again anti-Semitic and aggressively imperialist. 'We have shed the blood of our dead to make a nation and to create an empire ... We have a duty and a mission, the command of Gibraltar, African expansion and the permanence of a policy of unity.' In a flash of fascist existentialism, he declared: 'We want the hard, the difficult life, the life of virile peoples ... We offered five hundred thousand dead for the salvation and unity of Spain in the first European battle of the new order. We are not absent from the problems of the world ... Spain has two million warriors ...' He sang the praises of discipline and unity as the key to Spanish ambitions, and as the secret of Hitler's 'fantastic victories on the fields of Europe'.[144]

During the following day's Civil War Victory parade, there were carefully orchestrated demonstrations in favour of 'Gibraltar español' which caused Hoare and his wife ostentatiously to leave the diplomatic stand.[145] The Axis victory which Franco so enthusiastically took for granted in 1940 was not, as he later claimed, that over Communism since at this stage Germany and Russia were allies. As his speeches at the time made clear, the Axis's war was directed against the decadent democracies or 'plutocracies'. The press of both the Third Reich and Fascist Italy reported Franco's 17 July speech in the most enthusiastic terms. On the day after the speech, it was announced that Hitler had awarded Franco the highest order that the Third Reich could bestow upon a foreigner, the Grand Cross of Gold of the Order of the German Eagle.[146]

Ironically, this high point in his relationship with the Führer concealed the fact that Franco had not perceived the long-term significance of Hitler's armistice with France. He failed entirely to foresee that it had closed the door on his hopes of inheriting substantial parts of the French North African territories. Accordingly, his scavenging efforts to pick up an empire continued throughout the summer with attempts to whip up a tribal rebellion in French territory to justify a Spanish military intervention.[147] But nothing came of them. Even more humiliating were relations with London. Despite his pro-Axis gestures and rhetoric, Franco's policy was still subject to economic constraints. The British had been quietly putting pressure on Spanish fuel supplies which both inhibited possible war preparations and impeded distribution of the country's exiguous food supplies. Fuel shortages badly affected industry, ensured that houses, hospitals and schools faced a hard winter and gave rise to the appearance on Spain's roads of the *gasógeno*, a wood- or coal-burning device attached to the back of cars to produce combustible gases. Franco was reluctantly obliged to keep some options open. After all, little or no German help was materializing. Accordingly, on 24 July, he signed an agreement with Britain and Portugal for the exchange of goods through the sterling area.[148]

British resistance was forcing Hitler to adjust his priorities for Spanish entry into the war. The failure of the Luftwaffe to eliminate the RAF in the Battle of Britain was undermining his invasion plans, Operation Sea-lion. German thoughts turned to the idea of bringing down Britain by means other than frontal attack. On 15 August General Jodl suggested the intensification of U-boat warfare and the seizure of the nerve centres of her empire, Gibraltar and Suez, in a bid to give the Axis control of the Mediterranean and the Middle East. Already on 2 August, Ribbentrop had informed Stohrer that 'what we want to achieve now is Spain's early entry into the war'.[149]

Extremely serious consideration was given in Germany to the pros and cons of Spanish participation. As a result of his discussions with Ribbentrop, on 8 August Stohrer drew up in Berlin a long memorandum on the costs and benefits of a Spanish declaration of war on England. He recalled a statement on 3 August by Beigbeder to the effect that fuel shortages would limit a Spanish war effort without German assistance to one and a half months. This was a remarkably optimistic prediction. The major advantages of Spanish belligerency were perceived as the blow to English prestige, the curtailing of exports to England of Spanish ores and pyrites, the German acquisition of English-owned ore and copper mines and, above all, control of the Straits. The major disadvantages were seen as possible English counter-seizures of the Canary Islands, Tangier and the Balearic Islands and an extension of the Gibraltar zone, an English landing in Portugal, an English link with French forces in Morocco and the burden constituted by a Spanish drain on German and Italian supplies of food and fuel. Stohrer also drew attention to the enormous difficulties involved in transporting war material in Spain given her narrow roads and different railway gauge. He concluded that it was crucial to avoid too early a Spanish entry into the war for fear that the effort would be unendurable for Spain with consequent dangers for Germany.[150]

Equally pessimistic conclusions were reached by a report on Spanish military strength drawn up by the German High Command. It was judged that Spain had insufficient artillery to equip a wartime army, enough ammunition for only a few days of hostilities and armaments factories with a capacity far below wartime requirements. Fortifications on the Portuguese border were non-existent and those on the Pyrenean frontier inadequate in number and quality. 'Installations built around Gibraltar are of little value and essentially represent a waste of material.' The

Spanish high command was judged to be 'sluggish and doctrinaire', bogged down in the mentality of colonial war. The report concluded that 'without foreign help Spain can wage a war of only very short duration'.[151]

Nevertheless, the Germans began the process of ascertaining what exactly were Spain's essential civilian and military needs in terms of fuel, grain and other vital goods. The figures the Spaniards produced for civilian needs alone were substantial but realistic, that is to say not an invention to frighten off the Germans: 400,000 tons of gasoline, 6–700,000 tons of wheat, 200,000 tons of coal, 100,000 tons of diesel oil, 200,000 tons of fuel oil as well as large quantities of other raw materials, including cotton, rubber, wood pulp, hemp, jute and so on.[152] Admiral Canaris returned to Spain in the third week of July to make a reconnaissance of the area surrounding Gibraltar and to draw up plans for an attack on the rock. He had also been instructed to ascertain the details of the military equipment needed by Spain prior to any sort of belligerence at Germany's side.

Canaris was accompanied by Air General Wolfram von Richthofen, the one-time commander of the Condor Legion. They were met by General Vigón, who had replaced Yagüe as Air Minister on 27 June, and by General Martínez Campos, the Chief of the Spanish General Staff. Canaris took a gloomy view of Spain's military capabilities and told the German Chief of Staff, General Franz Halder, that 'Spain will not do anything against Gibraltar on her own accord'.[153] In contrast to German preoccupations, the acute problems of supplying a war machine were skated over in Madrid because of a widely held conviction in official circles that victory would be swift. So confident was Franco that he drew up the map of his 'African Empire' to be delivered to the Führer by General Eugenio Espinosa de los Monteros, his new Ambassador to the Third Reich. At the same time, as evidence of Spain's utility to the Axis, Beigbeder and Franco decided to inform Berlin that, thanks to the recent agreement with Lisbon, 'Portugal has been partially extracted from the British orbit and brought into ours'.[154]

Franco made his boast of Portugal's integration into the Spanish sphere of interest in a memorandum drawn up with Beigbeder which Stohrer finally sent to the Wilhelmstrasse on 21 August.[155] Apprehensive lest Berlin's silence with regard to his overtures could mean that Spain would not be invited to share the spoils and heartened by the massive stepping up of the Luftwaffe's attacks on Britain since 10 August, Franco had written a buoyant letter, on 15 August, to Mussolini from Madrid. In his letter, the Caudillo reminded the Duce of Spain's aspirations and claims in North Africa. He implied that a declaration of war was imminent and dependent only on the delivery of German supplies. Mussolini received the letter on 23 August and replied two days later in terms that were warmly effusive albeit non-commital on specific issues. He said that for Spain not to enter the war would be to 'alienate herself from European history, especially the history of the future, which the two victorious Axis powers will determine' and offered 'the full solidarity of fascist Italy' for Franco's aspirations. However, the Duce made it clear that if Franco waited until the end when Britain was irrevocably finished, he would have to relinquish the African prizes that he sought.[156]

Franco's reaction to the letter was revealing. At the time of its receipt at the Italian Embassy in Madrid, he was in his summer residence near La Coruña, the Pazo de Meiras. The recently arrived Ambassador Francesco Lequio had not yet presented his credentials. Unable without a breach of protocol to request an audience with Franco to deliver the letter personally, he simply informed Beigbeder of its arrival. Beigbeder telephoned Franco who broke with protocol and immediately issued an official invitation for Lequio to go to La Coruña as a guest of the Spanish Government. At the Pazo de Meiras, they were taken by the head of the military household, General Moscardó, to the Caudillo. The informal spontaneity of Franco's invitation and the solemnity of the reception were the consequence, Lequio was told by a functionary of the Caudillo's household, of Franco's 'profound, devoted admiration for the Duce'. Franco was thrilled to be on the equal terms with the Axis leaders suggested by the Duce's letter, exclaiming, in a voice choked with emotion, 'As always, the Duce is crystal clear. As always, he says what is essential. If they [the western Powers] had listened to him, we would not now be in the chaotic situation in which we find ourselves.' He then launched into a virulent attack on Britain and the United States, confiding in Lequio his total conviction that England was defeated and that her continued resistance would at best merely convert her into an American colony.[157]

XV

THE PRICE OF EMPIRE

Franco and Hitler, September–October 1940

IN THE early summer of 1940, enthusiasm for Spanish entry into the war had come from Madrid rather than Berlin. It was blatantly obvious that Franco, Serrano Suñer and even Beigbeder aspired to take part after the worst of the fighting was over but before the division of the spoils. But their offers had been brushed aside ungraciously by the Germans. By September, confident of an early German victory over Britain, Franco hastened to send Serrano Suñer to Berlin to clinch the conditions for Spain to be represented at the final conference table.

Franco's optimism about Spain's possible contribution to the Axis war effort was not shared by German military and economic experts. On 27 August, the Chief of the General Staff, General Halder, spoke of Spanish belligerence as one of Hitler's pipe-dreams. Halder's view was confirmed later on the same morning by Admiral Canaris who told him about the appalling food and fuel situation in Spain and the opposition of generals and senior clergymen to Franco. The Abwehr Chief remarked that 'Franco's policy from the start is not to come in until Britain is defeated, for he is afraid of her might.'[1] Göring declared that support to the extent requested by Franco was completely out of the question. Even small amounts were considered unlikely.[2] The only aid delivered by Germany was sixty-two tons of religious items looted from Poland sent to make good Civil War damage to Spanish churches.[3]

Seemingly unaware of the depth of German pessimism about Spanish military usefulness, Ambassador von Stohrer composed a preliminary draft of a protocol on Spanish entry into the war. Somewhat reworked, with the addition of further opinions from the *Oberkommando der Wehrmacht* (the Supreme High Command of the Armed Forces), this formed the basis of Ribbentrop's brief for discussions with Serrano Suñer who was due to arrive in Berlin in mid-September. By its terms, Spain would, in agreement with the Axis Powers, fix the time for her entry into the war. In return for the Reich supplying the necessary military equipment and raw materials, Spain would undertake to recognize her Civil War debts to Germany and pay them off through future deliveries of raw materials. Spain would also agree to the confiscation and transfer to Germany of French and English mining properties in Spain and Spanish Morocco. Spanish territory on the Gulf of Guinea was to be transferred to Germany. The Spanish economy would be integrated into a German-dominated European economy in which it would play only a subordinate role with her activities confined to agriculture, the production of raw materials and industries 'indigenous to Spain'.[4]

At the beginning of September, Stohrer was back in Madrid. In a ceremony held at El Pardo on 6 September, he bestowed upon Franco the Grand Cross of Gold of the Order of the German Eagle. It was Hitler's mark of appreciation to Franco for his decisive action in Tangier and for his offers of belligerence. It was obvious too from Stohrer's speech that the Führer was now ready to collect on those promises. An openly emotional Franco replied in terms of his faith in 'the triumph of our common ideals'. On the same day, the new Italian Ambassador Francesco Lequio presented his credentials and told Franco that he could rely on Italian support for Spain's legitimate aspirations.[5]

The publicity given to both ceremonies set the tone of camaraderie for Serrano Suñer's forthcoming mission to Berlin. That Serrano Suñer, who was after all Minister of the Interior, should be the emissary to Hitler reflected Franco's desire to use a person likely to be agreeable to the Germans. In typical style, Franco tried to derive additional profit from Serrano Suñer's journey. The negotiations with Vichy for the cession of the area near Fez were dragging. With the Italian and Spanish press trumpeting Serrano Suñer's visit as signifying Spanish membership of the Axis and the early satisfaction of her colonial ambitions in North Africa, Franco instructed his Ambassador to Vichy, Lequerica, to repeat Spanish complaints about alleged disorder in French Morocco. Lequerica was told to pass on to Pétain an unequivocal threat of Spanish intervention.[6] The clear hope was that Serrano Suñer's presence in Berlin, implying close Hispano-German friendship, might pressurize Vichy into territorial concessions.

The fears generated by Serrano Suñer's trip were reflected on 14 September in the action of the British Colonial Secretary, Lord Lloyd, who unofficially informed the Spanish Ambassador, the Duke of Alba, that he had advised Churchill to facilitate a Spanish occupation of French Morocco.[7] Churchill may have been using Lloyd to try to counter potential German offers to Franco. It was not just the British and the French who were assuming that Serrano Suñer's visit would result in Spanish entry into the war.[8] Two days before his arrival, Hitler had told General Halder of his 'intention to promise the Spaniards everything they want, regardless of whether the promise can be kept'.[9] Had he followed this intention through and perpetrated what he later called his 'grandiose fraud', then he might well have pulled Franco into the war on his side. In fact, the Führer, with his assault on Britain faltering, was concerned to retain the goodwill of Vichy. So, he did not react to the alarmist talk emanating from Madrid about disorders in Morocco, as Franco had hoped, by facilitating a Spanish occupation of French territory. Rather, to the evident chagrin of the Caudillo, he authorized the sending of Senegalese troops, armoured cars and aircraft to reinforce the French colonial army. Through Beigbeder, the Caudillo continued trying to persuade both the Germans and the Italians that Pétain was not to be trusted as the guardian of North Africa.[10] The same point was made by Serrano Suñer in Berlin.

It was obvious from his rapacious pressure on Vichy and from the Serrano Suñer mission that Franco would go to war if the Germans landed in England. However, his anxiety to climb aboard the German bandwagon was countered by alarming food shortages inside Spain, which had been exacerbated by the break-down of distribution networks dependent on imported fuel.* Without it signifying any change in his political allegiance, the Caudillo was forced to turn to the United States in search of economic assistance. Forgetting that in June he had arrogantly brushed off offers of help from both Weddell and Hoare, on 7 September 1940 Franco sent his Minister of Industry to ask Weddell for a credit of $100,000,000 to buy food, fuel and raw materials. Weddell felt, as did Hoare, inclined to take the gamble of benevolence towards Spain. There then followed a heated debate within the State Department over whether Franco was to be trusted with a credit, a debate fuelled by fears provoked by Serrano Suñer's visit to Berlin.[11] Eventually, a way out of the dilemma was proposed by Norman H. Davis, President of the American Red Cross and a close friend of the US Secretary of State, Cordell Hull. Davis

---

* The streets of the towns were inundated with beggars and there was a tenfold increase in the number of prostitutes. The numbers of both were inflated by war orphans. Intestinal diseases proliferated as a result of people eating potato and orange peelings and other scraps scavenged from rubbish bins.

suggested giving Spain aid from a special relief budget. It would signify American goodwill to the Spanish people yet be insufficient to encourage Franco in his war plans. Hull seized on the idea, hoping to demand in return assurances from Franco that Spain would remain at peace.[12]

Serrano Suñer arrived in Berlin on 16 September 1940 accompanied by a large party of Falangists, including Dionisio Ridruejo, his Director-General of Propaganda, to discuss Spain's contribution to the decisive blow against Britain. He was immensely impressed by the special train which the Germans sent to pick him up at Hendaye, by the discipline of the guard of honour and by the defeated look of the French. The Vichy authorities in turn were outraged that Serrano Suñer should progress through France as if he were one of the victors.[13] The *cuñadísimo*, for all that he may have been titillated by proximity to the victorious Wehrmacht, soon became bored by efforts to overwhelm him with demonstrations of German might in the form of visits to factories and military units. Nevertheless, Ramón Garriga, the representative in Berlin of the Spanish State news agency EFE, was given the clear message by members of the Spanish delegation that they had come to negotiate entry into the war.[14] One of them, Miguel Primo de Rivera, advocated sending a division of Falangist volunteers to help in the German assault on Britain.[15]

Operation Sealion (*Seeloewe*) for the invasion of England had been postponed temporarily on 14 September and was postponed indefinitely on 17 September as a result of the success of the RAF in the Battle of Britain. The Germans were less than honest with their Spanish guest about this. Indeed, at their first three-hour meeting on 16 September, Ribbentrop told Serrano Suñer that in England the situation was deteriorating and 'after a while there would be nothing left of London but rubble and ashes'. Serrano Suñer described the purpose of his visit as being formally, as a cabinet member and 'the personal agent of Spain', to take discussions on Spanish entry into the war beyond the earlier 'sporadic feelers'. He expressed surprise that the materials necessary for Spain's war effort had not yet arrived from Germany. Repeating the list of items required, he described French Morocco as belonging 'to Spain's *Lebensraum*'. In a further attempt to establish Spain's credentials as a ruthless member of the Axis club, he blatantly stated Spain's ambitions with regard to Portugal: 'Geographically speaking, Portugal really had no right to exist; she had only a moral and political justification for independence . . . Spain recognized this, but had to require that Portugal align herself with the Spanish group. '[16]

The harshness and affectation of Ribbentrop quickly provoked the intense dislike of Serrano Suñer.[17] At the 16 September meeting, Ribbentrop quibbled over the amounts of material requested by Spain but finally agreed that she would receive what was absolutely necessary to her. He then revealed the abyss which separated Franco and Hitler in their valuations of Spanish belligerence. Aware that the British would respond to the seizure of Gibraltar by taking the Canary Islands, the Azores or the Cape Verde Islands, the Führer wanted one of the Canary Islands for a German base, and further bases at Agadir and Mogador with 'appropriate hinterland'. He also demanded substantial economic concessions in terms of Civil War debt repayment and participation in mining interests in Morocco. The meeting ended with the overbearing Ribbentrop asking point blank when Spain could enter the war, to which Serrano Suñer replied that Spain would be ready the moment German heavy coastal artillery was installed near Gibraltar. Serrano Suñer had come expecting to be treated as a valued ally and instead he was being treated as the representative of a satellite state. Always touchy and fiercely patriotic, he regarded Ribbentrop's demands as intolerable impertinence and the lessons of his trip, when they sank in, were significantly to alter his attitude to the Third Reich and the question of Spain entering the war.[18]

That night, an RAF bombing raid obliged the Spanish delegation to descend into the air-raid shelters of their hotel, their awe-struck views on German invulnerability somewhat dented.[19] On the following day, Serrano Suñer was received by Hitler for a one-hour conversation. He began by transmitting a special message from Franco recording his gratitude, sympathy and high esteem for the Führer and his 'loyalty of yesterday, of today and for always'. He also brought with him a letter from Franco to Hitler, written in San Sebastián on 11 September 1940, which stated that the Serrano Suñer mission was a follow-up to the earlier offers of Spanish belligerence made by Vigón. It ended with an expression of the Caudillo's 'firm faith in your imminent and final victory and with the best wishes for your personal health and the happiness and welfare of the Greater German Reich'.[20]

Once the niceties had been completed, Serrano Suñer stated unequivocally that Spain was ready to enter the war as soon as her supply of foodstuffs and war material was secure and reiterated the request for coastal batteries near Gibraltar. Hitler asserted that heavy artillery would take months to install and that it would be more effective to station a group of Stukas in the area. He declared enthusiastically that the speedy capture of Gibraltar would be important, and easy, since it had already been the object of minute study by German experts. Hitler made only oblique reference to the Canaries and suggested that he meet Franco at the Franco–Spanish border. Shortly afterwards, Serrano again met Ribbentrop who pressed him hard for the cession of one of the Canary Islands and added that Germany wanted Spanish Guinea and the small Spanish islands of Central Africa in return for French Morocco. Serrano Suñer reacted negatively, asserting that, while Spain's youth clamoured for Gibraltar, it would be 'absolutely impossible' to agree to other amputations or limitations of Spanish territory. He suggested instead that Germany use Portuguese Madeira.[21]

As a result of his meeting with Serrano Suñer, Hitler wrote to Franco on 18 September. The problems with *Seeloewe* could be read between the lines particularly when the Führer stressed that the British blockade of Spain could only be broken by the expulsion of the English from the Mediterranean. This, he claimed, would 'be attained rapidly and with certainty through Spain's entry into the war' which would begin 'with the expulsion of the English fleet from Gibraltar and immediately thereafter the seizure of the fortified rock'. Thereafter, the defence of Spain's coasts should be entrusted to German dive bomber units. Since the loss of Gibraltar would impel Britain to try to seize one of the Canary Islands, Hitler urged Franco to permit the stationing there of Stuka or long-range fighter units. However, the merely relative importance which Hitler attached to Spanish entry was reflected in his closing words: 'Spain's entry into the fight will help show England even more emphatically the hopelessness of continuing the war and force her to give up once and for all her unjustified claims.'[22]

Despite the outrage of Franco and Serrano Suñer about specific German demands, it would be a long time before it slowly dawned on them, and on Franco in particular, that Spain's place in the new order would be that of a minor agrarian satellite. Hitler's colonial ambitions for a large central African empire with bases in the Canary Islands and Spanish Morocco as staging posts to it were of more importance to him than good relations with Franco.[23] In any case, Spanish belligerence would only be part of an indirect strategy against Britain. Hitler was not sufficiently interested in the southern flank to want to woo Franco. Such 'war on the periphery' was something in which Hitler would dabble while working on his grander strategies of annihilating Russia and encouraging Japan to attack the United States. Moreover, the cost of Spanish cooperation would have to be weighed against the requirements of both Italy and Vichy France.[24]

While Franco was still digesting Hitler's letter, Serrano Suñer was sent off on a tour of the battlefields of the western front. Between 19 and 20 September, Ribbentrop was in Rome to discuss with Mussolini the future direction of the war in the wake of the suspension of *Seeloewe*. Ribbentrop told Ciano in the car from the airport that Spanish intervention 'now seems to be assured and imminent'. He told Mussolini that 'Spain is ready to enter the war'. The Duce agreed that this was 'an event of great importance'.[25] Mussolini suggested that Spain join Italy and Germany in a Tripartite Pact, which be kept secret until Spanish entry into the war so as not to jeopardize the attack on Gibraltar. However, with his own North African ambitions in mind, the Duce tried to plant doubts in Ribbentrop's mind about Spanish military efficacy in Morocco.[26]

Before leaving Berlin, Serrano Suñer had sent Franco by special plane an account of his meetings with Ribbentrop. On his return to Berlin from Brussels, where his battlefield tour had ended, he was awaited by a long letter from his brother-in-law. The text demonstrated beyond doubt that, at that time, Franco believed blindly in the victory of the Axis and that he was fully decided to join in the war at its side. The Caudillo's tone oozed wide-eyed adulation of Hitler. 'One appreciates as always the sublimity and good sense of the Führer.' The disagreeable demands made on Serrano were put down to 'the selfishness and inflated self-regard of his underlings' who failed to see how the Spanish Civil War had facilitated Germany's victory over France. He urged Serrano Suñer to make the Germans realize that the Spanish conflict had helped Germany try out men, tactics and equipment which had been invaluable against France. Franco also referred obliquely to the way he had helped undermine the French position in the fifteen months after the war, 'constantly working in the shadows for the most rapid German success'. What Spain now offered Germany was 'a vast number of fighting men' (*una masa guerrera*), her geostrategic position and a way to split the South American Republics from the American bloc.

Franco shared Serrano Suñer's outrage at Ribbentrop's request for one of the Canary Islands, referring to 'what rightly provoked your indignation and which the pen refuses to write'. He then elaborated devices whereby Serrano Suñer might convince the Germans to reduce their demands. He remained, however, anxious to ensure that Spanish participation in the division of the spoils be clinched. There was no sense that Franco was astutely holding the Germans at bay. Rather he was trying to convince them that he was an ally to be trusted. To cede one of the

Canary Islands would be to create another Gibraltar. In a Spanish-German wartime alliance, 'the bases of one could become the bases of the other. If the Germans wanted Agadir, it could not be in perpetuity but on a ninety-nine year lease. German demands for raw materials from French Morocco could be satisfied as long as Spanish needs were met first. Franco saw German demands for control of British and French companies domiciled in Spain as economic imperialism. He was adamant that they could not express the true wishes of the Führer and attributed them to poor translations or the excessive zeal of Hitler's lesser functionaries.

'Such demands', he wrote optimistically, 'are incompatible with the existence of a mere treaty of friendship,' that is to say, let alone with the full-scale military alliance which we are talking about. 'It is all incompatible with the grandeur and independence of a nation.' With regard to the question of debt repayment, Franco suggested that German demands at least be reduced to the level found acceptable by Italy, 'a much poorer country'. 'Such a reduction would mean nothing to the Germans and if they refused, it would be taken amiss by the Spanish people'. When that particular point was made by Serrano Suñer to Hitler, the Führer was cut to the quick by the insinuation of his meanness, fumed for weeks, remarking on it to both Ciano and Mussolini. Franco then commented on his reaction to the letter from Hitler, 'which as always clears the horizon'. It confirmed his view that he and the Führer saw eye to eye and that all of the problems derived from German underlings. The Caudillo made specific reference at this point to the possible prolongation of the war. Far from perceiving this as a reason for not entering the war, he saw it as grounds for getting a better price. He suggested that the Spanish offer should be put into practice as soon as possible while the Germans thought that they still needed it.

Drawing on his own taste for war by attrition, he suggested that the conflict might not be as long as the Germans feared, 'because in war it sometimes happens that the victor does not realize that he is winning simply because the attrition which he suffers blinds him to the damage he is causing . . .' Franco's confidence in a relatively early end to the war, taken together with Spain's horrendous economic difficulties, led him to say 'it is in our interests to be inside [the Axis] but not to precipitate things'. He was confident that this would be possible, clutching at the straw that what Hitler had said to Serrano Suñer about an early attack on Gibraltar had been exaggerated by the interpreters. 'There is complete agreement between the Führer and ourselves and there only remains the technical evaluation of some factors which are not as crucial as he says.'[27]

The fact that Berlin was bombed by the RAF during his stay and the sight of tons of concrete being poured into German coastal fortifications convinced Serrano Suñer that it was going to be a long war. Nevertheless, Ramón Garriga of EFE found Serrano Suñer and his followers thrilled by the evidence of German might which they had seen on their journey.[28] On 24 September, Ribbentrop and Serrano Suñer were both back in Berlin for an extremely tough encounter at which they discussed Mussolini's proposal of a Tripartite Pact which had been passed on to the Spanish Minister during his stay in Brussels. By way of strengthening the Spanish claims to Morocco, Serrano Suñer said that he had just heard from Madrid that the British Ambassador had intimated that, after the war, England would not object to Spain getting French Morocco. (Churchill had indeed authorized Hoare to inform Beigbeder that Britain would be happy to see Hispano-French disputes in Morocco settled to the satisfaction of Spain – which he duly did on 21 September.)[29]

Commenting on Hitler's letter to Franco, Serrano Suñer declared that the Generalísimo 'had been distressed in a friendly way' because of the German claim for bases in Morocco. 'With great regret, he [Franco] had thought he recognized a certain sign of distrust toward Spain, and he would, therefore, like once more to re-emphasize solemnly that his attitude toward Germany was not a momentary opportunism, but an eternal reality.' In an alliance with Germany, all Spain's bases, ports and airports would be at her disposal. With regard to Germany's economic demands, they were seen in Madrid as unnecessarily impairing Spanish interests. A patronizing Ribbentrop pressed Serrano Suñer fiercely. He asked him point blank if Spain accepted the Duce's suggestion of a Tripartite Pact not to be published until the day on which Spain declared war with an attack on Gibraltar. Serrano Suñer responded with Franco's idea, expressed to him in the letter of 21 September, of a protocol containing three points: Spain's decision to participate in the war, the date still to be fixed; the assurance of German military and material aid to Spain and the recognition of Spain's territorial and national demands. This is more or less what would be signed by Franco and Serrano Suñer at the Hendaye meeting on 23 October.

Serrano Suñer continued to stone-wall on the issue of German bases, claiming that Spain could herself build up her defence capabilities in North Africa with the aid which she had requested from Germany. Ribbentrop then pressed him firmly on Franco's response to Hitler's requests for one of the Canary Islands, for the transfer of Spanish Guinea and

Fernando Po and for the bases in Morocco. After some prevarication, the *cuñadísimo* replied negatively in all cases.* Ribbentrop then raised the question of Spain's Civil War debts and demanded that English and French business assets in Spain be transferred to Germany. Serrano Suñer mounted a stout defence of Spanish interests. In the end, bases and debts were left pending until the arrival of Franco's reply to Hitler's letter of 18 September.[30] Commenting on the meeting, Ambassador von Stohrer accurately expressed the problem of the relations between Franco and Hitler: 'Spain cannot expect us to provide her with a new colonial empire through our victories and get nothing for it.'[31]

Franco's reply to Hitler was dated 22 September 1940 but did not leave Madrid until the following day because of delays in the translation into German.[32] It was accompanied by another long message to Serrano Suñer dated 23 September which constitutes an invaluable insight into how Franco's mind was working at the time of writing to the Führer. He was presumably influenced by Serrano Suñer's account of the talks so far, which have not survived. Serrano Suñer's letters to Franco were written by hand; he kept no copy and they have disappeared along with the majority of Franco's papers. As before, Franco remained convinced of Hitler's goodwill towards him and attributed all the difficulties in the negotiation to Ribbentrop.

Franco continued to think that the end of the war was nearer than did the Germans themselves. In this respect, he referred to an account given him by Captain Alvaro Espinosa de los Monteros, brother of the Ambassador in Berlin and himself naval attaché in Rome, of his recent lunch in Paris with Hermann Göring who had admitted to him that the bombing of England was not a success. Franco was not convinced by what Captain Espinosa told him: 'I believe that the bombing attacks are of immense efficacy and will eventually defeat the English.' He then recounted with obvious approval how his intimate, General Vigón, had said to Sir Samuel Hoare a few days previously 'You are defeated. Don't be stupid; make peace before things get worse.' He was convinced that the only thing standing in the way of a British surrender was London's distrust of German conditions which, he commented condescendingly, could be guaranteed by the Caudillo personally or Mussolini.[33]

Franco's letter to Hitler was delivered by Serrano Suñer on 25 September. The text mingled a sincerely obsequious tone with convol-

* This was no cunning plan to deceive the Germans. Serrano Suñer and Franco wanted to join the Axis to complete the job, as they saw, of restoring Spain's greatness, not to adopt satellite status.

uted arguments for not meeting Hitler's demands with regard to Agadir, Mogador, the Canary Islands and German air bases near Gibraltar. After all, such demands were incompatible with Franco's determination to rebuild imperial Spain. Nevertheless, there was nothing about the letter to suggest that Franco was not still totally committed to the Axis cause. He welcomed the proposed meeting on the Franco–Spanish border and made it clear that he regarded the settlement of civil war debts ('old matters') and 'the post-war exchange of commodities' as minor technical issues of little significance beside the great enterprise on which they were both about to embark.[34]

After writing the letter of 23 September to Serrano Suñer, Franco meditated overnight on the German economic demands which had been put on 18 September to Demetrio Carceller, a Falangist businessman, and to Colonel Tomás García Figueras, the Secretary-General of the Spanish High Commission in Morocco, who were both in Serrano Suñer's party.[35] The details were taken by special plane to Madrid by Garcia Figueras. The result of Franco's reflections was a further letter to Serrano Suñer. It opened with a reference to 'the new point which has cropped up' which might mean the demands brought by Colonel García Figueras or to the opinions of Captain Espinosa, who was working hard to convince Franco that the German navy was incapable of defeating the Royal Navy.[36] He may have finally heard about the German decision to postpone the attack on England. He certainly seems to have accepted finally that it was now going to be a long war.

The tone of the letter, while in no way suggesting a change in the underlying commitment to the Axis, was altogether less sanguine than Franco had been in the immediately preceding days. '*There is no doubt about the alliance. It is fully** expressed in my answer to the Führer and in the whole direction of our policy since our Civil War.' However, Franco now showed real concern about the prospect of a protracted war. Moreover, he was adamant in a way he had not been before about the need for adequate economic and military preparation. The scale of assistance required by Spain meant that 'everything needed to be specified in the protocol and, although there is no doubt about our decision, we have to think about the details of the agreement and the obligations of both parties'. The Pact with the Axis should remain secret until Madrid felt ready for war.[37]

In addition to evidence about German difficulties, there was opposition to Spanish belligerence building up within the higher reaches of the

* Underlined by Franco in the original.

Spanish army. The General Staff reported that the navy had no fuel, that there was no Air Force worthy of the name and no effective mechanized units, and that after the Civil War, the population would not tolerate more sacrifices. With tensions brewing between pro-British monarchists and pro-Axis Falangists, Franco latched onto the idea of the secret protocol with the Axis, which he hoped would guarantee his territorial ambitions yet still leave the precise date of Spanish entry up to him. However, the question of the date was never resolved because Hitler was neither able nor inclined to pay the Caudillo's double price of the prior German financing of Spanish military and economic preparation and the transfer to Spain of French North Africa. The harsh demands made by Hitler and Ribbentrop in their meetings with Serrano Suñer in Berlin on 16, 17, 24 and 25 September sowed the seeds of Franco's inclination to enter the war only if he was paid in advance.

After reading Franco's letter of 22 September, Hitler and Serrano Suñer confidently agreed that the various outstanding points of the negotiations could now be left until the Führer met the Caudillo. Serrano Suñer again added to Spain's existing demands the need to have Portugal in a subservient alliance. Despite the geographical absurdity of Portugal's existence, he declared arrogantly, Spain declined to absorb her and seven million 'weeping Portuguese'. The assumption that Spain was virtually aboard the Axis bandwagon was implicit in a memorandum presented by Serrano Suñer to Ambassador von Stohrer after this final conversation with the Führer. The memorandum reiterated Spain's 'readiness to conclude in the form of a tripartite pact a military alliance for ten years with Germany and Italy'. 'This secret protocol enters into force when, in accord with the other two Powers and with their aid, Spain has completed her military preparations and provided herself with the necessary raw materials, gasoline and foodstuffs.'[38]

Both sides regarded the visit of Serrano Suñer as something of a disappointment. The Germans thought that he had demanded too much; he thought that Hitler offered too little. On 27 September 1940, Ciano wrote in his diary, 'Generally speaking, Serrano Suñer's mission was not successful, and the man himself did not and could not please the Germans.'[39] Serrano returned to Spain via Italy.* Both he and the Caudillo

* While in Berlin, Serrano Suñer invited Heinrich Himmler to visit Madrid, ostensibly for a hunting trip but in fact to discuss security arrangements for the forthcoming Hitler-Franco summit and also to advise on the modernization of the Spanish secret police (Hoare, *Ambassador*, p. 76; Saña, *Franquismo*, p. 118). Curiously, on 27 September 1940, the Portuguese Ambassador reported that Himmler was in Madrid, (Pereira, *Correspondência*, II, p. 87).

saw the Italian connection as an important counter-weight to Germany. Accordingly, before Serrano Suñer had left Madrid for Berlin, Franco had formally requested that his minister be granted an audience with Mussolini and Ciano.[40] He was duly received on 1 October with great warmth by the Duce and Ciano. He spoke vehemently of his dislike for the blusteringly tactless Ribbentrop, which they both laughed off. At the same time, he said that Spain was preparing to take up arms to settle accounts with Britain – action he hoped would unite the squabbling factions behind Franco. Mussolini replied that he had always been convinced that Spain could not stand aside from the struggle. Therefore, he believed that Spain should accelerate her preparations and then a collective Axis decision taken about when she should intervene. The Duce suggested soothingly that the precise date should be at the least onerous moment for Spain and the most useful for the common cause. He made it clear that Italy had no spare resources with which to help Spain. Serrano Suñer took this to imply that Mussolini was not anxious for Spain to join in the war yet. It led him to suspect that the Duce desired to maintain his own position as Hitler's only Mediterranean ally.[41]

On 28 September, Hitler spoke with Ciano in Berlin and he made no secret of his impatience with the Spaniards, born of his experience during the Spanish Civil War. He declared that 'one could not make progress with Spaniards without quite concrete and detailed agreements'. He pointed to the startling imbalances of the agreement proposed by Franco and Serrano Suñer: Germany was to deliver grain, fuel, military equipment, all the troops and weapons necessary for the conquest of Gibraltar, all of Morocco and Oran in return for which Spain promised only her friendship. Understandably, the Führer expressed his doubts as to whether Spain had 'the same intensity of will for giving as for taking'. In fact, in the twelve days since Serrano Suñer had arrived in Berlin, both the tenor and the context of Hispano-German relations had changed dramatically. In particular, the talks themselves had opened Serrano Suñer's eyes to the harshness of the German position.

In addition, developments elsewhere had led to the Führer revaluing Spanish belligerence. Hitler's main concern was that any agreement on his part to meet Franco's Moroccan aspirations might leak out to the French and provoke an understanding between the Vichy defenders of French North Africa and de Gaulle, thereby permitting the English to establish themselves there. If the Spaniards were allowed to take over Morocco, at the first sign of English attack they would probably call for German and Italian help to hold it. 'Moreover,' declared the Führer, in a remark revealing a contempt for Franco's generalship which had festered since the Civil War, 'they would let the tempo of their Civil War prevail in their military measures.' In contrast to his sour memories of the Spanish war, Hitler, only a few days before, had been much impressed by the perform- ance of the Vichy garrison of Dakar in West Africa in beating off an Anglo- Free French naval attack on 23 September. This was the key to his attitude to Franco. Hitler now began to speculate on the possibility of incorporating Vichy into his alliance system as an enthusiastic participant.

Hitler remained outraged about Franco's outstanding Civil War debts. Considering the Nationalist cause in the Civil War to be a sacred crusade, Serrano Suñer had made it quite clear that he considered the German demand for a settlement of accounts to be a tactless confusion of economic and idealistic considerations. In consequence, Hitler expostulated to Ciano, 'as a German, one feels towards the Spanish almost like a Jew, who wants to make business out of the holiest possessions of mankind'. It was hardly surprising that Hitler should tell Ciano that he was opposed to Spanish intervention, 'because it would cost more than it is worth'.[42] This was a crucial admission. For months, the Spaniards would postpone their declaration of war as long as the Germans failed to deliver food and weapons. If Hitler had really wanted to sway Franco in his favour, it would have been easy enough to do so either by sending supplies or by taking a more generous line on Franco's imperial ambitions.

It was agreed during Ciano's visit to Berlin that Mussolini and Hitler would meet at the Brenner on 4 October 1940. When they met, Hitler made exactly the same points to the Duce as he had to Ciano. He dis- missed Franco's entry into the war as of strategic significance only in connection with the conquest of Gibraltar. On the basis of reports from Admiral Canaris, he believed that the Caudillo's military help would be nil. In any case, Hitler knew that the seizure of Gibraltar was secondary to the capture of Suez. If it took place before Suez was safely in Axis hands, it would merely provoke an English assault on the Canary Islands. Hitler told Mussolini that Franco had proposed leasing ports in the Canary Islands to the Germans. Hitler's real fear was that, if the French discovered that he was haggling with Franco over their empire, then they would simply abandon defence of their possessions or else French local forces in Africa would break away from Vichy.* In conclusion, Hitler

---

* Concerned as he was with his own claims on North Africa, the Duce said that agreement between Spain and France would be impossible if Spanish claims on Morocco were recog- nized. 'Spain demanded much and gave nothing'. His position with regard to Spain was 'wait and see'.

said that he planned to write to Franco to say that Oran could not be given to Spain.[43]

For the moment, the Caudillo remained excited about the prospect of securing French Morocco for Spain. Years later, Serrano Suñer described Franco's attitude as 'like that of an excited child intoxicated with what he had always wanted: the world in which he had made his way as a prominent soldier'.[44] His optimism was inflated by the idea that the Führer's vision, understanding and generosity were being undermined by the meanness of his subordinates. What is most striking about his views on the war and his attitude to Hitler in the autumn and winter of 1940 is their combination of narrow provincial naïvety and megalomanic complacency.

Mussolini was not the open-handed friend that the Caudillo and Serrano Suñer thought. The Duce, anxious that Franco should not get his hands on parts of North Africa coveted by Italy, encouraged the Führer to postpone Spanish belligerence and suggested that Spanish demands be met after the war. Hitler himself was trying to balance the conflicting demands of Franco, Pétain and Mussolini, something which he conceded was possible only through 'a grandiose fraud'.* When Ciano spoke to Serrano Suñer, who had remained in Rome to hear about the Brenner meeting, he was struck by the Spaniard's innocence. To Ciano's surprise, the *cuñadísimo* seemed blind to the fact that the Germans 'have had an eye on Morocco for a long time'.[45]

In an effort to derive benefit from Serrano Suñer's visit to Berlin, on 21 September, Beigbeder had told Hoare that Spain had been promised 'economic stability, Gibraltar and French Morocco' if she joined Hitler's continental block. He suggested that Britain try to prevent this by increasing economic aid to Spain and publicizing it. Beigbeder and Hoare agreed on the value of a British expression of sympathy for Spain's Moroccan ambitions.[46] On 29 September, Churchill minuted Halifax, 'I would far rather see the Spaniards in Morocco than the Germans, and if the French have to pay for their abject attitude, it is better that they should pay in Africa to Spain than in Europe to either of the guilty Powers. Indeed, I think you should let them know that we shall be no obstacle to their Moroccan ambitions, provided they preserve their neutrality in the war.'[47]

* That the Germans aimed to swindle Franco was revealed by Marshal Keitel, head of the OKW, and General Jodl, chief of Operations, to the Italian military attaché in Berlin. They told him that, in the light of Italian claims on Tunisia, Spain's aspirations in Morocco and Algeria could never be satisfied. Without referring to Germany's own desires, they also made it clear that something had to be left for Vichy France (Marras to Ministero della Guerra, 12 October 1940, *DDI*, 9, V, pp. 690–2).

With the Germans openly hostile to him, the volatile Beigbeder was displaying an open Anglophilia to Hoare. In the brief coexistence of Beigbeder's pro-British line with Serrano Suñer's pro-Axis stance perhaps lay the seeds of the tactic of playing off both sides which Franco was to use later in the war with varying degrees of crudity.

The British, and the Vichy French, were indeed toying with concessions to Spain which might counter German offers made to Serrano Suñer. On 30 September, hoping to give Franco a reason not to join the Axis in search of trophies in French Morocco, Vichy proposed the cession of the territory claimed by Spain in return for a renunciation of other claims. Not surprisingly, the offer was declined because Franco was reluctant to complicate the bigger deal on French Morocco which he was now hoping to clinch with Hitler.[48] Similarly, having already intimated to Alba and to Beigbeder their sympathy for a realignment of Moroccan borders in Spain's favour, the British were casting around for other ways of persuading Franco to stay out of the war. At a cabinet meeting on 2 October 1940, Lord Halifax again suggested a public statement that, after the war, Britain would be prepared to discuss the Gibraltar question. Once more, Churchill pointed out that, if Britain won, public opinion would not permit the return of Gibraltar and that, if she lost, there would be no choice. After further pressure, it was agreed to issue a general statement that 'all outstanding questions can be settled amicably between the two countries'.[49]

While pressuring the Vichy French for immediate concessions in Morocco, and despite his grandiose dreams of empire, the Caudillo still had to cope with the immediate food crisis in Spain. Distribution was collapsing and there were drastic bread shortages in some areas. With mass starvation now looming, the door to Anglo-American assistance had to be kept ajar. That was rendered difficult by American outrage at pro-Axis Spanish propaganda attacks on the USA directed at Latin America. The *Falange Exterior*, the Spanish equivalent of the Nazi *Auslandorganisation*, was the conduit for German anti-American agitation among the South American republics.[50] Nevertheless, in support of the British readiness to neutralize Franco by carefully doled-out assistance, Washington continued to give consideration to Norman Davis's notion of sending wheat to Spain through the Red Cross. On 30 September, Weddell told Beigbeder that American help depended on Spain staying out of the war. In the light of Serrano Suñer's continuing enthusiasm for belligerence, Beigbeder's reply must be seen either as an expression of his personal sentiments or more likely as an example of the incipient duplicity of

Franco. Beigbeder told the American Ambassador 'officially in the name of his Government that Spain would remain out of the European conflict unless and until she was attacked' and dismissed Serrano Suñer's trip as merely a courtesy visit.[51]

Whatever the US Ambassador thought of the assurances of Beigbeder, he must have found it difficult not to judge Franco's position in terms of the pro-Axis rhetoric and the spectacular military parades which accompanied the celebration of the '*Día del Caudillo*' on Tuesday 1 October. Numerous delegations from Fascist Italy and Nazi Germany set the tone. In Madrid, Franco was the object of an elaborate ceremony of adulation mounted in the royal Palacio de Oriente.[52] There was little sign of *hábil prudencia* in the swaggering vanity of the would-be emperor. To the contempt of the diplomatic corps, Franco received them in the throne room on a raised daïs. The assembled ambassadors were instructed to pass before him in a line and bow, a procedure never demanded even by the Kings of Spain.[53]

On 2 October, Beigbeder told Weddell in a dramatically solemn fashion that 'Your President can change the policy of Spain and of Europe by a telegram announcing that wheat will be supplied to Spain'.[54] The clear implication was that such a gesture might counteract the war-mongering of Serrano Suñer. After some consultation between Washington and London, it was decided to pursue the politically neutral device of food relief via the Red Cross. The British agreed 'on condition that American agents in Spain distributed the wheat, that none was re-exported, that publicity was given to the whole affair, and that wheat ships should go over singly and be stopped by us if anything went wrong'. Franco accepted these conditions on 8 October.[55] The State Department, however, remained reluctant to go ahead without further assurances from Franco about Spain's continued neutrality. However, in ignorance of just how close Franco was to entering the war, Roosevelt and Hull decided to make a gesture of generosity.[56]

Beigbeder's suggestions to Hoare and Weddell, and their subsequent advice, also lay behind a significant speech made by Churchill in the House of Commons on 8 October. He spoke of his government's readiness to adjust the blockade in order to meet Spanish needs and of the British desire to see Spain take her 'rightful place both as a great Mediterranean Power and as a leading and famous member of the family of Europe and of Christendom'. Although the speech was reported in Serrano Suñer's controlled press, the references to Spain were omitted.[57] That was an indication that Beigbeder's initiative, the American wheat and Churchill's friendly words would not be sufficient to dampen the pro-Axis enthusiasms of Franco's immediate circle. They were flaunted on 11 October when Mussolini sent Marshall De Bono to Madrid to confer upon Franco the Collar of the Order of the Annunziata. Clearly stirred, the Caudillo thanked De Bono effusively, speaking not as the prudent neutral but as the committed ally.[58] Similarly, when Franco sent a message to Salazar in an effort to play down the importance of his deepening links with Germany, he could not fail to reveal his conviction that the British Empire was finished and was about to be taken over by the United States.[59]

The rising fervour for the Third Reich was revealed most dramatically with the dismissal from Franco's government of the two most pro-Allied ministers. On 16 October 1940, Luis Alarcón de la Lastra was replaced as Minister of Industry and Commerce by the wily and unscrupulous Falangist businessman Demetrio Carceller Segura. Carceller was the architect of the economic policy whereby Spain exported food and raw materials to Germany in the hope of currying favour in time for the post-war share-out.[60] Beigbeder was replaced as Minister of Foreign Affairs by Serrano Suñer. Beigbeder learned of his dismissal from the morning newspapers.* He had been told nothing during a long session over dinner with Franco on the evening of 15 October.[61] Beigbeder believed that Stohrer had asked Franco to replace him because of his negotiations with Weddell over the wheat.[62] Serrano Suñer had already brought back from Berlin the message that Hitler regarded Beigbeder as unacceptably Anglophile. His dismissal came as a great shock to the British Embassy in Madrid and raised fears of an imminent Spanish declaration of war.[63] Once he had been sacked, the excitable Beigbeder drew even closer to Hoare. He was wildly indiscreet about his hostility to the Germans and spoke of Franco as 'the dwarf of the Pardo'.[64]

Serrano Suñer's elevation fed the long-circulating rumours that Franco might divest himself of the Presidency of the government and pass it to his brother-in-law.[65] Certainly the *cuñadísimo*'s accumulation of power was now considerable. Since no new Minister of the Interior was named, Serrano Suñer continued to control that Ministry as well as Foreign Affairs and, effectively, the Falange. Franco had asked his brother-in-law

---

* The delight of the pro-Axis camp in Madrid could be read in the article in *Arriba*, 17 October 1940, which attacked those who opposed the new direction of Spanish policy and cruelly goaded Beigbeder as 'the man who not only lacks a national sense but does not even have a Spanish name'.

to suggest a new Minister of the Interior. He suggested that the Caudillo assume the post himself. At first, Franco was hesitant until Serrano Suñer reminded him that Mussolini often held ministerial portfolios. The *cuñadísimo*'s game was clear. Franco would not have time to occupy himself with the day-to-day running of the Ministry which would therefore fall to the immensely efficacious Under-Secretary, José Lorente Sanz, who had been Serrano Suñer's nominee. When Serrano Suñer announced to his loyal group of confidants that he had been named Foreign Minister, he said 'at a delicate and serious moment, we are taking over Foreign Affairs and Lorente will stay here. That way we avoid a neutral coming here.' His faithful friends Ridruejo and Tovar still held the key posts in the press and propaganda section of the Ministry. Accordingly, through these henchmen, Serrano Suñer retained effective control of the Ministry of the Interior.[66]

Mussolini wrote to Hitler on 19 October 1940 that Franco's cabinet reshuffle 'affords us assurance that the tendencies hostile to the Axis are eliminated or at least neutralized'.[67] A further indication of the growing intimacy between Franco's Spain and the Third Reich came on Sunday 20 October 1940 when Reichsführer SS Heinrich Himmler began his three-day visit to Spain, in response to Serrano Suñer's invitation. He was accorded the highest conceivable honours. Greeted at the station by the Minister of Foreign Affairs and the top officials of the Falange, he was taken into Madrid through streets draped with swastika flags. He was received by Franco at the Pardo and attended a special bullfight in his honour.[68] In part, the purpose of his journey was to arrange security for the forthcoming Hitler-Franco meeting at Hendaye. However, longer term co-operation between the Gestapo and the Spanish police was also discussed. Liaison was provided by SS Sturmbannführer Paul Winzer, the police attaché at the German Embassy in Madrid. Winzer had helped train Franco's police towards the end of the Civil War. As a result of the agreement reached, greater facilities were granted for the Gestapo to pursue and interrogate the enemies of the Third Reich who fled to Spain.[69] Nevertheless, Himmler was taken aback by the scale of the post-Civil War repression – prisons were still overflowing with hundreds of thousands of prisoners and the silent executions of anonymous Republicans continued relentlessly. He believed that it made more sense to incorporate working class militants into the new order rather than annihilate them.[70]

During October 1940, the process of political revenge briefly became more public. On 21 October, the summary trial was held of a group of prominent Republicans,* who had been detained in occupied France at the behest of Lequerica and handed over to Francoist Spain by the Gestapo. They were all sentenced to death except one, Teodomiro Menéndez, thanks to the intervention of Serrano Suñer who appeared as a witness for the defence. Another, Julián Zugazagoitia, who as Republican Minister of the Interior had saved the lives of General Agustín Muñoz Grandes, Monsignor Escrivá de Balaguer, the founder of Opus Dei, Miguel Primo de Rivera and Raimundo Fernández Cuesta, was executed along with the journalist, Cruz Salido, on 9 November.[71] An equally notorious extradition was that of the Catalan President Lluis Companys. There have been acrimonious recriminations over responsibility for the Companys case.[72] After a summary trial on 14 October 1940 for 'military rebellion', he was shot on the following day.[73] Not afflicted by doubts about the guilt of his enemies, Franco did not pay much attention to the sheaves of death sentences put before him for signature.

Washington and London reasonably assumed the cabinet changes of 18 October and the imminent Franco-Hitler summit to be significant steps by the Generalísimo towards the Axis. However, both Hoare and Weddell advised that the talks about food relief for Spain should be kept open. Accordingly, Hull instructed Weddell to make it clear to Franco that help was dependent on his intentions.[74] Since his negotiations with Hitler were imminent, neither Franco nor his new Foreign Minister were to be available to discuss the American offer of grain until after their return from Hendaye.

Franco went to the historic meeting with Hitler at Hendaye on Wednesday 23 October 1940 hoping to get an appropriate reward for his frequently reiterated offers to join the Axis. His propagandists subsequently claimed that Franco held back the Nazi hordes at Hendaye and brilliantly kept a threatening Hitler at arm's length. In fact, an examination of the encounter suggests no inordinate pressure for Spanish belligerence on the part of Hitler. Nor does it diminish the conclusion that Franco remained as anxious in the autumn of 1940 as he had been in the early summer to be part of a future Axis world order. He went to Hendaye to derive profit from what he saw as the demise of the Anglo-French hegemony which had kept Spain in a subordinate position for over two centuries. Hitler went to the south of France in order to weigh up the respective costs of securing the collaboration in his European

---

* Julián Zugazagoitia, Francisco Cruz Salido, Teodomiro Menéndez, Cipriano Rivas Cherif, Carlos Montilla and Miguel Salvador.

block of Vichy France and of Franco's Spain. He saw Laval at Montoire-sur-Loire, a remote village railway station near Tours, on Tuesday 22 October, Franco on the Wednesday at Hendaye and Pétain on the Thursday again at Montoire.

Hitler did not intend to demand of Franco that Spain go to war immediately. That would have contradicted the exploratory nature of his journey. The Führer was preoccupied with the anxiety that Mussolini was about to get involved in a protracted and inconvenient Balkan war by attacking Greece. He was therefore ever more convinced that to hand French colonies over to the Spaniards was to make them vulnerable to British attack. He already believed after Dakar, as he later told Mussolini in Florence on 28 October, that the French should be left to defend French Morocco. Hendaye and Montoire were a reconnaissance to see if there was a way to make the aspirations of Franco and Pétain compatible and to help him decide his future strategy in south-western Europe.[75] The Führer was aware of the fact that his military and diplomatic advisers thought that he should not try to bring Franco into the war. His Commander-in-Chief, Brauchitsch, and his Chief of Staff, Halder, believed that 'Spain's domestic situation is so rotten as to make her useless as a political partner. We shall have to achieve the objectives essential to us (Gibraltar) without her active participation.'[76] State Secretary Weizsäcker wrote 'in my opinion, Spain should be left out of the game. Gibraltar is not worth it. Whatever England lost there would soon be made up with the Canary Islands. Spain today has neither bread nor petrol.' In his view, Spain joining the Axis had 'no practical worth'.[77]

To the contempt of the assembled Germans, Franco's train, which had had to come only a few kilometres, shuddered into the station shortly after 3.00 p.m., eight minutes late. According to the German Foreign Ministry official Dr Paul Schmidt, the train was one hour late although there is nothing to substantiate that assertion in reports of the time nor in the several accounts of Serrano Suñer.[78] It was later claimed by Franco's hagiographers, without any foundation other than a boast by the Caudillo made in 1958, that the alleged lateness was a skilful device by Franco to throw Hitler off balance.[79] Franco had no reason to want to do this. In fact, he was mortified by the small delay that his train suffered.* Feeling

* There was an unsuccessful attempt on the life of Franco made by Spanish anarchists in the course of the Hendaye meeting. They planned to throw hand grenades at Franco's train and the short delay in the Caudillo's arrival gave rise to rumours that the train had in fact been attacked. In the event, the enormous security precautions being taken guaranteed that the meeting would not be interrupted, (Eliseo Bayo, *Los atentados contra Franco* [Barcelona, 1977] pp. 54, 58–60).

that he was being diminished in the eyes of Hitler, he threatened to sack the lieutenant-colonel responsible for organizing his travel arrangements.[80] Substantial photographic evidence of the initial meeting on the platform at Hendaye station suggests that Franco was thrilled to be meeting the Führer. It was surely understandable that Franco's eyes should glisten with emotion since, for him, the meeting constituted an intensely historic moment.

In so far as it is possible to reconstruct the meeting which then followed in the parlour coach of the Führer's special train *Erika*, there is little to sustain the view that 'the skill of one man held back what all the armies of Europe, including the French, had been unable to do'.[81] An entirely accurate, minutely detailed, reconstruction is impossible despite the existence of several ostensibly eyewitness accounts. Six people took part – Hitler, Franco, Ribbentrop, Serrano Suñer, and the two interpreters, Gross and the Barón de las Torres. A seventh, Paul Schmidt, Ribbentrop's press secretary and interpreter, was hovering in the background. Four of the seven – Serrano Suñer, the Barón de las Torres, Ribbentrop and Schmidt – have left accounts of varying degrees of detail and reliability.[82] The fullest version is contained in the German Foreign Office record – produced by Schmidt. Just as other documents concerning the relations between Hitler and Franco are inexplicably missing, this record is incomplete.[83]

Hitler had previously been warned by Admiral Canaris that he would be disillusioned when he met Franco, 'not a hero but a little pipsqueak' (*statt eines Helden, ein Würstchen*).[84] The meeting began cordially enough at 3.30 p.m. To Franco's greeting 'I am delighted to see you, Führer', Hitler replied 'Finally, an old wish of mine is fulfilled, Caudillo'. Thereafter, rather than the conversation which Hitler might have expected to dominate, there were obliquely opposing monologues. Hitler gave the impression that he had bigger problems on his mind than any deal with Franco. Certainly, his behaviour was not that of someone about to deliver threats. He rambled around the point in a lengthy justification of Germany's present difficulties in the war, with particular emphasis on the role of the weather in the Battle of Britain. He surveyed his available military strength but did not, as was claimed in Franco's Spain, say 'I am the master of Europe and, as I have 200 divisions at my orders, there is no alternative but to obey'.*

---

* The threat is alleged in the Barón de las Torres' notes, dated 26 October 1940 and published in the Spanish press in 1989, *ABC, La guerra mundial* (Madrid, 1989) pp. 146–151. These and other notes by de las Torres were used in Ramón Serrano Suñer, *De anteayer y de hoy* (Barcelona, 1981) pp. 203–12. However, neither the tone nor some of

Hitler did explain laboriously and in convoluted terms why the fulfilment of Spain's Moroccan ambitions was problematic given the need for
the co-operation of the Vichy French. In this regard, he referred to his
conversation on the day before with Laval and his forthcoming encounter
with Pétain, in which his theme was that, if France came in with Germany,
then her territorial losses could be compensated with British colonies.
The bitter pill for Franco was Hitler's statement that 'If co-operation
with France proved possible, then the territorial results of the war might
perhaps not be so great. Yet the risk was smaller and success more readily
obtainable. In his personal view it was better in so severe a struggle to
aim at a quick success in a short time, even if the gain would be smaller
than to wage long drawn-out wars. If with France's aid Germany could
win faster, she was ready to give France easier peace terms in return.'

Franco can hardly have failed to notice that his hopes of massive
territorial gain at virtually no cost were being slashed before his eyes. He
had gone into the meeting naïvely convinced that Hitler, his friend,
would be generous. Accordingly, he tried to overwhelm the Führer with
a historical recital of Spanish claims in Morocco, the appalling economic
conditions in Spain, a list of supplies required to facilitate her military
preparations and a pompous assertion that Spain could take Gibraltar
alone. According to Schmidt, Franco irritated Hitler with his relentless
imperturbability and by droning on insistently 'in a quiet, gentle voice,
its monotonous sing-song reminiscent of the muezzin calling the faithful
to prayer'.[85] Hitler was especially infuriated when Franco repeated an
opinion which he had acquired from Captain Espinosa de los Monteros
to the effect that, even if England were conquered, the British Government and fleet would continue to fight the war from Canada with American support.[86] The outraged Führer jumped nervously to his feet, barking
that there was no point in further discussion. Hitler was frustrated with
what he saw as Franco's incorrigibly small-minded lack of vision in enter-
taining, and his bad taste in expressing, doubts about German victory
over England. However, he evidently thought better of breaking off the
meeting and sat down again.[87]

According to de las Torres, Hitler left the meeting muttering 'with

the details of de las Torres's account is convincing. References to the virility, patriotism and
realism with which Franco resisted Hitler's pressure are more redolent of the post-1945
propaganda exercise in rewriting the Caudillo's role in the war. The document makes,
among others, the dubious claim that Hitler, unprompted, offered Oran to a flabbergasted
Franco who refused on a point of honour. This is totally untenable given what is known
about Spanish ambitions in the area and Franco's pressure on Vichy in precisely that
direction during the preceding months.

this fellow, there is nothing to be done'('*mit diesem Kerl ist nichts zu machen*').[88] Clearly, had Hitler been threatening to use two hundred divisions against Spain, he would hardly have made a remark so redolent of impotence.* The interview ended at 6.05 p.m. and, after a short interval during which Serrano Suñer and Ribbentrop met, the party took dinner in Hitler's coach. According to Field-Marshal Keitel, who spoke briefly to Hitler during the dinner break, 'he was very dissatisfied with the Spaniards' attitude and was all for breaking off the talks there and then. He was very irritated with Franco, and particularly annoyed about the role played by Suñer, his Foreign Secretary; Suñer, claimed Hitler, had Franco in his pocket.'[89]

The two Foreign Ministers were then left to draw up a protocol.[90] It was significant that, in the conversation between Serrano Suñer and Ribbentrop which followed, the *cuñadísimo* 'noted at the outset that the Caudillo had not exactly understood the concrete questions dealt with in the conversations with the Führer'. In particular, he could not bring himself to accept that Hitler wished to collaborate with Pétain whom the Caudillo saw as finished.[91] Serrano Suñer expressed to Ribbentrop his surprise at Hitler's new line with regard to French Africa and his regret that 'this would render void Spain's maximum demands'. Nonetheless, consistent with the earlier proposals of Franco himself, he agreed to a secret protocol. Another Spanish aspiration which was not to be satisfied in the written agreement was a claim for a rectification of the Pyrenean frontier to give French Catalonia to Spain.[92]

The document had not been completed when the talks broke up. It is not known what the Caudillo and the Führer spoke about while their Foreign Ministers were negotiating. It would appear that, in the absence of Ribbentrop, Hitler had managed to rekindle Franco's enthusiasm for the Third Reich. The Caudillo's parting words to the Führer revealed his emotional commitment to the Axis: 'Despite what I've said, if the day ever arrived when Germany really needed me, she would have me unconditionally at her side without any demands on my part.' To the relief of Serrano Suñer, the German interpreter did not translate what he took to be merely a formal courtesy.[93]

With an astonishing mixture of naïvety and greed, Franco said to Serrano Suñer after the interview, 'These people are intolerable. They

---

* On 31 October, seven days after Hendaye, Serrano Suñer met the American Ambassador, Weddell, and repeated three times, that 'there had been no pressure, not even an insinuation on the part of either Hitler or Mussolini that Spain should enter the war'. *FRUS 1940*, II, p. 824.

want us to come into the war in exchange for nothing. We cannot trust them if they do not undertake, in whatever we sign, a binding, formal contract to grant to us now those territories which I have explained to them are ours by right. Otherwise, we will not enter the war now. This new sacrifice of ours would only be justified if they reciprocated with what would become the basis of our empire. After the victory, despite what they say, if they do not make a formal commitment now, they will give us nothing.'[94] What is striking about Franco's remarks was their implicit belief that 'a formal commitment' from Hitler would have been worth anything. This statement, and indeed the entire tenor of the meeting make a nonsense of the later claim by both Franco and Serrano Suñer that they were skilfully holding off Hitler. Their determination was not to hold on to neutrality, but to get the basis of a colonial empire. It was their good fortune that Hitler had other commitments and was unable to meet their imperial ambitions. Accordingly, neutrality became a kind of consolation prize.

After the meeting, when Franco's train finally drew off, it jolted so violently that only the intervention of General Moscardó prevented Franco tumbling off head-first onto the platform. It rained on the way back to San Sebastián and in the aged train once used by Alfonso XIII, known as the 'break de Obras Públicas', water leaked onto Franco and Serrano Suñer.[95] On returning to the Palacio de Ayete, Serrano Suñer and Franco worked on a text of the protocol between 2.00 and 3.00 a.m. The text prepared in advance by the Germans called upon Spain to join the war when the Reich considered it necessary. The Generalísimo and his brother-in-law sought in their text to find a less rigid formula which would still give them bargaining room. Before dawn, General Eugenio Espinosa de los Monteros, the Spanish Ambassador to Germany, appeared. In view of his account of German impatience, the text was sent back to Hendaye in his hands. Ribbentrop refused to accept small amendments to the protocol, although Serrano Suñer kept the news from Franco.[96] For all its vagueness, the protocol constituted a formal undertaking by Spain to join the war on the Axis side.[97]

Goebbels noted in his diary of the Hendaye talks 'The Führer has now had his projected meeting with Franco. I am informed by telephone that everything went smoothly. According to the information, Spain is firmly ours. Churchill is in for a bad time.'[98] Goebbels was not alone in receiving such a call. Ribbentrop also telephoned Ciano and expressed satisfaction with the meeting.[99] Both of these comments are entirely consistent with the fact that Hitler had been on something of a reconnaissance trip to compare the stances of Franco and Pétain. Franco had manifested an attitude of total loyalty to the Axis, albeit reserving his right to select the timing of Spanish participation in the war. In Madrid, it was assumed that Spain would soon be at war. There was panic in the diplomatic corps and the Portuguese Embassy was bombarded with requests for visas.[100] It was only later, as Spanish belligerence was interminably delayed, that Hitler came to regard the meeting as an outright failure.

However, that is not to say that he had enjoyed the encounter. After spending nine hours intermittently in Franco's company, Hitler told Mussolini later that 'Rather than go through that again, I would prefer to have three or four teeth taken out'.[101] Both Hitler and Ribbentrop were irritated by the fact that, oblivious of the needs of German policy towards Vichy, Franco relentlessly repeated what they saw as his ludicrously exaggerated imperial demands. On the drive away from Hendaye, Ribbentrop allegedly cursed Serrano Suñer as a 'Jesuit' and Franco as an 'ungrateful coward'.[102] Ribbentrop was exasperated by the difficulties he had encountered with Serrano Suñer who 'frequently revealed a lack of sufficient understanding for the fact that the realization of the Spanish aspirations depends exclusively on the military successes of the Axis Powers and that therefore these aspirations must be subordinated to the Axis policy of attaining final victory'.[103] Colonel Gerhard Engel, Hitler's Army Adjutant, reported that the Führer was infuriated (*wütend*) with the Hendaye meeting, ranting about 'Jesuit swine' and 'misplaced Spanish pride'.[104]

These contemptuous remarks have been cited by Francoist propagandists and by Serrano Suñer himself as proof that Hitler and Ribbentrop were apoplectic with rage because German might was held back by the skilful rhetoric of the Caudillo and his brother-in-law. In fact, such remarks completely undermine the claim that Franco blunted the German threat at Hendaye. If Hitler had really had two hundred divisions waiting to roll, nothing Franco or Serrano Suñer said would have made any difference. The insults are more indicative of Teutonic disdain for the self-regarding pretensions of the Caudillo and his pompous presumption of a status equal to the Führer's. The manifest pride and patriotism of both Franco and Serrano Suñer must have seemed infuriatingly misplaced in the representatives of a nation as economically and militarily weak as Spain.

In fact, Hitler's exasperation also derived from his frustration that he had not succeeded in deceiving the Spaniards over French Morocco by the seemingly frank admission that he could not give what was not yet his. He was, of course, confident of ultimately being able to dispose of the French colonial empire as he wished but had no intention of giving it to Franco. That was his 'grandiose fraud'. The Head of the German Foreign Office, Weizsäcker saw little importance in the fact that 'nothing concrete has been agreed concerning Spain's entry into the war' and thought Hendaye to be more of a failed 'conjuring trick'.[105] Serrano Suñer suggested years later that Hitler had not told a sufficiently big lie. In his view, Franco's *Africanista* obsession with Morocco was such that, if Hitler had offered it, he would have entered the war.[106] Franco himself has been quoted as telling the Civil Governor of León, Antonio Martínez Cattaneo, that 'it was Hitler who did not accept my conditions'.[107] While awaiting the arrival of Franco's train, Hitler himself had revealed the reasons why, for once, he was incapable of dabbling in full-scale untruth. Chatting with Ribbentrop on the platform at Hendaye, he had remarked that no firm promises of French territory could be given because, 'with these chattering Latins, the French are sure to hear something about it sooner or later'.[108]

It was fortunate for Franco that Hitler remained unwilling and indeed unable to pay his price. After all, one of his reasons for wanting Spain's participation was to be able to control North Africa and so preclude any increase in French resistance there. Yet Franco's price – the cession of French colonies – would almost certainly have precipitated an anti-German movement under de Gaulle that would pave the way to Allied landings. The Hendaye meeting came to a stalemate precisely on this problem. The protocol was signed, committing Spain to join the Axis cause at a date to be decided by 'common agreement of the three Powers' but after military preparations were complete. This effectively left the decision with Franco. Nevertheless, the Führer could have pushed him by beginning deliveries of food and military supplies. Hitler made firm promises concerning only Gibraltar and was imprecise about future Spanish control of French colonies in Africa. The vague promises which were made were not enough for the Caudillo. In the aftermath of Hendaye, Franco was obliged to recognize that his imperial pretensions were of little concern to Hitler and his emotional admiration for the Führer began to wither.

XVI

IN THE WINGS

Franco & the Axis November 1940–February 1941

FRANCO and Serrano Suñer came away from the Hendaye meeting reinforced in their sense of belonging to the Axis club. Their mood was reflected in Serrano Suñer's conversation with the Vichy Ambassador, Comte Renom de La Baume, on 26 October. In a haughtily cutting tone, he told the Comte that 'the possession of Morocco is crucial for the defence of the Peninsula' and said threateningly that good Franco–Spanish relations in the future depended on the solution of the 'African problem' and the 'France had some withdrawing to do' ('*des abandons à faire*').[1]

That such attitudes were built on empty bluster could be deduced from what Hitler had to say when he met Mussolini in Florence on 28 October. The Führer had travelled to Italy both to recount his meetings with Laval, Franco and Pétain, and to hear the full extent of Mussolini's adventure in Greece. He repeated his view that it made more sense to leave the Vichy French to defend Morocco than risk the consequences of giving it to Franco. It was clear from what Hitler said that his contemptuous remarks about Franco and Serrano Suñer after Hendaye derived from annoyance that their imperial aspirations stood in the way of his need to draw Laval and Pétain into some kind of subordinate alliance. Hitler told the Duce that Franco 'certainly had a stout heart, but only by an accident had he become Generalissimo and leader of the Spanish state. He was not a man who was up to the problem of the political and material development of his country.' Of Franco's demands, the Führer said that Spain 'could not get any more than a substantial enlargement of Spanish Morocco' and that the timing of Spanish entry into the war depended on the completion of her military preparations. At that point, they were in agreement that they and Franco meet in Florence to announce the Tripartite Pact with full publicity. Before that could materialize, the Führer's mind would be turning to Russia.[2]

Goebbels noted 'The Führer's opinion of Spain and Franco is not high. A lot of noise, but very little action. No substance. In any case, quite unprepared for war. Grandees of an empire that no longer exists. France, on the other hand, is a quite different matter. Where Franco was very unsure of himself, Pétain was clear and composed.' He also commented on reports about Spain from the *Auslandorganisation* which were circulating among the Nazi high command. They stressed that Franco was weak, not in control, Serrano Suñer very unpopular and Spain 'in a wild, almost anarchic state of disorder', the economy in ruins.[3] The rapid deterioration of Spain's economic position was to be the decisive obstacle to precipitate warlike action by Franco, bringing as it did an ever greater vulnerability to Anglo-American pressures and blandishments. As David Eccles wrote to his wife on 1 November 1940, 'The Spaniards are up for sale and it is our job to see that the auctioneer knocks them to our bid.'[4]

The economic power of the Anglo-Saxon powers would cause Franco to oscillate between the belligerents over the coming months but his essential loyalty remained with the Axis. Serrano Suñer said to the Portuguese Ambassador Pedro Theotonio Pereira on 6 November 1940 'the Generalísimo is a simple man. It is just as well he didn't speak much with Hitler.'[5] This was confirmed by Hitler who told Mussolini three months later that Franco 'had frequently agreed to German proposals and then been interrupted by Serrano Suñer, who had upset everything once more'.[6] Franco's diplomatic activities in early November also substantiated Serrano Suñer's judgement. The Generalísimo took several dangerous, and unnecessary, initiatives which can only be interpreted as indicating a readiness to join the war on the Axis side. If there were any disappointments at Hendaye for Franco, they were clearly superficial and shortlived.

On 30 October, Franco composed letters to both Mussolini and Hitler. To the Duce, he wrote strongly pressing Spanish claims to French Morocco and Oran while grudgingly acknowledging that the needs of the Axis-Vichy understanding meant that they could not be publicized.[7] The clear implication was that he had indeed been taken in by Hitler's 'grandiose fraud'. Indeed, both letters convey the pleasure that Franco felt as a result of belonging to the Axis club. His letter to Hitler was the more daring of the two. Nevertheless, the one directed to Mussolini was accompanied by a significant note from Serrano Suñer to Ciano. In it, the *cuñadísimo* explained why, ever since Hendaye, he had desperately been seeking a secret meeting in order to obtain Italian support to persuade the Germans not to put the interests of Vichy France before those of Spain.[8] He stated that out of loyalty to the Axis and in order to bring about the early conclusion of the war, Spain was prepared to accept Hitler's concessions to France, 'but not to be sacrificed in the interests of our eternal enemies'.[9]

The letter to Hitler was delivered by the Spanish Ambassador, General Eugenio Espinosa de los Monteros, on 3 November.[10] To the Führer, Franco promised faithfully that he would now carry out his verbal promises to enter the war and went on, at enormous and serpentine length, to repeat his territorial claims in French North Africa.[11] Both the letters to his fellow dictators and Serrano Suñer's note to Ciano leave no doubt that the Spanish response to Hendaye was one of disappointment that a firmer commitment had not been made to imperial prizes in return for Spanish belligerence. In an effort to show that Spain was a serious partner, on 3 November Franco dismantled the international administration of Tangier and incorporated the city into Spanish Morocco – and did so without informing either the Germans or the Italians. Stohrer believed that this reflected the Caudillo's disappointments at Hendaye and his desire to ensure 'Tangier at least'.[12] However, it is more likely that, as with his original move on Tangier in June, Franco intended to impress the Führer with his bellicosity. Together with the letter, it seemed to have the desired effect.

Understandably, Hitler was enthused by the letter and informed Generals Brauchitsch, Halder, Keitel and Jodl that he now wanted to hasten Spain's entry into the war and proceed to seize Gibraltar.[13] On 9 November, three copies of the secret German-Italian-Spanish protocol arrived in Madrid and were duly signed by Serrano Suñer and the German and Italian copies sent back by special courier.[14] An intensity in Hispano-German relations lacking during Serrano Suñer's visit to Berlin in September and at the Hendaye meeting was now generated by Italian reverses. Hitler was shaken by the British naval victory over the Italians at Taranto and by the turn of events in Greece which opened the way to a British offensive in the Balkans. To diminish the risk to Germany's Rumanian oil supplies, Hitler decided that he must close the Mediterranean.[15] Only now, for the first time, was he sufficiently keen on Spanish belligerence to force the pace and to put pressure on Franco. On 11 November, Ribbentrop invited Serrano Suñer to a meeting at the

Berghof on 18 November. The invitation was accepted once Serrano Suñer had discussed it with his brother-in-law.[16]

However, since writing his letter to Hitler on 30 October, the Caudillo had himself become more cautious. The Spanish economic situation was worsening by the day and the strength of British resistance had reactivated his deep-seated fears of a retaliatory strike against Spain or her overseas territories. With Hitler showing little sign of coming up with the specific undertakings of imperial profits that he sought, Franco was about to retreat to a more careful line than hitherto, that of postponing Spanish belligerence until British defeat was unmistakably imminent. Before Serrano Suñer went to Berchtesgaden, Franco held a meeting of the military Ministers in the government, General Varela (Army), General Vigón (Air Force) and Admiral Salvador Moreno (Navy), at which Spain's total inability to enter the war was discussed. On the table was an extremely realistic paper presented by Admiral Moreno, which had been drawn up by the naval staff – including his chief of operations, Franco's future *éminence grise*, the dour and plodding Captain Luis Carrero Blanco. Although taking for granted that Spain was on the Axis side, the paper shrewdly weighed up both Spanish maritime weakness in relation to the Royal Navy and the economic costs of Spanish belligerence. Its basic premise was that forced on Hitler himself by the Italian débâcle: the Axis needed to seize both Gibraltar and Suez, and capture of the former necessarily required Spanish intervention. However, the paper concluded that such an intervention would be impossible before the Axis capture of Suez. It was believed that the economic consequences of going to war against England while she still held Suez, with the inevitable Allied blockade, would be disastrous and, on balance, would be damaging to the Axis.[17] Carrero Blanco claimed years later that his part in drafting the paper took him in 1942 to the post of Under-Secretary of the Presidency whence he was to launch a career as Franco's influential political Chief of Staff that lasted over thirty years.[18]

Shortly after the Spanish High Command concluded that no initiative should be taken against Gibraltar while Britain still held Suez, proposals by Hitler's Chief of Staff, General Halder, for an assault on the Rock were converted in mid-November into detailed operational plans. Under the name Operation Felix, German troops would enter Spain on 10 January 1941 prior to beginning an assault on Gibraltar on 4 February.[19] German troops began to rehearse the assault near Besançon. The problem, as Canaris had reported and as Hitler's Chief Supply Planners quickly confirmed, was that Franco had not exaggerated when he had spoken of the prostrate condition of the Spanish economy. Troop movements would be rendered difficult by the different rail gauges on either side of the Franco-Spanish border and the notorious disrepair of Spanish track and rolling stock. Moreover, a disastrous harvest meant that Spain needed considerably more grain than specified in her earlier requests to the Germans. With famine conditions developing in many parts of the country, Franco had no choice but to seek to buy food in the Americas and that necessarily involved postponing a declaration of war.[20]

In any case, the postponing of Operation Sealion and a growing awareness of British naval strength had diluted Franco's euphoric confidence in German victory. As Canaris told Halder on 2 November, Franco was deeply anxious about the possibility of conflict with Britain and, particularly, about a British attack on the Canary Islands. Canaris's analysis of the Caudillo's domestic problems pointed to his remaining supine as far as the war was concerned. 'The internal administrative machinery has completely broken down. Franco cannot afford to take risks. Serrano Suñer is easily the most hated man in Spain.' German-Spanish understanding was handicapped by the 'unwarranted hauteur' and 'morbid sensitiveness' of both Franco and Serrano Suñer. 'To this must be added Franco's shyness.'[21] German military planners were fully cognizant of the weakness of the Spanish army. They were also inhibited with regard to a Spanish adventure because of 'the eastern operation'. There was not the remotest possibility that the Germans might attempt to seize Gibraltar against the will of the Spaniards since they had no desire to become embroiled in a difficult struggle and so delay even further the assault on the Soviet Union.[22]

At the same time as the Germans procrastinated over an attack on Gibraltar, the British Government, encouraged by Sir Samuel Hoare and Alexander Weddell, continued to advocate American food aid for Spain precisely to deprive Franco of the excuse to slip into the arms of the Axis. This was rendered difficult by the insistence of Secretary of State Cordell Hull that Franco publicly declare that he did not plan to change Spain's neutrality or envisage aid to Germany and Italy in the war against Great Britain. On 31 October, when Weddell had been received by Serrano Suñer, the Foreign Minister had flaunted Spain's political solidarity with Germany and Italy and then remarked that Franco was surprised that the Red Cross wheat had not arrived. Weddell reminded him that its delivery was conditional on Spain's international position. The meeting convinced Weddell – who was, of course, unaware of Franco's letter to Hitler written only the previous day – that food aid would contribute to Spain staying out of the war.[23]

Only two days after the bellicose expulsion of the remnants of the international administration from Tangier, the worsening grain crisis obliged Franco's cabinet on 5 November to agree to give publicity to the American offer. It was too late. Since the original offer had been made, Hendaye and Tangier had enraged American public opinion which was already incensed by reports of Franco's Axis sympathies and of the unending stream of executions of Republican prisoners.[24] On the other hand, Hoare and the British government were convinced that the best chance of preventing a desperate Spanish declaration of war was immediate food relief.[25]

While Serrano Suñer was again in Germany, Franco's cabinet decided, after some heartsearching, that the public declaration of neutrality required by Washington was an insuperable stumbling block. On 19 November, Weddell met, at their request, Carceller and the Minister without Portfolio, Pedro Gamero del Castillo, who was acting Secretary-General of the Falange and a close ally of Serrano Suñer. According to them, Franco could not declare neutrality for fear of provoking the hostility of a 'Germany at the frontier couched ready to spring'. This sly tack hardly accords with what is now known of Hispano-German relations in the immediate aftermath of Hendaye, the Tangier gesture and Franco's letter to Hitler. At the time, however, the argument, the fact that Serrano Suñer was at the time in Germany talking to Hitler and Spanish offers of private assurances from Franco eventually led to a softening of the State Department's position.[26]

In the meantime, the meeting between Hitler and Serrano Suñer took place at Berchtesgaden on 19 November 1940. There was a greater urgency now on Hitler's side and an element of prevarication on Serrano Suñer's. This was to be the first of his three visits to Germany during which he was to be put under real pressure. Although he tried to play down the consequences of Mussolini's 'mistake' in Greece, Hitler declared bluntly that 'it was imperative to act swiftly and decisively'. 'In the present circumstances it was absolutely necessary to shut off the Mediterranean', at Gibraltar and at Suez. Serrano Suñer pointed out that Spain needed English goodwill in order to import necessary food supplies, mentioning Franco's dismay at Hitler's failure to send either food or war material. Hitler replied that, if Spain became a belligerent, supplies would follow. Serrano Suñer said that, as a consequence of the greater Spanish solidarity with the Axis established at Hendaye, the United States had blocked a shipment of thirty thousand tons of wheat, a reference to the delayed Red Cross grain. Serrano Suñer's basic ideological sympathies were revealed when he remarked that 'above supply problems, however, stood history in which Spain wanted to participate this time too'.

Hitler insisted that Spain should enter the war as soon as possible. One month to six weeks of preparation would coincide with the best period for German troops to fight in Spain. Serrano Suñer hastened to remind Hitler of how disappointed he and the Caudillo were at the vagueness of the promises made in the secret Hendaye protocol concerning Spain's imperial demands. If the details were to become public, it would be believed in Spain that Germany had abandoned a 'friend of yesterday, today and for the whole future' in favour of a deal with Vichy. Serrano Suñer insisted that the vague promises of the protocol were worthless to which Hitler claimed that Spain would be satisfied in Morocco. The Führer then went on to reiterate why an early Spanish entry into the war was crucial. He offered artillery, ammunition and dive bombers to help secure the Canary Islands from English attack. Serrano Suñer replied that Spain had already attended to the defence of the islands. He took his leave, saying that he would use the remaining period of Spanish neutrality while military preparations were made to import as much Canadian, American and Argentinian wheat as possible into Spain. Hitler perhaps sensed, more than Ribbentrop, the difficulties over Spanish entry and, when he saw Ciano immediately after speaking to Serrano Suñer, he had suggested that Mussolini use his influence with Franco to clinch Spain's intervention.[27]

Ribbentrop now took it for granted that Spain was about to enter the war, and when he joined Serrano Suñer at his hotel later in the day, repeated what Hitler had said about the importance of acting quickly. Assuming that Serrano Suñer would speak to Franco as soon as he returned to Madrid, he suggested that the results of that conversation be communicated urgently to Berlin via Ambassador Stohrer in order for preparations to begin imediately. Ribbentrop asked for a Spanish representative to go to Berlin to arrange the details of German food and raw material deliveries. He was so anxious to move that he requested that Stohrer telegraph a code word to Berlin which would signify that Franco was prepared to enter the war. Ribbentrop gave a subjective overview of the war situation in which he dismissed American aid to England as vulnerable to U-boat attack and talked as if Stalin was about to join a German-led world coalition against the British. Serrano Suñer cut him short and said that he had heard that American aircraft deliveries to

England were rather substantial and was therefore glad to hear about the intensification of U-boat operations. The meeting was not a success and clinched the antipathy felt by Serrano Suñer towards Ribbentrop. [28]

The meeting between Hitler and Serrano Suñer had been followed in London with the greatest anxiety and intensified Allied concern to help resolve Spanish food problems. Churchill wrote to Roosevelt on 23 November: 'Our accounts show that the situation in Spain is deteriorating and that the Peninsula is not far from starvation point. An offer by you to dole out food month by month as long as they keep out of the war might be decisive. Small things do not count now and this is a time for very plain talk to them. The occupation by Germany of both sides of the Straits would be a grievous addition to our naval strain, already severe.'[29] Indeed, only the previous day, Mussolini had written to Hitler that the time was ripe to play 'the Spanish card' and had offered to meet Franco to put pressure on him to join the Axis.[30] Late in the evening of 25 November, Stohrer telegrammed Berlin to say that immediately after his return from Berchtesgaden, Serrano Suñer had spoken to Franco about his conversations with Hitler. Franco had endorsed his brother-in-law's position and immediately called the Armed Forces Ministers for a lengthy consultation. The Ministers had asked awkward questions about what Spain could expect to gain and reminded Franco about the country's appalling military weakness. Serrano Suñer told Stohrer that Franco's decision 'which would be given immediately after the conclusion of the military consultations, would naturally be in the affirmative and that our preparations could then begin', but, in the light of the questions asked by the military Ministers, 'an early reply to the last letter of the Caudillo to the Führer would be very desirable'.[31] Had a satisfactory reply to that letter been forthcoming, and had Hitler begun substantial food and weapons deliveries, it is likely that Franco would indeed have ordered military preparations. As it was, Hitler did not meet the demands made in the Caudillo's letter of 30 October and Franco sat tight.

On 28 November, Stohrer telegrammed Berlin to say that Serrano Suñer had just told him that Franco had agreed to start the preparations for war. In fact, this was not true and nothing happened. Stohrer, like Hitler and Ribbentrop a few days previously, seemed determined to see definite commitments where there was more than vague prevarication. It is not inconceivable that Serrano Suñer, to combat the growing opposition to his own position from within the military high command, was trying to curry favour with the Germans.[32] He had remarked during his visit to Berchtesgaden that 'Spanish participation in the war was the remedy for internal agitation'.[33] On 29 November, Serrano Suñer read out to Stohrer notes on Franco's position. They were significantly less definite than Stohrer's telegram on the previous day had implied. The notes contained Franco's agreement 'that the preparations for Spain's entry into the war are to be speeded up as much as possible'; but went on to say that 'the time required for this cannot be definitely determined' because of the need to be prepared for other military actions as a consequence of the attack on Gibraltar. Franco asked for the despatch of German military experts to liaise with the Spanish armed service ministries. Because of his fears of an English counter-attack on the Spanish Atlantic coasts, Franco suggested an attack on Suez to tie down the English Mediterranean fleet, a proposal indicating that, thanks to the meeting with his military Ministers, Franco's habitual caution had finally begun to reassert itself.[34]

Hitler responded to the news of Franco's barely cloaked desire to get Spain out of the firing line by sending Admiral Canaris to discuss the details.[35] As an indication of Franco's continuing support for the Axis, Serrano Suñer informed Stohrer that the Spanish government had agreed to German tankers being stationed in remote bays on the northern coast for the refuelling of German destroyers.[36] However, whatever the sympathies of Franco and Serrano Suñer for the Axis cause, their freedom of action was severely constrained by the increasingly desperate condition of the Spanish economy. The 'terrible picture' of British reports from Barcelona and Madrid inclined officials in the Foreign Office to believe that 'even the Germans might hesitate to "take over" a country in such a plight'.[37] On 21 November 1940, David Eccles had written to his wife 'As we walk to the Embassy in the morning we see an ever-increasing number of men, women and children picking over the dustbins and the slop pails standing on the kerb. As they spy a bit of potato peel among the filth, they eat it, and stuff into sacks garbage too horrible to describe.'[38] In this period of starvation and epidemics, the appalling privations of the many were in stark contrast to the decadence of the few who had access to the privileges of office and the black market.[39] In response to the appalling famine conditions in Spain, the Allies shrewdly pursued their policy of offering economic aid to Spain in return for her staying at peace.

At precisely the time that Serrano Suñer was reading to Stohrer Franco's reassurances about Spanish military preparations, the Caudillo himself was talking to the American Ambassador. With the economic crisis preying on his mind, he generated a mood of what, by his normally icy standards, approached warmth and cordiality. Weddell told him that, in the light of the recent visits by Serrano Suñer to Germany, the USA would find it difficult to proceed with the shipment of Red Cross wheat or to extend credits for raw material purchases unless there was some clarification of Spain's stance regarding the Axis. Franco mentioned his own and Serrano Suñer's recent conversations with Hitler without divulging their content. To Weddell's direct question, he denied that Spain had signed the Tripartite Pact – which Serrano Suñer had done three weeks earlier. Franco told him that US policy seemed to be based on the belief that Britain would win the war while his own policy was based on the reverse. In an attempt to curry sympathy and understanding, he obliquely referred to the threat he faced from the German divisions allegedly lying idle on the French-Spanish border.

When Weddell pushed Franco to say that Spain did not contemplate any departure from its present policy or aid to the Axis, he shiftily agreed to the first and to the second said that Spain could not help the Axis even if it wished and that no one could foretell what the future might bring. In Weddell's opinion, this latter was as near as the Caudillo was ever likely to go towards the declaration of neutrality required by Washington and, ignorant of the promises already made to Hitler, he urged the State Department to accept it as adequate for the release of the wheat. Washington remained suspicious of Franco not least because of the continued anti-US activities in Latin America of the *Falange Exterior*.[40] Frantic efforts were made by an unusually friendly Serrano Suñer to convince Weddell and Hoare that the wheat should now be sent. This no doubt reflected the fact that there were reports of bread riots and of violent attacks on bakeries. On 2 December 1940, by a supplementary agreement to the Anglo-Spanish trade agreement of 18 March 1940, the British had agreed to the transport of 150,000 tons of maize from South America and 100,000 tons of grain from Canada. In return, Spain had undertaken not to re-export certain raw materials to the Axis Powers and not to permit the transit of certain Portuguese goods to Germany and Italy. The State Department did not, however, agree to move the Red Cross wheat until 7 January 1941.[41]

An awareness of the food problem lay behind the beginning, in late 1940, of one of Franco's most important political friendships. It began in inauspicious circumstances. Franco summoned to his presence the Falangist Civil Governor of Málaga, José Luis Arrese, who was accused of plotting against him. Frenetically ambitious, Arrese had been condemned in 1937 as a supporter of Hedilla and had been rehabilitated by

Serrano Suñer. The oily Arrese not only managed to persuade Franco that he was not conspiring but also secured his favour by convincing him that he had original ideas for dealing with the regime's unpopularity. These ranged from designs for cheap housing to the extraordinary claim that the famine could be eased by feeding the people dolphin sandwiches (*bocadillos de carne de delfín*) and bread made from fish meal (*harina de pescado*). Arrese was to show himself a virtuoso of adulation over many years. Franco probably enjoyed the flattery but also recognized in Arrese someone who could be used. Arrese was the very picture of a Falangist, with a pencil moustache and hair sleeked back. His background as an Hedillista made him acceptable to the radical wing of the Falange which, together with his abject admiration for the Caudillo, made him the ideal agent to complete the process of taming the Falange. Franco would soon make Arrese Minister-Secretary General of FET y de las JONS.[42]

While Franco was coming face to face with the economic realities of his situation and the power of the Allies, he was in negotiation with Canaris about an attack on Gibraltar. Although he and Serrano Suñer played down their capitulation to Allied pressure, the Germans were fully aware of the concessions. On 11 December, Serrano Suñer denied Spanish promises of neutrality having been made to the Americans, but did admit that 'he unfortunately had to show the English some consideration here and there', which translated into an undertaking – not to re-export grain, phosphate and manganese ore – which he brushed off as 'vague, non-binding'. He denied that any undertakings had been made about Portuguese goods in transit and promised to continue deliveries to Germany, after Spain's own urgent needs had been met. Stohrer was convinced that Franco and his brother-in-law had said to the Allies only enough to maintain their freedom of action.[43]

The contradictions between Franco's conciliation of both the Allies and the Axis were soon made starkly clear. On 5 December, Hitler met with his High Command and decided to ask Franco for permission for German troops to cross the Spanish border on 10 January 1941 – hardly the action of a man with two hundred divisions poised to strike against Spain. It was planned for General Jodl to go to Spain to make the necessary arrangements for the attack on Gibraltar as soon as Canaris got Franco's agreement to the target date. The Führer wrote to Mussolini to say that, having doubts about the loyalty of Vichy forces in Africa, he regarded a final decision by Franco on his entry into the war as 'urgently necessary' and asked the Duce to intervene to clinch a date.[44] Two days later, on 7 December, Mussolini told the German Ambassador in Rome that, while Spain's active participation in the war was crucial regarding Gibraltar, in other ways, it 'was only a limited advantage'. His doubts were based on the disastrous economic situation in Spain and his own selfish fears that 'the Spaniards might later bring up inconvenient wishes in North Africa'.[45]

Canaris arrived in a freezing, snow-bound Madrid on 7 December. At 7.30 in the evening, he asked Franco, in the presence of General Vigón, to enter the war by permitting a German Army Corps with artillery to cross Spain to attack Gibraltar. Aware that the Germans had not established air superiority over the RAF and that Italian reverses in Albania were ever more severe, Franco told Canaris that Spain was simply unable, particularly in terms of food supplies, to meet Hitler's deadline. The deficit in foodstuffs was now estimated by Franco to be one million tons and it was compounded by appalling distribution difficulties on both roads and railways. Franco also repeated his fears that the seizure of Gibraltar would ensure that Spain would lose the Canary Islands and her other overseas possessions. Although couched in rhetoric about not wanting to be a burden on his German ally, this was his first admission of doubt about the prospects of early Axis victory. Franco may have been responding to hints from Canaris that Hitler might not win the war. Writing after the war, Weizsäcker claimed that Canaris had told him that he refused to be 'a pawn in the fraudulent game which was being carried on with the Spaniards'. Field-Marshal Keitel believed that Canaris 'did not make a serious effort to win over Spain for the operation but in fact advised his Spanish friend against it'. It is certainly possible that, believing Spanish entry to be a further complication in a war which was already lost, Canaris simply did nothing to persuade Franco to permit the attack on Gibraltar.[46]

In any case, Franco was not likely to be attracted by what was in effect an invitation to join the war on a one-issue basis. The Caudillo's shopping list remained enormous, including vast French imperial territories, and now Hitler was offering no more than to convert Gibraltar into a German base and, after the war, return it to Spain.[47] Hitler, with characteristic arrogance, saw this as Franco being offered a good deal – the eventual return of Gibraltar at no cost. Franco. however, seeing himself fobbed off with a vague promise in return for risking war with the British, needed no prodding from Canaris to reject the offer.

A puzzled Wilhelmstrasse telegrammed Stohrer to explain the 'flagrant contradiction' between what Franco was now saying on the one hand and the Hendaye discussions and the Hitler-Serrano Suñer meeting at

Berchtesgaden on the other. The Chief of the OKW then telegrammed Canaris on 8 December to extract the nearest deadline from Franco. Canaris replied that he had already pressed Franco without success. 'General Franco replied that he could not fix such a deadline, since it depended upon the further economic development of Spain, which could not be perceived today, as well as on the future development of the war against England. General Franco made it clear that Spain could enter the war only when England was about ready to collapse.'[48] Franco would later deny this interpretation of his words. Weizsäcker wrote in his diary, 'it is a poor consolation to have predicted Spain's position. They will only enter the war shortly before Axis victory, but what do we gain from that apart from some parasites?'[49]

Hitler was already deeply annoyed by the growing evidence that Franco was not prepared to fulfil the agreements made at Hendaye and in his subsequent letter. However, he had decided, if the Canaris mission failed, merely to recall those of his generals who were in Spain.[50] Throughout November, German special units had rehearsed the attack on Gibraltar but there was no question of an assault on the Rock without the acquiescence of Franco. A frontal sea attack was precluded by the fact that the German navy was over-committed to Operation Sealion against Britain, in abeyance but still pending, to protecting the French and Norwegian coasts and, above all, to continuing the Atlantic war. A land operation would involve the German troops in a march of 1,200 kilometres, carrying all their supplies along poor, often unmetalled roads, through narrow, winding mountain passes frequently affected by fog and ice, with no hope of living off the land or purchasing food and fuel as they went.*

On receiving Canaris's depressing report, with surprising equanimity, Hitler immediately ordered that Operation Felix be discontinued. He was convinced that to force his way through Spain would tempt the British to send troops to Spain to help local resistance and thereby open a new and unwanted theatre of operations.[51] The Führer decided that the dive-bombers to be used against Gibraltar were now needed in Southern Italy to attack British shipping between Sicily and North Africa, and were to be on hand in the event of an attack on Greece. In the short term, there was simply not the military capacity for an attack on Spain

---

* Churchill was confident that Hitler would not try to force his way through Spain. He wrote to General Ismay on 6 January 1941 that an invasion in winter was 'a most dangerous and questionable enterprise for Germany to undertake, and it is no wonder that Hitler, with so many sullen populations to hold down, has so far shrunk from it' – Winston S. Churchill, *The Second World War* III *The Grand Alliance* (London, 1950) p. 7.

once troops and equipment began to be gathered for the expeditionary force to reinforce the flagging Italians in Libya. In the longer term, the war against Russia was more important for Hitler than any actions on the periphery. Accordingly, preparations for a spring offensive to the East nudged aside plans for taking Gibraltar.[52]

It was entirely consistent with Franco's developing habit of playing both ends against the middle that he go on putting pressure on Vichy France for concessions in North Africa. Apparently seeing himself as part of the new order and therefore entitled to press for territorial adjustments, Franco received the new Vichy Ambassador, the elegant right-wing Anglophobe François Piétri*, on 7 December 1940 in a threatening manner. Piétri made an especially agreeable speech before Franco. Surrounded by his ministers and principal state dignatories, the Caudillo, in resplendent uniform, replied with what, in the diplomatic context, were coldly insulting references to Spanish claims on French Morocco. 'Friendship cannot exist without justice, and there are all to many injustices to repair for this friendship to become real.'[53]

The pro-Francoist argument that, from the winter of 1940, Franco skilfully fended off Hitler's blandishments and pressures is severely undermined by the fact that from that time, Spain had a low priority for Hitler. This is not say that the Führer was not disappointed with the Caudillo. In an end of year letter to Mussolini, he wrote that 'I fear that Franco is committing here the greatest mistake of his life.' Goebbels noted in his diary with equal disappointment that 'Franco is not pulling his weight. He is probably incapable of doing so. No backbone. And the domestic situation in Spain is anything but happy. The fact that we shall not have Gibraltar is a serious blow.'[54]

The famine and doubts about ultimate German victory led both Aranda and Kindelán to tell Hoare that they opposed Serrano Suñer's Germanofilia.[55] Kindelán said that he had come 'with the full knowledge and approval of General Vigón' which suggests that Franco knew, given his close relationship with Vigón. In that case, he might well have been ensuring that, in the event of seeking greater Allied economic aid, blame for pro-German actions could be diverted onto Serrano Suñer. Hardly had Hitler and the military High Command received Canaris's report when Stohrer informed Berlin that the worsening of famine conditions had taken precedence over every other issue including entry into the war.

---

* An intelligent Corsican and a distinguished fencer, the pliable Piétri had been a parliamentary deputy and minister under the Third Republic.

After several reductions in the bread ration, people could be seen fighting in the streets over crusts. Stohrer wrote on 11 December of *Madrileños* collapsing from lack of food and pointed out that the opposition of senior generals to Serrano Suñer was obliging Franco to tone down his enthusiasm for war.* Stohrer was convinced that his change of heart about entry into the war was entirely the result of the food crisis and his consequent fear for the safety of his regime. Stohrer reported that to overcome Franco's problems would entail economic support of 'tremendous proportions'.[56] During this time, Spanish foodstuffs were – astonishingly – being exported to Germany.[57]

Spain was now obliged to seek British and American permission to import Argentinian wheat. Stohrer wrote to the Wilhelmstrasse on 8 January 1941 that Franco had behaved correctly since it was only after his efforts to get German help had failed that he took up offers from England and America.[58] Stohrer's sympathy was well-founded. It was the famine which had caused Franco to pull back at the crucial moment. Indeed, Serrano Suñer told the Italian Ambassador Francesco Lequio on 8 January that 'If Spain had obtained from Germany the wheat necessary not for building up reserves but for her daily sustenance Spain would already be in the war at the side of the Axis. Unfortunately, that has not happened and the Spanish Government must contend with the odious blackmail of England and the United States. Tell him [Mussolini] that, despite all obstacles, Spain is seriously preparing in the military sphere to be ready for future trials.'[59]

On 9 January, an impatient Hitler decided that another effort must be made to bring Spain into the war. Stohrer was recalled to Berlin for instructions. In his absence, Gamero del Castillo, the Falangist Minister without Portfolio, spoke with Josef Hans Lazar, the sinister and powerful Press Attaché at the German Embassy. Gamero told him that a struggle was taking place between Franco and Serrano Suñer for control of the government. Franco, faced by the opposition of many senior generals and the Church, was reluctant to form an entirely pro-Nazi cabinet. Gamero and Serrano Suñer believed that 'an activist, homogeneous Serrano Suñer government should be formed as soon as possible'. Gamero wanted the Germans to intervene by letting it be known that the Third Reich wanted Serrano Suñer in charge.[60]

There is no doubt that by this time Hitler was totally disenchanted

---

* The complaints of the generals were reflected in efforts by the Falangist propaganda apparatus to deny any '*incomunicación*' between the Army and the Falange – *Arriba*, 10 December 1940.

with Franco. On 19 January 1941, he saw Mussolini and Ciano at the Berghof and roundly criticized the Caudillo for his failure to understand the fact that Axis control of Gibraltar and German bases in Spanish Morocco would entirely eliminate the problem of de Gaulle. He denounced Franco as 'only an average officer, who, because of the accident of circumstance, had been pushed into the position of Chief of State. He was not a sovereign but a subaltern in temperament.' Ribbentrop said that Stohrer had been instructed to make one last attempt to change Franco's mind. Hitler was not optimistic but asked the Duce to intervene with Franco. Mussolini doubtfully agreed to do so.[61]

On returning to Madrid from Salzburg, Stohrer had a long interview with Franco on Monday 20 January and put to him 'with ruthless candour' the views of Hitler and Ribbentrop. He laid special stress on the Führer's disappointment at Franco's reply to Canaris on 7 December and informed him that a Spanish entry into the war after the defeat of England was of no use to Germany. Stohrer said that Franco's attitude implied that he was no longer entirely convinced that the Reich would win the war. With the war practically won, 'for Spain, the historical hour had now struck'. He told him that Ribbentrop wanted to give him only forty-eight hours to decide since the Spanish situation was deteriorating and other dispositions had to be made for the troops currently in readiness to go into Spain.

Franco expressed astonishment at this pessimistic assessment and denied that the famine put his regime into jeopardy. He declared that his policy was unchanged, that 'his faith in the victory of Germany was also still the same'. Franco attributed his failure to enter the war as promised at Hendaye to the unexpected deterioration in the food situation when accurate estimates for the harvest had materialized in November. The scale of the deficit was such that to start a war would have been 'criminal'. It was the opinion of his military advisers that the diversion for military purposes of fuel and transport would only exacerbate the food crisis. However, Franco insisted that 'it was not a question at all of whether Spain would enter the war; that had been decided at Hendaye. It was merely a question of when.' He denied vehemently telling Canaris that Spain would enter the war only when England was already defeated: 'Spain intended to participate in the war fully and not obtain anything as a gift'. He claimed to have said rather that Spain intended to enter only when she would not be a burden on her allies.* Franco went on to

---

* That coincided with General Vigón's notes of the conversation, although Canaris's interpretation was hardly unreasonable.

insist that German help would be useless if it did not come until Spain had already entered the war. To the Caudillo's request for what Stohrer called 'pre-payment', the German Ambassador, having been previously authorized by Ribbentrop, stated that it could be considered if Franco gave assurances that Spain would enter the war at a time to be determined by the Reich. Franco requested time to consider the offer.[62]

That Franco's account of Spain's situation reflected genuine, albeit misplaced confidence, may be deduced from the astonishing fact that while Europe was being destroyed by war and Spain by starvation, the Generalísimo was writing a work of fiction. Entitled *Raza* (Race), a romanticized account of a Spanish family nearly identical to Franco's own, it was written in the form of a film-script although published as a novel. The plot relates the experiences of a Galician family, totally identifiable with Franco's own, from Spain's imperial collapse in 1898 to the Civil War. The pivotal character in the book is the mother figure Doña Isabel de Andrade. Alone, with three sons and a daughter to bring up, like Franco's mother Pilar Bahamonde, the pious Doña Isabel is a gentle yet strong figure. Pilar was abandoned by Francisco's dissolute, gambling, philandering father. In contrast, in the novel, the hero's father is a naval hero and Doña Isabel is widowed when he is killed in the Cuban war.

Raza was produced some time in the last months of 1940 and the first months of 1941. Franco dictated it, pacing up and down in his office. The text was then passed to the journalists Manuel Aznar and Manuel Halcón for the style to be corrected. Asked how he could spare the time in such tense moments to write fiction, Franco replied that it was merely a question of time-management, and that working to a timetable made everything possible. The fact of being able to write at all, almost as much as the intensely romantic style of the book, indicates the extent to which Franco was isolated from the real conditions of Spain at this time.

In the plenitude of his political power, Franco wrote a book in which he created a past worthy of the providential Caudillo. It was as if the fulfilment of many of his ambitions had made his past the more unacceptable. The novel exchanges the modest reality of Franco's family for the status of minor aristocracy, *hidalgos*. Similarly, the choice of pseudonym under which it was published, Jaime de Andrade, an ancient and noble family to which he was distantly connected through both of his parents, leaves little doubt about his social aspirations. *Raza* constitutes a revealing insight into Franco's egotistical drive to greatness. Not only does he romanticize his own parentage, childhood and social origins through the hero, José Churruca, but also manages to work in a reference to himself in all his own real glory as the all-seeing Caudillo.

The choice of title reflected Franco's current infatuation with Nazism. The internal logic of the title was that the hero and his family are considered to carry the essence of all that is valuable in the Spanish 'race' and so are able to save Spain from the foreign poisons of liberalism, freemasonry, socialism and Communism. This is what Franco considered himself to have done through the Civil War and through the subsequent relentless eradication of leftists. It is not difficult to see a link between Franco's fabrication of his own life and his dictatorial remodelling of the life of Spain between 1936 and 1975. In the book, he created the ideal family and father he did not have; in political life, he would rule over Spain as if he were the authoritarian father of a tightly united family. Since the early days of the Civil War, Franco had come to identify himself totally with Spain, or at least his version of Spain, which says more about his neurosis than his patriotism. Shortly after completion of the script, arrangements were made for it to be filmed, financed by the recently created *Consejo de la Hispanidad* and directed by José Luis Saenz de Heredia. With the resources of the State at his disposal, Saenz de Heredia was able to hire one of the leading romantic film actors of the day, Alfredo Mayo, to play the character based on Franco himself.[63] At the first private showing of the completed film, Franco cried profusely.[64] Over the next thirty years, he watched *Raza* many times.[65]

Raza's mood of belligerent imperialism is sufficiently near in tone to some of the things that Franco was saying at the time as to suggest that he was sincere in his claim that Spain would soon join Hitler in the war. Promises were not enough for Ribbentrop. He responded to Stohrer's account of his interview with Franco on 20 January with a blunt message which was to be read verbatim to the Caudillo. It stated harshly that 'Without the help of the Führer and the Duce there would not today be any Nationalist Spain or any Caudillo.' The Nazi Foreign Minister dismissed England's capacity to help Spain and emphasized that only Germany could provide effective aid. Asserting that the war was already won, he stated grudgingly 'The closing of the Mediterranean by the capture of Gibraltar would contribute toward an early end of the war and also open up for Spain the road to Africa with its possibilities. For the Axis, however, this action would be of strategic value only if it can be carried out in the next few weeks. Otherwise it will definitely be too late for it because of other military operations.' After condemning Spain's 'equivocal and vacillating attitude', Ribbentrop ended with an ultimatum:

'Unless the Caudillo decides immediately to join the war of the Axis Powers, the Reich Government cannot but foresee the end of Nationalist Spain.' Stohrer requested an appointment with Franco while trying vainly to persuade Ribbentrop to tone down the bullying tone of the message.[66]

The message was delivered on Thursday 23 January 1940. Franco 'very heatedly asserted that he had never vacillated and that his position was unswervingly on the Axis side, from gratitude and as a man of honour. He had never lost sight of entry into the war.' He made a lengthy, and apparently sincere, justification of his delay in terms of Spain's economic problems and, deeply stung by Ribbentrop's accusation, insisted that he had not deviated 'one millimetre from his Germanophile course' nor made any political concessions to the western Allies. Stohrer tried with difficulty to steer the conversation back to the central issue and finally managed to state that, subject to the well known requirement of an undertaking to enter the war at a date to be fixed by Germany, advance consignments of material were ready to be delivered. Franco said that he had instructed his 'Defence Council' to study the question. Serrano Suñer insisted that Germany must take responsibility for the fact that Spain was still not ready to fight by failing to send aid. Frustrated by his 'many digressions into details and non-essentials', Stohrer found the Caudillo irresolute in contrast to the apparently more decisive Serrano Suñer, but when he insisted on an early decision, Franco agreed.[67]

Ribbentrop replied to Stohrer's account of the 23 January 1941 meeting with an instruction for the Ambassador to read out to Franco an even stiffer message than the previous one. Since Franco was apparently unavailable until Monday 27 January, the message was delivered to Serrano Suñer. It stated that 'Only Spain's immediate entry into the war is of strategic value to the Axis and only by such a prompt entry can General Franco still render the Axis a useful service in return'. Ribbentrop insisted that Spain entrust to the Axis the determination of the date of her belligerence and demanded a 'final, clear answer'.[68]

The official Spanish reply was handed to Stohrer by Serrano Suñer on Saturday 25 January. Its text, which showed Franco's hand in many details, adopted a conciliatory tone, asserting that Germany misunderstood if it was thought that Spanish policy had changed. Spanish negotiations with England and the United States were presented as the fruit of necessity and a device to facilitate Spain's war plans. Part of the need to do so derived from the slow pace with which Germany examined Spain's economic difficulties. The note continued 'the date of our entry into the war is conditioned by very clear-cut and highly specific concrete requirements, which are not clumsy pretexts for delaying entry into the war until the moment when the fruits of a victory won by others can be garnered . . . Spain wishes to contribute materially to the victory, to enter the war, and to emerge from it with honours.'[69]

When he finally managed to see the Caudillo, in the presence of Serrano Suñer, on Monday 27 January, Stohrer said that the message was 'thoroughly unsatisfactory' since it contained no reference to a possible date for the war. Franco replied at inordinate length with a recital of the scale of Spain's economic difficulties. According to Stohrer, 'the only noteworthy item in this recital was that the Generalissimo did emphasize much more strongly than hitherto that Spain would undoubtedly enter the war'. He also requested the despatch of German economic experts to pronounce on Spain's needs and a military expert, perhaps Field Marshal Keitel, to do the same. Stohrer was convinced that Franco's feelings were genuinely injured by German lack of faith in his statements.[70] On the same day, Serrano Suñer made identical points to the Italian Ambassador.[71]

Both the German and Italian Ambassadors had seen enough of conditions in Spain to be immensely sympathetic to the plight of Serrano Suñer and Franco. Ribbentrop, however, was furious at Stohrer's understanding attitude and accused him of letting Franco divert him from his purpose. He demanded that Stohrer state precisely whether Franco had rejected Germany's desire for an immediate entry by Spain into the war.[72] Stohrer replied that whether Franco's stand constituted a 'clear and final rejection' of Germany's request depended on what was the latest date envisaged by the Reich for Spain's entry.[73] It is ironic that just as the Axis was despairing of Franco joining in the war, the Allies were growing more worried that the food crisis would drive him to do so. Barely ten days after Ribbentrop's 'final demand', Weddell expressed the fear that, without Allied food aid, 'Spain might under the stress of hunger engage in a mad African adventure hoping at one and the same time to secure food from French Morocco and also to gratify a territorial ambition'.[74]

On 5 February 1941, Hitler wrote to Mussolini, lamenting how a great opportunity to seal off the western end of the Mediterranean had been lost by Franco's lack of resolution. Faced with excuses from the Caudillo which meant a delay in Spanish belligerence until the autumn or the winter, the Führer again asked the Duce to try to persuade Franco to change his mind.[75] In fact, with the economic situation in Spain deteriorating daily, there was little possibility of that happening. German consuls were reporting that there was no bread at all in part of the country and there were cases of highway robbery and banditry. The Ambassador reported on dissent in military barracks and on political bitterness arising from 'the continued detention of 1–2 million Reds from the Civil War, who are poorly fed and whose families are starving'. The Caudillo received reports in similar vein from the Dirección General de Seguridad but, according to Serrano Suñer, he was indifferent to public feeling. The military was intensifying its pressure on Franco to get rid of Serrano Suñer, a prospect which Stohrer welcomed since he believed that the cabinet's inner divisions lay behind Franco's indecisiveness and so prevented Spanish entry into the war.[76]

Hitler himself made a further effort in a letter to Franco dated 6 February 1941. After reiterating the reasons why Spain should be linking arms with Germany and Italy, the Führer went on politely to demolish Franco's excuses for delay. There were no threats in the letter but rather a general invitation to join in an ideological war and an offer of supplies only after Spain declared war. In this regard, he commented in a way which could hardly entice Franco 'We are fighting a battle of life and death and cannot at this time make any gifts.' The Führer also compared Spain's 'very great territorial claims' in Africa with the 'very modest claims' made by Germany and Italy despite 'the most prodigious blood sacrifice'. It was as if Hitler had already decided that the Caudillo was a lost cause. When Franco received the letter on 8 February, he told Stohrer that he agreed with everything the Führer had said and undertook to reply on his return from a rapidly arranged visit to Italy. On the same morning, Franco had heard the news of the final annihilation of Marshal Graziani's army by the British at Bengazi.[77]

In the meanwhile, on 7 February 1941, Stohrer finally received a memorandum from the Spanish General Staff concerning the import requirements to remedy Spain's deficient military situation. General requirements for the country as a whole included fertilizers, gasoline, one million tons of grain, cotton, rubber, jute and other raw materials. The Army requested 3,750 tons of copper, radio, telephone and telegraph equipment, pharmaceutical and medical supplies, 90 four-gun batteries, 400 anti-aircraft guns, 8,000 trucks and considerable numbers of tractors. The Air Force requested Heinkel He-111 and Messerschmitt Bf-109 spares and material for the equipment of three squadrons of hydroplanes. The Navy requested torpedoes, mines, machine guns, depth charges, optical and other technical equipment all in considerable quantities. For the general transport requirements of the country, 180 locomotives and

16,000 railway wagons or 48,000 trucks were listed. Over and above military requirements, a further 13,000–15,000 trucks were requested. When the list was examined in Berlin, the Director of the Economic Policy Department concluded that the requests were 'so obviously unrealizable that they can only be evaluated as an expression of the effort to avoid entering the war under this pretext'. In his view, the list contained what was necessary to give Spain full economic and military striking power for a war – something that could only be achieved at the cost of serious damage to German interests.[78]

Mussolini having agreed to intercede with the Caudillo, a meeting between the two was arranged by Ciano for 12 and 13 February at Bordighera.[79] Franco dragged his feet about accepting the invitation and, with the more sycophantic of his ministers expressing horror that their Caudillo should travel at a time of international tension, eventually went convinced that he was doing the Duce a favour. After the military reverses suffered by Italy in the Balkans and in North Africa, Mussolini needed a propaganda event and Franco considered that Bordighera was it.[80] Mussolini had a rather different view. He wrote to the King three days before he was due to meet the Caudillo, 'I regard my journey to Bordighera as perfectly useless and I would willingly have avoided it. Franco will not say to me anything different from what he has already told the Führer'.[81] By the time that Franco met Mussolini, public opinion in Spain firmly opposed any intervention in the war. The Italian rout in Cyrenaica by a much smaller British force and the British naval bombardment of Genoa on 8 February caused some malicious anti-Italian merriment in Spain and even influenced the Caudillo himself.[82]

After arranging to make the journey overland through France, Franco and Serrano Suñer set off for Bordighera in a convoy of seventeen motor cars on the evening of 10 February.[83] When they met two days later, the senior official in the Italian delegation, Luca Pietromarchi, found Franco 'a chatterbox, disordered in his exposition, getting lost in minor details and giving free rein to long digressions on military matters' in contrast to the clarity of Serrano Suñer.[84] The Caudillo told a depressed and much-aged Mussolini of his faith in Axis victory and admitted candidly 'Spain wishes to enter the war; her fear is to enter too late'. He blamed his delays on German failure to deliver supplies and on Hitler's readiness, displayed at Hendaye, to court Vichy at the expense of his own imperial ambitions. Mussolini asked Franco if he would declare war if given sufficient supplies and binding promises about his colonial ambitions. Franco replied that Spain's military unpreparedness and famine conditions would still mean several months before she could join in the war but underlined that 'Spanish entry into the war depends on Germany more than on Spain herself, the sooner Germany sends help, the sooner Spain will make her contribution to the Fascist world cause'.[85]

Mussolini was inclined to stop trying to persuade Franco to join the Axis war effort in the short term. He said to his own staff 'how can you push into a war a nation with bread reserves for one day?' Instead, he told the Germans that the best policy with regard to the hesitant Caudillo should be simply to try to keep him within the Axis political sphere.[86] The five-hour meeting at Bordighera was shrouded in secrecy and the Spanish press merely hailed the consolidation of Italian-Spanish fraternity.[87] In the after-glow of meeting the Duce, Franco declared him to be 'the greatest political figure in the world. Whereas Hitler is a mystic, a diviner and very close to the mentality of the Slavs, Mussolini in contrast is human, clear in his ideas, never far from reality, in a word, "a true Latin genius".'[88] The Duce informed Hitler of Franco's desires at about the same time as the German Department of Economic Planning was reporting that Spanish demands could not be met without endangering the Reich's military capacity. Ribbentrop took the Bordighera meeting as signifying Franco's definitive refusal to join the war effort and therefore instructed Stohrer to take no further steps to secure Spanish belligerence.[89] When finally Hitler briefly contemplated forcing the issue, he had already committed his military machine to rescuing Italy from its disastrous involvement in the Balkans.[90]

En route back to Spain from Bordighera, Franco and Serrano Suñer had an inconclusive meeting at Montpellier with Pétain and his new prime minister, Admiral François Darlan.* Given Franco's determination to get control of as much as possible of France's North African empire, there was little basis for agreement with Pétain. The French received Franco with great pomp and ceremony and a lavish show of fine wines which reflected their hope of finding out if the Bordighera meeting threatened their North African territories. After the formal dinner, Franco and Pétain withdrew for private talks. Afterwards they appeared on the balcony of the Montpelleir prefecture, Franco raising his arms in the fascist greeting while Pétain confined himself to the military salute.[91] Serrano Suñer claimed plausibly that they had talked about 'the convenience of not irritating the Germans'. Pétain's *Chef de cabinet*, Henri Du Moulin

---

* This was Darlan's first official act since, on 10 February 1941, effectively becoming Pétain's heir apparent, taking over the Ministries of Foreign Affairs, Interior and Information.

de Labarthète recounted that the Marshal told him: 'He wanted me to support him with Hitler to prevent the passage of German troops through Spain. A funny thing to come to ask and something which I cannot decently undertake. But it's a pity. It would be in our interests.' Asked if Franco had changed since he last saw him, Pétain replied 'No, still the same, just as inflated, just as pretentious' (*Non, toujours le même: aussi gonflé, aussi prétentieux*).[92]

The Caudillo knew that Hitler's preference for Vichy had undermined his own ambitions at Hendaye. Given his unabated desire to gain control of French Morocco, it is likely that, if Franco did ask Pétain for help against the Germans, it was with a view to allaying Vichy suspicions.[93] Franco was playing a complex game with the French. He had realized that he could not expect to get French Morocco from the Germans without having to pay a high price. Accordingly, he had gone back to trying to bully the French into making concessions by playing on his own military strength. Throughout 1941 and 1942, virulent press campaigns and Falangist wall-daubings in favour of Spain's rights in French North Africa were accompanied by pressure on Ambassador Piétri, which, on the insistence of General Noguès, was ignored.[94]

Franco finally replied to Hitler's three-week-old letter on 26 February. With Yugoslavia and Greece falling to General Rundstedt and Rommel stiffening the Axis forces in North Africa, Franco was in a mood to open the bidding again but his price, if anything, had gone up. In addition to the recently submitted account of the material aid needed from Germany, Franco returned to his maximum territorial demands and also made it clear that Spanish entry into the war now required a prior closure of the Suez Canal by Axis forces, a strategic appreciation in fact shared by Admiral Raeder and Hitler's naval staff.[95] In fact, the letter reveals a degree of real enthusiasm for the Axis cause and a real anxiety for Spain to derive imperial profit from Hitler's successes, for which, in the Caudillo's opinion, Spain had paid in advance during the Civil War. There was an element of reproach that German supplies had not materialized while people were starving in Spain.[96]

Even before he had received the letter, Hitler wrote to Mussolini on 28 February 1941 'the gist of the long Spanish speeches and written explanations is that Spain does not want to enter the war and will not do so either. This is very regrettable since this eliminates, for the time being, the simplest possibility of striking at England at her Mediterranean position.'[97] The letter priced Spanish belligerence out of the market and Hitler described it to Ciano as 'a renunciation of the Hendaye agree-

ments'.[98] It is curious that Hitler accepted Franco's rebuff so calmly.*
Churchill speculated that 'Hitler was scandalised, but, being now set upon
the invasion of Russia, he did not perhaps like the idea of trying Napoleon's other unsuccessful enterprise, the invasion of Spain, at the same
time'.[99]

Franco was acting in the confidence that Hitler would not invade
Spain. Despite the later fabrications about German threats, there is no
evidence that, at the time, the self-opinionated and complacent Franco
feared that his honest excuses might provoke his friend Hitler, who owed
him so much, to contemplate invading Spain. In retrospect, for Churchill,
Franco's 'exasperating delay and exorbitant demands' were devices, the
'subtlety and trickery' by which he kept Spain out of the war.[100] Writing
in the late 1940s, Churchill had come to think more highly of the anti-
Communist Franco than he had done in 1941. He was certainly forgetting
the immense part played by British control of food and fuel supplies in
twisting Franco's arm. In any case, in early 1941, the urgency of seizing
Gibraltar was less for Hitler than it had been. German aircraft flying
from Libya were able to inflict substantial losses on British shipping
passing through the Sicilian Narrows. The value of Gibraltar was severely
diminished and the long route around the Cape of Good Hope had to
be used to supply Egypt and India.[101]

If Franco's attitude to Hitler had changed, it was not an ideological
conversion. He was not now skilfully fending off an enemy. His admiration for the Third Reich was undimmed but he had to face facts; the
war would be protracted, Spain's economic and military position would
not sustain a long war effort and the grudging price offered by Hitler
did not make the risks worth contemplating. Nevertheless, Franco's
enthusiasm for the Axis cause would never entirely die and would still
flare up dangerously from time to time.

---

* In mid-September 1940, Göring had brutally told Serrano Suñer that he believed that
the Führer should invade Spain. The Reichsmarschall would go on believing that the
failure to do so was Hitler's greatest blunder. In the summer of 1945, Göring told British
diplomats that the Führer was so infuriated that he was determined to show Franco that
he could go ahead without needing Spanish help of any kind – Creswell to Bowker, 25
July 1945, FO371/49550 XC/A/45932; Ivone Kirkpatrick, *The Inner Circle: Memoirs*
(London, 1959) pp. 193–5; Saña, *Franquismo*, p. 177.

XVII

TOWARDS A NEW CRUSADE

February 1941–January 1942

IF FRANCO could continue to believe complacently that he was still the valued friend of the Third Reich despite the economic difficulties which prevented him going to war, it was not an illusion shared in Berlin. Goebbels wrote bitterly in his diary that 'Franco is no more than a jumped-up sergeant-major'.[1] The changed tone of Hispano-German relations was marked at the end of February by German insistence on the repayment of Spain's Civil War debts which were agreed at 372 million Reichsmarks.[2] In the spring, Hitler said to Goebbels that if Gibraltar had been taken then Germany would now control Suez and the Middle East and England would be finished. The remark reflected Hitler's exasperation at the distance between Franco's pompous conviction of his value to the Axis and his failure to do the one thing requested of him.[3]

Hitler's impotent frustration* does not suggest that he was in a position to consider threats. In the course of the spring of 1941, the Germans became convinced that Britain would attempt a landing in Spain. Hitler told General Espinosa de los Monteros, the Spanish Ambassador, of his fears in this regard on 28 April.[4] The German High Command even drew up detailed contingency plans, code-named Operation Isabella, for invading Spain to repel a British attack and to attack Gibraltar.[5] After the war, much would be made of such plans by Franco's propagandists in order to give substance to the retrospective picture of his resistance

---

* 'The Führer reserves his harshest judgements for Franco and his lack of intelligence and courage. Even after hours of talk he had been unable to force him to an audacious decision. A clown! Conceited, arrogant and stupid. And that Serrano Suñer of his nothing but a Jesuit. Franco only rose to power on our backs. And that sort of thing never lasts. One must win power by one's own strength.' *The Goebbels Diaries 1939–1941* (London, 1982) p. 356.

to German pressure but these plans were not directed against Franco.

The Allies had little doubt where Franco's sympathies lay. In an effort to intensify pressure for Franco to remain neutral, Roosevelt's special envoy, Colonel William 'Wild Bill' J. Donovan, visited Madrid in late February. His efforts to meet Franco were frustrated, with the Caudillo's connivance, by Serrano Suñer.[6] However, when Serrano Suñer received Donovan on 28 February, he told him that 'we hope for and believe in the victory of Germany in the present conflict' but stated that Spain would remain aloof until her 'honour or interests or dignity' were at stake.[7] On 17 March 1941, Spain solemnly returned to the Third Reich the German consulate building in Tangier. As the only outstanding part of the Treaty of Versailles not overturned, the gesture was greeted with great delight in Berlin. More practically, the Germans quickly established their Tangier consulate as a major espionage and propaganda base in North Africa.[8]

The intensification of Serrano Suñer's pro-Axis fervour reflected the growing internal tensions between Army and Falange. The senior generals, with Kindelán at their head, were determined to bring down the *cuñadísimo*. Their confidence in mid-February that they were on the verge of success probably accounts for a recrudescence of Serrano Suñer's hostility to Portugal.[9] In the second half of February it had been rumoured in Madrid that Franco planned to meet Salazar. Serrano Suñer told the Italian Ambassador that the purpose was to set up a smokescreen to prevent the British discovering what had been agreed at Bordighera. He expressed himself in vehement terms about the cowardice of the Portugese in general and of Salazar in particular.[10] On 26 February, *Arriba* published a savage attack on Portugal which was believed by Ambassador Pereira to be the handiwork of Serrano Suñer. The Portuguese press was banned in Spain. Salazar concluded that Serrano Suñer was rabble-rousing within the Falange to secure Axis support for his domestic position.[11] That may have been so but it cannot be taken as implying any divergence between Franco and his brother-in-law over policy. Nicolás Franco assured Pereira that his brother 'knows everything and approves everything that his Minister of Foreign Affairs does. It is useless to think that he hides anything from him or proceeds without his support.'[12]

Serrano Suñer told Stohrer that he wanted to defer Franco meeting Salazar lest he be persuaded not to enter the war.[13] The Portuguese were cognizant of Falangist propaganda directed against them but not perhaps the full scale of the imperialist designs on Portugal nurtured by Serrano Suñer, Franco and other Spanish military figures.[14] The German Embassy in Madrid reported Spanish officers talking in terms of the improvement in Axis fortunes that could be expected 'when our western frontier reaches the Atlantic' or 'when German squadrons can fly from Portuguese bases which will be in Spanish hands'. The idea of absorbing his neighbour remained one of Franco's imperial dreams. In May 1941, one of the Generalísimo's ADCs, Major Navarro, told the German air attaché Colonel Kramer that a war against Portugal would be a useful diversion from Spain's internal political tensions. General Aranda also told both Kramer and Stohrer that he had been ordered to draw up preliminary plans for an attack on Portugal.[15]

With the sympathetic approval of Washington, on 7 April 1941, Britain granted Spain credits of £2,500,000.[16] However, German successes in the spring of 1941 wiped away the impact of the gesture. On 19 April, Alexander Weddell had a sharp exchange with Serrano Suñer which was deliberately blown up into a dramatic feud between them. In consequence, he would find his access to Franco systematically blocked for the next five months. The virtual freezing of relations with the USA would suit the Caudillo admirably while the war was still going in favour of the Axis since it permitted him to ignore American pressure to maintain strict neutrality.[17]

When the tense meeting between Weddell and Serrano Suñer took place, the Foreign Minister had just returned from seeing Franco, 'and seemed depressed and irritable'. The friction between the Caudillo and his brother-in-law, which was to erupt in a major political crisis in May 1941, was already coming to a head. There were rumours that the Army was planning a coup against Franco in order to get rid of Serrano Suñer.[18] Ambassador Weddell complained about the German censors interfering with the Spanish mail and about the way in which the controlled press seemed to be preparing the population for Spain's accession to the German-Italian-Japanese Tripartite Pact. He let Serrano Suñer know that his fervently pro-Axis speech at the recent German Press Exhibition had been noted.* Weddell also suggested that recent anti-American material read as if it had been translated from some foreign language, perhaps German, and that ready-made articles were being sent to newspapers by Hans Lazar, the German press attaché.[19]

---

* At the exhibition, Serrano Suñer said that, as head of the Spanish press for the previous three years, his policy had been to express friendship towards Germany. He boasted that, because no other country's press served the interests of friendship with Germany so unswervingly, this policy had been crowned with the personal thanks of the Führer, *Arriba*, 13 March 1941.

The irritability in Serrano Suñer discerned by Weddell was a reflection of the fact that the hostility of the most senior generals was again reaching fever pitch. The Portuguese Ambassador described Serrano Suñer as 'the most hated man in Spain'. Pereira thought Franco's own position to be desperate, reporting that 'all the generals who fought in the war say openly that he has failed, lacking any of the necessary qualities for a Chief of State'.[20] For his part, Franco did not bother to inform the council of ministers of his thoughts on foreign policy and had permitted discussion neither of the Hendaye nor the Bordighera meetings.[21] The essentially Anglophile Aranda had even gone so far as to seek help from the German Embassy in the power struggle against the Foreign Minister, attempting to curry favour with Berlin by suggesting that the High Command now desired Spanish entry into the war by early July.[22] The interventions of his senior colleagues had some impact on the Caudillo and no doubt were reflected in the meeting with Serrano Suñer that had preceded the latter's encounter with Weddell.

The German victories in North Africa, Yugoslavia and Greece convinced Franco that his underlying faith in an Axis victory was not misplaced. This was reflected in a speech that he made on 17 April 1941 on opening the Escuela Superior del Ejército. In a particularly bellicose mood, he declared that peace was merely a preparation for war, and the latter the normal condition of humanity.[23] Goebbels commented bitterly in his diary 'Then let him create normal conditions and fight alongside us. He is a totally conceited loudmouth.'[24] Goebbels' reaction was an understandable response to Franco's self-satisfied belief that he was the valued and valuable friend of the Third Reich. There is nothing about it which suggests that Franco was skilfully deceiving the Germans.

Nevertheless, with the Battle of the Atlantic raging throughout the spring of 1941, the possibility of a German take-over of the Iberian Peninsula began to worry Churchill. His anxiety was not centred only on the problems which would derive from a German ability to wage naval and air warfare from the Spanish and Portuguese coasts. He was equally concerned about the potential knock-on effects if the Germans thereby gained access to the Cape Verde and Canary Islands and the Azores. He wrote to Roosevelt on 24 April 1941 of his fears that Spain and Portugal had little capacity to resist German pressure. He had plans to respond to any German action against Gibraltar by seizing one of the Azores and one of the Cape Verde Islands.[25] Believing that there was intensifying German pressure for Spain to join the Axis war effort, Weddell too was anxious to be able to offer Franco some assurance that economic aid would be forthcoming provided he did not adopt an unfriendly attitude to the Allies. In fact, the main evidence of German pressure on Franco came from remarks made to British and American diplomats by the Caudillo himself and his staff as a device to screw more aid out of the western Allies. Involved in this shamefaced deception, Franco characteristically avoided contact with the American Ambassador. He was in any case preoccupied with resolving the growing political crisis involving Serrano Suñer. Moreover, in view of Axis successes, the Caudillo was reluctant to get too involved with the USA. Accordingly, Franco was simply not available for audience during the spring.[26]

In April 1941, reports from the *Auslandorganisation* and the German Embassy in Madrid stressed both Spain's continually deteriorating economic situation and the widespread dissatisfaction with Franco's government. On 22 April, Stohrer attributed the economic disaster partly to the inconsistent policies being pursued by Franco's Commerce and Finance Ministers and also to the mismanagement and corruption in the Falange for which Serrano Suñer was saddled with the blame. In Stohrer's opinion, Serrano Suñer was the most energetic minister but was inhibited by the machinations of his enemies in the cabinet and by Franco's reluctance to give him a free hand. Franco was deemed to be isolated and indecisive, bogged down in minutiae and making detailed decisions which often contradicted his general policies.

On his isolation, Stohrer made a comment which would be echoed by observers of the Caudillo over the next thirty-four years: 'There is increasing criticism that Franco sees fewer and fewer people and does not allow himself to be advised even by old friends.' More specifically, Stohrer predicted the serious crisis which was about to erupt, remarking that Franco's indecisiveness had caused friction with Serrano Suñer yet, to the outrage of the senior generals, he hesitated to drop him because of his reliance on his 'keen mind'. Stohrer also noted to Berlin that the population believed that the famine was the consequence of Spanish food exports to Germany.* Nevertheless, he remained convinced that both the military and Serrano Suñer were anxious for Spain's entry into the

---

* There were many reasons for the famine, ranging from autarchic policies, an overvalued peseta, incompetence in the Ministry of Agriculture and a collapse of the food distribution system in part because of the fuel shortage. However, in several ways, the Third Reich was plundering the Spanish economy. Manufactured goods paid for by Spanish customers were simply not delivered, essential fertilizer imports from Germany virtually dried up, and, in 1941, Spanish food exports to Germany were worth 94,186,000 pesetas as against imports of 1,784,000 pesetas. The difference of 92,402,000 pesetas was worth approximately £68,000,000 at 1992 prices – Viñas *et al.*, *Política comercial exterior*, I, pp. 389–412.

war. It is more likely that the enthusiasm for belligerence displayed to the Ambassador by some of the senior generals was merely a ploy to get German support in the power struggle against the *cuñadísimo*.[27] The same was true of Serrano Suñer's position. That was the view of the embittered Beigbeder.[28] And of Goebbels who considered Serrano Suñer 'the real fly in the ointment'.[29]

The overriding importance for Franco of not provoking the outright opposition of the generals was behind his efforts to meet some of their complaints in the small-scale, but crucially important, power struggle which broke out in May 1941. It began at the end of April when Vigón informed Franco that, if Serrano Suñer's power were not curtailed, the military ministers would resign *en bloc*.[30] The outcome of the crisis was one of the first, indirect, fruits of a British policy of bribing important elements in the Spanish High Command instituted nine months earlier as a result of which $13,000,000 had been distributed through Aranda.[31]

The crisis itself would centre on the vacant Ministry of the Interior. Since 17 October 1940, when Serrano Suñer had replaced Beigbeder at the Ministry of Foreign Affairs, Franco himself had formally taken over the Ministry although in practice the time-consuming task of administering its machinery was carried out on a day-to-day basis by the Under-Secretary, José Lorente Sanz. This permitted Serrano Suñer to continue to have enormous influence in many of the Ministry's functions, including press and propaganda. Franco had kept a distant eye on what went on in the Ministry through Colonel Valentín Galarza Morante, the head of the office of the president of the council of ministers (*Subsecretario de la Presidencia del Gobierno*). Given that Franco perceived political authority in the same terms as military command, *el mando*, the undersecretary to the Presidency was equivalent to a chief of the general staff for political affairs.[32] Serrano Suñer also had considerable authority in the Falange both through his position as president of the *Junta Política* and through his influence over the acting Secretary-General, Pedro Gamero del Castillo. As long as he thought Serrano Suñer to be selflessly at his service, Franco had no objection to his accumulation of posts. However, his suspicions that his brother-in-law might be building his own power base began to grow in early 1941.

In January, a book was published suggesting that Serrano Suñer was the real heir of José Antonio Primo de Rivera.[33] It has been alleged that the Caudillo's suspicions were further intensified by the disingenuous dinner table question of his fifteen year-old daughter Carmen, 'Who is in charge here? Papa or Uncle Ramón?'[34] The ever-distrustful Caudillo was alerted further by a move which virtually created an independent fascist press at the service of the *cuñadísimo*. A decree of 1 May 1941 exempted the Falangist press from censorship other than that exercised by its own *Delegación Nacional de Prensa y Propaganda de FET y de las JONS*.[35] On the following day, Serrano Suñer made a violent speech at Mota del Cuervo, attacking England and calling for a tightly knit Falange to assume the monopoly of power.[36] Mussolini and Ciano were delighted with the speech and its call for 'all power to fascism'.[37] Then Serrano Suñer suggested to Franco that Falangist representation in the cabinet should be increased by creating a Ministry of Labour for the young Valladolid fanatic José Antonio Girón de Velasco. On 3 May, Franco received a letter from Miguel Primo de Rivera resigning from his posts in the Falange in protest at the weakness of various Falangist organizations.* The accumulated signs that Serrano Suñer might be trying to turn the Falange into a fully fledged Nazi Party for his own purposes finally persuaded Franco that the insinuations of the military were correct. He agreed to the promotion of Girón but also took other measures to counter the surge of Falangist power.[38]

In a context of the already mounting crescendo of anti-Serrano Suñer criticism from his High Command and of clever insinuations from Vigón that, by holding the Ministry of the Interior himself, he was both diminishing his own prestige and permitting Serrano Suñer to exploit him, Franco acted. On 5 May 1941, he made Galarza Minister of the Interior. Two days later, he replaced him as undersecretary to the Presidency with the thirty-six year-old Chief of Operations of the Naval General Staff, Captain Luis Carrero Blanco.† His elevation would turn out to be the most important result of the crisis.‡ At the time, however, it seemed a minor aspect of a series of events which constituted a decisive battle in the war between the Falange and the military High Command. The scale of the changes to come was far-reaching and made it clear that the Caudillo had decided to clip Serrano Suñer's wings. Kindelán was named Captain-General of Catalonia and his predecessor, Orgaz, was made High Commissioner in Morocco. What the fiercely anti-Falangist Galarza now

---

* Miguel Primo de Rivera to Franco, 1 May 1941, *Documentos inéditos*, II-2, pp. 141–4.
† Short, with bushy eyebrows, the drab and hard-working Carrero Blanco shared all of Franco's political prejudices and, unlike the independent Serrano Suñer, was utterly devoted to him in a quietly servile way.
‡ Franco had originally offered the job to Serrano Suñer's friend José Lorente Sanz, under-secretary at the Ministry of the Interior, who refused on the grounds that the job carried too much responsibility (Lorente Sanz to Franco, 5 May 1941, *Documentos inéditos*, II-2, pp. 145–6).

did would not have been possible without detailed prior agreement with Franco. That being the case, it was hardly surprising that the Caudillo's decision to appoint Galarza was taken without his brother-in-law's knowledge. Indeed, Serrano Suñer and the other members of the government were given no warning of Galarza's appointment until they saw him seated at the table in the cabinet meeting of 5 May.[39]

The intentions of Galarza could be deduced from his speech on taking over the Ministry, in which he referred to 'certain defects' in its operation, and by his immediate replacement of Serrano Suñer's man, Lorente Sanz, with the traditionalist lawyer from Bilbao, Antonio Iturmendi. The appointment of Iturmendi was almost as significant as that of Carrero. Like Galarza himself, a major figure of the monarchist camp, he had been *Alcalde* (Mayor) of Bilbao during the Civil War and in years to come he would be a Francoist stalwart. The devastating anti-Serrano Suñer sweep was crowned by the removal of his henchman José Finat, the Conde de Mayalde, as Director-General de Seguridad.[40] He was replaced by Lieutenant-Colonel Gerardo Caballero Olébazar. In addition, a number of Civil Governors were replaced including Miguel Primo de Rivera in Madrid. Galarza also rescinded the decree exempting the Falangist press from censorship.[41]

A group of top Falangists, including Serrano Suñer, Dionisio Ridruejo, José Antonio Girón, José Luis de Arrese, Antonio Tovar, and Miguel Primo de Rivera met and decided to fight back against this apparent triumph of the military camp. Although the battle appeared to be between the Falange and Galarza, it was really a trial of strength between Franco and Serrano Suñer. Ridruejo wrote an article ridiculing Galarza, entitled '*Puntos sobre los íes: el hombre y el currinche*' ('Dotting the i's: the man and the new boy').[42] Galarza counter-attacked on two fronts. He commissioned an article by Juan Pujol in the newspaper *Madrid* on 12 May ridiculing the notion that the Falange should have any say in foreign policy and thus attacking Serrano Suñer. He also proceeded to the dismissal from his Ministry of the Falangists in charge of Press and Propaganda, including Dionisio Ridruejo and Antonio Tovar. The remaining members of the anti-Galarza group decided to resign their various posts, Girón who had just become Minister of Labour on 5 May, Arrese as Civil Governor of Málaga and Serrano Suñer as Minister of Foreign Affairs.[43]

There were clashes between the police and Falangists and the hostility between the military and the Falange reached boiling point. There were fatalities after fighting in León. This badly weakened Serrano Suñer's

standing in Franco's eyes. The Falangist press responded by disingenuously denouncing British machinations which, it was alleged, had
forged anti-military propaganda, presumably a reference to Ridruejo's
article.[44] One of the *cuñadísimo*'s staff was heard to say 'the Germans will
have to come and sort this out'.[45] After an intervention by von Stohrer,
Press and Propaganda were transferred from the Ministry of the Interior
to a new Vice-Secretariat of Popular Education within the Falange.
Franco was also annoyed that Galarza had permitted the press to pay too
much attention to the popular acclaim enjoyed by Orgaz on his departure
from Barcelona for Morocco. Serrano Suñer was alleged by Vigón to
have said to Franco 'here you have the results of taking the press away
from me'. Control of the press was thus restored to Serrano Suñer.[46]
There was an almost instantaneous outburst of pro-German fervour in
the newspapers which reflected Serrano Suñer's awareness of his need
for support in the ongoing power struggle.[47]

Many of the most senior monarchists in the Army were outraged by
what they saw as the intolerable antics of Serrano Suñer's clique. In
mid-May, Beigbeder, to the alarm of his British contacts, was rashly
talking of making a *pronunciamiento* in Morocco at the end of the month.[48]
More shrewdly, General Antonio Barroso suggested to Franco that he
deal separately with Arrese, Girón and Miguel Primo de Rivera.[49] In the
event, the crisis was finally resolved by a series of cabinet changes which
irrevocably undermined Serrano Suñer's position. The *cuñadísimo* had
tendered his resignation in the belief that he was acting in unison with
his friends from the top ranks of the Falange. He had cited the 12 May
article as evidence that Galarza was waging war against him. On 13 May,
Franco replied to resignation with a calming and friendly letter* in typically convoluted style.[50]

Franco's conciliatory stance derived from his fear that, if he removed
Serrano Suñer altogether, he would be left as the prisoner of the monarchist generals. It is also possible that the relationship between Carmen
and Zita Polo played a part. Vigón believed that a weak Franco had been
manipulated by the sisters. It is more likely that Franco acted on the

---

* 'Dear Ramón, I have received your letter along with a press cutting which I cannot
interpret in the way that you do in your letter. Put it in the hands of a learned and
objective person and I doubt that they could give it the same interpretation as you do. I
have done precisely that with negative results. I want you to meditate on the injustice and
the baselessness of your action before taking a decision which does so much for our
enemies and which, in these moments of confusion, could cause damage to Spain. I wish
with all my heart that God might illuminate and calm you. I expect you tomorrow at 4
o'clock so that we can talk calmly about all of this. I embrace you. Paco'

basis of his own calculations, although avoiding family frictions would have been a bonus.[51]

What Serrano Suñer did not discover until later was that several of his 'friends' had already met privately with Franco and accepted offers of senior posts. He found out in time and withdrew his resignation. In the cabinet reshuffle of 19 May, two additional Falangist ministers were appointed, Miguel Primo de Rivera as Minister of Agriculture and José Luis de Arrese as Minister-Secretary of the *FET y de las JONS*, while Girón remained as Minister of Labour. Only Serrano Suñer's faithful friends lost their posts.* At the time, the increase of Falangist representation in the cabinet made it seem as though Serrano Suñer had triumphed.[52] However, in retrospect, the May 1941 crisis would be seen to have been the beginning of his downfall. Franco had opened the crisis because he had discovered that Serrano Suñer was more loyal to his own ambitions for the Falange than to the Caudillo personally. In the course of its resolution, the Generalísimo learned that the Falange could be bought cheaply.[53]

Arrese was one of the first to benefit from Franco's susceptibility to flattery. Having now wormed his way into the Caudillo's affections, Arrese became more ambitious and set out to supplant Serrano Suñer as Franco's closest collaborator. On one occasion, he said to Serrano Suñer that Franco was as jealous of him as an ugly fiancé, hoping thereby either to ingratiate himself with Serrano or else to tempt the *cuñadísimo* into making some remark in agreement with this assessment that could be reported back to the Caudillo. As a servile lackey and for his utility in domesticating the Falange, Franco liked Arrese. By relaying to the Generalísimo all kinds of tittle-tattle about the ambitions of Serrano Suñer, he was to contribute over the next eighteen months to bringing down the *cuñadísimo*.[54]

Accordingly, the elevation of Arrese and the other Falangists did not represent, as many thought at the time, a victory for Serrano Suñer but rather the consolidation of Franco's own power over a docile section of the Falange. In fact, Serrano Suñer had not only challenged the Caudillo but, through the failure of the threatened mass resignations, he had also inadvertently revealed that he did not control the Falange. The entire affair taught Franco that it was easy for him to assume that control himself simply by exploiting the ambitions of many of its senior figures.[55] In that fact,

---

* Ridruejo and Tovar in the Press and Propaganda section of the Ministry of the Interior, Llorente Sanz as undersecretary of the Ministry, Mayalde as Director-General de Seguridad and Gamero del Castillo as acting Minister-Secretary of the Falange.

together with the assumption by Carrero Blanco of the position of under-secretary to the Presidencia del Gobierno, lay the real significance of the crisis. Henceforth, Franco was to be more receptive to criticisms of his brother-in-law and Carrero, like Arrese, was only too glad to supply them.*

Carrero Blanco was to be faithful to the point of self-abnegation yet also a fount of useful advice. His attraction to Franco was that he had no higher ambition than to serve him and that his basic ideas and obsession with the dangers posed by freemasonry, Communism and Jews were much closer to Franco's unspoken assumptions than those of Serrano Suñer. As the man who drew up the agenda of cabinet meetings and who served as the central channel of information received by Franco, Carrero's influence was immense and tended to confirm the Caudillo in his existing prejudices.[56] In contrast, the fundamental weakness of Serrano Suñer's position was the fact that his intelligence and political radicalism could be construed as dangerous ambition. He now had three influential and implacable enemies working against him, Varela and the military, Arrese and the so-called Francofalangists and Carrero Blanco.[57] The crisis worsened his health and provoked a recrudescence of a gastric ulcer from which he suffered acutely.[58]

With his political star waning, Serrano Suñer gave a convincing impression that he was genuinely anxious to see Spain enter the war but Ribbentrop set virtually no store by his bellicosity.[59] Both he and Franco were totally convinced that a British surrender was imminent. In view of Axis successes, the Caudillo was reluctant to get too involved with the USA and so remained unavailable for audience. Serrano Suñer also cancelled three consecutive meetings with Sir Samuel Hoare who told Weddell 'Suñer is doing everything he can to provoke us.' Weddell was convinced that, after the British evacuation of Crete in the last week of May, Franco believed that Suez would soon be in Axis hands.[60] According to Beigbeder, Franco's hope was that the war would end before there was any German intervention in Spain, thereby permitting him to declare his belligerency just as armistice negotiations were starting.[61] In view of their snubs of Weddell, it was made unmistakably clear to both Serrano Suñer and the Caudillo that requests for American economic and food assistance stood little chance of success.[62]

---

* Shortly after Carrero's appointment, Serrano Suñer told him that it would be part of his responsibility to protect Franco from sycophants. The manipulative way in which Carrero reported this conversation to the Caudillo was reflected in Franco's remark to Serrano Suñer 'I know that you had a go at me when you spoke to Carrero.' Serrano Suñer was convinced that Carrero would be no more than servile – Saña, *Franquismo*, pp. 261–2.

Franco's excuses for refusing to see the British and American Ambassadors were devious. His messenger was the affably cynical Demetrio Carceller, the Industry Minister involved in economic negotiations with Britain and the USA, who was used increasingly by Franco to appease the Allies while he reserved Serrano Suñer for dealings with the Axis. Carceller gave Hoare two reasons, both highly dubious. The first was that Franco had no wish to intervene in a merely personal quarrel between his brother-in-law and Weddell. The second was that he did not wish to be seen by his Axis friends to be too close to the Allies, having just resisted great pressure from Mussolini to join the Tripartite Pact.[63] That was in fact a gross mis-statement of the Duce's attitude at Bordighera and the subsequent correspondence between Ciano and Serrano Suñer. It was another example of Franco exaggerating Axis pressures in order to deceive the Allies and squeeze supplies from them.

The Italian approach was in fact altogether softer than the German although it was no more efficacious. On 11 June 1941, Serrano Suñer informed the German Ambassador formally that Ciano had written to him in the wake of the lengthy Führer-Duce conversations at the Brenner on 2 June, suggesting that Spain accede publicly to the Tripartite Pact. Franco, typically, had instructed Serrano Suñer to reply with a recital of the advantages and disadvantages of doing so. Although, equally typically, the reply did not draw conclusions, it was obvious that the possibility of accelerating a US entry into the war and the loss of grain and oil shipments currently en route to Spain meant that Spanish activation of the Pact signed after Hendaye would benefit neither the Axis nor Spain at this point. Ciano showed the letters to Ribbentrop on 15 June in Venice. The German minister remarked resignedly that the Spanish Foreign Minister's reasons had not changed in six months and that the Spaniards should be left liberty of action.[64]

Inhibited by his economic difficulties, Franco's belief in the ultimate victory of the Axis was inflamed anew by the Nazi invasion of the Soviet Union on Sunday 22 June 1941. Ironically, on the day before the invasion, in a long letter to Mussolini, Hitler commented 'Spain is irresolute and – I am afraid – will take sides only when the outcome of the war is decided.'[65] On being officially informed of the German attack on Russia, Serrano Suñer expressed great enthusiasm and informed Stohrer that he and Franco wished to send volunteer units of Falangists to fight,*

---

* This would take the form of the 18,000-strong expeditionary force known as the 'Blue Division' (*División Azul*) because of the blue shirts of the Falange.

'independently of the full and complete entry of Spain into the war beside the Axis, which would take place at the appropriate moment'.[66]

On the specific instructions of Serrano Suñer, the captive press rejoiced. The British Embassy was stormed by Falangists on 24 June, after Serrano Suñer had harangued them at the Falange headquarters in the Calle Alcalá. In the wake of his announcement that 'Russia is to blame for the Spanish Civil War' and 'History and the future of Spain require the extermination of Russia', the crowd moved on to the British Embassy where a truck-load of stones had been thoughtfully provided by the authorities. Efforts to break into the Embassy were fought off by the British guards backed by sixteen escaped British POWs. A German film crew was on hand to record the entire proceedings. When Hoare protested to Franco, the Caudillo dismissed the incident as a trivial matter concerning 'young hotheads'. It was announced that thousands of telegrams had been received from Falangists begging to be allowed to go and avenge Russian intervention in the Spanish Civil War. Arrese began a recruiting campaign within the Falange for volunteers to fight in Russia.* It was alleged that they were worried that they might not arrive in time, such was Spanish confidence in German military might.[67]

On the day after the riot, Serrano Suñer carefully parried Stohrer's request for a Spanish declaration of war against the Soviet Union by claiming that it would provoke an Allied blockade.[68] Nevertheless, Stohrer wrote to Berlin on 28 June 'The moves of Serrano Suñer in the last few days show even more clearly than hitherto that he is with clear aim preparing Spain's entrance into the war.' Serrano had told von Stohrer that the sending of the Blue Division was a ploy to commit Franco ('the easily influenced Chief of State') to the Axis. With an eye on his own part in the internal power struggle, he described military hostility to the sending of Falangist volunteers as reflecting the wider opposition of the generals to Spanish belligerence.[69]

Three days later, Spain moved towards the Axis when Serrano Suñer announced in a widely reproduced interview with *Die Deutsche Allgemeine Zeitung* on 2 July that non-belligerency would be replaced by what he described as 'moral belligerency'. Preparations were stepped up for the despatch of the Blue Division of Falangist volunteers. Serrano Suñer declared that 'there has been a surge of unrestrainable sympathy and

* Mussolini was upset at the announcement of the Blue Division, partly out of pique that he had not been consulted. Not wanting Franco to draw too near to Hitler without his own mediation, he would have liked to prevent the despatch of the Spanish volunteers to Russia – Ciano, *Diary 39–45*, p. 363.

admiration for the great German people, for its invincible army and its glorious Führer'.* He predicted that, after Russia's inevitable defeat, Britain would be forced to accept a dictated peace.[70] In a studied insult of the United States, which could hardly have taken place without Franco's assent, all the invited Spanish officials stayed away from Ambassador Weddell's summer reception.[71] The extent to which Serrano Suñer's star was now once again in the ascendent was reflected in the fact that the Spanish Ambassador to Germany, General Espinosa de los Monteros, was removed in late July and replaced by the *cuñadísimo*'s friend, the Conde de Mayalde. There was intense ill-feeling between Espinosa, a monarchist who, like many generals, was hostile to the *cuñadísimo*, and Serrano Suñer who wanted someone reliable in Berlin to put his case.[72]

In addition to the volunteer combatants, an agreement was made on 21 August 1941 between the *Deutsche Arbeitsfront* (German Labour Front) and the *Delegación Nacional de Sindicatos* (Falange union organization) for 100,000 Spanish workers to be sent to Germany. Trucks bedecked with posters and equipped with loud-speakers toured areas of high unemployment exhorting the unemployed to enlist. Theoretically 'volunteers', they were more often levies chosen by the Falange to fit the specific needs of Germany's industries.

In the event, the sending of the Blue Division was not a prelude to a declaration of war on Britain. Indeed, when Ribbentrop thanked Franco for the gesture and invited such a declaration, Franco refused on the entirely plausible grounds that his regime could not survive a full-scale Allied blockade. Nevertheless, he was keeping an iron in the fire, showing sufficient commitment to the Axis cause to have a say in the future division of the spoils. There can be no doubt of the sincerity and enthusiasm with which Franco participated in the anti-Communist struggle. The device of sending volunteers had two advantages. On the one hand, it meant that the feeding and arming of the Blue Division would be the responsibility of the Germans. On the other, volunteers were always formally deniable although the Allies were fully aware of the official patronage of the expedition. It was a high-risk strategy based on the confident assumption of eventual German victory. Franco was sailing near the wind. In mid-July it was revealed that the British knew that German submarines were being refuelled in the Canary Islands.[73]

---

* It is impossible to be precise about public opinion at this time. There was considerable ill-feeling against the Soviet Union among Falangists and Catholics but the hungry masses had no interest whatsoever in going to war. The overall balance of opinion seems to have favoured the Allies.

That the Caudillo believed the risks to be negligible may be deduced from his frequent assertions that the Allies had lost the war. When Sir Samuel Hoare remonstrated with him about the Blue Division, he replied that there were now two wars and that Spain could participate in a crusade against Russia without going to war with the western Allies.[74] On the fifth anniversary of the outbreak of the Spanish Civil War, 17 July 1941, Franco addressed the *Consejo Nacional* of the Falange.[75] Resplendent in the white summer uniform of *Jefe Nacional* of the Falange, the Caudillo was carried away by his enthusiasm for Hitler's Russian venture. Emulating Hitler's oratorical style, shouting and making aggressive gestures, he deliberately set out to offend the democracies.[76] He claimed mendaciously that the USA was holding back grain already purchased by Spain and stated provocatively that American offers of economic aid were a mask for political pressure 'incompatible with our sovereignty and with our dignity as a free people'. He denounced the fact that the United States was starting to put its economic power behind the British war effort. 'Gold ends by debasing nations as well as individuals. The exchange of fifty old destroyers for various remnants of an empire is eloquent in this regard.'*

Before an increasingly excited audience, Franco gave free rein to his pro-Hitlerian rhetoric. Linking the fate of Spain to the outcome of the war, he declared that he harboured no doubt about the result of the conflict. 'The die is cast. The first battles were fought and won on our battlefields.' He then gave a summary of the World War as an uninterrupted sequence of Axis triumphs. His enthusiasm could not have been more effervescent if the successes had been his own. After referring to 'the victorious campaign in Greece', he went on: 'The coasts of Norway, the waters of the Channel and the seas around Crete have been the theatre in which the Air Force has thrown back the enemy fleets.' He spoke of his contempt for 'plutocratic democracies', of his conviction that Germany had already won the war and that American intervention would be a 'criminal madness leading only to useless prolongation of the conflict

* That insulting reference to the arrangements made the previous autumn, whereby the United States had given fifty destroyers in return for bases in the British West Indies, provoked the ostentatious departure of both Hoare and Weddell (*FRUS 1941*, II, 908–11; Serrano Suñer, *Memorias*, pp. 348–9). The reference to the aged destroyers suggests some prompting from the Germans, since the fact that eighty per cent of the destroyers were unserviceable was not widely known at the time and Britain had officially declared them all to be 'in perfect condition' (David Wingeate Pike, 'Franco and the Axis Stigma', *Journal of Contemporary History*, vol. 17, no. 3, 1982, p. 374. Cf. Garriga, *España de Franco*, p. 203).

and catastrophe for the USA'. 'The Allies are on the wrong side in this war and they have lost it.' After a virulent diatribe against the Soviet Union, he spoke, in his ardent peroration, of 'these moments when the German armies lead the battle for which Europe and Christianity have for so many years longed, and in which the blood of our youth is to mingle with that of our comrades of the Axis as a living expression of our solidarity'.

Serrano Suñer was surprised by Franco's unrestrained fervour and alarmed by the enthusiasm of the crowd. Franco was mortified to overhear his brother-in-law muttering 'is this a bull-fight?' To the further annoyance of the Caudillo, who reproached Serrano Suñer later in the day for his failure to join in the general adulation, the *cuñadísimo* told him that he should leave such declarations to underlings who could always be disavowed if necessary.[77] Serrano Suñer's surprise at Franco's vehemence may have derived from the fact that Carrero Blanco provided the principal input for the speech. Nevertheless, the Foreign Minister's remonstrations seem to have had their effect. On the following day, in a broadcast address to the workers, his language was more prudent and aimed at labour issues. Both British and Portuguese observers saw the second speech as an appeal for 'red support' prior to a full-scale war effort. It was a view shared by an outraged Vigón.[78]

Soon afterwards, Serrano Suñer complained to Stohrer that Franco had opened the eyes of the English and Americans to 'the true position of Spain'. 'Previously', said Serrano Suñer, the English Government especially kept on believing that only he, the Foreign Minister, was pushing for war, while the "wise and thoughtful" Caudillo would preserve neutrality unconditionally. That illusion has now been taken from them.' Serrano Suñer was absolutely correct in his analysis. Oliveira Salazar, who had long assumed that Franco would remain neutral, was now impelled to think that he would join the war soon. The British Government took a provisional decision on 21 July to launch an attack on the Canary Islands. Vigón and other senior generals were furious at what they saw as Franco's irresponsibility. Serrano Suñer also complained to the German Ambassador that, after the bombshell of his speech, blithely ignoring its repercussions, Franco had left Madrid to go hunting ibex in the mountains.[79]

The job of minimizing the impact of the speech in Allied circles fell to others. The party line was that the speech had no implications for foreign policy. Carceller told Hoare that the speech was aimed at the Falange, in an attempt by Franco to steal the thunder of Serrano Suñer. Whatever the truth, Hoare wrote to Eden urging him to refrain from outright condemnation of Franco for fear of playing into Serrano Suñer's hands. Nicolás Franco told David Eccles in Lisbon that 'he thought his brother had gone too far, that his remarks were for internal consumption'.[80] In fact, there is little reason to believe that Franco's speech to the *Consejo Nacional* was anything but a sincere reflection of his eternal anti-Communism and his latent pro-Axis enthusiasm inflamed anew by the belief that the German invasion of Russia was the prelude to a rapid final victory.

In fact, Franco was immensely fortunate that Stalin did not choose to respond to the sending of the Blue Division with a declaration of war against Spain.[81] His declarations had provoked a turning point in Allied-Spanish relations. The American Under-Secretary of State, Sumner Welles, declared in response to the Caudillo's remarks, 'the only dignified course for this country is to withold further shipments of food and medical supplies to Spain.'[82] Thereafter, innumerable obstacles were put in the way of the export of vital American goods to Spain and the supply of oil dwindled to a trickle.[83] In London, Anthony Eden was henceforth determined to stand firm against Franco. He told Oliver Harvey that 'the argument that we can do nothing to annoy Franco or upset him because if we do the Germans will march in, applies no longer as the Germans are fully occupied in the East.' Like the Americans, Eden had long opposed the appeasement of Franco and now toyed briefly with trying to support the Spanish left against him. Eden was convinced that Franco would welcome the Germans if they chose to enter Spain.[84] In the House of Commons on 24 July, the Foreign Secretary declared that 'if economic arrangements are to succeed, there must be goodwill on both sides and General Franco's speech shows little evidence of such goodwill. His statement makes it appear that he does not desire further economic assistance for his country. If that is so, the British Government will be unable to proceeed with their plans and their future policy will depend on the actions and attitude of the Spanish Government'.[85] Churchill, however, drew back from his plans for a pre-emptive strike on the Canary Islands after coming round to the view that Franco's speech might have been for internal consumption.[86] After Eccles told Nicolás Franco on 6 August that the speech had shocked the Allies, he undertook to go to Madrid to tell his brother how damaging his speech had been.[87]

The senior generals were seriously disturbed by what they saw as the irresponsible adventurism involved in the creation and despatch of the *División Azul*. Among many junior officers, there was some enthusiasm for the chance to take revenge, in the company of the Wehrmacht, for the Soviet Union's participation in the Spanish Civil War. Accordingly, the military elders could only watch with dismay the departure of the Falangist and military volunteers under the command of the gaunt pro-Nazi General Agustín Muñoz Grandes. There was some disquiet at reports of the volunteers being obliged to swear an oath of loyalty to Adolf Hitler.[88] General Orgaz, now High Commissioner in Morocco, went so far as to discuss with civilian monarchists the possibility of military action against Franco. Along with four other key figures from the *Consejo Superior del Ejército* (Supreme Army Council), he was anxious to ensure that Spain stayed out of the war and to see Serrano Suñer's power diminished.* On 1 August, Orgaz informed Franco on behalf of all five generals that he should not make extreme pronouncements on issues of foreign policy such as his 17 July speech without first consulting them. Orgaz passed on to the Caudillo trenchant criticisms of Serrano Suñer and a demand for his early dismissal. In characteristic style, Franco avoided confrontation by agreeing to the request in principle. He then prevaricated by pointing out that, given the power of the Falange, to remove Serrano was more complicated than it appeared and would require time.[89]

It was a skilful response, which apparently aligned Franco with the generals in watchful suspicion of the Falange. It also exaggerated both the strength and the unity of the Falangist leadership which, since the May crisis, Franco knew to be malleable. As was to be expected, he then did nothing. In response to this inaction, on 12 August the senior generals sent General Aranda to reiterate the message in stronger terms. Franco responded in the same conciliatory manner in which he had greeted Orgaz. The generals involved had courted Hoare with a view to securing British support for a coup against Franco or for a government based in Morocco or the Canary Islands in the event of a German invasion.[90]

During the summer of 1941, Franco's government displayed an increasingly pro-German attitude, although on 25 August Hitler spoke with bitterness to Mussolini of his disappointment with Franco.[91] The controlled press frequently attacked England and the USA and glorified the achievements of German arms. In an interview given to Italian journalists, Serrano Suñer declared that once Germany had defeated Russia, Axis Europe would become an economic block which would strangulate the United States.[92] The staff of the British and American Embassies

* Kindelán, recently promoted to be Captain-General of the IV Military Region (Barcelona), General Saliquet, Captain-General of the I Region (Madrid), General Solchaga, Captain-General of the VII Region (Valladolid) and General Aranda, Director of the Escuela Superior del Ejército.

were treated coolly. As might have been expected after the declarations of Eden and Sumner Welles, imports of essential goods began to dry up as Spain found it harder to get American export licences and British navicerts (the certificates which permitted goods bound for Spain passage through British naval controls). When Willard Beaulac, Counselor at the US Embassy, called on Carceller on 6 August, the Minister explained away the Caudillo's rhetorical extremism as merely a device to keep the Germans out and suggested that anti-Americanism in the cabinet was confined to Serrano Suñer since 'Franco had strong democratic instincts'. Carceller also made the astonishing proposal that United States provide goods to be 'smuggled' into Germany to give the impression of co-operation.[93]

It was patently obvious that Franco was trying to play both ends against the middle and Carceller's Franco-inspired duplicity cut no ice in the State Department. On 13 September, Secretary Hull told the elegant Anglophile Spanish Ambassador Juan Francisco de Cárdenas who was just about to embark on a trip to Madrid, that 'in all of the relations of this Government with the most backward and ignorant governments in the world, this Government has not experienced such a lack of ordinary courtesy or consideration' and referred to 'the coarse and extremely offensive methods and conduct of Suñer in particular and in some instances of General Franco'. Hull said that he doubted that Cárdenas 'could make the slightest impression on Franco and Serrano Suñer since they were capable of adopting so unworthy and contemptible an attitude towards the United States Government without any cause whatever'. Cárdenas informed Franco of Hull's outrage and there was an improvement in the treatment of Weddell by both the Foreign Minister and the Caudillo.[94]

Shortages of coal, copper, tin, rubber and textile fibres presaged a breakdown of Spanish industry within a matter of months. In early September, Franco sent Carceller to seek help in Berlin with detailed instructions about what he was to say. Rather than direct economic assistance, fundamentally the Caudillo was seeking German acquiescence for a thaw in his attitude to the Anglo-Saxon powers in order to facilitate necessary imports. Carceller also had to explain in advance that the US Ambassador was to be granted the audience with Franco which had been denied for many months. Carceller revealed that his master intended to tell Weddell that he would maintain his present policy, implying that this meant a continuation of Spanish neutrality, while he really planned to continue to pursue his policy of 'unlimited support of Germany'. Carceller left his hosts with Franco's message that 'Spain was ready for everything no matter what was planned by the German side. Spain would, without further ado, accommodate herself into the framework of the all-European policy led by Germany; but then she should not be treated like Cinderella and left unnoticed, but should be included in the overall German economic planning."[95]

Franco had arrived almost imperceptibly at the position of pretending to be a friend to both sides although his heart lay with the Third Reich, as his more indiscreet speeches would reveal well into the 1950s. But military and economic realities obliged him to keep open the door to the Allies. His natural inclination in such a dilemma was to do nothing and watch. The Carceller mission did not constitute deception. Franco was simply explaining his strategy to his friends. He did not know that, in the first half of September 1941, German plans for Operation Felix against Gibraltar were again on the agenda. In May, the German Supreme High Command (the OKW) had called for the building of an auxiliary railway bridge at Hendaye and undertook to repair the existing one themselves. There was talk of withdrawing a division from the eastern front and preparations were being made for widening the Irun railway station.[96] Hitler's autumn review of the strategic situation on 8 September 1941 called for political and military co-operation with Spain to be intensified once more. However, military action in Spain was not expected until after the defeat of Russia which was anticipated to come in the spring of 1942 at the earliest.[97]

At the beginning of October 1941, under British pressure, Franco made the minor gesture of ordering the two German ships which refuelled U-boats engaged in the Battle of the Atlantic to be moved from the outer harbour of Las Palmas to the inner. The Germans protested that Franco was breaking a specific promise made in November 1939 to make submarine refuelling facilities available but given the scale of such operations elsewhere on Spanish territory, Stohrer was inclined not to push the point in this case.[98] Indeed, he remarked at the beginning of October to the newly arrived German military attaché, General Günther Krappe, that Germany did not want Spain to enter the war openly since Germany would lose her only outlet from the blockade ring. This was a reference to a complex triangular trading deception whereby Spain, behind the mask of neutrality, sent material it could ill-afford to Germany and had it replaced by Argentina.[99]

On 6 October, Weddell had his long-delayed interview with Franco, in the presence of the Foreign Minister. Commenting that wartime shortages were making American exports to Spain ever more difficult, Weddell underlined the need to place Hispano-American relations on a clear basis. Franco responded with a recital of Spain's problems in obtaining wheat, cotton and gasoline and said he wanted to see an improvement of economic relations with the USA. Weddell asked for clarification of a veiled threat made to him a week earlier by Serrano Suñer that difficulties over gasoline deliveries could 'strangle' Spain and put her neutrality in jeopardy. Before Franco could answer, Serrano Suñer intervened and said it had not been 'a threat but a reflection'.[100] It is instructive to compare this account, based on Weddell's report, with what Serrano Suñer told von Stohrer four days later. His account made it seem that Weddell had been eagerly seeking Spanish friendship and been rebuffed firmly. Serrano claimed that Weddell had offered loans and raw materials deliveries in return for neutrality but that Franco had replied that he merely desired the delivery of gasoline and other goods for which Britain had issued navicerts. The *cuñadísimo* proudly recounted that, when Weddell asked Franco if his statement that the 'economic thumbscrews' of England and America would drive Spain into the war was to be taken as a threat, he had intervened to say that it was merely a statement of fact and Franco had agreed.[101]

In the autumn of 1941, there was an intensification of the internal political crisis culminating in a tense meeting between Franco and Serrano Suñer in early October. Serrano Suñer complained to von Stohrer that his military opponents, particularly Aranda, accused him of doing great damage to Spain as a consequence of his pro-German policy. The military now believed that England and America would win the war and were already taking their economic revenge on Spain. He claimed that Aranda, egged on by Beigbeder, had been in contact with the British Ambassador and was trying to organize a military plot to change Spanish foreign policy. It is to be supposed that Serrano Suñer did not know the full extent of Aranda's commitments to the British nor of the scale of London's expenditure on bribing him and other senior generals. Serrano Suñer passionately asserted his conviction that the survival of both Franco and himself depended on the victory of the Third Reich. Stohrer shrewdly believed that Franco feared to dismiss Serrano Suñer lest it strengthen the monarchist camp.[102]

However, in domestic affairs, the drift of events was against Serrano Suñer. In November 1941, the Caudillo intensified the process of domesticating the Falange which had begun six months previously. In this, his instrument was to be the sycophantically docile Arrese, who set about buying off or removing 'old-shirt' zealots behind a smoke-screen of rhetorical radicalism. This could only spell the demise of the limited and pragmatic Falangist revolution favoured by Serrano Suñer. Those of doubtful loyalty to Franco were to be purged from the Falange as reds and radicals.[103] The press began to hammer home the twin notions that the Franco regime would survive no matter what the result of the war and that if the Caudillo fell, it would be the end of Spain.[104] It was a clear sign that Franco was wakening from his imperial dreams and addressing the practical problem of ensuring that he remained in power.

It was becoming clear to Franco, as to others, that Hitler had got himself into serious trouble in Russia. Those senior monarchist generals who acknowledged the possibility of ultimate Allied victory began to worry that the Eastern difficulties might impel Hitler to try to resolve the Mediterranean question with an attack on Gibraltar. They remained in touch with civilian monarchists about the possibility of sweeping Franco aside and imposing a monarchical restoration.* Believing a German invasion to be on the cards, they had elaborated plans for their own evacuation and the setting up of a military command in Morocco and a provisional civilian government with British backing in the Canary Islands. However, by the end of November 1941, as the danger of a German invasion receded, several of those involved began to withdraw. They were prepared reluctantly to plot to keep Spain out of the war but not if the objective were just to overthrow Franco.[105]

In the last week of November 1941, a meeting of the Anti-Comintern Pact Powers was held in Berlin and stimulated a revival of pro-Axis fervour in the Spanish press.[106] On 29 November, Serrano Suñer, Ciano, Ribbentrop, Stohrer and Hitler met to discuss the military situation. Serrano Suñer boasted that Spain 'performed every possible service for the Reich to the modest extent possible to her' but also suggested that the war would be long and difficult, a significant change from previous declarations of faith in a swift victory. He mentioned Franco's problems with monarchists, seditious generals and 'dormant reds' and claimed that news of his journey to Berlin had provoked the Americans into holding back two petrol tankers bound for Spain. Hitler was not impressed. In a report to the Duce, Ciano wrote 'Serrano has not yet discovered the

---

* In addition to Orgaz, Kindelán, Saliquet, Solchaga and Aranda, General Varela, the Army Minister and General Juan Vigón, the Air Minister, were also involved. Others implicated included General Ponte, who had moved from Morocco to take over as Captain-General of the II Military Region (Seville), the ex-Ambassador in Berlin and fervent enemy of Serrano Suñer, General Espinosa de los Monteros, now Head of Military Forces in the Balearic Islands and General Heli Rolando Tella, Military Governor of Burgos.

proper tone for speaking to the Germans, and does not even seem very anxious to find it. He says things with a brutality that makes one jump in one's chair.'[107]

On his return from Berlin, Serrano Suñer received a visit from Ambassador Weddell who quizzed him about an anti-Russian and anti-American speech that he had made in the German capital on 25 November. In it, as well as linking Russian and American Communism, he had declared that millions of Spaniards would fight to save Germany from Russia. Clearly injected with confidence by the meetings in Berlin, despite early evidence of German reverses in Russia, Serrano Suñer expressed great confidence in an Axis victory.[108]

At that point, Serrano Suñer still did not know that Washington had presented Cárdenas on 29 November with a list of the US conditions for continued trade with Spain. They included the demand that the supply of oil 'be subject to our supervision – to prevent diversion to the Axis'. Given that Spain was still supplying German U-boats, it was hardly surprising that Franco hesitated to comply with this condition until dwindling oil supplies forced him to do so. While agonizing over how to respond, Franco received a boost from an Axis success, being greatly heartened by the Japanese attack on Pearl Harbor on 7 December 1941. An official telegram of congratulation was sent from Madrid to Tokyo.[109] The Falangist press rejoiced in the blow to the United States and gleefully announced the imminent demise of England. Serrano Suñer was observed to be jubilant in the wake of the Japanese initiative. However, there were some misgivings when the Japanese invaded the Philippines and the majority of Latin American republics declared war on Japan.[110]

In the meanwhile, on 9 December, General Moscardó, both Falangist sports supremo (*Delegado Nacional de Deportes*), and Head of Franco's Military Household, was received by Hitler while in the course of a trip to visit the Blue Division. After passing on the Caudillo's good wishes and effusive belief in the final German triumph, Moscardó spoke of the Spanish anxiety to remove the dagger in the heart constituted by Gibraltar. Hitler replied that he was sorry not to be able to undertake anything in this respect at the moment and expressed regret that Franco had not seized the opportunity earlier in 1941. In fact, over a month before, the OKW had decided that the conditions did not exist for the intensification of military co-operation with Spain ordered by Hitler in preparation for an attack on Gibraltar.[111]

As his second flowering of pro-Axis enthusiasm withered in the winter of 1941, along with the fortunes of the German armies in Russia, Franco became cautious again. With the British victorious in North Africa, Spain's oil supplies running out and under some pressure from his own High Command, the Caudillo seems finally to have accepted that there were no real territorial compensations to be gained in return for taking a risk.*

The nature of the pressure that was being exerted on Franco by his top generals and his sinuous response to it were starkly revealed in the second week of December 1941. The *Consejo Superior del Ejército* met to discuss the grave internal and external political situations. After sessions involving Kindelán, Varela, Orgaz, Ponte, Saliquet and Dávila, a final session on 15 December 1941 was presided over by Franco himself at his El Pardo palace. The Caudillo's continuing confidence in an Axis victory was not shared by his senior generals despite the admiration which many of them felt for the Wehrmacht. At the beginning of the meeting, Franco did not take up the offer made by General Varela for individual generals to express their anxieties about the current situation. However, Kindelán took the initiative and presented a sternly critical account of Spanish politics, denouncing government incompetence and immorality and in particular the ineptitude and venality of the sprawling Falangist bureaucracy. Franco did not react visibly. Kindelán commented later that he could not tell if Franco listened with bored resignation or interest. He remained passive, nodding sagely as if he agreed entirely with Kindelán's courageous criticisms although his expression hardened noticeably when the exposition turned to the dented prestige of both Franco himself and the Army.

Kindelán claimed that the prestige of the Army was being severely damaged by the savage repression still taking place, with prisons overflowing, executions continuing weekly and labour battalions working in inhuman conditions. Just as alleged Civil War crimes had been deemed 'military rebellion' and were tried by court martial, so too any opposition to the regime had been made the responsibility of military tribunals by the *Ley de Seguridad del Estado* (Law of State Security) which had been introduced on 29 March 1941. Kindelán was also hostile to the use of military personnel in local administration, on supply commissions, as prosecutors and as tax collectors. He called upon Franco to abandon his

* The situation had changed to such an extent that Churchill wrote to Roosevelt on 16 December 1941 that Hitler would be chary of involving himself in 'guerrilla warfare with the morose, fierce, hungry people of the Iberian Peninsula. Everything possible must be done by Britain and the United States to strengthen their will to resist. The present policy of limited supplies should be pursued. Hope should be held out of an improvement of the Spanish/Moroccan frontier at the expense of France' – *Churchill & Roosevelt*, I, p. 298.

links with the Falange and to separate the posts of Head of State and Head of Government. Franco must have been outraged, yet he revealed nothing. He was far too canny to confront his assembled top brass for fear of provoking them into some rash action. As it was, Franco's self-control, and his ability to swallow, if not forgive, Kindelán's boldness, permitted him to weather a dangerous storm. He defused the situation with excuses about external dangers, the difficulties of filling important posts after the loss of so many good men in the civil war, and the material difficulties that Spain was undergoing. Kindelán was not satisfied and, with the assistance of the British Embassy, copies of his speech were distributed among monarchists, much to the annoyance of the German Embassy.[112]

Shortly afterwards, Kindelán repeated these views in a speech delivered at the Capitanía General in Barcelona on 26 January 1942 to commemorate the third anniversary of the Nationalist capture of the Catalan capital. Drawing attention to the attrition of the regime's prestige, he lamented the lack of any proper constitutional mechanisms for Franco's successsion and called for Franco to restore the monarchy as the only way to achieve the necessary 'conciliation and solidarity among Spaniards'. Committed as he was to his own survival in power, to the perpetuation of the divisive ideology of civil war hatreds and to an ever more elevated sense of his own mission, Franco was furious.[113] However, as befitted his characteristic caution, once again he did not react. To have done so now might have triggered off some concerted action from his military critics.

He preferred to wait. The wilder imperial ambitions of 1940 were giving way to a determination merely to stay in power. To do so would require skilful balancing acts between the Army and the Falange and between the Allies and the Axis. In the course of 1941, Franco had learned a lot both about domestic and international politics. Over the next three and a half years, he would learn even more.

XVIII

WATCHING THE TIDE TURN

January–December 1942

BY THE beginning of 1942, a euphoric Franco, seduced by the ever-expanding vistas of his own imperial greatness, was beginning to give way to a more astute and watchful politician. Harsh economic and military realities had forced him to draw in his horns a year earlier only to cast caution aside again when the German invasion of Russia made Axis victory seem imminent. Now, Hitler's difficulties in the East and the warnings of Kindelán at home forced him back to earth. Three years on from the end of the Civil War, the immediate post-war unity of his supporters was cracking and he gradually accepted that survival had to be a higher priority than empire. Imperial dreams were reluctantly shelved – if not yet forgotten – and he drew upon the natural reserves of cunning and slow-moving duplicity which had served him so well during his rise to power and in the process of unifying his coalition. Henceforth, in domestic and in foreign policy, he was to show the instinctive skill of the tight-rope walker, and some of the luck or *baraka* that had served him so well in Africa.

Throughout 1942, he was to need it. The policy of the United States was to bring Franco to heel by limiting food and fuel sales according to Spanish readiness to reduce its export of war supplies to Germany. That policy had been intensified after the Caudillo's rash speech of 17 July 1941. However, fearing that he might be thrown into the arms of Hitler, Churchill pushed for a more cautious approach. He wrote to Roosevelt on 5 January 1942, 'Please will you very kindly consider giving a few rationed carrots to the Dons to stave off trouble at Gibraltar? Every day we have the use of the harbour is a gain, especially in view of some other ideas we have discussed' – a reference to preparations for an invasion of North Africa (to be called Torch).[1]

What Churchill could not know was that the position of Serrano Suñer was now extremely fragile. After their meeting in mid-December 1941, the senior generals who made up the *Consejo Superior del Ejército* had met again on 9 January and launched another savage attack on the *cuñadísimo*. It was being rumoured in Madrid that Franco would relieve him of his post and send him to Rome as Ambassador.[2] In fact, Serrano launched a major fightback with a fierce editorial in *Arriba* on 13 January attacking the generals.[3] He survived the crisis, not least because Franco was not prepared to be railroaded by any one faction among his supporters. If the generals forced the demise of Serrano Suñer, Franco would be back to being the elected executive subject to the vigilance of those who elected him. However, the *cuñadísimo*'s days were numbered. The same was true of General Kindelán who had led the charge. Franco never forgave nor forgot attempts to undermine his power or force him into unwelcome decisions.

The tensions within the regime were reflected in the speeches made by Franco during a tour of Catalonia to celebrate the third anniversary of the capture of Barcelona from the Republicans. A massive exercise in orchestrated public adoration was mounted and the entire visit was choreographed as a great triumphal procession to present the Caudillo as the beloved leader of both the Army and the Falange. It was masterminded by Arrese, the time-serving Minister-Secretary of the Falange, who was keen to show the Caudillo his ability to generate popular enthusiasm. On his arrival in Barcelona on 26 January, Franco was greeted with the now usual pomp and ceremony by the Army, the Church and the Falange. After the release of three thousand doves, a fly-past of aircraft, artillery salutes and a military procession followed by a parade of twenty-four thousand Falangists, he was presented with the gold medal of the city. There were similar demonstrations over the next few days in Sabadell, Gerona, Tarragona and, en route back to the capital, in Zaragoza. On his return to Madrid, one hundred thousand Falangists lined the road. Fifty thousand copies of the speeches he had given in Catalonia were distributed among the crowd. For a month thereafter, there was lengthy exegesis of these speeches in the Falangist press which reached the conclusion that they were the words of a genius. Franco was delighted with the sheaves of cuttings which arrived each day on his desk and Arrese's stock rose ever higher.[4]

Through the cloudy rhetoric of these speeches glimmered two linked messages. The first was to prove to Spain and the world that, as his court jester Ernesto Giménez Caballero* put it, 'having won the war, he had now won the peace'.[5] Given the scale of repression in operation, with political executions still regularly taking place and left-wing prisoners numbered in the hundreds of thousands, rhetorical gestures to the Barcelona working class seemed singularly hollow. Nevertheless, Franco excused anarchist violence in the past as 'the virile expression of our race: explosions of rebellion against a decadent fatherland', and talked of social solidarity and the Falangist revolution.[6] Such rhetorical radicalism was aimed less at the genuine left, suffering in clandestinity, than at the Falange which, with the eager help of Arrese, Franco was trying to wean away from Serrano Suñer. Secondly, in words which must have meant little to the majority of his listeners, he harped on the need for unity. In Zaragoza, he also referred obliquely to the power struggle going on within his regime, 'what I ask of you is this, let us leave behind the petty resentments, the egoisms of *amour propre* and the cancer of envy; let us banish them, let us consider that for an enterprise as great as raising up Spain and leading her along the path of empire, three things are needed, a single command, a single discipline and a single obedience'.[7]

There can be little doubt that the continuing squabbles between Army and Falange, reports of which crossed his desk, were weighing heavily upon the Caudillo, all the more so as the Anglo-Soviet alliance stored up trouble for Hitler in Russia. Even if Franco calmly ignored Kindelán's demands for a monarchical restoration, military pressure and unavoidable evidence that there could now be no swift Axis victory forced him to contemplate indefinite postponement of Spanish entry into the war. Anxious to go on helping the Axis, in January 1942, the Caudillo agreed to continue Spanish exports of wolfram to Germany. Wolfram, as the ore from which tungsten is derived, was a crucial ingredient in the manufacture of high quality steel for armaments in general and particularly for machine tools and armour-piercing shells. Franco took this decision despite the fact that the Reich could not send equivalent amounts of German goods. Deliveries of machinery from Germany had simply failed to materialize and Spain was sitting on an increasingly large annual credit balance with the Reich.[8] Franco also agreed to Spain representing Japan's diplomatic interests in Latin America. Throughout early 1942, the press was unrestrained in its pro-Japanese fervour, an inclination which

---

* Ernesto Giménez Caballero was, along with Salvador Dalí and Luis Buñuel, one of the fathers of Spanish surrealism. He was one of the first Spanish fascists in the late 1920s and put his manic talents at the service of Franco during the Civil War, reaching delirious heights of sycophancy.

reflected the Caudillo's views as well as those of Serrano Suñer.[9]

For another three years, Franco's hopes were to be pinned on an Axis victory but now he had to make more of an effort to keep avenues open to the Allies. This was reflected in preparations* for a meeting with the Portuguese Premier and Foreign Minister, Antonio Oliveira Salazar.[10] Even at their 13 February 1942 Seville meeting, it was obvious where Franco's heart lay.[11] Despite the entry of the United States into the war, he told Salazar that an Allied victory was absolutely impossible and boasted that there was not the slightest danger of a German invasion of Spain as long as friendship with the Reich was maintained. In an excess of pro-Axis fervour, he proclaimed to Salazar that, if there were ever a danger of the Bolsheviks overrunning Germany, he would do everything possible to help and would send one million Spanish troops. Behind the courteous tone of the meeting there could be discerned Franco's long-nurtured ambition to wean Salazar from the Anglo-Portuguese alliance and into a dependent relationship with Spain. Salazar was informed by Serrano Suñer that the British intended to overthrow his regime and by Franco that a British landing on Portuguese territory would be taken as an act of aggression against Spain.[12]

Franco's mood of pro-Axis enthusiasm continued after Salazar's return to Lisbon. On 14 February 1942, the Generalísimo addressed high-ranking army officers at the Alcázar of Seville. Thrilled by the British disaster at Singapore on the previous day, he irresponsibly spoke in the eager voice of a friend of the certain victors. He seemed not have read the many reports from the Spanish Embassy in Berlin about the catastrophic situation of the German forces in Russia.[13] He expressed his astonishment that part of the world should be fighting Germany, 'to destroy the bulwark which held back the Russian hordes and defended western civilization'. Franco declared his 'absolute certainty' that Germany would not be destroyed. Fired with that confidence – and no doubt in the hope that his promise would never be put to the test – he publicly repeated what he had told Salazar, that 'if the road to Berlin were open, then it would not merely be one Division of Spanish volunteers but one million Spaniards who would be offered to help'.[14]

* Franco was dismayed by Salazar's insistence on a simple meeting without military parades, popular demonstrations or pomp. A decade and a half later, he would still be perplexed by Salazar's modesty. Interviewed by *Le Figaro* on 13 January 1958, he said that 'the most complete statesman, the one most worthy of respect, that I have known is Salazar. I regard him as an extraordinary personality for his intelligence, his political sense, his humanity. His only defect is probably his modesty' – *Discursos y mensajes del Jefe del Estado 1955–1959* (Madrid, 1960) pp. 478–9.

His words did not go unnoticed in either London or in Berlin.[15] A disgusted Goebbels wrote in his diary 'It would be far better if he declared war on Bolshevism. But what can you expect from that sort of general?' Earlier in the month, when Franco had asserted Spain's fidelity to the Catholic Church, Goebbels had written 'It would be far more fitting for Spain to remain faithful to the Axis. Franco, as we know, is a bigoted churchgoer. He permits Spain today to be practically governed not by himself, but by his wife and her father confessor. That's a nice revolutionary we placed on the throne!'[16]

Torn between his basic sympathy for the Axis cause and the growing economic pressure exerted by the Allies, Franco found the external tensions mirrored within his regime. On the one hand, oil supplies were so short that the Spanish refinery at Tenerife was forced to close down in February 1942. Franco was obliged grudgingly to accept Washington's demand that further US oil shipments be subject to supervisory inspections of their use.[17] Spain's parlous economic position led to frantic, and mendacious, public denials of the arrangements whereby Spain was supplying German submarines in the Canary Islands.[18] These humiliations led to vituperative anti-American comment in the press.[19] In fact, American suspicions about covert Spanish blockade-running were well founded. Von Stohrer's remarks to General Krappe in October 1941 about Spain's greater value as a blockade breaker than as a belligerent were echoed in February 1942. Ernst von Weizsäcker, the State Secretary, received the new military attaché in Tangier, Colonel Hans Renner, prior to him taking up his post. He told him that a quiet Spain and North Africa were in the interests of Germany. Weizsäcker stressed that it was necessary to keep Spain from coming out into the open in any way since 'we are receiving from Spain important support for our conduct of the war, and we must not lose it on any account'.[20]

At this moment, the Caudillo, burdened with difficult external and internal problems, heard of the death of his father on 24 February. There were telegrams of condolence from Mussolini and Marshal Pétain.[21] It may be supposed that, having to deal with the continuing squabbles between his own followers, he had little time or emotion to spare for grief, all the more so given his difficult relationship with Don Nicolás. Franco had spent a lifetime nurturing hatred of his father for the pain that he had caused his mother. Knowledge of that perhaps lay behind the fact that Don Nicolás had rarely lost an opportunity to express disdain for his second son. Certainly, Nicolás Franco Salgado-Araujo manifested neither pride nor pleasure in his son's political eminence and he often referred to him as 'inept' and ridiculed the flatterers who surrounded him. He found his son's obsession with the 'Jewish-masonic conspiracy' especially laughable, saying 'what could my son possibly know about freemasonry? It is an association full of illustrious and honourable men, certainly his superiors in knowledge and openness of spirit.'[22] He was often heard in Madrid bars when drunk loudly insulting his son's name even to the point of calling him 'a swine and a pimp' (*un cabrón y un chulo*). Since these outbursts led occasionally to him being arrested and detained until his identity was confirmed, the Caudillo was fully aware of his father's contempt.[23]

His reaction to the old man's death suggested that the desertion of the family and the later denigration of his achievements had eaten away at Franco's soul. During the eighty-four year-old Don Nicolás's extremely long death agony, afflicted with the consequences of a brain haemorrhage, his eldest son Nicolás travelled almost weekly to see him from Lisbon where he was Spanish Ambassador. The Caudillo, in contrast, ignored his dying father.[24] With vengeful harshness, he ordered that the body be taken from the home in Madrid's Calle Fuencarral that Nicolás Franco Salgado-Araujo had shared for thirty-five years with his common-law wife Agustina Aldana. She had lived with him during the greater part of his adult life and nursed him through his final illness. To Francisco, she was simply an immoral strumpet who had supplanted his mother and brought shame on his family. With a grief-stricken Agustina trying to hold on to the corpse, a squad of Civil Guards forcibly removed it. Nicolás Franco was buried in the full-dress uniform of a general of the pay corps with corresponding military honours, as if he had been a person fit to have been the Caudillo's father. Franco took further revenge against Agustina by refusing permission for her to attend the funeral. He also failed to take his farewell of his father, accompanying the cortège only as far as the gates of El Pardo.[25]

If he needed an excuse for his absence, the political situation was more than sufficiently time-consuming. The friction between the traditional right, represented by the generals, and the new right of the Falange now drove the Caudillo to take uncharacteristically decisive action. Military resentment of Serrano Suñer was growing daily.[26] An ever more prominent part was taken in machinations against the *cuñadísimo* by General Espinosa. Espinosa had been supplanted as Spanish Ambassador to the Third Reich in 1940 by Serrano Suñer's henchman, the Conde de Mayalde. Driven by resentment of the *cuñadísimo*, Espinosa was rumoured in March 1942 to be involved with Kindelán and Orgaz in preparations for an anti-Franco coup. On taking up the post of Captain-General of the VI Military Region (Burgos) in mid-April, he made a speech bitterly attacking 'the disloyalty and limitless ambition' of Serrano whom he had previously accused in private of treason. Franco's reaction was swift and Espinosa was sacked within a matter of days. However, his dismissal was balanced by that of Serrano's political secretary Felipe Ximénez de Sandoval who had been smeared by Arrese's followers as having been involved in alleged homosexual activities.[27] Kindelán and other senior generals were outraged by what happened to Espinosa and regarded his punishment as an imposition on the Generalísimo by Serrano Suñer. In fact, they over-estimated the power of Serrano Suñer who was increasingly isolated.[28]

After his initial enthusiasm for the Japanese assault on the United States, economic and political realism had prevailed with the Caudillo and relations had improved slightly with both London and Washington. He still believed of course that Britain and the United States were nests of conspiring Communists and freemasons and assured the Portuguese Ambassador at the end of March that the British and American secret services, hand-in-hand with Communism and freemasonry, planned to overthrow Salazar.[29] Nevertheless, despite his smouldering prejudices, in the spring of 1942, a significant change began gradually to take place in the economic role within Spain of the United States and Great Britain. Previously, Allied economic policy had been to control essential supplies to Spain in such a way as to restrain Franco's pro-Axis enthusiasms. However, in November 1941, the British had suggested to Washington a joint programme to buy up Spanish wolfram and other metals vital to the German war effort. In mid-March 1942, the American and British Embassies in Madrid began to implement a policy of pre-emptive (or as it was known 'preclusive') purchase. Washington also created the United States Commercial Corporation to supervise the American side of an operation which was soon to develop into a major instrument of economic warfare against Germany.[30] Eventually, this was to open up to Franco new possibilities of playing off the Allies and the Axis to the profit of his own regime.

Less anti-American material was appearing in the press. In some tiny measure that was a consequence of the fact that the astute and elegant Weddell, the object of the intense dislike of Serrano Suñer, had had to resign because of ill health. After his departure from Madrid in February 1942, the US Embassy was left in the hands of an extremely competent *Chargé d'Affaires*, Willard L. Beaulac. An affable and sensible man who

liked Spain and the Spaniards, Beaulac enjoyed less conflictive relations
with the Foreign Minister than had Ambassador Weddell. That in turn
was a reflection of the fact that both Franco and his brother-in-law were
coming to realize the importance of the USA's entry into the war. On
4 March 1942, Serrano Suñer told Beaulac that Spanish policy was to
keep out of the war and that required friendship with the Axis, engagingly
asking 'Would it do the Allies any good if Spain should begin to make
faces at Germany at this precise stage in human history?'[31] Meeting the
Vichy French Ambassador on 13 April 1942, Franco told him that in the
event of the Allies winning, the cause of Communism would triumph but
hinted that he was now determined to remain neutral.[32] Nevertheless, a
week later, Serrano Suñer made a speech affirming that Spain was at
Germany's side in its battle against Communism.[33] Moreover, confidential Spanish despatches being intercepted by the American secret services
convinced the Director-General William J. Donovan that the Spanish
Embassy in Washington was 'an enemy mission'.[34]

In the wake of the dismissal of Espinosa, the vehement Minister for
the Army, General Varela, had protested to Franco about the excesses
of the Falange, dwelling on the displeasure of the top brass at seeing
pimply, uneducated Falangists, with no respect for the Army, getting fat
salaries for artificial jobs. With an unaccustomed slyness, Varela criticized
Serrano Suñer by telling Franco in the last victory parade on 1 April he
had not been applauded as much as usual.[35] More disinterested if equally
outraged criticism came from the idealistic Falangist poet Dionisio Ridruejo who was invalided out of the División Azul at the end of April. On
being received by Franco in early May, he told him that, among his
comrades, there was much criticism of the corruption in Spain. The
Caudillo replied complacently that in other times, victors were rewarded
with titles of nobility and lands. Since to do so was now difficult, he
found it necessary to turn a blind eye to venality to prevent the spread
of discontent among his supporters.[36]

In May 1942, the ongoing ferment within regime circles spilled over
into street fights between monarchist Army officers and Falangist students
in Pamplona, Burgos and Seville.* There were violent clashes in the
Universities of Santiago and Madrid between Falangist and monarchist
students.[37] For the moment, Franco watched and waited.

It was fortunate for the increasingly beleaguered Caudillo that the new

* Goebbels noted 'never has a revolution yielded so few spiritual and political results as
Franco's' – Diaries, 1942–43, p. 167.

US Ambassador who arrived on 16 May 1942 was the Columbia University historian Professor Carlton Joseph Huntley Hayes who had been an energetic apologist for the Nationalist side during the Spanish Civil War. Unworldly, self-regarding and inexperienced in politics, Hayes resented the dominance established in the Madrid diplomatic corps by the more dashing Sir Samuel Hoare. Franco exploited the resulting friction, favouring Hayes with twice as many audiences as Hoare received.* Whereas Hoare remained convinced that Franco was pro-Axis, failing to join in the war only out of cautious expediency, Hayes thought the Caudillo pro-Allied. The tweedy, bespectacled Hayes liked Franco and appreciated his dry sense of humour, which he was one of the few people ever to find 'rather lively and spontaneous'. His ignorance about Franco led him in his memoirs to dismiss as a fabrication the Caudillo's offers to join the Axis in 1940. Franco in turn responded well to the plodding and pompous Hayes, while Serrano Suñer thought him a boring pedant.[38]

On 29 May 1942, Franco visited Medina del Campo to inaugurate the training school of the *Sección Femenina* of the Falange in the specially restored Castillo de la Mota. His penchant for regal ceremony and solemnity was indulged to the full. He was introduced by the monarchist-turned-Falangist José Pemartín who proclaimed that military and civic virtues far above 'the rights accruing from imperial bed-chambers' were the claim to kingship of 'the saviour of Spain'. These remarks gave rise to rumours that the Caudillo planned to proclaim himself king. The entire performance provoked the amusement of the diplomatic corps.[39] Since the castle had once belonged to Isabel la Católica, it was not surprising that Franco seized the opportunity to compare his triumphs with hers, identifying her misguided critics with those who foolishly criticized him. His interpretation of her achievements implied a clear alignment with Hitler. He praised her expulsion of the Jews, 'her totalitarian and racist policy' and her awareness of Spain's need for *Lebensraum* (*espacio vital*).[40]

It is possible that this speech had a domestic rather than an international purpose. If Franco was toying with the idea of dispensing with Serrano Suñer, he might well have wanted to clinch support within the Falange for himself first. It is more likely he was carried away by the historic and atmospheric location and his closeness to Arrese. Whatever his motives, the parlous state of the Army, without equipment or fuel, made

* Hayes had six audiences in four years (or one every eight months) as against Hoare's four in five years (or one every fifteen months).

even mere talk dangerous, especially when so aggressively adventurist.[41]

Throughout the summer of 1942, Franco began to distance himself from Serrano Suñer. Apart from complaints about the *cuñadísimo* from Varela and Vigón, the Caudillo was subject to constant poisonous insinuations about Serrano Suñer from the oily Arrese, and the dour Carrero Blanco. Cut off from the Pardo, the *cuñadísimo* himself made little secret of his weariness, disillusionment and irritation with the ongoing squabbles within the regime.[42] On 5 June, he dined with Pedro Teotónio Pereira and spent the entire evening contemptuously dismissing every name that came up in conversation. He said 'I have been an idiot for the last four years trying to resolve things partially. Nothing comes of that. I am waiting for the moment to clean the lot of them out.' Given his political weakness, the remark reflected more his bitterness and disillusion than any serious hopes of replacing Franco.[43]

On 10 June, Carlton Hayes presented his credentials and Franco received him for an unusually long and cordial interview of nearly half and hour during which his brother-in-law was a taciturn spectator. The Generalísimo tried out on Hayes a variant of his theory that there were two distinct wars going on. Whereas usually he claimed that there was a pointless Anglo-German war and a crucial anti-Communist war in which he admitted to having a stake, he now referred to a war in Europe against Russia and a war in the Pacific against Japan. Hayes did his best to convince the Caudillo that the overwhelming economic and military strength of the United States would ensure victory for the Anglo-Saxon powers. Franco remained unconvinced about the Americans' capacity to transport their armies and equipment across the Atlantic. With less than exquisite tact, he insisted that victory for the democracies would be a victory for Communism, adding that the British Labour Party was already Communist. Despite the blatant crudity of Franco's arguments, Hayes was impressed by his desire for good relations with the USA and also by his assurance of 'his sincere and earnest desire' to maintain Spanish neutrality.[44]

Hayes's confidence in Franco's good intentions was not shared by the British, nor by Allied supreme military command. In early 1942, British intelligence had intercepted and decoded German radio messages arising from an ambitious *Abwehr* operation, *Unternehmen Bodden*, for which Canaris had persuaded Franco to grant permission.* The *Abwehr* was

---

* The Bodden was a strait of water separating the Baltic island of Rügen from the German mainland. Its similarity to the Straits of Gibraltar, together with *Abwehr* activity in the vicinity of Algeciras, gave British intelligence the necessary clues to its purpose.

constructing, with the aid of the Spanish navy, a seabed sonic detection system across the Straits of Gibraltar and a chain of fourteen infra-red ship surveillance stations along the routes which would be used by the convoys for the projected Allied landings in North Africa (Operation Torch). With nine stations on the Spanish coast and a further five in Morocco, the system became fully operational on 15 April 1942. Information on Allied shipping thus gathered would be transmitted to U-boats in the Mediterranean and in the Atlantic within range of the Straits.

At one point, the Allies gave consideration to destroying the system by submarine action and commando raids. However, consistent with British policy towards Franco, a diplomatic solution was preferred. Supplied with details of the Bodden operation by Kim Philby, then head of the Iberian section of the British Secret Intelligence Service, Hoare put enormous pressure on Franco. On 27 May 1942, accompanied by the senior members of the British Embassy staff in full dress uniform, he showed Philby's dossier to Franco who had been assured by the Germans that the entire operation would be secret. Deeply embarrassed, and threatened with the curtailment of Allied oil shipments, he reluctantly responded with a promise to investigate. On 3 June, his staff admitted that the equipment was being installed by German technicians but claimed that it was for 'the defence of the coasts of Spain'. Typically, Franco stonewalled for months before asking Admiral Canaris to have his sonar and infra-red detection equipment dismantled. He ignored further British pressure in July and October and it was not until after the success of Torch that the system was eventually taken out of action.[45]

General Eisenhower and the staff preparing Operation Torch were intensely preoccupied by thoughts of the consequences if Franco were either to attack Gibraltar or permit the Germans to do so. The Governor and Commander-in-Chief of Gibraltar, Lieutenant-General Sir F. Mason-MacFarlane was anxious that Franco be neutralized by diplomatic reassurances that the Allies meant him no harm.[46] The anxious Allied military planners in England could not have known of the worries besetting the Spanish leadership.

Extremely tired, upset by the Ximénez de Sandoval affair, almost at a loose end, and certainly no longer at the centre of affairs, the *cuñadísimo* took a ten-day holiday in Italy from 15 to 25 June.[47] Politically, the visit had little point, although Serrano Suñer may have hoped to improve his deteriorating position.[48] It is difficult otherwise to understand why he had absented himself from Madrid just when the hostility of his enemies in both the Army and the Falange was reaching its peak.[49] Arrese was making a name for himself with a triumphant tour of Andalusia.[50] Gloomy and pessimistic, Serrano Suñer spoke to Ciano about Franco's failure either to establish a strong popular base for the Falangist revolution or to restore the monarchy. He attributed Franco's lack of adventure and imagination to the deadly influence of his wife and described the Caudillo as surrounded by nonentities who created in the Pardo an atmosphere which was a parody of the old Spanish court.[51] Ciano took him to lunch with King Vittorio Emanuele on 16 June. Serrano Suñer spoke of Franco 'as one speaks of a moronic servant. And he said this without caution, in front of everyone.'[52] The possibility that this was not reported to Franco is remote in the extreme.

For his part, Franco told Peña Boeuf, the Minister of Public Works, that his brother-in-law 'only does what he feels like' and complained to Carceller, the Minister of Industry and Commerce, that 'we have a Minister of Foreign Affairs who doesn't want to know anything about economic questions'.[53] Serrano Suñer was far from being Franco's only domestic concern. Just as the *cuñadísimo* played the German card in an effort to strengthen his position in the Spanish political maelstrom, there were also efforts by monarchist generals to get German support for a restoration. General Muñoz Grandes, Commander of the *División Azul*, had been asked by some of his peers to use his position to broach the subject of the Third Reich's acquiescence in the return of the monarchy.[54] In fact, they severely misjudged him since his sympathies lay, at this stage of his career, with radical Falangism.[55] Shortly afterwards, Vigón arranged a trip to Germany to seek support for a restoration under the subterfuge of seeking technical aid for the Air Force but was obliged to cancel his trip when Franco got wind of the real intention.[56] Franco was not slow to see that the independent and volatile Muñoz Grandes, with his operational command and his German contacts, was an infinitely more dangerous threat than the loyal conservative Vigón. When rumours reached El Pardo at the end of May 1942 that Muñoz Grandes was blaming Franco for the fact that Spain was in a mess, the Caudillo ordered that he be replaced by General Emilio Esteban Infantes, a friend from Moroccan days and one-time colleague at the Academia General Militar de Zaragoza.

Hitler, however, was coming to think that German interests would best be served by promoting the political career of Muñoz Grandes. In a mirror image of Franco's view that the goodwill of Hitler towards Spain was undermined by churlish subordinates, the Führer believed that the

Caudillo's pro-Axis destiny was being thwarted by the anti-Axis Serrano Suñer. Hitler received Muñoz Grandes on 13 July at the Wolf's Lair and listened enthusiastically to his diatribes against Serrano Suñer and his views on the need for a thorough-going fascist revolution. Muñoz Grandes said that he was prepared to implement a pro-Nazi policy as head of government with Franco relegated to the position of figurehead. Accordingly, to prepare the way for Muñoz Grandes' triumphant return to Spain, efforts were to be made to boost his popularity and he was to be decorated for his exploits with the *División Azul*. Hitler even toyed with giving Muñoz Grandes a prominent role in the expected capture of Leningrad and then sending him home with his men newly and lavishly equipped in order to tip the balance against Franco. The Führer arranged for obstacles to be put in the way of Esteban Infantes travelling from Berlin to the Eastern front and sent Canaris to ask a suspicious Franco to delay Esteban Infantes' assumption of his command. To the alarm of the Caudillo, the German military attaché in Spain also made contact with General Yagüe presumably to secure his support for the Muñoz Grandes operation against Franco.[57]

On 14 July 1942, the Caudillo received Stohrer for a long interview. The German Ambassador was discomfited by the growing Spanish credit balance in the current account with the Reich. He protested at recent Spanish demands for full payment in goods and about increasing delays for export permits for the raw materials which Germany needed. For the first time, Stohrer found himself having to remind Franco that such economic sacrifices were the undisputable duty of an ally of the Reich if Bolshevism was to be defeated. The Caudillo bristled at the suggestion that his anti-Communism might be wavering. He responded with a recital of the economic weapons which the Allies held over his head but nonetheless agreed to grant export licences for goods awaiting transport to Germany.[58]

Franco's views at this time were set out in his annual speech to the *Consejo Nacional de FET y de las JONS* on 17 July 1942. It began with a lengthy justification of the slowness of Spain's post-war economic recovery. There was an element of self-pity in his rhetorical question 'what do critical spirits know of the tense vigils in which a suffocating responsibility weighs on lonely shoulders?' The keynote, though, was an insistence on the unity of the Francoist coalition which indicates that he was genuinely preoccupied by the ever more open frictions between the Falange and the Army. He boasted of being able to mobilize three million fully equipped men on whom rested 'our security and the maintenance of our rights'. That gratuitous statement, as well as being highly questionable given the state of the Spanish armed forces at this point, was probably meant as a warning to the Allies and an inducement to the Axis. His views on the wider war situation had been modified little since the wild enthusiasms of the previous year's speech. He declared both that 'little will be saved of the liberal democratic system' and that 'in terms of war effort, the totalitarian regime has fully demonstrated its superiority; in economic terms, it is the only one which is capable of saving a nation from ruin'.

Perhaps hedging his bets, he went on to anticipate the creation of a non-representative Cortes to permit 'the contrasting of opinions, within the unity of the regime, the airing of aspirations'. It has been interpreted as a gesture to the Allies, which is unlikely if only because its inspirer was the pro-Axis Falange Secretary-General Arrese who was, in his turn, pirating the ideas of Serrano Suñer. The Cortes was in any case seen for what it was, a parliament of nominees. Franco accepted the project because he saw in it another institution with which to dilute the power of the Falange although *Arriba* happily commented on its similarity to the Italian Fascist Chamber of Corporations. It was also part of the process whereby the Caudillo surrounded himself with the trappings of the medieval Spanish monarchy.[59]

Those elements of Franco's speech which concerned the progress of the war caused considerable disquiet in the Allied camp. They seemed to be confirmed when American agents in Lisbon acquired the sealed war orders issued to the Spanish merchant fleet. A subsequent series of secret service burglaries of the Spanish Embassy in Washington in July, August, September and October 1942 showed that Franco remained undecided about whether to join the Axis.[60] It would not be until after the success of the Torch landings in North Africa in November that Allied concern about the Caudillo's intentions would diminish.*

In Madrid, it was felt that Serrano Suñer's journey to Italy may have shored up his position as the friend of Mussolini but it had put Franco on his guard against him.[61] Then, on 15 August, the press published a fervently pro-German article written by Serrano Suñer for *Wille und*

* At a private dinner in mid-July with the President and Director of EFE, the regime news service, and with Ramón Garriga, Carrero Blanco expressed his confidence that, by the end of 1942, the Wehrmacht would have conquered the Caucasus, Turkey, Iraq and Syria, and be on the point of taking Egypt from the East. Given the ever closer identification between Franco and Carrero, the other guests were convinced that they had heard an exact recital of what the Caudillo thought – Ramón Garriga, *Los validos de Franco* (Barcelona, 1981) pp. 235–9.

Macht, the newspaper of the *Hitlerjugend*. Its argument was that Spain was solidly behind Germany in her battle to establish a new order. It was in many respects a skilful piece, giving away nothing, asking the Third Reich to recognize services long since rendered. It stated explicitly that if it had not been for the Spanish Civil War, Germany would now be facing far greater difficulties in her fight with Russia since a Soviet Spain would be part of the equation.[62] It is difficult not to conclude from the timing of this no doubt sincere expression of pro-Axis sentiments that, in search of support in the power struggle inside Spain, Serrano Suñer was trying to show Berlin that he was as reliable a Germanophile as the rising star Arrese.

Ever since Serrano Suñer had arrived in Salamanca in early 1937, Franco had been learning the art of politics, especially international politics. The fifty year-old Caudillo was now ready to flex his muscles more independently. He had long perceived Serrano Suñer as a useful lightning conductor for complaints about the regime from his senior generals. He might have been happy to let him go on fulfilling that function had it not been for a series of factors. There was his resentment that Serrano Suñer was hogging the limelight, a feeling assiduously nurtured by the servile Arrese and Carrero Blanco.[63] Franco was deeply sensitive to stories that he was being upstaged by his brother-in-law. The same was true of his wife but more so. Both must have resented bitterly a story which went the rounds at the time. It was alleged that an old schoolfriend of the Polo sisters who had been in Latin America met Serrano Suñer's wife Zita and gushed 'How wonderful! I hear that you are married to the most important man in Spain. And whatever happened to your sister Carmen?' 'Poor thing', allegedly replied Zita, 'she ended up marrying a soldier.'[64] This was compounded by Señora Franco's anger at the fact that Madrid society knew that Serrano Suñer was two-timing her sister Zita with Consuelo (Sonsoles) de Icaza y León, the wife of Lieutenant-Colonel Francisco Díez de Rivera, the marqués de Llanzol. An American observer of the Spanish situation elegantly summed up Serrano Suñer's position: 'His personal conduct had bruised the intimacy within the Franco household.'[65] Time was running out for Serrano Suñer.

What provoked the final débâcle was the so-called Begoña affair in mid-August 1942 when the hostility of Traditionalists (or Carlists) and military supporters of Don Juan de Borbón against the Falange finally reached boiling point. The political turmoil of that crisis gave plausibility to suggestions made by Arrese to Franco during the summer at the Pazo de Meirás that a change was necessary and now possible. Arrese, through his contacts with Nazi Party elements in the German Embassy, had been able to reassure the Caudillo that Berlin would not be in the least distressed by the removal of Serrano Suñer.[66] It was to be the most serious internal crisis faced by Franco, certainly in the early 1940s, and possibly in the entire course of the dictatorship.

The depth of discontent between monarchists and Falange was brought home to Franco when the Traditionalist Minister of Justice, Esteban Bilbao, presented a letter of resignation in early August. Franco immediately sent him a flattering letter promising to look into his complaints about the Falange.[67] On 16 August 1942, the day on which Esteban Bilbao received this letter, the tension exploded during the annual ceremony held at the Santuario de la Virgen de Begoña, near Bilbao, to pray for the souls of the Carlist *Requetés* (militia volunteers) of the *Tercio de Nuestra Señora de Begoña* (Our Lady of Begoña Battalion) who had fallen during the Civil War. Don Juan de Borbón had been informed that Falangists would attempt to disrupt the occasion.[68]

The ceremony was presided over by the Anglophile General Varela who was well known for his Carlist sympathies and for his outspoken hostility to the Falange which he held responsible for the corruption and black marketeering which were flourishing in Spain.* After the service, as Carlists gathered outside the church shouting monarchist slogans and singing anti-Falangist jingles, a bloody incident was provoked by a group of Falangists, one of whom threw two bombs into the crowd. The first failed to explode but the other wounded nearly one hundred bystanders. The culprit was Juan Domínguez, the *Inspector Nacional* of the Falangist Students Union, the *Sindicato Español Universitario*. The fact that the Falangists had driven from Valladolid to Begoña and were carrying weapons including grenades, suggested a degree of premeditation.[69] Spurred on by his wife, the aristocratic Carlist Casilda Ampuero, Varela went beyond his immediate outrage and seized on the incident as an opportunity to strike a blow at the Falange in general and Serrano Suñer in particular.[70] He publicly interpreted the atrocity as a Falangist attack on the Army, sent a communiqué to that effect to the Captains-General and organized the court-martial of Domínguez. In this, he was seconded by the Minister of the Interior, Colonel Valentín Galarza, who sent telegrams to the civil governors of each province alleging that 'agents at

* Known as the *estraperlo*, after a roulette fraud of the 1930s, the black market exploited shortages and rationing in food, petrol, tobacco and all manner of consumer goods and services. In a context of grinding poverty and near starvation, spectacular fortunes were amassed, often by Falangists and government officials.

the service of a foreign power'* had tried to assassinate the Minister for the Army.[71]

The press maintained a deathly silence about Begoña but Franco's anxiety about the eruption of hostilities between the Army and the Falange spilled over into a series of speeches which he made during his annual holiday in Galicia. Significantly, throughout this tour, he had José Luis de Arrese at his side. By far the most revealing remarks were made on 24 August in La Coruña at a mass gathering described in the press as 'an act of confraternity between the Army and the Falange'. Franco praised the military spirit of the Falange and the Falangist virtues of the Army. The speech's attempt at conciliation between the two forces now at daggers drawn would have been obvious only to those caught up in the power struggle. Franco also made veiled references to foreign gold being used in Spain for subversive purposes 'to create Frenchified traitors to hand over our nation to the enemy' (*para crear afrancesados que entreguen nuestra nación al enemigo*). To underline further his closeness to the Germans, the Caudillo repeated the boasts, last heard in his 17 July speech, about Spain being able to muster three million men, who counted as six and 'show the worth of a race and a people which demands its place in the world'.[72]

Franco had quickly perceived that Varela's indignation cloaked a bid to make capital out of the incident. He bitterly resented any such attempt to usurp his position as ultimate political arbiter. In a long and tense telephone conversation on the same day, 24 August, Franco defended the Falangists involved while Varela stuck to his guns and denounced them as assassins.[73] Varela took a dangerous gamble. He gave the Generalísimo a letter of resignation in which he complained about the Falangist tone of Franco's speeches in Galicia. He said that he could continue in his post only if certain conditions were fulfilled. In addition to the punishment of those responsible, he demanded the expulsion from the single party of the instigators of the incident and the formation of a government 'of authority to rectify the errors of the past'. This clearly meant a cabinet in which monarchists would dominate and, presumably, begin the process of transition to a restored monarchy. Since Varela was backed by Galarza, Vigón, Admiral Salvador Moreno and Esteban Bilbao, Franco reacted cautiously. Partly to avoid antagonizing the military monarchists unnecessarily, and no doubt in part because he did not dissent from the judgement

* Domínguez's address book contained the names of prominent German diplomats in Spain. On the eve of his death, he was awarded the Cross of the German Eagle by Hitler.

on Domínguez, Franco acquiesced in his execution. He then tried to neutralize Varela and, with tears in his eyes, flattered him and begged him to stay. When Varela stuck to his guns, Franco, outraged at the threat to his own position implicit in the insubordination of both Varela and Galarza, accepted Varela's resignation and dismissed Galarza.[74] In doing so he was risking some reaction from the senior generals but banked on their fear of losing comfortable positions.

The crisis was far from over. It remained to be seen if the wider threat implicit in Varela's resignation would materialize. Franco began a frantic round of juggling. Bilbao, Vigón and Admiral Moreno believed that Varela had secured an adequate compromise from Franco and so did not feel obliged to emulate his intransigence. The *Secretario de la Presidencia del Gobierno*, Carrero Blanco, whispered in the Caudillo's ear that there had to be 'both victors and vanquished' after the crisis. Carrero found a ready audience for his suggestion that the dismissal of the two generals would be interpreted as meaning that Serrano Suñer was really in charge. However, it dramatically overestimates the role of Carrero Blanco to assume that he alone thereby convinced Franco that to balance matters he must exact comparable retribution from the Falange. It is inconceivable that, after the confrontation with Varela, Franco did not know that to survive the crisis there would have to be action against Serrano Suñer who as President of the *Junta Política* or executive committee of the Falange was its most powerful figure. It was less of a sacrifice than it once might have been. Franco had already turned against his brother-in-law sufficiently to permit Arrese, during the tour of Galicia, to convince him that the Begoña incident was a plot by Serrano Suñer to overthrow him.* He was heard to mutter that Serrano Suñer was 'a villain and a traitor' (*un malvado y un desleal*).[75]

In May 1941, Franco had not given in to military pressure to dismiss Serrano Suñer largely because to have done so would have made him the prisoner of his generals. Now it was a device to balance the blow to the generals constituted by Varela's departure.[76] At the beginning of September, a nervous Franco dismissed his brother-in-law as Foreign Minister. Looking extremely shifty, he said 'I want to speak to you about a serious matter, about an important decision that I have taken. After all that has happened, I'm going to replace you.' Serrano Suñer reacted with considerable aplomb and expressed relief that he could now take a rest.

---

* This accusation was given verisimilitude by the fact that the under-secretary of the Falange, Captain José Luna, a follower of Serrano Suñer, admitted having authorised the Falangist squad's journey to Begoña, albeit not the throwing of bombs.

When he proferred some papers which he had brought for signature, Franco said coldly 'I would prefer it if the new minister presented them to me.'[77] This was to be the sixty-six year-old General Count Francisco Gómez Jordana. Franco himself assumed the presidency of the *Junta Política* and therefore control of the Falange.

Jordana had served in General Primo de Rivera's Military Directory in the 1920s and as Franco's Foreign Minister from January 1938 to August 1939.[78] He was loyal to Franco, but enough of a monarchist for his appointment to be a sop to the high command and to make it seem as if Varela's gamble had nearly paid off.[79] It was proclaimed later by Franco's propagandists that the choice of Jordana as Serrano Suñer's replacement revealed a careful and prophetic shift in his foreign policy. At the time, however, the controlled press insisted that 'the substitution of certain persons in government positions does not and cannot produce the slightest variation of domestic or international policy'. It was merely a question of 'changing the guard'. Moreover, the other significant changes in the cabinet were the replacement of the Anglophile Varela as Minister for the Army with the notoriously pro-Axis General Carlos Asensio Cabanillas, and the monarchist Galarza with the Falangist lawyer, Blas Pérez González, hitherto Prosecutor of the Supreme Court. Accordingly, there is little reason to assume that Franco's cabinet changes were made with an eye on foreign affairs. The idea is rendered even less credible by the precipitate manner in which the Foreign Minister was dismissed and the way in which his successor was chosen.

Franco saw Jordana on Thursday 3 September. After giving him an account of the crisis, the Caudillo revealed that he was not first choice for the Ministry of Foreign Affairs but had originally been destined to replace Varela as Minister for the Army. Franco turned to Jordana after his first choice, an unnamed Admiral, had turned out to have links with the one-time President of the Second Republic, Niceto Alcalá Zamora, and his second, General Juan Vigón had, in solidarity with Varela, respectfully refused.[80] That alone suggests that Jordana was not chosen because of his Anglophilia. Serrano Suñer claimed convincingly some years later that Jordana's greatest advantage to Franco was the fact that he was a general conditioned to unquestioning military discipline who would never give him an argument, try to teach him lessons or put him in the shade.[81]

With Jordana heading for the Palacio de Santa Cruz, Franco had great difficulty finding a candidate for the Ministry of the Army. Varela was successful in persuading all his fellow Lieutenant-Generals not to replace him. Confidence that they would back him had been behind his letter of resignation. Franco was forced to go down to the level of Major-General before he could find a new minister in the person of General Carlos Asensio Cabanillas. A faithful Francoist with Falangist sympathies and hero of the Army of Africa's drive against Madrid in 1936, at first Asensio also refused the post to avoid difficulties with his immediate superiors. Franco overcame this reluctance simply by ordering Asensio to accept the appointment. He made it quite clear that he would resist to the death any challenges to his position telling Asensio 'I know that one day I'll leave here feet first.'[82]

A seriously dangerous crisis had been resolved and the undisputed winner was Franco. When Juan Vigón complained afterwards that it had all been unsatisfactory, meaning there had been no progress towards a restoration of the monarchy as he, Varela and other senior generals had hoped, Franco replied 'I thought that with the departure of Serrano Suñer you would be really pleased.'[83] The Army had been reinforced, gaining considerably over the Falange, although, in the final resolution, and in the promotion of Asensio, the dissident senior generals, led by Varela, Kindelán and Aranda, had lost ground to Franco. Nevertheless, the delight of the senior generals at the fall of the *cuñadísimo* could be perceived in the fact that, Kindelán aside, military dissent remained dormant for almost a year. The Falange was now in the hands of the most sycophantic elements concerned primordially with their own elevation within the regime. With Arrese now in the ascendant, it was more than ever the Franco-Falange.

There were interesting similarities between the Begoña crisis and the power struggle which had broken out in May 1941. Franco's acquisition in the earlier crisis of a shrewd but totally loyal and subservient servant in Carrero Blanco, was matched now by the appointment of the colourless forty-four year-old lawyer 'Comrade' Blas Pérez González as Minister of the Interior. A Major in the Army Juridical Corps, and a protégé of Lorenzo Martínez Fuset, Blas Pérez was to be one of Franco's most unconditionally faithful servants.[84] The two crises revealed the Caudillo's remarkable, and growing, talent for manipulating the component elites of the Francoist coalition. Perhaps most important in this respect was that what Franco was able to learn about the Army from the Begoña incident was roughly comparable to what he had learned about the Falange during the crisis of May 1941. It was made agreeably clear to him that the respectful restraint, not to say pusillanimity, of the majority of the anti-Falangist generals would consistently prevent serious efforts to overthrow him. Begoña was Franco's political coming of age. He would never again be as dependent on one man as he had been on Serrano Suñer. Franco now knew that his great political talent, and indeed the one on which his personal survival depended, was his ability to balance the internal forces of the Nationalist coalition. He would use it well.

Neither the Germans nor the Italians expressed much regret at Serrano Suñer's going, not least because he had increasingly been perceived as 'difficult'. *The Times* commented sagely that 'To regard Señor Suñer's dismissal as a political reverse for the Axis Powers is a tempting but unjustifiable exercise of the imagination.' Berlin and Rome had little cause for trepidation since the broad direction of Spanish policy did not change. Moreover, the Germans were delighted by the removal of Varela, whom they regarded as a dangerous Anglophile, and no less by the victory of Arrese over Serrano Suñer.[85] Indeed, the fall of Serrano Suñer and the promotion of the pro-Axis Asensio, together with the slowing down of German successes in Russia, took much of the steam out of Hitler's schemes to use Muñoz Grandes to impose a more pro-German policy on Madrid.[86] The Führer had grown sceptical about Muñoz Grandes' plans for overthrowing Franco which, with the Spanish Army again clearly under the Caudillo's thumb, he now dismissed as 'fantasy'. The dexterity with which Franco had managed the Begoña crisis had impressed Hitler and Ribbentrop was delighted, as were Göring and Himmler, that the demise of Serrano Suñer 'puts an end to his game of passing himself off as a friend of the Axis while preventing Spain joining the Axis coalition'.[87] Only two days after Serrano Suñer's dismissal, Hitler commented appreciatively 'Taking it all round, the Spanish press is the best in the world.'[88] There was again talk of Serrano Suñer being sent to Rome as Ambassador. However, for their different reasons, neither Jordana nor Franco had any desire to give him the chance. Ciano claimed to be relieved that Franco chose instead Raimundo Fernández Cuesta.[89] Serrano Suñer virtually disappeared from politics thereafter, rebuilding a successful career as a lawyer.* The relations between the two families became coolly polite,

---

* The fact that Serrano Suñer played little part in subsequent Falangist machinations goes some way to substantiating his frequent assertion that he had never been in politics out of personal ambition. He became a kind of elder statesman, writing occasionally in the conservative daily *ABC*, and from time to time urging Franco to open up the system. In later years, unlike many of his contemporaries, Serrano Suñer did not deny his own Fascist past nor try to turn himself into an instantaneous *demócrata de toda la vida* (lifelong democrat). He never repudiated his admiration for Mussolini, 'a real giant', and regularly attended masses in his memory. In 1959 and 1965 respectively, he wrote generous obituaries for Sir Samuel Hoare and Sir Winston Churchill – reprinted in Ramón Serrano Suñer, *Ensayos al viento* (Madrid, 1969) pp. 123–7, 141–50.

Zita Polo supplanted in Doña Carmen's affections by her friend Pura Huétor.[90]

In the wake of Serrano Suñer's going, Spanish policy began to change despite many public declarations to the contrary. Jordana told the Vichy French Ambassador on 9 September that the real maker of foreign policy was Franco and that he was no more than the Caudillo's 'docile executor'.[91] Nevertheless, Jordana's pro-Allied tendencies gradually had an impact despite the Generalísimo's persistent hopes for Axis success. The controlled press announced that continuity would not be interrupted.[92] The same message was explicit in an effusive letter which Franco sent to Mussolini on 18 September 1942, declaring that 'the changes carried out in the Spanish Government do not in the least affect our position in foreign affairs but are aimed rather at reinforcing our position in domestic politics.'[93] At that particular moment, there was no pressing need for Franco to write such an effusive letter – if anything, the reverse. A battle over the direction of foreign policy was fought out at a four-day cabinet meeting held on 17, 18, 19 and 21 September. The final communiqué reflected the conflict between Jordana and Arrese.[94] It linked Spanish policy to 'the imperatives of the new European order' yet made friendly references to Portugal and Latin America which signalled Jordana's desire to mend fences with Britain and America. That was only common sense in the light of the scale of the military preparations taking place at Gibraltar for 'Torch'. However, the bulk of the statement suggested the continuing strength of pro-Axis feeling in Franco's inner circle.[95]

Franco was beginning to be the silent arbiter of cabinet meetings, keeping his views to himself and letting others commit themselves. On the eve of Operation Torch, the Begoña reshuffle certainly favoured the Allied cause even if that was not Franco's intention. While in Italy in June, Serrano Suñer had told Ciano that, despite Spain's unpreparedness for war, she would certainly 'unsheathe her sword' in the event of an Allied landing in North Africa.[96] The tiny Jordana* had a reputation for courtesy, straightforward honesty and cautious common sense. He brought an element of prudence to Spanish foreign policy which it had lacked for the previous three years. His press office talked of Spanish military strength in terms of resisting invasion rather than of imperial conquest. Lisbon was delighted with his appointment.[97] Hayes was quick to see in Jordana's attentions a major shift towards the Allies. His choice

* Hoare described him as 'small to the point of insignificance' – 'When he sat in his chair, his feet did not touch the ground', (*Ambassador*, p. 175).

of the monarchist José Pan de Soraluce as Under-Secretary of Foreign Affairs was widely interpreted as a pro-Allied gesture.[98] Nevertheless, Jordana's appointment was not perceived in Berlin as an anti-German move.

Despite his admiration for the handling of the Begoña crisis, Hitler continued to resent the Caudillo – indeed since mid-1942 it had been forbidden for Franco's name to be mentioned in his presence. On 31 May 1942, the Caudillo had awarded the full honours of Captain-General to San Fuencisla, the patron saint of Segovia, in recognition of the miraculous defence of the city by General Varela.[99] When Hitler was informed, he commented 'I have the gravest possible doubts that any good can come of nonsense of this kind. I am following the development of Spain with the greatest scepticism, and I've already made up my mind that, though eventually I may visit every other European country, I shall never go to Spain.'[100] The Führer had begun to express regret for backing Franco in the Spanish Civil War: making his remark that 'The real tragedy for Spain was the death of Mola; there was the real brain, the real leader.' In the summer of 1942, he had said 'It is obvious that he is incapable of freeing himself from the influence of Serrano Suñer, in spite of the fact that the latter is the personification of the parson in politics and is blatantly playing a dishonest game with the Axis Powers.' Hitler was delighted by Serrano's down-fall, which further undermines the view that the replacement of the *cuñadísimo* was inspired by Franco's desire to distance himself from the Third Reich.[101]

In total contrast, Franco's admiration for Hitler was unabated. On 30 September 1942, in a bizarrely naïve variant of his two-war theory, the Caudillo assured Myron Taylor, President Roosevelt's personal representative to the Vatican, that Hitler was an honourable gentleman who had no quarrel with Great Britain nor any thought of impairing its independence.[102] However, in the course of October, when the Allied preparations for Operation Torch demonstrated that Britain was far from defeated, Franco behaved circumspectly. This has been interpreted as prophetic awareness of an eventual Allied victory. It was merely a reasonable short-term caution. The massing of force on his borders was hardly the best moment to cross swords with the Allies, particularly in the wake of Rommel's failure to conquer Egypt. Nevertheless, the military planners for Operation Torch were intensely aware of the damage that might be done either by Spanish hostility or else acquiescence in a German attack on Gibraltar. Accordingly, in mid-September the Foreign Office elaborated a scenario whereby the British and American Ambassadors would give Franco explicit reassurances that the operation did not threaten any Spanish territory. General Eisenhower was uneasy about this, telling his aide that, while military necessity might require such dealing with Franco, he thoroughly disliked his despotism and his contacts with Hitler and Mussolini.[103]

There can be no question that, in November 1942, Spain's attitude to Operation Torch was to have a major impact on the rest of the war. Thousands of Allied troops and tons of equipment were gathered in Gibraltar prior to the operation and eventually shipped through the Straits under Spanish guns on both sides of the Mediterranean. The heavily doctored post-1945 version of Franco's role in the war produced by his propaganda machine portrays the Caudillo as resisting German blandishments to cut Allied communications and so damage Operation Torch. To his admirers, this is proof of his benevolent service to the allies.[104] In fact, beyond an awareness of forthcoming Allied intervention somewhere in North Africa, neither Franco nor any of his ministers had any real inkling of what specifically was being prepared.[105] German pressure on Franco to oppose the Allied operation was curiously muted. It was however, not surprising that Churchill, in October 1944, in rejecting an offer by Franco to join in a post-war anti-Communist alliance, nevertheless commented on 'the supreme services' which Franco had rendered the Allied cause 'by not intervening in 1940 or interfering with the use of the airfield and Algeciras Bay in the months before Torch'. It is this (far from disinterested) testimonial from Churchill which lies behind the oft-propounded view of a canny Franco, foreseeing the eventual result and holding off by a charade of pro-Axis rhetoric an invasion of Spain by Hitler to seize Gibraltar.

However, Franco's caution was not born of any special perspicacity so much as intense awareness of the Allies' power of retaliation. He was especially worried that an attack on the Canary Islands might be in preparation. There were those in his Foreign Ministry, including the Director-General of Foreign Policy, José María Doussinague, who welcomed 'Torch' as an opportunity for Spain to draw closer to the Axis. They believed that now was the time to get food and arms in return for defending the Canary Islands and for granting Germany an unimpeded passage through Spain to North Africa. However, with sections of the American press calling for a break in diplomatic relations with Spain, Franco and Jordana were deeply apprehensive about a possible Allied attack on Spanish territory. So, on 26 October 1942, yet another entirely mendacious denial of the supply facilities for German submarines was issued.[106] In any case, Franco and Jordana were influenced by reports from the Ambassador in London, the Duque de Alba – eventually confirmed by assurances from both the British and American Ambassadors – that the Allies intended no hostility towards Spain. Hayes was told, when the time came, 'to give the most sweeping commitments that the United States will take no action which would in any way affect Spain or Spanish territories'. Hoare gave solemn assurances to both Jordana and Franco in late October as did Hayes on 3 November.[107]

The efforts of the Allied diplomats in Spain were crucial. Prior to giving these specific assurances of Allied benevolence, Sir Samuel Hoare had been working hard to persuade Franco of his dependence on British and American economic resources. He had spoken with the Caudillo on 19 October 1942 and thought his 'mind seemed more alert than usual'. Hoare assured him that there would be no British intervention in the internal affairs of Spain either during or after the war nor any British invasion or occupation of Spanish mainland or overseas territory. He reassured the Caudillo that Britain was giving no support to his Republican enemies and reminded him of the ease with which Spain was gaining navicerts for wheat. Lest Franco derive too much satisfaction from his conciliatory approach, Hoare also listed the anti-Allied activities taking place in Spain with official connivance, particularly the provisioning of German submarines in Vigo and asked him to account for five visits from Admiral Canaris in the previous six months. Franco said that he saw no reason why Spain, having stayed out of the war for three years, should not remain out until the end. He lightly brushed the submarine incidents aside as the result of 'inadvertence or corruption', and merely laughed when pressed about Canaris, which led Hoare to comment that 'Franco seemed throughout the interview more friendly and communicative than I have previously known him'. Nonetheless, in the Foreign Office, it was quickly noted that Franco 'was careful not to commit himself to give us satisfaction on any of our various complaints. In fact, he is still not sure that Germany will not win the war in the end and until he is sure we can expect little change in his present policy.'[108]

The German Embassy at this time was suggesting to Jordana that Spain inform London and Washington that an Allied landing in French Morocco would be considered a *casus belli*. On 27 October, Stohrer suggested that now was the moment for Spain to seize French Morocco. The idea of Spain thereby being able to interrupt any Allied advance in North Africa to the benefit of Rommel was attractive to Doussinague and other extremists in the Ministerio de Asuntos Exteriores.[109] However, again neither Franco nor Jordana were seriously tempted. If anything, they drew the conclusion that Stohrer's suggestion implied that the Germans were in no position to take more robust action of their own. At a tense cabinet meeting held on 4 November 1942 to discuss the international situation, the Allied assurances were thus crucial.[110] Even as the meeting was in session, Hull was telegraphing Hayes to the effect that Roosevelt wanted him to go beyond his statement of the previous day and to inform Franco that greater US economic assistance could be expected 'so long as Spain remains out of the conflict and does not permit her territory to be infringed by the Axis powers'.[111]

Roosevelt's anxiety derived from information received by the American military attaché from a Spanish army officer in early November 1942 that Franco had been requested by Hitler to permit the passage of German troops through Spain in the event of Allied military operations in north-west Africa.[112] This rumour was never substantiated and it seems more likely that German pressure did not go beyond Stohrer's insinuations of 27 October. Indeed, in October 1945, Serrano Suñer asserted that there had been no pressure at all.[113] Nevertheless, Roosevelt authorized Hayes to tell Franco that the United States would support him in resisting Axis aggression. Hayes gave that message to Jordana on the morning of Friday 6 November, suggesting that, to stay out of the war, Spain needed to make unequivocal declarations of its readiness to defend its neutrality against both sides. The fact that the Spanish Army was deployed to resist an attack from the south rather than from the north was deemed by Hayes to imply hostility to the Allied cause. The continuing enthusiasm for the Axis led Hayes to describe the Spanish press as 'an instrument of Axis political warfare'. However, there is no reason to doubt the normally truthful Jordana's categoric assertion that Germany had not requested Franco's consent for the passage of troops through Spain.

Nevertheless, Franco's hand could be seen in Jordana's clear attempt to be of service to the Axis by bluffing the Allies when he implied unequivocally to Hayes, as he had to Hoare on the previous day, that, if the Allies invaded French North Africa, then Spain would be forced to enter the war on Germany's side. This was exactly what Stohrer had requested exactly a week previously. Hayes countered Jordana's threat by assuring him that Allied intervention in the Peninsula would ensue if the Germans were permitted passage through Spain, a rather more direct threat than the British would have preferred. Hayes commented 'Following my previous conversation with the Minister in which he expressed gratitude for the assurances I gave him on behalf of our Government, Jordana spent the entire day of Tuesday [3 November] with General Franco and I consider that today he was interpreting Franco's attitude.'[114]

On Sunday 8 November 1942, Operation Torch began. Hayes and Hoare both saw Jordana in the course of the morning to reassure him that Spanish interests would be fully respected. Hayes had spread panic in Spanish official circles by demanding to see Jordana at 2 a.m. and insisting that the Foreign Minister arrange an immediate audience with Franco. A deeply anxious Jordana, in dressing-gown and pyjamas, received Hayes and Beaulac, dreading that he was about to be informed of an Allied assault on Spanish territory. Hayes was told that Franco, as so often, was absent on a hunting party. It has been suggested that Jordana said this merely to gain time and that Franco was actually with his military ministers reviewing the situation and then kept vigil through the dawn praying.[115] It is entirely possible, if he expected a declaration of war from the Allies, that Franco would feign absence in order to gain time. Hayes refused, for security reasons, to allay Jordana's fears by telling him the purpose of his demand. Finally, after the hour of the Torch landings had passed, he ended Jordana's misery by revealing to him the contents of the letter to Franco from Roosevelt which he had brought. Its tone was friendly, reassuring and ended with the words 'Spain has nothing to fear from the United Nations'. Hayes was eventually admitted to see Franco at 9 a.m. The Caudillo appeared calm, received him cordially and expressed his appreciation of the Allied guarantees. When relayed to Allied military leaders, Franco's assurance of his neutrality caused considerable relief. The Caudillo replied formally to Roosevelt on 10 November, accepting his assurances and expressing his 'intention of avoiding anything which might disturb our relations in any of their aspects'. Roosevelt's letter was printed in the Spanish press along with Franco's acceptance of its contents.[116]

The pro-Axis Minister for the Army, Asensio, along with the two senior Falangist Ministers, Girón and Arrese, believed that this was the ideal moment for a Spanish entry into the war on the German side. There were fierce clashes in the cabinet between them on the one hand and Jordana, Vigón and Moreno on the other.[117] Franco, having delayed discussion of the issue for a few days, remained the silent arbiter. In the interim, the Caudillo was the recipient of a disturbing visit on 11 November 1942, barely three days after the Allied landings. The most senior general on active service, the Captain-General of Barcelona, General Kindelán, travelled to Madrid to discuss the significance of the landings with the rest of the high command and with Franco himself. Kindelán informed the Caudillo in unequivocal terms that if he had committed Spain formally to the Axis then he would have to be replaced as Chief of State. In any case, he advised Franco to proclaim Spain a monarchy and declare himself regent. Franco gritted his teeth and responded in a conciliatory – and deceitful – way. He denied any formal commitment to the Axis, claimed that he had no desire to stay any longer than necessary in a post which he found every day more disagreeable and confided that he wanted Don Juan to be his ultimate successor.

Kindelán forcibly argued that the superior economic and industrial power of the Anglo-Saxon Allies would guarantee their eventual victory and that Spain must therefore remain neutral. He told Franco that it was not acceptable to the Army that its Commander should also be the head of a party, particularly one whose failure was as ignominious as that of the Falange. Since Kindelán could claim to be speaking for Generals Jordana, Dávila, Aranda, Orgaz, Juan Vigón and Varela, whom he had also seen on his trip, Franco simulated a cordial acceptance of what was said but his patience was wearing thin. On his return to Barcelona, Kindelán assembled at his home the generals and other senior officers of the Catalan military region. He told them that 'the ship of state is adrift in a sea of total misrule' and spoke of the incompetence and corruption of the Falangist bureaucracy. Asserting that no solution could be expected from the present regime, he called for a radical change of persons, methods of government and regime. Franco had squirmed long enough under Kindelán's criticisms and his assumption that the Caudillo's actions were subject to the approval of the generals who had voted for him in September 1936. After a cautious interval of three months, Franco replaced Kindelán* as Captain-General of Catalonia with the pro-Falangist Moscardó.[118]

It was not surprising that, with the military monarchists emboldened by the Allied landings, the Caudillo should chose this moment to rehabilitate Yagüe, making him, on 12 November 1942, commander of Spanish forces at Melilla.[119] Yagüe was still totally convinced of the imminence of Axis victory despite Operation Torch.[120] It was a remarkably clever posting. Franco knew that Yagüe was being courted by the Germans either as a possible direct replacement for him or else to bolster the claims of Muñoz

---

* In March 1943, Kindelán was made Director of the Escuela Superior del Ejército, where he would not have direct command of troops. Thereafter, he continued to agitate, in largely respectful terms, for Franco to begin preparations for the transition to an authoritarian monarchy.

Grandes. With monitoring the progress of the Allies to occupy him, the pro-Axis Yagüe was unlikely to become involved in plots against Franco although he was in any case too idealistic, and indeed loyal, to play the German game.[121] More significantly, in Melilla, Yagüe would be a valuable counterweight to the pro-Allied monarchist High Commissioner in Morocco, Kindelán's crony Orgaz. There was every chance that two potential rivals would neutralize one another.

By the time that the proposal of Asensio, Girón and Arrese for an early declaration of war on the Axis side was on the cabinet agenda, Allied successes were so spectacular as to inhibit any Spanish thoughts of independent hostile action. On 11 November, German troops had occupied Vichy, a reflection of a fear that events in North Africa might inspire a realignment of French loyalties or even be the prelude to an Allied landing in Mediterranean France. On the following day, Franco ordered a general mobilization to permit Spain to resist threats to her frontiers from either belligerent. This was interpreted by Hayes and in Washington, after secret service analysis of cable traffic between Madrid and Spain's Washington Embassy, as favourable to the Allies.[122] More unsubstantiated rumours circulated in mid-November in diplomatic and military circles that Hitler had demanded of Franco a free passage through Spain for his troops.[123] Jordana told Hayes that Spain had requested German guarantees that her territory would not be violated. Despite the hints to Hayes about German requests for the free passage of troops through Spain, it is clear from a letter written by Jordana to Alba in late November that there were no such pressures. Berlin gave verbal guarantees not to invade Spain and explicitly stated that the Third Reich had no intention of sending troops through Spanish territory. This was supported by the deployment of only light German forces in Southern France.[124] Equally, Jordana was adamant that the mobilization was not a move away from pro-Axis non-belligerency back towards a strict neutrality. That was contrary to messages being given by the Spanish Government to Washington.[125]

Imaginary Axis threats were inflated by Franco in order to ingratiate himself with the Allies and vice versa.[126] The fact that Anglo-American forces had entered precisely those French Moroccan and Algerian territories which he coveted can hardly have pleased Franco. His bitterness did not, however, blind him to the need to hedge his bets. That did not mean that he had lost his belief in an ultimate Axis victory.[127] That was quite clear in a letter which Jordana wrote in mid-November to the new Spanish Ambassador in Berlin, Ginés Vidal. Jordana systematically presented even the tiniest details of his work to Franco for approval.[128] His master's hand was visible in the instruction to seek for 'Spain (the only nation in the world which openly and sincerely professes her friendship for the Third Reich)' war material, free of charge, in order to resist the Allies and so relieve Germany of the need to defend an additional flank.[129] Given that Franco had accepted Roosevelt's guarantees both privately and publicly, the duplicity involved in the offer to Berlin was patent.

The emptiness of Franco's friendship was revealed less than a week later. On 14 November, Stohrer requested permission for German aircraft to use the Balearics as a base from which to rescue airmen shot down in the Mediterranean. When Jordana discussed the question with Franco at his regular Thursday meeting, on 19 November, the Caudillo refused on the grounds that to accede to the German request would provoke the hostility of the Allies.[130] The refusal was perhaps part of Franco's strategy of persuading the Germans that he must be given military help to permit him to stand up to the Allies. The Spanish Foreign Ministry drew up on 24 November 1942 a document entitled 'Grounds for Political Negotiations with Germany' which made it clear that Franco expected German weaponry without conditions, without payment, and without supervising officers or technicians.[131] It was a characteristic initiative, vague, ambiguous and opening up all kinds of possibilities. At face value, it was a genuine appeal to the Axis for help to strengthen an ally as the Allies massed near his frontiers. For all his sympathy with the Third Reich, Franco was trying to take advantage of Axis difficulties exactly as he was exaggerating German threats to squeeze benefits from the Allies.

The extent to which Franco was now coming in to his own, ready to pit his native cunning against both Axis and Allies, was soon made clear. At a cabinet meeting on Monday 23 November, he instructed Jordana to press the Americans on their promises to expand their economic assistance to Spain. Hayes optimistically took the consequent request for fuel and foodstuffs as a sign that 'Jordana and Franco probably are now contemplating eventual victory of the United Nations and can accordingly be counted on by us as potential friends rather than enemies'.[132] Hayes was right for the wrong reasons: an indication of Franco's mood at this time came during his fiftieth birthday celebrations. To Arrese and a Falange delegation, Franco declared that 'I match your faith and fanaticism with my own. I believe in Spain because I believe in the Falange.'[133] Thanking Hitler, who had sent a birthday telegram, as had King Vittorio

Emanuele III and Mussolini, Franco sent 'best wishes that victory accompanies your armies in the glorious enterprise of freeing Europe from the Bolshevik terror'.[134]

Three days later, in the Caudillo's address to the *Consejo Nacional* of the Falange on 7 December, there were expressions of his undiminished faith in the Axis. It was to be one of his most revealing and important speeches. However, there could also be discerned, in what was otherwise a remarkable display of complacency and myopia, the faint outlines of a formula for his own survival in the event of Axis defeat. In a disingenuous demonstration of his political thought, Franco took as his text the dramatic moments through which the world would pass after the war came to a close. He remained convinced that the liberal democracies were doomed: 'The liberal world is crumbling, the victim of the cancer of its own mistakes, and with it will fall commercial imperialism and finance capitalism with their millions of unemployed.' The cataclysm which Franco foresaw for Britain and America had already happened after the First World War in Italy, where 'the genius of Mussolini provided fascist solutions and outlets for everything that was just and humane in the rebellion of the Italian people'. Similar praise was found for Nazism.

With obvious pride the Caudillo unequivocally identified his regime with their 'youthful rebellion against the hypocrisy and inefficacy of the old liberal systems'. He predicted, in a display of near megalomania, that, whether the liberal democracies won or lost, they would have to face up to a revolutionary flood that could not be contained within bourgeois democracy and then they would have to look to the Spanish example. 'Because we know that we possess the truth and have worked for six years with that purpose, we watch events with serenity.' The speech was as nebulous as most of his public utterances. Yet, behind the confusions and the contradictions, could be seen the beginnings of what would eventually be inflated into the full-blown argument that Francoism was a truly original path, different from fascism. In the unlikely event of an Allied victory, the Caudillo was saying, I would be a valuable ally. Keeping open his options, as always, Franco also suggested that, when his interests demanded, he would contemplate the 'installation' of a new Falangist monarchy – not the restoration of the old constitutional monarchy.[135] The Caudillo was now, in the aftermath of Torch, seriously beginning to look for insurance.

XIX

THE HERO AS CHAMELEON

January 1943–January 1944

DESPITE HIS enduring sympathy with the Axis, Franco was now working hard to maintain good relations with both sides. On Thursday 26 November 1942, he had told Jordana that this would be a good moment to return the visit made by Oliveira Salazar to Seville nine months previously. The Spanish party set off on 18 December. After being received by the Portuguese President, General Carmona, Jordana and Salazar signed the treaty known as the Bloque Ibérico on 20 December 1942. It was hailed in both countries as a future bulwark of peace.[1] The visit was viewed with some hostility in Berlin.[2] But, if Franco had shelved his imperial designs on Portugal, he was far from slamming the door on the Third Reich. His posting of the pro-Axis General Yagüe to Morocco had obliged the Americans to assign large numbers of troops to guard against Spanish incursions into French Morocco.[3]

Yagüe was not the only pro-German in the Army. On leaving the Russian front, Muñoz Grandes was received again by Hitler, on 13 December 1942, and in recognition of his service at the head of the Blue Division, decorated with the much-coveted Oak Leaves to the Knight's Cross of the Iron Cross. Hitler still hoped to use him in some way – at the very least, to influence Franco. The austere Muñoz Grandes had seen enough in Nazi Germany to make him highly critical of the lack of social justice and efficient administration under Franco. Hitler's main concern, however, was to ensure that Spain would resist any invasion by the Allies and he made proposals about deliveries of German arms to Spain for that purpose. There was no question of German threats against Spain.[4] When Muñoz Grandes finally returned to Spain in February 1943, Franco promoted him to Lieutenant-General, awarded him the Falange's highest honour, the *Palma de Plata*, and named him Head of his Military Household in succession to Moscardó. They were typically astute moves. The promotion made Muñoz Grandes too senior to be able to command a mere division and thus prevented his return to Germany. The posting to El Pardo also denied Muñoz Grandes any operational command of troops, keeping him close and publicly displaying him as a courtier. Muñoz Grandes was happy to be near the Caudillo and confident that he could incline him in a more unreservedly pro-German direction.[5]

Ambassador Hayes had a faith in Franco's good intentions not shared by the US War Department, which observed that all Spain's military preparations were directed against the Allies. Intelligence reports noted that on the Spanish side of the French frontier there were no troops, just the usual frontier guards. In contrast, substantial reinforcements were being made in Morocco and pro-Allied Spanish officers were being replaced by known Axis sympathizers, of whom Yagüe was merely the most notable case.[6] Hitler might perhaps have welcomed Spanish help at this stage of the war although by now he could not afford to pay for it. In any case, Spanish prevarications in the course of the previous month can have done nothing to flesh out his withered faith in Franco. Nevertheless, when the new Spanish Ambassador, Ginés Vidal, presented his credentials in early December, he made new requests for help. The Führer asked for a list of Spanish requirements and stated without much conviction that he would do his best to meet them.[7]

In the Caudillo's suggestions to Germany that she *must* now supply food and arms, there were elements of low cunning and sheer rapacity, and a heartfelt desire not to see the balance of the World War tip too far against the Axis. Franco had not previously shown any distress about the massive current account imbalance between Spanish exports to the Third Reich, particularly of food, and deliveries to Spain of German goods.[8] Spain had not yet repaid her Civil War debts, but in 1941 German imports from Spain had been more than double her exports, 167 million marks to 82 million. In July 1942, Franco told Stohrer that he accepted that the imbalance had to be subordinate to the needs of the German war economy. Agreement was reached a few weeks later for a huge increase in Hispano-German trade in 1943 to 388 million marks worth of food and raw material from Spain and to 230 million marks worth of manufactured goods and machinery from Germany of which 130 million was to be in armaments.[9] After the Anglo-American landings, a new urgency was to be seen in Spanish requests for German armaments. Asensio, the Minister for the Army, told Stohrer that it was crucial to reinforce Spain's military potential before the Anglo-American forces in North Africa grew sufficiently strong to attack Spanish Morocco. General Orgaz made the same point to the German Consul General in Tangier.[10]

Hitler was sufficiently intrigued by the messages that he was receiving through both Ginés Vidal and Muñoz Grandes to send Canaris to Madrid to investigate.[11] On 29 December 1942, Jordana received the Head of the *Abwehr* and assured him that the Spanish mobilization order of 12 November had been provoked by the Allied landings in North Africa. In a near explicit admission of Franco's game of playing off both sides, he told Canaris that, if Germany could not help, then Spain would look elsewhere.[12] Enthusiasm for Spanish belligerence on the Axis side was still being expressed by Asensio, Yagüe and Muñoz Grandes. At dinner with them on New Year's Eve, Franco indulged his old fantasies and seemed to be in agreement.[13] The new drift of his policy was made clearer just over a week later. While he was making his overtures to Berlin via Vidal, Franco was simultaneously attempting to render his relations with the Allies more cordial. On 6 January 1943 at the annual Epiphany Banquet for the Diplomatic Corps, he was especially friendly to Sir Samuel Hoare, taking him aside and treating him to a personal resumé of his two-war theory and the need for an early peace treaty.[14] Franco was about to send a military commission to Berlin to discuss the arms deliveries that would be essential if Spain was to enter the war on the Axis side and the pro-German Arrese was about to take a message from the Caudillo to Hitler. The overture to Hoare was Franco's way of surrounding his plans with a smokescreen, and also taking out insurance lest the Axis connection not bear fruit.

In 1943, the international panorama in which Franco operated had changed dramatically. Not only had 'Torch' shifted the strategic balance, but many of the protagonists with whom Franco had to deal were changing too. Serrano Suñer had already gone. On 27 December 1942, Stohrer was recalled from Madrid. He was replaced partly because he had failed to predict the Torch landings and because Ribbentrop thought him insufficiently enthusiastic about the Nazi cause.[15] To a small extent, that was the work of Hoare who had deliberately pretended to have clandestine contacts with him – they lived in neighbouring houses – and praised his anti-Nazi sentiments. Madrid chortled to the joke that, when Hoare returned to London, he would be followed as British Ambassador by Stohrer.[16]

Stohrer was replaced by a scion of one of the great Prussian families, Hans Adolf von Moltke, who arrived in Madrid on 11 January 1943. Like von Stohrer, he was a diplomat of the old school, but much more enthusiastic about National Socialism. Reputed to be brutally arrogant, he had been German Ambassador in Poland until the Nazi invasion. Thereafter, he had been engaged in classifying the captured Polish archives. It was assumed in Madrid diplomatic circles that his appointment presaged the end of Stohrer's accommodating attitude.[17] Unlike Stohrer, Moltke neither spoke Spanish nor knew much about the internal situation of Spain. Accordingly, he came to rely, during his brief occupation of the post,* on the Embassy press secretary, the Nazi Hans Lazar. At the end of January 1943, Moltke's ignorance of the Spanish situation was nearly to provoke a serious rupture in German-Spanish relations when he allowed himself to be convinced by Nazi provocateurs in his Embassy that Franco had flown to Lisbon to meet Churchill to arrange Spanish entry into the war on the Allied side.[18]

That gaffe aside, Moltke made one significant contribution to Hispano-German relations. This was to be a secret protocol between the German and Spanish governments signed in Madrid on 10 February 1943. It had its origin in the Spanish document drawn up in November 1942 laying down the bases of Madrid's request for German military aid and the subsequent conversations of both Vidal and Muñoz Grandes with Hitler. On 13 January 1943, only two days after his arrival in Madrid, Moltke saw Jordana and requested guarantees that any arms sent would be used against the enemies of the Reich. Moltke got the reassurances but felt that Jordana was uncomfortable with the negotiation over arms – which suggested that it was the Caudillo's own initiative.[19]

That Franco was making the pace became more apparent when the new German Ambassador presented his credentials on 24 January. Moltke was surprised by the cordiality of Franco's reception. The Caudillo talked to him for an hour rather than the bare fifteen minutes demanded by protocol. Franco declared unequivocally that Germany was his friend; Britain, America and the 'Bolsheviks' his enemies. He swore that, within the limits of the possible, he was 'ready to support Germany in the struggle which destiny has imposed upon her'.† In a revealing comment on his two-war theory, Franco implied that one service which he could do for the Reich was 'to deepen the contradictions between England and the Soviet Union'.[20]

Five days later, Moltke told the Caudillo the conditions under which

* Moltke was to die in post from an acute appendicitis on 22 March 1943.
† Unaware of these dealings, the ever naïve Hayes wrote confidently to the State Department that 'the improvement in Spain's relations with the United Nations has been to the detriment of its relations with the Axis' – *FRUS 1943* (Washington, 1964) II, pp. 595–7.

Germany would give Spain the arms he had requested. There were those in the Spanish government, including Doussinague, who were excited by the prospect of a force of one million Spaniards armed by Germany fighting on behalf of the Third Reich. However, despite Franco's fervent protestations of a Spanish readiness to repel attack, little military equipment, other than eight aircraft, materialized in the next months.[21] The Secret Protocol,* dated 10 February, was signed on 12 February 1943.[22] In the event, the Secret Protocol was never activated but it constituted an irresponsible abandonment by Franco of Spain's freedom of action.[23] With narrow cunning, Franco was gambling that, armed with the pre-Torch assurances of benevolence towards Spain by both Roosevelt and Churchill, he could squeeze some profit from Hitler at low risk.

In the meanwhile, the *dramatis personae* continued to change at a giddying rate. On 5 February 1943, Mussolini dismissed Ciano and effectively took over the Foreign Ministry himself, working through the Under-Secretary Giuseppe Bastianini. Lequio had died in Rome in mid-January and his replacement Giacomo Paulucci di Calboli Barone presented his credentials on 20 April.[24] Hoping to get Franco to undertake to combat any possible allied landing in Spain, Mussolini had instructed Paulucci to suggest a meeting. Receiving Paulucci with great cordiality, Franco took the suggestion of a meeting with Mussolini to be only a personal initiative by the new Ambassador and brushed it aside with good wishes for Axis victory and references to his difficulties with monarchist opponents and his vulnerability to Allied pressures.[25]

The February Protocol was only one of several indications that Franco's faith in the Axis cause was dimmed but still alight. In its immediate aftermath, Jordana transmitted to the Americans the bare-faced lie that 'Franco has told the Germans not only that he will resist aggression from any side but that Spain will not even discuss possible military concessions to the Axis'.[26] The Caudillo's 1943 New Year message had been extremely pro-Hitler. The information which he received from his enthusiastically pro-Axis Ministers for the Army, General Asensio, and for the Air Force, General Vigón, and from the military attachés in Berlin was blindly optimistic. Indeed, the best-informed and most coldly realistic of the attachés, Lieutenant-Colonel Roca de Togores, displeased Asensio

---

* It declared simply that, in return for arms, war equipment of modern quality and in sufficient quantity, Spain would resist 'every entry by Anglo-American forces upon the Iberian Peninsula or upon Spanish territory outside of the Peninsula, that means, therefore, in the Mediterranean Sea, in the Atlantic and in Africa as well as in the Spanish Protectorate of Morocco, and to ward off such an entry with all means at its disposal'.

and was replaced on the technical pretext of having spent too long in Berlin.[27] The Falangist press also retained its enthusiasm for the Third Reich and denounced Roosevelt as 'guilty of provoking the Second World War'.[28] The German defeat at Stalingrad led at first only to demands that the Allies reconsider their mistakes before it was too late. It was then brushed off as a slight interruption on the road to inevitable German victory.[29] Axis publications formed a large proportion of the reading matter available to the Spanish public and cinema newsreels were largely Axis-originated. Hayes was finally driven to question how the freedom for psychological warfare granted the Axis could be compatible with Franco's assurance in his letter to Roosevelt of Spanish impartiality.[30]

In mid-January, the Minister-Secretary of the Movimiento, Arrese, along with a large group of Falangists, made an official visit to Berlin as the guest of the Nazi party. Arrese was hoping to get German support for his own political career and, in particular, to succeed where Serrano Suñer had failed in getting the Third Reich to grant Spain a North African empire. Franco hoped that the mission would counteract any German displeasure at Jordana's visit to Lisbon. He entrusted to Arrese a letter to Hitler requesting the arms which Muñoz Grandes had discussed with him. The Caudillo was insinuating that Arrese represented his own real pro-German feelings rather than the more conservative Jordana. It was an apparent repetition of the device whereby Serrano Suñer had bypassed Beigbeder in September 1941. Arrese played his part to the full despite being let down on occasions by his own lack of *savoir faire*. Determined to miss nothing granted to Serrano Suñer, he insisted on being received with full military honours despite not being a guest of the Government. The German authorities therefore arranged for him to be received in Hendaye and in Berlin by a detachment of the SS. He travelled to East Prussia where he had lunch with Ribbentrop and tea with Hitler on 19 January, as well as holding meetings with Goebbels and Bormann.[31]

Arrese made declarations in Germany which were so pro-Axis and contrary to Jordana's determination that Spain remain neutral that the Minister of Foreign Affairs offered Franco his resignation.[32] Franco neither accepted it nor reprimanded Arrese. On his return, Arrese gave vent to his pro-German sentiments in an enthusiastically belligerent speech to the Falange of Seville on 9 February 1943. Apparently oblivious to the defeat at Stalingrad three days earlier, he proclaimed that the Blue Division signified a commitment to fight Communism to the last.[33]

Franco was now conducting dual diplomacy, if not actively at least passively. He let Jordana be his mouthpiece to the Allies, Arrese to the

Axis. Whichever way the war went, he could always disown one or the other. This duality was matched by internal developments. To balance the apparent supremacy of the voluble Arrese, Franco established an annual requiem mass for all the kings of Spain to be said each year at the Escorial on 28 February, the anniversary of the death of Alfonso XIII. The Falange and the monarchy were each said to incarnate the permanent destiny of the Fatherland.[34] The clear implication was that the *Jefe Nacional* of the Falange was the modern equivalent of the great emperor-kings of the past.[35]

Throughout most of 1943, certainly up to the fall of Mussolini in the summer, Franco remained convinced that the Allies could not win and that their successes in Africa were of marginal importance. In a spirit of what seemed to Hoare to be 'impenetrable complacency', the Caudillo was convinced that he could eventually, after a long war, step in as broker between both sides. It was for that reason that he had floated the notion of two wars, one against Communism in which he was a belligerent and one in the west in which he was neutral. Hoare did not know what Franco had told Moltke on 24 January about his determination to drive a wedge between England and Russia. Already, on 6 January, Franco had broached with Hoare the Allies' 'grave error' of continuing to fight alongside Soviet Russia.[36] The Caudillo's views of the legitimacy of the German struggle against Russian Communism formed the basis of a lengthy exchange of written memoranda between Hoare and Jordana in late February 1943 and he assured the German Embassy that these initiatives helped the Third Reich by fostering the creation of a wide anti-Bolshevik front.[37]

On 12 February 1943, Ambassador Hayes arranged for a showing of *Gone with the Wind* at the American Embassy. It was a great propaganda coup and Franco asked for a private showing to be mounted at El Pardo. His personal cinema was a source of much pleasure to him. With his painting and writing activities in abeyance, films were his main diversion other than hunting and fishing which absorbed him more and more. Fresh-water fishing with his friend from La Coruña, Max Borrell, had resumed after the Civil War. They had started to fish too in the bays of the area in a small boat belonging to Borrell. Thoroughly enthused, Franco bought a larger boat from the Marqués de Cubas, called *Azorín*. Essentially a river boat, which had apparently once belonged to the Lord Mayor of London, its unsuitability for deep water expeditions eventually led to Franco commissioning a larger yacht, the *Azor*, which would become one of the passions of his life.[38]

Also in mid-February 1942, the Chaplain to the US Army, Archbishop

Francis J. Spellman of New York, stopped over in Madrid en route to the Vatican. He met Franco who did not miss the opportunity to pour into the Archbishop's sympathetic ear his notion that the war between the West and Germany was a regrettable mistake and that the real danger was Communism.[39]

In March, a Spanish armaments commission visited Berlin to arrange the details of the supplies agreed in the Secret Hispano-German Protocol. It was headed by General Carlos Martínez Campos, who was encharged by Franco with the further task of assessing the military capacity of the Third Reich in the wake of the defeat of Stalingrad. Armed with a list of Spanish needs in terms of aircraft and coastal defence batteries, Martínez Campos was received on 16 March by Marshal Keitel who was at pains to conceal the fact that Germany could not spare the arms in question. Two days later, at the Wolf's Lair, Hitler tried to convince him that it would be better to begin with some small deliveries of less sophisticated weapons. Taken off on a ten-day tour of the Nazi war industries, Martínez Campos was seduced by tales about the new wonder weapons with which the Third Reich would destroy Allied cities and armies and win the war easily. On his return to Madrid, he informed the Caudillo that the German war machine remained invincible.[40]

On 17 March 1943, Franco made his inaugural address to the newly fabricated pseudo-parliament, the Cortes. One third of the members were directly nominated by the Generalísimo. A further third were *ex officio* members – government ministers, members of the *Consejo Nacional*, the President of the Supreme Court, the *Alcaldes* of the fifty provincial capitals, rectors of universities and so on – all of whom had also been nominated to their posts by Franco or his ministers. Finally, the remaining third were 'elected' by the Falangist syndicates from carefully prepared lists of candidates. This was what Franco called 'organic' democracy. Although the 'representative' elements would be widened over the years, the Cortes met very rarely and always approved legislation submitted to it.* Ministers were responsible to the Caudillo, not to the Cortes.[41] His

* In private, Franco was contemptuous of the Cortes, inadvertently recognizing that it was merely a carefully constructed façade to cover his personal dictatorship. On one occasion in the early 1950s when his liberal Catholic Minister, Joaquín Ruiz Gimeñez, made some remark which suggested that he took seriously the farce of the Cortes, the Caudillo impatiently snapped 'and who do the Cortes represent?' – José María de Areilza, *Diario de un ministro de la monarquía* (Barcelona, 1977) pp. 73–6. On another, when one of his generals voted against a law in the Cortes, Franco was outraged, commenting 'if he doesn't like the project, he may abstain but never vote against since he owes his seat to me by direct nomination' – Franco Salgado-Araujo, *Mis conversaciones*, p. 214.

inaugural address echoed his constant refrain of the similarity between his rule and that of Spain's great medieval kings. The stress on historical precedent and Catholic elements in his social policy indicated that he was taking the first steps towards elaborating a unique Spanish way of politics which was authoritarian and hierarchical, similar to the Axis regimes, but sufficiently different to enable him to deny that similarity should the necessity arise.[42]

Franco's continuing hope of Axis success could be seen in his tacit acquiescence in the pro-German activities of Arrese and his entourage. Their anti-American propaganda constantly hindered Jordana's efforts to pursue a genuine neutrality.[43] Nevertheless, while still hoping for arms shipments from the Third Reich, a peace initiative was launched in April by Jordana at the celebrations held in Barcelona to commemorate the return from America of Christopher Columbus. The motivation was complex. Franco had taken on board enough of the significance of Stalingrad and El Alamein to sense the shift in the balance of military power. Now he hoped to contribute to slowing down things until Germany's wonder weapons came on stream. Even if he could do no more than foster negotiations which would permit the survival of the Third Reich, then his own future would be assured.[44] With one voice, the press echoed the Franco-Jordana initiative for a 'just and fraternal peace' while paying tribute to the 'peace-loving' Hitler's just struggle for independence.[45]

The Barcelona initiative was followed by Franco himself during a propaganda tour of Andalusia in early May. On 4 May at Huelva, after being presented with a 'victory sword', he described Spanish foreign policy as being inspired only by Christian spirit. He also developed the theme that Falangism was superior to both liberal democracy, 'the creator of modern slavery', and to Marxism, 'the annihilation of the individual'. At Jérez, he was made honorary *Alcalde*. On 7 May at Seville, after receiving the city's gold medal, he expressed fears as to how far the motorized hordes of Moscow might reach in their advance across Europe and also confidence in the ability of Spanish troops to halt them. On 8 May in Málaga, he linked up Germany's struggle in the East with the Nationalist cause in the Spanish Civil War, seeing both as conflicts between Christianity and barbarism. Finally, on 9 May 1943, in a speech to the Falange at Almería, Franco said 'We have reached a dead end in the struggle. Neither of the belligerents has the strength to destroy the other.' In a characteristic mix of the naïve and the hard-faced, he called for peace negotiations to achieve a united front against Communism and demanded a fairer distribution of the world granting Spain the place it deserved.[46]

The Caudillo placed sufficient importance on this speech to have it translated into English and printed as a pamphlet along with Jordana's earlier declarations in Barcelona.[47] There was outrage in Berlin since Franco's pamphlet was taken by the Allies as proof that the Axis was admitting defeat. After protests from Weizsäcker to Vidal, Franco hastened to inform the Americans that his peace feelers were not inspired by the Axis.[48] When he met Hayes to make this point on 11 May, Jordana pointed out that Spain could have damaged the Allies during Torch and did not do so despite what he called, with shameless exaggeration, 'tremendous German pressure'.

From the spring of 1943, Hayes noticed that he was especially favoured by Jordana, a reflection both of the fact that Franco's senior generals were becoming restless again and that his Ambassador to the Holy See, Domingo de las Bárcenas, had started sending alarming reports from Rome about Mussolini's increasingly precarious position, something to which the Ambassador to Italy, Fernández Cuesta, seemed oblivious.[49] An awareness of American strength was apparent when Franco spoke on 15 June with the new German Ambassador, Hans Heinrich Dieckhoff, who had arrived at the end of April after Moltke's sudden death in March. The arrival of the cosmopolitan Catholic Dieckhoff constituted a return to the more pliant ways of Stohrer.[50] The Caudillo told him that, since the Third Reich could never defeat both the USA and the British Empire, his peace feelers were in German interests.[51]

During the North African campaign, Allied policy towards Spain remained cautious because of the continuing threat that Franco might permit German troops passage to Gibraltar. However, by June 1943, with the Axis expelled from North Africa, the situation changed. A new sense of realism could be discerned in a lecture given by Carrero Blanco to the Real Sociedad Geográfica in Madrid on the importance of sea power in the present conflict. His remarks on the superiority of the Royal Navy over Axis navies suggested that, within the inner reaches of El Pardo, he at least had begun to doubt ultimate Axis victory.[52] Franco also actively sought to ingratiate himself with the Americans, although there is reason to believe that his motivation was concern about the growing pro-monarchist agitation within Francoist circles rather than any change of heart about the Axis.

Franco was working hard to consolidate military loyalty. On 5 June, an elaborately choreographed ceremony at the Alcázar de Toledo saw the 119 surviving fellow graduates of his own year (*promoción*) at the Toledo Military Academy pay homage to his present greatness. The master of ceremonies was Yagüe, commanding the two companies into which the now rather senior ex-cadets were formed under Generals Camilo Alonso Vega and Eduardo Sáenz de Buruaga. There was no spirit of comradely reunion but only of solemn pomp and adulation of the Caudillo. The entire gathering was contrived both as a reminder of the glories of the Francoist war effort and of the strength of military unity. Making a dramatic entry through the main door of the Alcázar, Franco inspected the two companies, heard mass and then received a specially struck medal from Yagüe.[53]

The need to demonstrate publicly the continuing loyalty of the military to his person was quickly revealed. Ten days later, a group of twenty-seven senior *Procuradores* (parliamentary deputies) from the Francoist Cortes, including the Duque de Alba, Antonio Goicoechea, the ex-ministers Alarcón de la Lastra and Valentín Galarza, and General Ponte, wrote an appeal to Franco, couched in respectful terms but containing a bombshell. Their manifesto called upon the Caudillo to settle the constitutional question by re-establishing the traditional Spanish Catholic monarchy before the war ended with an Allied victory. The clear implication was that only the monarchy could avoid Allied retribution for Franco's essentially pro-Axis position throughout the war. The signatories came from right across the Francoist spectrum, with representatives from the banks, the armed forces, monarchists and even Falangists. The Caudillo reacted swiftly to the challenge. Even before the manifesto was published, he had ordered the arrest of the Marqués de Eliseda who was collecting the signatures. As soon as it was published, showing how very little he was interested in his much-vaunted *contraste de pareceres* (contrast of opinions), he dismissed all the signatories from their seats in the Cortes immediately and sacked the five of them who were also members of the *Consejo Nacional*.[54] At the same time, he stepped up his efforts to cultivate his senior officers, spending time with them individually. In particular, he put great effort into winning over General Orgaz, the High Commissioner in Morocco. Jordana wrote that 'taming Orgaz has been one of the Generalísimo's greatest successes'.[55]

Prior to the manifesto of the *Procuradores*, there had been a notable softening of the approach towards the Allies. Unaware of the Spanish-German Protocol of February 1943 for the supply of German military equipment, the American Ambassador was pleased when Franco's government asked for Allied weaponry on the entirely spurious grounds that it was needed to allow Spain to reject German pressure to accept arms in payment of raw material purchases in Spain. There was no German pressure on Spain and certainly not to accept unwanted arms.[56]

In the second half of June, with Franco convinced that the Allies were behind the monarchist manifesto, Arrese lashed out with a furious pro-Axis onslaught by the Falangist press. Hayes was stung into protesting to Jordana.[57] Less convinced of Franco's sincerity, Hoare presented to Jordana in July a long list of unneutral acts by Spain including refuelling and repair of Axis submarines, the swift repatriation of Axis air crews who had made forced landings in Spain, sabotage attacks against Allied shipping from the south coast, the existence of observation and espionage networks, interference with British Embassy correspondence and the activities of the exclusively pro-Axis press.[58]

These complaints were ignored by Franco. Influenced by the pro-German Arrese, Girón and Blas Pérez, as well as Asensio and the younger generals, he still believed that Germany would defeat the Russians. Indeed, the attachés in Berlin were still sending reports that took for granted Allied defeat in Italy.[59] In that pro-Axis spirit, Franco made both public and private efforts to rally his forces. In his annual Civil War anniversary speech to the *Consejo Nacional de la FET y de las JONS* on Saturday 17 July, Franco moved to a last-ditch position. Merely ten days after the Allied landings in Sicily, he vehemently reasserted the hegemony of the Falange. Arriving with Arrese, both in the white summer uniform of the Falange, Franco saluted the waiting crowds fascist-style. With the monarchists having revealed themselves as untrustworthy, he was pinning his own survival on those who most depended on him, the Falangists. The Caudillo's speech slammed the door on internal political change and denounced the faint-hearted bourgeois and conservatives who failed to understand 'our revolution'. The continuing threat of Bolshevism was a reason to close ranks within the fortress of Falangism and not to risk the chaos of democratic institutions.[60]

Following the speech, there was the annual reception for the diplomatic corps at La Granja. Torrential rain forced the company indoors with the consequence that Franco's deliberate snubbing of both Hoare and Hayes was the more obvious. An outraged Hayes protested to Jordana about Franco's speech and demanded an audience with the Caudillo.[61] In fact, Jordana himself had been taken aback by the inopportune tone of Franco's declarations.[62] On the eve of his speech, Franco had drawn up, with the help of his faithful henchman, Carrero Blanco, an instruction to the eight *Capitanías Generales**. It was issued on 17 July 1943. Just as

---

* The military regions into which Spain was divided.

the speech had been an attempt to appeal to the Falangist rank-and-file, this document was meant to play on the reflexes of the most senior officers and provoke them into rallying around the regime. It claimed that an international masonic plot to drive a wedge between the Army and the Caudillo had been uncovered. Given the deeply engrained anti-masonic prejudices of Carrero Blanco and the Generalísimo, it is more than probable that they believed the contents of the document. To counter this non-existent conspiracy, the Franco-Carrero Blanco circular denounced the dangers involved in trying to re-establish a liberal monarchy on the grounds that it would in turn be only the first step to a return to pre-civil war anarchy and Communist domination.[63]

On Sunday 25 July 1943, dramatic news came from Italy. In the early hours of the morning, the Fascist Grand Council had passed a motion of no confidence in Mussolini and the King had seized the opportunity to have him arrested and replaced by Marshal Badoglio. There was panic in the Falange. Although there was no direct statement of the news in the press for two days, the consternation in the upper reaches of the regime could be deduced from *Arriba*'s Byzantine efforts to dissociate Falangism from fascism. Brave assertions were made that the Falange was not defeatist in the face of the Italian catastrophe.[64] Franco himself was desperately worried that if the Duce could fall so could the Caudillo. He wept as he recounted the events in Rome to the cabinet.[65] Despite the news black-out, copies were circulated of a letter from the secretary to the Spanish Ambassador in Rome graphically describing scenes of disorder and attacks on Fascist headquarters and speculating on the dangers of similar events occurring in Madrid. The Falangist Ambassador, Fernández Cuesta, was severely rebuked by Franco for permitting a dangerous act of defeatism. In public, the Caudillo vehemently asserted that there was no analogy between what was happening in Italy and conditions in Spain.[66]

His anxiety was firmly under control by the time that the audience demanded by Hayes took place on 29 July. The American Ambassador told the Caudillo that it was difficult to avoid the impression that his government was pro-Axis and not neutral. Hayes also pointed out that the fall of Mussolini presaged Allied victory and therefore difficulties for Franco himself. The Caudillo replied untruthfully that he had expected both Allied successes in North Africa and Italian defeat. He gave every impression of regarding his own regime as unassailable. He also affirmed his belief in German toughness, morale and ability to fight on. Hayes then made specific complaints regarding Spain's non-belligerent status, the anti-Allied activities of the Falange and the existence of the Blue Division and was treated to a virtuoso performance of obfuscation and lying. Franco brushed aside the first complaint with the assertion that non-belligerence was merely his way of showing that he was not indifferent in a struggle against Communism. He feigned surprise at Hayes' allegations about the Falange and its various press and propaganda agencies, then blamed underlings for failing to carry out his orders.

The truth made a brief appearance when Franco said that 'democratic propaganda was sometimes objectionable because it criticized the internal system in Spain', clearly something which was not a problem with Axis material. He justified the existence of the Blue Division in terms of Russian intervention in the Spanish Civil War. Hayes pointed out that Russian Communism had not been a problem for Franco when Germany and Russia were allies. The Caudillo also refined his two-war theory into a blatantly fanciful three-war theory, consisting of the war of the Anglo-Saxon powers against Germany in which Spain was neutral, of the war of the civilised nations against Japanese barbarism in which Franco would happily take part on the American side and of the war against Communism in which Spain was a belligerent. When Hayes pointed out in detail the absurdity of his arguments, Franco remained silent.[67]

Five days later, the US press attaché was informed that Franco had ordered Spanish press, radio and newsreel services to be placed on an impartial basis.[68] Nevertheless, the importance of Germany's fight against Russia remained a daily theme, although parts of the press did begin to show slightly more sympathy to the Allies.[69] Jordana informed Hayes on 7 August that, after consultation with his Ministers of the Army, Navy and Air, Franco had decided to seek a way of withdrawing the Blue Division and would make a declaration of neutrality at an early opportunity. Such a decision, if it had been taken, sat uncomfortably with the bizarre theories on international relations which Franco had expounded to Hayes on 29 July. Recruiting for the Blue Division was still taking place at the end of August.[70] Moreover, during the summer and autumn of 1943, the Germans tried to counter the inevitable impact in Spain of the fall of Fascist Italy by intensifying propaganda and stepping up anti-Allied sabotage activities in Spanish harbours.[71]

The changes in the international situation were exacerbating Franco's domestic problems. However, as so often, with his back to the wall, his cool fighting qualities came to the fore. One week after the fall of Mussolini, on 2 August 1943, Don Juan de Borbón telegrammed Franco, reminding him of the Duce's fate and asserting that the only way to avoid a catastrophe in Spain was the immediate restoration of the monarchy. The clear implication was that, if the Allies won the war and Franco was still in power, Spain would be punished as if she were one of the defeated Axis powers. Franco replied on 8 August 1943 with a telegram in which complacency and low cunning were equally balanced. Having confidently asserted that Spain could not suffer the fate of Italy thanks to the regime's success in keeping her out of the war, he went on to beg Don Juan not to make public any statement which might weaken the position of the regime internally or internationally.[72] It was at this time that one of Franco's close friends asked him how he would survive in the event of the Allies being victorious. He replied with astonishingly complacent equanimity 'send them my bill' (*pasar la cuenta*).[73]

Bravado aside, the build-up of anxiety among his formerly unconditional supporters can hardly have failed to preoccupy Franco. No doubt galvanized by the recent military developments in North Africa, and perhaps fancying himself as the Spanish Badoglio, General Orgaz took an uncharacteristic risk. He informed the ex-minister and inveterate monarchist conspirator, Pedro Sainz Rodríguez, that, by prior agreement with Aranda and other generals, he was ready to rise with one hundred thousand men to restore the monarchy, provided that immediate Allied recognition could be arranged by Don Juan's followers.[74] The Caudillo's anxiety must have been exacerbated when he was informed during his summer holiday at the Pazo de Meiras in La Coruña that his Lieutenant-Generals were meeting in Seville to discuss the situation and had composed a document calling upon him to take action.[75]

It was in the midst of these preoccupations that, on 20 August 1943, Hoare had a lengthy interview with Franco at the Pazo de Meirás. The meeting, which had been arranged nearly one month previously, was extremely well choreographed by the Spaniards, because Franco intended to make use of it to imply to the world that his relations with the British were excellent. The Spanish Air Minister, General Vigón, put a Douglas airliner at the Ambassador's disposal for the trip from a swelteringly hot Madrid to the Caudillo's summer retreat, a small fortified castle turned hunting lodge on a pleasant wooded slope a few miles outside La Coruña. Hoare hoped to puncture the Caudillo's 'incredible complacency' with an account of the invincibility of the Allies and with a series of complaints about Spain's 'obvious desire for an Axis victory'. Franco, however, mounted a soporific defence.

Self-possessed, calmly confident of his own position, he harped on the dangers to Europe of a victorious Russia. He showed utter indifference to the collapse of Mussolini, other than a quiet satisfaction that it proved his own superiority. Hoare later recalled that the rockets which he had hoped would shock Franco 'fizzled out in cotton wool'. At the time, both Hoare and senior Foreign Office officials were delighted that he had been able to sting the Caudillo into making the remarkable statement that Spain had now repaid its Civil War debt to the Axis and had given orders to the Falangist authorities that there was to be no more discrimination against Britain in the press. Both statements were soon proved to be empty. Franco was also apparently nonplussed by Hoare's question as to what would happen if Anglo-American bombers attacked the Blue Division. As Hoare left, the Spanish Chief of Protocol and interpreter, the Baron de las Torres, whispered in his ear, 'The Generalissimo is going to get rid of the Blue Division.' Despite his occasional discomfiture, Franco clearly hoped to exploit the publicity value of the meeting to the full and his Embassy in Washington announced that the talk had been 'friendly and satisfactory'. This obliged the British Foreign Secretary to state explicitly in the House of Commons that Britain remained gravely dissatisfied with Spain's continued flouting of neutrality.[76]

As the situation looked bleak for Franco, he fought back with an impressive display of nonchalance. Massive publicity was given to the celebration of the thousandth anniversary of the foundation of Castile. Franco had granted 500,000 pesetas* to its cost in the spring.[77] It had long been planned as yet another of those ceremonies so dear to Franco in which his valour, nobility and achievements were compared with those of the great royal warriors of Spain's glorious past. With the collapse of Italian fascism and the resurgence of monarchist opposition, the opportunity was seized to reaffirm Franco's greatness, his links with a monarchical past and the quintessential Spanishness of his regime. On Sunday 5 September, he attended Pontifical High Mass in Burgos Cathedral said by the Papal Nuncio. He was then presented with the Cruz de Alfonso VIII de Silos and a locket (relicario) said to have been carried into battle by Fernán González, the tenth-century warrior lord and founder of Castile. A medieval cavalcade processed before Franco and was followed by jousting. Then period dances were performed in honour of the reina de la fiesta, Franco's daughter Carmen, by dancers 'who pretended to be serfs' (que simulaban pertenecer al villanaje). On the following day, the Minister of National Education, José Ibáñez Martín, gushingly compared the exploits of the Caudillo in creating the 'New Spain' to those of Fernán

---

* About £360,000 at 1993 prices.

González in founding Castile. On 8 September, Arrese arrived in Burgos and tried to outdo Ibañez Martín's sycophancy by referring to Fernán González as *el caudillo rebelde*, an unmistakable comparison with Franco.[78]

In the wake of Italy's unconditional surrender, Franco faced rumblings of discontent from his own high command.[79] Unlike the pro-Allied Aranda and Kindelán, most Spanish generals favoured the Axis cause in the World War although they remained anxious for Spain to stay neutral. They had been prepared to see the question of the monarchist succession shelved until the result of the war was clear. By the late summer of 1943, however, the defeat of the Afrika Korps, the Allied invasion of Sicily and the collapse of Italian Fascism had convinced many of them that the moment had come for urgent consideration of the future. Like Kindelán, they believed that if the fruits of civil war victory were not to be swept away by the Allies turning against a pro-Axis Franco, then drastic measures had to be taken. Despite the fact that the tide was running in their favour, the reactions of the generals were extremely timid. Orgaz had finally refused to head a military coup against Franco, largely because he could not be sure of sufficient support. He and the other leading monarchist generals opted for the less dangerous measure of a petition to the Caudillo. Franco was kept fully informed by his intelligence services. Dated 8 September 1943, a letter was signed by eight Lieutenant-Generals, Kindelán, Varela, Orgaz, Ponte, Dávila, Solchaga, Saliquet and Monasterio, and handed to the Caudillo by General Varela on 15 September 1943.[80]

The implications of this letter were worrying for Franco. However, even before he accepted it, he unsettled Varela by a severe reprimand for carrying a swagger stick in his presence. Then, he simply accepted it without making any immediate response. There were a number of things that helped him bide his time calmly. He had had ample opportunity to watch the 'plot' develop and to consider its weaknesses. Apart from obliquely suggesting that Franco had remained in power for 'longer than the term originally foreseen', the respectful tone of the letter showed that the high command of the Army was more Francoist than monarchist. It did no more than ask Franco 'with loyalty, respect and affection, if he did not agree with them that the time had come to give Spain a monarchy'. Gil Robles (now a stalwart of the monarchist opposition) wrote in his diary of its 'vile adulation' and of his conviction that Franco would not pay it the slightest attention.[81] Franco was reassured by the fact that other senior generals, including Juan Vigón, García Valiño, Jordana, Muñoz Grandes, Yagüe, Serrador and Moscardó, did not sign. Moreover, he had every reason to be confident of the unconditional loyalty of his middle-rank officers who did not regard him merely as 'first among equals'.[82] Accordingly, he dealt with this crisis, as with others, with an extraordinary mixture of patience, self-confidence and outward calm.

The attitude of the bulk of the officer corps below the rank of Lieutenant-General accounts for the fact that Orgaz had so soon changed his mind about the possibility of a military action in favour of the monarchy. In addition, from early September 1943, Franco had on his desk a report accusing Orgaz of being involved in corrupt business deals in North Africa.[83] Its existence could have accounted for the diminution of Orgaz's readiness to plot in favour of the monarchy. Whatever his reasons, Orgaz informed Gil Robles in late September that a rising was most unlikely given that the younger generals and the entire officer corps from colonel downwards were committed to Franco. Indeed, Gil Robles, who was extraordinarily well informed, came to believe that the letter had had the effect of persuading other generals to close ranks around Franco.[84] Moreover, Franco was fully aware that the Allies had no desire to precipitate a change of government in Spain nor to intervene in her internal affairs. He had reason to believe that the guarantees still held which were given at the time of 'Torch' by both Churchill and Roosevelt that there would be no invasion of the Iberian Peninsula.[85] Perhaps too he shared some of the exultation shown by the Falangist press as a result of the German intervention in Italy and the daring rescue of Mussolini from captivity on 12 September by glider-borne SS commandos led by Colonel Otto Skorzeny.[86]

Nevertheless, to quell the timid rebellion of his senior generals, Franco made a small pro-Allied gesture. On 26 September, the withdrawal of the Blue Division was decided at a cabinet meeting but not made public. The value of this decision to the Allies was diminished by Arrese's proposals to permit volunteers to stay on in German units. The task of negotiating the withdrawal was undertaken secretly and with some success by Vidal. To the chagrin of both Vidal and of Jordana, many of the Spanish volunteers joined the SS in accordance with Arrese's stratagem. That Arrese should be given free rein in this way revealed both Franco's inveterate instinct for deceit and the difficulties with which Jordana had to cope.[87]

It was revealing that Franco had not followed the example of Salazar in moving openly towards the Allies after the fall of Mussolini. That the German reaction would necessarily have been muted was shown by the virtual silence which greeted the Portuguese decision to cede bases to the Allies in the Azores.[88] Instead of seeking a rapprochement with the Allies, Franco concentrated on consolidating support within Spain. Elaborate efforts were made to strengthen his position by an intensification of press adulation. Although officially sponsored, there was also an element of spontaneous desperation from those in the Falange who saw their own futures as totally tied to his survival.*

Franco marked his feast day on 1 October, the Day of the Caudillo, by addressing the *Consejo Nacional* of the Falange at the Palacio de Oriente. He began his forty-seven minute speech by describing Spain's role in the war as giving the world the greatest example of wisdom and serenity by saving her people from the horrors of war through a 'vigilant neutrality'. He went on to denounce the exiled Republicans who were encouraging the Allies to dismantle his dictatorship as soon as they had defeated Hitler. Lest such 'vile manoeuvres' bear fruit, he now set about distinguishing his regime from those of his erstwhile Axis partners. The *leitmotif* was his two-war theory. It was the unspoken premise on which he based his certainty that 'our truth' (*nuestra verdad*) and an allegedly advanced social programme rendered his regime superior to both Communism and the liberal democracy of the 'plutocracies'.[89] Later on the same day, at a reception for the diplomatic corps, Franco was resplendent, not in the uniform of *Jefe Nacional* of the Falange, which he had worn earlier in the day, but as Admiral of the Fleet. A reflection of the demise of Italy was the cordiality of his reception of Allied diplomats and, by comparison with previous years, an almost perfunctory greeting of the German Ambassador. For the first time, Franco also used the word 'neutrality' to describe Spain's position.[90]

The shift towards the Allies as a device to consolidate his position was matched by the announcement on the same day of the award of thirty-five military crosses and of the promotion of Yagüe to Lieutenant-General. Yagüe was given command of the VI Military Region, Burgos, as a counterbalance to the growing number of pro-Allied and pro-monarchist generals in the high command.[91] Franco also began to cultivate younger Falangist-inclined officers. He regarded the letter from the senior generals as an intolerable act of indiscipline but with the Allies closely monitoring the situation, his inclination to exact punishment was held in check. He had also taken note of Mussolini's error in confronting his enemies en masse. Accordingly, he adopted the divide-and-rule tactic of

* On 1 October 1943, '*Día del Caudillo*', *Arriba* hailed the seventh anniversary of his 'exaltation to the almost divine function of Governor' and praised his astonishing achievements in inspiring Spanish art, literature, music and architecture.

meeting each of the signatories to the letter in turn and assuring them that he had taken note of their request. He managed to persuade some of them that Hitler's secret weapons, of which he had been informed by Martínez Campos, could still win the war for the Axis. Kindelán, Orgaz and Ponte stood by what they had written. Others wavered and General Saliquet allegedly told Franco that he had been browbeaten into signing by the others.[92]

Gil Robles was astonished at the fact that the senior generals had seemed to expect Franco himself to take the initiative in bringing back the monarchy. He wrote privately in his diary that 'these "fervent monarchists", whose loyalty [to the Pretender] does not prevent them taking full advantage of the Francoist racket (tinglado), are the greatest enemy that the monarchy has'. At the end of September, he wrote a strong letter to the Minister for the Army, General Carlos Asensio, pointing out that a monarchist restoration granted by Franco would be worthless. It elicited only a polite acknowledgement. Needless to say, Franco was fully apprised of this exchange which was circulated among the higher ranks of the Army and the diplomatic corps.[93] By mid-October 1943, the storm had passed and Franco was able to begin an anti-monarchist offensive without worrying about opposition from his senior generals.

The feebleness of Franco's gestures in the direction of the Allies obliged Hayes, on 21 October 1943, to deliver to Jordana a stern letter remonstrating about the Spanish government's pro-German and anti-Russian stance. The letter contained the statement that 'in its own interest, Spain should without delay announce the withdrawal of the Blue Division'. Jordana told Hayes immediately that, as a result of measures then in train, after 25 October there would be no Spaniards on the Eastern front. That, of course, was simply not true as Washington well knew, even if Hayes did not. Then, after lengthy discussion between Jordana and Franco, a formal reply was drafted to Hayes's letter. Although signed by Jordana, it showed all the hallmarks of Franco's thinking. It was basically a vague and high-flown abstract defence of Spanish policy in terms of the struggle against Communism.[94]

The gradual shift in Franco's position was insufficient to please the Allies. As Hoare put it, 'Franco's obvious sympathies with the Axis and the impervious complacency with which he behaved towards the Allies were daily becoming more difficult to endure'. Nevertheless, British policy continued to be patient and tolerant of Franco's position in order to avoid provoking a crisis, despite incidents such as Falangist attacks on the British Vice-Consulate in Zaragoza and the American Consulate in

Valencia.[95] Since the summer, however, things were changing. The British Chiefs of Staff complained that Spanish troops were deployed against the Allies but not against the Germans and that 'we are forced by this disposition to maintain large forces ready to protect our lifeline through the Straits of Gibraltar and constantly to plan for immediate provision of additional forces to hold Gibraltar should Spain permit a German offensive through her territory'. It was felt that the time had come to insist that Franco shift the bulk of his forces from Morocco and Southern Spain to Northern Spain and cease military and economic aid to Germany.[96]

The USA, which was inclined to be rather harsher with Franco than was Britain, was restrained towards him while the outcome in Africa was undecided and it was still important to limit Spanish wolfram sales to Germany.[97] By mid-1943, however, the Germans were attempting to resolve the wolfram problem by buying up Spanish mines.[98] Another minor irritant in relations between the Allies and Franco was over the Spanish internment of Italian warships and merchant vessels. Such acts were now deemed to be hostile gestures against the Allies since they prevented the Italian ships being used in the Mediterranean to free American vessels for service in the Pacific.[99]

While a realignment of Spanish-Allied relations over the wolfram issue was taking place, the attitude to Franco of the Americans in particular was deeply soured in October 1943 by the so-called Laurel incident. On 18 October, Jordana sent a telegram of congratulation to José P. Laurel on his installation by the Japanese as puppet governor of the Philippines. Given Jordana's almost daily consultations with Franco and his own claim to submit every detail of foreign policy-making to him, it may be assumed that the telegram was sent with the Caudillo's knowledge and consent. Radio Tokyo made much of this message and a similar one from Berlin. Seeing the telegram as *de facto* recognition of Laurel's regime, the Allies were outraged. Jordana may have been pushed into this major ineptitude by pro-Axis officials in the Ministry of Foreign Affairs including Doussinague or he and Franco may have thought that it was a clever way of appeasing the Axis, without thinking through the implications of offending the Allies. Whatever the intention, the strength of American reaction seriously frightened the Spanish government. Jordana told Hayes on 5 November that he would resign over the issue. That statement, as no doubt it was intended to do, prompted Hayes to worry that Jordana might be replaced by someone less sympathetic to the Allies.[100]

Influential sections of the American press, the *New York Times*, the

Nation, the *New Republic*, columnists like Walter Winchell and Walter Lippman, and some powerful voices in Congress were advocating a tougher line with Franco in the wake of the Laurel affair. The US Army and the State Department began to discuss an embargo on oil shipments. On 6 November 1943, the new Under-Secretary at the State Department, Edward R. Stettinius Jr., instructed Hayes to request a complete and immediate embargo on wolfram exports and the removal of German agents from Tangier.[101]

Convinced that he could play off the Allies and the Axis, Franco responded to Hayes's demands by seeming to veer back towards the Germans. On 3 December, he spoke to the German Ambassador, Dieckhoff. When Dieckhoff complained that Spain was giving in to Allied pressure, particularly in the withdrawal of the Blue Division from Russia, Franco said that his own survival depended on an Axis victory and that an Allied triumph 'would mean his own annihilation'. Accordingly, he hoped with all his heart for German victory as soon as possible. It is significant that the Caudillo never made a similar statement of commitment to the Allied side to any British or American diplomat. He claimed that he had withdrawn the Blue Division before an Allied request to do so because of growing difficulties about recruiting volunteers and to avoid the humiliation of accepting an Allied ultimatum. Similarly, he pointed out that the internment of U-boat crews was an entirely symbolic ruse to deceive the Allies and that they would be released. All these things Franco considered to be trivial concessions. The crucial issue was that 'a neutral Spain which was furnishing Germany with wolfram and other products is at this moment of greater value to Germany than a Spain which would be drawn into the war'. At this point the Germans had reason to feel some satisfaction with their Spanish policy because Franco was paying off his Civil War debts with wolfram.[102]

Hoare wrote to the Foreign Office on 11 December, 'it is disturbing, though perhaps inevitable, that Franco should now be exploiting Allied patience, and the absence of effective Spanish opposition as evidence of the stability of his regime and the excellence of his relations with Great Britain and the United States. To me in Madrid this complacency is particularly galling . . . though we may succeed in restraining or stopping many unneutral acts, the present Spanish Government with Franco at its head is fundamentally hostile to the Allies and the aims for which we are fighting.' Although Hoare inclined to invoking a full-scale embargo on oil and rubber exports to Spain, he still believed that it was necessary to play Franco with a velvet glove for fear of losing control of the wolfram

market.[103] Nevertheless, Hoare's pressure on the Spanish Foreign Ministry was beginning to fray Jordana's nerves.[104]

Franco, in contrast, remained as imperturbable as ever. At the end of
1943, his intelligence services intercepted a letter from Don Juan to one
of his supporters in which he discussed a public break with the regime.
Franco wrote to the Pretender in an effort to prevent him taking such a
drastic step. The letter was written with the usual curious mixture of
cunning and naïvety. In denying that there was anything illegitimate
about his present position, he stated, in terms of arrogant self-confidence,
that 'among the rights that underly sovereign authority are the rights of
occupation and conquest, not to mention that which is engendered by
saving an entire society'. In a self-pitying tone, he made out that for him
the exercise of power was merely another duty.

In order to dent Don Juan's claims, he was at pains to make out that
the rising of 1936 was not specifically monarchist, but more generally
'Spanish and Catholic' and he cited Mola's expulsion of Don Juan himself
in August 1936 as proof. Accordingly, his regime had no obligation to
restore the monarchy. This sat ill with his own published reasons for
preventing Don Juan serving on the Nationalist side. It was quite obvious
that Franco was, as General Cabanellas had predicted in 1936, determined
not to give up power. In justification of his own legitimacy, he cited his
merits, accumulated during a life of intense service, his prestige among
all sectors of society and public acceptance of his authority. He went on
to state that what Don Juan was doing was the real illegitimacy and
claimed that his regime was moving towards a monarchical restoration
and that Don Juan's activities were impeding its arrival. The best policy
was to leave Franco to his self-appointed task of preparing the ground
for an eventual restoration.

Don Juan's reply was, by comparison, a masterpiece of clarity and not
without its ironic undertones. In response to Franco's insinuation that
he was out of touch with the real situation in Spain, the pretender pointed
out that in thirteen years of exile, he had learned more than he would have
living in a palace, where the atmosphere of adulation so often clouded the
vision of the powerful. With regard to their differing interpretations of
the international situation, Don Juan pointed out that Franco was one
of the very few people to believe in the long-term stability of the
National-Syndicalist State. He stated directly his conviction that Franco
and his regime could not survive the end of the war. To avoid a stark
choice between Franco's totalitarianism and a return to the Republic,
Don Juan appealed to the Caudillo's patriotism to restore the monarchy.

The pretender explained how he saw the monarchy as a regime for all Spaniards and how, for that reason, he had always refused Franco's invitations to express solidarity with the Falange.[105]

Don Juan's crystalline letter exuded the logic, common sense and patriotism that was lacking in Franco's convoluted effort. Yet it was the Caudillo who was to be proved right. With the 'blind faith' of his unquenchable optimism, he planned to sit tight in El Pardo, confident that the Allies had too many other things to worry about.

XX

'FRANCO'S VICTORY'

January 1944–May 1945

BY THE beginning of 1944, with the tide of war clearly turning, North
Africa secure and Italy out of the war, the USA was ever less inclined to
be patient with Franco. The American military staff was furious about
continued Spanish wolfram exports to Germany. Since the Laurel affair,
the Spanish Ambassador in Washington, Juan Francisco de Cárdenas,
had been unable to gain an audience with Roosevelt, and Franco's New
Year greetings to the President went unacknowledged.[1] On Friday
31 December 1943, Beaulac put the American position to the Under-
Secretary of the Spanish Foreign Ministry José Pan de Soraluce. Jordana
was spending the afternoon with Franco. On the following Monday,
3 January 1944, Jordana, fully primed by the Caudillo, went onto the
attack complaining to Hayes about what he described as the one-way
traffic of Spanish favours to the Allies. He asserted that Washington's
demands about wolfram would have economic effects far greater than
their impact on German arms production. In a daringly imaginative rein-
terpretation of reality, he complained that Washington was forgetting
that 'Spain has been actively and effectively co-operating with the Allies'
and was 'playing an important role as an impregnable barrier between
the Germans in the Pyrenees and Gibraltar and North Africa'.

Hayes took this brazen attitude as confirmation of the many reports
reaching him that Franco had been so impressed by German resistance
in Italy and by the German recovery of the Dodecanese Islands as to
conclude that the war would go on for some time and end in stalemate
in 1946. Hayes pointed out sharply to Jordana that Spain's policy was
conducted in Spanish and not Allied interests. Under pressure, Jordana
admitted that Franco had recognized the agent in Málaga of the German
puppet regime in Northern Italy, the Reppublica di Salò. Hayes now

inclined to believe that an oil embargo was necessary to stop Franco flirting with the Axis.[2] The hitherto benevolent attitude of Hayes towards Franco was soured at the Caudillo's annual Epiphany reception for the diplomatic corps on 6 January 1944. After keeping his guests waiting for over an hour, Franco made a regal entrance and then made a show of ignoring the British and American Ambassadors. An outraged Hayes, convinced that Franco was still in the hands of the Germans, commented 'this idiot is digging his own grave'.[3]

Both Hayes and Hoare were angered by the discovery that Demetrio Carceller had granted a credit of 425 million pesetas to Germany for the purchase of wolfram and that a deal was about to be struck whereby Spain's biggest wolfram mine would sell its entire production of 120 tons per month to the Third Reich.* Accordingly, they both agreed to recommend a suspension of petroleum exports to Spain.[4] Washington decided, in the light of Spanish internment of Italian warships, of the activities of German agents on Spanish soil and the continued presence of parts of the Blue Division in Russia, to refuse petroleum loadings from mid-February.[5]

The only efforts by the Franco regime to counter hard evidence of continued support for the Axis consisted of a press campaign aimed at distinguishing Francoism from fascism. Its basis was Franco's speech of 1 October 1943 lauding the originality of his political system.[6] Such assertions would eventually bear fruit in the atmosphere of the Cold War. Now, they were an irrelevance. Evidence of Franco's eager readiness to pay heed to the most alarmist information passed to him by the German Embassy was provided when he told his cabinet on 22 January that a British invasion of Portugal had already begun, an assurance which he was to repeat in mid-March and again in late April.[7]

London agreed with Hoare's reluctant advice that there should be an intensification of pressure on Franco. On 27 January 1944, the British

* Spain supplied Germany with other strategic raw materials. According to sources published in Germany in February 1944, 39.2 per cent of all Spain's exports went to Germany and 30 per cent went to German industries in occupied countries. Spanish armaments plants at Trubia and Reinosa produced gun barrels for the German army. In Valencia, hundreds of thousands of rifle cartridges were turned out daily for the Germans. In Barcelona, motors were built for German vehicles. Textile factories in Catalonia manufactured uniforms and parachutes for the German forces. Spain supplied Germany with ammonia, nitrogen and glycerine, as well as iron ore, pyrites, lead, zinc, nickel and wolfram. A complex shipping deception allowed vital Spanish material to be exported to Germany and replaced by substitute material from Argentina which thereby escaped the Allied blockade. (United Nations, Security Council, *Report of the Sub-Committee on the Spanish Question*, pp. 13–14.)

Ambassador visited the Caudillo at El Pardo ready to talk in terms of a total break-down of relations between Britain and Spain.[8] Again, Franco responded to Hoare's angry complaints with suffocating calm, speaking 'in the still small voice of a family doctor who wished to reassure an excited patient'. Verbally, at least, Franco accepted Hoare's charges and promised to take drastic action against German agents and to prohibit further exports of wolfram to Germany. On the following day, having been informed that the February 1944 petroleum deliveries to Spain had been suspended, Jordana made equally conciliatory remarks to Hayes. Within a matter of hours, the satisfaction of both the British and American Ambassadors was dissipated by the wide publicity given by the BBC and the American Office of War Information radio to the State Department press release that the United States had imposed an oil embargo on Spain in response to her various pro-Axis activities. Both Ambassadors now feared that Franco would dig in sullenly.[9]

The Allied curtailment of petroleum exports led to frantic Spanish diplomatic activity in both Madrid and Washington. Cárdenas talked about both Jordana and himself having to resign. Jordana told Hayes that the news of the embargo had come 'as a terrible shock to him and to the Spanish Government'.[10] At a cabinet meeting held on 2 February, Franco and Jordana were confident that a policy of stonewalling would enable them to exploit Anglo-American differences over the wolfram issue. The Spaniards had not failed to notice the mutual dislike between the pedantic Hayes and the quicksilver Hoare.[11] Franco's tactics make sense only in terms of his continuing hopes that the Third Reich could turn the tide, hopes encouraged by myopically optimistic reports from his military attachés in Berlin.[12] With dismay growing among the population at the spectre of Allied economic hostility, on 3 February 1944, Jordana offered immediate suppression of the German Consulate at Tangier and the expulsion of German agents from Spanish Morocco and eventually 'the energetic suppression of all German espionage and sabotage anywhere in Spanish territory'. He also offered total withdrawal of the Spanish legion and the remaining air squadron and other forces on the Russian front. With regard to Allied complaints about the interned Italian warships, however, he requested compensation in the form of armaments and aviation gasoline. He also suggested that the demands for a wolfram embargo could be resolved by negotiation.[13]

Despite frantic and emotional appeals by Cárdenas, Washington, however, was adamant that only 'a complete wolfram embargo' would be satisfactory.[14] With the agreement of Hoare, Hayes suggested a plan for a compromise solution on 4 February. However, after detailed examination of the question by the State Department and the British Embassy in Washington, Hull took the view that, having procrastinated for so long, Spain had provoked the petroleum ban and that it was up to Franco, and not Washington, to come up with a face-saving device.[15] On 11 February, Hayes was instructed to demand a total embargo on wolfram shipments to the Third Reich. Cárdenas begged US Under-Secretary Stettinius to restart oil deliveries in return for Spain undertaking to reduce wolfram supplies to Germany. Roosevelt was anxious for a hard line and Churchill was inclined, against Foreign Office advice, to go along with him. Both Hoare and Hayes in Madrid were readier to compromise. With a confidence, generated by reports from Alba, that the British were less inclined to intransigence, Franco played the crisis with more than his usual *sang froid* – hence the stonewalling of Jordana.[16]

There was an element of the crisis which was inflated and exploited with hypocritical cunning by Franco for his own narrow domestic purposes. The Spanish press did not reveal the reasons for the oil embargo and was thus able to portray the crisis as the consequence of external pressure to break the neutrality which the Caudillo worked so courageously to maintain. Official statements merely reiterated Spain's determination to maintain her neutrality at any price. There was also a tendency to write off any Allied grievances as the fruit of the machinations of exiled Republicans.[17] In retrospect, the wolfram crisis of 1944 might be seen as a dry run for the way in which Franco was to retain domestic support and survive the international ostracism of 1945–9. In other words, he invented an international siege and portrayed himself as the heroic national leader standing up for the independence of Spain against overwhelming odds.[18]

For fear that Franco might turn to Germany for oil, the British proposed that Spain cease exports for only six months but the State Department stood firm. American public opinion was hostile to oil being supplied to Franco and a presidential election was imminent.[19] Roosevelt and Cordell Hull held out and Hoare was instructed to act along the same lines as Hayes.[20]

One reason why the situation was so protracted was Franco's belief, encouraged by Carceller, that he would derive benefit from playing off both sides. Pro-Axis propaganda in the controlled press and radio networks was stepped up.[21] On Monday 14 February 1944, the Caudillo told the Duque de Alba that the war would last another six years and end in the total exhaustion of both sides. At that point, the Caudillo was confident that Spain would be in an important strategic position and needed by the democracies. 'Meanwhile, Spain needs a man like Franco to guide it.'[22] For this reason, no doubt, he instructed Jordana to stand firm, which he did. But, on 17 February, with Hoare, he did offer to reduce exports to 'an insignificant amount of no real military value to Germany'. This was regarded by both Hayes and Hoare as the basis of a compromise. Jordana also echoed Franco's view that 'Spain had done a great service to the Allies by not entering the war.' Hull telegrammed Hayes acerbically that 'it is not usual in the community of nations for a country to assume that it is rendering a great service to its neighbours by not attacking them'. He went on to state with stark clarity 'we cannot justify making sacrifices to support the Spanish economy in the absence of a willingness on the part of the Spanish Government to reciprocate our co-operative attitude; namely to take the step entirely compatible with Spanish neutrality of declaring a permanent embargo on the exportation of wolfram.'[23]

Hayes passed this on to Jordana on 21 February. The Spanish Foreign Minister made some more small detailed concessions. A similar set of proposals made by Jordana to Hoare met with the approval of London.[24] There was some backsliding by Jordana, who was increasingly isolated within Franco's cabinet. Most Ministers, with Carceller the loudest, regarded his efforts at compromise as a betrayal of national dignity. The Portuguese Ambassador was convinced that Franco paid considerable attention to Carceller because the Minister of Industry had helped him salt away money in Switzerland.*[25]

Although Allied goodwill was being gratuitously squandered, Franco, at home, retained his image as the man who did not vacillate in defence of national interests by dint of the press's partisan reporting of the crisis. However, it also reflected both his unsinkable self-confidence and his ability to impel those who surrounded him to share it. The scale, and unreality, of Franco's self-regard was revealed on 6 March when he was visited by Professor João Pinto da Costa Leite, the Portuguese Minister of Finance since 1940. In much the same way as he had patronized Calvo Sotelo in 1929, he treated the Portuguese economist to ninety minutes of pompous inanity on the subject of economics. More significantly, echoing his faith in the schemes proposed to him by Albert von Filek in 1940, he made it clear that he was not worried about the US petrol

* If Franco was taking precautions against having to go into exile, it was done with considerable discretion – Pereira's reference was virtually unique.

embargo because he believed that he had solved the problem of synthetic gasoline. He related two schemes, one for producing fuel from bituminous slate at a cost of 2,000 million pesetas and the other from hydrogenizing coal at a cost of 1,200 million pesetas, which he was convinced would make Spain self-sufficient in energy. He was apparently oblivious to the fact that, even if the schemes were technologically feasible, the cost was monumentally prohibitive. The Portuguese Ambassador commented to Salazar, 'it's sad. I prefer Don Quijote in the original version.'[26]

The reality of the situation was that the oil embargo was pushing the poverty-stricken Spanish economy further back towards the Middle Ages. Franco's 1 April 1944 victory parade had to take place without tanks or armoured cars.[27] The pro-Axis tenor of the press was maintained throughout the crisis.[28] However, by mid-April, the economic consequences of the petroleum embargo had forced Franco to turn down Dieckhoff's request for a resumption of wolfram shipments to Germany. After the most tedious negotiations, prolonged by the Caudillo's personal stubbornness, the Spaniards offered the Allies a dramatic restriction of monthly wolfram exports to a near token amount.[29] With the Germans offering oil, heavy machinery and foodstuffs in return for wolfram, Churchill persuaded Roosevelt, against the advice of Hull, to accept this compromise on the grounds that not to do so would slow down the cleaning up of German spy networks in Spain and also threaten British purchases of Spanish iron ore and potash. Hayes was furious, believing that Franco's resolve had been strengthened by reassurances from Hoare to Jordana on 27 April that oil could be supplied from British-controlled sources. Hull too felt that 'an absence of wholehearted British support' had denied Washington a full-scale victory over Franco.[30]

The eventual agreement with the Americans and the British signed on 2 May 1944 encompassed the resumption of petroleum exports to Spain in return for the reduction of wolfram exports to Germany to 20 tons in May, 20 tons in June and 40 tons per month thereafter, the closing down of the German Consulate in Tangier, the withdrawal of all remaining Spanish units from Russia and German spies and the expulsion of saboteurs from Spain.[31] Thanks to the stubborn hopes of Franco that Axis fortunes would improve, Spain had suffered a severe worsening of the food distribution crisis, a gratuitous hardship occasioned by the oil embargo. Yet, his ultimate grudging capitulation was interpreted with ineffable complacency as a sign of the capacity of his regime to uphold the prestige and liberty of Spain against the world. Avoiding any reference to German agents, Tangier or even wolfram, the Spanish climb-down was presented as a successful commercial agreement which proved that Franco enjoyed cordial relations with the Allies and would not be threatened by 'any contingency', a euphemism for the defeat of Hitler.[32]

Ribbentrop was furious and he berated Ginés Vidal in Berlin. Dieckhoff in Madrid protested energetically to both Franco and Jordana. The Caudillo shiftily told him that Germany had had its chance to stockpile wolfram and that now Spain could take no more risks.[33] At the same time, Jordana assured Hayes that he had issued the promised instructions for the removal of the German consulate in Tangier. However, throughout the rest of 1944, Hoare had reason to protest almost daily about the Spanish failure to expel the German agents and smuggling of wolfram to Germany continued on a small scale until the summer.[34] In late May too, there were Allied protests about attacks on American aircraft by anti-aircraft batteries in Spanish Morocco. It took until mid-June before the Spanish Government agreed to restrain their gunners.[35]

Despite the fact that the wolfram agreement was essentially a defeat for Franco and one that had caused considerable hardship to the Spanish people, he and Carrero Blanco had learned two valuable lessons from the entire process. In the first place, they had demonstrated the efficacy in domestic propaganda terms of presenting Franco as the heroic defender of national independence against the misguided and arrogant Allies. This was reflected in the unusually warm reception that Franco received some weeks later on visiting Bilbao.[36] Secondly, at a time when their forces were massing for the invasion of France and the Russians were advancing rapidly westwards, Britain and the United States had shown no inclination to crush Franco but rather had preferred to negotiate over wolfram. At this time, Franco was especially worried about the growing concentration of armed Republican exiles in the south of France.[37] But the wolfram negotiations gave him and Carrero the basis for hoping that the Allies would not be averse to the survival of the regime should Hitler be defeated.[38] This seemed to be confirmed unequivocally within a few weeks.

In a speech in the House of Commons on 24 May 1944, Churchill implicitly defended the negotiations which had taken place with Spain by what seemed to be praise of General Franco. Referring to the dangers from Spain in 1940, he paid tribute to the efforts of Hoare and Arthur Yencken, his able Embassy Counsellor, but indicated that 'the main credit is undoubtedly due to the Spanish resolve to keep out of the war'. With regard to Operation Torch, his gratitude to the Spanish government was even more fulsome. Churchill considered that, during Torch, Spain had made full amends for her earlier acts of assistance to Germany and concluded that 'as I am here today speaking kindly words about Spain, let me add that I hope she will be a strong influence for the peace of the Mediterranean after the war. Internal political problems in Spain are a matter for the Spaniards themselves. It is not for us – that is, the Government – to meddle in them.'[39]

Churchill's words certainly sprang from motives other than disinterested admiration for Franco. In the short term, he was trying to neutralize him during the forthcoming Normandy landings. He also had the longer term purpose of sanitizing Franco to be able to use him as a future bulwark of western Mediterranean policy. At the time, however, there was considerable furore in English and American political circles, and dismay within the anti-Franco opposition. The impact of the speech was intensified by the Madrid propaganda machine which presented it as a full-scale endorsement both of Franco's foreign policy and of his regime. Spanish newspapers were cruelly jubilant at the chagrin of the Republican exiles who had been looking to the Allies to dispose of Franco after defeating Hitler and Mussolini.[40] On his next visit to the Caudillo, Hoare tried in vain to disabuse him of the idea that Churchill had issued a declaration of unquestioning support for his regime.[41]

Churchill's speech was a hostage to fortune from which Franco was to squeeze the last ounce of benefit both domestically and internationally. Hugh Dalton thought it an ill-judged, romantic gesture, 'it was all totally unnecessary, but he made it up at 2.30 a.m. on the morning of his speech and the Foreign Office didn't see the draft until about an hour before it was to be delivered. They did their best to tone it down, but with hardly any success.'[42] Churchill wrote, in justification, to Roosevelt, 'I see some of your newspapers are upset at my references in the House of Commons to Spain. This is very unfair, as all I have done is to repeat my declaration of October 1940. I only mention Franco's name to show how silly it was to identify Spain with him or him with Spain by means of caricatures. I do not care about Franco but I do not wish to have the Iberian Peninsula hostile to the British after the war.'[43]

In fact, German observation posts, radio interception stations and radar installations were maintained in Spain until the end of the war. Influenced by partisan reports on the war situation which told him what he wanted to hear, Franco continued to prevaricate. From Vichy, Lequerica sent reports that Germany would soon be able to bomb New York. The Spanish military attaché in Berlin pinned his hopes on German use of the atomic bomb, even informing Madrid in July 1944 that Manchester had been totally annihilated by such means. Franco devoured optimistic predictions that the Germans were merely luring the Allied invaders to their doom and was irritated by Vidal's gloomily realistic reports from Berlin.[44] Throughout the Italian campaign and even after the Normandy landings, the Falangist press stuck doggedly to its belief in the invincibility of the Third Reich. As the Allied tide surged inexorably, the press gloated over the prospect of German retaliation through secret weapons. Reports were printed that most of southern England had been razed to the ground by flying bombs and London depopulated by large-scale evacuation.[45]

As the Allies advanced towards Germany from East and West, German difficulties were ingeniously reinterpreted, particularly by Franco's favourite columnist Manuel Aznar, as Hitler skilfully shortening supply lines, building up stocks of new weaponry and tempting his enemies into tank battles which they could not win.[46] Such press comment and the reports from his emissaries seem to have had their effect on the Caudillo himself. Throughout the summer of 1944 and even in the autumn, when Axis defeat loomed, he put faith in the possibility that the horrendous weapons of which Hitler boasted might yet reverse the outcome. He told the Duque de Alba that weapons such as the cosmic ray would be decisive in turning the tide and that, by landing in Normandy, the Allies had been drawn into a German trap – 'I follow the operations closely and I cannot account for eighty divisions which I believe we will see appear somewhere at any moment'.[47]

Whatever his innermost hopes, Franco's instinct for clinging to power was discernible in a greater readiness to tack to the Allied wind. That was an unmistakable feature of his annual speech to the Falange's *Consejo Nacional*, on Monday 17 July, the eighth anniversary of the military rising. Although sporting the white summer uniform of *Jefe Nacional*, he refrained from mentioning the Falange until the end of a rambling but revealing speech. The first half consisted of a self-adulatory catalogue of his regime's achievements in health, education and defence, which were referred to coyly as 'the Spanish peace' (*la paz española*) . His magnanimity was demonstrated by the announcement of the reduction of the sentences of political prisoners. The righteousness of the regime's policies was, he said, manifested by great popular demonstrations of support. These demonstrations, which he chose to see as entirely spontaneous, offended his government's natural modesty but were Spain's reply to the calumnies generated abroad by exiled Republicans. His calls for peace, which were aimed at pre-empting the total defeat of the Third Reich and at his own self-aggrandisement, were presented as a service to Europe and humanity.

With the threat of Communism louring, he declared Spain ready to collaborate in the post-war world as long as due respect was paid to the individuality of her political system. He rejected foreign criticism that his regime was not democratic on the grounds that the highest democracy lay in carrying out the teachings of the gospels. That, he declared, was precisely what the heroic Falangist, 'half-monk, half-soldier', was doing.[48]

The Spanish press was delirious in its praise.* Franco had set the world an example by creating in his Falange a synthesis transcending fascism and anti-fascism.[49] Unrestrained adulation for the Caudillo's prodigal generosity in showering the gifts of peace and prosperity upon Spain was based on the assertion that, from the beginning of the war, he had worked tirelessly to bring peace. Elaborate explanations were produced both of Franco's two-war theory and of the originality and uniqueness of Falangism.[50] Taken together, these two notions would, if accepted by the outside world, absolve Franco from his Axis stigma.†

On 3 August 1944, to the genuine regret of Hoare, Hayes and Pereira, Jordana died of angina.‡ Jordana had recognized the inexorable trend to ultimate Allied victory long before Franco. He had deferentially submitted every initiative to Franco for approval but, within his limits, was more pro-Allied than pro-Axis. Franco seemed absolutely unmoved by the death of this loyal and efficacious servant, neither expressing his sympathy to Jordana's family nor even attending the funeral.[51] Franco also missed totally the remarkable opportunity to diminish the hostility felt towards him in Allied circles by a clean break with his pro-Axis past. Instead, he sent immediately for José Félix Lequerica, the fiercely collaborationist Ambassador to Vichy.[52] Although it was not announced for some days, the decision was almost instantaneous and substantiated a remark by Carceller that Franco had planned to substitute Jordana anyway, presumably to remove an uncomfortable reminder of the truth of what had happened during the wolfram negotiations.[53]

Hayes described Lequerica's appointment as 'a tremendous blow'.[54] Years later, Serrano Suñer was still expressing astonishment that Franco should have appointed Lequerica, 'the Gestapo's man', with his role in

* A fake news item was manufactured purporting to be a favourable comment on Franco's speech by Anthony Eden in the House of Commons – *Arriba*, 23 July 1944.
† Significantly, after years of fulsome telegrams to Hitler on the occasions of his triumphs, the Caudillo refrained from congratulating him on his escape from the 20 July assassination attempt.
‡ Hoare wrote a generous tribute to Jordana's honesty and industry – 'I never saw a public man so obviously work himself to death' – in *The Times*, 8 August 1944; Pereira to Salazar, 6 August, 4 September 1944, *Correspondência*, IV, pp. 594, 608.

so many appalling events of the period.[55] On hearing the news of Jordana's death, Lequerica himself had told the Vichy correspondent of *La Vanguardia*: 'It can be anyone except me, since, after my pro-German activities in Vichy, I am doomed to disappear from the international scene as soon as Hitler is defeated.'[56] As well as appointing Lequerica, Franco also removed Jordana's pro-Allied under-secretary, Pan de Soraluce.

There was, as usual with Franco, a devious motive in choosing Lequerica and in maintaining in their posts notably pro-Axis ministers such as Asensio, Arrese, Girón and Blas Pérez. It had been widely assumed that Jordana would be replaced by Alba or Sangróniz but, if Franco had tacked to the prevailing wind by seeking more pro-Allied ministers, he would always have mistrusted them precisely because they were pro-Allied. In contrast, Lequerica and the backward-looking Falangists depended on him for their very survival and there was an element of desperation about their unconditional loyalty.

With Allied forces closing in on Paris, Franco resolved the now acute problem of his relations with Vichy by not sending a new Ambassador to substitute Lequerica. On 19 August, fighting against the Germans had begun in Paris. The city was liberated on 24 August by Free French Forces among whom anti-Francoist exiles played a prominent role. It is not known how Franco reacted to the news that armoured cars flying Spanish Republican flags and bearing names like Guadalajara and Teruel had been among the first into Paris. But his chagrin can be imagined. Indeed, it was not until after the German evacuation of Pétain to Belfort and the resignation of François Piétri, Vichy Ambassador to Spain, that the Caudillo broke off relations with Vichy. On 25 August, he recognized Jacques Truelle, the head of the Free French Liaison Mission in Madrid, as *Chargé d'Affaires* for all French interests.[57]

In August, Berlin recalled Ambassador Dieckhoff and did not replace him. With the Germans virtually defeated in France, Franco was going to have to come to terms with a new threat. Large numbers of armed Spanish Republicans who had fought with the French resistance were starting to drift southwards towards the Spanish border. Madrid was buzzing with rumours that Franco was about to be overthrown and that the Falange was to be dissolved. They were sufficiently pervasive for frantic official denials to be issued [58]

By the autumn of 1944, the need for the special missions in Madrid of both Sir Samuel Hoare and Carlton Hayes had diminished dramatically. Hayes had his last major audience with Franco on 11 September 1944. It was obvious that Franco, however reluctantly, was now courting the favour of the Allies. With total insincerity, the Caudillo told the US Ambassador of his 'relief' at the Allied military successes in France. This was in stark contrast to the anguish of his press. *Arriba* was now berating the Germans for the betrayal of their friends implicit in the defeats being suffered by the Wehrmacht. In fact, Franco's hopes for a German victory were maintained almost to the end. During the Ardennes counter-attack, he said 'now the Allies will see how the Germans surround them'. Hayes took Franco's declared Allied sympathies at face value and asked him to announce that Spain would not harbour Axis leaders who sought asylum. Typically, Franco avoided the issue by saying that, to his knowledge, no Axis leaders were contemplating seeking refuge in Spain and, if they were, they would be killed off by suicide or assassination or captured by the Allies before they could reach her borders.[59]

Hoare was to return to England to take up a seat in the House of Lords as Lord Templewood. He had made it clear to Lequerica that Franco's pro-Axis stance had diminished the possibilities of his being welcomed into the post-war community of nations.[60] At his farewell to Franco, Hoare planned to hand over a tough repetition of that message, which anticipated post-war allied policy towards Spain.[61] In fact, from October 1944, a series of frantic initiatives, both private and public, was begun by Franco to convince the Allies that he had never meant them any harm and that his links with the Axis had been aimed only against the Soviet Union. Before, during and after the celebrations on 1 October 1944 of the eighth anniversary of Franco's assumption of power, the press harped on his untiring efforts since 1939 as a peacemaker. Any suggestion that Spain had favoured the Axis was the malicious work of the exiled 'red scum'. In an outrageous distortion of the facts, Franco was praised for not exploiting French weakness in the summer of 1940.[62]

In mid-September 1944, Carrero Blanco had produced a memorandum for Franco on the post-war world. Its theme was that Britain, in fighting Germany, had chosen the wrong enemy and contributed to the emergence of godless Russia as a superpower. Since Germany's miracle weapons were likely to gain Hitler no more than a negotiated peace, Spain should mediate now with Britain to prevent the total destruction of the Third Reich.[63] Accordingly, on 18 October 1944, the Caudillo wrote a letter to the Duque de Alba the contents of which the Ambassador was asked to pass on to Churchill. Its text showed that he had given some consideration to Carrero Blanco's memo. In it, he proposed a future Anglo-Spanish anti-Bolshevik alliance. In Franco's Hitlerian analysis, 'after the terrifying test Europe has gone through, those who have shown themselves strong and virile among the nations great in population and resources are England, Spain and Germany'. However, Germany, along with France and Italy, was incapable now of standing up to Russia. American domination of Europe would be disastrous. Accordingly, Britain and Spain together should work to destroy Communism. He dismissed his own pro-Axis activities as 'a series of small incidents'. The only obstacle, he claimed, with astounding myopia, to better Anglo-Spanish relations during the previous years had been British interference in Spain's internal affairs, in particular the activities of the British Secret Service. A sly conclusion suggested that any help given to exiled anti-Francoists would only play into Russian hands.* The letter would not be delivered to the British Foreign Secretary until 21 November.[64]

Talk of the struggle against Communism was not rhetoric as far as Franco was concerned. In October 1944, any rebellious thoughts among the senior military monarchists were banished by the invasion of the Val d'Aran in the Pyrenees by Spanish Republicans who had fought in the French resistance. In fact, the majority of those involved were Communists, although the Moscow-based leadership of the PCE was desperately trying to stop the invasion attempt. The repulsion of the incursions and the subsequent guerrilla war came as a Godsend to Franco. As Hoare wrote to London, 'the reckless movement of a few hundred Spanish adventurers on the frontier has given him the chance of posing as the champion of Spain against a Red invasion. It has also provided him with a pretext for arresting and executing a formidable number of his political opponents.'[65] They made possible the revival of the Civil War mentality, gave the Army something to do and generally reunited the officer corps around Franco.[66] As Captain-General of Burgos, Yagüe played a crucial role in repelling the guerrilla incursion.

In fact, this minor military side-show against Communists may have stimulated the sympathy of Churchill to the ideas expressed in the Caudillo's letter. In any case, as early as 23 April 1944, Churchill had telegrammed the British Ambassador in Moscow in connection with Russian accusations that Spain had been a German supply base since 1939: 'it was a very good thing that Franco did not let the Germans through to attack Gibraltar and get across into North Africa. This has to be con-

---

* Hoare wrote 'It is difficult to say whether the effrontery of Franco's arguments or the naïvety of his method of expressing it was the more remarkable.' He saw the letter as one more example of the 'complacency that dispensed him from the conventional rules of speech and action and caused him to believe in his own infallibility of judgement and knowledge although a volume of controversial evidence challenged him at every turn'.

sidered too and you might remind our friends, as opportunity served, that at that time we were absolutely alone in the world (and the Soviets were feeding Germany with essential war munitions). So don't let's all be too spiteful about the past.'[67]

Churchill's stark realism in defence of British interests was a powerful influence in favour of Franco. Other Ministers held different views. In the wake of a highly critical memorandum from Hoare, the Deputy Prime Minister, Clement Attlee, produced a note for the War Cabinet on 4 November 1944. In it, he ruled out direct intervention in Spain but suggested efforts to undermine the dictatorship: 'We should use whatever methods are available to assist in bringing about its downfall. We should especially in the economic field work with the United States and France to deny facilities to the present regime.' Attlee believed that Britain was 'running into the danger of being considered to be Franco's sole external support'.[68] Attlee's hostility to Franco was shared by the Foreign Secretary, Eden.

Franco may well have known what was happening in Whitehall. Churchill had long since made his views public and the Duque de Alba was extraordinarily well-informed about what went on in high places in London. Accordingly, while his press continued to give free rein to Nazi enthusiasms, Franco presented himself as a friend of the Allies with a virtuoso display of hard-faced cynicism. He gave an interview to the Director of the United Press Foreign Service, A. L. Bradford, which was published on 7 November and widely reproduced. It took the form of written replies to a previously submitted questionnaire for which he had had the assistance of Doussinague.[69] The published version presented a disingenuous, not to say shamelessly mendacious, account of his policy during the previous five years. Forgetting his appeals to Hitler for the dismemberment of French Morocco at the time of the French defeat in 1940, he described his attitude to France as one of friendship and *hidalguía* (noble generosity). His most outlandish statement was that the *División Azul* 'implied no idea of conquest or passion against any country' and that 'when the Spanish government realized that the presence of these volunteers could affect its relations with those Allied countries with which it had friendly relations, it took the necessary steps to make those volunteers return home'.

He explained away the dictatorial nature of his regime with the patronizing statement that 'certain peculiarities of the Spanish temperament' made it impossible to sustain democratic institutions. Projecting himself as a firm father overseeing a recalcitrant family, he stated that democracy invariably ended up unleashing violence among Spaniards. He boasted of his regime's Catholic principles, its 'organic democracy' and its 'spirit of justice'. Among a series of remarks aimed at currying favour with the American public, he claimed that the discovery of America had given Spain 'an American character'. Asserting that Spain's internal regime was no obstacle, he brazenly demanded a place at the post-war peace conference on the grounds of his 'serene and dispassionate understanding of what is and is not just'. With an eye on Anglo-Saxon public opinion, Franco made vague promises of forthcoming elections and extremely confusing hints about the installation of a new monarchy. A delirious Spanish press responded with reports of the 'universal expectation' and awed admiration with which the world had perceived the 'transcendental importance' of Franco's remarks.[70] It is not known how the Caudillo reacted to reports from his ambassadors around the world that his declarations had been received with near universal hostility.[71]

Such blatantly insincere overtures had entirely counter-productive effects in the capitals of the Great Powers. In London, the Foreign Secretary was outraged.[72] A British Government spokesman told the House of Commons that there was 'no reason why any country which has not made a positive contribution to the United Nations' war effort should be represented at the peace conference'. The British press recalled some of Franco's more unrestrained pro-German speeches.[73] In Washington, Roosevelt and Cordell Hull were firmly opposed to Franco's membership of the United Nations.[74] Fortunately for Franco, perhaps, Hull was about to retire in poor health to be replaced by his Under-Secretary, Edward Stettinius.

Eden proposed instructing the Ambassador in Washington, Lord Halifax, to ask the US government to support a British plan to give Franco a solemn warning that Spain could not expect to play a full part in the post-war world. This would be backed up by an oil embargo. Eden's draft telegram was written in the strongest terms. He described Franco's mood as one of 'smug complacency' based on his mistaken belief that his regime enjoyed the approval of both the British and the American Governments. 'In consequence, he believes that he will be able to maintain a double policy of totalitarianism in Spain and friendly relations with the victorious Allies.' Eden was especially indignant about Franco's 'insolent suggestion that Falangist Spain has a right to a seat at the peace conference'.[75] In response to Eden's proposal, Churchill wrote 'I am no more in agreement with the internal government of Russia than I am with that of Spain, but I certainly would rather live in Spain than in

Russia.' Convinced, rightly, that Franco would fight to the death rather than give up power, the Prime Minister wrote: 'What you are proposing to do is little less than stirring up a revolution in Spain. You begin with oil: you will quickly end in blood. Should the Communists become master of Spain, we must expect the infection to spread very fast through Italy and France.'[76] The telegram was not sent.

The tension between these two attitudes would continue to mark British policy towards Franco until the late 1940s. Out of the tension came inaction and that would suit the Caudillo very well. On 27 November 1944, the War Cabinet discussed various documents concerning Spain including Hoare's intensely critical memorandum of 15 October, and Franco's recently delivered letter. Churchill admitted that it was likely that Franco had been encouraged to write the letter by a misinterpretation of his speech in the House of Commons on 24 May. Anthony Eden got Churchill to agree to reply in terms which would leave the Caudillo in no doubt that the future world organization would not be aligning Spain and Britain against Russia.[77]

On 2 December 1944, Churchill wrote to Eden 'I shall try my best to write your insulting letter to Franco over the weekend, but I cannot give any guarantee. The relations between England and Spain have undergone many vicissitudes and variations since the destruction of the Spanish Armada and I cannot feel that a few hours more consideration on my part of the letter for which I am to be responsible are likely markedly to affect the scroll of history.'[78] Hoare had already written on 13 November a fierce draft of what he hoped to say in his farewell to Franco. He told Churchill that 'Nothing short of high explosive will have any effect on General Franco's complacency.'

Churchill, however, was too busy to give much time to Franco and, when Hoare had his final interview with the Caudillo on 12 December 1944, the Prime Minister's letter was still unwritten. Sir Samuel told Franco that his recent advances to Churchill were unlikely to be successful, which failed entirely to dent the Caudillo's self-satisfaction. A 'complacent and unruffled' Franco listened in near silence to Hoare's criticisms, although he did deny that the Falange was in any way similar to the Nazi or Fascist parties. When the British Ambassador referred to the large numbers of executions still taking place in Spain, Franco said that he planned to make reforms. To Hoare's annoyance, he 'showed no signs of being worried about the future of Spain'. Only on taking his leave did Hoare notice 'a sign that the wind had begun to blow in this unventilated shrine of self-complacency. Photographs of the Pope and

President Carmona [of Portugal] had taken the place of honour previously held on his writing table by Hitler and Mussolini.' Hayes had already noted the change-over in July 1944.[79]

Franco's changing ideological predilections were no doubt influenced by the rumours flying around in December to the effect that the Allies were planning to replace him with a government headed by the conservative Republican Miguel Maura.[80] There was more than a hint of whistling in the dark about a spectacular ball mounted at El Pardo on 23 December 1944 to celebrate the eighteen year-old Carmen Franco's *presentación en sociedad* (coming out). The event also reflected the Caudillo's doting affection for Nenuca, as he called his only daughter, and his, and more particularly his wife's, penchant for royal pomp. There was a sumptuous banquet for two thousand guests. It had been decided that no one from the diplomatic corps would be invited. However, at the last minute, as part of his efforts to curry favour with the Americans, Hayes, his wife and daughter were invited. Otherwise, military friends of the Generalísimo and Francoist courtiers abounded. The old aristocracy conspicuously stayed away. Afterwards there was dancing in the eighteenth-century salons and entertainment from the most prominent figures of Spanish showbusiness. On the following morning, in regal style and accompanied by the Bishop of Madrid-Alcalá, Carmen visited an old people's home.[81]

The event was no doubt meant to announce to those who anticipated an early transition to the monarchy that Franco was staying put, serene as always in his 'blind faith' in his own good fortune. But, while the regime revellers danced the night away, armed secret policemen arrested a large number of well-known establishment figures. Personages above suspicion of subversion, like the distinguished conservative intellectual, Dr Gregorio Marañón, and the ex-CEDA Minister of Justice, Cándido Casanueva, received sinister visits in the early hours of the morning. Lequerica later tried bizarrely to explain away these embarrassing incidents as the work of 'Communist elements' in the police trying to bring the regime into disrepute. The real crime of those arrested was to have had contact with Gil Robles or Miguel Maura.[82] Franco reacted implacably when under threat.

Nevertheless, the faith publicly displayed at Nenuca's coming-out ball was not entirely misplaced. On 11 December, Churchill had written again to Eden about the draft of the reply to Franco prepared by the Foreign Office, 'I do not think the balance of help and hindrance given us by Spain in the war is fairly stated . . . Therefore I should like to see the passages reciting our many grievances somewhat reduced . . .'[83] Churchill eventually approved a somewhat diluted version which was sent on 20 December 1944 and not delivered until early January 1945. A copy was sent to Stalin.

While acknowledging that Spain stayed out of the war in June 1940 and during Operation Torch in 1942, Churchill's letter reminded Franco of the extent of German influence in Spain and of his own many declarations that the defeat of the Allies was both 'desirable and unavoidable'. The Prime Minister wrote unequivocally that 'it is out of the question for His Majesty's Government to support Spanish aspirations to participate in the future peace settlements. Neither do I think it likely that Spain will be invited to join the future world organisation.' With regard to Franco's anti-Russian statements, Churchill wrote 'I should let your Excellency fall into serious error if I did not remove from your mind the idea that His Majesty's Government would be ready to consider any *bloc* of Powers based on hostility to our Russian allies, or on any assumed need of defence against them.'[84]

A copy of Churchill's letter to Franco was sent to the US Department of State. It was accompanied by a letter which laid out British policy towards the Caudillo. The text made it clear that, fundamentally, London would like to see him and the Falange removed as 'an unfortunate anomaly' but regarded any attempt to dislodge him by force as undesirable since it would lead to another civil war.[85] That policy was to remain unchanged for many years. Franco was to derive considerable benefit from its essential contradiction. In particular, he would be able to foment popular feeling within Spain against a British rhetoric of hostility behind which lay an innocuous policy of non-intervention. It was a variant on British policy during the Spanish Civil War and would have roughly the same effect on Franco in favouring his interests at the same time as it intensified his resentment of 'perfidious Albion'.

On receiving Churchill's letter, Franco was unabashed and continued his efforts to rewrite history in a way now becoming familiar. His reply to Churchill's letter simply ignored its content and chose to interpret it as a friendly plea for cordial and close relations between Britain and Spain.[86] As he had done since the autumn of 1944, the Caudillo continued to claim that he had been the Allies' secret friend throughout the war. In this regard, the monumentally cynical Lequerica was the ideal man to tell the lie with a straight face. He had assured Hoare at the beginning of October that Franco was anxious to break away from Nazi and Fascist influences and follow the line of Great Britain in foreign policy.[87] He peddled the same line with increasing frequency but never more blatantly than in assuring the Associated Press correspondent Charles Foltz, as Franco had done already to the United Press, of Spain's American destiny.[88]

From December 1944 onwards, there was a build-up of anti-Bolshevik rhetoric in the Spanish press to give substance to Franco's claim to have been hostile only to Communism and not to the democracies.[89] Goebbels commented 'there is no serious political move behind this. Franco is a pompous ass.'[90] The trend of Franco's policy in the first half of 1945 was to trim for the future. He knew that if the Allies wanted to overthrow him, it would not be too difficult since most of his erstwhile collaborators would desert in droves. In an effort to dissuade them, monarchist plotters were receiving short shrift. Well-known figures of the pre-war extreme right* were arrested and sent into exile.[91] Their fate was gentle by comparison with that reserved for the Caudillo's enemies on the left. Frequent executions of 'Communists' still took place.[92]

Nevertheless, Franco knew that he could not survive on repression alone if the international situation was adverse. He pinned his hopes on the Americans regarding him as a better bet for anti-Communist stability in Spain than either the Republican opposition or Don Juan de Borbón. His confidence on that score was intensified by the indiscreet assurances of Carlton Hayes that Roosevelt would never under any circumstances contemplate intervention in Spain.[93] Franco showed his gratitude to Hayes when he left Madrid to return to the United States in mid-January 1945. Contrary to precedent, Mrs Hayes and her daughter took tea with Franco's wife Doña Carmen and his daughter Nenuca while Hayes had his final interview with the Caudillo. The Ambassador was presented with a portrait of himself by the fashionable artist Zuloaga, the fee for which was met by the Spanish government.[94]

The British *Chargé d'Affaires*, James Bowker, referred to Franco 'making Hayes while the sun shines'. The exaggeration of American friendship had the obvious domestic purpose of inflating Franco's international stature to maintain the morale of his supporters and deflate that of his enemies. Together with wildly unfavourable comparisons in the press between Hayes and Hoare, it was also meant to foment divisions between the British and American Embassies in Madrid.[95] The Falangist doormen in every apartment block were ordered to arrange for American and Spanish flags to decorate the building when Hayes's successor,

---

* Including the Marqués de Quintanar, the Marqués de Eliseda, Alfonso García Valdecasas and Lieutenant-Colonel Juan Ansaldo.

Norman Armour, arrived to present his credentials. Franco regarded the United States as a hotbed of dangerous freemasonry and he told a friend of his unease about playing the American card and becoming dependent on 'America's political hysteria'.[96]

Franco stifled his prejudices and made a great effort to flatter Armour. The effort was unsuccessful in personal terms but every last drop of propaganda was squeezed from it. When Franco invited Armour to dinner at El Pardo, the press printed photographs of him at the same table as the Minister-Secretary of the Falange. The implication that Franco had secured American support was of immense help to him in maintaining the loyalty of his generals.[97] In fact, American policy was rather more actively hostile to Franco than British although informed opinion in Washington was uneasy about economic sanctions or intervention to remove Franco in the absence of any well-defined successor regime. Like the British, the Americans were fearful of sparking off another civil war.[98] Accordingly, in the spring of 1945, the Foreign Office and the State Department reached broad agreement on policy towards Franco.[99]

The Yalta Conference, held between 4 and 11 February 1945, had called for free elections in the liberated countries. Franco was shrewd enough to assume that Stalin was unlikely to permit elections in Poland or Hungary and that he too might get away without fatal political changes. On 11 February, Japanese troops had perpetrated a massacre in Manila in the course of which many people taking refuge in the Spanish Consul-ate-General were murdered. Throughout March, the press throbbed with outraged speculation that Spain was about to declare war on Japan.[100] Only after two months of deliberation did Franco sever diplomatic relations with Japan.[101] Had he done so a year earlier in the wake of the Laurel incident, it would have been meaningful. Now, a distant Japan provided a painless, and empty, anti-Axis gesture.*

Hitler was disgusted by Franco's barely concealed desire 'to gain a good mark from America'. The Führer commented percipiently that: 'At heart, Franco realizes that it is now no good playing the British card and he is relying more on America.' Hitler returned to the same theme on more than one occasion, saying that 'the Americans are working actively in the background to cheat not only the Soviets but also the British out of the international game. This seems to have been noticed by Franco.' The Führer was convinced that 'Franco is trying by every conceivable

* Franco's Falangist poodle Arrese added to the long list of his inanities by telling an official of the American Embassy that he was ready to lead a new División Azul against Japan – Emmet John Hughes, *Report from Spain* (London, 1947) pp. 210–11.

means to take a hand in the great game and having failed with Britain –
and Britain, moreover, having too little power at present to afford him
the necessary protection – he is now making a renewed attempt in the
United States.'[102]

The difficulties facing the Caudillo in attempting to draw a veil over
his recent passions could be discerned in the letter which, on 10 March
1945, Roosevelt wrote to his new Ambassador in Madrid. The text was
the fruit of the newly agreed Anglo-American policy towards Spain. In
unequivocal language, it made clear the President's view that memories
of neither Spain's war record nor the activities of the Falange could be
erased by last-minute changes in Spanish policy. While reiterating that
'it is not our practice in normal circumstances to interfere in the internal
affairs of other countries unless there exists a threat to international
peace', Roosevelt stated that 'I can see no place in the community of
nations for governments founded on fascist principles'.[103] Although
Roosevelt's letter was not released to the press until 26 September 1945,
at the first opportunity Armour made its contents known to Lequerica.
Washington and London were united at this stage in hoping that Franco
would shortly be succeeded by a regime 'based on democratic principles,
moderate in tendency, stable and not indebted for its existence to any
outside influences'.[104]

The imminent collapse of the Axis caused Franco profound anxiety.
When Armour presented his credentials on 24 March 1945, the Caudillo
was still peddling his two-war theory. Although he reluctantly recognized
that Nazism was doomed, he took the opportunity to point out what that
meant in terms of the dangers of Communism. Armour spoke unequivocally along the lines of Roosevelt's letter, telling Franco that, to the
American people, the Falange was the symbol of his collaboration with
the Axis. The Caudillo disingenuously claimed that the Falange was not
a political party 'but rather a grouping together of all those having a
common interest, an objective – the welfare of Spain, the maintenance
of order, the development of the country along sound religious, cultural
and economic lines et cetera. It was open to anyone to join and included
representatives from all walks of life.' Even more astounding were his
outrageous boasts about the regime's achievements 'in rebuilding the
devastation caused by the civil war and in healing the wounds arising out
of the bitterness the conflict had engendered'. When Armour expostulated about the thousands of political prisoners in Spanish jails and the
continuing executions, Franco expressed his pain at such infamies. There
were only twenty-six thousand political prisoners, he said with satisfac-

tion.* Franco's appalling figures were almost certainly understated, they did not include those in labour battalions and other categories.[105]

The pressure from the victorious Allies inevitably boosted the confidence of Franco's exiled and internal enemies. However, the Caudillo could count on the murderous divisions among the exiled Republicans and the growing tensions among the Allies. Nevertheless, rumblings among the senior generals were getting louder, with the brightest and most ambitious of them, Rafael García Valiño and Antonio Aranda, beginning to show their hands. Fully aware of what was going on, Franco had announced a number of important postings on 4 March by way of seizing the initiative. Varela became High Commissioner in Morocco much to the chagrin of his predecessor Orgaz who was made Head of the General Staff. The crucial posting was that of the gloomily chain-smoking pro-Nazi Muñoz Grandes to be Captain-General of Madrid, the linchpin in terms of the Caudillo's political security. It was a shrewd move since Muñoz Grandes was not only one of the best of Franco's generals, but, having burned his boats politically, now had no option but to support the regime to the last. His post as Head of the Caudillo's Military Household was filled once more by the faithful mediocrity General José Moscardó who was replaced as Captain-General of Barcelona by the dourly loyal General José Solchaga. These cunning promotions ensured that no one group in the Army could count on sufficient key positions to threaten Franco.[106]

Franco felt seriously threatened when Don Juan, encouraged by General Kindelán and his civilian advisors, issued his so-called Lausanne Manifesto on 19 March 1945. In it, the Pretender denounced the totalitarian nature and the Axis connections of the Francoist regime and called upon Franco to make way for a moderate, democratic, constitutional monarchy.[107] Although not reproduced by the Spanish press, it was broadcast by the BBC. A group of senior monarchists was set up, consisting of the Duque de Alba and Generals Aranda, Alfonso de Orleáns and Kindelán, to oversee the expected transition. They went so far as to draft the text of a decree law announcing the monarchy and composed a provisional government.†[108]

---

* Two years earlier, on 15 March 1943, the outgoing Minister of Justice, Estéban Bilbao, admitted to seventy-five thousand political prisoners, a figure which referred only to those in prison and not those in labour battalions and military prisons, *Faro de Vigo*, 19 March 1943. The figure did not include many political prisoners classified as common criminals. Since Republican prisoners were still heavily involved in work on Franco's pet project, the Valle de los Caídos, there is no question of him being ignorant of their existence.
† Kindelán was to be President, Aranda Minister of National Defence, Varela Minister for the Air Force and General Juan Bautista Sánchez González Minister for the Army.

The Lausanne Manifesto was accompanied by an instruction to prominent monarchists to resign from their posts in the regime. The first to do so was Alba who abandoned the London Embassy.[109] He was quickly followed by General Alfonso de Orleans, effective head of the Air Force. Their examples were not replicated. As a rebellion, it was a damp squib. Franco responded by ordering General de Orleans confined to his estates near Cádiz.[110] Two prominent Catholics, Alberto Martín Artajo, President of Catholic Action, and Joaquín Ruiz Giménez were sent to tell Don Juan that the Church, the Army and the bulk of the monarchist camp remained loyal to Franco. They had no need to tell him that the Falange was deeply opposed to a restoration.[111] Without the military support of the Allies or the prior agreement of the Spanish Army and Church, Don Juan was naïvely depending on Franco withdrawing in a spirit of decency and good sense. Alba, bitterly disillusioned with Franco, told Martín Artajo, 'all he wants is to stay in power for ever; he is infatuated and haughty. He is a know-all and is recklessly banking on the international situation.'[112] Franco himself commented to Kindelán, 'as long as I live, I'll never be a Queen Mother'.[113]

It was an indication of just how seriously Franco took the Lausanne Manifesto that he immediately mounted an operation to neutralize any resurgence of monarchist sentiment in the high command. He summoned a meeting of the *Consejo Superior del Ejército* for 20 March. It remained in session for three days until 22 March and, unusually, Franco himself took the chair. He brushed aside a call from Kindelán in favour of a restoration and made an enormous effort to justify himself. The scale of the lies which he told the meeting suggest that his natural tendency to wishful thinking was intensified by sheer desperation. He informed his generals that Spain was so orderly and contented as to ensure that foreign countries including the United States would soon imitate and adopt Falangist principles. Declaring that Britain was finished, he backed up this judgement with the fanciful claim that Churchill had confessed to him that he was the prisoner of freemasons. He tried to undercut any monarchist conspiracy among his generals by brandishing the danger of Communism for which he blamed Britain, the most active supporter of the monarchists. He claimed that he was on the most cordial terms with the United States and cited a mythical personal assurance from the President of support for his government – conveniently forgetting Roosevelt's letter to Ambassador Armour about which he already knew. Referring to General Mola's republicanism, he alleged, with bare-faced cheek, that it had taken his own efforts to put the monarchist restoration on the agenda.

Only Kindelán challenged the absurdity of some of these affirmations for which he was rewarded by Franco with kindly derision. Many of the other generals, however, seemed satisfied by what they had heard.[114]

The annual Civil War victory celebrations were choreographed with both eyes on the mounting opposition. The press praised Franco who had saved the Spanish people from 'martyrdom and persecution', the fate, it was implied, to which the failures of the monarchy had exposed them.[115] The 1 April parade to mark the sixth anniversary was reviewed by the Caudillo on horseback. The press devoted rather more space even than usual to the event in order to pay tribute to Franco's victory over the 'thieves', 'assassins' and Communists of the Second Republic. The barely veiled message was that these same criminals were even now plotting their return with the help of the Allies. The exiled Republicans were taken to task for using the argument that the Spanish Civil War was the first battle in the present world war.[116] It was conveniently forgotten that Franco used exactly the same argument as his justification for receiving imperial crumbs from Hitler's table.

It was at this stage that Franco came up with one of the more blatantly crude of his devices to wipe out the past. At a cabinet meeting lasting several days in the first half of April 1945, he discussed the idea of adopting a 'monarchical form of government'. This was a public relations exercise meant to counter the activities of Don Juan and to deflect the hostility of the Allies to a regime organized along fascist lines. A *Consejo del Reino* (Council of the Kingdom) would be created to determine the succession – a gesture somewhat diluted by the announcement that Franco would continue as Head of State and that the King designated by the *Consejo* would not assume the throne until Franco either died or abandoned power himself. The forthcoming pseudo-constitution known as the *Fuero de los Españoles* (Spaniards' Charter of Rights) was also announced.* Foreign press correspondents were to be freed of censorship. The death penalty for offences committed during the civil war was allegedly to be abolished, although in fact there would be an execution on such grounds as late as 1963.[117]

More energy was devoted to driving a wedge between the United States and Britain.[118] There was a distinct difference in the receptions accorded the new ambassadors, London's nomination on 22 April of Sir

---

* Franco showed his cynicism about the proposed *Fuero de los Españoles* being prepared by Arrese and his reliance on a strong apparatus of state terror when he said 'Arrese, don't hurry, it's all the same to me to govern with the constitution of 1876'– José Luis de Arrese, *Una etapa constituyente* (Barcelona, 1982) p. 70.

Victor Mallet being given a press coverage far less effusive than that accorded to the arrival of Norman Armour five weeks earlier.[119] The death of President Roosevelt on 12 April and his replacement by Harry S. Truman did little for Franco's hopes in this regard. The plain-speaking Truman loathed Franco's deviousness, his repressive regime, his religious bigotry and his denunciations of freemasonry, liberalism and democracy. Nevertheless, a whispering campaign was mounted in Spain to give the impression of firm American support for Franco. It took the form of entirely fabricated rumours that Washington had begged him to keep the Falange in existence to help combat the Russian menace.[120]

There was no doubt that the changes being contemplated by the Caudillo were aimed at persuading the western Allies that sufficient reforms were under way to counteract Soviet demands for action against Franco. The feebleness of the reforms was deviously justified as necessary to maintain stability and avoid civil war in Spain. Both Asensio and Lequerica suggested to Armour that 'the Russians in their plan to dominate Europe would attempt to make use of the not inconsiderable elements in Spain favorable to them in order to bring about a violent upheaval'.[121]

It was by now as much out of fears for the future, as from ideological solidarity, that the Franco regime collectively continued to hope against hope that the defeat of Hitler might be avoided. The last German garrisons in the south of France were supplied with food and ammunition from Spanish ports on the Bay of Biscay, something which required official connivance at the very highest level.[122] As Allied forces stumbled across the horrendous sights of the extermination camps, the British at Belsen, the Americans at Buchenwald and the Russians at Auschwitz, Nazi officials were being given certificates of Spanish nationality.[123] The Francoist press played down the horrors of the holocaust as the unavoidable consequence of wartime disorganization.[124] When Berlin fell, the press printed tributes to the inspirational presence of Hitler in the city's defence and to the epoch-making fighting qualities of the Wehrmacht. *Informaciones* declared that Hitler had preferred to sacrifice himself for Europe rather than unleash his secret weapons. Allied victory was seen as the triumph of materialism over heroism. A burst of praise for the Franco regime as a peaceful bulwark against anarchy was a veiled announcement that there would be no sudden political change in Spain. The Caudillo appeared regularly in Falangist uniform and many major figures of the regime called at the German Embassy to express their condolences for the death of the Führer.[125] Franco did not break off diplomatic relations with the Third Reich until 8 May, VE Day.[126]

The end of the war in Europe was greeted with the most extreme eulogies to the 'Caudillo of Peace' for the wisdom and firmness which had enabled him to bestow the gift of peace upon Spain. According to *Arriba*, the end of the war was 'Franco's Victory'. *ABC* carried a front page picture of the Caudillo with the statement 'he appears to have been chosen by the benevolence of God. When everything was obscure, he saw clearly . . . and sustained and defended Spain's neutrality.'[127] Franco was now free to devote all his energies to regularizing his position with the victorious Allies. He was playing a dangerous double game, which was made possible only by his total control of the media in Spain. Part of the game consisted in rallying support around himself on the pretext that Spain was the object of a ruthless international siege. That line in his propaganda was to assume a central role after the end of the war but it had taken shape during the wolfram crisis. The Civil War was now presented as an example of a victorious Spanish unity against foreign interference and the Falange, or *Movimiento* as it was increasingly called, its institutional guarantee.[128]

Spanish neutrality in the Second World War was hailed for the next thirty years as Franco's greatest achievement. However, Franco ultimately avoided war not because of immense skill or vision but rather by a fortuitous combination of circumstances to which he was largely a passive bystander: the disaster of Mussolini's entry into the war which made the Führer wary of another impecunious ally, then Hitler's inability to pay the high price sought by the Caudillo for his belligerence and, throughout, the skilful use made by Allied diplomats of British and American food and fuel resources in an economically devastated Spain. Under such circumstances, it was hardly surprising, as von Stohrer had remarked to General Krappe in October 1941, that the Führer should conclude that Spain was more useful to Germany under the mask of neutrality as her only outlet from the British blockade.[129] Above all, Franco's neutrality rested on the appalling economic and military plight of a Spain shattered by the Civil War, a disaster from which the Caudillo thus derived enormous benefit.

XXI

THE HERO BESIEGED

1945–1946

AT THE end of the war, building on Churchill's 1944 praise in the Commons, the weight of Franco's propaganda machine was thrown into the task of rewriting the history of his role in the Second World War. For the rest of his life, he would assert that he had never even contemplated entering the war.[1] However, the Caudillo was aware in the spring of 1945 that some stormy seas would have to be navigated in the months to come. His first priority was to start gathering a dedicated crew. To this end, in the last months before the end of the war, as well as reshuffling the Captains-General in the interests of security, he promoted a number of militants from the officer training corps of the Falangist students' union, the *Sindicato Español Universitario*, to commissions in the Army and created the *Guardia de Franco*, a paramilitary formation of Falangist zealots. He also ensured that senior Juanista monarchists were kept away from positions from which they could endanger his position and strengthen that of Don Juan.

Franco had seen opportunities open up to him in seemingly endless profusion after 1936. The collapse of the Axis signified the end of what had seemed to be an ever onward and upward progress from rebel to Generalísimo to Chief of State to would-be emperor. With that optimistic serenity which always inspired his followers, whether in Moroccan skirmishes or Civil War battles, Franco gave no sign of dismay at the collapse of his dreams. Perhaps because there was always an element of fantasy about what he did, he was able, without a backward glance, to create a new goal, his own political survival, which he interpreted and projected publicly as a life and death struggle for the very soul of Spain. The monumental egotism that lay at the heart of his being enabled him to shrug off the demise of his erstwhile benefactors Hitler and Mussolini as matters of little significance relative to his own providential mission.

The Caudillo's strategy for survival was simple in theory, if complex in execution. With an eye on his domestic position, he would work hard to consolidate the loyalty of the triple pillars of the regime, the Church, the Army and the Falange. At the same time, for foreign consumption, he would emphasize the Catholic and monarchical elements, and play down the fascist ones, in what would be presented as a uniquely Spanish polity. The central objective was that Franco remain in power. After nearly a decade of exposure to daily adulation, he was now sufficiently messianic in his self-perceptions to see no contradictions between his own political needs and those of Spain. It is clear that, just as, in his early military career, he had had no difficulty creating and living the persona of the reckless desert hero, now he believed in, openly proclaimed and acted out to the full the role of the providential steersman navigating the storm-tossed ship of Spain. This image vied with that of the divinely inspired commander of a besieged fortress. As so often before, the unreal and naïve elements of these inflated beliefs were married to a cynical and ruthless political sense.

His thinking during the difficult months of the collapse of the Third Reich and the subsequent reconstruction of the international order was revealed in early April 1945 to the chubby forty year-old Catholic lawyer Alberto Martín Artajo, President of Catholic Action in Spain and a prominent member of the powerful pressure group, the *Asociación Católica Nacional de Propagandistas*. On behalf of a group of influential Catholics worried about the international opprobrium directed against Spain, Martín Artajo suggested that he eliminate 'the external signs' of Axis links and consider permitting an independent press. Franco was not disposed to accept such advice. He dismissed international criticism as the work of a masonic conspiracy in which he bizarrely saw the hands of the monarchist Sainz Rodríguez and a dissident Falangist, Santiago Montero Díaz. His response to the idea of changing of personnel or institutions was altogether more realistic and astute: 'we must make sure that people don't get disillusioned; they have served faithfully. There is nothing to be gained by ceding ground and it might be taken for weakness.'[2]

With his uncanny eye for valuable collaborators, Franco saw in the ambitious Martín Artajo someone whom he could use. In the first instance, he sent him to Lausanne as an emissary to Don Juan. On his return, on 1 May, Franco spoke to him for two and a half hours. Martín Artajo offered him the collaboration of the Catholics whom he represented and of the powerful Catholic press networks which could put the regime's case abroad. In return, they wanted the regime to survive with a more Catholic and less Falangistic face and, sooner rather than later, evolve towards a monarchist restoration. Still smarting from the Lausanne Manifesto, Franco was not open to talk either of compromise with Don Juan whom he dismissed as only 'a Pretender' or of changes to the Falange which he praised as 'an efficacious instrument'. He described the Falange as 'a bulwark against subversion', a safety valve – 'it gets the blame for the government's errors' – and a machinery for educating and mobilizing the masses who greeted him on his journeys around Spain.[3] Nevertheless, by insinuating to him that some political evolution was possible, Franco would turn Martín Artajo into a convincing advocate of his cause to the rest of the world.

While there was panic in Falangist circles about the consequences of the collapse of the Axis, Franco himself remained cool. He outlined a clear and confident vision of how he would survive. He would produce a law which turned Spain into a kingdom but that would not necessarily mean bringing back the Bourbons. In language which must have shocked the rather prim Martín Artajo, he summed up his view of the decadence of the constitutional monarchy by reference to the notorious immorality of the nineteenth-century Queen Isabel II. He said 'the last man to sleep with Doña Isabel cannot be the father of the King and what comes out of the belly of the Queen must be examined to see if it is fit'. Clearly, Franco did not regard Don Juan de Borbón as fit to be king – 'he has neither will nor character'. A monarchical restoration would take place, declared Franco, when he decided and the Pretender had sworn an oath to uphold the fundamental laws of the regime.[4]

Franco was still feeling his way in a difficult situation but he intended to admit only enough apparent change to neutralize his monarchist opponents and to buy time from his foreign enemies. His expectation that he would survive derived from his belief that the alliance between the democracies and the Soviet Union was a monstrous aberration.[5] The only tactic available to him was to sit out the international ostracism until the natural antagonism between the Communist and capitalist *blocs* crystallized. Then the geopolitical advantages of Spain would buy him out of his original sin and into the western bloc. It was to be a much easier task than Franco and his henchmen expected at the time or later made out.

There were only superficial reasons for believing that the Allies intended to 'finish the job' and remove Franco. Admittedly, at Yalta, on 12 February 1945, the Great Powers had promised free elections to the peoples of former Axis satellites – a move with implications for Spain. And so, with no apparent regrets, Franco had not hesitated to start rewriting the story of his links with the Axis. He declared in Madrid on 2 April 1945 that while the Republic had received massive aid from the Soviet Union and from international Communism, the Nationalists had been helped only by 'a few hundred Irish Catholics and several thousand other foreigners who came to demonstrate symbolically that Spain was not isolated'. He also made, for domestic consumption, the preposterous claim that current foreign hostility to Spain was the fruit of jealousy of her new-found strength.[6]

During the period from 1945 to 1950, Franco convinced himself that he and Spain were under deadly siege. As both Cabanellas and Casado had prophesied, he had come to regard Spain, in military terms, as the position which it was his task to defend until death. With the opposition re-emerging and hoping for backing from the Allies, many of the Caudillo's followers wavered during what has been called 'the black night of Francoism'.[7] Franco lost no sleep. He interpreted the situation as a siege and applied to it his experiences in the Legion. Numerous declarations over the next few years would reveal his dogged determination to fight to the end but none more than remarks made to his brother Nicolás in August 1945. Nicolás was on one of his habitual visits to El Pardo. Franco showed him two photographs, one of the corpses of Mussolini and Clara Petacci, the other of Alfonso XIII stepping ashore in Marseille on the first stage of his exile. He told his brother, 'if things go badly I will end up like Mussolini because I will resist until I have shed my last drop of blood. I will not flee like Alfonso XIII.'[8] Franco was privately affected by the fate of the Duce which apparently triggered off memories of the Annual disaster when those Spanish officers who surrendered to the Moors ended up dead, their corpses mutilated.[9] His response was, to use Garriga's graphic metaphor, 'to behave like the captain of a ship who, in order to guarantee the loyalty and discipline of the crew, scuttles the life-boats and destroys the life-jackets to show that either everyone is saved or else everyone goes down together'.[10] In other words, as he had indicated to Martín Artajo, far from shedding the Falange, he would draw it round him as a praetorian guard.

Franco's ship did not have to wait long for stormy seas. The founding conference of the United Nations was held at San Francisco between 25 April and 26 June 1945. On 19 June, the Mexican delegation proposed the exclusion of any country whose regime had been installed with the help of the armed forces of the States that had fought against the United

Nations. The Mexican resolution, drafted with the help of exiled Spanish Republicans, could apply only to Franco's Spain. It was approved by acclamation.[11] The anti-Franco exiles now took it for granted that the Caudillo's days were numbered. Even his erstwhile admirer and supporter, the Cardinal Archbishop of Toledo, Plá y Deniel, taking for granted that there would now be negotiations for the restoration of the monarchy, offered to act as intermediary.[12] However, there were glimmers of light behind the black clouds. In Washington, there was already some anxiety that too hard a line on Franco might encourage Communism in Spain. In fact, Franco saw the part played by his exiled enemies in generating international ostracism as simply proving the rightness of his own position.[13] Nevertheless, on the day after the Mexican resolution, he broadcast to Latin America his first attempt to discount 'the campaigns of defamation' and instead to present a Spain which was 'holy, warlike, artistic, generous, honourable and marvellous'.[14] On 22 June, Lequerica denied that the San Francisco resolution had any bearing on Spain on the mendacious grounds that the Franco regime was created without Axis help. [15]

As part of the long and arduous effort to dissociate himself and his regime from the Axis, Franco made two important declarations for both international and domestic consumption in the summer of 1945. He gave an interview to a representative of the British United Press in which he mixed stark truths with blatant lies. He gave, however, as so often before, every sign of believing every word that he said. He confided in his interviewer that 'You can tell the world that the Falange wields no political power in Spain and takes no political decisions because all political power and decisions depend absolutely on the Government and on no other entity whatsoever. There are some members of my Government who happen to be members of the Falange Party, just as they might happen to be anything else, but first and last, they are members of my cabinet.' At one level, it was a disingenuous remark since all members of the Government were technically members of the *Falange Española Tradicionalista y de las JONS* and there were no other legally permitted political parties in Spain. At the same time, Franco was accepting a vague and shifting distinction which had always been implicit in the composite nature of the united *FET y de las JONS* and would become ever more common within the regime. This was between the *FET y de las JONS* as a broad umbrella for all Francoists, whether they were predominantly monarchists, Catholics, Carlists, soldiers or militant Falangists, and the Falange as the most radical among them.

Franco went on to deny that Spain had been allied in any way to Germany or Italy. 'It is true', he said, 'that when Germany seemed to be winning the war, some members of the Falange tried to identify Spain with Germany and Italy, but I immediately dismissed all persons so inclined. I never had the slightest intention of taking Spain into the war.' It is representative of the difficulties of interpreting Franco that this virtuoso display of amnesia reflected at the same time both cold, hard-faced cynicism and the fact that Franco now sincerely believed his own flattering reconstruction of his recent past. Another sleight of hand was served up which hinted at political change: 'With the exception of certain relatively short periods in our history, Spain and the Government have been traditionally monarchist. Therefore to provide for eventualities we have already decided upon the creation of a Council of State which, whenever the necessity arises, would decide on the problem of succession as regards the throne . . . and the necessary prerequisites of fitness to ascend the throne.'[16]

One month later, as the Potsdam Conference was beginning, in his annual speech to the *Consejo Nacional* of the Falange, the Generalísimo publicly began the process of fabricating a new and internationally acceptable façade for his regime. The necessary concession to the changed situation was a pseudo-constitution in the form of the '*Fuero de los Españoles*' (Spaniards' Charter of Rights). Superficially, the decree guaranteed the civil liberties of Spaniards. In the small print, freedom of expression, for instance, did not extend to opposing the 'fundamental principles of the State'. Equally, the freedom of association stopped short at any grouping interpreted as undermining 'the spiritual, national and social unity of the fatherland' – in other words, political parties or trades unions.

The speech, written by Franco himself with the assistance of Carrero Blanco, began with self-congratulations for the achievement of remaining neutral during the war. He offered Spain a bleak choice: 'the order, the peace and the joy which makes Spain one of the few peoples still able to smile in this tormented Europe' or Communism with its bloody consequences. 'They are mistaken', he declared, dashing the hopes of those expecting change, 'who believe that Spain needs to import anything from abroad.' The clear message for both foreign and domestic audiences was that the Franco regime was unique, not comparable to the Axis regimes on which it had in fact been modelled, and, if dissimilar to democracy, capable of coexisting with the western powers. The idea that the removal or even modification of Francoism would open the door to Communism was to be the constant excuse in this period for avoiding change of any but the most superficial kind. The only thing which Franco had to say about the future was that his regime could be succeeded only by the traditional, medieval monarchy. There was no question of immediate change nor of a modern constitution. His final words made it perfectly clear that he intended to ride out the storm of Axis defeat and Allied opprobrium: 'like the good captain, we must keep the ship firmly on course, adjusting to the storms which might lash her'.[17]

Intensely aware of the menace of Potsdam, the good captain quickly hastened to renew the crew of his ship. The flag of the Axis, if not that of the Falange, was hauled down and replaced by that of a heavily conservative variant of Christian Democracy. On 18 July 1945, the more obviously Axis-tainted ministers were dropped without any signs of regret. Lequerica was replaced, on the advice of Carrero Blanco, by Martín Artajo; Asensio by the anodyne General Fidel Dávila; Arrese was removed and not replaced as *Ministro-Secretario del Movimiento*. In fact, Franco had not planned to eliminate the party ministry from the cabinet until Martín Artajo told Carrero Blanco that he would not join the cabinet otherwise. Franco did not usually let himself be pushed into decisions but he and Carrero Blanco were convinced that the clever but pliable Martín Artajo was crucial to their strategy for survival. Although prominent Catholics like José María Gil Robles and Manuel Giménez Fernández opposed collaboration with the regime, the presence of Martín Artajo and others, such as the ex-Cedista José María Fernández Ladreda as Minister of Public Works, gave credibility to Franco's new image as an authoritarian Catholic ruler.[18]

Falangists remained in the cabinet, but less prominently.* Shortly, control of the press would pass from the Falange to the grey and stolid Francoist José Ibáñez Martín, who remained as Minister of Education. The equally loyal and politically colourless Blas Pérez González remained as Minister of the Interior. Franco's lifelong friend Juan Antonio Suanzes was Minister of Commerce and Industry. Since it was Franco's practice to make his own broad policy while leaving ministers freedom only in their narrower departmental interests, ministerial changes were necessarily cosmetic. Given the international context and Franco's primordial concern for foreign policy, the key change was the introduction of Martín Artajo. Nevertheless, he had carefully assembled a cabinet of fiercely loyal

---

* Raimundo Fernández Cuesta became Minister of Justice. José Antonio Girón remained as Minister of Labour. Carlos Rein Segura became Minister of Agriculture. These three constituted Franco's token public commitment to the social rhetoric of the Falange.

Francoists, a crew which would put aside partisan loyalties in the common aim of keeping the ship afloat.[19]

Martín Artajo, a one-time member of the CEDA and protégé of Angel Herrera, the *éminence grise* of Spanish Christian Democracy, had met Carrero Blanco in 1936 when they had both taken refuge in the Mexican Embassy in Madrid.[20] He accepted the post after consultation with the Primate, Cardinal Plá y Deniel, and both were naïvely convinced that he could play a role in smoothing the transition from Franco to the monarchy.[21] Franco was happy to let them think so, just as simultaneously, he was reassuring Falangists that nothing would change. Having already announced that he would soon initiate the process for Spain to become a monarchy once more, empty legislation such as the *Fuero de los Españoles* would give the impression that the Caudillo presided over a *sui generis* democracy and was not dictator. His overall strategy was clear. Catholics, rather than Falangists, would carry the burden, with the support of the Vatican, of blunting the enmity of the victorious democracies.

Franco astutely permitted Martín Artajo to think that he could push him into liberalizing the regime, knowing that his assurances to foreign governments would be the more convincing. He had Martín Artajo's measure all along, conceding nothing. The Caudillo maintained an iron control over foreign policy, yet used him to the full as the acceptable face of his regime for the international community. Artajo told José María Pemán that he spoke on the telephone for at least one hour every day with Franco and used special earphones to leave his hands free to take notes. Pemán cruelly wrote in his diary 'Franco makes international policy and Artajo is the minister-stenographer'. In the first meeting of the new cabinet team, on 21 July, Franco told his ministers that concessions would be made to the outside world only in non-essentials and out of convenience.[22]

Franco also made long-term plans to neutralize Don Juan and his supporters during the storms to come. On the advice of Carrero Blanco, on whom he relied more and more, he adopted a two-pronged strategy to tame the threat of a democratic monarchy. First there would be an attempt to wean Don Juan away from his more radical advisers such as Gil Robles, Saínz Rodríguez and Eugenio Vegas Latapié, and then pressure to ensure that Franco would have a say in the education of Don Juan's eight year-old son, Juan Carlos. Accordingly, although he had already decided after the Lausanne Manifesto that Don Juan could never be King, Franco opted for a policy of reducing friction with him while encouraging Francoist monarchists to get close to the royal camp. The visits of opportunists such as José María de Areilza* to see Don Juan were dutifully reported to the British Embassy to give the impression that Franco was negotiating the terms of a restoration and so buy him more time.[23]

Shortly after his ministerial changes, Franco's fate came onto the international agenda at the Potsdam Conference. Stalin tried to get British and American backing for breaking off all relations with Spain and giving support to 'the democratic forces'. His initiative was cautiously welcomed by Truman who was inhibited by fear of sparking off another civil war. Stalin then suggested, if breaking relations was 'too severe a demonstration', that the Big Three adopt some other more flexible means to show the Spanish people their sympathy. Stalin's suggestion was resisted by Churchill who pointed out both that Britain did not want to risk her valuable trading relationship with Spain and also that interference in the internal affairs of other states was contrary to the United Nations Charter.[24] Arguing that however odious Franco might be, it was not for the Three Powers to dictate to 'the smaller enemy countries the kind of government they must have', the British delegation was seeking 'some fairly anodyne form of anti-Franco resolution'.[25] When Churchill was replaced at Potsdam on 28 July, after the Conservative defeat in the British general elections two days earlier, the Russian proposal of a total break of relations with the Franco regime was rejected by his successor Clement Attlee and his Foreign Minister, Ernest Bevin. The Mexican resolution accepted at San Francisco was ratified by the three Great Powers in their final communiqué on 2 August 1945 which reiterated Spain's exclusion from the United Nations on the grounds of the origins, nature, record and Axis links of the Franco regime.[26] Franco, however, could take heart from the absence of any suggestion of intervention against him.

Shortly before the Potsdam declaration, Franco received the new British Ambassador, Sir Victor Mallet, for the first time. When Mallet told him that his regime was associated in British minds 'with friendship towards Fascists and Nazis and [that] both deeds and speeches during the war will not be forgotten', the Caudillo put on a virtuoso performance of ingratiating affability. He beamed complacently as Mallet tried to bring home the extent of his isolation and 'smilingly insisted that relations

---

* A Basque monarchist linked to the Falange in the 1930s, Areilza – who, through his marriage was Conde de Motrico – had been *Alcalde* of Bilbao after its capture. In 1941, he wrote, with Fernando María Castiella, the ferociously imperialist text *Reivindicaciones de España* and aspired to be Ambassador to Fascist Italy. After the war, he moved back to the pro-Francoist monarchist camp, was Ambassador to Buenos Aires and Paris, before becoming a full-scale supporter of Don Juan and opponent of Franco in the 1960s.

would improve'. The Caudillo dismissed the Ambassador's references to his wartime support for the Axis with lengthy interruptions aimed at proving that his pro-Germanism had been much exaggerated. He lightly brushed off the Blue Division as a 'mere drop of water'. With the British election in mind, he coolly went on to tell Mallet that his programme of social reform was closer to the ideals of the Labour Party than to those of the Conservatives.[27]

Franco was vexed by the Potsdam declaration but quickly perceived that it was less hostile than he and Martín Artajo had feared.[28] The brazen Spanish reply to the communiqué showed the hand of Franco himself. Issued on 5 August 1945, it declared that Spain was not begging for a place in any international organization and would certainly accept only a position commensurate with her historical importance, size of population and her services to peace and culture. The note went on: 'Similar reasons made her, under the monarchy, leave the old League of Nations.'* Inevitably, Spanish neutrality during the war was praised as an 'outstanding record of nobility' (*destacada ejecutoria* – one of Franco's favourite expressions).[29]

Despite their own divisions, the exiled leaders of the Spanish Left were optimistically convinced that the San Francisco and Potsdam declarations about Spain, when considered alongside the arrival in power of the British Labour Party, signified the imminent doom of Franco. In anticipation of his demise, a government-in-exile was formed in late August under the presidency of José Giral.[30] The exiles' conviction that Franco was finished was shared in regime circles in Madrid. The great wit, Agustín de Foxá, remarked 'what a kick they're going to give Franco in our arse'.[31] Franco in contrast gave every sign of being blithely unconcerned. He made his stand on his anti-Communism and in the confidence that circumstances would change. In numerous speeches and declarations, he took credit for avoiding Spanish involvement in the world war and portrayed Spain as a happy, unified oasis of peace in a troubled world in which the Communist hordes were ceaselessly on the prowl.[32]

That he was ploughing in fertile fields soon became obvious. Ernest Bevin, in his first speech to the House of Commons as Foreign Secretary on 20 August 1945, signalled clearly that the western powers would not take any action against Franco when he said 'the question of the regime in Spain is one for the Spanish people to decide'. He thereby confirmed

---

* In referring only to the brief temporary withdrawal of Spain in 1928, Franco conveniently drew a veil over his own withdrawal from the League of Nations on 8 May 1939 in solidarity with the Axis.

that the Spanish policy of the new British Labour government was as innocuous as that of its Conservative predecessor. Bevin can only have given Franco enormous comfort when he said that the British government would 'take a favourable view if steps are taken by the Spanish people to change their regime, but His Majesty's Government are not prepared to take any steps which would promote or encourage civil war in that country'.[33]

The Soviet Ambassador in London, Feodor Tarasovitch Gousev, was shocked by what he perceived as an abandonment of the Potsdam agreements. He told Bevin four days later that the speech 'would be read by Franco as meaning that no action was intended against him: he was evidently trying to consolidate his position and throw dust in the eyes of the Allies by announcing that he intended to hold elections'.[34] Gousev was right. Franco had been examining secret police reports on the fears provoked among regime forces by the Potsdam declaration. Not only had the Left been heartened, but anti-Falangist sentiments had been voiced in Catholic and military circles. Monarchists had assumed that the restoration of Don Juan was imminent.[35] The gloom which had descended on regime circles after Potsdam lightened appreciably in the wake of Bevin's speech. Martín Artajo openly expressed his relief to Mallet.[36] 'Spain', he said, 'has only to sit waiting at her door to see the funeral procession of the enemies which she defeated in 1939.'[37]

It was at precisely this time that Franco's ever more influential assistant, the future Admiral Carrero Blanco, drew up a long report on the regime's survival. It was a deeply cynical document which combined the most narrow-minded provincialism with brilliant perspicacity. In terms which mirrored the Caudillo's own attitudes, and skilfully flattered his own view of his achievements, Anglo-Saxon ostracism was dismissed outright as jealousy 'because Spain is now independent, politically free, vigorous and on the way up, because she is a Spain unknown since the plundering of Utrecht [a reference to the loss of Gibraltar] and that irritates and hurts them'. At the same time, Carrero Blanco's astute conclusions were that, after Potsdam, Britain and France would never risk opening the door to Communism in Spain by supporting the exiled Republicans. Accordingly, 'the only formula possible for us is order, unity and hang on for dear life. Good police action to foresee subversion; energetic repression if it materializes, without fear of foreign criticism, since it is better to punish harshly once and for all than leave the evil uncorrected'.[38]

Franco continually referred to foreign pressure for democratic change as 'the masonic offensive'. He assured a cabinet meeting on 8 September that there were fifteen million freemasons in England who all voted Labour.[39] He bought time by letting Martín Artajo assure foreign diplomats that he would hand over to Don Juan 'within the next two years'.[40] However, the limits of the 'new' post-war Franco, and the primacy of his personal role in foreign policy, were made clear when he rejected Martín Artajo's recommendation of the liberal Dr Gregorio Marañón as Ambassador to London. He distrusted him, as he had the Duque de Alba, as someone who would be loyal to higher ideals than the survival in power of Franco. He chose instead the Francoist Domingo de las Bárcenas, who had been his trusted Ambassador to the Vatican during the Second World War. Similarly, Martín Artajo wanted to send a Christian Democrat – either José Larraz or Luis García Guijarro – to Washington, but Franco insisted on Lequerica. London hesitated before finally granting the *agrément* to de las Bárcenas; Washington refused outright to accept a man regarded as a fascist.[41]

Even in such bleak international and domestic contexts, Carrero Blanco's memorandum had proposed an entirely reasonable strategy. Churchill and Bevin had made it clear that the British would never intervene in Spain. The nearest that London came to action was a series of subtle snubs which were barely noticed by Franco, who was described by a Foreign Office official as having 'a skin like a rhinoceros'.[42] France, given intense public hostility to Franco, was potentially the most militantly anti-Francoist of the three western powers. However, both the moderate Catholic Foreign Minister, Georges Bidault, and the President of the Council of Ministers, General de Gaulle, were hostile to action against Franco. While Bidault temporized, de Gaulle sent a secret message to Franco to the effect that he would resist left-wing pressure and would maintain diplomatic relations with him.[43]

In the summer of 1945, internal regime opposition centred on Generals Aranda and Kindelán. They tried to put pressure on the new Minister of War Fidel Dávila to give Franco an ultimatum to go. Given Dávila's unquestioning loyalty to Franco, their efforts were in vain. In fact, Aranda and Kindelán were becoming increasingly lone wolves, talking to members of the left-wing opposition and to any foreign diplomats ready to listen to them.[44] Their impact was insignificant by comparison with the massive propaganda barrage being mounted by the Caudillo to create an image of the regime's permanence. On 25 August 1945, Franco sacked Kindelán as Director of the Escuela Superior del Ejército for a fervently royalist speech predicting that the Pretender would soon be on the throne with the full support of the Army. Ironically, Kindelán, one of the few

Spanish generals to foresee eventual Allied victory, was replaced by Juan Vigón, who, like the Caudillo, had sustained his faith in the triumph of the Third Reich until late in the day.[45]

That he could deal so relatively leniently with Kindelán was an indication of Franco's determination not to offend the Army. The exiled Republicans were bitterly divided and had little support within a Spain traumatized by the Civil War and beaten into political apathy by the ongoing repression. It was the Great Powers and influential elements among his own supporters which gave Franco cause for concern. To consolidate the fidelity of the Army, he lost no opportunity to appeal for military vigilance in defence of the unity of the *Patria*, which effectively meant in defence of his position. The dismissal of Kindelán was followed by a vehement plea for loyalty at the Escuela Superior del Ejército on 15 October.[46]

Franco's attitude to the Falange was altogether more complicated and devious, as his earlier conversations with Martín Artajo had revealed. On 3 September 1945, Serrano Suñer wrote a letter to Franco, proposing that he use the breathing space provided by Bevin's speech to proceed to the demobilization (*licenciamiento*) of the Falange and a renovation of political personnel, with a national government including such figures as Gregorio Marañón, José Ortega y Gasset and Francesc Cambó. In the margin of the letter, Franco wrote 'No' alongside the suggestion and 'Ha, Ha, Ha' by the names.[47] He made only the most superficial of changes. At the cabinet meeting of 7 September, it was decided, in response to British demands, to withdraw Spanish forces from Tangier. Franco commented realistically 'it's not much to lose if we can't defend it'. Spain's only imperial conquest in the Second World War was thus erased. At the same meeting, the fascist salute was abolished, much to the chagrin of the Falangist Ministers.[48]

Such cosmetic devices changed nothing in Franco's relationship with the Falange. Its political value to him was revealed during the spectacular commemoration of the *Día del Caudillo* on 1 October 1945, the ninth anniversary of his exaltation to supreme power. The ceremonies were orchestrated as a defiant demonstration to the outside world of the strength of popular and institutional support for Franco. The Church participated fully. Franco, accompanied by his ministers, the leaders of the Falange and the high command of the three armed services attended a choral mass with orchestra, officiated by the fiercely Francoist bishop of Madrid/Alcalá, Leopoldo Eíjo y Garay, in the Church of San Francisco el Grande, the nearest to a cathedral in the capital. That was followed by a solemn Te Deum officiated by the Papal Nuncio, Monsignor Cicognani. The assembled dignatories were then treated to a sumptuous reception in the throne room of the Palacio Real presided over by the Caudillo. Outside in streets lavishly decorated with Spanish flags, the Falange youth front, the *Frente de Juventudes*, marched. There were similar celebrations all over Spain.[49]

The celebrations, and the reaction to Serrano Suñer's letter, merely confirmed what Franco had revealed to Martín Artajo in the spring. He had made a shrewd cost/benefit analysis of the relative utility of maintaining or abolishing the Falange. He realized that changes to the Falange, or Movimiento as it was increasingly called, would do little to modify the attitude of the western democracies towards him. In any case, the Falange was *his* movement and his every inclination was to keep in existence a machine which provided him with constant adulation. To dismantle the Falange would create two major problems. The first would be how to deal with the hundreds of thousands of hangers-on who lived off the Movimiento and its sprawling bureaucracy. The second would be the extent to which the removal of the Movimiento would open the way to a wide range of opposition groups. In contrast, to maintain the Movimiento was to retain a gigantic apparatus staffed by people who knew that their existence had been threatened, who had nowhere else to go and would therefore give unquestioning loyalty to the Caudillo. By doing nothing, Franco consolidated the fervent support of hundreds of thousands of Falangists.

Indeed, while appearing to be merely sitting tight, Franco was working hard to defend his position. He kept his cabinet in session for four full days between 3 and 11 October. Behind a smokescreen of rhetoric about an increase of 'popular participation in the tasks of the State', which was to be implemented by periodic referendums, Franco made it obvious to his Ministers that he intended to permit little change. A return to political parties was unthinkable, since in Britain it had brought the country to socialism, and he insinuated, to the verge of Communism. Since the West did not want to see revolution or Communism in Spain, he would pursue a strategy of blunting international criticism with cosmetic changes while awaiting a revaluation of the regime when the West and the Soviet Union eventually fell out. Franco suggested a propaganda emphasis on the *Fuero de los Españoles* and a law to establish the eventual succession. When Artajo suggested a partial amnesty for political prisoners, Franco replied 'we do not wipe the slate clean' (*nosotros no borramos*). When municipal elections were proposed as a way of giving an impression of democracy, Franco

suggested that elections be announced and then delayed indefinitely. He
agreed to a referendum and talked of 'wearing a democratic suit as an
insurance policy' and declared optimistically 'we are on the verge of the
miracle'.[50]

The remarks made during the cabinet meeting suggested that a world
of experience separated the cautious and cunning Franco of 1945 from
the eager imperialist of 1940. That was clear too from the way he put
into practice his awareness of the Anglo-American anxiety not to provoke
further civil war. On 26 October, he demonstrated publicly that to remove
him would mean fighting and defeating the Army. The occasion was
another reunion of his surviving fellow cadets from the Toledo Infantry
Academy. A massively choreographed public homage was mounted with
the participation of the leaders of the coalition that had won the Civil
War, the Army, the Church and the Falange. With an escort of generals,
he entered the Cathedral under a canopy and was blessed by the Primate,
Archbishop Pla y Deniel. After being named 'protector of the city' and
presented with an elaborate baton of office (vara), inlaid with gold and
silver, he proceeded to the Alcázar. Surrounded by the symbolism of
Nationalist resistance under siege, he declared defiantly that the Army
would block the evil designs of political parties and freemasons.[51] Lavish
press coverage of the occasion constituted a kind of mass blackmail.
Franco knew from the reports of his own secret services that most Spaniards were desperately determined never again to undergo the horrors of
civil war.[52] It was also the view of the US Embassy that 'it is the will of
almost all categories of Spaniards to avoid more bloodshed; and no single
fact plays more directly into the hands of General Franco than the argument that precipitate change means another 1936'.[53]

Contradictory reports about Franco's mood filtered out of El Pardo
to the western embassies. Mallet and Armour would be told by eager
informants one day that the Caudillo had decided that the game was up
and on the next that he would resist to the last in the wreckage of his
palace.[54] The situation was tense but his 'blind faith' and unflappable
optimism did not desert him. He revealed his bunker mentality when he
told General Martínez Campos that, 'I will not make the same mistake
as General Primo de Rivera. I don't resign. For me, it's straight from here
to the cemetery.'[55] Franco's growing confidence exasperated Norman
Armour.[56] But the American Ambassador had a small satisfaction in early
October when the US Embassy bulletin published Roosevelt's March
letter to Franco. Ninety thousand copies were quickly distributed and
queues of those anxious to get a copy formed in front of the Embassy.

Franco's annoyance was reflected in a Spanish Government protest to Armour.[57]

A deeply frustrated Armour left Madrid at the beginning of December 1945. Armour's retirement, after thirty-three years in the US foreign service was routine but it was interpreted in the American press as an official rebuke to Franco. This impression was confirmed by the fact that he was not replaced at ambassadorial level.[58] When Armour took his leave on 29 November, he told Franco of his disappointment at the slowness of political change. With 'complete confidence and self-righteousness', Franco merely counselled patience.[59] For two hours, he expatiated on the dangers of moving too fast and assured Armour that all criticism of his regime abroad was Communist-inspired.[60] Thereafter, the United States Embassy would be in the hands of a *Chargé d'Affaires* until 1951. Franco realized that such humiliating gestures went no further than rhetoric. He was quick to see that they could, in fact, be used to his advantage in generating a siege mentality within Spain. Moreover, as Mallet realized, the chances of him being removed depended on the Army and Franco knew how to cement its support by hints that, if he disappeared, 'the days of fat living' would end.[61]

On the day after taking his leave of Franco, Armour visited Martín Artajo who assured him with apparent conviction that the Caudillo would eventually make way for a constitutional monarchy. He repeated the feeble excuses produced by Franco for his delay in relinquishing power. One was his doubt that, if he handed power back to the generals from whom he had received it in 1936, they would relinquish it to a civilian government. Another was that, unlike Alfonso XIII who had abandoned his throne in 1931 under the 'erroneous' impression that the country was against him, Franco would not leave under the mistaken view that the people did not love him. Artajo pointed out plausibly that Franco was not a man who reacted well to rough treatment or direct pressure. However, his own frustration at the Caudillo's slow political evolution glimmered through his curious request to Armour that the new Secretary of State, James F. Byrnes or Ernest Bevin or both send a message to Franco urging faster political change.[62]

How little the Caudillo had to fear from British anti-Franco rhetoric was starkly revealed in early December 1945. The arrival in London of his new Ambassador, Domingo de las Bárcenas, provoked a debate in the House of Commons on 5 December. The left Labour MP Ian Mikardo asked Bevin if he was aware that the trial of twenty-two Spanish anti-fascists, suspended in October after British protests, was about to be resumed and if he intended to approach the Spanish government about it. To Bevin's anodyne answer that the British Ambassador would watch the situation, Mikardo came back with a mischievously perceptive question: 'Can the right honourable gentleman explain to the House how it comes about that General Franco, who was so worried about his position at the end of July, has now regained confidence to such an extent as to thumb his nose in this way at His Majesty's Government?' Bevin could only reply: 'The mercurial habits of this dignitary are very difficult to explain.' The anti-Francoist Captain Noel-Baker then asked if Bevin was aware that the recent arrival of a new Ambassador was being interpreted in Spain and elsewhere as implying His Majesty's Government's support for the Franco regime. Referring to his speech on 20 August, Bevin stated that 'Our attitude to Franco was made quite clear . . . it has not changed; we detest the regime'.[63] In fact, on the same day, Bevin had told de las Bárcenas of his 'very great regret that the Franco regime was still being maintained' and asked him to inform Franco that British patience was 'becoming exhausted'.[64]

Franco was indifferent to such verbal antipathy. In mid-December, Martín Artajo told José María Pemán that Franco actually believed in the Falange and treated the Falangist ministers as if they were his special favourites, like members of the family. Pemán wrote in his diary, 'if they had told me that Franco had a lover it would have seemed bad, not to say strange, but this is worse: he has got a conviction.' 'I thought that in Africa he had learned to have horses shot from under him and stay unharmed. This was bad for the horses but excellent for him and it could be now for Spain. But now there is a horse to which he feels so close that he is prepared to go down with it.'[65] In fact, the normally shrewd Pemán was wrong. Franco may have had an emotional commitment to the Falange but it did not undermine his capacity for ruthless calculation. He had in fact worked out that there was more benefit to be derived from keeping the Falange. Not only was it a massive bulwark of support but international criticism of it also helped him capitalize on mass resentment of foreign 'interference'.

International hostility aside, at the end of 1945, Franco was threatened internally on two fronts. On the one hand, there was the pressure from monarchists within the Francoist coalition for him placate the western powers by making way for Don Juan. Outmanoeuvring the monarchists would require all his cunning. He would use the same device to deal with both international ostracism and monarchist sentiment, changing the name of his regime to 'kingdom' without altering its substance and making himself regent for life. Secondly, he faced the opposition of the defeated republicans. Whereas Franco dealt with the opposition of the monarchists with subtlety and duplicity, the Left faced only the most implacably brutal repression. Imprisonment, executions, torture and exile had taken a savage toll and made fear a way of life for those who opposed the dictator. Hunger and the difficulty of getting work without safe-conducts for travel and certificates of political reliability diminished the combative capacity of the Left. Nevertheless, the defeat of the Axis had allowed many of the Spanish *maquisards* who had played a key role in the French resistance to return to Spain.

By the end of 1945, a full-scale guerrilla war against the regime was beginning to build up in the north and east. Dominated by Communists, but including also Socialists and anarchists, the so-called Spanish *maquis* would threaten the regime until 1947. Franco reacted by keeping his regime on a war footing and not hesitating to evacuate entire areas in order to pursue scorched earth tactics. The guerrilla opposition would not be entirely eradicated until 1951.[66] It was hardly surprising then that, in the State budget for 1946, forty-five per cent was dedicated to the apparatus of repression, the police, the Civil Guard and the Army.[67] At cabinet meetings, the subject of repression swamped any efforts by Martín Artajo to open a discussion on political evolution. The meetings went on late into the night, dragged out by long rambling speeches from Franco about masonic conspiracies or the possibility of reviving the Spanish film industry with big productions of his favourite *zarzuelas* (operettas). When the parlous state of the economy was discussed, he dismissed the inflation which crippled it as the invention of bankers, 'typical of credulous simple-tons in economics' (*propio de los papanatas en lo económico*).[68]

Early in 1946, in a widely publicized interview, Franco again denied that he had ever supported the Axis. Yet since the closing stages of the war, his regime had been giving succour to many escaped Nazis, fascists and supporters of Vichy France. German Government and Nazi Party property in Spain which was supposed to have been placed under embargo was spirited away with official connivance. By the device of granting Spanish nationality to war criminals, it was possible to deny that they were given asylum. The Caudillo personally connived at the escape of Leon Degrelle, the Belgian SS general. The Italian General Gambara and other members of the one-time Corpo di Truppe Volontarie found a welcome in Spain as did the Nazi special operations ace, Otto Skorz-eny.[69] A report was placed before the United Nations Security Council claiming that between two and three thousand German Nazi officials, agents and war criminals were living in Spain in addition to tens of thousands of ex-members of the Vichy militia. It was calculated by the US Government that the financial holdings of ex-Nazis in Spain amounted to $95,000,000.* The Polish Government alleged that ex-Gestapo agents had found positions in the Spanish secret police and military intelligence although the British and American Governments knew of no cases. Spaniards previously employed by the Gestapo, however, were believed to have been incorporated into Franco's security services.[70]

In early 1946, there was some debate between London and Washington as to the possible effects on Franco's position of the publication of captured German documents revealing the extent of his collaboration with the Axis.† The general opinion in the Foreign Office was that it would do him little harm since his domestic propaganda about his own 'hábil prudencia' had already had its effect. According to Mallet, 'there are many here who consider Franco's foreign policy in keeping Spain out of the war to be his only solid achievement'.[71] Nevertheless, Washington published some of the more damning evidence in a booklet entitled *The Spanish Government and the Axis* in early 1946, which provoked an immediate denial from Martín Artajo.[72]

Franco turned the international criticism of the regime to his own advantage by portraying it as a the work of a Communist-masonic conspiracy dedicated to destroying Spain. His arguments were bizarre and feeble yet, paradoxically, effective. In September 1945, he briefed the religious advisers of the Falange's *Sección Femenina*. He told them that the Civil War had been undertaken to combat the 'satanic machinations' of perverted freemasons and now Spain was coming under attack from 'the masonic super State' which controlled the world's press and radio stations as well as many key politicians in the western democracies. Using the Crusade rhetoric of the Civil War, he declared that Spain was subjected to this because she had carried the gospel to the world and her men were the soldiers of God.[73] He had found another *persona* – the captain of the besieged Numantine fortress.‡ There was a long tradition in Spain of rallying national unity by inventing sinister foreign enemies. Franco carried this message to the people on punishing tours around the

* Approximately £428,000,000 at 1993 prices.
† British Naval Intelligence supplied the United Nations with a report entitled 'Use of Spanish Ports by Axis with Connivance of local Spanish Authorities' (Paper for UNO Subcommittee, prepared by British Naval Attaché Madrid, FO371/60332, Z6254/8/G41).
‡ The siege of Numantia by the Romans in 154 BC was the national symbol of heroic last-ditch resistance.

country at which his speeches were hailed by crowds mobilized by the Falange.[74]

Franco was infuriated by the fact that his efforts to cope with international hostility were made more difficult by Don Juan de Borbón. Ever since the end of the war, as part of his efforts to present his regime to the outside world as monarchist, Franco had been suggesting that the heir to the throne take up residence in Spain.[75] The Caudillo had been convinced by the monarchist José María Oriol that Don Juan now regretted the Lausanne manifesto. In fact, Don Juan was determined not to return until Franco left and, at the beginning of February 1946, he took up residence in the fashionable Portuguese resort of Estoril near Lisbon. The Caudillo's hopes of controlling him through his brother, Nicolás, his ambassador in Portugal, came to nought. The Pretender's presence in the Iberian peninsula set off a wave of monarchist enthusiasm which was expressed in various ways. Most worryingly from Franco's point of view, there was a collective letter of greeting known as 'el saludo' signed by 458 of the most important figures of the Spanish establishment, including twenty ex-ministers, the presidents of the country's five biggest banks, many aristocrats and prominent university professors. It expressed their wish to see the restoration of the monarchy, 'incarnated by Your Majesty'.[76]

When it was published on 13 February, Franco was livid. He reacted, as so often, as if he was faced with a mutiny by subordinates. He told a cabinet meeting held on 15 February, 'This is a declaration of war, they must be crushed like worms.' Again seeing a masonic conspiracy, he announced that he would put all the signatories in prison without trial. He backed down only after General Dávila and others had pointed out the damaging international repercussions of such a move. He then went through the list of signatories listing appropriate ways of punishing them, by withdrawing passports, tax inspections or dismissal from their posts. Nevertheless, still not satisfied, the Caudillo said that he was determined to make a scapegoat in the affair, named Kindelán as the ringleader and ordered him imprisoned immediately. Only after an appeal by Dávila on the grounds of Kindelán's age and poor health did he agree to Kindelán being exiled to the Canary Islands. Knowing that there were many in the Army who supported Kindelán's views, even if they did not have his courageous readiness to speak out, Franco had long responded cautiously to secret police reports of his activities. Now, the scale of support for a restoration implied by the saludo led him to make an example of Kindelán. It was an effective measure which provoked no more than mutterings among a few monarchist generals.[77] Few wanted to share his debilitating and humiliating punishment.[78]

A proposal from the Caudillo that Don Juan come to Spain for a private meeting was snubbed. The Pretender fobbed off Nicolás Franco with the excuse that he needed time to prepare himself. Don Juan's advisers realized that Franco wanted to use such a meeting to stress the Pretender's subordinate position.[79] The snub enraged Franco anew. He instructed elements of the Falangist *Sindicato Español Universitario* to disrupt the classes of the professors who had signed the *saludo* and sent a note to Don Juan in which he announced that relations between them were broken on the grounds that Estoril was fomenting monarchist conspiracy against him. Franco acted out of pique, but there was a strong element of calculation in his reaction. The more daring monarchists now began to seek contacts on the Left but many of the more opportunistic conservatives who had signed the *saludo* scuttled back to Franco.[80]

However threatened Franco may have felt by the activities of the *Juanista* monarchists, by the guerrilla war or even by the Republican government-in-exile, the propaganda campaign to present him as the champion of a beleaguered Spain was paying off. Right across the country, in the first months of 1946, there were demonstrations in favour of Franco organized by the Movimiento.[81] In the meanwhile, the repressive apparatus kept the mass opposition of the Left at bay. Moreover, international hostility was abating somewhat.

In mid-January, the American Under-Secretary of State, Dean Acheson, told the British Minister in Washington, John Balfour, that Anglo-American policy towards Franco would have to move on from mere statements of distaste. He suggested a joint declaration from France, the United States and Britain that for Spain to be accepted into the international community, the Spanish people would have to remove Franco and set up a caretaker government to organize elections.[82] The idea was discussed by Byrnes, Bevin and Bidault. But the notion was gaining ground in Washington that the Soviet Union actively wanted to see civil war fomented in Spain, to secure a victory for Communism, a flank position relative to Italy and France and a bridgehead to Latin America.[83] After the British Ambassador in Washington, Lord Halifax, made similar points to Acheson, American pressure diminished.[84]

Ironically, while British policy in fact aimed at restraining the French and the Americans from taking precipitate action against Franco, the rhetoric of the Foreign Secretary gave a different message to the Caudillo. In order to cope with the repugnance felt within the Labour Party for

Franco, Bevin continued to express abhorrence for his regime while stressing that the dictatorship should be overthrown 'by the activities of the Spanish people themselves'.[85] On 9 February 1946, the General Assembly of the United Nations reminded members of the resolutions of the San Francisco and Potsdam conferences and recommended that they conduct their relations with Spain in line with the letter and spirit of those resolutions.[86]

This declaration was regarded by many in France as entirely inadequate. Since the resignation of General de Gaulle as president of the French provisional government on 20 January 1946, left-wing pressure for action against Franco had been more effective. On 21 February, one of the leaders of the anti-Franco guerrilla movement and a hero of the French resistance, Cristino García, was executed along with nine others, after the most cursory of trials.[87] By ostentatiously ignoring pleas for clemency from the French Government and proceeding with the executions, Franco was making a deliberate gesture of defiance to his enemies abroad. His timing also reflected the fact that he was touring Spain addressing Falangist rallies on his determination to fight Communism. More significantly, it was a warning to those in Spain who hoped that he might withdraw without bloodshed. A few days later, thirty-seven members of the Partido Socialista Obrero Español were given heavy prison sentences for trying to reorganize the party.[88]

On 26 February 1946, in response to the waves of public outrage provoked by the execution of the anti-Franco militants, the French cabinet decided to close the frontier with Spain and break off economic relations. It also proposed that the Spanish question be discussed by the Security Council of the UN Bidault called for Anglo-American support for these measures which would have meant an economic blockade against Spain.[89] Moreover, both London and Washington were reluctant to let the Spanish question be put before the Security Council where the USSR would have an opportunity to influence the direction of events. Both the British Chiefs of Staff and Foreign Office officials were already inclined to the American view that the Russians wanted to see a civil war provoked in Spain. They had their view confirmed by the influential exiled diplomat and writer, Salvador de Madariaga. It is ironic that Madariaga, loathed by Franco precisely because of his acute criticisms of the regime, should have thus done the dictator a service in confirming the Foreign Office inclination to caution.[90]

With considerable British and American misgivings, and as a device to head off the French desire to put the Spanish question on the agenda of the Security Council, on 4 March 1946, a Tripartite Declaration of the United States, Great Britain and France announced that 'As long as General Franco continues in control of Spain, the Spanish people cannot anticipate full and cordial association with those nations of the world which have, by common effort, brought defeat to German Nazism and Italian Fascism, which aided the present Spanish regime in its rise to power and after which the regime was patterned.' However, the limits of the Declaration lay in the statement that: 'There is no intention of interfering in the internal affairs of Spain. The Spanish people themselves must in the long run work out their own destiny.' The pious hope was expressed that, without risking civil war, 'leading patriotic and liberal-minded Spaniards may soon find the means to bring about a peaceful withdrawal of Franco, the abolition of the Falange and the establishment of an interim or caretaker Government'. The Tripartite note was thus milder than the Potsdam Declaration and the exclusion of the Soviet Union from the declaration emphasized that fact.[91] In cabinet meetings, Franco fumed about 'those bandits', complained that France was 'the Quisling of Russia' and that President Truman was a 'rough freemason from the south'.[92] Not surprisingly, he and the beneficiaries of his regime ignored the invitation to vacate power and face the prospect of trial as war criminals.

Franco realized that the Allies would not be intervening in Spain to implement their suggestions. The British and American governments were more discomfited by the Tripartite note than was the Caudillo himself. The Anglo-American policy of non-intervention aimed to put Spain into cold storage. It acknowledged the impossibility of bringing Spain into the concert of nations at the moment but it was also based on a determination that the Iberian Peninsular should not fall under Soviet influence. This effectively meant abandoning the cause of the anti-Franco democrats and conniving at the Caudillo's survival by blocking calls for action on Spain. The Republicans, who were already perceived as left-wing and pro-Communist, were branded as Soviet lackeys when they were left with no choice but to accept the backing of the USSR and her East European satellites. Thus, non-intervention in post-war Spain benefited Franco in the same way as it had during the Civil War.[93]

Toothless international ostracism had the effect within Spain of con-firming the regime's projection of itself as inevitable and immoveable. Falangists and others tied to Franco by networks of corruption and com-plicity in the machinery of repression were ever more persuaded that their futures were safe with Franco. The Caudillo had already foreseen this when he spoke to Martín Artajo at the end of the world war.

If Franco suffered any anxiety about the Tripartite Declaration, it must have been short-lived. On the following day, 5 March, Churchill once more came to his rescue in the celebrated 'Iron Curtain' speech at Fulton, Missouri. Franco took the speech to mean that it was only a matter of time before his value to the West would be recognized. He was quick to take the hint and proclaim ever louder that the Bolshevik threat was at Spain's door. As long as the exiled opposition remained divided, only foreign military pressure or foreign economic sanctions could have worried him and the Tripartite note had reassured him on that score.[94]

Three days after the note, Franco presided at the opening of new exhibition halls at the Museo del Ejército (Army Museum). The entire occasion was a glorification of the Nationalist cause in the Civil War, a reminder to Franco's supporters that their best defence against the return of a vengeful Left was unity. Speaking of international hostility, he declared 'it should surprise us least of all, since we never heard of anything but sacrifices and discomfort, of austerity and long vigils, of service and sentry duty. But in such service, you can occasionally rest. I cannot; I am the sentry who is never relieved, the one who receives the unwelcome telegrams and dictates the solutions; the one who is watchful while others sleep.' With a skilful rhetoric that justified the ongoing repression, he claimed that he had accepted the posts of Generalísimo and Head of State only on the condition that he was also to be allowed, after the victory, to undertake the long-term task of 'eliminating the causes of so much misfortune'.

Lest anyone might think that he enjoyed power, he was at pains to stress the personal costs of his selfless dedication. Forgetting his hunting and fishing trips, his golf and his long holidays, he told his audience of senior military men that, unlike him, they could forget their cares and preoccupations. 'I, as Chief of State, see my private life and my hobbies severely limited; my entire life is work and meditation.' The self-glorification with a touch of self-pity was typical. Equally illuminating was the anecdote which he then recounted – his audience of comrades-in-arms tempting him to uncharacteristic public levity. 'During the early days of the Crusade, bad news outweighed good and a staff officer with a long face would bring one bit of bad news after another. I had to smile and try to cheer him up. Then one day, he took ill and another officer, Captain Medrano, stood in for him. Medrano had to bring me one of the worst pieces of news of the entire war and he came in smiling and optimistic and I asked him: "What's up, Medrano?" With a smile on his lips, he replied: "nothing much General. I've just got a little report for you here". I read it and said: "Very good. From now on you are always going to bring me the reports." And the fact is you have to put a good face on bad news. The worse the news, the happier the face.' In this unexpected burst of frankness, the Generalísimo had given his recipe for survival.[95]

It was a recipe which would have to serve for some time yet. The French remained determined to step up pressure on him. On 12 March, Bidault proposed to Britain and the United States that joint economic sanctions be imposed upon Spain. Both Secretary Byrnes and the British Ambassador reiterated to Bidault their respective Governments' commitment to non-intervention. Britain opposed economic sanctions as futile without the co-operation of many nations including Argentina, Franco's staunchest ally.[96] The Foreign Office communicated to the State Department its impatience belief that the French had 'an erroneous belief in the possibilities of effective outside action to hasten the fall of General Franco' and that the closure of the frontier was seen by Spaniards 'as Communist-inspired and as an unwarranted attempt to interfere by means of outside pressure in Spanish internal affairs. It has in consequence only served to strengthen General Franco's position.' The State Department was in substantial agreement with this assessment.[97]

The Tripartite Declaration did have some minor impact within the Army. General Ponte, Captain-General of Seville wrote on 12 March 1946 to General Varela, at the time High Commissioner in Morocco, that the high command should co-ordinate its position. Ponte suggested that since the Army put Franco into power, it should not flinch from removing him. Varela replied on 23 March that nothing should be done which might create divisions in the Army or give any hint of such military disunity to the outside world.[98] Ponte's letter was the last spasm of serious anti-Franco activity from a general with an active command until the mid-1950s. Varela's response was proof that Franco had been skilful in his promotions and could feel confident in the loyalty of the Army.

Even more confidence could he feel in the Movimiento. On 1 April, the annual parade to commemorate the Nationalist victory in the Civil War turned into a massive demonstration of support for Franco, culminating in his appearance on the balcony of the Palacio Real to acccept the cheers of the multitude.[99] On 6 April, the Falangist Minister of Labour, José Antonio Girón de Velasco, led a delegation of Civil War ex-combatants to present Franco with fifty albums containing 300,000 signatures affirming their loyalty to him at a moment when he was under attack from 'the hired assassins of the forces of evil' (*una banda de sicarios de las fuerzas del mal*). A delighted Franco told them that the masonic and Communist plots masterminded by the scum of the Republican exile meant only that 'we exist, we are not dead and our flag flies in the wind'.[100]

The abandonment by the Allies of the Spanish democratic opposition to Franco was soon confirmed. On 17 April 1946, the Polish delegate on the Security Council, Oscar Lange, proposed the immediate suspension of diplomatic relations with Spain on the grounds that she was a danger to world peace, a claim supported by the absurd allegation that atomic bombs were being manufactured there by escaped Nazis. As a non-Communist academic who had been exiled in the USA, Lange was an excellent spokesman to give the impression that Poland was acting independently of the Soviet Union. Behind the scenes, the British and American representatives at the United Nations, Sir Alexander Cadogan and Edward R. Stettinius Jr., joined forces to deflect the Russian and Polish manoeuvre and the 'Spanish Question' was referred on 29 April 1946 to a five-man subcommittee to be chaired by the representative of Australia, Dr Herbert V. Evatt. It was to report to the General Assembly one month later.[101]

The publicity given to the Polish accusations led to Franco making declarations which combined outrageous untruth with sincere belief, cunning with naïvety. On 14 May 1946, he spoke to the Cortes for more than two hours. He told the anxious *Procuradores* that 'there is talk beyond our frontiers of the problem of Spanish politics. I deny that in Spain there is any problem to solve. We solved our political problems with our blood and our effort.' Frequently interrupted by the cheers of his hand-picked listeners, he went on to affirm that accusations that he was a dictator were stupid and malicious. He denied that he had come to power with Axis assistance, referring to Italian and German assistance during the Civil War as 'a drop in the ocean'. He inadvertently revealed his own duplicity when he tried to explain away his erstwhile sympathy for the Axis. 'Others try to present us to the world as nazifascists and antidemocrats. There was a time when we did not mind that mistake given the prestige which such regimes enjoyed in the world. Today, however, when so many insulting accusations of cruelty and ignominy have been piled upon the defeated, in justice we must underline the very different characteristics of our State.'

This was the prelude to a defence of his particular form of rule, different from fascism but providing the authority necessary to keep in check the Spanish tendency to 'egoism and anarchy'. This near perfect polity was now under attack only because Spain had suppressed freemasonry and defeated Communism. He then neatly implied that only he stood between Spain and the anarchy sought by his foreign detractors and appealed to Catholics everywhere to work to put an end to the persecution of Spain. By presenting international ostracism as directed against Spain and not against himself, Franco ceased to be the cause of her ills and became her champion against ancient enemies. Referring to the war, with outrageous cheek, he praised himself for the generosity with which he dealt with France in the summer of 1940, claimed that Spain saved Britain from defeat in 1940 and, forgetting the question of wolfram, asserted that he had shown only goodwill to the United States.[102]

The subcommittee set up after Polish allegations against Spain had investigated, and resoundingly confirmed in its report of 31 May 1946, the Axis-assisted origins and fascist nature of the Franco regime, its pro-Axis conduct during the Second World War, its ongoing support for Nazi war criminals and foreign fascist organizations, the disproportionate scale of the Spanish armed forces, its uranium and armament production, the execution, imprisonment and repression of its political opponents, and the profascist activities of the Falange. However, it was unable to recommend that the Security Council interfere in Spain since Franco had committed no act of aggression nor threatened international peace. Nonetheless, despite serious British misgivings, the subcommittee did conclude that Franco's Spain, although not an immediate threat, represented 'a potential threat to international peace and security'. The final recommendation was that the Security Council support the Tripartite Declaration of March by recommending that the General Assembly call on all its members to break off all relations with Spain.[103]

On 5 June 1946, the Spanish Government issued a long and fiercely indignant reply to the subcommittee's report on which Franco's own stamp was clear. Throughout the document, the Spanish people and Franco were assumed to be one and the same. Despite underlining the 'supreme indifference' with which 'the Spanish people' viewed the opinion of those who had no right to judge their conduct, the document went on to protest at the implied interference of the subcommittee in Spain's internal affairs. With feigned high-mindedness, the statement claimed that this protest was being made on behalf of all the middle-sized powers, neutral nations and small countries who were similarly threatened by the arrogance of the great powers.[104]

More effective, for Franco's purposes, than the official Spanish reply was a speech made on the same day by Winston Churchill. In an ironic commentary on the inefficacy of the United Nations policy towards Spain, he congratulated Attlee's Government on 'a wise restraint, or, at least, a marked lack of enthusiasm, in not interfering in the internal affairs of Spain. None of us like the Franco regime, and, personally, I like it as little as I like the present British Administration, but, between not liking a Government and trying to stir up civil war in a country, there is a very wide interval.' The Conservative leader asserted that French economic pressure 'has only had the result of giving Franco a new lease of life' and declared that 'anything more silly than to tell the Spaniards they ought to overthrow Franco, while, at the same time assuring them that will be no military intervention by the Allies can hardly be imagined'. Denouncing the Polish intervention before the UNO as Soviet-inspired, he said that 'there is as much freedom in Spain under General Franco's reactionary regime, and a good deal more security and happiness for ordinary folk, than in Poland at the present time'.*[105]

Despite the sympathy for his regime emanating from conservatives and Catholics in both Europe and America, Franco's propaganda machinery was working frantically to persuade the Spanish people that Spain was the victim of an international siege (cerco internacional).[106] The disastrous economic performance of his economic system of autarky was blamed on the siege.† In fact, to share the benefits of post-war economic reconstruction, Franco would have had to pay the price of political reform – and he was not prepared to do that. The idea of the 'siege' was a convenient screen for almost every failing of the regime. Martín Artajo, with his hopes of liberalization and a monarchical restoration, became a kind of exile in a cabinet dominated by the Falangists.[107]

The report of the subcommittee on Spain was discussed at the 44th, 45th, 46th and 49th sessions of the Security Council held in New York on 6, 13, 17 and 26 June 1946. Its recommendations were toned down

---

* The comparison of life in Spain with the lot of the citizens of Poland after successive Nazi and Soviet invasions and occupations was hardly a fair one. In the light of Franco's own boasts about the peace and prosperity which he had bestowed upon neutral Spain, the scale of hunger and repression in the country remained startling. Indeed, the American Chargé d'Affaires, Philip Bonsal, believed that a spiral of economic difficulties and political disorders would soon create insuperable problems for Franco (FRUS 1946, V, p. 1077).
† In fact, the system, which permitted Spanish manufacturers to import crucial raw materials and machinery only with government licences, was both corrupt and incompetent. Permits could be obtained for anything, at a price, and much of Spain's scarce foreign currency went on imported luxury goods. At the same time, resources were squandered on massive prestige projects – such as Franco's astronomically expensive schemes for energy self-sufficiency, hydroelectric plants, naval building programmes, the creation of chemical and metal industries – the results of which would not be seen for decades.

by both the United States and British representatives but it was decided rather feebly that, since the subcommittee had shown that the Franco regime constituted a potential threat to world peace, the Spanish question should be subject to constant vigilance by the Security Council. This effectively signified a recognition that no measures were likely to be taken against Franco. Efforts by the Russian delegation led by Andrei Gromyko to harden the declaration against Franco were unsuccessful and ultimately strengthened the Caudillo's position by giving a semblance of credibility to his claim to be a bulwark of western defence.[108]

In this context of international disagreement over Spain, an ever more confident Franco was beginning to relax somewhat. After a 'tranquil summer', he expressed to Artajo his delight that 'the world squabbles and leaves us in peace'.[109] He announced in an interview in *Arriba* that the Spanish people 'know what to expect from abroad and what History shows is that the hatred of Spain isn't something invented today or even yesterday. Spain lives in truth and sincerity and the rest of the world in perpetual hypocrisy.'[110] The tenth anniversary of his formal assumption of power as Head of State, the *Día del Caudillo* on 1 October, gave rise to even more spectacular celebrations than those of previous years. Franco was in Burgos where he attended a Te Deum in the Cathedral and then received a shield of gold and platinum encrusted with rubies, diamonds and emeralds, the gift of the authorities of the fifty Spanish provinces.[111]

At the beginning of November 1946, the United Nations Security Council passed the Spanish question to the General Assembly. It meant renewed publicity for the international opprobrium directed against Franco but less likelihood of concerted concrete action. Proposals for action against Franco were sent to the President of the General Assembly by the representatives of thirteen European and American countries. These included a call from the United States for Franco to cede power to a representative provisional government and from the Soviet Socialist Republic of Bielorussia for the breaking of economic relations with Spain.[112] As far as Franco was concerned, the United Nations criticisms were 'arbitrary and unjust' and clear proof that there was a Soviet-inspired plot to isolate Spain.[113] A massive propaganda campaign was initiated to give the impression of total national unity around Franco, including a staged rally of 'workers' acclaiming him 'the first worker of Spain' in the Plaza de Oriente.[114]

The Spanish question was discussed by Committee 1 of the General Assembly (Political and Security Questions) between 2 and 4 December 1946 at Lake Success in New York State. Numerous representatives condemned the Franco regime as fascist while only a few voices, including those of Paraguay and El Salvador, spoke out against a general rupture of diplomatic relations with Spain. The American and British representatives, Warren Austin and Sir Hartley Shawcross, while acknowledging the repugnant nature of the Franco regime, argued against outside interference lest it provoke civil war in Spain. Eventually, a subcommittee was named to draft a resolution to be put to the General Assembly. It took as its text the proposed resolution of the United States, which had been submitted by Senator Tom Connally.[115] Léon Jouhaux, for France, had ridiculed the idea that another condemnation of Franco would somehow encourage the Spanish people to take their destiny in their own hands. But the so-called 'Connally resolution' was yet another denunciation of the Axis links of Franco followed by an appeal to the Spanish people to 'give proof to the world that they have a government which derives its authority from the consent of the governed' and an invitation to Franco to 'surrender the powers of Government'.[116]

In anticipation of some international action against Spain, Franco's cabinet discussed 'manifestations of national irritation'. The Falangist Ministers Girón and Fernández Cuesta were charged with organizing the campaign.[117] The fruit of their efforts was a massive demonstration in the Plaza de Oriente in front of the Palacio Real on 9 December 1946. Shops were ordered to close for the day. Thronged by the Falangist Syndicates, the Youth Front and civil war veterans organizations, the Plaza was alleged by the Francoist police to have held 700,000 people. Contemporary photographs show the square and surrounding streets packed to overflowing. However, the Plaza – 46,600 square metres and well-populated with bushes and statues – and the adjoining streets are likely to have held fewer than that number. Nevertheless, it was an immensely impressive turn-out. Hundreds of banners carried slogans attacking the Russians, the French and foreigners in general. There were insistent and deafening chants of '¡Franco sí, comunismo no!'.

An understandably delighted Franco appeared on the balcony of the Palace at 12.30 p.m. Immediately dropping into Civil War rhetoric, he addressed himself to 'Combatientes, ex-cautivos y españoles todos' (wartime combatants, ex-prisoners [of the Republic] and Spaniards one and all). Constantly interrupted by frenetic applause, he denounced 'those abroad who speculate with your loyalty and our domestic peace'. He declared that 'what is happening at the United Nations should not surprise us Spaniards. When a wave of Communist terror is laying Europe to waste with total impunity, along with rape, murder and persecution of the kinds that many of you saw [in the Spanish Civil War], we can hardly be surprised when the sons of Giral and of La Pasionaria find an atmosphere of tolerance and support from the official representatives of those unfortunate countries.' When he stated that 'no one has the right to interfere in the private matters of other nations', the cheers and clapping drowned out his words. He invited his supporters 'to unite the force of our righteousness [*razón*] to the fortress of our unity'. He ended with the boast that 'the proof of Spain's resurgence is the fact that the rest of the world is dangling from our feet'. For over an hour after his speech, he and Doña Carmen stood on the balcony listening to the chants of 'Franco! Franco! Franco!'.[118]

The final agreed resolution on Spain was adopted by a plenary session of the General Assembly on 12 December 1946. It excluded Spain from all its dependent bodies, called upon the Security Council to study measures to be adopted if, within a reasonable time, Spain still had a government lacking popular consent; and called on all member nations to withdraw their ambassadors. The resolution was passed by thirty-four votes, including France, Britain, the Soviet Union and the United States, to six, all Latin American countries, with thirteen abstentions.[119] At the cabinet meeting on 13 December, Franco crowed that the United Nations was 'fatally wounded'.[120] Four days later, he went to Zaragoza to take part in a ceremony which celebrated Spanish resistance against the Napoleonic invasion. Evoking the spirit of that era, he affirmed the superiority of his system and claimed that, in moral standing and social evolution, Spain was ten years ahead of other nations. He also authorized a new coinage to be minted on which would appear his bust and the words 'Caudillo by the grace of God'.[121]

XXII

A WINNING HAND

1947–1950

DESPITE THE Francoist campaign to imply that Spain was the victim of international aggression, the United Nations had effectively endorsed the Anglo-American policy of non-intervention: the measures voted on 12 December 1946 were exclusively diplomatic and did not extend to economic or military sanctions.* Indeed, the United States had accepted only with reluctance the inclusion of the phrase 'within a reasonable time'.[1] Franco was probably sincere when he later claimed to have been relieved by the United Nations' adoption of the tactics of ostracism because it enabled him to place himself in the long Spanish historical tradition of heroic struggles against overwhelming odds, from the Romans through the Moorish to the Napoleonic invasions. They had given him a winning hand to play. The great popular demonstration in the Plaza de Oriente enraptured him because, he believed and his press insistently claimed, it encompassed both his friends and his enemies spontaneously united in the eternal Spanish response to those foreigners who try to tamper with their independence.[2]

Franco's personal, albeit covert, reply to the United Nations resolution was to begin to write for *Arriba* an occasional series of articles denouncing freemasonry in general and the United Nations Secretary-General, the Norwegian Trygve Lie, and the President of the General Assembly, the Belgian Paul-Henri Spaak, in particular. Lie and Spaak were, in Franco's opinion, freemasons under the orders of Moscow. The articles, published under the pseudonym Jakim Boor (the two pillars of the masonic temple), ran until May 1951. Their central thesis was that freemasonry, which

---

* Most countries chose to withdraw their ambassadors but continued to run their embassies under a *Chargé d'Affaires*. Spain merely withdrew her ambassadors from those countries but maintained the rest of her staff in otherwise fully functioning embassies.

Franco did not distinguish from liberal democracy, was engaged in a conspiracy with Communism to destroy Spain. Freemasonry, 'one of the most repugnant mysteries of the modern age', was the instrument by which the British had destroyed the Spanish empire. Now, freemasonry was to democratic parties as Marxism was to Communist parties. These views were shared vehemently by Carrero Blanco. The articles were written in a fresher and livelier style than other contemporary writings and speeches of the Caudillo. It has been suggested that Franco's friend, the writer Joaquín Arrarás, assisted him in the composition of the articles,[3] although the prose style may reflect a lack of inhibition bestowed by the use of a pseudonym. Certainly it permitted Franco to indulge his vanity to the extent of writing about himself in the third person underlining the worldwide masonic hatred of 'our Caudillo' and the fact that the people of Spain were 'with Franco to the death'.[4] To strengthen his cover, it was announced in the press that Franco had received 'Jakim Boor' in an audience.[5] The articles were collected as a book in 1952 and, for the rest of his life, Franco remained convinced that all the copies had been bought up by freemasons to prevent it being read.[6]

At the same time, Franco and Carrero Blanco put considerable effort into making his regime acceptable to the same western democracies excoriated in the articles. On 31 December 1946, Carrero drew up a memorandum urging Franco to exploit the popular support manifested in the Plaza de Oriente by institutionalizing his regime as a monarchy and then giving it a veneer of 'democratic' legitimacy with a referendum. The central idea was that the 'biological inadequacies' of any hereditary monarch could be neutralized by Franco remaining as Head of State and the King being subject to the advice of a *Consejo del Reino* (council of the kingdom), made up of loyal Movimiento figures and controlled by Franco. The Caudillo knew that it was even simpler never to restore the monarchy in his lifetime. Carrero Blanco's memorandum was thus refined further in another working paper presented on 22 March 1947, which suggested that Franco name his own royal successor.[7] These ideas for a Francoist 'monarchy' were given urgency by the fact that Don Juan was attracting support from the Carlists on the extreme right to Socialists on the democratic Left.[8]

A reflection of Franco's continuing optimism was given in early January 1947 when he received a delegation of senior generals. His old friend, the venerable Andrés Saliquet, took the liberty of saying 'Paco, we are all worried about the United Nations decision.' The Caudillo replied cynically 'there's no need to worry. What's the matter? Isn't your soap factory doing well?' It was a neat, if cruel, reminder to them all that he was fully aware that many senior generals enjoyed directorships in companies happy to pay for their influence in the quest for rare raw materials or electrical power. In the embarrassed silence which followed, Franco explained his conviction that the growing Russo–American antagonism guaranteed that he would soon be courted by Washington.[9] Franco had appreciated better than any of them that the United Nations decision meant the end for the Republican government-in-exile. The Great Powers were not going to restore the Republic and the discredited and disillusioned Giral resigned. As the pragmatic Socialist leader Indalecio Prieto perceived, the only option now was for the Spanish Left to try to build a broad front with the monarchist opposition to Franco. A more moderate government-in-exile was set up to this end under the leadership of the Socialist Rodolfo Llopis.[10]

The prospects for Prieto's strategy were not bright: diplomatic reports reaching Franco reassured him that British Conservatives and important elements in the Pentagon saw his utility as a bulwark against Soviet advances.[11] Neither the British nor the Americans gave concrete assistance to plots hatched by Aranda and Beigbeder to replace Franco with a provisional government representing various monarchist and moderate left-wing forces.[12] Franco was fully informed by his secret police about these conspiracies and he had on his desk incriminating dossiers on both men.[13] Nevertheless, the recall of ambassadors led to a frantic effort in the Spanish Ministry of Foreign Affairs to establish contacts with anyone who would listen. After the anti-Francoist Paul T. Culbertson was sent to Madrid as American *Chargé d'Affaires* in June 1947, clumsy gestures contrary to all protocol were made to cultivate him and his family.[14] Any kind of visit, by passing bishop or sheikh, was magnified in the hope of diminishing the impact of the isolation. In this regard, Martín Artajo was especially dexterous in using his network of contacts with the Vatican. The regime wit, Agustín de Foxá, called the Ministry the '*Monasterio de Asuntos Exteriores*' and Martín Artajo 'the Prior'.[15]

In mid-February 1947, Franco gave a long interview to Constantine Brown of the Washington *Evening Star*. Fundamentally a defence of Spain's role in the Second World War, it provided a revealing picture of the Caudillo's self-perception at this time. His belief in his own propaganda that he was the reluctant instrument of a divine mission was made clear by the mixture of megalomania and false humility of his opening words. 'I am a man who never harboured any ambitions of command or power. Ever since my youth, life subjected me to hard tests by obliging me to undertake positions of command and responsibility far beyond my years; but I have a concept of responsibility and of the fulfilment of my duties, and duty is a fact consubstantial with the conscience of each person. If I believed that the interest of my *Patria* lay in my resigning my command, have no doubt that I would do it without hesitation and with joy since command constitutes for me both a duty and a sacrifice.'[16]

Franco had reasonable grounds for such confidence. Even if the economic sanctions sought by the French and others had been applied rigorously, there is every likelihood that the regime would have survived. Although there were significant food shortages and insufficient electrical power for the basic needs of industry, the British estimated that food supplies would last from four months to indefinitely and oil supplies for at least six months.[17] Franco had established a lifeline by clinching friendly relations with a number of countries, including Salazar's Portugal and, above all, with the sympathetic regime of the pro-Axis Juan Domingo Perón in Argentina.[18] With obvious relief, Martín Artajo spoke of the 'breathing spell' provided by Argentinian help.[19] Franco's optimism was also boosted by the formalization of western resistance to Communist expansion. On 12 March 1947, in response to British inability to sustain military aid to Greece and Turkey, the Truman doctrine of support for 'free peoples to work out their own destinies in their own way' was announced.[20]

As the international context changed, Franco accelerated his plans to camouflage his regime with the trappings of acceptability. Carrero Blanco's ideas formalized in a draft text of the 'Law of Succession' were discussed in a cabinet meeting on 28 March 1947 and made public three days later. The first article declared that 'Spain, as a political unit, is a Catholic, social and representative state which, in keeping with her tradition, declares herself constituted as a kingdom'. The second article declared that 'the Head of State is the Caudillo of Spain and of the Crusade, Generalísimo of the Armed Forces, Don Francisco Franco Bahamonde'. The Falange's docility as its cherished 'revolution' was consigned to the rubbish heap was testimony to the fear eating at the souls of Franco's closest supporters. The unfortunate company kept by the regime between 1936 and 1945 was to be forgotten and the fascist tendencies shown during that period were simply to be replaced with a monarchical façade. In any context other than that of the Cold War, the palpable deceit would have been laughable. The straight-faced declaration that Franco would govern until prevented by death or incapacity, the Caudillo's right to name his own royal successor, the lack of any indication that the royal family had any rights of dynastic succession, the statement that the future King must uphold the fundamental laws of the regime and could be removed if he departed from them – all this showed that nothing but the label had changed. However, when the Korean War broke out three years later, that repackaging was virtually all that would be needed to put an end to international ostracism and open the way to incorporation into the western community.

The new law was part of an elaborate show aimed at convincing the western powers and Spanish monarchists that the regime was evolving towards a restoration. The stage-management of the production required Don Juan to speak the right lines and not denounce the scheme. That part of the show was handled with notable clumsiness. On the day before the *Ley de Sucesión* was promulgated, Carrero Blanco arrived in Estoril. He carried a message to Don Juan which had already been conveyed in various forms by Nicolás Franco and by Alberto Martín Artajo. It was more or less 'identify yourself with the regime, trust in Franco, be patient and reconcile yourself to being Franco's heir'. The specific purpose of Carrero's mission was to inform the Pretender of the project that would become law on 31 March. He had been ordered by Franco to seek an audience for precisely 31 March, in order to deny Don Juan the possibility of doing anything to impede the plans.

When they met in the late morning, Carrero delivered a long and flowery catalogue of Franco's achievements. He spoke about the new law in terms which gave Don Juan the impression that he was being consulted about a draft project. To Carrero's discomfort, Don Juan pointed out that Franco could not present himself as the restorer of the monarchy when he prohibited monarchist activities. The Pretender's desire to be King of all Spaniards provoked Carrero into a candid statement of the Francoist view of politics. 'In Spain in 1936 a trench was dug; and you are either on this side of the trench or else opposite . . . You should think about the fact that you can be King of Spain but it must be of the Spain of the *Movimiento Nacional*: Catholic, anti-Communist, anti-liberal and violently free of any foreign influence in its policies.' Before leaving, on Franco's instructions, Carrero handed Don Juan files bulging with denunciations against his followers. Don Juan promised to read the text of the *Ley de Sucesión* and meet Carrero on the following day to give him his opinion.

After the audience was over, Carrero took his leave and Don Juan retired to his rooms. Only then did Carrero slip back unobtrusively to the Villa Giralda and leave a message with an official of the royal household that Franco would be going on national radio that night to announce the definitive text of the new law. He left hastily before Don Juan was given the message and realized that they had not been discussing a proposed draft. The deception inclined Don Juan and his advisers to strengthen their links with the left-wing anti-Franco opposition. On 7 April 1947, Don Juan issued the 'Estoril Manifesto' denouncing the illegality of the succession law which proposed to alter the nature of the monarchy without consultation with either the heir to the throne or the people. Franco, Artajo and Carrero were agreed that Don Juan had thereby eliminated himself as a suitable successor to the Caudillo. On 13 April, the *Observer*, the BBC and the *New York Times* published declarations by Don Juan to the effect that he was prepared to reach an agreement with Franco as long as any such agreement was limited to the details of the peaceful and unconditional transfer of power. The manifesto and the press interview unleashed a furious press campaign against Don Juan as the tool of international freemasonry and Communism.[21]

The Great Powers were not fooled by the *Ley de Sucesión*. The US Acting Secretary of State, Dean Acheson, wrote to the US Ambassador in London, Lewis Douglas, that 'as long as Franco, or a successor appointed in accordance with the new decree, continues in power there can be no real improvement of economic stagnation in Spain. We will continue to be blocked from providing the effective assistance which would make possible the economic reconstruction of that country and thereby build an effective barrier to civil strife and Communist domination.' Acheson voiced the growing suspicion in both London and Washington that 'Moscow not only is interested in keeping Franco in power until political and economic distress in Spain reaches the point of revolution, but also derives considerable propaganda advantages from the present situation by placing the western powers on the defensive as defenders of fascism and of reaction.' Acheson was still convinced 'that Franco and any regime perpetuating the principles of his control must go' but had no practical policy for making that happen. He ruled out force but suggested that Franco himself be offered safe conduct out of Spain and that economic assistance be promised but given only after his departure.[22]

The British remained in a contradictory position. To placate the Labour Party, Bevin maintained a public hostility to Franco.[23] At the same time, he agreed with the Foreign Office view that active measures against the dictator such as economic sanctions were likely to be ineffective and have disproportionately high costs for Britain, who would have to organize a blockade at the risk of a deterioration of relations with Portugal and Argentina, lose allegedly irreplaceable imports* and suffer reprisals against British businesses in Spain.[24] The British *Chargé* in Madrid, Douglas Howard, dampened any optimism about Acheson's proposal to invite Franco to leave with a convincing report that the military high command would remain loyal to him both as the best defence against Communism and to protect their own material interests.[25] By 25 April, Bevin, who was in Moscow at a conference of foreign ministers, had concluded that the plan for a joint approach to Franco was 'ill-considered and based on wishful thinking' and was too dangerous because it set a precedent for interference in the internal affairs of Spain which might be used by the Soviet Union elsewhere. Bevin now suggested that nothing more be done at the United Nations and that 'defensive lobbying' begin to ensure support for such inaction.[26] Fortunately for Franco, the State Department had no real commitment to intervention of the kind implicit in Acheson's suggestion.[27]

In response to the appalling living conditions of the working class,† and despite heavy police repression, industrial unrest finally erupted at the beginning of May 1947. The series of strikes which broke out across Spain was largely concentrated in the Basque Country but there were others in Catalonia, Madrid and in the shipyards of El Ferrol. That the strikes were as widespread as they were, despite a decade of brutal repression, was a reflection of the plummeting living standards of working-class Spain.[28] The response of the regime was immediate and harsh. Units of the Legion and of the Civil Guard were sent to Bilbao and an additional 2,500 police were sent to the city. Employers were ordered to sack strikers 'without a second thought'. Those who did not were imprisoned. The Basque government-in-exile hoped that the strikes would help convince the Great Powers that Franco presided over a deeply hated, repressive and fascist regime.[29] The Basque President José Antonio de Aguirre saw the strike as being precisely the popular action against the regime which the United Nations condemnation of December 1946

---

* Most of Britain's crucial imports from Spain could be substituted from other sources of supply – iron ore from Sweden and Algeria, potash fertilizers from Chile and Morocco, oranges from Palestine and South Africa. Sherry of course was another matter. See Qasim Bin Ahmad, *The British Government*, pp. 239–85.

† Spanish industrial cities were flooded at this time by rural labourers fleeing the countryside. Despite police controls at railway stations, large numbers inflated the ranks of the unemployed and kept wages down. For those with jobs, despite long working hours, their low wages were insufficient to feed a family since wages had fallen to half their pre-Civil War level and prices risen by more than 250 per cent. In consequence, begging, petty pilfering and prostitution increased in the towns.

had foreseen.[30] He proclaimed the strikes as 'the greatest victory obtained by popular forces against the Franco regime', which, given the scale of the repressive machinery deployed, was an extreme exaggeration.[31] Aguirre's claim was echoed by the French and Spanish Communist Parties.[32] However, it was typical of Franco's good fortune that the strike wave just convinced London and Washington of the need to shore up his position as a bulwark against what was seen as Communist-inspired mischief.

General George C. Marshall had replaced James F. Byrnes as US Secretary of State on 21 January 1947.* The Marshall Plan for the economic reconstruction of western Europe was launched in a speech at Harvard on 5 June 1947. Shortly afterwards, at the suggestion of the French Foreign Minister, Georges Bidault, Spain was excluded from the Paris conference called for 12 July 1947 to examine the economic needs of Europe. The exclusion was described as provisional, and would be lifted if Spain changed its government. The Spanish government responded by publishing in Washington a series of pamphlets aimed at demonstrating that, without Spain, the Marshall Plan was doomed to failure.[33] At one level, the European Recovery Programme, as the Marshall Plan was known formally, even with Spain excluded, favoured Franco's survival. Moscow's refusal to permit its Eastern European dependents to accept aid was a major step towards the division of Europe into two blocks, a division which implicitly increased Spain's strategic value to the West.

Help from the Argentinian populist dictator Perón was crucial in bridging the time gap between the exclusion of Spain from the Marshall Plan and the change in US attitudes. It began with an agreement of 30 October 1946 for the delivery of wheat on credit.[34] Argentina had defended Spain in the United Nations General Assembly in December 1946 and, flouting the UN resolution on the recall of ambassadors, had sent a new envoy, Pedro Radío in January 1947. His arrival was greeted by orchestrated demonstrations and euphoric press coverage.[35] Even more spectacular propaganda was made out of the visit to Spain by the glamorous María Eva Duarte de Perón (Evita) in the summer of 1947.[36] The visit coincided with the referendum being organized by Blas Pérez, the Minister of the Interior, to ratify the *Ley de Sucesión*. Franco would have preferred to receive Perón himself. However, Perón was either too shrewd to associate himself further with Franco, who was hated throughout much of Latin

---

* During Byrnes' long absences from Washington at the United Nations, Dean Acheson had been acting Secretary of State. Marshall kept Acheson on until June 1947 when he returned to his law practice and was replaced by Robert A. Lovett.

America, or else too cautious to risk a prolonged absence from Buenos Aires. As it was, Evita insisted on being treated as a major Head of State and Franco was sufficiently anxious for the visit and the consequent publicity to agree to the pantomime.[37]

A special Iberia aircraft was sent to collect her. During the last section of her flight, the aircraft was escorted by Spanish Air Force fighters. At Barajas, the bejewelled Evita was received by Franco and Doña Carmen, the cabinet, leading figures of the Falange, the Army and the Church hierarchy. Franco bent double to kiss her hand. Huge crowds greeted the cavalcade of cars when it passed through Madrid en route to El Pardo. On a sweltering 9 June, schools were closed, government officials were given a day off work and the syndical machine went into action to guarantee a massive pro-Franco rally when the Caudillo bestowed on Evita the Gran Cruz de Isabel la Católica. Both the Generalísimo, in army uniform, and his guest, in a somewhat inappropriate mink coat, gave the fascist salute to the chanting Falangists in the Plaza de Oriente. Throughout the visit, a fashion duel was fought out between Evita and Doña Carmen, in which the most often flourished weapons were extravagant hats. The victory went to the Argentinian.[38]

Not since the visit of Heinrich Himmler in October 1940 had the welcome mat been put out for a foreign dignitary. The visit of Evita did the regime more good, at least within Spain. On the eve of the referendum, appearances alongside the beautiful Evita, the rallies and the publicity were extremely useful for Franco. Calls for a 'yes' vote appeared opposite coverage of the Sra Perón tour. Her speeches contained lavish praise of the Caudillo.[39] In the frenzy of her visit, the Spanish press omitted to mention that she was also visiting Portugal, Italy, the Vatican, Switzerland and France.* In response to Spain's exclusion from the Marshall Plan, the Franco-Perón Protocol was signed by which further credit was granted to Spain and wheat deliveries guaranteed until 1951.[40]

As the day of the referendum came near, pro-Franco propaganda grew more frenetic. Spaniards were told to vote 'yes' if they were Catholics and if they did not want to see their fatherland in the hands of Communists and to vote no if they wanted to abandon Catholicism, to betray those who died on the Nationalist side in the Civil War and to help international Marxism destroy Spain's prosperity.[41] The full power of the Church was mobilized. In some places, local Falangist officials insisted

* Great publicity was also generated for the visit to Buenos Aires made in the following October by Martín Artajo – *ABC*, 10, 11, 12, 14, 17, 20 October 1948.

that ration cards would not be valid unless presented and stamped at the polling booths. A relatively large turn-out was inevitable. In many country areas, people simply did not believe that the ballot was secret. Everywhere, a heavy police presence at voting stations gave credence to that view. According to the official figures, for which there was no independent scrutiny, in the referendum held on Monday 6 July 1947, out of a qualified electorate of 17,178,812, 89 per cent of those eligible voted – 15,219,565 votes. There were 14,145,163 or 93 per cent 'yes' votes cast – the remaining 7 per cent was made up of 4.7 per cent (722,656) 'no' votes and 2.3 per cent (351,746) blank or spoiled ballot papers. There was sufficient abstention in the big cities to cast doubt on the published figures. Nevertheless, for all of the pressure, intimidation and falsification, the results showed that Franco now enjoyed considerable popular support.[42]

Franco's personal reaction to the referendum was described by General Kindelán in a letter to Don Juan de Borbón. 'Franco is in a state of total euphoria. He is a man in the enviable position of believing everything that pleases him and forgetting or denying that which is disagreeable. He is, moreover, arrogant and intoxicated by adulation and drunk on applause. He is dizzy from height; he is sick with power, determined to hold on to it come what may, sacrificing whatever is necessary and defending his powers with beak and claws. Many think that he is perverse and evil, but I don't think so. He is crafty and cunning (*taimado y cuco*), but I believe that he operates in the conviction that his destiny and that of Spain are consubstantial and that God has placed him in the position which he occupies for great things.'[43]

Although totally sceptical of the democratic validity of the referendum, by mid-July 1947, the official policies of both Britain and American were coming to reflect a growing acceptance that Franco would be around for some time to come. The British had hardened their belief that it would be counter-productive to put pressure on Franco to leave voluntarily.[44] The Caudillo, however, guided by his lifelong distrust of the British and mistakenly believing that the Foreign Office was more determined to depose him than the State Department, continued to try to drive a wedge between the two Allies, as he had done in the days of Hayes and Hoare and of Armour and Mallet. On 27 July 1947, Franco took the extraordinary step of making arrangements for a secret meeting with the US *Chargé d'Affaires*, Paul Culbertson. Apart from resentful complaints about British duplicity, his central theme was that the Allies should be deeply grateful for his neutrality during the Second World War and for his anti-Communism. He ignored efforts by Culbertson to press him on the lack of political liberties in Spain. Culbertson came away convinced that Franco sincerely believed that his policies were in the best interests of Spain and were raising economic and social standards.[45]

On 18 July 1947, Franco showed that, in the wake of the *Ley de Sucesión*, he would act as sovereign in the newly proclaimed Spanish kingdom. The throne was empty but he assumed royal prerogative to the extent of bestowing titles of nobility. General Mola, José Calvo Sotelo and José Antonio Primo de Rivera were all given posthumous dukedoms. General Moscardó became Conde del Alcázar de Toledo. Over the next twenty-five years, he was to bestow thirty-nine titles.[46] As always with Franco, there were several motives to any action. The usurpation of a royal privilege was a warning to the Pretender. It reminded his supporters of the unlimited preferment in his gift and posed a cruel dilemma for the monarchists among them. If they refused a title, they were openly declaring their enmity; if they accepted, they were betraying their monarch. At the same time, the ennoblement in the early 1950s of many wartime figures, like Saliquet, Dávila, Queipo de Llano, Yagüe and Varela served as a reminder of the Civil War and announced that national reconciliation was still far from the Generalísimo's thoughts. In the case of General Alfredo Kindelán, whom he named marqués in 1961, there was a cruel sarcasm in Franco choosing to bestow 'nobility' on someone who had played an active, but ultimately unsuccessful, role in the campaign to restore the monarchy under Don Juan.

Franco continued to do everything possible to cultivate influential American opinion whether by granting press interviews or receiving visitors. In October 1947, a group of three Senators, and eight Congressmen of the Smith-Mundt Committee in transit from Rome were picked up at Barajas and whisked first to the Ministry of Foreign Affairs and then to El Pardo to meet Franco. That he would make himself available at an hour's notice indicates the importance that he gave to the contact. When he received the call from the Ministry to explain both the urgency of his meeting them, Franco was wearing military uniform. He called for a grey civilian suit saying to his aide 'We mustn't give these Americans the feel of a military regime.' Then he changed his mind, saying 'It's too great a concession. An Admiral's day uniform would be better; when all is said and done, it's a blue suit. And they love anything to do with the navy.' When they arrived, he spoke to them of the Communist danger and the threat posed by the Soviet Union. They went away delighted with the Caudillo and his uniform.[47]

The signs of a change in the American attitude were unmistakable.

The signals coming from London indicated the opposite. In October 1947, Gil Robles was in contact with Indalecio Prieto under the auspices of the British Foreign Secretary Ernest Bevin.[48] Alarmed reports from his embassy in London led an outraged Franco to take action. A note of protest about what was seen as interference in Spanish affairs was sent to the British government.[49] In fact, Bevin sponsored these talks only to placate anti-Franco sentiment in the Labour Party. His own view, and the official British Foreign Office line, was that nothing could be done to remove Franco and that the lesser evil was to prevent the Spanish question coming up at the Security Council in a way which might benefit Russia.[50] Culbertson spoke with José Erice, Director-General of Foreign Policy, at the Spanish Ministry of Foreign Affairs on 23 October. From the report of their conversation, Franco gleefully concluded that the Department of State disapproved of Bevin's action and, in opposition to the British, had decided to eliminate the Spanish question from the United Nations.[51] Franco complacently saw approval of his person and politics on the basis of a wishful thinking which dramatically exaggerated differences between British and American policy. Nevertheless, in practical terms he was right.

After his meeting of 23 October, although personally critical of Franco, Culbertson advised Washington against upsetting the 'applecart here regardless of [the] number of rotten apples in [the] cart'.[52] Both Britain and the United States were caught between a dislike of the Spanish dictatorship and a realistic awareness of his value in the Cold War. In fact, Washington was about to abandon any thoughts of removing him from power. On 24 October 1947, the US Policy Planning Staff, under the direction of George F. Kennan, had sent General Marshall and Dean Acheson a report recommending a rapid normalization of US economic and political relations with Spain. Marshall and Secretary of Defense James Forrestal approved Kennan's recommendations immediately. The consequent change in American policy was visible when next Franco's position came before the United Nations.[53]

Between 10 and 12 November 1947, the Spanish question was again discussed by the General Assembly's Committee 1 (Political and Security Questions) at Lake Success. These sessions marked a major turning point in terms of Franco's international position. The Polish delegate Oscar Lange convincingly demolished the flimsy democratic elements of Franco's recent *Ley de Sucesión* and produced evidence that the Generalísimo continued to grant asylum to large numbers of Nazi war criminals. The Czechoslovak representative, Jan Masaryk, protested that Franco continued to keep thousands of his opponents in inhuman prison conditions. Nevertheless, there was little support for proposals to apply full-scale economic sanctions. Both Britain and the United States played down their earlier condemnations of Franco and were more hostile than ever to interference in the internal affairs of Spain. Accordingly, when the General Assembly at Flushing Meadow on 17 November voted on Committee 1's mild resolution, the United States voted against the paragraph reaffirming the resolution of 12 December 1946, which failed to gain the necessary two thirds majority. The message was clear.

On 18 December 1947, Culbertson received instructions to adopt a friendly attitude towards Spain. With ambassadors drifting back to Madrid, it was hardly surprising that Franco should claim a great victory.[54] At the beginning of January 1948, the trend was confirmed when Franco received a telegram from his staff in Washington informing him that State Department officials had expressed their desire to see an American Ambassador in Madrid.[55] Effectively, the Caudillo had survived the worst. The United Nations Resolution of December 1946, the peak of the international ostracism, had failed. The Communist take-over in Czechoslovakia in February 1948 and the Berlin blockade from 24 June 1948 to 4 May 1949 would do the rest. That what had happened was not a brilliant personal achievement of Franco could not have been deduced from his smugly triumphant broadcast to the nation on 31 December which oozed self-righteous delight at the turn of events.[56]

Although the United States had no intention of sending an Ambassador to Madrid in the foreseeable future, the State Department was now moving ahead of its British counterpart in readiness to 'normalize' relations with the Caudillo. This reflected the extent to which the Pentagon had accepted the recommendations of George F. Kennan's October 1947 paper. Theodore C. Achilles, the Chief of the State Department's western European Affairs Division, outlined Washington's new position when he wrote to Culbertson: 'international pressure to "kick-Franco-out-now" has failed and has served only: (1) to strengthen his resistance to any liberalization under foreign pressure; (2) to increase support for him in Spain among those who would like a more democratic government but object to foreign pressure or fear renewed disorders; and (3) to give the Communists everywhere one more chance to cause trouble and embarrassment.' The British regarded the change in American attitude as disastrous and likely to cause embarrassment to her other European partners.[57]

Franco could now confidently sit tight in the knowledge that the tide was turning rapidly in his direction. This was reflected in the amount of time that he felt able to devote to his pleasures. A fishing holiday in Asturias at Easter was becoming a fixture in his calendar and, since the delivery of his yacht *Azor*, a real passion for deep-sea fishing in the Atlantic with Max Borrell had developed and occupied ever more of the summer.[58] His golf had come on since his early efforts in the Canary Islands and, at one point in this period, he had the audacity to explain to a speechless Duque de Alba – James Fitz-James Stuart y Falcó, he was also the Duke of Berwick – how to make a golf-course.[59] He could always find time too for hunting. At sea, he spent considerable time playing cards (*mus* and *tresillo*) and dominoes with his inner circle of military friends, General Camilo Alonso Vega, Admiral Pedro Nieto Antúnez and General Pablo Martín Alonso.

In political terms, Franco's growing confidence permitted him to block with deliberate incomprehension Culbertson's efforts to get even token gestures of liberalization in both economic management and the political repression.[60] When the British *Chargé*, Douglas Howard, protested about Nazi war criminals in Spain, the persecution of Protestants and the trials of leftists, his complaints were confidently brushed aside by Martín Artajo who then subjected Howard to an aggressive dressing down, something which could not have happened without Franco's acquiescence.[61] Indeed, far from evincing the slightest inclination even to meet the Americans half-way, Franco instructed Martín Artajo to push for a more penitent approach from Washington. Martín Artajo told Culbertson on 9 March 1948 that, despite the US attitude at the United Nations in November, Spain was not satisfied that her good name had been vindicated and therefore wished the USA to take the lead in righting the injustice.[62] Franco knew that the change in the State Department's attitude was running behind that of the American military and financial establishments.[63] At the end of March, the US Joint Chiefs of Staff had expressed their interest in having three airfields in Spain equipped to handle the heaviest American bombers.[64]

Spain's inclusion in the Marshall Plan was approved by Congress as part of the Foreign Assistance Bill on 30 March 1948. This was in response to an amendment sponsored by Congressman Alvin O'Konski, who argued that 'to exclude Spain is a shameful and stupid appeasement of the reds of Moscow and the reds of our own State Department and Department of Trade'. O'Konski's success reflected the fact that, shortly before voting took place, news of Russian demands to control traffic into and out of Berlin had begun to spread around the House of Representatives. However, Truman blocked the inclusion of Spain by pointing out that it was up to the members of the European Recovery Programme to decide whether to admit new members. Truman was motivated both by disgust at the lack of religious freedom in Franco's Spain and by a willingness to adjust to popular opinion in Britain and France.[65]

Franco's perception that the western powers were no longer seriously interested in displacing him was intensified after the Berlin blockade. The German crisis also consolidated support for the Generalísimo among moderates within Spain who had no love for him but believed that it was no moment to contemplate a change of regime and destabilize Spain.[66] The consequent boost to his confidence freed Franco to give more time and energy both to recouping his position internationally and to the internal battle with Don Juan and his supporters. In early August 1948, he sent his old *Africanista* friend General Eduardo Sáenz de Buruaga, to see the Governor of Gibraltar, Sir Kenneth Anderson, in a vain effort to loosen the deadlock in Anglo-Spanish relations. Anderson sought advice from the Madrid Embassy and was told to inform Saenz de Buruaga that there could be no normal relations with Britain while the 'detested and notorious Falange' kept its stranglehold on Spain.[67] The rebuff would be instrumental in pushing the Caudillo to concentrate his efforts at seduction of a western Power on the United States.

Franco's initiative in the direction of Don Juan was altogether more satisfying for him. The tension between the Caudillo and the Pretender was not in the interests of either. Franco, however, held most of the cards. In early January 1948, Culbertson had told two of Don Juan's advisers, one of whom, José María de Oriol, was also a regular visitor to El Pardo, that the United States saw no point in provoking the fall of Franco by means of economic blockade since the benefit would not go to the monarchists but to the Left. Culbertson recommended that they tell Don Juan to seek some agreement with Franco.[68] The Caudillo had been concerned by reports from his secret services on the growing links between the monarchists and the Left but, fully informed by Oriol of Culbertson's remarks, he was able to react with equanimity.[69]

Franco went to great lengths to arrange a meeting with Don Juan on his yacht, the *Azor*. After fending off various discreet invitations passed to him by courtiers who were also in close contact with Franco, the Pretender agreed to meet the Caudillo in the Bay of Biscay, on 25 August 1948.[70] Don Juan made the decision to see him without informing his own close political advisers, including Gil Robles and Pedro Sainz Rodríguez. Don Juan had insisted that Franco join him first on the yacht *Saltillo* which belonged to his friend Pedro Galíndez. Either out of a desire for Don Juan to be seen as coming to his 'territory' or because he was frightened of seeming ridiculous in moving in and out of boats on a choppy sea, Franco refused. When Don Juan came aboard the *Azor*, Franco greeted him effusively and, almost on cue, wept profusely. They then spoke alone in the main cabin for three hours. Apart from the short official account given to the Spanish press, the only detailed information derives from Don Juan's various accounts.

The Pretender had arrived, it seems, feeling emotional and nervous. However, he quickly sensed that the Caudillo believed him to be an idiot, entirely in the hands of embittered advisers and totally ignorant of Spain. A voluble Franco, who barely allowed him to get a word in edgeways, counselled patience and asserted that he was in splendid health and expected to rule Spain for at least another twenty years. He boasted that, under his own guidance, Spain would soon be one of the richest countries in the world. He spoke of his devotion to Alfonso XIII and again cried. Franco alleged that there was no enthusiasm in Spain either for a monarchy or for a republic although he claimed that he could, if he wanted, make Don Juan popular in a fortnight. He was completely thrown when Don Juan asked him why, if it was so easy to manufacture popularity, he constantly cited popular hostility as a motive for not restoring the monarchy. The only reason that the Caudillo could cite was his fear that the monarchy would not have the firmness of command (*mando*) necessary. In contrast to what he must have supposed to be Don Juan's practice, he declared 'I do not permit my ministers to answer me back. I give them orders and they obey.' In the course of the interview, Franco chortled sardonically about the fact that he had not informed his ministers about the meeting. Reviewing his generals in cruelly dismissive terms, he commented that Solchaga was 'an idiot' and Yagüe 'a raving lunatic' (*un loco rematado*). He also assured Don Juan that 'anyone can be bought'.

Franco's real purpose in arranging the meeting finally became apparent when he showed immense interest in the Pretender's ten year-old son Juan Carlos completing his education in Spain. Juan Carlos in Spain would be a hostage to justify Franco's indefinite assumption of the role of regent and an instrument to control the political direction of any future monarchical restoration. Franco spoke with a mixture of cunning and prejudice of the dangers run by princes under foreign influence (*príncipes extranjerizados*). Don Juan pointed out that it would be impossible for his son to go to Spain while it remained an offence to shout '¡Viva el Rey!' (long live the King) and monarchist activities were subjected to fines and police surveillance. Franco offered to change all that. No firm arrangements were made about Juan Carlos. Over lunch, Franco started to talk about his hunting prowess and tactlessly mentioned a recent hunting party on the royal estate at Gredos. Don Juan asked if it was true that wild goats had been killed with machine guns. The embarrassed Caudillo admitted that this was so although he excused it with the claim that only wounded goats were thus treated. Don Juan reproved him for his lack of sportsmanship at which Franco changed the subject to speak of salmon fishing. On leaving, Franco suggested that they remain in touch via the Duque de Sotomayor, making the astonishing remark that 'I can trust no one since all my staff are very indiscreet.'[71]

Whatever superiority Don Juan might have felt over Franco in terms of good taste, *savoir faire* or intellectual power, he had agreed to the meeting because he had already reached the conclusion that the Caudillo would survive and that a future monarchical restoration was feasible only with his approval. Don Juan told an official of the American Embassy that prior to the meeting he was making no progress in his relations with Franco and that now he had got 'his foot in the door'. Gil Robles, in contrast, was convinced that the meeting had set back the cause of the monarchy for many years simply by demonstrating to Franco that Don Juan was not the malleable playboy that he had previously assumed him to be. The Duque de Sotomayor, Don Juan's representative in Spain, and Julio Danvila, acting as intermediaries from Franco, pressed him for a decision about Juan Carlos's education. Don Juan replied that he was fully aware that any announcement would be used by Franco to imply that he had abdicated. When he stood firm, Sotomayor resigned as his representative. In his heart, however, Don Juan was convinced that there would be no restoration against the will of Franco.[72]

The benefits of the rapprochement between the dictator and the would-be King were one-sided. One of the threats which had impelled Franco to initiate the contact was now dissipated. The negotiations between monarchists and Socialists resulted in the so-called Pact of St Jean de Luz in October 1948 but they were now shorn of much of their efficacy. To the consternation of the moderate Socialists and Republicans, the contact between Don Juan and Franco completely discredited the democratic monarchist option for which they had broken with the Communist Party and the Socialist Left.[73]

Confident enough to give free rein to his sentimental streak, Franco treated Don Juan with superficial respect. On the occasion of Franco's twenty-fifth wedding anniversary, and recalling that Alfonso XIII had been *padrino*, albeit *in absentia*, at his wedding, Don Juan sent a message of congratulation. The Caudillo responded with a hand-written letter of thanks which began 'My Prince' and stated what a great honour it had been for him and his wife to have received his greeting and that he owed his happiness to Don Juan's father. The letter ended 'with loyalty and affection'.[74] The hard significance of the *Azor* meeting was, however, revealed when Franco had news leaked that Juan Carlos would be educated in Spain. With no more concessions from Franco than a promise that the monarchist daily *ABC* could function freely and that restrictions on monarchist activities would be lifted, Don Juan capitulated and agreed to send Juan Carlos to Spain where he arrived on 9 November. With his habitual caution, Franco refused to permit the young prince to use the title *Príncipe de Asturias*, normally given to heirs to the Spanish throne. A group of teachers of firm pro-Francoist loyalty was arranged for the young prince. The recently arrived Juan Carlos was immediately received by Franco at El Pardo. The publicity given to the visit was handled in such a way as to give the impression that the monarchy was subordinate to the dictator. That, along with the torpedoing of the monarchist-Socialist negotiations, had been one of the principal objectives behind the entire *Azor* operation.[75]

At virtually no cost, Franco had left the moderate opposition in embittered disarray and driven a wedge between Don Juan and his most fervent and loyal supporters.[76] He had created a situation in which many influential members of the conservative establishment who had wavered since 1945 would incline again towards his cause. The controlled press was ordered to keep references to the monarchy to a minimum. In international terms, the Caudillo had cleverly bought his regime more time. In the widely publicized report of a conversation with the British Labour MP for Loughborough, Dr M. Follick, Franco declared that it was his intention to restore the monarchy although he sidestepped the question of when.[77] In a context of ever greater international tension, any apparent 'normalization' of Spanish politics was eagerly greeted by the western powers. Within less than a year, a deeply disillusioned Don Juan would order an end to the policy of conciliation.[78] By then it would be too late, Franco having squeezed every drop of benefit out of the apparent closeness between them.

In March, Franco had sent the consummate cynic, José Félix de Lequerica – who had been refused by the United States as Ambassador in 1945 – to Washington with the specious title of 'Inspector of Embassies and Legations' to create a Spanish lobby. Large amounts of money were spread in pursuit of this goal. Influential lawyers hired at enormous cost concentrated their efforts on political, military, religious and financial targets. Lequerica could already count on the support of Senator Pat McCarran and Republican Congressman Alvin E. O'Konski. The deeply Catholic McCarran was Democratic Senator for Nevada and Chairman of the Senate Appropriations Committee, a position which gave him immense influence in Washington and which he used effectively. In the words of Dean Acheson, 'the Senator was not a person who in the eighteenth century would have been termed a man of sensibility'.[79] Lequerica himself was busy wining and dining an enthusiastially pro-Franco lobby of influential American Catholics, anti-Communists, military planners, anti-Truman Republicans and businessmen with interests in Spain.[80]

Having established cordial relations with Cardinal Spellman, Lequerica, with barefaced cheek, told his interlocutors that support for the Spanish lobby meant the votes of thirty million American Catholics. He encouraged the Caudillo to make statements to the American press which he began to do regularly.[81] The growing warmth of relations with certain sectors of the American establishment was reflected in visits to Spain made by James Farley, a prominent American Catholic and one of the presidents of the Coca Cola Corporation. On 24 September, Artajo was informed by Culbertson that the US delegation to the United Nations could now count on the necessary two thirds majority to permit the repeal of the December 1946 resolution.

On 30 September 1948, a US military mission headed by Senator Chan Gurney visited El Pardo. Gurney was Republican Senator for South Dakota and Chairman of the Senate Armed Forces Committee. It was rumoured that the mission had come to discuss military bases in Spain. After Culbertson had introduced the party to the Caudillo, there was a long embarrassing pause. Finally, Franco, who remained standing throughout the forty-minute interview, launched into a denunciation of France as the weak link of western defence. He claimed to know from French military contacts that, in the event of Soviet attack, the French army would be unable to defend French soil and be forced to fall back on the Pyrenees.[82] The coincidence of Gurney's trip with the annual *Día del Caudillo* enabled the Spanish press to present the visit as an American endorsement of Franco's rule. On his return to Washington, Senator Gurney made statements in favour of Spanish inclusion in the Marshall Plan and in the United Nations. He also recommended military aid for Spain. Franco's press presented this as causing a sensation in the world's newspapers. Lequerica immediately telegrammed Franco that he should make a statement to the prominent American columnist Cyrus L. Sulzberger of the *New York Times*. He agreed to do so.[83]

Meanwhile, in a ceaseless effort to maintain domestic popular support, Franco continued to tour Spain. In the autumn of 1948, he visited Andalusia. All his public appearances had an international as well as a domestic dimension. The rapturous demonstrations and the cheering crowds were orchestrated to show the outside world that it was pointless to contemplate dividing the Spanish people and their beloved Caudillo. The apparent unity thus displayed was also intended, as were the fiercely anti-Communist speeches, to be an advertisement for Francoist Spain's utility in the Cold War. On 11 October, he spoke to the local military authorities in Seville at the Capitanía General. In the course of a hyperbolic speech about Spanish military prowess, he made the astonishing statement that the atomic bomb would never overcome the capacity of the Spaniards to resist with guerrilla tactics. He also quoted with pride his notorious speech of 14 February 1942 to Army officers in Seville when he had declared his 'absolute certainty' that Germany would not be destroyed and had rashly offered 'one million Spaniards' to defend Berlin. With that characteristically seamless combination of naïvety and duplicity which enabled him to propound the most outrageous untruths with genuine conviction, he now resurrected that statement as a prophetic and courageous anticipation of the Cold War.[84]

As part of Franco's efforts to create a bridgehead into the western community through links with Latin America, the *Día de la Raza*, as Columbus Day was known in the Falangist calendar, was given over to a spectacular ceremony of Pan-Hispanic solidarity. 1948 was also the seventh centenary of the foundation of the Castilian navy. On 12 October, twenty-eight warships lay at anchor in the estuary of the River Odiel at Huelva. A salute was fired from a Spanish gunboat and the Caudillo, accompanied by Doña Carmen and his daughter 'Nenuca', passed the assembled vessels in review. There were ships from the navies of Argentina, Brazil, Colombia, the Dominican Republic, Peru and Spain. Afterwards, at the monastery of La Rábida where Christopher Columbus kept vigil on the night before setting out from Palos de Moguer on his historic voyage, Franco was invested with the title of Lord High Admiral of Castile. It was a lifetime's ambition fulfilled and his delight could be discerned both in his jauntily enthusiastic speech about Spanish naval tradition and in his beaming face.[85] He had long been presented by the propaganda machine of his regime as the El Cid of the twentieth century;

now he had recovered from his brother Ramón the title of the Christopher Columbus of the twentieth century.

In the course of the Andalusian tour, an incident took place which underlined that Franco's perception of himself as a near royal personage was totally sincere and not just part of his propaganda repertoire. It was a revealing commentary on the sincerity of his discussions with Don Juan about a future return of the monarchy. As part of the seventh centenary celebrations of the conquest of Seville from the Moors, a monument was being erected to the Sacred Heart of Jesus some way from the provincial capital. Franco was to attend and there was to be an official banquet after the ceremony of inauguration. An official of Franco's household went to the Archbishop's palace to discuss protocol and proposed that one table be headed by the Caudillo with Cardinal Segura at his right hand and the other headed by Doña Carmen. Segura refused on the grounds that he should preside at the second table. He pointed out that, according to the statutes of the Sacred College of Cardinals, a cardinal could give up his place only to a King, a Queen, a Head of State or a crown prince. There was consternation among the officials since Franco insisted that his wife be given identical treatment to his own. If Franco was adamant, Segura offered three solutions, that Doña Carmen did not attend, that he the Archbishop did not attend or that there be no banquet at all. The third solution was adopted. Francisco and Carmen were furious and machinations began which culminated five years later with Segura losing power in his own diocese.[86]

Such minor embarrassments aside, flushed with the pleasure which a thorough immersion in popular adulation always gave him, Franco was now convinced that the worst was over. Don Juan was tamed. The military representatives of the United States were already beginning to beat a path to his door. On 4 October 1948, General Marshall in Paris for the first part of the Third Session of the UN General Assembly, told Bevin and Dr Robert Schuman, the French Foreign Minister, that the recognition of Franco presented no problem for the United States. They made it clear to him that public opinion in Britain and France was not yet ready to tolerate normal relations with the Caudillo.[87] A proposal to support Spanish entry into the United Nations was discussed by the French cabinet on 10 November. After an acrimonious debate, it was decided not to do so but not to oppose Spanish entry into the technical dependencies of the UNO. On 18 November, Spain was invited to participate in the International Statistical Commission. Among the votes in favour were those of Britain and the USA.[88]

The Caudillo's confidence was reflected both in declarations to the *New York Times* and to *Newsweek* and in his end of the year broadcast. Franco conveniently forgot years of contemptuous insults about the 'masonic super-state' and the mindless materialism of the Americans. The Caudillo treated the correspondent of the New York daily to a virtuoso display of the most servile pro-Americanism and made offers of Spanish participation in a US-Spanish alliance, which were not published in Spain.[89]

Five days after the publication of his interview in the *New York Times*, anxious to consolidate a favourable position, the Caudillo committed an uncharacteristically precipitate act. Through one of his officials, he offered the American *Chargé* in Madrid a bilateral economic arrangement with the USA outside the Marshall Plan. To clinch such an arrangement, Spain would accept any 'reasonable conditions' which the Americans might suggest and also make available bases in the Canary and Balearic Islands and on the Spanish mainland. Culbertson did not take up the offer.[90] Nevertheless, in an effort to get American public opinion on his side, Franco made a hard-faced effort to drive a wedge between Washington and London. In an interview published in *Newsweek* on 22 November 1948, he assured the American people that British selfishness was depriving them of the peace of mind that a Spanish alliance could bring, clearly a reference to the British rebuff for the overture he had made in August through General Sáenz de Buruaga.[91]

When Franco said 'the British', he could only have meant the Labour Government. His natural readiness to think ill of perfidious Albion led him to generalize. In fact, there were voices in the Conservative Party, including that of Churchill, echoing the calls of some American Senators and Congressmen for the resumption of full diplomatic relations with Spain. Churchill's advocacy of Franco's incorporation into western defence reached its peak on 10 December 1948 in the House of Commons. He dramatically exaggerated Franco's benevolence during the Second World War, declaring 'No British or Americans were killed by Spaniards and the indirect aid we received from Spain during the war was of immense service'. Not only had he forgotten the wolfram war and the innumerable acts of aid to Axis forces countenanced by Franco, but, as Captain Noel-Baker reminded him, he seemed to have forgotten that Britain was, at the time, allied to the Soviet Union. Taking up Churchill's cynical deletion of the Blue Division, the Labour Parliamentary Under-Secretary of State for Foreign Affairs, Christopher Mayhew, said 'the record of Franco Spain during the war is a serious reason why we cannot consider the welcoming of Franco Spain into the Community of western Europe'.[92]

With the Labour Party in power, Franco considered that there was little to be done for his cause in Britain. Accordingly, he concentrated his diplomatic efforts on the United States and the Vatican. The piously Catholic Joaquín Ruiz Giménez was despatched as Ambassador to the Holy See at the end of November 1948. His mission was to pave the way to a Concordat which Franco wanted as a public seal of divine approval for his regime. Privately, he already assumed himself to have that approval. In his broadcast to the nation on 31 December 1948, he thanked God for giving 'a pleasant wind and a calm sea to the ship of the *Patria*'. In the main, however, the speech was a complacent anthology of self-congratulation. He declared that 'we have come through the worst years' and then went on to a hymn of self-praise in an entirely surrealistic account of the great economic progress of Spain under his prescient guidance.[93]

Franco's complacency now that he had ridden the worst of the storm generated a stultifying political atmosphere. Serrano Suñer was stung out of his private existence to make a daring plea for a new stage in Spanish political life. Complaining of the 'dangerous boredom' which afflicted Spain, he criticized those 'without vision' who could not see that Spain could not permit herself the luxury of trying to live isolated in the world.[94] Franco was outraged. 'He'll find out about boredom. I'll close *ABC* for three months and I'll exile this arrogant Serrano to the Canary Islands.' About this time, talking with the Colombian poet Eduardo Carranza, Franco explained why no opening-up (*apertura*) was possible. When Carranza cited the attitude of Serrano Suñer as an indication that evolution might be necessary, the Caudillo replied 'if it's necessary to shoot Ramón, he will be shot too'.[95]

Franco's self-congratulation suggested that he was oblivious to the fact that, in working class districts of major towns, people in rags could be seen hunting for scraps. Outside Barcelona and Málaga, many lived in caves. Most major cities had shanty towns on their outskirts made of cardboard and corrugated iron huts where people lived in appallingly primitive conditions. The streets were thronged with beggars. State medical and welfare services were virtually non-existent other than the soup kitchens provided by the Falange. Hardship, malnutrition, epidemics, the growth of prostitution, the black market, corruption were consequences of his regime's policies which inevitably did not figure in the Caudillo's optimistic survey. He was concerned only with 'his' Spain, not that of the left-wing workers who belonged to the 'other' anti-Spain.[96]

In any case, Franco had his mind on other things. On 7 January, it had been announced that the shrewd Dean Acheson would return as Secretary of State to replace General Marshall, who was retiring because of ill health.[97] Carrero Blanco drew up a memorandum on how to derive benefit from the emergence of NATO and how best to react to the arrival of Acheson. In his fevered imaginings, no doubt shared by Franco, this change 'marked a new attempt by Truman at an understanding with the USSR' – which meant that Acheson made good relations with Britain and France a higher priority than normalizing relations with Franco. However, according to Carrero Blanco, despite the misfortune of Acheson's appointment, Franco could be confident that the Pentagon and the senior military staff of the United States would not permit Truman to go on appeasing Stalin. The American military fully perceived the strategic and ideological value to western defence of Franco's Spain. In a total identification with Franco's way of thinking, Carrero proposed that, since 'they need us and will call us', Spanish interest in joining the new defence system be carefully concealed. Links with Portugal should be strengthened so that Spain would be seen as an indispensable part of the Iberian block. When the Allies came courting, Carrero Blanco suggested, the price for Spain's participation in NATO should the return of Gibraltar.[98]

Franco was eager to join Nato.[99] But, in response to Carrero Blanco's memo, he played hard to get. Interviewed in the *Daily Telegraph*, he was cool about Spain wanting to join the United Nations and the 'western Union' [i.e. NATO]. There was both cynicism and sincerity in his words, the self-interest of increasing his price for military co-operation and a resentment at Spain's economic plight (his own responsibility for which he conveniently forgot): 'If there are eight hungry men on a desert island and a ship arrives bringing food for seven of them, imagine the sentiments of the eighth. Well, we in Spain happen to be that eighth man'. That admission of Spain's economic desperation was not published in Spain. His dignified rejection of the United Nations was. Citing the 'many injustices done' to Spain, he dismissed NATO of no interest without a full and specific statement of the 'advantages, guarantees, rights and obligations' involved.[100]

On the day that his interview was published, there was an embittered debate in the House of Commons in which Conservative critics savaged the Labour Government's continued commitment to excluding Franco from the western community. Captain Noel-Baker quoted Franco's rejection of membership of the United Nations and NATO in his *Daily*

Telegraph interview and commented 'as far as closer relations with the western democracies are concerned, the Spanish dictator has himself given his own answer, a blunt, peevish and ill-mannered 'NO' to proposals never made to him, and has given a very good exhibition of a bad-mannered refusal to attend a party to which he had never been invited'. But Labour indignation about Franco was increasingly being overtaken by world events and Spain's value to western defence. Under pressure, Christopher Mayhew, the Parliamentary Under-Secretary of State for Foreign Affairs, admitted that the British Government would not necessarily oppose moves to reinstate ambassadors.[101]

Seizing on such unmistakable clues, Franco hoped to sell his collaboration dearly to the Allies. The Spanish economic situation was so desperate that the Minister of Industry and Commerce, Juan Antonio Suanzes had predicted that, if American financial aid were not secured, collapse could be anticipated within six months. Persistent drought was causing severe electricity restrictions which hit industrial production. Calculations of the size of the wheat harvest were consistently being scaled down. The bread ration was reduced to 150 grammes per day after Perón refused to ship more wheat to Spain without fulfilment of Spanish commitments to Argentina.[102]

Franco's unsinkable optimism in such appalling conditions was boosted by the success of the Spanish lobby among the American military elite. With anxiety for Spanish bases growing, one of the lobby's first triumphs was the announcement, on 8 February 1949, that the Chase Manhattan and National City Banks of New York had made a loan of $25,000,000 to the Spanish government. As collateral, Spain had given twenty-six tons of gold deposited in London. Since the loan had needed the approval of the State Department, it was an indication of the changing image of Franco in the United States.[103]

On 31 March 1949, Franco broadcast to the nation on the eve of the massive celebrations of the tenth anniversary of his victory in the Civil War. In the knowledge that four days later the Atlantic Pact would be signed and NATO created without Spanish participation, his speech was directed at the western Powers in a tone of outraged resentment that they had not included him and thus failed to recognize his massive contribution to the defence of western civilization. 'The situation of the world throws light on the fact that we were right.' His vengeful domestic policies were described complacently as 'the generosity with which we have administered our victory'. Attributing Spain's neutrality during the Second World War to his vigilance as captain while his war-weary people rested, he described his exercise of power as a burdensome sacrifice in the interests of Spain. 'My certificate of nobility (*ejecutoria*) is written with one word: duty, the concept of duty. Anything comfortable, agreeable or to do with my feelings has been sacrificed to the dictates of that one word.'[104] There is no reason to supect that Franco, who in the late 1940s was becoming ever more obsessive in taking his hunting, shooting and fishing pleasures, did not believe every word of his speech.

Such rhetoric, however, could not prevent Spain being excluded from participation in the North Atlantic Treaty while Salazar's Portugal was not. That reflected the indispensable strategic value of the Azores, the continuing hostility towards Franco of public opinion in most European countries and the fact that Salazar had handled his wartime neutrality with infinitely greater subtlety than Franco.[105] Nevertheless, there were hopeful signs for Franco. When Committee 1 (Political and Security Questions) of the General Assembly of the United Nations met in New York in early May 1949, there were two draft resolutions on Spain: one from Poland strengthening the resolution of December 1946 and prohibiting commercial treaties with Spain; the other from Brazil and three other Latin American countries attempting to restore diplomatic relations. The Polish proposal, which accused Franco of being the puppet of the United States, was roundly defeated. The Latin American proposal was approved for forward transmission to the General Assembly. Since Washington now wanted the return of ambassadors to Madrid but did not wish to cause insoluble difficulties for Britain and France, the United States abstained from the votes on Spain when the General Assembly of the United Nations met.[106] After bitter arguments on 11 and 16 May, the Latin American resolution came within four votes of reaching the required two thirds majority. Britain and France joined the USA among the abstentions.[107]

Two days later, Franco delivered to the Cortes a speech intended largely for international consumption. He was still smarting from a statement made at the United Nations on 16 May 1949 by the British Minister of State at the Foreign Office, Hector McNeil, in which he had declared that to provide arms to Franco was 'like putting a gun into the hands of a convicted murderer'.[108] The Caudillo's speech was two-edged, both supplicatory and boastful, with a smear of anti-British venom. In its endless justifications of the past, it showed Franco still angling to ensure a prime place in the international community. At the same time, in its self-congratulation, the speech revealed Franco laying out his stall before those who would be expected to pay for his participation in western defence. One of its themes was that his Spain was as democratic as those countries whose leaders expected him to liberalize. He also boasted that his regime was at the forefront of world developments, its social achievements distinguishing it from both liberal capitalism and Marxist materialism. The speech inevitably contained an account of the alleged economic achievements of recent years.

Clearly addressing himself to Washington, he set out to show the United States that he constituted an altogether sounder ally than perfidious Albion by denouncing Socialism as every bit as evil as Communism, a dig at the British Labour government. In contrast, he underlined 'our title of nobility (*ejecutoria de nobleza*), our well-proven chivalry (*hidalguía*) and our unselfish generosity (*desinterés*)'. He then went into a long and convoluted defence of his 'noble' and 'elegant' behaviour during the Second World War in which he gave the game away by saying 'we had every right to be what we would have liked to be [presumably belligerents on the Axis side] since we had no reason to feel any gratitude to one side [the western Allies], yet we were neutral'. He claimed mendaciously that, in return for not allowing German troops through the Iberian Peninsula, Britain had offered to help make Spain the most powerful nation in the Mediterranean.[109] He was blithely unaware that his resentment over what he perceived as a bribe unpaid rather undermined his self-image as an idealistic and chivalrous neutral and showed his role in the war for what it had been, that of a rapacious opportunist prepared to sell neutrality for an empire on the cheap. Oblivious to the fact that, even if Churchill had tricked him as alleged, he had done so in order to undermine the relationship between Hitler and one of his satellites, Franco exposed his own continuing failure to reassess the role of the Axis in the Second World War. To his grudge over his interpretation of unfulfilled British offers, he added a dose of bile over the way he was treated after the Laurel incident and during the wolfram crisis of 1944. Most galling to him was the fact that Churchill had chosen not to accept his far-sighted offer of anti-Soviet co-operation.

This lengthy catalogue of injuries suffered and benevolent offers churlishly rejected was by way of hinting why he might not be overly anxious now to join the western alliance. 'We find the states of Europe so clumsy, so old and so divided, and their politics so riddled with Marxisms, passions and malice, that inadvertently they impel us to where our heart calls us, to greater closeness and understanding with the peoples of our stock; America again attracts the historic destiny of Spain.' He did not mean just Latin America: his 'policy of friendship and understanding with the peoples of Hispanic origin necessarily led Spanish foreign policy to a greater understanding with the entire American continent, in which North America, by dint of her wealth and power, has come to occupy the leading place.' Any difficulties remaining between Spain and the USA were the fault of others concerned with their spheres of influence, an unmistakable reference to Britain and France.[110] It was ironic that Franco should attack Britain at a time when pressure for a more friendly policy towards Spain was building up in London in the press, parliament and the City.[111]

On 20 July 1949, a solemn investiture ceremony took place in Madrid at which Franco was given the title of First Journalist of Spain and presented with press card no.1 by a commission of newspapermen.[112] Barely three weeks later, Lequerica had to find a way of telling the First Journalist that his articles on freemasonry were likely to damage Spanish-American relations. Franco was keen on an English translation of his series and Lequerica had to advise him that 'in some countries, there is little taste for authentic, deep truths'.[113]

On all fronts, things were going well for Franco. When Don Juan intimated that he would not permit his son's return to Spain to continue his studies, Franco instructed his brother Nicolás to inform him that the Cortes would pass a law specifically excluding him from the throne. Don Juan, influenced by the fiercely collaborationist Danvila, responded to the threat as the Caudillo had hoped.[114] The sporadic guerrilla opposition which had flickered since 1945 in the remote sierras of the south and the bleak mountains of Asturias was petering out. The announcement by Truman on 23 September 1949 of the successful explosion of a Soviet atom bomb in August intensified pressure within the United States for a rapprochement with Spain in order to secure both air and naval bases.[115] A powerful advocate of a military alliance with Franco was Admiral Forrest Sherman who was convinced of the geostrategic importance of Spain to the USA. As Commander-in-Chief of the US Sixth Fleet, he visited many Spanish ports. By chance, his son-in-law, Lieutenant-Commander John Fitzpatrick, had been appointed Assistant Naval Attaché in Madrid in 1947. When Sherman and his wife visited their daughter in the Spanish capital, they were cultivated by the authorities.[116]

Admiral Richard L. Conolly, Commander of US Naval Forces in the Eastern Atlantic and the Mediterranean, was also keen to secure bases in Spain. In September 1949, a squadron of the United States Eastern Atlantic Fleet put in at El Ferrol and stayed for five days. Conolly and some of his senior officers visited Franco at the Pazo de Meirás. American sailors were allowed to visit Madrid. It was a clear indication of things to come and reflected the interests of a powerful group in the US defence establishment, led by Louis A. Johnson, Secretary of Defence. Having initiated an economy programme which had cut plans for a big expansion of the US aircraft-carrier fleet, Johnson was particularly interested in land bases for American bombers. He was not the only one. Lequerica's Washington lobby could count on the support of several influential political and military figures, including Joseph McCarthy, Republican Senator for Wisconsin. Senator Pat McCarran sponsored three attempts to facilitate a loan to Spain through the United States Export-Import Bank. Thanks to 'the autumn plan' (an expensive diplomatic initiative dreamed up by Lequerica and personally authorised by Franco) increasing numbers of Senators and Congressmen paraded through Madrid at Spain's expense. One of the junketers, James J. Murphy, Democratic Congressman for New York, called Franco 'a very, very lovely, and loveable, character'.[117]

In subsidising these trips through the 'autumn plan', Franco seems to have believed that he was merely facilitating 'Spain's truth making its way in the world'. Certainly, the visitors convinced the Caudillo that he had been right all along and need make no changes in his political system – much to the chagrin of Culbertson who was trying to prod him in the direction of economic and political liberalization.[118] At the beginning of October 1949, Mao Tse-Tung had established the Chinese People's Republic. Although he was not the stooge of Moscow, in the West it seemed as if another huge area of the world had fallen into the orbit of the Soviet Union. The temperature of the Cold War dropped several degrees to the benefit of Franco.

In the meanwhile, on 22 October 1949, in an attempt to get nearer to North Atlantic Treaty via Salazar, Franco visited Portugal and was delighted to be treated with great solemnity and pomp. His attitude contrasted with the austerity manifested by Salazar on his visit to Spain in 1942. The Caudillo travelled by road to Vigo where he went on board the battlecruiser *Miguel de Cervantes* which put to sea at the head of a flotilla of eleven warships. At the mouth of the Tagus, it was met by four Portuguese destroyers and escorted to Lisbon. There he was met by the President of Portugal Marshal Carmona and Salazar himself. There followed a fly-past of Hurricanes and Spitfires and a parade of fifteen thousand Portuguese soldiers. Installed in the Palace of Queluz, he was joined by Doña Carmen who had made the journey from Madrid by train. He had hoped during his visit to Portugal to demonstrate to the world his dominance over Don Juan. Via his brother Nicolás, he demanded Don Juan's presence at Queluz on 22 October in a courtesy visit. Don Juan refused. It was the only blot on an otherwise brilliant propaganda triumph for the Caudillo. The entire operation was a skilfully mounted affirmation of the utility of the Iberian Peninsula to the western Alliance, and, to ensure the maximum efficacy for the domestic dimension of the Portuguese trip, a huge propaganda climax to the visit was arranged. 27 October was declared a public holiday in Spain and large numbers of Falangists and peasants bussed in from the Castilian provinces lined the streets to greet the returning Caudillo.[119]

Such *divertissements* did little to distract attention from the seriousness of Spain's food crisis. A concerned Varela visited Franco shortly after Christmas 1949. When he mentioned the desperate shortage of wheat, Franco said complacently that he could easily get all the foreign credits necessary to solve the problem but was not prepared to pay the political price of the changes required by the western powers. He preferred to wait certain that 'they will bend their heads, because the world needs Spain more than Spain needs the world'. Varela also complained that the regime's corruption was facilitated by the lack of press freedom and the fact that the Cortes had no real power. Franco acknowledged that more power for the Cortes and a freer press might help clean up the corruption but asserted that the negative consequences would be worse. Corruption mattered little to him when his own permanence in power was at stake. The interview ended with Franco telling Varela that 'I will not give Spain any freedom in the next ten years. Then, I will open my hand somewhat'.[120]

That he could speak with such confidence reflected his awareness of developments in the Anglo-Saxon world. Churchill had mocked the Labour Government on 17 November 1949: 'Fancy having an ambassador in Moscow but not having one in Madrid. The individual Spaniard has a much happier and freer life than the individual Russian or Pole or Czechoslovak.'[121] In Washington too, calls for a return of ambassadors to Madrid came from Senator Arthur H. Vandenburg, the Republican leader, from the Democratic Senator from Texas, Tom Connally, Chairman of the Senate Foreign Relations Committee, and Judge John Kee, Chairman of the House Committee on Foreign Affairs.[122] President Truman, who loathed Franco, was about to find himself obliged to capitulate before the growing Congressional demand for a rapprochement with Spain. Acheson had been threatened by McCarran that, until policy towards Spain was changed, the State Department's appropriations would

be examined with a fine-tooth comb.[123] Admitting that the 1946 resolution had failed, on 18 January 1950 Acheson stated in a widely reproduced letter to Connally that the United States would be prepared to
vote for a resolution permitting member nations to send ambassadors to
Madrid and Spain admitted to international technical agencies. Referring
to the political origins of the regime, Acheson indicated that fuller integration into western Europe, including presumably NATO, would
require political liberalization in Spain.[124] However, despite those stric-
tures, the writing was on the wall. Acheson's letter provoked enormous
delight in Madrid.[125]

When Acheson's letter was published in Spain, it was taken as proof
that the United States recognized that the Caudillo had been right all
along. At the same time, Acheson's comment on his fascist connections
was denounced as an impertinent interference in Spain's internal
affairs.[126] The fact that the controlled press should thus rebuff what was
a painful gesture for Acheson reflected Franco's hope of pushing up his
price as a military ally.[127] His confidence could be seen in a savage attack
on Britain published in *Arriba* under the pen-name Macaulay, behind
which hid the First Journalist himself. The use of a pseudonym was a
flimsy cover. The British Ambassador was fully aware of the identity of
the author.*[128]

It is to be doubted that Franco cared. He demonstrated his confidence
even further when, on 22 February 1950, he had a number of prominent
monarchists arrested in night-time swoops by the secret police and
imprisoned for 'conspiracy to restore the monarchy'.[129] There could be
no doubt now that he knew that he had come through.

* Franco's authorship was proudly claimed by the newspaper after his death – *Arriba*,
20 November 1975.

XXIII

THE SENTINEL OF THE WEST

1950–1953

THE POLITICAL atmosphere in Spain during the years of international ostracism was suffocating. The controlled press denigrated foreigners and Spaniards who called for political change as the dupes of Communism and freemasonry. The brutal repression of the regime's enemies was portrayed in heroic and moralistic terms. Behind the smokescreen of adulation of the Caudillo's even-handed patriotism, appalling living standards for the defeated working class co-existed with fat-living and corruption among the victorious elite. Cynicism pervaded this Francoist elite as memories of Civil War sacrifices receded. Varela, Yagüe, Muñoz Grandes and some of the other more austere generals protested. However, as Franco's remarks to Saliquet had indicated, austerity in the higher echelons was the exception and contrasted with the behaviour of the Caudillo and his family.[1]

Franco considered himself to be of exemplary austerity. Certainly, he did not womanize, did not smoke, drank wine in moderation at meal times and did not gamble beyond a small-scale flutter on the national lottery or when playing cards with friends – and later, on the pools. However, the entire resources, antiques and art-works, palaces and estates of the one-time royal patrimony were at the exclusive disposal of his family, a privilege of which he took full advantage particularly for hunting. The expense of his hunting and fishing expeditions was enormous. Deep-sea fishing required the year-round maintenance of the yacht *Azor* and the supply of naval escorts when he chased tuna and whales far into the Atlantic. Both hunting and fresh-water fishing involved moving large retinues around Spain. Moreover, in addition to the hidden costs of the neglect of government business, since not only the Caudillo but also several of his Ministers would be involved, there was also the effort that went into ensuring that the Caudillo had successful trips. This would involve baiting large areas of the sea over lengthy periods and feeding deer and other game at strategic points on hunting reserves. Both the 'stage-management' of Franco's hunting triumphs and his own equanimity were revealed in an incident which took place in February 1950. Franco was hunting in the mountains and anxious to shoot an especially fine stag which had disappeared. The stag was found in a particularly inaccessible part of the mountains. Franco was brought there and was 'with great difficulty, propelled, hoisted, pushed and shoved up'. The entire party was in a great hurry lest the stag might have moved on. When they reached the top, the stag was standing not far away. To the astonishment of his entourage, Franco declared that the exertion had tired him and calmly sat down and rested. After some time had elapsed, he took his rifle and shot the stag.[2]

The burdens of government and the hobbies aside, Franco's greatest concern was his only child 'Nenuca' (Carmen). The apple of his eye, she often accompanied him on his hunting jaunts. On 10 April 1950, Nenuca married a minor society playboy from Jaén, Dr Cristóbal Martínez Bordiu, soon to be Marqués de Villaverde. A satirical ditty sung in Madrid summed up the popular attitude. It ran: 'the girl wanted a husband, the mother wanted a marquis, the marquis wanted money, now all three are happy' (*La niña quería un marido/la mamá quería un marqués/el marqués quería dinero/¡ya están contentos los tres!*).[3] The preparations and the accumulation of presents were such that the press was ordered to say nothing for fear of provoking unwelcome contrasts with the famine and poverty which afflicted much of the country.[4] Those wishing to ingratiate themselves with Doña Carmen took advice from her inseparable companion, the Marquesa de Huétor,* on the most appropriate presents. The wedding was on a level of extravagance that would have taxed any European royal family. Guards of honour, military bands, hundreds of guests including all members of the cabinet, the diplomatic corps and a glittering array of aristocrats, took part in a full-scale State occasion. The ceremony, which took place in the chapel at El Pardo, was reported, but the press failed to mention the gifts. Editorial comments praising the austerity of the occasion were, however, laughably at odds with the cover-

---

* The portly Pura, Marquesa de Huétor de Santillán, was the wife of Ramón Diez de Rivera y Casares, Marqués de Huétor de Santillán, the Head of Franco's Civilian Household. Increasingly, she became the filter through which Doña Carmen, and often Franco himself, learned about the outside world. Her malicious gossip could make or break those in the closed circles of El Pardo. She was particularly hostile to the Serrano Suñer family.

age, on other pages, of the banquet offered at El Pardo for 800 people.

Popular attention was caught by the beautiful jewellery worn by the bride and by the bridegroom's recently acquired Ruritanian outfit, that of a Knight of the Holy Sepulchre, complete with sword and crested helmet. The nuptial mass was said by the Bishop of Madrid-Alcalá, Monseñor Leopoldo Eijo y Garay. The sermon was delivered by Cardinal Pla y Deniel who crowned years of outrageous adulation by suggesting that the newly-weds model their family life on that of 'the family of Nazareth' or that of 'the exemplary Christian home of the Chief of State'. The Caudillo, dressed in the ornate dress uniform (*de gran gala*) of Captain-General of the Armed Forces, gave away his daughter. Franco displayed an intensity of sentiment entirely appropriate for the wedding of his beloved daughter. His emotion, however, did not prevent him leaving the *prie-dieu* allotted to him as *padrino* of the bride in order to take the one at the side of the altar reserved for the Head of State. This gesture was interpreted by a slavishly sycophantic press as evidence of his self-sacrifice, unable even at such a moment to abandon the cares of State.[5]

Nenuca's marriage was to change Franco's life. Between 1951 and 1964, she would give Franco seven grandchildren on whom he would lavish an indulgent affection hitherto absent from his life.* Martínez Bordiu abandoned the old motorbike on which he had visited his fiancée for a series of Chrysler and Packard convertibles and was soon known by Madrid wags as the Marqués de *Vayavida* (what a life). He also lost little time in taking advantage of his link to the dictator's family to foster his business interests. Together with the Marqués de Huétor de Santillán, the head of Franco's household (*casa civil*), Martínez Bordiu made a fortune from various sources. One derived from their acquisition of the exclusive licence for importing Vespa motor-scooters from Italy, at a time when Spain had little foreign currency for imports. The machines were in a standard green and, in Madrid, his nickname changed to the Marqués de *Vespaverde*.† A so-called Villaverde clan emerged, headed by Martínez Bordiu's uncle and godfather José María Sanchiz. Soon they controlled considerable banking interests. Sanchiz made a fortune for the Villaverde clan with property speculations and import-export licences and he helped Franco buy, and then acted as administrator of, a substantial estate at

---

* All seven were born in El Pardo: María del Carmen on 26 February 1951, María de la O on 19 November 1952, Francisco on 9 December 1954, María del Mar on 6 July 1956, Cristóbal on 10 February 1958, María Aránzazu on 16 September 1962, Jaime on 8 July 1964.

† It was said by Madrid wits that VESPA stood for *Villaverde Entra Sin Pagar Aduana* (Villaverde enters without paying customs duties).

Valdefuentes, near Móstoles on the Extremadura road out of Madrid. The Marqueses de Huétor also exploited their connections with Franco. Eventually, the numerous Villaverde clan came to displace the families of Franco's brother Nicolás and sister Pilar at the Pardo.[6]

It was only after the connection with the Villaverdes was sealed that Doña Carmen gave free rein to her passion for antiques and jewellery. In this, she was egged on by the Marquesa de Huétor de Santillán who assured her that everyone in Spain with a high standard of living owed everything to the Caudillo.[7] The meanness and acquisitiveness of *La Señora* became legendary. It has been claimed that the jewellers of Madrid and Barcelona set up unofficial insurance syndicates to indemnify themselves against her visits. She was equally fond of antiques. In La Coruña and Oviedo, jewellers and antique-dealers often shut up shop when it was known that she was in town.[8] Daily contact with sycophants in search of official favour provided ample opportunities for the acquisition of desirable pieces. Unwanted gifts were exchanged for more desirable ones. Doña Carmen's courtiers, led by the Marquesa de Huétor, gave advice on acceptable items to would-be donors.[9] What Franco thought of Martínez Bordiu is not known although he never released him from the obligation to call him *Excelencia* and to use the formal *Usted* mode of address. He was concerned mainly that his daughter be happy. In general, he turned a blind eye to corruption since it kept the elite loyal to him. In any case, in 1950, he still had more important things on his mind.

By the beginning of 1950, the United States military establishment was intensifying its efforts to incorporate the fervently anti-Communist Franco into its defensive orbit. Portugal was pressing for Spanish inclusion in the North Atlantic Treaty but American policy was inhibited by fear of alienating Britain and France.[10] In the House of Commons, the Conservatives pressed for a renewal of diplomatic relations with Spain but Ernest Davies, Parliamentary Under-Secretary of State for Foreign Affairs, stated that Britain would abide by the 1946 United Nations resolution because the Franco regime remained 'as repugnant to us now' as then.[11] Truman's attitude to Franco had not changed. He declared at the end of March 1950 that he could see no difference between the USSR, Hitler's Germany and Franco's Spain since all were police states.[12] However, anti-Franco sentiments were being overtaken by bigger events.

An indignant Franco attributed British and American hostility to masonic plots. He still produced regular diatribes against freemasonry, often with anti-Semitic overtones.[13] His prejudices were confirmed by despatches from Lequerica who, telling his master what he wanted to hear, explained Truman's attitude as dictated by his need to appease American freemasons.[14] The Caudillo naïvely thought that the pseudonym Jakim Boor permitted him to give free rein to his views of freemasonry as an evil conspiracy with Communism while publicly expressing admiration for all things American. The fact that such articles on the crimes of freemasonry were published in *Arriba* was interpreted in the the *New York Times* as signifying official sanction at the highest level for the views expressed therein. The White House received thousands of telegrams of protest. Washington was fully aware that the articles were the work of Franco.[15]

Culbertson was sourly disillusioned. Worn down by Franco's sanctimonious self-satisfaction, he wrote to Acheson a despatch which reflected his disgust at Franco's failure 'to evolve in the direction of some democratic concept and toward a government which does not rest solely on the power and life expectancy of one man'. Of the Caudillo, Culbertson commented 'he thinks he knows better than anyone else what is best for Spain and the Spaniards today. He listens to what he wants to hear, shuts his mind and ear to all other'. Referring to Franco's absolute refusal to contemplate political reform, he wrote 'Franco is the kind of Spaniard who likes to get into the movie without buying a ticket.' Nevertheless, since the elimination of Franco was not a realistic policy and given the value placed on Spain by the defence establishment, Culbertson sadly concluded that the United States should work towards the return of ambassadors to Madrid.[16]

The announcement of the Soviet atomic bomb and the triumph of Mao Tse-Tung in China were followed by several espionage scandals including the cases of the senior State Department official Alger Hiss, and of Ethel and Julius Rosenberg, who were accused of belonging to the spy-ring associated with the British German-born atomic scientist Klaus Fuchs. A paranoiac belief grew that the recently triumphant United States was now threatened because the secrets of its greatest weapons had been given away.[17] The witch-hunting Senator Joseph E. McCarthy fomented the view that Communists were the cause of all America's problems. With doubts growing about the West's defensive capability, the Joint Chiefs of Staff pressed for an alliance with Spain in order to be able to use Iberia as 'the last foothold in continental Europe' without which a re-entry into a Soviet-held Europe might not be possible. At first, Truman regarded the demands of General Omar Bradley, chief of the Joint Chiefs of Staff, as politically unrealistic.[18] Within a little over a week, his views would change dramatically as a result of an event which swept away the doubts of many of Franco's liberal critics in the United States.

South Korea, under United States control since the end of the Second World War, was invaded by North Korean troops on 24 June 1950. With wild speculation about Soviet imperialist intentions rampant, the Truman administration passed from a strategy of containment to a more aggressive response to Soviet expansionism. The Caudillo offered to send troops. It was not surprising that Franco's stock was rising rapidly despite Spain's military debility. It was widely assumed that a third world war was now imminent. Certainly there was no question now of any action which might increase instability in Spain. There were still those in London who felt that military assistance against Communism should be better sought 'in ex-Nazi Germany than in still-fascist Spain' but the prevailing view in the Foreign Office that the failure of ostracism should be recognized and Spain brought back into the international community, 'despite Franco'. However, after lengthy consideration, Bevin decided that there should be no change.[19]

Franco laboured to take advantage of the international situation. He told the correspondent of the Washington *Evening Star* that he had proof that Russia was about to invade Europe with twelve paratroop divisions. France would offer no resistance and the Russians would sweep through Spain towards Gibraltar and North Africa. In contrast, Spain had half a million men who would resist if only they could be armed.[20] It is unlikely that Franco fully believed these scaremongering stories. Nevertheless, in the context of Lequerica's well-paid Spanish lobby, they worked. On 26 September 1950, with American troops committed in Korea, the Initiatives Commission of the United Nations voted to reconsider diplomatic relations with Spain. While Lequerica spread bribes to influential journalists and politicians in Washington, the Generalísimo milked the changing situation for every political advantage possible. Moreover, not all Franco's friends were on the pay-roll. The Portuguese delegate to the United Nations made an impassioned appeal for Spanish inclusion in NATO.[21] Between 25 and 27 September 1950, Dr Antonio Oliveira Salazar returned the visit which Franco had made to Portugal a year earlier.[22]

For domestic consumption, Franco underlined his strength of purpose by intensifying anti-British measures. Difficulties were created for British subjects going in and out of Gibraltar. To drive a wedge between Britain and the USA, the Generalísimo told the Italian newspaper *Roma* that there was no difference between 'the Socialist imperialism of London' and 'the Communist imperialism of Moscow'.[23] His views on Britain were developed on 11 October during an audience with the Earl of Bessborough, chairman of the Rio Tinto Corporation. On being ushered into the dictator's presence, Lord Bessborough was startled to find Franco looking at him in complete silence. Bessborough was informed by the interpreter, the Marqués de Prat, that he was expected to open the conversation. When he said that not all Englishmen shared the anti-Franco views of the Labour Government, the Caudillo launched into a disquisition on the defects of the Conservative Party and the links between Socialism and Communism. With an astonishing blend of patronizing complacency and ignorance of the social realities of both post-Beveridge Britain and post-Civil War Spain, he said that the rich in England must make sacrifices for the benefit of the working classes, just as the rich were doing in Spain. Franco also stressed that Spain and England should join in the defence of their common western civilization. When Bessborough said that rapprochement would be helped if the Spanish press toned down its attacks on Britain, the Caudillo replied disingenuously that the press in Spain was free to say what it wished and that he was powerless to intervene unless the head of a friendly state were insulted. Bessborough, who was an admirer of Franco, left El Pardo stunned by the Caudillo's total lack of acquaintance with the outside world.[24]

On 31 October 1950, the Special Ad Hoc Political Committee meeting at Lake Success, New York, voted to drop the December 1946 resolution on withdrawal of ambassadors.[25] On 2 November, Truman announced that it would be 'a long, long time' before the United States named an ambassador to Franco – though it was to be done within weeks. The remark was not published in Spain but was broadcast on the BBC Spanish Service. Paradoxically, it delighted the Falangist hard-liners who feared that Franco would dump them once the ostracism was over.[26] On 4 November, the General Assembly of the United Nations, meeting at Flushing Meadow, voted to authorize the return of ambassadors to Madrid by thirty-eight votes to ten, with twelve abstentions. Moreover, on the grounds that recognition of a regime implied no judgement of its internal policies, Spain was admitted to its Food and Agriculture Organization. Britain and France abstained while the United States voted in favour.[27] Despite the fact that the condemnatory preamble to the original resolution remained in place, the mood in Madrid was euphoric and the decision was loudly proclaimed as a 'Spanish Victory' and Franco himself portrayed it as full-scale international endorsement.[28]

Lequerica telegrammed Martín Artajo with an account of the expenses he had incurred in fees and bribes, urgently requesting $237,000 for retainers to politicians and journalists. To pave the way for the 4 November vote, he had distributed $44,350 and had to borrow a further $57,500, all of which he considered to have been well spent.[29] Some months earlier, Theodore Achilles, the Director of the US State Department's Office of western European Affairs had predicted to a senior Spanish official that 'Lequerica would undoubtedly claim and probably receive credit for the eventual return of Ambassadors but I thought it only fair that his government should know that this and any other step in improving relations would be taken despite Lequerica's efforts rather than because of them.'[30] Lequerica did indeed claim the credit which was owed less to sponsored junketing than to Spain's strategic value and to the consistency of Franco's anti-Communism. In the unlikely event of the Spanish official passing on Achilles' prediction, it would have carried no weight with Franco. In spending the money, Lequerica, whose shameless sycophancy was usually highly successful with the Caudillo, was merely confirming his master's belief that 'anyone can be bought'.[31]

Franco was not satisfied with the victory and relentlessly pushed for more. Aware of the domestic political value of interpreting hostility against Franco as the international persecution of Spain, he presented the November 1950 resolution as merely the overdue rectification of an injustice. He now expected substantial compensation for the economic hardships of recent years, which he blamed on the international ostracism.[32] In fact, economic distress derived less from the lack of ambassadors than from Franco's refusal to countenance political reform. That had cut off Spain from international aid while Suanzes' policies of autarky and high exchange rates had made it more difficult to repair the devastation of the Civil War.

As part of his endless manoeuvring in search of indemnification for the injustice he had suffered, Franco began to press for the return of Gibraltar. It was a characteristic way of signalling that his readmission to international society was not taking place on suffrance but represented an act of condescension on his part. He was emboldened to make difficulties by Egyptian demands for the withdrawal of British troops from the Suez Canal and hoped to enlist the support of anti-colonial sentiment in the United States. The recovery of Gibraltar, as the last part of Spain in alien hands, was an ambition understandably cherished by a Caudillo always anxious to emulate the great warrior leaders of Spanish history. Just as the Catholic Kings had expelled the Moors, he would like to expel the infidel freemasons of perfidious Albion. The campaign of 1950 was mounted on Franco's personal initiative, in collusion with the Falangist leader Raimundo Fernández Cuesta and against the advice of Martín Artajo.[33] The campaign built up momentum through December and took the form of articles placed in the controlled press, including one by Carrero Blanco writing as 'Juan de la Cosa', another by the Caudillo himself, a press interview with Franco, virulent articles by regime hacks and the organization of student demonstrations against Britain.[34]

Franco did want to recover Gibraltar but knew that he would not get it with student agitation. The demonstrations were a bellicose gesture to draw Falangist attention away from his obsequiousness to the United States. He would get his economic reward although it would not be given to right any injustices. The conviction in Washington that Franco's Army needed rearming accounted for the authorization by the Truman administration, on 16 November 1950, of a $62,500,000 loan to Spain.*[35] On the following day, Truman secretly agreed to the appointment of an Ambassador to Franco.[36] At the end of the month, 200,000 Chinese troops joined in the Korean War and pushed the United Nations troops back into the south. To begin the process of bringing Spain formally into the anti-Soviet bloc, the appointment of Stanton Griffis as the new American Ambassador to Madrid was made public on 27 December.[37]

It was hailed in Spain as another victory for Franco. In fact, there had never been anything resembling a full-scale *cerco internacional* (international siege) and, when Franco was welcomed to the bosom of the western allies, it was because they wanted him not because he had manipulated them. Nevertheless, his euphoria shone through his end-of-year message to the people on 31 December 1950. The broadcast was a virtuoso set of variations on the theme of 'I told you so' in which 'Franco' and 'Spain' were again indistinguishable. Ignoring the catastrophic state of the economy, he claimed that great social and economic advances had been achieved against the obstacle of an international conspiracy to keep Spain weak. With notable exaggeration, he listed the achievements of autarky and, as so often before, the providential gift of unnamed mineral discoveries which would soon transform the Spanish economy. He then asked a rhetorical question in which historical ignorance was mixed with megalomania. 'What Spanish regime, throughout history, has been more productive in its tasks and created for the Nation, in any area, a wealth comparable to that which we have created so far?'

---

* Spain thus became the only country in Europe to receive aid through the European Co-operation Administration without belonging to the Marshall Plan.

Commenting ironically on the Cold War revaluation of the defeated Axis powers, Franco inadvertently revealed his undiminished sympathies for Fascist Italy, Nazi Germany and imperial Japan. 'Who could have foreseen that those German armies which victoriously burst onto Europe with unstoppable impetus would soon have to retrace their steps and find themselves captive and at the mercy of their enemies? How could it have been foreseen that the Imperial Italy forged in North Africa would so quickly succumb under the crisis of the last war? Who could predict that yesterday's powerful victors in the Pacific would immediately see themselves opposed by the same peoples that they had liberated? How can one explain that, after so few years, it is now deemed necessary to raise up again in Europe and Asia the two countries which were destroyed with such vicious brutality?'[38]

Franco had come out of the years of ostracism with his domestic power undisputed. By exploiting international moral opprobrium as if it were an ruthless siege of Spain, bent on unleashing the horrors of civil war, he had consolidated his popular support considerably. He had tamed the monarchist opposition, crushed the guerrilla resistance, and seen the Church and the Army become more Francoist in their loyalties. A fearsome apparatus of repression remained in place. As he had told Varela less than twelve months earlier, he had no intention of liberalizing his rule. His confidence was shown when he gave his Foreign Minister a small lesson in the exercise of arbitrary power. Martín Artajo now hoped to see a moderate Catholic named as Spanish Ambassador to the United States. Anticipating this, the slimy Lequerica wrote a sinuous plea to the Generalísimo implying that Martín Artajo's plans were less than loyal to the Caudillo. He argued that, if he himself were not permitted to present his credentials as ambassador, it would be said that 'people were simply burnt up in the service of Your Excellency and of the regime. Real Francoists, it would be said, are fine for bad times and emergencies, but when things improve, they are sent home precisely because they are Francoists. And those who are cold, neutral or even hostile to Franco will return to abandon or betray the regime to the United States.'[39]

The Caudillo named Lequerica Ambassador. The choice reflected not gratitude for services rendered but a desire to settle the grudge Franco had borne since Truman's rejection of Lequerica as Ambassador in 1945. Franco was demonstrating to Truman and Acheson that they, not he, had changed. At the same time, as in so many other choices of key personnel, Franco saw in Lequerica someone who had nowhere else to go and whose loyalty, whether sincere or cynical, could therefore be relied upon. When Lequerica presented his credentials on 17 January 1951, President Truman manifested his distaste for the Caudillo's representative by barely shaking hands and despatching him in record time.

The new American Ambassador, the sixty-four year-old Stanton Griffis arrived in Spain on 19 January 1951. He was not a professional diplomat but an investment banker with interests in the entertainment business, including Paramount Pictures and the Madison Square Garden boxing arena. He had spent some time in Spain in early 1943 as an emissary of Colonel William Donovan of the Office of Strategic Services and had been hankering to be US Ambassador to Spain since early 1948.[40] Previously posted in Poland and Argentina, he had been criticized by the American press for praising Perón.[41] He would soon make equally warm remarks about Franco. The exchange of ambassadors was the beginning of a process which would see Spain admitted to UNESCO on 17 November 1952; sign a Concordat with the Vatican on 27 August 1953; sign the Pact of Madrid with the United States on 26 September 1953, and be admitted to the United Nations in December 1955.

Keen to stress that the British, like the Americans, had changed while he remained immutable, he made for London, as for Washington, a provocative choice of Ambassador. He could easily have upgraded the status of the current *Chargé*, the Duque de San Lúcar la Mayor, who was a competent diplomat, a monarchist and liked in establishment circles in London. The last two, if not the first, of those reasons accounted for Franco's determination to replace him. At this time, the Caudillo was venting his spleen against England through the radio broadcasts of Carrero Blanco who, as 'Juan de la Cosa', ranted against the infiltration of the British establishment by Marxists and freemasons. It had been assumed in the Spanish Foreign Ministry that either San Lúcar or Sangróniz, now Ambassador to Italy, would come to London.[42] When Franco proposed sending Fernando María Castiella, his Ambassador to Peru, the *agrément* was refused. As a volunteer in the *División Azul*, Castiella had sworn allegiance to Hitler and had been awarded the Iron Cross. He was the co-author, with José María de Areilza, of *Reivindicaciones de España*, a vehement statement of Spanish imperial aspirations during the Second World War.

When he was rejected, Franco immediately proposed the brother of José Antonio Primo de Rivera, the playboy Miguel, once Falangist Minister of Agriculture and now *Alcalde* (mayor) of Jérez. He was banking successfully on London not refusing the *agrément* twice and was able thereby to oblige the British to accept a symbol of the unalloyed

Falangism that they had ostracized for five years.[43] In contrast, the new British representative, Sir John Balfour, was a career diplomat. A brilliant linguist, familiar with the Hispanic world, he had served in the Madrid Embassy earlier in his career and came now from Argentina where he had been Ambassador.[44]

Although the British military establishment, like its American counterpart, had concluded that there were significant advantages in Spanish membership of the Atlantic Pact, the bulk of European political opinion remained unremittingly hostile to Franco and would ensure that Spain would never enter NATO while he was in power. Accordingly, encouraged by a policy memorandum from Carrero Blanco, Franco inclined still more to seek a bilateral relationship with the United States.[45] With that in the bank, European disdain would count for little. On 13 February 1951, he made a statement to the Hearst press chain aimed at demonstrating to America Spain's superiority as an ally as against Britain or France. His fundamental prejudices gleamed brightly through the fog of obsequiousness. He boasted that, unlike all other European governments, he had eliminated Communism from his country and alleged that for this reason Britain, France and America had ostracised Spain – something which would make sense only if all three countries, as he secretly thought, were infiltrated by Communists and freemasons. Despite American mistakes, however, he graciously admitted to feeling admiration for the USA's greatness, power, talent for organization, progress and industrial dominance. Making a virtue of necessity, he was cool about NATO, knowing an invitation to join was not likely to materialize. He stated that 'altogether less complicated, much better and more satisfactory would be a direct arrangement for collaboration with North America'.[46]

Such flannel convinced only those in the United States who wanted to be convinced. He exploited his exclusion from NATO for domestic purposes, re-running 'the international siege' propaganda with the Americans left out. Spaniards could thus feel good about the links with the USA and resentful about the British and French who, they were told, meanly refused to share American aid. Franco retaliated against British opposition to his membership of NATO by arranging demonstrations in Barcelona on 19 February in favour of the return of Gibraltar. He decreed that 4 August of every year would be the *Día de Gibraltar*, with rallies to be held by the Falange's Youth Front to mark 'the pain suffered by Spain as a result of a foreign occupation'.[47]

When Stanton Griffis presented his credentials on 1 March, a spectacular ceremonial display was mounted to impress the Ambassador and the

American press with the standing of the Caudillo and his friendship for the United States. Three red and gold eighteenth-century coaches each drawn by six horses carried the Ambassador and his staff to the Palacio de Oriente. They were escorted by two hundred lancers of Franco's Moorish Guard in red tunics and white cloaks, astride black horses, a platoon of buglers in orange uniforms and white cloaks, mounted on white horses, and a squadron of armed police. A delighted Griffis acknowledged the cheers of the crowd which lined the streets. At the Palacio de Oriente, the procession was greeted by a full military band and a large military and diplomatic reception committee. Griffis was then treated to a private audience with the Caudillo and Martín Artajo.[48] The impact of the lavish reception for Griffis was demonstrated on the following day when the eager Ambassador paid fulsome public tribute to the Generalísimo, to the anxiety of Balfour. Knowing Griffis from Buenos Aires, Balfour feared that he would be enmeshed by Franco in his schemes to play off British and American policy on Spain.[49] Significantly, when Balfour presented his credentials two weeks later, it was a much more low-key affair. He cheerfully described himself to Griffis as 'an ambassador in the doghouse'.[50]

Confident that the American connection was in the bag, a euphoric Franco now tried to get the British where he wanted them. He was led to believe by the naïve Miguel Primo de Rivera that a planted question by a Conservative MP in the House of Commons would lead to the new Parliamentary Under-Secretary at the Foreign Office, Ernest Davies, making a pro-Spanish statement, along the lines of 'let us first consolidate our friendship and then we can talk about military alliances'.[51] The Caudillo was convinced that Whitehall would be anxious to appease him after his recently unveiled capacity to mount at will 'popular' demonstrations for the return of Gibraltar. In fact, the choice of a Falangist ambassador and the playing of the Gibraltar card had exactly the opposite effect. The previously choreographed incident took place after midnight on 20 February. Peter Smithers, Conservative MP for Winchester, made a long speech about British strategic interests, the theme of which was that 'whatever Government there is in Spain, the important thing is that it should be on our side'. He ended by asking 'what the Government propose to do to enable the Spanish people to play their part in the great effort which is being made by western Europe to defend itself?'

Davies replied that the inclusion of Spain in NATO would not benefit western defence since equipping Spanish forces would hinder the rearmament of the existing signatories of the Treaty. Politically, 'the moral basis of NATO might be weakened rather than strengthened by the inclusion of forces as opposed to the democratic way of life as is Communism itself'. Franco was outraged by Davies' 'injurious' quotation of the preamble to the Atlantic Treaty '"to safeguard the freedom, common heritage, and civilization of their peoples, founded on the principles of democracy, individual liberty and the rule of law". Those things are absent from Franco Spain at the present time.' The Caudillo regarded the inclusion of Portugal, while he was excluded, as proof of the democracies' outright hypocrisy and a humiliation for Spain.[52]

Whatever setbacks his policy towards Britain might suffer, Franco could take consolation in the strengthening of links with the USA. By April 1951, the Commander-in-Chief of US Naval Forces in the Eastern Atlantic and Mediterranean had been instructed by the US Joint Chiefs of Staff to make contact with the appropriate Spanish military authorities to lay down the basis for future co-operation and for the establishment of American air and naval bases on Spanish territory. In view of Spain's combination of military weakness and strategic importance, and given the political objections to her inclusion in NATO, the British Chiefs of Staff agreed that the only solution was a separate US-Spanish bilateral agreement. The continuing repugnance provoked by Franco both within the Labour Party and in many European countries made it impossible for the US policy to be endorsed openly.[53]

In bleak contrast to Franco's complacent statements about his foreign and domestic achievements, by the late 1940s the deficiencies of his rule had become starkly obvious. Spain paid the economic price for Franco's survival. Isolation from the international economy and exclusion from the Marshall Plan did less damage to the Spanish economy than an arti-ficially high exchange rate for the peseta, maintained for reasons of honour. A succession of mediocre ministers in the economic and agricul-tural fields had made little progress in recovery from the devastation of the Civil War. In this regard, the British *Chargé* had written to London 'Franco desperately needs a Minister of Finance who is capable of work and not stone deaf!'[54] Despite his pride in his own economic expertise, the Caudillo had little real interest in the subject and nothing to offer in terms of policies. However, food shortages, inflation, growing internal pressure for industrial development and disturbing signs of reviving labour militancy were inclining him reluctantly to contemplate the modi-fication of autarky.

Per capita meat consumption in Spain in 1950 was only half of what it had been in 1926 and bread consumption only half of what it had been in 1936. Even according to massaged official statistics, prices had risen twice as fast as working-class wages since the Civil War. In any case, working-class families had to buy food on the black market where prices were more than double the official rate.[55] Agricultural inefficiency meant that Spain had to import food with her dwindling foreign currency reserves. Raw materials and energy were also in short supply. The logic of the situation demanded Spain's integration into the international economy and more American credit. However, the liberalization insisted upon by the Americans, and the worldwide rise in raw material prices stimulated by the Korean War, led to inflation in the cost of basic necessities. Bread, potatoes and rice had rocketed in price. Power cuts were leaving factories idle and effectively slashing workers' wages.

For Franco, labour unrest was simply a law and order problem generated by Communist agitators. Nevertheless, he was forced to react to reports of growing tension in working class districts as a consequence of worsening living conditions. Thus, he made a surprisingly realistic speech to the National Congress of Workers on 11 March 1951. The Caudillo's sombre words made a nonsense of most of his previous pronouncements on the economy. It is significant that such harsh realism should be employed when the audience was the working class. 'The national economy has its limitations and its demands. It is only possible to share out what is produced.' 'We must wipe from the conscience of the Spaniards the puerile mistake that Spain is a rich country, rich in natural products. No sir, there are rich nations and poor nations and Spain is not one of the rich ones.'[56]

On the following day, the crisis in working-class living standards provoked a serious challenge to the regime in Barcelona. Despite the fact that social tensions in the city had been at boiling point for some time, the attention of the hated Civil Governor of Barcelona, the Falangist Eduardo Baeza Alegría, was distracted by his relationship with a local cabaret artist. Baeza Alegría had been appointed by the Caudillo personally, as a reward for an enthusiastic reception that he had organized for him in 1946 on a visit to Zaragoza. Confident of the Caudillo's favour, and without considering the impact on local living costs, Baeza had blithely authorized an increase in tram fares in February. This had led at the beginning of March to a boycott of public transport and the stoning of trams. Baeza blamed the crisis on 'professional agitators at the service of political ideologies of sad memory'.[57]

However, by 12 March, the boycott had escalated into a general strike involving 300,000 workers. The Civil Governor's facile theories were undermined by the participation of local Falangists along with activists of the Catholic workers' organization, HOAC – the *Hermandad Obrera de Acción Católica* (the Workers Fraternity of Catholic Action) – and members of the middle class. The British Embassy even claimed to have 'irrefutable evidence' that the strike was initially planned by hard-line radicals of the group of veteran Falangists known since 1949 as the '*Vieja Guardia*'. Baeza Alegría requested troops when some cars and buses were overturned. That firmness and his explanation that the unrest was the work of agitators appealed to Franco's basic prejudices. Nevertheless, when Luis Galinsoga, director of *La Vanguardia Española*, and one of Franco's most sycophantic admirers, telephoned Pacón with alarmist reports that events were getting out of hand, the Caudillo reacted quickly. He rang the Minister of the Interior, Blas Pérez, and ordered that 'public order must be maintained without panic on the part of the forces of order. If the Civil Governor feels unable to maintain public tranquillity, he should delegate control to the Captain-General.' Franco ordered three destroyers and a minesweeper to the port of Barcelona and marines marched through the streets. However, the austere monarchist General Juan Bautista Sánchez, Captain-General of Barcelona since 1949, believed that it was wrong for the Army to repress disorder which had been provoked by the Civil Governor's incompetence. One of the few generals whom Franco could not manipulate easily, Bautista Sánchez prevented large-scale bloodshed by calmly confining the garrison to barracks.[58]

For losing control of the situation, Baeza was dismissed on 17 March. The first choice to replace him was the Conde de Mayalde, once Ambassador in Berlin. On asking Blas Pérez if he would be able to release supplies of bread and olive oil, Mayalde was told to forget that sort of approach and to make use of the Civil Guard. Mayalde suggested therefore that a more appropriate choice would be a soldier.[59] Baeza was replaced by the hard-line General Felipe Acedo Colunga, a military prosecutor, notorious for his part in the trial of the Socialist leader Julián Besteiro in 1939.[60]

Franco was fortunate that, in the deepening atmosphere of the Cold War, his repressive policies were taken as testimony to his fierce anti-Communism. Fortunately too, his repressive labour legislation, fostering as it did high profit margins, made Spain attractive to foreign investors. Clearly, labour dissent and its repression were not issues of the highest priority for Stanton Griffis. On 14 March 1951, in the midst of the crisis in Barcelona, Franco received the American Ambassador who had requested a special audience to discuss 'a specific problem' which President Truman had asked him to try to solve. The 'problem', which Griffis said could undermine Hispano-American relations, was the religious big-otry which an outraged Truman saw in Spanish discrimination against non-Catholic religious denominations. Protestant funerals, for example, were regularly prohibited by officious Civil Governors.[61] Protestant prayer meetings were often broken up by the police. Griffis pointed out that Truman was about to set the budget for the next year, in which credits for economic and military aid to Spain would be included, and begged Franco to make a gesture such as instructing Civil Governors to respect the rights theoretically enshrined in the *Fuero de los Españoles*. Franco blamed intransigent clerics and undertook to raise the matter at the next cabinet meeting, implying that he was as subject to cabinet decisions as any democratic head of government. When Griffis left Spain ten months later, the scale of official religious intolerance would not have changed at all.

Having broached the religious matter, Griffis then asked Franco directly if he was prepared to join NATO. The Caudillo replied that he thought a bilateral pact with the United States more appropriate. Griffis, aware of the views of the United States' other allies, replied that separate negotiations with Spain would be difficult. He then asked Franco if he would be prepared to send them Spanish troops to fight with American and other NATO forces beyond the Pyrenees. After some prevarication, and pressure from Griffis, Franco said that he would collaborate in a wider defence effort. Griffis then pushed even further, asking Franco outright if, in the event of conversations between the Spanish and American general staffs, Spain might put her air, land and naval bases at the disposal of the USA. Franco replied that the two world wars had shown that all nations belonged to great coalitions and, that being the case, the military bases would be made available to the western allies although they would remain Spanish.[62]

On 23 April 1951, 250,000 men began a forty-eight hour strike in the shipyards, steelworks and mines of the Basque Country. Falangists and members of HOAC joined in alongside leftists and Basque Nationalists. The regime denounced the strike as the work of foreign agitators. The employers, aware of the problem of the cost of living, refused to imple-ment official orders for mass sackings. Despite savage police beatings of strike leaders, many of whom were rounded up and taken to a concen-tration camp near Vitoria, industrial action continued sporadically into May.[63] The high cost of living had been discussed at a cabinet meeting on 5 April. Franco blamed the economic situation on Spain's foreign enemies and dismissed the labour discontent arising from it as mutiny.[64]

On 12 May, speaking to the *Hermandad de Labradores y Ganaderos* (the fraternity of farmers and stock-breeders) the Falangist rural syndicate, he declared that 'the strike is a crime' and related internal unrest to the siege mounted by international Communism and freemasonry.

To counter the blow to his prestige, Franco tried to divert attention from the strikes with a foreign policy circus. At the beginning of May, he claimed to have documents proving that false news was broadcast by the BBC on masonic orders.[65] Now a press campaign including contributions by Carrero Blanco, writing under the pseudonym 'Ginés de Buitrago', denounced the strikes as the work of French and British freemasons.[66] The Caudillo kept up anti-British propaganda over the Gibraltar issue as a crude diversion from domestic tensions.[67]

It was indicative of the growing strength of Franco's position that, on 10 July 1951, Don Juan wrote a letter in which he threw away years of opposition to the regime. Aware of the growing closeness between the Caudillo and Washington, Don Juan clearly felt the need to mend his fences. He suggested that the regime had suffered considerable erosion as a result of the recent strikes, which he rightly attributed to the economic situation and government corruption. He suggested a negotiated transition to the monarchy, offering Franco a way to consolidate the principles to which he was committed within the stability of a monarchy which would unite all Spaniards. While he renounced none of his rights, Don Juan was abandoning his past championship of a democratic monarchy and accepting the Movimiento. Franco's reply, which he disdainfully delayed for two months, ignored the offer and denied that there was any administrative corruption in Spain or that the economic situation was anything but entirely favourable.[68]

Ultimately, Franco's efforts at driving a wedge between the United States and its British and French allies had less impact than simple geopolitical realities. On 14 June 1951, Ambassador William D. Pawley, political adviser to General Omar Bradley, visited Madrid and was received by Martín Artajo and the Director of American Policy at the Spanish Ministry of Foreign Affairs, the Marqués de Prat de Nantouillet. Just before leaving for Madrid, Pawley had spent some hours in Paris with General Eisenhower, NATO C-in-C, discussing possibile military aid for Spain and Spanish participation in western defence. Pawley openly declared that the Pentagon and the Combined General Staff favoured Spanish rearmament and blamed political obstacles for delay. Martín Artajo responded with an indication that Franco would permit the Spanish Army, armed and equipped by the USA, to fight beyond the Pyrenees.

Pawley then asked, on behalf of Eisenhower, about Spain's position if the USA could overcome British and French opposition to Spanish membership of NATO. Artajo replied with the standard line that a bilateral Hispano-American treaty would be better.[69]

Eleven days later, Griffis informed Artajo that a US military mission could shortly be sent to Spain to negotiate a bilateral pact. Within twenty-four hours, Franco responded with an agreement in principle.[70] In fact, the British, like the French, Scandinavian, Dutch, Belgian and Italian governments, remained opposed to the inclusion of Spain in western defence. Acheson, Marshall, General Omar Bradley and Admiral Sherman discussed the views of the Europeans on 10 July and concluded that military necessity outweighed political sentiment.[71]

Truman remained hostile to Franco, telling Admiral Sherman, now the Chief of Naval Operations, 'I don't like Franco and I never will but I won't let my personal feelings override the convictions of you military men'. So, with Sherman arguing forcefully in favour of a military alliance with Spain, Truman reluctantly acquiesced.[72] On 13 July, Franco received a group of Senators from the Senate Foreign Relations Committee. His quiet style and lack of bombast impressed them enormously and convinced them that he was not like 'typical European and Latin American dictators'. They were delighted to hear Franco say that he would let Spanish troops fight beyond the Pyrenees. They left convinced of Griffis's opinion that the Caudillo should not be replaced by Don Juan.[73] On 16 July 1951, Sherman and officers of his general staff unexpectedly visited the Caudillo for what turned out to be preliminary discussions about the leasing of bases in Spain.[74]

Sherman went bluntly to the point, listing American needs in terms of air bases and anchorage facilities for aircraft-carriers. Spanish airfields would have to made capable of handling high-powered long-distance heavy bombers. The United States would foot the bill. Sherman wanted from Franco agreement in principle: military and economic missions would follow to work out the details. Franco must have been delighted. The long wait was over. Now he was being courted by the most powerful nation on earth and he had little difficulty in convincing himself that it would be a dialogue of equals. His ingratiating response to Sherman cunningly suggested that any delay might diminish the efficacy of the proposed co-operation. The implication was that American aid should be abundant and fast.

He told Sherman that the cession of bases to the USA would provoke an immediate attack from the Soviet Air Force and claimed that the

Spanish forces needed to be brought to a point at which they could resist the Russians. Giving a hint that his price would be high, he said that it was one thing to collaborate in peacetime, another to seek alliances in wartime. Just how high the price might be was indicated when Franco revealed that the Spanish armed forces had no radar and were short of aircraft, heavy tanks, anti-aircraft and anti-tank equipment. While Sherman was mainly concerned to see Spain's airports adapted for powerful bombers and her harbours made ready to deal with American aircraft-carriers, Franco was keen to extend the terms of American assistance. He told Sherman that there was little point in the USA seeking military collaboration if Spaniards did not have enough to eat. In terms that recalled his offers of belligerence to Hitler, Franco pointed out that Spain had insufficient stocks of fuel, wheat and other commodities to be able to go to war. Sherman promised that, on his return to Washington, the General Staff and the Department of Defence would seek credits from Congress.

Franco's desire to squeeze the highest price possible was overridden by his feverish anxiety to clinch a deal. When Sherman asked him for a date on which the military mission could come to Spain to start work, he replied 'Immediately. From the 20th of this month, since tomorrow is the eve of a fiesta. The 18th is the fiesta and on the 19th I am changing my cabinet but from the 20th onwards they can come anytime.' When Griffis asked whether Spanish officers left Madrid in summer, Franco replied 'when there are urgent matters to resolve, there are neither summers nor vacations for Spanish officers'. In a blatant attempt to increase his value in the eyes of the Americans, Franco expressed worries about the French.[75] The press hailed the Caudillo as the valued ally of the United States while reporting British Labour hostility.[76] The way was open to detailed negotiations and Franco had made it clear that everything was negotiable. It was now up to Washington to decide what it wanted and how much it was prepared to pay. Within a month, high-powered American military and economic study groups were in Spain.

Two days after the Sherman interview, Franco remodelled his cabinet. On 18 July, his Ministers attended the Caudillo's annual reception for the diplomatic corps at the gracious royal palace of La Granja near Segovia. On returning home, some of them found letters informing them that they had been replaced in the reshuffle to be announced on the next morning.* Like its predecessor, the new team was a response to changed

* Simultaneously timid and cruel, Franco was rarely able to tell Ministers to their faces that they were being replaced. It was common for them to learn of their fate from a letter sent by motorcycle despatch rider or even from the morning newspapers.

international circumstances. Since May, Martín Artajo had been pressing Franco to include more moderate Catholics. Franco took little notice of Artajo. The only new Catholic face that Franco wanted to see was Fernando María de Castiella as Minister of Education. However, to the astonishment of both Artajo and the Caudillo, Castiella refused and Franco turned instead to the deeply Catholic Joaquín Ruiz Giménez. Castiella was sent to Rome as Ambassador to the Holy See to negotiate the Concordat, a difficult task given the Vatican's desire to see Franco replaced by a moderate monarchy under Don Juan.[77]

The only political sense in which this was the more liberal government that the Americans had hoped for was in the appointment of Ruiz Giménez as Minister of Education. Otherwise, in the context of the Korean War, the Caudillo felt confident enough to reassert the Falangist tone of his regime. The Ministry of War went to General Agustín Muñoz Grandes who had led the Blue Division against the Russians and been decorated by Hitler with the Iron Cross. Now he would be responsible for negotiating the military agreement with the Americans. Similarly, the newly created Ministry of Information and Tourism, which would be responsible for selling that agreement to the nation, was given to Gabriel Arias Salgado, who for much of the Second World War had run the controlled press in the interests of the Third Reich. Neither had been chosen just for their impeccable anti-Communist credentials, since all Franco's ministers were anti-Communists, but as a further reminder to the Americans that they and not the Caudillo had changed.

It had been rumoured earlier in the year that the next cabinet would be the one to hand over power to Don Juan.[78] By reinforcing the presence of Falangists, Franco was putting the lid on that possibility. Moreover, the political obsolescence of the Falange meant that Franco could count absolutely on the loyalty of those who had nowhere else to go. He knew that by making Falangists accomplices in the surrender of sovereignty to the United States he could diminish any possible nationalist backlash. The tame Raimundo Fernández Cuesta was brought back as Minister-Secretary General of the *Movimiento* and the irascible José Antonio Girón remained as Minister of Labour. With the same bare-faced cheek with which he had peddled his two and three war theories in the war, he was embarking on a quest to become an ally of the world's most powerful democracy with a team dominated by virulent enemies of liberalism.[79]

It was surprising, in some respects, that Carrero Blanco remained a fixture and was, indeed, promoted to ministerial rank, 'to save me' as Franco put it, 'having to tell you every Friday what went on in cabinet meetings'.[80] In the late autumn of 1950, there had been rumours that marital difficulties had caused Carrero Blanco to fall from favour in El Pardo. Doña Carmen was adamant about such matters. A reunion with Lorenzo Martínez Fuset and his wife during a visit by Franco and Doña Carmen to the Canary Islands in late October 1950 led to speculation in informed circles that the ex-juridical assessor was about to return from obscurity and take over from Carrero. But Martínez Fuset refused to go back on his decision not to return to politics.[81] And Carrero managed to patch up his marital difficulties. It was said that he had been helped by a young Catholic law professor called Laureano López Rodó. In the new cabinet composed, like that of 1945, of men of unquestioning loyalty to Franco, the most faithful of them all, Carrero Blanco, now free of moral taint, became chief of the political general staff.

If the cabinet was backward-looking in political terms, in the economic sphere, it marked one of the major turning points of the regime. With the Americans pressing for economic liberalization, Juan Antonio Suanzes, Franco's lifelong friend and the architect of autarky, was dropped as Minister of Industry and Commerce. He was replaced by two Ministers, at Commerce the quick-witted economist Manuel Arburúa and at Industry Joaquín Planell, an artillery general who had fought at Alhucemas in 1925, had been military attaché in Washington in the 1930s and until 1951 was vice-president of INI. The Caudillo revealed his lingering commitment to autarky by keeping on Suanzes as President of INI where he still had influence within economic policy as a whole. The seventy-three year-old Joaquín Benjumea, of whom it was said that he was 'tired, deaf and ill and has begged to be allowed to resign for the last two years', was finally relieved of the Ministry of Finance* which he had held since 1939.[82] He was replaced by Francisco Gómez y Llano, a grey functionary.

The new cabinet was to make the first tentative steps towards opening up the economy to external market forces. Franco had to make himself more acceptable to the United Nations but could not dismantle his dictatorship without committing political suicide. The price to be paid for American support was the sacrifice of autarky. The ultimate rewards would be massive: in the short term, American friendship; in the long, economic growth.[83] In that sense, it saw the beginning of a growing distance between Franco and his regime. The time was rapidly approaching when highly trained technocrats rather than old military chums would

---

* Benjumea was rewarded with the post of Governor of the Bank of Spain which he held until his death in 1963 aged eighty-five.

be required to run the economy. It was getting beyond Franco's comprehension. His bewilderment coincided, especially after 1953 and the alliance with the United States, with a feeling that he deserved, and could
risk, a period of rest.

In London and Paris, it was believed that, by negotiating with Franco,
the Americans undermined the moral superiority of the western block.
The French, in particular, feared that an alliance with Spain meant that,
in the event of Soviet attack, the Americans would abandon France and
dig in behind the Pyrenees. Franco himself was quick to seize the opportunity to play off the western allies against one another. In August 1951,
he gave an unashamedly cynical interview to *Newsweek*. Skilfully couched
in anti-colonialist terms which he hoped would appeal to an American
audience, his remarks implied that the reactionary imperialist prejudices
of the British Right and the dangerous passions of the British Socialists
were standing in the way of the crucial anti-Communist alliance between
Spain and the United States. His American readers might well have been
puzzled by his claim that 'because we have a fifteen-year lead in confronting the political, social and economic problems of our time, we are
nearer to solving them than all the other European countries'.[84]

With Labour in power in Britain, Franco could insinuate to the Americans that Britain's leaders were but a step away from full-blown Communism. However, he was delighted when the Conservatives returned to
power in Britain in October 1951. Having both Atlantic powers interested
in his strategic contribution, his position was immeasurably more secure.
The hypocrisy of the pro-Americanism expressed to *Newsweek* was
revealed on 7 November 1951, when he once more granted an audience
to the Earl of Bessborough who brought an informal message from Sir
Anthony Eden, once more Foreign Secretary, to the effect that he looked
forward to maintaining correct and friendly relations with Spain. The
Caudillo beamed with delight and said that Spain and England were old
friends.* He commented conspiratorially that it was a great pity that the
United States had no settled policy. When Bessborough asked him if
he thought that Truman and Acheson were now making a successful
contribution to world affairs, he sniggered and said that they must be
getting tired of being a laughing-stock.[85]

* Like many others, Bessborough was struck by Franco's quietly amiable manner: 'I
reflected how unlike the Caudillo is to one's idea of the typical dictator. He speaks so
simply, courteously and naturally and in an unaffected and quiet voice. The mind of the
dictator only appears in his evidently complete conviction that every opinion he expresses
is incontrovertible and the last word on the subject; but in justice it should be added that
he listens with great patience and good humour to what is said to him.'

If this deep contempt for the United States regularly bobbed to the surface, so too there were severe limits to any desire on Franco's part to mend fences with London. Secure now with the Americans, his main priority with the British was the Gibraltar issue. In late November 1951, the Caudillo gave an interview to the *Sunday Times*. In terms of apparent common sense, he claimed that the days were long gone when Britain needed Gibraltar as an invulnerable nest for its fleet. Britain could no longer go it alone but now was part of wider defence associations. He offered a rent-back scheme whereby, after being returned to Spanish sovereignty, Gibraltar could remain as a free port and as a British naval base.[86] The British were not interested, not least in the light of the fate of the contractual arrangements which underlay their presence at Suez.[87] The Minister of Education, Joaquín Ruiz Giménez, told the British Ambassador that the Gibraltar agitation was meant as a sop to 'neutralist' opinion in Spain which was hostile to the negotiations with the United States. Pamphlets also circulated in Madrid accusing Franco of using the Rock to distract attention from food shortages.[88]

In January 1952, Stanton Griffis was replaced as United States Ambassador to Spain by the career diplomat Lincoln McVeagh.[89] Lequerica and Franco had been delighted with the uncritical Griffis and found him an amiable collaborator, 'the best ambassador imaginable'.[90] Truman, however, was less enthusiastic than his Ambassador. At a press conference on 7 February 1952, as negotiations with Spain were beginning, the President embarrassed the State Department and the emissaries to Franco by commenting that he was still 'not fond of' the Caudillo. He was infuriated by the undiminished scale of religious intolerance in Spain which, somewhat unfairly, he attributed to Franco personally. No doubt Franco could have done something to mitigate discrimination. Nevertheless, incidents such as the burning of books at a Protestant church in Badajoz, and the subsequent arson at a British Protestant Church in Seville had deeper origins than Franco's prejudices.[91]

On his return to the United States, Griffis advocated massive economic and military aid to Spain.[92] With the US Navy and Air Force still enthusiastic about bases in Spain, further negotiations were left in the hands of the military.[93] However, the American teams sent to investigate Spain's economy and military preparedness had presented deeply pessimistic reports about the appalling condition of both. Franco's exaggerated expectations that American money would put everything right caused the negotiations to drag on. In any case, the urgency of the matter had been diminished somewhat for the Americans by the establishment of bases in

North Africa. Franco set his price high because Lequerica had claimed to be able to 'fix anything in Washington' – something that he found plausible given Lequerica's considerable expenditure on the 'Spanish lobby'.[94]

Eventually the slowness of progress began to worry Franco who wrote, at the suggestion of Lequerica, a conciliatory letter to Truman in late February. Before it was sent, a virulent anti-Protestant pastoral by Cardinal Segura, Archbishop of Seville, accused Franco of bargaining away Spanish Catholicism for American dollars, and an attempt was made to burn down an Evangelical Church in Seville. Re-dated 17 March 1952, the letter expressed Franco's thanks for American financial help and hopes that the alliance negotiations would prosper. Amongst the platitudes was an explanation of the religious frictions as the consequence of malicious exaggerations by 'enemies of our understanding' and a proud claim that the private practice of other religions was guaranteed by his legislation.[95] It did not occur to him that, to Harry Truman, the fact that the public practice of other religions was not protected by law indicated discrimination. The letter remained unanswered for more than four months because Truman's advisers considered that, whatever he said, it would be twisted by Franco to his own benefit. After a frantic intervention by Lequerica, the President sent a bare acknowledgement of receipt.[96]

In his speech on the opening of the Cortes on 17 May 1952, the Caudillo announced that the negotiations would bring economic and military aid 'without the slightest diminution of our sovereignty'. To emphasize his guardianship of traditional Spanish rights, and to emphasize the permanence of his rule, the ceremonial of the occasion was mounted with more regal pomp and circumstance than ever. He was escorted by two regiments of infantry, a squadron of cavalry and a battery of armoured vehicles. In the manner of Alfonso XIII, the Generalísimo advanced to the entrance door behind the rostrum through a line of halberdiers, attended by heralds in historic costume.[97]

In the speech to the Cortes, Franco heaped praise on what he saw as his uniquely democratic system. His words made a cruel commentary on the brutal repression that was taking place in the wake of the strikes of the previous year. The leader of the Catalan Communist Party, Gregorio López Raimundo, was made the scapegoat for the strikes in Barcelona and court-martialled. Massive international attention saw the prosecutor's demand for a sentence of twenty years' imprisonment reduced to four. Fourteen men held responsible for the strikes in the Basque Country were also sentenced to twenty-year prison terms. Forty men were given fifteen-year sentences for abetting the Communist Party in Galicia. Efforts to crush the clandestine anarcho-syndicalist union, the *Confedera-ción Nacional del Trabajo*, saw trials of large numbers of its militants in Barcelona and Seville and the passing of two death sentences. Barely a day went by without arrests, police beatings and eventual courts martial at which punitive sentences were meted out. Men were shot 'trying to escape' and several prisoners died from the injuries received in custody.[98]

At this time of intensifying repression, Franco was more concerned than ever with his carefully cherished image. In particular, his role as defender of the faith had been put in question by Segura and by mutter-ings in the hierarchy that a deal with the United States would open up Catholic Spain to the pernicious doctrines of Protestantism. Franco viewed the Church in the manner of a medieval king, regarding it as the legitimizing agent of his own divine purpose. Accordingly, he was careful to cultivate a public image of total identification between himself and the ecclesiastical hierarchy. On 28 May 1952, he arrived in Barcelona to attend the International Eucharistic Congress. In the uniform of Admiral of the fleet (*Capitán General de la Armada*), he and Doña Carmen sailed into the port on board the battle-cruiser *Miguel de Cervantes* escorted by the other warships of the Mediterranean fleet and a squadron of aircraft in formation.[99]

In an elaborately choreographed religious ceremony on Sunday 1 June, presided over by the Papal Delegate Cardinal Tedeschini, Franco, in the manner of a medieval crusading king, consecrated Spain to the Eucharist. In his speech, he declared that 'the history of our nation is inseparably united to the history of the Catholic Church, its glories are our glories and its enemies our enemies'. Father José María Bulart, Franco's personal chaplain, sang the praises of 'the Eucharistic devotion of the Caudillo' and spoke of his late-night rosaries.[100] Franco threw himself into the role of champion of the Church with delight, basking in the hysterical adu-lation of the 300,000 faithful present. To appear as being united in spirit with his Catholic people would do him no harm in the eyes of American and European Catholics and take him nearer the coveted Concordat with the Vatican.[101] There could be no more effective mask for his savagely repressive rule.

Franco had come a long way since the insecurities of 1945 and 1946. The efficacy of his repressive apparatus meant that he had few concerns about threats from the Left. Nevertheless, he still needed the moral approval of the Vatican and the United States to undermine the hopes of the monarchists. Above all, he needed American economic aid and so followed the 1952 US presidential election campaign with avid interest. His hopes lay with the Republicans, assuming that the favourite, Eisenhower, would regard him with considerably more sympathy than Truman. The Caudillo gave an interview to the *Washington Post* on 7 September, receiving the correspondent at the Palacio de Ayete in San Sebastián dressed in a civilian suit. He was at his most reasonable and affable to demonstrate that here was no bombastic military dictator of a banana republic. He attributed the slowness of the negotiations over the bases to the thoroughness of the military and economic experts who were responsible for the fine print. With an eye on his Spanish audience, he stressed that the United States had come to him and not the other way around. For the Pentagon's consumption, he suggested that, in the event of war with the Soviet block, the West would have to dig its last trenches in the Iberian Peninsular which would necessitate a full rearmament of Spain and Portugal.[102]

After Eisenhower's victory, Franco rammed home the message with an interview reproduced by large numbers of newspapers and radio stations in the United States and many other countries. Again quiet and friendly, he reverted now to a general's uniform, thereby stressing both his common ground with Eisenhower and also that he was an ally ready to fight. Putting aside his private views about the Americans, he loosed a torrent of pro-Americanism criticizing the rest of the world for failing to recognize 'the splendid sacrifice that the United States are making in Korea'. So heartfelt was Franco's desire to ingratiate himself with the Americans, that, in an echo of his gesture to Hitler in 1941, he offered to send a division of volunteers to fight in Korea under Spanish regular officers, 'although technically this is a war of the United Nations and the United Nations has excluded Spain'. His gushing rhetoric suggested that he was worrying about the delays in the bases negotiations and the consequent financial credits.[103]

After his inauguration in January 1953, President Eisenhower signalled the importance which he gave to an understanding with Franco when he replaced Ambassador McVeagh with James C. Dunn, at the time Ambassador in Paris, a man keen to see an agreement clinched with Spain. On 9 April 1953, Dunn had a long meeting with Franco and Martín Artajo. Afterwards, the Ambassador issued a statement which can only have delighted the Caudillo. He stated starkly 'we want the bases' and affirmed that the United States intended 'to strengthen the cordial relations existing between our countries'.[104] Negotiations should now have accelerated. However, they did not because Franco had been led by

Lequerica into thinking that the Americans were more anxious for a deal with him than was in fact the case.[105] Martín Artajo was marginalized from the negotiations and the tone was set by Franco himself and his Chief of Staff, Lieutenant-General Juan Vigón, who took instructions directly from him.[106]

In the last resort, however, when the Americans pressed for the agreement to be finalised, Franco was forced, rather than lose it, to drop his extreme demands and accept what was virtually an American text. The prospect of his ultra-nationalist regime having to concede bases to a foreign power and diminish national sovereignty worried Franco. It was a stroke of luck that the British now announced that the recently crowned Queen Elizabeth II would visit Gibraltar in 1954. News of the Coronation preparations and travel agency advertisements for trips to London had been banned from the Spanish press. British citizens in Spain who celebrated the Coronation were harassed.[107] Franco seized the opportunity provided by the announcement of the royal visit to the Rock to give a militant interview to *Arriba*.

He declared that: 'Just because we don't talk about it does not mean that that shameful disgrace does not exist.' His explanation of how the 'disgrace' had lasted so long was wide-ranging in its display of prejudice. It was the result of 'the policy of foreigners of weakening our *Patria*, creating problems for our Nation, undermining and influencing the ruling classes, fomenting insurrection in the colonies and fomenting revolutionary movements from masonic lodges and left-wing internationals'.[108] The unsheathing of the Gibraltar issue was a cheap and effective way of whipping up nationalist support by stressing the sinister intentions of imperialist Britain, thereby maintaining in a small way the spirit of 1945–50 and diverting attention from the costs of the agreement with the USA.[109] It did nothing, however, to hasten the return of Gibraltar nor to incline London to view the regime more favourably. Eventually, evidence that the American Government had no intention of supporting his claim would lead to Franco realistically toning down his own attitude to Gibraltar, although he always kept public agitation on the matter as a useful residual device to divert attention from other issues.[110]

At the end of August, the lengthy Concordat negotiations with the Vatican were successfully concluded. The delay had reflected Vatican doubts about the stability and international acceptability of Franco's dictatorship.[111] While significantly less important than the regime was to make out, the Concordat was a major step towards international recognition for the Caudillo.[112] In return, he gave the Church a pre-eminent voice in education and social morality as well as the exclusive right to proselytize as the official State religion.[113] Franco wanted the Papal seal of legitimacy for his semi-monarchical rule, to justify the coins which were stamped 'Caudillo by the grace of God' and his arrogation of royal status in the ecclesiastical sphere.* With the Concordat, the Caudillo got what he wanted although that did not mean that there would no conflicts in his dealings with the Church. In particular, the regal right to 'presentation of bishops', by which he could choose from three names presented to him by the Nuncio, would come to have increasing political significance, especially in areas of strong local nationalisms such as Catalonia and the Basque Country.[114]

Franco's relationship with Catholicism was never simple. As a young man, he had displayed conventional piety when in El Ferrol with his mother and manifested a bluff soldierly rejection of religion when in Africa. In 1936, he had quickly perceived the political value of Church endorsement of his position and had relished playing the role of 'defender of Christianity'. In the subsequent decades, with age and his wife's influence, his piety returned and he said the rosary daily. That was in private. Publicly, though delighted to have ecclesiastical endorsement, he behaved like any strong medieval king, ready to impose his will upon the Church. On 21 December 1953, Pius XII granted Franco, 'our beloved son', the highest Vatican decoration, the Supreme Order of Christ. The Concordat, and the appointment of a compliant Papal Nuncio, Monsignor Ildebrando Antoniutti, permitted Franco to clip the wings of his Grand Inquisitor, Cardinal Segura. Bishop José María Bueno Monreal was sent to Seville as apostolic administrator with rights of succession to Segura. The Cardinal remained Archbishop but with increasingly reduced powers and followed wherever he went by secret policemen.[115]

The Concordat could not have the practical political importance of the deal with the United States, on which the Caudillo's hopes were much more tightly focused. In early September, Dunn took the text to Washington for President Eisenhower's agreement.[116] Many details remained to be worked out when the integration of Spain into the western camp was formalized by the signing on 26 September 1953 of the Defence Pacts with the USA. The unresolved details concerned the conditions of American utilization of the bases in wartime and ultimate Spanish command over them.[117] The ambiguities and grey areas in the final agreement

---

* He continued to enter and leave Churches under the canopy previously reserved for the Kings of Spain, a privilege rarely used by Alfonso XIII.

favoured the Americans although Franco claimed that he had not ceded any national sovereignty in the negotiations.

In practice, the defender of national independence had renounced a large tranche of national sovereignty. A wartime emergency, in which there would be only minutes in which to get fighters airborne, would not permit any further negotiations. This was effectively recognized in secret additional clauses to the treaty whereby, in the case of a Soviet aggression, the United States had to do little more than 'communicate the information at its disposal and its intentions' to the Government in Madrid. It was not stated even if such a communication had to be in writing. There was little reciprocity. Any American commitments were subordinate to prior commitments to NATO from which Spain would continue to be excluded. In the event of Spain being attacked by a non-Communist aggressor, the USA was under no obligation to come to her aid. Indeed, large areas of Spain remained without adequate defence coverage. Similarly, the priority given to American undertakings within NATO meant that Spain had to accept surplus and second-rate equipment. In that sense, Franco had sold the pass.[118]

In subordinating Spain to wider American defence needs, if not entirely accepting satellite status, Franco showed just how high a price he was prepared to pay to keep himself in power.[119] In the final stages of the haggling, he had told his negotiators, 'in the last resort, if you don't get what you want, sign anything that they put in front of you. We need that agreement.'[120] The abandonment of the traditional Spanish policy of neutrality was indeed a high price, although Franco had been prepared to pay it to the Third Reich in 1940 when he had also regarded the rewards as sufficiently tempting. The way the agreements were presented to the Spanish people, however, permitted Franco to bask in the self-invented glory of being the equal partner of the President of the world's greatest military power. Photomontages of Franco and Eisenhower together were accompanied by articles to the effect that the nations of the world were reeling with amazement and delight at this latest triumph of the Caudillo.[121]

The mutual defence pact brought $226 million in military and technological assistance from the USA. General economic aid was limited to projects of an infrastructural kind with military sigificance, the building of roads, ports and defence industries. Other American commitments meant that deliveries of military equipment would be confined largely to equipment surplus to the general NATO arms build-up, weapons, aircraft and vehicles already used in the Second World War and/or Korea. In return, Franco permitted the establishment of American air bases at Tor-rejón near Madrid, Seville, Zaragoza and Morón de la Frontera and a small naval base at Rota in Cádiz, as well as an enormous range of smaller Air Force installations and naval refuelling facilities in Spanish ports. More was offered than the Americans were ever able to take up. Equally, American military personnel stationed in Spain were exempt from Spanish law and tax systems. The Caudillo had bargained away neutrality and sovereignty without distinguishing between the good of Spain and the good of Francisco Franco. In particular, the siting of bases next to major cities constituted an act of sheer irresponsibility.[122]

Franco had got what he wanted: the end of international ostracism, a massive consolidation of his regime and the right to present himself publicly as the valued ally of the United States. The price was a dimin-ution of sovereignty and the danger of war in the atomic age. The immedi-ate benefits for the regime were the integration of Spain into the western system, the transfer of military expenses out of the general budget and the neutralization of military discontent about resources. Franco's hope of securing general economic aid were frustrated in that the great flood of Marshall Aid to Europe was already past its peak. Economic aid came with conditions which meant that, in the medium to long term, he would have to admit changes in the very nature of his regime.[123] Beyond the high-sounding rhetoric, there were practical conditions, in terms of estab-lishing a realistic exchange rate for the peseta, balancing the state budget, restoring confidence in the financial system, all of which struck at the very existence of his cherished system of autarky. Previous internal efforts to introduce greater flexibility into government financial mechanisms had been abortive.[124]

Initially, Franco paid no attention to these detailed conditions, perhaps assuming that the United States put too high a value on his strategic services to want to invoke the small print of the agreement. The desired changes would not be implemented until the end of the decade. In the event, it would not be American pressure but the collapse of the Spanish economy which would oblige him to permit economic liberalization. Franco continued to cling to autarky for another six years, finally aban-doning it both reluctantly and uncomprehendingly. Paradoxically, the agreements provided the economic stimulus which would expose the structural rigidities of Francoist autarky. In that sense, they constituted another step in the process of economic and social development which would eventually render the Caudillo an obsolete irrelevance.

XXIV

YEARS OF TRIUMPH AND CRISIS

1953–1956

ON 1 October 1953, Franco presented the texts of the bases agreement to the Cortes as the logical culmination of a selfless policy pursued since 1936 in defence of western civilization. He denied that there was any question of selling territory for economic and military aid: he had initiated negotiations purely out of concern for the defence of the West and now collaboration with the USA simply facilitated the necessary Spanish rearmament. Praising his own perspicacity, he criticized Churchill for myopically rejecting his offer of an alliance in October 1944, an error he patronizingly attributed to the natural reluctance of a defeated imperial power to share its privileges. He declared that the lack of a strong and determined foreign policy like his own was the cause of Spanish decadence in the previous two hundred years, for once forgetting his usual conviction that masonic conspirators were to blame.[1]

At the time, coming in the wake of the recently concluded Concordat, the agreements were hailed as a monumental achievement by the Caudillo. The American ambassador, James Dunn, was photographed with him, the camera angle suggesting that a supplicant was being graciously received. The Falange and its dependent organizations, the Syndical and Student Organizations, the *Frente de Juventudes* and the *Sección Femenina* worked feverishly to mount on the same day, the *Día del Caudillo*, another great rally in the Plaza de Oriente. Shops were shut and workers and peasants were bussed in from all over Spain, with a day's pay and a packed lunch. A visibly delighted Franco received the popular homage on the balcony of the royal palace. Although the bases agreements provoked little international press comment and some bitter criticism in the United States, *Arriba* described the rest of the world as speechless with admiration.[2]

Franco claimed, entirely erroneously, that the pacts constituted a full-blown alliance between equals when, in fact, they were no more than agreements on a specific issue. He had stated to the Cortes, and his press had echoed the point, that with the pacts, he had given back to Spain an international prestige that she had not had since the days of Philip II.* Not having read the secret clauses of the pacts, the Francoist editorialists declared that Spain was again in control of her own destiny. Neither they nor Franco mentioned the fact that Spain was drastically more vulnerable to Soviet aggression than before.[3] Having survived the dark days of the international ostracism, it is probable that Franco could have stayed in power without American backing.[4] With the internal opposition crushed, there was an air of permanence about his rule. However, by becoming an ally, however subordinate, of the United States, Franco made things immeasurably easier for himself. He opened the way to entry into the United Nations and full international recognition – a deeply demoralizing blow to the already divided opposition in exile.

Franco consolidated the propaganda triumph of his American link when he presented the Concordat to the Cortes on 26 October 1953. Even by his own standards, the speech was arrogantly vainglorious: 'My spirit is overcome with innermost satisfaction, which I hope you share, at having been able to render the Nation and our Holy Mother the Church the most important service of our times.' The Caudillo made no reference to the fact that the delays in securing the Concordat had reflected misgivings in the Vatican about certain aspects of his rule and explained at length how his rule was entirely in accordance with the wishes of the Church. Indeed, such was the identification between Church and State that he described that many of the *Procuradores* must have wondered why he needed a Concordat at all.[5]

The possibility that the creation of 'new Gibraltars' in the form of American bases and the fervent embrace of the Catholic Church might offend Falangist sensibilities led to a public demonstration by Franco of his commitment to the *Movimiento*. On 29 October 1953, the twentieth anniversary of the foundation of Falange Española by José Antonio Primo de Rivera, Franco addressed about 125,000 blue-shirted 'Falangists' at the Real Madrid football stadium at Chamartín. According to the corre-

---

* From the premise that Franco had been right all along, it was but a short leap to the claim that now the entire western world was falling in behind his leadership. Luis de Galinsoga proclaimed him 'Caudillo of the West', the only truly great man of the twentieth century, a giant by the side of such dwarves as Churchill and Roosevelt (*La Vanguardia Española*, 1 October 1953).

spondent of the French daily *Le Monde*, Jean Créac'h, eighty per cent of those present were peasants or unemployed agricultural labourers who had been bussed in from the provinces and paid a day's wage.[6] Franco appeared in the black uniform of *Jefe Nacional*. To pre-empt Falangist criticism of friendship with the power which took the last remnants of the Spanish empire and had recently been vilified for its part in the 'siege' of Spain, Franco triumphantly presented the agreements as part of a greater Spanish service to the West: 'The battle that we have won in these years of difficult peace is our second victory over Communism.'[7]

Within a few years, however, Franco would come to realize some of the disadvantages of the bases agreement. He complained of the dangers posed by American installations to nearby cities and of the disappointing scale of American economic help, significantly less than that received by Yugoslavia, Turkey, Greece and Brazil.[8] His resentment that he had not got the best of the bargain surfaced years later, when he said privately that 'according to what Don Camilo [Alonso Vega] told me, the best thing that the Americans did for us was empty the Madrid bars and cabarets of whores, since they almost all marry American sergeants and GIs'. He went on to say that 'it causes me some anxiety to see the world in the hands of the North Americans. They are very childish. The aplomb of the English reassures me more.'[9] Such ruminations, however, should not obscure the fact that, at the time, the bases agreement and the Concordat made 1953 a pinnacle in Franco's life.

The link-up with the United States reduced pressure on Franco from inside and outside the regime, both by stemming a further decline in living standards and permitting a revival of anti-Communist crusade propaganda which helped keep alive the spirit of the Civil War. There can be little doubt that after the signing of the Pact of Madrid, Franco felt that he had definitively established his regime. The guerrilla war was over save for sporadic outbursts of urban resistance, mainly the work of isolated anarchists in Barcelona.[10] The huge investment in State terror made between 1939 and 1945 was paying off in the political apathy of the bulk of the population. Franco's opponents had learned their lesson and torture, prisons and occasional executions served as a reminder for those who forgot. The Civil Guard, the Armed Police and the secret police did their gruesome work, tirelessly dismantling clandestine efforts to rebuild parties and unions.

The Caudillo's confidence that he could now rest on his laurels was obvious in the increasing amount of time that he devoted to his hobbies – hunting, fresh water and deep-sea fishing with Max Borrell, golf, watching endless westerns in the private cinema in El Pardo, painting, and developing his large estate at Valdefuentes, where he grew wheat, potatoes and even tobacco. Having at its disposal the services, manpower and machinery of the Ministry of Agriculture, it became immensely profitable. When he was in residence at El Pardo, Franco would go to Valdefuentes most afternoons to take the air after a late lunch.[11]

In 1954, Lequerica said to Pacón that 'to be a minister of Franco is to be a little king who does whatever he feels like without restraint from the Caudillo' and that 'the ministers who cause him trouble are those who don't know how to run their departments'.[12] Franco had always left his ministers to get on with the technical side of their ministries, to make fortunes if they wished, or merely to be efficient or even incompetent, while he dictated the broad lines of policy, especially foreign policy. But, after 1953, there became apparent a readiness to leave the drudgery of day-to-day government to others. He turned a blind eye to corruption, whether committed by his political servants or by his extended or family 'clan', as long as absolute uncritical loyalty was maintained.*

Pacón reflected that 'the Caudillo is effusive with those who dominate him and with the flatterers (*pelotillas*) who swamp him with gifts and lavish hospitality, but as cold as an ice flow with the majority of those of us who are not sycophants, are serious in our conduct and speak to him loyally, whether he likes it or not.' The Caudillo may have rewarded the adulators or ignored their corruption, but flatterers fared no better than the faithful Pacón when it came to human warmth.† José Antonio Girón, the long-serving Falangist Minister of Labour, complained 'he is cold with that coldness which at times freezes the soul'.[13] Franco was ill at ease with those who served him. Even at hunting parties, he could be cold and distant. Neither a genuine friend of forty years' standing, Max Borrell, nor his political Chief of Staff for thirty-five years, Carrero

---

* Franco showed no interest in putting a stop to graft as opposed to using knowledge of it to increase his power over those involved. He often repaid those who informed him of corruption not by taking action against the guilty but by letting them know who had informed on them (Serrano Suñer, *Memorias*, p. 230; Franco Salgado-Araujo, *Mis conversaciones*, pp. 19, 37, 56–8, 83, 178).

† In 1963, José María Sanchiz, uncle and godfather of Nenuca's husband Cristóbal, administrator of Franco's estate at Valdefuentes, and a hunting companion of ten years' standing, ventured to ask him: '*¿No le parece que hemos llegado al punto en que nos podríamos tutear?*' (isn't it time we used the intimate form of address?) to which Franco replied glacially '*El trato que me corresponde es "Excelencia"*' (the correct mode of address for me is "Your Excellency"). Franco also cultivated the illusion of regal distance by the device of keeping his hand near his waist so that visitors were obliged to bow in order to shake his hand.

Blanco, were ever released from the obligation to address Franco as *Excelencia*.[14]

The tone of Franco's rule after the clinching of the American agreements was significantly more relaxed than before. The frenzied anti-foreign propaganda of earlier periods gave way to a more routine glorification of the Caudillo and his works. This was apparent with regard to Gibraltar. In January 1954, he authorized a demonstration by students of the Falangist student syndicate, the *Sindicato Español Universitario*, outside the British Embassy in Madrid in protest at Queen Elizabeth II's visit to the Rock. Young Falangists had been subjected by Franco and the Falangist leadership to ultra-nationalist, xenophobic and imperialist calls for a virile and patriotic reaction to the insult constituted by the Gibraltar visit. Official support for the demonstration included the provision of a truck-load of stones of a size convenient for throwing. At mid-point, however, fearful that it might get out of hand, the police charged and the students used the stones against the authorities. The students' indignation at their treatment led to a series of minor university disturbances.[15] Franco felt the greatest personal indignation about Queen Elizabeth II's visit to Gibraltar and wrote several virulent articles under the pen-name 'Macaulay' in *Arriba*. However, he muted the louder public protests, partly to diminish the ferment in the universities and in response to a clear statement from the US Ambassador that the position of Britain and Gibraltar within NATO ensured American support for London over the Rock.[16]

There was a notable cooling of Franco's public ardour over Gibraltar thereafter. Although some of his Ministers would be fiercely militant on the subject from time to time, he himself would always calmly take the longest of long-term views. That is not to say that it did not concern him. On 22 August 1954, he spoke at the Palacio de Ayete about Gibraltar with the then Minister of State at the British Foreign Office and future Foreign Secretary, Selwyn Lloyd. An otherwise affable encounter cooled only when Selwyn Lloyd told the Caudillo that 'no British Government could discuss the sovereignty over Gibraltar'. The British Minister found the sixty-two year-old Caudillo 'much sprucer, slimmer, fitter and younger than I expected ... I had the feeling that nothing I said made any impression on him but I am told that this is common form. When he was not talking himself, his mouth set in rather an obstinate, discontented expression. But I gained the impression of a mentally alert and competent man, master of himself and of those about him, but firm in his own opinions to the extent of obstinacy.'[17]

Having clinched the relationship with the United States, the urgency with which Franco had had to attend to foreign affairs since 1936 diminished. There were occasional big moments such as the visit in June 1954 of Rafael Leónidas Trujillo, the corrupt dictator of the Dominican Republic, who was received by an evidently delighted Franco.[18] However, the visits of such minor potentates lost their value to Franco once the United States, France and Britain decided to support Spanish entry into the United Nations.[19] This did not satisfy his desire to integrate Spain into NATO and a diminution of British and French hostility was not enough to open the doors and so he looked to the United States as the key.[20] He was sufficiently taken by the implication of a General being President of the United States to countenance a long-term publicity effort comparing him with Eisenhower, the first fruit of which was the publication of photographs of him playing golf and painting. He had first claimed an interest in painting in the 1920s, and had revived it in the 1940s, aware of Hitler's claims to be an artist. Now, the fact of Churchill's skill and Eisenhower's interest made it a respectable hobby again. He was photographed at his easel wearing, somewhat implausibly, a pin-striped suit and a large hat.[21]

Only a small selection of Franco's pictures have ever been published, most having been destroyed in a fire in 1978. Assuming that they are genuinely his work, they show a very competent amateur whose work is ultimately of more interest to the psychiatrist than to the art critic. The subject matter suggests a conservative, petty bourgeois taste. The influences are unmistakably those of the Goya tapestry cartoons and the great age of seventeenth-century Spanish painting in landscapes, still-lives of dead game and guns, a bloodthirsty portrayal of a bear being attacked by a pack of dogs.* One interesting exception is a Modigliani-like portrait of his daughter Carmen.

There were other pleasures and accolades in the wake of the deal with the USA. On 8 May 1954, the year of its seven hundredth anniversary, Franco was given an honorary doctorate of law at the ancient University of Salamanca. Ironically, for a man who usually manifested contempt for

---

* It is noteworthy that the few well-known pictures show a remarkable similarity of subject matter to paintings by Carrero Blanco. Apart from still-lives of game and guns, among Carrero's canvases was one of a bull being attacked by a pack of dogs. The differences lie in the fact that Carrero Blanco's paintings are almost invariably copies of classic Spanish and Dutch painters whereas Franco, although conservative and derivative in style, is somewhat more imaginative in his choice of subjects. (See the pictures reproduced in Julio Rodríguez Martínez, *Impresiones de un ministro de Carrero Blanco* [Barcelona, 1974] pp. 144–7.)

intellectuals, his speech suggested that he was moved by the honour. Nevertheless, apart from some token words of modesty, his speech was otherwise all self-glorification. Referring to 'those of us who make history', he compared his own achievements to those of 'the royal caudillos who in the thirteenth century, while resting from their victorious Reconquest, laid the foundations on which the glorious University of Salamanca was to be raised'.[22] Franco sat unresponsively throughout the proceedings looking only at Doña Carmen who did not take her eyes off him. Although stonily impassive, he was clearly insecure out of his normal element. He gave the impression of being terrified of making a mistake in front of the assembly of what he took to be men of wisdom. The security measures surrounding the Caudillo were notable. As one of the professors reached for his tobacco pouch, he was surrounded by police despite the fact that only persons of the highest loyalty and eminence had been invited to the ceremony and one of them could have attempted to kill Franco only at the cost of sacrificing his own life. At the dinner which followed, Franco barely spoke and when he decided to leave, he simply stood up abruptly and moved off without a word.[23]

1954 also saw the completion of the crypt at the Valle de los Caídos. It had been an obsession, 'the other woman', ever since the inauguration of the works in 1940. More than any other legacy of his regime, it mirrored Franco's conception of himself as an historic figure on a par with Philip II. He had had it doubled in size from the original conception and the crypt was finished on 31 August. It had been a colossal undertaking, dug out from solid granite, 262 metres long, and 41 high at the cross. Many of the great building companies of the Francoist boom got their start there, Banús, Agromán, and, particularly, Huarte, who got the contract to build the cross which was not finished until September 1956. Weighing 181,620 metric tons, the cross was 150 metres high, with arms 46 metres long and wide enough to hold two saloon cars.[24]

The only remaining thorn in Franco's side was constituted by Don Juan and the muted opposition of the monarchists. Franco did not like to be reminded of his unfulfilled promise to restore the monarchy. In 1954, under the influence of Gil Robles, Don Juan had begun to stand up to Franco again. Juan Carlos had finished the secondary education imparted by his private tutors and, on 16 July 1954, Don Juan sent a *note verbale* to the Caudillo to say that it was time for his son to begin his university education at Louvain. It arrived just as Franco was putting the finishing touches to his own scheme for Juan Carlos to enter the military academy at Zaragoza for a period, followed by time at the naval and air academies, the social science and engineering faculties of Madrid University and then some practice in the art of government 'at the side of the Caudillo'. Franco wrote a reply to Don Juan coldly stating that those who hoped to govern Spain should be educated in Spain. The dismissive implication was that Don Juan did not figure in his plans for any future monarchical restoration. Franco's letter also threatened that if Don Juan did not accept the programme for Juan Carlos, he would be 'closing the natural and viable road that could be offered for the installation of the monarchy in our *Patria*'. The word 'installation' meant, in Franco's jargon, that there would be no restoration of the legitimate Bourbon line but rather the imposition of a Francoist monarchy whose incumbent had to be chosen and trained to continue the traditions of his regime. Don Juan realized the dangers of linking the monarchy to a political system which was far from enjoying universal acceptance, but he was too frightened by the prospect of breaking totally with Franco to back down. To Franco's delight, this provoked the resignation of Gil Robles.[25]

Perhaps inevitably, the very complacency which Franco and his followers experienced in the wake of the Concordat and the Pact of Madrid had the effect of loosening the unity-under-siege which the regime had experienced in the period 1945–53. Franco had, of course, always had to deal with political, economic and religious pressure groups, managing with a skill verging on artistry the competition of individuals for preferment and important posts. He had handled the rivalry of the Falange and the monarchists in the Army up to 1945 and of the Falange and the Catholic monarchists in the post-war period. But generations were coming to maturity which had not fought in the Civil War. Despite, or because of, exposure to the Francoist education system, they were blasé about the Caudillo's achievements as the 'saviour' of Spain. A more open jockeying for position began to emerge among groups which he did not control totally and younger men somewhat less bewitched by the magic of the Caudillo.

The apparent tranquillity of Franco's delicately balanced system was challenged by a curious amalgam of collaborationist followers of Don Juan and members of Opus Dei.* The group was proclaimed to be the *Tercera Fuerza* (third force) by their self-appointed theorist, Rafael Calvo Serer, a regime intellectual connected with Opus Dei. To Franco, Calvo

* Opus Dei was an increasingly powerful Catholic secular order – a conservative elite whose members were enjoined to perform their apostolic task by excelling at their chosen profession.

Serer's vision smacked dangerously of political parties. The 'third force' saw itself as ploughing a middle furrow within the regime, between the 'left', constituted by the Falange, and the right, made up of Martín Artajo's conservative Catholics, or self-proclaimed Christian Democrats. It consisted of about thirty prominent bankers, lawyers and professors. Some of the leading lights were figures connected with Opus Dei while others, like the lawyer Joaquín Satrústegui and General Jorge Vigón, were simply supporters of Don Juan. They were committed to the eventual restoration of a traditional monarchy under Don Juan, albeit within the context of the *Movimiento*. In an article published in Paris in September 1953, and widely circulated within the Francoist establishment, Calvo Serer claimed that the Falangists and the regime Catholics had lost their way. For suggesting that only the new group could renovate the regime, liberalize the administration and modernize the economy, Calvo Serer was dismissed from his posts in the *Consejo Superior de Investigaciones Científicas* (the higher council of scientific research).[26]

The existence of the *Tercera Fuerza* was a reflection of concern among Franco's supporters for the future. In February 1954, he had received a visit from several generals, including Juan Bautista Sánchez. To his alarm they openly broached the subject of what would happen after his death and pressed him to make arrangements for the monarchist succession.[27] Franco saw danger in any hint of support for the Pretender. In the autumn of 1954, the coming out (*la puesta de largo*) of Don Juan's daughter, the Infanta María Pilar, motivated fifteen thousand applications for passports from Spanish monarchists who wished to travel to Portugal to pay homage to the royal family. In the event, three thousand made the journey for the celebrations held on 14 and 15 October. Charabancs and cars brought not just the wealthy and Army officers but significant numbers of the more modest middle classes. Nicolás Franco was present at the spectacular ball given at the Hotel Parque in Estoril and reported back to his brother about the warmth and spontaneous enthusiasm which had greeted the appearance of Don Juan.[28] Franco was deeply displeased by the social event in Estoril and talked of removing the privilege of a diplomatic passport enjoyed by the highest ranking nobility, the *grandes de España* 'because they use it conspire against the regime'. Carrero Blanco consoled him with the thought that the popularity of the monarchy was the work of freemasons.[29]

It was with this background of potential challenge to Franco that the *Tercera Fuerza* was put to the test. The occasion was the holding of limited municipal elections in Madrid on 21 November 1954, the first since the Civil War. They were presented by the regime as genuine elections because one third of the municipal councillors would be 'elected' by an electorate of 'fathers of families' and married women over the age of thirty. Sponsored by the newspaper *ABC*, the four monarchist candidates were subjected to intimidation by Falangist thugs and by the police. They were Joaquín Satrústegui, Joaquín Calvo Sotelo, a prominent playwright and brother of the assassinated José, Juan Manuel Fanjul, son of the general, and Torcuato Luca de Tena, of the family that owned *ABC* and one-time director of the paper. Regarding these elections as a kind of referendum, the controlled press mounted a huge propaganda campaign in favour the four *Movimiento* candidates put up by the Minister of the Interior Blas Pérez and the Minister Secretary, Raimundo Fernández Cuesta. It was a crass error on the part of Pérez and Fernández Cuesta to label their candidates in such a way, since it exposed the farcicality of Franco's claim that all Spaniards were part of the *Movimiento*. Together with Carrero Blanco and Gabriel Arias Salgado, the Minister of Information, Blas Pérez and Fernández Cuesta had agreed some days before that they would resort to electoral falsification (*pucherazo*) rather than risk defeat. Monarchist publicity material was destroyed and voting urns were spirited away to prevent monarchist scrutiny of the count. Inevitably, official results gave a substantial victory to the Falangist candidates. Nevertheless, the monarchists claimed to have received over sixty per cent of the vote.[30]

At first, Franco was prepared to take the word of his Minister of the Interior, Blas Pérez, that the municipal elections constituted a manifestation of popular acclaim for him. However, within a week, influential monarchists were calling for the newspapers to rectify their erroneous accounts of the elections, backing up their demands with threats of resorting to the courts. Franco then received requests for audiences from two important figures. The first was Joaquín Calvo Sotelo, who wrote and complained that 282 monarchists had been arrested during the campaign. Calvo Sotelo was put off until mid-January but his request had the effect of convincing Franco that the Falange had lied to him. The second was from the Minister of Justice, the traditionalist Antonio Iturmendi, who presented his resignation. It did not take Franco much effort to talk him out of it, but the impact of the gesture made its mark on the Caudillo.

Iturmendi's complaints were as nothing by the side of what General Juan Vigón, now Chief of the General Staff, had to tell the Caudillo. Military Intelligence Services had gathered information which showed that the bulk of the Madrid garrison had voted for the monarchy. Vigón told Franco that 'the regime lost the elections of 24 November'. This constituted a threat serious enough to compel Franco to take action to neutralize the resurgence of military monarchism. After delivering a severe dressing-down to Blas Pérez and the Director-General of Security whom he accused of tricking him, he immediately instructed his brother Nicolás to let it be known in Estoril that he would be pleased to meet Don Juan.[31]

No longer concerned about his international position and safe in the knowledge that his repressive apparatus kept the working class and the left-wing opposition at bay, Franco had few political problems. Henceforth, his most substantial concerns would be the neutralization of the monarchists and simultaneously the consolidation of his own plans for the succession. His inclination was always to check monarchist independence by encouraging successive revivals of the Falange. Moreover, Franco was determined that if the monarchy were to return, it would have to be a Falangist monarchy. However, the difficulty now about a strategy which had served him well throughout the 1940s was that the Falange was increasingly anachronistic while the monarchist option seemed more in tune with the outside world. Above all, the autarkic policies favoured by both Franco and the Falange were being exposed as incapable of coping with Spain's economic problems. In retrospect, 1953 may be seen as the high point of Franco's political career, a moment of triumph with the forces of the Nationalist coalition united around him. Before the end of the decade, while his survival would hardly be threatened, he would find himself no longer entirely in control, forced to abandon the Falange and leave the detailed management of economics and, by extension, politics to expert technocrats.

Franco's objectives in arranging to meet Don Juan in December 1954 were, as with the *Azor* meeting in 1948, to convince royalists inside Spain of his own good faith as a monarchist. Any impression that they might be discussing ways of hastening a restoration was entirely erroneous. Franco had left little room for doubt that he would hand over only on his death or total incapacity and then only to a king who was committed to the unconditional maintenance of the dictatorship. Franco had written to Don Juan on 2 December 1954 a letter which made it clear that he saw the education of Juan Carlos in such terms. Describing himself as 'identified with the feelings of great sectors of the nation', he wrote, 'I believe that it is indispensable that the education of the Prince take place not just on our territory but also within the principles which inspire the *Movimiento Nacional*.' The letter ended with Franco upbraiding Don Juan for the behaviour of his supporters in Madrid in running as candidates in the municipal elections against the *Movimiento*.[32]

It is possible that Franco was influenced slightly in writing that letter by an imminent event within his own family – the arrival on 9 December, five days after the Caudillo's sixty-second birthday, of his first grandson. On the day of the birth, Cristobal Martínez Bordiu was talking about changing the baby's name by reversing his matronymic and patronymic. As Francisco Franco Martínez-Bordiu, the new arrival was a potential heir to his grandfather. The formal agreement was given by a servile Cortes on 15 December which gave rise to rumours that Franco planned to establish his own dynasty.[33] Whether the prospect of his own heir contributed to the stiffening of his attitude to Don Juan is impossible to say. In any case, by the time the Pretender received the letter of 2 December, Franco had had the communications from Joaquín Calvo Sotelo, Antonio Iturmendi and Juan Vigón which demonstrated to him that the monarchist challenge was stronger than his lackeys in the *Movimiento* had led him to believe. Accordingly, there was some hard negotiating with Don Juan's representative, the Conde de los Andes, about the agenda to be discusssed in the forthcoming meeting.[34]

Franco secretly left Madrid at 8.00 a.m. on the morning of 29 December 1954 accompanied by Admiral Pedro Nieto Antúnez. His Cadillac and its convoy of guards headed for Navalmoral de la Mata in the province of Cáceres in Extremadura. The meeting – at Las Cabezas, the estate of the Conde de Ruiseñada, Juan Claudio Güell, the Pretender's representative in Spain – lasted from 11.20 a.m. to 7.30 p.m. with a late lunch break. The ever-affable Don Juan set a cordial atmosphere. He felt confident, telling Franco that he had received thousands of messages of support from Spain including telegrams from four Lieutenant-Generals.

Such hints about the current debate on the monarchist succession did not matter to Franco other than as a reference to a far distant and theoretical future. This became clear when he began to talk of the possibility of separating the functions of Head of State and Head of Government. He would do so only, he said, 'when my health gives out, or I disappear or because the good of the regime, with the evolution of time, needs it; but, as long as I have good health, I don't see any advantages in change.' With a remarkably frank display of narrow-mindedness, he said 'In confidence, I will tell Your Highness that I see disadvantages; because with a head of government, if I remain as Head of State, public opinion will blame me for everything bad that happens while everything good will be credited to the head of government.'

Clearly at his ease, talking without pause or even a sip of water, he proceeded to give Don Juan an interminable history lesson. Efforts by Don Juan to get a word in edgeways and turn the discussion to the timing of the transition to the monarchy and the terms of the post-Franco future met with a frosty response. Franco took the opportunity to criticize many prominent monarchists, accusing Pedro Sainz Rodríguez of being a freemason. When Don Juan praised Sainz Rodríguez as a faithful counsellor, in whom he had complete confidence, Franco replied 'I never placed my complete confidence in anyone.' Disinterring a hitherto deeply buried admiration for José Antonio Primo de Rivera, Franco's eulogies about the achievements of Falangist vertical syndicalism were a warning that he would call on the Falange should Don Juan try to mobilize his monarchists inside Spain.

Don Juan's suggestions for freedom of the press, an independent judiciary, social justice, trade union freedom and political representation convinced Franco that the Pretender was the puppet of dangerous aristocratic meddlers who were probably freemasons. Through what seemed impenetrable verbiage glimmered the Caudillo's message: if Don Juan did not accept that his son Juan Carlos should be educated under his tutelage, he would consider it as a renunciation of the throne. Don Juan thus agreed that Juan Carlos be educated at the three service academies, at the university and at Franco's side. However, the Pretender made it quite clear that none of this constituted a renunciation of his dynastic rights. With the greatest reluctance, Franco accepted a joint communiqué whose terms implicitly, if not explicitly, recognized the hereditary rights to the throne of the Bourbon dynasty.[35]

With an engaging lack of awareness of his own experience, Franco patronizingly warned Don Juan against sycophants. He ended with some malice, 'we have many years in which to discuss these questions.' On parting, Don Juan suggested that they each name two persons of unquestionable loyalty to keep in constant contact on the various issues discussed. Franco expressed surprise that Don Juan should have as many as two people he could trust. Don Juan said that he could produce one hundred such names. 'Well, I could not,' replied the Caudillo. The interchange revealed as much about the cheerfully trusting Don Juan as it did about Franco. For all his distrust of his collaborators, the Caudillo felt able to boast that 'I don't find governing an onerous task' and 'Spain is easy to govern.' But Don Juan found him disillusioned, resentfully complaining that all that he had done for Spain was not adequately appreciated although he claimed that he was 'loved by his people'.[36]

The joint communiqué aside, Franco had made no real concessions about a future restoration, or rather installation as he called it. Nevertheless, the theatrical gesture of meeting Don Juan had, for the moment, drawn the sting of the monarchists and gave the impression that progress was being made. On 30 December, he told Pacón that he would not give up active rule and retire to the Headship of State, since 'my role would just be decorative and it would not be so easy to direct politics and orient things in the form I want and which I consider beneficial for the nation.'[37] In his end of year message on 31 December 1954, he made it quite clear that he had conceded nothing to Don Juan. Using the royal 'we', he declared that, 'if . . . we took from our traditions the form of a kingdom, which gave unity and authority to our Golden Age, this does not mean under any circumstances the resuscitation of the vices and defects which in the last centuries ruined it.' In Francoist code, this meant that there would be no restoration of the Bourbon dynasty. In the wake of the Las Cabezas meeting, the Caudillo was publicly affirming that he did not renounce the rights enshrined in the *Ley de Sucesión* to choose a successor who would guarantee the continuity of his authoritarian regime. He denounced any move towards political liberalization and attributed calls for reform to bad Spaniards in the service of sinister foreign enemies.[38]

The communiqué issued after the Las Cabezas meeting gave rise to monarchist-inspired rumours that the Caudillo was now actively preparing an early transition to the monarchy. Shortly afterwards, there were mutterings of protest from hard-line Falangists about such a prospect. Franco responded with a widely reproduced interview which dispelled hopes of his early departure. 'Although my magistracy is for life,' he declared pompously, 'it is to be hoped that there are many years before me, and the immediate interest of the issue is diluted in time.' Franco made it clear that the monarchy would be a Falangist one in no way resembling that which fell in 1931.[39] It was a question now of defusing Falangist opposition to what seemed a swing towards the monarchy. Banking on the readiness of the Falangist hierarchy to swallow whatever he put on their plates, Franco was asking them to postpone the 'pending revolution' even longer in return for a Francoist future under a Francoist king.[40]

Nonetheless, in February 1955, he authorized Raimundo Fernández Cuesta to draft laws to block loopholes in the *Ley de Sucesión* and irrevocably tie any royal successor to the *Movimiento*.[41] To make that more acceptable to his monarchist supporters, the Falangist edges of the *Movimiento* were blurred. On 19 June 1955, Fernández Cuesta declared in

Bilbao that to ensure the survival of the regime after Franco's death, judicial, political and institutional guarantees would be necessary. The role of the *Movimiento* would be to sustain the monarchy which succeeded Franco and to keep it on the straight and narrow path of Francoism. It was the formal recognition by the Falange of the inevitability of a monarchical succession and a redefinition of the *Movimiento* in terms wider than *FET y de las JONS*. There can be no doubt that the speech was made at the behest of Franco. In return for accepting a monarchist succession and its own definitive domestication, the Falange's functionaries would be guaranteed a major institutional role, well-paid jobs and sinecures, and a commitment to the one-party state and the corporative syndicates by the regime present and future.[42]

For their part, the monarchists had to accept that the monarchy would be restored only within the *Movimiento*. Julio Danvila, a friend of both Don Juan and Franco, anxious to bring the two together and to further the establishment of a Francoist monarchy, had concocted the text of an 'interview' with Don Juan giving royal approval of Fernández Cuesta's speech. Franco agreed to the text, which Danvila then took to Estoril where, it was rumoured, Don Juan had the greatest reservations about it being published. Danvila then told the Caudillo that the Pretender had accepted the 'interview', at which point Franco amended the text to bring it even more into line with his own thinking and obliged *ABC* and *Ya* to publish it on 24 June 1955. Although outraged, Don Juan did not protest since a public break between him and Franco would have encouraged the anti-monarchical machinations of the extremist elements of the Falange.[43]

Franco's remark to Don Juan about how easy it was to govern Spain was entirely sincere. The ease of his shoddy manoeuvre with the Don Juan interview proved his point. More importantly, in terms of mass acceptance of the regime, the political apathy generated by years of carefully applied state terror made Spain 'easy to govern'. The Caudillo was beginning to delegate ever more and to feel free to spend increasing amounts of time hunting and fishing. The pursuit of large tuna was becoming a passion. As Nenuca's family grew, Franco took greater pleasure in his grandchildren. Gradually, his intimates began to notice a reluctance to give attention to day-to-day political developments.[44] There would still be crises to overcome but, to an extent, they became crises in part precisely because he let politics be a smaller drain on his time. A large proportion of his official business consisted of receiving delegations of one kind or another at El Pardo which took up all of Tuesday and Wednesday. These were cold, arid occasions. Thursdays were devoted to receiving ambassadorial credentials, Fridays to cabinet meetings. There was little or no time to meditate on the general drift of state problems, of which the most acute continued to be the economy paralysed by continuing inflation and stagnation.[45]

The central reason for the neglect of politics was that Franco rarely refused an invitation to a hunt. By the end of 1954, he was spending Saturdays, Sundays and Mondays hunting in the season and occasionally entire weeks at a time. His shooting was improving significantly. Hunting parties were organized around his presence and became the occasion for much wheeling and dealing. Ministers developed an interest in hunting because they could not afford to be absent from what was perceived as the Caudillo's charmed circle. Their consequent neglect of government business seemed not to bother him. While hunting, he was subjected to adulation and to malicious gossip about those who were not present, as well as constant requests for favours of one kind or another. Businessmen sponsored costly hunting parties in order to be able to get near to ministers. The corruption which surrounded these jaunts was notorious but Franco did not abstain from attending because he derived an obsessive pleasure from his skill with a shotgun. Even in the most difficult moments of the Second World War, he had often abandoned his daily tasks to indulge in the royal sport of hunting. His main objective seemed to be to kill as much as possible, suggesting that hunting, like soldiering before it, was the outlet for the sublimated aggression of the outwardly timid Franco. Many of the hunts were exhausting and his doctor complained that he often fired as many as six thousand cartridges in a day. It was a matter of pride with him never to admit to having been tired by a day's hunting or fishing. To flatter his prowess as a hunter, his kills were facilitated by the device of stags being fed every day at certain spots, where he could come across them 'casually'.[46]

An indication of Franco's self-perception in the mid-1950s was given in a speech that he made in Burgos on 24 July 1955. The occasion was the inauguration of a statue of El Cid. He mocked Joaquín Costa's celebrated phrase 'lock the tomb of El Cid with seven keys', which had been a call for Spain's violent, imperial traditions to be abandoned in recognition of her humbler economic position and narrower horizons in the early twentieth century. Franco had long seen himself as a warrior hero analogous to El Cid, as a man who had revived the sleeping beauty of Spain from its long centuries of slumber in mediocrity. In the 'normality' of recent years, he had had fewer opportunities to inflate his ego with such metaphors. Now he seized his chance without apparent irony or embarrassment. Referring to the 'great fear [of cowardly liberals] that El Cid might arise from his tomb and be incarnated in the new generations', he claimed that 'the great service of our Crusade, the virtue of our *Movimiento* is to have awakened an awareness of what we were, of what we are and what we can be'. With an implicit comparison between the great hero of the past and the great hero of the present, Franco cited El Cid as the symbol of the new Spain: 'in him is enshrined all the mystery of the great Spanish epics: service in noble undertakings; duty as norm; struggle in the service of the true God'. He was talking about himself.[47] When Franco first began to see himself as a modern-day El Cid, it was partly a response to the adulation of eager sycophants but also made plausible by his background as the dashing hero of the desert war in Africa and as the energetic and determined Generalísimo of the Nationalist struggle in the Civil War. But, isolated in El Pardo, surrounded by flattering time-servers, the complacent, and increasingly narrow-minded Caudillo retained little that was heroic.

For all his much-vaunted physical stamina, there is little doubt that the Generalísimo wanted to spend less time on politics and more with his family and at his pleasures. However, despite his ostrich-like assumption that all ills were the work of Satanic minorities radio-controlled from masonic lodges and left-wing internationals abroad, the municipal elections of November 1954 and the meeting at Las Cabezas had put the post-Franco succession firmly on the agenda. After the Pact with the USA and the Concordat, Franco could have stood down with considerable prestige and handed over to Don Juan but he was not remotely prepared to contemplate such an idea.

With the young Juan Carlos being educated in Spain, the problem of monarchist opposition seemed to be under control. Franco's health was good, permitting rigorous days of hunting or working days in which he would often remain standing from 7.00 a.m. to midnight without needing to rest. He would preside over cabinet meetings for nine hours at a time, taking neither food nor drink, nor breaking to relieve himself.[48] He had no inclination to give up power. He could handle routine business, and still manipulate his servants and play them off against one another. However, he was out of touch with the broader reality of social change and the aspirations of substantial sections of the Spanish population. The view of good and bad Spaniards, of victors and vanquished, of Francoists and anti-Francoists, which had served him well since 1939 was being rendered irrelevant by generational change. He remained convinced by the vision of the world provided for him by Carrero Blanco and others, a world in which he was the beloved father of his people protecting them from freemasons and Communists. Inevitably, his sharpness was blunted by daily immersion in flattery. Moreover, his appetite for maintaining a minute control of the political currents within the regime must have been sated by twenty years in the eye of the storm.

As Franco showed signs of wanting to retreat from the daily business of politics, he and his wife were assuming the distant air of royal personages except when they were with immediate family and friends. The annual summer party held at the elegant royal palace of La Granja had all the appearance of a royal occasion. Surrounded by the diplomatic corps, the military and religious authorities, members of the government and top functionaries and Falangists, Franco and Doña Carmen would receive homage. The annual departure to the Palacio de Ayete in San Sebastián with the government recalled the traditional summer custom of Alfonso XIII and his court. Often morose, Franco could become animated when talking about his deep-sea fishing or hunting triumphs. It was while holidaying on the *Azor* near San Sebastián, and spending huge amounts of State funds in long sea expeditions in search of tuna, that Franco revealed his isolation from the reality of daily life in mid-1950s Spain when he commented without irony that 'one is happier living austerely'.[49] Doña Carmen did not want her husband bothered with disagreeable news but 1955 saw the beginnings of intense crisis on two fronts and he would be forced in 1956, with discernible signs of reluctance, to join the fray.

The first problem concerned the Moroccan colony for which he had fought in his youth. Morocco, at this time, was still of primordial importance to military honour in general and to Franco in particular. Franco had entrusted the key role of High Commissioner to General Rafael García Valiño, one of the youngest and most brilliant tacticians among the Nationalist generals during the Civil War. García Valiño was regarded in some regime circles as a potential rival to the Caudillo. He had once declared that on the day that Franco died, he would turn up at El Pardo to take over. Franco's intimate crony General Camilo Alonso Vega, in particular, regarded him as 'ambitious and dangerous'.[50]

At a time when his French counterpart, General Guillaume, was intensifying the repression of Moroccan nationalists, García Valiño was pursuing, with the covert encouragement of Franco, an actively anti-French policy. He authorized local political parties, gave the Spanish zone a degree of autonomy and secretly encouraged the rebels in the French zone with arms and money. Franco permitted García Valiño's irresponsibility for a number of reasons. To an extent, there was little choice. The lamentable condition of the Spanish army was hardly such as to allow it to fight a major colonial war with any hope of success. The French empire was crumbling in both the Arab world and in the Far East, so Spain could not hope to fare better. Moreover, the rise of Nasser had encouraged militant Arab nationalism. Accordingly, Franco hoped to derive benefit from French discomfort and to make the best of Spanish weakness. A desire to see the Spanish Moroccan empire aggrandised at the expense of the French had been a constant feature of his African policy since 1939. Now, by allowing García Valiño to encourage local aspirations, he thought to ingratiate himself with the Arab world and perhaps secure Arab votes in the United Nations for Spanish membership. [51]

Subsequently, Franco was to maintain that García Valiño had been out of control and acting on his own initiative. This was simply not true. Indeed, Franco wrote a newspaper article under the pseudonym *Hispanicus* in favour of García Valiño's policy. What did annoy the Caudillo was the high-handed way in which his High Commissioner conducted internal Moroccan affairs. In particular, he was inclined against García Valiño by tittle-tattle emanating from his wife's crony, the Marquesa de Huétor de Santillán, about the disdain with which he treated members of the Franco family when they visited the zone.[52] That aside, he fully endorsed his High Commissioner's stance.

In August 1953, the French had deposed the Sultan Mohammed V. On 21 January 1954, García Valiño, speaking to a large crowd, declared his solidarity with the victims of French repression. Five days later, Franco granted a pardon to all Moroccan political prisoners. In early February, he received a delegation of Moroccan nationalists and denounced the French. Throughout 1954, the French repression intensified and García Valiño declared Spanish support for 'the evolution of the Moroccan people' and continued to abet the anti-French liberation movement. Finally, in August 1955, under pressure in both Vietnam and Algeria, the French decided to cut their losses in Morocco and lifted martial law. In November 1955, the Sultan was brought back. García Valiño congratulated a delighted multitude in Tetuan. Both he and the Caudillo seemed to believe that the deterioration of the French position had no relevance for the Spanish zone. With a blind and patronizing racism, they were confident, as were most Spanish *Africanistas*, that the Moroccans loved their Spanish rulers.[53]

The Caudillo made token references to future independence, but on 30 November 1955, he confidently predicted that the Moroccans would not be ready for this for twenty-five years. With the French beginning to talk seriously to the Moroccans, at the beginning of 1956, García Valiño sent a frantic telegram to Franco saying that, unless concrete promises of independence were made and a major programme of public works initiated to soak up local unemployment, the nationalist movement would soon turn against Spain. Franco telephoned García Valiño on 9 January 1956 and proposed that he issue vague statements about future independence. The local nationalists reacted to the Spanish procrastination by using the same violent methods which had been successful against the French. García Valiño now denounced his erstwhile nationalist friends as Communist subversives, closed down their newspapers and arrested prominent militants. When, on 2 March 1956, the French announced independence for Morocco, the Caudillo was left stranded. On 6 March, violent nationalist riots broke out in the Spanish zone.

Franco was obliged on 15 March to free the recently arrested nationalists and to announce that Spain would relinquish its own protectorate. On 5 April, Franco received Mohamed V in Madrid and was treated with the kind of icy disdain that he normally dispensed to others. In the unpleasant negotiations that took place on 4 April, he finally showed a sense of realism. After years of confident assertions about the special friendship with Morocco, his policy had shown no insight at all. Hunting and fishing had occupied a lot of his time during the final colonial crisis and he had tended to leave things to García Valiño. However, in the last resort, he knew that there was no question of fighting to keep the protectorate. The declaration of independence was signed on 7 April 1956.[54] To soften the blow, he intensified pressure on Britain over Gibraltar.

While Franco was occupied with the Moroccan crisis, domestic political problems arose which convinced him of the risks involved in permitting any kind of political interplay that resembled the activities of parties. Ever since his meeting with Don Juan at Las Cabezas in late 1954, he had been trying to ignore sporadic evidence of discontent within the Falange. In fact, the discontent went back further. Under the collaborationist leadership of Arrese and Fernández Cuesta, the Falange had accepted the regular postponement of its 'pending revolution'. However, new generations which had not fought in the Civil War were impatient with the endless compromises and the Falange's status as the Caudillo's claque. Their frustrations had been brutally underlined in January 1954 when the police had crushed the SEU demonstration over Gibraltar. In February 1955, the extreme Falangist militia, the *Guardia de Franco*, chanted insulting slogans against Prince Juan Carlos and were reported to have called Franco a traitor for his dallyings with Don Juan.[55] What was disturbing from Franco's point of view was the fact that these incidents had revealed an erosion of the unquestioning loyalty to his person that had previously been a central feature of the *Movimiento*.

Various liberalizing initiatives by the Minister of Education, Joaquín Ruiz Giménez in the universities had exacerbated tensions within the *Movimiento*. One early symptom was the publicity which surrounded the death and burial of the philosopher José Ortega y Gasset in October 1955. He was denounced by many in the regime, but there were those who paid homage to him as a free-thinker, using him as symbol with which to express their discontent with the stifling mediocrity of regime culture. A well-attended meeting in his memory held in the Madrid Faculty of Philosophy and Letters caused Ruiz Giménez considerable anxiety.[56] The students knew little about Ortega but he symbolized critical thought and the free interplay of ideas, things ruthlessly suppressed under Franco.

In fact, ferment in the universities was not the only sign that things were moving behind the repressive facade of regime uniformity. Working-class and left-wing opposition could be taken for granted by Franco as an annoying reality, which he dismissed as the work of sinister foreign Communist and masonic elements to be dealt with by harsh repression. The rivalry between military monarchists and senior Falangists had also been easily absorbed into his world picture. Believing that everyone could be bought, he had set about buying them or cajoling them or deceiving them. The rumblings of the mid-1950s were something different altogether and much more intractable for Franco. Spanish students of this period, even left-wing and liberal ones, were almost exclusively from comfortable middle-class families. Like the young Falangists who expressed a different kind of dissatisfaction with the regime, they could not simply be subjected to the savage repression casually dispensed against working-class strikers. Moreover, Franco had neither the time nor the flexibility to learn about these new forces. He desired to enjoy the fruits of power uninterrupted and to reap the rewards of 'saving an entire society' as he put it. The Moroccan crisis interferred with those ambitions but he did not take off excessive amounts of time to deal with it. If he had a domestic political concern, now that his own survival was comfortably assured, it was to ensure that a form of Francoism would exist after his 'lifelong magistracy' eventually came to an end.

Accordingly, he did not take seriously either the student unrest or the Falangist rejection of the slide into conservative monarchism. At the

November 1955 rally in El Escorial to commemorate the anniversary of the death of the Falange's founder, José Antonio Primo de Rivera, Franco confirmed the fears provoked among Falangists by his apparent rapprochement with Don Juan at Las Cabezas. Eschewing the usual black uniform and blue shirt of the *Jefe Nacional*, he distanced himself from the Falange by arriving for the ceremony in the uniform of a Captain-General. From the ranks of guard of honour, a voice called out 'we want no idiot kings'. It has also been alleged that a cry of 'Franco traitor' was heard. There were other minor incidents reflecting Falangist discontent with the complacency of the regime which Franco dismissed as of little consequence.[57] As an assessment of the political weight of such elements, it was a realistic response. However, Franco seriously misread the student unrest incidents as a symptom of the fact that Spanish society was beginning to move in a different direction from the regime. More than ever, the comforting assumption that any opposition was of Communist or masonic inspiration was inadequate.

An important indication of changes in Spanish middle class society of which Franco had virtually no inkling was provided by the Rector of the Universidad Complutense de Madrid, the liberal Pedro Laín Entralgo, a repentant Falangist and a nominee of Ruiz Giménez. In the wake of the student disturbances of January 1954, Laín Entralgo had begun to study the attitudes of Spanish youth. His report suggested that there was a widespread dissatisfaction with the stultifying atmosphere of the Francoist university. It implied too that students had severe misgivings about the moral standing of the regime and its servants. Laín claimed that what university students expressed today, the rest of society would feel tomorrow. His report was a plea for the windows of the regime to be opened before Marxism began to grow in its fetid atmosphere. Laín requested an audience with Franco at the end of December 1955 in order to give him the first bound copy of the report. Franco seemed insecure when faced with a topic about which he had no first-hand knowledge and he let Laín do most of the talking. The interview ended inconclusively but Laín Entralgo believed that the Caudillo later read the text. Both Martín Artajo and Ruiz Giménez mentioned to Laín that the Generalísimo had begun to use phrases in cabinet meetings which were suspiciously similar to the style of the report. Shortly afterwards, the sociologist José Luis Pinillos made a study of student attitudes and concluded that a large majority regarded the political and military authorities as incompetent and deeply immoral.[58]

Franco referred to the university tensions in his end of year broadcast on 31 December 1955. The speech was his reply both to Laín Entralgo's report and to the extent to which the poll held by Pinillos had indicated opposition to the dictatorship among the sons of Franco's most influential supporters.[59] Franco seemed finally to have realized that, nearly two decades after he became Head of State, the political atmosphere in Spain was changing. After the diplomatic triumphs of 1953, the artificial maintenance of unity as a response to international siege was no longer realistic. So, instead of the usual resumé of his great domestic and international achievements, Franco devoted this annual message to the dangers of subversion. The implication was that, as a result of the success of his 'captaincy', Spaniards were becoming too complacent and were therefore easy prey to the foreigners who wanted to divide them. He referred to the libertinage of the airwaves. His festering contempt for Ortega and the liberal intelligentsia was revealed when he spoke of the 'liberal after-tastes which are sometimes noticed and, like whited sepulchres, have brilliance and charm, but up close give off the masonic stink and stench which characterized our sad years'. He called upon the loyal intelligentsia to combat the subversion.[60]

Franco's defensive appeal disappointed many of his supporters.[61] It found an echo only in the most reactionary sections of the Falange. Hard-liners of various Falangist organizations and pressure groups, the *Vieja Guardia* (old guard), the *Asociación de ex-cautivos* (association of ex-prisoners of the Republic), the *Guardia de Franco*, the *Frente de Juventudes* and wartime veterans or *ex-combatientes* began to mobilize. Without the fabricated external threat of the 'international siege' to make Falangists huddle together around the Caudillo, the conservative mediocrity of the regime could no longer be ignored. The great Falangist revolution had not been made and the sight of the Caudillo consorting with the Anglophile Don Juan was an unpleasant reminder that, for all the symbols and the rhetoric, the regime was not really Falangist. In 1956, there would be increasing signs of Falangist indiscipline albeit under the banner of extreme Francoism. That Franco permitted such activities reflected both his essential sympathy with the negative, anti-liberal, anti-masonic, anti-Communist, elements of their rhetoric and his need for a bogy to use in the ongoing tug-of-war with the monarchists. Acquiescence in the antics of Falangist hotheads also reflected the extent to which, given the attractions of hunting and fishing, the sixty-four year-old Franco did not want to be bothered with intra-regime squabbles.

In fact, the cracks which appeared in 1956 went beyond the tantrums of younger Falangists. The discontent had many facets, ranging from internal feuding amongst the regime forces to working-class discontent at appalling housing conditions and living standards. At the beginning of 1956, the new British Ambassador Sir Ivo Mallet commented that, in the wake of the end-of-year broadcast, the view had taken hold that Franco was 'a complete cynic, interested only in keeping power as long as he lives, and indifferent to what may happen when he dies. He is said to keep two folders on his desk, one marked "problems which time will solve" and the other "problems which time has solved", his favourite task being, it is said, to transfer papers from one folder to the other.' The Caudillo must have reflected that any shift of power towards the monarchists would weaken his own position, since their loyalties lay elsewhere. Accordingly, as monarchists grew more confident, his natural reaction was always to incline back to the Falange which depended on him for its very existence.[62]

Franco's complacency had left him unprepared for the crisis that was imminent. At the beginning of 1956, the disquiet of Franco's more moderate and passive supporters took the form of alarmist rumours, including speculation about the Caudillo's physical degeneration. Rumours that Franco needed an operation on the prostate gland had been rife for some months. After some investigation, Mallet concluded that 'the probable truth is that he may have a growth in the bladder necessitating treatment at present and perhaps an operation later. Outwardly he appears in his portly way to be in the pink of health and continually receives delegations, pays visits and goes shooting.' Mallet was convinced that Franco would respond to this potential crisis as to earlier ones by doing nothing and just sitting tight.[63]

Franco was fortunate that the resurgence of discontent on several fronts paradoxically inclined army officers, Falangists and monarchists to rally to his cause. In 1956, the university tensions which had emerged in January 1954 and again after the death of Ortega y Gasset broke out into the open. Left-wingers and liberals were pushing for an opening-up of the system. Clashes took place in the old Law Faculty in San Bernardo in the centre of the city between the progressives and the militant Falangists who lashed out as they saw their 'pending revolution' apparently being postponed for ever. Organized bands of Falangists rampaged through the University on 8 February beating up students and destroying offices and lecture rooms. The conflict intensified on the following day. A group of armed thugs from the extremist *Guardia de Franco* returning from a ceremony in memory of Matías Montero, a Falangist killed during the Second Republic, clashed with some progressive students. One of the Falangists,

Miguel Alvarez Pérez, was shot and seriously hurt, either by a policeman or by the accidental discharge of a gun carried by one of his own companions. The symbolic similarity between Alvarez and Montero allowed hard-liners to invoke the spirit of the pre-Civil War Falange.

On the evening of 9 February, wild rumours circulated that the Falange was planning a bloody revenge not least as a means of reasserting its political position. It was said that Tomás Romojaro, Fernández Cuesta's under-secretary of the Movimiento, had authorised the arming of Falangist squads. Black lists of 'traitors' were drawn up. Ruiz Giménez was warned by both the Minister of Labour José Antonio Girón and the Minister of the Interior Blas Pérez that his life was in danger.[64] In fact, the Captain-General of Madrid, General Miguel Rodrigo Martínez, had made it unmistakably clear to the Falangists that he would tolerate no violence. He, the Minister for the Army, General Muñoz Grandes, and General Carlos Martínez Campos, tutor to Prince Juan Carlos, visited the Caudillo on the morning of 10 February and expressed their displeasure at the activities of the *Guardia de Franco*.[65]

Franco had been kept fully informed of developments by the police during the night of 9 February. His initial response was not to take things too seriously, partly because of a gut sympathy with the Falangists and partly out of an instinctive habit of underplaying crises. The controlled press blamed the incidents on Communist agitators.[66] When Muñoz Grandes, Rodrigo Martínez and Martínez Campos appeared at El Pardo before breakfast to ask Franco, in the name of the Army, what he planned to do to control the Falange, he replied, with his customary insouciance, that he thought that the threats would come to nothing. However, when Muñoz Grandes told him that, if any of the names on the 'black list' were harmed, then the Army would take over Madrid, Franco allegedly promised to order the arrest of the Falangist conspirators.[67]

At a cabinet meeting held later on the same day, the rights 'enshrined' in the pseudo-constitution the *Fuero de los Españoles* were suspended for the first time. Nevertheless, Franco was sufficiently unruffled by the incidents to set out immediately on a large hunting party along with Muñoz Grandes, Arburúa, and a group of aristocrats and businessmen. He was back in time for another cabinet meeting, on 13 February. Attempts by Martín Artajo to suggest reasonably that the shot which hit Miguel Alvarez might have come from the Falangists or from the police was brusquely cut short by the Caudillo who accused him of swallowing information which came from 'the enemy' – a reference to the BBC which he remained convinced was the mouthpiece of international free-masonry.[68]

Franco believed that Ruiz Giménez's liberalizing tendencies had per-mitted left-wing elements to come to the surface in the universities. Equally, it was felt in El Pardo that Fernández Cuesta had failed to check the emergence of anti-Franco tendencies within the *Movimiento*.[69] In fact, students, who regarded Franco as a political fossil, would agitate against his regime with increasing frequency and intensity throughout the 1960s. Despite his usual inclination not to be forced into precipitate action, and his reluctance to replace ministers once he had got used to them, the resurgence of hostility between the Falange and the military high com-mand impelled the Caudillo to seek scapegoats in the form of the two ministers. Ruiz Giménez was summoned for an audience with Franco on 14 February and offered his resignation. Franco did not reply directly but, in typically crab-like fashion, he said that, as a result of the crisis, the Minister of Education and the Minister-Secretary of the *Movimiento* were 'going to leave'. He also said that he did not have time for a full-scale cabinet reshuffle.[70]

If anything that remark suggested a concern for his hunting commit-ments rather than a perception of the real depth of the political difficulties facing him. The fact that the two victims of the crisis came from two of the main regime groups represented in the cabinet does not mean that Franco was undertaking a subtle balancing act. The dismissals were a knee-jerk response to an unexpected problem. At the time of the riots, Fernández Cuesta was abroad on official visits to Brazil representing Franco at the inauguration of Juscelino Kubitschek and to the Dominican Republic returning Trujillo's 1954 visit to Spain. He was immediately summoned back from Washington where he had stopped off *en route* from Latin America. Shortly after arriving in Madrid on 14 February, he was received by Franco at El Pardo. He gave the Caudillo a gift of a spectacle case which he had brought him from New York. Franco did not acknowledge the gift and then listened impassively as Fernández Cuesta reported on his trip. When he began, in his capacity as Minister-Secretary of the *Movimiento*, to comment on the recent crisis, Franco icily told the perplexed Fernández Cuesta that it was not his concern since he was no longer Minister-Secretary.[71]

On the same morning, Franco sent for Arrese who had been in political retirement since 1945. When he arrived at El Pardo at 6.30 p.m. on the evening of 14 February, Franco painted a dramatic picture of liberal threats and Falangist indiscipline. He made it clear that he was looking for someone who could reimpose discipline without appearing to be crushing the Falange. Arrese had done the same job after the crisis of 1942 and there was a certain predictability about his being called in again as the reliable fireman. Believing that much of the present crisis derived from the lack of an exclusively Falangist government – he regarded the existing cabinet as a coalition – Arrese skilfully insinuated that Fernández Cuesta had been the victim of his own benevolence. Franco indicated to him that he could take over the programme of constitutional preparations for the post-Franco future entrusted to Fernández Cuesta a year earlier and, after token resistance, Arrese accepted the job.[72] Both Fernández Cuesta and Ruiz Giménez were officially replaced on 16 February.

Franco brought back Arrese to give his special varnish of compliance to the Falange, although his ambitious plans were quickly to provoke an alarming polarization of the Francoist coalition. Ruiz Giménez was replaced by another Falangist, Jesús Rubio García-Mina, a conservative professor whose view on the recent disturbances was 'students should study' (*estudiantes a estudiar*).[73] Unlike the cabinet changes of 1945 and 1951, these were not deliberate and considered changes of direction but rather botched emergency repairs along the road. Neither was the right man for the situation, Arrese because he was dangerously ambitious, Rubio because he was too unimaginative. Franco had little choice but to cling to the Falange. Doing so in 1945 and 1951 had seemed daring, when concessions to the monarchists had been considered inevitable. There was nothing daring about the partial reshuffle of 1956. Franco could not abandon the Falange without putting his fate into the hands of those senior Army officers who wanted an earlier rather than a later restoration of the monarchy. However, Arrese was not an option for the long term. The Falangist violence of February 1956 was a symptom of a death agony rather than of youthful vitality. Franco's reflex reaction to the crisis of 1956 was the first sign that he was beginning to be less dominant a figure. Preoccupied by the Moroccan problem and underestimating the seriousness of the crisis, he did not control events but was driven by them.[74]

The instinctive response of reasserting Falangist pre-eminence was understandable given the mutual dependence of Franco and the Falange. Franco was responding not only to the immediate crisis of February 1956 but also to evidence of hard-line Falangist discontent which had been building up since his rapprochement with Don Juan a year earlier. At the end of 1955, leaders of the Falange in Madrid had presented Franco with a memorandum demanding the swift implementation of the *revolución pendiente* under the exclusive control of the Falange. It was effectively a blue-print for a more totalitarian one-party State structure.[75] The appointment of Arrese was in large part a response to this sentiment. As Minister-Secretary, Arrese would try to implement many of the memorandum's recommendations with the acquiescence of Franco. However, in the changing Spain of the 1950s, slowly integrating into western capitalism, totalitarian Falangism was not a serious long-term option.

The Caudillo's inadequate response to the crisis of February 1956 was the consequence of the fact that, at a time when he wanted to sit on his laurels and enjoy his pre-eminence, he had to cope simultaneously with too many pressures. Franco's cousin recounted various meetings at this time with a silent and morose Caudillo at lunch in the Pardo, failing to respond to conversational gambits and chewing on toothpicks which he then left in a pile on the table.[76] There was no public indication of the horror felt by Franco the *Africanista* at the prospect of the decolonization of Morocco but the loss of his habitual serenity in private was understandable in a man who had once said 'Without Africa, I can scarcely explain myself to myself.'[77]

Franco was too preoccupied in March 1956 to think through the full implications of Arrese's scheme to draft a set of 'Fundamental Laws', a kind of Francoist constitutional reform. Arrese's ideas for the '*refalangistización*' or '*totalitarización*' of the regime owed much to the model of the Third Reich and envisaged a massive increase of power for the single party. In April, in the aftermath of the loss of Morocco, Franco grasped at Arrese's scheme as a way of revitalizing his rule. Instead of calming spirits within the regime after the February crisis, Arrese's plans caused intense polarization. They were perceived by traditionalists, Juanista monarchists and Catholics as a neo-Nazi scheme to block any future liberalization under a restored monarchy and perpetuate the Falangist domination of the regime.

It is a measure of Franco's preoccupation that he failed to be suspicious of the scale of Arrese's ambition and then was himself seduced by a scheme presented to him in the most sycophantic wrapping. Arrese had discussed with Girón and other top Falangists the wisdom of trying to persuade Franco to step down as head of government. That remained a medium-term aim. In the meanwhile, Arrese's priority was to toughen the legislative framework within which the post-Franco succession would have to take place. With his inimitably obsequious style, he told Pacón, presumably in the hope that it would be passed on to the Caudillo, that nobody could replace or succeed Franco.[78] Arrese had then told Franco in person on 27 February 1956 that all his powers could not possibly be transmitted to his royal successor because the Caudillo was 'unrepeatable'. Since the decadence of liberal monarchy was one of Franco's favourite themes, it was easy for Arrese to persuade him that safeguards were needed to prevent the risk of democratic reform under a weak king. Arrese was thus given the green light to eliminate any loophole in the set of laws which made up the Francoist 'constitution' that might permit a future King to uncouple himself and Spain from the *Movimiento*. Arrese's plan was announced in Valladolid at the twenty-second anniversary ceremony on 4 March 1956 of the union of *Falange Española* and the *Juntas de Ofensiva Nacional Sindicalistas*. In a rhetorically violent speech, Arrese talked of smashing Communism and liberalism with 'fists and guns' and declared that the first objective was to 'capture the street'.[79]

At first, Arrese's plans were elaborated while Franco was distracted throughout March not only with the imminent decolonization of Spanish Morocco but also with rumblings of economic and social discontent. The cost of living index had risen by fifty per cent in the course of the previous twelve months. With wages effectively frozen, the economic crisis was borne by the working class. Faced by the threat of a repeat of the Barcelona strike of 1951, Franco's cabinet met on 3 March to discuss rising working-class militancy. The urgent advice of José Antonio Girón, the Falangist Minister of Labour, supported by Arrese, was for the government to decree across-the-board wage rises of twenty-three per cent. There was a bitter argument between Girón and the Minister of Commerce, Manuel Arburúa, who pointed out the inflationary consequences of such a strategy. In the short term, Girón won the battle. It was naïvely announced that the pay increase would have no repercussions on prices.[80] The wage rises did not come in time to delay a series of strikes which began in the shoe-manufacturing industry in Pamplona in April and then spread through the Basque steel industry and into the Asturian coalfields.[81] Even after the strike wave temporarily abated, the personal conflict between Girón and Arburúa, and between their different conceptions of the role of the state in the economy, continued to be a problem for Franco.*

The Caudillo's greatest skill was always to let every section of the Francoist coalition believe that, if only they remained loyal, they could get what they wanted. Only absorption in other problems explains why,

---

* The urgency of the labour situation occupied an emergency ten-hour cabinet meeting on 29 April in the Alcázar of Seville, where the Caudillo had gone for the annual *Feria* – *The Times*, 30 April 1956.

for once, Franco should permit Arrese to close off all options for the succession but the Falangist one. Although he did not show it publicly, Franco was devastated by the loss of the colony for which he had fought in his youth. It was also the end of the dream of African empire for which he had so nearly gone to war on Hitler's side in 1940.[82] For the first time since he had come to power, Franco had suffered a humiliation that he could not turn to his advantage or rewrite in flattering terms. The decolonization of Morocco also meant that Franco had to disband his *Guardia Mora*, one of the most characteristic symbols not only of his semi-royal, imperial status but also of the terror on which his rule was built.

The *Guardia Mora* had served as a reminder of the fear generated by the Army of Africa in the Civil War. Its disappearance did not mean the end of the division of Spain into victors and vanquished but it did represent a further step towards the drab ordinariness of Franco's rule. The defiance of 1946 seemed as far away as the imperial ambitions of 1939. The Caudillo himself was looking to enjoy the present and to secure the future. The days of glorious struggle and comparisons with El Cid were gone. In response to his slide into comfortable routine, the regime forces would make their own dispositions for their futures. Paradoxically, as his rule lost dynamism and direction and they squabbled, they needed him the more as the arbiter who held the system together. Franco, of course, was happy to go on being indispensable.

XXV

LEARNING TO DELEGATE

Homo Ludens, 1956–1960

ARRESE'S speech of 4 March had alerted other regime forces to the danger of a Falangist bid for an iron monopoly of the Francoist future. The new Minister-Secretary was behaving with an ebullience which suggested that he had the Caudillo's encouragement. His confidence blinded him to the strength of the forces which quickly ranged against him. The first to mobilize were the supporters of Don Juan, although traditionalists and Catholics within the Francoist elite were not far behind. In the spring of 1956, Don Juan's representative, the Conde de Ruiseñada, gave General Juan Bautista Sánchez, the Captain-General of Barcelona, a plan aimed at blocking Arrese's schemes by means of an early restoration of the monarchy. The Ruiseñada plan envisaged Franco being obliged to withdraw from active politics to the position of regent. The day-to-day running of the government would be assumed by Bautista Sánchez until the King was restored. The involvement of Bautista Sánchez – the most respected professional in the armed forces – helped secure the support of other monarchist generals against Arrese.[1]

That Franco was anything but neutral in the competition for the future being contested by Falangists and monarchists was made starkly obvious in the spring of 1956. His views were proclaimed publicly in the course of a propaganda tour of Andalusia during which he was accompanied by Arrese. After the gloom of recent months, Franco was energetic and enthusiastic on his journey. His mood suggested that, with Morocco irrevocably gone, a weight had been lifted from his shoulders. As at so many other times in his life, he revealed his remarkable capacity to shrug off disasters and press forward. In the militant Falangism that he was to display in Andalusia, he sought a terrain in which to flex his muscles and compensate for the humiliation at the hands of the Sultan. His rhetorical excesses also served to show the Falange that despite the imperial setback, the fervour of their *Jefe Nacional* was undiminished. Opening a shipyard in Seville on 24 April, he declared that his regime bore comparison with 'the best regimes ever known or even imaginable'.[2] As the tour progressed, Arrese rekindled the fighting spirit of harder times. There is little doubt that Franco liked the always beaming Arrese and responded well to his particular kind of adulation. Arrese boasted that he was 'the minister who had the closest relationship with Franco'.[3] When they went on tour together, as they had in Catalonia in January 1942 and Galicia in August 1942, Arrese seemed to inspire Franco with enthusiasm for his role as *Jefe Nacional*.

Apart from skilfully preparing the enthusiastic receptions given to the Caudillo by crowds of delirious Falangists, Arrese would work on him during the long car journeys persuading him that such demonstrations reflected mass enthusiasm for a more strongly Falangist line. He was delighted to hear Franco give a tone of 'superfalangism and aggression' to his speeches. In Huelva on 25 April, the Caudillo delighted his audience with an unmistakable and insulting reference to the monarchists and to Juan Carlos. He declared that 'we take no notice of the clumsy plotting of several dozen political intriguers nor their kids. Because if they got in the way of the fulfilment of our historic destiny, if anything got in our way, just like in our Crusade, we would unleash the flood of blue-shirts and red berets which would crush them.'[4] Addressing twenty-five thousand Falangists in Seville on 1 May 1956 with palpable emotion, Franco passionately denounced the enemies of the Falangist revolution, the liberals and the political wheeler-dealers, working in the interests of masonic lodges and Communist internationals. Referring to his own near-monarchical status, he announced that Spain was constitutionally 'a monarchy without royalty' and worked himself up to the declaration that 'the Falange can live without the monarchy but what could not survive is a monarchy without the Falange.'[5] Franco's belligerence suggested that he was aware of the plan being discussed by Ruiseñada and General Bautista Sánchez.

It is striking that Franco, whose greatest talent was his ability to maintain the political balance and shroud his intentions in nebulous vagueness, should have gone so far. Looking to make up for the Moroccan humiliation, on the long drives between the provincial capitals of Andalusia, Franco had let himself be enthused by Arrese's talk of a glorious Falangist future into explicit declarations disturbing to many monarchists who were happy to go along with the regime as long as it was ideologically undefined.[6] In addition to the disquiet generated by Franco's speeches, alarm was provoked by the arrogant presumption with which Arrese went around the Francoist elite making consultations about possible cabinet changes and constitutional amendments. There was even talk of the rehabilitation of Hedilla and, on the assumption that the predominance of the Falange was assured, many prominent figures began to cultivate Arrese as the man of the hour.[7] Fernández Cuesta told Sir Ivo Mallet that the aim of the new constitution was to give the Falange a position comparable to that of the Communist Party in Russia.[8]

At least two of the members of the committee helping to draw up Arrese's proposed laws were bothered by the fact that the drafts made no mention of the monarchy in their detailed provisions for the post-Franco succession. Carrero Blanco drew up notes in which he suggested that what Spain needed was 'a traditional monarchy for the present day'. The Minister of Justice, the traditionalist Antonio Iturmendi, was also following Arrese's preliminary efforts with some hostility. He had commissioned one of his brightest collaborators to go through Arrese's project with a critical eye. The man undertaking the job was the Catalan monarchist and professor of administrative law, Laureano López Rodó.[9] A deeply religious member of Opus Dei, the austere López Rodó, who would soon rise to dizzying eminence, was the very model of an Opus Dei militant, quietly confident, hard-working and efficient.

Despite the early misgivings of Carrero Blanco and Iturmendi, on 20 June, when Arrese discussed his plans with the Caudillo, he came away with the impression that he could still rely on his full support.[10] However, other influences were building up around Franco. General Antonio Barroso, his wartime chief of operations and now Director of the *Escuela Superior del Ejército* was deeply alarmed by his 1 May speech. On the eve of taking up the post of Head of the Caudillo's Military Household, on 1 July 1956, he protested to Franco about the Arrese plan. Along with two other monarchist generals, one of whom may well have been Bautista Sánchez, he is alleged to have broached the Ruiseñada plan to the Caudillo. They suggested that a military directory take over and hold a plebiscite on the issue of monarchy or republic, in the confident expectation that such a plebiscite would produce support for the monarchy.[11] Faced with a plan to eliminate him politically, Franco was superficially non-committal but his concern for the Army's loyalty was reflected in the fact that one month earlier he had decreed massive pay rises for officers.[12] His reaction to the visit by Barroso may be deduced from a sudden and perceptible change in his attitude to Arrese. Two days after it, Franco received Arrese at El Pardo and alarmed him by expressing noticeably less enthusiasm for his schemes without a hint as to why.[13]

Until that moment, Arrese had felt himself to be 'the golden boy (*el niño mimado*) of El Pardo' and dreamed of an all-Falangist government. Arrese was friendly with Doña Carmen and spent considerable time showering obsequious attention upon her, converting her to his vision of the Falangist future.[14] His success might be measured from the fact that, throughout June 1956, the normally secretive Franco had discussed with him a possible ministerial reshuffle to be announced on 18 July, the fifth anniversary of his last cabinet change. The Minister-Secretary had then provoked hostility in the Francoist elite by speculating openly about the removal of Martín Artajo and the creation of an all-Falangist cabinet to be dominated by himself as secretary to the president, the key strategic post from which to influence Franco. Whispers in the Pardo that Arrese was too ambitious and the political anxieties provoked by his constitutional plans soon led Franco to see him as 'a wild horse which had to be reined in'. The Caudillo dropped his plans for a cabinet reshuffle.[15]

One of the immediate consequences of the arrival of Arrese in the government was a fifty per cent reduction in the number of cabinet meetings. Pointing out that the interminable meetings were not efficacious, and often ran into the early hours of the following morning without doing useful business, Arrese managed to persuade Franco to reduce them from weekly to fortnightly. He claimed to be anxious to conserve Franco's energies for Spain but it was also part of his plan to strengthen the role of the Falange. He hoped vainly that some cabinet functions might be taken over by the executive committee of the Falange, the *Junta Política*.[16]

Criticisms of Arrese's plans from Barroso, Iturmendi and Carrero Blanco account for the cooling of Franco's support for his scheme. Nonetheless, when the Caudillo came to make his speech to the *Consejo Nacional de FET y de las JONS* on 17 July 1956, the twentieth anniversary of the military uprising, he gave little sign of being about to abandon Arrese, at whose suggestion the occasion was mounted. That Franco should go along with his initiative was a significant gesture of support for the Falange. The *Consejo Nacional* had not met since Arrese's departure from government in 1945 and it formed a crucial part of his plans for the future as watchdog of the ideological purity of Franco's successor. Arrese helped Franco sketch out the plan of his speech and gave him notes on the proposed 'fundamental laws' in order, in his own phrase, 'to ensure that he did not say anything, either under the influence of other sectors of the *Movimiento* or to calm liberal and monarchist anxieties, which might put us in an embarrassing situation later on'.[17]

The Caudillo, drawing on Arrese's notes, confirmed the central role of the *Movimiento* in the plans for the succession.* A long hymn of praise to his achievements in power, the speech calmed Falangist fears that a future monarchist successor might use his absolute powers to bring about a transition to democracy. Franco stated, still following Arrese's notes, that the unique combination 'in my magistracy of a series of providential circumstances' was unrepeatable and the powers of any future Head of State would be carefully defined by the *Movimiento* which would have 'a permanent mission, a constant task, to guarantee the permanence of the principles for which people fought and died'.[18]

Assuming that Franco was aware of the implications of what he was reading out, he seemed to be announcing his support for Arrese's draft of the 'constitutional' framework which would guarantee the *Movimiento*'s political monopoly. A deeply gratified Arrese took Franco's words to be an endorsement of his position. More militant Falangists were disappointed with what they perceived as a limitation on their ambitions. In self-justification before those who felt that his regime was becoming too conservative, Franco tried to demonstrate that his regime had fulfilled the ambitions of José Antonio Primo de Rivera. His words were taken as meaning that there was no 'pending revolution', that the regime had made as much of a revolution as it was ever going to make. Girón commented to Arrese that the disappointment of the *Consejeros* could be heard in the sad tone with which they joined Franco in singing the Falangist anthem *Cara al sol* at the end of the session.[19]

Martín Artajo was thoroughly alarmed by the *Jefe Nacional*'s speech. In praising the Falangist system, Franco lauded Fascist Italy and Nazi Germany and made sneering reference to the post-war democratic systems 'imposed' on the defeated Axis powers by frightened and envious western allies. Franco's words reflected his deeply held conviction that prosperity was impossible in a democracy. He had recently confided to his cousin Pacón his belief that the Allies had obliged Germany, Italy and Japan to adopt democracy in order to guarantee their economic prostration.[20] In response to a plea from Martín Artajo, the anti-democratic remarks were omitted from the published version of the speech.[21]

---

* The notes had been drafted for Arrese by Emilio Lamo de Espinosa, recently named Director of the Falange's Instituto de Estudios Políticos.

Spending August in San Sebastián, Franco was made aware that the tensions over the plans for the future were dwarfed by the conflict raging between his ministers about how to cope with spiralling inflation and intensifying social problems. The squabbling affected several of them but the gladiatorial champions of the two opposed views were Girón, who pushed for more spending, and Arburúa, who counselled austerity, particularly where the state holding company, the INI or Instituto Nacional de Industria, was concerned.[22] Franco seemed indifferent. Having already put off a major reshuffle on 18 July, he had no desire to face up to the problem in the midst of his summer holidays. Conflicts surged again in September, by which time he had moved the entire court to Galicia. At a cabinet meeting held in the Pazo de Meirás on 14 September, considerable heat was generated by the issue of Spain's stance on the Suez crisis. Arrese was keen to associate Falangist Spain with Nasser's Egyptian nationalism as a gesture of virile solidarity and as a blow against perfidious Albion. Franco's sympathies lay with Arrese and he had authorized the sale of weapons to Colonel Nasser, a decision which subsequently was to undermine his efforts to secure membership of NATO. However, with Martín Artajo arguing the need for Spain not to be out of step with her American ally, the Caudillo restrained Arrese.[23]

In the middle of the holidays, General Barroso had replaced Franco Salgado-Araujo as Head of the Caudillo's Military Household. Pacón had served with his cousin in one capacity or another for more than forty years and indeed now stayed on as his military secretary. Nevertheless, he was deeply hurt by the fact that, having reached the obligatory retirement age, the coldly indifferent Franco did nothing to mark the occasion, neither thanking him for his years of service nor even mentioning it. On taking up his post, Barroso confided a number of worries to Pacón. Ultimately loyal to the Caudillo despite his monarchist sympathies, Barroso was concerned by complaints that Franco was beginning to put off important business. His greatest anxiety was the succession since he believed that, if Franco died soon, it would simply be resolved by the decisive action of the most daring general, probably García Valiño or Muñoz Grandes.

Barroso believed that Franco was losing touch with the military hierarchy and was particularly disturbed – as were Pacón and the Minister for the Army, Muñoz Grandes – by the increasing ostentation of the Franco family which was causing disquiet within the generally austere high command. Since Nenuca's marriage to the playboy Cristóbal Martínez de Bordiu in 1950, the dictator's wife had plunged into high society and given ever freer rein to her penchant for jewellery and antiques. This had led to her acquiring the popular nickname *Doña Collares* (Doña Necklaces).[24] Pacón was outraged by Carmen's efforts in the summer of 1956 to doctor her husband's past. In early August 1956, the Alcalde of El Ferrol had requested that the Franco family home in the Calle María be converted into a museum. Franco agreed but before the house was handed over to the municipal authorities, his wife had it restructured and refurnished. The house and its modest furnishings had been a reflection of the life of a middling naval officer with four children. By stocking it with antiques and expensive porcelain, albeit of the appropriate period and in exquisite taste, Carmen Franco set out to create an upper middle-class or semi-aristocratic past for her husband.[25]

The Caudillo's springtime fervour for a revived Falange withered in the late autumn. His enthusiasm was badly battered in the storm provoked when Arrese launched his plans publicly. To coincide with the twentieth anniversary of Franco's elevation to the Headship of State, Arrese organized on 29 September a huge Falangist rally in Salamanca to celebrate the meeting of generals twenty years earlier which had elected Franco Head of the Government of the Spanish State. The original wooden cabin (*barracón*) where the generals met had rotted away but Arrese arranged for a replica to be erected. Martín Artajo was unable to attend because of prior ministerial commitments which led Franco to comment indignantly to Arrese 'what business can Artajo have that is so important that he can't leave it for me?' After inspecting twenty thousand Falangists on the famous wartime airfield, the Caudillo made a speech which gave an unashamedly egoistic panorama of the previous twenty years.

Franco claimed that the elevation of 29 September 1936 had been forced upon him. The task, which the generals at the time had seen as provisional while the war lasted, he recalled as being to oversee both 'the long and painful struggle and, after it, the indispensable total transformation of our *Patria* in order that the blood spilled might bear fruit, the tireless sacrifice of my entire life in the service of the nation'. His sacrifice had not been in vain: 'If politics is the art of serving the common good, I doubt that there could have been a policy which could better have served the collective interest of the Spanish people. It goes without saying that we certainly cannot find a better one in the entire history of our nation.' The Falangist ministers present were delighted but Franco had made no mention of the 'fundamental laws' in preparation. The message perceptible between the lines was that Arrese was being abandoned to face alone the furore to come.[26]

Before Franco spoke, and to the annoyance of Carrero Blanco and Iturmendi, Arrese had announced that the draft of his *Leyes Fundamentales* had been given to members of the *Consejo Nacional* for their final views. Although the text recognized Franco's absolute powers for life, it left his successor at the mercy of the *Consejo Nacional* and of the Secretary-General of the Falange, a position which Arrese envisaged for himself. When the text was distributed, there was uproar in the Francoist establishment. Monarchists, Catholics, archbishops and generals joined in opposing a text which proposed giving the *Movimiento* totalitarian control over all aspects of Spanish life. Esteban Bilbao, the President of the Cortes, and the Conde de Vallellano, the Minister of Public Works, compared the project to Soviet totalitarianism.[27] There was outrage in the Army at what seemed like an attempt to block the return of the monarchy. And, on 12 December 1956, three of the four Spanish Cardinals, the Primate, Enrique Plá y Deniel, Archbishop of Toledo, Benjamín Arriba y Castro, Archbishop of Tarragona, and Fernando Quiroga Palacios, Archbishop of Santiago de Compostela sent Franco a letter denouncing Arrese's text for flouting Papal encyclicals such as *Non abbiamo bisogno* and *Mit brennender Sorge* in its similarity to Nazism, Fascism and Peronism.[28]

The impact of the Cardinals' protest was made clear when Franco saw Arrese on 18 December. With their document in his hand, the Caudillo said to Arrese, 'I have something very disagreeable and very serious here' and made it clear that he would not be confronting the Church hierarchy. Arrese offered his resignation. Franco said that a better solution would be for him to amend his text.[29] Franco remained sympathetic to Arrese's plans and to Arrese himself but instructed Carrero Blanco to persuade him to amend the schemes in such a way as to make them acceptable to their opponents. Arrese regarded this as the 'castration' of his plans. However, he acquiesced after three long interviews with Franco on 7, 8 and 9 January 1957 in the course of which he was told by the Caudillo that ministers resigned only when he wanted them to. Franco gave Arrese the impression that he still supported his ideas, but had his hands tied by military and clerical opposition. An extremely watered-down text was eventually produced.[30] Ever flexible in his own interests, Franco had given way before the pressure of the forces hostile to Arrese. Nonetheless, he was reacting to events, not controlling them.

Between the two extremes of Ruiseñada's negotiated transition to Don Juan and a retreat into Arrese's fortress Falangism, there emerged the middle option favoured by Carrero Blanco and ultimately adopted by Franco. This consisted of an attempt to create the legislative framework for an authoritarian monarchy to guarantee the continuity of Francoism after the death of the Caudillo. The technician encharged with the job of producing a blueprint was the administrative lawyer Laureano López Rodó. Carrero Blanco was immensely impressed with the critique of Arrese's text that López Rodó had drawn up for Iturmendi. Recognizing his talent and capacity for hard work, at the end of 1956 Carrero Blanco asked him to set up a technical secretariat in the *Presidencia* to prepare plans for a major administrative reform.[31] As secretary of the *Presidencia* (the office of the president of the council of ministers), the doggedly loyal Carrero Blanco was Franco's political Chief of Staff. As Franco began to relax his hold on day-to-day politics, Carrero Blanco, who shared all of Franco's political prejudices and some of his political cunning, was gradually assuming some of the tasks of a prime minister. López Rodó, in his turn Carrero's Chief of Staff, would consolidate that tendency by creating an administrative machine to confront the complex technical problems of a modern economy. This inevitably marginalized Franco.

López Rodó had a long-term plan for a gradual evolution towards the monarchy. He constituted, in Francoist terms, a sanitized, less risky, version of Rafael Calvo Serer's *Tercera Fuerza* middle way between a Falangist Left and a Christian Democrat Right.[32] Unlike Calvo Serer, who was a Juanista, López Rodó would work towards a restoration in the person of Prince Juan Carlos. The partisans of Don Juan were less patient. Bautista Sánchez was trying to consolidate support for Ruiseñada's plan to sideline Franco and place Don Juan upon the throne. Since the Caudillo suspected the fervently Catholic and monarchist Bautista Sánchez of being a freemason, he was under constant surveillance by the intelligence services.[33] In December 1956, a meeting of those military and civilian monarchists involved in the Ruiseñada plan was to take place under the cover of a hunting party at one of the estates of Ruiseñada, 'El Alamín' near Toledo. Bautista Sánchez decided against attending after Muñoz Grandes reminded him that, as a *Procurador*, he must attend a meeting of the Cortes. Not to attend because of military duties in Catalonia was one thing, not to do so in order to attend a conspiratorial meeting near Madrid was altogether more dangerous.[34]

Things came to a head in mid-January 1957, when another transport users' strike broke out in Barcelona. While it was not as dramatic nor violent as that of 1951, it was linked with anti-regime demonstrations in the university organized on the pretext of solidarity with the uprising in Hungary.[35] The Civil Governor, General Felipe Acedo Colunga, used considerable force in evacuating the university and stopping demonstrations in favour of the strikers. Bautista Sánchez was critical of Acedo Colunga's harsh methods, counselled caution and was therefore considered in some circles to have given moral support to the strikers.[36] Franco was displeased by the Captain-General's failure to help Acedo. At the same time, he was sufficiently concerned about pro-Arrese demonstrations by Falangists to authorize Blas Pérez, the Minister of the Interior, to have Arrese's house watched and his telephone tapped.[37]

There were rumours flying around Madrid that Bautista Sánchez was planning a coup. Franco himself seems to have toyed with the bizarre notion that the Captain-General was fostering the strike in order to provide the excuse for a coup in favour of the monarchy. The plan hatched with Ruiseñada was more than enough to rouse the ire and suspicions of the Caudillo. However, as far as military action is concerned, it is likely that the rumours were based on the wishful thinking of prominent monarchists. The conversations of royalist plotters with the Pretender's household in Portugal were being tapped by the security services. The Caudillo, ever cautious, reacted as if their optimistic speculations merited some anxiety.[38]

To be on the safe side, Franco sent two regiments of the Legion to join in manoeuvres being supervised by Bautista Sánchez in Catalonia. The Lieutenant-Colonel commanding the regiments informed Bautista Sánchez that he could take direct orders only from Franco himself.[39] Muñoz Grandes also appeared in the course of the manoeuvres and had a tense interview with Bautista Sánchez in which he apparently informed him that he was being relieved of the command of the Capitanía General de Barcelona. On the following day, 29 January 1957, Bautista Sánchez was found dead in his room in a hotel in Puigcerdá. The most dramatic and bizarre rumours that he had been murdered quickly ran around Spain.[40] What is most likely is that, having long suffered poor health, Bautista Sánchez had died of a heart attack after the shock of his painful interview with Muñoz Grandes.[41] The large numbers of mourners at his funeral were testimony to the hopes that had been placed in him. Franco commented to Julio Danvila 'death has been kind to him. Now he won't have to fight the temptations that tormented him so much in his last days. We were very patient with him, helping him to avoid the scandal of treachery that he was about to commit.'[42] He told his cousin that he was relieved not to have had to sack Bautista as Captain-General of Barcelona.[43]

After the ferment of internal regime opposition generated by Arrese's schemes, the monarchist opposition headed by Bautista Sánchez, the

Barcelona strike and serious economic problems, Franco concluded that a major cabinet reshuffle was now unavoidable. Since the beginning of October 1956 when Arburúa had painted a stark picture for the cabinet, he had been aware of the crumbling economic situation, with rocketing inflation matched by a disastrous balance of payments situation. Most of the problems were the consequence of the ineptitude of his cabinets, for which the Caudillo cannot escape responsibility. Many difficulties derived from Franco's attachment to autarky and the central role of the Instituto Nacional de Industria directed by his friend Suanzes. The INI's show-piece projects made demands on scarce resources of capital and materials. Franco's loose, not to say non-existent, control of ministers encouraged overspending. The consequent shortages of government funds were met by resort to the printing press. At the same time, there was little in the way of monetary and fiscal policies to regulate demand. In addition to these inflationary pressures, the previous year's wage rises had increased industrial and agricultural costs by more than forty per cent.[44]

Franco's reluctance to initiate cabinet changes was a symptom of a lifelong caution which was turning into a noticeable distaste for con-fronting new problems. He liked the routine and the familiar and, as February 1956 had shown, he was less agile in a crisis than he had once been. In particular, he was loath to initiate the reshuffle because he feared that there could be no suitable replacement for his Minister of Com-merce, Arburúa. His worry was based on a belief that so few men were versed in the arcane secrets of international trade and finance that the wizardry of Arburúa was virtually unique.[45] In confiding these anxieties to his cousin, he was inadvertently hinting at his own obsolescence. The damaging rigidity of bureaucratic autarky and Spain's need for highly sophisticated economic techniques were things which Franco perceived only dimly.[46] He would reluctantly take advice on this and acquiesce in Spanish integration into the Organization for European Economic Co-operation and the International Monetary Fund. However, in doing so, and in the cabinet changes about to be announced, he would be relinquishing further his own close control over events.

The cabinet reshuffle of February 1957 was to be one of the great watersheds of Franco's political career. It marked the beginning of his transition from active politician to symbolic figurehead. The details were worked out in close collaboration with Carrero Blanco whose influence was growing by the day. Lequerica called Carrero Blanco the 'duque de Olivares', a reference to Philip IV's all-powerful minister. Eleven years younger than the Caudillo to whom he was devoted, the tireless Carrero was no more of an economist than his master. However, he in his turn relied increasingly on the highly talented Laureano López Rodó who, at thirty-seven years of age, had become technical secretary-general of the Presidencia.[47] His relationship to Carrero Blanco mirrored that of Carrero with Franco. The long-term strategic implications of the cabinet changes advised by Carrero and López Rodó went beyond anything anticipated at the time by Franco.

While it is clear that the Caudillo deliberately undertook that part of the reshuffle which implied the political disarmament of the Falange, there were would be other consequences of the changes which he did not foresee. Over the next two years, it would become clear that the new appointments had meant the abandonment of every economic idea that the Caudillo had ever held dear and the uninhibited embrace of modern capitalism. That in its wake would bring huge foreign investment, massive industrialization, population migration, urbanization, educational expansion, the social consequences of which were to turn Franco and Falangism into historical anachronisms. In the event, he would take the credit for economic development, as he had for wartime neutrality and for surviving the Cold War. At the time, however, his control over events extended only to the immediate political balancing act. If anyone had a grand design in the cabinet changes of February 1957, it was not Franco but López Rodó.

Franco announced his changes almost as an afterthought at nearly midnight on 22 February just as he was closing the day's cabinet meeting. The reasons he gave were the attrition suffered by ministers, the fact that some had asked to be relieved and because 'people are tired of always seeing the same faces in the papers'.[48] In fact, having decided to acquiesce in Catholic and monarchist opposition to Arrese's scheme, Franco now used the cabinet changes to begin a long process of completely emasculating the Falange. It would always remain useful to him but he was determined that it would never get out of hand again as it had during 1956. Arrese was removed as Minister-Secretary and was replaced by an even more collaborationist Falangist, the ambitious and flexible José Solís Ruiz, head of the Falangist syndicates. Franco had originally thought of the hard-liner José Antonio Elola, but was dissuaded by Carrero Blanco after López Rodó had pointed out that Elola was a virulent anti-monarchist. López Rodó recommended Solís to Carrero. Lacking Arrese's grandiose visions, the loquaciously amiable Solís had no greater concern than to remain in Franco's favour. Solís was anxious to keep his salary as *Delegado Nacional de Sindicatos* in addition to his pay as Minister-Secretary, which

Franco permitted him to do. It ensured that he would be too busy to try to follow in Arrese's footsteps as a reforming Minister-Secretary. It also signalled that the acceptable terrain for the Falange was social security and labour legislation.[49] Girón, a powerful personality now completely discredited by the economic consequences his 1956 wage rises, was replaced as Minister of Labour by an altogether more manipulable Falangist, the colourless Fermín Sanz Orrio. The equally grey Jesús Rubio was kept on at the Ministry of Education.

Arrese mistakenly went on thinking that he was still an important part of the Francoist inner circle for some time after Franco had decided to dump him. Expressing his self-interest in terms of sycophantic loyalty, Arrese had told Franco that he would make the sacrifice of staying on in the cabinet to prevent it being thought that his departure implied an outbreak of hostilities between the Falange and its Caudillo. In fact, he was kept on, as a sop to the Falange, in the innocuous position of Minister of Housing where he could find a social outlet for his ideological zeal. In the days preceding the announcement, while Franco had worked on his new team, Arrese had made desperate efforts to persuade him to include a stronger Falangist presence. Franco toyed with him for more than a week. As Arrese produced names of prospective ministers, Franco would laugh and say 'you're getting cold' or 'you're getting warm', before finally ignoring his advice as he had intended all along. Arrese's ambition to remain in the cabinet entirely suited Franco since it provided a lightning conductor for Falangist anger at the defeat of the constitutional project. As he put it to Carrero Blanco, 'I don't want him to leave waving the banner of his *Leyes Fundamentales*. I need him to cool down first in the Ministry of Housing.'[50]

To guarantee public order in the wake of the previous year's student disorders, Falangist indiscipline and labour unrest, the Caudillo appointed a man he once said was too hard and inflexible to be Minister of the Interior.[51] His lifelong friend and close collaborator, the sixty-eight year-old General Camilo Alonso Vega, replaced Blas Pérez who was the victim of a whispering campaign by a jealous Carrero Blanco.[52] Alonso Vega's iron control of law and order would be crucial during the period of economic upheaval that was to follow the decision to float the Spanish economy on international waters. Franco offered Blas Pérez a newly created Ministry of Health and, to the Caudillo's astonishment and suspicion, he refused. As one of the minority of Franco's ministers not to have lined his pockets while in office, Blas Pérez wanted to return to his private law practice. Unable to believe such a simple explanation, Franco assumed some sinister motive and had his collaborator of fifteen years' standing watched by the secret police.[53]

Muñoz Grandes, partly because of his involvement with Bautista Sánchez and partly because he was deeply unpopular in the Army, was removed as Minister for the Army. Franco compensated him by a symbolic promotion to the rank of Captain-General – the nearest Spanish equivalent to Field-Marshal. This exalted rank, as opposed to the title of Captain-General carried by the head of a military region, had previously been held only by Franco himself and, posthumously, by General Moscardó. The Caudillo saw Muñoz Grandes as the man who, in an emergency, should take over from him. General Barroso, whom Franco had known as a loyal collaborator since 1936, despite his reputation as a liberal monarchist, became Minister for the Army. This was an attempt to neutralize monarchist sentiment in the high command. Burdened with the difficult and unpopular job of reducing the size of the Army in the wake of the agreement with the USA and the loss of Morocco, Barroso would be unable to make the position a power base for monarchist conspiracy.[54]

Martín Artajo, after twelve years as Foreign Minister was replaced by Fernando María de Castiella. Having had a meeting with Franco on the evening of 21 February, Martín Artajo was deeply hurt that the Caudillo had not warned him about the dry announcement which he made at the following day's cabinet session. Franco had ample reason to be grateful to Martín Artajo for the services rendered in the international arena since 1945 but dropped him without a second thought. The one-time Falangist and now Christian Democrat Castiella was ideologically indistinguishable from Martín Artajo but an astute choice. Franco believed that the leader of the Christian Democrat faction would have to go in order to soften the blow for the Falange of Arrese's humiliation. Castiella's record as author of the bible of Falangist imperialism *Reivindicaciones de España* and as a combatant in the *División Azul* were ideal credentials in the eyes of even the most militant Falangists.[55]

Even more important than the political neutralization of the Falange was the inclusion in the new cabinet of the 'technocrats' who would soon undertake the modernization of the economy. The choice of men interested in integrating Spain into the world economy signified the end of Falangist economics. The new Minister of Finance, Mariano Navarro Rubio, was a Catholic lawyer and a member of Opus Dei. He was the quintessential Francoist functionary, a officer in the military juridical corps who had held senior posts in the Falangist syndicates and had been Under-Secretary at the Ministry of Public Works. Competent and hard-working, he was also a member of the board of directors of the Opus Dei-controlled Banco Popular.[56] The new Minister of Commerce, Alberto Ullastres Calvo, was a university professor of economic history and was also a member of Opus Dei. The fact that López Rodó was a member too led to speculation that the three constituted a sinister block at the orders of a secret society. The enormous influence that the three would exercise over the next few years, laying the basis for the regime's survival through its economic and political transformation, would fuel a belief within the displaced Falange that they had hijacked the Caudillo and the *Movimiento*. Falangist resentment, combined with a readiness to believe in sinister masonic conspiracies, led to the emergence of the idea of the Opus as a Catholic freemasonry or mafia. Other criticisms came from the *Asociación Nacional Católica de Propagandistas*, the Catholic pressure group which was itself not dissimilar to Opus Dei.[57]

The arrival of the technocrats has been interpreted variously as a planned take-over by Opus Dei and a clever move by Franco to 'fill vacant seats in the latest round of musical chairs'.[58] In fact, the arrival of the technocrats was neither sinister nor cunning but rather a piecemeal and pragmatic response to a specific set of problems. By the beginning of 1957, the regime faced political and economic bankruptcy. Franco and Carrero Blanco were looking for new blood and fresh ideas. To be acceptable, new men had to come from within the *Movimiento*, be Catholic, accept the idea of an eventual return to the monarchy and be, in Francoist terms, apolitical. López Rodó, Navarro Rubio and Ullastres were ideal. López Rodó was the nominee of Carrero Blanco.[59] The dynamic Navarro Rubio was the Caudillo's choice. Franco had known him since 1949. He was a *Procurador en Cortes* for the *Sindicatos* and had been highly recommended by the outgoing Minister of Agriculture, Rafael Cavestany.[60] Both López Rodó and Navarro Rubio suggested Ullastres.[61] Without being a monolithic unit, López Rodó, Navarro Rubio and Ullastres worked together as a team, despite occasional frictions, to push for the administrative and economic modernization of the regime.[62]

Although he did not have cabinet rank, the influence of López Rodó was to be immense and ultimately to have substantial repercussions on Franco's political life. And, although it was far from obvious at the time, a decree which was published on the same day that the new cabinet was announced was to hasten the transformation of Franco's role from that of active president of the council of ministers to something resembling a figurehead. He would still be much more than a ceremonial Head of State but the Caudillo would have less to do with the daily machinery of government. The Decree-Law of the Juridical Regime of the Administration of the State of 25 February 1957, which was the work of López Rodó, laid down the basis for a reorganization of the government. It constituted as big a leap forward in the 'normalization' of procedures as had the move instituted in the course of 1937 by Serrano Suñer from the battlefield improvisation of the Burgos Junta to a formal cabinet.

The decree, ratified by the Cortes in mid-July 1957, set up a ministerial department, known as the Presidencia del Gobierno, a full-scale prime minister's office which would initiate, draft and programme legislation. It was a sign of the growing complexity of day-to-day administration. In 1956, the rambling cabinet meetings with interminable conflicts between 'economic' and 'social' (Falangist) ministers had led to Arrese's efforts to persuade Franco to reduce their frequency. Loosely presiding over the squabbles between the various factions in his cabinet, Franco often lost control of the discussion.[63] Henceforth, such conflicts would be resolved by interministerial committees. The committees could be chaired either by the Caudillo or by the Minister of the Presidencia. An Office of Economic Co-ordination and Planning was created to provide the technical services for the committees. In addition to his crucial post as technical secretary to the Presidencia, López Rodó was named director of the new office. Crucially, the autonomy of the Falange was curtailed since the previously independent budget of the Minister-Secretary now fell under the control of the Presidencia. The business of government became less 'political' and more austerely 'administrative'. In practical terms, given Franco's hunting and fishing passions, that meant that henceforth strategic policy would be much more likely to be made by Carrero Blanco and López Rodó than by the Caudillo.[64]

Following the arrival of the technocrats in the new cabinet, there was inevitably a period of what has been called 'disorientation', during which public debt, inflation and balance of payments problems continued.[65] This was largely the legacy of autarky. Believing that it was possible to impose price stability by decree, Franco had accepted Girón's claim in 1956 that strikes could be avoided by massive wage rises without there being any impact on prices. In fact, Girón's strategy had unleashed a major inflationary spiral. By the spring of 1957, the pressure on living standards saw the beginning of a new and more militant strike wave. It is an indication of how little Franco understood what was happening in economic terms that he perceived industrial unrest as either perversity or the work of outside agitators manipulated by Communists and free-masons. He believed that talk of inadequate wages and hunger had no basis other than foreign propaganda. 'They tell the people that they should work less and that they should get more pay; but they don't tell them that that is what makes the cost of living rise and destroys the economy of a nation.'[66]

That the marginalization of Franco was implicit in the activities of the technocrats may be perceived in some astonishingly frank remarks made shortly after the cabinet changes by López Rodó to the Conde de Ruiseñada. He told him that the '*Tercera Fuerza*' plans of Opus Dei members like Rafael Calvo Serer and Florentino Pérez Embid were doomed to failure since 'it is impossible to talk to Franco about politics because he gets the impression that they are trying to move him from his seat or preparing the way for his replacement'. He then made the revealing comment that 'the only trick is making him accept an administrative plan to decentralize the economy. He doesn't think of that as being directed against him personally. He will give us a free hand and, then, once inside the administration, we will see how far we can go with our political objectives, which have to be masked as far as possible.'[67] In May 1957, López Rodó outlined to Dionisio Ridruejo, the Falangist poet who had broken with the regime, his far-reaching political plans. Concerned about the fragility of a system dependent on the mortality of Franco, López Rodó wanted to replace it with a more secure structure of institutions and constitutional laws. On the premise that 'the personal power of General Franco has come to an end', López Rodó hoped to have Juan Carlos officially proclaimed royal successor. Until 1968, when the Prince would reach thirty, the age at which the *Ley de Sucesión* permitted him to assume the throne, Franco would remain as Head of State. To prevent the Caudillo suffering unnecessary political attrition, the post of prime minister would then be created.[68]

In his hopes of being able to engineer an early transition to the monarchy, López Rodó's optimism was analogous to that of Martín Artajo twelve years earlier. He was forced to slow down in November 1957 when Franco took umbrage at the fact that the decrees emanating from the *Secretaría General Técnica de la Presidencia del Gobierno* were tending to limit his powers.[69] Given the delicacy with which his hidden agenda for political change had to be put before the Caudillo, and the existence of anti-monarchism within the *Movimiento*, the realization of his programme would take another twelve years.

In fact, there were many within the regime who suspiciously perceived the plans for political transformation behind the ostensible objective of economic liberalization. After the change in the exchange rate from five pesetas to the US dollar to the more realistic forty-two pesetas and Ullastres' announcement in August 1957 of his determination to free price controls, there was considerable alarm.[70] Franco, however, remained cool despite misgivings about the devaluation. Navarro Rubio found him deferential and respectful, the humble layman in the presence of the arcane expert.[71] If he was uninvolved in the dramatic economic changes underway, he was even less exercised by the problem of the succession. López Rodó had been encharged by Carrero Blanco with the elaboration of a set of constitutional texts which would permit the eventual installation of the monarchy yet still be acceptable to those who wanted the *Movimiento* to survive after the 'biological fact', as the death of Franco was coming to be called. The issue in general, and López Rodó's draft texts in particular, were discussed interminably in the cabinet, but Franco seemed to be in no hurry nor even especially engaged by a process which he regarded as simply fine-tuning the *Ley de Sucesión*.

Ruiseñada and López Rodó tried, throughout the summer of 1957, to arrange an interview between Franco and Don Juan. The Pretender refused since he could see no sign of progress or reform in the regime. López Rodó himself explained his scheme for gradual evolution to Don Juan on 17 September. He was in Lisbon as part of an economic delegation and took the opportunity to reassure the Pretender that things were moving, albeit slowly. In a conversation lasting more than three hours, López Rodó claimed that Franco wanted to put an end to the uncertainty surrounding his succession but was obsessed with the fear that when he died his life's work might simply be overthrown by his successor. Thus, in accordance with the *Ley de Sucesión*, whoever was chosen would have to accept the basic principles of the Francoist State. Don Juan made it clear that for him to take the first step would be 'like being forced to take a purgative'.[72]

In November 1957, Franco had to face another phase of Spain's lingering crisis of decolonization and his handling of it confirmed that he was gradually losing his capacity to react flexibly to problems. One of Spain's few remaining colonial territories, Ifni, on the Atlantic coast of North Africa, was the object of a territorial claim by Morocco in August 1957 and a subsequent encroachment by irregular Moorish forces. In fact, reports had been coming in to Franco for months about anti-Spanish activities in Morocco and hostile infiltrations of Ifni.[73] The Governor-General of Spanish West Africa, General Mariano Gómez Zamalloa, recommended a pre-emptive strike from Villa Cisneros, the Spanish base in the southern Sahara, to break up the gathering invaders. Franco, for whom procrastination had now become an unbreakable habit rather than a sign of cunning, failed to react despite his conviction that the Soviet Union was behind the entire operation. He was anxious not to undermine his pro-Arab policy which had paid dividends in building up votes at the United Nations and he was aware of American support for Mohammed V, the King of Morocco. While he hesitated, guerrilla incursions intensified throughout the summer and early autumn. On 23 November 1957, Moroccan guerrillas attacked the principal town Sidi Ifni.[74]

Franco was in Northern Spain but hurried back to Madrid. The American weaponry received as part of the bases agreement could not be used against another ally of the United States. In an ironic reversal of history, Spanish troop reinforcements had to be sent across the Straits in Second World War-vintage Junkers and Heinkel bombers similar to those which had taken part in Franco's airlift of Moorish mercenaries to Spain in 1936. Barroso and other generals were furious with Franco for the neglect and complacency which they believed had permitted the situation to arise. They believed that an early pre-emptive attack would have cooled Moroccan ardour. Under fire from Spanish Messerschmitt 109s, the Moroccan advance faltered. Before the generals in Madrid could relax, another assault was mounted near El Aaiun, principal city of Spanish Sahara. Franco finally decided to dump his pro-Arab policy and agreed to joint operations with the French against the Moroccan liberation forces. In the last resort, however, American fears of Morocco being pushed into the Soviet orbit led to pressure from Washington for a peace settlement. Franco accepted and an uneasy agreement was made in June 1958.[75]

The Moroccan crisis coincided with a new strike wave in the Asturian coal-mines and in Catalonia in the spring of 1958. Increasingly divorced from everyday politics, Franco took little interest other than to denounce the strikes as the work of foreign agitators and to accuse the working class of laziness.[76] Without Arrese to enthuse him to make contact with the Falange rank-and-file, Franco's tours and public appearances became more infrequent. The new cabinet was much less prone to conflict than its predecessor. He left his ministers to get on with their business and so found more time for hunting and fishing. Moreover, since the composition of a technocratic cabinet and the subsequent creation of the Presidencia, Franco's political life consisted more of routine audiences than of cunning arbitration of squabbling factions.

The first fruit of López Rodó's work as head of Carrero Blanco's secretariat of the Presidencia was the 'Declaration of the Fundamental Principles of the *Movimiento Nacional*'. It was unveiled in the Cortes on 17 May 1958 by Franco who made sententious reference to 'my responsibility before God and before History'. The gradual reform hinted at by López Rodó to Ruiseñada and Don Juan could be discerned in the formal decoupling of the regime from Falangism. The twelve principles were an innocuously vague and high-minded statement of the regime's Catholicism and commitment to social justice. The seventh principle stated that 'the political form of the Spanish State, within the immutable principles of the *Movimiento Nacional* and the *Ley de Sucesión* and the other Fundamental Laws, is the traditional, Catholic, social and representative monarchy'.[77] Of the *Movimiento Nacional* understood as *Falange Española Tradicionalista y de las JONS*, he had nothing to say.

It appeared as if he was edging crab-like towards a restoration and many monarchists interpreted the speech in that spirit. Franco had asked Carrero Blanco to prepare his speech and he in turn had passed on the task to López Rodó. The Caudillo did not discuss the text in cabinet before delivering it and several ministers had revealed their dismay at its departure from Falangism by ostentatiously failing to applaud.[78] Now sixty-five, Franco was increasingly concerned about the future and the continuity of the regime after his death. He often asked friends older than himself what it was like to be seventy. Admiral Salvador Moreno replied that it felt no different except that one made more mistakes than before, a phrase which Franco began to repeat frequently.[79] Two days after the Cortes meeting, he spoke of making Agustín Muñoz Grandes chief of the general staff to ensure that his wishes would be carried out in the event of him dying before the constitutional process was complete. He did so on 6 June 1958. The promotion of Muñoz Grandes and remarks to the effect that he would make a good regent made it unequivocally clear that Franco had no intention of handing over to any successor before he died.[80]

On 10 June 1958, with Spain's foreign exchange reserves dwindling dramatically, Navarro Rubio presented to the cabinet a report which was the basis of the harsh monetary stabilization programme which was to be the foundation for Spain's future economic development. It constituted, as López Rodó had hinted to Ruiseñada twelve months previously, a reversal of twenty years of Francoism. Franco seemed unaware, or unconcerned, of just how far-reaching the changes were. He expressed surprise that things were as bad as Navarro Rubio painted them but 'had no objection to things being tidied up'. He was, however, sufficiently aware of the political implications to insist that, although Navarro Rubio's programme be implemented, his report remain secret.[81] Three weeks later, Franco commented to Pacón that Navarro Rubio 'is very theoretical and an extremely cold man'.[82]

During the summer of 1958, Franco spent every moment that he could fishing. Immediately after the speech to the Cortes, he had disappeared for two weeks fishing in Asturias. He returned delighted to have caught nearly sixty salmon, some of more than thirty pounds. In September, during his stay in Galicia, he was thrilled to catch a whale weighing twenty tons. He told Pacón 'it cost me twenty hours of struggle until at the end it gave up. I hope to catch an even bigger one. I really enjoy this sport and it is a great relief from my work and worries.' What was striking was Franco's passion for records and ever larger kills. In 1959, he was to boast to his cousin that on one shooting expedition in a few days he had shot nearly five thousand partridges.[83]

On 9 October 1958, Pope Pius XII died and was replaced three weeks later by Cardinal Angelo Roncalli. As Pope John XXIII, Roncalli instituted a liberalization of the Catholic Church which would cause severe problems for Franco. It was a change which underlined the anachronistic nature of his political survival. Reforms in the strategic direction of Vatican policy coincided with increasing involvement by Spanish priests in the labour movement. Both as a result of the success of the Catholic workers' organization, the *Hermandades Obreras de Acción Católica*, and of the increasing involvement of individual clerics in the worker priest movement, the pro-Francoism of the Church began to crumble.

Disputes with the Church were not yet on the Caudillo's immediate agenda. The most acute problem at the end of 1958 was the collapse of the Spanish economy amidst rocketing inflation and growing working-class discontent, a situation which could not have been perceived from Franco's smug end-of-the-year message for 1958. It was largely a scissors-and-paste concoction of paragraphs provided by his ministers. A statistical review of the regime's triumphs tried to obscure the economic difficulties. At the same time, the speech was concerned principally to stress that any future succession had to be within the confines of the *Movimiento*. Denouncing the 'frivolity, lack of foresight, neglect, clumsiness and blindness' of the Bourbon monarchy, he claimed that anyone who did not recognize the legitimacy of his regime was suffering from 'personal egoism and mental debility'.[84]

The speech suggested that Franco was having doubts about the wisdom of having authorized a monarchist gathering scheduled to take place one month later. On 29 January 1959, liberal supporters of Don Juan would be meeting at a dinner to be held at the Hotel Menfis in Madrid to launch an association known as *Unión Española*. It was the brainchild of the liberal monarchist lawyer and industrialist, Joaquín Satrústegui. Although Gil Robles was present, he did not make a speech. Those who did made it clear that the monarchy, to survive, could not be installed by a dictator but had to be re-established with the popular support of a majority of Spaniards. Satrústegui denied the Caudillo's claim made in the end-of-year message to the effect that the *Cruzada* was the fount of the regime's legitimacy, pointing out that the monarchy could not be based on the victory of one side in a civil war. Franco was furious when he heard the details of the Hotel Menfis after-dinner speeches and fined Satrústegui the not inconsiderable sum of fifty thousand pesetas. That the penalties were not more severe, comparable for instance to those meted out to left-wing opponents, derived from the fact that Franco did not want to be seen to be persecuting the followers of Don Juan.[85] Shortly afterwards, he said about Don Juan to his cousin Pacón: 'he is entirely in the hands of the enemies who want to wipe out the regime and the sweeping victory that we gained.'[86]

The emergence of *Unión Española* was merely one symptom of unrest within the Francoist coalition. Left-wing and regionalist opposition in the universities and the labour movement was to be expected. However, now there were several focuses of opposition on the Right. Gil Robles was organizing a group called *Democracia Social Cristiana*. His one-time ministerial colleague under the Second Republic, Manuel Giménez Fernández, was organizing a more liberal Christian Democrat opposition grouping, *Izquierda Democrática Cristiana* (the Christian Democratic Left). Carlists and Juanista monarchists were openly organizing. Falangists who opposed the trend of the regime organized to defend the essence of the 'pending revolution' as the *Vieja Guardia* and the *Círculos José Antonio*.[87] Their emergence goes some way to explaining why Franco was prepared to put his faith in López Rodó, Navarro Rubio and Ullastres. He saw them as bringing new blood and ideas which might give a fresh lease of life to an exhausted system without changing it. In the long run, it was a vain hope but, in the medium term, it was to help secure another fifteen years.

The extent to which an out-of-touch Franco seemed happy to leave matters to his technocrats was revealed at the beginning of 1959 when

an International Monetary Fund mission began to investigate the difficulties of the Spanish economy. Navarro Rubio pushed for the free con-
vertibility of the peseta as a further step towards the necessary integration
of the Spanish economy into the international system. Given the likely
opposition from several ministers, Franco simply omitted to put the
measure before the cabinet for full discussion.[88] However, in February,
he opposed a further devaluation of the peseta in response to the advice
of the International Monetary Fund. The Caudillo's bewilderment about
what was going on became apparent when he told Navarro Rubio that
he could not see why the rate should be changed and so permit Americans
to buy more with a dollar in Spain than Spaniards could buy in the United
States. On 18 February, he refused Ullastres permission to accept an
offer from the International Monetary Fund to elaborate a stabilization
plan for the Spanish economy. Navarro Rubio was appalled and requested
an immediate audience. Despite Spain's near bankruptcy, Franco told
Navarro Rubio to reject IMF offers because he distrusted the good intentions of foreigners and believed that Spain could solve her problems
alone. To little avail, Navarro Rubio bombarded Franco with evidence
of Spain's parlous financial state. Finally, in desperation, he asked him
'if we have to return to ration cards, what will we do if the orange harvest
is hit by frost?' An alarmed Caudillo looked at him, unable to answer as
he repeated the question. Eventually, faced with the spectre of the return
of gas-powered motor-cars, Franco stood up, shrugged his shoulders and
authorized the opening of formal talks with the IMF.[89]

Franco remained far from convinced that it was the right thing to do.
Some days later, Carrero Blanco told López Rodó that 'he is not happy;
he's deeply suspicious'.[90] It was as if he was locked into the idea of the
'international siege' of 1945–50. He suspected that once economically
dependent on international goodwill, he might be pressured into political
reform or even resignation.[91] He could not see the stabilization plan as
a short-term device to get enough dollars to survive.[92] However, once
the links with the IMF were established, the pressure for a further devaluation of the peseta from forty-two to the US dollar to sixty became
irresistible. There was considerable opposition in the cabinet and all eyes
were on Franco.

Navarro Rubio believed that, having agreed to IMF involvement in
the stabilization programme, Franco would have to accept the financial
consequences. When the issue was put before the cabinet, he was instinctively hostile until Ullastres revealed just how near Spain was to bankruptcy.[93] The stabilization plan adopted by the cabinet on 6 March 1959,

the new devaluation of the peseta and the reduction of public spending all had social consequences. Many businesses were forced to close and unemployment began to rise which led to protests in cabinet. The growth of working-class militancy provoked Alonso Vega into especially vehement diatribes in cabinet while limits on public spending eventually provoked the resignation of Arrese from the Ministry of Housing in 1960. Cabinet meetings became, for Navarro, 'Fridays of sorrow' as he had to combat the complaints of his ministerial colleagues as their budgets were reduced. Franco let the arguments run but, when he did intervene, it was to put his weight behind the new policy.[94]

Having made the decision to leave the running of the economy to Navarro Rubio and Ullastres, Franco characteristically stood by it. Outside the cabinet, he never put any but the most timid objections to either and always backed down if they showed firmness.[95] He could easily be convinced by technical argument, largely because the issues were beginning to be of a complexity that was way beyond his own crude home-spun economics. On one occasion when Arrese explained to him the ramifications of a scheme for subsidized rents, he said 'I don't understand a word but I believe you.'[96] He protected Navarro Rubio after he said in the Cortes in July 1958 that the military budget was an unproductive drain on the economy and the Minister for the Army, Barroso, had wanted to set up a *tribunal de honor* to judge him. Nevertheless, the Generalísimo did draw the line at support for Navarro Rubio's plans for a massive reduction of the military budget by means of a transition to a small professional army.[97] Franco harboured some hopes that, after the stabilization plan had fulfilled its objectives, it would be possible to return to the kind of autarky favoured by Arrese and Suanzes yet he had to accept the resignations of both.[98]

To dampen the speculation about his future, which had fuelled the emergence of a conservative opposition, Franco permitted the monarchists in his cabinet to elaborate their own constitutional scheme for the post-Franco succession. The first draft was given to Franco by Carrero Blanco on 7 March 1959 together with a sycophantic note urging the completion of the 'constitutional process': 'If the king were to inherit the powers which Your Excellency has, we would find it alarming since he will change everything. We must ratify the life-time character of the magistracy of Your Excellency who is Caudillo which is greater than King because you are founding a monarchy.'[99] Franco was hesitant, confiding in Pacón one week later that Don Juan and Prince Juan Carlos must accept that the monarchy could be re-established only within the *Movimiento*, because a liberal constitutional monarchy 'would not last a year and would cause chaos in Spain, rendering the Crusade useless'.[100] Unwilling to do anything that might hasten his own departure, he did nothing with the constitutional draft for another eight years. To increase his freedom of action and to put pressure on Don Juan, Franco began quietly to cultivate Alfonso de Borbón-Dampierre, the son of Don Juan's brother Don Jaime. Solís and other Falangists seized the opportunity to promote the cause of a possible *príncipe azul* (a Falangist prince).[101]

The question of the Caudillo's mortality was raised directly by an event which gave him intense satisfaction. The official inauguration of the Valle de los Caídos on 1 April 1959, the twentieth anniversary of the end of the Civil War, was an occasion to rival the 1939 victory celebrations. The entire cabinet, the *Procuradores* of the Cortes, the full membership of the *Consejo Nacional*, representatives of all regime institutions, military and civilian authorities from every province, two Cardinals and a large panoply of archbishops and bishops, and the diplomatic corps filled the huge basilica. The Generalísimo, in the uniform of a Captain-General, and Doña Carmen, dressed in black with a mantilla and high comb, walked up the centre aisle under a canopy to their special thrones near the high altar. Thousands of workers were given a day off with pay and a packed lunch and coaches brought them to Cuelgamuros free of charge.[102]

The construction had cost the equivalent of £200,000,000. Franco equated the Valle de los Caídos with El Escorial, the one the symbol of the past greatness of the era of Philip II, the other the symbol of the greatness of his era.[103] His speech, about the heroism of 'our fallen' in defence of 'our lines', was triumphant and vengeful. He gloated over the enemy that had been obliged 'to bite the dust of defeat' and showed not the slightest trace of desire to see reconciliation between Spaniards. The controlled press described the inauguration as the culmination of his victory in 1939.[104] If not before, certainly by the time of inauguration, Franco was talking about himself being buried in the basilica. Diego Méndez, the architect who replaced Muguruza on the latter's death, assuming that this was what Franco wanted, was planning a tomb on the opposite side of the great altar from José Antonio's, sited so that the deceased Caudillo 'would be the master of the house, who receives others into his home'. It was a good guess on his part. On the day of the inauguration, Franco was strolling round with Méndez and pointed precisely at the spot which Méndez had chosen and said '*Bueno, Méndez, y en su día yo aquí, ¿eh?*' (when the time comes, me here).[105]

Such ceremonies were nostalgic flashbacks to a Spain that was about to disappear. The stabilization plan, worked out under the supervision of the IMF and the OECE, aimed to cut domestic consumption by the massive devaluation of the peseta, fierce credit restrictions and cuts in public spending. It was intended that the surplus production would go to export and bring in hard currency which could then be used for imports of capital goods. It would be a major step towards the economic modernization of Spain. The inevitably high social costs were visited upon the working class in terms of a wage freeze, rising unemployment and shortages of basic consumer goods. This was part of the gradual process of transition from a paternalist centralized economy to a free-market one, a process which would never be completed under Franco. Unemployment kept down the number of strikes, but nonetheless, the late 1950s saw the beginnings of a considerable revival of clandestine trade union activity, organized by Catholic groups as well as by the Communist Party and other left-wing organizations.[106] The Caudillo chose to believe, as so often before, that all social problems were a reflection of the fact that Spain was under siege again from international Communism and freemasonry.[107]

If the opening of the Valle de los Caídos had been the apotheosis of Franco's domestic career, it was followed six months later by the culmination of his international career. While President Eisenhower was in London in August Castiella gave him an invitation from Franco to visit Madrid.[108]Eisenhower had been interested in a meeting since the establishment of US bases in Spain. As C-in-C of NATO forces, he had been unable to do so given the hostility to Franco of other NATO members. Now, he considered it proper to stop in Madrid for a brief discussion of practical co-operation in the running of the bases as part of a world tour due to start at the end of November 1959. On 21 December, Eisenhower landed at the American base at Torrejón de Ardoz near Madrid. On descending from the aircraft, he shook Franco's hand rather stiffly and made a short statement. Franco, on the other hand, made an obsequious speech of welcome. 'Permit me to express, on behalf of the Spanish people and of myself,' he gushed, 'our humble admiration for the task to which you have devoted yourself with such personal courage, our gratitude for having come to visit us and to inform us about your momentous journey, and, finally, our firm hope that your immense effort and the historic mission of your great country will be crowned with the prize of a just and lasting international order.' Eisenhower's mood changed as he drove into the city with Franco. The cavalcade was received with a massive popular welcome, partly spontaneous but largely orchestrated by the Falange which had brought thousands of its followers in trucks.

Franco was overcome with emotion at the personal visit of Eisenhower. Just as his eyes had glistened with the emotion of being the Führer's equal now too he cried at the State banquet mounted at the Palacio de Oriente, made another fawning speech and was visibly thrilled to be on familiar terms with the President of the United States. On the following morning, there was a meeting over breakfast at which the atmosphere was coldly uncomfortable until Franco began to talk about partridge shooting. Eisenhower was delighted to find out that the Caudillo shared his passion for bird shooting. In the course of the conversation, the Generalísimo tried to secure Eisenhower's support for his position in Morocco by attributing the anti-Spanish movement there to Soviet agitators.[109] Ironically, Franco's embarrassment favourably impressed Eisenhower. Like many American visitors before him, Eisenhower liked Franco's quiet, modest air and lack of bombast. The President wrote later 'I was impressed by the fact that there was no discernible mannerism or characteristic that would lead an unknowing visitor to conclude that he was in the presence of a dictator.' The President naïvely speculated later whether Franco might not win free elections in the unlikely event of his ever holding any.[110]

After the breakfast meeting, US Army Air Force helicopters took Eisenhower and Franco to Torrejón. On his departure, Eisenhower effusively embraced Franco. The Caudillo, for his part, was completely captivated, talking about nothing else for weeks on end. Indeed, the scale of his new-found admiration for the United States led his sister to comment 'if only Hitler and Mussolini could have heard him'.[111]

Despite Franco's delight in the glory of being visited by Eisenhower, he nevertheless used his end-of-year message on 31 December 1959 to assert the individuality of his regime, a sop perhaps to nationalist elements in the Falange. He denounced freemasonry and 'formalist inorganic democracy', by which he meant democracy as practised in the United States and Northern Europe. In contrast, he praised the system of organic democracy to be found within the *Movimiento Nacional* and proclaimed that his regime was not a dictatorship since there was nothing provisional about it. As he read out sections of the speech provided by his economics ministers, he asserted that it was his autarkic policies which had laid the foundations for the achievements of Ullastres and Navarro Rubio.[112] In both the political and economic spheres, it was the speech of a regal figurehead.

In fact, there were many reasons for the economic success which began
to be apparent after 1960. Franco, and his supporters, attributed an 'economic miracle' to his genius and foresight. However, despite an alleged
special interest in economics, most of Franco's recorded statements about
economic problems reveal, rather than genius, at best, an unsophisticated
common sense; at worst, a naïve gullibility. Far from being the consequence of his long-term vision, growth began only after Franco had
passed responsibility for economic policy making to Navarro Rubio and
Ullastres. Their policies were the fruit of liberal economics and integration into the international capitalist system. Both economic liberalism
and international capitalism had been excoriated by Franco since 1939.
Long after the February 1957 cabinet changes, indeed long after the
introduction by the 'technocrats' of measures of liberalization and
integration into international economic organizations, both Franco and
Carrero Blanco were still committed to autarky.[113]

Development thus occurred with entirely different policy objectives
and instruments than those in which Franco had put his faith between
1939 and 1959. Moreover, it is conveniently forgotten by Francoists that
Spain's boom took place as the world as a whole was undergoing a period
of sustained growth. That context accounts largely for the fact that Spain
was able to export excess labour, largely to Northern Europe. The
migrant workers then remitted their earnings back in foreign currency.
Disposable income in the pockets of German, French and British workers
accounted for the valuable foreign currency earned by the consequent
tourist boom. Nevertheless, it cannot simply be argued that Spain leap-
frogged to prosperity on the backs of her Northern neighbours. Spain
experienced growth rates which were comparable to those of Japan, albeit
starting from a significantly lower statistical base. If Franco bears any
personal credit for this, it lies not in his conscious economic policy-
making or long-term vision but in the inadvertent contribution of other
policy decisions: the fact that his anti-Communism eventually brought
American aid in the mid-1950s; the attraction to foreign investors of an
authoritarian regime with repressive labour legislation. Like Japan, Italy
and Germany, Spain had undergone a fascist experience which had
destroyed the organized working-class movement and traumatized the
work force. In that sense, the Franco regime's contribution to economic
growth was the decades of hardship which left Spanish workers willing
to work long hours for low wages. Combined with the repression of strike
activity and good facilities for the repatriation of profits, this made Spain
an obvious target for foreign capital in the early 1960s.

Franco nurtured hopes that the stabilization plan did not herald the death of Falangist economics and the maintenance by the State of three hundred thousand of his most fervent supporters. However, his instinctive perception that the future lay with the technocrats was demonstrated by the manner of the final departure of Arrese. On 27 February 1960, inaugurating the National Council of Housing, Architecture and Urbanism, Arrese made a speech openly criticizing the stabilization plan. In front of a group of ministers, including Navarro Rubio, Arrese unveiled a proposal to build one million dwellings. He denounced the 'meanness' of a policy by which spending on social projects, once based on massive borrowing, was now to be limited to what the technocrats coldly judged that the state could afford. As the speech continued, the barometer of regime sentiment, the sibylline head of the *Movimiento*, José Solís, showed symptoms of deep unease. Convinced that he had done everything possible to facilitate Arrese's spending plans, Navarro Rubio was furious at what he saw as a propaganda stunt to brand the technocrats as the main obstacle to social justice.

Franco was not present at the ceremony but his stance quickly became apparent. After a war of press releases from the Ministries of Housing and Finance, the power struggle moved to the cabinet when Arrese distributed copies of a letter he had written to Navarro Rubio. In pious terms of abject loyalty to the Caudillo, Arrese expressed his regret that 'the deepest economic depression experienced by the regime' was preventing the fulfilment of the hopes of millions of Spanish families. In reply, Navarro Rubio gave the cabinet figures showing that Arrese had funds to do much of what he proposed. The Minister of Housing stood his ground. Unable to produce a compromise in full session of the cabinet, Franco summoned Arrese and Navarro Rubio to his office on 17 March 1960. Arrese threatened, in blustering terms, to resign if he did not get a public apology from Navarro Rubio, who naturally refused. After quietly listening to Arrese's harangue, the Caudillo shrugged his shoulders and accepted his resignation. Arrese was replaced by an anodyne Falangist crony of Solís, José María Martínez y Sánchez Arjona. López Rodó found him a cooperative colleague and Navarro Rubio, equally keen to 'de-politicize' politics and turn it into technical administration, saw him as 'a normal minister who spoke about numbers.'[114]

The entire episode not only marked a significant triumph for the technocrats, but also revealed much about Franco. It showed yet again that he never permitted personal loyalty to stand in the way of realistic political judgements. He liked Arrese, probably much more than he ever did the busily efficient Navarro Rubio. He certainly felt more sympathy for the departing minister's political philosophy. In putting his weight behind the technocrats, he was breaking with the past. The consequence was a step further in the gradual process whereby Franco was being reduced to a figurehead, ready to take credit for what was happening in economic policy but ever further removed from the practical implementation of policies which he no longer entirely understood.

XXVI

INTIMATIONS OF MORTALITY

1960–1963

STIMULUS for Francoists to think about the future was provided by rumours of the Caudillo's failing health as he neared the end of his seventh decade. In the spring of 1960, rumours in Madrid that he had suffered a heart attack were so widespread that an official denial was issued by his household. It also leaked out, belatedly, that, on returning in his Rolls Royce from a hunting party in Jaén on 25 January 1960, a fault in the heating system led to the interior being filled with carbon monoxide fumes from the exhaust. Noting his drowsiness, Doña Carmen had the presence of mind to order the car stopped before any serious harm was done and he suffered only a severe headache. There were rumours that the car had been tampered with.[1]

With such reports of Franco's mortality current, special significance was read into Franco's third meeting with Don Juan, their second at Las Cabezas, on 29 March 1960.[2] Las Cabezas had been inherited, on the Conde de Ruiseñada's death, by his son, the Marqués de Comillas. The encounter had originally been scheduled to take place one week earlier in the Parador at Ciudad Rodrigo near the Portuguese border. But, when news of the meeting was leaked in the press and gossip spread that Franco planned to hand over power to Don Juan, he cancelled it. In the wake of the Unión Española meeting at the Hotel Menfis, Franco was anything but well-disposed towards the Pretender. He told Pacón that 'Don Juan is hopeless and he gets less trustworthy all the time' and was particularly annoyed by the Pretender's aspiration to be the King of all Spaniards, of both left and right: he commented with engaging frankness, 'Don Juan ought to understand that for that it would not have been necessary to have a bloody civil war.' His contempt for Don Juan's liberalism was balanced by a belief in the greater suitability of Juan Carlos, because of his education in Francoist Spain. He was beginning to hope that Don Juan would abdicate in favour of his son although, as he told Pacón, 'as long as I have my health and my mental and physical faculties, I will not give up the Headship of State'.[3]

When the rearranged encounter took place, a grey-suited Franco was accompanied by the head of his civilian household, the Marqués de Casa Loja and the second-in-command, General Fernando Fuertes de Villavicencio. It was a significantly shorter meeting than its predecessor six years earlier. In general, Franco showed even less interest in bringing Don Juan around to his point of view, having already eliminated him as a possible successor.[4] He astutely refrained from making that decision public, convinced that if he did so, Juan Carlos would side with his father. If that happened, Franco would lose great swathes of monarchist support.*

Franco's often mendacious remarks to Don Juan at Las Cabezas revealed his ongoing obsession with freemasonry and his resentment at the Pretender's independent stance. Don Juan complained about a book, *Anti-España 1959*, published in Madrid by an obsessive regime propagandist, Mauricio Carlavilla, who was also a secret policeman. The book denounced the monarchist cause as the stooge of freemasonry and a smokescreen for Communist infiltration, as well as insinuating that Don Juan himself was a freemason.[5] Franco, who could plausibly have feigned ignorance, resentfully justified the book as a tit-for-tat reply to the memoirs of the monarchist aviator Juan Antonio Ansaldo. He made it clear that, if he had not inspired Carlavilla's book, he was nonetheless delighted by it. Ansaldo's *¿Para qué . . .* (For What?) had attacked Franco's failure to restore the monarchy as a betrayal of the war against the Republic. Don Juan pointed out that, since Ansaldo's book was banned in Spain, there was no need for a reply. He went on to complain about attacks on the monarchy in the press which Franco shiftily attributed outrage over the 1945 Lausanne Manifesto on the part of journalists – a matter on which he still harboured a grudge. Franco underlined his identification with Carlavilla's line of argument when he criticized Don Juan's reliance on Pedro Sainz Rodríguez whom he again accused of being a freemason. After suffering a twenty-minute diatribe on this point, Don Juan replied that nothing that he had heard altered his high opinion

---

* In the event of ever needing to organize a rapid succession process, Franco planned to offer the throne to Juan Carlos and simultaneously ask Don Juan to abdicate, confident that the Pretender would agree rather than risk a public break with his son – Franco Salgado-Araujo, *Mis conversaciones*, pp. 304, 334.

of Sainz Rodríguez. A rattled Franco replied spitefully that he knew of other masons in Don Juan's circle including his uncle, General Alfonso de Orleans, and the Duque de Alba.[6]

Don Juan gave Franco the text of a proposed communiqué which stated that the talks had taken place in an atmosphere of cordiality and that the education in Spain of Juan Carlos did not prejudge the question of the succession nor prejudice 'the normal transmission of dynastic obligations and responsibilities'. It closed with the statement that 'the interview ended with the strengthened conviction that the cordiality and good understanding between both personalities is of priceless value for the future of Spain and for the consolidation and continuity of the peace and the work carried out so far'. Franco read the text and discussed it at length with Don Juan, arguing it point by point and protesting at a reference to Juan Carlos as Príncipe de Asturias, the usual title given to the heir to the throne. Don Juan conceded that. However, Franco accepted the text with alacrity when Don Juan said that, if he wanted to take longer, he was happy to keep Juan Carlos with him for an academic term. On the following day, Don Juan's staff went ahead and issued the agreed text in good faith. However, to their astonishment, the version published in Madrid contained significant variants from Don Juan's text. On arriving at El Pardo late on 29 March, Franco had unilaterally amended the agreed communiqué.[7]

He added a reference to himself as Caudillo, a title never acknowledged by Don Juan. To the the phrase about the transmission of dynastic responsibilities, he added 'in accordance with the *Ley de Sucesión*', a law still repudiated by Don Juan. In the last sentence, he removed the reference to 'both personalities' to ensure that he and Don Juan should not be seen to be on an equal footing. Finally, he added to the sentence a reference to the *Movimiento Nacional* implying that future relations between himself and Don Juan would take place in that context.[8] He told his cousin that he had felt no need to consult with Don Juan since these additions merely reflected the legal situation in Spain and 'I knew that he would have to agree'.[9] The Spanish censorship machinery not only blocked attempts from Estoril to have rectifications published but also permitted the Spanish press to print accusations that Don Juan had omitted the references which Franco had in fact added. Don Juan was understandably annoyed at this underhand dealing by Franco.[10] Immediately after Franco had taken his leave, his advisers got the strong impression that he would never again meet the Pretender to discuss politics.

A twofold process was taking place which was moving Franco imperceptibly from the centre of political life. On the one hand, in response to economic modernization, a society was emerging with problems and concerns that meant nothing to a Franco locked into the mental set of the Civil War and the 1940s. On the other, partly as a result of his inclination to devote himself to his hobbies and partly in response to the growing complexity of government business, he was increasingly the silent chairman of the cabinet, leaving the detail of government to Carrero Blanco and the technocrats.

This was apparent during a lengthy tour of Catalonia in May 1960. As so often on such visits, he arrived by sea, on the battle-cruiser *Galicia*.[11] During his stay, the annual Civil War victory parade was held in Barcelona.[12] On 20 May, he held a cabinet meeting at which Navarro Rubio presented crucial legislation. In response to protests from both the Church and the Syndical Organization about the social consequences of the stabilization plan, Navarro Rubio had elaborated a Law of Social Funds for spending on four specific social projects which had implications for the Ministries of Labour, Education, Interior and the *Movimiento*. Although the *Presidencia del Gobierno* was fully apprised of the fact that the project would be presented, neither Franco nor several ministers had been warned. Implicitly, Navarro Rubio's project shifted the politically crucial power of social patronage within the *Movimiento* away from the Falange. When Franco saw the affected Ministers individually before the opening of the session, both the Minister-Secretary of the *Movimiento*, Solís, and the Minister of Education, Jesús Rubio, told him that the new law was a blow against the Falange. Fermín Sanz Orrio, the Minister of Labour, and Camilo Alonso Vega, the Minister of the Interior, expressed their satisfaction at getting more funds. Paralysed with indecision, Franco left the issue for a month and it was eventually accepted at a cabinet meeting on 10 June 1960 but, when it went to the Cortes in July, the Falangist *Procuradores* effectively neutralized it with amendments.[13]

The Law of Social Funds, and the way in which it was elaborated and introduced, emphasized the extent to which Franco was presiding over a machine whose inner workings were becoming a mystery to him. Symptomatically, during the extended visit to Catalonia and to the Balearic Islands in early May, his speeches as he inaugurated public works were soporifically anodyne.[14] With Spaniards concerned about living standards, anxious to forget the Civil War and the Cold War, he was reading out speeches about the regime's commitment to economic development but the refrains about past triumphs made it difficult to avoid the impression that the entire process was beginning to pass him by. The political elite still needed Franco as ultimate arbiter but less and less as day-to-day ruler.

The distance which now separated the sixty-eight year-old Franco from much of Spanish society was illustrated by a curious incident at the end of May 1960. He intervened personally to cancel both legs of the football match between Spain and the Soviet Union in the quarter finals of the first-ever European Nations Cup. Franco had insisted that both legs be played on neutral ground in part because of complaints by Camilo Alonso Vega and Carrero Blanco that there were combatants of the *División Azul* still detained in Russian concentration camps. He also had on his desk reports from the police predicting popular demonstrations in favour of the Soviet team. When the Russians refused to comply, the matches were cancelled. At a time when both Real Madrid and the Spanish national team were at the peak of their glory, the decision merely caused international ridicule and domestic unpopularity. To make matters worse, news of the withdrawal of the national team from the competition was not published in the Spanish press. It was merely announced that the USSR had qualified for the semi-finals. Then, it was stated that the reason for Spain's elimination was that the USSR had refused to play on neutral ground.[15] Franco told Pacón that he had also been reacting against Communist radio broadcasts which claimed that the Russian team would receive a colossal popular reception in Madrid as an expression of hostility to him. This, combined with the Russian request for the Soviet anthem to be played and for the Soviet flag to be flown, was altogether too much. When Pacón asked why the reasons for Spain's withdrawal were not published, a puzzled Franco replied, 'the bosses of the Football Federation know the reasons already'.[16]

The position of the technocrats was significantly consolidated as a result of the relative success of the stabilization plan. The political and economic changes implicit in their policies provoked a sense of desperation within the Falange. From time to time, this erupted. On 20 November 1960, the twenty-fourth anniversary of the death of José Antonio Primo de Rivera, Franco attended a ceremony at the Valle de los Caídos. He entered the basilica under a canopy. As the lights went down and the priest raised the host, a loud voice was heard to shout 'Franco, you are a traitor.' Franco heard but showed no sign of doing so. Immediately, guards began to go along the ranks in search of the culprit. It was not the first time that Falangist frustration had been expressed in such a way. A year earlier when the ceremony had been presided over by Carrero Blanco because Franco had influenza, a voice had shouted 'Carrero go home!' On another occasion, as Franco left the basilica surrounded by ministers, low voices muttered 'get rid of those who surround you, they do more harm than good.' Then, no action had been taken. However, now, on an occasion in which he was being glorified in his own basilica, he was not prepared to overlook a direct insult. The Falangist responsible was Román Alonso Urdiales, a young radical outraged by the corruption of the regime. At the time, he was doing military service. He was arrested and tried by court martial on 20 December 1960. It was rumoured that Franco wanted to have him shot but eventually refrained in order not to make a martyr. Urdiales was sentenced to twelve years' imprisonment and was sent to a punishment battalion in the Sahara desert.[17]

On 19 December 1960, Navarro Rubio announced the first *Plan de Desarrollo* or Development Plan which had been elaborated in collaboration with the World Bank. The gradual liberalization of the Spanish economy under the technocrats had little impact on Franco's own views. At a cabinet meeting on 19 January 1961, when Ullastres complained of Falangist attacks on his policies, Franco responded by launching a lengthy defence of autarky. Later in the meeting, Franco expressed his anxieties about the World Bank mission encharged with advising on the Development Plan. He was convinced that Sir Hugh Ellis Rees, its leader, an Irish Catholic, was a freemason. He also expressed his belief that the main object of the Plan should be to create jobs. It took Navarro Rubio some time to convince him that the level of employment could not be assured in isolation since it was dependent on numerous other factors. Only with the greatest reluctance did Franco agree to the arrival of the World Bank mission and to the Development Plan.[18]

As he clung to the ideas of the Civil War, the Caudillo would find himself increasingly isolated from some of his ministers, many of whom were twenty to thirty years his junior. He remained obsessed with freemasonry.[19] And he still admired both Hitler and Mussolini for the 'energy, authority and patriotism' with which they crushed Communism and revived their nations. His attitude to the United States remained a curious mixture of awe and contempt. He believed that American government was totally in the hands of freemasons who would open the door to Communism. He did not trust the newly elected President, John F. Kennedy, because he was 'surrounded by leftists and enemies of the Spanish regime'.[20]

At the beginning of 1961, Franco had commissioned Carrero Blanco to produce a report on the likely consequences for his regime of Kennedy's arrival in the White House. Submitted on 23 February 1961, it stated that the world was dominated by the three internationals, the Communist, the socialist and the masonic, which shared a determination to destroy the Franco regime. It warned against those who called for the legalization of political parties, because they were motivated only by the desire to weaken Spain. 'We must be ready to defend our unity within the most tightly clenched intransigence.' In addition to warning that any American pressure for political liberalization must be resisted, Carrero Blanco also advocated a harder line when it came to renewing the bases agreement. Reports that the new President was making contact with Spanish Republican exiles and was hostile to the bases in Spain inclined Franco to accept the report in its entirety.[21]

From 20 April to 6 May 1961, Franco made a triumphal tour of Andalusia. In the midst of considerable acclamation, some genuinely popular and some artifically mounted by the *Movimiento*, of undiluted adulation from local officials and of visits to new buildings and public works projects which normally made up such tours, he had a novel experience. The Civil Governor and Jefe Provincial of Seville, the liberal monarchist Hermenegildo Altozano Moraleda, astonished the Caudillo by taking him to see the appalling shanty towns on the outskirts of Seville. Franco was taken aback by the inhuman conditions in which he observed people living. At the end of the tour, the impact on him of what he had seen was reflected in a speech he made in Córdoba. 'On this journey', he said, 'as on others, I have become aware of the persistence of many social injustices and of greatly irritating differences.' However, his response was merely to appeal to the paternalism of the Andalusian rich. Blithely unaware of the savage social tensions in southern Spain, he appealed 'to the nobility of Andalusia, to the generosity of the men of this land, to those whose possessions and goods we saved, to the businessmen, that they might collaborate in a Christian spirit in the creation of social justice and assist with good faith our social legislation, and I trust that, in return, the workers and labourers might pay them back with their hard work and enthusiasm'.[22]

During the return journey, he was notably discomfited when Alonso Vega commented that the cheering Falangists who had greeted him were probably paid to do so. Don Camilo pointed out that their blue shirts could never have been worn before since they still bore the creases of their boxes.[23] Once back in El Pardo, Franco did nothing about the misery he had witnessed beyond asking himself rhetorically why the Seville authorities did not appeal to the State for funds to remedy the problem. Such a suggestion in itself revealed his limited comprehension of the austerity imposed on public spending by the stabilization plan. More significantly, Franco did not prevent Solís, as Minister-Secretary of the *Movimiento*, from trying to hound Altozano from his job in punishment for the dual crimes of his monarchist sentiments and his gesture of upsetting the Caudillo with the harsh reality of Andalusian society. Franco was astonished to hear from Pacón that there were similar shanty towns around Madrid and Barcelona and declared 'I am prepared to throw myself into this business to try to resolve it.'[24] It was a good resolution soon forgotten.

1961 was the twenty-fifth anniversary of the military uprising of 1936. It was also the year in which the sacrifices of the stabilization plan began to bear fruit. The contrast between the violent past and the prosperous present was increasingly perceptible in the dual tone of Franco's speeches. The parts written by his ministers were full of statistics, politically colourless. The parts written by the Caudillo himself were nostalgic, belligerent, far removed from the domestic and international reality of the 1960s. He talked as if his enemies were those of 1936, but many of them were dead. His real enemies were young workers ready to strike against factory conditions in which hours were long and safety regulations ignored, students protesting against the stifling atmosphere of the Francoist universities, young Basque and Catalan priests denouncing the repression of nationalist sentiment. With Eisenhower as President of the United States committed to a defensive anti-Communism, Franco had not seemed so out of place. In the world of Kennedy, determined to beat Communism by a more aggressive projection of the benefits of capitalism, the Caudillo seemed a fossilized survivor.

On 3 June, on inaugurating the new legislature of the Cortes, Franco made a speech nearly two hours long. The tone was entirely triumphalist, a boastful survey of the achievements of the past twenty-five years contrasted with a bleak survey of the calamities of the previous centuries. There were references to a possible *Ley Orgánica del Estado* to complete the Francoist constitution and guarantee the future succession. He even mentioned the way in which the development of the European Common Market was beginning to determine the direction and scope of Spanish trade. However, both of these hints of possible preparations for the future came only after a justification of the present, a hymn of praise to *Caudillaje* and a denunciation of political parties and the liberty which divides or merely leads to compromise.

There was still no talk of forgiveness: the Civil War remained the Crusade fought by good Spaniards against bad. He denied that his regime was a dictatorship at the same time as he railed against democracy. His self-congratulation was not limited to his achievements in Spain but also on the world stage, claiming that 'Without our victory, Spain in its entirety would be Communist and the Iberian peninsula would have constituted over the last twenty-five years the most efficacious and stimulating factor for the extension of Communism to Spanish America and a launching pad for international Marxism throughout Africa' and that the allies owed him a colossal debt for his comportment during the Second World War.[25]

1961 was choreographed as yet another high point in Franco's career. One week before the anniversary of the military rebellion, Don Juan wrote to congratulate him on twenty-five years in power and to proclaim the ties between the monarchy and the uprising of 1936. Franco replied with a letter in which his delight was interleaved with cunning references to his triumphant tour of Andalusia and to his popularity in general. The hint was unmistakable: he had no need to consider stepping down.[26] On 17 July 1961, the annual victory parade was led by a contingent of fifty thousand ex-combatants from the Civil War. The main celebrations of the year, however, were centred not on the anniversary of the uprising but, as might have been anticipated, on that of Franco's elevation to supreme power.

On 1 October 1961, a carefully orchestrated set of ceremonies began in Burgos. Speaking from the balcony of the *ayuntamiento* (town hall), as he had twenty-five years previously on receiving power from the hands of General Cabanellas, Franco displayed understandable pride and emotion. With characteristically blinkered arrogance, he substituted the reality of his careful scramble for power with a tale of heroism, self-sacrifice and vision. He suggested that even then, in 1936, he had been planning the future as it had subsequently developed: 'From the first moment, I was aware of the burden which was thrown onto my shoulders.' Forgetting the effort that he and his brother Nicolás had put into ensuring that he won the power struggle, he declared 'in such conditions, command could not be desirable. Only a high concept of duty and confidence in God and in the rightness of our cause made the fulfilment of my duty more bearable'. He omitted all reference to the political assistance of Serrano Suñer in building his State apparatus and mentioned only with vengeful satisfaction the hundreds of thousands of wrecked lives on which his regime was built.[27]

In between the nostalgic and self-congratulatory speech at the *ayuntamiento*, and a similar one to the *Consejo Nacional* on the following day, there was a much more practical interlude. Later in the day on 1 October, Franco made a speech to representatives of the three armed services in which he commented that the development of nuclear weapons had devalued the conventional weapons acquired as part of the 1953 Pact with the USA. 'For that reason,' he stated, 'with four fifths of the time agreed already gone, our agreements need to be examined again and renewed in accordance with the new situation.'[28] It was the opening shot in the negotiations for the renewal of the bases agreement which would begin one year later. In fact, the days of substantial economic and military aid in return for the bases had gone but Franco's words had sufficient impact in Washington for the Secretary of State Dean Rusk to visit El Pardo before the end of the year.[29]

As part of the anniversary celebrations, a meeting of the cabinet took place in the Palacio de la Isla, Franco's wartime headquarters in Burgos. López Rodó and Carrero Blanco had hoped to see some symbolic progress towards resolving the succession. López Rodó had persuaded both Carrero Blanco and Alonso Vega to press Franco to announce that the constitutional law which he had drafted in 1957, the *Ley Orgánica del Estado*, would be submitted to the Cortes. But Franco shrank from committing himself irrevocably to a particular option for the succession and ignored suggestions made by Carrero and Don Camilo during the short session.[30]

That the Caudillo did not want his moment of glory diluted by mention of a distant future was underlined on 2 October 1961, when he addressed the *Consejo Nacional de FET y de las JONS* at the Monasterio de las Huelgas, where the first ever session had taken place in 1937. The occasion – organized by the rising star of the *Movimiento*, its Vice-Secretario General, Fernando Herrero Tejedor – was choreographed to praise Franco's past triumphs.[31] To emphasize the Falange's abject subordination to the Caudillo, José Solís, as Minister Secretary of the *Movimiento*, began the ceremony by addressing Franco simply as '*Señor*', a form of address reserved for kings.

The tone of Franco's own speech was as usual one of arrogant self-congratulation and reaffirmation of the values of the Nationalist side in the Civil War. He began by praising his far-sighted vision twenty-five years previously and went on, by taking sentences out of context from his speeches since 1939, to prove that the recent economic surge had been skilfully planned all along. Despite using some of José Antonio Primo de Rivera's most well-known phrases, Franco ostentatiously failed to mention the Falange and spoke throughout of the *Movimiento Nacional*.

It was an exercise in ideological sleight-of-hand aimed at reconciling the Falange of autarky and 'the pending revolution' with the conventionally capitalist Spanish economy emerging under the technocrats.[32] At the end, the Falangist anthem *Cara al sol* was sung by the *Consejeros* most of whom raised their right arms in the fascist salute. For once Franco did not join them but stood to attention with his arms at his side as did Carrero Blanco and López Rodó. A deeply embarrassed Solís was obliged to do the same. Significantly, Prince Juan Carlos was not present.[33]

The distance between the Falange and the technocrats was to be the greatest source of tension within the regime in Franco's last years. There was a battle, in a phrase attributed to López Rodó, to 'furnish Franco's head with ideas'.[34] However, the extent to which Franco perceived this conflict as a problem should not be exaggerated. Militant Falangists committed to making 'the pending revolution' were fewer by the day. The *Movimiento* had been thoroughly domesticated under Arrese, Fernández Cuesta and Solís. When Franco talked of the *Movimiento* rather than of the Falange, it may have upset the hard-core militants but for many it reflected a certain truth. Bright, hard-working functionaries were emerging who were more concerned to get top jobs in the state apparatus than to implement the ideology of Falangism. That was entirely true of men like López Rodó and Navarro Rubio who were labelled as being primarily of Opus Dei but were more accurately seen as part of what came to be called the 'bureaucracy of number ones' (*la burocracia de los número uno*), those who had won competitive civil service examinations or university chairs while very young. Other prominent administrators of Francoism in the 1960s, like Manuel Fraga and Torcuato Fernández Miranda, were usually described as Falangists. Similarly, among the best brains of the next generation, Rodolfo Martín Villa was thought of as primarily a Falangist; José María López de Letona and Fernando Herrero Tejedor were both linked to Opus Dei – yet they could all more properly be identified as meritocratic functionaries. This is not to say that there were no longer factional rivalries in the Francoist coalition but rather that for a decade they were considerably muted. Ironically, in the early 1960s, there was more tension between López Rodó and Navarro Rubio than between López Rodó and Fraga.[35]

After the collapse of Arrese's plans for Falangist revival in 1956, the nature of political competition under Franco changed. Between 1957 and 1973, the new technically competent functionaries held sway. They regarded themselves as 'apolitical' – by which they meant that their central concern was the technical task of state administration and that they belonged in broad terms to the *Movimiento* rather than to a particular faction. Their competence in a complex world marginalized Franco from the day-to-day running of government yet, paradoxically, strengthened his position in two ways. First, he took the credit for the economic improvements which took place under their administration. They were happy to let Franco take the honour because, in the gossip-ridden world of El Pardo, to fall short of adulation would have been to risk their jobs. Secondly, because – unlike the factions of the past – the technocrats had no political clientèle, they owed their well-remunerated prominence to Franco and never thought of challenging him. To have done so would have been dangerous and simply impelled their rivals to rally to him.

As Franco came in the 1960s to be less active, he also became more impregnable. His strength lay in his position as the keystone of the Francoist arch. In the period between 1938 and 1953, international hostility to the Caudillo encouraged rival factions and he had played them off with great skill and cunning. After 1953, he felt much more secure and even over-confident. In consequence, he had had to weather the crisis of 1956, the outrage provoked by Arrese's constitutional plans and the collapse engendered by his autarkic economic policies. In 1957, he had been fortunate to find the technocrats to get him out of a jam. With their policies working, no one Francoist faction would take the risk of destabilizing a situation from which all derived benefit. Each hoped to get his support against the others.

Thus, while López Rodó and Carrero Blanco offered Franco efficiency and relief from the cares of government, Solís sought to offer him more spectacular, if less substantial, gifts. On 13 October 1961, he and Herrero Tejedor organized at the Valle de los Caídos a massive fascist rally gathered as the European Assembly of Ex-Combatants. Extreme rightists from all over Europe, the defeated of 1945, came to pay homage to the only victorious general of the pre-war right. Alongside Italian Fascists and Nazis, Eastern European fascist generals paraded with the massed ranks of Civil War veterans. Franco was careful not to attend but he sent a message of welcome and congratulation via his friend General Pablo Martín Alonso.[36]

The rally reflected Falangist fears of their own obsolescence. Living off the State bureaucracy, they could feel only the greatest anxiety about a future without Franco. The military had more reason for confidence but also saw Franco as the key to a politically acceptable regime. Even Don Juan and the monarchists felt that, in outright opposition to Franco, they could do little to hasten the re-establishment of the throne. Moreover, most factions at the heart of the Francoist coalition tended to huddle together as the social turmoil unleashed by economic change brought a surge of opposition in the factories, the universities and the regions. But those at the political extremes of the coalition leaked away – the Christian Democrat liberals on its left to create their own opposition groupings and the Falangist old guard on the right to coagulate around the pressure group known as the *Guardia de Franco* led by the Falangist General Tomás García Rebull. In fact, neither would seriously threaten Franco whose position, during the prosperity of the 1960s, came to seem immutable other than by death.

The fearful rumours about the Caudillo's mortality which had circulated in the spring of 1960 were overshadowed by the sheer panic provoked at the end of 1961. On a cold and wet Christmas eve, Franco followed a suggestion from José Sanchiz, the administrator of his Valdefuentes estate, and went pigeon shooting in the hills behind El Pardo. At 5.15 p.m., he was hurt in a serious accident when his Purdey shotgun exploded, badly damaging his left hand.[37] The press played down the extent of his injuries – fractures of his index finger and his second metacarpal bone. Much of the muscle and nerve tissue of the hand was destroyed.[38] He was taken to the Air Force Hospital in calle Princesa. The radiologist called to attend to him, unaware of the real identity of his patient, remarked to a nurse how much he looked like the Caudillo, at which Franco commented 'people say that' (*eso dicen algunos*).[39] For Franco to be given a general anaesthetic was a matter of some moment yet he used none of the complex mechanisms established in the *Ley de Sucesión*. Instead, he telephoned Carrero Blanco and ordered him to inform only the military ministers and the General Staff of the Army. Then, he sent for his lifelong friend Camilo Alonso Vega, the hard-line Minister of the Interior. He made no provision for a diminished ability to govern other than to ask Don Camilo, somewhat cryptically, to 'keep an eye on things' (*ten cuidado de lo que ocurra*). Franco was justifiably confident that Alonso Vega, his Director-General of Security, Carlos Arias Navarro, and the Director-General of the Civil Guard could between them maintain public order.

Franco was considerably incapacitated by the consequent pain and unable to sleep at night for the first months of 1962 despite taking painkillers.[40] Even after the plaster was removed in April, he gingerly held his left arm to his chest. Only extensive physiotherapy over the next three months returned him to near normal mobility.[41]

Exhaustive tests by military armourers and by the English manufacturers proved that the accident could not be attributed to a defective gun. This led to speculation that Franco had been the target of an assassination attempt with Alonso Vega especially convinced that the ammunition had been tampered with.[42] This was unlikely since Franco took his ammunition at random from sealed stock specially prepared for El Pardo and no other defective cartridges were found. With many guns loaded from the same stock, an assassin would have had little chance of targeting Franco by introducing one defective cartridge. Finally, it was concluded that smaller calibre ammunition belonging to his daughter had found its way into his pocket, been loaded into the gun, failed to detonate, become lodged in the barrel and exploded at his next shot.[43] Nonetheless, the incident impelled many Francoists to turn their attention to the succession.[44] Their anxieties for the future would be intensified when large-scale strikes swept across the industrial north in the spring of 1962.

At the time of the accident, Franco himself gave no public hint of preoccupation. In his end-of-the-year broadcast, he had described himself as the captain of the ship of Spain and the people as the 'crew and the beneficiaries' of his rectitude, virtues and skill as a navigator.[45] But where the navigator was steering Spain was not clear. All that could be discerned of his intentions was a commitment to some eventual but far-distant royal succession. The trust he placed in Alonso Vega after the shooting accident raised the possibility of a Francoist regency. In such a scenario, a hard-liner would guarantee that the eventual monarch would not stray from the authoritarian path. The role of watch-dog of Francoist ideals would be entrusted first to Muñoz Grandes and then to Luis Carrero Blanco, although in fact Franco would outlive them both. In the meanwhile, despite Franco's close relationship with Carrero, who was committed to the monarchy of Juan Carlos, there remained considerable room for speculation and manoeuvre regarding the selection of a royal candidate.

While still suffering acutely from the after-effects of the accident, Franco agreed on 25 January 1962 to Carrero Blanco's suggestion that López Rodó be made head of the Commissariat for the Development Plan, a central planning body suggested by the World Bank advisers. Following the model of the French Development Plan, the *Comisario* would be the delegate of the *Presidencia del Gobierno* in each of the economic ministries with power to create inter-ministerial committees.[46] It constituted a massive increase of power and responsibility for López Rodó. Navarro Rubio was understandably annoyed that the Commissariat had not been set up in the Ministry of Finance. He believed that the Development Plan should be forged within budgetary limits which only his Ministry could set. Also regarding his ministerial salary as inadequate, Navarro Rubio, after some hesitation, presented his resignation. Franco agreed but asked him to delay announcing it. In the meantime, he arranged for Navarro Rubio to be offered the highly paid post of Governor of the Bank of Spain.[47] The Caudillo remained unshaken in his belief that all men had their price. In fact, Navarro Rubio was not to be able to take up his new post for another three years.

Navarro Rubio was not the only opponent of López Rodó's elevation. At the cabinet meeting held on 26 January 1962, Solís tried to get the Development Commissariat located in the Syndical Organization. Franco guillotined the debate by stating bluntly that López Rodó was the best candidate. Solís continued to whisper in the Caudillo's ear that the appointment constituted the triumph of a plot by Opus Dei to take over the economy. Solís was actually fighting a rearguard action. In the nomination of López Rodó as *Comisario*, the Falange had lost a decisive battle in the war for the post-Franco future. However, López Rodó actually took on the job for the obvious reason of personal ambition and because he believed that adminstrative and economic reform were better guarantees for the survival of the system than the *inmovilismo* (resistance to change) manifested in Franco's reluctance to resolve the succession. López Rodó considered that the integration of Spain into the dynamic European economy would necessarily have political consequences in Spain. His views were summed up in the phrase, erroneously attributed to him, that only when per capita income in Spain reached $1,000 per annum could there be democratization.[48]

Now in his seventies, Franco had not agonized over the appointment. He saw López Rodó simply as one of the brightest and best of his functionaries. Decreasingly active in the daily tasks of government, Franco would be distanced further from the centre of gravity by López Rodó's elevation. López Rodó reported directly to Carrero Blanco or to Franco. Since neither Carrero nor the Caudillo understood the full complexity of what he was doing, the *Comisario* had considerable autonomy. Moreover, over the next few years, the hard-working López Rodó came to dominate the cabinet sub-committee for economic affairs (*La Comisión Delegada de Asunto Económicos*) formally presided over by Franco. The committee soon came to be a mini-council of ministers. It was attended at first by the various ministers with responsibility for economic matters. However, as it became the real locus of power, issues other than economics were discussed and other ministers sought excuses to attend. Just as previously, as Secretary to the Presidencia, López Rodó had had a primordial role in preparing and prioritizing cabinet business, now he had a similar, and even more powerful role, in initiating and co-ordinating economic policy. Franco became aware of the issues only when they came, pre-wrapped, to the sub-committee.[49] Since he had no reason to question the loyalty of Carrero Blanco's protégé, he was delighted to be relieved of irksome economic detail.

Despite the accident, Franco's interest in hunting remained all-consuming. As soon as he could shoot again, he spent every weekend hunting in the south.[50] That aside, there were signs of a slowing down and about this time, he started for the first time in his life to rely on a daily siesta. Having been delighted by the introduction of television into Spain, he also spent increasing numbers of hours watching the many sets placed around the Pardo.[51] His favourite programmes were movies and sport, particularly football. The revival of Spanish football since the arrival of Hungarian refugees like Ladislao Kubala, Ferenc Puskas and Sandor Kocsis had enthused Franco who saw the triumphs of Real Madrid and of the Spanish national team as somehow his own.[52] He began to do the pools (*quiniela*) every week, for a time signing his coupon (*boleto*) Francisco Cofran. He won twice. It is difficult somehow to imagine Hitler or Mussolini doing the pools.[53]

To maintain the momentum of economic reform initiated by the technocrats, Castiella and the economic ministers suggested that Spain petition to join the European Economic Community. The Caudillo agreed with the greatest reluctance. On 9 February 1962, the official request to open negotiations was presented by the Spanish Ambassador in Paris, José María de Areilza, to the French Prime Minister Maurice Couve de Murville, at the time President of the EEC. A major diplomatic campaign was mounted in the capitals of the member countries. The application was doomed to be shipwrecked on Franco's fiercely negative attitude to democracy. He had already shown, in relation to the IMF mission, that he was suspicious of the economists of the democracies. He was even more bleakly convinced that an approach to the Community would allow his enemies to try to blackmail Spain into political liberalization. He and Carrero Blanco regarded the EEC as 'a fief of freemasons, liberals and Christian-Democrats'. Categorically determined to accept no political conditions, Franco began to talk of decoupling the Spanish economy from northern Europe and orientating her trade towards the Communist bloc. Apart from revealing yet again his ignorance of economic realities, the notion also showed that his obsession with freemasons was more virulent even than his anti-Communism. In the event, the EEC agreed to begin negotiations for some form of economic agreement but made it clear that major constitutional changes would be necessary before any form of political link could be contemplated.[54]

Franco's political views had not evolved in response to the social and economic changes now under way. His reaction to anything which displeased him was to retreat into Civil War rhetoric. At the beginning of 1962, he was still indignant at Don Juan's proclaimed desire to be King of all Spaniards which, he claimed contemptuously, included 'all the defeated, Basque and Catalan separatists, Communists, anarchists, socialists, all kinds of republicans and terrorists as well, why not? they're all Spaniards'.[55] When Juan Carlos married Princess Sofía of Greece in Athens on 14 May 1962, a ceremony attended by large numbers of Spanish monarchists, Franco ordered that the occasion be given as little publicity as possible in Spain and that Don Juan should not figure in any photographs published. As his official representative, he sent his Minister of the Navy, Admiral Felipe Abarzuza, in the battle-cruiser *Canarias* – an unmistakable symbol of his Civil War victory. When Juan Carlos and Sofía made a special trip to Madrid at the beginning of their honeymoon in order to pay their respects to Franco, he made a point of also granting a long audience to the President of the *Comunión Tradicionalista*, José María Valiente, the advocate of the Carlist pretender, Carlos Hugo, as successor.[56]

The Community's refusal to open political negotiations merely convinced him that Spain was still surrounded by hostile forces determined to bring him down. This belief was reinforced shortly after by the outbreak of a wave of industrial unrest. Throughout April and May 1962, there were strikes in the Asturian mines and the Basque steel industry. Despite the massive and brutal deployment of the Civil Guard and the armed police, the strikes spread to Catalonia and Madrid. Stopped not by repression but by wage increases, the strikes marked the begining of the end for the Falangist vertical syndicates and the emergence of a new clandestine working class movement.[57]

The strikes were economic rather than political in motivation. In the economic revival which followed the harsh austerity of the stabilization plan from 1959 to 1961, the workers were determined to improve wage levels. Their victory showed that the State-owned enterprises and private sector industrialists were prepared to pay to avoid interrupting valuable production. Franco did not see things in such terms but crudely attributed labour unrest to outside agitators. He was infuriated by the many declarations of solidarity with the strikers received from France, Italy, Germany

and Britain and was perplexed because many priests had expressed support
for the workers, particularly in the Basque Country. The activities of the
JOC (*Juventud Obrera Católica* – Catholic Workers Youth) and the HOAC
(*Hermandad Obrera de Acción Católica* – Workers' Fraternity of Catholic
Action) were in his view 'not apostolic' but rather 'open the way to
Communism'.[58] Privately and in public, he reverted to explanations in
Civil War terms of the 'enemy' and foreign Communist and masonic
agitators.[59]

His public analysis of the strikes on 27 May 1962 was given at the
Cerro Garabitas, a battleground of the Civil War siege of Madrid, a
choice of venue fraught with aggressive symbolism. To an audience consisting of the Falangist war veterans of the *Hermandad de Alfereces Provisionales* (Brotherhood of Provisional Officers), he dismissed the strikes as
unimportant, taken seriously abroad only because of the wild statements
of 'the odd Basque separatist priest or the clericalist errors of some exalted
priest'. Declaring that the Civil War was still being fought, he attacked the
unnamed 'enemy' for trying to make capital out of 'minor malfunctions of
our labour relations' ('*pequeños fallos en nuestras relaciones laborales*') and
declared that Spanish workers should know that 'no one can go further
than the Spanish State in the work of social justice'. Significantly, he
ended his speech to regime hard-liners with remarks meant to calm any
anxieties they might have about the future as a result of the shotgun
accident. 'There are those who speculate clumsily about my age. I can
only say to them that I feel young, as you do, that after me everything
will be left well tied down and guaranteed by the will of the great majority
of Spaniards, among whom, along with the *Movimiento*, you constitute
the sinews and the essence and by the faithful and insuperable guard of
our Armies. Our work is the mandate of our dead.'[60]

The regime's abortive overtures to the EEC and the 1962 strikes
stimulated a resurgence of sympathy in Europe for the anti-Franco opposition. To capitalize on this, monarchists, Catholics and repentant Falangists from inside Spain met exiled Socialists and Basque and Catalan
nationalists in Munich at the IV Congress of the European Movement
from 5 to 8 June 1962. The participation of the delegates from the interior
had been openly announced in advance. The final communiqué of the
meeting was a moderate and pacific call for evolution in Spain. On the eve
of the Congress, Franco had received a warning from the extreme-right
Falangist group, *Vieja Guardia*, of a conspiracy to undermine the regime
by freemasons, Jews and Catholics. This, not unnaturally, struck a
chord.[61] As the Congress closed, his cabinet remained in session dis-

cussing its implications until 3.00 a.m. in the morning of 9 June. It was decided to suspend the flimsy constitutional guarantees of the *Fuero de los Españoles*. The Caudillo was furious at what he saw as plot to torpedo his regime's efforts to secure an association with the European Community.[62] Many of the Spanish delegates, including Dionisio Ridruejo and José María Gil Robles, were arrested and sent into exile for their part in what was denounced as the 'filthy Munich cohabitation' ('*contubernio*'). Franco's reaction severely damaged the Spanish case for entry into Europe.[63]

Franco urged Arias Salgado, the Minister of Information, to unleash a violent response in the press. The barrage of hysteria even extended to blaming Don Juan for what had happened.[64] This reflected both Franco's personal outrage and a generalized fear about the future within the *Movimiento*. Not only had the Caudillo 's shooting accident raised the spectre of his death, but the strike wave had undermined the myth of the regime's invulnerability. There was now a disturbing plausibility about Communist claims that their policy of 'national reconciliation', adopted in 1956, was about to bear fruit in a wide front of anti-Franco forces.[65] Moreover, there had been glimmers of conflict with the Catholic Church since Pope John XXIII's convocation of the Second Vatican Council in 1959. After it began its work in October 1962, the rifts would open ever wider.[66] The Pope's encyclical of 1961 *Mater et Magistra* had alarmed the hard-liners in Franco's cabinet with its talk of just wages and humane conditions for industrial and agricultural labourers, redistributive taxes and trade union rights. At Munich, Catholics and monarchists had consorted with exiled democrats and groups sponsored by the Church were at the heart of the reviving internal opposition.

Franco's view about the Church's increasingly liberal stance had been made clear at Garabitas. Now he gave vent to his feelings about Munich in a series of speeches that he made shortly after in Valencia. On 16 June, he denounced foreign criticism of his regime and, indicating his cheering followers – many of whom had, as usual, been brought in from the surrounding countryside and paid a day's wage to come, declared 'here are my powers; the closest union with my people'. Accusing Europe of wanting to see only failure in Spain, he proudly surveyed his own 'exemplary and noble record [*ejecutoria ejemplar*]'. In denouncing the press of Europe as the lackey of world Communism, Franco undid much of the patient labour of his technocrats and diplomats, and also provoked ridicule at his own expense.[67] On the following day, he derided liberalism as weak, useless and rotten. On the day after, he attacked those who went to Munich 'as the wretched ones who conspire with the reds to take their miserable complaints before foreign assemblies'.[68]

To protest at the punishment of Spanish participants in the Munich Congress, a delegation of the International Executive Committee of the European Movement visited Madrid. They spoke to Franco on 6 July for one hour and ten minutes. They pointed out that the Spaniards at Munich had done no more than make made a moderate declaration in favour of non-violent political evolution. Told that his reaction to Munich was an implicit condemnation of the EEC and undermined his own government's application to join, Franco remained unperturbed. With his usual hard-faced cheek, he blandly denied their criticisms, asserting that he accepted the right of Spaniards to participate in the European Movement and objected only to the fact that the invitations were limited only to those who opposed his regime and that the interior delegates had been able to meet with his exiled enemies. This was true – only those with some commitment to democracy had been invited.[69]

Despite his defiance in Valencia, Franco realized that his reaction to Munich had been a serious error. On top of the remarks made by the European delegation, he was amply informed by Castiella and Areilza of the damage caused in the capitals of Europe. His Ambassador in Washington, Antonio Garrigues, sent reports of widespread criticism of the regime in the United States.[70] He might not have understood the finer points of economic theory, but his sensitivity to threats to his survival was undiminished. Accordingly, he overcame the inertia born of his reluctance to change ministerial teams and instituted a major cabinet reshuffle on 10 July.[71] The new economic line was consolidated. Ullastres and Navarro Rubio remained in place and were joined by the dynamic thirty-eight year-old naval engineer Gregorio López Bravo, another Opus Dei member, as Minister of Industry. The handsome and dashing López Bravo had been brought in not as part of some Opus Dei plot but because he had been recommended to Franco by his friend from El Ferrol, the Falangist Admiral Pedro Nieto Antúnez, who became Minister for the Navy. Other more 'progressive' technocratic elements introduced at the expense of Falangists were the Opus Deista Professor of Chemistry, Manuel Lora Tamayo, as Minister of Education and Jesús Romeo Gorría as Minister of Labour, both on the recommendation of López Rodó. Romeo's predecessor, Fermín Sanz Orrio, was replaced in punishment for failing to prevent the strikes of April and May. Solís, on the other hand, was kept on, both as a sop to the Falange and as a reward for the energy that he had displayed during the strikes.

The dourly authoritarian Minister of Information, Gabriel Arias Salgado, despite enjoying Franco's special favour, was saddled with the blame for the grotesquely miscalculated press reaction to Munich and replaced by Manuel Fraga Iribarne. Usually labelled a Falangist, Fraga had come to prominence as an associate of the Catholic Joaquín Ruiz Giménez. In fact, he was more a versatile and flexible apparatchik of the *Movimiento* than a militant Falangist although he had been recommended by both Solís and Nieto Antúnez. Possessing inexhaustible energy and ambition, the forty year-old Fraga had come first in every public competitive examination in which he had taken part, been a full Professor of Law before he was thirty, written many books and held a wide range of posts. With his short-cropped hair and natty suits, the can-do style and appearance of a busy American entrepreneur, Fraga was seen as someone capable of resolving the intractable problems of maintaining press censorship in a vertiginously changing society. In a sense, his partial liberalization of the press would make him one of the grave-diggers of the regime.

As an acknowledgement of the fact that he was nearly seventy years old, Franco introduced for the first time a vice-president of the council of ministers in the person of the sixty-six year-old General Agustín Muñoz Grandes, who was, and remained, Chief of the General Staff. The appointment was a belated precaution in the light of the warning provided by the shooting accident and reassured Falangists jealous of the westernizing Opus Dei technocrats. Muñoz Grandes was an extreme right-winger who shared and encouraged Franco's instinctive hard-line response to the growing opposition. In the event, the rigid and austere Muñoz Grandes was never used as deputy for Franco. Indeed, he was a poor politician and let Carrero Blanco carry out most of the functions of vice-president and so accumulate even more power. Franco's need for familiar faces as well as the dynamic young men was also reflected in the fact that the seventy-three year-old Alonso Vega was kept on as Minister of the Interior and in the appointment of the sixty-four year-old Admiral Nieto Antúnez as Minister for the Navy.[72]

The incorporation into the cabinet of Fraga and López Bravo was to be a major triumph of public relations. Both contributed by their youth and energy to a renovation of the image of the regime. The apparent liberalization implicit in the cabinet changes was, however, accompanied by harsher police measures against the left-wing opposition. Arrests, torture and trials of leftist militants were still commonplace.[73] A new strike wave in Asturias and Catalonia during August and September was countered by ferocious police measures.[74]

To secure his own permanence in power, Franco was pinning his hopes on a dual policy of brutal repression combined with an intensified bid for growth. With Fraga as Minister of Information, the censorship was loosened and the image of Franco and his government was much more skilfully manipulated. The first fruits of the change could be perceived in his end of year broadcast on 30 December 1962. The tone was much less anachronistic, more mellow even, than before. The usual tirades against freemasons and other enemies of the true Spain were, for once, absent and the stress was on economic achievements. Much of the speech had been written by Fraga and López Rodó.[75] Talk of 'the Spanish miracle' and lengthy statistical comparisons between 1936 and 1962 in everything from fertilizer production to irrigation schemes was tendentious but a more modern and effective way of making points previously expressed in an obsolete rhetoric of vengeful triumphalism.

The claim that the present economic growth had been planned all along was not new. What was entirely novel was the veneer of awareness of social tensions. The impact of the strikes of 1962 and the Fraga face-lift were both apparent in the astonishing statement that the Government intended to resolve problems which he attributed to 'the enormous effort of growth being made by the country'. If such rhetorical humility was far from the triumphalism of the past, even more remarkable was a short-lived departure from the language of victors and vanquished. The impact of labour unrest was also visible in his announcement of the introduction for the first time of a minimum wage. Although the hopes this raised were to be dashed soon enough – not least by the pitifully low minimum wages introduced on 1 January 1963 – the speech suggested that the cabinet changes of 1962 had marked even more of watershed than those of 1957.[76]

Having survived the crisis of 1962, Franco began to enjoy ever more a political life appropriate to a royal personage. His daily work consisted increasingly of receiving delegations of admirers, representatives of certain industries or cultural activities, giving prizes and medals, opening dams and other public works, and receiving the credentials of ambassadors. Noticeably more time went into leisure pursuits, hunting and fishing as ever, but now more golf, and a growing absorption in watching sport and films on television, some painting, doing the pools and very long holidays divided between the Pazo de Meirás and the Palacio de Ayete in San Sebastián in the summer and at La Piniella in Asturias at Easter. The length and number of hunting and fishing trips increased. At the beginning of October every year he would make a long hunting trip to the Sierra de Cazorla in the province of Jaén.[77]

While Franco left the administration of daily politics and the economy to his ministers, several of them were doing more than just managing the present. The new cabinet contained two groups with different kinds of plans for the future. Carrero Blanco and the technocrats, backed by conservative monarchists like Camilo Alonso Vega and the Minister of Public Works, Jorge Vigón, were united in wanting to restore the monarchy. The technocrats regarded the modernization of the economy as a prior condition, and saw political stability as a necessary context. The more conservative members of the group had no desire to see political change and hoped to use economic prosperity as a device for buying off political dissent.[78] Castiella, Fraga, Solís and Nieto Antúnez, on the other hand, were keener on political modernization and less ready to concede primacy to the economy. Fraga was anxious to open up the regime by means of a more liberal information policy.[79] Solís was behind talk of a kind of limited pluralism within the *Movimiento* by the introduction of 'political associations'. The main focus of tension was between Solís and Carrero Blanco. Solís, having failed to secure the Commissariat for the Development Plan for the Syndical Organization, began a guerrilla campaign of attacking it. The defence was assumed by Carrero Blanco, with López Rodó in the background. Franco let the battle go on, stirring himself only rarely to try to make peace.[80]

While dynamic ministers like Fraga and López Bravo worked hard to give a modern image to the government, Franco moved into the past. The Vatican Council convinced him that the Curia was infiltrated by freemasons and Communists. He confided in Pacón at the beginning of 1963 that he continued to collate information from his secret service on what happened in masonic lodges and socialist and Communist meetings around the world. 'Nothing will catch me by surprise; it is necessary to be prepared for the struggle.' He drew up lengthy notes on the links between the masonic danger and Catholic liberalization. So worried was he that any king who succeeded him might open the doors to freemasonry and Communism that he toyed with appointing a regent to guarantee the continuity of the regime. He still waxed indignant at Don Juan's determination to be King of all Spaniards. 'It is inconceivable', he said, 'that the victors of a war should cede power to the defeated'.[81]

In March 1963, Franco called together the one hundred and sixty members of the *Consejo Nacional*. In 1957, Franco had emasculated the Falange politically, deprived it of access to the levers of the economy and pushed it into labour relations. The strikes of 1962 had exposed the inadequacy of the Falangist Syndical Organization. How tarnished and useless it had become for Franco could be seen in the fact that his speech about regime labour legislation, which was largely written for him by Fraga, again failed to mention the Falange. He spoke only of the *Movimiento Nacional* which he described as 'a common enterprise of all Spaniards' and called upon the *Consejeros* to help perfect the instruments of state. By launching a debate, Franco hoped to give Europe the impression that he was contemplating liberalization. He also stimulated speculation in Spain that such concern for the future might signify that he saw 1964, the twenty-fifth anniversary of the end of the Civil War, as a suitable moment at which to step down.[82]

The heightened political expectations and the misery of social dislocation which accompanied economic progress fuelled strikes and unrest and the inevitable Francoist response was repression. Franco himself, however, was genuinely convinced that he presided over a paradise of individual freedom. 'In Spain', he told Pacón, 'there have never existed the freedoms that exist today; every Spaniard does what he feels like and thinks as he pleases, being able to participate in public life through syndical elections, muncipal elections and those for the elected section of the Cortes. The press today has freedom of speech and no Spaniard is punished for having ideas different from those of the regime or even for defending such ideas in the company of his friends.' Denying that his rule was dictatorial, he asserted that 'in Spain today, one governs through the popular will'.[83]

The barbaric nature of the regime in general and of Franco in particular was unmasked by the trial and execution of the Communist Julián Grimau García in 1963. A senior Communist Party official, Grimau had been arrested in Madrid on 7 November 1962. Horribly beaten and tortured, he was thrown out of a window of the *Dirección General de Seguridad* (national police headquarters) by interrogators attempting to conceal what they had done. Despite his appalling injuries, he was then tried on 18 April by court martial. He was condemned to death for 'military rebellion', an indictment which covered crimes allegedly committed during the Civil War. Grimau was merely one of more than one hundred members of the opposition tried by court martial in the first months of 1963.[84] Prior to the trial, in a press conference, Fraga declared that Grimau was a repellent murderer. In addition to the minister's gaffe, the trial itself was marked by serious legal flaws.[85]

A wave of demonstrations against Franco spread through the major

capitals of Europe and America. It reflected considerable political ineptitude on the part of Franco's servants and the Caudillo's own increasing
neglect of the detail of day-to-day politics that the Grimau trial should
coincide with a major step in the process of Catholic reform initiated by
Pope John XXIII. His liberal encyclical *Pacem in Terris* (the follow-up to
Mater et Magistra) had been published on 11 April. It was hailed by the
opposition as an attack on the regime. Advocating wider human rights
such as freedom of association, of political participation and of expression
for ethnic minorities, it was seen in El Pardo as a stab-in-the-back by a
once trusted ally. Franco simply attributed the changes introduced by
John XXIII to the successful infiltration of the Vatican by freemasons
and Communists.[86] Published while the Grimau trial was taking place,
the encyclical acquired even greater political significance.

Pleas for clemency for Grimau were made by ecclesiastical dignitaries
from around the world, and from political leaders including Nikita
Khrushchev, Willy Brandt, Harold Wilson, and Queen Elizabeth II.
Franco was unperturbed. His growing distrust of the Church was reflected
in his rejection of an appeal on behalf of Grimau from Cardinal Giovanni
Battista Montini, Archbishop of Milan.* The cabinet met on Friday
19 April to discuss the encyclical and the sentence on Grimau. His sentence occupied most of the discussion. Aware of the international repercussions, Castiella argued firmly in favour of clemency but Franco was
adamant that Grimau must die and a majority of ministers agreed.[87]
Grimau was executed by firing squad on 20 April.[88]

The Ambassador in Paris, José María de Areilza, on a visit to Madrid,
had pleaded with Castiella to intervene to stop the execution. Castiella
told him that he had already tried only to find himself confronting Franco
and a united cabinet. The consequent wave of international revulsion
severely set back the regime's efforts to improve its image. In particular,
the popular outrage felt in France sabotaged General de Gaulle's plans
to hasten a closer association of Spain with the European Economic
Community. Valéry Giscard d'Estaing, the French Minster of Finance,
cancelled a proposed audience with the Caudillo.[89] In partial, and
extremely clumsy response, it was decided, at the cabinet meeting of
3 May 1963, to create the *Tribunal de Orden Público*, by which political
offences would henceforth be treated as civilian crimes rather than as

---

* Franco had brusquely dimissed an earlier appeal for clemency made by Montini during
the trial in October 1962 of a group of anarchists. Cardinal Montini became Pope on
18 June 1963 two weeks after the death of John XXIII. It did not bode well for Franco's
relationship with the Catholic Church.

military rebellion.[90] Four months later, after an indecently hurried trial, two anarchists, Francisco Granados Gata and Joaquin Delgado Martinez, were executed by the barbaric method of strangulation (*garrote vil*) for alleged implication in a bombing incident at Madrid police headquarters. Although more muted than in the case of Grimau, the international clamour was considerable.[91]

The world's reaction to the Grimau case, coming after the outcry which followed Munich, recalled for Franco the international ostracism of 1945 and 1946. However, there were now crucial differences. With Spain increasingly integrated into the world economy, and with Spaniards getting used to the consequent prosperity, there could be no retreat into fortress Francoism. In any case, the forces which huddled round the Caudillo during the Cold War had evolved. The Army was still faithful but the rest of the coalition was increasingly shaky. Even the Falange was less reliable: its senior militants were geriatric and corrupt; its young men cynical functionaries simply alive to the main chance. Franco suspected monarchists of wanting to open the door to democracy and Communism. Above all, the position of the Catholic Church was changing at an alarming rate.

Franco had worrying reports on his desk from his Ambassadors to Italy, Alfredo Sánchez Bella, and to the Holy See, José María Doussinague. They suggested that there were officials in the Curia hostile to Franco and that the Vatican was anxious to see the legalization of a Christian Democrat Party in Spain under the leadership of Joaquín Ruiz Giménez who was about to launch what was to be an immensely influential liberal Catholic journal, *Cuadernos para el Diálogo*. Like many other Catholics who had previously supported the regime, Ruiz Giménez was moving nearer the opposition in the light of the liberalization of the Church being implemented through the Vatican Council.[92] News that Cardinal Montini had been elected as Pope Paul VI reached Franco during a cabinet meeting held in Barcelona on 21 June 1963. He exclaimed bitterly 'A jug of cold water'.[93] This did not prevent him from immediately sending a telegram of filial good wishes to the newly elected Pope.[94]

Eight days later, in Tarragona on 29 June, he made a speech of militant Catholicism which might be taken as his reply to *Pacem in Terris*, as his comment on recent efforts by the Vatican to re-establish relations with the Communist regimes of the East and as a challenge to the new Pope. He flaunted his credentials as Defender of the Faith in belligerent terms. Against 'the crudest materialism which is invading the countries of the world', he declared that 'Spain still stands firm to defend spirituality' and that 'we fight on the side of God'.[95]

It was an ironic twist of fate that exactly ten years after the dual triumph of the American bases agreement and the Concordat that frictions should emerge in Franco's relations with both Washington and the Vatican. Just how serious he was about fighting on the side of God would be put to the test when the renewal of the bases agreement was negotiated. At the beginning of 1963, he had been inclined to take a tough line over the forthcoming negotiations. Aware that the bases made Spain a target for Russian hostility and that the Spanish military had only obsolete weaponry, Franco's stance was that for the agreement to be renewed there would have to be some compensation acceptable to Spanish public opinion.[96] Accordingly, following the Caudillo's instructions, Castiella pressed for Spain to be given a place under the American nuclear umbrella as well as $300,000,000-worth of military equipment. But, while Franco overvalued Spain's importance to the United States, in contrast, the cost of foreign military aid was seriously worrying the Kennedy administration. Spanish demands therefore appeared outrageous. Moreover, the development of Polaris submarines and intercontinental ballistic missiles was diminishing the value to the Pentagon of the land bases in Spain.

In fact, Washington was trying to meet the costs of global defence by getting her allies to buy more American equipment through a programme known as Off-Set. The first Off-Set delegation to Spain, under William Bundy, had arrived in Madrid in mid-January. Contrary to Castiella's demand, Bundy offered $75,000,000 in US aid with Spain purchasing $175,000,000-worth of US arms. Bundy was brushed off and, at a cabinet meeting on 25 January 1963, it was decided that the bases agreement be renewed only if the conditions were significantly improved. A statement was issued in Madrid to the effect that the agreement would not be renewed. A second delegation, led by Deputy Secretary of Defence, Roswell Gilpatric, set out to visit Tokyo, Bonn and Madrid at the beginning of February. When Gilpatric reached Japan, he was awaited by a telegram from Madrid stating that the relevant ministers, Muñoz Grandes, Castiella and Navarro Rubio, would be unable to see him because they were hunting with Franco. It is not clear if the Caudillo appreciated the relatively high status of Gilpatric. In Washington, this was taken as an offensive rebuff. In this as in so many things, Franco was demonstrating either a dangerous complacency or a lack of attention to detail.

Time was not on Franco's side. On 20 March 1963, the House Foreign

Affairs Committee issued a report on foreign aid which criticized aid to Spain as 'excessive'. Nevertheless, on 5 April, the Spanish cabinet again decided to stand firm in the negotiations. At this point, the Ambassador in Washington, Antonio Garrigues,* a personal friend of John Kennedy, intervened with a lengthy memorandum to Franco requesting *carte blanche* to get the best deal possible. The deterioration of his relationship with the Vatican may have made Franco think twice about also risking his friendship with the United States. He backed down and placed negotiations in the hands of Garrigues. With his Minister-Counsellor, Nuño Aguirre de Carcer, Garrigues worked out a formula for agreement. However, in early June, talks reached breaking point when Garrigues stormed out of the Pentagon declaring that the US terms would be accepted only by some other Ambassador. He then flew to Madrid to report.

On arrival, he offered his resignation to Castiella who was deeply alarmed about the implications of breaking with the United States. However, when Franco received Garrigues, he rejected the resignation and said that he was not worried, that the Ambassador had done what he had to do, and that, if a break came, it came and that was that. This seems to have been less a reflection of the Caudillo's legendary coolness than an irresponsible lack of concern. At the end of August, Garrigues travelled in the company of Castiella, Nieto Antúnez and Muñoz Grandes to the Pazo de Meirás to inform Franco of the draft formula he had elaborated. To the astonishment of Garrigues, Franco paid virtually no attention to what he had to say, treating him to a long and rambling disquisition on the subject of crop rotation in Galicia. On 19 September, when Garrigues explained the proposed formula for renewal of the agreements to a special cabinet meeting, the American offer was regarded as derisory. However, when he argued that, if the offer were rejected, Spain would face international isolation, Franco, aware of crumbling relations with the Vatican, supported him and overruled cabinet opposition. The agreements were renewed for another five years.

In practical terms, the United States had secured the better deal – although without Garrigues there might have been no deal at all. The United States was now permitted to base a squadron of Polaris submarines at Rota. Garrigues had secured improved economic terms: Spain would receive $100 million in military aid and purchase $50 million-worth of US equipment under the Off-Set scheme. There was to be no economic

* Garrigues was a first rank international lawyer who had represented many American firms in Spain. During the Civil War, he had become friendly with Kennedy's older brother, Joseph, who was later killed in the Second World War.

aid other than the offer of $100 million in Export-Import Bank loans. The most significant improvement for Spain was the inclusion for the first time of a joint declaration that 'a threat to either country would be a matter of common concern to both countries'. This was still a less extensive commitment than those given by the USA to her NATO partners – which did not prevent the Francoist press hailing the deal as a full-scale alliance.[97]

Thanks to Garrigues, one of the pillars of Franco's international prestige remained in place. Had the negotiations broken down, a pall would have been cast over the 1964 celebrations of the anniversary of the end of the Civil War. Under the slogan invented by Manuel Fraga, 'Twenty-Five Years of Peace', 1964 was to be converted into the apogee of Franco's rule. It would have been an appropriate moment to announce his choice of successor but Franco would ignore it. Over the next five years, confident that his technocrats could be left to provide increased prosperity and efficient administration, Franco's main concern, beyond his hobbies, was the future. The 'depoliticized' administration being mounted by López Rodó constituted a firm foundation for the continuity of the regime after his death, or as he usually put it 'when I'm no longer around' (*cuando yo falte*). Franco would take an inordinately long time to make a relatively simple decision. That was partly because he was losing sharpness and concentration. The hesitations also sprang from a distaste for any contemplation of death or of the abandonment of power. They also reflected an instinctive fear that to resolve the succession in a particular direction would provoke resentments and opposition from those who were disappointed. Nevertheless, that Franco could spend half a decade musing on the topic was a measure of the extent to which his position was secure. Within five years, that situation was to change dramatically.

XXVII

PREPARING FOR IMMORTALITY

1964–1969

ON 23 January 1964, Fraga gave Franco two documents – a broad survey of the political situation concluding with a call for reform and a draft press law. When they met five days later to discuss them, Franco expressed doubts about liberalizing information policy. He was worried about what a freer press might make of something like a recent incident in which the bishop of Calahorra had been caught in a hotel bedroom with a woman. Then, Franco had simply imposed silence. Now, despite his qualms, he felt that Fraga was chalking up sufficient successes to be given a hearing and he authorized him to bring the project to a cabinet meeting.[1]

Franco was especially delighted with Fraga's campaign to mark the anniversary of the end of the Civil War. Every town and village in Spain was bedecked with commemorative posters. In contrast, Franco was deeply annoyed when the Church hierarchy's official journal *Ecclesia* published an editorial on the anniversary arguing that peace and order were all well and good but not enough. Franco commented to Fraga on 30 March 'you should go and cover the Primate's palace with your posters', to which Fraga replied that it would not do for the Minister of Tourism to be caught vandalizing Toledo.[2] The celebrations of 'Twenty-Five Years of Peace' began officially with a solemn Te Deum in the basilica at the Valle de los Caídos. That ceremony, together with an interview given by Franco to *ABC* made it clear that, for the Caudillo, what was being celebrated was 'twenty-five years of victory' rather than of peace.[3]

The anniversary revels confirmed Franco's belief in his own immense popularity. Among the activities mounted by the Ministry of Information were itinerant exhibitions about his achievements, prizes for literary works best reflecting the spirit of the Franco age, as well as innumerable press articles and television programmes. There was also a limited amnesty for some of Spain's several thousand political prisoners. The celebrations would close at the end of the year with the launching of the film hagiography *Franco, ese hombre* (Franco, that man) written by José María Sánchez Silva and directed by José Luis Saenz de Heredia. It was a skilful piece of work, a reverential corporate video for the Caudillo, 'Franco, that man who forged twenty-five years of peace with his spirit of steel on the anvil of his life'. The picture it presented was of a hero who saved a country in chaos from the hordes of Communism, then saved it again from the hordes of Nazism and later became the benevolent father of his people. It was a considerable box-office success but Franco himself did not like the film, commenting only 'too many parades'.[4]

Prior to the première of the film, the culmination of the official rejoicing, there were many other events in which Franco played a direct role. On 9 April 1964, He gave the *Consejo Nacional* his self-congratulatory interpretation of the peace, attributing the present economic development to his own foresight. He described the social achievements of his regime as being far in advance of recent Papal encyclicals. Turning to Europe, he warned his audience of plots and sectarianism, of 'secret machinations, subversive action and the power of occult forces'. This reflected his annoyance at the opposition to Spanish entry into the EEC being mounted by Holland and Belgium and the issue of Gibraltar which kept alive his resentment of Great Britain. There was not much in the speech for the would-be reformers in his cabinet. Although he made much of his efforts to provide for the continuity of the regime after his disappearance, he made no specific announcement about the future. Declaring that 'the powers which come together in my person are by their very nature not transferable', he talked once more of the need to regulate the functions of the Head of State, of the Government and of the *Movimiento*.[5]

For the reformist elements in the cabinet who hoped that the euphoria of the twenty-five years of peace might provide a suitable context in which Franco might promulgate the *Ley Orgánica del Estado* and name his successor, it was a disappointment. Watching him read out the speech, Fraga was struck by how the Caudillo seemed to have aged. A week later, face-to-face in a private audience with him, the impression that Franco was fading fast was overwhelming.[6] (He was probably manifesting the first symptoms of the Parkinson's disease which would intermittently afflict him in his last years.) On 30 April 1964, Franco was presented with a medal to commemorate the twenty-five years of peace. In his speech of thanks, he said that he looked forward to a similar ceremony in another twenty-five years time.[7]

In the spring of 1964, the celebratory spirit was marred by a resurgence of tension in the Asturian mines. The immediate cause of the strikes which broke out in April was a new labour law that the miners rejected because it failed to deal adequately with the appalling problems of silicosis. As the strike spread, the government lashed out with a savage repression. Men were dismissed and strikers arrested, many of whom would languish in prison until 1970.[8] On 8 May 1964, in a meeting of the *Comisión Delegada de Asuntos Económicos*, there was a violent argument about industrial policy between the Minister of Industry, Gregorio López Bravo, and the Minister of Labour, Jesús Romeo Gorría. Romeo accused López Bravo of being too ready to buy off strikers. Franco intervened on the side of López Bravo, commenting sourly that the Ministry of Labour and the Syndical Organization were infiltrated by Communists.[9] In a mixture of paternalism and paranoia, he told Pacón that 'many mine-workers obey hidden powers which threaten them with sanctions if they do not obey strike calls from abroad'.[10]

Alonso Vega wanted to intensify the repression, but Fraga, Castiella and the reformists in the cabinet managed to persuade Franco that more violence would be counter-productive.[11] The scale of the opposition, and the artificiality of the regime's division of Spaniards into victors and vanquished was exposed by the arrest on 26 April 1964 of a Communist militant who turned out to be the son of General José Lacalle Larraga, the Minister of Aviation. For once, Franco was remarkably understanding, commenting to Pacón that such things happen in the best families, quoting the case of his own brother Ramón. There were rumours of the case being the occasion for a cabinet reshuffle. Franco, always reluctant to initiate a ministerial change and basking in the 'Twenty-Five Years of Peace' celebrations, dismissed suggestions that Lacalle resign.[12]

In private, Franco spoke of his hopes that Prince Juan Carlos would officially accept the *Ley de Sucesión* and swear as King to fulfil the principles of the *Movimiento*.[13] Significantly, on 24 May 1964, at the annual Civil War victory parade, Prince Juan Carlos took the salute along with the Caudillo. However, Franco gave no indication of his plans for the succession other than such hints. The pleasure that he derived from the 1964 celebrations increased both his reluctance to plan for the future and his conviction that he was indispensable. During the summer of 1964, he cited the applause which greeted him on trips to Seville and Bilbao as his principal argument when his ministers suggested change. He was similarly affected by the frenetic scenes – planned by Solís – which greeted his arrival, on 21 June, at the Santiago Bernabeu football stadium at Chamartín in Madrid for the final of the European Nations Cup.

After the great international *faux pas* of Franco's prohibition of the European Nations Cup football match against the USSR in 1960, the *Movimiento*'s sports section, the *Delegación Nacional de Deportes*, had gone to enormous lengths to make amends to FIFA and had secured the privilege of hosting the championship in Spain in 1964. Held on 21 June, the final was between Spain and the Soviet Union. Until only a few days before, there had been some doubt as to whether Franco would attend since there was the danger of his having to present the trophy to the Russian captain. A senior Falangist in the Spanish Football Federation even proposed, unsuccessfully, that the Soviet team be doped to protect the Caudillo from any such embarrassment. On entering the stadium along with Doña Carmen, General Muñoz Grandes and other ministers, the Caudillo was greeted with well-orchestrated chants of 'Franco! Franco! Franco!' from Falangists and gradually the bulk of the rest of the 120,000 fans then joined in. He was delighted with the crowd, with Spain's 2–1 victory and with the words of the Spanish team coach, Major Villalonga, who said after the match 'we offer up this victory first of all to the Caudillo who came to honour us this evening with his presence and to inspire the players'. Since the match and the popular reception were broadcast to fifteen European countries, his delight was understandable. The press hailed the victory as the logical culmination of Franco's victory in the Civil War. Reluctant to turn his back on such adulation, Franco grew noticeably cool to the notion of reform.[14]

The warmth of popular acclaim made Franco the more aware of foreign criticism. The rising tide of strikes, student demonstrations and agitation in the regions found a ready echo in the press of Europe and, most hurtfully for the Caudillo, in some Catholic publications. The contrast between perceived popularity at home and criticism abroad naturally confirmed Franco in his view that subversion in Spain was the work of sinister foreign forces. However, he was not so easily able to shrug off the implicit criticisms of his rule emanating from the Second Vatican Council. In the main, he simply did not understand them, any more than did many of Spain's aged bishops. His belief in his own divine purpose remained unshaken, confirmed frequently by extravagant praise from some sections of the Spanish Church hierarchy.[15] He clung to his deeply traditionalist and reactionary notion of Catholicism (*nacional-Catolicismo* in regime jargon) within which the Civil War was a religious crusade. Just how threatening he found the changing face of the Church could be seen in his speech to the Cortes on 8 July 1964. Together with his speech to the *Consejo Nacional* three months earlier, it constituted a defiant attempt to show that his variant of Catholicism was at the pinnacle of religiosity and social justice. With its pedagogic tone, the speech was a patronizing attempt to guide the new Pope, benevolently offering him the Francoist example of Catholicism as he took up the baton of renovating the universal Church.

The Caudillo harped on the debt the Church owed him by contrasting Spain's return 'to the road of religious faith' with the Godless countries of the East where the Church suffered under the yoke of Communism and those of the West where crime and suicide were rife. At a time when the Vatican Council was trying to decouple the Church from politics, he praised the harmony and social peace allegedly achieved under his Catholic rule. The growing Catholic unease that this peace had been built on the imprisonment, torture, exile and even execution of the regime's enemies, together with the growth of the Catholic workers' organization HOAC, was seen by the Caudillo as proof of 'the progressive infiltration of Communism in some Catholic organizations'. The reformers of the regime were again disappointed as Franco denounced liberal democracy as an exhausted system repudiated by the masses.[16]

The extent to which Franco was increasingly out of touch with contemporary Catholicism was exposed starkly in the autumn. On 10 September, at a cabinet meeting held in the Pazo de Meirás, Castiella, supported by Fraga, presented a draft project on religious freedom for non-Catholics. Prior to presenting it, Castiella had obtained the approval of the Church hierarchy. Carrero Blanco was furious and presented Franco with a memorandum accusing Castiella and Fraga of 'opening up to the Left'. The religious triumphalism and siege mentality of Carrero Blanco and Franco were on a collision course with the humanist and pluralist renewal of Catholicism being elaborated by the bishops assembled in Rome. The clash between Carrero Blanco and Castiella brought out into the open the underlying tension between the so-called *inmovilistas* (intransigent conservatives) and the *aperturistas* (those prepared to contemplate some opening up). Fraga was surprised by Franco's reaction which was simply to do nothing. Although it may be supposed that he sympathized with Carrero, he preferred to wait and see what would happen in Rome and left the two opposing forces to neutralize one another. While a fierce battle raged among his ministers, the Caudillo appeared calmly unflappable. How much he appreciated what was going on is impossible to say.[17]

The issue of freedom of religion was only one of several areas of friction with the Vatican. In general, the new orientation of the Conciliar Church struck at the heart of Franco's reactionary variant of Catholicism (*integrismo*) and heralded the reversal of the clerical legitimization of the regime on which Franco had been able to rely since 1936. It also gave impetus to opposition against the regime in Catalonia and the Basque Country where Catholic nationalists would soon be able to argue that Franco's dictatorial rule violated the teachings of the Church. Freedom of religion was linked by the Second Vatican Council to the question of the Church's independence of political structures.[18] On 18 September 1964, the Council approved a resolution requesting States to renounce their privileges of intervening in the nomination of bishops. Franco refused categorically to negotiate, using the blatantly dishonest argument that, since this was a traditional privilege of the Kings of Spain, only a King had the right to renounce it. He resisted giving up what he saw as the only clause of the 1953 Concordat which was favourable to the Spanish State. Moreover, he was appalled by the idea that the Papal Nuncio might name bishops in the interests of the communities in which they would work rather than as instruments of national unity.

Despite such tensions, Franco devoted even more time to hunting. Partridges and deer were his favourite prey.* When the press referred to Franco, it was usually to commend his astonishing feats as a hunter or his capacity to withstand interminable hours of work. However, those close to him saw a different picture. At the end of October 1964, Fraga was alarmed to note on meeting him for a work session that the Caudillo was unable to keep his eyes open. A few days later, Franco returned from a hunting trip with a heavy cold and looking older than ever. It may be surmised that these were signs of the development of his Parkinson's disease. That would be consistent with the fact that the decline of his health was intermittent – as the year wore on, he seemed to liven up. Parkinson's disease isolated Franco from the world around him. A further contributing factor to that isolation was the manic enthusiasm of both Franco and his wife for the television.[19] On public occasions, he spoke less frequently and for shorter time. Apart from big setpiece occasions, such as the opening of the Cortes or the end-of-year broadcast, the long harangues were a thing of the past. Indeed, public appearances would

---

* On 1 February 1964, Fraga took part in a hunt with the Caudillo and accidentally shot Nenuca in the bottom. Franco commented caustically 'those who don't know how to hunt should not be here' – Fraga, *Memoria breve*, p. 99; Peñafiel, *El General*, pp. 69–70.

diminish as it became harder to conceal the symptoms of the disease – a rigid stance, an unsure walk and a vacant, open-mouthed facial expression.

On 30 December 1964, in his annual broadcast, Franco's worries about the work of the Vatican Council were to the fore. He still misunderstood its purpose – which is hardly surprising, since many of the senior Spanish bishops with whom he occasionally talked were as old as or older than himself.[20] Admitting that the reasons for the Council's convocation escaped him, his interpretation of what was happening was that the Council was like a retreat from which the bishops would emerge inflamed with evangelical fervour. With characteristic self-satisfaction, he described the social recommendations emanating from the Council as especially pleasing since they merely reflected what he had long been doing.[21]

Throughout 1963 and 1964, Franco had been bombarded from all sides of the regime with suggestions for the *Ley Orgánica*. Both Fraga and Solís produced draft constitutions. Solís was pushing his idea of 'political development' through organized currents of opinion within the *Movimiento* to be known as 'associations'.[22] The urgency of the matter was underlined on 14 January 1965 when Camilo Alonso Vega took advantage of his lifelong friendship with Franco to broach the topic of the succession. 'The country follows you and loves you,' he said, 'it will say yes to whatever you do. You must name a President of the Government and define a political system which will guarantee the future. The other ministers think the same as I do, but I am not speaking to you as their representative. If they don't say it to you, it is because they don't know you as well as I do. I can remember you in short pants and we've played together. If I can't speak to you like this, who can? Or maybe it's not allowed to say such things to you. People are worried about the future.' Franco listened with a smile, made a joke about their age and said that he was working on the *Ley Orgánica del Estado* which would be produced 'before you think'.[23]

Despite the remark to Don Camilo, Franco's clock was marked, as always, in months and years rather than minutes and hours. In the spring of 1965, there were serious university disturbances in Madrid and Barcelona. The students, like their counterparts elsewhere in Europe, were inspired by new currents of left-wing thought. For the first time ever, at a cabinet meeting on 5 March 1965, the political difficulties facing the regime were discussed openly. Just as the Church was beginning to pull away from the regime, many one-time Francoist intellectuals were following Dionisio Ridruejo and Joaquín Ruiz Giménez into the opposition. When several university professors, including Enrique Tierno Galván,

Agustín García Calvo and José Luis López Aranguren, were removed from their posts for complicity in the student unrest, the one-time Falangist Pedro Laín Entralgo wrote to Franco to announce his rejection of the regime. Carrero Blanco claimed implausibly at the 5 March cabinet meeting that unrest was just the consequence of uncertainty about the post-Franco future and proposed that the *Ley Orgánica del Estado* be propounded as soon as possible. The entire cabinet jumped on the bandwagon and spoke in support of Carrero. Franco complained of the difficulty of finding a solution which would please everyone, by which he was hinting at his own reluctance ever to close off options. However, he ended the debate saying 'I have undertaken to do it and I will do it.'[24]

One week later, on entering the Caudillo's study, Carrero Blanco found him working on the *Ley Orgánica*. Franco commented that he did not think that it would be necessary to maintain the *Secretaría General del Movimiento* – in fact, it would not be dismantled until a year after his death. On the same day, Navarro Rubio spoke with Franco about the new constitution. In the course of their conversation, the seventy-two year-old Caudillo revealed that he anticipated ruling for some time to come. He said that he would have much preferred to leave the *Ley Orgánica* until later because the longer he left it the more in tune it would be with the future. However, he reluctantly recognized that he had no choice but to start the process now.[25] Leading regime figures paraded before Franco to press him to make his dispositions. On 25 March 1965, the Minister of Education, Lora Tamayo also attributed difficulties in the universities to the uncertainty about the future. 'Do you think that the future doesn't worry me?' Franco asked, to which Lora Tamayo replied that the longer he left it, the harder it would be and stated that, if he died without resolving the problem, there would be chaos. Franco disagreed, stating 'No, because the good people would come out into the street.' When his alarmed Minister spluttered 'that would mean another civil war', Franco was unconcerned, making the astonishing comment that the costs of the Spanish Civil War had been low.[26]

On 1 April 1965, Franco read to Carrero Blanco a near final draft of the *Ley Orgánica*. On the same day, Muñoz Grandes was diagnosed as having renal cancer. After an operation to remove a kidney, his iron constitution kept him alive until 1970, but he could no longer fulfil the role of guarantor of the post-Franco succession. At a cabinet meeting held on 2 April, under the shadow of the Vice-President's illness, the subject of the future was again raised openly by Navarro Rubio, closely seconded by Castiella and Fraga. As the debate got more heated, in response to pressure from Fraga, Franco abandoned his usual distant passivity and began to give explanations about how difficult it was and said that he needed more time. Fraga pressed him, saying 'there's no time to spare and I beg you to make use of what we've got'. Franco exploded 'do you think I don't realize, do you think I am a circus clown?' He was irritated to be pressured into adopting a rate of advance faster than his instinctive snail's pace and the storm passed. For the rest of the meeting, Franco beamed craftily at his ministers which led his fellow *gallego*, Camilo Alonso Vega, to guess that he already had a draft of the law.[27] Yet, despite the apparent promise made in the cabinet meeting, within three weeks Franco had retreated into apathy. At a meeting of the Comisión Delegada on 23 April, he avoided all attempts to raise the question. Seeming more indecisive by the day, he was in no mood for progress on the constitutional question and resented any pressure.[28]

In July 1965, after interminable doubts and hesitations, Franco reshuffled the cabinet. When Carrero Blanco pointed out that there was some urgency about the matter, the Generalísimo had said that he would leave it until October. Only when Carrero reminded him that he had been putting off changes since the summer of 1964 did he start to examine the list of new ministers elaborated by his *Ministro-Secretario de la Presidencia*. This cabinet has been described as 'the last of the classic cabinet balancing acts of Franco'.[29] In fact, rather than any equilibrium skilfully concocted by Franco, the cabinet was provided virtually prepackaged by Carrero Blanco. At best, the list suggested that Franco was sufficiently pleased with the job done by the technocrats to confirm their predominance. At worst, it showed him surrendering important issues to Carrero. In theory, Muñoz Grandes remained as Vice-President but the decline of his health pushed Carrero nearer the forefront.

Navarro Rubio's resignation was finally accepted and he was replaced by his most senior functionary, Juan José Espinosa San Martín.[30] It was a mark of the greater priority being given to relations with the EEC that Alberto Ullastres became Spanish Ambassador to the Community and was replaced as Minister of Commerce by Faustino García Moncó. Like Espinosa, García Moncó was a member of Opus Dei. López Rodó remained as *Comisario del Plan de Desarrollo* and became Minister without portfolio. Carrero's battle with Solís would henceforth be assumed directly by López Rodó. The seventy-two year-old Jorge Vigón was replaced as Minister of Public Works by Federico Silva Muñoz, at the suggestion of López Rodó. The grey technocrat Silva would prove himself an efficient reformer of Spain's transport system. Cirilo Cánovas was replaced as Minister of Agriculture by Adolfo Díaz Ambrona, another López Rodó suggestion. Within the narrow confines of the Francoist system, it was an almost apolitical government. As Nieto Antúnez commented, 'the new government has a good general staff, but it has not got the people with it'.[31] The composition of the cabinet in fact suggested a consolidation of the technocrats' position but the Falangists were far from conceding defeat. Surprised that Franco had not discussed the imminent changes with him, Nieto Antúnez told Fraga 'I thought the Caudillo was closer to me than that'.[32]

During a summer in which Franco did much hunting and fishing, the cabinet meeting of 13 August saw the opening of discussion about Fraga's Press Law. Fraga had previously given the Caudillo several drafts which he had subjected to close criticism. Franco was adamant that liberty should not become the *libertinaje* (licentiousness) which he saw in democratic countries. He insisted that newspaper editors be held responsible for what they published and run the risk of the closing or confiscation of their papers. He also listed the areas which would be untouchable, which included the Church and the *Movimiento*. Carrero Blanco and Alonso Vega expressed serious reservations against which Fraga energetically defended his proposal. Franco closed the debate saying 'I don't believe in freedom of the press, but it is a step which many important reasons force us to take.' Over subsequent months, the text was debated, with fierce opposition coming from Alonso Vega and other reactionary elements who tried to persuade Franco that it threatened the very foundations of the regime. Nevertheless, by February 1966, it was ready for submission to the Cortes for rubber-stamping. Franco commented cynically to Fraga, 'let us not be too good-natured. Like everyone else, let us use indirect means of control.'[33]

From the mid-1960s, implausible efforts were made to present a picture of a superhumanly fit Caudillo. He derived great satisfaction from the operation, spending at one point two weeks baiting an area of the Spanish coast in order to attract tuna. When he was satisfied with his preparations, he went out in the *Azor* with Nieto Antúnez and Solís and caught a tuna weighing 375 kilos, a European record. At one time, he had been capable of pursuing a sperm whale for days. Max Borrell commented that 'to see him chase a whale is to understand all the successes of his political and military careers', 'his perseverance is such that he would chase a whale to Russia'.[34] It was announced on 20 August 1966 that he had caught a twenty-five ton whale in the Atlantic near Vigo and on 7 September that he had harpooned thirty-six whales. In the summer of 1968, aged seventy-six, he was reported to have caught a twenty-two ton whale.[35] It may be supposed that the crew of the *Azor* played some part in these achievements.

Despite the time spent fishing for whales and hunting ibex and deer, Franco still took seriously the task of presiding over cabinet meetings, although he did little to impose a particular direction on government policy. He would begin by receiving ministers individually and they would tell him what they wanted to do. Then when they had assembled and ceremonially filed in, he would give a resumé of the main domestic and international political events since the last meeting. As the decade wore on, and Franco's health deteriorated, these surveys got shorter and were eventually omitted altogether. Indeed, his interventions of any kind grew rarer. His control over cabinet debates remained loose. It was a mini-parliament in which, within certain rules of deference to him, and civility to one another, ministers had considerable freedom to say what they liked. He rarely cut the rambling interventions of ministers, especially Alonso Vega and Castiella. Occasionally, he would make lightly sarcastic remarks about Alonso Vega whom he liked to rib,* threatening to put an egg-timer in front of him and once, when he finished his intervention quickly, commenting that 'several delegations have arrived to congratulate you'.[36]

Franco was a passive chairman who did not initiate business. It did not seem to bother him that throughout 1965 and 1966 there were increasingly acrimonious conflicts between Solís and Romeo Gorría on the one hand and López Bravo and López Rodó on the other. In open opposition to López Rodó's partisanship for Juan Carlos, Solís favoured the claim to the post-Franco throne of Alfonso Borbón-Dampierre, son of Don Juan's brother Jaime.† Franco shrank from taking action because of a combination of apathy, a deep reluctance to disturb ministers once they had settled into their posts and a sentimental attachment to the Falange. He told López Rodó that he needed the Falange as the Pope needed the clergy.[37] On 24 February 1966, he made a remark to Carrero

---

* On a hunting party at the Sierra de Gredos in Avila in 1958, a female goat (*capra*) had been shot by accident, contrary to all the rules of conservation. With Franco's knowledge, Alonso Vega, an expert shot and an unlikely culprit, was blamed and mercilessly taunted by Franco for the alleged offence to Don Camilo's near apoplexy – Vaca de Osma, *Paisajes*, p. 190.
† Since Don Jaime had voluntarily renounced his rights to the throne in 1933 in recognition of his disablement and then contracted a morganatic marriage, the claim to the throne of his son Alfonso was highly questionable. That would not have mattered to most Francoists had Franco decided arbitrarily to name Alfonso his heir. Since the early 1950s, the ever-impoverished Don Jaime had been covertly encouraged by Franco to press his own and his son's rights to the throne – Bardavio, *La rama trágica*, pp. 62–71.

Blanco which revealed that his habitual indecision and evasion had turned into impotence: 'I was ready to accept the resignation of the Minister of Labour and I provoked him to make him offer it but he didn't. I don't like his policy.' In May 1966, López Rodó complained to Franco that the government's economic policy was being attacked by Solís's *Movimiento* press. The Caudillo feebly tried to defend Solís, then agreed with López Rodó that the situation was intolerable and finally did nothing.[38] A report on the political situation drawn up by Girón at the end of September and given to Pacón for transmission to Franco complained of the Caudillo's loosening grip: 'the dictatorship of one man has become that of eighteen ministers'. In the early 1960s, cabinet meetings were still interminable and often went on from 10.00 a.m. until late at night. However, by the middle of the decade, the late sessions became infrequent and, by its end, even afternoon sessions would be rare.[39]

One of the many figures of the Francoist establishment to propose ideas for the *Ley Orgánica* was the liberal Catholic lawyer Antonio Garrigues, who had been sent in February 1964 as Ambassador to the Holy See in the hope that he might repeat in the Vatican his success in Washington. In early September 1965, Garrigues spoke to Franco at the Palacio de Ayete in San Sebastián about his ideas for the future. After listening with a silent smile, the Caudillo asked if he knew what the *Movimiento* meant to him. When Garrigues admitted his ignorance, a broadly grinning Franco explained 'well, it's like this, for me the *Movimiento* is like the claque. Have you never noticed that when there is a big group of people, it only needs a few to start clapping for the others to follow them and join in? Well, that, more or less, is what I see as the purpose of the *Movimiento*.'[40] Franco had come a long way in the decade since Arrese had enthused him with his abortive plans to assert the eternal dominance of the Falange over Spanish politics.

At a cabinet meeting on 19 November 1965, the new Minister of Justice, Antonio Oriol, spoke of the recent denunciation of dictatorship by the Second Vatican Council whose proceedings were drawing to a close. Franco said complacently 'I do not take the reference to dictators as being directed at me, although the statement could cause problems for some countries in Latin America.'[41] On 25 November 1965, Manuel Fraga gave an interview to *The Times*, in which he stated that Spaniards were now convinced that the successor to Franco would be Juan Carlos. He could hardly have made such a statement without the authorization of Franco. Senior Falangists responded by intensifying their partisanship for the rival royal pretender, Alfonso de Borbón-Dampierre.[42]

The question of the future continued to be the most divisive issue in the council of ministers. On 20 January 1966, Carrero Blanco suggested to Franco that, on age grounds, he replace Muñoz Grandes as Chief of the General Staff. Committed to the technocrats' strategy of an authoritarian regime built on economic prosperity and crowned by Juan Carlos as successor to Franco, Carrero Blanco feared that Muñoz Grandes would favour a more radical Falangist option. Franco procrastinated, believing that he could not release Muñoz Grandes lest he become a rallying point for discontented Falangists, but reassured Carrero that 'he is sick, he won't last'.[43] On 9 February 1966, López Rodó followed up this initiative in a long conversation with Franco, pressing him to resolve the succession. He used the same arguments as Lora Tamayo earlier in the year, pointing out that without clear plans for the future, he would be succeeded by chaos. Franco's eyes filled with tears and he said, 'Yes, it would be chaos, it would be chaos' but then complained of the difficulties arising from the existence of so many competing candidates. Apart from Don Juan and Juan Carlos, there were Alfonso de Borbón-Dampierre, and the Carlist Carlos Hugo de Borbón Parma, in whose favour his father Don Javier de Borbón-Parma had abdicated his own claim.[44] The proliferation of candidates was quietly encouraged by Franco as a convenient excuse to keep his options open. In fact, he had privately eliminated the claims of Don Juan as too liberal and of Javier and Carlos Hugo as foreigners ineligible to be Kings of Spain.[45]

Franco's continuing delays about naming a successor derived in part from a reluctance to admit that his rule would ever end. He knew that once a successor was named, there would be rush of opportunists eager to ingratiate themselves with the nominee, something which could only diminish his own power. In addition, he was inhibited by the fact that he had on his desk secret police reports on Juan Carlos's contacts with progressive elements which suggested that, as King, he would not oppose the restoration of a multi-party system.[46] Franco would not proceed without the reassurance that Juan Carlos would swear to be bound by the principles of the *Movimiento*. This was ironic, given the widespread view on the Left and within the Falange that the tall, handsome twenty-eight year-old Prince was an empty-headed mediocrity comfortably installed as Franco's stooge. His diffidence and reserve did nothing to dispel that image.

Of one thing the Caudillo remained certain, as he told Fraga at the beginning of June: Don Juan was out of the question because he would divide Francoist forces.[47] Franco was furious because Don Juan, despairing of Franco ever making him his successor, had established a secretariat, a virtual shadow cabinet, headed by José María de Areilza, the Conde de Motrico. In October 1964, Areilza had resigned as Franco's Ambassador in Paris, convinced by the international storms over Munich and Grimau and by an icy interview with the Caudillo, that the regime was in a cul-de-sac. In early 1964, he had spoken to Franco about the need for reform and been heard in glacial silence. Since it had long been assumed that Areilza was the natural successor to Castiella, his gamble in going over to the monarchist opposition was a considerable blow to the Caudillo. Given the scale of his ambition, the ex-ambassador was a sensitive barometer of the political climate.[48]

Alonso Vega headed growing opposition to the new press law within the cabinet, but Franco continued to give Fraga his support. Censorship would no longer be carried out before publication but after, leaving newspaper editors and journalists to guess what they could get away with. Franco, the Falange, the Army and the principles of the regime could not be criticized but, for all the limitations, the law constituted a real change and the most reactionary elements in the regime were furious at the implications. Fraga was subjected to pressure from Raimundo Fernández Cuesta and from Franco's son-in-law, the Marqués de Villaverde, who said bitterly 'neither you [the reformists] nor that man [Franco] are governing with energy'.[49]

In the spring of 1966, Muñoz Grandes commented to the Civil Governor of Avila, 'the Generalísimo forgets to ring me nowadays. He is in low form and lets people get away with too much (*deja hacer demasiado*) and there are plenty of fly types ready to take advantage.'[50] On occasion, however, Franco could be as lucid and decisive as he ever had been. On 9 March 1966, a democratic student union was established in a Capuchin convent in Sarriá in Barcelona. At the following day's cabinet meeting, Alonso Vega gave a watered down version of events. Franco, who appeared completely distracted, suddenly interrupted and painted a more detailed and blacker picture. He ordered Alonso Vega to leave the session and give orders for the occupiers to be evicted from the convent. Don Camilo stood to attention and requested permission to leave. He returned twenty minutes later, requested permission to speak and reported: 'General, the orders have been carried out without casualties on either side.'[51]

By this time, Franco was playing a sufficiently small part in daily government for rumours to be rife that he was ill. When Pacón told him, he said 'the only illness that I have is my seventy-three years, and that's certainly enough to give me no illusions about living for many more

years'. Distanced from the day-to-day responsibilities and power
struggles, he seemed more relaxed. The decline of his health was unmistakable although the evidence of senility alternated with long periods of
fitness.[52] After his annual spring fishing holiday in Asturias, he returned
sunburnt and boasting of his long hours in the icy rivers of the region.[53]
A few days later, he presided over the annual victory parade, standing
erect in torrential rain for an hour and a half, saluting rigidly.[54]

But his age and the succession were on his mind. On 13 June 1966,
he gave Carrero Blanco the final draft of the *Ley Orgánica del Estado*. He
then went on a tour of Catalonia which was meant to counter student
and regionalist unrest and came back in mid-July totally exhausted. He
had no interest in the problems of government and was merely looking
forward to his summer holiday and a long rest. It had been agreed that
the text of the *Ley Orgánica* would be presented to the Cortes at the
beginning of October. In fact, even after the holidays, he showed little
interest in discussing it. That, together with hesitations about whether
to permit debate on the text in the Cortes, ensured that it would not be
presented until 22 November.[55] It was decided that there would be no
discussion of the complex law. It would be submitted first to the Cortes
and then to the Spanish people without any public examination of its
advantages and disadvantages, or even much in the way of explanation.
Franco was in no mood for concessions, by now seriously irritated by
the unwelcome consequences of Fraga's press law. On 4 November, he
commented 'I am getting fed up with the fact that the press wakes up
each day asking itself what shall we criticize today?'[56]

In his speech on the occasion of the presentation of the *Ley Orgánica
del Estado* to the Cortes on 22 November 1966, a bespectacled Franco
read a speech which managed to be valedictory in tone while proclaiming
that he had no intention of retiring. It covered much familiar ground: his
view of Spanish history, his belief in his providential mission, his acceptance
of the deification to which he had been subjected by his own propaganda
machine. He attributed Spanish neutrality in the Second World War to
the fact that 'God gave us the necessary strength and the requisite clairvoy-
ance'. Surveying the thirty years of his rule, he declared with pride and
self-pity, 'during these last thirty years, I have dedicated my life to the cause
of Spain. And the distance was so great between the point of departure and
the goal we had set ourselves that only faith and the help of God gave me
strength to accept the high and grave responsibility of governing the
Spanish people. Whoever takes on such a responsibility can never be
relieved nor rest, but must burn himself out in finishing the task.'

To put his achievements into perspective, he called on his listeners to 'think of the *Patria* that I received, and that from that anarchic and impoverished Spain has arisen a political and social order through which we have achieved a transformation of our structures, reaching a rhythm of perfection and progress never equalled . . . Night after night it was my job to keep watch at the death-bed of the invalid [Spain] who was dying, who had been led to war, to ruin and to hunger, who was surrounded by the Great Powers like birds of prey.' He justified the survival of Francoism beyond the grave. Denouncing political parties as a threat to national unity, he offered in their stead what he called 'the legitimate contrast of opinions'. The new *Ley Orgánica del Estado* was seen as crowning a constitutional process begun with the *Fuero del Trabajo*, 'a veritable Magna Carta of social justice in Spain'. The long and complex text was then read out by the President of the Cortes, Antonio Iturmendi. Without any debate of its ten sections, sixty-six articles and many additional clauses, the Caudillo then called on the *Procuradores* to give their assent to the new law which they did by acclaim.[57]

Newsreel records and eyewitness accounts of the scene that day in the Cortes expose a stark contrast between the ringingly triumphant rhetoric and the weak voice with which an aged, infirm Franco read it. After the lengthy applause which greeted his appearance, he looked around him unseeingly. At the beginning, he used both hands to emphasize points but after a while only the right. Having then settled into a fast mumbled monotone, his head dropped and he did not look up from the text that he was reading. In the last half hour or so, although individual words were being read out intelligibly, the overall rhythm and intonation had gone.[58] These were symptoms of the inexorable Parkinson's disease. Its effects were increasingly noticeable but there was a tacit conspiracy within the regime elite and the Spanish media not to notice them.[59]

Three weeks later, he spoke to the nation on television and radio to seek a 'yes' vote in the forthcoming referendum. In line with the official slogan '*¡Franco sí!*', the Caudillo made the referendum a vote of confidence in him personally. He expressed regret that there were those who dreamed of dressing in foreign fashions, oblivious to the fact that democracy was a fiction. For their benefit, he cited foreign hostility as proof of international admiration of his regime. With a 'yes'-vote, Spaniards could repay him for the sacrifice of his life for their well-being. He appealed to all the generations: 'You all know me. The oldest, since the times in Africa, when we fought for the pacification of Morocco; the more mature, when, in the midst of the disasters of the Second Republic, you placed your hopes in my captaincy for the defence of the threatened peace; the combatants of the Crusade, because you will never be able to forget the emotional times of those shared efforts in the victory over Communism; those who suffered under the yoke of the red domination, because you will always remember the infinite joy of your liberation; those who since then have stayed loyal to my captaincy, because you are part of that victory over all the conspiracies and sieges that were laid against Spain; those who have lived these twenty-seven years of peace, encouraging our people with your songs of faith and hope, because you all know only too well how I have always kept my word'.

There was little hope for the future here, just a heartfelt request for payment for the past. 'I was never motivated by ambition for power. Ever since I was young, they put on my shoulders responsibilities which were greater than my years and my rank. I would have liked to enjoy life like so many ordinary Spaniards, but the service of the *Patria* took over my every hour and occupied my life. I have been governing the ship of State for thirty years, saving the nation from the storms of the contemporary world; but, despite everything, here I remain, still in harness with the same spirit of service as in my early years, using what remains to me of useful life in your service. Is it too much to ask that I in turn ask your support for the laws which for your exclusive benefit and for that of the nation are about to be submitted to a referendum?'[60]

Franco's speech was merely one part of a massive campaign mounted by Fraga with the full power of the media directed to securing a 'yes' vote. The streets and highways were plastered with gigantic placards of a beaming, benevolent patriarch. A 'no' vote was described as a vote for Moscow. On 14 December 1966, eighty-eight per cent of the possible electorate voted in the referendum on the *Ley Orgánica* of whom less than two per cent voted 'no'. The validity of the vote was questionable. There had been no discussion of the virtually incomprehensible new law: the opposition had been intimidated and silenced. With policemen looking on, people placed their open voting slips into glass urns – there were neither envelopes nor cubicles. There were cases of multiple voting by the same individuals and, in some places, official efficiency and enthusiasm led to Franco getting a 'yes' vote from as much as 120 per cent of the local electorate, a phenomenon explained away by the concept of 'voters in transit' (*transeuntes*). It appeared as if large sections of the population had suddenly become nomads, although no municipalities registered drops in their electorate sufficient to account for the large votes elsewhere by alleged absentees.[61]

The referendum was, nonetheless, in general terms, a victory for Franco. Many had voted 'yes' in gratitude for the past and for growing prosperity but many did so also in the hope that they would be bringing nearer the transition from Franco's dictatorship to the monarchy. Franco was delighted and particularly pleased with the tireless and inventive Fraga.

With the arrangements for the succession now in place, there was little left for Franco to do by the beginning of 1967. Now seventy-four years-old, he seemed at times barely a shadow of his former self. In newsreels, there was an ever more noticeable rigidity in movement and a lack of energy in his speeches. The difficulties of filming his broadcasts meant that a number of television technicians were aware of it. His ministers knew but chose not to recognize his decline. After all, even after the *Ley Orgánica*, there was great uncertainty about what would happen after his departure. Rather than risk losing everything, many of the regime elite tacitly agreed to behave as if Franco were still entirely in control. The Caudillo did function normally for much of the time although he was out of touch for long periods too.

The machinery of government was in the hands of Carrero Blanco and López Rodó. The years of state terror banked between 1936 and 1944 had paid off handsomely in mass political apathy. The central issue of day-to-day politics was now the future and that gave rise to a process of jockeying for position in which Franco necessarily played an extremely marginal role. He was no longer major player in the game of Spanish politics. That was reflected in the fact that he was photographed more often playing with his grandchildren, hunting or fishing. Smiling in timid manner, Franco was now the distant patriarch. That he had time on his hands was reflected in the fact that, in May 1967, Franco won the substantial sum of one million pesetas* on the pools (*quiniela*) with a coupon signed 'Francisco Franco' and his address given simply as 'El Pardo, Madrid'.[62] His good fortune no doubt convinced him of the essentially democratic nature of the regime.

Franco also gave some indications, in his foreign policy, of having mellowed, or at least of having become even more cautious. In both cabinet meetings and numerous conversations with his cousin Pacón, he showed a calm rationality about Gibraltar and relations with the United States which contrasted with the obsessive aggressiveness of Castiella and other ministers. In February 1966, he had said 'the British won't give in

* About £6,000 at the time or just under £50,000 today.

easily, the fruit is not yet ripe and perhaps we won't see it fall; but I am sure that one day the Rock will return to Spain'. Eight months later, with the case before the United Nations and Spanish proposals being given short shrift in London, Franco said in cabinet that aggressive propaganda was a mistake and that it was pointless to try to humiliate the British. He firmly squashed a proposal by Castiella to fly barrage balloons around Gibraltar to obstruct British air access to the Rock. Despite United Nations resolutions demanding an end to Gibraltar's colonial status, he said that nothing could be done without first convincing British public opinion that the Rock belonged to Spain.

Even after the plebiscite organized by the British in September 1967 in which the Gibraltarians voted nearly unanimously to remain British, he was adamant that an aggressive stance by Spain would be entirely counter-productive. Two months later, he was saying calmly 'it should not be thought that we are going to get anywhere any faster by using violence'.[63] He stated in a cabinet meeting in 1967, 'I see no point in trying to trip up the strong.'[64] At the end of November 1968, he told López Rodó that the Soviet Union's advocacy of the Spanish case over Gibraltar would simply consolidate American support for Britain. Accordingly, he ordered Castiella's propaganda campaign to be stopped.[65] However, on 7 June 1969, with Franco's approval, the frontier with Gibraltar was finally closed. He declared that it should remain closed until the negotiations for the return of the Rock were satisfactorily concluded.[66]

In March 1967, Solís gave Franco the text of a proposed *Ley Orgánica del Movimiento*, which authorized 'political associations' within the strict confines of the single party. Throughout April and May, Franco received hostile advice on the project from various ministers. Alonso Vega, Carrero Blanco, López Rodó, Silva Muñoz and others preferred the present vague *Movimiento* as an umbrella covering all Spaniards. It fitted in with the technocrats' concept of apoliticism, a prosperous economy ruled over by an efficient administration. Solís's plan was a diluted version of Arrese's 1956 schemes. Franco, convinced that the referendum had confirmed his personal power, liked the idea of constitutional obstacles to political parties and refused to squash Solís's project.[67]

While Solís and López Rodó squabbled over the future, Franco's relationship with the Church declined rapidly. The successive Papal Nuncios, Antonio Riberí and Luigi Dadaglio, selected the most progressive candidates for vacant bishoprics. Franco refused to relinquish his *derecho de presentación*, the privilege of selecting the names of bishops from lists provided by them. In practice, he never contradicted the Nuncios' recommendations. Accordingly, the Caudillo's obstinacy derived from his conception of his semi-royal status and also from an innately cautious reluctance to give up a defensive weapon in the event of the Vatican succumbing altogether to the masonic and Communist influences which he discerned behind its post-conciliar liberalism. 1967 was a crucial year since there were to be twelve episcopal replacements because of age. It was the opportunity for which Pope Paul VI had been waiting to start bringing the deeply conservative Spanish hierarchy into line with the more progressive ordinary clergy.

Antonio Garrigues appealed to the Caudillo to make the gesture of unilaterally giving up the rights enshrined in the Concordat. Franco refused. He was inclined, on the basis of secret service reports on his desk, to attribute progressivism in the Church to the sexual degeneration of individual clerics.[68] In fact, more than ever before, Franco accepted the most sinister theories about his enemies. A report from his Ambassador in Rome, Alfredo Sánchez Bella, about American financing of Socialist parties was behind a curious updating of the prejudice which blamed everything on international freemasonry. On 13 March 1967, Franco told Pacón 'I believe that all the activities which have been carried out in the western world against us have been carried out by organizations which receive funds from the CIA above all with the intention of establishing in Spain an American-style political system on the day that I cease to be around.'[69]

In February 1967, Muñoz Grandes had told Fraga that he and Franco were drifting ever further apart: 'we are fed up with arguing'. On 21 July, Franco said to him 'I know that you are unhappy. I won't oblige you to stay any longer'. However, having made the decision, he ordered Carrero Blanco to delay publication of the news in the *Boletín Oficial del Estado* until the political elite was away from Madrid on holiday. Muñoz Grandes had been the great hope of those elements within the Falange which opposed the transition to the monarchy and Franco seems to have feared their hostile reaction. It had been hoped by the technocrats that, within the terms of the *Ley Orgánica*, the Caudillo would nominate Carrero Blanco as President of the Council of Ministers. The doggedly faithful Carrero, however, did not want the post, believing that no one could do the job better than Franco. For two months, the Caudillo did nothing, to let time take any heat out of the situation. The ambitious Fraga aspired to the post but, since Franco did not trust him, and Carrero Blanco even less, there was not the slightest possibility. Finally, after a cabinet meeting in San Sebastián, as they travelled by car to an official function, Franco casually told Carrero Blanco that he was to be vice-president of the council of ministers. It was announced on 21 September 1967.[70] It was the logical move. He had served the Caudillo loyally since 1941 and their views were almost indistinguishable.

For most Francoists, Carrero Blanco was a guarantee of untrammelled Francoism. However, in the upper reaches of the regime, his commitment to the cause of Juan Carlos made him the object of jealous suspicion. It was rumoured that Franco had had a cerebral haemorrhage in September 1967 and for that reason had handed over to Carrero Blanco.[71] There is no indication of this in the memoirs of those who dealt with him on a regular basis. Nevertheless, anxiety that an incapacitated Franco was now entirely in the hands of López Rodó and the Opus Dei group inclined unreconstructed Francoists to make their own dispositions. They feared that, in backing Juan Carlos, Franco might be opening the way to a liberal monarchy, and certainly one which would put an end to the monopoly of privilege previously held by the Falange/Movimiento. The public war against the Opus was carried on in the mid-1960s through the *Movimiento* press network. Franco commented 'the only newspapers which don't say what their owners tell them to are those of the *Movimiento*'.[72]

There was a more insidious private struggle carried on by a circle of right-wingers who gathered in El Pardo in an attempt to mobilize an increasingly decrepit Franco to the cause of *inmovilismo*. Consisting of Cristóbal Martínez Bordiu, Doña Carmen and hard-line Falangists like Girón, they had close links with military hard-liners who saw the Army as the praetorian guard of the regime. They could derive satisfaction from the fact that, by the late 1960s, the most prominent of the so-called 'blue'* or Falangist generals (*generales azules*), such as Alfonso Pérez Viñeta, Tomás García Rebull, Carlos Iniesta Cano and Angel Campano López, were reaching key operational positions. In his last years, largely because of his disease and the drugs taken to mitigate the symptoms, Franco became a passive shuttlecock between these groups. He was funda-mentally committed to the Carrero Blanco/López Rodó vision of the transition to an authoritarian monarchy. However, as he grew older, his instincts made him more prone to listen to the alarmist accounts of what was happening put his way by this *camarilla* (clique). He showed less political energy, and gave fewer signs of reading the press or even of knowing what his ministers had done. The extent to which he was out of touch was revealed when he asked Fraga, Solís and other ministers for

---

* A reference to the *camisa azul* (blue shirt) of the Falangist uniform.

names to help him make senior appointments: 'you who are out in the world can help me. I have been locked up here for so long that I don't know anyone anymore.'[73] When not away on hunting trips, he watched television and movies. He was captivated by the arrival of colour television.[74]

The new Cortes which met on 17 November 1967 had had one third of its *Procuradores* elected by heads of families. This was not a significant liberalization: all *Procuradores* were members of the *Movimiento* and nearly half were state functionaries. In any case, as Franco made a point of telling one of his ministers, the Cortes was not sovereign. Only the Caudillo could sanction laws.[75] On the day before the State opening, he told Pacón that he had no intention of using the occasion to announce changes in the government. Playing down the significance of the appointment of Carrero as vice-president, he acknowledged that the 'constitution' provided for the appointment of a prime minister in the event of his health making it necessary but boasted that fortunately there was no need.[76] In his speech of inauguration, he mocked those who wanted to bring back liberal democracy. There were hints that there would be some kind of *apertura* (opening) although he was at pains to stress the narrow limits within which that might happen.[77]

On his seventy-fifth birthday, Franco told his cousin that he felt strong but that he had no illusions about living into his nineties.[78] His political reflexes were getting sclerotic. Infuriated by the university disturbances of early 1968, Franco was unwavering in his conviction that university unrest was the work of foreign agitators and that radical priests were merely Communists in disguise.[79] He was delighted at the energetically violent repression of left-wing and liberal priests and university students by General Pérez Viñeta, the Captain-General of Barcelona.[80]

He spoke of the monarchist daily *ABC* as an 'enemy'. His determination to prevent Don Juan coming to the throne was based less on hostility to his present views than on a simmering resentment of the 1945 Lausanne Manifesto. His attitude to Don Juan and his doubts about the monarchical succession came to the surface in the course of the celebrations which followed the birth of Juan Carlos's son Felipe on 30 January 1968. Franco refused to go to Barajas Airport to meet Queen Victoria Eugenia, widow of Alfonso XIII, mother of Don Juan and grandmother of Juan Carlos, who came from Nice on 7 February for the baptism. 'You must realize, Your Highness,' he said to Juan Carlos, 'that I cannot compromise the State with my presence there.' Juan Carlos drily reminded him that he had already committed the State to the monarchy in the *Ley Orgánica del Estado*. Franco refused to receive Don Juan. He was furious that several ministers, Castiella, Espinosa San Martín and Lora Tamayo went to the airport without asking his permission. He expressly prevented Alonso Vega from going. Only Antonio María Oriol, the Minister of Justice with formal responsibility for relations with the royal family, was there with Franco's blessing. At the baptism ceremony, with Don Juan, Prince Juan Carlos and the baby Felipe together, Victoria Eugenia cornered Franco and said: 'Well, Franco, you've got all three Bourbons in front of you. Decide.' He did not reply.[81]

With age, there was no inclination on Franco's part to reconciliation with his enemies. Although he enjoyed being told that he was Caudillo of all Spaniards, he wished to be leader only of 'good' Spaniards. When he heard of suggestions that the mutilated war veterans of the Republican forces receive state pensions, he was furious at the idea of putting 'the dregs of Spanish society' on the same level as gentlemen and heroes.[82] Franco's prejudices came to the fore when López Rodó commented to him on the assassination of Robert Kennedy on 4 June 1968. López Rodó took the opportunity to press the Caudillo to restore the monarchy, saying that, if such things happened in the course of a presidential election in the United States, then elections for the president of a Spanish Republic could always throw up an insane would-be assassin. Franco coolly rejected the idea, remarking categorically 'there are more lunatics in the United States'.[83]

On 29 April 1968, Pope Paul VI wrote to Franco asking him to give up his privilege of choosing bishops from a list of three names provided by the Papal Nuncio. The Pope undertook to let him have the names of bishops in advance in order that he might make known any objections. Dropping his earlier claim to be unable to renounce a royal privilege not his to relinquish, Franco now replied with a letter of Byzantine cynicism, refusing on the equally spurious grounds that only the Cortes could modify the present situation and that Spanish public opinion would not stand for a unilateral concession. By so doing, Franco clinched the hostility of the Vatican to the regime.[84] In the 1970s, Rome would use two devices for asserting its position. One was to submit only one name for vacant bishoprics, as opposed to the three from which Franco would normally choose. The other was to avoid the issue and simply to appoint auxiliary bishops. Technically only caretakers, their appointment did not require the Caudillo's approval.

Franco was bewildered by the growing liberalism of the Catholic Church and by the activities of some bishops, like José María Cirarda of

Santander, who denounced the repressive activities of the police. He was convinced that he had saved the Church and that, if Communism were ever successful in Spain, churches would be burnt and bishops murdered.[85] More and more priests were getting actively involved in outright support of the labour and regionalist opposition to the regime. In the summer of 1968 Franco authorized his Minister of Justice, Oriol, to set up a special prison for priests at Zamora. To the immense embarrassment of both the Church hierarchy and Franco, more than fifty priests were imprisoned.[86] The consequence of the leftwards move of the Church was the emergence within the Francoist coalition of ultra-right-wing anti-clericalism, strongest in Blas Piñar's neo-Nazi political association *Fuerza Nueva* and its armed terror squads, the *Guerrilleros de Cristo Rey* (the warriors of Christ the King).

Franco seems to have taken only sporadic note of the fact that his cabinet had been virtually paralysed for an entire year by the hostility between the Falangists and the technocrats. Throughout the summer and autumn of 1968, Carrero Blanco, Fraga and others had been urging him to renew the government but he simply ignored them. Carrero Blanco wrote Franco a long letter on 11 July outlining the disagreements between ministers. He also denounced the policy of Fraga at the Ministry of Information for destroying public morality by permitting bookshops to be flooded with 'Marxist works and novels of the most unrestrained eroticism'. Presumably unaware of the real contents of Spanish bookshops at the time, Franco believed such horror stories but failed to react. It would be more than a year before he got around to the desired reshuffle and Carrero Blanco was driven to comment 'how slow this man is to give birth'.[87]

Symptoms of his decreasing energy and sense of urgency were impossible to ignore. Much was made in political circles of an incident alleged to have taken place during the summer of 1968 in Santander. After a meeting to report on ministerial business, a member of his own cabinet asked him to sign a photograph in which they both appeared along with other ministers. Franco agreed, put on his glasses, picked up a pen and then hesitated, peered at the minister and asked him who he was.[88] Throughout the autumn, Franco received from Oriol, López Bravo and Silva Muñoz reports urging him to name a president of the council of ministers.[89] It would be another five years before he did so. On the other hand, he did intervene to block an effort by Solís to attack López Rodó's Second Development Plan as an instrument of American colonialization.[90]

Anti-Americanism was part of the rhetorical armoury of the Falange. In September 1968, the bases agreement which had been renewed in 1963 expired once more. The choice was between simply accepting its automatic extension or terminating it and thereby opening up a six-month consultation period during which the agreement could be renegotiated. Castiella made an unsuccessful opening bid in July 1968, demanding $1 billion in aid in return for the continued presence of American bases on Spanish soil. Carrero Blanco and the military ministers, backed up by reports from the general staff, valued the American link too highly to want to risk it and wanted automatic renewal. Castiella, Fraga, Solís and most of the rest of the cabinet favoured cancelling the agreement in the hope of negotiating better terms. Franco backed the latter view, his acquisitiveness never far from the surface. After a lengthy and acrimonious debate on 24 September 1968, the cabinet agreed to invoke the termination procedure of the 1953 agreement and issued a statement calling for the removal of the US air base at Torrejón outside Madrid.

Franco drew up a document on the basis of which the renewal was to be negotiated.[91] Its overpricing, a typical Franco tactic, was a serious error of judgement. Within days, he was seriously worried, after receiving further protests from Muñoz Grandes, still Chief of the General Staff, that without the American friendship the Spanish armed forces would be condemned to impotence and that the chances of recovering Gibraltar would be dramatically diminished. When he received Dean Rusk for a long and difficult meeting on 18 November, three days after the election of Richard Nixon to the Presidency, a barely lucid Franco did no more than grunt monosyllables. Later in the month, with the ecstatic support of the *Movimiento* press, Castiella proposed the withdrawal of the US Sixth Fleet from the Mediterranean. Franco was deeply irritated because he had not been consulted and because Castiella's move was naïvely inappropriate at a moment, after the Soviet invasion of Czechoslovakia, when anti-Communism was again a negotiable asset. Already suspected of being too Vaticanist, Castiella had effectively put an end to his ministerial career although, characteristically, it would be nearly a year before Franco took action against him.[92]

In the wake of the Vietnam War, there was growing Senate opposition to existing American commitments in Spain let alone to massive increases thereof. Accordingly, when Richard Nixon assumed the presidency at the beginning of 1969, Franco obliged Castiella to agree in principle to a five-year renewal of the bases agreement and lower Spain's price to $300,000,000. However, under the leadership of Senators Fulbright and

Symington, Senate opposition to any deal with Franco was growing. To avoid serving a notice to quit on the Americans, Castiella had to announce on 26 March 'an agreement in principle' and so create an extension for further negotiations. In May, with Senate opposition being marshalled by Senator Fulbright, Castiella made a revised offer of a one-year extension in return for $50,000,000 in military aid and $25,000,000 in credits for arms purchases. This averted the threat of a complete breakdown of Hispano-American relations.[93]

There was little hint of these problems in Franco's lifeless end-of-year broadcast on 30 December 1968. While condemning the continuing violence in the universities, his tone was one of quiet satisfaction.[94] But if Franco showed every sign of planning to remain in power, the attrition of years in power was having its effect. Cabinet meetings were now fortnightly, alternating with sessions of the Comisión Delegada de Asuntos Económicos. Although ministers had regularly left cabinet meetings to smoke a cigarette or visit the lavatory, Franco himself never did.* When he did so for the first time ever on 6 December 1968, it was taken as a serious indication of his deteriorating health. Carrero Blanco took the chair. The Caudillo had to leave the meeting again on 5 January 1969.[95] It was a symbolic anticipation of his more general absence during the final six years of his rule.

The Caudillo's private hesitations about the succession were largely resolved by the autumn of 1968.[96] On 20 December 1968, he ordered the Carlist pretender Don Hugo-Carlos de Borbón-Parma expelled from Spain in reprisal for his political machinations. Shortly afterwards, Juan Carlos clinched his position. On 8 January 1969, the Prince was interviewed by the official news agency EFE and declared his unreserved commitment to the idea of monarchical installation rather than restoration. The declarations had been drafted by Fraga. The emphasis on the Prince's loyalty to Franco and to the *Movimiento* delighted the Caudillo, as was evident in his comments to both Fraga and López Rodó later in the same day. When Fraga came out of Franco's study, López Rodó said to him merely 'the perfect crime'. When López Rodó went in, he said to Franco 'the Prince has burned his boats. Now all that is lacking is Your Excellency's decision.' On 15 January, Franco more or less told Juan Carlos that he would be naming him as successor before the end of the year. Franco said to him: 'You need not worry, Your Highness. Don't

---

* His ability to remain impassive throughout long meetings to the chagrin of his ministers was described by one of them as the triumph of the continent over the incontinent.

let anything lead you astray now. Everything is done.' Juan Carlos replied 'rest assured, *mi general*, I have learned much from your *galleguismo* (Galician craftiness).' As they both laughed, Franco complimented him, 'Your Highness does it very well.'[97]

However, with university agitation reaching new peaks, Alonso Vega, backed by Carrero Blanco, Nieto Antúnez and Solís, proposed a state of emergency at a cabinet meeting on 24 January 1969. It was a massive overkill, typical of the nearly eighty year-old authoritarian nicknamed by students 'Don Camulo' because of his mule-like obduracy. When Franco indicated his approval of the measure, it was symptomatic of the political bankruptcy of the government that the rest of the cabinet hastened to agree. In private, López Rodó, Silva Muñoz and other technocrats mocked the measure which they had not opposed in cabinet. Five days after the decision was taken, Silva wrote López Rodó a note in which he said 'you don't have to be psychic to foresee the failure of the repression'. Others spoke of 'killing flies with artillery'.[98] However, the main concern of López Rodó was not the absurdity of the regime dinosaurs trying to stamp out the consequences of social change but that, with a state of emergency in place, Franco would delay yet again the nomination of Juan Carlos as successor.

At the cabinet meeting of 21 March 1969, Alonso Vega, who had previously been briefed by López Rodó, called for the state of emergency to be lifted. Solís opposed this, precisely because a successor could not be named while it remained in force. The technocrats used the argument that the regime should not approach its thirtieth anniversary in such conditions. Fraga argued that it would severely damage the tourist trade. Finally, Franco closed the debate saying 'given that the Minister of the Interior requests it, the state of emergency should be lifted'. Aware of the way Oliveira Salazar's collapse had caught that ruler unawares in August 1968, the technocrats pressed the Caudillo to make a decision on the succession while he was still able. On 7 May, Fraga spoke to Franco about his age and the growing political vacuum. He listened politely, said nothing, then left for a ten-day salmon-fishing expedition in Asturias.[99]

Before leaving for his fishing holiday on 7 May, Franco received from Carrero Blanco a long report on the political situation. In it, he dealt with a series of problems. The first was the efforts of Solís and others in the *Movimiento* to create an independent power base through a new Syndical Law which Carrero compared to the Arrese proposals of 1956. He recommended that they be shelved. On relations with the Church, he denounced Castiella as defending the interests of the Vatican rather than those of Spain and also accused him of mishandling the negotiations with Washington. He then turned to the emergent threat of the Basque revolutionary separatist organization *Euzkadi ta Askatasuna* (Basque Homeland and Liberty). Carrero Blanco pointed out that the extirpation of ETA required enormous delicacy if the operation were not to damage relations with the Basque Country and with the Church. Finally, he respectfully called upon the Caudillo to resolve the anxieties of the Spanish people about the future by naming a successor.[100] The months which followed would show just how far Franco was now reliant on Carrero Blanco. The Syndical Law would be blocked, a deal made with the United States and Castiella removed. In large part because Franco was not fully on top of events, the struggle against ETA would be left to hard-line elements in the Army and Carrero's pessimistic predictions would come true.

On 28 May, the eve of his own eightieth birthday, Camilo Alonso Vega had a long conversation with Franco. He asked to be replaced as Minister of the Interior, reminded the Caudillo of the need to settle the succession in favour of Juan Carlos and suggested that the moment had come to take a well-earned rest by naming Carrero Blanco as president of the government. Franco said little but, on the following day, he told Carrero Blanco that he would name Juan Carlos successor before the summer.[101] Then, pressed by Falangists to do no such thing, he hesitated again, fearful as he told Carrero Blanco of deserting his loyal followers. With the technocrat ministers getting more impatient, Franco told Carrero Blanco on 26 June that he would make the announcement before 18 July 1969. However, on 30 June, when Antonio Iturmendi, the President of the Cortes asked him when the ceremony would be, he said indecisively 'it could be either before or after the summer'. However, he picked 17 July as the day for the announcement.

He said nothing to Juan Carlos who was about to visit his father in Portugal. On his return, on 12 July, the Caudillo gave him the news. When Juan Carlos asked him why he had said nothing before, Franco cunningly replied 'because I would have asked you to give me your word of honour not to reveal the secret and, if your father had asked you, you would have had to lie. And I preferred that you should not lie to your father.' In fact, what his apparently high-minded duplicity achieved was a rupture between father and son, something which, in political terms at least, he had wanted to do ever since the first agreement on the *Azor* that the Prince would be educated in Spain. After not telling his father that he was about to be proclaimed *Príncipe de España*, Juan Carlos was assumed by Don Juan to have betrayed him. Relations between them were strained for some time afterwards.[102] Franco wrote to Don Juan asking him to accept the designation of his son as 'the coronation of the political process of the regime'. Don Juan dissociated himself with dignity from what had happened. Stressing his belief that the monarch should be King of all Spaniards, above groups and parties, based on popular support and committed to individual and collective liberties, he implicitly denounced a monarchy which was irrevocably linked to the dictatorship. [103]

A beaming Franco announced his decision to the council of ministers on 21 July 1969. He said 'the years pass. I am 76, I will soon be 77. My life is in God's hands. I wanted to confront this reality.' He quoted Don Juan's reaction as proof that he was entirely useless as a Francoist monarch. The hard-line Falangist opponents of Juan Carlos hastened to mount a last-ditch defence against his nomination. The device proposed by Solís in the cabinet was a secret Cortes vote. He hoped to use his control of *Movimiento* patronage to engineer an unfavourable vote which could be flatteringly interpreted to Franco as a recognition of Juan Carlos's inadequacy as a successor to such a providential figure. Franco, backed by Carrero Blanco, Alonso Vega and the technocrats, would have none of it. He wanted to see each *Procurador* vote.[104]

The designation of Juan Carlos as *Príncipe de España*, and not *Príncipe de Asturias*, the traditional title of the heir to the throne, was Franco's way of breaking with both the continuity and the legitimacy of the Borbón line. The new monarchy would be his and his alone. This was recognized by Don Juan who obliged his son to return to him the insignia (*placa*) of *Príncipe de Asturias*.[105] In his speech to the Cortes on the following day, Franco mocked those who 'speculate with the crisis of the day when my captaincy may be lacking' and took pride in the precision of the instruments now created for the succession.[106] The *Procuradores*, no doubt relieved to have their doubts finally resolved, interrupted his speech constantly to applaud and, at the end, stood and chanted 'Franco! Franco! Franco!'. The Prince swore fidelity to the principles of the *Movimiento*, having first been privately assured by his counsellor Torcuato Fernández Miranda that his oath would not prevent a future process of democratic reform.[107]

Franco clearly had great faith in Juan Carlos and had come to like and respect him over the years. It was a faith shared neither by the left-wing opposition nor by many on the Falangist wing of the regime, nor even, in the late 1960s, by Juan Carlos himself. Despite the support of López Rodó and Carrero Blanco, his reserved and melancholic manner gave the impression that he was 'a guest who wasn't sure if he was meant to stay to dinner'. In fact, though in private conversation and public speeches Franco had made it obvious that he expected his successor to continue his work, he had never given any explicit instructions to Juan Carlos. He had obliged the Prince to spend periods in the three armed forces, in the university and in the finance ministries, but he had not indoctrinated him politically. Whenever Juan Carlos asked him for advice, he would reply: 'What's the point in expecting me to tell you? You will not be able to govern like me.' Aware that Juan Carlos would face problems that he could not foresee, Franco trusted him sufficiently to give him a free hand.[108] It is difficult to avoid the conclusion that the Prince, having learned from his mentor how to keep his cards close to his chest, planned all along to deceive him by working for the transition to democracy after his death.

XXVIII

THE LONG GOODBYE

1969–1975

FRANCO's belief that, with the succession resolved, a trouble-free future could now be enjoyed was to be rudely shattered in the second half of 1969. ETA was a threatening black cloud on the horizon. More immediately, however, in mid-August 1969, there erupted the political volcano known as the Matesa scandal. Matesa (*Maquinaría Textil del Norte de España Sociedad Anónima*) was a company which manufactured textile machinery in Pamplona. Under its director, Juan Vilá Reyes, Matesa had developed a shuttleless loom which it was exporting to Europe, Latin America and the USA. The apparent successes of Vilá Reyes made him the toast of the technocrats. In order to qualify for export credits, subsidiary companies were set up in Latin America which ordered large numbers of looms. Financial irregularities were discovered in late 1968, and it was alleged that the subsidiaries and their orders were a fraudulent device to qualify for the credits.[1] A detailed report by the Director-General of Customs had been sent to both the Minister of Finance, Espinosa San Martín, and to Franco. The Caudillo was not too concerned, having been convinced in January 1969 by Vilá Reyes that the company was merely bending archaic regulations to boost much needed exports.[2]

Vilá Reyes claimed that it was only the accusations themselves which caused genuine orders to his subsidiaries to be cancelled. Whatever the truth of the matter, the company's problems intensified to the extent that they were discussed at a cabinet meeting on 14 August at the Pazo de Meirás. The *Movimiento* press unleashed a stream of attacks on the technocrats with *Arriba* denouncing what it called 'a national disaster'.[3] With Juan Carlos now named as successor, Solís was frantically seizing the last chance to break the hegemony of the Opus Dei group before the post-Franco future began. In part also, the scandal-mongering of the

Movimiento press represented an effort to revive declining circulations with sensationalist mud-slinging. Whatever his motives, Fraga, as Minister of Information, let it happen, but the anti-Opus Dei tactic of Solís backfired badly. Franco did not believe that the ministers linked to Opus Dei acted as a sinister independent block and delighted in their absolute personal loyalty to him, commenting that 'they are perfect gentlemen'.[4] They had solved problems and caused him no embarrassment.* When the Matesa scandal was raging, Franco refused to let it be mentioned in his presence.[5] According to his sister, he did not regard the offences as of significant gravity.[6]

On 11 September, in San Sebastián, Silva Muñoz gave the Caudillo a bulky dossier of press cuttings and a report on the media campaign against the technocrats. The report alleged that there was an orchestrated attempt to provoke a major political scandal by the *Movimiento* press and through the Ministry of Information's press agencies. Franco was infuriated particularly because of the foreign publicity.[7] In the event, the two ministers who had jurisdiction over the issues at stake, Espinosa San Martín in Finance and García Moncó in Commerce, chose to resign although Franco seems to have had no doubts of their honesty.[8] It is unlikely that Franco, now often heavily medicated for Parkinson's disease, followed all the ramifications of the Matesa affair. Given his belief that *el mando* (authority), in this case the government's, should not be undermined, it was easy enough to persuade him that the 'crime of scandal', imputable to the Falangists who gave the issue publicity, was more serious than any original fraud.

Carrero Blanco was determined that the two ministers implicated in the generation of the scandal, Fraga at Information and Solís as *Ministro-Secretario del Movimiento*, would also have to go. To make sure, on 16 October, he read Franco a report accusing Solís of trying to build an independent power base in the *Sindicatos* and Fraga of subjecting the Matesa case to 'scandalous politicization'. It was a reactionary denunciation of the Press Law for permitting attacks 'on the Spanish way of being and on public morality'. Fraga's crime in Carrero Blanco's eyes was to have permitted the media to reflect the reality of Spain in the 1960s: he alleged that the press was so negative that readers might get the impression that Spain was a country which was 'politically stagnant, economically monopolistic and socially unjust. The press exploits por-

---

* Allegedly, when Solís and other Falangists complained at length about the technocrat ministers, Franco abruptly terminated the audience saying 'What have you got against the Opus? Because while they work you just fuck about.' (Peñafiel, *El General*, p. 102.)

nography as a commercial instrument. In literature, the theatre and the cinema, the situation is equally serious in political and moral terms. Bookshops are full of Communist and atheist propaganda; theatres put on works which prevent decent families attending; cinemas are plagued with pornography. To encourage cheap tourism, *streap-tesse* [sic] is protected in *play-voy* [sic] clubs.' Such surrealistic nonsense went straight to the Caudillo's heart. The report also repeated in more detail Carrero's early denigration of Castiella's misguided policies towards the United States, Gibraltar and the Vatican.[9]

Franco's dwindling energy and his unquestioning acceptance of Carrero's report were reflected in the cabinet changes of 29 October 1969. There was nothing even resembling a balancing act and the cabinet was once again Carrero Blanco's more than his own. Carrero's list was supplied largely by López Rodó and Silva Muñoz.[10] The Caudillo suggested to Carrero Blanco that now was the time for him to take over as prime minister but his vice-president refused, believing himself to be incapable of controlling ministers.[11] Castiella was dropped after twelve years in the post. The dynamic Gregorio López Bravo, for whom Franco had developed a paternal affection, was given his choice of ministries. To the chagrin of Silva Muñoz, who had wanted the post for himself, he chose the Ministry of Foreign Affairs.[12]

The new Minister Secretary-General of the *Movimiento*, the diminutive, dapper and sinuously intelligent Torcuato Fernández Miranda, also a member of Opus Dei, was close to Juan Carlos and could be expected to continue the job of de-Falangizing the *Movimiento*. Both the new Minister of Industry, José María López de Letona, and of Finance, Alberto Monreal Luque, were López Rodó protégés from the Commissariat of the Development Plan. Also part of the technocrat circle were the new Minister of Commerce, Enrique Fontana Codina, of Housing, Vicente Mortes Alfonso, and of Agriculture, Tomás Allende y García Baxter. Camilo Alonso Vega was finally allowed to retire and his replacement, the military lawyer, Tomás Garicano Goñi, was also suggested by López Rodó.

Since all its members came from the two conservative Catholic pressure groups, Opus Dei and the *Asociación Católica Nacional de Propagandistas*, and were all committed to Juan Carlos, it was known as the *gobierno monocolor* (monochrome government). Even the Army Minister, General Juan Castañón de Mena – who was close to both the Caudillo and Prince Juan Carlos – was also a sympathizer of Opus Dei and party to Carrero Blanco's schemes for a modified Francoist monarchy.[13] Franco gave the

ABOVE: Franco and Pétain,
Montpellier, 13 February 1941.
Behind Franco can be seen General
Moscardó in the dark glasses, Admiral
Darlan, Serrano Suñer and a smiling
Pacón

RIGHT: Serrano Suñer greeted
by Hitler, watched by Ciano,
29 November 1941

LEFT: Franco greets Evita Perón under the watchful eye of Doña Carmen, June 1947

BELOW: Franco's efforts to rewrite his role in the Second World War as seen from London

Hullo, Hollywood! I want Spanish victory film, please, to show how Spaniards won war for Allies!

ABOVE: Franco receives from Admiral Bastareche the symbolic ring of Lord High Admiral of Castile, Huelva, 12 October 1948

BELOW: Franco and Doña Carmen preside at the Holy Week bullfights in Seville, c.1948

LEFT: Franco the fisherman, on board his yacht *Azor* with a swordfish, 1949

BELOW: Franco receives a group of US Senators led by Owen Brewster, 30 September 1949

ABOVE: General Franco and a regally attired Doña Carmen during their state visit to Portugal, Palacio de Ajuda, Lisbon, 26 October 1949

RIGHT: Franco, watched by Doña Carmen, makes a speech to the young men of the Falangist rural fraternities (*Hermandades*), El Pardo, *c.*1950

OPPOSITE ABOVE: Franco addresses fifty thousand Nationalist veterans, urging them to keep alive the values of the Civil War, at the Alto de los Leones de Castilla, a wartime battle-site to the north of Madrid, 19 October 1952

OPPOSITE BELOW: Franco places the *birreta* on three new cardinals resident in Spain. In the picture, the Apostolic Nuncio Monsignor Cicognani, watched by Doña Carmen and (*far left*) the Barón de las Torres. Madrid, 19 January 1953

RIGHT: Carmen Franco (Nenuca) with her daughters María del Carmen (*standing*) and María de la O (*in her lap*), 1953

BELOW: Franco meets the American Ambassador, James C. Dunn, at El Pardo, 9 April 1953, with Alberto Martín Artajo and the Barón de las Torres

LEFT: Franco, accompanied by his Minister of Education, Joaquín Ruiz Giménez, in the cloisters of the University of Salamanca to receive an honorary doctorate of law, 8 May 1954

LEFT: Carrero Blanco, Rafael Trujillo, Señora de Trujillo and Franco visit the newly built monastery at the Valle de los Caídos, 9 June 1954

ABOVE: Franco and Doña Carmen surrounded by the singers and dancers taking part in the annual fiesta to celebrate the outbreak of civil war on 18 July 1936, Palacio de la Granja, 18 July 1954

RIGHT: Doña Carmen and Franco in the Pazo de Meirás, summer 1955

Franco in his study at the Pazo de Meirás (c.1955) surrounded by the books of Emilia Pardo Bazán, the novelist who previously owned the estate

Franco the hunter, Sierra de Cazorla (Jaén), c.1959

RIGHT: Franco embraces
President Eisenhower at
Madrid Airport, December
1959

BELOW: Franco greeted by
Doña Carmen, Nenuca and
his grandchildren after
returning from hospital
after his hunting accident,
28 December 1960

ABOVE: Franco fishing in Asturias, 20 April 1968

LEFT: Franco and his horse 'Zegri' in the hills (*montes*) of Toledo, *c*.1968

BELOW: Franco the golfer, 1972

ABOVE: Franco and Doña
Carmen watch the regatta in
La Coruña, *c.*1969

RIGHT: A distracted Franco
greets Fraga, La Coruña,
.1969

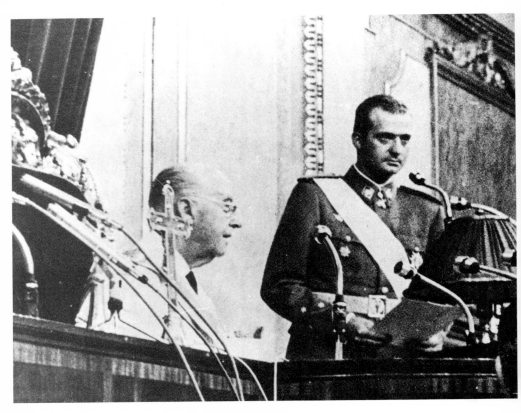

ABOVE: Franco with Juan Carlos as he accepts designation as successor, the Cortes, 23 July 1969

LEFT: Richard Nixon meets Franco and his grandchildren, Madrid, 2 October 1970

ABOVE: Franco and Juan
Carlos, *c.*1971

RIGHT: Laureano López Rodó
(*centre*) introduces the
President of a Commission to
Franco, El Pardo, 3 November
1971

ABOVE: Prime Minister Carlos Arias Navarro visits Franco at the Ciudad Sanitaria Francisco Franco, July 1974

LEFT: Franco leaves hospital, July 1974

Ministry of Information* to the deeply conservative Catholic, Alfredo Sánchez Bella, his Ambassador in Rome, as a reward for a series of assiduous reports on Castiella's relations with the Vatican.[14] The monochromatic character of Carrero Blanco's cabinet reflected his political simplicity. He was happy to have a technocratic team provided by López Rodó. However, as soon as it ran into problems, his essentially reactionary instincts would come back into play.

It was ironic that just when Franco thought that the future had been settled, his dreams should be shattered by Falangist tantrums. The intra-regime squabbles over Matesa went far beyond the immediate isssue. In part, it was a question of competition for the spoils of power. However, it also reflected growing unease about labour, student and regionalist unrest. Franco's supporters were beginning to break up into factions which reflected not the traditional divisions into Falangists, monarchists, Catholics and so on but rather differing, and kaleidoscopically changing, perceptions of how best to survive the imminent disappearance of Franco. The technocrats hoped that prosperity and efficient administration would permit a painless transition to a Francoist monarchy under Juan Carlos. Others, like Fraga and Solís, saw the scale of opposition as requiring political reform of the system. Others still came to believe that it was modernization which, like the sorcerer's apprentice, had opened the floodgates to opposition and so sought a return to hard-line Francoism. Despite having sponsored the domination of the technocrats, when faced with the crises of the early 1970s, Franco and Carrero too instinctively returned to the siege mentality of the 1940s.

Franco's outrage at the behaviour of key elements in the *Movimiento* accounts for his acceptance of an all-technocrat cabinet. On 28 October 1969, a grovelling Solís had hastened to El Pardo and tried to persuade the Caudillo to step back from his cabinet changes. He even begged Juan Carlos to intervene with Franco to stop what he called 'a *coup d'état*'. Fraga wept as he handed over to Sánchez Bella and there were furious comments when Fernández Miranda wore a white shirt at the ceremonial handing-over of office by Solís. By spurning the Falangist blue, he was declaring the *Movimiento* to be bigger than the Falange.[15]

It was symptomatic of Franco's loosening grasp on Spain's political and social reality that he sanctioned the creation of a government which

---

* There had been a possibility that the Ministry of Information would go to Adolfo Suárez, the fast-rising protégé of Fernando Herrero Tejedor. Like Fernández Miranda, Suárez was a man of the *Movimiento* linked to Opus Dei, and together they would play crucial roles in the post-Franco transition to democracy.

excluded other factions of the regime. Blind to the fact that the political instruments of his dictatorship were not adequate to cope with a dramatically different Spain, he assumed that this cabinet would be capable of resolving the serious problems already on the agenda. López Rodó's strategy of depoliticized adminstrative efficiency and economic prosperity was now in place, but it faced insuperable opposition from inside as well as outside the regime. Inevitably, and soon, the inability of the *monocolor* team to settle the ferment of Spanish society would open the way, with Franco's approval, to a return of the repressive brutality of the post-Civil War period.

In December 1969, at a dinner attended by Fraga, Nieto Antúnez, Solís and the ex-Minister of Agriculture, Adolfo Díaz Ambrona, an ironic comment on Franco's limited role in the crisis was coined and went the rounds: 'In Franco's day, this kind of thing would never have happened.' Within a month, a tearful Franco would insinuate to Fraga that the reshuffle had been stage-managed against his will.[16] In his end-of-year message on 30 December 1969, however, the Caudillo had confidently declared, in what was to become the nautical catch-phrase of his twilight years, that 'all is lashed down and well lashed down' (*todo ha quedado atado, y bien atado*). He ended with the promise that 'while God gives me life, I will be with you, working for the *Patria*.'[17]

In fact, all was not so well lashed down. Outright opposition in the universities, factories and regions continued to intensify. There were twenty thousand miners on strike in Asturias. As the year progressed, there would be major disputes in the shipyards, the Granada and Madrid construction industries and the Madrid metro, all of which would be met by police violence. Since the strikes were often supported by the clergy, parallel terror squads began to carry out the work of repression that the government did not wish to be seen doing. The squads, working under the name *los Guerrilleros de Cristo Rey*, included paid thugs as well as young Falangist militants. They were organized by Carrero Blanco's more or less private intelligence service, the *Servicio de Documentación de la Presidencia del Gobierno*.[18] The *Guerrilleros* were linked to the neo-fascist political association *Fuerza Nueva* (New Force) led by Blas Piñar, a member of the *Consejo Nacional* and a friend of Carrero Blanco. The cabinet acquiesced in this violence because the existence of a wild extreme right let the government present itself as somehow belonging to the centre.[19]

The more progressive Francoists, like Fraga, began to work for reform within the system. The regime's loyalists were divided between the grey technocrats, known as *continuistas*, and the intransigent 'ultras' or *inmovil-istas* whose readiness to fight progress to the last led to them being known in Hitlerian terms as the *bunker*. The *bunker* could count on sympathy among hard-line Falangists of the older generation, the gilded youth that made up the terror squads and many extreme right-wing officers in the Armed Forces with the *generales azules* at their head.

At the end of the decade, the *bunker* mounted a two-pronged assault in El Pardo against the planned Francoist monarchy under Juan Carlos. First, through sympathizers in Franco's family circle, they started a whispering campaign against Opus Dei and the cabinet. Then, they began to push the cause of Don Alfonso de Borbon-Dampierre, son of Don Juan's elder brother Jaime and soon to be fiancé of Franco's eldest grandchild, María del Carmen Martínez Bordiu, a great favourite of Doña Carmen. Relying on the phrase in the *Ley de Sucesión* about 'the prince with the best rights', they could point to Alfonso de Borbón-Dampierre's enthusiastic Francoism and his friendship with Cristóbal Martínez-Bordiu.[20]

It is unlikely that Franco himself gave much thought to the possibility of establishing a royal dynasty. However, the cause of Alfonso, the *príncipe azul* (blue prince), was to be much favoured by the extreme right and especially by Franco's wife and his son-in-law. Ultras in the top echelons of the *Secretaría General del Movimiento* put pressure on provincial governors to play down visits from Juan Carlos and to inflate those by Alfonso de Borbón. In general terms, the dynastic question provided a focus for the intrigues which increasingly occupied, and divided, the Francoist 'families'. By having Juan Carlos at his side each year in the annual victory parade, Franco made a gesture to the monarchists: otherwise, in order not to alienate those Falangists who favoured Alfonso, he stood studiously aloof.[21]

Franco's decline was reflected in an increasingly noticeable withdrawal from political tasks. On 2 June 1969, he had alarmed Silva Muñoz, his Minister of Public Works, on a journey from Madrid to Córdoba to inaugurate new projects. During the long car drive, Franco drooped apparently unconscious onto Silva's shoulder. Although it was just the effect of the Caudillo's medication (probably dopomine), Silva was scared that the end was imminent.[22] When in El Pardo, his daily routine was still to rise at eight, and undergo a session of massage and physiotherapy with his doctor Vicente Gil. He would then breakfast with the family and browse superficially through the newspapers. He had until nearly seventy played a few games of tennis, often with Vicente Gil, or else gone riding in the woods near the palace. Those activities were no longer feasible. On Fridays, there would be still be meetings of either the council of ministers or the economic committee. By the end of the 1960s, they were drastically reduced to morning sessions only and then only fortnightly. His opening surveys of international and domestic events were also a thing of the past and he now rarely broke his silence during the proceedings.[23]

On Tuesdays, he held military audiences, on Wednesdays, civilian audiences. He would receive his visitors standing up. Once seated with the light behind him, it was difficult for them to know where he was looking. He had a device for judging the importance of what they had to say. He would interrupt with a quiet but irrelevant question. If the interlocutor let himself be side-tracked by the question and did not return to his ostensible business, Franco would conclude that it had merely been a pretext to secure an audience and indulge his vanity.[24] On audience days, he would lunch very late, often as late as five or six in the afternoon. Normally, he would lunch at 2.00 p.m. On Sundays, if not away on one of the still frequent hunting parties, he would start the day with mass and then go fishing at La Granja or shooting in the grounds of El Pardo.

Franco was reckoned by the family to be a solid trencherman (*un gran hambrón*, in the words of his grandson Francisco). In these last years, whenever his weight went over ninety kilos, (fourteen stones) Vicente Gil would put him on a strict diet and nag him constantly. Franco responded by calling him *gruñón* (misery guts) and eating clandestine snacks.* After lunch, he would take a walk, paint, play a round of golf or go to inspect his farm at Valdefuentes. He might then return to his office for three or four hours. In the evenings, he would watch television or else play cards, *mus* and *tresillo*, with his military friends. After a late, light dinner, he would say the rosary with Doña Carmen then fall asleep reading, usually biographies of great men or magazines.[25]

Franco had long invested in his sporting activities the dogged determination which had characterized his political triumphs. The press continued to use his hunting and fishing exploits as signs of his endless vitality, but the trembling hands must have had an effect on his aim and the tendency to doze off must have diminished his concentration during long sea chases. Whatever his prowess, he remained assiduous as a hunter and a fisherman. He certainly spent even more time fishing, especially at Easter in the Asturian rivers Narcea, Sella and Cares, using either La

* To the fury of Dr Gil, on hunting trips, those who wanted to ingratiate themselves with Franco would offer him delicious but fattening tit-bits.

Piniella or the Hotel Pelayo in Covadonga as his base. In the summer of 1971, fishing at Puentedeume between La Coruña and El Ferrol, he was alleged to have caught 196 *reos*, a small river salmon. He would happily fish in the most inclement weather and if any of his entourage ever complained he would simply say, 'Well, I'm not cold.' He still disappeared for several long hunting trips and was photographed playing golf during the summer at the La Zapateira club in La Coruña – allegedly staying on the green for hours on end without a break irrespective of the weather.[26]

Such dedication to pleasure suggests that the Caudillo was oblivious to the fact that, within the regime, positions were being taken up for the aftermath of his demise. Within the *Movimiento*, various options appeared ranging from the fascist extreme right of Blas Piñar's *Fuerza Nueva* to progressive *aperturistas* like Fraga. In July 1970, a tearful Doña Carmen begged 'Pedrolo' Nieto Antúnez to speak to Franco about the drift of events. Nieto told Fraga that he found Franco more solitary and preoccupied than ever before. Castiella spoke of him only as '*el cansado*' (the tired one).[27] His supporters were deeply preoccupied by the intensifying agitation in the universities and the labour movement and, even more so, by ETA. Its terrorist activities in the late 1960s shattered the regime's myth of invulnerability. Extreme rightists in the Army, the so-called *generales azules* convinced Franco to reply with a show trial of sixteen Basque prisoners, including two priests. That their narrowly vengeful views prevailed was a symptom of the decadence of the regime, Franco's declining judgement and the lack of political sensibility of Carrero Blanco.

The repercussions started even before the trial had begun and they affected Franco directly. On 18 September 1970, while the Caudillo was presiding over the world *jai alai** championships at the San Sebastián *Frontón* (court), Joseba Elósegi, a member of the Partido Nacionalista Vasco set fire to himself and jumped from the wall of the frontón in front of Franco shouting *Gora Euzkadi askatasuna* (Long live free Euskadi).[28] He was carried away badly burned while an unperturbed Franco continued to watch the game. Elósegi had been in command of the only military unit in Guernica on the day that it was bombed on 26 April 1937. He thus drew international attention to the Basque cause and its continued persecution by the Franco dictatorship. He wrote in his diary on 28 August 'I do not intend to eliminate Franco, I want only for him to feel on his own flesh the fire that destroyed Guernica.'[29] The Elósegui incident

* A game not unlike squash played with wooden balls and scoops.

severely undermined the efforts of the globe-trotting new Foreign Minister, López Bravo, to modernize the regime's image.

One of the problems inherited by López Bravo was the damage done to the relationship with the United States by Castiella's unsuccessful attempt to hold Washington to ransom over the renewal of the American bases agreement. Castiella's revised offer of a one-year extension in return for $50,000,000 in military aid and $25,000,000 in credits for arms purchases remained to be resolved. At the end of September 1970, Richard Nixon landed in Madrid en route from visiting another aged autocrat, Marshal Tito. He was accompanied by Henry Kissinger, head of the National Security Council, who found Franco's Spain 'as if suspended, waiting for a life to end so that it could rejoin European history'. The United States was still interested in Spain strategically and was anxious to see a moderate evolution after Franco's death. US policy was to maintain a working relationship with his regime while extending contacts within the moderate opposition. There was discreet American pressure to persuade him to hand over to Juan Carlos before incapacity deprived him of control of the transition. Fortunately for Franco, with the Middle East in turmoil and other American bases in the Mediterranean in jeopardy, Washington saw the maintenance of bases in Spain as the highest priority in its Spanish policy.

In personal terms, Nixon was anxious that the crowds which greeted him should at least equal and ideally exceed those for Eisenhower eleven years before. The popular reception was warm as Franco and Nixon drove from Barajas into Madrid flanked by mounted lancers. Franco cleverly flattered Nixon by telling him how difficult it was to get the press to accept a figure as plausible once the crowd exceeded several hundred thousand. When Nixon and Kissinger met Franco for what were meant to be 'substantive talks', they knew that any allusion to the post-Franco transition would be disastrous. However, they were taken aback to discover that the seventy-eight year-old dictator had been exhausted by the motorcade and was starting to doze off even as the President began to talk. Soon the Caudillo and Kissinger were snoozing gently while Nixon talked to López Bravo.[30]

Between the meeting with Nixon in September and the trials of the Basque militants two months later, Franco had travelled thirty years back in time. The trials began in December at Burgos, headquarters of the military region in which the Basque Country lay. Even before they had started, Franco's brother Nicolás had written to him on 6 November about the death sentences demanded by the prosecutors: 'Dear Paco,

Don't sign these sentences. It's not in your interests. I'm telling you because I love you. You are a good Christian and afterwards you will regret it. We're getting old. Listen to my advice, you know how much I love you.'[31] After violent clashes between the police and anti-regime demonstrators in Madrid, Barcelona, Bilbao, Oviedo, Seville and Pamplona, on the morning of 14 December, four Captain-Generals visited Franco to tell him that the Army wanted more energetic government.* The Caudillo then held an emergency cabinet meeting on that afternoon at which the Minister of the Interior, General Garicano Goñi, and the three military ministers called for the suspension of *habeas corpus*. Franco went along with them.[32]

On the morning of 17 December, the press and the radio called on people to go to the Plaza de Oriente in Madrid. Rural labourers were bussed in from all over Old and New Castile. Large crowds gathered outside the Palacio de Oriente shouting for Franco. 'Pedrolo', Pacón and Vicente Gil, among others, contacted El Pardo and urged the Caudillo to come. It was a mark of Franco's waning strength that he could be manipulated by ultra elements keen to see the regime return to its hard-line origins. Not having expected to attend, a bewildered Franco and Doña Carmen immediately set out for Madrid in civilian clothes. While Doña Carmen gave the fascist salute, he acknowledged the chants of the crowd by raising both hands. Many of the banners attacked the 'weakness' of the *monocolor* cabinet. Dr Gil, Franco's manically devoted personal physician, an ultra with a passion for boxing, violently berated the Minister of Information, Alfredo Sánchez Bella, for failing to give the fascist salute. This provoked only a mild reprimand from the Caudillo.[33]

The trials ended with three of the ETA militants found guilty of two capital charges each and given two death sentences each. On the evening before the cabinet meeting on 30 December 1970 at which the sentences were to be reviewed, López Bravo saw Franco and attempted to make him see the negative impact on Spain's international image if they were confirmed. The Caudillo listened to him for an hour, then said 'López Bravo, you have not convinced me.' López Rodó and Carrero Blanco were agreed that it would be politically disastrous for Franco to approve the death sentences. On the following day, at the meeting, López Bravo spoke first and at length in favour of commutation of the sentences. He was followed by the other ministers not all of whom favoured pardons.

* The delegation consisted of General Joaquín Fernández de Córdoba of Madrid, Tomás García Rebull of Burgos, Alfonso Pérez Viñeta of Barcelona and Manuel Chamorro of Seville.

Franco had arrived at the meeting convinced that the death penalties should be confirmed but finally permitted himself to be persuaded. After listening to the cabinet, he said nothing and, only after a meeting of the *Consejo del Reino** which also recommended clemency, did he announced his decision to commute the death penalties to prison sentences.[34]

In his end-of-year message televised on 30 December 1970, Franco explained the international protests against the Burgos trials in familiar terms: 'The peace and the order that we have enjoyed for more than thirty years awaken the hatred of those forces which were always the enemy of the prosperity of our people.' He described the pardons as symptoms of the regime's strength. He ended with a reassuring declaration: 'The firmness and the strength of my spirit will not let you down while God gives me life to continue ruling over the destinies of our *Patria*.'[35]

The Burgos trials were a disaster for the regime in that they dramatically altered the balance of forces in Spain. The regime's clumsiness had united the opposition as never before, the Church was deeply critical and the more progressive Francoists were beginning to abandon what they saw as a sinking ship. Under pressure, both Franco and Carrero Blanco inclined to the *inmovilista* cause which boded ill for the technocrats. The pardons may have been manifestations of strength but to have held the trials at all was a symptom of Franco's loosening grip.

In early January 1971, the Spanish state news agency, EFE, announced that Franco, some of his ministers and members of his family had taken part in a weekend hunting trip in Ciudad Real at which nearly three thousand partridges were shot.[36] This post-Burgos event was almost certainly stage-managed to give an impression of serenity and optimism in the face of foreign opprobrium. On this as on other hunting trips, just as at El Pardo, Franco was subjected by the family clique to criticisms of the technocrats.[37] Franco's interest in hunting had not diminished along with his physical strength. Indeed, the persistent publicity given to the trips makes it all the more difficult to isolate exact landmarks in the deterioration of his health. However, by the early 1970s, the symptoms of Parkinson's disease, unsteady hands, stiff movements, vacant expression, were becoming more and more unmistakable. In February 1971, General Vernon A. Walters, deputy chief of the CIA, visited Madrid on behalf of President Nixon. His mission was to ask the Caudillo what would happen in Spain after his death. Franco told Walters that the

---

* The Council of the Kingdom was made up of the *Movimiento*'s 'great and good'.

succession to Juan Carlos would take place without any disorder and that 'the Army would never let things get out of hand'. Walters found Franco 'old and weak. His left hand trembled at times so violently that he would cover it with his other hand. At times he appeared far away and at others he came right to the point.'[38]

At the last week of January 1971, Juan Carlos and Princess Sofía visited Washington. The Prince gave some press interviews about the future which inclined American policy-makers to support him. He was quoted as saying 'I believe that the people want more freedom. It is all a question of knowing how fast.' On his return to Spain, Juan Carlos, expecting Franco to be furious, hastened to El Pardo to see him. How much Franco assumed Juan Carlos to be of his way of thinking was revealed in his unexpected reaction. To the Prince's surprise, Franco spoke in terms which recalled his own double-dealing with the western powers during the years of international ostracism: 'There are things which you can and must say outside Spain and things which you must not say inside Spain.'[39]

In so far as Franco inhabited the political world of his own regime, he was increasingly trapped in a narrow space between the grey technocrats and the ultra-right of the *bunker* which, with increasing frequency, openly denounced the 'weakness' of the technocrats. Franco's isolation was symbolized by a declaration of the Joint Assembly of Bishops and Priests on 13 September 1971. Chaired by the Primate, Cardinal Vicente Enrique y Tarancón, the Assembly rejected Franco's triumphalist division of Spain into victors and vanquished. The declaration begged forgiveness of the Spanish people for the clergy's failure to be 'true ministers of reconciliation'.[40] It was hardly surprising that Franco should feel beleaguered. He responded by harking back nostalgically to the triumphs of the 1930s and 1940s. He became more susceptible to the whispers of the El Pardo clique and the hard-line Falangists, among whom Girón maintained easy access to the Caudillo.

The two dominant influences on the Caudillo, the technocrats and the *bunker*, could be seen at work on 1 October 1971. On that day, to celebrate the thirty-fifth anniversary of his elevation to power, Franco announced a pardon which would apply to most of those on trial for the Matesa affair. He said: 'If, for political reasons, I have had to pardon the ETA assassins, why can I not do the same with good collaborators who have simply made a mistake or been a bit negligent?'[41] He also addressed a multitude of the faithful from the balcony of the Palacio de Oriente. The event had been in preparation for some time, with large placards calling people to attend the '35 years homage to Franco . . . this time, just for the hell of it' (*esta vez porque sí*) or 'one day for an entire life' (*un día por toda una vida*). He was being paraded by a government which had lost its way but, with the blue-shirted fanatics bussed in from all over Spain chanting 'Franco! Franco! Franco!', it was just like old times. Thousands of soldiers attended in civilian clothes. It was rumoured that the *Movimiento* had issued cassocks to Falangist militants to make it appear that many priests supported the Caudillo against the Vatican. Unaware that the *Movimiento* had manufactured the demonstration, Franco was deeply moved and his delight glimmered throughout a speech in which the old clichés were mixed with conviction that the future was secure. The frenetic reception revitalized him and confirmed his decision to stay on. *Arriba* commented, without apparent irony, 'The living, and also the dead, shouted and cheered with us.'[42]

On the occasion of the opening of the Cortes on 19 November 1971, he referred to the demonstration of support in the Plaza de Oriente. The 'clamorous confirmation of the people' was taken as full endorsement of his thirty-five years in power. His complacent tone suggested that he had not begun to ask why his regime needed ultra-rightist terror squads to hold back the rising tide of labour militancy and clerical opposition.[43] While admitting the possibility of the *contraste de pareceres* (contrast of opinions) through associations and *hermandades* (fraternities) kept strictly within the *Movimiento*, he slammed the door firmly on anything that might lead to political parties. He explained recent strikes in the state-owned SEAT car plant in terms of the ever present international siege.[44]

The stench of decadence at Franco's court was intensified in 1972. His brother, Nicolás, was implicated in one of the greatest financial scandals of the dictatorship, the so-called *aceite de Redondela* affair. Four million kilos of olive oil, held as a State reserve stock in tanks belonging to a fat and edible oil refining company, REACE (*Refinerías del Noroeste de Aceites y Grasas S.A.*), was found to be missing. Not anticipating that the stocks would be called upon, the company had been speculating with the oil. Nicolás Franco was a major shareholder in the company. In the course of the judicial investigation into the affair, six people met violent deaths. A major cover-up was mounted to silence the links between Nicolás and those accused of fraud. The Caudillo was deeply preoccupied by the affair, first uncharacteristically bad-tempered then sinking into a depressive silence.[45]

The decadence was revealed in a different way when, on 18 March 1972, his eldest granddaughter, María del Carmen Martínez Bordiu y Franco married Alfonso de Borbón-Dampierre, the eldest son of Don

Jaime and first cousin of Juan Carlos. The link with a direct descendant of Alfonso XIII inflamed the ambitions of both the Marqués de Villaverde and Doña Carmen who issued wedding invitations which referred to *Su Alteza Real el Príncipe Alfonso* (His Royal Highness Prince Alfonso), a title to which he had no right. The wedding reception was even more spectacularly lavish than that of the bride's parents in 1952. Two thousand guests assembled at El Pardo and witnessed Cristóbal Martínez-Bordiu once more indulge his taste for Ruritanian uniforms. Imelda Marcos, a friend of the bride's parents, was one of the few foreigners invited who chose to attend. At the behest of the family, the Caudillo took the place of the bride's father as *padrino*, in order to enhance the status of the occasion. He cut a pathetic figure, his eyes watering, his mouth open and his hands trembling. As soon as the newly-weds had returned from their honeymoon, Doña Carmen insisted that her granddaughter be treated as if she were a princess, curtseying formally when María Carmen entered a room and issuing instructions to guests and servants that she be called 'Your Highness'.*[46]

With the support of the El Pardo clique, Alfonso had tried to get Franco to give him a title which matched that of Juan Carlos. The idea was for him to be Príncipe de Borbón and entitled to be called Royal Highness. Don Juan de Borbón, as the head of the royal family, opposed this on the reasonable grounds that only the first son of the King, the Príncipe de Asturias, had the right to be called a prince. To minimize the acrimony, he agreed instead that Alfonso be given the title of Duque de Cádiz. The El Pardo clique easily convinced Franco that Don Juan's objections were aimed at his family: he told the Minister of Justice, Oriol, 'Don Alfonso had the title of prince and now, because he is marrying my granddaughter, they want to take it away from him.' In the end, Franco accepted the legal reports on the issue prepared by the Ministry of Justice. Inevitably, relations between the Villaverde family and Estoril were severely soured in a way which had negative repercussions on Franco's attitude to Juan Carlos. At the time, Franco was motivated by no more than the desire to see his granddaughter become a princess. However, Alfonso de Borbón's presence in El Pardo inflamed the ambitions of the Villaverde clan, just as his ultra-rightist views reinforced their reactionary influence.[47]

Franco's interminable hours of work when not on long holidays and

---

* After a son, Francisco, was born to the couple, Doña Carmen was often heard to ask the household staff 'has Sire had his bottle? (*¿Le han dado ya el biberón al señor?*).

FRANCO

hunting and fishing trips had given way to hours in front of the television. The Caudillo needed substantial siestas. In cabinet meetings and important audiences, he said virtually nothing, and often dozed off. In the family rooms in El Pardo, he was morose and and showed no interest in anything but the television. Vicente Gil found him one day lost to the world reading aloud the label of a bottle of after-shave lotion. When awake, in audiences, his hands were seen to tremble uncontrollably and his sight was deteriorating. For the 1972 victory parade, on 20 May, a hidden shooting stick was rigged up for him to sit on. His own health worried him less than that of his wife and he told Juan Carlos that Doña Carmen had been diagnosed as suffering an incurable heart disease.

The Caudillo himself suffered from fungal infections in the mouth and from related pains in the leg which prevented him, in the autumn of 1972, going on his annual hunting trip to the south. Shortly afterwards, the annual 1 October reception was shortened because he was unable to stand for a long period. The medication for Parkinson's disease caused him to become increasingly indecisive. Vicente Gil wanted him to cut out hunting trips, especially those in rough country. However, it was impossible to resist the political pressure to accept invitations.[48]

By this time family gatherings wider than Franco himself, Doña Carmen and the Villaverdes took on an air of protocol and were awkward occasions. Even in the close family circle, difficult subjects, the political mistakes or the corruption of Franco's close subordinates, were taboo and, if raised, irritably dismissed as inventions. Whenever Francisco's loquacious sister Pilar mentioned cases of corruption she was reprimanded for believing rumours. Conversation became animated only when it turned to 'traitors' or 'the ungrateful' which tended to mean the technocrats.[49] The immediate family clique, increasingly hostile to the Carrero Blanco/López Rodó/Juan Carlos option, pressed the Caudillo 'to sort things out'. An anxious Doña Carmen complained to Vicente Gil that her friends constantly asked her what would become of them? 'And Paco doesn't want to do anything. They all think I have influence over him and I have none.' As Franco grew more infirm, Doña Carmen became more outspoken. She regarded the Minister of the Interior, Garicano Goñi, and the Foreign Minister, López Bravo, as weak and disloyal, telling Carrero Blanco in February 1973 that something ought to be done about them.[50]

On 4 December 1972, Franco reached his eightieth birthday. His ADC revealed that his legs were swollen and that to get him away from the television to play a round of golf or just to take the air was increasingly difficult.[51] The recording of his end-of-year speech had to be interrupted several times for him to rest. Even so, when it was broadcast on 30 December 1972, he appeared decrepit and noticeably older than a year previously. In a voice which broke and occasionally faded into inaudibility, he assured the viewers that he would hang on indefinitely: 'Here you will have me, with the same firmness as many years ago, for as long as God wants to let me go on serving the destinies of the *Patria* with efficacy.'[52]

Such declarations seemed unreal given the symptoms of Parkinson's disease which could no longer be concealed from visitors.[53] The fears provoked by Franco's health within the inner circles of the regime were exacerbated by an atmosphere of ever fiercer social and political tensions. In April 1973, a striker was killed by the police near Barcelona. Carrero Blanco had lost confidence in the technocrats and was secretly encouraging the activities of the ultra-rightist terror squads of *Fuerza Nueva*. Franco himself was uneasy that the government was not doing enough to combat the activities of ETA.[54] The belief in El Pardo that events were slipping out of control came to a head on 1 May 1973 when a policeman was stabbed to death during a May Day demonstration. At the funeral of the murdered officer, 'ultra' policemen and Falangist war veterans howled for repressive measures. There were mass arrests of leftists, and Garicano Goñi, disappointed at the lack of will for reform and alarmed at the growing influence of extreme rightists, resigned on 2 May. The El Pardo clique finally convinced Franco that the cabinet had failed in the primordial task of maintaining public order. On 3 May, he again told an unwilling Carrero Blanco that he was going to be made president of the council of ministers and should start drawing up his cabinet.

At the beginning of June, the decision was formalized and Carrero Blanco's cabinet list approved. It went some way to reversing the technocrat dominance of the cabinet of 1969. The Caudillo's favourite López Bravo was dropped, presumably to please Doña Carmen. López Rodó lost his crucial influence in domestic policy and was exiled to the Ministry of Foreign Affairs where he was used to put a moderate veneer on an essentially reactionary cabinet. As vice-president and Minister-Secretary General of the *Movimiento*, Carrero Blanco chose Torcuato Fernández Miranda. Two hard-line Falangist followers of Girón, José Utrera Molina at Housing and Francisco Ruiz-Jarabo at Justice, reflected the influence of the El Pardo clique as did Julio Rodríguez, the new Minister of Education.*

* As Rector of the Universidad Autónoma de Madrid, Julio Rodríguez had been notorious for his violent methods which even included personally joining in police charges against left-wing students in the University.

It was a sign both of his faith in Carrero Blanco's reactionary instincts and of his own waning strength that the Caudillo accepted the list with only one change. Carrero had wanted Fernando de Liñán as Minister of the Interior. Having been convinced by the El Pardo clique that the government was too soft, Franco insisted on the inclusion of Don Camilo's one-time Director-General of Security, Carlos Arias Navarro, a tough law-and-order man who had started his career as a prosecutor during the repression of Málaga in 1937. He was *Alcalde* of Madrid and a favourite of Doña Carmen.* Carrero Blanco instead made Liñán Minister of Information, a post which had been originally intended for the man of the future, Adolfo Suárez.

It was a deeply depressing moment for all those inside and outside the regime with hopes for progressive change. Nevertheless, the new cabinet also constituted Franco's rejection of the more ambitious aspirations of the group of ultra-Falangists who hovered around El Pardo. He would allow Fernández Miranda to begin exploring the possibility of 'political associations' as a way of permitting currents of opinion within the *Movimiento* while still holding the line against political parties. He had given power to Carrero Blanco for five years. Carrero was seventy years old, had neither popular nor military support. His authority depended entirely on the continued existence of his master. Madrid wits called it 'the funeral cabinet'.[55]

Fraga, en route to London as Ambassador, went to take his leave of Franco and got the overwhelming impression that he was 'beyond the physical and mental demands of his great responsibility'.[56] If Franco died first, it is difficult to imagine that Carrero would have been able to rule for long thereafter since he lacked the will, the authority and the ideas. As it was, by the end of the year, his cabinet was adrift in a sea of industrial unrest in Catalonia, Asturias and the Basque Country which had been provoked by austerity measures taken to stem inflation. With the first energy crisis brewing and Spain heavily dependent on imported energy, the technocrat strategy of buying off political discontent with rising prosperity was doomed. Carrero Blanco's only response was intensified repression.

On Thursday 20 December 1973, the *proceso 1001*, the show trial of ten leaders of the underground trade union, Comisiones Obreras, was

---

* When Arias went to thank the new president, Carrero said drily 'you don't need to thank me. I took no part in your appointment. You know the Generalísimo's handwriting. You can see in this list, among the amendments and crossings-out, your name written by the Caudillo.'

due to commence. It was to be a public demonstration of the regime's determination to crush the clandestine unions. Shortly before 9.30 a.m., a squad of ETA activists assassinated Carrero Blanco by detonating an explosive charge in the street under his car as he returned from daily mass.[57] For two hours after the first news, Franco was left believing that he had died in a gas explosion. Torcuato Fernández Miranda, the vice-president, telephoned El Pardo at midday and informed Franco that it had been a political assassination. Franco was reluctant to believe that the explosion had not been a coincidence. Fernández Miranda then went to El Pardo and was received by Franco, ill with flu, wearing a dressing gown. His immediate reaction was to totter several steps murmuring over and over 'these things happen'. His only instructions were that the cabinet should maintain its serenity. He did not make an appearance to pay his respects to the corpse at the *capilla ardiente* (improvised chapel of repose) set up in the *Presidencia del Gobierno*. He seemed completely over-whelmed.[58] He was unable to eat and shut himself in his study.[59]

The relationship between Franco and Carrero Blanco had been immensely close yet somehow very distant. The deeply respectful Carrero had always called Franco *mi general* and used the formal *Usted* mode of address. He had referred to him in the third person either as *el Caudillo* or *el Generalísimo*. Franco for his part had always called him Carrero.[60] Yet, the Admiral had become his *alter ego* and even for as cold and withdrawn a character as Franco, the loss cut deep. His plans for with-drawing from political responsibilities were shattered. Infirm, he was more vulnerable now to the El Pardo clique than he had been six months earlier.

Fernández Miranda automatically took over as interim prime minister. However, the Director-General of the Civil Guard, Carlos Iniesta, issued an order for his men to repress subversives and demonstrators energeti-cally 'without restricting in any way the use of firearms'. He explicitly ordered them to go beyond their rural jurisdiction and keep order in urban centres. It was a gross abuse of his authority. Cooler heads pre-vailed. After taking advice from the Chief of the General Staff, Manuel Díez Alegría, a triumvirate consisting of Fernández Miranda, the Minister of the Interior, Arias Navarro, and the senior military minister, Admiral Gabriel Pita da Veiga, acted to prevent a bloodbath. Within less than an hour, Iniesta was obliged to withdraw his telegram.[61]

On the day after the assassination, tempers ran high: at a mass for Carrero, the liberal Cardinal-Archbishop of Madrid, Vicente Enrique y Tarancón, was jostled and insulted by extreme right-wingers. Franco received short visits from Gerald Ford and Marcelo Caetano. He then chaired a cabinet meeting. After shaking hands with each minister, he began to speak of 'the horrendous crime which has taken the life of our president' then broke down in tears and stared at Carrero's empty chair. He quickly composed himself and opened the meeting proper. The only business transacted was the posthumous conferment on the murdered prime minister of the title Duque de Carrero Blanco.

At 8.00 a.m. on the next day, 22 December, a red-eyed Franco told one of his aides, the naval Captain Antonio Urcelay, that he had not been able to sleep during the previous night and commented desolately: 'Urcelay, they have cut my last link with the world.' Franco then attended another funeral mass at the Church of San Francisco el Grande. He cried and moaned quietly throughout the ceremony. At the end, he walked over to Carmen Pichot, Carrero Blanco's wife, and weeping profusely, took her hand. It had been hoped that he would appear on television but it was decided that it would counter-productive for him to be seen so depressed. During the mass, Carrero's deranged Minister of Education, Julio Rodríguez, ostentatiously turned his back on Cardinal Tarancón and, immediately afterwards, drove to the Dirección General de Seguridad to offer to lead a hit-squad to go into France to hunt down and kill the Admiral's assassins.[62]

When the first shock passed, it was clear to many in the regime that Carrero Blanco's death reopened many options, both reactionary and progressive. ETA had chosen him as a target precisely because of his pivotal role in the Caudillo's plans for the continuity of the regime. Franco was again subjected to pressure from his immediate circle of courtiers. It was a measure of his ever greater physical and mental decline that the final outcome was more of their making than of his. When the search began for a successor to Carrero Blanco, the views of Doña Carmen and the Marqués de Villaverde played a major part in blocking the promotion of Fernández Miranda, because of his known commitment to the cause of Juan Carlos. They were able to persuade Franco that the cession of power to Carrero Blanco had been a mistake because it had opened the way to Juan Carlos who harboured secret liberal plans.[63] It is likely that during this crisis, the Parkinson's disease was so advanced as to leave Franco somewhat distanced from events.*

---

* His sister Pilar spoke with him about lists of possible successors to Carrero being discussed in political circles. When he asked what names were on them, her son read them out. His only reaction, when pressed, was to say 'one or two ring a bell' (*alguno me suena*) – Baón, *La cara humana*, p. 145.

Franco's inclination, once the automatic substitution of Carrero by Fernández Miranda had been discounted, was to appoint his old friend Admiral Pedro Nieto Antúnez. In theory, he could not just appoint anyone but had to choose the prime minister from a *terna* (a list of three names) to be presented to him by the *Consejo del Reino*. But there was no possibility that the *Consejo* would not include in the *terna* the name that he intended to pick. 'Pedrolo' was an apparently safe choice, a personal friend and fishing and card-playing companion, a senior military Francoist, not so selfless as Carrero Blanco but the nearest off-the-peg substitute. On 22 December, Franco told him that he would be president and he reluctantly accepted the nomination as an order, immediately contacting potential collaborators including Fraga, whom he invited to be vice-president, and López Bravo, whom he invited to be Minister of Foreign Affairs. Unaware of this and of the fact that Franco regarded him as a treacherous liberal, Fernández Miranda was briefly confident that he would be president. However, when he broached the subject on 24 December, the Caudillo, lucid once more, asked brutally 'are you insinuating that I include you in the *terna* of the *Consejo del Reino*?'

On 26 December, Franco went through the formal motions of asking Juan Carlos for his opinion on the next prime minister and the Prince suggested Fraga or Fernández Miranda, but he had no intention of paying any heed to the views of the Prince. Later in the day, he had a long meeting with the President of the *Consejo del Reino* and of the Cortes, the Falangist Alejandro Rodríguez Valcárcel, and went through the charade of discussing twenty-two names. They reduced them to twelve ranging from Arrese and Girón on the Falangist right to López Rodó. On the morning of 27 December, Franco told Rodríguez Valcárcel that he had reduced the list overnight to five: Fernández Miranda, Fraga, Nieto Antúnez, Arias Navarro and the relatively liberal Minister of Finance, Antonio Barrera de Irimo. In the evening, he told Rodríguez Valcárcel that Nieto Antúnez was his choice. For the regime 'ultras' and the El Pardo clique, this was not unalloyed good news. Born in 1898, 'Pedrolo' was only six years younger than the Caudillo himself and five years older than Carrero Blanco. He provided no guarantee that the problem of a replacement would not arise again in the near future. Moreover, although sufficiently Francoist in his credentials, he was likely to rely on the now openly reformist Fraga in the same way as Carrero Blanco had relied on López Rodó. All in all, the El Pardo clique had to stop his nomination.

Accordingly, on the night of 27 December, Franco was subjected to intense pressure to change his mind in favour of the tough Arias Navarro.

A friend of the Franco family, particularly of Doña Carmen, Arias was one of Franco's card cronies and a friend of Vicente Gil. Renowned for his hard line on public order matters, he was regarded as the natural heir to Alonso Vega whose protégé he had been.[64] The pressure came from Doña Carmen and from Dr Gil. They were backed up by General José Ramón Gavilán, the second-in-command of Franco's military household, and his adjutant, Captain Urcelay. The influence of Girón could be perceived in the arguments put before Franco. Doña Carmen is alleged to have opened the operation by saying to her husband: 'They are going to kill us all like Carrero Blanco. We need a hard president. It has to be Arias. There is no one else.' After the overnight pressure, on the morning of 28 December, Franco announced to Rodríguez Valcárcel that Arias Navarro would be his prime minister, saying 'Pedrolo is nearly as old as I and has the same problems with his memory.' Later on the same day, the *Consejo del Reino* 'elected' the *terna* which duly included Arias, as well as José Solís and José García Hernández, a dour *Movimiento* apparatchik.[65]

In his end-of-year message on 30 December 1973, Franco paid tribute to Carrero Blanco in a rather off-hand manner. Dismissing the assassination as the work of a tiny minority controlled by foreigners, he took pride in the functioning of the Francoist institutions during the crisis. Calling for unity, he offered to go on indefinitely. 'After thirty-seven years at the head of the State, here you have me with you, with the same vocation of service to the *Patria* that I always had, conscious that authority can never be a privilege, but rather a duty which demands fidelity and sacrifice.' The words '*no hay mal que por bien no venga*' (it's an ill wind that blows nobody any good), were added in his own handwriting to the typed text of the message. This was assumed in the inner circles of the regime to be an acknowledgement that Franco now saw the Carrero Blanco period as a mistake.[66] The choice of a replacement for Carrero Blanco, taken under pressure, was Franco's last major political decision. Thereafter, the transition to democracy had begun and he would be a spectator of its political struggles. According to López Rodó, 'Franco, without Carrero, was another Franco'.[67]

Arias's government was a curious rag-bag of hard-liners and progressives. The liberal wing included Antonio Barrera de Irimo, seen as representative of the more dynamic elements of Spanish business, and followers of Fraga, like Antonio Carro, as Minister of the Presidencia, and Pío Cabanillas, as Minister of Information. The most reactionary elements were two fanatical Falangists, survivors from Carrero's cabinet and supporters of Girón: José Utrera Molina as Minister-Secretary of the *Movimiento* and Francisco Ruiz Jarabo as Minister of Justice. José García Hernández, the Minister of the Interior, and Vice-President in charge of internal security, was, like Arias himself, a one-time assistant to Alonso Vega.[68] When Arias proposed Fraga himself as his Minister of Foreign Affairs he was vetoed by Franco who wanted López Rodó to continue. Arias was adamant that he did not want him and they compromised on the diplomat Pedro Cortina Mauri, another El Pardo favourite.[69]

Arias had wanted Fernando Herrero Tejedor to head the *Movimiento* but, such was Girón's influence in El Pardo, that he was obliged by Franco to appoint Utrera Molina. Utrera, a doctrinaire Falangist, told Arias that he had no intention of letting the *Movimiento* be the equivalent of a herd of political sheep. At his ceremonial induction into the post, he was surrounded by a galaxy of *bunker* celebrities – Arrese, Fernández Cuesta, Solís and Girón from the Falange and *generales azules* like Iniesta Cano and García Rebull.[70] Arias did not have the courtesy to consult with Juan Carlos over his proposed new cabinet. The El Pardo clique approved of the marginalization of the Prince, but Arias Navarro would not be what they hoped for. The structural problems of the regime would oblige an uncomprehending Arias to go much much further than Carrero Blanco had ever done in the direction of change. Franco did not understand Arias and missed the symbiotic relationship with Carrero.

Arias's instincts were authoritarian and repressive but he was sufficiently vain to be concerned with his public face. And the more liberal or *aperturista* members of his team, particularly Pío Cabanillas, the Minister of Information, persuaded him that to defend the essences of Francoism it was necessary at least to change its image. Accordingly, he lent himself to reading out a declaration of progressive intent on 12 February 1974. It included the statement that responsibility for political innovation could no longer lie only on the shoulders of the Caudillo.* Arias seems to have appreciated neither the full implications of the speech nor the problems that what came quickly to be called the 'spirit of 12 February' would bring him with the *bunker*.[71] It only became gradually apparent that the *bunker* would wheel out the Caudillo in an effort to block any progressive initiative. The process was facilitated by the fact that, as he

---

* The text was prepared in Antonio Carro's Ministry of the Presidencia by two of his subordinates, Gabriel Cisneros and Luis Jaúdenes, members of the Catholic reformist group known collectively as *Tácito*. Carro introduced many of the *Tácito* group into the government as under-secretaries in various ministries.

aged and regressed, Franco was highly susceptible to accusations that some of Arias's ministers were freemasons.

The minister with whom Franco had closest contact other than Arias was Utrera with whom he soon established a paternal relationship.* In January 1974, when Utrera told a delighted Caudillo that he intended to undertake the ideological rearmament of the *Movimiento*, he replied 'we have committed the error of lowering our guard' but reassured his Minister that there was time to make good the mistake. To Arias, Franco gave no indication of what he wanted to happen but made violent interventions against what he did not. After Arias's speech, the Caudillo asked Utrera to explain to him 'the spirit of 12 February'. Thoroughly alarmed by that explanation, Franco said 'if the regime allows its doctrinal essence to be attacked and its defenders fail to defend what is fundamental, we will be driven to think that some are contemplating a cowardly suicide'.[72]

With prices rocketing in the wake of the energy crisis, worker militancy intensified in early 1974. Arias oscillated between some toleration for the moderate opposition and harsh repression of labour and student unrest. His chances of being able to adjust the regime's creaking structures to the changes in Spanish society were diminished by the louring presence of the Caudillo. Paradoxically, Franco's first intervention was to restrain Arias's reactionary instincts. The prime minister was on the verge of expelling the Bishop of Bilbao, Monsignor Antonio Añoveros, from Spain. Añoveros's crime was to have permitted the publication on 24 February of homilies in defence of ethnic minorities. Unwilling to risk the excommunication of his prime minister, Franco obliged Arias to back down.[73] However, Franco refused to commute the death sentences passed against a Catalan anarchist Salvador Puig Antich and a common criminal, Heinz Chez. Despite an international outcry reminiscent of the Burgos and Grimau trials, both were executed by *garrote vil* (strangulation) on 2 March 1974.

The reactionary instincts of the El Pardo clique were provoked by the fall of the dictatorship in Portugal. On 28 April 1974, just three days after the cataclysm in Portugal, Girón launched a broadside against Arias in *Arriba*. Franco made it clear to Utrera that he was anything but displeased by this so-called *Gironazo*.[74] As part of the same operation, the military *bunker* set about establishing control of the crucial sectors of the Army. While Girón and other civilian ultras attacked Arias and his cabi-

---

* Their contact – which bypassed the prime minister – could be justified to Arias by the fact that Utrera, as Minister Secretary, was Vice-President of the *Consejo Nacional de FET y de las JONS* of which Franco, as *Jefe Nacional*, was President.

net, the retired General García Rebull denounced political parties as 'the opium of the people' and politicians as 'vampires'. The military scheme was for Iniesta to side-step his imminent retirement as Director-General of the Civil Guard and to replace the liberal Manuel Díez Alegría as Chief of the General Staff. General Angel Campano would take over the Civil Guard and officers suspected of liberalism would be purged. The scheme enjoyed the support of the El Pardo clique although the failing Caudillo was not told. Informed about the plot by the Minister for the Army, General Francisco Coloma Gallegos, Arias hastened to see the Caudillo and alarmed him with a threat of resignation. Franco, who regarded military regulations and seniority procedures as sacrosanct, backed Arias and Iniesta was forced to retire on schedule on 12 May 1974.[75]

On 26 June, on a visit to Franco, Fraga found him tired and distant – 'he listened but heard nothing'.[76] He went into the Francisco Franco Hospital on 9 July on the advice of Vicente Gil for treatment for phlebitis in the right leg. Gil attributed the problem to a combination of repeated pressure from his fishing rod which he supported on the leg and the fact that he had sat through every televised football match during the 1974 World Cup. Dressed in a lounge suit and carpet slippers, unaware of those who greeted him, the doddering Franco appeared now anything but a ruthless dictator. The illness caused Franco to miss a cabinet meeting, on 11 July 1974, for only the second time in his career.* Gil's choice of hospital annoyed the Marqués de Villaverde who was in the Philippines attending the 1974 Miss World Contest as the guest of Ferdinand and Imelda Marcos. On his return, Villaverde criticized Dr Gil's handling of the crisis, no doubt by way of drawing attention away from his own absence and the seamy reasons for it.

The treatment of the phlebitis was complicated by the fact that the medication that Franco was taking to alleviate the symptoms of Parkinson's disease was causing gastric ulcers. The anticoagulants to treat the blood clot associated with the phlebitis were incompatible with the treatment for his stomach problem. On 18 July, upset by a TVE film about his life and by the fact of having to miss the traditional reception for the diplomatic corps and the political elite at La Granja, Franco's general condition worsened. On 19 July, Arias and the President of the Cortes appeared with the necessary papers for Franco to sign to implement article 11 of the *Ley Orgánica del Estado* whereby he would stand down

* The first had been on 19 November 1959 when he had had severe influenza.

and Juan Carlos would take over as interim Head of State. Arias and Gil urged him to sign, which he duly did. Doña Carmen and Villaverde were furious. The Marqués allegedly said to Gil: 'What a lousy trick you have played on my father-in-law! What a favour you have done for that kid Juanito.' It was Cristóbal's recognition that the claims of Alfonso de Borbón-Dampierre were finally dead.

Juan Carlos, who had no wish to be tarnished by the actions of a government which he had not chosen, was extremely reluctant to accept. He planned to reform the regime but could not do so if he were Head of State only on an interim basis. He hoped vainly that Franco would hand over to him on a permanent basis. Villaverde was anxious that Franco remain in charge and even told his chaplain, Father José María Bulart, to leave because 'the presence of a priest makes people nervous'. In fact, to the later fury of Villaverde, Bulart secretly gave Franco extreme unction. Throughout his time in hospital, Franco behaved with total docility and never complained.[77]

The tension between the pugnacious Gil and the self-regarding Villa-verde got worse by the day and came to a head on 21 July with an unseemly scuffle in the corridor outside the Caudillo's room. For a fight to break out between the two ultra-rightists supposedly charged with the care of Franco's health mirrored the decomposition of the regime. By 24 July, the Caudillo had improved and Dr Manuel Hidalgo Huerta, Director of the Provincial Hospital of Madrid, held a press conference and declared that he was better and could go on holiday whenever he wanted. At the end of July, the ill-feeling between Gil and Villaverde led to Gil being replaced as Franco's personal physician by Dr Vicente Pozuelo Escudero. Gil was shattered to be told by an off-hand Doña Carmen, 'there are plenty of doctors, but only one son-in-law'. He was even more flabbergasted when, in recognition of his forty years of devoted service to Franco, she sent him a television set – which compared well with the packet of cigarettes that, as first lady, she gave to the doctor who had saved Franco's life in 1916 at El Biutz or the two cases of wine she sent to Dr Hidalgo Huerta after he operated on Franco during his final illness.[78]

On taking over his responsibilities on 31 July, Dr Pozuelo was received by Franco clad in dressing gown, pyjamas and slippers. His voice was so feeble that Pozuelo, who had not been permitted to consult Gil, immediately deduced that he had Parkinson's disease. At 5.00 p.m. on the same afternoon, nine of the doctors of the large medical team responsible for the Caudillo's health held a council. For the first time, it was officially announced that Franco was suffering from Parkinson's disease as a result of sclerosis of the blood vessels. In consequence, a programme of exercise therapy and rehabilitation was to be organized under the overall responsibility of Dr Pozuelo. One of the things which Pozuelo noticed was the extent to which Franco watched television at every opportunity and without fail if sport was on. He was still a fervent Real Madrid supporter and never missed a match.

The doctor's main preoccupation was how to raise the morale of his eighty-two year-old patient. He came up with the device of playing the Caudillo tapes of the military marches used by the Spanish Legion. On the first occasion that he heard 'Soy valiente y leal legionario' (I am a brave and loyal legionaire), Franco's eyes shone, his lips tightened, he threw back his shoulders and he began to smile, once more a *novio de la muerte* (bridegroom of death). Thereafter, one of the exercises to which he was subjected was to march to the tunes of his youth. Another was to try to recount his memories to a tape-recorder. He applied himself to the physical exercise programme with military discipline and Pozuelo was struck as well by the efforts he made to conceal his suffering from others. To improve his image, Pozuelo began to rehearse public appearances with him, both in terms of physical movements, going up aircraft steps and so on, and speeches and conversations in audiences.[79] After a programme of exercise and a more varied diet, the Caudillo's health improved and, on 16 August, he was able to fly to Galicia for his annual holiday at the Pazo de Meirás. He strolled around the estate incessantly and was filmed apparently playing nine holes of golf.

As a result of the whispering campaign of the El Pardo clique, Franco had begun to distrust Juan Carlos, fearing that he might recall his father to Spain as King. Accordingly, the Caudillo had been in as little hurry to see Juan Carlos take over as the Prince was to do so. The reason was simple. While Franco was alive, he could do nothing that would not immediately be reported negatively to the Caudillo. Indeed, on assuming the Headship of State, Juan Carlos telephoned his father who had immediately returned to Estoril, having been sailing off the southern coast. Secret service reports of the conversation raised Franco's fears, no doubt fanned by the family, that Juan Carlos was in league with Don Juan. On 9 August, as interim Head of State, Juan Carlos presided at a cabinet meeting in El Pardo. When it was over, one of Franco's aides said to Utrera, 'I know you are loyal, so I want to warn you that something is being cooked up. Be careful'. Utrera took it to mean a plan to have Franco declared incapable of returning to the Headship of State. As the

ministers went into the gardens to greet a convalescent Franco, the Marqués de Villaverde behaved as if he had assumed, in all its ramifications,
the role of head of the family. He was rude to Juan Carlos and treated
his own son-in-law, Alfonso, as if he were the senior royal personage.
Villaverde was reported to have travelled to Marbella to consult with
Girón about how best to block Arias's drift into reformism (*apertura*).[80]

Franco kept himself informed through Utrera Molina and the El Pardo
clique. He received the Minister Secretary on 28 August and they went
for a stroll in the grounds of the Pazo de Meirás. Utrera painted Franco
a vivid, and alarming, picture of the frenetic political wheeling-and-
dealing that was going on in anticipation of his death. He warned him
of the Trojan Horse of those – Antonio Carro, Pío Cabanillas and the
Directores Generales (junior ministers) belonging to the liberal Catholic
Tácitos group – who hoped to use the notion of 'political associations' as
a bridge to parties and the dissolution of the *Movimiento*. Utrera urged
him to pick a new cabinet and repeated the warning which he had been
received earlier about those who wished to have him declared incapacitated. This inspired Franco to a disquisition on the subject of resentment
and ingratitude. He then told Utrera that he was thinking of resuming
his powers. 'I am not a dictator who clings on in order not to lose
prerogatives', he declared, 'but it is not the first time that Spain has
demanded my sacrifice. After a prudent interval when I have made the
changes which can no longer be put off, I will reconsider my decision.'
They went on talking of the liberal threat and Franco used the language
of the *bunker*, declaring that 'the Army will defend its victory'.[81]

Juan Carlos presided at another cabinet meeting on 30 August. To
stress the provisional nature of his headship of State, the session was held
at the Pazo de Meirás: once more Franco received the ministers in the
garden. The Minister of the Interior, José García Hernández, said to him
'*Mi general*, it is time for you to lighten your duties and leave the helm
in other hands.' Franco looked fixedly at him and said 'You know that is
not possible.'[82] On 31 August, the medical team decided that Franco was
restored to full health. He took the initiative of asking his doctors to
draw up a communiqué (*parte médico*) announcing that his period of conva-
lescence was at an end. When Juan Carlos asked Franco's daughter Nenuca if her father had any intention of resuming as Head of State, she
replied that his health made that completely impossible. The Caudillo
said nothing to the Prince about the communiqué but, as soon as it was
ready, he took the decision to resume his powers. In the most precipitate
manner, a delighted Villaverde telephoned Arias Navarro at his holiday

home at Salinas and the Prince, who was in Mallorca. The decision was made public on 2 September. Such an unwise step, with Franco clearly incapable of resuming major executive responsibilities, has provoked plausible speculation about the machinations of Villaverde and Doña Carmen each of whom was desperately anxious about the future.[83]

On Franco's return to Madrid, there took place a revealing incident. Pozuelo noticed during a routine examination that Franco had a large callus on the little toe of his right foot. On its removal, Pozuelo found traces of an abscess to which he attributed the thrombophlebitis that Franco had suffered during the summer. The callus had been caused by the cheap heavy shoes which Franco normally wore. When Pozuelo told him that the greater sensitivity of his skin with age indicated that he should wear a lighter shoe, Franco protested on the grounds that he had lots of pairs of his usual ones, given him free by the manufacturer which, he explained, caused him pain 'only until I get used to them'.* Pozuelo pointed out that doctors believed that the shoe should adjust to the foot and not the other way around. 'You people like an easy life' (*Ustedes son unos comodones*), replied Franco.[84]

The victory of the El Pardo clique over Juan Carlos implicit in Franco's return to power was a victory for the *bunker*. It was soon followed by an assault on the most liberal minister in the Arias cabinet, Pío Cabanillas. Franco was given a dossier of Spanish magazine pages containing advertisements for beach-ware and camping equipment featuring bikini-clad models skilfully interleaved with pages of foreign soft pornography to give the impression that such material was published in Spain. If six years previously, he had believed Carrero Blanco that Fraga was opening the door to Marxism and erotic subversion, he was now even more easily convinced that Pío Cabanillas must go. He was especially irritated by evidence in the dossier that, like Fraga with the Matesa scandal, Pío Cabanillas was permitting the press to publicize the *aceite de Redondela* case which had finally come to trial. He was alleged to have exploded 'what is the use of everyone saying that Cabanillas is so clever if he hasn't been able to keep my brother's name out of the press? I don't want to see Cabanillas again at a cabinet meeting.'

On 24 October, Arias had his weekly meeting with Franco and was ordered to remove Cabanillas. In solidarity, Antonio Barrera de Irimo

---

* One day in the early 1960s, Franco admired a pair of shoes worn by his brother-in-law and private secretary, Felipe Polo. On being told that they were imported from England and what they cost, Franco said, 'I couldn't afford to pay that much' – Manuel Vázquez Montalbán, *Los demonios familiares de Franco* (Barcelona, 1978) pp. 90–1.

resigned. So that the victory of the *bunker* should not be too evident, Arias and his vice-president Antonio Carro proposed to Franco that the balance which he had always held up as an ideal would be best served by the removal of Utrera and Francisco Ruiz Jarabo. Franco refused categorically on the grounds that they were both 'very loyal'.[85]

By late 1974, Franco was showing even more marked signs of senility. His mouth gaped in a permanent yawn. Dark glasses hid his constantly watery eyes. His gestures were jerky and indecisive and he appeared to be unaware of what was going on around him. Those who spoke to him noticed that he had lost the capacity for logical thought. Occasionally, he would tune into normality but the general impression was of impenetrable distraction.[86] Yet, in this last year of his life, Franco pursued, or for political reasons was pushed into, a daily programme which was the despair of his doctors. His continued obsession with hunting and fishing, or else the desperate desire of the El Pardo clique and other senior Francoists for him to be seen to be active, led to him joining tiring expeditions in inclement weather. During the winter of 1974–75, there were several shooting parties, in open country, in wet blustery weather, with temperatures often near or below zero. During the first such excursion of 1975, at the beginning of January in the Sierra Morena, the need to stand still for long periods so as not to disturb the prey worried Dr Pozuelo that Franco would end up with nephritis (inflammation of the kidneys) or prostatitis, especially when he complained that he couldn't shoot straight because his hands were so cold. On the following night, 4 January, he was seized with an uncontrollable shiver which led Pozuelo to give him antibiotics. When his urine was analysed, there were signs of albuminuria and haematuria, symptoms of kidney infection.[87]

From the end of 1974, Franco's health began to deteriorate rapidly with the greatest distress being caused him by dental trouble.[88] In his end-of-year broacast on 30 December 1974, he had given thanks for his 'complete recuperation' from the illness of the previous summer and took pride in the solidity of his institutions and the way they functioned during his indisposition.[89] He seemed unaware of the disintegration of the Francoist coalition. The first skirmish of the new year took place in February. Arias was infuriated when the *Movimiento* press failed to note the anniversary of his 12 February speech and he ordered Utrera to dismiss the editor of *Arriba*. When Utrera demurred, Arias shouted down the telephone that he would soon find out who was in charge in Spain. Utrera scurried to inform Franco only to find a feeble and fearful Caudillo telling him to do what the President ordered so as to avoid trouble.[90] Utrera took

Franco documents revealing Arias's plans to dissolve the *Movimiento* and tapes of Arias saying 'Franco is old' and 'the only one with any guts here is me' (*aquí no hay más cojones que los míos*). When Utrera said 'Arias is a traitor', Franco began to cry and said only 'Yes, yes, Arias is a traitor, but don't tell anyone. We must work with caution.'[91] The medication for Parkinson's disease had left him timorous.

Arias's Minister of Labour, Licinio de la Fuente, resigned on 24 February in protest at obstacles put in the way of his plans to recognize the right to strike. At last, Arias was impelled to fight back against the *bunker*. He visited Franco on 26 February and said that he would like not just to replace the Minister of Labour but also other ministers. They met again 3 March and Franco opposed any change. However, Arias, by alleging that Utrera had fabricated evidence against him and by threatening to resign, intimidated a weak and nervous old man. Franco was thus obliged to permit a ministerial reshuffle in which both Ruiz Jarabo and Utrera were removed. Fernando Herrero Tejedor, the chief prosecutor of the Supreme Court, arrived as the great new liberal promise in the post of Minister-Secretary General of the *Movimiento*. Arias's hope was that Herrero would be able to make something of the project for political associations. Franco accepted him because of the highly competent investigative report that he had produced on the death of Carrero Blanco.[92]

There was considerable wishful thinking as well as nostalgia when Arias declared on television on 27 February 1975 that anyone who harboured doubts, felt discouraged or was lukewarm in their enthusiasm 'should go to the Palace of El Pardo, and, although it might be from afar, contemplate the light permanently glowing in the Caudillo's study where the man who has consecrated his entire life to the service of Spain continues, without mercy for himself, firm at the helm, steering the course of life so that Spaniards may arrive at the safe haven that he desires for them'.[93] When Utrera went to El Pardo on 11 March to say his goodbyes to the Caudillo, there took place a scene which symbolized the end of an era. Franco praised Utrera's loyalty and as the audience came to an end, he asked him never to change, 'a loyalty like yours is not common'. Carried away by emotion, Utrera promised to stay at his post until his last breath. At that, Franco broke down, embraced him and wept copiously. When Franco let him go, Utrera stepped back, stood to attention and, asked permission to say his last goodbye in the manner of a *centurion* of the Falangist Youth. With his arm raised in the fascist salute, he barked '*Caudillo, a tus órdenes, ¡Arriba España!*'. Franco stood pathetically with his own trembling arm raised in response.[94]

Franco was deeply disturbed by the events in Portugal which he saw as a pointless destruction of the achievements of Salazar. At this time, Fraga gave Nieto Antúnez the draft programme for a political association which he hoped to register. The view of Arias and other ministers was that it needed toning down. Franco himself, after reading it, sarcastically asked Nieto 'for what country is Fraga writing these projects?'[95] The continuing leftwards trend of events in the neighbouring country was to have less repercussion on his health than the determination of Hassan II to remove the last traces of Spanish presence from Morocco. Franco found incidents in Spanish Sahara particularly distressing.[96]

On 31 May, President Gerald Ford arrived in Spain for a two-day visit. He received a significantly less rapturous reception than either Nixon or Eisenhower before him. That reflected both Franco's own incapacity and also the fact that State Department policy was inclining away from the Caudillo and towards the future in the form of Juan Carlos. President Ford spent rather more time with the Prince than with Franco.[97]

On 23 June 1975, the new *Ministro Secretario General del Movimiento*, Fernando Herrero Tejedor died in a road accident near Villacastín in the province of Valladolid. Franco was told while attending a bull-fight. He was much affected by the news which he took as a providential sign that the experiment with associations did not have divine approval.[98] The logical successor should have been Herrero's ambitious second-in-command, Adolfo Suárez. However, Suárez, like Herrero Tejedor, was too compromised by *aperturismo*. Instead, Franco insisted to a baffled Arias that the new Minister-Secretary should be José Solís. In doing so, the Caudillo was reflecting the belief in regime circles that, with the enemy grouping for an assault, he should surround himself with his reliable old guard. The El Pardo clique, in close touch with Girón and Alejandro Rodríguez Valcárcel, the President of the Cortes, persuaded Franco to extend the life of the present Cortes by six months. They hoped thereby to have gained the time necessary to push Arias out and secure the elevation of Solís, Rodríguez Valcárcel or even Girón, to the presidency. The Martínez-Bordiu clique was the transmission belt to Franco of the views of beleaguered Francoism.[99]

Franco continued to believe in his own divine mission unaware that his ultra supporters were increasingly frantic with anxiety about their own futures.[100] To do what they planned, they needed Franco alive. Franco was exhausted and anxious to rest, even talking of retiring to a monastery to die, in emulation of the last days of Charles V. To the end, he maintained the comparisons with the great Spanish monarchs of the past. It has been suggested that his panic-stricken wife persuaded him not to abandon politics just as his equally worried son-in-law later kept him alive electronically.[101] Franco alive was an appeal of last resort for the bunker. In his final months, the *bunker* worked on his fears and prejudices, their efforts facilitated by his undying conviction of the sinister threat of freemasonry.[102] On 15 July, Franco received in El Pardo a delegation of the *Hermandad Nacional de Alféreces Provisionales* (National Fraternity of Provisional Lieutenants), a *bunker* stronghold, led by the reactionary Marqués de la Florida. On the previous day, a policeman had been murdered by FRAP terrorists.* Franco dismissed the delegation's highly charged denunciations of the Left: 'I believe that you pay too much attention to dogs that bark. In reality, they are tiny minorities which demonstrate our vitality and which put to the test our *Patria's* strength and capacity to resist'. Yet, in the belligerent language of the 1940s, he urged Florida and his *Hermandad* to defend to the death the Civil War victory.[103]

In the summer of 1975, the sense of the regime crumbling was all-pervading. While Franco was on his annual holidays in Galicia, rumours spread that, on his return, he would replace Arias with Solís. A cabinet meeting held at the Pazo de Meiras on 22 August introduced a fierce new anti-terrorist law, whose blanket provisions covered all aspects of opposition to the regime.[104] The first fruits of the law were a series of trials which would lead to the final black episode in Franco's life. On 28 August, a court martial in Burgos sentenced to death two members of ETA and, on 19 September, another in Barcelona passed a third death penalty. In between, two more courts martial on 11 and 17 September, held at a military base near Madrid, sentenced eight members of FRAP to death. A worldwide wave of protests greater even than that occasioned by the trial of Grimau, provoked Franco's indignation. Fifteen European governments recalled their ambassadors. There were demonstrations and attacks on Spanish embassy buildings in most European countries. At the United Nations, the President of Mexico, Luis Echevarría, called for the expulsion of Spain. Pope Paul VI appealed for clemency as did all the bishops of Spain. Don Juan sent an appeal through his son. Similar requests came from governments around the world.[105] Franco ignored them all.

At the three and a half hour cabinet meeting held on 26 September,

---

* Frente Revolucionario Antifascista y Patriótica (the Revolutionary Antifascist and Patriotic Front) was a Maoist group which emerged in the late 1960s and had been infiltrated by police *agents provocateurs*.

presided over by an extremely infirm Caudillo, five death sentences were confirmed. At dawn on the following day, the condemned were shot. The international protests intensified with the Pope in the forefront. The Spanish embassy in Lisbon was sacked.[106] If, as the Caudillo himself had claimed, the pardons after the 1970 Burgos trials were a sign of the regime's strength, the executions of 27 September 1975 were a symbol of terminal decline. ETA was more of a threat than five years previously but the difference between 1970 and 1975 was the influence exercised over Franco by the ultra right.*

By now, Franco was losing weight and having trouble sleeping. On 1 October 1975, the thirty-ninth anniversary of his elevation to the Headship of State, he appeared before a huge crowd at the Palacio de Oriente. Buses had brought representatives of the *Movimiento* from all over Spain. In the previous days, the television had been urging viewers to attend and offices, factories and shops were officially closed to facilitate this. On this last appearance in public, the now diminutive, hunched Caudillo had evident difficulty in breathing as he croaked out the same paranoiac clichés as always. Spain's problem was, he declared 'a masonic left-wing conspiracy within the political class in indecent concubinage with Communist-terrorist subversion in society'. He took his leave of the crowd weeping and with both hands raised.[107]

Exposure to the stabbing autumn winds of Madrid on 1 October set off the escalation of medical crises which ended in his death. After a day of nose-blowing on 14 October and other symptoms of influenza, the first of these began in the early hours of the morning of 15 October. Franco awoke with pains in his chest and shoulders: he had suffered a heart attack. Despite this, he refused to suspend his work programme, holding eleven formal audiences on Thursday 16 October and watching films in the evening.[108] Desperate to maintain an image of normality, the Marqueses de Villaverde spent these first days of Franco's illness at a long hunting party in Ciudad Real in the course of which Cristóbal Martínez-Bordiu gave voice to the panic running through Francoist circles. During a break in the proceedings, he seized the machine-pistol of one of the Civil Guards who were escorting the party and, shouting 'We have to be ready. They'll be coming for us but, as far as I'm concerned...,' had started wildly pumping bullets into a nearby rock.[109]

Against the advice of his doctors, Franco insisted on chairing a cabinet meeting on Friday 17 October. He refused to have the ministers come

---

* It is possible too that Franco wanted revenge for the death of Carrero.

to his bedroom or to go to the meeting in a wheelchair. His alarmed doctors conceded only on condition that he wore electrodes connected to a heart monitor. During the session, a minister recounted a visit made by Prince Juan Carlos to La Mancha. When he mentioned that the crowd had chanted '*¡Franco! ¡Franco! ¡Franco!*', the Caudillo's heart began to thump to such an extent that the doctors in the adjoining room were convinced that the end had come. In the course of the meeting, news came in of the Moroccan 'green march' on Spanish Sahara which caused him to have a relapse.[110]

On Saturday 18 October, Franco got up and worked in his study for the last time, probably writing his last will and testament. On Sunday 19 October, he heard mass and took communion. At 11 p.m. on the night of 20 October, he had another mild heart attack. Although he was able to watch a film on Wednesday 22 October, his condition began to deteriorate badly from that evening. Unable to sleep, he complained of fierce pains in the shoulders and the lumbar region. He had had a third heart attack. He was, however, able to whisper to Arias that he send Solís as special emissary to Morroco 'to play the gypsy' (*gitanear*) with Hassan II and gain time. Franco's death was accidentally announced on ABC News in Washington.[111]

Franco suffered another bout of cardiac insufficiency on 24 October. His dental problems flared up again and he began also to suffer abdominal distension as a result of stomach haemorrhage. On Saturday 25 October, he was given extreme unction. On Sunday 26 October, after a further internal haemorrhage, it was widely assumed that the end was nigh and several radio stations played suitably lugubrious music. By 29 October, he was receiving constant blood transfusions. Throughout this time, he was in acute pain. By 30 October, there were signs of peritonitis. On being told of the heart attacks and the serious intestinal complications, Franco said, 'Article 11, implement article 11'. Martínez-Bordiu and Arias, thrown together in alliance, hoped to get Juan Carlos to accept an interim position, as he had reluctantly done a year previously, but now he refused. Franco was no longer Head of State. Sections of the press began to build up the image of Juan Carlos and to talk of Franco in the past tense.[112]

By the night of 2–3 November, Franco's intestinal haemorrhage was intensifying. The bed, the carpet and a nearby wall were soaked in blood. To stem it, the twenty-four specialists now in attendance decided on an emergency operation. With no time to get him to a properly equipped hospital, Franco was pushed on a trolley to an improvised operating theatre in the first-aid post of the guard at El Pardo. A copious trail of blood marked his route. In the course of a three-hour operation, supervised by Dr Manuel Hidalgo Huerta, the medical team discovered an ulcer which had opened an artery. Franco survived the operation but was now found to be suffering uraemia (a morbid condition of the blood due to the retention of urinary matter normally eliminated by the kidneys).[113] He had to have dialysis. It was decided to move him to a properly equipped hospital, the Ciudad Sanitaria La Paz, where he was taken in a military ambulance.

Three days later, with the uraemia intensifying, at 4.00 p.m. on 5 November, another operation began; it lasted four and a half hours and saw two thirds of his stomach removed.[114] Thereafter, he was kept alive by a massive panoply of life-support machines, regaining consciousness occasionally to murmur 'how hard it is to die'. The hospital was besieged by journalists. Enormous sums were offered for photographs of the dying dictator. Dr Pozuelo indignantly rejected fabulous offers only to discover later that the Marqués de Villaverde had already made full use of his own camera.[115] On 15 November a further massive haemorrhage began. Franco's stomach was inflated as a result of the peritonitis. A third operation began in the early hours of the morning, after which Hidalgo Huerta's team remained deeply pessimistic.[116] The determination of the El Pardo entourage to keep Franco alive despite his intense suffering was not unrelated to the fact that the term of office of Alejandro Rodríguez Valcárcel as President of the *Consejo del Reino* and of the Cortes was due to end on 26 November. If Franco could recover sufficiently to renew Rodríguez Valcárcel's mandate, the clique would have a key man in a position to ensure that the president of the council of ministers chosen by Juan Carlos would be 'reliable'.[117]

Franco was alive but only just, barely conscious, and entirely dependent on the complex life-support machinery. Finally, his daughter Nenuca insisted that he be allowed to die in peace. At 11.15 p.m. on 19 November, the various tubes connecting him to the machines were removed on the instructions of Martínez-Bordiu. He probably died shortly afterwards. The official time of death was given as 5.25 a.m. on 20 November 1975, the official cause as endotoxic shock brought about by acute bacterial peritonitis, renal failure, bronchopneumonia, cardiac arrest, stomach ulcers, thrombophlebitis and Parkinson's disease.[118]

EPILOGUE

'No enemies other than the enemies of Spain'

AS SOON as he had realized that he was dying, Franco had written his political testament to the Spanish people. It was read out on television by a tearful Carlos Arias Navarro at 10.00 a.m. on 20 November. Describing himself as a 'faithful son of the Church' who 'wanted to live and die as a Catholic', the Caudillo had written: 'I beg forgiveness of everyone, just as with all my heart I forgive those who declared themselves my enemies even though I never thought of them as such. I believe that I had no enemies other than the enemies of Spain.' And he warned 'Do not forget that the enemies of Spain and of Christian civilization are on the alert.'[1]

As the news of his death was flashed to every corner of Spain, many mourned and many rejoiced. Outfitters and haberdashers ran out of black ties and black cloth for armbands yet it is said that people danced in the streets of some Basque towns. The novelist Manuel Vázquez Montalbán captured the atmosphere of the time in Barcelona: 'throughout 20 November 1975, the city filled with silent passers-by, walls reflected in their eyes, their throats dried by prudent silence. Up the Rambla and down. As ever. Security guards, police and paramilitaries observed the muted demonstration while with their sixth sense they heard the 'Hymn of Joy' sung by the hidden soul of the 'Rose of Fire' [Barcelona], by the cautious soul of the widowed city, by the wise soul of the occupied city. Above the skyline of the Collserola mountains, champagne corks soared into the autumn twilight. But nobody heard a sound. Barcelona was, after all, a city which had been taught good manners. Silent in both its joy and its sadness.'[2]

Franco's body lay in state on a dais in the Sala de Columnas of the Palacio de Oriente. During the fifty hours that the Sala was open to the public, queues formed several kilometres long. Between 300,000 and 500,000 people filed past his body and not just to reassure themselves that he was dead.[3] For some days before his death, the functionaries at the Valle de los Caídos had been trying to find the monumental stone which matched the one that covered the grave of José Antonio Primo de Rivera. When it was finally located, several days of rehearsal were needed before its enormous weight could be handled with the dexterity required for the funeral ceremony. On 23 November, while the funeral cortège was still *en route* between Madrid and Cuelgamuros, one of the waiting mourners fell into the grave and was rendered unconscious. Considerable difficulty was experienced getting him out before he could be taken to hospital. At 1.00 p.m., the cortège arrived at the Basilica, and the coffin was carried by the Marqués de Villaverde, his son Francisco Franco Martínez-Bordiu, Alfonso de Borbón and representatives of the Army, Navy and Air Force.[4] No significant Head of State, other than Franco's admirer the Chilean dictator, General Augusto Pinochet, attended the funeral.

Juan Carlos put no pressure on Doña Carmen to leave El Pardo and she remained in residence for another two and a half months. Each morning Franco's personal standard (*guión*) was run up the flagpole and lowered in the evening. A company of soldiers continued to act as a guard of honour. Crates of jewellery, antiques, pictures and tapestries were packed, loaded onto lorries, along with the Caudillo's papers, and either distributed around the various family properties in Spain or else spirited off to safe foreign havens.*[5] At 6.10 p.m. in the evening of a cold 31 January 1976, a tearful Doña Carmen left El Pardo for the last time. After inspecting a guard of honour with military band, she departed escorted by the Marqueses de Villaverde and several 'ultra' ex-ministers including Girón and Utrera Molina. Groups of ultras lined the road outside the palace chanting 'Franco! Franco! Franco!' and singing the Falangist anthem *Cara al sol*.

Doña Carmen moved into Madrid and stayed at the apartment of her daughter and son-in-law in the building that she owned at Hermanos Becquer 8 until, in 1978, her own apartment in the same building was ready. Despite the loss of Franco, the family remained immensely rich.[6] And, in addition to her accumulated wealth, Doña Carmen received lavish pensions from the state.[7] At first, she attended some ultra-rightist commemorations of her husband's death but quickly withdrew into silence.[8] In contrast, Nenuca and her husband continued to join Girón and Blas Piñar in nostalgic demonstrations in the Plaza de Oriente.[9]

---

* There have been claims that some of the priceless items properly belonged to the nation but that there was no vigilance by officials of the *Patrimonio Nacional* which was, in any case, headed by the Head of Franco's Household, General Fernando Fuertes de Villavicencio.

Shorn of the protecting layers of special privilege, the family came under attack from the press and was also involved in a series of unexplained accidents. A debate began in the media as to whether Franco's papers were the property of his family or of the State. Before the issue could be resolved, a huge collection of thirty-nine years' worth of State papers, private notes and secret reports, along with the Caudillo's own paintings, were said to have disappeared when a mysterious fire broke out at the Pazo de Meirás on 18 February 1978. A fabulous treasure trove of antiques and *objets d'art* saved from the flames by the firemen provoked speculation about how much belonged to the *Patrimonio Nacional*. Barely two months later, on 25 April 1978, some of Franco's possessions were stolen during a robbery at Valdefuentes.[10] Then, on 12 July 1979, Doña Carmen and the Marqués de Villaverde narrowly escaped death when the Hotel Corona de Aragón in Zaragoza was set ablaze. With many other senior Francoists, they were staying at the hotel in order to see the cadets at the military academy, including Villaverde's son, Cristóbal Martínez-Bordiu Franco, receive their commissions. The family's hopes that Cristóbal Jr. might follow in his grandfather's footsteps came to nothing when, shortly after, he declared that the military life did not suit him and left the Army.

The Martínez-Bordiu family was also involved in a number of scandals. In 1979, María del Carmen Martínez-Bordiu, Duquesa de Cádiz, abandoned her husband Alfonso de Borbón-Dampierre and her two sons to take up with a Parisian antique dealer, Jean Marie Rossi. Her elder son Francisco was killed in a car accident in 1984 after which Alfonso de Borbón was tried and sentenced to a short prison sentence for dangerous driving. After efforts to establish a claim to the throne of France, he died in a bizarre skiing accident in 1989.[11] Within five years of Franco's death, the estate at Valdefuentes had fallen into neglect. Under the management of Franco's oldest grandson, Francisco Franco Martínez-Bordiu, its prosperity trickled away to nothing and it became the location for horror and pornographic films. An attempt to make his fortune in Latin America ended when he was charged with fraud in Chile.[12] His mother Nenuca was also the object of press attention. On 7 April 1978, now Duquesa de Franco but travelling as Señora Martínez, she was stopped at the frontier trying to take some gold coins out of the country, allegedly to have a special clock made in Switzerland.[13]

If the eclipse of the Franco family was predictable, what was altogether more surprising was the relative silence about the dictatorship which followed Franco's death. By tacit national consent, the regime was relegated to oblivion. The Caudillo had occupied much of his last fifteen years planning for the perpetuation of Francoism after his death but his schemes came to nothing. Collectively rejecting those plans, a broad spectrum of Spaniards co-operated in what came to be known as the *pacto del olvido* (the pact of forgetfulness). In order to ensure a bloodless transition to democracy, the victims of the repression renounced their desires for revenge, demanded no settling of accounts. There were no purges of the executioners, the torturers, the jailers, the informers or of those close to Franco who had enriched themselves during the years of the dictatorship. By the same token, large numbers of Franco's more moderate and far-sighted supporters forgot their own pasts, some collaborating sincerely in building the democratic consensus, others merely fabricating new autobiographies as '*demócratas de toda la vida*' (life-long democrats).

The fact that the urge for revenge was kept in check did not mean that Franco himself was forgotten. The years following his death saw a publishing boom fuelled by an insatiable national appetite for intimate gossip about the life of a man previously cocooned by layers of adulation. To the chagrin of unrepentant Francoists, the memoirs of those who knew him presented a less than edifying picture. His sister Pilar, his cousin Pacón, his niece Pilar Jaraiz, his grandson Cristóbal, the widow of his brother Ramón, his brother-in-law Ramón Serrano Suñer, his medical advisers Dr Ramón Soriano, Dr Vicente Gil, Dr Vicente Pozuelo, and numerous ex-ministers, Pedro Sainz Rodríguez, Manuel Fraga, Laureano López Rodó, José María de Areilza, José Utrera Molina and many others wrote highly successful books. The portrait which they presented, inadvertently or otherwise, was of the astonishing personal mediocrity which characterized 'a sphinx without a secret'.[14]

However, even more than these revelations, it was the unexpected ease with which the mass of Spaniards opted for democracy and simply sidestepped the Caudillo's schemes for the future of Spain that provided the most telling commentary on his place in history. That is not to say that Franco achieved nothing but it does underline the extent to which his triumphs were sectarian and personal. For himself and his supporters, there was no contradiction between the good of Spain and the good of Franco. He never made any bones either about identifying himself with Spain or indeed about the fact that, when he spoke of Spain, his definition was narrowly partisan. For decades, Franco ridiculed Don Juan de Borbón for his patriotic desire to be King of all Spaniards and, to his dying day, vengefully tried to maintain the Civil War divisions of victors and vanquished. Spain as defined by Franco, the Spain that won the Civil War, the postwar Spain that was maintained by an apparatus of repression, had virtually ceased to exist by 1975. A massive social and economic revolution had taken place since the late 1950s during the period in which Franco had been a symbolic rather than an active leader, increasingly cut off from reality. For the majority of the population which had been born since the Civil War, Franco's Spain was not so difficult to forget.

Franco will be remembered first for his ruthless conduct of the Nationalist war effort between 1936 and 1939 and the determination with which he pursued the systematic annihilation of his enemies on the left and secondly for the sheer duration of his survival thereafter. His hallmark was the instinctive cunning and hard-faced, unflappable *sang froid* with which he staged-managed the rivalries between the various regime forces and easily defeated challenges by those – from Serrano Suñer to Don Juan – who were his superiors in intelligence and integrity. Franco's achievements were not those of a great national benefactor but of a skilful manipulator of power who always looked to his own interests. As Salvador de Madariaga wrote: 'The highest interest of Franco is Franco. The highest interest of de Gaulle is France.'[15]

The Spanish ruling classes abdicated power to Franco and the other generals in 1936, just as their Italian counterparts had done with Mussolini and the Fascists in 1922 and the German ruling classes with Hitler and the Nazis in 1933, convinced that once working-class challenges to the existing system were crushed, power could be taken back. Franco, with his reverence for the bourgeoisie, the aristocracy and the crown, seemed a good choice, a certainty to restore the monarchy at the earliest opportunity. That for thirty-nine years after his elevation to the provisional headship of the wartime state he blocked the return of the rightful heir to the throne and managed to remain in power is the measure of his remarkable political skill.

That skill was exercised in relation to members of his own political coalition. Against his enemies, he was ruthless in the use of state terror, the effects of which reverberated for decades after its scale had been significantly reduced. It was a kind of political investment, a bankable terror, which accelerated the process of Spain's depoliticization, pushing the mass of Spaniards into political apathy. Franco presided over the entire process from afar. The story is told that when asked by his friend General Alonso Vega about the fate of an old comrade from the Moroccan wars, Franco replied '*le fusilaron los nacionales*' (the Nationalists shot him), for all the world as if it had nothing to do with him at all.[16] Nonetheless,

in his speeches, he made no secret of his belief in the necessity of blood
sacrifices and, since he was the supreme authority within the system of
military justice, there can be no dispute as to his full knowledge and
approval of executions.

Franco considered that he was merely the Solomonesque dispenser of
justice just as, on numerous occasions, he denied that he was a dictator.
In March 1947, he told Edward Knoblaugh of the International News
Service that there was no dictatorship in Spain: 'I am not free, as it is
believed abroad, to do what I want.' And, in June 1958, he assured a
French journalist that 'to describe me as a dictator is simply childish'.[17]
Regarding himself as the saviour of Spain universally beloved by all but
the sinister agents of occult powers, Franco could think of himself in
such benevolent terms with total sincerity. He believed the one-party
state, the censorship, the prison camps and the apparatus of terror were
somehow balanced by a readiness to let his ministers talk interminably
in cabinet meetings – which reflected only poor chairmanship. In fact,
the device of leaving his subordinates great leeway made him appear less
despotic but was an effective means of absolutist control. Those who
became enmeshed in corruption or the repression became ever more
dependent on his good will.

Franco's conviction that he was not a dictator was characteristic of his
lack of critical self-perception. The series of masks behind which he hid
– gallant desert hero in Africa, twentieth century El Cid during the Civil
War, would-be imperial leader in the early 1940s, commander of the
besieged garrison of Spain in the late 1940s – were deeply gratifying to
him and gave him a conviction which made him impervious to discourage-
ment. After the international triumphs of 1953, the mask which served
him for the rest of his days was that of the benevolent and beloved
patriarch of the Spaniards. Like its predecessors, it was a persona in
which he believed totally and which derived its strength in part from the
undeniable fact that it met a need among his admirers and supporters.
After all, Franco did not rule by repression alone: he enjoyed a considerable popular support. There were those who, for reasons of wealth,
religious belief or ideological commitment, actively sympathized with the
values of the Nationalist war effort between 1936 and 1939. Then there
was the passive support of those who had been conditioned into political
apathy by political repression, the controlled media and an appallingly
inadequate state education system. Then, from the late 1950s onwards,
there was the support of those who were simply grateful for rising living
standards.

However, the economic boom – so assiduously claimed by his propagandists as the greatest achievement – was, like wartime neutrality, little to do with Franco. His stewardship in the economic sphere – even if judged solely by his own standards and objectives – was lamentable. During the Civil War and until the late 1950s, with boundless pride in his own economic competence, he clung to the fascist notion of autarkic central control, not least because the idea of a command economy fitted well with his military mentality. In October 1939, his rigidly autarkic plan for national reconstruction together with his rejection of British and American offers of credits, when Spain was crippled by shortages of food and raw materials, had catastrophic effects. His policies led to incalculable hardship. Black-marketeering and corruption remained features of the Spanish economy well into the second half of the 1950s. Had Franco had the magnanimity and patriotism to make way for Don Juan de Borbón in 1945, Spain would have had a constitutional monarchy in time to enjoy the benefits of the Marshall Plan and of early membership of NATO and the EEC. As it was, he remained in power, blaming the disastrous consequences of autarky on international malevolence, and *per capita* income did not regain the levels of 1936 until the mid-1950s.[18]

After 1959, the Spanish economy underwent a profound transformation because the policies espoused by Franco were abandoned. By the early 1950s, with economic stagnation threatening the stability of his dictatorship, he was obliged, through his technocrats, to seek salvation in the international capitalist order which had been loudly denouncing since 1936. Stabilization and development plans in accordance with the recommendations of international financial institutions and French planning models made a nonsense of the previous twenty years of autarky and constituted an economic U-turn comparable to the political abandonment of fascism. Far from masterminding the process, Franco grudgingly acquiesced in changes which he did not understand in order to remain in power.

Development was ultimately the fruit of the combination of domestic capital accumulation born of the repressive labour legislation of the 1940s, the receipts from emigrant workers and tourism, and foreign investment attracted by an anti-Communist, anti-union regime. Only in the sense that his repressive regime created stability and a docile labour force which made Spain attractive to foreign investors, did Franco contribute to economic growth – but, as his commitment to autarky starkly demonstrated, the way in which it came about was not his objective.

Franco seized upon growing material affluence as a source of political legitimation but his regime found itself rendered obsolete by the very pace of social and economic change. By the end of the 1960s, many in the industrial, banking and business fraternities found themselves frustrated by the Falangist syndicates' paternalistic regulation of the labour market and by the political ostracism which kept Spain out of the EEC. At the same time, particularly after the first energy crisis of 1974, a working class still deprived of political rights could no longer be bought off by constant increases in living standards. A far-reaching consensus between the progressive elements of the Right and the Left was to lie at the heart of the national rejection of Franco's plans for the post-Franco future and was to underwrite the transition to democracy.

After Franco's death, the immediate succession mechanisms worked well and Juan Carlos became King within the terms of the Caudillo's pseudo-constitution. That fact neutralized those in the *Movimiento* who suspected that he might share the liberal inclinations of his father. At the same time, Franco's long hesitations over the nomination of a successor and his flirtation with Alfonso de Borbón had left Juan Carlos sufficiently distanced from the regime to enjoy an initial tolerance from the democratic opposition. From mid-1976, the new King played a central role in the complex process of dismantling the Francoist apparatus and in the creation of a democratic legality. It had been Franco's intention to install an entirely Francoist monarchy to perpetuate his regime. Even that part of his scenario came to nought when, with the process of democratization well under way, on 14 May 1977, one month before the first elections since 1936, Don Juan de Borbón renounced his rights to the throne and thus gave his son full dynastic legitimacy.

Judged in terms of his ability to stay in power, Franco's achievement was remarkable. However, the human cost in terms of the executions, the imprisonments, the torture, the lives destroyed by political exile and forced economic migration points to the exorbitant price paid by Spain for Franco's 'triumphs'. Scarred by the horrors of the Civil War and the postwar repression, Spaniards rejected both political violence and Franco's idea that, by right of conquest, one half of the country could rule over the other. During the transition to democracy, they collectively displayed a political maturity which contradicted Franco's belief that they were incapable of living under a democratic system.

By mid-1977, Franco's most cherished ambitions for the future lay in ruins. The Movimiento was dismantled, trade unions were legalized, political parties, including the hated *Partido Comunista de España*, were permitted. Having set out to eradicate Communism from Spain, Franco had left the PCE with dramatically greater strength than it had had before the Civil War. Within a year of his death, the PCE had over 200,000 members, its Secretary-General Santiago Carrillo played a major role in the transition and the party gained ten per cent of the vote in the 1977 elections.[19] Equally Franco's determination to eliminate separatism had left more powerful regionalist movements in the Basque country and Catalonia than had ever existed before 1936 along with nascent nationalist movements in Andalusia, Galicia, the Valencian Region and even areas such as the Rioja, León and Castille. The democratic constitution of 1978 enshrined rights of regional autonomy which overturned the rigid centralism for which, in part, the Francoists had fought the Civil War.

Inevitably, the most dramatic difficulties encountered by Spain's new-born democracy were the direct legacy of Franco's rule. His intransigent centralism and its brutal application to the Basque Country were to bear fruit in ETA terrorism and the popular support which it enjoyed until the end of the 1970s. Equally, his politics of postwar revenge had been fostered most vehemently in the military academies where officer cadets were conditioned to associate democracy with disorder and regional separatism. As the dictatorship was rapidly dismantled, it was hardly surprising that some of its senior military defenders found themselves isolated from the massive political consensus in favour of democratization. That did not, of course, inhibit them from endeavouring at several moments in the late 1970s and, most dramatically, in the attempted coup of Colonel Tejero on 23 February 1981, to impose the bunker's view of what Spain's political destiny should be.[20] Courageous mass demonstrations involving millions of people were the popular response to Tejero's attempt to turn back the clock to Franco's time.

In the many national, regional and municipal elections that have been held in Spain since 1977, parties openly espousing Francoist values have never gained more than two per cent of the vote. Spaniards have embraced democracy and rejected Franco's blueprint for their future. Every year on 20 November, a band of his most fervent supporters meet and feebly chant '*Franco resucita, el pueblo te necesita*' (Franco rise up, the people need you). Every year, the ageing Francoist stalwarts are fewer and fewer. Soon there will be none.

NOTES

PROLOGUE

1 Carlos Fernández, *El general Franco* (Barcelona, 1983) pp. 311–20.

2 Rogelio Baón, *La cara humana de un Caudillo* (Madrid, 1975) p. 91.

3 Amando de Miguel, *Franco, Franco, Franco* (Madrid, 1976) p. 117.

4 Almirante Carrero Blanco, *Discursos y escritos 1943–1973* (Madrid, 1974) p. 32.

5 Enrique González Duro, *Franco: una biografía psicológica* (Madrid, 1992) pp. 265–6.

6 Testimony of P. Bulart, María Mérida, *Testigos de Franco: retablo íntimo de una dictadura* (Barcelona, 1977) p. 36.

7 Jaime de Andrade, *Raza anecdotario para el guión de una pelicula* (Madrid, 1942). For an acute commentary, see Román Gubern, *"Raza" (un ensueño del General Franco)* (Madrid, 1977).

8 José María Pemán, *Mis encuentros con Franco* (Barcelona, 1976) p. 53; testimony of Pilar Franco Bahamonde, Mérida, *Testigos*, p. 92.

9 Juan Antonio Ansaldo, *¿Para qué...? (de Alfonso XIII a Juan II)* (Buenos Aires, 1951) p. 51.

10 John Whitaker, 'Prelude to World War: A Witness from Spain', *Foreign Affairs*, Vol. 21, No. 1, October 1942, p. 116.

11 Cantalupo to Ciano, 24 March 1937, ASMAE, SFG, b. 38, T.657/320.

12 Pemán, *Mis encuentros*, p. 20.

13 Pemán, *Mis encuentros*, p. 9.

CHAPTER I

1 Joseph M. Levinson, 'Is General Franco a Jew?' *Israel's Messenger*, 3 April 1939; Ramón Garriga, *Ramón Franco, el hermano maldito* (Barcelona, 1978) pp. 20–3; S. F. A. Coles, *Franco of Spain* (London, 1955) pp. 103–6; José María Fontana, *Franco: radiografía del personaje para sus contemporáneos* (Barcelona, 1979) pp. 37–40; Sir Robert Hodgson, *Spain Resurgent* (London, 1953) p. 109.

2 Pilar Franco Bahamonde, *Nosotros los Franco* (Barcelona,

1980) p. 15; Ramón Garriga, *Nicolás Franco el hermano brujo* (Barcelona, 1980) p. 13; Luis Suárez Fernández, *Francisco Franco y su tiempo*, 8 vols (Madrid, 1984) I, pp. 57–60.

3   Pilar Jaraiz Franco, *Historia de una disidencia* (Barcelona, 1981) pp. 17–20; Franco, *Nosotros*, p. 80; George Hills, *Franco: The Man and his Nation* (London, 1967) p. 19.

4   Franco, *Nosotros*, p. 76; Joaquín Arrarás, *Franco* 7ª edición (Valladolid, 1937) p. 14; Suárez Fernández, *Franco*, I, p. 66.

5   Pedro Teotonio Pereira, *Memórias* 2 vols (Lisbon, 1973) II, p. 63.

6   Franco, *Nosotros*, pp. 21, 24.

7   Carmen Díaz, Viuda de Franco, *Mi vida con Ramón Franco* (Barcelona, 1981), p. 22; Franco, *Nosotros*, pp. 26–7, 70; Garriga, *Ramón Franco*, pp. 14–15; Garriga, *Nicolás Franco*, pp. 14–17.

8   Francisco Salva Miquel & Juan Vicente, *Francisco Franco (historia de un español* (Barcelona, 1959) p. 33.

9   *ABC*, 5 February 1926.

10 Jaraiz Franco, *Historia*, pp. 25–32; Franco, *Nosotros*, pp. 20, 25; *Mi vida*, pp. 22, 60, 94; Guillermo Cabanellas, *Cuatro Generales* (Barcelona, 1977) 2 vols, I, p. 59.

11 Jaraiz Franco, *Historia*, pp. 47, 86; Francisco Franco Salgado-Araujo, *Mi vida junto a Franco* (Barcelona, 1977) p. 14.

12 Federico Grau, 'Psicopatología de un dictador: entrevista a Carlos Castilla del Pino', *El Viejo Topo*, Extra No. 1, 1977, pp. 18–22.

13 Jaraiz Franco, *Historia*, pp. 37, 59–60; Franco, *Nosotros*, pp. 27, 72, 74, 233; Vicente Pozuelo, *Los 476 últimos días de Franco* (Barcelona, 1980) pp. 87–8; Garriga, *Ramón Franco*, pp. 17–20.

14 Pozuelo, *Los 476 últimos dís*, p. 204.

15 Ricardo de la Cierva, *Francisco Franco: biografía histórica* 6 vols (Barcelona, 1982) I, p. 25.

16 Comandante Franco, *Diario de una bandera* (Madrid, 1922) p. 21.

17 *La Vanguardia Española*, 18 July 1941.

18 Jaraiz Franco, *Historia*, p. 37; Hills, *Franco*, pp. 24–5.

19 Hills, Franco, p. 29.

20 Franco, *Nosotros*, p. 18, 24, 70–6; Hills, *Franco*, p. 29.

21 Franco, *Nosotros*, pp. 73, 77.

22 Isidre Molas, *Lliga Catalana: un estudi de Estasiologia* 2 vols (Barcelona, 1972) I, pp. 42–5.

23 José Alvarez Junco, *El emperador del Paralelo: Lerroux y la demagogia popular* (Madrid, 1990) pp. 347–8.

24 Joaquín Romero Maura, *The Spanish Army and Catalonia: The "Cu-Cut! Incident" and the Law of Jurisdictions, 1905–1906* (London, 1976) pp. 13–29; Alvarez Junco, *Lerroux*, pp. 317–18.

25 Franco Salgado-Araujo, *Mi vida*, pp. 16–17; Suárez Fernández, *Franco*, I, pp. 76–7; Hills, *Franco*, pp. 29–30.

26 Baón, *La cara humana*, p. 188, says that Nicolás Franco Salgado-Araujo applied for a transfer to Madrid.

27 For sympathetic portraits of Nicolás Franco Salgado-Araujo, see Jaraiz Franco, *Historia*, pp. 53–61, and Franco, *Nosotros*, pp. 16, 72. Both deny that he gambled or drank seriously.

28 Franco, 'Transcript of taped memoirs', Pozuelo, *Los 476 últimos días*, pp. 89–90.

29 Andrade, *Raza*, p. 66.

30 Grau, 'Psicopatología', pp. 18–22.

31 Franco, 'transcript of taped memoirs', Pozuelo, *Las últimas 476 días*, pp. 93, 97.

32 De la Cierva, *Franco*, I, p. 53; Hills, *Franco*, p. 62.

33 Hills, *Franco*, p. 62.

34 Suárez Fernández, *Franco*, I, pp. 97–9; Hills, *Franco*, p. 63.

35 Franco, 'Transcript of taped memoirs', Pozuelo, *Los 476 últimos días*, p. 100.

36 Angel Ossorio, *Mis memorias* (Buenos Aires, 1946) pp. 92–6; Joaquín Romero Maura, *"La rosa de fuego" El obrerismo barcelonés de 1899 a 1909* (Barcelona, 1975) pp. 501–42; Alvarez Junco, *Lerroux*, pp. 374–9; Joan B. Culla i Clarà, *El republicanisme Lerrouxista a Catalunya (1901–1923)* (Barcelona, 1986) pp. 205–14; Xavier Cuadrat, *Socialismo y anarquismo en Cataluña (1899–1911): los orígenes de la CNT* (Madrid, 1976) pp. 368–401; Joan Connelly Ullman, *The Tragic Week: A Study of Anti-Clericalism in Spain 1875–1912* (Harvard, 1968) pp. 129–297.

37 Franco Salgado-Araujo, *Mi vida*, p. 20; Suárez Fernández, *Franco*, I, p. 92; Franco, 'Transcript of taped memoirs', Pozuelo, *Los 476 últimos días*, p. 102.

38 Suárez Fernández, *Franco*, I, p. 97; De la Cierva, *Franco*, I, p. 55; Hills, *Franco*, p. 77.

39 Franco Salgado-Araujo, *Mi vida*, p. 23.

40 Faustino Moreno Villalba, *Franco, héroe cristiano en la guerra* (Madrid, 1985) p. 27.

41 Franco Salgado-Araujo, *Mi vida*, p. 26.

42 Arrarás, *Franco*, p. 21.

43 Fernando Reinlein García-Miranda, 'Del siglo XIX a la guerra civil', in Colectivo Democracia, *Los Ejércitos . . . más allá del golpe* (Barcelona, 1981) pp. 13–33; Stanley G. Payne, *Politics and the Military in Modern Spain* Stanford, 1967) pp. 87–9.

44 David S. Woolman, *Rebels in the Rif: Abd el Krim and the Rif Rebellion* (Stanford, 1969) pp. 4–5; J. A. S. Grenville, *Lord Salisbury and Foreign Policy: The Close of the Nineteenth Century* (London, 1964) pp. 431–3; Richard Shannon, *The Crisis of Imperialism 1865–1915* (London, 1974) pp. 342–4.

45 Woolman, *Rebels*, pp. 8–14.

46 Suárez Fernández, *Franco*, I, p. 105.

47 'Declaraciones de S.E. a Manuel Aznar', 31 December 1938, *Palabras del Caudillo 19 Abril*

1937 – 31 Diciembre 1938
(Barcelona, 1939) p. 314.

48 Vicente Gracia & Enrique
    Salgado, *Las cartas de amor de
    Franco* (Barcelona, 1978)
    pp. 28–97.

49 Arturo Barea, *La forja de un
    rebelde* (Buenos Aires, 1951)
    pp. 295–6, 320–1.

50 Georges Rotvand, *Franco means
    Business* (London, n.d.)
    pp. 8–9.

51 Coles, *Franco*, pp. 26, 123.

52 Arrarás, *Franco*, p. 135.

53 Suárez Fernández, *Franco*, I,
    pp. 111–14; Hills, *Franco*,
    pp. 95–6; Franco Salgado-
    Araujo, *Mi vida*, p. 30.

54 Arrarás, *Franco*, pp. 131–3.

55 Fontana, *Franco*, pp. 32–3,
    Garriga, *La Señora*, pp. 58–9.

56 Pedro Sainz Rodríguez,
    Testimonio y recuerdos (Barcelona,
    1978) pp. 334–5.

57 Suárez Fernández, *Franco*, I,
    p. 115 makes the entirely
    disingenuous suggestion that
    Franco had not done his case any
    good by his lack of repressive
    zeal during the repression of the
    general strike in Asturias in the
    summer of 1917. See also
    pp. 137–8.

58 Arrarás, *Franco*, p. 33.

59 Arrarás, *Franco*, p. 34.

60 Arturo Barea, *The Struggle for
    the Spanish Soul* (London, 1941)
    p. 23.

61 Interview with Carmen Polo,
    Estampa, 29 May 1928; Jaraiz
    Franco, *Historia*, pp. 37–40;
    Franco, *Nosotros*, p. 81.

62 Suárez Fernández, *Franco*, I,
    p. 119, affirms that Franco was
    attracted not by her money but

by 'an attraction towards that
sector of society which has
inherited the cultivation of
dignity which derives from the
minor nobility'.

63 Sainz Rodríguez, *Testimonio*,
    p. 323.

64 Ramón Garriga, *La Señora de El
    Pardo* (Barcelona, 1979)
    pp. 37–8.

65 '50 años de matrimonio', *La Voz
    de Asturias*, 21 October 1973;
    Juan Cueto Alas, *Guía secreta de
    Asturias* (Madrid, Editorial Al-
    berak, 1975) pp. 139–40.

66 Adrian Shubert, *The Road to
    Revolution in Spain: The Coal
    Miners of Asturias 1860–1934*
    (Urbana, 1987) pp. 46–118;
    Antonio L. Oliveros, *Asturias en
    el resurgimiento español (apuntes
    históricos y biográficos* (Madrid,
    1935) pp. 113–77.

67 Carolyn P. Boyd, *Praetorian
    Politics in Liberal Spain* (Chapel
    Hill, 1979) pp. 84–5, 286;
    Enrique Moradiellos, *El Sindicato
    de los Obreros Mineros Asturianos
    1910–1930* (Oviedo, 1986)
    pp. 58–9; Juan Antonio
    Lacomba Avellán, *La crisis
    española de 1917* (Madrid, 1970)
    pp. 269–4. Burguete tried to
    join the Socialist Party in 1931
    and his application was refused,
    Oliveros, *Asturias*, p. 120. He
    joined the Communist Party
    during the Civil War, Suárez
    Fernández, *Franco*, I, p. 131.

68 Francisco Aguado Sánchez, *La
    revolución de octubre de 1934*
    (Madrid, 1972) p. 193; Luis
    Galinsoga & Francisco Franco-
    Salgado, *Centinela de occidente
    (Semblanza biográfica de Francisco*

Franco (Barcelona, 1956) pp. 35–6; Brian Crozier, *Franco: A Biographical History* (London, 1967) p. 50.

69 Discurso ante los mineros en Oviedo, 19 May 1946, Francisco Franco, *Textos 1945–1950* (Madrid, 1951) pp. 417–25.

70 Hills, *Franco*, pp. 104–5. Cf. Galinsoga & Franco Salgado, *Centinela*, p. 36; Crozier, *Franco*, p. 51.

71 Llaneza, letters from prison, published in *El Minero de la Hulla*, August and September 1917, reprinted in Manuel Llaneza, *Escritos y discursos* (Oviedo, 1985) pp. 206–14.

72 Boyd, *Praetorian Politics*, p. 160.

73 Gabriel Cardona, *El poder militar en la España contemporánea hasta la guerra civil* (Madrid, 1983) pp. 70–1; Boyd, *Praetorian Politics*, p. 286; Woolman, *Rebels*, p. 64.

74 Woolman, *Rebels*, pp. 64–5.

75 José Millán Astray, 'Prólogo', Franco, *Diario de una Bandera*, p. 7; General Millán Astray, *Franco, el Caudillo* (Salamanca, 1939) pp. 9–12.

76 Franco Salgado-Araujo, *Mi vida*, p. 42.

77 *Estampa*, 29 May 1928.

78 Barea, *Forja*, pp. 315–16; Franco, *Diario*, pp. 18–19.

79 Franco Salgado-Araujo, *Mi vida*, pp. 50–2.

80 Millán Astray, *Franco*, p. 16.

81 Barea, *Spanish Soul*, pp. 29–31.

82 On Castro Girona and Xauen, see Barea, *Forja*, pp. 305, 323–5; Franco, *Diario*, pp. 51–52.

83 Millán Astray, *Franco*, pp. 11–12; Franco Salgado-Araujo, *Mi vida*, p. 46; Hills, *Franco*, pp. 111–15; Suárez Fernández, *Franco*, I, pp. 145–9.

84 Garriga, *La Señora*, p. 40.

85 José Martín Blázquez, *I Helped to Build an Army: Civil War Memoirs of a Spanish Staff Officer* (London, 1939) p. 302; Herbert R. Southworth, *Antifalange: estudio crítico de "Falange en la guerra de España: la Unificación y Hedilla" de Maximiano García Venero* (Paris, 1967) pp. xxi–xxii; Guillermo Cabanellas, *La guerra de los mil días* 2 vols (Buenos Aires, 1973) II, p. 792.

86 Franco, *Diario*, p. 177.

87 Sainz Rodríguez, *Testimonio*, p. 272.

88 Francisco Franco Salgado-Araujo, *Mis conversaciones privadas con Franco* (Barcelona, 1976) pp. 184–5.

89 Garriga, *La Señora*, p. 33; Joan Llarch, *Franco: biografía* (Barcelona, 1983) pp. 158–9.

90 Barea, *La forja*, pp. 408–9.

91 Woolman, *Rebels*, pp. 83–8.

92 Franco, *Diario*, pp. 99–103.

93 Woolman, *Rebels*, pp. 90–5.

94 Franco, *Diario*, pp. 109–13.

95 Franco, *Diario*, pp. 118–21.

96 Franco Salgado-Araujo, *Mi vida*, p. 49.

97 'El "As" de la Legión', *ABC*, 22 February 1922; Franco, *Diario*, pp. 145–51.

98 Franco, *Diario*, pp. 280–5.

99 Franco Salgado-Araujo, *Mi vida*, pp. 52–3; Franco, *Diario*, pp. 155–9.

100 Compare the contrasting

sentiments expressed in Franco, *Diario*, pp. 177 & 199–200.

101   Franco, *Diario*, p. 228.

102   Franco, *Diario*, pp. 243–50.

103   *El Correo Gallego*, 20 April 1922.

104   Sainz Rodríguez, *Testimonio*, p. 323.

105   *El Correo Gallego*, 10 October 1921.

106   *El Carbayón*, 23 February 1922.

CHAPTER II

1   The Sanjurjo remark is quoted in the adulatory profile, 'El "As" de la Legión', *ABC*, 22 February 1921.

2   Suárez Fernández, *Franco*, I, pp. 162–3; Hills, *Franco*, p. 130.

3   The most widely publicised accusation against the King was that by Vicente Blasco Ibáñez, *Alfonso XIII Unmasked* (London, 1925) pp. 78–83.

4   Dictamen de la Minoría Socialista, *El desastre de Melilla: dictamen formulado por Indalecio Prieto como miembro de la Comisión designada por el Congreso de los Diputados para entender en el expediente Picasso* (Madrid, 1922) p. 31

5   Woolman, *Rebels*, pp. 106–8; Barea, *Forja*, pp. 413–21.

6   Suárez Fernández, *Franco*, I, pp. 161–3.

7   *Nuevo Mundo*, 26 January 1923. On the rumour of ghost-writing, see Stanley G. Payne, *The Franco Regime 1936–1975* (Madison, 1987) p. 72.

8   Woolman, *Rebels*, pp. 113–14.

9   Garriga, *La Señora*, p. 44.

10   *La Voz de Asturias*, 10 June 1923.

11   *La Voz de Asturias*, 10 June 1923.

12   Niceto Alcalá Zamora, *Memorias* (Barcelona, 1977) p. 116.

13   De la Cierva, *Franco*, I, pp. 214–15.

14   María Teresa González Calbet, *La dictadura de Primo de Rivera: el Directorio militar* (Madrid, 1987) pp. 55–63.

15   Javier Tusell, *Radiografía de un golpe de Estado: el ascenso al Poder del General Primo de Rivera* (Madrid, 1987) pp. 110–12.

16   José Luis Gómez Navarro, *El régimen de Primo de Rivera: reyes, dictaduras y dictadores* (Madrid, 1991) pp. 126–9; González Calbet, *La Dictadura*, pp. 81–94.

17   Franco Salgado-Araujo, *Mi vida*, p. 66; Barea, *Forja*, pp. 449, 459–61.

18   Franco Salgado-Araujo, *Mi vida*, pp. 64–6; De la Cierva, *Franco*, I, pp. 207–9, 223; Hills, *Franco*, p. 132.

19   *El Carbayón*, 23 October; *ABC*, 23, 25 October 1923; *Región*, 23 October 1923; *La Voz de Asturias*, 24 October 1923, 21 October 1973.

20   Jaraiz Franco, *Historia*, pp. 40–2.

21   *Estampa*, 29 May 1928; Franco, *Nosotros*, p. 81.

22   *El Carbayón*, 23 October 1923.

23   *El Carbayón*, 23 October 1923.

24   *Asturias*, 15 November 1923.

25   *Mundo Gráfico*, 31 October 1923.

26   Suárez Fernández, *Franco*, I, p. 170.

27   Franco Salgado-Araujo, *Mis conversaciones*, pp. 62–3,

377–8; Hills, *Franco*, pp. 133–5. Franco's account to Hills, understandably after forty years, confuses the chronology of his meeting with Primo.

28 Payne, *Politics and the Military*, p. 209.

29 Franco, *Diario*, p. 278.

30 It is possible to trace the magnification of Franco's role through his biographies, Arrarás (1937), *Franco*, pp. 113–14; Galinsoga & Franco Salgado (1956), *Centinela*, pp. 89–91; Crozier (1967), *Franco*, p. 83; Hills (1967), *Franco*, p. 135. Cf. Woolman, *Rebels*, p. 187.

31 Teniente Coronel Franco, 'Pasividad e inacción', *Revista de Tropas Coloniales*, April 1924, reprinted in *Revista de Historia Militar* Año XX, No. 40, 1976, pp. 166–7.

32 Suárez Fernández, *Franco*, I. p. 171.

33 Suárez Fernández, *Franco*, I, p. 172.

34 Franco to Pareja, 5 July 1924, reproduced in De la Cierva, *Franco*, I, pp. 236–7.

35 Barea, *Forja*, pp. 472–3; Payne, *Politics and the Military*, p. 211.

36 José Calvo Sotelo, *Mis servicios al Estado* (Madrid, 1931) pp. 238–9; Arrarás, *Franco*, pp. 100–1; Galinsoga, *Centinela*, pp. 89–91; General Francisco Javier Mariñas, *General Varela (de soldado a general)* (Barcelona, 1956) pp. 35–6; Franco Salgado-Araujo, *Mis conversaciones*, pp. 137–8.

37 De la Cierva, *Franco*, I. pp. 235, 238–9.

38 Franco to Pareja, undated (late July 1924), reproduced in De la Cierva, *Franco*, I, pp. 240–1.

39 Gonzalo Queipo de Llano, *El general Queipo de Llano perseguido por la dictadura* (Madrid, Javier Morato, 1930) p. 105, Cf. De la Cierva, *Franco*, I, pp. 245–6.

40 Suárez Fernández, *Franco*, I, pp. 175–80; Galinsoga & Franco Salgado, *Centinela*, pp. 93–100; Payne, *Politics and the Military*, pp. 214–7. Millán Astray, *Franco*, p. 14, claimed, implausibly, that the entire operation had been masterminded by Franco.

41 Coronel Franco, 'Xauen la triste' in *Revista de Tropas Coloniales*, July 1926, reprinted in *Revista de Historia Militar*, Año XX, No. 40, 1976, pp. 174–8.

42 Carlos Fernández, *El almirante Carrero* (Barcelona, 1985) pp. 23–6.

43 Dated 1 March 1925, the letter was reproduced in *Estampa*, 29 May 1928.

44 *ABC*, 9 September 1925; Francisco Franco, diary entry for 8 September 1925 'Diario de Alhucemas', *Revista de Historia Militar*, No. 40, 1976, p. 229; Franco's testimony to Vicente Pozuelo, Pozuelo, *Los últimos 476 días*, p. 41.

45 *ABC*, 27 May 1926; Woolman, *Rebels*, pp. 191–3.

46 The 1970 censored version in *Revista de Historia Militar*, No. 40, 1976, pp. 227–47, and in the original in Francisco Franco Bahamonde, *Papeles de la*

guerra de Marruecos (Madrid, 1986) pp. 203–27.

47   Franco, diary entry for 22 September 1925, *Revista de Historia Militar*, No. 40, 1976, pp. 232–3.

48   Ramón Soriano, *La mano izquierda de Franco* (Barcelona, 1981) p. 144.

49   *El Eco de Santiago*, 4 February 1926.

50   *Hoja de servicios del Caudillo de España* edited by Esteban Carvallo de Cora (Madrid, 1967) pp. 115–18.

51   Barea, *The Struggle*, p. 32.

52   J. W. D. Trythall, *Franco* (London, 1970), p. 31.

53   *La Voz de Galicia*, 29, 31 January, 2, 3, 4, 5, 6, 7, 9. 10, 11 February; *ABC*, 11 February 1926; Ramón Franco & Julio Ruiz de Alda, *De Palos al Plata* (Madrid, 1926) *passim*; Díaz, *Mi vida*, pp. 28–51.

54   *El Compostelano*, 2 February; *La Voz de Galicia*, 6, 7 February 1926.

55   Sainz Rodríguez, *Testimonio*, p. 333.

56   Arrarás, *Franco*, pp. 139–40.

57   *La Voz de Galicia*, 9, 10, 11, 12, 13 February; *El Compostelano*, 12 February 1926.

58   Eugenio Vegas Latapie, *Memorias políticas: el suicidio de la monarquía y la segunda República* (Barcelona, 1983) p. 38.

59   Hills, *Franco*, pp. 154–5.

60   *El Carbayón*, 17 September 1926. Her birth certificate is reproduced in Soriano, *La mano*, p. 104.

61   Soriano, *La mano*, p. 103.

62   Díaz, *Mi vida*, pp. 21, 157.

63   Franco, *Nosotros*, pp. 84–5. Curiously, two of Franco's biographers have also attributed Carmencita's birth to 1928, Hills, *Franco*, p. 157; Crozier, *Franco*, p. 92.

64   Franco Salgado-Araujo, *Mi vida*, p. 71.

65   Franco Salgado-Araujo, *Mi vida*, p. 74; Jaraiz Franco, *Historia*, p. 79; Fontana, *Franco*, p. 80.

66   José María Gárate Córdoba, ' "Raza", un guión de cine', *Revista de Historia Militar*, No. 40, 1976, p. 59.

67   On the spare time, Soriano, *La mano*, p. 61; on the reading, Francisco Franco, *"Apuntes" personales sobre la República y la guerra civil* (Madrid, 1987) p. 6.

68   Franco, *Apuntes*, p. 6; Soriano, *La mano*, p. 61.

69   Franco, *Apuntes*, p. 6; Soriano, *La mano*, pp. 61–2, 154–5.

70   Michael Alpert, *La reforma militar de Azaña (1931–1933* (Madrid, 1982) pp. 106–9, 120; Shlomo Ben-Ami, *Fascism from Above: The Dictatorship of Primo de Rivera in Spain 1923–1930* (Oxford, 1983) pp. 356–8; Payne, *Politics and the Military*, pp. 240–2.

71   Emilio Mola Vidal, *Obras completas* (Valladolid, 1940) p. 1024.

72   General E, López Ochoa, *De la Dictadura a la República* (Madrid, 1930) pp. 78–116; Francisco Hernández Mir, *La Dictadura ante la Historia: un crimen de lesa patria* (Madrid, 1930) pp. 259–72; Vicente Marco Miranda, *La conspiraciones*

contra la Dictadura (1923–1930) relato de un testigo 2ª edición (Madrid, 1975) pp. 53–68.

73 López Ochoa, *De la Dictadura*, pp. 109–24; Gabriel Maura, *Bosquejo histórico de la Dictadura* 2 vols (Madrid, 1930) I, pp. 317–22, 325–37, 360–77; Payne, *Politics and the Military*, pp. 236–40.

74 Ben-Ami, *Fascism from Above*, pp. 361–3; Duque de Maura & Melchor Fernández Almagro, *Por qué cayó Alfonso XIII* (Madrid, 1948) p. 395.

75 The difference between the *juntero* and the *Africanista* attitude to Primo's military policy is clear in the respective accounts of López Ochoa, *De la Dictadura*, pp. 78–124 and Mola, *Obras*, p. 395.

76 Mola, *Obras*, p. 1026; Julio Busquets, *El militar de carrera en España* 3ª edición (Barcelona, 1984) pp. 78–80.

77 Calvo Sotelo, *Mis servicios*, p. 239.

78 Carlos Blanco Escolá, *La Academia General Militar de Zaragoza (1928–1931)* (Barcelona, 1989) pp. 45–2, 59–66.

79 Barea, *La forja*, p. 409.

80 Franco Salgado-Araujo, *Mi vida*, pp. 71, 82–5, 108, 113, 122, 142.

81 Jaraiz Franco, *Historia*, p. 78.

82 Franco Salgado-Araujo, *Mi vida*, p. 153.

83 *Estampa*, 29 May 1928.

84 Discurso de apertura de la Academia General Militar, *Revista de Historia Militar*, Año XX, No. 40, 1976, pp. 333–4.

85 Arrarás, *Franco*, pp. 150–1; Carlos de Silva, *General Millán Astray, el legionario* (Barcelona, 1956) pp. 143–7; Blanco Escolá, *Academia*, pp. 102–6, 111–17.

86 Franco Salgado-Araujo, *Mi vida*, pp. 78, 82.

87 Franco Salgado-Araujo, *Mi vida*, pp. 81–4; Hills, *Franco*, pp. 155–7; Mariano Aguilar Olivencia, *El Ejército español durante la segunda República* (Madrid, 1986) pp. 119–29.

88 Peter Kemp, *Mine Were of Trouble* (London, 1957) p. 115.

89 'Discurso de despedida en el cierre de la Academia General Militar', *Revista de Historia Militar*, Año XX, No. 40, 1976, pp. 335–7.

90 *The Morning Post*, 20 July 1937.

91 Mola, *Obras*, p. 1027; Cabanellas, *Cuatro generales*, pp. 140, 142; Coronel Jesús Pérez Salas, *Guerra en España (1936–1939)* (México D.F., 1947) pp. 85–7; Antonio Cordón, *Trayectoria (recuerdos de un artillero)* (Paris, 1971) pp. 192–4.

92 Blanco Escolá, *Academia*, pp. 127–48.

93 Arrarás, *Franco*, pp. 145–6; Franco Salgado-Araujo, *Mi vida*, pp. 77–8; Busquets, *El militar*, pp. 117–39; Blanco Escolá, *Academia*, pp. 69–72, 149–98; Garriga, *La señora*, p. 57.

94 Franco Salgado-Araujo, *Mis conversaciones*, p. 106; De la Cierva, *Franco*, II, pp. 18–19.

95 On this subject, I am immensely grateful to Herbert R. Southworth who kindly permitted me to consult his

work-in-progress on the Entente.

96 Crozier, *Franco*, p. 92. Crozier's book was published in Spain in Franco's lifetime and his account of what Franco had told him was neither altered by the censorship machinery nor repudiated subsequently. Cf. Hills, *Franco*, p. 157; Suárez Fernández, *Franco*, I, pp. 197–8.

97 Franco Salgado-Araujo, *Mi vida*, p. 82.

98 Franco Salgado-Araujo, *Mi vida*, p. 84.

99 Garriga, *La señora*, p. 62.

100 Vegas Latapie, *Memorias*, p. 101; Franco Salgado-Araujo, *Mi vida*, p. 89.

101 Díaz, *Mi vida*, p. 24.

102 Ramón's account in Comandante Franco, *Aguilas y garras: historia sincera de una empresa discutida* (Madrid, n.d. [1929]).

103 Calvo Sotelo, *Mis servicios*, p. 245.

104 *ABC*, 2 July 1929.

105 Díaz, *Mi vida*, pp. 64–91; Garriga, *Ramón Franco*, pp. 127–54.

106 Reprinted in Garriga, *Ramón Franco*, pp. 173–4.

107 Reprinted in Garriga, *Ramón Franco*, pp. 175–8.

108 Shlomo Ben-Ami, *The Origins of the Second Republic in Spain* (Oxford, 1978) pp. 76–94.

109 On Ramón's revolutionary activities, see Comandante Franco [Ramón], *Madrid bajo las bombas* (Madrid, 1931) pp. 83–114; Díaz, *Mi vida*, pp. 94–153.

110 Mola, *Obras*, pp. 389–95,

408–12, 454–5; Franco, *Madrid bajo las bombas*, pp. 117–46; Garriga, *Ramón Franco*, pp. 186–94.

111 Arrarás, *Franco*, pp. 155–7; Franco Salgado-Araujo, *Mi vida*, p. 90.

112 Alejandro Lerroux, *La pequeña historia: apuntes para la historia grande vividos y redactados por el autor* (Buenos Aires, 1945) pp. 568–9.

113 Franco Salgado-Araujo, *Mi vida*, pp. 91–2.

114 Graco Marsá, *La sublevación de Jaca; relato de un rebelde* (Paris, 1931) *passim*; José María Azpíroz Pascual & Fernando Elboj Broto, *La sublevación de Jaca* (Zaragoza, 1984) pp. 27–36.

115 Franco Salgado-Araujo, *Mi vida*, p. 92; Azpíroz Pascual & Elboj Broto, *Jaca*, pp. 81–93.

116 Azpíroz Pascual & Elboj Broto, *Jaca*, pp. 109–17.

117 Franco, *Madrid bajo las bombas*, pp. 171–5.

118 Garriga, *Ramón Franco*, p. 204.

119 Díaz, *Mi vida*, p. 138; Garriga, *Ramón Franco*, pp. 209–11.

120 Azpíroz Pascual & Elboj Broto, *Jaca*, p. 66.

121 *ABC*, 14, 15 February 1931.

122 *ABC* 18, 19 February 1931; Dámaso Berenguer, *De la Dictadura a la República* (Madrid, 1946) pp. 320–33.

123 Azpíroz Pascual & Elboj Broto, *Jaca*, pp. 144–9; Berenguer, *De la Dictadura*, p. 344; Suárez Fernández, *Franco*, I, p. 206.

124 Franco Salgado-Araujo, *Mi vida*, p. 93.

125 Pemán, *Mis encuentros*, pp. 14–16; Saña, *El franquismo*,

p. 42; Garriga, *La señora*, pp. 57–9.

126 Fernando García Lahiguera, *Ramón Serrano Suñer: un documento para la historia* (Barcelona, 1983) p. 41; Ramón Serrano Suñer, *Entre el silencio y la propaganda, la Historia como fue. Memorias* (Barcelona, 1977) pp. 54–6.

CHAPTER III

1 Franco Salgado-Araujo, *Mi vida*, pp. 96–7.

2 Berenguer, *De la Dictadura*, pp. 355–6.

3 Franco Salgado-Araujo, *Mis conversaciones*, pp. 88, 498.

4 Lerroux, *La pequeña historia*, pp. 83–4; Marqués de Hoyos, *Mi testimonio* (Madrid, 1962) p. 59; Mola, *Obras*, p. 631.

5 Serrano Suñer, *Memorias*, p. 20; Garriga, *Franco-Serrano*, p. 16.

6 Franco Salgado-Araujo, *Mi vida*, pp. 96–7; Franco Salgado-Araujo, *Mis conversaciones*, pp. 450–2.

7 Berenguer, *De la Dictadura*, pp. 357–8; Julián Cortés Cavanillas, *La caída de Alfonso XIII* 7ª edición (Madrid, n.d., but 1933) pp. 217–18.

8 Berenguer, *De la Dictadura*, pp. 349–51, 394–7.

9 Mola, *Obras*, p. 867.

10 Arrarás, *Franco*, pp. 159–60. On Franco's aversion for the Republic, Franco, *Nosotros*, p. 90.

11 Franco Salgado-Araujo, *Mi vida*, pp. 98–100; Suárez Fernández, *Franco*, I, p. 211, confuses General Gómez

Morato with the wartime air ace Joaquín García Morato.

12 Franco, *Apuntes personales*, pp. 7–9.

13 Franco Salgado-Araujo, *Mis conversaciones*, p. 88.

14 Alfredo Kindelán Núñez del Pino, 'Semblanza político-militar del general Kindelán', prologue to Alfredo Kindelán Duany, *Mis cuadernos de guerra* 2ª edición, (Barcelona, 1982) p. 40.

15 *ABC*, 18, 21 April 1931; Azaña, diary entry for 27 January 1932, *Obras*, IV, p. 315.

16 Ramón Salas Larrazábal, *Historia del Ejército popular de la República* 4 vols (Madrid, 1973) I, pp. 7, 14, 22–3; Santos Juliá, *Manuel Azaña: una biografía política, del Ateneo al Palacio Nacional* (Madrid, 1990) pp. 98–106.

17 Antonio Cordón, *Trayectoria (recuerdos de un artillero)* (Paris, 1971) p. 196; Salas Larrazábal, *Ejército popular*, I, pp. 5–6; Juliá, *Azaña*, p. 106.

18 Alpert, *La reforma militar*, pp. 125–31; *ABC*, 24 April; *La Época*, 24 April 1931.

19 Maximiano García Venero, *El general Fanjul: Madrid en el alzamiento nacional* (Madrid, 1967) pp. 189–93.

20 José Antonio Vaca de Osma, *Paisajes con Franco al fondo* (Barcelona, 1987) pp. 18–21.

21 Mola, *Obras*, pp. 1056–8; Alpert, *La reforma militar*, pp. 133–50; Aguilar Olivencia, *El Ejército*, pp. 65–75.

22 Alpert, *La reforma militar*, pp. 150–74. There is

23 considerable debate over the figures, see Salas Larrazábal, *Ejército popular*, I, pp. 8–13.

23 Franco Salgado-Araujo, *Mi vida*, pp. 101–2.

24 Berenguer, *De la Dictadura*, p. 407.

25 Mola, *Obras*, pp. 879–80; José María Iribarren, *Mola, datos para una biografía y para la historia del alzamiento nacional* (Zaragoza, 1938) pp. 39–40.

26 Carolyn P. Boyd, ' "Responsibilities" and the Second Republic, 1931–1936' in Martin Blinkhorn, ed., *Spain in Conflict 1931–1939: Democracy and its Enemies* (London, 1986) pp. 14–39.

27 Franco Salgado-Araujo, *Mis conversaciones*, p. 464.

28 Franco Salgado-Araujo, *Mi vida*, pp. 100–2.

29 Franco Salgado-Araujo, diary entry for 28 February 1955, *Mis conversaciones*, pp. 88–9.

30 Mola, *Obras*, pp. 1062–3; Aguilar Olivencia, *El Ejército*, pp. 147–57.

31 Alpert, *La reforma militar*, pp. 216–28; Azaña, diary entry for 20 July 1931, *Obras*, IV, p. 35.

32 *La Correspondencia Militar*, 18 June, 17, 31 July 1931; Mola, *Obras*, pp. 1045–65. Cf. Cordón, *Trayectoria*, p. 194.

33 The speech is reproduced in full in Eduardo Espín, *Azaña en el poder: el partido de Acción Republicana* (Madrid, 1980) pp. 323–34. See p. 330. On the injustice of the accusation, see Alpert, *La reforma militar*, pp. 293–7, whose account is slightly flawed by his acceptance of the misquoted version of the speech given in Miguel Maura, *Así cayó Alfonso XIII*, (México D.F., 1962) p. 227. Payne, *Politics & the Military*, p. 275 accepts that Azaña said the offending phrase. Hugh Thomas, *The Spanish Civil War* 3rd edition (London, 1977) p. 92, accepts Maura's version.

34 Salas Larrazábal, *Ejército popular*, I, pp. 14, 78; Aguilar Olivencia, *El Ejército*, pp. 77–83.

35 Cordón, *Trayectoria*, pp. 192–3, 197; Juliá, *Azaña*, pp. 101–2.

36 Crozier, *Franco*, p. 92.

37 Díaz, *Mi vida*, p. 159; Arxiu Vidal i Barraquer, *Esglesia i Estat durant la segona República espanyola 1931–1936* 8 vols, (Montserrat, 1971–1990) I, p. 85; Garriga, *Ramón Franco*, p. 232.

38 Franco, *Apuntes*, p. 4.

39 Salas Larrazábal, *Ejército popular*, I, p. 19; Cordón, *Trayectoria*, pp. 192–3; Arrarás, *Franco*, pp. 166–7; Mola, *Obras*, p. 1027; Franco Salgado-Araujo, *Mi vida*, pp. 104–6; Franco, *Nosotros*, pp. 90–2.

40 Azaña, diary entry for 20 July 1931, *Obras*, IV, p. 35.

41 'Discurso de despedida en el cierre de la Academia General Militar', *Revista de Historia Militar*, Año XX, No. 40, 1976, pp. 335–7.

42 Franco Salgado-Araujo, 25 May 1964, *Mis conversaciones*, p. 425; Franco Salgado-Araujo, *Mi vida*, p. 122.

43 Franco Salgado-Araujo, *Mi vida*, pp. 11, 104.

44 Azaña, 16, 22 July 1931, *Obras*, IV, pp. 33, 39. See also 9 December 1932, *Memorias íntimas de Azaña* (Madrid, 1939) pp. 307–8. Franco's service record in *Hoja de Servicios*, pp. 82–3.

45 Franco to General Gómez Morato, Vª División, 24 July, Gómez Morato to Minister of War, 28 July, Archivo Azaña, Ministerio de Asuntos Exteriores, RE. 131–1.

46 Testimony of Ramón Serrano Suñer to the author; Garriga, *La señora*, p. 70.

47 Azaña, 22 July 1931, *Obras*, IV, pp. 40–2; Joaquín Arrarás, *Historia de la segunda República* 4 vols (Madrid, 1956–1968) I, pp. 153–8.

48 Azaña's diaries aside, there is little evidence that Franco was engaged in active subversion at this time. See Azaña, 25 July, 15, 16 September 1931, *Obras*, IV, pp. 46, 129, 131; Joaquín Arrarás, *Historia de la segunda República española* 4 vols (Madrid, 1956–1968) I, p. 470.

49 Azaña, 12, 13, 14 August 1931, *Obras*, IV, pp. 79–80, 83.

50 Azaña, 21 August 1931, *Obras*, IV, pp. 95–6.

51 Arrarás, *Franco*, p. 166.

52 Boyd, ' "Responsibilities" ', pp. 22–3.

53 Azaña, diary entry for 2 September 1931, *Obras*, IV, pp. 115–16.

54 *El Debate*, 16, 18, 20, 23 October, 1, 3, 10, 12 November; *ABC* 11, 14, 16 October 1931.

55 Azaña, 5 July, 2 September 1931, *Obras*, pp. 12, 115; Franco, *Madrid, bajo las bombas*, p.v.; Díaz, *Mi vida*, pp. 154–5; Garriga, *Ramón Franco*, pp. 234–44; Alcalá-Zamora, *Memorias*, p. 201.

56 Suárez Fernández, *Franco*, I, pp. 232–7.

57 Boyd, ' "Responsibilities" ', pp. 32–3.

58 *La Voz de Galicia*, 5, 28 February 1932. Franco Salgado-Araujo, *Mi vida*, p. 107.

59 Juliá, *Azaña*, p. 176.

60 See the submissions of the defence lawyers at the trial of the villagers, Luis Jiménez de Asúa, Juan-Simeón Vidarte, et al., *Castilblanco* (Madrid, 1933).

61 *ABC*, 1, 2, 3 January 1932; Azaña, 5 January 1932, *Obras*, IV, pp. 293–6.

62 Paul Preston, *The Coming of the Spanish Civil War: Reform, Reaction and Revolution in the Second Republic 1931–1936* (London, 1978) pp. 67–9.

63 *El Socialista*, 2, 5 January; *El Heraldo de Madrid*, 5, 6, 9 January 1932; *Diario de Sesiones de Cortes*, 5, 6 January 1932.

64 Azaña, 1, 3, 4 February 1932, *Obras*, IV, pp. 322–4; *La Voz de Galicia*, 5 February 1932.

65 Ismael Saz Campos, *Mussolini contra la II República: hostilidad, conspiraciones, intervención (1931–1936)* (Valencia, 1986) p. 39.

66 Azaña, 8, 11, 12, 17 January 1932, pp. 299–301, 306–10; Lerroux, *La pequeña historia*, pp. 143–6; Diego Martínez

Barrio, *Memorias* (Barcelona, 1983) pp. 105–6.

67 Jaraiz Franco, *Historia*, p. 79; Borrell in Mérida, *Testigos*, pp. 218–27; Vicente Gil, *Cuarenta años junto a Franco*, p. 68.

68 Franco, *Apuntes* (Barcelona, 1981) p. 9; Franco Salgado-Araujo, *Mi vida*, p. 108; Suárez Fernández, *Franco*, I, pp. 246–7.

69 Sainz Rodríguez, *Testimonio*, pp. 325–6; Julián Lago, *Las contramemorias de Franco* (Barcelona, 1976) pp. 137–8.

70 Ansaldo, *¿Para qué . . . ?*, p. 51.

71 José María Gil Robles, *No fue posible la paz* (Barcelona, 1968) p. 235.

72 Franco Salgado-Araujo, 29 June 1965, *Mis conversaciones*, p. 452.

73 Azaña, 11 July 1932, *Obras*, IV, pp. 433–4; *El Sol*, 12 July 1932. On Lerroux and military prejudices, see Alvarez Junco, *El emperador del Paralelo*, pp. 320–3. On his involvement with Sanjurjo, Nigel Townson, 'The Collapse of the Centre: The Radical Republican Party during the Second Spanish Republic'. Unpublished doctoral thesis, University of London, 1991, chapter 4.

74 Galinsoga & Franco Salgado, *Centinela*, p. 157.

75 Franco Salgado-Araujo, *Mis conversaciones*, pp. 498–9.

76 Vegas Latapie, *Memorias*, p. 184.

77 110 right-wing newspapers were closed down for four months. *ABC*, 30 November 1932; Alfonso Senra, *Del 10 de agosto a la sala sexta del Supremo*

(Madrid, 1933) pp. 32–5. On the coup, see Ansaldo, *¿Para qué . . . ?*, pp. 32–45; Arrarás, *Cruzada*, I, p. 435–529; Santiago Galindo Herrero, *Los partidos monárquicos bajo la segunda República* 2nd ed. (Madrid, 1956) pp. 156–66; Emilio Estéban Infantes, *La sublevación del general Sanjurjo* (Madrid, 1933) *passim*.

78 Azaña, diary entry for 12 August 1932, *Memorias íntimas*, pp. 204–5.

79 Franco Salgado-Araujo, *Mi vida*, pp. 108–9; Franco Salgado-Araujo, *Mis conversaciones*, p. 452.

80 *Acción Española*, tomo XVIII, (Burgos) March 1937, pp. 17–19; Vegas Latapie, *Caminos*, p. 79. The subscription was a gift from the Marqués de la Vega de Anzó.

81 Ansaldo, *¿Para qué . . . ?*, pp. 47–50.

82 Azaña, diary entry for 29 August 1933, quoted General Jorge Vigón, *General Mola (el conspirador)* (Barcelona, 1957) p. 79, presumably from the stolen portion of his diary which was apparently available to top Francoists.

83 Ansaldo, *¿Para qué . . . ?*, p. 50.

84 *La Voz de Galicia*, 18, 20 September; *El Eco de Santiago*, 19, 20, 21, 22 September; *El Compostelano*, 19, 20 September 1932; Franco Salgado-Araujo, *Mi vida*, p. 109; Galinsoga & Salgado-Araujo, *Centinela*, pp. 158–9; Suárez Fernández, *Franco*, I, p. 249.

85 It has been claimed that at this

86 Baón, *La cara humana*, p. 110.

87 Ansaldo, *¿Para qué . . . ?*, p. 51; Antonio Cacho Zabalza, *La Unión Militar Española* (Alicante, 1940) p. 14; Vicente Guarner, *Cataluña en la guerra de España* (Madrid, 1975) pp. 64–6; Julio Busquets, 'La Unión Militar Española, 1933–1936', Historia 16, *La guerra civil* 24 vols (Madrid, 1986) III, pp. 86-99.

time, he bent to prevailing winds by trying to become a freemason, his application being turned down by officers close to Azaña. Ferrer Benimelli, 'Franco contra la masonería', pp. 43–4.

88 Azaña, 8 February 1933, *Memorias íntimas*, p. 310; Alpert, *La reforma militar*, pp. 223–8.

89 Franco, *Apuntes*, p. 9.

90 Azaña, 8 February 1933, *Memorias íntimas*, p. 310.

91 Franco, *Apuntes*, p. 9.

92 Azaña, 24 February 1933, *Memorias íntimas*, p. 310.

93 Azaña, 1 March 1933, *Obras*, IV, p. 447.

94 *ABC*, 12 September 1933.

95 Arrarás, *Franco*, pp. 168–71. It is curious that Gil Robles, who misses no opportunity to stress his links with Franco, makes no mention of this in his memoirs.

96 Suárez Fernández, *Franco*, I, p. 262.

CHAPTER IV

1 Enrique Montañés, *Anarcosindicalismo y cambio político: Zaragoza, 1930–1936* (Zaragoza, 1989) pp. 98–100; Arrarás, *Historia de la segunda República*, II, pp. 249–56.

2 Aguilar Olivencia, *El Ejército*, pp. 371–2.

3 Diego Hidalgo Durán, *Un notario español en Rusia* (Madrid, 1929); Diego Hidalgo, *¿Por qué fui lanzado del Ministerio de la Guerra? Diez meses de actuación ministerial* (Madrid, 1934) pp. 38, 103–4; Concha Muñoz Tinoco, *Diego Hidalgo, un notario republicano* (Badajoz, 1986) pp. 19, 87–9.

4 Hidalgo, *¿Por qué fui lanzado?*, pp. 105–12; Muñoz Tinoco, *Hidalgo*, pp. 89–92; Elsa López, José Alvarez Junco, Manuel Espadas Burgos & Concha Muñoz Tinoco, *Diego Hidalgo: memoria de un tiempo difícil* (Madrid, 1986) pp. 153–62; Cardona, *El poder militar*, pp. 197–8; Carlos Seco Serrano, *Militarismo y civilismo en la España contemporánea* (Madrid, 1984) p. 408.

5 *ABC*, 28, 30 March 1934; Franco Salgado, *Mi vida*, p. 114.

6 Interview with Diego Hidalgo in *The Sunday Express*, 15 May 1938.

7 Hidalgo, *¿Por qué fui lanzado?*, pp. 77–9; *The Sunday Express*, 15 May 1938; Franco Salgado, *Mi vida*, p. 114.

8 Antonio Lizarza Iribarren, *Memorias de la conspiración* (Pamplona, 1953) pp. 22–6; *How Mussolini Provoked the Spanish Civil War: Documentary Evidence* (London, 1938) *passim*; Saz, *Mussolini contra la II República*, pp. 66–82; John F.

Coverdale, *Italian Intervention in the Spanish Civil War* (Princeton, 1975) pp. 50–4.

9 Franco to the Secretary of the Entente, 16 May 1934, reproduced in Bureau Permanent de l'Entente Internationale Anticommuniste, *Dix-sept ans de lutte contre le bolchévisme 1924–1940* (Geneva, 1940) p. 35; further correspondence in *Documentos inéditos para la historia del Generalísimo Franco* I (Madrid, 1992) pp. 11–12.

10 Southworth, manuscript on Entente; Suárez Fernández, *Franco*, I, pp. 268–9; Hills, *Franco*, p. 207.

11 Franco Salgado, *Mi vida*, pp. 112–14; Jaraiz Franco, *Historia*, p. 94; Franco, *Nosotros*, pp. 18–19.

12 Jaraiz Franco, *Historia*, pp. 82, 95.

13 Rafael Salazar Alonso, *Bajo el signo de la revolución* (Madrid, 1935) pp. 50–73; Preston, *CSCW*, pp. 108–112.

14 Salazar Alonso, *Bajo el signo*, pp. 141ff.; Preston, *CSCW*, pp. 112–17.

15 *El Debate*, 11 September; *CEDA*, 15 September 1934; Gil Robles, *No fue posible*, pp. 127–30.

16 Gil Robles, *No fue posible*, p. 131; Salazar Alonso, *Bajo el signo*, pp. 319–20.

17 *CEDA*, nos 36–7, December 1934.

18 *El Carbayón*, 18, 23, 29 September 1934.

19 Hidalgo, *¿Por qué fui lanzado?*, pp. 79–81; Arrarás, *República*, II,

pp. 440–1; Manuel Ballbé, *Orden público y militarismo en la España constitucional (1812–1983)* (Madrid, 1983) p. 374.

20 García Venero, *Fanjul*, p. 196.

21 *Diario de las sesiones de Cortes*, 4, 7 November 1934.

22 Salazar Alonso, *Bajo el signo*, Ricardo de la Cierva, *Historia de la guerra civil española* (Madrid, 1969) pp. 302–3.

23 Primo de Rivera to Franco, 24 September 1934, José Antonio Primo de Rivera, *Textos de doctrina política* 4th ed. (Madrid, 1966) pp. 297–300; Serrano Suñer, *Memorias*, pp. 54–6; Franco, *Apuntes*, p. 9.

24 Santos Juliá, 'Fracaso de una insurrección y derrota de una huelga: los hechos de octubre en Madrid', *Estudios de Historia social*, No. 31, October–December 1984, pp. 37–47; Francisco Aguado Sánchez, *La revolución de octubre de 1934* (Madrid, 1972) pp. 351–456.

25 Frederic Escofet, *Al servei de Catalunya i de la República* 2 vols (Paris, 1973) Vol. 1 *La desfeta 6 d'octubre 1934*, pp. 109–44; J. Costa i Deu & Modest Sabaté, *La veritat del 6 d'octubre* (Barcelona, 1936) *passim*; Enric Ucelay da Cal, *La Catalunya populista. Imatge, cultura i política en la etapa republicana (1931–1939)* (Barcelona, 1982) pp. 208–20.

26 The most detailed reconstruction of the events in Asturias is Pablo Ignacio Taibo II, *Octubre 1934: el ascenso* and

Octubre 1934: la caída vols 7 &
8 of *Historia general de Asturias*
(Gijón, 1978).

27 Hidalgo, *¿Por qué fui lanzado?*,
pp. 79–81; Muñoz Tinoco,
Hidalgo, p. 93; López et al.,
Hidalgo, pp. 171–2.

28 General López Ochoa,
*Campaña militar de Asturias en
octubre de 1934 (narración táctico-
episódica)* (Madrid, 1936)
pp. 11–12, 26–9; Gil Robles,
No fue posible, pp. 140–1; César
Jalón, *Memorias políticas:
periodista, ministro, presidiario.*
(Madrid, 1973) pp. 128–31;
Juan-Simeón Vidarte, *El bienio
negro y la insurrección de Asturias*
(Barcelona, 1978) pp. 358–9;
Aguado Sánchez, *La revolución*,
pp. 188–93; Cardona, *El poder
militar*, pp. 203–5.

29 Alcalá Zamora, *Memorias*,
p. 296; Vidarte, *El bienio negro*,
pp. 290–1.

30 Ballbé, *Orden público*, pp. 371–2.

31 Hidalgo, *¿Por qué fui lanzado?*,
pp. 79–81; Franco Salgado, *Mi
vida*, p. 115.

32 Serrano Suñer, *Memorias*, p. 52;
Aguado Sánchez, *La revolución*,
p. 192.

33 Franco, *Apuntes personales*,
pp. 11–12.

34 Franco Salgado-Araujo, *Mi
vida*, pp. 114–16; Arrarás,
Franco, p. 189.

35 Claude Martin, *Franco, soldado y
estadista* (Madrid, 1965)
pp. 129–30.

36 Jalón, *Memorias*, pp. 139–40;
Franco Salgado, *Mi vida*,
pp. 116–21; Aguado Sánchez,
La revolución, pp. 193–6, 257–8;
Manuel D. Benavides, *La

revolución fue así (octubre rojo y
negro) reportaje* (Barcelona,
1935) p. 330; Bernardo Díaz
Nosty, *La comuna asturiana:
revolución de octubre de 1934*
(Bilbao, 1974) pp. 164, 240–4,
314. On Franco's relationship
with Ricardo de la Puente and
his execution, see Jaraiz
Franco, *Historia*, pp. 73–85.

37 Arrarás, *Historia de la segunda
República española* II, pp. 614,
637–8; Joaquín Arrarás, *Historia
de la cruzada española* 8 vols
(Madrid, 1939–1943) II, t.7°,
p. 259; Franco, *Apuntes*, p. 12.

38 On the repression, see *ABC*, 13
October 1934; Ignacio Carral,
Por qué mataron a Luis de Sirval
(Madrid, 1935) pp. 37–60; Díaz
Nosty, *La comuna*, pp. 355–72;
José Martín Blázquez, *I Helped to
Build an Army: Civil War
Memoirs of a Spanish Staff Officer*
(London, 1939) pp. 12–33.

39 Suárez Fernández, *Franco*, I,
p. 285, denies that Franco had
anything to do with Doval's
appointment. Cf. Hidalgo, *¿Por
qué fui lanzado?*, pp. 91–3;
Ballbé, *Orden público*, pp. 372–3;
Serrano Suñer, *Memorias*, p. 52;
Aguado Sánchez, *La revolución*,
pp. 308–9; La Cierva, *Historia*,
p. 448.

40 *Diario de las sesiones de Cortes*, 6
November 1934; *ABC*, 6, 7, 10,
12, 13, 16, October 1934.

41 For López Ochoa's account of
the atrocities committed by the
Legion and by Doval's men, see
Vidarte, *El bienio negro*,
pp. 359–62.

42 Escofet, *La desfeta*, pp. 147–93;
Juan Antonio Sánchez y García

Sauco, *La revolución de 1934 en Asturias* (Madrid, 1974) pp. 146–8.

43   Celesia to Mussolini, 17 October 1934, Archivio Storico e Diplomatico del Ministero degli Affari Esteri, Rome (henceforth ASDMAE) Politica Spagna, b.9, R.3316/1717; Franco Salgado-Araujo, *Mi vida*, p. 123.

44   Arrarás, *Cruzada* II, p. 277; Gil Robles, *No fue posible*, pp. 141–8.

45   Ansaldo, *¿Para qué . . . ?*, pp. 92–3.

46   *Diario de las sesiones de Cortes*, 6, 7, 14, 15 November 1934.

47   Lerroux, *La pequeña historia*, pp. 344–7, 354.

48   Franco, *Apuntes personales*, p. 13.

49   *Documentos inéditos*, I, p. 12.

50   Franco, *Apuntes personales*, p. 13.

51   Suárez Fernández, *Franco*, I, p. 287.

52   Manuel Azaña, *Mi rebelión en Barcelona* (Madrid, 1935) *passim* but especially pp. 11–22; Paul Preston, 'The Creation of the Popular Front in Spain', in Helen Graham & Paul Preston, editors, *The Popular Front in Europe* (London, 1987) pp. 84–105.

53   Gil Robes, *No fue posible*, p. 235.

54   Serrano Suñer, *Memorias*, pp. 50, 52.

55   Franco, *Apuntes*, pp. 13–14; Franco Salgado-Araujo, *Mi vida*, pp. 122–4. In Seville in the first weeks of the Civil War, Franco was still talking about the sloppy preparation of the *Sanjurjada*, Pemán, *Mis encuentros*, p. 56.

56   Franco Salgado-Araujo, *Mi vida*, p. 122; Garriga, *La señora*, p. 83.

57   Gil Robles, *No fue posible*, pp. 234–43; Antonio López Fernández, *Defensa de Madrid* (México D.F., 1945) pp. 40–3.

58   Iribarren, *Mola*, p. 44; De la Cierva, *Franco*, II, p. 162.

59   Gil Robles, *No fue posible*, pp. 243–64; Arrarás, *Franco*, pp. 193–4; Alcalá Zamora, *Memorias* pp. 334–5.

60   *Documents on German Foreign Policy* Series C, vol. IV (London, 1964) pp. 641–50; Gerhard L. Weinberg, *The Foreign Policy of Hitler's Germany: Diplomatic Revolution in Europe, 1933–1936* (Chicago, 1970) p. 285; Angel Viñas, *la alemania nazi y el 18 de Julio* 2ª edición (Madrid, 1977) pp. 104–13.

61   Franco, *Apuntes*, pp. 14–15.

62   Franco, *Apuntes*, pp. 18–19.

63   Vegas Latapie, *Memorias*, p. 248.

64   H. R. Southworth, 'Entente', ms, pp. 69–75, 82–91; *Documentos inéditos*, I, pp. 13–23.

65   Gil Robles interview in Armando Boaventura, *Madrid-Moscovo da ditadura à República e à guerra civil de Espanha* (Lisbon, 1937) p. 191; Joaquín Chapaprieta Torregrosa, *La paz fue posible: memorias de un político* (Barcelona, 1971) pp. 292–332.

66   Gil Robles, *No fue posible*, pp. 358–67; Boaventura, *Madrid-Moscovo*, p. 192.

67   Franco to Gil Robles, April 1937, quoted Jaime del Burgo,

Conspiración y guerra civil
(Madrid, 1970) pp. 228–9;
Documentos inéditos, I, p. 28.

68   Ansaldo, *¿Para qué?*,
pp. 110–11.

69   *El Debate*, 15 December 1935;
Gil Robles, *No fue posible*,
pp. 375–6; Arrarás, *Franco*,
p. 198.

70   For Moscardó's accounts, see
Maximiano García Venero,
*Falange en la guerra de España: la
Unificación y Hedilla* (Paris, 1967)
p. 66 and Benito Gómez
Oliveros, *General Moscardó*
(Barcelona, 1956) p. 104.
Fernández Cuesta's version
appears in a letter to Felipe
Ximénez de Sándoval, 9
February 1942, reproduced in
Gil Robles, *No fue posible*, p. 367,
and Raimundo Fernández
Cuesta, *Testimonio, recuerdos y
reflexiones* (Madrid, 1985)
pp. 52–3. For a commentary,
Southworth, *Antifalange*,
pp. 91–4.

71   Franco, *Apuntes*, pp. 21–2.

72   Franco, *Apuntes*, pp. 23–4;
Suárez Fernández, *Franco*, I,
p. 301.

73   Manuel Portela Valladares,
*Memorias: dentro del drama
español* (Madrid, 1988)
pp. 168–9.

74   Franco, *Apuntes*, pp. 24–5,
35–6.

75   *The Daily Mail*, 29 January 1936.

76   Hills, *Franco*, pp. 209–10;
Arrarás, *Franco*, pp. 206–7;
Franco Salgado-Araujo, *Mi
vida*, pp. 126–7.

77   Serrano Suñer, *Memorias*, p. 56.
As part of the operation
undertaken in 1937 to link the
names of Franco and José
Antonio, Arrarás places this
interview in early March and
portrays the two protagonists as
decisive collaborators in the
preparations for the rising,
Arrarás, *Franco*, p. 228.

78   Javier Tusell, *Las elecciones del
Frente Popular* 2 vols (Madrid,
1971) I, p. 13. Cf. José Venegas,
Las elecciones del Frente Popular
(Buenos Aires, 1942) p. 65.

79   Portela, *Memorias*, pp. 175–8;
Gil Robles, *No fue posible*,
pp. 492–3; Alcalá Zamora,
Memorias, p. 347.

80   Gil Robles, *No fue posible*,
pp. 492–3; Franco, *Apuntes*,
pp. 25–8. Arrarás, *Cruzada*, II,
p. 439, gives a different
chronology with Franco talking
to Pozas and Molero *before* Gil
Robles saw Portela. Gil Robles'
version is more plausible and is
confirmed by Franco's own
sketchy account. Arrarás's
reversal of events is accounted
for by Francoist resentment of
Gil Robles.

81   Portela, *Memorias*, pp. 183–4;
Alcalá Zamora, *Memorias*,
p. 347; Franco, *Apuntes*,
pp. 28–30.

82   Franco, *Apuntes*, p. 30; De la
Cierva, *Historia*, I, p. 640.

83   Franco, *Apuntes*, p. 30; Arrarás,
Cruzada, II, p. 440; Gil Robles,
No fue posible, pp. 494–5.
Arrarás implies that Portela
had personally told Franco
about the decree. This is counter
to Portela's assertion in his
memoirs that he had never met
Franco before the evening of 17
February, and his claim at the

time to General Núñez de Prado that he did not tell Franco about the decree, Portela, *Memorias*, p. 184; Juan-Simeón Vidarte, *Todos fuimos culpables: testimonio de un socialista español* (México D.F., 1973) p. 49. It is in any case highly unlikely that the Minister of the Interior would call upon the Chief of the General Staff to do a job which corresponded to the head of the Civil Guard.

84 Franco, *Apuntes*, pp. 28–30; Servicio Histórico Militar, *Historia de la guerra de liberación* (Madrid, 1945) I, p. 421; Vidarte, *Todos fuimos culpables*, p. 48.

85 *El Sol*, 19 February 1936; Portela, *Memorias*, pp. 184–5, 190; Arrarás, *Cruzada*, II, p. 441; Franco interview in Boaventura, *Madrid-Moscovo*, pp. 207–8; B. Félix Maíz, *Alzamiento en España: de un diario de la conspiración* 2nd ed. (Pamplona, 1952) p. 37.

86 *El Socialista*, 19 February 1936; Manuel Goded, *Un "faccioso" cien por cien* (Zaragoza, 1939) pp. 26–7; Azaña, 19 February 1936, *Obras*, IV, p. 563; Diego Martínez Barrio, *Memorias*, (Barcelona, 1983) pp. 303–4; Gil Robles, *No fue posible*, pp. 497–8; Vidarte, *Todos fuimos culpables*, pp. 40–2, 48–9.

87 Arrarás, *Cruzada*, II, p. 443.

88 Portela, *Memorias*, pp. 186–7; Gil Robles, *No fue posible*, pp. 500–1.

89 Portela, *Memorias*, pp. 192–3; Gil Robles, *No fue posible*, pp. 499–500.

CHAPTER V

1 Ministerio de la Guerra Estado Mayor Central, *Anuario Militar de España año 1936* (Madrid, 1936) p. 150.

2 Franco, *Apuntes*, p. 23; Franco Salgado-Araujo, *Mi vida*, p. 131.

3 The version which recounts what Franco said in the most outspoken form is Arrarás, *Cruzada*, III, p. 58. See also Arrarás, *Franco*, pp. 228–9; Franco Salgado-Araujo, *Mi vida*, p. 131.

4 José María Iribarren, *Con el general Mola: escenas y aspectos inéditos de la guerra civil* (Zaragoza, 1937) p. 14; Franco Salgado-Araujo, *Mi vida*, p. 132.

5 There is some confusion over the participants at the meeting in Delgado's house, Gil Robles, *No fue posible*, pp. 719–20; Arrarás, *Cruzada*, II, p. 467; Franco, *Apuntes personales*, p. 33; Maíz, *Alzamiento*, pp. 50–1; Iribarren, *Mola*, pp. 45–6; Iribarren, *Con el general Mola*, pp. 14–15; Felipe Bertrán Güell, *Preparación y desarrollo del alzamiento nacional* (Valladolid, 1939) p. 116; Kindelán, *Mis cuadernos*, p. 81; Kindelán, 'La aviación en nuestra guerra' in *La guerra de liberación nacional* (Zaragoza, 1961) pp. 354–5.

6 Franco Salgado, *Mi vida*, pp. 134–6. Galinsoga & Franco Salgado, *Centinela*, pp. 195–6.

7 Franco Salgado-Araujo, *Mi vida*, p. 136–7.

8 *The Morning Post*, 20 July 1937; Arrarás, *Cruzada*, III, p. 56;

Franco Salgado-Araujo, *Mi vida*, p. 142.

9   Hodgson, *Spain Resurgent*, p. 103.

10 Ramón Garriga, *Los validos de Franco* (Barcelona, 1981) pp. 16–18.

11 Franco Salgado-Araujo, *Mi vida*, pp. 146–50; Arrarás, *Franco*, pp. 257–9.

12 Chilton to Eden, 22 April 1936, PRO FO 371/20521 W3720/62/41.

13 Pedrazzi to Ciano, 28 April 1936, ASDMAE, Politica Spagna, b.9, TE1248/548; Saz Campos, *Mussolini*, pp. 179–80.

14 For more details, see Preston, *CSCW*, pp. 182–3.

15 Gil Robles, *No fue posible*, pp. 561–2; García Venero, *Fanjul*, pp. 208–12.

16 *Ya*, 23 April 1936.

17 Gil Robles, *No fue posible*, pp. 563–7; García Venero, *Fanjul*, pp. 226–8; Serrano Suñer, *Memorias*, pp. 56–8.

18 Gil Robles, *No fue posible*, pp. 563–4.

19 Basilio Alvarez, *España en crisol* (Buenos Aires, Editorial Claridad, 1937) p. 69.

20 Arrarás, *Franco*, p. 231.

21 Arrarás, *Cruzada*, II, p. 488.

22 Franco, speech of 9 November 1947, *Textos 1945–1950*, p. 109.

23 Franco, *Apuntes*, pp. 34–5.

24 Arrarás, *Historia de la segunda República española*, IV, pp. 165–6.

25 Indalecio Prieto, *Convulsiones de España* 3 vols (México D.F., 1967–69) I, pp. 387–403.

26 Gil Robles, *No fue posible*, pp. 567–72.

27 Iribarren, *Mola*, p. 71.

28 Iribarren, *Mola*, pp. 79–87.

29 Bertrán Güell, *Preparación*, pp. 119–24; B. Félix Maíz, *Mola aquel hombre* (Barcelona, 1976) p. 121; Iribarren, *Mola*, p. 47.

30 Iribarren, *Mola*, pp. 52, 58–62; Juan José Calleja, *Yagüe: un corazón al rojo* (Barcelona, 1963) pp. 75–7.

31 Arrarás, *Cruzada*, III, p. 61, claims that Franco wrote 30 coded letters on the forthcoming uprising to Galarza from the Canary Islands. Cabanellas, *Cuatro generales*, I, p. 445, affirms that Franco wrote only three letters connected with the rising, which is altogether more likely.

32 Franco Salgado Araujo, *Mi vida*, pp. 139, 145; Soriano, *La mano*, p. 145.

33 H. Edward Knoblaugh, *Correspondent in Spain* (New York, 1937) p. 21; Pemán, *Mis encuentros*, p. 56.

34 Iribarren, *Mola*, p. 53; Vigón, *Mola*, pp. 91–2.

35 Serrano Suñer, *Memorias*, p. 53.

36 Serrano Suñer, *Memorias*, p. 53.

37 Calleja, *Yajüe*, pp. 75–6; Arrarás, *Cruzada*, II, p. 523; Maíz, *Alzamiento*, p. 153.

38 Vigón, *Mola*, pp. 93–4; Iribarren, *Mola*, pp. 55–6.

39 Frederic Escofet, *Al servei de Catalunya i de la República* 2 vols, II, *La victoria 19 de juliol 1936* (Paris, 1973) pp. 39–42, 151, 455–7.

40 The first published reference to the letter was in *The Times*, 7 September 1936. A fuller text

was published in Spain in Arrarás, *Franco*, pp. 233–7. A slightly different text was printed in Galinsoga & Franco Salgado, *Centinela*, pp. 203–6. For the postings, see Arrarás, *Cruzada*, II, pp. 522–4.

41   Arrarás, *Franco*, p. 233; Arrarás, *Cruzada*, II, pp. 523–4; Suárez Fernández, *Franco*, II, pp. 37–40; Maíz, *Alzamiento*, p. 191. See the analysis by Southworth, *Antifalange*, pp. 99–101.

42   Franco to Gil Robles, 12 April 1937, Burgo, *Conspiración*, pp. 228–9; Iribarren, *Con el general Mola*, pp. 16–17.

43   Franco Salgado-Araujo, *Mi vida*, p. 147.

44   Iribarren, *Con el general Mola*, p. 17; *Mola*, p. 65.

45   Testimony of Ramón Serrano Suñer to the author.

46   Sainz Rodríguez, *Testimonio*, p. 247; Ansaldo, *¿Para qué . . . ?*, p. 121.

47   Pemán, *Mis encuentros*, p. 32.

48   Sainz Rodríguez, *Testimonio*, p. 340.

49   Gil Robles, *No fue posible*, p. 780; Ramón Garriga, *El general Yagüe* (Barcelona, 1985) pp. 61–8.

50   Maíz, *Mola*, pp. 217, 238, 160; José Ignacio Luca de Tena, *Mis amigos muertos* (Barcelona, 1971) p. 162; Torcuato Luca de Tena, *Papeles para la pequeña y la gran historia: memorias de mi padre y mías* (Barcelona, 1991) pp. 204–10; Antonio González Betes, *Franco y el Dragón Rapide* (Madrid, 1987) pp. 83–94; Luis Bolín, *Spain: the Vital Years*

(Philadelphia, 1967) pp. 10–15.

51   Douglas Jerrold, *Georgian Adventure* (London, 1937) pp. 367–73; Bolín, *Spain*, pp. 16–30; Arrarás, *Cruzada*, III, pp. 62–4; González Betes, *Dragón Rapide*, pp. 96–121.

52   Kindelán, 'Prólogo', *Mis cuadernos*, 2ª ed., p. 42.

53   Alfredo Kindelán, *La verdad de mis relaciones con Franco* (Barcelona, 1981) pp. 173–4; Saña, *Franquismo*, pp. 48–9; Vegas Latapie, *Memorias*, p. 276; Serrano Suñer, *Memorias*, pp. 120–1.

54   Arrarás, *Cruzada*, III, p. 61.

55   Franco Salgado-Araujo, *Mi vida*, p. 150; Serrano Suñer, *Memorias*, p. 120; Saña, *Franquismo*, p. 49; González Betes, *Dragón Rapide*, pp. 122–3.

56   Garriga, *Franco–Serrano Suñer*, p. 42; Jaraiz Franco, *Historia*, pp. 105–30 and especially pp. 116–17; Franco, *Nosotros*, pp. 41, 50, 94.

57   *The Morning Post*, 20 July 1937.

58   Franco Salgado-Araujo, *Mi vida*, p. 151; González Betes, *Dragón Rapide*, pp. 127–30; Arrarás, *Cruzada*, III, pp. 64–5; Arrarás, *Franco*, p. 250. Sangróniz had written a book about Morocco in the 1920s – José Antonio Sangróniz, *Marruecos. Sus condiciones físicas, sus habitantes y las instituciones indígenas* (Madrid, Sucesores de Rivadeneyra, 1921).

59   *The Morning Post*, 20 July 1937.

60   Federico Bravo Morata, *Franco y los muertos providenciales*

(Madrid, 1979) pp. 19–47; González Betes, *Dragón Rapide*, pp. 132–4.

61 Arrarás, *Franco*, p. 261; Arrarás, *Cruzada*, III, p. 66; Millán Astray, *Franco*, p. 21; Suárez Fernández, *Franco*, II, p. 50; Franco Salgado-Araujo, *Mi vida*, pp. 145, 152.

62 Bravo Morata, *Franco y los muertos*, pp. 38–44; Payne, *Politics and the Military*, p. 341; Thomas, *Civil War*, p. 212.

63 Franco Salgado-Araujo, *Mi vida*, pp. 152–4; González Betes, *Dragón Rapide* pp. 134–7, 142–7; Arrarás, *Cruzada*, III, pp. 66–7.

64 Arrarás, *Cruzada*, III, pp. 17–44.

65 Arrarás, *Cruzada*, III, pp. 67–9; Franco Salgado-Araujo, *Mi vida*, pp. 154–5; Luis Romero, *Tres días de Julio (18, 19 7 20 de 1936)* 2ª edición (Barcelona, 1968) pp. 11–13.

66 Eugenio Vegas Latapie, *Los caminos del desengaño: memorias políticas II 1936–1938* (Madrid, 1987) p. 80.

67 Arrarás, *Cruzada*, III, pp. 70–3; Suárez Fernández, *Franco*, II, pp. 53–4.

68 Franco Salgado-Araujo, *Mi vida*, pp. 155–6.

69 Franco Salgado-Araujo, *Mi vida*, pp. 158–60; Arrarás, *Cruzada*, III, pp. 74–5; Franco, *Nosotros*, p. 108; Garriga, *La Señora*, p. 92.

70 González Betes, *Dragón Rapide*, pp. 159–64; Franco Salgado-Araujo, *Mi vida*, pp. 160–1.

71 Gil Robles, *No fue posible*, p. 782.

72 Franco Salgado-Araujo, *Mi vida*, p. 161; Arrarás, *Cruzada*, III, p. 75; Bolín, *Spain*, p. 45; Hills, *Franco*, p. 232; Crozier, *Franco*, p. 185; González Betes, *Dragón Rapide*, p. 167.

73 Franco Salgado-Araujo, *Mi vida*, p. 161; Arrarás, *Cruzada*, III, p. 75; Bolín, *Spain*, p. 47. According to González Betes, *Dragón Rapide*, p. 156, the emergency pilot who accompanied Franco's party, Lieutenant Villalobos, claimed to have shaved Franco in Casablanca.

74 Cabanellas, *Guerra*, I, p. 657.

75 Bolín, *Spain*, pp. 50–2; Franco Salgado-Araujo, *Mi vida*, pp. 160–4; Arrarás, *Franco*, pp. 263–71; González Betes, *Dragón Rapide*, pp. 164–78; Arrarás, *Cruzada*, III, pp. 73–8.

76 *Daily Express*, 26 June 1938; Bolín, *Spain*, pp. 52–3; González Betes, *Dragón Rapide*, pp. 186–9.

CHAPTER VI

1 Franco Salgado-Araujo, *Mi vida*, p. 165; Arrarás, *Cruzada*, III, pp. 80–2.

2 Franco Salgado-Araujo, *Mi vida*, pp. 166–7.

3 Arrarás, *Cruzada*, III, pp. 49–51, 82.

4 Abd el Hajid Ben Jellon, 'La participación de los mercenarios marroquíes en la guerra civil española (1936–1939)', *Revista Internacional de Sociología* Vol. 46, no. 4, October–December 1988, pp. 527–41.

5   *The Times*, 26 August 1926;
    Whitaker, 'Prelude' pp. 105–6;
    María Rosa de Madariaga,
    'Imagen del moro en la
    memoria colectiva del pueblo
    español y retorno del moro en la
    guerra civil de 1936', *Revista
    Internacional de Sociología*
    Vol. 46, no. 4, October–
    December 1988, pp. 590–6.

6   Franco, *Diario*, p. 177;
    Southworth, *Antifalange*,
    pp. xxi–xxii, 112–13,
    photographs no. 44a, 44b.

7   Cabanellas, *Guerra*, I, pp. 569–71.

8   *The Times*, 20 July 1936.

9   Escofet, *La victoria*, pp. 427–30.

10 Testimony of Carlos Marín de
    Bernardo, Mérida, *Testigos*,
    pp. 189–90.

11 Vaca de Osma, *Paisajes*, pp. 35–6.

12 Arrarás, *Cruzada*, III, pp. 308–21.

13 Arrarás, *Cruzada*, III,
    pp. 156–210; Antonio
    Bahamonde y Sánchez de
    Castro, *Un año con Queipo:
    memorias de un nacionalista*
    (Barcelona, 1938) pp. 23–9;
    Guzmán de Alfarache, *¡18 de
    julio! Historia del alzamiento
    glorioso de Sevilla* (Seville, 1937)
    pp. 51–179; Cabanellas, *Guerra*,
    I, pp. 401–2; Ian Gibson,
    *Queipo de Llano: Sevilla, verano
    de 1936* (Barcelona, 1986)
    pp. 157–8.

14 The best account is in García
    Venero, *Fanjul*, pp. 296–345.
    See also, Luis Romero, *Tres días
    de Julio* 2ª edición (Barcelona,
    1968) pp. 453 ff.

15 Goded, *Un "faccioso"*,
    pp. 43–67; Escofet, *La victoria*,
    pp. 200–400; Romero, *Tres
    días*, pp. 341–53, 409–10;

    Francisco Lacruz, *El alzamiento,
    la revolución y el terror en
    Barcelona* (Barcelona, 1943)
    pp. 13–116.

16 Oscar Pérez Solís, *Sitio y defensa
    de Oviedo* 2ª edición (Valladolid,
    1938) pp. 5–28; Julián
    Zugazagoitia, *Guerra y
    vicisitudes de los españoles*, 2ª
    edición. 2 vols, (Paris, 1968) I,
    pp. 50–4; Arrarás, *Cruzada*, IV,
    pp. 141–3.

17 García Venero, *Fanjul*,
    pp. 286–91; Martínez Barrio,
    Memorias, pp. 358–63.

18 Martínez Barrio, *Memorias*,
    pp. 363–4; Maíz, *Alzamiento*,
    p. 304; Iribarren, *Con el general
    Mola*, pp. 65–6.

19 Arrarás, *Cruzada*, III, pp. 444–9.

20 Franco Salgado-Araujo, *Mis
    conversaciones*, p. 526.

21 *ABC*, Seville, 22 July; *The Times*,
    24 July 1936; Arrarás, *Cruzada*,
    III, p. 84.

22 Jaraiz Franco, *Historia*, pp. 73,
    82–5; Franco Salgado-Araujo,
    Mi vida, p. 167.

23 Kindelán, *Mis cuadernos*, p. 82.

24 Iribarren, *Con el general Mola*, p. 97.

25 Iribarren, *Con el general Mola*, p. 73.

26 Vegas Latapie, *Los caminos*,
    pp. 23–4; Ansaldo, *¿Para qué?*,
    pp. 137–8; Sainz Rodríguez,
    Testimonio, p. 249.

27 Ansaldo, *¿Para qué?*,
    pp. 138–44; José Antonio Silva,
    Cómo asesinar con un avión
    (Barcelona, 1981) pp. 40–58.

28 Tom Gallagher, *Portugal: A
    Twentieth Century Interpretation*
    (Manchester, 1983) pp. 86, 108,
    n.5.

29 General Vicente Rojo, *¡Alerta
    los pueblos! estudio político-militar*

del período final de la guerra española 2ª edición (Barcelona, 1974) pp. 185–6; Luigi Longo, *Le brigate internazionali in Spagna* (Rome, 1956) pp. 14–16; Thomas, *Civil War*, p. 328.

30 Arrarás, *Cruzada*, III, p. 83.

31 *The News Chronicle*, 29 July, 1 August 1936.

32 Alpert, Michael. *La guerra civil española en el mar* (Madrid, 1987) p. 86; Calleja, *Yagüe*, p. 88; Iribarren, *Con el general Mola*, p. 167.

33 Kindelán, *La verdad*, pp. 173–4; *Mis cuadernos* pp. 82–3.

34 Kindelán's son claimed that the ideas for the airlift and for a blockade-breaking convoy emanated from his father, 'Prólogo', *Mis cuadernos*, p. 45.

35 Arrarás, *Cruzada*, III, pp. 211–13.

36 Kindelán, *Mis cuadernos*, pp. 83–5; Gerald Howson, *Aircraft of the Spanish Civil War 1936–1939* (London, 1990) pp. 116–17, 145–6; Servicio Histórico Militar, (Colonel José Manuel Martínez Bande) *La campaña de Andalucía* 2ª edición (Madrid, 1986) pp. 51–3; Coronel José Gomá, *La guerra en el aire* (Barcelona, 1958) pp. 70–1; Jesús Salas Larrazabal, 'El puente aereo del Estrecho', *Revista de Aeronáutica y Astronáutica*, September 1961, pp. 747–50.

37 Franco Salgado-Araujo, *Mi vida*, p. 173–5.

38 Gomá, *La guerra en el aire*, pp. 70; Howson, *Aircraft*, pp. 76–8, 191.

39 Gomá, *La guerra en el aire*, pp. 67; Jesús Salas Larrazabal, *La guerra de España desde el aire* 2ª

edición (Barcelona, 1972) p. 87.

40 Arrarás, *Cruzada*, III, pp. 118–19; Franco Salgado-Araujo, *Mi vida*, pp. 181–2; Kindelán, *La verdad*, 176–7; Martinez Bande, *La campaña de Andalucia*, pp. 55–8; Martin, *Franco*, pp. 191–2.

41 Maiz, *Alzamiento*, pp. 307–11.

42 *FRUS 1936*, II, p. 449; *ABC*, Seville, 26 July 1936; Iribarren, *Con el general Mola*, pp. 106–7, 122.

43 Arrarás, *Cruzada*, III, p. 513, IV, p. 218.

44 Suárez Fernández, *Franco*, II, p. 77; Fernando & Salvador Moreno de Reyna, *La guerra en el mar (Hombres, barcos y honra)* (Barcelona, 1959) p. 42.

45 Julio de Ramón-Laca, *Baja la férula de Queipo: como fue gobernada Andalucía* (Seville, 1939) pp. 5–45.

46 Franco Salgado-Araujo, *Mi vida*, p. 180; Cabanellas, *Guerra*, I, pp. 642–3.

47 Bolín, *Vital Years*, pp. 159–66; Franco Salgado-Araujo, *Mi vida*, pp. 175–6.

48 Renzo de Felice, *Mussolini il duce: lo stato totalitario 1936–1940* (Turin, 1981) p. 363.

49 De Felice, *Il duce*, pp. 364–5.

50 Saz, *Mussolini*, pp. 181–4; Roberto Cantalupo, *Fu la Spagna Ambasciata presso Franco* (Milan, 1948) p. 63; Coverdale, *Italian Intervention*, pp. 69–72; Bolín, *Vital Years*, pp. 70–2.

51 Saz, *Mussolini*, pp. 180, n. 97, 243; Bolín, *Vital Years*, pp. 167–9; Coverdale, *Italian Intervention*, pp. 71–2.

52 Saz, *Mussolini*, pp. 183–5. The

telegrams are reproduced on
pp. 247–51.

53 José Ignacio Escobar, *Así empezó*
(Madrid, 1974) pp. 55–60; José
Gutiérrez-Ravé, *Antonio
Goicoechea* (Madrid, 1965)
pp. 35–7; Saz, *Mussolini*,
pp. 186–8; Arrarás, *Cruzada*, III,
p. 126; Coverdale, *Italian
Intervention*, pp. 72–4; Viñas,
La Alemania nazi, pp. 308–11;
Saz, *Mussolini*, p. 191.

54 *FRUS 1936*, II, p. 453.

55 Doubt has been thrown on the
claim that Mussolini insisted
that the aircraft be paid for in
cash before their delivery.
Money provided by the
millionaire smuggler, Juan
March, often alleged to have
been used to pay for the initial
Savoia bombers, was used for
other aircraft purchases. For the
claim, see Gutiérrez-Ravé,
Goicoechea, p. 37; Coverdale,
Italian Intervention, p. 74. For its
refutation, see Saz, *Mussolini*,
pp. 189–90.

56 *FRUS 1936*, II, p. 451.

57 *Documents on German Foreign
Policy* [henceforth *DGFP*]
Series D, Volume III, (London,
1951) pp. 15, 17; Saz, *Mussolini*,
p. 185; Howson, *Aircraft*, p. 274.

58 Saz, *Mussolini*, pp. 191–3.

59 Franco, *Apuntes*, p. 39; Viñas,
Alemania nazi, pp. 274–6.

60 *DGFP*, D, III, pp. 4–5, 7; Viñas,
Alemania nazi, pp. 264–9,
293–4, 323.

61 On Bernhardt and Langenheim,
see Viñas, *Alemania nazi*,
pp. 276–92.

62 *DGFP*, D, III, pp. 7–8; Viñas,
Alemania nazi, pp. 316–29;

Howson, *Aircraft*, pp. 206–7;
Suárez Fernández, *Franco*, II,
pp. 60–8.

63 *DGFP*, D, III, pp. 10–11; Viñas,
Alemania nazi, pp. 330–8.

64 Viñas, *Alemania nazi*,
pp. 330–42. Göring conflated
this meeting and a later one in
his evidence at Nuremberg,
International Military Tribunal,
Trial of Major War Criminals
(Nuremberg, 1947) IX,
pp. 280–1.

65 Joaquim von Ribbentrop, *The
Ribbentrop Memoirs* (London,
Weidenfeld and Nicolson,
1954) p. 59.

66 Viñas, *Alemania nazi*,
pp. 341–52; *DGFP*, D, III,
pp. 113–14; Angel Viñas et al.,
*Política comercial exterior en
España 91931–1975)* 2 vols
(Madrid, 1979) I, pp. 146–9;
Denis Smyth, 'The Moor and
the Money-lender: Politics and
Profits in Anglo-German
Relations with Francoist Spain
1936–1940' in Marie-Luise
Recker, *Von der Konkurrenz zur
Rivalität: Das britische-deutsche
Verhältnis in den Ländern der
europäischen Peripherie
1919–1939* (Stuttgart, 1986)
pp. 146–7.

67 Heinz Höhne, *Canaris* (London,
1979) pp. 230–1; André
Brissaud, *Canaris* (London,
1973) pp. 8, 36–42; Weinberg,
Foreign Policy, p. 288; Viñas,
Alemania nazi, p. 342; *DGFP*,
D, III, p. 34, 40, 442–3.

68 Angel Viñas, 'Los espías nazis
entran en la guerra civil' in
*Guerra, dinero, dictadura: ayuda
fascista y autarquía en la España*

de Franco (Barcelona, 1984) pp. 51–3; Höhne, *Canaris*, p. 234; Klaus A. Maier, *Guernica 26.4.1937: Die deutsche intervention in Spanien und der "Fall Guernica"* (Freiburg, 1975) pp. 24–5; Coverdale, *Italian Intervention*, pp. 87, 103, 106, 119–20.

69 On Mola's contacts, see Viñas, *Alemania nazi*, pp. 274–5.

70 Escobar, *Así empezó*, pp. 75, 81–118.

71 *DGFP*, D, III, pp. 9, 16.

72 Howson, *Aircraft*, pp. 206–7; Maier, *Guernica*, pp. 21–2.

73 Franco's telegram is quoted in José Manuel Martínez Bande, 'Del alzamiento a la guerra civil verano de 1936: correspondencia Franco/Mola', *Historia y Vida*, No. 93, 1975, p. 21, where a misprint attributes it wrongly to 20 July although internal evidence makes it clear that it had to be either 29 or 30 July, cf. Suárez Fernández, *Franco*, II, p. 79. Iribarren, *Con el general Mola*, p. 157, cites the telegram arriving on 29 July.

74 *Hitler's Table Talk*, p. 687.

75 Franco, *Apuntes*, p. 40; 'Correspondencia Franco/Mola', p. 22; Escobar, *Así Empezó*, p. 243. On Mola's pessimism, Vigón, *Mola*, pp. 176–201.

76 Iribarren, *Mola*, p. 149; 'Correspondencia Franco/Mola', p. 21.

77 *The Times*, 30 July 1936; Franco Salgado-Araujo, *Mi vida*, p. 180.

78 Franco Salgado-Araujo, *Mi vida*, pp. 181–2, seems to be conflating the 3 August meeting with that held on 20 July.

79 *DGFP*, D, III, pp. 26–7.

80 'Correspondencia Franco/Mola', p. 22.

81 Franco Salgado-Araujo, *Mi vida*, pp. 182–3; Martínez Bande, *La campaña de Andalucía*, pp. 57–63; Arrarás, *Cruzada*, III, pp. 129–39; Arrarás, *Franco*, pp. 281–3; Alpert, *La Guerra en el mar*, pp. 89–95; Gomá, *La guerra en el aire*, pp. 74–81; 'Almogávares en el aire', *Revista de Aeronáutica* August 1942, pp. 97–100.

82 Kindelán, 'La aviación', p. 365; *The Times*, 15 August 1936; Suárez Fernández, *Franco*, II, p. 80.

83 Francisco Bonmati de Codecido, *El Príncipe Don Juan de España* (Valladolid, 1938) pp. 224–37; Iribarren, *Mola*, pp. 166–7; Vegas Latapie, *Caminos*, pp. 37–44; Cabanellas, *Guerra*, I, pp. 636–7.

84 Pemán, *Mis encuentros*, pp. 188–93.

85 *The Times*, 10 August 1936.

86 Howson, *Aircraft*, p. 123.

87 Franco Salgado-Araujo, *Mi vida*, p. 184.

88 Carlos Asensio Cabanillas, 'El avance sobre Madrid y operaciones en el frente del centro', *La guerra de liberación nacional* (Zaragoza, 1961) pp. 160–5; Servicio Histórico Militar (José Manuel Martínez Bande), *La marcha sobre Madrid* (Madrid, 1968) pp. 24–34; Manuel Sánchez del Arco, *El sur de España en la reconquista de*

Madrid (Seville, 1937)
pp. 62–81; Calleja, *Yagüe*,
pp. 90–1, 94–6; Salas
Larrazabal, *La guerra desde el
aire*, p. 64.

89 Mijail Koltsov, *Diario de la
guerra de España* (Paris, 1963)
pp. 88–9; Whitaker, 'Prelude',
pp. 105–6.

90 Cardozo, *The March*,
pp. 160–2.

91 Martínez Bande, *La marcha sobre
Madrid*, pp. 165–70.

92 Pemán, *Mis encuentros*, p. 64.

93 *DGFP*, D, III, pp. 86–7.

94 'Correspondencia Franco/
Mola', pp. 22–3.

95 *DGFP*, D, III, p. 40.

96 Escobar, *Así Empezó*,
pp. 119–24.

97 Calleja, *Yagüe*, pp. 100–2.

98 On the strategic error, Thomas,
Civil War, p. 373. Arrarás,
Cruzada, VII, pp. 24–6; Aznar,
Historia militar, p. 103;
Martínez Bande, *La marcha sobre
Madrid*, pp. 34–5 all attribute
the decision to Yagüe. Luis
María de Lojendio, *Operaciones
militares de la guerra de España*
(Barcelona, 1940) pp. 141–2,
avoids comment. Calleja, *Yagüe*,
p. 97 makes it clear that there
was full consultation with
Franco.

99 Martínez Bande, *La marcha
sobre Madrid*, pp. 35–41;
Sánchez del Arco, *El sur*,
pp. 82–91; Calleja, *Yagüe*,
p. 105.

100 Mário Neves, *La matanza de
Badajoz* (Badajoz, 1986) pp. 13,
43–5, 50–1. See also Jay Allen,
'Blood flows in Badajoz', in
Marcel Acier, editor, *From
Spanish Trenches: Recent Letters
from Spain* (London, 1937)
pp. 3–8; Whitaker, 'Prelude',
pp. 104–6; Calleja, *Yagüe*,
pp. 99–109. On the pro-
Francoist propaganda campaign
to deny what happened in
Badajoz, see Herbert Rutledge
Southworth, *El mito de la
cruzada de Franco* (Paris, 1963)
pp. 217–31.

101 *FRUS 1936*, II, p. 456–7;
DGFP, D, III, pp. 54–5; Iva
Delgado, *Portugal e a guerra civil
de Espanha* (Lisbon, n.d.)
pp. 30–9.

102 Manuel Aznar, *Historia militar
de la guerra de España
(1936–1939)* (Madrid, 1940)
p. 99; Martínez Bande, *La marcha
sobre Madrid*, pp. 18–19, 30.

103 Gibson, *Queipo*, pp. 101–5. I am
grateful to Ian Gibson who
made available to me copies of
documents relating to the trial
and execution of Campins,
including a telegram from his
wife begging for news ten days
after his death.

104 Franco Salgado-Araujo,
Mi vida, pp. 185–8,
348–53.

105 Franco, *Nosotros*, p. 88.

106 Gibson, *Queipo*, pp. 104–5;
Cabanellas, *Guerra*, II,
pp. 872–3.

107 Vegas Latapie, *Caminos*, p. 32;
Cabanellas, *Guerra*, I, pp. 638–9.

108 Iribarren, *Mola*, p. 11.

109 Iribarren, *Con el general Mola*,
pp. 164–5, 243.

110 Arthur Koestler, *Spanish
Testament* (London, 1937)
pp. 29–30; H. R.
Knickerbocker, *The Seige of the*

Alcazar: A War-Log of the Spanish Revolution (London, n.d. [1936]) pp. 27–8, 41.

111 Whitaker, 'Prelude', p. 116.
112 Harold G. Cardozo, *The March of a Nation* (London, 1937) p. 141. Cf. William Foss & Cecil Gerahty, *The Spanish Arena* (London, 1938) p. 62; Rotvand, *Franco*, p. 20; Coles, *Franco*, p. 71.
113 Cabanellas, *Cuatro generales*, II, p. 327.
114 *DGFP*, D, III, pp. 42–3.
115 'Correspondencia Franco/Mola', p. 22.
116 Franco Salgado-Araujo, *Mi vida*, pp. 189–90; Iribarren, *Con el general Mola*, pp. 262–4.
117 Iribarren, *Con el general Mola*, p. 271.
118 Iribarren, *Con el general Mola*, p. 263.
119 Iribarren, *Mola*, pp. 149–53.
120 *Dez anos de política externa (1936–1947) a a nação portuguesa e a segunda guerra mundial* III (Lisbon, 1964) p. 156; *The Times*, 11, 17 August 1936.
121 'Correspondencia Franco/Mola', p. 23.
122 Viñas, 'Los espías', pp. 50, 56–8; Robert H. Whealey, *Hitler and Spain: The Nazi Role in the Spanish Civil War* (Lexington, Kentucky, 1989) p. 7.
123 Martínez Bande, *La marcha sobre Madrid*, pp. 45–56; Calleja, *Yagüe*, pp. 111–12; Sánchez del Arco, *El sur*, pp. 94–114.
124 'Correspondencia Franco/Mola', pp. 23–4. Martínez

Bande's claim, *ibid*, pp. 23–4, 28–9.

CHAPTER VII

1 Vegas Latapie, *Caminos*, p. 71.
2 Letter of Ramón Garriga to the author, 1.XI.90; Vegas Latapie, *Caminos*, pp. 72, 89; Iribarren, *Mola*, pp. 193, 211; Arrarás, *Franco*, p. 287; Franco Salgado-Araujo, *Mi vida*, pp. 192–3; Suárez Fernández, *Franco*, II, p. 95.
3 José María Pemán, *Mis almuerzos con gente importante* (Barcelona, 1970) p. 138.
4 Franco Salgado-Araujo, *Mi vida*, p. 197; Baón, *La cara humana*, p. 125.
5 Coverdale, *Italian Intervention*, pp. 102–4, 119–20; Höhne, *Canaris*, p. 235; Weinberg, *Diplomatic Revolution*, p. 292; Maier, *Guernica*, pp. 24–5.
6 Raymond Carr, *The Spanish Tragedy: The Civil War in Perspective* (London, 1977) p. 152; Whitaker, 'Prelude', pp. 104–5.
7 Servicio Histórico Militar (Coronel José Manuel Martínez Bande), *Nueve meses de guerra en el norte* (Madrid, 1980) pp. 64–93; Aznar, *Historia militar*, pp. 127–43; Thomas, *Civil War*, pp. 376–80.
8 *DGFP*, D, III, pp. 16, 28.
9 Iribarren, *Con el general Mola*, p. 320; Suárez Fernández, *Franco*, II, p. 97.
10 Arrarás, *Cruzada*, VII, p. 152; Franco Salgado-Araujo, *Mi vida*, pp. 197–8.

11 Koltsov, *Diario*, pp. 94–103.

12 On the siege, there is an enormous and largely romanticised literature. For eye-witness accounts, see Comandante Benito Gómez Oliveros, *General Moscardó (sin novedad en el Alcázar)* (Barcelona, 1956) pp. 129–218; Comandante Alfredo Martínez Leal, *El asedio del Alcázar de Toledo: memorias de un testigo* (Toledo, n.d.); Moscardó's diary and the improvised newspaper *El Alcázar*, reprinted in Arrarás, *Cruzada*, VII, pp. 143–90. Among the epic chronicles, see Aznar, *Historia militar*, pp. 183–211; Knickerbocker, *Seige*, pp. 171–85; D. Muro Zegri, *La epopeya del Alcázar* (Valladolid, 1937); Major Geoffrey McNeill-Moss, *The Epic of the Alcazar* (London, 1937); Alberto Risco S.J., *La epopeya del Alcázar de Toledo* (San Sebastián, 1941) and Cecil Eby, *The Seige of the Alcázar* (London, 1966). For critical accounts, Antonio Vilanova, *La defensa del Alcázar de Toledo (epopeya o mito)* (Mexico, 1963); Luis Quintanilla, *Los rehenes del Alcázar de Toledo* (Paris, 1967); Southworth, *El mito*, pp. 92–116; Herbert L. Matthews, *The Yoke and the Arrows* (London, 1958) pp. 173–6.

13 Martínez Bande, *La marcha sobre Madrid*, pp. 56–71; Garriga, *Yagüe*, pp. 111–12.

14 Whitaker, 'Prelude', pp. 105–6; Webb Miller, *I Found No Peace* (London, n.d. [1937]) pp. 328–9.

The massacre was denied by Harold G. Cardozo, *The March of a Nation: My Year of Spain's Civil War* (London, 1937) pp. 73–4, 111–14.

15 Kindelán, *Mis cuadernos*, p. 85; Calleja, *Yagüe*, pp. 113–15.

16 Payne, *Falange*, p. 130; Cabanellas, *Guerra*, I, p. 647, n. 19.

17 Thomas, *Civil War*, p. 412; Salas Larrazabal, *Ejército Popular*, I, p. 475.

18 Arrarás, *Cruzada*, VII, pp. 41–3.

19 Vegas Latapie, *Caminos*, p. 72.

20 Kindelán, *Mis cuadernos*, pp. 85–6; Serrano Suñer, *Memorias*, p. 132; Garriga, *Yagüe*, p. 113.

21 *DGFP*, D, III, pp. 95–6; *Ciano's Diplomatic Papers* ed. Malcolm Muggeridge (London, 1948) pp. 53–4.

22 Kindelán, *Mis cuadernos*, p. 103; Franco Salgado-Araujo, *Mi vida*, pp. 207–8.

23 José Luis Alcofar Nassaes, *C.T.V. Los legionarios italianos en la guerra civil española 1936–1939* (Barcelona, 1972) pp. 51–2.

24 Pemán, *Mis encuentros*, p. 63.

25 Messerschmidt report, 8 September 1936, *DGFP*, D, III, pp. 85–9.

26 Garriga, *Nicolás Franco*, pp. 101–2, an account based on information given to him in Salamanca shortly afterwards by Bernhardt's secretary, (letter of Ramón Garriga to the author, 30 April 1991). See also the memorandum of Count Du Moulin-Eckart, 8 October 1936, to the effect that the

nomination of Franco as Generalísimo and Head of State had been accelerated in order to take advantage of the fact that, with the battle for Madrid still pending, there would be little resistance to Franco's elevation. 'From intimations, I gathered that a conference with German agencies had not been without influence in this decision.' *DGFP*, D, III, p. 107.

27 Kindelán, *Mis cuadernos*, pp. 103–4.

28 On Cabanellas, Cabanellas, *Cuatro generales*, I, pp. 231, 425–8; on Queipo, Escobar, *Así Empezó*, p. 133; Cabanellas, *Guerra*, pp. 196, 305–6; García Venero, *Falange/Hedilla*, p. 180.

29 Kindelán, *Mis cuadernos*, p. 105.

30 Iribarren, *Mola*, pp. 232–3.

31 Sainz Rodríguez, *Testimonio*, pp. 248–9.

32 Vegas Latapie, *Caminos*, p. 87.

33 Sainz Rodríguez, *Testimonio*, p. 272.

34 Suárez Fernández, *Franco*, II, p. 100.

35 Vigón, *Mola*, p. 248.

36 Martínez Bande, *La marcha*, p. 72; Salas Larrazabal, *Ejército popular*, I, p. 475.

37 Iribarren, *Mola*, p. 232.

38 Vaca de Osma, *Paisajes*, p. 55; Escobar, *Así empezó*, p. 148.

39 Kindelán, *Mis cuadernos*, p. 105; Vegas Latapie, *Caminos*, p. 85; Escobar, *Así empezó*, pp. 149–50; Garriga, *Nicolás Franco*, p. 102; Jean Créac'h, *Le coeur et l'épée: chroniques espagnoles* (Paris, 1958) pp. 179–80.

40 Hills, *Franco*, pp. 124–5, 250.

41 Boaventura, *Madrid-Moscovo*, p. 212.

42 Martínez Bande, *La marcha*, pp. 74–5.

43 For the military aspects of the capture of Toledo, Martínez Bande, *La marcha sobre Madrid*, pp. 71–5; Lojendio, *Operaciones*, pp. 154–60. On the exclusion of the press, Knickerbocker, *Seige*, pp. 172–3; Miller, *I Found No Peace*, pp. 329–30; Matthews, *The Yoke*, p. 176. For the reprisals, Peter Wyden, *The Passionate War: The Narrative History of the Spanish Civil War, 1936–1939* (New York, 1983) pp. 142–5; Geoffrey Cox, *Defence of Madrid* (London, 1937) p. 54; Bowers, *My Mission*, p. 313; Whitaker, 'Prelude', p. 106; Miller, *I Found No Peace*, pp. 336–8.

44 De la Cierva, *Franco*, III, pp. 8–9; Calleja, *Yagüe*, p. 116.

45 Kindelán, *Mis cuadernos*, pp. 105–8. The dates given by Kindelán, 30, 31 September are incorrect.

46 Kindelán, *Mis cuadernos*, pp. 108–10; De la Cierva, *Franco*, III, p. 10.

47 Luciano González Egido, *Agonizar en Salamanca: Unamuno julio-diciembre 1936* (Madrid, 1986) p. 109.

48 Cabanellas, *Guerra*, I, pp. 654–5.

49 Garriga, *España de Franco*, p. 73 and letter to the author, 30 April 1991; García Venero, *Falange/Hedilla*, p. 167; Charles Foltz, Jr., *The Masquerade in Spain* (Boston, 1948) p. 178;

Créac'h, *Le coeur*, p. 182;
Serrano Suñer, *Memorias*,
p. 163. See also Thomas, 1st
edition.

50 *ABC*, Seville, 30 September; *The
Times*, 2 October 1936; Burgo,
Conspiración, p. 491.

51 Cabanellas, *Guerra*, I, p. 652; cf.
the reaction of the monarchist
intellectual Pedro Sainz
Rodríguez, Vegas Latapie,
Caminos, p. 79.

52 Rafael Martínez Nadal, *Antonio
Torres y la política española del
Foreign Office (1940–1944)*
(Madrid, 1989) p. 231; cf.
Cardozo, *The March*, p. 153.

53 *The Times*, 2 October 1936.

54 Antonio Montero Moreno,
*Historia de la persecución religiosa
en España 1936–1939* (Madrid,
1961) pp. 688–708; María
Luisa Rodríguez Aisa, *El
Cardenal Gomá y la guerra de
España: aspectos de la gestión
pública del Primado 1936–1939*
(Madrid, 1981) pp. 109–25.

55 La Cierva, *Franco: un siglo*, I,
p. 513.

56 Simon Haxey, *Tory M.P.*
(London, 1939) pp. 210–20;
Richard Griffiths, *Fellow
Travellers of the Right: British
Enthusiasts for Nazi Germany
1933–39* (London, 1980)
pp. 260–64.

57 *ABC* (Seville), 2 October; *The
Times*, 3 October 1936; *FRUS
1936*, II, p. 534; *Palabras de
Franco* (n.p., 1937) pp. 9–18;
Cabanellas, *Guerra*, I,
pp. 657–8; Franco Salgado-
Araujo, *Mi vida*, pp. 210–11;
Burgo, *Conspiración*, pp. 491–2.

58 Kindelán, *Mis cuadernos*, p. 111;

Serrano Suñer, *Memorias*,
pp. 53–4, 220–2; Cabanellas,
Guerra, I, p. 659. General
Cabanellas died on 14 May
1938.

59 *DGFP*, D, III, p. 103–7; Franco
to Hess, 2 October, 'Hispanicus',
Foreign Intervention in Spain
(London, 1937) p. 44.

60 Testimony of Ramón Serrano
Suñer to the author, Madrid, 21
November 1990. The priest in
question seems to have been
Don Manuel Fidalgo Alonso,
who according to the parish
records of San Cucao, was the
parish priest between 1928 and
1932. Since the Asturian
diocesan records were destroyed
during the Civil War, it is
impossible to be sure.

61 Stanley G. Payne, *Falange: A
History of Spanish Fascism*
(Stanford, 1961) pp. 148–9;
García Venero, *Falange/
Hedilla*, pp. 307–9.

62 Mérida, *Testigos*, p. 31.

63 Garriga, *La señora*, p. 105.

64 Suárez Fernández, *Franco*, II,
p. 121; Crozier, *Franco*, p. 219.

65 For the texts of both telegrams,
see Rodríguez Aisa, *Gomá*,
pp. 32, 61.

66 Admiral Cervera, quoted
Cabanellas, *Guerra*, I, p. 660.

67 González Egido, *Unamuno*,
p. 120.

68 Cabanellas, *Guerra*, II,
pp. 937–9.

69 Escobar, *Así empezó*, pp. 155–6;
Pereira, *Memórias*, II, p. 18;
Jaraiz Franco, *Historia*, p. 152;
testimony of Ramón Serrano
Suñer to the author; Pemán, *Mis
encuentros*, p. 18; Franco

Salgado-Araujo, *Mi vida*, p. 221.

70 Broadcast of 4 October 1936, reprinted in Millán Astray, *Franco*, pp. 41–4.

71 Luis Moure Mariño, *La generación del 36: memorias de Salamanca y Burgos* (La Coruña, 1989) pp. 69–71; Vegas Latapie, *Caminos*, p. 175.

72 Knickerbocker, *Siege*, pp. 23, 27, 136–7; Kemp, *Trouble*, p. 49.

73 Herbert R. Southworth, *Guernica! Guernica!: A Study of Journalism, Propaganda and History* (Berkeley, 1977) pp. 45–50; Koestler, *Spanish Testament*, pp. 27–8, 219–31; Sir Peter Chalmers Mitchell, *My House in Málaga* (London, 1938) pp. 274–91; Letter of Ramón Garriga to the author, 1.XI.90.

74 Kemp, *Trouble*, p. 50; Foltz, *The Masquerade*, p. 116; Whitaker, 'Prelude', p. 108; Southworth, *Guernica!*, pp. 50–2.

75 Virginia Cowles, *Looking for Trouble* (London, 1941) pp. 86–90.

76 González Egido, *Unamuno*, pp. 125–7.

77 Escobar, *Así empezó*, p. 167; Vegas Latapie, *Caminos*, p. 113.

78 The local press said nothing of the incident, *La Gaceta Regional*, 13 October 1936. For an eye-witness account and Unamuno's own letters on the incident, see Moure Mariño, *Memorias*, pp. 73–9, 87–95. See also González Egido, *Unamuno*, pp. 127–58; Elías Díaz, 'Miguel de Unamuno y la guerra civil' in *La voluntad del humanismo: homenaje a Juan Marichal*, ed. B.

Ciplijauskaité & C. Maurer, (Barcelona, 1990) pp. 209–21.

79 Unamuno to Quintín de Torre, 1, 13 December 1936, in González Egido, *Unamuno*, pp. 210–12, 226–8.

80 García Venero, *Falange/Hedilla*, pp. 320–3.

81 Moure Mariño, *Memorias*, pp. 105–6; Franco Salgado-Araujo, *Mis conversaciones*, pp. 430–1.

82 On the 'Fifth Column', see Cabanellas, *Guerra*, II, pp. 684–6. On the massacres, see Ian Gibson, *Paracuellos: cómo fue* (Barcelona, 1983) pp. 73–134; Carlos Fernández, *Paracuellos de Jarama: ¿Carrillo culpable?* (Barcelona, 1983) pp. 60–83.

83 See Gibson, *En busca*, pp. 161–70; Southworth, *Antifalange*, pp. 144–8.

84 *FRUS 1936*, II, p. 568.

85 Viñas, *Guerra*, pp. 69–78; taped testimony of Hans Joachim von Knobloch to Sheelagh Ellwood, 14 January 1978, García Venero, *Falange/Hedilla*, pp. 200–3.

86 Communication of Folke von Knobloch to the author, 26 March 1991; *DGFP*, D, III, pp. 114–15; taped testimony of von Knobloch; Viñas, *Guerra*, pp. 78–97; García Venero, *Falange/Hedilla*, pp. 203–4.

87 Taped testimony of von Knobloch; Garriga, *Franco-Serrano Suñer*, pp. 29–30; García Venero, *Falange/Hedilla*, pp. 205–7.

88 Ramón Garriga, *La España de Franco: las relaciones con Hitler*,

2ª edición (Puebla, Mexico, 1970) pp. 18–19.

89 García Venero, *Falange/Hedilla*, p. 255.

90 García Venero, *Falange/Hedilla*, pp. 255–8; Southworth, *Antifalange*, pp. 164–6.

91 Serrano Suñer, *Memorias*, p. 169; see the comments by Dionisio Ridruejo in Ronald Fraser, *Blood of Spain: The Experience of Civil War 1936–1939* (London, 1979) p. 316.

92 Serrano Suñer, *Memorias*, p. 170.

93 Facsimile of letter from José Antonio Primo de Rivera to Serrano Suñer and Fernández Cuesta, 19 November 1936, Angel Alcázar de Velasco, *Serrano Suñer en la Falange* (Madrid/Barcelona, 1941) between pp. 166–7; Zugazagoitia, *Guerra*, I, pp. 256–64; Prieto, *Convulsiones*, I, pp. 130–53; Southworth, *Antifalange*, p. 203; Serrano Suñer, *Memorias*, pp. 483–4.

94 Saña, *Franquismo*, pp. 51–3.

95 Garriga, *Ramón Franco*, pp. 272–9.

96 The letter is reproduced in facsimile in Díaz, *Mi vida*, p. 182; Garriga, *Ramón Franco*, pp. 280–2.

CHAPTER VIII

1 *The Times*, 6 October 1936; Cardozo, *The March*, p. 157.

2 Alcofar Nassaes, *C.T.V.*, pp. 42–6.

3 Martínez Bande, *La marcha*, pp. 81–95; Lojendio, *Operaciones militares*, pp. 162–70; Aznar,

Historia militar, pp. 277–9; Mariñas, *Varela*, p. 129.

4 Kindelán, *Mis cuadernos*, pp. 93–7.

5 Calleja, *Yagüe*, p. 120, speaks without apparent irony of Franco's 'deferential condescension' in allowing Mola to take over.

6 Franco Salgado-Araujo, *Mi vida*, pp. 211–12.

7 Cox, *Defence of Madrid*, p. 19.

8 *ABC* (Seville), 8 October 1936; Julio Alvarez del Vayo, *Freedom's Battle* (London, 1940) p. 37.

9 Salas Larrazábal, *Ejército popular*, I, pp. 532–6; Howson, *Aircraft*, p. 15.

10 Koltsov, *Diario*, p. 182.

11 Vigón, *Mola*, p. 279.

12 Whitaker, 'Prelude', p. 114.

13 Calleja, *Yagüe*, pp. 122–3; Mariñas, *Varela*, pp. 129–30; Vigón, *Mola*, p. 280.

14 Martínez Bande, *La marcha*, pp. 104–109; Calleja, *Yagüe*, pp. 123–7.

15 García Venero, *Falange/Hedilla*, pp. 253–4.

16 Aznar, *Historia militar*, p. 281.

17 Howson, *Aircraft*, pp. 192–201; Andrés García Lacalle, *Mitos y verdades: la aviación de caza en la guerra española* (México D.F., 1973) pp. 173–9, 561–7.

18 *Ciano's Diplomatic Papers* ed. Malcolm Muggeridge (London, 1948) pp. 53–4.

19 Neurath to German Embassy in Italy, 27 October, Hassell to Wilhelmstrasse, 28, 29 October 1936, *DGFP*, D, III, pp. 121–2; U.S. Ambassador to Italy, Phillips, to Hull, 31 October

1936, *FRUS 1936*, II, p. 551.

20 *DGFP*, D, III, pp. 124–5.

21 *FRUS 1936*, II, pp. 558, 576. Whealey, *Hitler and Spain*, p. 8 puts the figure at nearer 4,000 in all.

22 Koltsov, *Diario*, pp. 183–6; Zugazagoitia, *Guerra*, I, pp. 178–84; Francisco Largo Caballero, *Mis recuerdos* (México D.F., 1954) pp. 187–9.

23 Koltsov, *Diario*, pp. 189–90; Julio Aróstegui & Jesús A. Martínez, *La Junta de Defensa de Madrid* (Madrid, 1984) pp. 43–80; López Fernández, *Defensa*, pp. 80–92; General Vicente Rojo, *Así fue la defensa de Madrid* (México D.F., 1967) pp. 45–8.

24 Gibson, *Queipo*, pp. 240, 263–4, 279.

25 Zugazagoitia, *Guerra*, I, p. 198; Gregorio Gallego, *Madrid, corazón que se desangra* (Madrid, 1976) pp. 220–2; Largo Caballero, *Mis recuerdos*, pp. 188–9.

26 López Fernández, *Defensa*, pp. 109–110; Lázaro Somoza Silva, *El general Miaja: biografía de un heroe* (México D.F., 1944) p. 185.

27 Claude G. Bowers, *My Mission to Spain* (London, 1954) p. 320; Southworth, *El mito*, p. 90; *FRUS 1937* (Washington, 1954) I, pp. 279–80.

28 Henry Buckley, *Life and Death of the Spanish Republic* (London, 1940) p. 261.

29 Rojo, *Así fue*, pp. 36–45; report of Miaja to Junta de Defensa, 7 November 1936, *Junta de Defensa*, p. 292.

30 Robert G. Colodny, *The Struggle for Madrid: The Central Epic of the Spanish Conflict 1936–1937* (New York, 1958) p. 43; Louis Fischer, *Men and Politics: An Autobiography* (London, 1941) p. 373; General Vicente Rojo, *España heroica: diez bocetos de la guerra española* 3ª edición (Barcelona, 1975) pp. 43–4.

31 López Fernández, *Defensa*, pp. 93–105; Koltsov, *Diario*, pp. 202–4; Zugazagoitia, *Guerra*, I, pp. 189–202; Rojo, *España heroica*, pp. 44–9.

32 Rojo, *España heroica*, pp. 49–55; Colodny, *The Struggle*, p. 24; Jef Last, *The Spanish Tragedy* (London, 1939) pp. 28–30.

33 Zugazagoitia, *Guerra*, I, p. 191; Colodny, *The Struggle*, p. 24.

34 Buckley, *Life and Death*, pp. 262–3; Somoza Silva, *Miaja*, pp. 142–8.

35 Aznar, *Historia militar*, pp. 282–5; Lojendio, *Operaciones militares*, pp. 170–1; Vigón, *Mola*, pp. 281–2; Whitaker, 'Prelude', p. 115.

36 Colodny, *The Struggle*, pp. 76–82; Martínez Bande, *La marcha*, pp. 113–45.

37 Buckley, *Life and Death*, pp. 264–8; Colodny, *The Struggle*, pp. 84–8.

38 Thomas, *Civil War*, p. 486; Bolín, *Vital Years*, p. 228; Kindelán, *Mis cuadernos*, p. 91.

39 *FRUS 1936*, II, p. 569.

40 Rojo, *Así fue*, pp. 55–103; Colodny, *The Struggle*, pp. 52–91; Thomas, *Civil War*, pp. 473–82.

41 Franco Salgado-Araujo, *Mi*

vida, p. 215; Vigón, *Mola*, pp. 286–7.

42 This is clear from the apologetic tone of Francoist historians, Lojendio, *Operaciones militares*, pp. 173–6; Aznar, *Historia militar*, pp. 285–9; Suárez Fernández, *Franco*, II, p. 136.

43 Aznar, *Historia militar*, pp. 285–6.

44 Whitaker, 'Prelude', p. 115; *FRUS 1936*, II, pp. 582, 601.

45 *FRUS 1936*, II, pp. 566, 579, 582.

46 *DGFP*, D, III, pp. 125–6, 131–3; *Manchester Guardian*, 19 November 1936.

47 *Daily Telegraph*, 19 November 1936.

48 *DGFP*, D, III, p. 134.

49 *The Times*, 1 December 1936.

50 Ismael Saz & Javier Tusell, editors, *Fascistas en España: la intervención italiana en la guerra civil a través de los telegramas de la "Missione Militare Italiana in Spagna" (15 diciembre 1936 – 31 marzo 1937)* (Madrid/Rome, 1981) p. 25. Henceforth cited as MMIS, *Telegramas*.

51 *DGFP*, D, III, pp. 139–40, 143–4, 147–8; Coverdale, *Italian Intervention*, pp. 122–3, 157–8.

52 *FRUS 1936*, II, p. 561.

53 *DGFP*, D, III, pp. 154–5, 159–62; *Die Weizsäcker-Papiere 1933–1950* (Frankfurt/M, 1974) pp. 104–5.

54 Verbale riunione a Palazzo Venezzia del 6 dicembre 1936–XV, ASMAE, Gabinetto, Spagna, b.3031; Hassell to Wilhelmstrasse, 28 November, 1, 17 December, Dieckhoff to

Hassell, 2 December, Dieckhoff memorandum, 11 December, *DGFP*, D, III, pp. 143–4, 146, 149–50, 165, 169; De Felice, *Mussolini il Duce*, pp. 383–4; MMIS, *Telegramas*, p. 27; Coverdale, *Italian Intervention*, pp. 160–4; Weinberg, *Diplomatic Revolution*, pp. 296–8.

55 Alcofar Nassaes, *C.T.V.*, pp. 32–3, 54; MMIS, *Telegramas*, pp. 20–1; Coverdale, *Italian Intervention*, pp. 165–6.

56 Martin Blinkhorn, *Carlism and Crisis in Spain 1931–1939* (Cambridge, 1975) pp. 273–5.

57 Del Burgo, *Conspiración*, pp. 687–92; García Venero, *Falange*, pp. 291–3; Faupel to Wilhelmstrasse, 14 April 1937, *DGFP*, D, III, p. 268; Blinkhorn, *Carlism*, pp. 275–7.

58 Don Juan de Borbón to Franco, 7 December 1936, Pedro Sainz Rodríguez, *Un reinado en la sombra* (Barcelona, 1981) p. 347.

59 Franco to Don Juan de Borbón, 12 January 1937, Sainz Rodríguez, *Un reinado*, p. 347.

60 *ABC*, Seville, 18 July 1937.

61 Servicio Histórico Militar (Coronel José Manuel Martínez Bande), *La lucha en torno a Madrid* (Madrid, 1968) pp. 43–51.

62 Faupel to Neurath, 10 December 1936, *DGFP*, D, III, pp. 159–62.

63 Cantalupo, *Fu la Spagna*, p. 108; Franco, *Apuntes*, p. 42; Suárez Fernández, *Franco*, II, p. 162;

Alcofar Nassaes, *C.T.V.*,
pp. 53–4.

64 MMIS, *Telegramas*, pp. 28–9;
Coverdale, *Italian Intervention*,
pp. 167–8, 170–1.

65 Weizsäcker, *Papiere*, p. 103;
'Hispanicus', *Foreign
Intervention*, pp. 138–40;
Weinberg, *Diplomatic
Revolution*, pp. 297–8.

66 Martínez Bande, *La campaña de
Andalucía*, pp. 184–6.

67 De Felice, *Mussolini il Duce*,
p. 385.

68 *DGFP*, D, III, p. 222.

69 MMIS, *Telegramas*, pp. 29, 67,
94.

70 Martínez Bande, *La lucha en
torno a Madrid*, p. 37.

71 Martínez Bande, *La lucha en
torno a Madrid*, pp. 179–80;
Lojendio, *Operaciones militares*,
pp. 182–95; Aznar, *Historia
militar*, pp. 326–8.

72 MMIS, *Telegramas*, p. 79.

73 Carlos de Arce, *Los generales de
Franco* (Barcelona, 1984) p. 186;
Martínez Bande, *La lucha en
torno a Madrid*, pp. 51–69;
Rojo, *Así fue*, pp. 103–18;
Aznar, *Historia militar*,
pp. 329–33.

74 Rodríguez Aisa, *Gomá*,
pp. 85–93.

75 Rodríguez Aisa, *Gomá*,
pp. 93–100.

76 *DGFP*, D, III, pp. 169, 173–4,
178, 191, 206, 219, 222; De
Felice, *Mussolini il Duce*,
pp. 390–1; Alcofar Nassaes,
C.T.V, pp. 58–61.

77 *DGFP*, D, III, p. 191.

78 MMIS, *Telegramas*, pp. 93–4.

79 Verbale della riunione a Palazzo
Venezia del 14 gennaio 1937-

XV, ASMAE, Gabinetto,
Spagna, b.3031; *DGFP*, D, III,
pp. 225–6; De Felice, *Mussolini
il Duce*, pp. 389–90; Coverdale,
Italian Intervention, pp. 171–3.

80 De Felice, *Mussolini il Duce*,
pp. 389–90.

81 Comunicazione fatta al
Generale Franco a nome del
Governo italiano e del Governo
tedesco il 23 gennaio 1937-XV
decisa a Roma nella riunione a
Palazzo Venezia col Generale
Göring il 14 gennaio, ASMAE,
Gabinetto, Spagna, b.3031.

82 Anfuso to Ciano, 27, 28 January
1937, ASMAE, Gabinetto,
Spagna, b.3031, and MMIS,
Telegramas, pp. 112–18.

83 Nota verbal del Gobierno del
General Franco, 24 January
1937, ASMAE, Gabinetto,
Spagna, b.3031.

84 Anfuso to Ciano, 23, 27, 28
January, Roatta to U.S., 30
January, 14 February 1937,
MMIS, *Telegramas*, pp. 108, 112,
114–16, 121, 140; Faupel to
Wilhelmstrasse, 27 January
1937, *DGFP*, D, III, p. 236;
Alcofar Nassaes, *C.T.V.*, pp. 54,
65; Suárez Fernández, *Franco*,
II, p. 163.

85 Mussolini to Roatta, 18
December 1936, MMIS,
Telegramas, p. 69.

86 Kindelán, *Mis cuadernos*, p. 113;
Martínez Bande, *La campaña de
Andalucía*, pp. 184–6.

87 Martínez Bande, *La campaña de
Andalucía*, pp. 153–65.

88 MMIS. *Telegramas*, pp. 69, 108.

89 MMIS, *Telegramas*, p. 72, 76–8.

90 MMIS, *Telegramas*, pp. 43, 78,
92–5; Emilio Faldella, *Venti*

mesi di guerra in Spagna
(Florence, Le Monnier, 1939)
pp. 233–5.

91 Richthofen diary, quoted Maier,
Guernica, pp. 35–6.

92 Faldella, *Venti mesi*, pp. 239–43.

93 Franco Salgado-Araujo, *Mi
vida*, pp. 216–19.

94 Martínez Bande, *La campaña de
Andalucia*, pp. 170–211;
Faldella, *Venti mesi*, pp. 243–7;
Aznar, *Historia militar*,
pp. 339–55; Ramón Garriga,
Guadalajara y sus consecuencias
(Madrid, 1974) pp. 52–4;
Roatta to U.S., 8 February
1937, MMIS, *Telegramas*,
p. 130.

95 Koestler, *Spanish Testament*,
pp. 186–204, 216; Chalmers
Mitchell, *Málaga*, pp. 238–42,
251–4; Manuel Azaña, *La
velarda en Benicarló* in *Obras*, III,
p. 400.

96 Bowers to Hull, 12 April 1937,
FRUS 1937, I, pp. 279–80.

97 Martínez Bande, *La campaña de
Andalucía*, pp. 210–11; Alcofar
Nassaes, *C.T.V.*, p. 70;
Southworth, *Antifalange*,
pp. 159–60; Southworth, *El
mito de la cruzada*, pp. 274–5. For
eye-witness accounts, see T. C.
Worsley, *Behind the Battle*
(London, 1939) pp. 179–208;
Chalmers Mitchell, *Málaga*,
pp. 266–7; Bahamonde, *Un año
con Queipo*, pp. 126–36;
Cantalupo, *Fu la Spagna*,
pp. 130–45.

98 Franco Salgado-Araujo, *Mi
vida*, p. 219.

99 Cantalupo to Mussolini, 29
March 1937, ASMAE, SFG,
b.38, T.709/345, p. 3; Faldella,

Venti mesi, pp. 193–4; Aznar,
Historia militar, pp. 353–5;
Martínez Bande, *La campaña de
Andalucía*, pp. 169–70, 214–15.

100 Largo Caballero, *Mis recuerdos*,
pp. 200–1; Dolores Ibárruri, *El
único camino* (Paris, 1964)
pp. 373–7; Burnett Bolloten,
*The Spanish Civil War:
Revolution and Counterrevolution*
(Chapel Hill, 1991) pp. 343–6;
Franz Borkenau, *The Spanish
Cockpit* 2nd edition (Ann Arbor,
1963) pp. 223–4.

101 Kindelán, *Mis cuadernos*,
pp. 113–14; Alcofar Nassaes,
C.T.V., pp. 70–2.

102 MMIS, *Telegramas*, pp. 129,
131–2; Faldella, *Venti mesi*,
pp. 249–50.

103 MMIS, *Telegramas*, pp. 88–9,
96, 106.

104 Cabanellas, *Guerra*, II,
pp. 718–19.

105 *ABC*, Seville, 18, 20 February
1937.

106 Di Febo, *La Santa*, pp. 66–9.
The relationship between
Franco and the Saint is dealt
with ironically in Francisco
Umbral, *Leyenda del César
visionario* (Barcelona, 1991)
p. 113.

107 Rodríguez Aisa, *Gomá*, p. 154;
Di Febo, *La Santa*, pp. 69–71;
Franco Salgado-Araujo, *Mi
vida*, p. 219.

108 Franco Salgado-Araujo, *Mi
vida*, p. 220.

109 Cantalupo, *Fu la Spagna*,
pp. 76–7; Coverdale, *Italian
Intervention*, pp. 187–8.

110 Cantalupo to Ciano, 17
February 1937, ASMAE,
Spagna Fondo di Guerra, b.38,

no.287/137; Cantalupo, *Fu la Spagna*, pp. 108–12.

111 Faldella to Franco, 13 February 1937, MMIS, *Telegramas*, pp. 209–10. See *Ibid*, pp. 47–8; Olao Conforti, *Guadalajara: la prima sconfitta del fascismo* (Milan, U.Mursia C., 1967) pp. 27–30; Faldella, *Venti mesi*, p. 254; Coverdale, *Italian Intervention*, pp. 213–15, 254.

112 Conforti, *Guadalajara*, pp. 30–2; Coverdale, *Italian Intervention*, p. 215.

113 Cantalupo to Ciano, 17 February 1937, ASMAE, SFG, b.38, no.287/137.

114 Conforti, *Guadalajara*, p. 33.

115 Franco's note to Faldella, dated 14 February, is reproduced in MMIS, *Telegramas*, pp. 211–13; Conforti, *Guadalajara*, pp. 33–4; Coverdale, *Italian Intervention*, pp. 216–17.

116 Cantalupo to Ciano, 1 March 1987, ASMAE, SFG, b.38, no.392/193.

117 Saz/Tusell, 'Presentación', MMIS, *Telegramas*, p. 49.

118 Martínez Bande, *La lucha en torno a Madrid*, pp. 76–111; Rojo, *España heroica*, pp. 57–68; Rojo, *Así fue*, pp. 152–69.

119 Conforti, *Guadalajara*, pp. 34–5; Faldella, *Venti mesi*, p. 255; Saz/Tusell, 'Presentación', MMIS, *Telegramas*, pp. 49–50, 191–95; Cantalupo, *Fu la Spagna*, pp. 185–6; Martínez Bande, *La lucha en torno a Madrid*, pp. 121–31; Coverdale, *Italian Intervention*, p. 217–18.

CHAPTER IX

1 Rodríguez Aisa, *Gomá*, pp. 204–13.

2 Cantalupo, *Fu la Spagna*, pp. 124–5; Vegas Latapie, *Caminos*, pp. 157–8.

3 Cantalupo, *Fu la Spagna*, p. 131.

4 MMIS, *Telegramas*, pp. 154, 158; Cantalupo, *Fu la Spagna*, pp. 131–42; Coverdale, *Italian Intervention*, p. 192.

5 Anastasio Granados, *El Cardenal Gomá: Primado de España* (Madrid, 1969) pp. 145–6; Rodríguez Aisa, *Gomá*, pp. 61–5; José M. Sánchez, *The Spanish Civil War as Religious Tragedy* (Notre Dame, Indiana, 1987) pp. 79–81.

6 *The Daily Mail*, 1 March 1937.

7 Southworth, *Antifalange*, p. 202; *La España de Franco*, pp. 7–8; Garriga, *Los validos*, pp. 42–3, 72–3; Franco Salgado-Araujo, *Mi vida*, pp. 232, 239; Saña, *Franquismo*, pp. 116–17; Galinsoga & Franco Salgado, *Centinela*, p. 302.

8 Thomas, *Civil War*, p. 514.

9 Testimony of Franco to Dr Soriano, Soriano, *La mano izquierda*, pp. 146–7.

10 Serrano Suñer, *Memorias*, pp. 243–52; Saña, *Franquismo*, pp. 115–20; Garriga, *Franco-Serrano Suñer*, p. 57.

11 Saña, *Franquismo*, pp. 119–20.

12 Serrano Suñer in an interview with the author in 1979; Ridruejo quoted in Lago, *Las contramemorias*, p. 121.

13 Iribarren, *Con el general Mola*, pp. 210–11; García Venero, *Falange/Hedilla*, pp. 232–3;

Cabanellas, *Guerra*, II,
pp. 849–50.

14 Cantalupo, *Fu la Spagna*,
pp. 146–58; Coverdale, *Italian Intervention*, pp. 190–1.

15 Cantalupo, *Fu la Spagna*,
pp. 101–10; Coverdale, *Italian Intervention*, pp. 218–19;
Martínez Bande, *La lucha en torno a Madrid*, pp. 117–19.

16 MMIS, *Telegramas*, pp. 157,
193–4; Franco to Roatta,
undated, probably 1 March
1937, in Martínez Bande, *La lucha en torno a Madrid*, p. 125,
n. 137.

17 Franco to Roatta, 5 March
1937, in Martínez Bande, *La lucha en torno a Madrid*,
pp. 131–2; Roatta's summary
of the letter, MMIS, *Telegramas*,
pp. 193–4.

18 Martínez Bande, *La lucha en torno a Madrid*, p. 130; MMIS,
Telegramas, p. 161.

19 Martínez Bande, *La lucha en torno a Madrid*, pp. 197–209.

20 Coverdale, *Italian Intervention*,
pp. 222–4.

21 Martínez Bande, *La lucha en torno a Madrid*, pp. 130–3;
Zugazagoitia, *Guerra*, I,
pp. 248–9; Colodny, *Struggle*,
p. 132.

22 MMIS, *Telegramas*, pp. 161–83;
Faldella, *Venti mesi*,
pp. 255–66; Conforti,
Guadalajara, pp. 51–178;
Martínez Bande, *La lucha en torno a Madrid*, pp. 133–46;
Cantalupo, *Fu la Spagna*,
pp. 182–91; Longo, *Le Brigate Internazionali*, pp. 285–314;
Coverdale, *Italian Intervention*,
pp. 225–38, 256–60; Alcofar

Nassaes, *C.T.V.*, pp. 91–5; Salas
Larrazabal, *Ejército popular*, I,
pp. 861–80; Colodny, *Struggle*,
pp. 128–43; Aznar, *Historia militar*, pp. 369–83.

23 MMIS, *Telegramas*, pp. 162,
164, 166, 169–70, 193–5;
Martínez Bande, *La lucha en torno a Madrid*, pp. 160–1.

24 Cantalupo to Ciano, 16 March
1937, ASMAE, SFG, b.38,
T.559/267; MMIS, *Telegramas*,
p. 165; Cantalupo, *Fu la Spagna*,
p. 191, 196–7.

25 Mariñas, *Varela*, p. 151; Inés
García de la Escalera, *El general Varela* (Madrid, 1959) p. 24.

26 MMIS, *Telegramas*, p. 194.

27 MMIS, *Telegramas*, pp. 56–8,
174–5, 192, 217–19;
Cantalupo, *Fu la Spagna*,
pp. 192–3; Coverdale, *Italian Intervention*, pp. 238–40;
Franco Salgado-Araujo, *Mi vida*, p. 222.

28 Franco to Mussolini, 19 March
1937, quoted in Ufficio Spagna,
'Indagine sull'azione di
Guadalajara', Archivio Centrale
dello Stato, Segretaria
Particolare del Duce,
Carteggio riservato, f.463/R,
pp. 47–9; MMIS, *Telegramas*,
pp. 58–9; Wolfram von
Richthofen, entry for 19 March
1937, 'Spanien-Tagebuch', in
Maier, *Guernica*, p. 75;
Coverdale, *Italian Intervention*,
pp. 241–2; Colodny, *The Struggle*, p. 142.

29 Martínez Bande, *La lucha en torno a Madrid*, pp. 146–60;
Faldella, *Venti Mesi*,
pp. 266–75; Conforti,
Guadalajara, pp. 178–324;

Kindelán, *Mis cuadernos*, p. 118;
Koltsov, *Diario*, p. 357;
Cantalupo, *Fu la Spagna*,
pp. 200–2; Coverdale, *Italian
Intervention*, pp. 240–48;
Alcofar Nassaes, *C.T.V.*,
pp. 95–100; Aznar, *Historia
militar*, pp. 383–7; Salas
Larrazabal, *Ejército popular*, I,
p. 880; Vigón, *Mola*, p. 308.

30 Cantalupo, *Fu la Spagna*,
pp. 192–6; *DGFP*, D, III,
pp. 258–60.

31 Cantalupo to Ciano, 24 March
1937, ASMAE, SFG, b.38,
T.657/320; Cantalupo, *Fu la
Spagna*, pp. 207–9; *FRUS 1937*,
I, pp. 268–9; Garriga,
Guadalajara, pp. 173–4; Salas
Larrazabal, *Ejército popular*, I,
pp. 882–4; Antonio Ruiz
Vilaplana, *Doy fe . . . un año de
actuación en la España
nacionalista* (Paris, n.d. [1938])
pp. 237–48; Colodny, *Struggle*,
p. 141.

32 Cantalupo, *Fu la Spagna*, p. 155.

33 MMIS, *Telegramas*, pp. 52, 179,
193; Cantalupo, *Fu la Spagna*,
pp. 187–8; Rojo, *Así fue*, p. 186.

34 MMIS, *Telegramas*, p. 189;
Franco Salgado-Araujo, *Mi vida*,
pp. 224–5.

35 Conforti, *Guadalajara*, p. 360;
Coverdale, *Italian Intervention*,
pp. 255–6.

36 Coverdale, *Italian Intervention*,
pp. 253–5.

37 Cantalupo to Ciano, 17
February 1937, ASMAE, SFG,
b.38, no.287/137.

38 Ufficio Spagna, 'Indagine
sull'azione di Guadalajara',
Archivio Centrale dello Stato,
Segretaria Particolare del Duce,

Carteggio riservato, f.463/R,
p. 67; MMIS, *Telegramas*,
pp. 53–5; Rojo, *Así fue*, p. 164.

39 Cantalupo, *Fu la Spagna*, p. 197;
Martínez Bande, *La lucha en torno
a Madrid*, pp. 160–1, 166–7.

40 De Felice, *Mussolini il Duce*,
pp. 405–11.

41 Cantalupo to Mussolini, 29
March 1937, ASMAE, SFG,
b.38, T.709/345, pp. 2–5;
Cantalupo, *Fu la Spagna*,
pp. 210–14.

42 *Documents on the Italian
Intervention in Spain* (n.p.
[London], 1937).

43 Rojo, *Así fue*, pp. 187–9;
Thomas, *Civil War*, pp. 604–5.

44 Cantalupo, *Fu la Spagna*, p. 196.

45 Cantalupo, *Fu la Spagna*,
pp. 222–8; Coverdale, *Italian
Intervention*, pp. 272–5.

46 Garriga, *Guadalajara*,
pp. 208–9.

47 Kindelán, *Mis cuadernos*,
pp. 120–3; Vigón, *Mola*,
pp. 303–4.

48 Franco Salgado-Araujo, *Mi
vida*, p. 225.

49 Kindelán, *Mis cuadernos*,
pp. 115–16.

50 *DGFP*, D, III, pp. 267–70;
Payne, *Franco Regime*,
pp. 131–6.

51 Kindelán, *Mis cuadernos*,
pp. 119–20; José Manuel
Martínez Bande, *Vizcaya*
(Madrid, 1971) pp. 13–17;
Maier, *Guernica*, pp. 44–5;
Vigón, *Mola*, p. 311; Angel
Viñas, 'La responsibilidad de la
destrucción de Guernica' in
Viñas, *Guerra*, p. 99.

52 Williamson Murray, *German
Military Effectiveness*

(Baltimore, 1992) pp. 104–5.

53   Maier, *Guernica*, pp. 45–6; Richthofen, diary entries for 24, 26 March 1937, 'Spanien-Tagebuch', pp. 77–81; Viñas, 'La responsibilidad', pp. 99–102.

54   Alcofar Nassaes, *C.T.V.*, pp. 112–15; Maier, *Guernica*, p. 48; Howson, *Aircraft*, pp. 209; Murray, *Military Effectiveness*, p. 148.

55   *DGFP*, D, III, p. 266; Coverdale, *Italian Intervention*, pp. 275–80; Alcofar Nassaes, *C.T.V.*, pp. 107–9.

56   *DGFP*, D, III, pp. 125–6.

57   Viñas, 'La responsibilidad', p. 106.

58   Richthofen, 'Spanien-Tagebuch', diary entries for 24, 28 March 1937, pp. 79, 82.

59   G. L. Steer, *The Tree of Gernika: A Field Study of Modern War* (London, 1938) p. 159; *FRUS 1937*, I, p. 291; Aznar, *Historia militar*, p. 398.

60   Steer, *Gernika*, pp. 160–70; Southworth, *Guernica!*, pp. 368–9; Salas Larrazabal, *La guerra desde el aire*, pp. 187–8.

61   Martínez Bande, *Vizcaya*, p. 84.

62   It is not clear if Sperrle and Richthofen saw Mola together or separately. Maier, *Guernica*, pp. 49–52; Richthofen, diary entry for 2 April 1937, 'Spanien-Tagebuch', pp. 86–7; Viñas, 'La responsibilidad', pp. 102–3.

63   Martínez Bande, *Vizcaya*, p. 84.

64   *The Times*, 9 April 1937; Thomas, *Civil War*, p. 616; Aznar, *Historia militar*, pp. 398–401.

65   Richthofen, diary entry for 2 April, 'Spanien-Tagebuch', p. 87; Viñas, 'La responsibilidad', pp. 103–4.

66   Martínez Bande, *Vizcaya*, pp. 82–92; Aznar, *Historia militar*, pp. 401–6; Lojendio, *Operaciones militares*, pp. 269–76; Carr, *Spanish Tragedy*, pp. 184–6.

67   Cantalupo, *Fu la Spagna*, pp. 229–30.

68   Cantalupo to Mussolini, 29 March 1937, ASMAE, SFG, b.38, T.709/345, pp. 8–9.

69   Cantalupo to Mussolini, 29 March 1937, ASMAE, SFG, b.38, T.709/345, p. 3; Cantalupo, *Fu la Spagna*, p. 230.

70   Cantalupo, *Fu la Spagna*, p. 231.

71   Cantalupo, *Fu la Spagna*, pp. 232–3.

72   Richthofen, diary entry for 18 April 1937, 'Spanien-Tagebuch', pp. 96–7; Viñas, 'La responsibilidad', pp. 106–8.

73   Richthofen, diary entry for 23 April 1937, 'Spanien-Tagebuch', p. 101.

74   In part because of the politically motivated withdrawal of battalions of the anarcho-syndicalist CNT, Steer, *Gernika*, pp. 213–33.

75   Richthofen, diary entry for 25 April 1937, 'Spanien-Tagebuch', pp. 101–3.

76   Richthofen, diary entry for 26 April 1937, 'Spanien-Tagebuch', p. 103–4; Peter Monteath, 'Guernica Reconsidered: Fifty Years of Evidence', *War & Society*, Vol. 5, No. 1, May 1987, pp. 97–8.

77   Howson, *Aircraft*, pp. 136, 175,

182, 209, 272; Richthofen, diary entry for 26 April, 'Spanien-Tagebuch', p. 103; Ramón Hidalgo Salazar, *La ayuda alemana a España 1936–1939* (Madrid, 1975) pp. 142–5; Salas Larrazabal, *Ejército popular*, II, p. 1436, n.47.

78 Alberto Onaindía, *Hombre de paz en la guerra* (Buenos Aires, 1973) pp. 229–45; Joseba Elosegui, *Quiero morir por algo* (Bordeaux, 1971) pp. 145–59; Steer, *Gernika*, pp. 236–45; Southworth, *Guernica!*, pp. 353–70; Alberto Reig Tapia, 'Guernica como símbolo' in Carmelo Garitaonandía & José Luis de la Granja, editors, *La guerra civil en el país vasco* (Bilbao, 1987) pp. 149–50; Thomas, *Civil War*, pp. 624–9; Salas Larrazabal, *Ejército popular*, II, pp. 1384–92, III, 2861–6.

79 Richthofen, diary entries for 27 & 30 April 1937, 'Spanien-Tagebuch', pp. 106, 109; Maier, *Guernica*, pp. 59–64; Southworth, *Guernica!*, pp. 276–7; Monteath, 'Guernica', pp. 90–1, 102–3, n.48; Viñas, 'La responsibilidad', pp. 114–22.

80 *The Times*, 28 April 1937.

81 Southworth, *Guernica!*, pp. 239–325; Monteath, 'Guernica', pp. 79–85. In *The Guardian*, 27 May 1991, Brian Crozier denied that Guernica had been destroyed by German bombers.

82 Salas Larrazabal, *Ejército popular*, II, p. 1390; Hills, *Franco*, p. 277. For a thorough

demolition of the idea, see Southworth, *Guernica!*, pp. 263–7.

83 *Palabras del Caudillo 19 abril 1937 – 31 diciembre 1938* (Barcelona, 1939) p. 273.

84 Similar claims that Mola was upset are in contradiction with his own prior bloodthirsty threats, Vicente Talón, *Arde Guernica* (Madrid, 1970) pp. 115–18; Del Burgo, *Conspiración*, p. 862.

85 Foltz, *The Masquerade*, pp. 54–5; Del Burgo, *Conspiración*, p. 862.

86 Darstellung des Obersten Freiherr von Beust, in Maier, *Guernica*, p. 157.

87 Both telegrams in Servicio Histórico Militar, Archivo de la Guerra de Liberación/D.N./A.7/L.368/43 quoted Reig Tapia, 'Guernica', pp. 133–4, who attributes the reply to Franco. Talón, *Arde Guernica*, pp. 112–13, attributes it to an unnamed lieutenant colonel of Franco's general staff.

88 Southworth, *Guernica!*, pp. 301–2, 373–5; Viñas, 'La responsibilidad', pp. 122–35, where it is suggested that an ulterior motive of Franco's telegram was to hide his collusion with Sperrle and throw the responsibility for Guernica lower down the command chain onto Vigón and Richthofen.

89 *Palabras del Caudillo 19 abril 1937 – 19 abril 1938* (n.p., 1938) pp. 116, 120, 133; *Palabras del Caudillo 19 abril 1937 – 31 diciembre 1938*, pp. 137;

Francisco Franco, *Textos de doctrina política: palabras y escritos de 1945 a 1950* (Madrid, 1951) pp. 675, 687.

90   Salas Larrazabal, *La guerra desde el aire*, p. 190.

91   Maier, *Guernica*, pp. 65–6.

92   A point still denied by Francoist historians – see Salas Larrazabal, *Ejército popular*, II, pp. 1390–2; Martínez Bande, *Vizcaya*, pp. 107–8; Hidalgo Salazar, *La ayuda alemana*, pp. 142–5.

93   Carr, *Spanish Tragedy*, pp. 186–8.

94   Southworth, *Guernica!*, pp. 188–9, 383–4; Steer, *Gernika*, p. 260.

CHAPTER X

1   Cantalupo to Ciano, 17 February 1937, ASMAE, Spagna Fondo di Guerra, 287/137, b.38.

2   Cantalupo to Ciano, 1 March 1937, ASMAE, SFG, 392/193, b.38.

3   Rodríguez Aisa, *Gomá*, p. 153.

4   Sheelagh Ellwood, 'La crisis de Salamanca: la Unificación', *Historia 16*, No. 132, April 1987, p. 12.

5   MMIS, *Telegramas*, pp. 37, 88.

6   Vegas Latapie, *Caminos*, p. 295; Franco Salgado-Araujo, *Mi vida*, p. 202. See also Koestler, *Spanish Testament*, pp. 23–5.

7   'Correspondencia Franco/Mola', p. 22.

8   Del Burgo, *Conspiración*, pp. 207–8.

9   Report of Cantalupo, Salamanca, 6 March 1937, ASMAE, SFG, b.38, N.464/

226; Cabanellas, *La guerra*, I, pp. 622–3; González Egido, *Agonizar en Salamanca*, pp. 95–6, 105, 228, 245.

10   Koestler, *Spanish Testament*, p. 33.

11   Vegas Latapie, *Caminos*, p. 204.

12   Testimony of Gil Robles to the author in Madrid in 1970.

13   Franco in 8 October 1947, quoted Armando Chávez Camacho, *Misión de prensa en España* (México D.F., 1948) pp. 437–8. He repeated the story in 1961, Soriano, *La mano*, p. 152.

14   See his letter to *The Universe*, 22 January 1937.

15   Del Burgo, *Conspiración*, pp. 221–2.

16   Blinkhorn, *Carlism*, pp. 280–4; Angel Alcázar de Velasco, *Siete días de Salamanca* (Madrid, 1976) pp. 108–19.

17   A. de Lizarza (pseudonym of Andrés María de Irujo), *Los vascos y la República española* (Buenos Aires, 1944) pp. 124–8.

18   Serrano Suñer, *Memorias*, pp. 126–54. See also Faupel's assessment, *DGFP*, D, III, pp. 285–6.

19   Saña, *Franquismo*, pp. 57–8.

20   Ramón Serrano Suñer, *Entre Hendaya y Gibraltar* (Madrid, 1947) pp. 17–22.

21   Franco, *Nosotros*, p. 42; Garriga, *Nicolás Franco*, p. 111; Garriga, *Franco-Serrano Suñer*, p. 41.

22   García Lahiguera, *Serrano Suñer*, pp. 89–90.

23   Testimony of Ramón Serrano Suñer to the author, 21.XI.90.

24   Serrano Suñer, *Entre Hendaya y*

Gibraltar, pp. 22–3; Saña, *Franquismo*, pp. 63–6.

25 Saña, *Franquismo*, p. 86.

26 Maximiano García Venero, *Historia de la Unificación (Falange y Requeté en 1937)* (Madrid, 1970) pp. 176–8; Ernesto Giménez Caballero, *Memorias de un dictador* (Barcelona, 1979) pp. 100–1.

27 For a brilliant analysis of Franco's attitude to power, see Dionisio Ridruejo, *Casi unas memorias* (Barcelona, 1976) p. 115.

28 Escobar, *Así empezó*, p. 151–5; María Jesús Cava Mesa, *Los diplomáticos de Franco: J.F. de Lequerica, temple y tenacidad (1890–1963)* (Bilbao, 1989) p. 128.

29 Hedilla to Serrano Suñer, 17 March 1947, *Cartas cruzadas entre D. Manuel Hedilla Larrey y D. Ramón Serrano Suñer* (Madrid, 1947) reprinted in Luis Ramírez, *Nuestros primeros veinticinco años* (Paris, 1964) pp. 136.

30 Foltz, *The Masquerade*, p. 84. On Hedilla's violence, see García Venero, *Falange/Hedilla*, pp. 70–1; Southworth, *Antifalange*, pp. 10, 94.

31 Serrano Suñer, *Memorias*, pp. 158–9; Saña, *Franquismo*, p. 80.

32 Franco Salgado-Araujo, *Mi vida*, p. 220.

33 Saña, *Franquismo*, pp. 67–70.

34 Saña, *El franquismo*, pp. 22–4; Vegas Latapie, *Caminos*, p. 80; Alfredo Kindelán Núñez del Pino, 'Semblanza político-militar del general Kindelán' in

Kindelán, *Mis cuadernos*, pp. 50, 60; Kindelán, *La verdad*, pp. 16–17; Serrano Suñer, *Memorias*, p. 165.

35 García Venero, *Falange/Hedilla*, pp. 338–40; Alcázar de Velasco, *Siete días*, pp. 20–1; Southworth, *Antifalange*, pp. 182–4, 204–8.

36 Southworth, *Antifalange*, pp. 206–7.

37 García Venero, *Falange/Hedilla*, pp. 312–13.

38 García Venero, *Falange/Hedilla*, pp. 296–7; Vegas Latapie, *Caminos*, p. 190.

39 Escobar, *Así empezó*, pp. 160–1.

40 García Venero, *Falange/Hedilla*, pp. 313–14; Serrano Suñer, *Entre Hendaya y Gibraltar*, pp. 24–5; Alcázar de Velasco, *Siete días*, p. 127.

41 Vicente de Cadenas y Vicent, *Actas del último Consejo Nacional de Falange Española de las JONS (Salamanca, 18–19.IV.1937) y algunas noticias referentes a la Jefatura nacional de prensa y propaganda* (Madrid, 1975) p. 94; Southworth, *Antifalange*, pp. 191–2.

42 García Venero, *Falange/Hedilla*, pp. 287–91, 338–41, 348–54.

43 Payne, *Falange*, pp. 162–3; Alcázar de Velasco, *Siete días*, pp. 19–21; García Venero, *Falange/Hedilla*, pp. 297, 349–50, 376–6.

44 García Venero, *Falange/Hedilla*, pp. 212, 221, 238-51.

45 Southworth, *Antifalange*, p. 179.

46 Cantalupo, *Fu la Spagna*, pp. 150–1; Cantalupo, 'La situazione politico-militare

834   FRANCO

della spagna nel nono messe di guerra civile', ASMAE, SFG, b.38, R.3428.

47 Danzi to MAE, 1 March 1937, ASMAE, SFG, b.38, T.151; García Venero, *Falange/Hedilla*, p. 321.

48 Cantalupo to Ciano, 6 March, ASMAE, SFG, b.38, N.464/226; 16 March, ASMAE, SFG, b.38, T.559/267; 23 March 1937, ASMAE, SFG, b.38, T.644/313.

49 Cantalupo to Mussolini, 29 March 1937, ASMAE, SFG, b.38, T.709/345, p. 12.

50 García Venero, *Falange/Hedilla*, p. 321.

51 Cantalupo, *Fu la Spagna*, pp. 164–6, 196–8.

52 García Venero, *Falange/Hedilla*, p. 296, 341–4; Alcázar de Velasco, *Siete días*, pp. 187–9; Southworth, *Antifalange*, pp. 184–5.

53 Payne, *Falange*, p. 151; García Venero, *Falange/Hedilla*, pp. 348–9; Southworth, *Antifalange*, p. 185.

54 Alcázar de Velasco, *Siete días*, pp. 36, 44–50, 276–7.

55 Danzi to Ciano, 12 April 1937, ASMAE, SFG, b.38, N.332.

56 García Venero, *Falange/Hedilla*, pp. 355–6; García Venero, *Historia de la Unificación*, pp. 189–92.

57 *DGFP*, D, III, pp. 268–9.

58 Cantalupo, 'La situazione politico-militare della Spagna nel nono messe di guerra civile' 9 aprile 1937–XV, ASMAE, SFG, b.38, R.3428.

59 Alcázar de Velasco, *Siete días*, pp. 79, 92–4.

60 Alcázar de Velasco, *Siete días*, p. 156, 193–202.

61 García Venero, *Falange/Hedilla*, pp. 356–7; Alcázar de Velasco, *Siete días*, pp. 87–8; Payne, *Falange*, p. 163.

62 Vegas Latapie, *Caminos*, pp. 192–3; García Venero, *Falange/Hedilla*, p. 358; Alcázar de Velasco, *Siete días*, p. 128.

63 Ridruejo, *Memorias*, pp. 92, 94; Alcázar de Velasco, *Siete días*, p. 249.

64 García Venero, *Falange/Hedilla*, pp. 359–63; Alcázar de Velasco, *Siete días*, pp. 130 ff., Southworth, *Antifalange*, pp. 187–8; Payne, *Falange*, pp. 164–5.

65 Hedilla to Carrero Blanco, 24 March 1947, *Cartas cruzadas*, pp. 141–2; García Venero, *Falange/Hedilla*, p. 371; Alcázar de Velasco, *Siete días*, pp. 166–8; Luis Pagés Guix, *La traición de los Franco ¡Arriba España!*, (Madrid, n.d. [1937]) (in Southworth, *Antifalange*, pp. 245–58) p. 249.

66 Southworth, *Antifalange*, p. 192; Alcázar de Velasco, *Serrano Suñer*, p. 68; García Venero, *Historia de la Unificación*, p. 201.

67 Pagés Guix, *La traición*, p. 250.

68 García Venero, *Falange/Hedilla*, p. 370; Vegas Latapie, *Caminos*, p. 196; Ridruejo, *Memorias*, p. 93.

69 Alcázar de Velasco, *Siete días*, pp. 169–73, 231; García Venero, *Historia de la Unificación*, p. 202.

70 Southworth, *Antifalange*, pp. 196–8; Alcázar de Velasco, *Siete días*, pp. 174–87.

71 Foltz, *The Masquerade*, pp. 84–5; Southworth, *Antifalange*, p. 197; Alcázar de Velasco, *Siete días*, p. 211.

72 The account of López Puertas in Alcázar de Velasco, *Siete días*, pp. 228–40; *Cartas cruzadas*, pp. 150–3. Alcázar de Velasco suggests that Dávila killed Goya, *Siete días*, p. 2246.

73 Alcázar de Velasco, *Siete días*, pp. 240–2; García Venero, *Falange/Hedilla*, p. 381.

74 See *Cartas cruzadas*, pp. 150–3. Hedilla and his friends all mistakenly give the date as 14 April. This is accepted by Payne, *Falange*, pp. 165–6. The same interpretation is to be found in García Venero, *Falange/Hedilla*, pp. 372–5, 381. The complications of reconstructing the events of 16 April are outlined in Southworth, *Antifalange*, pp. 219–26.

75 Vegas Latapie, *Caminos*, p. 200.

76 Southworth, *Antifalange*, pp. xxi, 7.

77 García Venero, *Falange/Hedilla*, pp. 376–7, 395; Alcázar de Velasco, *Siete días*, pp. 226–7.

78 *Cartas cruzadas*, p. 136, 147; Southworth, *Antifalange*, pp. 180–1.

79 Vegas Latapie, *Caminos*, p. 200; García Venero, *Falange/Hedilla*, pp. 381–2.

80 Cadenas y Vicent, *Actas*, pp. 89–107; Alcázar de Velasco, *Serrano Suñer*, pp. 62–76; Alcázar de Velasco, *Siete días*, pp. 257–73; García Venero, *Falange/Hedilla*, pp. 384–7; Payne, *Falange*, pp. 167–8.

81 Danzi to Ciano, 18 April 1937, ASMAE, SFG, b.38, T.1090.

82 García Venero, *Falange/Hedilla*, p. 387; Southworth, *Antifalange*, pp. 209–10.

83 *Palabras de Franco I año triunfal* (Bilbao, 1937) pp. 21–9. Confusion of the chronology at this time is accentuated by Hedilla's deficient memory and the fact that in later editions of Franco's speeches, the speech of 18 April is attributed to the following day, see *Palabras del Caudillo* (Barcelona, 1939) p. 9.

84 *ABC* (Sevilla), 20 March 1937; Alcázar de Velasco, *Siete días*, pp. 297–8; García Venero, *Falange/Hedilla*, p. 387.

85 Vegas Latapie, *Caminos*, pp. 201–2; García Venero, *Falange/Hedilla*, pp. 388–9.

86 Serrano Suñer, *Entre Hendaya y Gibraltar*, pp. 23–31.

87 Testimony of Serrano Suñer and Giménez Caballero to the author; Giménez Caballero, *Memorias*, pp. 98–100; Alcázar de Velasco, *Siete días*, pp. 284–8; Saña, *Franquismo*, pp. 77–8; García Venero, *Falange/Hedilla*, p. 392.

88 García Venero, *Falange/Hedilla*, p. 394.

89 García Venero, *Falange/Hedilla*, pp. 395–6.

90 Rodríguez Aisa, *Gomá*, pp. 158–9.

91 García Venero, *Falange/Hedilla*, p. 396.

92 Vegas Latapie, *Caminos*, p. 208; García Venero, *Falange/Hedilla*, p. 401.

93 Alcázar de Velasco, *Siete días*, p. 22.

94 José María Fontana, *Los catalanes en la guerra de España* 2ª edición (Barcelona, 1977) p. 298.

95 *DGFP*, D, III, pp. 277–8, 281–3; Serrano Suñer, *Entre Hendaya y Gibraltar*, pp. 41–2; Pierre Broué & Emile Témime, *The Revolution and the Civil War in Spain* (London, 1972) pp. 428–32.

96 *Cartas cruzadas*, pp. 136, 138, 142; García Venero, *Falange/Hedilla*, pp. 397, 404–5; Southworth, *Antifalange*, p. 210; Ridruejo, *Memorias*, p. 93.

97 García Venero, *Falange/Hedilla*, pp. 405–8, 411–20; Vegas Latapie, *Caminos*, pp. 211–14; Danzi? (signature unclear) to Ciano, 11 May 1937, ASMAE, SFG, b.38, T.1266/589; Ridruejo, *Memorias*, pp. 94–6; Cardozo, *The March*, pp. 308–9; Foss & Gerahty, *Arena*, pp. 6333–4. One of those charged with these 'crimes' does not deny that they took place, Angel Alcázar de Velasco, *La gran fuga* (Barcelona, 1977) pp. 29–44. Cf. Southworth, *Antifalange*, pp. 217–18, 226–38.

98 García Venero, *Falange/Hedilla*, pp. 419–20.

99 Ridruejo, *Memorias*, pp. 96–7; Serrano Suñer, *Memorias*, pp. 173–4.

100 Ridruejo, *Memorias*, p. 99.

101 Faupel to Wilhelmstrasse, 1 May 1937, *DGFP*, D, III, p. 277.

102 Faupel to Wilhelmstrasse, 9 June 1937, *DGFP*, D, III, pp. 312–13.

103 Hedilla's letter, *Documentos inéditos*, I, p. 124; Vegas Latapie, *Caminos*, pp. 214–16.

104 Saña, *Franquismo*, pp. 318–20; García Venero, *Historia de la Unificación*, pp. 222–3.

105 Danzi? to Ciano, 11 May 1937, ASMAE, SFG, b.38, T.1266/589.

106 Alcázar de Velasco, *Siete días*, pp. 12–14, 21–5; García Venero, *Falange/Hedilla*, pp. 438–67.

107 *Cartas cruzadas*, pp. 147–9; García Venero, *Falange/Hedilla*, p. 445; Suárez Fernández, *Franco*, II, pp. 206–8.

108 García Venero, *Falange/Hedilla*, pp. 402–3.

109 Southworth, *Antifalange*, p. 176.

110 *DGFP*, D, III, pp. 281–2; Payne, *Falange*, p. 169.

111 Danzi? to Ciano, 11 May 1937, ASMAE, SFG, b.38, T.1266/589.

112 Serrano Suñer, *Entre Hendaya y Gibraltar*, pp. 44–5.

113 Burgo, *Conspiración*, p. 809.

114 Burgo, *Conspiración*, pp. 233–4.

115 Alcázar de Velasco, *Serrano Suñer*, p. 98; Escobar, *Así empezó*, p. 62.

116 Blinkhorn, *Carlism*, pp. 290–1; Burgo, *Conspiración*, pp. 797–815.

117 *DGFP*, D, III, pp. 294–5; Rodríguez Aisa, *Gomá*, pp. 160–8; Sánchez, *Religious Tragedy*, pp. 126–7.

118 Rodríguez Aisa, *Gomá*, pp. 233–269; 411–18, 442–54; Sánchez, *Religious Tragedy*, pp. 92–5.

119 Montero Moreno, *La persecución religiosa*, pp. 726–41;

Rodríguez Aisa, *Gomá*, pp. 254–69.

120 Rodríguez Aisa, *Gomá*, pp. 245–51.

121 Rodríguez Aisa, *Gomá*, pp. 271–89.

122 Vegas Latapie, *Caminos*, pp. 307–8; Rodríguez Aisa, *Gomá*, pp. 298–300.

CHAPTER XI

1 Vegas Latapie, *Caminos*, p. 259, 279–80.

2 Francesc Cambó, *Meditacions: dietari (1941–1946)* (Barcelona, 1982) p. 1449.

3 Escobar, *Así empezó*, pp. 233–4.

4 Alberto Onaindía, *El "Pacto" de Santoña, antecedentes y desenlace* (Bilbao, 1983) pp. 33–50; José Antonio de Aguirre y Lecube, *De Guernica a Nueva York pasando por Berlín* 3ª edición (Buenos Aires, 1944) pp. 30–9; Cantalupo, *Fu la Spagna*, pp. 225–8; Coverdale, *Italian Intervention*, pp. 284–7; José María Garmendia, 'el Pacto de Santoña', in Carmelo Garitaonandia & José Luis de la Granja, editors, *La guerra civil en el País vasco 50 años después* (Bilbao, 1987) pp. 157–61; Rodríguez Aisa, *Gomá*, pp. 219–20. In fact, for some days before then, it was known in diplomatic circles that such a negotiation was about to start, *DAPE*, IV (Lisbon, 1965) pp. 274–5.

5 Azaña, *Obras*, IV, pp. 588, 655–6; Pablo de Azcárate, *Mi embajada en Londres durante la guerra civil española* (Barcelona,

1976) pp. 64–7; Andrés Saborit, *Julián Besteiro* (Buenos Aires, 1967) pp. 272–3; Ignacio Arenillas de Chaves, *El proceso de Besteiro* (Madrid, 1976) p. 239.

6 *DGFP*, D, III, pp. 289–92.

7 *DGFP*, D, III, pp. 293–5.

8 Rodríguez Aisa, *Gomá*, pp. 165–71.

9 Testimony of Ramón Serrano Suñer to the author, 21.XI.90.

10 Cabanellas, *Guerra*, II, pp. 994–5; Arrarás, *Cruzada*, III, p. 449.

11 Testimony of Ramón Serrano Suñer to the author, 21.XI.90; Saña, *Franquismo*, pp. 94–5.

12 *ABC*, Seville, 4 June 1937.

13 Howson, *Aircraft*, p. 38; Silva, *Como asesinar*, pp. 78–90; Bravo Morata, *Los muertos providenciales*, pp. 149–81.

14 Testimony of Ramón Serrano Suñer to the author, 21.XI.90. It was a political/poetic licence which led Serrano Suñer to declare in Bilbao on 19 June 1938 that 'General Franco cried – we saw him – full of emotion for his comrade-in-arms', Ramón Serrano Suñer, *Siete discursos* (Bilbao, 1938) p. 48.

15 Faupel to Wilhelmstrasse, 9 July 1937, *DGFP*, D, III, p. 410.

16 Vegas Latapie, *Caminos*, pp. 291–2; Escobar, *Así empezó*, pp. 160–4.

17 Rodríguez Aisa, *Gomá*, p. 171.

18 Ruiz Vilaplana, *Doy fe . . .*, pp. 121–2.

19 *Hitler's Table Talk*, p. 608.

20 Garriga, *Franco-Serrano Suñer*, pp. 50–1.

21 Franco Salgado-Araujo, *Mi vida*, pp. 226–7.

22 Franco Salgado-Araujo, *Mi vida*, pp. 227–8, 238.

23 Martínez Bande, *Vizcaya*, pp. 149–60; Lojendio, *Operaciones militares*, pp. 286–90; Aznar, *Historia militar*, pp. 418–24.

24 Elosegui, *Quiero morir*, pp. 208–13; Martínez Bande, *Vizcaya*, pp. 175–97; Manuel González Portilla & José María Garmendia, *La guerra civil en el País Vasco* (Madrid, 1988) pp. 32–9; Southworth, *Guernica!*, pp. 384–6; Monteath, 'Guernica', p. 98; Hills, *Franco*, p. 278.

25 Onaindía, *El "Pacto"*, pp. 50–6; Fraser, *Blood of Spain*, pp. 408–9; Coverdale, *Italian Intervention*, pp. 286–8.

26 Juan de Iturralde, *La guerra de Franco: los Vascos y la Iglesia* 2 vols (San Sebastián, 1978) II, pp. 285–99; Cabanellas, *Guerra*, II, p. 861; Manuel Tuñón de Lara, *La España del siglo XX* (Paris, 1973) p. 458.

27 *DGFP*, D, III, pp. 412–13.

28 *DGFP*, D, III, pp. 408–10.

29 Maier, *Guernica*, p. 73.

30 W. von Oven, *Hitler und der Spanische Bürgerkrieg. Mission und Schicksal der Legion Condor* (Tubingen, 1978) p. 403.

31 *DGFP*, D, III, p. 411.

32 Kindelán, *Cuadernos*, p. 129.

33 Franco Salgado-Araujo, *Mi vida*, pp. 228–9.

34 De la Cierva, *Franco*, III, p. 213.

35 Salas Larrazábal, *La guerra desde el aire*, pp. 235–44; Howson, *Aircraft*, p. 233.

36 Kindelán, *Cuadernos*, pp. 131–7.

37 Franco Salgado-Araujo, *Mi vida*, p. 229; Lojendio, *Operaciones militares*, p. 332; Aznar, *Historia militar*, pp. 460–1; Garriga, *Yagüe*, pp. 134–5.

38 On the battle and its strategic significance, see Faldella, *Venti mesi*, p. 357; Rojo, *España heroica*, pp. 91–101; Salas Larrazábal, *Ejército popular*, II, pp. 1215–64; Aznar, *Historia militar*, pp. 430–63; Lojendio, *Operaciones militares*, pp. 331–44; Thomas, *Civil War*, pp. 710–16.

39 Salas Larrazábal, *Ejército*, II, p. 1248.

40 Servicio Histórico Militar (Coronel José Manuel Martínez Bande), *El final del frente norte* (Madrid, 1972) pp. 13–36.

41 Garriga, *Yagüe*, p. 135.

42 Kindelán, *Cuadernos*, pp. 136–7; Aznar, *Historia militar*, p. 460; Luca de Tena, *Mis amigos muertos*, pp. 205–6.

43 Franco Salgado-Araujo, *Mi vida*, p. 232; *Palabras de Franco I ano triunfal*, pp. 37–45.

44 *Palabras de Franco, I año triunfal*, pp. 49–61.

45 Martínez Bande, *El final del frente norte*, pp. 41–89; Lojendio, *Operaciones militares*, pp. 290–303; Aznar, *Historia militar*, pp. 465–83; Thomas, *Civil War*, pp. 717–19; Franco Salgado-Araujo, *Mi vida*, p. 238.

46 *Ciano's Diary*, edited by Malcolm Muggeridge, (London, 1947) p. 15; Onaindía, *El "Pacto"*, pp. 57–81; Coverdale, *Italian Intervention*, pp. 289–90.

47 Onaindía, *El "Pacto"*, pp. 108–9.

48 Onaindía, *El "Pacto"*, pp. 110–50; Garmendia, 'El Pacto', pp. 162–76.

49 Onaindía, *El "Pacto"*, pp. 153–64; Garmendia, 'El Pacto', pp. 177–8; Steer, *Gernika*, pp. 386–94; Martínez Bande, *El final del frente norte*, pp. 89–98; Coverdale, *Italian Intervention*, pp. 291–4; P. M. Heaton, *Welsh Blockade Runners in the Spanish Civil War* (Newport, 1985) pp. 68, 101–2; Iturralde, *La guerra de Franco*, II, pp. 301–10; *FRUS 1937*, I, p. 433, 465–6.

50 Onaindía, *El "Pacto"*, p. 171.

51 The fullest account of the frictions between the Basques and the Republican Government is *El Informe del Presidente Aguirre al Gobierno de la República sobre los hechos que determinaron el derrumbamiento del frente del norte (1937)* (Bilbao, 1978) *passim*. See also González Portilla & Garmendia, *La guerra civil en el País Vasco*, pp. 40–4.

52 Fraser, *Blood*, p. 412.

53 *Ciano's Diary 1937–1938* (London, 1952) pp. 17–18; Coverdale, *Italian Intervention*, pp. 317–18.

54 Smyth, 'The Moor and the Moneylender', pp. 147–8; Viñas *et al.*, *Política comercial exterior*, I, pp. 149–50; *DGFP*, C, III, pp. 499–503.

55 Serrano Suñer, *Entre Hendaya y Gibraltar*, pp. 48–9.

56 *DGFP*, D, III, pp. 413–14, 417, 421–2.

57 *DGFP*, D, III, pp. 434–5.

58 *DGFP*, D, III, pp. 496–542; Coverdale, *Italian Intervention*, pp. 324–6; Smyth, 'The Moor and the Moneylender', pp. 149–55.

59 Rojo, *España heroica*, pp. 103–15; Servicio Histórico Militar (Coronel José Manuel Martínez Bande), *La gran ofensiva sobre Zaragoza* (Madrid, 1973) pp. 78–167; Lojendio, *Operaciones militares*, pp. 346–9; Salas Larrazábal, *Ejército*, II, pp. 1287–1330; Aznar, *Historia militar*, pp. 499–516; Thomas, *Civil War*, pp. 722–8; Franco Salgado-Araujo, *Mi vida*, pp. 241–2.

60 Angel Viñas, 'La Legión Cóndor en Asturias' in *Guerra, dinero, dictadura*, p. 147; Martínez Bande, *El final del frente norte*, pp. 109–75; Salas, Larrazábal, *Ejército*, II, pp. 1470–99; Lojendio, *Operaciones militares*, pp. 303–26; Aznar, *Historia militar*, pp. 517–29; Thomas, *Civil War*, pp. 728–31.

61 Thomas, *Civil War*, p. 733; Aznar, *Historia militar*, pp. 528–9.

62 Serrano Suñer, *Entre Hendaya y Gibraltar*, p. 29.

63 Serrano Suñer, *Entre Hendaya y Gibraltar*, pp. 52–4.

64 Garriga, *España de Franco*, pp. 47–9; Serrano Suñer, *Memorias*, p. 172.

65 Payne, *Falange*, p. 184.

66 Escobar, *Así empezó*, pp. 230–42.

67 *Palabras del Caudillo 19 abril 1937 – 19 abril 1938*, pp. 186-7.

68 Ridruejo, *Memorias*, p. 121;

Serrano Suñer, *Memorias*, p. 172; Vegas Latapie, *Caminos*, pp. 427–30; Saña, *Franquismo*, p. 148.

69   Vegas Latapie, *Caminos*, pp. 515–16.

70   Zugazagoitia, *Guerra*, II, p. 44.

71   Ciano, *Diary 1937–1938*, 17 October 1937, p. 22; *DGFP*, D, III, p. 521.

72   Servicio Histórico Militar (Coronel José Manuel Martínez Bande), *La batalla de Teruel* 2ª edición (Madrid, 1990) pp. 16–26; Cabanellas, *Guerra*, II, p. 1008.

73   Aznar, *Historia militar*, p. 536.

74   Kindelán, *Mis cuadernos*, pp. 140–51; Franco Salgado-Araujo, *Mi vida*, p. 248; Lojendio, *Operaciones militares*, pp. 361–3; Aznar, *Historia militar*, pp. 535–41.

75   Rojo, *España heroica*, pp. 117–19; Aznar, *Historia militar*, pp. 543–4.

76   Rojo, *España heroica*, pp. 119–125; Martínez Bande, *Teruel*, pp. 52–64; Aznar, *Historia militar*, pp. 545–54; Salas Larrazábal, *Ejército*, II, pp. 1637–49; Lojendio, *Operaciones militares*, pp. 365–7.

77   Martin, *Franco*, p. 293; Aznar, *Historia militar*, pp. 551, 622; Garriga, *Yagüe*, pp. 139–40.

78   Rojo, *España heroica*, p. 125, 128–9.

79   Ciano, *Diary 37–38*, p. 46.

80   Martínez Bande, *Teruel*, p. 89; Kemp, *Mine Were of Trouble*, p. 125.

81   Franco Salgado-Araujo, *Mi vida*, p. 250; Lojendio, *Operaciones militares*, p. 369.

82   Rojo, *España heroica*, pp. 127–8; Franco Salgado-Araujo, *Mi vida*, pp. 248–50; Martínez Bande, *Teruel*, pp. 70–120; Salas Larrazábal, *Ejército*, II, pp. 1649–50; Lojendio, *Operaciones militares*, pp. 369–80; Aznar, *Historia militar*, pp. 554–65.

83   Ciano, *Diary 37–38*, p. 50.

84   Franco Salgado-Araujo, *Mi vida*, pp. 248–53; Martínez Bande, *Teruel*, pp. 123–61; Eloy Fernández Clemente, *El coronel Rey d'Harcourt y la rendición de Teruel: historia y fin de una leyenda negra* (Terual, 1992) *passim*.

85   Ciano, *Diary 1937–1938*, 14 January 1938, pp. 64–5.

86   *DGFP*, D, III, p. 576.

87   Martínez Bande, *Teruel*, pp. 165–209; Lojendio, *Operaciones militares*, pp. 380–95; Aznar, *Historia militar*, pp. 569–85; Salas Larrazábal, *Ejército*, II, pp. 1672–1704.

88   *DGFP*, D, III, pp. 556–7; Howson, *Aircraft*, pp. 20–8.

89   *The Times*, 4 March 1938; *DGFP*, D, III, pp. 613–14.

90   *DGFP*, D, III, pp. 581–2, 588–9.

91   Stohrer to Wilhelmstrasse, 14 February, 9 March, Weisäcker Memorandum, 28 February 1938, *DGFP*, D, III, pp. 589, 607–8, 615; Coverdale, *Italian Intervention*, pp. 335–9.

92   Serrano Suñer, *Entre Hendaya y Gibraltar*, pp. 60–4; Equipo Mundo, *Los 90 Ministros*, pp. 19–67.

93   For an excellent account of the

various ministers and their
political interests, see Serrano
Suñer, *Entre Hendaya y Gibraltar*,
pp. 60–4. See also Ridruejo,
Memorias, p. 122.

94 Ciano, *Diary 37–38*, p. 37;
Serrano Suñer, *Memorias*,
pp. 255–60; Saña, *Franquismo*,
p. 88; Escobar, *Así empezó*,
pp. 276–9; Franco, *Nosotros*,
pp. 51–3. Garriga, *Nicolás
Franco*, pp. 160–1, 168–70,
suggests financial improprieties.

95 Serrano Suñer, *Entre Hendaya y
Gibraltar*, pp. 57–70; *España de
Franco*, pp. 40–3.

96 *Palabras del Caudillo 19 abril
1937 – 19 abril 1938*,
p. 201.

97 See the interview of Franco with
Manuel Aznar, 31 December
1938, *Palabras del Caudillo 19
abril 1937 – 31 diciembre 1938*,
pp. 300–10.

98 Hodgson to Eden, 1 February
1938, FO425/415, W2122/29/41.

99 Hodgson to Eden, 11 February
1938, FO425/415, W2127/29/41.

100 Coverdale, *Italian Intervention*,
pp. 342–5; García Lahiguera,
Serrano Suñer, p. 129; Saña,
Franquismo, pp. 99–100.

101 Ridruejo, *Memorias*, pp. 122,
195–6; Serrano Suñer,
Memorias, p. 262; Lago, *Las
contra-memorias*, p. 119.

102 Franco Salgado-Araujo, *Mi
vida*, pp. 258–9.

CHAPTER XII

1 Calleja, *Yagüe*, pp. 146–7;
Garriga, *Yagüe*, 142–3; Aznar,
Historia militar, p. 611.

2 Franco Salgado-Araujo, *Mi

vida, pp. 257–8; Alpert, *Guerra
en el mar*, pp. 334–7.

3 Servicio Histórico Militar
(Coronel José Manuel
Martínez Bande), *La llegada al
mar* (Madrid, 1975) pp. 13–14.

4 Ribbentrop Memorandum 21
March, Stohrer to
Wilhelmstrasse, 23, 24, 26
March 1938, *DGFP*, D, III,
pp. 622–8; Ciano, *Diary
1937–1938*, 20 March 1938,
p. 91.

5 Thomas, *Civil War*, p. 798.

6 Lojendio, *Operaciones militares*,
pp. 445–7.

7 Enrique Castro Delgado,
Hombres made in Moscú
(Barcelona, 1965) p. 560;
Aznar, *Historia militar*,
pp. 612–13.

8 Calleja, *Yagüe*, p. 148; Martínez
Bande, *La llegada*, pp. 42–5;
Lojendio, *Operaciones militares*,
pp. 448–51; Aznar, *Historia
militar*, pp. 623–9.

9 Martínez Bande, *La llegada*,
p. 72; Servicio Histórico
Militar (Coronel José Manuel
Martínez Bande), *La ofensiva
sobre Valencia* (Madrid, 1977)
pp. 13–14.

10 Cordón, *Trayectoria*,
pp. 376–80; Martínez Bande, *La
llegada*, pp. 44–64; Garriga,
Yagüe, p. 143.

11 Martínez Bande, *La llegada*,
pp. 75–126; Lojendio,
Operaciones militares,
pp. 460–79; Salas Larrazábal,
Ejército popular, II, pp. 1742–57;
Aznar, *Historia militar*,
pp. 641–63.

12 Kindelán, *Mis cuadernos*,
pp. 157–63.

13 *DGFP*, D, III, p. 628.
14 Salas Larrazábal, *Ejército popular*, II, p. 1818.
15 Martínez Bande, *La ofensiva*, pp. 16–18.
16 Jaime Martínez Parrilla, *La fuerzas armadas francesas ante la guerra civil española (1936–1939)* (Madrid, 1987) pp. 184–92; David Wingeate Pike, *Les français et la guerre d'Espagne 1936–1939* (Paris, 1975) pp. 296–7; *DGFP*, D, III, pp. 620–2.
17 See the interview with Franco, *The Times*, 4 March 1938.
18 Azaña, *Obras*, III, p. 537.
19 Garriga, *Yagüe*, pp. 145–6; Rojo, *¡Alerta los pueblos!*, pp. 40, 46–50, 54–5.
20 García Venero, *Falange/Hedilla*, pp. 436–7; Garriga, *Yagüe*, pp. 147–8; Rafael Abella, *La vida cotidiana durante la guerra civil 1) La España Nacional* (Barcelona, 1978); Yagüe to Franco, 9 December 1938, *Documentos inéditos*, I, pp. 278–9.
21 Martínez Bande, *La llegada*, pp. 141–73; Lojendio, *Operaciones militares*, pp. 479–89; Aznar, *Historia militar*, pp. 671–5.
22 *ABC* (Seville), 16 April 1938.
23 Canaris to Weizsäcker, 5 April, Weizsäcker Memorandum, Stohrer to Wilhelmstrasse, 8 April 1938, *DGFP*, D, III, pp. 630–1, 640–2.
24 Martínez Bande, *La ofensiva*, pp. 45–68; Lojendio, *Operaciones militares*, pp. 493–506.
25 *DGFP*, D, III, p. 647, 651–4, 671, 675–81.
26 The resignation caused considerable delight in Rome. Ciano, *Diary 1937–1938*, 21 February 1938, p. 78.
27 Coverdale, *Italian Intervention*, p. 355.
28 Martínez Bande, *La ofensiva*, pp. 69–95.
29 Martínez Bande, *La ofensiva*, pp. 95–6.
30 Lojendio, *Operaciones militares*, pp. 506–18; Aznar, *Historia militar*, pp. 677–99; Martínez Bande, *La ofensiva*, pp. 99–135.
31 *FRUS 1938*, I, pp. 200–5; Ciano, *Diary 1937–1938* 9, 28 June 1938, pp. 126, 132; Thomas, *Civil War*, pp. 826–9.
32 Thomas, *Civil War*, p. 828.
33 *DGFP*, D, III, pp. 674, 682–5; *FRUS 1938*, I, pp. 208–9.
34 *DGFP*, D, III, pp. 681–5; Ciano, *Diary 1937–1938*, 28, 29 June 1938, p. 132; *Ciano's Diplomatic Papers*, pp. 219–20.
35 *DGFP*, D, III, p. 684.
36 *FRUS 1938*, I, pp. 211, 215–16.
37 *DGFP*, D, III, pp. 700–3.
38 *DGFP*, D, III, pp. 703–4.
39 *DGFP*, D, III, pp. 696–8, 703–8. The Generalísimo's closer relationship with the Germans was evident too in a request made, on his orders, by Orgaz of General von Brauchitsch, for the Spanish Army to adopt the training regulations of the Germany Army and for Spanish war industries to be developed in close connection with those of the Third Reich, which was

tantamount to making Spain a military satellite of Germany.

40 *DGFP*, D, III, pp. 700–1, 713, 725–6.

41 Phillimore to Halifax, 12 July, P.M.'s Office to Caccia, 14 July 1938, Halifax Private Papers, PRO FO800/323. On Phillimore, see Griffiths, *Fellow Travellers*, p. 260; Haxey, *Tory M.P.*, p. 216.

42 *ABC* (Seville), 2 October 1938.

43 Cabanellas, *Guerra*, II, p. 939.

44 Franco, *Palabras del Caudillo, 19 abril 1937 – 31 diciembre 1938*, pp. 131–41.

45 Fernando González, *Liturgias para un Caudillo: manual de dictadores* (Madrid, 1977) pp. 84–7.

46 Servicio Histórico Militar (Coronel José Manuel Martínez Bande), *La batalla del Ebro* 2ª edición (Madrid, 1988) pp. 73–5; Calleja, *Yagüe*, pp. 163–6; Garriga, *Yagüe*, p. 150; Cabanellas, *Guerra*, II, p. 1028.

47 Martínez Bande, *Ebro*, pp. 87–113; Juan Modesto, *Soy del quinto regimiento (notas de la guerra española)* (Paris, 1969) pp. 175–92; Lojendio, *Operaciones militares*, pp. 518–26; Aznar, *Historia militar*, pp. 707–17, 723–35.

48 Franco Salgado-Araujo, *Mi vida*, pp. 262–4.

49 Galinsoga & Franco-Salgado, *Centinela*, pp. 305–6; Martínez Bande, *Ebro*, p. 127.

50 Lojendio, *Operaciones militares*, pp. 420–1; Galinsoga & Franco-Salgado, *Centinela*, p. 307.

51 Franco Salgado-Araujo, *Mi vida*, p. 264; Aznar, *Historia militar*, pp. 739–70.

52 Kindelán, *Mis cuadernos*, p. 173.

53 Manuel Tagüeña Lacorte, *Testimonio de dos guerras* (México D.F., 1973) p. 230; Martínez Bande, *Ebro*, p. 168.

54 Kindelán, *Mis cuadernos*, p. 184.

55 Franco Salgado-Araujo, *Mi vida*, p. 264; Kindelán, *Mis cuadernos*, pp. 171, 186, 205.

56 *DGFP*, D, III, pp. 742–3.

57 *DGFP*, D, III, pp.736–7.

58 Ciano, *Diary 1937–1938*, p. 148.

59 *DGFP*, D, III, pp. 739–41; Suárez Fernández, *Franco*, II, p. 319; Ciano, *Diary 1937–1938*, p. 159.

60 *DGFP*, D, III, pp. 741–2, 746–8; *DAPE*, I (Lisbon, 1961) pp. 449–51; Garriga, *España de Franco*, pp. 115–16.

61 *DGFP*, D, III, pp. 747–8.

62 Hodgson to F.O., 23 September 1938, FO371/22698, W13084/12909/41.

63 Mounsey to Cadogan, 28 September 1938, FO371/22698, W13118/12909/41.

64 *DGFP*, D, III, pp. 749–50, 752; *DAPE*, I, pp. 456–7, 460–1.

65 Ciano, 26 September 1938, *Diary 1937–1938*, p. 163.

66 *DAPE*, II (Lisbon, 1973) p. 166; report of Spanish military attaché in Berlin, 17 November 1938, *Documentos inéditos*, I, p. 225.

67 Williamson Murray, *The Change in the European Balance of Power, 1938–1939* (Princeton, 1984) pp. 195–274; Donald Cameron Watt, *How War Came: The Immediate Origins of the*

Second World War 1938–1939
(London, 1989) pp. 27–9.

68 Alba to Halifax, 3 October 1938,
    FO371/22698, W13345/
    12909/41.

69 Murray, *European Balance*,
    p. 215.

70 *DGFP*, D, III, pp. 753–7;
    Suárez Fernández, *Franco*, II,
    p. 318.

71 *FRUS 1939* II (Washington,
    1956) pp. 715–16.

72 *DAPE*, I, pp. 443–4; *DGFP*, D,
    III, pp. 753–7.

73 *DGFP*, D, III, pp. 760–1,
    767–8, 775–9, 782–8, 802.

74 *DGFP*, D, III, pp. 795–6, 808.

75 Franco Salgado-Araujo, *Mi
    vida*, pp. 265–5; Martínez
    Bande, *Ebro*, pp. 252–68.

76 Thomas, *Civil War*, p. 854.

77 *DGFP*, D, III, p. 745.

78 Chetwode to Halifax, 14
    November 1938, Halifax Private
    Papers, PRO FO800/323.

79 Díaz, *Mi vida*, pp. 17–19,
    216–26; Franco, *Nosotros*,
    pp. 200–13; Giménez
    Caballero, *Memorias*,
    p. 80.

80 *DGFP*, D, III, pp. 800–1.

81 *Palabras del Caudillo 19 abril
    1937 – 31 diciembre 1938*,
    pp. 284–5. See Paul Preston,
    *The Politics of Revenge: Fascism
    and the Military in the 20th
    Century Spain* (London, 1990)
    pp. 30–47.

82 Servicio Histórico Militar
    (Coronel José Manuel Martínez
    Bande), *La campaña de Cataluña*
    (Madrid, 1979) p. 15.

83 Martínez Bande, *Cataluña*,
    pp. 41–60; Thomas, *Spanish
    Civil War*, pp. 867–8;

Coverdale, *Italian Intervention*,
p. 375.

84 Garriga, *La Señora*, pp. 123–6.
    Fenosa stood for Fuerzas
    Eléctricas del Nor-Oeste
    Sociedad Anónima, the name of
    Barrié's electrical generating
    company.

85 *DGFP*, D, III, p. 828.

86 Jaraiz Franco, *Historia*,
    pp. 97–8, 143.

87 *FRUS 1939*, II, pp. 722–3;
    Martínez Bande, *Cataluña*,
    pp. 60–92.

88 Ciano, 25, 28, 31 December
    1938, *Diary 1937–1938*,
    p. 209.

89 *Palabras del Caudillo 19 abril
    1937 – 31 diciembre 1938*,
    pp. 295–315.

90 Martínez Bande, *Cataluña*,
    pp. 100–22.

91 Martínez Bande, *Cataluña*,
    pp. 265–6.

92 *Ciano's Diary 1939–1943*
    (London, 1947) p. 5; Coverdale,
    Italian Intervention, pp. 376–80;
    Martínez Bande, *Cataluña*,
    pp. 148–74.

93 *DGFP*, D, III, p. 844.

94 Estado Español, Ministerio de
    la Gobernación, *Dictamen de la
    comisión sobre ilegitimidad de
    poderes actuantes en 18 de julio de
    1936* (Barcelona, 1939);
    Ministerio de Justicia, *Causa
    general. La dominación roja en
    España. Avance de la información
    instruida por el ministerio público*
    (Madrid, 1944); Payne, *Franco
    Regime*, pp. 220–8.

95 Cabanellas, *Guerra*, II,
    pp. 1075–6.

96 *FRUS 1939*, II, pp. 751, 763;
    Documentos inéditos, I, pp. 292–3,

323–4; José Manuel Martínez Bande, *El final de la guerra civil* (Madrid, 1985) pp. 296–314.

97 Saña, *Franquismo*, pp. 90–1.

98 Ciano, *Diary 1939–1943*, 28 March 1939, p. 57.

99 Gil, *Cuarenta años*, pp. 37–8; Franco Salgado-Araujo, *Mi vida*, p. 276; *DGFP*, C, III, pp. 888–9.

CHAPTER XIII

1 In January 1939, Franco had signed an agreement on cultural and spiritual collaboration between Germany and Spain. After the Papal Nuncio, Monsignor Gaetano Cicognani, and Cardinal Gomá complained that this gave unlimited facilities for Nazi propaganda to the detriment of the Church, Franco did not ratify the agreement, Rodríguez Aisa, *Gomá*, pp. 503–9.

2 Rodríguez Aisa, *Gomá*, pp. 315–18; Gomá-Franco letters, Cardenal Gomá, *Por Dios y por España 1936–1939* (Barcelona, 1940) pp. 537–40.

3 Iturralde, *La guerra de Franco*, II, pp. 550–2; Rodríguez Aisa, *Gomá*, p. 318.

4 Garriga, *La Señora*, p. 182. Tusquets had published *Orígenes de la revolución española* (Barcelona, 1932); *La Francmasonería, crimen de lesa patria* (Burgos, n.d. [1937]) (freemasonry, crime of treason); *Masonería y separatismo* (Burgos, 1937).

5 José A. Ferrer Benimelli, 'Franco contra la masonería', *Historia 16*, año II, no. 15, julio de 1977, pp. 37–51; Suárez Fernández, *Franco*, III, pp. 92–100.

6 E. Allison Peers, *Spain in Eclipse 1937–1943* (London, 1943) pp. 98–100.

7 *DGFP*, D, III, pp. 838, 841–2, 845–6, 852.

8 *DGFP*, D, III, pp. 856–8.

9 *The Times*, 13, 16 February 1939; *DGFP*, D, III, pp. 854–5.

10 Testimony of Doña María Cristina de Borbón, *Hola*, No. 2478, 6 February 1992.

11 Whitaker, 'Prelude', p. 116.

12 *DGFP*, D, III, pp. 865–6; *Ciano's Diary 1939–1943*, p. 43.

13 *DGFP*, D, III, pp. 874–6.

14 Ciano, *Diary 1939–1943*, pp. 32–3; *DGFP*, Series D, Vol.III, pp. 880–1. Cf.Martin Wight, 'Spain and Portugal' in Arnold Toynbee & Frank T. Ashton-Gwatkin, *Survey of International Affairs 1939–1946: The World in March 1939* (London, 1952) pp. 145–6.

15 *DGFP*, D, VI, p. 301.

16 On Peterson's appointment, see *The Times*, 4, 31 March 1939. The conversation with Jordana, Peterson to Halifax, 12 April 1939, FO371/24150, W6173/824/41. Cf.Peterson, *Both Sides*, pp. 182–3.

17 *DAPE*, V (Lisbon, 1967) pp. 725–8.

18 On negotiations for close cooperation on cultural affairs, mining interests, police matters, military training and the future development of the Spanish Army, and facilities for German citizens to live and work in

Spain, Spanish support for German claims in Tangier and privileges for German merchant shipping, *DGFP*, D, III, pp. 632–4, 761–4, 803–4, 811–14, 827–8, 884–6.

19 Carlos Ruiz Ocaña, *Los ejércitos españoles: las fuerzas armadas en la defensa nacional* (Madrid, 1980) p. 113; DGFP, D, X, pp. 461–4; Payne, *Politics and the Military*, p. 421.

20 Matthieu Séguéla, *Pétain-Franco: les secrets d'une alliance* (Paris, 1992) pp. 25–9; Philippe Simonnot, *Le secret de l'armistice* (Paris, 1990) p. 97.

21 *Documents Diplomatiques Français 1932–1939*, 2ᵉ Série (1936–1939) XV (Paris, 1981) pp. 54–5.

22 *The Times*, 25 March 1939; Pétain to Bonnet, 27 March 1939, *DDF*, 2ᵉ Série, XV, n.1, p. 234; François Piétri, *Mes années d'Espagne 1940–1948* (Paris, 1954) p. 49; Séguéla, *Pétain-Franco*, pp. 31–7.

23 *The Times*, 12, 14 April 1939; Bonnet to Pétain, 15 April, Pétain to Bonnet, 20, 22 April 1939, *DDF*, 2ᵉ, XV, pp. 656, 721–2, 770–2.

24 Peterson to Halifax, 12 April 1939, FO371/24150, W6173/824/41; *The Times*, 12 April 1939.

25 *DGFP*, D, III, pp. 902–15; *The Times*, 20 April, 12 May 1939.

26 Francesc Cambó, *Meditacions: dietari (1936–1940)* (Barcelona, 1982) p. 563.

27 *The Times*, 17, 18, 25 April, 4, 19 May 1939.

28 *The Times*, 19, 20 May; *ABC*, 20 May; *The Daily Telegraph*, 20 May 1939. For a vivid description of the parade, see Daniel Sueiro & Bernardo Díaz Nosty, *Historia del franquismo*, 2 vols, 2nd edition (Barcelona, 1985) I, pp. 22–4. On the *señoritos'* cavalry, see Pemán, *Mis encuentros*, p. 90.

29 *Palabras del Caudillo 19 abril 1937 – 7 diciembre 1942* (Madrid, 1943) pp. 117–22.

30 *ABC*, 21 May; *The Times*, 21, 22 May 1939.

31 Gomá-Franco letters in Gomá, *Por Díos*, pp. 537–44.

32 *The Times*, 24 May 1939.

33 Viola to Ciano, 25, 29 May, *I Documenti Diplomatici Italiani*, 8ª serie, vol.XII (Rome, 1952) pp. 17–18, 48.

34 He criticized the atheistic excesses of Nazism although he hinted that he wanted to improve his German contacts, *DGFP*, D, III, pp. 916–17; *The Times*, 14 June 1939; *Ciano's Diary 1939–1943*, pp. 99–100; Serrano Suñer, *Entre Hendaya y Gibraltar*, pp. 91–118.

35 *DGFP*, Series D, Vol.VI, pp. 695–7; Ciano, *Diary 1939–1943*, pp. 97, 102.

36 Roncalli to Ciano, 16 June, Viola to Ciano, 24 June 1939, *DDI*, 8ª, XII, pp. 220–1, 268–9. On the Jordana-Serrano Suñer rivalry, see the quotations from Jordana's diary in Xavier Tusell & Genoveva García Queipo de Llano, *Franco y Mussolini: la política española durante la segunda guerra mundial* (Barcelona, 1985) pp. 21–2, 39–40, 44.

37 Pétain to Bonnet, 9 June 1939, *Documents Diplomatiques*

Français, 2ᵉ Série (1936–1939) Tome XVI (Paris, 1983) pp. 747–50; *DGFP*, D, III, pp. 847–8.

38 Ciano, *Diary 1939–1943*, 5 June 1939, p. 100; Peterson, *Both Sides*, pp. 223–4; Sir Samuel Hoare, *Ambassador on Special Mission* (London, 1946) pp. 56–8. In contrast, he assured the Vichy French Ambassador that he felt no resentment towards France, Piétri, *Mes années*, p. 35.

39 *Arriba*, 20 January 1940.

40 *DGFP*, D, VI, pp. 697–8, 719–21, 830, 882.

41 Ciano, *Diaries 1939–1942*, p. 102; *Ciano's Diplomatic Papers*, edited by Malcolm Muggeridge (London, 1948) p. 295.

42 Oliveira Salazar had assumed it to be certain six weeks earlier, telegram to Monteiro (London), 2 May 1939, *DAPE* II (Lisbon, 1978) p. 332.

43 Serrano Suñer, *Entre Hendaya y Gibraltar*, pp. 99, 122; Saña, *Franquismo*, p. 189.

44 *The Times*, 6 June 1939; *Palabras del Caudillo 1937–1942*, pp. 135–45.

45 Roncalli to Ciano, 5 June 1939, *DDI*, 8ª, XII, pp. 101–3.

46 *The Times*, 13 July 1939; Charles R. Halstead, *Spain, the Powers and the Second World War* unpublished Ph.D. thesis, University of Virginia, 1962, pp. 75–80.

47 Ambassade d'Espagne au Départment, 17 March 1939, *DDF*, 2ᵉ Série, XV, pp. 51–3; Juan Avilés Farré,

'L'Ambassade de Lequerica et les rélations hispano-françaises 1939–1944', *Guerres Mondiales et Conflits Contemporains* No. 158, April 1990, pp. 65–78, and 'Lequerica, embajador franquista en París', *Historia 16*, No. 160, August 1989, pp. 12–20; Marc Ferro, *Pétain* (Paris, 1987) pp. 51–2; Garriga, *España de Franco*, pp. 147–9.

48 Peterson, *Both Sides*, pp. 175, 203.

49 Pétain to Bonnet, 9 June 1939, *DDF* 2ᵉ S., T.XVI, pp. 749–50; Hoare, *Ambassador*, p. 56.

50 *DGFP*, D, VI, pp. 830–2, VIII, p. 24.

51 Viola to Ciano, 4, 5 July 1939, *DDI*, 8ª, XII, pp. 345, 362–4.

52 Gorostarzu to Guy la Chambre, 20 July 1939, *DDF* Tome XVII (Paris, 1984) p. 435.

53 Serrano Suñer had written a fulsome letter of invitation to Ciano in mid-June, and requesting a signed photograph of the Duce for Franco, Serrano Suñer to Ciano, undated; Ciano to Serrano Suner, 25 June 1939, *DDI*, 8ª, XII, pp. 277–8, 631–2.

54 *The Times*, 1, 4, 11 July 1939.

55 Pétain to Bonnet, 17 July 1939, *DDF*, 2ᵉ S., T.XVII, p. 380.

56 Ciano, *Papers*, pp. 290–5; Mussolini to Franco, 6 July 1939, *DDI*, 8ª, XII, p. 368; Tusell & García Queipo de Llano, *Franco y Mussolini*, pp. 38–9.

57 Pereira to Salazar, 11 July 1939, *Correspondência de Pedro Teotónio Pereira para Oliveira Salazar, I*

(1931–1939) (Lisbon, 1987) pp. 179–80.

58 Franco Salgado-Araujo, *Mis conversaciones*, p. 327; Serrano Suñer, *Memorias*, pp. 215–16; Cabanellas, *Cuatro generales*, II, pp. 439, 443; Sainz Rodríguez, *Testimonio*, pp. 272–4.

59 Beigbeder to Franco, 11 May 1939, *Documentos inéditos*, I, pp. 412–13; Suárez Fernández, *Franco*, II, pp. 422–3.

60 *The Times*, 22, 25, 26 July 1939; Cabanellas, *Cuatro generales*, pp. 438–9; Ciano, *Diary 1939–1945*, pp. 117, 119, 294–5; Tusell & García Queipo de Llano, *Franco y Mussolini*, pp. 41–2; Garriga, *España de Franco*, pp. 64–5.

61 Memorandum of conversation between Canaris and the Head of Italian Naval Intelligence Distribution, Lais, 22 July 1939, *DDI*, 8ª, XII, pp. 485–7.

62 *The Times*, 7, 8, 9 August 1939.

63 *Boletín Oficial del Estado*, 9 August; *Arriba*, 9 August; *Ya*, 9 August 1939.

64 Ciano to Roncalli, 17 August, Roncalli to Ciano, 19 August 1939, *DDI* 8ª, XIII, pp. 50, 74.

65 Gazel to Charvériat, Pétain to Franco, 3 August 1939, *DDF*, 2ᵉ S., T.XVII, pp. 696, 715.

66 *DGFP*, D, VII, pp. 388–9, 290; Roncalli to Ciano, 30 August 1939, *DDI*, 8ª, XIII, pp. 276–7.

67 *La Vanguardia*, 11 August 1939.

68 José M. Doussinague, *España tenía razón (1939–1945)* (Madrid, 1949) pp. 19–21; Tusell 7 García Queipo de

Llano, *Franco y Mussolini*, p. 40.

69 Gambara to Ciano, 9 August 1939, *DDI*, 8ª, XII, p. 607.

70 Ciano to Gambara, 8, 9 September 1939, *DDI*, 9ª, I, pp. 64, 72.

71 François-Poncet to Bonnet, 17 August 1939, *Documents Diplomatiques Français 2ᵉ Série (1936–1939)*, Tome XVIII (Paris, 1985) p. 145. In conversation with Franco Salgado-Araujo in 1955, Franco swore that Beigbeder had been a Germanophil, Franco Salgado-Araujo, *Mis conversaciones*, p. 66

72 Saña, *Franquismo*, p. 222. Beigbeder was a manic enthusiast for Franco until late 1940 and then turned against him.

73 *The Times*, 26 July, 11 August 1939; Ciano, *Diary 1939–1943*, 9 June 1939, p. 102; Saña, *Franquismo*, pp. 221–2.

74 Halstead, *Spain*, pp. 128–9, 132–3; Halstead, 'A "Somewhat Machiavellian" Face: Colonel Juan Beigbeder As High Commissioner in Spanish Morocco, 1937–1939', *The Historian*, Vol. 37, 1974, pp. 46–66; Halstead, 'Un "Africain" méconnu: le colonel Juan Beigbeder', *Revue d'Histoire de la deuxième guerre mondiale*, Vol. 21, No. 83, 1971, pp. 31–60.

75 The later rivalry between them does not justify the claim that the appointment of Beigbeder marked the beginning of Franco's tactic of dual conduct

of foreign affairs, Halstead, *Spain*, pp. 133–4.

76 Serrano Suñer, *Entre Hendaya y Gibraltar*, pp. 123–32.

77 Saña, *Franquismo*, p. 97; Garriga, *España de Franco*, p. 172–3; Serrano Suñer, *Memorias*, p. 233.

78 José Luis de Arrese, *Una etapa constituyente* (Barcelona, 1982) p. 131.

79 Kindelán, *La verdad*, pp. 90–9.

80 Kindelán, *La verdad*, pp. 30–2.

81 Kindelán, *La verdad*, pp. 116–18.

82 *The Halder War Diary 1939–1942* edited by Charles Burdick and Hans-Adolf Jacobsen (London, 1988) p. 29.

83 Pereira to Salazar, Monteiro to Salazar, 30 August 1939, *DAPE*, II, pp. 513, 510; Pétain to Bonnet, 26 August 1939, *DDF 2ᵉ S.*, T.XIX (Paris, 1986) pp. 60–2; Allison Peers, *Spain in Eclipse*, pp. 143–4.

84 Saña, *Franquismo*, pp. 142–3.

85 Pereira to Salazar, 25 August 1939, *Correspondência*, I, pp. 190–5.

86 Pereira to Salazar, 31 August 1939, *DAPE*, II, pp. 518–23; Corbin to Bonnet, 31 August 1939, *DDF*, 2ᵉ S., T.XIX, p. 297.

87 Gambara to Ciano, 7 September 1939, *DDI*, 9ª serie, vol.I (Rome, 1954) pp. 56–9.

88 Tusell y García Queipo de Llano, *Franco y Mussolini*, pp. 46–8. This contradicts the claims made by Doussinague, *España tenía razón*, pp. 14–15, 30–5. Cf.Katherine Duff, 'Spain between the Allies and the Axis',

in Arnold & Veronica Toynbee, editors, *Survey of International Affairs 1939–1946: The War and the Neutrals* (London, 1956) pp. 264–5.

89 For descriptions of Beigbeder, David Eccles, ed., *By Safe Hand: Letters of Sybil and David Eccles 1939–1942* (London, 1983) pp. 75–6; Hoare, *Ambassador*, pp. 50–1.

90 *DGFP*, D, VII, pp. 501–2; Peterson, *Both Sides*, pp. 191–2.

91 On Lazar, see *The Times*, 6 January 1943; Hoare, *Ambassador*, pp. 54–5. On Falangist dominance of the official press in Spain, see Serrano Suñer, *Entre Hendaya y Gibraltar*, p. 132; Javier Terrón Montero, *La prensa de España durante el régimen de Franco* (Madrid, 1981) pp. 41–54.

92 Paul Reynaud, *Au coeur de la Mêlée 1930–1945* (Paris, 1951) p. 919; Peterson, *Both Sides*, pp. 191–5.

93 Gomá, *Por Dios*, pp. 224–302; Iturralde, *La guerra*, II, pp. 534–5; Rodríguez Aisa, *Gomá*, pp. 322–9.

CHAPTER XIV

1 *Boletín Official del Estado*, 4 September 1939. Cf.Serrano Suñer, *Entre Hendaya y Gibraltar*, p. 89.

2 Mussolini to Franco, 6 September, Gambara to Ciano, 17 September 1939, *DDI*, 9ª, I, pp. 37, 174.

3 Franco to Mussolini, 21 August, 3 September, García Conde to Ciano, 3 September 1939, *DDI*,

8ª, XIII, pp. 85, 388. Rumania refused to join Franco's call for neutrality on the grounds that it would hinder the efforts of those powers which wanted to aid Poland, Ghigi (Italian Minister in Bucharest) to Ciano, 4 September 1939, *DDI*, 9ª, I, p. 19.

4 Serrano Suñer, *Entre Hendaya y Gibraltar*, pp. 133–5, 142–3.

5 Pétain to Bonnet, 3 September 1939, *DDF*, 2ᵉ S., T.XIX, p. 422.

6 *Arriba*, 5, 7 October; 19 November 1939 (on German support).

7 Bova Scopa (Geneva) to Ciano, 16 September 1939, *DDI*, 9ª, I, p. 155.

8 *DGFP*, D, VIII, pp. 181–2; *Documentos inéditos*, I, pp. 603–10; Manuel Espadas Burgos, *Franquismo y política exterior* (Madrid, 1988) p. 97.

9 *Arriba*, 28, 29 September; *La Vanguardia*, 5 October 1939.

10 Pereira to Salazar, 28 September 1939, *Correspondência*, I, pp. 201–5.

11 Alba to Beigbeder, 2 October 1939, MAE, R 1371/3B.

12 *Arriba*, 22 October, 1 November; *ABC*, 2, 5 November 1939.

13 Francisco Franco Bahamonde, 'Fundamentos y directrices de un Plan de saneamiento de nuestra economía, armónico con nuestra reconstrucción nacional', *Historia 16*, No. 115, noviembre de 1985, pp. 44–9.

14 Viñas et al., *Política comercial exterior*, I, pp. 268–81; Manuel Jesús González, *La política económica del franquiismo*

(1940–1970) (Madrid, 1979) pp. 46–7; Payne, *Franco Regime*, pp. 249–53; Sima Lieberman, *The Contemporary Spanish Economy: A Historical Perspective* (London, 1982) pp. 165–76; Joseph Harrison, *The Spanish Economy in the Twentieth Century* (Beckenham, 1985) p. 121.

15 Rafael Abella, *La vida cotidiana bajo el régimen de Franco* (Barcelona, 1985) pp. 49–60; Rafael Abella, *Por el Imperio hacia Díos* (Barcelona, 1978) pp. 101–32.

16 *La Vanguardia Española*, 1 October; *Arriba*, 3 October 1939.

17 *ABC*, 19 October 1939.

18 Séguéla, *Pétain-Franco*, p. 43.

19 Garriga, *La Señora*, pp. 126–8, 145; Plenn, *Wind*, pp. 74–5.

20 Samuel Ros & Antonio Bouthelier, *A hombros de la Falange: Historia del traslado de los restos de José Antonio* (Barcelona, 1940) *passim*; Ian Gibson, *En busca de José Antonio* (Barcelona, 1980) pp. 246–8; Sueiro & Díaz Nosty, *Franquismo*, I, pp. 176–82.

21 Hamilton, *Appeasement's Child*, pp. 91–2.

22 Monteiro to Salazar, 23 November 1939, *DAPE*, VI (Lisbon, 1970) p. 199.

23 *ABC*, 1 January 1940. Omitted from subsequent compilations of Franco's speeches, the complete text was published at the time as *Mensaje del Caudillo a los españoles: discurso pronunciado por S.E. el Jefe del Estado la noche del*

31 de diciembre de 1939 (Madrid, n.d., but 1940).

24 *The Goebbels Diaries 1939–1941*, edited by Fred Taylor (London, 1982) 3 January 1940, p. 85.

25 *Arriba*, 24 December; *La Vanguardia*, 24 December 1939.

26 *Mensaje 31.XII.1939*, p. 27; Garriga, *España de Franco*, pp. 58, 126.

27 *La Voz de Galicia*, 8 February 1940; *La Vanguardia Española*, 21 January, 8 February 1940; Foltz, *The Masquerade*, pp. 258–60; Ansaldo, *¿Para qué . . . ?*, pp. 254–6.

28 Peterson, *Both Sides*, p. 207.

29 Hamilton, *Appeasement's Child*, p. 206; Halstead, *Spain*, pp. 157–9, 179.

30 *DGFP*, D, VIII, pp. 324–5; IX, p. 588; X, pp. 291, 299–300; XI, p. 48, 185; Ferro, *Pétain*, pp. 51–2.

31 Ansaldo, *¿Para qué . . . ?*, pp. 230–3.

32 Pereira to Salazar, 23 April 1940, *DAPE*, VI, pp. 543–4.

33 See undated letter of Kindelán to Varela, Kindelán, *La verdad*, pp. 117–18; Tusell 7 García Queipo de Llano, *Franco y Mussolini*, pp. 97–98.

34 Gambara to Ciano, 13 August, 11, 28 October, 21, 28, 29 December 1939, 30 January, 19 March 1940, Ciano to Gambara, 14 August, 27, 31 October 1939, 13, 23 February 1940, *DDI*, 8ª, XIII, pp. 17–18, 25; 9ª, I, p. 442; 9ª, II, pp. 20, 28, 41–2, 519–20, 576–7; 9ª, III, (Rome, 1959) pp. 190–1, 253–4, 310, 508.

35 Gambara to Ciano, 5 April 1940, *DDI*, 9ª, III, p. 613.

36 Mussolini to Franco, 8 April 1940, *DDI*, 9ª, III, pp. 623–4.

37 Gambara to Ciano, 13 April 1940, *DDI*, 9ª, IV (Rome, 1960) pp. 52–3.

38 *Arriba*, 11 April 1940.

39 *Informaciones*, 4 May 1940.

40 *DGFP*, D, VIII, pp. 190–2.

41 *Arriba*, 13 March, 5, 9 June; *ABC*, 9 June; *El Alcázar*, 21 June 1940.

42 *Arriba*, 12, 13 April 1940.

43 *Arriba*, 2 April 1940; Antonio Marquina Barrio, *La diplomacia vaticana y la España de Franco (1936–1945)* (Madrid, 1983) pp. 243–53, 514–18; Garriga, *Segura*, pp. 271–4.

44 *ABC*, 1, 2, 3 April 1940; Sueiro, *Valle*, p. 12.

45 Millán Astray, *Franco*, pp. 11–12.

46 Sueiro, *Valle*, pp. 8–24, 44–73, 118–43, 184–205. On the psychological significance of the monument, see Grau, 'Psicopatolgía', p. 22.

47 Pozuelo, *Los 476 últimos días*, p. 166.

48 Marquina, *La diplomacia vaticana*, pp. 253–62; Serrano Suñer, *Memorias*, p. 274; Garriga, *Segura*, pp. 274–6.

49 Franco to Mussolini, 30 April 1940, *DDI*, 9ª, IV, pp. 210–11; diary entry for 3 May 1940, *Ciano's Diary 1939–1943*, p. 243.

50 Pereira to Salazar, 23 April 1940, *DAPE*, VI, pp. 543–8.

51 For a sympathetic portrait of Weddell, see Charles R. Halstead, 'Diligent Diplomat:

Alexander W. Weddell as American Ambassador to Spain, 1939–1942', *The Virginia Magazine of History and Biography*, Vol. 82, No. 1, January 1974, pp. 3–38.

52   *Foreign Relations of the United States 1940*, Vol II (Washington, 1957) pp. 794–6.

53   *Arriba*, 11, 17 May 1940; *DGFP*, D, IX, p. 315.

54   Simonnot, *Le secret*, p. 118.

55   Interview with Franco in *Arriba*, 25 February 1951, reprinted in Francisco Franco, *Discursos y mensajes del Jefe del Estado 1951–1954* (Madrid, 1955) pp. 38–42; interview with Fanco in *Le Figaro*, 13 June 1958, reprinted in Francisco Franco, *Discursos y mensajes del Jefe del Estado 1955–1959* (Madrid, 1960) p. 488.

56   Séguéla, *Pétain-Franco*, p. 45.

57   Ferro, *Pétain*, pp. 7, 43; Reynaud, *Au coeur de la Mêlée*, pp. 919–20.

58   Simonnot, *Le secret*, pp. 212–14.

59   *DGFP*, Series D, p. 396; Pereira to Salazar, 27 May 1940, *Correspondência de Pedro Teotónio Pereira para Oliveira Salazar II (1940–1941)* (Lisbon, 1989) p. 39.

60   Herbert Feis, *The Spanish Story: Franco and the Nations at War* (New York, 1948) [page references to the New York, 1966 paperback edition] pp. 28–31.

61   Hoare to Eden, 11 July 1941, Templewood Papers, XIII–19.

62   Eccles, *By Safe Hand*, pp. 19, 26, 49, 51–3.

63   Peterson, *Both Sides*, pp. 228–33.

64   Hoare, *Ambassador*, pp. 14–18, 30–2; J. A. Cross, *Sir Samuel Hoare: A Political Biography* (London, 1977) pp. 322–8; Kenneth Young, editor, *The Diaries of Sir Robert Bruce Lockhart*, Vol. 2 *1939–1965*, (London, 1980), diary entry for 19 May 1940, pp. 56–7; David Dilkes, editor, *The Diaries of Sir Alexander Cadogan 1938–1945* (London, 1971) 19, 20 May 1940, pp. 286–7.

65   Cross, *Hoare*, pp. 328–9; Sir John Lomax, *The Diplomatic Smuggler* (London, 1965) pp. 84–9; Carlton J. H. Hayes, *Wartime Mission in Spain* (New York, 1945) p. 35; Eccles, *By Safe Hand*, pp. 100–2, 294; Pereira, *Memórias*, II, pp. 205–6.

66   *Arriba*, 24, 26, 29, 31 May 1940; Ramón Serrano Suñer, 'Prólogo', in David Jato, *Gibraltar decidió la guerra* (Barcelona, 1971) p. 15.

67   *DGFP*, D, pp. 509–10.

68   Saña, *Franquismo*, pp. 170–1. Franco's official biographer, Suárez Fernández, *Franco*, III, p. 121, sees the letter as proof that Franco had no intention of entering the war.

69   Lequerica to Beigbeder, 19 May, Mussolini to Franco, 20 May, *Documentos inéditos*, II–1, pp. 191–2; Zoppi to Ciano, Ciano to Zoppi, 20 May 1940, *DDI*, 9ª, IV, pp. 399–401.

70   Lequerica to Beigbeder, 29 May 1940, MAE/R 2295, leg. 4, no. 769; Simonnot, *Le secret*, pp. 214–48.

71 Eccles, *By Safe Hand*, pp. 14–16, 54–7, 78, 84–5, 90–3.

72 *Arriba*, 1, 2, 4, 5, 9 June; *ABC*, 9 June 1940.

73 Mussolini to Franco, 9 June 1940, *DDI*, 9ª, IV, p. 620.

74 Ciano to Serrano Suñer, 3, 8 June 1940, Ciano, *Papers*, pp. 370–1; Espadas, *Franquismo*, p. 104; Calleja, *Yagüe*, pp. 197–8.

75 Tusell & García Queipo de Llano, *Franco y Mussolini*, pp. 99–102.

76 *FRUS 1940*, II, p. 796.

77 Franco to Mussolini, 10 June 1940, *DDI*, 9ª, IV, p. 630; Saña, *Franquismo*, pp. 167–8.

78 Anfuso to Zoppi, 12 June 1940, *DDI*, 9ª, V (Rome, 1965) p. 6.

79 Zoppi to Ciano, 13, 15 June 1940, *DDI*, 9ª, V, pp. 14, 24.

80 *DGFP*, Series D, Vol. IX, p. 560.

81 *Arriba*, 13 June, *ABC*, 13 June 1940.

82 Monteiro to Salazar, 13 June 1940, *DAPE*, VII, pp. 131–2; *The Times*, 14 June 1940; Pereira, *Memórias*, II, pp. 214–15.

83 Pereira to Salazar, 23 April, 14 May, 10 June 1940, *DAPE*, VI, pp. 544–5; VII, pp. 30–1, 115–17.

84 Notes by Teixeira de Sampaio, 13 June 1940, *DAPE*, VII, pp. 135–6.

85 Zoppi to Ciano, 15 June 1940, *DDI*, 9ª, V, pp. 23–5.

86 Doussinague, *España tenía razón*, pp. 41–2.

87 Hoare, *Ambassador*, p. 45.

88 Hoare, *Ambassador*, pp. 24–30, 45–8; *FRUS 1940*, II, pp. 796–9; Hayes, *Wartime Mission*, p. 40.

89 Doussinague, *España tenía razón*, pp. 22, 103–5, 328. Jordana told the Vichy Ambassador on 9 September 1942 that he was only 'the docile executor' of Franco's policy, Piétri, *Mes années*, p. 65.

90 Pereira, *Memórias*, II, pp. 223–3.

91 *DGFP*, D, IX, p. 542; Serrano Suñer, *Entre Hendaya y Gibraltar*, pp. 159–60.

92 Naval Intelligence Division (Admiralty), 'Spanish Assistance to the German Navy', 10 January 1946, FO371/60331, Z594/8/G41; *DGFP*, Series D, Vol. IX, pp. 449–53; Vol. XI, p. 445; *Documents secrets du Ministère des Affaires Etrangères d'Allemagne: Espagne* (Paris, 1946) pp. 103–4; Charles B. Burdick, ' "Moro": The Resupply of German Submarines in Spain, 1939–1942', *Central European History* Vol. 3, No. 3, 1970, pp. 256–84; Duff, 'Spain', p. 268; cf. Doussinague, *España tenía razón*, pp. 73–4.

93 United Nations, Security Council, Official Records, First Year: 2nd Series, Special Supplement, *Report of the Sub-Committee on the Spanish Question* (New York, June 1946) pp. 13, 83–9.

94 Serrano Suñer, *Entre Hendaya y Gibraltar*, pp. 151–2; Doussinague, *España tenía razón*, pp. 43–4.

95 Halder, *Diary*, 31 July 1940, pp. 244–5; Alan Clark,

Barbarossa: The Russian–German Conflict 1941–1945 (London, 1965) pp. 17–26. I am grateful to Professor Brian Bond for clarifying this point.

96 Reynaud, *Au coeur de la mêlée*, pp. 855–6; Séguéla, *Pétain-Franco*, p. 80–1.

97 Zoppi to Ciano, 17 June 1940, *DDI*, 9ª, V, pp. 32–3.

98 Sir Llewellyn Woodward, *British Foreign Policy in the Second World War* 5 vols (London, 1970–1976) I, p. 435–6; conversation of Serrano Suñer with Charles Favrel of Paris Presse, 21 October 1945, FO371/49663, Z13272/11696/41.

99 *Bulletin of Spanish Studies*, Vol. XVII, no. 68, October 1940, pp. 220–2; Saña, *Franquismo*, pp. 168–70.

100 Foltz, *The Masquerade*, p. 152.

101 Memorandum of conversation between Vigón and Ribbentrop, Château Ciergnon, 15 May 1940, among captured German documents, FO371/60332, Z2727/8/41.

102 *DGFP*, D, IX, pp. 585–8.

103 Garriga, *España de Franco*, pp. 170–1.

104 Lequerica to Beigbeder, 17 June 1940, *Documentos inéditos*, II-1, pp. 209–10; Paul Baudouin, *The Private Diaries of Paul Baudouin (March 1940–January 1941)* (London, 1948) pp. 118–123; François Charles-Roux, *Cinq mois tragiques aux affaires étrangères 21 Mai–1er Novembre 1940* (Paris, 1949) pp. 52–4; *DGFP*, D, IX, p. 590; Ansaldo, *¿Para qué . . . ?*, p. 214. See also Cava Mesa, *Lequerica*, pp. 167–74.

105 The reason given to Lequerica was 'the high regard in which Pétain held Franco', Lequerica to Beigbeder, 17 June 1940, *Documentos inéditos*, II-1, p. 209.

106 *ABC*, 22 June 1940. It is not clear where the phrase arose. Pétain is alleged to have used a slight variant, '*l'épée la plus pure au monde*' on 17 July 1940 in his first message to Franco after assuming power – Séguéla, *Pétain-Franco*, p. 56.

107 Cf. Simonnot, *Le secret*, pp. 216, 232.

108 Halstead, *Spain*, pp. 259–60; Feis, *Spanish Story*, pp. 35–6.

109 Simonnot, *Le secret*, pp. 220–9; Séguéla, *Pétain-Franco*, p. 51–5.

110 Zoppi to Ciano & Spanish Embassy (Rome) to Anfuso, 18 June 1940, *DDI*, 9ª, V, pp. 33–4, 41–2.

111 Charles-Roux, *Cinq mois tragiques*, pp. 193, 224–8; Simonnot, *Le secret*, pp. 244–6; Séguéla, *Pétain-Franco*, pp. 81–7.

112 *DGFP*, D, IX, pp. 620–1; Gerhard L. Weinberg, *World in the Balance: Behind the Scenes of World War II* (Hanover, New Hampshire, 1981) p. 111.

113 Feis, *Spanish Story*, pp. 33–4; Halstead, *Spain*, pp. 271–2.

114 Hoare, *Ambassador*, p. 48.

115 Zoppi to Ciano, 17 June 1940, *DDI*, 9ª, V, p. 33.

116 *The Times*, 21, 26, 29, 31 July 1940.

117 ABC, 31 July 1940; Feis, *Spanish Story*, pp. 34–53.
118 *DGFP*, D, X, pp. 15–16.
119 Halstead, *Spain*, p. 274; *DGFP*, D, X, pp. 97–9.
120 *DGFP*, D, X, pp. 97–99; *Arriba*, 28 June 1940; Suárez Fernández, *Franco*, III, pp. 144–8; Garriga, *Yagüe*, pp. 181–4; Garriga, *España de Franco*, pp. 171–4; Hoare, *Ambassador* (London, 1946) pp. 52–3; Payne, *Falange*, pp. 213–15; Klaus-Jörg Ruhl, *Franco, Falange y III Reich* (Madrid, 1986) pp. 61, 317; Denis Smyth, *Diplomacy and Strategy of Survival: British Policy and Franco's Spain, 1940–41* (Cambridge, 1986) pp. 33–6.
121 ABC, 25, 26 June; *Arriba*, 29 June 1940.
122 *El Diario Vasco*, 30 June 1940, gave full photographic coverage to the encounters. See also Espadas Burgos, *Franquismo*, pp. 105–7; Hoare, *Ambassador*, pp. 52–3.
123 Hoare, *Ambassador*, pp. 50–2.
124 Serrano Suñer, 'Prólogo' to Jato, *Gibraltar*, pp. 16–17; Saña, *Franquismo*, p. 164.
125 Beigbeder to Franco, 25, 26 June 1940, *Documentos inéditos*, II-1, pp. 215, 226–7; *DGFP*, D, X, pp. 2.
126 *DGFP*, D, X, pp. 9, 187–9, 199–200, 276–7, 283, 290–1, 317–18, 366–7, 376–9, 397–401, 409–10; Bova Scoppa (Lisbon) to Ciano, 1, 2 August 1940, *DDI*, 9ª, V, pp. 311, 324–5; Walter Schellenberg, *The Schellenberg Memoirs: A Record of the Nazi Secret Service*

(London, 1956) pp. 126–43; Mariano González-Arnao Conde-Luque, '¡Capturad al duque de Windsor!', *Historia 16*, No. 161, September 1989; Michael Bloch, *Operation Willi: The Plot to Kidnap the Duke of Windsor July 1940* (London, 1984) *passim*. For Bermejillo's reports, see Suárez Fernández, *Franco*, III, pp. 143–4, 149.
127 Bruce Lockhart, *Diaries*, entry for 28 June 1940, p. 63.
128 Stohrer to Wilhelmstrasse, 12 July 1940, *DGFP*, D, X, p. 200.
129 Hoare, *Ambassador*, pp. 34–8; Smyth, *Diplomacy*, pp. 32–40.
130 At PRO CAB 65/7, W.M. 171(40)5 and CAB 65/7, W.M. 172(40)7, the record is censored. This account is based on copies of the relevant minutes kindly supplied by Mr John Costello. In fact, the essence of the meeting is fully conveyed by Woodward, *British Foreign Policy*, I, p. 436. See also Smyth, *Diplomacy*, pp. 42–3.
131 Rafael Rodríguez Miñino Soriano, *La misión diplomática del XVII Duque de Alba en la Embajada de España en Londres (1937–1945)* (Valencia, 1971) p. 63.
132 Churchill, *Finest Hour*, p. 564; Smyth, *Diplomacy*, pp. 43–4.
133 Stohrer to Wilhelmstrasse, 2 July 1940, *DGFP*, D, X, pp. 97–8.
134 *DGFP*, D, X, pp. 82, 148–9.
135 *DGFP*, D, IX, pp. 605–6, 608–11.
136 Karl Heinz Abshagen, *Canaris* (London, 1956) pp. 79, 111–13; Hugh Trevor-Roper, *The Philby*

Affair (London, 1968)
pp. 111–16; Garriga, *España de Franco*, pp. 71–3.

137 Zoppi to Ciano, 6 July 1940, *DDI*, 9ª, V, pp. 185–6. The minutes of a meeting between Canaris, Beigbeder and Vigón are included in Beigbeder to Franco, 30 June 1940, *Documentos inéditos*, II-1, pp. 238–9.

138 Pereira to Salazar, 6 July 1940, *Correspondência*, II, pp. 57–60.

139 Pereira to Salazar, 6 July 1940, *DAPE*, VII, pp. 216–19; Halstead, *Spain*, pp. 297–8; Charles R. Halstead, 'Peninsular Purpose: Portugal and its 1939 Treaty of Friendship and Non-Aggression with Spain', *Il Politico: Rivista Italiana di Scienze Politiche* (Pavia) XLV, No. 2, 1980, pp. 287–311.

140 *Fotos*, 13 July 1940.

141 *DGFP*, D, X, pp. 105–6, 224–5; Pereira to Salazar, 28 February 1941, *DAPE*, VIII (Lisbon, 1973) p. 150; Pereira, *Memórias*, II, pp. 227–9.

142 Pereira to Salazar, 12 July, notes of Salazar's conversation with Nicolás Franco, 13 July 1940, *DAPE*, VII, pp. 247–8, 250–2.

143 *DGFP*, D, X, p. 515; *DAPE*, VII, pp. 294–7, 323–5, 336–9; *DDI*, 9ª, V, pp. 296–7; Pereira, *Memórias*, II, pp. 230–3; Hoare, *Ambassador*, pp. 58–9; Halstead, *Spain*, pp. 298–300; Halstead, 'Consistent and total peril from every side: Portugal and its 1940 Protocol with Spain', *Iberian Studies*, Vol. III, No. 1, Spring 1974, pp. 15–29; Tusell & García Queipo de

Llano, *Franco y Mussolini*, p. 90.

144 *The Times*, 18 July, *La Vanguardia Española*, 18 July; *Arriba*, 18 July 1940.

145 *The Times*, 18 July, *La Vanguardia Española*, 19 July 1940; Hoare, *Ambassador*, pp. 48–9; Smyth, *Diplomacy*, pp. 41–2; François Mirandet, *L'Espagne de Franco* (Paris, 1948) pp. 63–5.

146 *La Vanguardia Española*, 19 July; *The Times*, 19 July 1940; Garriga, *España de Franco*, pp. 151–3.

147 Beigbeder to Asensio, 25 July 1940, *Documentos inéditos*, II-1, pp. 273–6; *ABC*, 30 July; *Arriba*, 1 August 1940; Charles-Roux, *Cinq mois*, pp. 194, 222–41; Ridruejo, *Memorias*, p. 214; Séguéla, *Pétain-Franco*, pp. 87–92.

148 Hoare, *Ambassador*, pp. 62–3.

149 *DGFP*, D, X, p. 396; Hoare, *Ambassador*, p. 44; Serrano Suñer, *Entre Hendaya y Gibraltar*, p. 65; Churchill, *Finest Hour*, p. 463; Bloch, *Willi*, pp. 50, 241–3.

150 Memorandum by Stohrer, 8 August 1940, *DGFP*, D, X, pp. 442–5.

151 Note of the High Command, 10 August 1940, *DGFP*, D, X, pp. 461–4.

152 *DGFP*, D, X, pp. 466–7, 499–500.

153 Halder, 9 August 1940, *Diary*, p. 247; Macgregor Knox, *Mussolini Unleashed 1939–1941: Politics and Strategy in Fascist Italy's Last War* (Cambridge, 1982) p. 184; André Brissaud,

Canaris (London, 1973)
pp. 191–4.

154 Beigbeder to Franco, 10, 14
August 1940, *Documentos inéditos*,
II-1, pp. 285–7; Suárez
Fernández, *Franco*, III,
pp. 171–2.

155 Stohrer to Wilhelmstrasse, 21
August 1940, *DGFP*, D. X,
pp. 521–2.

156 Franco to Mussolini, 15 August,
Mussolini to Franco, 25 August
1939, *DDI*, 9ª, V, pp. 403–5,
478–9; Ciano, *Diary
1939–1943*, p. 285; Serrano
Suñer, *Entre Hendaya y Gibraltar*,
pp. 103–4; Weinberg, *World in
the Balance*, p. 120.

157 Lequio to Ciano, 2 September
1940, *DDI*, 9ª, V, pp. 521–3.

CHAPTER XV

1 Halder, *Diary*, 27 August 1940,
pp. 251–2.

2 *DGFP*, D, X, p. 561.

3 Mirandet, *Franco*, p. 87.

4 *DGFP*, D, X, pp. 561–5, XI,
pp. 37–40, 81–2.

5 *Ya*, 7 September; *The Times*, 8,
9 September 1940; Arthur F.
Loveday, *Spain 1923–1948:
Civil War and World War*
(Bridgewater, 1949) p. 172.

6 Charles-Roux, *Cinq mois
tragiques*, pp. 241–2; Lequio to
Ciano, 8 September 1940, *DDI*,
9ª, V, pp. 557–8.

7 Doussinague, *España tenía razón*,
pp. 233–4. On this, and other
British insinuations of sympathy
for Spanish aspirations, see Duff,
'Spain', p. 272.

8 *The Times*, 26, 29, 30 September
1940.

9 Halder, *Diary*, 14 September
1940, p. 259.

10 Lequio to Ciano, 16 September,
Ciano to Zamboni (Berlin) 19
September 1940, *DDI*, 9ª, V,
pp. 576–7, 594–5.

11 *FRUS 1940*, II, pp. 805–08;
Feis, *Spanish Story*, pp. 55–7.

12 *FRUS 1940*, II, pp. 808–10;
Feis, *Spanish Story*, pp. 57–9.

13 Charles-Roux, *Cinq mois
tragiques*, p. 243.

14 Garriga, *España de Franco*,
p. 180.

15 The Portuguese Ambassador
believed that Serrano Suñer
passed on the suggestion to the
Germans, Pereira to Salazar, 30
August, 1 October 1940,
Correspondência, II, pp. 79–80,
88.

16 *DGFP*, D, XI, pp. 83–7;
Serrano Suñer, *Entre Hendaya y
Gibraltar*, p. 165–71. In
November 1940, Serrano Suñer
told the Italian Ambassador
Lequio that the attitude of the
Portuguese Government was
'vile and servile with regard to
England and doubtful as far as
Spain is concerned', and in May
1942 that Portugal should be
part of a greater Spain, Lequio
to Ciano, 24 November 1940,
26 May 1942, *DDI*, 9ª, VI,
p. 232, VIII, p. 618. On
Falangist ambitions to annex
Portugal, with a benevolent
view of Franco's stance, see
Hugh Kay, *Salazar and Modern
Portugal* (London, 1970) p. 156.

17 Ernst von Weizsäcker, *Memoirs
of Ernst von Weizsäcker* (London,
1951) pp. 239, 266; Saña, *El
franquismo*, pp. 176, 180;

Doussinague, *España tenía razón*, p. 46; Schellenberg, *Memoirs*, pp. 135, 143.

18 *DGFP*, D, XI, pp. 87–91. On British plans for the seizure of the Canary Islands, 'Operation Bugle', cf. Martin Gilbert, *Finest Hour: Winston S. Churchill 1939–1941* (London, 1983) p. 567.

19 Garriga, *España de Franco*, pp. 182–4.

20 *DGFP*, Series D, Vol. XI, pp. 62–3.

21 *DGFP*, Series D, Vol. XI, pp. 93–102; Serrano Suñer, *Entre Hendaya y Gibraltar*, pp. 175–83. Serrano Suñer's 1947 version is more forthright than that recorded by the Germans at the time, substituting 'absolutely impossible' by 'criminal and monstrous'.

22 *DGFP*, D, XI, pp. 106–8.

23 Weinberg, *World in the Balance*, p. 122.

24 Knox, *Mussolini Unleashed*, pp. 183–5; Smyth, *Diplomacy*, pp. 93–4.

25 *Ciano's Diary 1939–1943*, p. 291; Ciano, *Papers*, pp. 389–93.

26 *DGFP*, D, XI, pp. 134–6, 150–2.

27 Franco to Serrano Suñer, 21 September 1940, Serrano Suñer, *Memorias*, pp. 331–40; *Entre Hendaya y Gibraltar*, p. 183.

28 Serrano Suñer, *Entre Hendaya y Gibraltar*, pp. 188–91; Garriga, *España de Franco*, pp. 188–9.

29 Smyth, *Diplomacy*, pp. 50–1, 95–6.

30 *DGFP*, D, XI, pp. 166–74.

31 *DGFP*, D, XI, pp. 183–4.

32 Letter of Serrano Suñer, *ABC*, 3 November 1990; Hernando Espinosa de los Monteros to Serrano Suñer, 5 November 1990 (Serrano Suñer archive); Hernando Espinosa de los Monteros to the author, 15 January 1991; testimony of Ignacio Espinosa de los Monteros to the author, 16 February 1991; Serrano Suñer, *De anteayer*, pp. 213–15.

33 Franco to Serrano Suñer, 23 September 1940, Serrano Suner, *Memorias*, pp. 341–2.

34 *DGFP*, D, XI, pp. 153–5.

35 *DGFP*, Deries D, Vol. XI, pp. 155, note 5; 214–19.

36 Serrano Suñer, *De anteayer*, p. 214.

37 Franco to Serrano Suñer, 24 September 1940, Serrano Suner, *Memorias*, pp. 342–8.

38 *DGFP*, Series D, Vol. XI, pp. 199–204.

39 *Ciano's Diary 1939–1943*, p. 294.

40 Lequio to Ciano, 13 September 1940, *DDI*, 9ª, V, pp. 572–3.

41 Conversation Mussolini–Serrano Suñer, 1 October 1940, *DDI*, 9ª, V, pp. 639–40; Serrano Suñer, *Entre Hendaya y Gibraltar*, pp. 195–8.

42 *DGFP*, D, XI, pp. 211–14; *Ciano's Diary 1939–1943*, p. 294.

43 *DGFP*, D, XI, pp. 245–59. See the shorter account in *DDI*, 9ª, V, pp. 655–8 and the version given by Keitel and Jodl to the Italian military attaché, Marras to Ministero della Guerra, 12 October 1940, *DDI*, 9ª, V,

pp. 690–1. Cf. Halder, *Diary*, 30 September, 15 October 1940, pp. 260–6.

44 Saña, *Franquismo*, p. 185.

45 *Ciano's Diary 1939–1943*, pp. 294–6; Ciano, *Papers*, pp. 393–6; Knox, *Mussolini Unleashed*, pp. 189, 196.

46 Hoare, *Ambassador*, pp. 67–71, where the meeting is wrongly placed on 27 September; see Smyth, *Diplomacy*, pp. 95–6.

47 Gilbert, *Finest Hour*, p. 816.

48 Charles-Roux, *Cinq mois tragiques*, pp. 244–8.

49 *The Second World War Diary of Hugh Dalton 1940–1945* edited by Ben Pimlott (London, 1986) 2 October 1940, p. 87; Smyth, *Diplomacy*, pp. 98–9.

50 Halstead, *Spain*, p. 358. The most thorough accounts of the Falange Exterior are Allan Chase, *Falange: The Axis Secret Army in the Americas* (New York, 1943) and Ovidio Gondi, *La hispanidad franquista al servicio de Hitler* (México D.F., 1979). See also Plenn, *Wind*, pp. 231–46.

51 *FRUS 1940*, II, pp. 81–-12; Feis, *Spanish Story*, pp. 59–61.

52 *La Vanguardia Española*, 1, 2 October; *Arriba*, 1, 2, 3 October 1940.

53 Pereira to Salazar, 1 October 1940, *Correspondência*, II, pp. 88–9.

54 *FRUS 1940*, II, p. 812.

55 *FRUS 1940*, II, pp. 812–15; Dalton, *War Diary*, 7 October 1940, p. 89 (where Beigbeder's remarks to Weddell are misconstrued as emanating from Franco himself).

56 Cordell Hull, *Memoirs* 2 vols (London, 1949) I, p. 876; Feis, *Spanish Story*, p. 61.

57 365 H.C.DEB., 5s, c.302; *The Times*, 11 October 1940.

58 *Palabras del Caudillo 1937–1942*, pp. 189–90.

59 Notes of conversation, Salazar/Nicolás Franco, 16 October 1940, *DAPE*, VII, pp. 518–21.

60 The consequence was that Spain received less in terms of German goods than did Sweden, Switzerland and Turkey, Garriga, *España de Franco*, pp. 194–6.

61 Pereira to Salazar, 18 October 1940, *Correspondência*, II, p. 97.

62 Espadas, *Franquismo*, p. 110.

63 *FRUS 1940*, II, pp. 820–2; Lequio to Ciano, 18 October 1940, *DDI*, 9ª, V, p. 715.

64 Serrano Suñer, 'Prólogo', Jato, *Gibraltar*, pp. 17–18; Hoare, *Ambassador*, pp. 66–73; Pereira, *Memórias*, II, p. 222; Escobar, *Diálogo íntimo*, p. 29; Mirandet, *L'Espagne*, p. 69.

65 Mirandet, *L'Espagne*, p. 70; Espadas, Franquismo, p. 108.

66 Testimony of Ramón Serrano Suñer to the author, 21.XI.90; Saña, *Franquismo*, p. 160; Hamilton, *Appeasement's Child*, p. 93. On Lorente Sanz, see Ridruejo, *Memorias*, p. 134.

67 Mussolini to Hitler, 19 October 1940, *DDI*, 9ª, V, pp. 720–2; translation from *DGFP*, D, XI, pp. 331–4.

68 *Arriba*, 20, 23 October 1940; *ABC, La II guerra mundial* (Madrid, 1989) p. 160.

69 United Nations, Security Council, *Report of the Sub-Committee on the Spanish*

Question, p. 14; Bloch, *Willi*, p. 150.

70 Hoare, *Ambassador*, p. 76; Mirandet, *Franco*, pp. 72–3. However, cf. Burdick, *Strategy*, pp. 50–1. On Himmler's reaction to continued repression, see Garriga, *España de Franco*, pp. 208–9.

71 Cipriano Rivas Cherif, *Retrato de un desconocido: vida de Manuel Azaña* (Barcelona, 1980) pp. 496–500; Saña, *Franquismo*, pp. 118–19. See the special supplement *Julián Zugazagoitia: protagonista y víctima*, *El Sol*, 10 November 1990.

72 Serrano Suñer blamed Lequerica, Saña, *Franquismo*, pp. 124–7. See also Gregorio Morán, *Los españoles que dejaron de serlo: Euskadi, 1937–1981* (Barcelona, 1982) pp. 114–15. According to Cava Mesa, *Lequerica*, pp. 184–9, Lequerica did no more than obey orders emanating from the Director General de Seguridad, José Finat.

73 Angel Ossorio y Gallardo, *Vida y sacrificio de Companys* 2ª edición (Barcelona, 1976) pp. 285–92; *Catalunya sota el règim franquista: Informe sobre la persecució de la llengua i la cultura de Catalunya pel règim del general Franco* (Paris, 1973) pp. 388–91.

74 *FRUS 1940*, II, pp. 822–3; Feis, *Spanish Story*, pp. 63–6.

75 Norman Rich, *Hitler's War Aims: Ideology, the Nazi State, and the Course of Expansion* 2 vols (London, 1973–4) I, pp. 169–70.

76 Halder, *Diary*, 11 October 1940, p. 262.

77 Weizsäcker, diary entry for 21 October 1940, *Papiere*, p. 221.

78 Paul Schmidt, *Hitler's Interpreter: The Secret History of German Diplomacy 1935–1945* (London, 1951) p. 193; *La Vanguardia Española*, 24 October 1940; *The Memoirs of Field-Marshal Keitel* (London, 1965) p. 125; Serrano Suñer, *Memorias*, pp. 289–90; testimony of Serrano Suñer, 21.XI.90.

79 José María Sánchez Silva & José Luis Saenz de Heredia, *Franco . . . ese hombre* (Madrid, 1975) p. 139. Franco's declarations to *Le Figaro*, 13 June 1958 reproduced in Francisco Franco, *Discursos y mensajes del Jefe del Estado 1955–1959* (Madrid, 1960) p. 479. Cf. Trythall, *Franco*, p. 171.

80 Testimony of Ramón Serrano Suñer to the author, 21.XI.90.

81 Silva & Saenz de Heredia, *Franco*, p. 139; Suárez Fernández, *Franco*, III, p. 119.

82 For a valiant attempt to reconstruct in minute detail what was said, see Ignacio Espinosa de los Monteros y Bermejillo, *Hendaya y la segunda guerra mundial (vivencias y razones)* undated, unpublished manuscript (Madrid, n.d.) pp. 72–108. Ramón Garriga, Berlin representative of the official Spanish news agency EFE, produced a perceptive pro-Serrano Suñer account in his *España de Franco*, pp. 209–17.

83 That record was printed

incomplete as Document No. 8 in *The Spanish Government and the Axis: Official German Documents* (Washington, Department of State, 1946). Further sheets of the document were found subsequently and a somewhat longer version appears in *DGFP*, Series D, Vol. XI, pp. 371–6. It is followed by a record of the later meeting between Serrano Suñer and Ribbentrop, signed by Schmidt, *DGFP*, Series D, Vol. XI, pp. 376–9. Schmidt's own account in Schmidt, *Hitler's Interpreter*, pp. 192–7. Serrano Suñer's version is in *Memorias*, pp. 283–305. On the post-1945 destruction of documents, see Saña, *El Franquismo*, pp. 200–1. See also Carlos Rojas, 'En torno a la entrevista de Hendaya', *ABC*, 28 February 1976. On Franco's attempts to get the testimony of prominent Germans in support of his own interpretation of his neutrality, see Suárez Fernández, *Franco*, IV, p. 43.

84 *Hitler's Table Talk*, 7 July 1942, pp. 567–9.

85 Schmidt, *Hitler's Interpreter*, p. 196; Saña, *Franquismo*, pp. 190–3.

86 Espinosa de los Monteros, *Hendaya*, pp. 66–83.

87 Schmidt, *Hitler's Interpreter*, p. 196; *The Ribbentrop Memoirs* (London, 1954) p. 30.

88 In a letter from De las Torres to Serrano Suñer, 21 November 1972, reproduced in facsimile, Serrano Suñer, *De anteayer*, p. 203. He originally wrote incorrectly *mit diesem Kerle* and in trying to make sense of his own notes altered it first to *mit diesem Kerl* and later to *mit diesen Kerlen*. For an analysis of De las Torres' error, see Espinosa de los Monteros, *Hendaya*, pp. 82–4.

89 Keitel, *Memoirs*, p. 126.

90 *Arriba*, 24 October; *La Vanguardia Española*, 24 October 1940; Brissaud, *Canaris*, pp. 204–9 makes the unlikely claim that Canaris had visited Franco at an unspecified date in September 1940 and rehearsed him on how to resist Hitler's demands. There is no evidence to support this.

91 Saña, *Franquismo*, p. 191.

92 Ciano, *Papers*, pp. 401–2.

93 Serrano Suñer, *Memorias*, p. 298.

94 Serrano Suner, *Memorias*, p. 299.

95 *Hitler's Table-Talk*, p. 569; Serrano Suñer, testimony 21.XI.90, and *Memorias*, pp. 299–300.

96 Serrano Suñer, *Memorias*, pp. 300–1; Saña, *Franquismo*, pp. 195–8; Espadas, *Franquismo*, pp. 117–18. For the text of the protocol, see *DGFP*, Series D, Vol. XI, pp. 466–7.

97 Serrano Suñer, *Memorias*, p. 284.

98 Goebbels, *Diaries*, 25 October 1940, p. 153.

99 Ciano, *Diary 1939–1943*, p. 300.

100 Pereira to Salazar, 25 October 1940, *Correspondência*, II, p. 100.

101 Ciano, *Papers*, p. 402; Ulrich
     von Hassell, *The Von Hassell
     Diaries 1938–1944* (London,
     1948) p. 144.
102 Schmidt, *Hitler's Interpreter*,
     p. 197.
103 *DGFP*, Series D, Vol. X,
     pp. 392–3.
104 Halder, *Diary*, 1 November
     1940, p. 272; *Heeresadjutant bei
     Hitler 1938–1943.
     Aufzeichnungen des Majors Engel*
     (Stuttgart, 1974) p. 88.
105 Weizsäcker, diary entry for 25
     October 1940, *Papiere*, p. 221;
     Weizsäcker, *Memoirs*, p. 239.
106 Saña, *Franquismo*, pp. 192–3.
107 Jato, *Gibraltar*, p. 53.
108 Schmidt, *Hitler's Interpreter*,
     pp. 193–4.

CHAPTER XVI

1   Charles-Roux, *Cinq mois
     tragiques*, p. 148.
2   Robert O. Paxton, *Vichy France:
     Old Guard and New Order
     1940–1944* (London, 1972)
     pp. 74–5; Knox, *Mussolini
     Unleashed*, pp. 226–30; *DGFP*,
     Series D, Vol. XI, pp. 411–22;
     Ciano, *Papers*, pp. 401–4; Von
     Hassell, *Diaries*, p. 147.
3   Goebbels, *Diaries*, 1, 3, 5
     November 1940, pp. 159–64.
4   Eccles, *By Safe Hand*, p. 180.
5   Pereira to Salazar, 7 November
     1940, *DAPE*, VII, pp. 580–1.
     Serrano Suñer would make a
     similar point two months later
     in conversation with the Italian
     Ambassador, Lequio to Ciano,
     27 January 1941, *DDI*, 9ª, VI,
     p. 506.
6   *DGFP*, D, XI, pp. 1129–33.

7   Franco to Mussolini, 30
     October 1940, *DDI*, 9ª, VI
     (Rome, 1986) pp. 10–12.
8   Lequio to Ciano, 27, 29, 30
     October, 8 November; Ciano
     to Lequio, 7, 9 November 1940,
     DDI, 9ª, C, pp. 758, 9ª, VI, pp. 3,
     8, 43, 47, 52.
9   Serrano Suñer to Ciano, 31
     October 1940, *DDI*, 9ª, VI,
     pp. 15–16.
10   *DGFP*, D, XI, p. 452. Like other
     key documents relating to
     Franco's commitments to
     Hitler, this letter was not found
     among the captured German
     papers but was published after
     the Caudillo's death by Serrano
     Suñer.
11   Serrano Suñer, *Memorias*,
     pp. 301–05.
12   *DGFP*, Series D, Vol. XI,
     p. 465.
13   Halder, *Diary*, 4 November
     1940, pp. 277–80.
14   Ribbentrop to Stohrer, 6
     November 1940, *DGFP*, D, XI,
     pp. 478–9.
15   Rich, *Hitler's War Aims*, I, p.
     171.
16   Ribbentrop to Stohrer, 11
     November 1940, *DGFP*, Series
     D, Vol. XI, pp. 513–14; Saña,
     Franquismo, p. 203.
17   Serrano Suñer, *Entre Hendaya y
     Gibraltar*, pp. 233–5. The
     paper, attributed exclusively to
     Carrero Blanco, is presented in
     full in *ABC, La II guerra
     mundial*, Vol. 11, pp. 163–7. A
     polemic over the authorship of
     the document broke out after the
     death of Franco, see Serrano
     Suñer, *De anteayer*, p. 215;
     Fernández, *Carrero*, pp. 68–70.

18   The claim was made by Carrero Blanco at a dinner held on 6 May 1971 to celebrate the thirtieth anniversary of his appointment as Subsecretario de la Presidencia, Laureano López Rodó, *El principio del fin: Memorias* (Barcelona, 1992) pp. 180–1, 592–3; Cf. Joaquín Bardavío, *La crisis: historia de quince días* (Madrid, 1974) pp. 26–7.

19   Halder, *Diary*, 11, 22, 24 October, 4 November, pp. 262, 267–8, 275–9; Directive No. 18, 12 November 1940, *Hitler's War Directives 1939–1945*, edited by H. R. Trevor-Roper (London, 1966) pp. 81–7.

20   *DGFP*, D, XI, pp. 528–30, 574–6, 581–2; Charles B. Burdick, *Germany's Military Strategy and Spain in World War II* (Syracuse, 1968) pp. 77ff.

21   Halder, *Diary*, 2 November 1940, p. 275.

22   Halder, *Diary*, 13, 18, 27 November 1940, pp. 280–1, 284, 288.

23   *FRUS 1940*, II, pp. 824–7; Feis, *Spanish Story*, pp. 99–101.

24   Bert Allan Watson, *United States–Spanish Relations, 1939–1946* Unpublished Ph.D. dissertation, George Washington University, 1971, pp. 51–7.

25   *FRUS 1940*, II, pp. 828–33; Feis, *Spanish Story*, pp. 101–02.

26   *FRUS 1940*, II, pp. 836–38.

27   *DGFP*, D, XI, pp. 598–606; Serrano Suñer, *Entre Hendaya y Gibraltar*, p. 235–49; Ciano, *Papers*, pp. 409–11; Halder, *Diary*, 18 November 1940, p. 285; *The Times*, 19 November 1940.

28   *DGFP*, D, XI, pp. 619–23; Serrano Suñer, *Entre Hendaya y Gibraltar*, p. 250–7. Later on in the war, Goering and Goebbels came to agree that Ribbentrop had played a large part in the German failure to draw Spain into the war, *The Goebbels Diaries* (January 1942–December 1943) edited by Louis P. Lochner (London, 1948) 2 March 1943, p. 201.

29   *Churchill & Roosevelt: The Complete Correspondence* 3 vols (Princeton, 1984) I, p. 86; Smyth, *Diplomacy*, pp. 121–3.

30   Mussolini to Hitler, 22 November 1940, *DDI*, 9ª, VI, pp. 157–8 and in translation in *DGFP*, D, XI, pp. 671–2.

31   *DGFP*, D, XI, pp. 705–6.

32   Halder, *Diary*, 28 November 1940, p. 289.

33   Weizsäcker, diary entry for 28 November 1940, *Papiere*, p. 227.

34   *DGFP*, D, XI, pp. 725, 739–41.

35   *DGFP*, D, XI, p. 782; Brissaud, *Canaris*, p. 223.

36   *DGFP*, D, XI, pp. 787–8.

37   Memorandum of Vice-Consul in Barcelona, Yencken to Halifax, 5 November minute of Roberts, 12 November 1941, FO371/24509, C12016/40/41.

38   Eccles, *By Safe Hand*, p. 206.

39   For a sombre account of both see Hamilton, *Appeasement's Child*, pp. 112–46. The orgiastic frenzy of Madrid high society at this time may be discerned in Aline, Countess of Romanones, *The Spy Wore Red* (London, 1987) *passim*.

40   *FRUS 1940*, II, pp. 839–41, 845–50.

41   *FRUS 1940*, II, pp. 839–44; Feis, *Spanish Story*, pp. 104–08.

42   Serrano Suñer, *Memorias*, pp. 190–3; Saña, *Franquismo*, pp. 146–7.

43   *DGFP*, D, XI, pp. 777–8, 842–3, 975–6, Vol. XII, p. 165; Alfieri (Berlin) to Ciano, 3 January, Lequio to Ciano, 12 December 1940, 7 January 1941, *DDI*, 9ª, VI, pp. 266–7, 390–1, 413.

44   Hitler to Mussolini, 5 December 1940, *DDI*, 9ª, VI, pp. 236–8.

45   *DGFP*, D, XI, pp. 802–3.

46   Halder, *Diary*, 8 December 1940, p. 300; Abshagen, *Canaris*, pp. 212–13; Weizsäcker, *Memoirs*, p. 239; Keitel, *Memoirs*, p. 133.

47   Keitel, *Memoirs*, p. 133.

48   *DGFP*, D, XI, pp. 812, 816–7, 8522–3; Brissaud, *Canaris*, pp. 224–6; Serrano Suñer, *Entre Hendaya y Gibraltar*, pp. 258–9.

49   Weizsäcker, diary entry for 8 December 1940, *Papiere*, p. 228; Weizsäcker, *Memoirs*, p. 239.

50   Alfieri (Berlin) to Ciano, 9 December 1940, *DDI*, 9ª, VI, p. 252.

51   Keitel, *Memoirs*, p. 133.

52   Halder, *Diary*, 9, 13 December 1940, pp. 301–5; *Hitler's War Directives*, 13, 18 December 1940, pp. 90–4.

53   *Arriba*, 8, 12 December 1940; Piétri, *Mes années*, pp. 36–7; Séguéla, *Pétain-Franco*, pp. 120–3; Mirandet, *Franco*, p. 80.

54   Hitler to Mussolini, 31 December 1940, *DGFP*, D, XI, pp. 990–4; Goebbels, *Diaries*, 19 December 1940, p. 210.

55   Hoare to Eden, 3, 14 December 1940, FO371/24059 C13128/40/41 & C13541/40/41.

56   *DGFP*, D, XI, pp. 824–5, 847–50; John Lukacs, *The Last European War: September 1939 – December 1941* (London, 1976) p. 114.

57   Memorandum of conversation between Serrano Suñer and Bernard Malley, undated, in the autumn of 1945, FO/371/49663 Z13272/11696/41.

58   Stohrer to Wilhelmstrasse, 8 January 1941, *DGFP*, D, XI, pp. 1054–5.

59   *DGFP*, D, XI, p. 1056; Lequio to Ciano, 10 January 1941, *DDI*, 9ª, VI, pp. 424–5.

60   *DGFP*, D, XI, pp. 1069–70. On Lazar, see Hamilton, *Appeasement's Child*, pp. 174–7 and Garriga, *España de Franco*, pp. 86–93. For a colourful account of Lazar as Himmler's man in Spain, see Romanones, *The Spy*, *passim*. It was a view shared by *The Times*, 6 January 1943. Ramón Garriga later came to believe that Lazar had been a double agent ultimately loyal to the British.

61   *DGFP*, D, XI, pp. 1129–33; Ribbentrop-Ciano conversation, 19 January 1941, *DDI*, 9ª, VI, p. 470; Ciano, *Papers*, pp. 417–19.

62   Stohrer to Wilhelmstrasse, 20 January 1941, *DGFP*, Series D, Vol. XI, pp. 1140–3.

63   Gárate Córdoba, '"Raza"',

pp. 61–2; Gubern, *"Raza"*, P. 7.

64 Baón, *La cara humana*, p. 89.

65 Jaime Peñafiel, *El General y su tropa. Mis recuerdos de la familia Franco* (Madrid, 1992) p. 92.

66 Ribbentrop to Stohrer, 21 January 1941, *DGFP*, D, XI, pp. 1157–8.

67 *DGFP*, D, XI, pp. 1171–5.

68 Ribbentrop to Stohrer, 24 January 1941, *DGFP*, D, XI, pp. 1183–4.

69 *DGFP*, D, XI, pp. 1188–91.

70 *DGFP*, D, XI, pp. 1208–10.

71 Lequio to Ciano, 27 January 1941, *DDI*, 9ª, VI, pp. 506–8.

72 Ribbentrop to Stohrer, 28 January 1941, *DGFP*, D, XI, pp. 1217–18.

73 *DGFP*, D, XI, pp. 1222–378.

74 *FRUS 1941*, II, pp. 880–1.

75 Hitler to Mussolini, 5 February 1941, *DGFP*, D, XII (London, 1962) p. 30.

76 *DGFP*, D, XII, pp. 36–7; DGS report, 16 January 1941, *Documentos inéditos* II-2, pp. 19–20.

77 *DGFP*, D, XII, pp. 37–42, 58.

78 *DGFP*, D, XII, pp. 51–3, 78–9.

79 Ciano to Serrano Suñer, 22 January, Lequio to Ciano, 26, 27, 28 January 1941, *DDI*, 9ª, VI, pp. 485, 504, 506, 510.

80 Lequio to Ciano, 29 January 1941, *DDI*, 9ª, VI, p. 518; Serrano Suñer, *Entre Hendaya y Gibraltar*, pp. 262–3.

81 Mussolini to Vittorio Emanuele, 9 February 1941, *DDI*, 9ª, VI, p. 558.

82 Hoare, *Ambassador*, pp. 95, 104.

83 Pereira to Salazar, 10 February 1941, *Correspondência*, II,

pp. 162–3; Séguéla, *Pétain-Franco*, pp. 135–41.

84 Tusell & García Queipo de Llano, *Franco y Mussolini*, p. 120.

85 Serrano Suñer, *Entre Hendaya y Gibraltar*, pp. 261–4; Mussolini-Franco conversation, 12 February 1941, *DDI*, 9ª, VI, pp. 568–76; Ciano, *Papers*, pp. 421–30; Halder, *Diary*, 15 February 1940, p. 319; Tusell & García Queipo de Llano, *Franco y Mussolini*, pp. 120–2.

86 Mussolini to Cosmelli (Berlin) 14 February 1941, *DDI*, 9ª, VI, pp. 582–3; *DGFP*, D, XII, pp. 96–7; Weizsäcker, diary entry for 16 February 1941, *Papiere*, p. 237; *Documents secrets*, pp. 72–3; Tusell & García Queipo de Llano, *Franco y Mussolini*, p. 122.

87 *Arriba*, 13, 15 February 1941.

88 Lequio to Ciano, 18 February 1941, *DDI*, 9ª, VI, p. 604.

89 *DGFP*, D, XII, pp. 131–2.

90 Burdick, *Military Strategy*, pp. 103 ff.

91 *The Times*, 14, 15 February 1941, commented on the silence of the Spanish press. This account draws heavily on Piétri, *Mes années*, pp. 53–6.

92 H. Du Moulin de Labarthète, *Le temps des illusions: souvenirs, juillet 1940–avril 1942* (Geneva, Éditions du Cheval Ailé, 1946) pp. 229–30; Serrano Suñer, *Entre Hendaya y Gibraltar*, pp. 265–6; Séguéla, *Pétain-Franco*, pp. 147–60.

93 *DGFP*, D, XII, pp. 113–14.

94 Piétri, *Mes années*, pp. 56, 102–10.

95 Some authors have speculated,

without any proof, that, on this point, Franco had been coached by Canaris, something denied by Serrano Suñer – William L. Shirer, *The Rise and Fall of the Third Reich* (London, 1959) pp. 812–19; Brissaud, *Canaris*, pp. 230–1. See Saña, *Franquismo*, p. 207 & Garriga, *España de Franco*, pp. 85–6.

96 *DGFP*, D, XII, pp. 176.

97 *DGFP*, D, XII, p. 197.

98 Conversation Hitler-Ciano, 25 March 1941, *DDI*, 9ª, VI, p. 745.

99 Churchill, *Second World War*, II, *Their Finest Hour*, p. 468. Cf. David Eccles' letter to Norman Davis, Eccles, *By Safe Hand*, pp. 270–1.

100 Churchill, *Finest House*, pp. 467–8

101 Arthur Bryant, *The Turn of the Tide 1939–1943* (London, 1957) pp. 271, 278–9, 286; Hamilton, *Appeasement's Child*, p. 211.

CHAPTER XVII

1 Goebbels, *Diaries*, 6 March 1941, p. 257.

2 *DGFP*, D, XII, pp. 194–5.

3 Goebbels, *Diaries*, 9 May 1941, pp. 355–6.

4 *DGFP*, D, XII, pp. 731–3.

5 *DGFP*, D, XII, pp. 731–3.

6 That is the only possible conclusion to be drawn from Nicolás Franco's remarks to Pereira on 31 March 1941, *DAPE*, VIII, p. 248.

7 Hoare, *Ambassador*, pp. 107–110; Weddell to Hull, 1 March 1941, *FRUS 1941*, pp. 881–5.

8 Garriga, *España de Franco*, pp. 243–4.

9 Pereira to Salazar, 17, 19, 25 February 1941, *Correspondência*, II, pp. 172, 176, 192.

10 Lequio to Foreign Ministry, 18 February 1941, *DDI*, 9ª, VI, pp. 604–5.

11 Pereira to Salazar, 28 February, 3, 14 March, Salazar to Pereira, 6 March 1941, *DAPE*, VIII (Lisbon, 1973) pp. 148–50, 157–8, 169–70, 199–200.

12 Pereira to Salazar, 3, 31 March 1941, *Correspondência*, II, pp. 214–16, 246; *DAPE*, VIII, p. 248.

13 Pereira to Salazar, undated [26 February 1941], *Correspondência*, II, pp. 197–8; Salazar to Pereira, 9 April 1941, *DAPE*, VIII, p. 267; *DGFP*, D, XII, pp. 119–20, 589–90.

14 Memorandum of Alexandre do Amaral Mimoso Ruiz, 28 February 1941, *Correspondência*, II, pp. 212–13.

15 Stohrer to Wilhelmstrasse, 7 May 1941, *Documents secrets*, pp. 72–8.

16 *FRUS 1941*, II, pp. 886–7.

17 Charles R. Halstead, 'The dispute between Ramón Serrano Suñer and Alexander Weddell', *Rivista di Studi Politici Internazionali*, No. 3, 1974, pp. 445–71; Escobar, *Diálogo íntimo*, pp. 52–3. Halstead bends over backwards to understand Serrano Suñer's position as, over forty years after the event, did Beaulac, *Silent Ally*, pp. 108–12. In fact, it is clear that Weddell's protests were deemed to give

offence only long after they had occurred and when it suited the Franco regime. Cf. Weddell to Hull, 23 September 1941, *FRUS 1941*, II, pp. 917–19.

18 *The Times*, 25 April 1941.
19 *FRUS 1941*, Vol. II, pp. 888–90; *DGFP*, D, XII, pp. 590–1; Hoare, *Ambassador*, p. 106. Serrano Suñer, *Entre Hendaya y Gibraltar*, pp. 274–6, gives a Jesuitical account of this interview, wrongly dating it to the day of the Japanese attack on Pearl Harbour.
20 Pereira to Salazar, 6, 7 March 1941, *Correspondência*, II, pp. 221, 226–7.
21 Pereira to Salazar, 25, 31 March, 1 April 1941, *Correspondência*, II, pp. 235, 242, 249.
22 *DGFP*, D, XII (London, 1962) pp. 36–7, 613–14. Cf.Ruhl, *III Reich*, p. 66; Marquina, 'Aranda', p. 24.
23 *Informaciones*, 17 April 1941.
24 Goebbels, *Diaries*, 19, 20 April 1941, p. 323–4.
25 *Churchill & Roosevelt*, I, pp. 173–4.
26 Weddell to Hull, 21, 25 April, 3 May 1941, *FRUS 1941*, II, pp. 891–7.
27 Indeed, Aranda, who was by some margin the general most closely linked to the British, had again approached the German Embassy in such terms in mid-April 1941 – *DGFP*, D, XII, pp. 611–15. Cf. *The Times*, 25 April 1941.
28 Beigbeder, who was pessimistic about any real military opposition to Franco, revealed this to Vice-Admiral J. F.

Somerville on 23 April 1941, FO371/26982, C5625/4298/41, and again to the British military attaché Brigadier Torr on 28 May, Beigbeder's notes, Torr's report & Hoare to Eden, 30 May 1941, Templewood Papers, XIII-21.
29 Goebbels, *Diaries*, 20 May 1941, p. 373.
30 Pereira to Salazar, 1 May 1941, *Correspondência*, II, pp. 286–7.
31 Denis Smyth, 'Les Chevaliers de Saint-George: la Grande-Bretagne et la corruption des généraux espagnols (1940–1942)', *Guerres mondiales et conflits contemporains*, No. 162, Avril 1991, pp. 29–54; Anthony Cave Brown, *The Last Hero: Wild Bill Donovan* (London, 1982) pp. 225–6.
32 Carrero Blanco to Franco, 25 August 1941, *Documentos inéditos*, p. 316.
33 Angel Alcázar de Velasco, *Serrano Suñer en la Falange* (Madrid, 1941).
34 Garriga, *Franco–Serrano Suñer*, pp. 120, 134.
35 Memorandum of conversation between Serrano Suñer and Bernard Malley, autumn 1945, undated, FO371/49663, Z13272/11696/41. For the decree, *Boletín Oficial del Estado*, 4 May 1941.
36 *Arriba*, 3 May 1941.
37 Ciano to Serrano Suner, 4 May 1941, Ciano, *Papers*, pp. 440–1; Ciano, *Diary 1939–1943*, 5 May 1941, p. 339.
38 Stanley G. Payne, *The Franco Regime 1936–1975* (Madison, 1987) pp. 286–6.

39 Testimony of Ramón Serrano Suñer to the author, 21.XI.90; Pereira to Salazar, 16 May 1941, *Correspondência*, II, p. 314; Saña, *Franquismo*, p. 161.

40 Bernard Malley, of the British Embassy Press Office, described Mayalde as 'a blackguard of the worst type', undated memorandum of conversation with Serrano Suner, autumn 1945, FO371/49663 Z13272/11696/41.

41 *ABC*, 6, 8 May; *El Alcázar*, 6 May; *Arriba*, 10, 11 May; *Boletín Oficial del Estado*, 10 May 1941. There is some confusion as to whether Mayalde, Miguel Primo de Rivera and the provincial governors were dismissed or resigned. The press merely records their replacement. Serrano Suñer speaks of other resignations on 12 and 13 May, a week after their replacement. Given the determination with which officials clung to their posts, dismissal seems more likely. However, see Tusell & García Queipo de Llano, *Franco y Mussolini*, pp. 131–2.

42 *Arriba*, 8 May 1941. See also Serrano Suner to Franco, 10 May 1941, *Documentos inéditos*, II-2, pp. 148–50.

43 Saña, *Franquismo*, pp. 160–1.

44 *Arriba*, 9, 11, 14 May 1941; Pereira to Salazar, 11 May 1941, *Correspondência*, II, p. 308.

45 Pereira to Salazar, 11 May 1941, *Correspondência*, II, p. 310.

46 *DGFP*, D, XII, pp. 795–6; Pereira to Salazar, 18 June 1941, *Correspondência*, II, p. 366;

Hamilton, *Appeasement's Child*, pp. 95–7; Hoare, *Ambassador*, p. 112.

47 *Arriba*, 14, 18, 22 May; *ABC*, 17, 19, 22 May, *La Vanguardia Española*, 14 May 1941; *DGFP*, D, XII, pp. 774, 781.

48 Report of conversation with Colonel Beigbeder by Wing-Commander A. H. James M.P., & minute by Roger Makins, 14 May 1942, FO371/26982, C5141/4298/41.

49 Testimony of Ramón Serrano Suñer to the author, 21.XI.90.

50 A facsimile of the letter is reproduced in Serrano Suñer, *Memorias*, p. 180.

51 Pereira to Salazar, 18, 20, 22 May, 9 June 1941, *Correspondência*, II, pp. 314–16, 321–3, 327–8, 349–50.

52 *Arriba*, 21, 22 May 1941.

53 Serrano Suñer, *Memorias*, pp. 200–1.

54 Serrano Suñer, *Memorias*, pp. 190–3; Saña, *Franquismo*, pp. 259–68.

55 On the relationship between Franco and Arrese, see Garriga, *España de Franco*, pp. 322–6.

56 See Carrero Blanco to Franco, 25 August 1941, *Documentos inéditos*, II-2, pp. 326–7.

57 Garriga, *Franco–Serrano Suner*, pp. 133–4; Payne, *Franco Regime*, pp. 291–3.

58 Bruce Lockhart, *Diaries*, 17 June 1941, pp. 103–4.

59 *DGFP*, D, XII, p. 930.

60 *FRUS 1941*, II, pp. 891–903; *Arriba*, 31 May' *ABC*, 31 May, 6, 9 June 1941.

61 Beigbeder's notes of conversation with Brigadier

Torr, 28 May 1941,
Templewood Papers, XIII-21.

62 *FRUS 1941*, II, pp. 904–6, 908.
63 *FRUS 1941*, II, pp. 907–8.
64 Record of conversation between
Hitler and Mussolini, 3 June
1941, *DGFP*, D, XII,
pp. 940–51, 1007–8; Ciano,
Papers, pp. 440–6.
65 *DGFP*, D, XII, pp. 1067.
66 *DGFP*, D, XII, pp. 1080–1;
Pereira to Salazar, 25 June
1941, *Correspondência*, II,
pp. 378–9.
67 *Arriba*, 24, 25 June; *El Alcázar*,
26, 27, 28 June 1941; *The
Times*, 26 June 1941; Hoare,
Ambassador, pp. 114–15.
68 *DGFP*, D, XIII (London, 1964)
pp. 16–17.
69 *DGFP*, Series D, Vol. XIII,
pp. 38–9.
70 *Arriba*, 3 July 1941; *The Times*,
4 July 1941.
71 Feis, *Spanish Story*, p. 136.
72 For an account of the
differences between Espinosa
and Serrano Suñer, see Espinosa
to Franco, 25 January 1941,
Documentos inéditos, II-2,
pp. 49–55. In defence of his
position, Espinosa claimed to
have been recalled because he
was too pro-German, *DGFP*, D,
XIII, pp. 353–4, 357–8.
Serrano Suñer seized on this
later when trying to use the
dismissal as proof of his own
distance from the Axis, Saña, *El
franquismo*, pp. 205–6.
73 *DGFP*, D, XIII, pp. 168–9.
74 Hoare, *Ambassador*, pp. 138–9.
75 *La Vanguardia Española*, 18 July
1941; *The Times*, 19 July 1941.
The text is omitted from all the
collected editions of Franco's
speeches. It is re-printed in full
in *Bulletin of Spanish Studies*,
Vol. XVIII, no. 72, October
1941, pp. 210–17.
76 Pereira to Salazar, 21 July 1941,
Correspondência, II, pp. 388–9.
77 Serrano Suner, *Memorias*,
p. 349; Saña, *Franquismo*,
pp. 252–4.
78 The text is also omitted from all
the collected editions of Franco's
speeches. It may be found in full
in *Bulletin of Spanish Studies*,
Vol. XVIII, no. 72, October
1941, pp. 217–19. Pereira to
Salazar, 21, 25 July 1941,
Correspondência, II, pp. 387–91;
Minute of M. S. Williams, 7
August 1941, FO371/26891,
C8744/3/41.
79 Pereira to Salazar, 25 July 1941,
Correspondência, II, p. 391;
DGFP, D, XIII, pp. 222–4;
Salazar to Tovar (Berlin) &
Portuguese Minister in Rome,
23, 28 July 1941, *DAPE*, IX
(Lisbon, 1974) pp. 87, 99;
Smyth, *Diplomacy*, pp. 230–6.
80 Hoare to Eden, 2 August 1941,
Templewood Papers, XIII, 21;
Campbell to Eden, 6 August
1941, Eccles, *By Safe Hand*,
pp. 304–5.
81 Garriga, *España de Franco*,
p. 305.
82 *The Times*, 22 July 1941.
83 Feis, *Spanish Story*, pp. 138–9.
84 *The War Diaries of Oliver Harvey*
edited by John Harvey (London,
1978), entries for 18 & 19 July
1941, pp. 21–2.
85 373 H.C.DEB., 5s, cc.1074–5;
The Times, 25 July 1941.
86 Smyth, *Diplomacy*, pp. 232–7.

87 Hoare to Eccles, 22 July 1941;
Sir Ronald Campbell,
Ambassador in Lisbon, to Eden,
6 August 1941, Eccles, *By Safe
Hand*, pp. 301, 304–5.

88 *Arriba*, 13, 15, 29 July, 1, 2
August; *ABC*, 19 July, 1 August
1941; Ruhl, *III Reich*, pp. 22–6;
Gerald R. Kleinfeld & Lewis A.
Tambs, *Hitler's Spanish Legion:
The Blue Division in Russia*
(Carbondale, 1979) pp. 1–17.

89 Hoare telegram, 5 August 1941,
FO371/26891, C8744/3/41;
Gil Robles, diary entry for 10
June 1941, *La monarquía*, p. 17.

90 Hoare telegram, 13 August
1941, FO371/26891 C9154/3/
41; Hartmut Heine, *La Oposición
política al franquismo*
(Barcelona, 1983) pp. 253–4;
Smyth, *Diplomacy*, p. 210.

91 Ciano, *Papers*, p. 450.

92 *Arriba*, 31 July 1941.

93 *FRUS 1941*, II, pp. 911–13.

94 *FRUS 1941*, II, pp. 913–25;
Hull, *Memoirs*, II, pp. 1187–8;
Feis, *Spanish Story*, p. 9.

95 *DGFP*, D, XIII, pp. 444–6,
459–60.

96 Kramarz to Ritter, 28 May
1941, *Documents secrets*,
pp. 78–9; Eisenlohr to Stohrer,
11 September 1941, *DGFP*, D,
XIII, pp. 478–9.

97 Warlimont to Ritter, 10
September 1941, *DGFP*, D,
XIII, pp. 498–9.

98 *DGFP*, D, XIII, pp. 628–30,
647–8.

99 United Nations, Security
Council, *Report of the Sub-
Committee on the Spanish
Question*, pp. 12, 85.

100 Weddell to Hull, 30 September,
1, 6 October, *FRUS 1941*, II,
pp. 924–9.

101 *DGFP*, D, XIII, pp. 633–4.

102 Pereira to Salazar, 24
September 1941;
Correspondência, II, p. 435;
DGFP, D, XIII, pp. 630–2; on
Aranda and the British, see Cave
Brown, *Last Hero*, p. 225.

103 *El Alcázar*, 25 November;
Arriba, 27, 28, 29 November
1941; Sheelagh M. Ellwood,
Spanish Fascism in the Franco Era
(London, 1987) pp. 66–70;
Payne, *Franco Regime*,
pp. 288–9.

104 *Arriba*, 19 November; *ABC*, 23
November 1941.

105 Minute of M. S. Williams, 7
December 1941, and telegram of
Hoare to Foreign Office, 27
December 1941, and minute of
M. S. Williams, 1 January 1942,
FO371/27005, C13543/13543/
41 and C14329/13543/41;
Pereira to Salazar, 5 November
1941, *Correspondência*, II, p. 457;
Heine, *La oposición*, pp. 255–6.

106 *Arriba*, 2 December' *ABC*, 26
November 1941.

107 Memorandum of conversation
between Hitler, Ciano and
Serrano Suñer, 30 November
1941, *DGFP*, D, XIII,
pp. 904–6; Ciano, *Diaries
1939–1943*, p. 402; Ciano,
Papers, pp. 461–2.

108 Weddell to Hull, 2 December
1941, *FRUS 1941*, II, pp. 932–4.

109 Feis, *Spanish Story*, pp. 147–51;
Saña, *Franquismo*, pp. 243–4.

110 *Arriba*, 11, 13, 14 December
1941; Pereira to Salazar, 17
December 1941; *DAPE*, X
(Lisbon, 1974) pp. 297–8;

Bulletin of Spanish Studies,
Vol. XIX, nos. 73–4,
January–April 1942, pp. 1–2;
Espadas, *Franquismo*, p. 124.

111   *DGFP*, D, XIII, p. 774.

112   Note verbale from German
Embassy, 13 December 1941,
Documentos inéditos, II-2,
pp. 424–6; Kindelán, *La verdad*,
pp. 46–50; Pereira to Salazar,
17 December 1941, *DAPE*, X,
pp. 297–300; Ruhl, *III Reich*,
p. 95; Suárez Fernández, *Franco*,
III, p. 323.

113   For the text of the speech, see
Kindelán, *La verdad*, pp. 120–2.

CHAPTER XVIII

1   *Churchill & Roosevelt*, I,
p. 313.

2   Lequio to Ciano, 9, 10, 12
January 1942, *DDI*, 9ª, VIII
(Rome, 1988) pp. 113, 116–17,
123–4; Pereira to Salazar, 28
January 1942, *Correspondência de
Pedro Teotónio Pereira para
Oliveira Salazar III (1942)*
(Lisbon, 1990) p. 27.

3   *Arriba*, 13 January 1942.

4   *Arriba*, 27, 28, 29, 31 January, 1
February; *ABC*, 29 January 1942.
Arrese to Franco, 2 February
1942, *Documentos inéditos*, III,
pp. 206–9. For the role of
Arrese during and after the
Catalan visit, see Garriga,
España de Franco, pp. 373–4.

5   *Arriba*, 1 February 1942.

6   *El Caudillo en Cataluña* (Madrid,
1942) pp. 18–19, 31, 33–9.

7   *El Caudillo en Cataluña*,
pp. 24–6, 55; *Arriba*, 30
January, 6 February 1942.

8   Feis, *Spanish Story*, pp. 186–8.

9   *ABC*, 18 January' *Arriba*, 14, 21,
24 February, 5 March 1942.

10   Pereira to Salazar, 24, 28
January, 31 March 1942,
Correspondência, III, pp. 15–16,
27–8, 46.

11   *The Times*, 13 February; *ABC*,
13 February 1942.

12   Lequio to Ciano, 19 February,
Fransoni (Lisbon) to Ciano, 21
February 1942, *DDI*, 9ª, VIII,
pp. 322–3, 335–8; Stohrer to
Wilhelmstrasse, 19 February
1942, *Documents secrets*,
pp. 86–95; *FRUS 1942*, III,
pp. 281–3. Testimony of
Serrano Suner to the author and
to García Lahiguera, *Serrano
Suñer*, p. 205.

13   Garriga, *España de Franco*,
pp. 345–6.

14   *Palabras del Caudillo abril 1937 –
diciembre 1942*, pp. 235–7.

15   *The Times*, 16 February
1942.

16   Goebbels, *Diaries*, 1, 16
February 1942, pp. 28, 52.

17   Feis, *Spanish Story*, pp. 155–7.

18   *Arriba*, 26 February; *Ya*, 27
February 1942.

19   *Arriba*, 18, 20 March 1942.

20   United Nations, Security
Council, *Report of the Sub-
Committee on the Spanish
Question*, p. 85.

21   *Ya*, 25, 26 February 1942.

22   Jaraiz Franco, *Historia*,
pp. 58–60.

23   *Interviu*, No. 383, 14–20
September 1983.

24   Notes of conversation between
Salazar and Nicolás Franco, 16
October 1940, *Dez Anos de
Política Externa (1936–1947) A
Nação Portuguesa e a segunda*

guerra mundial, Vol. VII (Lisbon, 1971) p. 518.

25 Jaraiz Franco, *Historia*, pp. 60–1; Franco, *Nosotros*, pp. 29–31; Garriga, *La Señora*, pp. 170–2.

26 Pereira to Salazar, 1 April 1942, *Correspondência*, III, p. 53.

27 Hoare to F.O., 20 April 1942, FO371/31235, C4198/220/41; *Arriba*, 27 March, 21 April 1942; Garriga, *España de Franco*, pp. 357–9; Heine, *La oposición*, p. 257; Sainz Rodríguez, *Un reinado*, p. 147.

28 Pereira to Salazar, 8, 25 April 1942, *Correspondência*, III, pp. 62–3, 80–1.

29 Pereira to Salazar, 31 March 1942, *Correspondência*, III, pp. 47–50.

30 Feis, *Spanish Story*, pp. 160–79.

31 *FRUS 1942*, III, pp. 275–8, 281–3; Beaulac, *Career Ambassador*, pp. 154, 183.

32 Piétri, *Mes années*, p. 62.

33 *FRUS 1942*, III, pp. 286–7.

34 Cave Brown, *Last Hero*, p. 224.

35 José María Toquero, *Franco y Don Juan: La oposición monárquica al franquismo* (Barcelona, 1989) pp. 49–50.

36 Ramón Garriga, *Franco–Serrano Suñer: un drama político* (Barcelona, 1986) p. 178.

37 Lequio to Ciano, 27 April 1942, *DDI*, 9ª, VIII, pp. 522–4; Heine, *La oposición*, p. 258.

38 Cross, *Hoare*, pp. 338–9; Serrano Suñer, *Entre Hendaya y Gibraltar*, p. 276; Hayes, *Wartime Mission*, pp. 15, 26, 31, 40, 65–6, 293. For contrasting portraits of Hayes by Embassy colleagues, see the devastating attack by the press officer Plenn, *Wind*, pp. 247–69 and the sympathetic remarks of the Counseller Beaulac, *Career Ambassador*, p. 184. For a view similar to Plenn's, see Chase, *Falange*, pp. 245–7.

39 Yencken to Eden, 11 September 1942, FO371/31237, 8984/220/41; Pereira to Salazar, 4 June 1942, *Correspondência*, III, p. 10.

40 *Arriba*, 30 May 1942.

41 Hayes to Hull, 1 June 1942, *FRUS 1942*, III, pp. 288–9.

42 Piétri, *Mes années*, p. 63.

43 Pereira to Salazar, 6 June 1942, *Correspondência*, III, pp. 114–16.

44 *FRUS 1942*, III, pp. 290–90; Hayes, *Wartime Mission*, p. 31; *The Times*, 11 June 1942.

45 Cadogan, *Diaries*, pp. 489–90; Kim Philby, *My Silent War* (London, 1968) pp. 40–1. The best scholarly account of the affair is Denis Smyth, 'Screening "Torch": Allied Counter-Intelligence and the Spanish Threat to the Secrecy of the Allied Invasion of French North Africa in November 1942' in *Intelligence and National Security*, Vol. 4, No. 2, April 1989, pp. 339–44. See also F. H. Hinsley et al. *British Intelligence in the Second World War* Vol. 2 (London, HMSO, 1981) pp. 719–21; Woodward, *British Foreign Policy*, IV, pp. 8–9; Anthony Cave Brown, *The Secret Servant: The Life of Sir Stewart Menzies, Churchill's Spymaster* (London, 1988) pp. 424–5; Trevor-Roper, *Philby*, p. 119.

46 Diary entries for 15, 16, 22, 23, 24, 25 August 1942, Captain Harry C. Butcher, *My Three Years with Eisenhower* (New York, 1946) pp. 58–60, 68–74.

47 Lequio to Ciano, 14 May, Serrano Suñer to Ciano, 19 May, Ciano to Serrano Suñer, 25 May, Ciano to Lequio, 2 June 1942, *DDI*, 9ª, VIII, pp. 591, 599–600, 612, 645; Ciano, *Diary 1939–1943*, 26 May 1943, p. 473.

48 Lequio to Ciano, 5 June, Buti (Paris) to Ciano, 13 June, Chef de Cabinet Lanza d'Ajeta to Ciano, 14 June 1942, *DDI*, 9ª, VIII, pp. 651–4, 670–1.

49 Pereira to Salazar, 17 June, 17 July 1942, *Correspondência*, III, pp. 126, 131.

50 *ABC*, 23 June 1942.

51 Conversations Ciano–Serrano Suner, Livorno, 15–19 June 1942, *DDI*, 9ª, VIII, pp. 690–2. Cf. Mackensen to Ribbentrop, 17 June 1942, *Documents secrets*, pp. 106–8.

52 Ciano, *Diary 1939–1945*, 15–23 June, 4 September 1942, pp. 481–3, 501.

53 Pereira to Salazar, 17 June 1942, *Correspondência*, III, pp. 124–5.

54 Stohrer to Wilhelmstrasse, 8 May 1942, *Documents secrets*, pp. 96–101.

55 He told his German contacts that Don Juan de Borbón was nothing but 'the son of an English woman who has grown up in England which is Spain's bitter enemy' – Kleinfeld & Tambs, *Hitler's Spanish Legion*, p. 195.

56 Stohrer to Wilhelmstrasse, 29 May, 11 June 1942, *Documents secrets*, pp. 101–3, 105–6; Lequio to Ciano, 29 May, Chef de Cabinet Lanza d'Ajeta, 14 June 1942, *DDI*, 9ª, VIII, p. 625, 677.

57 *Hitler's Table Talk*, 7 July, 5 September 1942, pp. 568–70, 691–4; *Akten zur deutschen auswärtigen Politik 1941–1945*, Series E, Vol. III (Göttingen, 1969) pp. 140–1; Kleinfeld & Tambs, *Hitler's Spanish Legion*, pp. 191–7, 204; Garriga, *Yagüe*, pp. 210–11; Payne, *Franco Regime*, p. 301.

58 Feis, *Spanish Story*, p. 188.

59 *Arriba*, 18, 24 July 1942; *The Times*, 18 July, 4 August 1942; Pereira to Salazar, 30 July 1942, *Correspondência*, III, pp. 146–7; Saña, *Franquismo*, pp. 262–3.

60 Donald Downes, *The Scarlet Thread: Adventures in Wartime Espionage* (London, 1953) pp. 91–103; Cave Brown, *Last Hero*, pp. 226–30.

61 Fracassi to Ciano, 30 June 1942, *DDI*, 9ª, VIII, pp. 725–7.

62 *Arriba*, 15 August 1942; *The Times*, 17 August 1942. Suárez Fernández, *Franco*, III, pp. 343–4, insinuates that the publication of a Germanophil article was timed to precede the following day's bomb incident at Begoña.

63 Saña, *Franquismo*, pp. 259–60.

64 Hamilton, *Appeasement's Child*, p. 90; Garriga, *Franco–Serrano Suñer*, p. 134.

65 For speculation on Serrano Suñer's sentimental entanglements, see Hamilton, *Appeasement's Child*, p. 97; Pike,

'Stigma', pp. 384, 402; Feis, *Spanish Story*, p. 181; Fernández, *Tensiones militares*, pp. 41–3.

66 Garriga, *España de Franco*, pp. 357–67, 373–8; Toquero, *Franco y Don Juan*, p. 55.

67 Toquero, *Franco y Don Juan*, p. 51.

68 Sainz Rodríguez, *Un reinado*, p. 290.

69 Madrid Embassy to HMPSSFA, 17 August 1942, FO371/31236, C8368/220/41, and enclosure, Carlist pamphlet, 'Crimen de la Falange en Begoña: Un régimen al descubierto' (the Falange's crime at Begoña: a regime unmasked); Pereira to Salazar, 20 August 1942, *Correspondência*, III, p. 166.

70 Yencken to Eden, 4 September 1942, FO371/31237, C8740/220/41.

71 On Domínguez's German links, see García Lahiguera, *Serrano Suñer*, pp. 227–8; Hoare, *Ambassador*, p. 165.

72 *Arriba*, 22, 23, 25 August 1942; *The Times*, 22, 24, 25 August 1942. *Palabras del Caudillo 1937–1942*, pp. 267–80. This speech was never reproduced in the post-1945 compilations of the Caudillo's utterances.

73 For more or less Falangist versions of the Begoña affair, see Serrano Suñer, *Memorias*, pp. 364–71; Saña, *El franquismo*, pp. 263–9; García Lahiguera, *Serrano Suñer*, pp. 215–29; Payne, *Falange*, pp. 216, 219–20, 234–6. For a Francoist version laying the blame on the Germans, see Doussinague, *España tenía razón*, p. 130. For the transcript of Franco's telephone conversation with Varela, see Laureano López Rodó, *La larga marcha hacia la monarquía* (Barcelona, 1977) pp. 503–7. See also Antonio Marquina Barrio, 'El atentado de Begoña', *Historia 16*, No. 76, August 1982, pp. 11–19.

74 Yencken to F.O., 27, 29 August 1942, FO371/31236, C8350/220/41 & C8427/220/41; Pereira to Salazar, 29 August, 1 September 1942, *Correspondência*, III, pp. 175–6, 187–9; Toquero, *Franco y Don Juan*, p. 54.

75 Yencken to F.O., 3 September 1942, FO371/31236, C8555/220/41; Pereira to Salazar, 3 September 1942, *Correspondência*, III, pp. 192–3; Toquero, *Franco y Don Juan*, p. 55; López Rodó, *La larga marcha*, pp. 28–30.

76 Yencken to Eden, 4 September 1942, FO371/31237, C8740/220/41.

77 See Serrano Suñer's own rather benevolent account in Saña, *El franquismo*, pp. 271–6. Ellwood, *Spanish Fascism*, pp. 84–90; Hoare, *Ambassador*, pp. 140, 164–71; Doussinague, *España tenía razón*, pp. 130–1; Marquina, 'Begoña', pp. 11–19.

78 *La Vanguardia Española*, 4 September 1942.

79 Testimony of Serrano Suñer to the author in 1979; Serrano Suñer, *Memorias*, p. 371; Ruhl, *III Reich*, pp. 118–19.

80 Jordana, diary entries for 3, 6

September 1942, quoted by Tusell & García Queipo de Llano, *Franco y Mussolini,* pp. 168–9.

81 Serrano Suñer, *Memorias,* p. 358. For a contrary view of Jordana's independence, see Tusell & García Queipo de Llano, *Franco y Mussolini,* p. 171.

82 *Arriba,* 4 September; *La Vanguardia Española,* 4, 5 September 1942; Saña, *El franquismo,* pp. 267, 271–3; Serrano Suñer, *Memorias,* pp. 370–2; Garriga, *España de Franco,* pp. 377–80.

83 Toquero, *Franco y Don Juan,* p. 55.

84 *Arriba,* 4 September; *La Vanguardia Española,* 5 September 1942. On Blas Pérez's early career, see Garriga, *Los validos,* pp. 126–63.

85 *The Times,* 4, 5 6 September; *Daily Telegraph,* 4 September 1942; Serrano Suñer, *Entre Hendaya y Gibraltar,* pp. 211–18; Von Hassell, *Diaries,* pp. 239–48.

86 Kleinfeld & Tambs, *Hitler's Spanish Legion,* pp. 207–8.

87 Ruhl, *III Reich,* pp. 118–19.

88 *Hitler's Table Talk,* p. 694.

89 *Ciano's Diaries,* pp. 501, 511; Pereira to Salazar, 16 September, 21 October 1942, *Correspondência,* III, pp. 197, 200, 223.

90 Serrano Suñer, *Memorias,* pp. 447–55.

91 Piétri, *Mes Années,* p. 65.

92 *Arriba,* 4 September 1942.

93 Javier Tusell, 'Franco no fue neutral', *Historia 16,* no. 141, 1988, p. 19.

94 Javier Tusell, 'Un giro fundamental en la política española durante la segunda guerra mundial: la llegada de Jordana al Ministerio de Asuntos Exteriores', José Luis García Delgado, editor, *El primer franquismo: España durante la segunda guerra mundial* (Madrid, 1989) p. 284.

95 *La Vanguardia Española,* 22 September; *Ya,* 22 September; *The Times,* 23 September 1942.

96 Tusell, 'Franco no fue neutral', p. 18.

97 Pereira to Salazar, 16, 22 September, 8 October 1942, *Correspondência,* III, pp. 198, 204–5, 212–14; Tusell, 'Un giro fundamental', p. 286.

98 *FRUS 1942,* III, pp. 295–8; Beaulac, *Career Ambassador,* pp. 181, 186. See also Javier Tusell, 'La etapa Jordana (1942–1944)', *Espacio, Tiempo y Forma* (UNED, Madrid) Serie V., tomo 2, 1989, pp. 169–89.

99 *Bulletin of Spanish Studies,* Vol. XIX, no. 76, October 1942, p. 200.

100 *Hitler's Table Talk,* 5 June 1942, pp. 514–15.

101 *Hitler's Table Talk,* 7 July, 1 August, 5 September 1942, pp. 567–9, 607–8, 691–4.

102 Hayes, *Wartime Mission,* pp. 71–2.

103 Diary entry for 18 September 1942, Butcher, *My Three Years,* pp. 111–12.

104 José María de Areilza, *Embajadores sobre España* (Madrid, 1947) pp. 4–5, 57–8; Hills, *Gibraltar,* pp. 436–8.

105 Smyth, Screening "Torch" ', *passim.*

106   Doussinague, *España tenía razón*, pp. 69–87, and especially pp. 71–3.

107   *FRUS 1942*, III, pp. 299–302.

108   Hoare to Foreign Office, 19 October 1942, and minute of M. S. Williams, 22 October 1942, FO371/31230, C10035/175/41. See also Hoare, *Ambassador*, pp. 173–4.

109   Doussinague, *España tenía razón*, pp. 98–101.

110   Doussinague, *España tenía razón*, pp. 82–6.

111   *FRUS 1942*, III, p. 303.

112   *FRUS 1942*, III, pp. 301–2.

113   Conversation of Serrano Suñer with Charles Favrel, 21 October 1945, PRO FO371/49663, Z13272/11696/41; Doussinague, *España tenía razón*, pp. 98–105.

114   *FRUS 1942*, III, pp. 303–6; *Churchill & Roosevelt* I, pp. 664–6; Cadogan, *Diaries*, 6 November 1942, pp. 489–90; Hoare, *Ambassador*, p. 163, 176–9.

115   Baón, *La cara humana*, p. 140. See Pereira to Salazar, 9 November 1942, *Correspondência*, III, pp. 244–5.

116   Three of the four present during the dramatic meeting at Jordana's apartment wrote accounts which fundamentally coincide. See Hayes, *Wartime Mission*, pp. 86–94; Beaulac, *Career Ambassador*, pp. 187–8; Doussinague, *Espana tenía razón*, pp. 90–2. Jordana's version was given to Pereira, Pereira to Salazar, 9 November 1942, *Correspondência*, III, pp. 243–5; Butcher, *My Three Years*, diary

entry for 9 November 1942, p. 179; Roosevelt's letter to Franco and Franco's reply, *FRUS 1942*, III, pp. 306, 308.

117   Pereira to Salazar, 18 November 1942, *Correspondência*, III, pp. 254–5.

118   Kindelán, *La verdad*, pp. 32–6, 50–9, 125–7.

119   *El Alcázar*, 13 November 1942.

120   Carlos Iniesta Cano, *Memorias y recuerdos* (Barcelona, 1984) pp. 149–50.

121   Ruhl, *III Reich*, pp. 178–82; Garriga, *Yagüe*, pp. 222–3.

122   *FRUS 1942*, III, pp. 308–11; Cave Brown, *Last Hero*, p. 233.

123   Diary entry for 15 November 1942, Butcher, *My Three Years*, p. 193.

124   *FRUS 1942*, III, pp. 313–14.

125   Jordana to Alba, 27 November 1942, reprinted in Doussinague, *España tenía razón*, pp. 103–5; *FRUS 1942*, III, p. 311.

126   Hoare, *Ambassador*, pp. 176–82; Hayes, *Wartime Mission*, pp. 92–6.

127   Garriga, *España de Franco*, pp. 407–8.

128   Doussinague, *España tenía razón*, pp. 207.

129   Jordana to Vidal, 12 November 1942, Doussinague, *España tenía razón*, p. 204.

130   Doussinague, *España tenía razón*, pp. 106–7.

131   Doussinague, *España tenía razón*, p. 205.

132   Hayes to Hull, 24 November 1942, *FRUS 1942*, III, pp. 312–13.

133   Palabras del Caudillo 1937–1942, p. 597.

134 *ABC*, 6 December 1942.

135 *Palabras del Caudillo 1937–1942*, pp. 599–607.

CHAPTER XIX

1 *ABC*, 18, 19, 20 December; *Arriba*, 18, 19, 20 December; *Diário de Lisboa*, 17, 19, 20 December 1942; Doussinague, *España tenía razón*, pp. 116–26.

2 Ramón Garriga, *La España de Franco: de la División Azul al pacto con los Estados Unidos (1943 a 1951)* (Puebla, 1971) p. 9.

3 Diary entries for 3 December 1942, 10 April 1943, Butcher, *My Three Years*, pp. 211, 283.

4 Pereira to Salazar, 4 February 1943, *Correspondência de Pedro Teotónio Pereira para Oliveira Salazar IV (1943–1944)* (Lisbon, 1991) pp. 22–3.

5 Pereira to Salazar, 3 March 1943, *Correspondência*, IV, p. 58; Garriga, *De la División Azul*, pp. 10–13; Kleinfeld & Tambs, *Hitler's Spanish Legion*, pp. 229–35.

6 *FRUS 1942*, III, pp. 313–17.

7 Doussinague, *España tenía razón*, p. 205.

8 Fernando Eguidazu Palacios, 'Factores monetarios y de balanza de pagos en la neutralidad española', *Revista de Estudios Internacionales*, no. 2, abril–junio 1984, pp. 355–83.

9 Report of Director of Commercial Policy Service to Ribbentrop, 9 November 1942, *Documents secrets*, pp. 116–20.

10 Stohrer to Wilhelmstrasse, 18 December, Ribbentrop to Ritt,

22 December 1942, *Documents secrets*, pp. 121–4.

11 Ribbentrop to Heberlein (Madrid), 28 December 1942, *Documents secrets*, pp. 124–7.

12 Doussinague, *España tenía razón*, p. 205.

13 Kleinfeld & Tambs, *Hitler's Spanish Legion*, pp. 234–7.

14 Doussinague, *España tenía razón*, pp. 143–4; Hoare, *Ambassador*, pp. 184–5.

15 Curt Prüfer, *Rewriting History: The Original and Revised World War II Diaries of Curt Prüfer, Nazi Diplomat* (Kent, Ohio, Kent State University Press, 1988) pp. 23, 39, 166.

16 Lomax, *Diplomatic Smuggler*, p. 86; Cross, *Hoare*, pp. 183, 337; Doussinague, *España tenía razón*, pp. 128–9.

17 Pereira to Salazar, 27 December 1942, *Correspondência*, III, pp. 280–1; Piétri, *Mes années*, p. 86.

18 Ruhl, *III Reich*, pp. 49–50; Hayes, *Wartime Mission*, p. 96; Doussinague, *España tenía razón*, pp. 131–3.

19 Moltke to Wilhelmstrasse, 13 January 1943, *Documents secrets*, pp. 127–30; Garriga, *De la División Azul*, p. 30.

20 Moltke to Wilhelmstrasse, 24 January 1943, *Documents Secrets*, pp. 131–4.

21 Doussinague, *España tenía razón*, p. 206.

22 Moltke to Wilhelmstrasse, 9, 12 February, *Documents secrets*, pp. 140–2; Secret Protocol between the German and Spanish Governments, 10 February 1943, USA

Department of State, *The Spanish Government and the Axis* (Washington, 1946) p. 35.

23 Garriga, *De la División Azul*, p. 31.

24 He had been named in February, *Ya*, 28 February, 21 April 1943.

25 Tusell y García Queipo de Llano, *Franco y Mussolini*, pp. 191–3.

26 *FRUS 1943*, II, p. 597.

27 Garriga, *De la División Azul*, pp. 43–6.

28 *Arriba*, 5, 8, 19, 21, 24, 27 January, 19, 21, 27 February; *Informaciones*, 7, 9, 13, 25 January 1943.

29 *Arriba*, 10, 25, 27 February 1943.

30 *FRUS 1943*, II, p. 598–601.

31 *La Vanguardia Española*, 19, 20, 24 January; *Arriba*, 24 January 1943; Kleinfeld & Tambs, *Hitler's Spanish Legion*, pp. 239–40.

32 Garriga, *De la División Azul*, pp. 19–27; Suárez Fernández, *Franco*, III, p. 370.

33 *ABC*, 10 February 1943. The full text may be found in José Luis de Arrese, *Escritos y discursos* (Madrid, 1943) pp. 183–92.

34 *ABC*, 28 February 1943.

35 *El Alcázar*, 1 March 1943.

36 An account of this interview was given to the Germans, *Documents secrets*, pp. 135–6.

37 Hoare, *Ambassador*, pp. 184–96; Moltke to Wilhelmstrasse, 24 February, 3 March 1943, *Documents secrets*, pp. 143–9.

38 Borrell interview in Mérida, *Testigos*, pp. 218–19.

39 Hayes, *Wartime Mission*,

pp. 97–8; Francis J. Spellman, *Action This Day: Letters From The Fighting Fronts* (London, 1944) pp. 20–1.

40 Kleinfeld & Tambs, *Hitler's Spanish Legion*, pp. 310–13.

41 Bernardo Díaz-Nosty, *Las Cortes de Franco* (Barcelona, 1972) pp. 15–32.

42 *Arriba*, 18, 19, 20, 21 March 1943; *The Times*, 18 March 1943.

43 Pereira to Salazar, 10, 22 February 1943, *Correspondência*, IV, pp. 31, 45–7; Garriga, *De la División Azul*, pp. 48–50.

44 *ABC*, 17 April 1943; Pereira to Salazar, 18 April 1943, *Correspondência*, IV, pp. 118–23; Doussinague, *España tenía razón*, pp. 180–91; Hayes, *Wartime Mission*, pp. 129–30.

45 *ABC*, 18, 20 April; *La Vanguardia Española*, 20 April; *El Alcázar*, 19 April 1943.

46 *ABC*, 2, 5, 7, 8, 9 May; *Arriba*, 2, 5, 12 May 1943; Doussinague, *España tenía razón*, pp. 207–9.

47 *Statements made by H.E. General Franco and Count Jordana on the International Policy of Spain* (Madrid, 1943).

48 Garriga, *De la División Azul*, pp. 66–7.

49 Suárez Fernández, *Franco*, III, p. 409.

50 Piétri, *Mes années*, pp. 86, 144.

51 Dieckhoff to Ribbentrop, 1 May 1943, *Documents secrets*, pp. 155–7; Hayes, *Wartime Mission*, pp. 99–104, 129–30; Doussinague, *España tenía razón*, pp. 209–10, 215–19. On Dieckhoff's appointment see, *ABC*, 18 April 1943.

52 *ABC*, 1 June; *The Times*, 3 June 1943.

53 Garriga, *Yagüe*, pp. 229–31.

54 *The Times*, 7 July 1943; Pereira to Salazar, 24 June 1943, *Correspondência*, IV, pp. 213–14.

55 López Rodó, *La larga marcha*, pp. 36–8; Suárez Fernández, *Franco*, III, 403.

56 Pereira to Salazar, 4, 10 February 1943, *Correspondência*, IV, pp. 23, 31.

57 *FRUS 1943*, II, pp. 609–11; Hayes, *Wartime Mission*, p. 150.

58 Hoare, *Ambassador*, pp. 197–204.

59 Pereira to Salazar, 15 July 1943, *Correspondência*, IV, p. 230; Garriga, *De la División Zaul*, pp. 83–6.

60 *Arriba*, 18 July; *La Vanguardia Española*, 18, 20 July; *The Times*, 18 July 1943.

61 Pereira to Salazar, 22 July 1943, *Correspondência*, IV, pp. 234–9.

62 Jordana, diary entry for 23 July 1943, quoted by Tusell, 'La etapa de Jordana', p. 180.

63 López Rodó, *La larga marcha*, pp. 39–41.

64 *La Vanguardia Española*, 27, 28, 29 October; *Arriba*, 27, 28, 29, 30 July, 1 August 1943.

65 Gil Robles, diary entries for 6, 11 August 1943, *La monarquía*, pp. 51, 53; testimony of José Antonio Girón de Velasco, Mérida, *Testigos*, p. 117.

66 Tusell & García Queipo de Llano, *Franco y Mussolini*, pp. 208–9; Raimundo Fernández Cuesta, *Testimonio, recuerdos y reflexiones* (Madrid, 1985) pp. 221–2; Hoare, *Ambassador*, pp. 211–12.

67 *FRUS 1943*, II, pp. 611–17; Hoare to Foreign Office, 30 July 1943, FO371/34755, C88732/24/41; Hayes, *Wartime Mission*, pp. 157–62; Doussinague, *España tenía razón*, p. 229.

68 Hayes, *Wartime Mission*, p. 163.

69 *Arriba*, 13, 14 August; *ABC*, 12, 13 August; *Ya*, 13, 22, 26 August 1943.

70 Hayes, *Wartime Mission*, p. 165; *FRUS 1943*, II, pp. 617–19, 621.

71 *FRUS 1943*, II, pp. 619–20, 629–30.

72 López Rodó, *La larga marcha*, pp. 515–19; Gil Robles' diary entry for 25 August 1943, refers to Franco's 'impertinent refusal', Gil Robles, *La monarquía*, p. 55.

73 Emmet John Hughes, *Report from Spain* (London, 1947) pp. 224–5; Duff, 'Spain', p. 306. Cf. Arthur P. Whitaker, *Spain and the Defence of the West: Ally and Liability* (New York, 1961) p. 22.

74 Gil Robles, diary entry for 23 August 1943, *La monarquía*, p. 55; Sainz Rodríguez, *Un reinado*, p. 161.

75 It has been claimed that the report was carried to Galicia by the young Catholic monarchist intellectual, Rafael Calvo Serer. Cf. Ricardo de la Cierva, *Historia del franquismo: I orígenes y configuración (1939–1945)* (Barcelona, 1975) pp. 265–70.

76 Hoare to Foreign Office, 18, 21 August 1943, FO371/34755, C9473/24/41, C9602/24/41; Hoare, *Ambassador*, pp. 218–22. Cf. Jordana to

Alba, 30 August 1943, MAE, R1371/3B.

77 *Arriba*, 8 April 1943.

78 *Ya*, 7 September; *ABC*, 7, 10 September; *The Times*, 8 September; *Arriba*, 7, 8, 9 September 1943; Arrese, *Escritos y discursos*, pp. 213–31.

79 Fernández, *Tensiones militares*, p. 91; La Cierva, *Franquismo*, I, p. 264.

80 Suárez Fernández, *Franco*, III, p. 431.

81 For the full text, see López Rodó, *La larga marcha*, pp. 43–4. For Gil Robles' comment, diary entry for 18 September 1943, Gil Robles, *La monarquía*, p. 60.

82 José Fortes & Restituto Valero, *Qué son las Fuerzas Armadas* (Barcelona, 1977) p. 45; Tusell & García Queipo de Llano, *Franco y Mussolini*, p. 224–5.

83 Suárez Fernández, *Franco*, III, p. 432.

84 Gil Robles, diary entries for 26 September & 31 October, 1943, *La monarquía*, pp. 61, 67.

85 Woodward, *British Foreign Policy*, II, pp. 353–56, IV, pp. 8–17; Suárez Fernández, *Franco*, III, p. 433; Gil Robles, *La monarquía*, p. 53.

86 *Arriba*, 16, 18 September; *Informaciones*, 13, 14, 15 September 1943.

87 Garriga, *De la División Azul*, pp. 101–4.

88 Garriga, *De la División Azul*, pp. 104–6.

89 *La Vanguardia Española*, 2 October 1943.

90 *FRUS 1943*, II, pp. 620–2; Hayes, *Wartime Mission*, p. 175.

91 *La Vanguardia Española*, 2 October 1943.

92 Suárez Fernández, *Franco*, III, pp. 431–2; Gil Robles, diary entries for 3 October & 17 November 1943, *La monarquía*, pp. 62, 68; Saña, *Franquismo*, p. 282; Payne, *Military*, p. 434; Heine, *La oposición*, p. 261.

93 Gil Robles, diary entries for 25 August & 31 October 1943, and letter to General Asensio, 28 September 1943, Gil Robles, *La monarquía*, pp. 55, 67, 360–6; enclosure to Pereira to Salazar, 20 November 1943, *Correspondência*, IV, pp. 326–30; Suárez Fernández, *Franco*, III, p. 432.

94 *FRUS 1943*, II, pp. 622–7; Jordana to Hayes, 29 October 1943, Doussinague, *España tenía razón*, pp. 246–56.

95 *Bulletin of Spanish Studies*, Vol. XXI, no. 82, April 1944, p. 85; Hoare, *Ambassador*, pp. 239–40.

96 Draft memorandum by the British Chiefs of Staff, 19 August 1943, PRO FO371/34755, C10232/24/41.

97 *FRUS 1943*, II, p. 633. On U.S. Spanish economic relations, see *FRUS 1943*, II, pp. 632–711; Feis, *Spanish Story*, pp. 191–262; James W. Cortada, *Relaciones España–USA 1941–45* (Barcelona, 1973) pp. 23–126; Beaulac, *Career Ambassador*, pp. 163–72; Hayes, *Wartime Mission*, pp. 181–7.

98 Hayes to Hull, 24 June 1943, *FRUS 1943*, II, pp. 634–5.

99 See exchange of telegrams between the Madrid Embassy

and the State Department, September–December 1943, *FRUS 1943*, II, pp. 711–22.

100 *FRUS 1943*, II, pp. 631–2, 722–38, 727–31; Cortada, *España–USA*, pp. 40–53. Blaming Jordana, Doussinague, *España tenía razón*, pp. 88–9, 280–90, claimed that the telegram was non-commital about recognizing Laurel and its effusive tone was a device to protect Spanish citizens from Japanese reprisals.

101 *FRUS 1943*, II, pp. 648–9; Hayes, *Wartime Mission*, pp. 191–3.

102 *The Spanish Government and the Axis*, pp. 34–7; Hoare, *Ambassador*, p. 258.

103 Hoare, *Ambassador*, pp. 246–7.

104 Jordana to Alba, 27 December 1943, MAE, R1371/3B.

105 Franco to Don Juan, 6 January, Don Juan to Franco, 25 January 1944, reprinted in Sainz Rodríguez, *Un reinado*, pp. 359–62; López Rodó, *La larga marcha*, pp. 520–3; Suárez Fernández, *Franco*, III, pp. 478–81.

CHAPTER XX

1 *The Diaries of Edward R. Stettinius, Jr., 1943–1946*, edited by Thomas M. Campbell & George C. Herring (New York, 1975) entry for 20 January 1944, pp. 21–2.

2 *FRUS 1944*, IV, (Washington, 1966) pp. 297–301; Hayes, *Wartime Mission*, pp. 210–11.

3 Pereira to Salazar, 10, 15 January 1944, *Correspondência*, IV, pp. 364–6, 371–2.

4 Hayes, *Wartime Mission*, pp. 211–12.

5 Hull to Hayes, 18, 25, 27 January 1944, *FRUS 1944*, IV, pp. 301–4.

6 *Arriba*, 23 January; *Ya*, 23 January; *The Times*, 25 January 1944.

7 Pereira to Salazar, 27 January, 16 March, 22 April 1944, *Correspondência*, IV, pp. 388–9, 463, 498–500.

8 Pereira to Salazar, 27 January 1944, *Correspondência*, IV, p. 389.

9 Hoare, *Ambassador*, pp. 249–56; *FRUS 1944*, IV, pp. 305–8; *The Times*, 29, 30 January 1944; Hayes, *Wartime Mission*, pp. 212–14.

10 *FRUS 1944*, IV, pp. 309–14; *The Times*, 1, 2 February 1944.

11 Doussinague, *España tenía razón*, pp. 305–8; Pereira to Salazar, 9 February 1944, *Correspondência*, IV, p. 409.

12 Garriga, *De la División Azul*, pp. 161–2.

13 *FRUS 1944*, IV, pp. 319–25; *The Times*, 3 February 1944.

14 *FRUS 1944*, IV, pp. 325–7.

15 *FRUS 1944*, IV, pp. 332–6; *The Times*, 8 February 1944.

16 Hayes, *Wartime Mission*, pp. 215–20.

17 *ABC*, 2, 19, 25 February; *El Alcázar*, 2 February' *Arriba*, 2, 3 February, 5 March 1944.

18 Pereira to Salazar, 19 February 1944, *Correspondência*, IV, p. 432.

19 Churchill to Roosevelt, 13 February 1944; Roosevelt to Churchill, 15, 23 February 1944, *Churchill & Roosevelt*, II,

pp. 725–6, 728, 751; Cadogan, *Diaries*, 28 January, 3, 4 February 1944, pp. 602–3; Stettinius, *Diaries*, 22 February 1944, pp. 28–9; Hoare, *Ambassador*, pp. 257–62.

20 *FRUS 1944*, IV, pp. 344–5.
21 *ABC*, 10, 12 February 1944.
22 *FRUS 1944*, IV, pp. 338–9.
23 *FRUS 1944*, IV, pp. 339–44; Pereira to Salazar, 16 February 1944, *Correspondência*, IV, pp. 427–8.
24 *FRUS 1944*, IV, pp. 346–53.
25 Pereira to Salazar, 1, 7, 12, 23 March, 18 October 1944, *Correspondência*, IV, pp. 447, 452–3, 455–6, 469, 642.
26 Pereira to Salazar, 7, 12 March 1944, *Correspondência*, IV, pp. 453–4, 456.
27 *The Times*, 3 April 1944.
28 *ABC*, 25, 26 March, 23 April; *Arriba*, 24 March, 16 April 1944.
29 *FRUS 1944*, IV, p. 400; Pereira to Salazar, 23 March 1944, *Correspondência*, IV, p. 469.
30 Churchill to Roosevelt, 30 March, 17, 22 April, Roosevelt to Churchill, 21, 25 April 1944, *Correspondence*, III, pp. 66–8, 99, 106–8, 114; *FRUS 1944*, IV, pp. 353–408; Cadogan, *Diaries*, 20, 22 April 1944, pp. 622–3; Hoare, *Ambassador*, pp. 262–6; Hayes, *Wartime Mission*, pp. 217–25; Dean Acheson, *Present at the Creation: My Years in the State Department* (New York, 1969) p. 95.
31 *FRUS 1944*, IV, pp. 408–14; *The Times*, 3 May 1944.
32 *ABC*, 3, 7, 9 May; *Arriba*, 10 May 1944.
33 Pereira to Salazar, 5 May 1944, *Correspondência*, IV, p. 531.
34 Hayes, *Wartime Mission*, p. 229; Hoare, *Ambassador*, pp. 267–8.
35 *FRUS 1944*, IV, pp. 414–23.
36 Pereira to Salazar, 22 June 1944, *Correspondência*, IV, pp. 576–7.
37 Piétri, *Mes Années*, pp. 73–4.
38 Garriga, *De la División Azul*, pp. 177–9.
39 400 H.C.DEB., 5s, cc.768–72; Hoare, *Ambassador*, p. 267.
40 *ABC*, 25, 26 May; *Ya*, 26, 30 May; *Arriba*, 31 May 1944.
41 Martínez Nadal, *Antonio Torres*, pp. 183–6; Suárez Fernández, *Franco*, III, p. 495.
42 Dalton, *War Diary*, 9 June 1944, pp. 755–6.
43 Churchill to Roosevelt, 4 June 1944, *Correspondence*, III, pp. 162–3.
44 Garriga, *De la División Azul*, pp. 189, 191–6.
45 *Arriba*, 17 May, 14, 17 June, 11 July, 17 August; *ABC*, 17 June, 7 July; *Ya*, 20 June, 21 July 1944.
46 *Arriba*, 30 July, 1, 5, 9, 13, 27 August; *ABC*, 25 August 1944.
47 Serrano Suñer, *Memorias*, p. 358; Saña, *Franquismo*, p. 281. In an interview at the end of the war, Serrano Suñer himself admitted to having been impressed by what he had heard about the V-1 and V-2 and the work of German nuclear scientists, interview with Charles Favrel, FO371/49663, Z13272/11696/41.
48 *La Vanguardia Española*, 18 July 1944.
49 *Arriba*, 18, 21 July 1944.
50 *ABC*, 18, 22, 23 August; *Arriba*, 19, 24, 25 August 1944.

51 Hoare, *Ambassador*, pp. 269–72; Garriga, *De la División Azul*, pp. 201–3.
52 *Arriba*, 13, 15 August 1944.
53 Pereira to Salazar, 9 August 1944, *Correspondência*, IV, pp. 601–2.
54 Pereira to Salazar, 4 September 1944, *Correspondência*, IV, p. 608.
55 Saña, *Franquismo*, p. 321.
56 Garriga, *De la División Azul*, p. 204; Cava Mesa, *Lequerica*, pp. 208–13.
57 Piétri, *Mes années*, pp. 253–6; Hayes, *Wartime Mission*, pp. 253–4; Hughes, *Report*, pp. 211–12.
58 *ABC*, 26 September 1944.
59 *Arriba*, 6 September 1944; Hayes, *Wartime Mission*, pp. 259–62. For the remark about the Ardennes, Saña, *Franquismo*, p. 281.
60 Hoare to Eden, 3 October 1944, FO371/39671, C13318/23/41.
61 Hoare memorandum, 15 October 1944, FO371/39671, C14492/23/41.
62 *Arriba*, 30 September; *ABC*, 30 September 1944.
63 Suárez Fernández, *Franco*, III, pp. 543–4.
64 Hoare, *Ambassador*, p. 283. The full text is given on pp. 300–04.
65 Hoare memorandum, 16 November 1944, PRO PREM 8/106.
66 *Arriba*, 11, 26 October 1944.
67 Martin Gilbert, *Road to Victory: Winston S. Churchill 1941–1945* (London, 1986) p. 749.
68 Attlee to War Cabinet, 4 November 1944, PRO PREM B/106.

69 Pereira to Salazar, 22 November 1944, *Correspondência*, IV, p. 703.
70 *ABC*, 7, 8 November; *Ya*, 7 November; *Arriba*, 8 November 1944; *Franco ha dicho ... (Madrid, 1947)* pp. 239–45. As late as 1984, Franco's official bibiographer could praise his restraint towards France in 1940, Suárez Fernández, *Franco*, I, p. 17.
71 Suárez Fernández, *Franco*, III, p. 551.
72 Eden memorandum, 18 November 1944, PRO PREM 8/106.
73 *Bulletin of Spanish Studies*, Vol. XXII, no. 85, January 1945, p. 47.
74 Hull, *Memoirs*, II, pp. 1334–5.
75 Eden to Halifax, draft telegram, 9 November 1944, PRO PREM 8/106.
76 Churchill to Eden, 10 November 1944, PRO PREM 8/106.
77 War Cabinet Conclusions, 27 November 1944, PRO PREM 8/106.
78 Gilbert, *Road to Victory*, p. 1071.
79 Hoare to Eden, 12 December 1944, FO371/39672, Z17266/23/41; Hoare, *Ambassador*, pp. 283–4; Cross, *Hoare*, pp. 343–4; Hayes, *Wartime Mission*, p. 243.
80 Torr memorandum, 30 January 1945, FO371/49587, Z1595/233/41; *Bulletin of Spanish Studies*, Vol. XXII, no. 86, April 1945, pp. 86–7.
81 Pereira to Salazar, 28 December 1944, *Correspondência*, IV, p. 738; *La Vanguardia Española*, 24

December 1944; Garriga, *La Señora*, pp. 197–8.

82 Bowker to Eden, 29 December 1944, FO371/49587, Z381/233/41.

83 Churchill minute PRO PREM 8/106; Winston S. Churchill, *The Second World War*, VI, *Triumph and Tragedy* (London, 1954) pp. 616–17.

84 PRO PREM 8/106; Gilbert, *Road to Victory*, p. 1071; Hoare, *Ambassador*, pp. 304–6.

85 F.O. to State Department, 15 January 1945, FO371/39672, C18083/23/41.

86 Franco to Churchill, 25 February 1945, reproduced in Alberto J. Lleonart y Anselem & Fernando María Castiella y Maiz, *España y ONU I (1945–46)* (Madrid, 1978) pp. 10–11.

87 Hoare to F.O., 3 October 1944, FO371/39671/C13318/23/41.

88 *ABC*, 23 January; *Ya*, 24 January 1945.

89 *Arriba*, 6 December 1944, 9 March 1945; *ABC*, 9 December 1944; *Ya*, 17 December 1944.

90 *Final Entries 1945. The Diaries of Joseph Goebbels* edited, introduced and annotated by Hugh Trevor-Roper (New York, 1978) 1 March 1945, p. 17.

91 Bowker to Eden, 20 January 1945, FO371/49857, Z1359/233/41; Ansaldo, *¿Para qué . . . ?*, pp. 313–14.

92 Bowker to Eden, 2 March 1945, FO371/49587, Z3353/233/41.

93 Florentino Portero, *Franco aislado: la cuestión española (1945–1950)* (Madrid, 1989) pp. 75–6.

94 Bowker to Eden, 16 January 1945, FO 371/49581, Z904/118/41.

95 *Ya*, 21 January 1945.

96 Bowker to Eden, 26 January 1945, FO371/49581, Z1559/118/41.

97 *FRUS 1948*, III, pp. 679–80; Bowker to Eden, 12 June 1945, FO371/49589, Z7338/233/41; Garran minute, 3 September 1945, FO371/49613, Z10105/537/41.

98 Halifax to Eden, 2 February 1945, FO371/49581, Z1907/118/41.

99 *FRUS 1945*, V, pp. 672–3; Halifax to Eden, 7 April 1945, FO371/49611, Z4450/537/41.

100 *Arriba*, 13, 17, 18, 23, 24, 25 March; *ABC*, 18, 24, 25 March 1945.

101 *ABC*, 12 April 1945.

102 Goebbels, *Final Entries*, 21, 23, 24, 25 March 1945, pp. 199, 212, 220, 227; Doussinague, *España tenía razón*, pp. 346–8.

103 *FRUS 1945*, V, (Washington, 1967) pp. 667–8.

104 *FRUS 1945*, V, pp. 672–3.

105 *FRUS 1945*, V, pp. 668–71. On the maintenance of civil war divisions, see Paul Preston, *The Politics of Revenge: Fascism and the Military in 20th Century Spain* (London, 1990) chapter 2.

106 Brigadier Torr Memorandum, 30 January 1945, FO371/49587, Z1595/233/G41; Suárez Fernández, *Franco*, IV, p. 21; La Cierva, *Historia del franquismo*, I, p. 294.

107 López Rodó, *La monarquía*, pp. 48–50; Suárez Fernández, *Franco*, IV, pp. 18–19.

108 Kindelán, *La verdad*, p. 89.

109 Bowker to Eden, 27 March 1945, FO371/49629, Z4138/1484/41.

110 *ABC*, 6 April 1945; Kindelán, *La verdad*, pp. 229–36.

111 Bowker to Eden, 25 April 1945, FO371/49589, Z5249/1484/G41.

112 Javier Tussell, *Franco y los católicos: la política interior española entre 1945 y 1957* (Madrid, 1984) p. 60.

113 Vázquez Montalbán, *Los demonios*, p. 105.

114 Bowker to Eden, 27 March 1945, FO371/49587, Z4137/233/41. For a list of those present, see Ricardo de la Cierva, *Francisco Franco: un siglo de España* 2 vols (Madrid, 1973) II, p. 406.

115 *Arriba*, 28 March 1945.

116 *Ya*, 1, 3 April; *ABC*, 3 April 1945.

117 *FRUS 1945*, V, pp. 673–6.

118 *Ya*, 31 January, 18 February, 1, 14 March; *Arriba*, 14 February 1945.

119 *ABC*, 14 March, 23 April 1945.

120 Bowker to Eden, 12 June 1945, FO371/49589, Z7338/233/41.

121 Armour to Stettinius, 1 May 1945, *FRUS 1945*, V, pp. 676–8.

122 Bowker to Eden, 8, 14 May 1945, FO371/49550, Z6008/2/41 & *Ibid*, Z6421/7/41.

123 Bowker to Eden, 31 May 1945, FO371/49550, Z7213/7/41.

124 *ABC*, 26 April; *Mundo*, 29 April 1945.

125 *Arriba*, 3, 5, 10 May; *ABC*, 3, 11 May; *Informaciones*, 3, 7 May; *The Times*, 3 May 1945.

126 *The Times*, 11 May 1945. The Falange connived at asset-stripping of the Embassy by Nazi officials, Foltz, *Masquerade*, pp. 279–81.

127 *Arriba*, 8 May; *ABC*, 8 May 1945.

128 Speeches in Madrid, 2 April and Valladolid, 20 May 1945, Francisco Franco, *Textos de doctrina política: palabras y escritos de 1945 a 1950* (Madrid, 1951) pp. 5–8, 611–13.

129 United Nations, Security Council, *Report of the Sub-Committee on the Spanish Question*, p. 12.

CHAPTER XXI

1 Soriano, *La mano*, p. 159; Borrell interview, Mérida, *Testigos*, p. 225.

2 Tussell, *Franco y los católicos*, pp. 54–5.

3 Tusell, *Franco y los católicos*, pp. 56–8.

4 Tusell, *Franco y los católicos*, pp. 58–9.

5 Suárez Fernández, *Franco*, I, p. 19.

6 Franco, *Textos de doctrina política*, p. 613; *The Times*, 4 April 1945.

7 Max Gallo, *Spain Under Franco: A History* (London, 1973) pp. 153–9.

8 Garriga, *Nicolás Franco*, p. 266.

9 Garriga, *De la División Azul*, p. 311.

10 Garriga, *De la División Azul*, p. 424.

11 Lleonart, *España y ONU*, I, pp. 30–3.

12 Bowker to Churchill, 26 June 1945, FO3711/49589, Z7876/233/41.

13 Pemán, *Mis encuentros*, p. 185.
14 Franco, *Textos 1945–1950*, pp. 9–12.
15 Lleonart, *España y ONU*, I, pp. 34–5.
16 *The Times*, 18 June 1945.
17 *La Vanguardia Española*, 18 July; *The Times*, 18 July 1945; Franco, *Textos 1945–1950*, pp. 15–25.
18 Tusell, *Franco y los católicos*, pp. 63–77; Suárez Fernández, *Franco*, IV, p. 44; Gil Robles, diary entry for 21 July 1945, *La monarquía*, pp. 126–7; Garriga, *De la División Azul*, pp. 334–5.
19 Equipo Mundo, *Los 90 Ministros de Franco* (Barcelona, 1970) pp. 185–202; Payne, *Franco Regime*, pp. 350–1.
20 Testimony of Alberto Martín Artajo, Mérida, *Testigos*, p. 197.
21 Tusell, *Franco y los católicos*, pp. 61–2.
22 Tusell, *Franco y los católicos*, pp. 84–94, 118; Portero, *Franco aislado*, pp. 106–10.
23 Bowker to Churchill, 5 June 1945, FO371/49589, Z7168/233/41; López Rodó, *La larga marcha*, pp. 54–5; Gil Robles, diary entry for 11 October 1945, *La monarquía*, pp. 134–5.
24 FO371/49612, Z9049/537/G41; Duff, 'Spain', p. 310; Lleonart, *España y ONU*, I, pp. 37–41; Churchill, *Second World War*, VI, p. 566; Harry S. Truman, *Memoirs, Year of Decisions 1945* (London, 1955) pp. 272–3, 284–5.
25 Hoyer Millar to Cadogan, 19 July 1945, FO371/49612, Z94049/537/G41.
26 *FRUS 1945*, II, pp. 1499, 1509;

Qasim Bin Ahmad, *The British Government and the Franco Dictatorship, 1945–1950* Ph.D., University of London, 1987, pp. 65–70.
27 Mallet to F.O., 27 July 1945, FO371/49617, Z8861/829/41.
28 Tusell, *Franco y los católicos*, pp. 96–7.
29 Lleonart, *España y ONU*, I, pp. 42–4; *FRUS 1945*, V, p. 683.
30 José María del Valle, *Las instituciones de la República en exilio* (Paris, 1976) pp. 113–31; Heine, *La oposición*, pp. 157–74.
31 Garriga, *De la División Azul*, p. 354.
32 Speeches, 20 May, 20 June, 2, 17 July 1945, Franco, *Textos 1945–1950*, pp. 5–25.
33 413 H.C.DEB, 5s, col.296; Alan Bullock, *Ernest Bevin: Foreign Secretary 1945–1951* (London, 1983) pp. 71–2.
34 Bevin to Clark Kerr (Moscow), 24 August 1945, FO371/49613, Z9949/537/41.
35 Suárez Fernández, *Franco*, IV, pp. 55–7.
36 Mallet to F.O., 22 August 1945, FO371/49613, Z9844/537/41; Mallet to Bevin, 25, 26 August 1945, FO371/49613, Z9941/537/41, Z10132/537/41; Garriga, *De la División Azul*, pp. 359–61.
37 Angel Viñas, *Los pactos secretos de Franco con Estados Unidos: bases, ayuda económica, recortes de soberanía* (Barcelona, 1981) p. 27.
38 López Rodó, *La larga marcha*, pp. 57–59; Tusell, *Franco y los católicos*, pp. 99–100.

39 Tusell, *Franco y los católicos*, p. 100.

40 Mallet to Bevin, 22 September, 6 October 1945, FO371/49590, Z10932/233/41, Z11432/233/41.

41 Garran minute, 15 September, FO371/49580, Z10630/110/41; Mallet to Bevin, 15 September, FO371/49580, Z10685/110/41; Mallet to Bevin, 21 September 1945, FO371/49580, Z10918/110/41; Portero, *Franco aislado*, pp. 111–12.

42 Hoyer Millar minute, 3 September 1945, FO371/49613, Z10105/537/41.

43 Portero, *Franco aislado*, pp. 133–8; Randolph Bernard Jones, *The Spanish Question and the Cold War 1944–1953* (Unpublished Ph.D. thesis, University of London, 1987) pp. 49–51.

44 Gil Robles, diary entry for 16 August 1945, *La monarquía*, p. 131.

45 Ansaldo, *¿Para qué... ?*, pp. 332, 336; Garriga, *De la División Azul*, p. 295.

46 See also speeches to the high command, 7 January 1946, to the general staff, 16 February 1946, Franco, *Textos 1945–1950*, pp. 539–49.

47 On the letter, Garriga, *De la División Azul*, pp. 382–6; Saña, *Franquismo*, pp. 289–92, 301–3; on the annotations, Suárez Fernández, *Franco*, IV, pp. 58–9.

48 Tusell, *Franco y los católicos*, pp. 100–2; Suárez Fernández, *Franco*, IV, pp. 52–3.

49 *La Vanguardia Española*, 2 October 1945; *Arriba*, 2 October 1945.

50 Tusell, *Franco y los católicos*, pp. 102–6; *ABC*, 12 October 1945.

51 *ABC*, 27 October; *Arriba*, 27 October 1945; Trythall, *Franco*, pp. 203–4. The full text of the speech was not published until 1960, Francisco Franco, *Discursos y mensajes del Jefe del Estado 1955–1959* (Madrid, 1960) pp. 739–53.

52 Suárez Fernández, *Franco*, IV, p. 102.

53 *FRUS 1946* (Washington, 1969) V, p. 1039.

54 Mallet to Bevin, 3 December 1945, FO371/49629, Z13504/1484/G41.

55 Kindelán, *La verdad*, p. 287.

56 *FRUS 1945*, V, pp. 684–7.

57 Ministerio de Asuntos Exteriores to U.S. Embassy, 10 October 1945, *FRUS 1945*, V, pp. 690–2.

58 Halifax to Bevin, 22 November 1945, FO371/49581, Z1296/118/41.

59 *FRUS 1945*, V, pp. 694–5.

60 Mallet to Bevin, 1 December 1945, FO371/49614, Z13350/537/41.

61 Mallet to Bevin, 3 December 1945, FO371/49629, Z13504/1484/G41.

62 *FRUS 1945*, V, pp. 695–7; Mallet to Bevin, 1 December 1945, FO371/49614, Z13350/537/41.

63 416 H.C.DEB, 5s, cols.2314–2315; Garriga, *De la División Azul*, pp. 410–11.

64 Bevin to Mallet, 5 December

1945, FO371/49614, Z13392/537/41.

65 Tusell, *Franco y los católicos*, pp. 111–12.

66 The bibliography on the guerrilla is huge but largely anecdotal. Overviews from the regime's point of view are Tomás Cossías, *La lucha contra el "maquis" en España* (Madrid, 1956); Francisco Aguado Sánchez, *El Maquis en España: su historia* (Madrid, 1975); Francisco Aguado Sánchez, *El Maquis en España: sus documentos* (Madrid, 1976). On the left, see Andrés Sorel, *Búsqueda, reconstrucción e historia de la guerrilla española del siglo XX, a través de sus documentos, relatos y protagonistas* (Paris, 1970); Carlos J. Kaiser, *La guerrilla antifranquista: historia del maquis* (Madrid, 1976); Eduardo Pons Prades, *Guerrillas españolas 1936–1960* (Barcelona, 1977); Hartmut Heine, *A guerrila antifranquista en Galicia* (Vigo, 1980); Daniel Arasa, *Años 40: los maquis y el PCE* (Barcelona, 1984).

67 Garriga, *De la División Azul*, p. 415.

68 Tusell, *Franco y los católicos*, pp. 111, 113.

69 The British had complained in January 1945, FO371/49548, Z1360/7/41; Foltz, *Masquerade*, pp. 278–85; Saña, *Franquismo*, pp. 305–8; Suárez Fernández, *Franco*, IV, pp. 108, 111.

70 United Nations, Security Council, *Report of the Sub-Committee on the Spanish Question*, pp. 16–21.

71 Mallet to F.O., 25 February 1946, FO371/60331, Z1827/8/41.

72 *Réplica del Gobierno español a la publicación hecha por el Departamento de Estado* (Madrid, 1946) reprinted in Lleonart, *España y ONU*, I, pp. 67–80.

73 Franco, *Textos 1945–1950*, pp. 334–5.

74 *The Times*, 20 May 1946.

75 Suárez Fernández, *Franco*, IV, pp. 53, 62–3.

76 Mallet to Bevin, 15 February 1946, FO371/60373, Z2125/41/41; López Rodó, *La larga marcha*, p. 62; Xavier Tusell, *La oposición democrática al franquismo 1939–1962* (Barcelona, 1977) pp. 114–16.

77 Torr memorandum, 20 February 1946, FO371/60373, Z1741/41/41; Kindelán, *La verdad*, pp. 128–30, 254; Tusell, *Franco y los Católicos*, pp. 150–1.

78 Suárez Fernández, *Franco*, IV, pp. 127–32, 153–7, 301.

79 Gil Robles, diary entry for 9 February 1946, *La monarquía*, pp. 161–2.

80 Tusell, *La oposición democrática*, pp. 114–16; Gil Robles, diary entries for 15, 28 February 1946, *La monarquía*, p. 163, 1168–9.

81 *Arriba*, 5 January, 17, 19, 20 February 1946; *ABC*, 5, 9 January, 22, 23 February 1946; *The Times*, 11 February, 20 May 1946; Gil Robles, diary entry for 27 February 1946, *La monarquía*, p. 168.

82 *FRUS 1946*, V, p. 1030.

83 *FRUS 1946*, V, pp. 1033–6, 1038–42, 1044–5.

84 Bevin to Halifax, 26 January 1946, FO371/60349, Z60111/36/41; Halifax to Bevin, 28 January 1946, FO371/60349, Z882/36/41.

85 418 H.C.DEB, 5s, c.142.

86 Lleonart, *España y ONU*, I, pp. 57–9.

87 Gregorio Morán, *Miseria y grandeza del Partido Comunista de España 1939–1985* (Barcelona, 1986) pp. 103, 107; David Wingeate Pike, *Jours de gloire, jours de honte: le Parti Comuniste d'Espagne en France depuis son arrivée en 1939 jusque'à son départ en 1950* (Paris, 1984) p. 59.

88 Hoyer-Millar memorandum, 3 March 1946, FO371/603352, Z210/36/41.

89 *FRUS 1946*, V, pp. 1043–4.

90 Hoyer–Millar memorandum, 3 March 1946, FO371/60352, Z210/36/41.

91 *The Times*, 5 March 1946; Whitaker, *Spain*, pp. 25–7; Portero, *Franco aislado*, pp. 151–5.

92 Tusell, *Franco y los católicos*, p. 115.

93 Qasim Bin Ahmad, *The British Government*, pp. 34–44.

94 *The Economist*, 9 March 1946.

95 *Arriba*, 8 March 1946; *ABC*, 8 March 1946.

96 *FRUS 1946*, V, pp. 1049–58.

97 F.O. to Washington Embassy, 30 March 1946, FO371/60354, Z2886/36/41; *FRUS 1946*, V, pp. 1062–4.

98 López Rodó, *La monarquía*, pp. 69–70.

99 *Arriba*, 2 April 1946.

100 *Arriba*, 7 April 1946; Franco, *Textos 1945–1950*, pp. 551–2.

101 Stettinius, *Diaries*, 15 April 1946, pp. 466–9; Lleonart, *España y ONU*, I, pp. 81–3.

102 Franco, *Textos 1945–1950*, pp. 31–59; *The Times*, 15 May 1946.

103 United Nations, Security Council, *Report of the Sub-Committee on the Spanish Question*; *FRUS 1946*, V, pp. 1072–4; Portero, *Franco aislado*, pp. 174–6.

104 'Nota de réplica del Ministerio de Asuntos Exteriores al Informe del Subcomité del Consejo de Seguridad' reprinted in Lleonart, *España y ONU*, I, pp. 109–20; Bonsal to Byrnes, *FRUS 1946*, V, pp. 1075–7.

105 423 H.C.DEB. 5s, cols.2016–2017.

106 Agustín del Río Cisneros, *Política internacional de España: El caso español en la ONU y en el mundo* (Madrid, 1946) *passim*.

107 Tusell, *Franco y los católicos*, p. 153.

108 Lleonart, *España y ONU*, I, pp. 104–9, 130–96.

109 Tusell, *Católicos*, p. 116.

110 *Arriba*, 18 July 1946, reproduced in *Textos 1945–1950*, pp. 61–6.

111 *Arriba*, 1, 2 October' *ABC*, 1, 2 October; *La Vanguardia Española*, 1, 2 October 1946; *The Times*, 2 October 1946.

112 Lleonart, *España ONU*, I, p. 215.

113 *Arriba*, 14 November 1946.

114 *Arriba*, 30 November 1946.

115 Lleonart, *España y ONU*, I, pp. 240–94.

116 *FRUS 1946*, V, pp. 1080–2.
117 Tusell, *Franco y los católicos*,
     p. 154.
118 *Arriba*, 10 December; *La
     Vanguardia Española*, 10
     December 1946.
119 *ABC*, 10, 11, 12, 13 December
     1946; Lleonart, *España y ONU*,
     I, pp. 310–89.
120 Tusell, *Franco y los Católicos*,
     p. 154.
121 *The Times*, 17 December 1946;
     Garriga, *De la División Azul*,
     p. 469.

CHAPTER XXII

  1 *FRUS 1946*, V, pp. 1085–8.
  2 Pemán, *Mis encuentros*,
     pp. 118–19.
  3 Payne, *Franco Regime*, p. 403,
     has suggested that Franco's
     friend, the writer Joaquín
     Arrarás, assisted him in the
     composition of the articles. La
     Cierva, *Franquismo*, I, p. 384,
     attributes them exclusively to
     the pen of Franco. On Carrero
     Blanco's views on freemasonry,
     see Garriga, *Los validos*,
     pp. 228–30.
  4 *Arriba*, 14 December 1946;
     Suárez Fernández, *Franco*, IV,
     pp. 139–40; J. Boor, *Masonería*
     (Madrid, 1952) pp. 8–9.
  5 Baón, *La cara humana*, p. 99.
  6 Franco told his Minister of
     Finance in June 1969 that this
     was the case, Bayod, *Franco visto*,
     p. 158.
  7 López Rodó, *La larga marcha*,
     pp. 73, 529–32; Gil Robles,
     diary entries for 5 November
     1945, 15 March 1946, *La
     monarquía*, pp. 138, 173–4.

  8 Tusell, *La oposición democrática*,
     pp. 117–120.
  9 Garriga, *De la División Azul*,
     pp. 472–4.
 10 Del Valle, *República en exilio*,
     pp. 224–31.
 11 Portero, *Franco aislado*, p. 182.
 12 *FRUS 1947* (Washington, 1972)
     III, pp. 1056–60.
 13 Suárez Fernández, *Franco*, IV,
     pp. 137, 145–6, 151–7.
 14 Vaca de Osma, *Paisajes*,
     pp. 137–41.
 15 Vaca de Osma, *Paisajes*, p. 132.
 16 Franco, *Textos, 1945–1950*,
     pp. 219–26.
 17 Economic Intelligence
     Department, report on Spanish
     vulnerability to sanctions, 23
     April 1947, FO371/67868,
     Z4313/3/41G; Howard to
     Bevin, 7 July 1947, FO371/
     67869, Z6479/3/41G.
 18 *The Times*, 14 October 1946.
 19 *FRUS 1946*, V, pp. 1079–80.
 20 Kenneth O. Morgan, *Labour in
     Power 1945–1951* (Oxford,
     1984) pp. 251–3; Acheson,
     Present, pp. 294–301; Herbert
     Feis, *From Trust to Terror: The
     Onset of the Cold War 1945–1950*
     (London, 1970) pp. 191–8.
 21 Carrero's report to Franco is
     reproduced in López Rodó, *La
     larga marcha*, pp. 75–89; Don
     Juan's reaction, *Ibid.*, pp. 89–99;
     Gil Robles, diary entries for 31
     March–15 April 1947, *La
     monarquía*, pp. 206–14,
     388–93; Tusell, *La oposición
     democrática*, pp. 161–9; Tusell,
     Franco y los católicos, pp. 161–2.
 22 *FRUS 1947*, pp. 1066–8.
 23 On the neutralization of Labour
     hostility to Franco, see Qasim

Bin Ahmad, *The British Government*, pp. 286–343.

24 Bevin memorandum 'Economic Sanctions Against Spain', 3 January 1947, FO371/67867, Z270/3/41G.

25 *FRUS 1947*, III, pp. 1068–73; Howard to Hoyer Miller, 15 April 1947, FO371/67867, Z3740/3/41.

26 Bevin to Sargent, 25 April 1947, FO371/67868, Z4093/3/41; Douglas to Marshall, 1 May 1947, *FRUS 1947*, III, pp. 1068–74.

27 *FRUS 1947*, III, pp. 1078–80; Qasim Bin Ahmad, *The British Government*, pp. 125–34.

28 *Mundo Obrero*, 8 May 1947; *El Socialista*, 16 May 1947; José María Lorenzo Espinosa, *Rebelión en la Ría. Vizcaya 1947: obreros, empresarios y falangistas* (Bilbao, 1988) pp. 17–69.

29 Juan Carlos Jiménez de Aberasturi & Koldo San Sebastián, *La huelga general del 1° de mayo de 1947 (artículos y documentos)* (San Sebastián, 1991) pp. 48–61.

30 *Le Monde*, 9 May 1947.

31 Jiménez, *La huelga*, p. 61.

32 *Nuestra Bandera*, Nos. 17, 18, April–May, June 1947; *L'Humanité*, 13 May 1947.

33 Viñas, *Guerra, dictadura, dinero*, pp. 265–87; A. J. Lleonart y Anselem, *España y ONU II (1947)* (Madrid, 1983) p. 117.

34 *Arriba*, 1 November 1946.

35 *Arriba*, 14 January; *ABC*, 14, 15, 16, 17 January 1947.

36 *Arriba*, 28 March 1947; Franco Salgado-Araujo, *Mis conversaciones*, p. 321.

37 José María de Areilza, *Memorias exteriores 1947–1964* (Barcelona, 1984) p. 28.

38 *Arriba*, 9, 10 June; *ABC*, 10 June; *The Times*, 9 June 1946; *Observer*, 13 June 1947; Garriga, *La Señora*, pp. 211–12.

39 *Arriba*, 24 June; Areilza, *Memorias*, pp. 216–18.

40 Whitaker, *Spain*, p. 25; Areilza, *Memorias*, pp. 216–18.

41 *Ya*, 6 July 1947; *Arriba*, 5, 6 July 1947.

42 Tusell, *Franco y los católicos*, pp. 163–5; Payne, *Franco Regime*, p. 375; Trythall, *Franco*, pp. 203–6.

43 Kindelán, *La verdad*, p. 344.

44 *FRUS 1947*, III, pp. 1084–5; F.O. to British Embassy (Washington) FO371/67869, Z7004/3/41.

45 *FRUS 1947*, III, pp. 1085–7.

46 *La Vanguardia Española*, 18 July 1948; González, *Liturgias*, pp. 162–5.

47 Vaca de Osma, *Paisajes*, pp. 133–6; Suárez Fernández, *Franco*, IV, p. 188–9.

48 Gil Robles, 15, 16, 17, 18, 19, 20 October 1947, *La monarquía*, pp. 240–2; Garriga, *De la División Azul*, pp. 499–504.

49 Howard to F.O., 23 October 1947, FO371/67870, Z9291/3/41; *The Times*, 23, 24 October 1947.

50 Allen (British Embassy) to Reber, 28 July 1947, *FRUS 1947*, III, pp. 1087–8.

51 Suárez Fernández, *Franco*, IV, pp. 191–9.

52 *FRUS 1947*, III, pp. 1088–90.

53 *FRUS 1947*, III, pp. 1091–5; Walter Millis, ed., *The Forrestal*

Diaries (New York, Viking Press, 1951) p. 328; Walter LaFeber, *America, Russia and the Cold War 1945–1975* 3rd ed. (New York, 1976) pp. 66–7.

54 *FRUS 1947*, III, pp. 1096–7; Lleonart, *España y ONU*, II, pp. 230–313; Suárez Fernández, *Franco*, IV, pp. 212–14.

55 Suárez Fernández, *Franco*, IV, p. 226.

56 Franco, *Textos, 1945–1950*, pp. 111–14.

57 *FRUS 1948* (Washington, 1974) III, pp. 1017–20; Harvey to Balfour, 24 November 1947, FO371/67871, Z9953/3/41.

58 Coles, *Franco*, p. 63; Franco Salgado-Araujo, *Mi vida*, p. 319.

59 Federico Sopeña, *Escrito de noche* (Madrid, 1985) p. 133.

60 *FRUS 1948*, III, pp. 1020–5.

61 Howard to Crosthwaite, 11 February 1948, FO371/73333, Z1458/84/41.

62 *FRUS 1948*, III, pp. 1026–7.

63 Suárez Fernández, *Franco*, IV, p. 237.

64 The State Department was uneasy about the impact on public opinion of any deal about bases but suggested using civilian capital to get the airfields created, *FRUS 1948*, III, pp. 1034–5, 1039–40.

65 *La Vanguardia Española*, 31 March 1948; *The Times*, 1, 2 April 1948; Portero, *Franco aislado*, pp. 309–13; Suárez Fernández, *Franco*, IV, pp. 239–40.

66 *FRUS 1948*, III, pp. 1028–30.

67 Anderson to Howard, 9 August, Howard to Anderson, 2 September, minute by Crosthwaite, 23 September 1948, FO371/73336, Z7447/84/41.

68 Gil Robles, diary entries for 13 January, 15 February 1948, *La monarquía*, pp. 253, 255.

69 Suárez Fernández, *Franco*, IV, pp. 243–9.

70 Tusell, *La oposición democrática*, pp. 197–202; Suárez Fernández, *Franco*, IV, pp. 249–51.

71 Don Juan gave his account to José María Gil Robles on 1 September 1948 (Gil Robles, *La monarquía*, pp. 265–73), to Pedro Sainz Rodríguez at about the same time (Sainz Rodríguez, *Un reinado*, pp. 220–2), to an unnamed person (either Gil Robles or Sainz Rodríguez) 'close to Don Juan' who informed the U.S. Ambassador in Lisbon, Lincoln MacVeagh; and directly to Theodore Xanthaky, Special Assistant to MacVeagh (*FRUS 1948*, III, pp. 1050–1, 1059–63; *FRUS 1949*, IV, p. 755). Cf. Ramón de Alderete, . . . *y estos borbones nos quieren gobernar* (Paris, 1974) pp. 56–8; *The Times*, 28 August; *ABC*, 29 August 1948.

72 Gil Robles, diary entries for 1, 25 September, 5 October 1948, *La monarquía*, pp. 272–5.

73 Tusell, *La oposición democrática*, pp. 203–5.

74 *FRUS 1948*, III, p. 1062.

75 *ABC*, 10 November 1948; Gil Robles, diary entries for 8, 10, 11, 27 October, 19 December 1948, *La monarquía*, pp. 276–81, 286.

76 Testimony to the author of Eugenio Vegas Latapié and José María Gil Robles, Madrid, 1970; Gil Robles, diary entry for 6 August 1949, *La monarquía*, p. 302.

77 *The Times*, 13 November 1948; Follick in 460 *H.C.DEB.* 5s, c.1757; Gil Robles, diary entries for 12, 13 October, 2, 4, 10, 12, 13 November, *La monarquía*, pp. 278–83.

78 Gil Robles, diary entries for 4, 8, 30 June, 6, 18, 27 July 1949, *La monarquía*, pp. 298–301.

79 Acheson, *Present*, p. 151.

80 Cava Mesa, *Lequerica*, pp. 265–310; Theodore J. Lowi, 'Bases in Spain' in Harold Stein, ed., *American Civil-Military Decisions: A Book of Case Studies* (Birmingham, Alabama, 1963) pp. 675–6; R. Richard Rubottom & J. Carter Murphy, *Spain and the United States Since World War II* (New York, 1984) pp. 10–11; Viñas, *Guerra, dictadura, dinero*, pp. 284–7; Whitaker, *Spain*, pp. 32–4.

81 Suárez Fernández, *Franco*, IV, p. 261.

82 Johnston to F.O., 2 October 1948, FO371/73337, Z7957/84/41.

83 *Arriba*, 2 October 1948; *ABC*, 2 October 1948; Suárez Fernández, *Franco*, IV, pp. 266–9.

84 *Arriba*, 13 October 1948; *The Times*, 13 October 1948; Franco, *Textos 1945–1950*, pp. 571–4. For the 1942 speech, see *Palabras del Caudillo abril 1937–diciembre 1942*, pp. 235–7. On 9 October in Cordoba, he declared that even the old, children and women would join in the battle against 'the Asiatic hordes', *ABC*, 10 October 1948.

85 *ABC*, 13 October; *Arriba*, 13, 14 October; *The Times*, 14 October 1948; Franco, *Textos 1945–1950*, pp. 601–2.

86 Garriga, *Segura*, pp. 294–6.

87 Marshall memorandum, 4 October 1948, *FRUS 1948*, III, pp. 1053–4.

88 Suárez Fernández, *Franco*, IV, pp. 273–5.

89 *The New York Times*, 12 November 1948; Franco, *Textos, 1945–1950*, p. 269; Whitaker, *Spain*, pp. 35–6.

90 *FRUS 1948*, III, p. 1063.

91 *The New York Times*, 12 November 1948; Franco, *Textos, 1945–1950*, p. 269; Qasim Bin Ahmad, *The British Government*, pp. 150–1; Whitaker, *Spain*, pp. 35–6.

92 459 *H.C.DEB.* 5s, cc.718–19.

93 Franco, *Textos, 1945–1950*, pp. 131–6.

94 *ABC*, 1 January 1949; Serrano Suñer, *Memorias*, p. 330.

95 Garriga, *Franco-Serrano Suner*, p. 182.

96 Bartolomé Barba Hernández, *Dos años al frente del Gobierno Civil de Barcelona y varios ensayos* (Madrid, 1948) pp. 45–50; Rafael Abella, *Por el Imperio hacia Dios: crónica de una posguerra (1939–1950)* (Barcelona, 1978) pp. 101–32; Qasim Bin Ahmad, *The British Government*, pp. 277–8.

97 Acheson, *Present*, pp. 331–3.

98 Laureano López Rodó, *Testimonio de una política de*

Estado (Barcelona, 1987) pp. 197–200.

99 Portero, *Franco aislado*, pp. 316–17.

100 *The Daily Telegraph and Morning Post*, 1 February 1949; Franco, *Textos, 1945–1950*, pp. 277–81.

101 2 February 1949, 460 *H.C.DEB.* 5s, cc.1750–94.

102 *FRUS 1949*, IV, pp. 729–30; Hankey to Bevin, 'Spain: Annual Report for 1949', 27 January 1950, FO371/89479, WS1011/1. The draught would last until 1956, with rainfall ranging from one third to one half of the historical average.

103 Whitaker, *Spain*, pp. 34–5; Gil Robles, diary entry for 11 February 1948, *La monarquía*, pp. 291–2.

104 *ABC*, 1 April 1949; *The Times*, 1 April 1949; Franco, Textos, 1945–1950, pp. 137–43.

105 Whitaker, *Spain*, pp. 36–7.

106 *FRUS 1949* (Washington, 1975) IV, pp. 721–4, 730–5.

107 A. J. Lleonart y Anselem, *España y ONU III (1948–1949): La cuestión española* (Madrid, 1985) pp. 54–8, 148–372; *FRUS 1949*, IV, pp. 742–3.

108 Cadogan to F.O., 21 May 1949, FO371/79710, Z3843/1054/41; Russell minute, 20 May. Mallet minute, 24 May 1949, FO371/ 79711, Z4321/1054/41.

109 A lengthy Foreign Office investigation revealed that the conversations in question could in no sense be interpreted in the transactional significance given them by Franco. See Russell minute 20 May, I. Mallet minute, 24 May, & I.

Mallet to Howard, 16 June 1949, all in FO371/79711, Z4321/1054/41, & Maltby (P.S. to Eden) to Reddaway, 23 June 1949, FO371/79711, Z4604/ 1054/41.

110 Franco, *Textos, 1945–1950*, pp. 147–73.

111 Portero, *Franco aislado*, pp. 343–7.

112 *The Times*, 22 July 1949.

113 Suárez Fernández, *Franco*, IV, p. 365.

114 Gil Robles, diary entries for 25, 26 September 1949, *La monarquía*, pp. 304–6.

115 *FRUS 1949*, IV, p. 761.

116 Benjamin Welles, *Spain: The Gentle Anarchy* (London, 1965) pp. 286–7; Lowi, 'Bases', p. 692.

117 Cava Mesa, *Lequerica*, pp. 310–12; Lowi, 'Bases', pp. 677–80; Whitaker, *Spain*, pp. 23, 36–7; Viñas, *Los pactos*, pp. 43–4; Suárez Fernández, *Franco*, IV, pp. 366–7; Garriga, *De la División Azul*, pp. 548–9, 563–7.

118 *FRUS 1949*, IV, p. 761.

119 *The Times*, 22, 24, 25, 28 October 1949; *Arriba*, 22, 23, 25, 26, 27 October 1949; Gil Robles, diary entries for 19, 21, 23 October 1949, *La monarquía*, pp. 308–12; Franco Salgado-Araujo, *Mi vida*, pp. 327–8.

120 Gil Robles, diary entry for 17 January 1950, *La monarquía*, pp. 318–19.

121 469 *H.C.DEB.* 5s, c.2225.

122 Qasim Bin Ahmad, *The British Government*, pp. 177–82.

123 Lowi, 'Bases', p. 683.

124 *FRUS 1950*, III, pp. 1549–55;
Franks to F.O., 19 January 1950,
FO371/89496, WS10345/3,
WS10345/4; Hoyer-Miller to
Bevin, 13 February 1950,
FO371/89496, WS10345/13.

125 Hankey to Shuckburgh, 25
January 1950, FO371/89496,
WS10345/9.

126 *Arriba*, 24, 25 January
1950.

127 Hankey to Bevin, 1 February
1950, FO371/89480, W1013/3.

128 *Arriba*, 19 February 1950;
Madrid Embassy to F.O., 19
February 1950, FO371/89487,
WS1021/8.

129 Hankey to F.O., 1 February
1950, FO371/89480,
W1013/4.

CHAPTER XXIII

1 Preston, *Politics of Revenge*,
pp. 141–2.

2 Hankey to Russell, 1 February
1950, FO371/89484,
WS1017/1.

3 Sánchez Soler, *Villaverde*,
pp. 36–7, 52–3.

4 Madrid Embassy to F.O., 16
April 1950, FO371/89487,
WS1021/15.

5 *La Vanguardia Española*, 11, 12
April 1950; *Arriba*, 11, 14 April
1950; *ABC*, 11, 12 April 1950;
testimony to the author of
Fernando Serrano-Suñer Polo,
cousin of Nenuca and a guest at
the wedding. See also Garriga,
La Señora, pp. 222–4.

6 Franco Salgado-Araujo, *Mis
conversaciones*, pp. 9, 17–18, 189;
Franco, *Nosotros*, pp. 144–6,
215–20; Garriga, *La Señora*,
pp. 224–7, 243–4; Sánchez
Soler, *Villaverde*, pp. 56–70,
76, 110–16.

7 Garriga, *Franco-Serrano Suñer*,
p. 179.

8 Franco Salgado-Araujo, *Mis
conversaciones*, p. 195; Peñafiel, *El
General*, pp. 140–1; Vaca de
Osma, *Paisajes*, p. 189; Pilar
Jaraiz Franco, *Historia de una
disidencia* (Barcelona, 1981)
p. 41.

9 Garriga, *La Señora*, pp. 225–7;
Fernández, *Franco*, pp. 271–3.

10 *FRUS 1950*, III, pp. 1557–60.

11 27 March 1950, 473 H.C.DEB,
5s, cc.40–1.

12 Suárez Fernández, *Franco*, IV,
pp. 408–9.

13 Jakim Boor, *Masonería*,
pp. 137–41.

14 Viñas, *Los pactos*, p. 59.

15 Jakim Boor, *Masonería*, 121–9;
Suárez Fernández, *Franco*, IV,
pp. 431–3.

16 *FRUS 1950*, III, pp. 1563–7.

17 David Caute, *The Great Fear:
The Anti-Communist Purge under
Truman and Eisenhower*
(London, 1978) pp. 58–69,
566–7.

18 Memorandum of Chairman of
the Joint Chiefs of Staff to the
Secretary of Defense, 3 May,
Truman to Acheson, 16 June
1950, *FRUS 1950*, III,
pp. 1560–2.

19 Younger to Bevin, 3 August,
FO371/89502, WS1031/39;
W. I. Mallet to Bevin, 2 August,
FO371/89502, WS1051/39;
W. I. Mallet to Hankey, 11
September 1950, FO371/89503,
WS1051/63.

20 Burrows (Washington) to

Young, 4 September 1950, FO371/89503, WS1051/58.

21 Jebb to F.O., 17 September 1950, FO371/89503, WS1051/69.

22 *FRUS 1950*, III, pp. 1574–6.

23 Hankey to Attlee, 9 October 1950, FO371/89504, WS1051/82.

24 Bessborough memorandum, 11 October 1950, with Hankey to Bevin, 18 October 1950, FO371/89504, WS1051/86. A slightly fuller account, Bessborough, 'Interview with General Franco' in Rio Tinto Zinc Archives. I am grateful to Antonio Gómez Mendoza for providing me with a copy of this latter document.

25 A. J. Lleonart y Anselem, *España y ONU IV (1950). La "cuestión española"* (Madrid, 1991) pp. 215–62.

26 Hankey to Bevin, 8 November 1950, FO371/89506, WS1051/115; Lowi, 'Bases', p. 688.

27 Lleonart, *España y ONU*, IV, pp. 269–310; Qasim Bin Ahmad, *The British Government*, pp. 191–5.

28 *Arriba*, 5 November 1950.

29 Suárez Fernández, *Franco*, IV, p. 448, V, p. 24; Viñas, *Los pactos*, pp. 31–2.

30 *FRUS 1950* (Washington, 1977) III, p. 1556.

31 *FRUS 1949*, IV, p. 755.

32 *ABC*, 7 November 1950.

33 Hankey to Bevin, 6 December 1950, FO371/89523, WS1082/4; Hankey to F.O., 10, 16 December 1950, W. I. Mallet minute, 14 December 1940, FO371/89508, WS1051/159;

W. I. Mallet minute, 19 December 1950, Hankey to Young, 20 December 1950, FO371/89509, WS1051/170; Young minute, 5 January 1951, FO371/96173, WS1051/6.

34 *Arriba*, 25 (Franco), 26 November, 3 (Carrero Blanco), 10 December 1950; *ABC*, 1 December 1950; Franco, *Textos, 1945–50*, pp. 325–7; Young minute, 5 January 1951, FO371/96194, WS1081/4.

35 *FRUS 1950*, III, pp. 1573–4.

36 Franks to F.O., 18 November 1950, FO371/89507, WS1051/129.

37 Viñas, *Los Pactos*, pp. 59–60; Garriga, *De la División Azul*, pp. 577–85.

38 Francisco Franco, *Discursos y mensajes del Jefe del Estado 1951–1954* (Madrid, 1955) pp. 7–20.

39 Lequerica to Franco, 25 October 1950, Suárez Fernández, *Franco*, IV, pp. 440–1.

40 Forrestal, *Diary*, p. 445.

41 Stanton Griffis, *Lying in State* (New York, 1952) pp. 73–96, 107–8, 249; Cava Mesa, *Lequerica*, pp. 319–21.

42 Hankey to Young, 15 November 1950, FO371/89507, WS1051/132.

43 *The Times*, 1 February 1951; Suárez Fernández, *Franco*, V, pp. 7–9.

44 Griffis, *Lying in State*, pp. 284–5; Qasim Bin Ahmad, *The British Government*, pp. 350–2.

45 Suárez Fernández, *Franco*, IV, p. 413.

46 Franco, *Discursos y mensajes 1951–1954*, pp. 33–7; Trythall, *Franco*, p. 211.

47 Suárez Fernández, *Franco*, V, p. 11; Garriga, *De la División Azul*, pp. 616–18.

48 *ABC*, 2 March 1951; Griffis, *Lying in State*, pp. 283–4.

49 Balfour to F.O., 4 March 1951, FO371/96172, WS10345/1 & WS10345/2.

50 Balfour to F.O., 15 March 1951, FO371/96174, WS1051/34; *The Times*, 16 March 1951; Griffis, *Lying in State*, pp. 284–5.

51 Suárez Fernández, *Franco*, V, pp. 11–12.

52 484 H.C.DEB, 5s, cc.1249–58, 20 February 1951; Suárez Fernández, *Franco*, V, p. 12.

53 Qasim Bin Ahmad, *The British Government*, pp. 217–38; Jones, *Spanish Question*, pp. 195–211, 215–26.

54 Hankey to Young, 20 December 1950, FO371/89509, WS1051/170.

55 Sebastian Balfour, *Dictatorship, Workers, and the City: Labour in Greater Barcelona since 1939* (Oxford, 1989) pp. 20–2.

56 Franco, *Discursos 1951–1954*, pp. 43–8.

57 *La Vanguardia Española*, 3 March 1951.

58 Balfour to F.O., 12 March 1951, FO371/96156, WS1016/17; Balfour to Morrison, 21 March 1951, and enclosure, report from British Consulate, 14 March 1951, FO371/96156, WS1016/19; *La Vanguardia Española*, 13, 14 March 1951; Félix Fanés, *La vaga de tramvies del 1951* (Barcelona, 1977)

pp. 50–157; Balfour, *Dictatorship*, pp. 22–30; Franco Salgado Araujo, *Mi vida*, pp. 329–30. On *Vieja Guardia*, see Emilio Romero, *Los papeles reservados* 2 vols (Barcelona, 1985) I, pp. 119–30.

59 Garriga, *De la División Azul*, pp. 595–6.

60 Garriga, *De la División Azul*, p. 600.

61 Hankey to Bevin, 12 March 1951, FO371/96156, WS1016/16.

62 Griffis, *Lying in State*, pp. 269–70, 287–9; Balfour to Young, 28 March 1951, FO371/96183, WS1071/36; Viñas, *Los pactos*, pp. 73–9.

63 Balfour to F.O., 24 April 1951, FO371/96157, WS1016/29; Tewson (TUC) to Morrison, 3 May 1951, FO371/96157, WS1016/38; Balfour to Morrison, 16 May 1951, FO371/96158, WS1016/41; Manuel González Portilla & José María Garmendía, *La posguerra en el País Vasco: política, acumulación, miseria* (San Sebastián, 1988) pp. 275–84.

64 Balfour to Morrison, 23 May 1951, FO371/96158, WS1016/56/51.

65 Franco, *Discursos 1951–1954*, pp. 50–1, 57.

66 *Arriba*, 13, 15, 19 (Carrero Blanco) May 1951.

67 Balfour to Bevin, 8 March 1951, FO371/96174, WS1051/27; Franco, interviewed by Ward Price, *The Daily Mail*, 29 May 1951.

68 López Rodó, *La larga marcha*, pp. 112–13; 550–4.

69 Viñas, *Los pactos*, pp. 87–90.

70 Viñas, *Los pactos*, pp. 90–1.

71 F.O. to Washington, 7 July, Francks to F.O., 12 July 1951, FO371/96185, WS1071/69G & WS1071/71.

72 Welles, *Gentle Anarchy*, p. 287; Lowi, 'Bases', p. 692.

73 Antonio Marquina Barrio, *España en la política de seguridad occidental 1939–1986* (Madrid, 1986) pp. 420–22; Viñas, *Los pactos*, pp. 92–4.

74 *ABC*, 17 July 1951; *The Times*, 19 July 1951.

75 Viñas, *Los pactos*, pp. 95–102; Griffis, *Lying in State*, pp. 294–5; Marquina, *España*, pp. 422–4; Lowi, 'Bases', pp. 692–5; Cava Mesa, *Lequerica*, pp. 323–5.

76 *ABC*, 17 July 1951; *Ya*, 17 July 1951.

77 Perowne to Younger, 16 June, 11 July 1950, FO371/89483, WS1016/31, WS1016/37; Tusell, *Franco y los católicos*, pp. 220–5, 250–1.

78 Pemán, diary entry for 20 January 1951, quoted by Tusell, *Franco y los católicos*, p. 287.

79 Equipo Mundo, *Los 90 ministros de Franco*, pp. 203–46; Garriga, *De la División Azul*, pp. 624–9.

80 Garriga, *Los validos*, p. 274.

81 Garriga, *Los validos*, pp. 122–4; Jesús Ynfante, *La prodigiosa aventura del Opus Dei: Génesis y desarrollo de la Santa Mafia* (Paris, 1970) pp. 177–8.

82 Hankey to Bevin, 12 March 1950, FO371/9615, WS1016/16.

83 Viñas et al., *Política comercial*, I, pp. 635–9, 671–2.

84 *Newsweek*, 27 August 1951;

85 Balfour to Eden, 12 November 1951, FO371/96179, WS1051/90.

86 *The Sunday Times*, 25 November 1951; Franco, *Discursos 1951–1954*, pp. 107–8.

87 Balfour to Young, 28 November 1951, Balfour to Harrison, 11 December 1951, FO371/96192, WS1071/203; FO371/96179, WS1051/104.

88 Balfour to Eden, 10 December 1951, FO371/96194, WS1081/31.

89 Griffis, *Lying in State*, pp. 297–9.

90 Cava Mesa, *Lequerica*, p. 333.

91 Lowi, 'Bases', p. 694; Viñas, *Los pactos*, p. 141.

92 Griffis, *Lying in State*, pp. 302, 307.

93 Lowi, 'Bases', pp. 694–6; Viñas, *Los pactos*, pp. 125–33.

94 Balfour to Eden, 10 July 1952, Steel to Cheetham, 22 August, Murray to Cheetham, 28 August 1952, FO371/1020222, WS1102/21, WS1102/24G, WS1102/25; Viñas, *Los pactos*, pp. 120–1, 177; Franco Salgado-Araujo, *Mis conversaciones*, p. 56.

95 Viñas, *Los pactos*, pp. 143–4; Garriga, *Segura*, pp. 301–4; Suárez Fernández, *Franco*, V, pp. 90–2.

96 Viñas, *Los pactos*, pp. 145–7.

97 Balfour to Eden, 29 May 1952, FO371/102000, WS1015/2; Franco, *Discursos 1951–1954*, p. 173.

98 Eden to Tewson, 26 March 1952, FO371/102002,

WS1016/25; Madrid Embassy to F.O., 27 March, 3 April 1952, FO317/102004, WS1016/69, FO371/101999, WS1013/5; Balfour to F.O., 1 April 1952, FO317/102004, WS1016/73; McClelland to Hepple, 9 July 1952, FO371/102006, WS1016/115; Murray to Cheetham, 18 September 1952, FO371/102007, WS1016/129.

99 *La Vanguardia Española*, 29 May 1952.

100 *La Vanguardia Española*, 31 May, 3 June 1952.

101 José Chao Rego, *La Iglesia en el franquismo* (Madrid, 1976) pp. 88–90; Franco, *Discursos 1951–1954*, pp. 199–200.

102 Franco, *Discursos 1951–1954*, pp. 221–4.

103 Franco, *Discursos 1951–1954*, pp. 275–80.

104 Lowi, 'Bases', pp. 696–7.

105 Makins (Washington) to Eden, 18 February 1953, FO371/107687, WS1073/1.

106 Viñas, *Los pactos*, pp. 165–9, 183–93, 252; Marquina, *España*, pp. 498–554.

107 Balfour to Cheetham, 29 April, 24 May, Bellotti to F.O., 26 June 1953, FO371/107682, WS1051/9 & WS1051/19, FO371/107686, WS10/2/6.

108 Franco, *Discursos 1951–1954*, pp. 360–4.

109 Balfour to Cheetham, 31 July, Balfour to Young, 11 December 1953, FO371/107682, WS1051/24, FO371/107690, WS1081/50.

110 Young memorandum, 8 December, Harrison to Minister of State, 17 December, Balfour memorandum, 17 December 1953, FO371/107682, WS1051/38.

111 Balfour to Eden, 21 March 1953, FO371/107731, WS1782/2.

112 Tusell, *Franco y los Católicos*, pp. 258–82; Guy Hermet, *Les Catholiques dans l'Espagne Franquiste* 2 vols (Paris, 1980–1) II, pp. 204–18; Joaquín L. Ortega, 'La Iglesia española desde 1939 hasta 1976' in Ricardo García-Villoslada, editor, *Historia de la Iglesia en España* Vol. V (Madrid, 1979) pp. 671–8.

113 José Angel Tello, *Ideología política: la Iglesia católica española (1936–1959)* (Zaragoza, 1984) pp. 111–16.

114 Chao Rego, *La Iglesia*, pp. 93–102; Norman B. Cooper, *Catholicism and the Franco Regime* (Beverly Hills, 1975) pp. 16–18; Rafael Gómez Pérez, *El franquismo y la Iglesia* (Madrid, 1986) pp. 66–70; Feliciano Blázquez, *La traición de los clérigos en la España de Franco: crónica de una intolerancia (1936–1975)* (Madrid, 1991) pp. 103–5.

115 Garriga, *Segura*, pp. 311–20.

116 Lowi, 'Bases', pp. 696–7; Viñas, *Los pactos*, p. 252.

117 Viñas, *Los pactos*, pp. 165–9, 183–93.

118 Hood minute, 5 November 1953, FO371/107686, WS1072/43; Viñas, *Los pactos*, pp. 195–202, 313–14.

119 De la Cierva, *Franquismo*, II, p. 114.

120 José María de Areilza, *Diario de un ministro de la Monarquía* (Barcelona, 1977) p. 45.

121 *ABC*, 27, 29, 30 September 1953; *Arriba*, 27, 29, 30 September 1953.

122 Viñas, *Los pactos*, pp. 181–2, 203–50, 292; Lowi, 'Bases', pp. 697–8.

123 Viñas, *Los pactos*, pp. 261–75.

124 Viñas, *Los pactos*, pp. 263–4; Viñas et al., *Política comercial*, I, pp. 497–501, 532–45.

CHAPTER XXIV

1 Franco, *Discursos 1951–1954*, pp. 376–84.

2 *Arriba*, 1 October; *The Times*, 1 October 1953; Viñas, *Los pactos*, pp. 277–84.

3 *ABC*, 27, 30 September 1953; *Arriba*, 27, 29 September 1953; Viñas, *Los pactos*, pp. 277–84.

4 Matthews, *The Yoke*, p. 108.

5 Franco, *Discursos 1951–1954*, pp. 388–409.

6 *The Times*, 30 October 1953; Créac'h, *Le coeur*, pp. 319–20.

7 Franco, *Discursos 1951–1954*, pp. 414–15.

8 Viñas, *Los pactos*, pp. 315–16.

9 Soriano, *La mano*, p. 70.

10 Antonio Téllez, *La guerrilla urbana. 1: Facerias* (Paris, 1974) pp. 101–262; Antonio Téllez, *La guerrilla urbana en España: Sabaté* (Paris, 1973) pp. 65–113.

11 Sánchez Soler, *Villaverde*, pp. 63–8; Franco Salgado-Araujo, *Mis conversaciones*, pp. 9, 90–2, 111, 132.

12 Franco Salgado-Araujo, *Mis conversaciones*, pp. 50, 80.

13 Franco Salgado-Araujo, *Mis conversaciones*, pp. 159–60, 395.

14 See the interviews with Máximo Rodríguez Borrell recounted in Mérida, *Testigos*, pp. 217–27 and S. F. A. Coles, *Franco of Spain* (London, 1955) pp. 13–15. Vaca de Osma, *Paisajes*, p. 186.

15 Balfour to Eden, 3 February 1954, FO371/113041, WS1081/44; Pablo Lizcano, *La generación del 56: La Universidad contra Franco* (Barcelona, 1981) pp. 95–99; Pedro Laín Entralgo, *Descargo de conciencia* (Barcelona, 1976) pp. 404–5.

16 Makins (Washington) to Eden, 18 February 1954, FO371/113042, WS1081/61; Balfour to Eden, 19 May 1954, FO371/113043, WS11081/92; *Arriba*, 16 May 1954.

17 Minister of State's interview with General Franco, enclosure to Mallet to Ward, 25 April 1955, FO371/117870, WS1051/22G.

18 Franco Salgado-Araujo, *Mi vida*, pp. 337–8.

19 Young minute, 4 August 1955, FO371/117870, RS1051/23; F.O. to Madrid, 27 October 1955, FO371/117872, RS1051/32; Mallet to Macmillan, 5 November 1955, FO371/117872, RS1051/40.

20 Scott to Hood, 1 April, Madrid to F.O., 21 April, Pilcher to Laskey, 31 May, Steel to Macmillan, 5 July 1955, FO371/117873, RS1071/3, RS1071/5, RS1071/8, RS1071/19.

21 On the outfit, see 'Brass Sombreros', *The Sunday Times*,

25 July 1954; Pozuelo, *Los 476 últimos días*, p. 166, and the paintings reproduced between pp. 176–7, and in *Interviu*, No. 39, 21–27 December 1983.

22 Franco, *Discursos 1951–1954*, pp. 456–67.

23 Enrique Tierno Galván, *Cabos sueltos* (Barcelona, 1981) pp. 180–4.

24 Sueiro, *Valle de los Caídos*, pp. 123–43.

25 Gil Robles, diary entries for 13 May, 21, 22 June, 25 July, 7 September 1954; Don Juan to Franco, 16 July, Franco to Don Juan, 17, 20 July 1954, *La monarquía*, pp. 327–8, 411–18; López Rodó, *La larga marcha*, pp. 115–17, 554–5.

26 Rafael Calvo Serer, *Franco frente al Rey: el proceso del régimen* (Paris, 1972) pp. 29–30; Créac'h, *Le coeur*, pp. 317–18.

27 Créac'h, *Le coeur*, p. 332.

28 *ABC*, 20 October 1954; Créac'h, *Le coeur*, pp. 335–7; Suárez Fernández, *Franco*, V, p. 157.

29 Franco Salgado-Araujo, *Mis conversaciones*, pp. 18, 23.

30 *Arriba*, 23 November 1954; Calvo Serer, *Franco frente al Rey*, pp. 29–30; Franco Salgado-Araujo, *Mis conversaciones*, p. 30; López Rodó, *La larga marcha*, p. 117; Créac'h, *Le coeur*, pp. 338–9; Toquero, *Franco y Don Juan*, pp. 253–5.

31 Créac'h, *Le coeur*, pp. 339–40; Suárez Fernández, *Franco*, V, p. 159.

32 Franco to Don Juan, 2 December 1954, Sainz Rodríguez, *Un reinado*, pp. 383–4.

33 Franco Salgado-Araujo, *Mis conversaciones*, pp. 45, 48, 65.

34 Franco Salgado-Araujo, *Mis conversaciones*, pp. 52–3.

35 Mallet to Eden, 11 January 1955, FO371/117914, RS1942/4; Sainz Rodríguez, *Un reinado*, pp. 222–35; Créac'h, *Le coeur*, pp. 341–5.

36 Stirling to Macmillan, 19 April 1955, FO371/117914, RS1942/15; Franco Salgado-Araujo, *Mis conversaciones*, pp. 59–64; Pemán, *Mis encuentros*, p. 232; Garriga, *La Señora*, pp. 236–7; Tusell, *La oposición*, pp. 235–6.

37 Franco Salgado-Araujo, *Mis conversaciones*, p. 63.

38 Franco, *Discursos 1951–1954*, pp. 551–3.

39 *Arriba*, 23, 27 January 1955; *ABC*, 1 March 1955.

40 Mallet to Eden, 26 January 1955, FO371/117914, RS1942/6.

41 Franco Salgado-Araujo, *Mis conversaciones*, pp. 89–90.

42 *Arriba*, 20 June 1955; Mallet to Macmillan, 5 July 1955, FO371/117914, RS1942/21.

43 *ABC*, 24 June 1955; Mallet to Macmillan, 5 July 1955; Stirling to Macmillan, 26 July 1955; Balfour memorandum, 7 September 1955, FO371/117914, RS1942/21, RS1942/25, RS1942/27; Créac'h, *Le coeur*, pp. 353–4.

44 Salva Miquel & Vicente, *Franco*, pp. 293–7; Franco Salgado-Araujo, *Mis conversaciones*, pp. 84–5.

45 Franco Salgado-Araujo, *Mis conversaciones*, pp. 23, 32–3, 71–2; Carlos Rein Segura

interview in Bayod, *Franco visto*, p. 78.

46 Franco Salgado-Araujo, *Mis conversaciones*, pp. 32–3, 36–7, 126; Garriga, *La Señora*, pp. 249–53. For an analysis of Franco's passion for hunting, see González Duro, *Franco*, pp. 314–16.

47 Franco, *Discursos 1955–1959*, pp. 88–9.

48 Créac'h, *Le coeur*, p. 394.

49 Franco Salgado-Araujo, *Mis conversaciones*, pp. 125, 127–8, 132–5.

50 Manuel Fraga Iribarne, *Memoria breve de una vida pública* (Barcelona, 1980) p. 287; Franco Salgado Araujo, diary entry for 30 October 1954, *Mis conversaciones*, p. 22. The remark was made when Valiño was Captain-General of Valladolid.

51 Miguel Martín, *El colonialismo español en Marruecos (1860–1956)* (Paris, 1973) pp. 219–23.

52 Suárez Fernández, *Franco*, V, pp. 176–80, 183–5, 192–207; Franco Salgado Araujo, *Mis conversaciones*, pp. 110, 116–17, 168, 223.

53 Franco Salgado-Araujo, Mi vida, p. 27.

54 *The Times*, 5 April 1956; Martín, *El colonialismo*, pp. 227–39; Whitaker, *Spain*, pp. 328–9; Suárez Fernández, *Franco*, V, pp. 193–205; La Cierva, *Historia*, II, pp. 138, 146; Franco Salgado-Araujo, *Mis conversaciones*, pp. 170–3.

55 Chancery (Madrid) to Southern Department, 18 February 1955, FO371/117914, RS1942/10.

56 Tusell, *Franco y los católicos*, pp. 375–8.

57 Calvo Serer, *Franco frente al Rey*, p. 14; Franco Salgado-Araujo, *Mis conversaciones*, pp. 146–7; Créac'h, *Le coeur*, p. 358.

58 Laín Entralgo, *Descargo*, pp. 414–18; Créac'h, *Le coeur*, pp. 359–60. The reports are printed in Roberto Mesa, ed., *Jaraneros y alborotadores: documentos sobre los sucesos estudiantiles de febrero de 1956 en la Universidad Complutense de Madrid* (Madrid, 1982) pp. 45–53, 58–64.

59 Washington Chancery to Madrid Chancery, 14 January 1956, FO371/124127, RS1015/4.

60 Francisco Franco, *Discursos y mensajes del Jefe del Estado 1955–1959* (Madrid, 1960) p. 136.

61 Mallet to Macmillan, 10 January, Mallet to Lloyd, 17 January 1956, FO371/124127, RS1015/3, RS1015/6.

62 Mallet to Lloyd, 17 January 1956, FO371/124127, RS1015/2.

63 Mallet to Ward, 27 January 1956, FO371/124127, RS1015/10; *The Times*, 5 October 1955.

64 Mesa, ed., *Jaraneros*, pp. 109–12; Laín Entralgo, *Descargo*, pp. 418–23; Lizcano, *La Universidad*, p. 142; Franco Salgado Araujo, *Mis conversaciones*, pp. 163–4; Tusell, *Franco y los católicos*, p. 382; Franco Salgado-Araujo, *Mi vida*, p. 343; Créac'h, *Le coeur*, pp. 362–3.

65 Pilcher to Young, 12 October 1956, FO371/124128, RS1015/43; Payne, *Politics and the Military*, p. 443.

66 *Arriba*, 9, 10 February 1956; Mallet to Lloyd, 11 February 1956, FO371/124127, RS1015/12.

67 Testimony to the author of Rafael Calvo Serer, London, 1975; Créac'h, *Le coeur*, pp. 364–5.

68 Créac'h, *Le coeur*, pp. 364–5; Tusell, *Franco y los Católicos*, pp. 382–3; Franco Salgado-Araujo, *Mi vida*, p. 343.

69 Franco Salgado-Araujo, *Mis conversaciones*, p. 159.

70 Tusell, *Franco y los Católicos*, pp. 383–4.

71 Fernández Cuesta, *Testimonio*, pp. 241–5.

72 Arrese, *Una etapa*, pp. 16–22.

73 Equipo Mundo, *Los 90 Ministros*, pp. 249–53.

74 Mallet to Lloyd, 17, 18, February 1956, FO371/124127, RS1015/13, RS1015/14.

75 Madrid Chancery to Southern Department, 24 February 1956, FO371/124127, RS1015/18.

76 Franco Salgado-Araujo, *Mis conversaciones*, p. 167.

77 'Declaraciones de S.E. a Manuel Aznar', 31 December 1938, *Palabras del Caudillo 19 Abril 1937–31 diciembre 1938*, p. 314.

78 Franco Salgado-Araujo, *Mis conversaciones*, p. 166.

79 Madrid Chancery to Southern Department, 10 March 1956, FO371/124127, RS1015/21; Arrese, *Una etapa*, pp. 34–8; Suárez Fernández, *Franco*, V, pp. 264–5; *Arriba*, 6 March 1956.

80 Arrese, *Una etapa*, pp. 32–3.

81 Mallet to Lloyd, 29 May 1956, FO371/124128, RS1015/30; Llibert Ferri, Jordi Muixí & Eduardo Sanjuan, *Las huelgas contra Franco (1939–1956)* (Barcelona, 1978) pp. 226–38; Faustino Miguélez, *La lucha de los mineros asturianos bajo el franquismo* (Barcelona, 1976) pp. 94–5.

82 Franco, *Nosotros*, pp. 147–8.

CHAPTER XXV

1 Sainz Rodríguez, *Un reinado*, p. 163; Suárez Fernández, *Franco*, V. pp. 153, 266.

2 Franco, *Discursos 1955–1959*, pp. 158–9.

3 Arrese, *Una etapa*, p. 104.

4 Franco, *Discursos 1955–1959*, pp. 163–5.

5 Arrese, *Una etapa*, pp. 42–5; Franco, *Discursos 1955–1959*, pp. 181–90; Madrid Chancery to Southern Department, 5 May 1954, FO371/124128, RS1015/23.

6 Arrese, *Una etapa*, pp. 64, 66.

7 Arrese, *Una etapa*, pp. 45–8.

8 Mallet to Young, 21 June 1956, FO371/124128, RS1015/34.

9 Laureano López Rodó, *Memorias* (Barcelona, 1990) pp. 51–2; López Rodó, *La larga marcha*, pp. 124–30; Arrese, *Una etapa*, pp. 71, 80.

10 Arrese, *Una etapa*, p. 81; López Rodó, *Memorias*, pp. 58–9.

11 Payne, *Military*, pp. 443; Whitaker, *Spain*, pp. 141–2.

12   Payne, *Military*, p. 534.

13   Arrese, *Una etapa*, pp. 82–3.

14   Arrese, *Una etapa*, pp. 87, 235.

15   Arrese, *Una etapa*, pp. 86–93.

16   Arrese, *Una etapa*, pp. 32–3.

17   Arrese, *Una etapa*, pp. 98–103.

18   Franco, *Discursos 1955–1959*, pp. 214–15; Mallet to Lloyd, 20 July 1956, FO371/124128, RS1015/39A.

19   Franco, *Discursos 1955–1959*, pp. 201–30; Arrese, *Una etapa*, pp. 102–4.

20   Franco Salgado-Araujo, *Mis conversaciones*, p. 67.

21   Franco, *Discursos 1955–1959*, pp. 201–30; Suárez Fernández, *Franco*, V, p. 293; Arrese, *Una etapa*, pp. 102–4.

22   Arrese, *Una etapa*, pp. 110–11.

23   Madrid Chancery to Southern Department, 22 March 1957, FO371/130345, RS1071/6; Arrese, *Una etapa*, pp. 122–3.

24   Ramón Garriga, *La Señora de El Pardo* (Barcelona, 1979) p. 11.

25   Franco Salgado-Araujo, *Mis conversaciones*, p. 174–6, 178–9.

26   *The Times*, 1 October 1956; Franco, *Discursos 1955–1959*, pp. 233–8; José Luis de Arrese, *Treinta años de política* (Madrid, 1966) pp. 1146–52; Arrese, *Una etapa*, pp. 124–31; Madrid Chancery to Southern Department, 6 October 1956, FO371/124128, RS1015/42; López Rodó, *Memorias*, pp. 64–5; López Rodó, *La larga marcha*, pp. 132–3.

27   Arrese, *Una etapa*, pp. 132–5, 144–92; López Rodó, *Memorias*, pp. 65–77; López Rodó, *La larga marcha*, pp. 133–5; Créac'h, *Le coeur*, pp. 386–7.

28   Tusell, *Franco y los católicos*, pp. 409–25; Suárez Fernández, *Franco*, V, pp. 306–12.

29   Arrese, *Una etapa*, pp. 212–18.

30   Mallet to Lloyd, 15 January 1957, FO371/130325, RS1015/3; Arrese, *Una etapa*, pp. 234–42, 253–65; Suárez Fernández, *Franco*, V, pp. 314–15; Tusell, *Franco y los católicos*, pp. 426–8.

31   Laureano López Rodó, *Memorias* (Barcelona, 1990) pp. 66–9; López Rodó, *La larga marcha*, pp. 120–1.

32   An opinion held by Calvo Serer, testimony to the author, London, 1975.

33   Franco Salgado-Araujo, *Mis conversaciones*, p. 184.

34   Calvo Serer, *Franco frente al Rey*, p. 36; Sainz Rodríguez, *Un reinado*, p. 164; López Rodó, *La larga marcha*, pp. 123–4; Toquero, *Franco y Don Juan*, p. 266; Suárez Fernández, *Franco*, V, pp. 319–20.

35   Madrid Chancery to Southern Department, FO371/130325, RS1015/5; Créac'h, *Le coeur*, pp. 387–8.

36   Luis Ramírez, *Nuestros primeros veinticinco años* (Paris, 1964) pp. 111–12; Franco Salgado-Araujo, *Mis conversaciones*, p. 200; Jaume Fabre, Josep M. Huertas & Antoni Ribas, *Vint anys de reistència catalana (1939–1959)* (Barcelona, 1978) pp. 208–11.

37   Arrese, *Una etapa*, p. 234, 244; Tusell, *Franco y los católicos*, pp. 428–9.

38 Franco Salgado Araujo, *Mis conversaciones*, pp. 176, 195–8; Suárez Fernández, *Franco*, V, pp. 269, 319; López Rodó, *La larga marcha*, p. 124; Sainz Rodríguez, *Un reinado,* p. 166.

39 Calvo Serer, *Franco*, p. 37; La Cierva, *Franquismo*, II, p. 155 who quotes, without naming him, a minister.

40 Franco Salgado-Araujo, diary entry for 6 April 1957, *Mis conversaciones*, p. 209 comments on the rumours having been picked up by the Cuban press. For the more outlandish versions of what happened, see Ramírez, *25 años*, p. 117 and Busquets, *Pronunciamientos*, pp. 140–1; cf. Serrano Suñer, *Memorias*, p. 238.

41 Franco Salgado-Araujo, diary entry for 17 May 1955, *Mis conversaciones*, pp. 107–10.

42 Sainz Rodríguez, *Un reinado*, p. 166.

43 Franco Salgado-Araujo, *Mis conversaciones*, p. 198.

44 Mallet to Lloyd, 16 January 1957, FO371/130349, RS1106/1; Créac'h, *Le coeur*, pp. 369–72.

45 Franco Salgado-Araujo, *Mis conversaciones*, p. 191.

46 Joseph Harrison, 'Towards the liberalization of the Spanish economy, 1951–1959' in Colin Holmes & Alan Booth, editors, *Economy and Society: European Industrialization and Its Social Consequences* (Leicester, 1990) pp. 102–15.

47 López Rodó, *Memorias*, pp. 89–99; Suárez Fernández, *Franco*, V, pp. 320–1.

48 Arrese, *Una etapa*, pp. 282–3.

49 Mallet to Lloyd, 1 March 1957, FO371/130325, RS1015/9; López Rodó, *Memorias*, pp. 93–4; Welles, *Gentle Anarchy*, p. 127; Arrese, *Una etapa*, p. 281.

50 Arrese, *Una etapa*, pp. 275–82; López Rodó, *Memorias*, pp. 92–3; Payne, *Falange*, pp. 261–2.

51 Franco Salgado-Araujo, *Mis conversaciones*, p. 201.

52 Payne, *Franco Regime*, p. 453.

53 Garriga, *Los validos*, pp. 207–8; Franco Salgado-Araujo, *Mis conversaciones*, p. 201.

54 Franco Salgado-Araujo, diary entries for 23 February, 15 July 1957, *Mis conversaciones*, pp. 201, 246–7.

55 Tusell, *Franco y los Católicos*, pp. 429–32; Payne, *Franco Regime*, p. 452.

56 Mariano Navarro Rubio, *Mis memorias: Testimonio de una vida política truncada por el Caso MATESA* (Barcelona, 1991) pp. 64–5; Ynfante, *Santa Mafia*, pp. 233–5.

57 Daniel Artigues, *El Opus Dei en España 1928–1962: su evolución ideológica y política de los orígenes al intento de dominio* (Paris, 1971) pp. 181–95; Ynfante, *Santa mafia*, pp. 163–207. On the ACNP, see A. Sáez Alba (pseud. Alfonso Colodrón) *La otra "cosa nostra": la Asociación Católica nacional de Propagandistas y el caso de El Correo de Andalucía* (Paris, 1974) pp. lxxiii–lxxxii.

58 Ynfante, *Santa mafia*, pp. 178–9; Payne, *Falange*, pp. 262–3.

59 López Rodó, *Memorias*, p. 66; Ynfante, *Santa Mafia*, pp. 177–8.

60 Navarro Rubio, *Mis memorias*, pp. 59–74.

61 López Rodó, *Memorias*, p. 91; Navarro Rubio, *Mis memorias*, p. 74; Interview of Alberto Ullastres, *Diario 16, Historia del Franquismo*, ed. Justino Sinova, 2 vols (Madrid, 1985) II, p. 471.

62 Navarro Rubio, *Mis memorias*, pp. 78–9.

63 Navarro Rubio, *Mis memorias*, p. 240.

64 López Rodó, *Memorias*, pp. 80–8, 96–108; Artigues, *Opus Dei*, pp. 185–7.

65 Manuel Jesús González, *La economía política del franquismo (1940–1970): Dirigismo, mercado y planificación* (Madrid, 1979) pp. 134–7.

66 Franco Salgado-Araujo, *Mis conversaciones*, pp. 203, 228.

67 Toquero, *Franco y Don Juan*, p. 267.

68 López Rodó, *Memorias*, pp. 105–6; Ynfante, *Santa Mafia*, pp. 178–9.

69 Calvo Serer, *Franco frente al Rey*, pp. 88–9.

70 Mariano Navarro Rubio, 'La batalla de la estabilización' in *Anales de la Real Academia de Ciencias Morales y Políticas*, No. 53, 1976, pp. 175–8; Suárez Fernández, *Franco*, VI, p. 8.

71 Navarro Rubio, *Mis memorias*, p. 78.

72 López Rodó, *La larga marcha*, pp. 145–8; Toquero, *Franco y Don Juan*, pp. 267–70.

73 Rafael Casas de la Vega, *La última guerra de Africa (campaña de Ifni-Sáhara* (Madrid, 1985) pp. 41–105; Suárez Fernández, *Franco*, VI, pp. 25–7.

74 Casas de la Vega, *La última guerra*, pp. 127–249.

75 Franco Salgado-Araujo, *Mis conversaciones*, pp. 218–23; Welles, *Gentle Anarchy*, pp. 238–44; Suárez Fernández, *Franco*, VI, pp. 28–40.

76 Mallet to FO, 28 March 1958, FO371/136711, RS2183/1; Franco Salgado-Araujo, *Mis conversaciones*, p. 228.

77 López Rodó, *Memorias*, pp. 139–44.

78 Franco Salgado-Araujo, *Mis conversaciones*, p. 236.

79 López Rodó, *Memorias*, p. 83.

80 Franco Salgado-Araujo, *Mis conversaciones*, pp. 236, 266.

81 Navarro Rubio, 'La batalla de la estabilización', pp. 178–86.

82 Franco Salgado-Araujo, *Mis conversaciones*, p. 244.

83 Franco Salgado-Araujo, *Mis conversaciones*, pp. 248, 270. On the psychological basis of this passion, see González Duro, *Franco*, pp. 313–18.

84 Franco, *Discursos 1955–1959*, pp. 557–68. Cf. Chancery to Southern Department, 2 January 1959, FO371/144927, RS1015/1.

85 Report from Bank of London & South America, Madrid, 3 March 1959, FO371/144927, RS1015/9; Toquero, *Franco y Don Juan*, pp. 297–300; Suárez Fernández, *Franco*, VI, pp. 78–82.

86 Franco Salgado-Araujo, *Mis conversaciones*, p. 259.

87 Tusell, *La oposición democrática*,

pp. 314–36, 340–57; Calvo Serer, *Franco frente al Rey*, pp. 55–8; Javier Tusell & José Calvo, *Giménez Fernández: precursor de la democracia española* (Madrid, 1990) pp. 269–80; Ellwood, *Prietas las filas*, pp. 220–8.

88　Navarro Rubio, 'La batalla de la estabilización', pp. 188–96; Mallet to FO, 5, 11 February 1959, FO371/144927, RS1015/4, FO371/144926, RS1013/1; Mallet to Selwyn Lloyd, 17 February 1959, FO371/144950, RS1102/1.

89　Navarro Rubio, 'La batalla de la estabilización', pp. 196–9; Navarro Rubio, *Mis memorias*, pp. 124–6; Report from Bank of London & South America, Madrid, 3 March 1959, FO371/144927, RS1015/9; Calvo Serer, *Franco frente al Rey*, p. 79.

90　López Rodó, *Memorias*, p. 184.

91　Interview with Navarro Rubio, Bayod, *Franco*, p. 89.

92　Calvo Serer, *Franco frente al Rey*, p. 79.

93　Navarro Rubio, 'La batalla de la estabilización', pp. 201–2.

94　Navarro Rubio, *Mis memorias*, pp. 140–1; Navarro Rubio interview in Bayod, *Franco*, p. 89.

95　Ullastres interview in Sinova, *Historia del franquismo*, II, p. 473; Arrese interview in Bayod, *Franco*, pp. 59–61.

96　Arrese interview in Bayod, *Franco*, p. 58.

97　Navarro Rubio, *Mis memorias*, pp. 141–8; Mariano Navarro Rubio, 'La batalla del desarrollo' in *Anales de la Real Academia de Ciencias Morales y Políticas*, No. 54, 1977, pp. 198, 205–7; Franco Salgado-Araujo, *Mis conversaciones*, pp. 246–7.

98　Testimony of Rafael Calvo Serer to the author, London, 1976; Cf. Calvo Serer, *Franco frente al Rey*, pp. 77–9.

99　Suárez Fernández, *Franco*, VI, p. 96.

100　Franco Salgado-Araujo, *Mis conversaciones*, p. 259.

101　Joaquín Bardavío, *La rama trágica de los Borbones* (Barcelona, 1989) pp. 68–9.

102　*The Times*, 2 April 1959.

103　Franco, *Discursos 1955–1959*, pp. 625–6.

104　*Arriba*, 2 April 1959; Franco, *Discursos 1955–1959*, pp. 593–9.

105　Sueiro, *El Valle de los Caídos*, pp. 208–9.

106　José Maravall, *El desarrollo económico y la clase obrera* (Barcelona, 1970) pp. 91–2; Javier Domínguez, *Organizaciones obreras cristianas en la oposición al franquismo (1951–1975)* (Bilbao, 1985) pp. 47–66.

107　*The Times*, 1 July 1959; Franco, *Discursos 1955–1959*, pp. 641–3.

108　Dwight D. Eisenhower, *The White House Years: Waging Peace 1956–1961* (London, 1965) p. 423.

109　Franco, *Discursos 1955–1959*, pp. 699–705; memorandum of Eisenhower–Franco conversations, Suárez Fernández, *Franco*, VI, pp. 140–52; Welles, *Gentle Anarchy*, pp. 247–52.

110 Eisenhower, *The White House Years*, pp. 509–10.
111 Franco, *Nosotros*, p. 115.
112 Franco, *Discursos 1955–1959*, pp. 707–35; López Rodó, *Memorias*, pp. 204–5.
113 Joan Clavera, Joan M. Esteban, María Antonia Monés, Antoni Montserrat & Jacint Ros Hombravella, *Capitalismo español: de la autarquía a la estabilización* 2 vols (Madrid, 1973) I, pp. 78–90; Juan Muñoz, Santiago Roldán & Angel Serrano, *La internacionalización del capital en España 1959–1977* (Madrid, 1978) pp. 17–43.
114 Navarro Rubio, *Mis memorias*, pp. 142–5, 278–9; López Rodó, *Memorias*, pp. 217–8; Arrese interview in Bayod, ed., *Franco visto*, pp. 59–61; Suárez Fernández, *Franco*, VI, p. 188.

CHAPTER XXVI

1 *The Times*, 29 March 1960; Franco Salgado-Araujo, *Mis conversaciones*, p. 279; Peñafiel, *El General*, pp. 70–2.
2 *The Times*, 30 March 1960.
3 Franco Salgado-Araujo, *Mis conversaciones*, pp. 277, 280.
4 Franco to Don Juan, 12 March 1960, reprinted in Sainz Rodríguez, *Un reinado*, pp. 400–1; *Ibid.*, pp. 236–7.
5 Mauricio Carlavilla, *Anti-España 1959: autores, cómplices y encubridores del communismo* (Madrid, 1959) pp. 117–24.
6 Toquero, *Franco y Don Juan*, pp. 280–3; Franco Salgado-

Araujo, *Mis conversaciones*, pp. 280–1.
7 *The Times*, 31 March 1960; Toquero, *Franco y Don Juan*, pp. 280–4.
8 Sainz Rodríguez, *Un reinado*, pp. 238–9; López Rodó, *Memorias*, pp. 280–4.
9 Franco Salgado-Araujo, *Mis conversaciones*, p. 286
10 Don Juan to Franco, 11 April 1960, Franco to Don Juan, 27 April 1960, reprinted in Sainz Rodríguez, *Un reinado*, pp. 402–3.
11 *La Vanguardia Española*, 1 May 1960.
12 *La Vanguardia Española*, 10 May 1960.
13 Navarro Rubio, *Mis memorias*, pp. 170–80.
14 Francisco Franco, *Discursos y mensajes del Jefe del Estado 1960–1963* (Madrid, 1964) pp. 13–46.
15 Carlos Fernández Santander, *El futbol durante la guerra civil y el franquismo* (Madrid, 1990) pp. 171–5; Duncan Shaw, *Fútbol y franquismo* (Madrid, 1987) p. 168; Suárez Fernández, *Franco*, VI, p. 201.
16 Franco Salgado-Araujo, *Mis conversaciones*, pp. 290–1.
17 Sueiro, *Valle*, pp. 223–30; Franco Salgado-Araujo, *Mis conversaciones*, pp. 302–3.
18 López Rodó, *Memorias*, pp. 257–9.
19 Suárez Fernández, *Franco*, VI, pp. 202–3.
20 Franco Salgado-Araujo, *Mis conversaciones*, pp. 307, 311–12, 324.
21 Suárez Fernández, *Franco*, VI,

pp. 261–5; Marquina, *España*, pp. 746–9; Calvo Serer, *Franco frente al Rey*, pp. 58–66.

22 Franco, *Discursos 1960–1963*, pp. 199–201.

23 Calvo Serer, *Franco frente al Rey*, p. 86.

24 Franco Salgado-Araujo, *Mis conversaciones*, pp. 317–19; López Rodó, *Memorias*, p. 271; Navarro Rubio, *Mis memorias*, pp. 223–6.

25 Franco, *Discursos 1960–1963*, pp. 207–53.

26 Don Juan to Franco, 10 July 1961, Franco to Don Juan, 22 July 1961, reprinted in full in López Rodó, *Memorias*, pp. 698–702.

27 Franco, *Discursos 1960–1963*, pp. 291–306.

28 Franco, *Discursos 1960–1963*, pp. 307–16.

29 Marquina, *España*, pp. 750, 754–9.

30 López Rodó, *La larga marcha*, pp. 189–90, 198–9.

31 Gregorio Morán, *Adolfo Suárez: historia de una ambición* (Barcelona, 1979) pp. 140–1.

32 Franco, *Discursos 1960–1963*, pp. 317–41.

33 López Rodó, *La larga marcha*, pp. 190–1.

34 Morán, *Suárez*, pp. 141, 199.

35 López Rodó, *Memorias*, pp. 262–3; López Rodó, *La larga marcha*, p. 199.

36 Morán, *Suárez*, pp. 142–3.

37 *ABC*, 26 December 1961; Gil, *Cuarenta años*, p. 131.

38 *ABC*, 27 December 1961; *The Times*, 27 December 1961; Soriano, *La mano*, pp. 14–20.

39 Soriano, *La mano*, p. 15.

40 Franco Salgado-Araujo, *Mis conversaciones*, p. 331.

41 Franco Salgado-Araujo, diary entries for 8, 15 January 1962, *Mis conversaciones*, pp. 331–2; Soriano, *La mano*, pp. 26–8, 47.

42 López Rodó, *La larga marcha*, pp. 195–6.

43 Soriano, *La mano*, pp. 29–35.

44 José María de Areilza, *Crónica de libertad* (Barcelona, 1985) pp. 36–7; López Rodó, *La larga marcha*, pp. 195–8; López Rodó, *Memorias*, pp. 301–2.

45 Franco, *Discursos 1960–1963*, pp. 357–9; López Rodó, *La larga marcha*, pp. 196–7.

46 López Rodó, *Memorias*, pp. 306–10.

47 Navarro Rubio, *Mis memorias*, pp. 227–30.

48 López Rodó, *Memorias*, pp. 307–11; López Rodó, *La larga marcha*, pp. 199–201.

49 López Rodó, *Memorias*, pp. 312–15, 538.

50 Soriano, *La mano*, pp. 88–93; Gil, *Cuarenta años*, p. 132.

51 Franco Salgado-Araujo, *Mi vida*, p. 345; Soriano, *La mano*, pp. 87–8; Pozuelo, *Los últimos 476 días*, pp. 35, 109, 178.

52 Shaw, *Fútbol*, pp. 145–80.

53 Vicente Gil, *Cuarenta años junto a Franco* (Barcelona, 1981) pp. 84–5; Fernández, *El fútbol*, pp. 196–7.

54 Franco Salgado-Araujo, *Mis conversaciones*, p. 322; López Rodó, *Memorias*, pp. 325–17; Areilza, *Memorias*, pp. 169–70.

55 Franco Salgado-Araujo, *Mis conversaciones*, p. 332.

56 Peñafiel, *El General*, pp. 121–8.

57 *Mundo Obrero*, 1 May 1962; *The Times*, 12 May 1962; Ignacio Fernández de Castro & José Martínez, *España hoy* (Paris, 1963) pp. 67–97, 103–28, 140–92; Parti Communiste Français, *Dos meses de huelgas* (Paris, 1962) pp. 41–95; Miguelez, *La lucha*, pp. 103–13.

58 Franco, manuscript notes on clerical participation in labour matters, May 1962, Fundación Francisco Franco, Legajo 53, Nos. 12–18, in *Manuscritos de Franco*, doc. 39bis.

59 Franco Salgado-Araujo, *Mis conversaciones*, pp. 337–41.

60 *Arriba*, 27 May 1962; Franco, *Discursos, 1960–1963*, pp. 389–97.

61 Suárez Fernández, *Franco*, VI, p. 377.

62 *ABC*, 9 June 1962; Franco Salgado-Araujo, *Mis conversaciones*, p. 343; Soriano, *La mano*, pp. 151–2; López Rodó, *Memorias*, pp. 335–6; Suárez Fernández, *Franco*, VI, p. 357.

63 Areilza, *Memorias exteriores*, pp. 170–82; Calvo Serer, *Franco frente al Rey*, pp. 112–13.

64 *Arriba*, 9, 10, 12 June 1962; *ABC*, 9, 11, 12 June 1962.

65 Partido Comunista de España, *Declaración por la reconciliación nacional, por una solución democrática y pacífica del problema español* (Paris, 1956) pp. 3, 5, 29–31, 37–40; Paul Preston, 'The PCE's Long Road to Democracy 1954–1977' in Richard Kindersley, editor, *In Search of Eurocommunism* (London, 1982) pp. 36–65.

66 Hermet, *Les catholiques*, II, pp. 287–97.

67 *La Vanguardia Española*, 17 June 1962; *The Times*, 18 June 1962; Franco, *Discursos 1960–1963*, pp. 399–404; Calvo Serer, *Franco frente al Rey*, p. 112.

68 Franco, *Discursos 1960–1963*, pp. 412, 423–4, 427.

69 *Le Monde*, 7 July 1962.

70 Suárez Fernández, *Franco*, VI, pp. 394–5.

71 Franco Salgado-Araujo, *Mis conversaciones*, pp. 343–4.

72 *ABC*, 11 July 1962; *Le Monde*, 11 July 1962; López Rodó, *Memorias*, pp. 339–47; Franco Salgado-Araujo, *Mis conversaciones*, p. 344; Fraga, *Memoria breve*, pp. 29–32, 43; Mérida, *Testigos*, pp. 68–70; Welles, *Gentle Anarchy*, pp. 88–99.

73 *ABC*, 22 September 1962.

74 *España hoy*, pp. 293–303.

75 Fraga, *Memoria breve*, p. 59; López Rodó, *Memorias*, pp. 359–60.

76 *España hoy*, p. 307. Franco, *Discursos 1960–1963*, pp. 471–99.

77 Baón, *La cara*, p. 70; Fraga, *Memoria breve*, p. 52; Franco Salgado-Araujo, *Mis conversaciones*, pp. 382, 397; Gil, *Cuarenta años*, pp. 107–36.

78 Fraga, *Memoria breve*, pp. 33, 41–2.

79 Manuel Fernández Areal, *La libertad de prensa en España 1938–1971* (Madrid, 1971) pp. 69–75; Terrón Montero, *La prensa*, pp. 166–75.

80 López Rodó, *Memorias*, pp. 364–5, 518–19.

81 Franco, typescript notes on freemasonry, 1963, Fundación Francisco Franco, Legajo 246, No. 4, in *Manuscritos de Franco*, doc. 45; Franco Salgado-Araujo, *Mis conversaciones*, pp. 366–9; Jaraiz Franco, *Historia*, p. 191.

82 Franco, *Discursos 1960–1963*, pp. 505–39; Welles, *Gentle Anarchy*, pp. 103–6; López Rodó, *Memorias*, pp. 375–6; Fraga, *Memoria breve*, p. 65.

83 Franco Salgado-Araujo, *Mis conversaciones*, p. 376.

84 *Le Monde*, 13, 18, 19 April 1962; Amandino Rodríguez Armada & José Antonio Novais, *¿Quién mató a Julián Grimau?* (Madrid, 1976) pp. 17–103, 113–14; *Julián Grimau: crimen y castigo del general Franco* (Buenos Aires, 1963) pp. 9–15. The regime's reply to criticism was *¿Crimen o castigo? documentos inéditos sobre Julián Grimau García* (Madrid, 1963).

85 Rodríguez Armada & Novais, *Grimau*, pp. 110–11; Ballbé, *Orden público*, p. 425; Alfonso Armada, *Al servicio de la Corona* (Barcelona, 1983) pp. 72–3, 76–7.

86 Lannon, *Privilege*, pp. 246–9; testimony of Joaquín Ruiz Giménez, Mérida, *Testigos*, p. 235; Franco Salgado-Araujo, *Mis conversaciones*, pp. 381–2.

87 Franco Salgado-Araujo, *Mis conversaciones*, pp. 380–1; López Rodó, *Memorias*, p. 379; Fraga, *Memoria breve*, pp. 69–70; testimony of Manuel Fraga to author.

88 *ABC*, 28 April 1962; Rodríguez Armada & Novais, *Grimau*, pp. 109–59.

89 Areilza, *Memorias exteriores*, pp. 164–5.

90 Ballbé, *Orden público*, pp. 420–7; Fraga, *Memoria breve*, p. 71.

91 Octavio Alberola & Ariane Gransac, *El anarquismo español y la acción revolucionaria 1961–1974* (Paris, 1975) pp. 107–12; Edouard de Blaye, *Franco and the Politics of Spain* (Harmondsworth, 1976) p. 221.

92 Suárez Fernández, *Franco*, VII, pp. 88–91; Calvo Serer, *Franco frente al Rey*, pp. 132–5.

93 Fraga, *Memoria breve*, p. 77.

94 Franco, *Discursos 1960–1963*, p. 561.

95 Franco, *Discursos 1960–1963*, pp. 565–7.

96 Franco, manuscript notes on negotiations for the renewal of the 1953 agreements (May 1963), Fundación Francisco Franco, Legajo 241, Nos 21, 13, 25, in *Manuscritos de Franco*, doc. 44.

97 Fraga, *Memoria breve*, pp. 60, 65, 67; Welles, *Gentle Anarchy*, pp. 294–308; Marquina, *España*, pp. 761–78; Antonio Garrigues y Díaz Cañabate, *Diálogos conmigo mismo* (Barcelona, 1978) pp. 98–9; Rubottom & Murphy, *Spain*, pp. 79–84; Suárez Fernández, *Franco*, VII, pp. 101–9; Blaye, *Franco*, pp. 222–3.

CHAPTER XXVII

1 Fraga, *Memoria breve*, p. 99.

2 Fraga, *Memoria breve*, p. 106.

3 *ABC*, 1 April 1964.

4 Sánchez Silva & Saenz de Heredia, *Franco . . . ese hombre*,

p. 14; Fraga, *Memoria breve*,
p. 120; Franco Salgado-Araujo,
Mis conversaciones, pp. 428–9,
431; Suárez Fernández, *Franco*,
VII, p. 150.

5 Francisco Franco, *Discursos y
mensajes del Jefe del Estado
1964–1967* (Madrid, 1968)
pp. 19–40.

6 Fraga, *Memoria breve*, p. 107;
López Rodó, *Memorias*,
pp. 458–9.

7 Franco, *Discursos 1964–1967*,
p. 43.

8 Miguélez, *La lucha*, pp. 121–6.

9 López Rodó, *Memorias*, p. 456;
Fraga, *Memoria breve*, p. 108.

10 Franco Salgado-Araujo, *Mis
conversaciones*, p. 424.

11 Fraga, *Memoria breve*, p. 110.

12 Franco Salgado-Araujo, *Mis
conversaciones*, pp. 423–4; Calvo
Serer, *Franco frente al Rey*,
p. 144.

13 Franco Salgado-Araujo, *Mis
conversaciones*, p. 421.

14 Fraga, *Memoria breve*, pp. 112,
115; Franco Salgado-Araujo, *Mis
conversaciones*, p. 426;
Fernández, *El Futbol*, pp. 185–7;
Calvo Serer, *Franco frente al Rey*,
p. 145.

15 Gómez Pérez, *El franquismo*,
pp. 104–6.

16 Franco, *Discursos 1964–1967*,
pp. 51–92.

17 Fraga, *Memoria breve*,
pp. 117–16; López Rodó,
Memorias, pp. 475–8.

18 Blázquez, *La traición*, pp. 158–64;
Lannon, *Privilege*, pp. 250–1.

19 Franco Salgado-Araujo, *Mis
conversaciones*, p. 407; Fraga,
Memoria breve, pp. 89, 99, 103,
123–5.

20 Gómez Pérez, *El franquismo*,
pp. 111–13.

21 Franco, *Discursos 1964–1967*,
pp. 119–35; Blázquez, *La
traición*, pp. 160–1.

22 López Rodó, *La larga marcha*,
pp. 208, 217–25; Payne, *Franco
Regime*, pp. 508–10.

23 López Rodó, *Memorias*, p. 498.

24 López Rodó, *La larga marcha*,
p. 226; Fraga, *Memoria breve*,
p. 133.

25 López Rodó, *La larga marcha*,
pp. 226–7.

26 López Rodó, *Memorias*, p. 512;
López Rodó, *La larga marcha*,
pp. 227–8.

27 López Rodó, *Memorias*,
pp. 519–20; López Rodó, *La
larga marcha*, pp. 229–30;
Fraga, *Memoria breve*, p. 135.

28 Fraga, *Memoria breve*, pp. 136,
138.

29 Payne, *Franco Regime*, p. 511.

30 Espinosa San Martín interview
in Bayod, *Franco*, pp. 149–50.

31 López Rodó, *Memorias*,
pp. 532–9; López Rodó, *La larga
marcha*, pp. 235–6; Fraga,
Memoria breve, p. 142.

32 Fraga, *Memoria breve*, p. 142.

33 Franco, typescript notes
borrador de Ley de Prensa,
1964, Fundación Francisco
Franco, Legajo 157, No. 1, in
Manuscritos de Franco, doc. 46;
Fraga, *Memoria breve*,
pp. 144–5, 151, 158–9.

34 Baón, *La cara humana*,
pp. 64–7, 73, 160.

35 *ABC*, 28 August 1968; Garriga,
La Señora, pp. 278–80, 296.

36 Espinosa San Martín interview
in Bayod, *Franco*, p. 151.

37 Laureano López Rodó,

Memorias: Años decisivos
(Barcelona, 1991) p. 43; Joaquín
Bardavio, *La rama trágica de los
borbones* (Barcelona, 1989)
pp. 111–18.

38 López Rodó, *Memorias años
decisivos*, pp. 22, 33–4.
39 López Rodó, *Memorias*,
pp. 539–43, 564; Franco
Salgado-Araujo, *Mis
conversaciones*, pp. 455–6;
Fraga, *Memoria breve*,
pp. 159–60; Espinosa San
Martín interview in Bayod,
Franco, pp. 150–1.
40 Garrigues, *Diálogos*, pp. 58–9;
Cambio 16, No. 278, 10 April
1977, p. 13. Neither source
gives a date, but cf. López
Rodó, *Memorias*, p. 559.
41 López Rodó, *Memorias*, p. 579;
Fraga, *Memoria breve*, p. 151.
42 Toquero, *Franco y Don Juan*,
pp. 334–6; Bardavío, *La rama
trágica*, pp. 102–6, 111–14;
López Rodó, *La larga marcha*,
pp. 237–9; Fraga, *Memoria
breve*, pp. 150–1.
43 López Rodó, *Memorias años
decisivos*, p. 93; López Rodó, *La
larga marcha*, p. 238.
44 López Rodó, *Memorias años
decisivos*, pp. 18–20; López
Rodó, *La larga marcha*,
pp. 239–43. On the Carlist
claimants, see Javier Lavardín,
*Historia del último pretendiente a
la corona de España* (Paris, 1976)
passim.
45 Franco Salgado-Araujo, *Mis
conversaciones*, pp. 465, 506,
514.
46 Suárez Fernández, *Franco*, VII,
p. 328–9; López Rodó, *Memorias
años decisivos*, pp. 41–2.

47 Fraga, *Memoria breve*, p. 172.
48 Areilza, *Crónica*, pp. 19–21,
42–4; Toquero, *Franco y Don
Juan*, pp. 343–8; Suárez
Fernández, *Franco*, VII,
pp. 171–2.
49 Fraga, *Memoria breve*,
pp. 166–9.
50 Vaca de Osma, *Paisajes*, p. 211.
51 Emilio Attard, *La Constitución
por dentro* (Barcelona, 1983)
p. 115; López Rodó, *Memorias
años decisivos*, p. 26.
52 Fraga, *Memoria breve*, p. 64.
53 Franco Salgado-Araujo, *Mis
conversaciones*, p. 469.
54 Fraga, *Memoria breve*, p. 170–2.
55 Fraga, *Memoria breve*,
pp. 174–5; López Rodó, *La larga
marcha*, p. 248.
56 Calvo Serer, *Franco frente al Rey*,
pp. 169–70; Fraga, *Memoria
breve*, p. 183.
57 Franco, *Discursos, 1964–1967*,
pp. 219–51.
58 Crozier, *Franco*, pp. 486–7.
59 Payne, Franco Regime, p. 495.
60 Franco, *Discursos, 1964–1967*,
p. 259.
61 *Cuadernos de Ruedo Ibérico*,
No. 10, December
1966–January 1967, pp. 27–63;
Blaye, *Franco*, pp. 236–8.
62 Gil, *Cuarenta años*, pp. 84–5;
Fernández Santander, *El futbol*,
pp. 196–7.
63 López Rodó, *Memorias años
decisivos*, p. 138; Franco Salgado-
Araujo, *Mis conversaciones*,
pp. 465, 495, 504–5; Espinosa
San Martín interview in Bayod,
Franco, p. 158; Hills, *Rock of
Contention*, pp. 455–66.
64 Espinosa San Martín interview
in Bayod, *Franco*, p. 153.

65 López Rodó, *Memorias años decisivos*, pp. 370–1.

66 Espinosa San Martín interview in Bayod, *Franco*, p. 158.

67 López Rodó, *La larga marcha*, pp. 262–3; Suárez Fernández, *Franco*, VII, pp. 377–9.

68 Suárez Fernández, *Franco*, VII, pp. 380–90.

69 'Franco: los archivos secretos IV', *Tiempo*, 14 December 1992; Franco Salgado-Araujo, *Mis conversaciones*, pp. 497–8.

70 De la Cierva, *Franquismo*, II, pp. 250–1; Fraga, *Memoria breve*, p. 194; López Rodó, *La larga marcha*, pp. 263–5; López Rodó, *Memorias años decisivos*, p. 207.

71 Calvo Serer, *Franco frente al Rey*, p. 171.

72 Espinosa San Martín interview in Bayod, *Franco*, p. 154.

73 Fraga, *Memoria breve*, p. 215.

74 Franco Salgado-Araujo, *Mis conversaciones*, pp. 530, 533, 537; Fraga, *Memoria breve*, pp. 216, 243.

75 Espinosa San Martín interview in Bayod, *Franco*, p. 158.

76 Franco Salgado-Araujo, *Mis conversaciones*, p. 511.

77 Franco, *Discursos 1964–1967*, pp. 291–333.

78 Franco Salgado-Araujo, *Mis conversaciones*, p. 511.

79 Franco Salgado-Araujo, *Mis conversaciones*, pp. 513–14.

80 *Le Monde*, 14 March 1969; Franco Salgado-Araujo, *Mis conversaciones*, pp. 540–1, 547.

81 López Rodó, *La larga marcha*, pp. 267–70.

82 Franco Salgado-Araujo, *Mis conversaciones*, pp. 515–17, 525–6, 530–1.

83 López Rodó, *Memorias años decisivos*, p. 307.

84 Franco, manuscript draft of letter to Paul VI, 1968, Fundación Francisco Franco, Legajo 17, No. 2, in *Manuscritos de Franco*, doc.49; De la Cierva, *Franquismo*, II, p. 258.

85 Franco Salgado-Araujo, *Mis conversaciones*, pp. 538–9.

86 Francesc Amover, *Il carcere vaticano: Chiesa e fascismo in Spagna* (Milan, 1975) pp. 28–47; Fernando Gutiérrez, *Curas represaliados en el franquismo* (Madrid, 1977) *passim*.

87 López Rodó, *Memorias años decisivos*, pp. 308–18, 325.

88 Rafael Calvo Serer, *La solución presidencialista* (Barcelona, 1979) p. 39.

89 López Rodó, *Memorias años decisivos*, pp. 346–9, 355–7.

90 López Rodó, *Memorias años decisivos*, pp. 362–4.

91 López Rodó, *Memorias años decisivos*, pp. 339–41, 370–1; Fraga, *Memoria breve*, pp. 229–30; Marquina, *España*, pp. 810–14.

92 López Rodó, *Memorias años decisivos*, pp. 366–7; Marquina, *España*, pp. 814–21.

93 Marquina, *España*, pp. 820–36; Rubottom & Murphy, *Spain*, pp. 87–91.

94 Francisco Franco, *Discursos y mensajes del Jefe del Estado 1968–1970* (Madrid, 1971) pp. 52–69.

95 López Rodó, *Memorias*, p. 542; Fraga, *Memoria breve*, pp. 234,

241; interview with López Rodó
in Bayod, *Franco*, p. 167;
Raymond Carr, 'The legacy of
Francoism' in José L. Cagigao,
John Crispin and Enrique
Pupo-Walker, editors, *Spain
1975-1980: The Conflicts and
Achievements of Democracy*
(Madrid, 1982) p. 136; Coles,
Franco, p. 29.

96 López Rodó, *Memorias años
decisivos*, pp. 358-9.

97 López Rodó, *La larga marcha*,
pp. 279, 291-3, 301; López
Rodó, *Memorias años decisivos*,
pp. 381-4; Fraga, *Memoria
breve*, pp. 236-7; Suárez
Fernández, *Franco*, VIII,
pp. 66-72.

98 López Rodó, *Memorias años
decisivos*, p. 386.

99 López Rodó, *La larga marcha*,
pp. 303-11; Espinosa San
Martín interview in Bayod,
Franco, p. 160; Fraga, *Memoria
breve*, pp. 245-6; Franco
Salgado-Araujo, *Mis
conversaciones*, pp. 544-5.

100 López Rodó, *Memorias años
decisivos*, pp. 423-6.

101 López Rodó, *La larga marcha*,
pp. 320-5.

102 Joaquín Bardavío, *Los silencios del
Rey* (Madrid, 1979) p. 35;
López Rodó, *La larga marcha*,
pp. 325-36.

103 López Rodó, *Memorias años
decisivos*, pp. 456-66.

104 Bardavío, *Los silencios*, pp. 35-6;
López Rodó, *Memorias años
decisivos*, pp. 471-6; López
Rodó, *La larga marcha*,
pp. 362-5.

105 Sainz Rodríguez, *Un reinado*,
p. 276.

106 Franco, *Discursos 1968-1970*,
pp. 85-97.

107 Bardavío, *Los silencios*,
pp. 49-50.

108 Bardavío, *Los silencios*, pp. 27,
50-2.

CHAPTER XXVIII

1 Accounts favourable to Matesa
in Navarro Rubio, *Mis memorias*,
pp. 345-431 and López Rodó,
Memorias años decisivos,
pp. 494-521, 553-63.

2 Suárez Fernández, *Franco*, VIII,
pp. 158-9.

3 *Arriba*, 24, 27 August 1969.

4 Franco Salgado-Araujo, *Mis
conversaciones*, pp. 527, 530.

5 Jaraiz Franco, *Historia*, p. 204.

6 Franco, *Nosotros*, p. 158.

7 López Rodó, *Memorias años
decisivos*, pp. 507-9, 682-90.

8 Espinosa San Martín interview
in Bayod, *Franco*, pp. 161-3.

9 Carrero Blanco to Franco, 16
October 1969, reprinted in
López Rodó, *La larga marcha*,
pp. 654-9; Franco Salgado-
Araujo, *Mis conversaciones*,
p. 549; Fraga, *Memoria breve*,
pp. 252-3; De la Cierva,
Franquismo, II, pp. 324-16;
López Rodó, *Memorias años
decisivos*, pp. 499-505.

10 López Rodó, *La larga marcha*,
pp. 390-5; López Rodó,
Memorias años decisivos,
pp. 520-3, 534-7.

11 Navarro Rubio, *Mis memorias*,
p. 245.

12 López Bravo interview, Bayod,
Franco, p. 120; López Rodó,
Memorias años decisivos, pp. 522,
535.

13 *ABC*, 29 October 1969; Equipo Mundo, *Los 90 Ministros*, pp. 420–500; Calvo Serer, *La dictadura*, pp. 166, 168; Armada, *Al servicio*, pp. 68, 72, 78, 93–4, 100–1, 119, 121, 135; José Ignacio San Martín, *Servicio especial: a las órdenes de Carrero Blanco* (Barcelona, 1983) pp. 198, 253; López Rodó, *La larga marcha*, p. 200.

14 Morán, *Suárez*, pp. 198–9, 204–5.

15 López Rodó, *Memorias años decisivos*, pp. 536–9.

16 Fraga, *Memoria breve*, pp. 261–3.

17 Franco, *Discursos 1968–1970*, pp. 107–21.

18 Many journalists and opposition figures in Spain at the time believed this to be the case. It is impossible to prove although there are hints in the memoirs of the officer in charge of the SDPG, San Martín, *Servicio especial*, pp. 23–42.

19 'Luis Ramírez' (pseudonym Luciano Rincón), 'Morir en el bunker', *Horizonte Español 1972* 3 vols (Paris, 1972) I, pp. 1–20; Preston, *Politics of Revenge*, pp. 165–74.

20 Garriga, *La Señora*, pp. 235, 289–92, 297–301; Joaquín Giménez Arnau, *Yo, Jimmy: mi vida entre los Franco* (Barcelona, 1981) p. 26.

21 López Rodó, *La larga marcha*, pp. 274–5, 286–9; López Rodó, *Memorias años decisivos*, p. 307.

22 Fraga, *Memoria breve*, p. 263; López Rodó, *Memorias años decisivos*, p. 436.

23 López Rodó, *Memorias*, pp. 83–4.

24 Morán, *Suárez*, p. 184.

25 Baón, *La cara humana*, pp. 30–1, 37–8; Fraga, *Memoria breve*, p. 263.

26 Baón, *La cara humana*, pp. 64–7, 73, 160; Franco Salgado-Araujo, *Mis conversaciones*, p. 555.

27 Fraga, *Memoria breve*, pp. 268, 272.

28 *Le Monde*, 20–21 September 1970.

29 Testimony of Elosegi to the author in 1985; Elosegi, *Quiero morir*, pp. 37, 40; Southworth, *Guernica!*, p. 309.

30 Henry Kissinger, *The White House Years* (London, 1979) pp. 930–2; Laureano López Rodó, *El principio del fin: Memorias* (Barcelona, 1992) pp. 84–5.

31 Nicolás Franco to Francisco Franco, 6 November (no year given). The letter is reproduced in facsimile in Jaraiz Franco, *Historia*, p. 203 and transcribed in Franco, *Nosotros*, p. 62. See also Garriga, *Nicolás Franco*, pp. 339–40. The fact that the date on letter did not include the year led *Diario 16, Historia de la transición* (Madrid, 1984) I, p. 144, and Payne, *Franco Regime*, p. 615, to assume that it related to the executions of 27 September 1975. It would have made little sense for Nicolás to request the commutation of executions carried out ten days earlier.

32 *Le Monde*, 16, 18 December 1970.

33 *Le Monde*, 18, 19, 21 December 1970; Gil, *Cuarenta años*,

pp. 98–103; Blaye, *Franco*, pp. 310–11; Franco Salgado-Araujo, *Mis conversaciones*, p. 560; López Rodó, *El principio*, pp. 113–15.

34 *Le Monde*, 29, 30 December 1970; López Bravo & Garicano Goñi interviews in Bayod, *Franco*, p. 124, 201–2; López Rodó, *La larga marcha*, pp. 405–6; López Rodó, *El principio*, pp. 122–9, 579–82.

35 Franco, *Discursos 1968–1970*, pp. 167–78.

36 Garriga, *La Señora*, pp. 252–3.

37 Franco Salgado-Araujo, *Mis conversaciones*, p. 559.

38 Vernon A. Walters, *Silent Missions* (New York, 1978) pp. 555–6.

39 Bardavío, *Los silencios*, pp. 53–4; López Rodó, *El principio*, p. 146.

40 Cooper, *Catholicism*, pp. 39–40.

41 Fraga, *Memoria breve*, pp. 280–1.

42 *Arriba*, 2 October 1971; *ABC*, 2 October 1971; *Mundo Obrero*, 15 October 1971; *Cuadernos de Ruedo Ibérico*, nos 33–5, October 1971–March 1972, pp. 3–18.

43 Paul Preston, *The Triumph of Democracy in Spain* (London, 1986) pp. 40–2.

44 *Pensamiento político de Franco* 2 vols (Madrid, 1975) I, pp. 3–17.

45 Garriga, *Nicolás Franco*, pp. 311–17; Gil, *Cuarenta años*, pp. 87, 93; De la Cierva, *Franquismo*, II, pp. 355–6.

46 Bardavío, *La rama trágica*, pp. 163–72, 192; Peñafiel, *El General*, pp. 166–8, 203–10.

47 Garriga, *La Señora*, pp. 332–3; López Rodó, *La larga marcha*, pp. 411–29; Suárez Fernández, *Franco*, VIII, pp. 273–82; Bardavío, *La rama trágica*, pp. 153–7, 181–92.

48 Gil, *Cuarenta años*, pp. 42–3, 60, 87–8, 91; López Rodó, *La larga marcha*, pp. 323, 419, 435; López Rodó, *El principio*, pp. 280–1; Fraga, *Memoria breve*, pp. 285, 287–92; Calvo Serer, *La solución*, pp. 38–9.

49 Jaraiz Franco, *Historia*, pp. 156, 162–3, 174, 205.

50 Gil, *Cuarenta años*, pp. 50–1; Fraga, *Memoria breve*, pp. 288–9; Payne, *Franco Regime*, p. 585.

51 López Rodó, *El principio*, p. 325.

52 Franco, *Pensamiento político*, I, p. 34; López Rodó, *El principio*, pp. 336–8.

53 López Rodó, *El principio*, pp. 335–6.

54 López Rodó, *El principio*, p. 345.

55 *Le Monde*, 4, 5–6, 7 August 1973; Ismael Fuente, Javier García & Joaquín Prieto, *Golpe mortal: asesinato de Carrero y agonía del franquismo* (Madrid, 1983) p. 164; Carlos Arias interview in Bayod, *Franco*, p. 308; Bardavío, *Los silencios*, pp. 61–2; López Rodó, *La larga marcha*, pp. 440–53.

56 Fraga, *Memoria breve*, p. 298; López Rodó, *El principio*, pp. 541–2; De la Cierva, *Franquismo*, II, p. 321.

57 Julen Agirre (pseud. Eva Forest), *Operación Ogro: cómo y porqué ejecutamos a Carrero Blanco* (Hendaye/Paris, 1974) *passim*; Joaquín Bardavío, *La*

crisis: historia de quince días
(Madrid, 1974) pp. 47–56.

58 Fuente et al., Golpe mortal,
p. 172; José Utrera Molina, Sin
cambiar de bandera (Barcelona,
1989) pp. 70–4; De la Cierva,
Franquismo, II, p. 389.

59 Franco, Nosotros, p. 150.

60 Fuente et al., Golpe mortal,
pp. 54–7.

61 Pueblo, 22 December 1973;
Iniesta Cano, Memorias,
pp. 218–22; San Martín,
Servicio especial, pp. 90–114;
Bardavío, La crisis, pp. 111–16;
Fuente et al., Golpe mortal,
pp. 184–7.

62 Le Monde, 25, 30–31 December
1973, 5–8 January 1974; Sunday
Times, 23 December 1973; Julio
Rodríguez Martínez,
Impresiones de un ministro de
Carrero Blanco (Barcelona, 1974)
pp. 54–8, 80, 86; Utrera, Sin
cambiar, pp. 76–8; Fuente et al.,
Golpe mortal, pp. 195–201; San
Martín, Servicio especial,
pp. 99–102.

63 Jaraiz Franco, Historia, p. 208;
De la Cierva, Franquismo, II,
pp. 391–2.

64 Fuente et al., Golpe Mortal,
p. 288. Rafael Borrás Bertriu et
al., El día en que mataron a
Carrero Blanco (Barcelona,
1974) pp. 252–6; Mundo, 5
January 1974.

65 Gil, Cuarenta años, pp. 139–63;
Utrera, Sin cambiar, pp. 83–5;
Bardavío, Los silencios, pp. 65–9;
Fuente et al., Golpe mortal,
pp. 172–3, 282–301; López
Rodó, La larga marcha,
pp. 459–61; Rodríguez,
Impresiones, p. 96; Fraga,

Memoria breve, pp. 309–10.

66 Franco, Pensamiento, I,
pp. 35–8; Bardavío, Los silencios,
p. 74.

67 López Rodó, El principio, p. 152.

68 The Times, 4 January 1974; Le
Monde, 4 January 1974.

69 Fuente, Golpe mortal, p. 283.

70 Utrera, Sin cambiar, pp. 85–92; De
la Cierva, Franquismo, II, p. 395.

71 The Times, 13 February 1974; Le
Monde, 14 February 1974;
Carlos Arias Navarro, Discurso
del Presidente del Gobierno a las
Cortes Españolas, 12.II.1974
(Madrid, 1974); Interview with
Carro Martínez, Bayod, Franco,
pp. 348–9.

72 Utrera, Sin cambiar, pp. 98, 103;
De la Cierva, Franquismo, II,
pp. 395–6.

73 Le Monde, 26 February, 5, 9
March 1974; El Alcázar, 7, 8
March 1974.

74 Arriba, 28 April 1974; ABC, 30
April 1974; Cambio 16, 13 May
1974; Utrera, Sin cambiar,
pp. 116–22.

75 Le Monde, 15 May 1974;
Financial Times, 29 May 1974;
Manuel Gutiérrez Mellado, Un
soldado para España (Barcelona,
1983) pp. 47–9; Preston,
Triumph, pp. 60–2.

76 Fraga, Memoria breve, p. 330.

77 Gil, Cuarenta años, pp. 167–90;
López Rodó, La larga marcha,
pp. 463–5; Bardavío, La rama
trágica, pp. 203–4; Baón, La
cara humana, pp. 37, 217–20.

78 Gil, Cuarenta años,
pp. 193–202, 209, 212;
Pozuelo, Los últimos 476 días,
pp. 22–3; Peñafiel, El General,
pp. 155–6, 160.

79 Pozuelo, *Los últimos 476 días*, pp. 29–30, 35, 37–9, 47, 51.

80 Bardavío, *Los silencios*, pp. 95–101; Utrera, *Sin cambiar*, p. 147.

81 Utrera, *Sin cambiar*, pp. 155–60.

82 Utrera, *Sin cambiar*, p. 163.

83 *ABC*, 3 September 1974; Pozuelo, *Los últimos 476 días*, pp. 67–71; López Rodó, *La larga marcha*, pp. 467–8; Garriga, *La Señora*, pp. 343–6; Bardavío, *Los silencios*, pp. 100–2.

84 Pozuelo, *Los 476 últimos días*, pp. 121–5.

85 *Le Monde*, 4, 8, 18, 30, 31 October 1974; José Oneto, *Arias entre dos crisis 1973–1975* (Madrid, 1975) pp. 149–53; Carro Martínez interview, Bayod, *Franco*, pp. 354–6; Sánchez Soler, *Villaverde*, p. 100; Utrera, *Sin cambiar*, pp. 173–5; De la Cierva, *Franquismo*, II, p. 402.

86 Bardavío, *Los silencios*, p. 102.

87 Pozuelo, *Los 476 últimos días*, pp. 126–9, 133–6, 141–7.

88 Pozuelo, *Los 476 últimos días*, pp. 133, 177–8.

89 Franco, *Pensamiento político*, I, pp. 39–43.

90 Utrera, *Sin cambiar*, pp. 226–33.

91 Javier Figuero & Luis Herrero, *La muerte de Franco jamás contada* (Barcelona, 1985) pp. 19–21.

92 *Le Monde*, 5 March 1975; Carro Martínez interview, Bayod, *Franco*, pp. 356–7; Utrera, *Sin cambiar*, pp. 248–59; Morán, *Suárez*, pp. 286–7.

93 *Arriba*, 28 February 1975.

94 Utrera, *Sin cambiar*, pp. 266–73.

95 Fraga, *Memoria breve*, pp. 346–9.

96 Pozuelo, *Los 476 últimos días*, pp. 131–2, 158.

97 Rubbotom & Murphy, *Spain and the U.S.*, pp. 113–14.

98 *Cambio 16*, 23–29 June 1975; Pozuelo, *Los 476 últimos días*, pp. 178–80.

99 Pedro J. Ramírez, *El año que murió Franco* (Barcelona, 1985) pp. 51–2, 68–9; Morán, *Suárez*, pp. 295–6.

100 Pozuelo, *Los 476 últimos días*, p. 157.

101 Franco, *Nosotros*, pp. 236–7.

102 Pozuelo, *Los 476 últimos días*, p. 187; Fraga, *Memoria breve*, p. 363.

103 *Arriba*, 16 July 1975.

104 Ramírez, *El año*, pp. 112, 118–21.

105 *Ya*, 30 September 1975; *Diario 16, Historia de la transición*, pp. 130–7.

106 Pozuelo, *Los 476 últimos días*, pp. 209–10; Ramírez, *El año*, pp. 204–6.

107 *Arriba*, 2 October 1975; *Cambio 16*, 6 October 1975; Pozuelo, *Los 476 últimos días*, pp. 210–12.

108 Pozuelo, *Los 476 últimos días*, pp. 215–16; Baón, *La cara humana*, p. 227.

109 Peñafiel, *El General*, pp. 47–52.

110 Franco, *Nosotros*, pp. 167–8; Pozuelo, *Los 476 últimos días*, pp. 218–21; Figuero & Herrero, *La muerte*, p. 26.

111 Baón, *La cara humana*, p. 230.

112 *ABC*, 2, 7 November; *Ya*, 29, 30 October, 9, 14, 18 November 1975.

113 Manuel Hidalgo Huerta, *Cómo y porqué operé a Franco* (Madrid, 1976) pp. 18–34.

114 Hidalgo Huerta, *Cómo y porqué*, pp. 35–55.

115 Peñafiel, *El General*, pp. 29–35; Ramírez, *El año*, p. 255.

116 Hidalgo Huerta, *Cómo y porqué*, pp. 59–69.

117 *Arriba*, 14, 18 November 1975; Figuero & Herrero, *La muerte*, pp. 35–6, 50–1.

118 *Arriba*, 20 November 1975; *Ya*, 20 November 1975; Pozuelo, *Los 476 últimos días*, pp. 224–41; Baón, *La cara humana*, pp. 226–50; Figuero & Herrero, *La muerte*, pp. 102–12.

EPILOGUE

1 Franco, *Pensamiento político*, I, p. xix.

2 Manuel Vázquez Montalbán, *Barcelonas* (London, 1992) p. 175.

3 *El País*, 20 November 1985.

4 Sueiro, *El Valle*, pp. 216–17.

5 Peñafiel, *El General*, pp. 132–3, 175.

6 Giménez Arnau, *Yo Jimmy*, pp. 164ff.

7 Inmaculada G. Mardones, 'Sin Franco no viven peor', *El País*, 20 November 1985.

8 Sánchez Soler, *Villaverde*, pp. 110–16, 120–7, 131–6.

9 Sánchez Soler, *Villaverde*, pp. 150–3.

10 *Cambio 16*, 5 March 1978; Sánchez Soler, *Villaverde*, pp. 176–7.

11 Bardavío, *La rama trágica*, pp. 208–44.

12 Sánchez Soler, *Villaverde*, pp. 70–5.

13 Peñafiel, *El General*, pp. 175–80; Sánchez Soler, *Villaverde*, pp. 154–171.

14 Javier Tusell, 'Una esfinge sin secreto y el largo purgatorio de un país', *Diario 16*, 4 December 1992.

15 Salvador de Madariaga, *General, márchese usted* (New York, 1959) p. 11.

16 Baón, *La cara humana*, p. 143. The first comment on this came from the pen of Miguel de Unamuno in a letter written on 13 December 1936 to Quintín de Torre, reproduced in Luciano González Egido, *Agonizar en Salamanca: Unamuno julio–diciembre 1936* (Madrid, 1986) pp. 226–8.

17 Franco, *Textos 1945–1950*, p. 229; Franco, *Discursos 1955–1959*, pp. 496–7.

18 José Luis García Delgado, 'Un legado económico ambivalente', *El País*, 3 December 1992; Ramón Tamames, 'Franco y la economía', *El Mundo*, 2 December 1992.

19 Carlos Elordi, 'El PCE por dentro', *La Calle*, No. 95, 15 January 1980; Jorge de Esteban & Luis López Guerra, *Los partidos políticos en la España actual* (Barcelona, 1982) pp. 139, 144–5; Richard Gunther, Giacomo Sani & Goldie Shabad, *Spain After Franco: The Making of a Competitive Party System* (Berkeley, Cal., 1986) pp. 65–70.

20 Preston, *Politics of Revenge*, pp. 165–202.

SOURCES

I. PRIMARY SOURCES

(i) Unpublished Sources

1. OFFICIAL ARCHIVES

A. Ministerio de Asuntos Exteriores (Madrid)

Archivo General: Serie de Archivo Renovado (MAE/R files)
Archivo Azaña

B. Public Record Office (London)

(i) CABINET OFFICE

CAB 65 War Cabinet Minutes
CAB 66 War Cabinet Memoranda
CAB 128 Cabinet Conclusions, 1945–50
CAB 129 Cabinet Papers, 1945–50
PREM 8 Correspondence and Papers of Prime Minister's Office 1944–1945

(ii) FOREIGN OFFICE

FO371 General Correspondence
FO425 Confidential Prints

C. Rome

(i) ARCHIVIO STORICO E DIPLOMATICO DEL MINISTERO DEGLI AFFARI ESTERI

Politica Spagna
Spagna Fondo di Guerra

(ii) ARCHIVIO CENTRALE DELLO STATO

Segretaria Particolare del Duce

2. PRIVATE COLLECTIONS

Serrano Suñer papers, Madrid
Templewood Papers, University Library, Cambridge

(ii) Published sources

1. PARLIAMENTARY DEBATES

Diario de las sesiones de Cortes, Congreso de los Diputados, comenzaron el 8 de diciembre de 1933 17 Vols, (Madrid, 1933–1936)
Diario de las sesiones de Cortes, Congreso de los Diputados, comenzaron el 16 de marzo de 1936 3 Vols, (Madrid, 1936)
Diario de sesiones de las Cortes Constituyentes de la República española, comenzaron el 14 de julio de 1931 25 vols (Madrid, 1931–1933)
Hansard, *House of Commons Debates* Fifth Series

2. OFFICIAL DOCUMENTS AND CORRESPONDENCE

Actas del último Consejo Nacional de Falange Española de las JONS (Salamanca, 18–19.IV.1937) y algunas noticias referentes a la Jefatura nacional de prensa y propaganda edited by Vicente de Cadenas y Vicent, (Madrid, 1975)
Actas de las sesiones de la Junta de Defensa de Madrid edited Julio Aróstegui & Jesús A. Martínez, (Madrid, 1984)
[Aguirre y Lecube, José Antonio], *El Informe del Presidente Aguirre al Gobierno de la República sobre los hechos que determinaron el derrumbamiento del frente del norte (1937)* (Bilbao, 1978)
Akten zur deutschen auswärtigen Politik 1941–1945, Series E, Vol. III (Göttingen, 1969)
Arxiu Vidal i Barraquer, *Esglesia i Estat durant la segona República espanyola 1931–1936* 8 vols, (Montserrat, 1971–1990)
Cartas cruzadas entre D. Manuel Hedilla Larrey y D. Ramón Serrano Suñer (Madrid, 1947)
Catalunya sota el règim franquista: Informe sobre la persecució de la llengua i la cultura de Catalunya pel règim del general Franco (Paris, 1973)
Churchill & Roosevelt: The Complete Correspondence edited by Warren F. Kimball, 3 vols (Princeton, 1984)
Correspondência de Pedro Teotónio Pereira para Oliveira Salazar, I (1931–1939) (Lisbon, 1987); *II (1940–1941)* (Lisbon, 1989); *III (1942)* (Lisbon, 1990); *IV (1943–1944)* (Lisbon, 1991)
Dez anos de política externa (1936–1947) a nação portuguesa e a segunda guerra

mundial vol.I (Lisbon, 1961); vol. II (Lisbon, 1978); vol. III (Lisbon, 1964); vol. IV (Lisbon, 1965); vol. V (Lisbon, 1967); vol. VI (Lisbon, 1970); vol. VII (Lisbon, 1971); vol. VIII (Lisbon, 1973); vol. IX (Lisbon, 1974); vol. X (Lisbon, 1974); vol. XII (Lisbon, 1985)

Dictamen de la Minoría Socialista, *El desastre de Melilla: dictamen formulado por Indalecio Prieto como miembro de la Comisión designada por el Congreso de los Diputados para entender en el expediente Picasso* (Madrid, 1922)

I Documenti Diplomatici Italiani, 8 serie, vol. XII (Rome, 1952); vol. XIII (Rome, 1953); 9 serie, vol. I (Rome, 1954); vol. II (Rome, 1957); vol. III, (Rome, 1959); vol. IV (Rome, 1960); vol. V (Rome, 1965); vol. VI (Rome, 1986); vol. VIII (Rome, 1988)

Documentos inéditos para la historia del Generalísimo Franco 3 vols [to date] (Madrid, 1992)

Documents Diplomatiques Français 1932–1939, 2ᵉ Série (1936–1939) XV (Paris, 1981); Tome XVI (Paris, 1983); Tome XVII (Paris, 1984); Tome XVIII (Paris, 1985); Tome XIX (Paris, 1986)

Documents on British Foreign Policy 2nd Series, Vol. XVII (London, 1979)

Documents on German Foreign Policy Series C, vol. IV (London, 1964); Series D, vol. III (London, 1951); vol. VI (London, 1956); vol. VII (London, 1956); vol. VIII (London, 1954); vol. IX (London, 1956); vol. X (London, 1957); vol. XI (London, 1961); vol. XII (London, 1964); vol. XIII (London, 1964)

Documents on the Italian Intervention in Spain (n.p. [London], 1937)

Documents secrets du Ministère des Affaires Etrangères d'Allemagne: Espagne (Paris, 1946)

Estado Español, Ministerio de la Gobernación, *Dictamen de la comisión sobre ilegitimidad de poderes actuantes en 18 de julio de 1936* (Barcelona, 1939)

Foreign Relations of the United States 1936 Vol. II (Washington, 1954)

Foreign Relations of the United States 1937 Vol. I (Washington, 1954)

Foreign Relations of the United States 1938 Vol. I (Washington, 1955)

Foreign Relations of the United States 1939 Vol. II (Washington, 1956)

Foreign Relations of the United States 1940, Vol. II (Washington, 1957)

Foreign Relations of the United States 1941 Vol. II (Washington, 1959)

Foreign Relations of the United States 1942 Vol. III (Washington, 1961)

Foreign Relations of the United States 1943 Vol. II (Washington, 1964)

Foreign Relations of the United States 1944 Vol. IV (Washington, 1966)

Foreign Relations of the United States 1945 Vol. V (Washington, 1967)

Foreign Relations of the United States 1946 Vol. V (Washington, 1969)

Foreign Relations of the United States 1947 Vol. III (Washington, 1972)

Foreign Relations of the United States 1948 Vol. III (Washington, 1974)

Foreign Relations of the United States 1949 Vol. IV (Washington, 1975)

Foreign Relations of the United States 1950 Vol. III (Washington, 1977)

'Hispanicus', *Foreign Intervention in Spain* (London, 1937)

[Hitler, Adolf], *Hitler's War Directives 1939–1945*, edited by H.R. Trevor-Roper, (London, 1966)

How Mussolini Provoked the Spanish Civil War: Documentary Evidence (London,

1938) Lleonart y Anselem, Alberto J. & Castiella y Maiz, Fernando María. *España y ONU I (1945–46)* (Madrid, 1978)

Lleonart y Anselem, A.J. *España y ONU II (1947)* (Madrid, 1983)

Lleonart y Anselem, A.J. *España y ONU III (1948–1949): La "cuestión española"* (Madrid, 1985)

Lleonart y Anselem, A.J. *España y ONU IV (1950): La "cuestión española"* (Madrid, 1991)

Mesa, Roberto, editor. *Jaraneros y alborotadores: documentos sobre los sucesos estudiantiles de febrero de 1956 en la Universidad Complutense de Madrid* (Madrid, 1982)

Ministerio de Justicia, *Causa general. La dominación roja en España. Avance de la información instruida por el ministerio público* (Madrid, 1944);

Ministerio de la Guerra Estado Mayor Central, *Anuario Militar de España año 1936* (Madrid, 1936)

[Missione Militare Italiana in Spagna]. *Fascistas en España: la intervención italiana en al guerra civil a través de los telegramas de la "Missione Militare Italiana in Spagna" (15 diciembre 1936–31 marzo 1937)* edited by Ismael Saz, & Javier Tusell, (Madrid/Rome, 1981)

Nuremberg, International Military Tribunal, *Trial of Major War Criminals* (Nuremberg, 1947)

Partido Comunista de España, *Declaración por la reconciliación nacional, por una solución democrática y pacífica del problema español* (Paris, 1956)

United Nations, Security Council, Official Records, First Year: Second Series, Special Supplement, *Report of the Sub-Committee on the Spanish Question* (New York – June 1946)

USA Department of State, *The Spanish Government and the Axis* (Washington, 1946)

[Weizsäcker, Ernst von], *Die Weizsäcker-Papiere 1933–1950* (Frankfurt/M, 1974)

3. NEWSPAPERS AND PERIODICALS

ABC

ABC (Seville)

Acción Española

El Alcázar

Arriba

Asturias

Boletín Oficial del Estado

Bulletin of Spanish Studies

Cambio 16

El Carbayón (Oviedo)

CEDA

El Compostelano (Santiago de Compostela)

El Correo Gallego

La Correspondencia Militar
Cuadernos de Ruedo Ibérico
The Daily Express
The Daily Mail
The Daily Telegraph
The Daily Telegraph and Morning Post
El Debate
Diario 16
Diário de Lisboa,
El Diario Vasco
El Eco de Santiago (Santiago de Compostela)
The Economist
La Época
Estampa
El Faro de Vigo
Fotos
La Gaceta Regional
El Heraldo de Madrid
Hola
L'Humanité
Informaciones
Interviú
Israel's Messenger
Manchester Guardian
Le Monde
The Morning Post
Mundo
Mundo Gráfico
Mundo Obrero
The News Chronicle
Newsweek
The New York Times
Nuestra Bandera
Nuevo Mundo
The Observer
Opinión
El País
La Prensa (Gijón)
Región
El Socialista
El Sol
The Sunday Express
The Sunday Times
Tiempo
The Times

La Vanguardia Española
La Voz de Asturias
La Voz de Galicia
The Universe
Ya

4. WRITINGS & SPEECHES OF FRANCISCO FRANCO

Andrade, Jaime de. [pseudonym Francisco Franco Bahamonde], *Raza anecdotario para el guión de una pelicula* (Madrid, 1942)

Boor, Jakim. [pseudonym Francisco Franco Bahamonde), *Masonería* (Madrid, 1952)

Franco Bahamonde, Francisco. *"Apuntes" personales sobre la República y la guerra civil* (Madrid, 1987)

[Franco Bahamonde, Francisco]. *El Caudillo en Cataluña* (Madrid, 1942)

Franco, Comandante. *Diario de una bandera* (Madrid, 1922)

Franco, Francisco *Discursos y mensajes del Jefe del Estado 1951–1954* (Madrid, 1955)

Franco, Francisco. *Discursos y mensajes del Jefe del Estado 1955–1959* (Madrid, 1960)

Franco, Francisco. *Discursos y mensajes del Jefe del Estado 1960–1963* (Madrid, 1964)

Franco, Francisco. *Discursos y mensajes del Jefe del Estado 1964–1967* (Madrid, 1968)

Franco, Francisco. *Discursos y mensajes del Jefe del Estado 1968–1970* (Madrid, 1971)

[Franco Bahamonde, Francisco]. *España y Francisco Franco, XXV aniversario de la exaltación a la Jefatura del Estado* (Madrid, 1961)

[Franco Bahamonde, Francisco]. *Francisco Franco, escritor militar*, Número especial, *Revista de Historia Militar*, Año XX, No.40, 1976

[Franco Bahamonde, Francisco]. *Franco ha dicho . . . recopilación de las más importantes declaraciones del Caudillo desde la iniciacion del Alzamiento Nacional hasta el 31 de diciembre de 1946* (Madrid, 1947)

[Franco Bahamonde, Francisco]. *Franco ha dicho. Primer apéndice (contiene de 1 enero 1947 a 1 abril 1949)* (Madrid, 1949)

Franco Bahamonde, Francisco. 'Fundamentos y directrices de un Plan de saneamiento de nuestra economía, armónico con nuestra reconstrucción nacional', *Historia 16*, No.115, noviembre de 1985

Franco, facsimile typescript and manuscript notes, Fundación Francisco Franco, *Manuscritos de Franco* (Madrid, 1986)

[Franco Bahamonde, Francisco]. *Hoja de servicios del Caudillo de España*, Excmo. Sr. Don Francisco Franco Bahamonde y su genealogía, edited by Esteban Carvallo de Cora, (Madrid, 1967)

[Franco Bahamonde, Francisco]. *Las cartas de amor de Franco*, edited by Vicente Gracia & Enrique Salgado, (Barcelona, 1978)

[Franco Bahamonde, Francisco]. *Mensaje del Caudillo a los españoles, discurso pronunciado por S.E. el Jefe del Estado la noche del 31 de diciembre de 1939* (Madrid, 1939)

Franco Bahamonde, Francisco. *Papeles de la guerra de Marruecos* (Madrid, 1986)

[Franco Bahamonde, Francisco]. *Palabras del Caudillo 19 abril 1937 – 31 diciembre 1938* (Barcelona, 1939)

[Franco Bahamonde, Francisco]. *Palabras del Caudillo 19 abril 1937 – 19 abril 1938* (n.p., 1939)

[Franco Bahamonde, Francisco]. *Palabras del Caudillo 19 abril 1937–7 diciembre 1942* (Madrid, 1943)

[Franco Bahamonde, Francisco]. *Palabras de Franco: I año triunfal* (Bilbao, 1937)

[Franco Bahamonde, Francisco]. *Pensamiento político de Franco*, edited by Agustín del Río Cisneros, 2 vols (Madrid, 1975)

[Franco Bahamonde, Francisco]. *Statements made by H.E. General Franco and Count Jordana on the International Policy of Spain* (Madrid, 1943)

Franco Bahamonde, Francisco. *Textos de doctrina política: palabras y escritos de 1945 a 1950* (Madrid, 1951)

5. DIARIES, MEMOIRS, LETTERS & SPEECHES OF OTHER PROTAGONISTS

Acheson, Dean. *Present at the Creation: My Years in the State Department* (New York, 1969)

Agirre, Julen (pseud. Eva Forest). *Operación Ogro: cómo y porqué ejecutamos a Carrero Blanco* (Hendaye/Paris, 1974)

Aguirre y Lecube, José Antonio de. *De Guernica a Nueva York pasando por Berlín* 3 edición (Buenos Aires, 1944)

Alcalá Zamora, Niceto. *Memorias* (Barcelona, 1977)

Alcázar de Velasco, Angel. *Serrano Suñer en la Falange* (Madrid/Barcelona, 1941)

Alcázar de Velasco, Angel. *Siete días de Salamanca* (Madrid, 1976)

Alderete, Ramón de. *. . . y estos borbones nos quieren gobernar* (Paris, 1974)

Alvarez, Basilio. *España en crisol* (Buenos Aires, Editorial Claridad, 1937)

Alvarez del Vayo, Julio. *Freedom's Battle* (London, 1940)

Ansaldo, Juan Antonio. *¿Para qué . . . ? (de Alfonso XIII a Juan III)* (Buenos Aires, 1951)

Areilza, José María de. *Crónica de libertad* (Barcelona, 1985)

Areilza, José María de. *Diario de un ministro de la monarquía* (Barcelona, 1977)

Areilza, José María de. *Embajadores sobre España* (Madrid, 1947)

Areilza, José María de. *Memorias exteriores 1947–1964* (Barcelona, 1984)

Areilza, José María de, & Castiella, Fernando María. *Reivindicaciones de España* (Madrid, 1941)

Arenillas de Chaves, Ignacio. *El proceso de Besteiro* (Madrid, 1976)

Carlos Arias Navarro, *Discurso del Presidente del Gobierno a las Cortes Españolas, 12.II.1974* (Madrid, 1974)

Armada, Alfonso. *Al servicio de la Corona* (Barcelona, 1983)

Arrese, José Luis de. *Una etapa constituyente* (Barcelona, 1982)

Arrese, José Luis de. *Escritos y discursos* (Madrid, 1943)

Arrese, José Luis de. *La revolución social del nacional-sindicalismo* (Madrid, 1959)

Arrese, José Luis de. *Treinta años de política* (Madrid, 1966)

Attard, Emilio. *La Constitución por dentro* (Barcelona, 1983)

[Azaña, Manuel]. *Memorias íntimas de Azaña* (Madrid, 1939)

Azaña, Manuel. *Mi rebelión en Barcelona* (Madrid, 1935)

Azaña, Manuel. *Obras completas* 4 vols (Mexico D.F., 1966–1968)

Azcárate, Pablo de. *Mi embajada en Londres durante la guerra civil española* (Barcelona, 1976)

Bahamonde y Sánchez de Castro, Antonio. *Un año con Queipo: memorias de un nacionalista* (Barcelona, 1938)

Barba Hernández, Bartolomé. *Dos años al frente del Gobierno Civil de Barcelona y varios ensayos* (Madrid, 1948)

Baudouin, Paul. *The Private Diaries of Paul Baudouin (March 1940-January 1941)* (London, 1948)

Bayod, Angel. *Franco visto por sus ministros* (Barcelona, 1981)

Beaulac, Willard L. *Career-Ambassador* (New York, 1951)

Beaulac, Willard L. *Franco: Silent Ally in World War II* (Carbondale, 1986)

Berenguer, Dámaso. *De la Dictadura a la República* (Madrid, 1946)

Boaventura, Armando. *Madrid-Moscovo da ditadura à República e à guerra civil de Espanha* (Lisbon, 1937)

Bolín, Luis. *Spain: the Vital Years* (Philadelphia, 1967)

Bowers, Claude G. *My Mission to Spain* (London, 1954)

[Bruce Lockhart, Sir Robert]. *The Diaries of Sir Robert Bruce Lockhart*, Vol. 2 *1939–1965*, edited by Kenneth Young, (London, 1980)

Bureau Permanent de l'Entente Internationale Anticommuniste, *Dix-sept ans de lutte contre le bolchévisme 1924–1940* (Geneva, 1940)

Burgo, Jaime del. *Conspiración y guerra civil* (Madrid, 1970)

Butcher, Captain Harry C. *My Three Years with Eisenhower* (New York, 1946)

[Cadogan, Sir Alexander]. *The Diaries of Sir Alexander Cadogan 1938–1945* edited by David Dilkes, (London, 1971)

Calvo Serer, Rafael. *Franco frente al Rey: el proceso del régimen* (Paris, 1972)

Calvo Serer, Rafael. *La dictadura de los franquistas: el "affaire" del "Madrid" y el futuro político* (Paris, 1973)

Calvo Serer, Rafael. *La solución presidencialista* (Barcelona, 1979)

Calvo Sotelo, José. *Mis servicios al Estado. Seis años de gestión: apuntes para la Historia* (Madrid, 1931)

Cambó, Francesc. *Meditacions: dietari (1936–1940)* (Barcelona, 1982)

Cambó, Francesc. *Meditacions: dietari (1941–1946)* (Barcelona, 1982)

Cantalupo, Roberto. *Fu la Spagna. Ambasciata presso Franco. Febbraio–Aprile 1937* (Milan, 1948)

Carrero Blanco, Almirante. *Discursos y escritos 1943–1973* (Madrid, 1974)

Castro Delgado, Enrique. *Hombres made in Moscú* (Barcelona, 1965)

Chalmers Mitchell, Sir Peter. *My House in Málaga* (London, 1938)

Chapaprieta Torregrosa, Joaquín. *La paz fue posible: memorias de un político* (Barcelona, 1971)

Charles-Roux, François. *Cinq mois tragiques aux affaires étrangères 21 Mai–1er Novembre 1940* (Paris, 1949)

Churchill, Winston S. *The Second World War* 6 vols (London, 1948–1954).

[Ciano, Galeazzo], *Ciano's Diary 1937–1938* (London, 1952)

[Ciano, Galeazzo], *Ciano's Diary 1939–1943* (London, 1947)

[Ciano, Galeazzo], *Ciano's Diplomatic Papers* ed. Malcolm Muggeridge (London, 1948)

Cordón, Antonio. *Trayectoria (recuerdos de un artillero)* (Paris, 1971)

Cowles, Virginia. *Looking for Trouble* (London, 1941)

[Dalton, Hugh]. *The Second World War Diary of Hugh Dalton 1940–1945* edited by Ben Pimlott (London, 1986)

Díaz, Carmen, (Viuda de Franco). *Mi vida con Ramón Franco* (Barcelona, 1981)

Doussinague, José M. *España tenía razón (1939–1945)* (Madrid, 1949)

Downes, Donald. *The Scarlet Thread: Adventures in Wartime Espionage* (London, 1953)

Eccles, David, editor, *By Safe Hand: Letters of Sybil and David Eccles 1939–1942* (London, 1983)

Eisenhower, Dwight D. *The White House Years: Waging Peace 1956–1961* (London, 1965)

Elosegui, Joseba. *Quiero morir por algo* (Bordeaux, 1971)

[Engel, Gerhard]. *Heeresadjutant bei Hitler 1939–1943: Aufzeichnungen des Majors Engel* (Stuttgart, 1974)

Escobar, Adrián C. *Diálogo íntimo con España: memorias de un embajador durante la tempestad europea* (Buenos Aires, 1950)

Escobar, José Ignacio. *Así empezó* (Madrid, 1974)

Escofet, Frederic. *Al servei de Catalunya i de la República* 2 vols, Vol. 1 *La desfeta 6 d'octubre 1934*, Vol. II, *La victoria 19 de juliol 1936* (Paris, 1973)

Estéban Infantes, Emilio. *La sublevación del general Sanjurjo* (Madrid, 1933)

Faldella, Emilio. *Venti mesi di guerra in Spagna* (Florence, Le Monnier, 1939)

Feis, Herbert. *The Spanish Story: Franco and the Nations at War* 2nd edition (New York, 1966)

Fernández Cuesta, Raimundo. *Testimonio, recuerdos y reflexiones* (Madrid, 1985)

Fischer, Louis. *Men and Politics: An Autobiography* (London, 1941)

Fontana, José María. *Los catalanes en la guerra de España* 2 edición (Barcelona, 1977)

[Forrestal, James]. *The Forrestal Diaries* edited by Walter Millis, (New York, Viking Press, 1951)

Fraga Iribarne, Manuel. *Memoria breve de una vida pública* (Barcelona, 1980)

Franco, Comandante [Ramón]. *Aguilas y garras: historia sincera de una empresa discutida* (Madrid, n.d. [1929])

Franco, Comandante [Ramón]. *Madrid bajo las bombas* (Madrid, 1931)

Franco, Ramón, & Ruiz de Alda, Julio. *De Palos al Plata* (Madrid, 1926)

Franco Bahamonde, Pilar. *Nosotros los Franco* (Barcelona, 1980)

Franco Salgado-Araujo, Francisco. *Mis conversaciones privadas con Franco* (Barcelona, 1976)

Franco Salgado-Araujo, Francisco. *Mi vida junto a Franco* (Barcelona, 1977)

García Lacalle, Andrés. *Mitos y verdades: la aviación de caza en la guerra española* (México D.F., 1973)

Garrigues y Díaz Cañabate, Antonio. *Diálogos conmigo mismo* (Barcelona, 1978)

Gil, Vicente. *Cuarenta años junto a Franco* (Barcelona, 1981)

Gil Robles, José María. *La monarquía por la que yo luché: páginas de un diario 1941–1954)* (Madrid, 1976)

Gil Robles, José María. *No fue posible la paz* (Barcelona, 1968)

Giménez Arnau, Joaquín. *Yo, Jimmy: mi vida entre los Franco* (Barcelona, 1981)

Giménez Caballero, Ernesto. *Memorias de un dictador* (Barcelona, 1979)

Giménez Caballero, Ernesto. *España y Franco* (Cegama, 1938)

[Goebbels, Joseph], *Final Entries 1945. The Diaries of Joseph Goebbels* edited, introduced and annotated by Hugh Trevor-Roper (New York, 1978)

[Goebbels, Joseph], *The Goebbels Diaries 1939–41*, edited by Fred Taylor, (London, 1982)

[Goebbels, Joseph], *The Goebbels Diaries* [January 1942-December 1943] edited by Louis P. Lochner, (London, 1948)

Gomá, Cardenal. *Por Dios y por España 1936–1939* (Barcelona, 1940)

Griffis, Stanton. *Lying in State* (New York, 1952)

Guarner, Vicente. *Cataluña en la guerra de España* (Madrid, 1975)

Gutiérrez Mellado, Manuel. *Un soldado para España* (Barcelona, 1983)

[Halder, Franz], *The Halder War Diary 1939–1942* edited by Charles Burdick and Hans-Adolf Jacobsen (London, 1988)

[Harvey, Oliver]. *The War Diaries of Oliver Harvey* edited by John Harvey (London, 1978)

Hassell, Ulrich von. *The Von Hassell Diaries 1938–1944* (London, 1948)

Hayes, Carlton J.H. *Wartime Mission in Spain* (New York, 1945)

Hidalgo Durán, Diego. *¿Por qué fui lanzado del Ministerio de la Guerra? Diez meses de actuación ministerial* (Madrid, 1934)

Hidalgo Durán, Diego. *Un notario español en Rusia* (Madrid, 1929)

Hidalgo Huerta, Manuel. *Cómo y porqué operé a Franco* (Madrid, 1976)

[Hitler, Adolf]. *Hitler's Table Talk 1941–1944* (London, 1953)

Hoare, Sir Samuel. *Ambassador on Special Mission* (London, 1946)

Hodgson, Sir Robert. *Spain Resurgent* (London, 1953)

Hoyos, Marqués de. *Mi testimonio* (Madrid, 1962)

Hull, Cordell. *Memoirs* 2 vols (London, 1949)

Ibárruri, Dolores. *El único camino* (Paris, 1964)

Iniesta Cano, Carlos. *Memorias y recuerdos* (Barcelona, 1984)

Iribarren, José María. *Con el general Mola: escenas y aspectos inéditos de la guerra civil* (Zaragoza, 1937)

Iribarren, José María. *Mola, datos para una biografía y para la historia del alzamiento nacional* (Zaragoza, 1938)

Jalón, César. *Memorias políticas: periodista. ministro. presidiario.* (Madrid, 1973)

Jaraiz Franco, Pilar. *Historia de una disidencia* (Barcelona, 1981)

Jerrold, Douglas. *Georgian Adventure* (London, 1937)

Jiménez de Asúa, Luis, Vidarte, Juan-Simeón, et al., *Castilblanco* (Madrid, 1933)

[Keitel, Wilhelm]. *The Memoirs of Field-Marshal Keitel* (London, William Kimber, 1965)

Kemp, Peter. *Mine Were of Trouble* (London, 1957)

Kindelán Duany, Alfredo. *La verdad de mis relaciones con Franco* (Barcelona, 1981)

Kindelán Duany, Alfredo. *Mis cuadernos de guerra* 2 edición, (Barcelona, 1982)

Kirkpatrick, Ivone. *The Inner Circle: Memoirs* (London, 1959)

Koestler, Arthur. *Spanish Testament* (London, 1937)

Koltsov, Mijail. *Diario de la guerra de España* (Paris, 1963)

Laín Entralgo, Pedro. *Descargo de conciencia* (Barcelona, 1976)

Largo Caballero, Francisco. *Mis recuerdos* (México D.F., 1954)

Lerroux, Alejandro. *La pequeña historia: apuntes para la historia grande vividos y redactados por el autor* (Buenos Aires, 1945)

Liddell Hart, Captain. *Britain and Spain* (London, 1938)

Lizarza Iribarren, Antonio. *Memorias de la conspiración* (Pamplona, 1953)

Llaneza, Manuel. *Escritos y discursos* (Oviedo, 1985)

Lomax, Sir John. *The Diplomatic Smuggler* (London, 1965)

Longo, Luigi. *Le brigate internazionali in Spagna* (Rome, 1956)

López Fernández, Antonio. *Defensa de Madrid* (México D.F., 1945)

López Ochoa, General E. *Campaña militar de Asturias en octubre de 1934 (narración táctico-episódica)* (Madrid, 1936)

López Ochoa, General E. *De la Dictadura a la República* (Madrid, 1930)

López Rodó, Laureano. *La larga marcha hacia la monarquía* (Barcelona, 1977)

López Rodó, Laureano. *Memorias* (Barcelona, 1990)

López Rodó, Laureano. *Memorias: años decisivos* (Barcelona, 1991)

López Rodó, Laureano. *Política y desarrollo* (Madrid, 1970)

López Rodó, Laureano. *El principio del fin: Memorias* (Barcelona, 1992)

López Rodó, Laureano. *Testimonio de una política de Estado* (Barcelona, 1987)

Luca de Tena, José Ignacio. *Mis amigos muertos* (Barcelona, 1971)

Luca de Tena, Torcuato. *Papeles para la pequeña y la gran historia: memorias de mi padre y mías* (Barcelona, 1991)

Madariaga, Salvador de. *Memorias (1921–1936) amanecer sin mediodía* (Madrid, 1974)

Madariaga, Salvador de. *General, márchese usted* (New York, 1959)

Maíz, B. Félix. *Alzamiento en España: de un diario de la conspiración* 2nd ed. (Pamplona, 1952)

Maíz, B. Félix. *Mola aquel hombre* (Barcelona, 1976)

Marco Miranda, Vicente. *Las conspiraciones contra la Dictadura (1923–1930) relato de un testigo* 2 edición (Madrid, 1975)

Marsá, Graco. *La sublevación de Jaca; relato de un rebelde* (Paris, 1931)

Martín Blázquez, José. *I Helped to Build an Army: Civil War Memoirs of a Spanish Staff Officer* (London, 1939)

Martínez Barrio, Diego. *Memorias* (Barcelona, 1983)

Maura, Gabriel. *Bosquejo histórico de la Dictadura* 2 vols (Madrid, 1930)

Maura, Miguel. *Así cayó Alfonso XIII*, (México D.F., 1962)

Mérida, María. *Testigos de Franco: retablo íntimo de una dictatdura* (Barcelona, 1977)

Millán Astray, General José. *Franco, el Caudillo* (Salamanca, 1939)

Modesto, Juan. *Soy del quinto regimiento (notas de la guerra española)* (Paris, 1969)

Mola Vidal, Emilio. *Obras completas* (Valladolid, 1940)

Moulin de Labarthète, H. Du. *Le temps des illusions: souvenirs, juillet 1940-avril 1942* (Geneva, 1946)

Moure Mariño, Luis. *La generación del 36: memorias de Salamanca y Burgos* (La Coruña, 1989)

Navarro Rubio, Mariano. *Mis memorias: Testimonio de una vida política truncada por el Caso MATESA* (Barcelona, 1991)

Neves, Mário. *La matanza de Badajoz* (Badajoz, 1986)

Onaindía, Alberto. *El "Pacto" de Santoña, antecedentes y desenlace* (Bilbao, 1983)

Onaindía, Alberto. *Hombre de paz en la guerra* (Buenos Aires, 1973)

Ossorio, Angel. *Mis memorias* (Buenos Aires, 1946)

Pagés Guix, Luis [pseudonym]. *La traición de los Franco ¡Arriba España!*, (Madrid, n.d. [1937])

Pemán, José María. *Mis almuerzos con gente importante* (Barcelona, 1970)

Pemán, José María. *Mis encuentros con Franco* (Barcelona, 1976)

Pereira, Pedro Theotonio. *Memórias postos em que servi e algumas recordações pessoais* 2 vols (Lisbon, 1973)

Peterson, Maurice. *Both Sides of the Curtain* (London, 1950)

Pérez Salas, Coronel Jesús. *Guerra en España (1936–1939)* (México D.F., 1947)

Pérez Solís, Oscar. *Sitio y defensa de Oviedo* 2 edición (Valladolid, 1938)

Philby, Kim. *My Silent War* (London, 1968)

Piétri, François. *Mes années d'Espagne 1940–1948* (Paris, 1954)

Portela Valladares, Manuel. *Memorias: dentro del drama español* (Madrid, 1988)

Pozuelo, Vicente. *Los últimos 476 días de Franco* (Barcelona, 1980)

Pradera, Victor. *El Estado nuevo* 3rd edition (Madrid, 1941)

Prieto, Indalecio. *Palabras al viento* 2ª edición (México D.F., 1969)

Prieto, Indalecio. *Palabras al viento* 2 edición (México D.F., 1969)

Primo de Rivera, José Antonio. *Textos de doctrina política* 4th edition (Madrid, 1966)

Prüfer, Curt. *Rewriting History: The Original and Revised World War II Diaries of Curt Prüfer, Nazi Diplomat* (Kent, Ohio, 1988)

Queipo de Llano, Gonzalo. *El general Queipo de Llano perseguido por la dictadura* (Madrid, Javier Morato, 1930)

Reynaud, Paul. *Au coeur de la Mêlée 1930–1945* (Paris, 1951)

Ribbentrop, Joaquim von. *The Ribbentrop Memoirs*, (London, 1954)

Richthofen, Wolfram von. 'Spanien-Tagebuch', in Maier, Klaus A. *Guernica 26.4.1937. Die deutsche Intervention in Spanien und der 'Fall Guernica'* (Freiburg, 1975)

Ridruejo, Dionisio. *Casi unas memorias* (Barcelona, 1976)

Rodríguez Martínez, Julio. *Impresiones de un ministro de Carrero Blanco* (Barcelona, 1974)

Rojo, General Vicente. *¡Alerta los pueblos! estudio político-militar del período final de la guerra española* 2 edición (Barcelona, 1974)

Rojo, General Vicente. *Así fue la defensa de Madrid* (México D.F., 1967)

Rojo, General Vicente. *España heroica: diez bocetos de la guerra española* 3 edición (Barcelona, 1975)

Romanones, Aline, Countess of. *The Spy Wore Red* (London, 1987)

Romero, Emilio. *Los papeles reservados* 2 vols (Barcelona, 1985)

Ruiz Vilaplana, Antonio. *Doy fe . . . un año de actuación en la España nacionalista* (Paris, n.d. [1938])

Sainz Rodríguez, Pedro. *Un reinado en la sombra* (Barcelona, 1981)

Sainz Rodríguez, Pedro. *Testimonio y recuerdos* (Barcelona, 1978)

Salazar Alonso, Rafael. *Bajo el signo de la revolución* (Madrid, 1935)

San Martín, José Ignacio. *Servicio especial: a las órdenes de Carrero Blanco* (Barcelona, 1983)

Saña, Heleno. *El franquismo sin mitos: conversaciones con Serrano Suñer* (Barcelona, 1982)

Schellenberg, Walter. *The Schellenberg Memoirs: A Record of the Nazi Secret Service* (London, 1956)

Schmidt, Paul. *Hitler's Interpreter: The Secret History of German Diplomacy 1935– 1945* (London, 1951)

Serrano Suñer, Ramón. *De anteayer y de hoy* (Barcelona, 1981)

Serrano Suñer, Ramón. *Ensayos al viento* (Madrid, 1969)

Serrano Suñer, Ramón. *Entre el silencio y la propaganda, la Historia como fue. Memorias* (Barcelona, 1977)

Serrano Suñer, Ramón. *Entre Hendaya y Gibraltar* (Madrid, 1947)

Serrano Suñer, Ramón. *Semblanza de José Antonio, joven* (Barcelona, 1958)

Serrano Suñer, Ramón. *Siete discursos* (Bilbao, 1938)

Sopeña, Federico. *Escrito de noche* (Madrid, 1985)

Soriano, Ramón. *La mano izquierda de Franco* (Barcelona, 1981)

Spellman, Francis J. *Action This Day: Letters From The Fighting Fronts* (London, 1944)

[Stettinius, Edward R. Jr]. *The Diaries of Edward R. Stettinius, Jr., 1943–1946*, edited by Thomas M. Campbell & George C. Herring (New York, 1975)

Tagüeña Lacorte, Manuel. *Testimonio de dos guerras* (México D.F., 1973)

Tierno Galván, Enrique. *Cabos sueltos* (Barcelona, 1981)

Truman, Harry S. *Memoirs. Year of Decisions 1945* (London, 1955)

Tusquets, P. Juan. *La Francmasonería, crimen de lesa patria* (Burgos, n.d. [1937])

Tusquets, P. Juan. *Masonería y separatismo* (Burgos, 1937)

Tusquets, P. Juan. *Orígenes de la revolución española* (Barcelona, 1932)

Utrera Molina, José. *Sin cambiar de bandera* (Barcelona, 1989)

Vaca de Osma, José Antonio. *La larga guerra de Francisco Franco* (Madrid, 1991)

Vaca de Osma, José Antonio. *Paisajes con Franco al fondo* (Barcelona, 1987)

Vegas Latapie, Eugenio. *Caminos del desengaño: memorias políticas II 1936–1938* (Madrid, 1987)

Vegas Latapie, Eugenio. *Memorias políticas: el suicido de la monarquía y la segunda República* (Barcelona, 1983)

Vidarte, Juan-Simeón. *El bienio negro y la insurrección de Asturias* (Barcelona, 1978)

Vidarte, Juan-Simeón. *Todos fuimos culpables: testimonio de un socialista español* (México D.F., 1973)

Walters, Vernon A. *Silent Missions* (New York, 1978)

Weizsäcker, Ernst von. *Memoirs of Ernst von Weizsäcker* (London, 1951)

Worsley, T.C. *Behind the Battle* (London, 1939)

Zugazagoitia, Julián. *Guerra y vicisitudes de los españoles*, 2ª edición, 2 vols, (Paris, 1968)

II. SECONDARY SOURCES

(i) Books and unpublished theses

ABC, La II guerra mundial (Madrid, 1989)

Abella, Rafael. *Por el Imperio hacia Dios: crónica de una posguerra (1939–1950)* (Barcelona, 1978)

Abella, Rafael. *La vida cotidiana durante la guerra civil 1) La España Nacional* (Barcelona, 1978)

Abella, Rafael. *La vida cotidiana en España bajo el régimen de Franco* (Barcelona, 1985)

Abshagen, Karl Heinz. *Canaris* (London, 1956)

Aguado Sánchez, Francisco. *El Maquis en España: sus documentos* (Madrid, 1976)

Aguado Sánchez, Francisco. *El Maquis en España: su historia* (Madrid, 1975)

Aguado Sánchez, Francisco. *La revolución de octubre de 1934* (Madrid, 1972)

Aguilar Olivencia, Mariano. *El Ejército español durante la segunda República* (Madrid, 1986)

Alarcón Benito, Juan. *Francisco Franco y su tiempo* (Madrid, 1983)

Alberola, Octavio, & Gransac, Ariane. *El anarquismo español y la acción revolucionaria 1961–1974* (Paris, 1975)

Alcofar Nassaes, José Luis. *C.T.V. Los legionarios italianos en la guerra civil española 1936–1939* (Barcelona, 1972)

Alfarache, Guzmán de. *¡18 de julio! Historia del alzamiento glorioso de Sevilla* (Seville, 1937)

Alpert, Michael. *El ejército republicano en la guerra civil* 2 edición (Madrid, 1989)

Alpert, Michael. *La guerra civil española en el mar* (Madrid, 1987)

Alpert, Michael. *La reforma militar de Azaña (1931–1933)* (Madrid, 1982)

Alpuente, Moncho. *Hablando francamente* (Barcelona, 1990)

Alvarez Junco, José. *El emperador del Paralelo: Lerroux y la demagogia popular* (Madrid, 1990)

Amover, Francesc. *Il carcere vaticano: Chiesa e fascismo in Spagna* (Milan, 1975)

Arasa, Daniel. *Años 40: los maquis y el PCE* (Barcelona, 1984)

Arce, Carlos de. *Los generales de Franco* (Barcelona, 1984)

Arrarás, Joaquín. *Franco*, 7 edición (Valladolid, 1939)

Arrarás, Joaquín. *Historia de la Cruzada española* 8 vols, 36 tomos, (Madrid, 1939–43)

Arrarás, Joaquín. *Historia de la segunda República española* 4 vols (Madrid, 1956–1968)

Artigues, Daniel. *El Opus Dei en España 1928–1962: su evolución ideológica y política de los orígenes al intento de dominio* (Paris, 1971)

Avni, Haim. *España, Franco y los judíos* (Madrid, 1982)

Aznar, Manuel. *Franco* (Madrid, 1975)

Aznar, Manuel. *Historia militar de la guerra de España (1936–1939)* (Madrid, 1940)

Azpíroz Pascual, José María & Elboj Broto, Fernando. *La sublevación de Jaca* (Zaragoza, 1984)

Balfour, Sebastian. *Dictatorship, Workers, and the City: Labour in Greater Barcelona since 1939* (Oxford, 1989)

Ballbé, Manuel. *Orden público y militarismo en la España constitucional (1812–1983)* (Madrid, 1983)

Baón, Rogelio. *La cara humana de un Caudillo* (Madrid, 1975)

Bardavío, Joaquín. *La crisis: historia de quince días* (Madrid, 1974)

Bardavío, Joaquín. *La rama trágica de los borbones* (Barcelona, 1989)

Bardavío, Joaquín. *Los silencios del Rey* (Madrid, 1979)

Barea, Arturo. *La forja de un rebelde* (Buenos Aires, 1951)

Barea, Arturo. *The Struggle for the Spanish Soul* (London, 1941)

Barreiro Fernández, Xosé Ramón. *Historia contemporánea de Galicia* 4 vols (La Coruña, 1982)

Bayo, Eliseo. *Los atentados contra Franco* (Barcelona, 1977)

Benavides, Manuel D. *La revolución fue así (octubre rojo y negro) reportaje* (Barcelona, 1935)

Ben-Ami, Shlomo. *Fascism from Above: The Dictatorship of Primo de Rivera in Spain 1923–1930* (Oxford, 1983)

Ben-Ami, Shlomo. *The Origins of the Second Republic in Spain* (Oxford, 1978)

Bertrán Güell, Felipe. *Preparación y desarrollo del alzamiento nacional* (Valladolid, 1939)

Blanco Escolá, Carlos. *La Academia General Militar de Zaragoza (1928–1931)* (Barcelona, 1989)

Blasco Ibáñez, Vicente. *Alfonso XIII Unmasked* (London, 1925)

Blaye, Edouard de. *Franco and the Politics of Spain* (Harmondsworth, 1976)

Blázquez, Feliciano. *La traición de los clérigos en la España de Franco: crónica de una intolerancia (1936–1975)* (Madrid, 1991)

Blinkhorn, Martin. *Carlism and Crisis in Spain 1931–1939* (Cambridge, 1975)

Blinkhorn, Martin, ed., *Spain in Conflict 1931–1939: Democracy and its Enemies* (London, 1986)

Bloch, Michael. *Operation Willi: The Plot to Kidnap the Duke of Windsor July 1940* (London, 1984)

Bolloten, Burnett. *The Spanish Civil War: Revolution and Counterrevolution* (Chapel Hill, 1991)

Bonmati de Codecido, Francisco. *El Príncipe Don Juan de España* (Valladolid, 1938)

Borkenau, Franz. *The Spanish Cockpit* 2nd edition (Ann Arbor, 1963)

Borràs Betriu, Rafael et al., *El día en que mataron a Carrero Blanco* (Barcelona, 1974)

Boyd, Carolyn P. *Praetorian Politics in Liberal Spain* (Chapel Hill, 1979)

Bravo Morata, Federico. *Franco y los muertos providenciales* (Madrid, 1979)

Brenan, Gerald. *The Spanish Labyrinth* (Cambridge, 1943)

Brissaud, André. *Canaris* (London, 1973)

Broué, Pierre, & Témime, Emile. *The Revolution and the Civil War in Spain* (London, 1972)

Buckley, Henry. *Life and Death of the Spanish Republic* (London, 1940)

Bullock, Alan. *Ernest Bevin: Foreign Secretary 1945–1951* (London, 1983)

Burdick, Charles B. *Germany's Military Strategy and Spain in World War II* (Syracuse, 1968)

Busquets, Julio. *El militar de carrera en España* 3ª edición (Barcelona, 1984)

Cabanellas, Guillermo. *Cuatro generales* 2 vols (Barcelona, 1977)

Cabanellas, Guillermo. *La guerra de los mil días* 2 vols (Buenos Aires, 1973)

Cacho Zabalza, Antonio. *La Unión Militar Española* (Alicante, 1940)

Cagigao, José L., Crispin, John & Pupo-Walker, Enrique, editors. *Spain 1975–1980: The Conflicts and Achievements of Democracy* (Madrid, 1982)

Calleja, Juan José. *Yagüe: un corazón al rojo* (Barcelona, 1963)

Cameron Watt, Donald. *How War Came: The Immediate Origins of the Second World War 1938–1939* (London, 1989)

Cardona, Gabriel. *El poder militar en la España contemporánea hasta la guerra civil* (Madrid, 1983)

Cardona, Gabriel. *El problema militar en España* (Madrid, 1990)

Cardozo, Harold G. *The March of a Nation: My Year of Spain's Civil War* (London, 1937)

Carlavilla, Mauricio. *Anti-España 1959: autores, cómplices y encubridores del communismo* (Madrid, 1959)

Carr, Raymond. *Spain 1808–1975* (Oxford, 1982)

Carr, Raymond. *The Spanish Tragedy: The Civil War in Perspective* (London, 1977)

Carr, Raymond, & Fusi, Juan Pablo. *Spain: Dictatorship to Democracy* (London, 1979)

Carral, Ignacio. *Por qué mataron a Luis de Sirval* (Madrid, 1935)

Casas de la Vega, Rafael. *La última guerra de Africa (campaña de Ifni-Sáhara)* (Madrid, 1985)

Caute, David. *The Great Fear: The Anti-communist Purge under Truman and Eisenhower* (London, 1978)

Cava Mesa, María Jesús. *Los diplomáticos de Franco: J.F. de Lequerica, temple y tenacidad (1890–1963)* (Bilbao, 1989)

Cave Brown, Anthony. *The Last Hero: Wild Bill Donovan* (London, 1982)

Cave Brown, Anthony. *The Secret Servant: The Life of Sir Stewart Menzies, Churchill's Spymaster* (London, 1988)

Chao Rego, José. *La Iglesia en el franquismo* (Madrid, 1976)

Chase, Allan. *Falange: The Axis Secret Army in the Americas* (New York, 1943)

Chase, Gilbert. *The Music of Spain* 2nd ed. (New York, Dover, 1959)

Chávez Camacho, Armando. *Misión de prensa en España* (México D.F., 1948)

Cierva, Ricardo de la. *Francisco Franco: biografía histórica* 6 vols (Barcelona, 1982)

Cierva, Ricardo de la. *Francisco Franco: un siglo de España* 2 vols (Madrid, 1973)

Cierva, Ricardo de la. *Hendaya. Punto final* (Barcelona, 1981)

Cierva, Ricardo de la. *Historia del franquismo: I orígenes y configuración (1939– 1945)* (Barcelona, 1975)

Cierva, Ricardo de la. *Historia del franquismo: II aislamiento, transformación, agonía (1945–1975)* (Barcelona, 1978)

Cierva, Ricardo de la. *Historia de la guerra civil española* (Madrid, 1969)

Cierva, Ricardo de la. *La derecha sin remedio (1801–1987)* (Barcelona, 1987)

Clark, Alan. *Barbarossa: The Russian–German Conflict 1941–1945* (London, 1965)

Clavera, Joan; Esteban, Joan M.; Monés, María Antonia; Montserrat, Antoni & Ros Hombravella, Jacint. *Capitalismo español: de la autarquía a la estabilización* 2 vols (Madrid, 1973)

Colectivo Democracia, *Los Ejércitos . . . más allá del golpe* (Barcelona, 1981)

Coles, S.F.A. *Franco of Spain* (London, 1955)

Colodny, Robert G. *The Struggle for Madrid: The Central Epic of the Spanish Conflict 1936–1937* (New York, 1958)

Conforti, Olao. *Guadalajara: la prima sconfitta del fascismo* (Milan, 1967)

Cooper, Norman B. *Catholicism and the Franco Regime* (Beverly Hills, 1975)

Cortada, James W. *Relaciones España–USA 1941–45* (Barcelona, 1973)

Cortés Cavanillas, Julián. *La caída de Alfonso XIII* 7 edición (Madrid, n.d., [1933])

Cossías, Tomás. *La lucha contra el "maquis" en España* (Madrid, 1956)

Costa i Deu, J. & Sabaté, Modest. *La veritat del 6 d'octubre* (Barcelona, 1936)

Coverdale, John F. *Italian Intervention in the Spanish Civil War* (Princeton, 1975)

Cox, Geoffrey. *Defence of Madrid* (London, 1937)

Créac'h, Jean. *Le coeur et l'épée: chroniques espagnoles* (Paris, 1958)

Cross, J.A. *Sir Samuel Hoare: A Political Biography* (London, 1977)

Crozier, Brian. *Franco: A Biographical History* (London, 1967)

Cuadrat, Xavier. *Socialismo y anarquismo en Cataluña (1899–1911): los orígenes de la CNT* (Madrid, 1976)

Cueto Alas, Juan. *Guía secreta de Asturias* (Madrid, Editorial Al-berak, 1975)

Culla i Clarà, Joan B. *El republicanisme Lerrouxista a Catalunya (1901–1923)* (Barcelona, 1986)

Delgado, Iva. *Portugal e a guerra civil de Espanha* (Lisbon, n.d.)

Diario 16, Historia de la transición edited by Justino Sinova. 2 vols (Madrid, 1984)

Diario 16, Historia del Franquismo, edited by Justino Sinova. 2 vols (Madrid, 1985)

Díaz Nosty, Bernardo. *La comuna asturiana: revolución de octubre de 1934* (Bilbao, 1974)

Díaz Nosty, Bernardo. *Las Cortes de Franco* (Barcelona, 1972)

Domínguez, Javier. *Organizaciones obreras cristianas en la oposición al franquismo (1951–1975)* (Bilbao, 1985)

Eby, Cecil. *The Seige of the Alcázar* (London, 1966)

Equipos de Estudio, *Noticia, rumor, bulo: la muerte de Franco: ensayo sobre algunos aspectos del control de la información* (Madrid, 1976)

Equipo Mundo, *Los noventa ministros de Franco* (Barcelona, 1970)

Ellwood, Sheelagh M. *Spanish Fascism in the Franco Era* (London, 1987)

Espadas Burgos, Manuel. *Franquismo y política exterior* (Madrid, 1988)

Espín, Eduardo. *Azaña en el poder: el partido de Acción Republicana* (Madrid, 1980)

Espinar Gallego, Ramón *et al. El impacto de la II guerra mundial en Europa y en España* (Madrid, 1986)

Espinosa de los Monteros y Bermejillo, Ignacio. *Hendaya y la segunda guerra mundial (vivencias y razones)* unpublished (Madrid, n.d.)

Fabre, Jaume; Huertas, Josep M. & Ribas, Antoni. *Vint anys de resistència catalana (1939–1959)* (Barcelona, 1978)

Fanés, Félix. *La vaga de tramvies del 1951* (Barcelona, 1977)

Febo, Giuliana di. *La santa de la raza: un culto barroco en la España franquista (1937–1962)* (Barcelona, 1988)

Feis, Herbert. *From Trust to Terror: The Onset of the Cold War 1945–1950* (London, 1970)

Felice, Renzo de. *Mussolini il duce: lo stato totalitario 1936–1940* (Turin, 1981)

Felice, Renzo de. *Mussolini l'alleato I. L'Italia in guerra 1940–1943 1. Dalla guerra "breve" alla guerra lunga* (Turin, 1990)

Felice, Renzo de. *Mussolini l'alleato I. L'Italia in guerra 1940–1943 2. Crisi e agonia del regime* (Turin, 1990)

Fernández, Julio L. *Los enigmas del Caudillo: perfiles desconocidos de un dictador temeroso e implacable* (Madrid, 1992)

Fernández Areal, Manuel. *La libertad de prensa en España 1938–1971* (Madrid, 1971)

Fernández de Castro, Ignacio, & Martínez, José. *España hoy* (Paris, 1963)

Fernández Clemente, Eloy. *El coronel Rey d'Harcourt y la rendición de Teruel: historia y fin de una leyenda negra* (Teruel, 1992)

Fernández Santander, Carlos. *Antología de 40 años (1936–1975)* (La Coruña, 1983)

Fernández Santander, Carlos. *El almirante Carrero* (Barcelona, 1985)

Fernández Santander, Carlos. *El futbol durante la guerra civil y el franquismo* (Madrid, 1990)

Fernández Santander, Carlos. *El general Franco* (Barcelona, 1983)

Fernández Santander, Carlos. *Paracuellos de Jarama: ¿Carrillo culpable?* (Barcelona, 1983)

Fernández Santander, Carlos. *Tensiones militares durante el franquismo* (Barcelona, 1985)

Ferrer Benimeli, José Antonio. *Masonería española contemporánea* 2 vols (Madrid, 1980)

Ferri, Llibert; Muixí, Jordi, & Sanjuan, Eduardo. *Las huelgas contra Franco (1939–1956)* (Barcelona, 1978)

Ferro, Marc. *Pétain* (Paris, 1987)

Figuero, Javier, & Herrero, Luis. *La muerte de Franco jamás contada* (Barcelona, 1985)

Foltz, Charles Jr., *The Masquerade in Spain* (Boston, 1948)

Fontana, José María. *Franco: radiografía del personaje para sus contemporáneos* (Barcelona, 1979)

Fortes, José, & Valero, Restituto. *Qué son las Fuerzas Armadas* (Barcelona, 1977)

Foss, William, & Gerahty, Cecil. *The Spanish Arena* (London, 1938)

Fraser, Ronald. *Blood of Spain: The Experience of Civil War 1936–1939* (London, 1979)

Fuente, Ismael; García, Javier, & Prieto, Joaquín. *Golpe mortal: asesinato de Carrero y agonía del franquismo* (Madrid, 1983)

Fusi, Juan Pablo. *Franco: autoritarismo y poder personal* (Madrid, 1985)

Galindo Herrero, Santiago. *Los partidos monárquicos bajo la segunda República* 2nd ed. (Madrid, 1956)

Galinsoga, Luis, & Franco-Salgado, Francisco. *Centinela de occidente (Semblanza biográfica de Francisco Franco* (Barcelona, 1956)

Gallagher, Tom. *Portugal: A Twentieth Century Interpretation* (Manchester, 1983)

Gallego, Gregorio. *Madrid, corazón que se desangra* (Madrid, 1976)

Gallo, Max. *Spain Under Franco: A History* (London, 1973)

García, P. *Los chistes de Franco* (Madrid, 1977)

García de la Escalera, Inés. *El general Varela* (Madrid, 1959)

García Lahiguera, Fernando. *Ramón Serrano Suñer: un documento para la historia* (Barcelona, 1983)

García-Villoslada, Ricardo, editor. *Historia de la Iglesia en España* Vol. V (Madrid, 1979)

García Venero, Maximiano. *El general Fanjul: Madrid en el alzamiento nacional* (Madrid, 1967)

García Venero, Maximiano. *Falange en la guerra de España: la Unificación y Hedilla* (Paris, 1967)

García Venero, Maximiano. *Historia de la Unificación (Falange y Requeté en 1937)* (Madrid, 1970)

Garriga, Ramón. *El Cardenal Segura y el Nacional-Catolicismo* (Barcelona, 1977)

Garriga, Ramón. *La España de Franco: de la División Azul al pacto con los Estados Unidos (1943 a 1951)* (Puebla, México, 1971)

Garriga, Ramón. *La España de Franco: las relaciones con Hitler* 2 edición (Puebla, México, 1970)

Garriga, Ramón. *Franco-Serrano Suñer: un drama político* (Barcelona, 1986)

Garriga, Ramón. *El general Yagüe* (Barcelona, 1985)

Garriga, Ramón. *Guadalajara y sus consecuencias* (Madrid, 1974)

Garriga, Ramón. *Nicolás Franco, el hermano brujo* (Barcelona, 1980)

Garriga, Ramón. *Ramón Franco, el hermano maldito* (Barcelona, 1978)

Garriga, Ramón. *La Señora de El Pardo* (Barcelona, 1979)

Garriga, Ramón. *Los validos de Franco* (Barcelona, 1981)

Garitaonandía, Carmelo, & Granja, José Luis de la. editors, *La guerra civil en el país vasco* (Bilbao, 1987)

Georgel, Jacques. *El franquismo: historia y balance 1939–1969* (Paris, 1970)

Gibson, Ian. *The Assassination of Federico García Lorca* (London, 1979)

Gibson, Ian. *En busca de José Antonio* (Barcelona, 1980)

Gibson, Ian. *Federico García Lorca* (London, 1989)

Gibson, Ian. *Paracuellos: cómo fue* (Barcelona, 1983)

Gibson, Ian. *Queipo de Llano: Sevilla, verano de 1936* (Barcelona, 1986)

Gilbert, Martin. *Finest Hour: Winston S. Churchill 1939–1941* (London, 1983)

Gilbert, Martin. *Road to Victory: Winston S. Churchill 1941–1945* (London, 1986)

Goded, Manuel. *Un "faccioso" cien por cien* (Zaragoza, 1939)

Gomá, Coronel José. *La guerra en el aire* (Barcelona, 1958)

Gómez Navarro, José Luis. *El régimen de Primo de Rivera: reyes, dictaduras y dictadores* (Madrid, 1991)

Gómez Oliveros, Comandante Benito. *General Moscadó (sin novedad en el Alcázar)* (Barcelona, 1956)

Gómez Pérez, Rafael. *El franquismo y la Iglesia* (Madrid, 1986)

Gondi, Ovidio. *La hispanidad franquista al servicio de Hitler* (Mexico D.F., 1979)

González, Fernando. *Liturgias para un Caudillo: manual de dictadores* (Madrid, 1977)

González, Manuel Jesús. *La economía política del franquismo (1940–1970): Dirigismo, mercado y planificación* (Madrid, 1979)

González Betes, Antonio. *Franco y el Dragón Rapide* (Madrid, 1987)

González Calbet, María Teresa. *La dictadura de Primo de Rivera: el Directorio militar* (Madrid, 1987)

González Calleja, Eduardo, & Limón Nevado, Fredes. *La Hispanidad como instrumento de combate: raza e imperio en la prensa franquista durante la guerra civil española* (Madrid, 1988).

González Duro, Enrique. *Franco: una biografía psicológica* (Madrid, 1992)

González Egido, Luciano. *Agonizar en Salamanca: Unamuno julio–diciembre 1936* (Madrid, 1986)

González Portilla, Manuel, & Garmendia, José María. *La guerra civil en el País Vasco* (Madrid, 1988)

González Portilla, Manuel, & Garmendía, José María. *La posguerra en el País Vasco: política, acumulación, miseria* (San Sebastián, 1988)

Graham, Helen & Preston, Paul, editors. *The Popular Front in Europe* (London, 1987)

Granados, Anastasio. *El Cardenal Gomá: Primado de España* (Madrid, 1969)

Grenville, J.A.S. *Lord Salisbury and Foreign Policy: The Close of the Nineteenth Century* (London, 1964)

Griffiths, Richard. *Fellow Travellers of the Right: British Enthusiasts for Nazi Germany 1933–39* (London, 1980)

[Grimau García, Julián]. *¿Crimen o castigo? documentos inéditos sobre Julián Grimau García* (Madrid, 1963)

[Grimau García, Julián]. *Julián Grimau: crimen y castigo del general Franco* (Buenos Aires, 1963)

Gubern, Román. *«Raza» (un ensueño del General Franco)* (Madrid, 1977)

Gutiérrez, Fernando. *Curas represaliados en el franquismo* (Madrid, 1977)

Gutiérrez-Ravé, José. *Antonio Goicoechea* (Madrid, 1965)

Halstead, Charles R. *Spain, the Powers and the Second World War* unpublished Ph.D. thesis, University of Virginia, 1962

Hamilton, Thomas J. *Appeasement's Child: The Franco Regime in Spain* (London, 1943)

Harrison, Joseph. *The Spanish Economy in the Twentieth Century* (Beckenham, 1985)

Haxey, Simon. *Tory M.P.* (London, 1939)

Heaton, P.M. *Welsh Blockade Runners in the Spanish Civil War* (Newport, 1985)

Heine, Hartmut. *A guerrila antifranquista en Galicia* (Vigo, 1980)

Heine, Hartmut. *La oposición política al franquismo* (Barcelona, 1983)

Hermet, Guy. *Les Catholiques dans l'Espagne Franquiste* 2 vols (Paris, 1980–1)

Hermet, Guy. *L'Espagne de Franco* (Paris, 1974)

Hernández Mir, Francisco. *La Dictadura ante la Historia: un crimen de lesa patria* (Madrid, 1930)

Hidalgo Salazar, Ramón. *La ayuda alemana a España 1936–1939* (Madrid, 1975)

Hills, George. *Franco: The Man and His Nation* (New York, 1967)

Hills, George. *Rock of Contention: A History of Gibraltar* (London, 1974)

Hinsley, F.H. et al. *British Intelligence in the Second World War* 4 Vols in 5 parts (London, 1979–1990)

Höhne, Heinz. *Canaris* (London, 1979)

Howard, Michael. *British Intelligence in the Second World War V: Strategic Deception* (London, 1990)

Howard, Michael. *Grand Strategy IV: August 1942-September 1943* (London, 1970)

Howson, Gerald. *Aircraft of the Spanish Civil War 1936–1939* (London, 1990)

Hughes, Emmet John. *Report from Spain* (London, 1947)

Iturralde, Juan de. *La guerra de Franco: los vascos y la Iglesia* 2 vols (San Sebastián, 1978)

Jackson, Gabriel. *The Spanish Republic and the Civil War* (Princeton, 1965)

Jato, David. *Gibraltar decidió la guerra* (Barcelona, 1971)

Jellinek, Frank. *The Civil War in Spain* (London, 1938)

Jiménez de Aberasturi, Juan Carlos, & San Sebastián, Koldo. *La huelga general del 1 de mayo de 1947 (artículos y documentos)* (San Sebastián, 1991)

Jones, Randolph Bernard. *The Spanish Question and the Cold War 1944–1953* (Unpublished Ph.D. thesis, University of London, 1987)

Juliá, Santos. *Manuel Azaña: una biografía política, del Ateneo al Palacio Nacional* (Madrid, 1990)

Kaiser, Carlos J. *La guerrilla antifranquista: historia del maquis* (Madrid, 1976)

Kay, Hugh. *Salazar and Modern Portugal* (London, 1970)

Kleinfeld, Gerald R. & Tambs, Lewis A. *Hitler's Spanish Legion: The Blue Division in Russia* (Carbondale, 1979)

Knickerbocker, H.R. *The Seige of the Alcazar: A War-Log of the Spanish Revolution* (London, n.d. [1936])

Knoblaugh, H. Edward. *Correspondent in Spain* (New York, 1937)

Knox, Macgregor. *Mussolini Unleashed 1939–1941: Politics and Strategy in Fascist Italy's Last War* (Cambridge, 1982)

Lacomba Avellán, Juan Antonio. *La crisis española de 1917* (Madrid, 1970)

Lacruz, Francisco. *El alzamiento, la revolución y el terror en Barcelona* (Barcelona, 1943)

LaFeber, Walter. *America, Russia and the Cold War 1945–1975* 3rd ed. (New York, 1976)

Lago, Julián. *Las contramemorias de Franco* (Barcelona, 1976)

Lannon, Frances. *Privilege, Persecution, and Prophecy: The Catholic Church in Spain 1875–1975* (Oxford, 1987) pp. 215–16

Lannon, Frances, & Preston, Paul, editors, *Elites and Power in Twentieth-Century Spain: Essays in Honour of Sir Raymond Carr* (Oxford, 1990)

Last, Jef. *The Spanish Tragedy* (London, 1939)

Lavardín, Javier. *Historia del último pretendiente a la corona de España* (Paris, 1976)

Lieberman, Sima. *The Contemporary Spanish Economy: A Historical Perspective* (London, 1982)

Little, Douglas. *Malevolent Neutrality: The United States, Great Britain, and the Origins of the Spanish Civil War* (Ithaca, 1985)

Lizarza A. de (pseudonym of Andrés María de Irujo). *Los vascos y la República española* (Buenos Aires, 1944)

Lizcano, Pablo. *La generación del 56: La Universidad contra Franco* (Barcelona, 1981)

Llarch, Juan. *Franco biografía* (Barcelona, 1983)

Lojendio, Luis María de. *Operaciones militares de la guerra de España* (Barcelona, 1940)

López, Elsa, Alvarez Junco, José, Espadas Burgos, Manuel & Muñoz Tinoco, Concha. *Diego Hidalgo: memoria de un tiempo difícil* (Madrid, 1986)

Lorenzo Espinosa, José María. *Rebelión en la Ría. Vizcaya 1947: obreros, empresarios y falangistas* (Bilbao, 1988)

Loveday, Arthur F. *Spain 1923–1948: Civil War and World War* (Bridgewater, 1949)

Lukacs, John. *The Last European War: September 1939-December 1941* (London, 1976)

McNeill-Moss, Major Geoffrey. *The Epic of the Alcazar* (London, 1937)

Maier, Klaus A. *Guernica 26.4.1937: Die deutsche intervention in Spanien und der "Fall Guernica"* (Freiburg, 1975)

Maravall, José. *El desarrollo económico y la clase obrera* (Barcelona, 1970)

Mariñas, General Francisco Javier. *General Varela (de soldado a general)* (Barcelona, 1956)

Marquina Barrio, Antonio. *España en la política de seguridad occidental 1939–1986* (Madrid, 1986)

Marquina Barrio, Antonio. *La diplomacia vaticana y la España de Franco (1936–1945)* (Madrid, 1983)

Marquina Barrio, Antonio, & Ospina, Gloria Inés. *España y los judíos en el siglo 20XX* (Madrid, 1987)

Martin, Claude. *Franco, soldado y estadista* (Madrid, 1965)

Martín, Miguel. *El colonialismo español en Marruecos (1860–1956)* (Paris, 1973)

Martínez Leal, Comandante Alfredo. *El asedio del Alcázar de Toledo: memorias de un testigo* (Toledo, n.d.)

Martínez Nadal, Rafael. *Antonio Torres y la política española del Foreign Office (1940–1944)* (Madrid, 1989)

Martínez Parrilla, Jaime. *La fuerzas armadas francesas ante la guerra civil española (1936–1939)* (Madrid, 1987)

Matthews, Herbert L. *The Yoke and the Arrows: A Report on Spain* (London, 1958)

Maura, Duque de & Fernández Almagro, Melchor. *Por qué cayó Alfonso XIII* (Madrid, 1948)

Miguel, Amando de. *Franco, Franco, Franco* (Madrid, 1976)

Miguel, Amando de. *Sociología del Franquismo* (Barcelona, 1975)

Miguélez, Faustino. *La lucha de los mineros asturianos bajo el franquismo* (Barcelona, 1976)

Mirandet, François. *L'Espagne de Franco* (Paris, 1948)

Molas, Isidre. *Lliga Catalana: un estudi de Estasiologia* 2 vols (Barcelona, 1972)

Montañés, Enrique. *Anarcosindicalismo y cambio político: Zaragoza, 1930–1936* (Zaragoza, 1989)

Montero Moreno, Antonio. *Historia de la persecución religiosa en España 1936–1939* (Madrid, 1961)

Moradiellos, Enrique. *El Sindicato de los Obreros Mineros Asturianos 1910–1930* (Oviedo, 1986)

Moradiellos, Enrique. *Neutralidad benévola: el Gobierno británico y la insurrección militar española de 1936* (Oviedo, 1990)

Morales Lezcano, Victor. *Historia de la no-beligerancia española durante la segunda guerra mundial* (Las Palmas, 1980)

Morán, Gregorio. *Adolfo Suárez: historia de una ambición* (Barcelona, 1979)

Morán, Gregorio. *Los españoles que dejaron de serlo: Euskadi, 1937–1981* (Barcelona, 1982)

Morán, Gregorio. *Miseria y grandeza del Partido Comunista de España 1939–1985* (Barcelona, 1986)

Moreno de Reyna, Fernando & Salvador. *La guerra en el mar (hombres, barcos y honra* (Barcelona, 1959)

Moreno Villalba, Faustino. *Franco, héroe cristiano en la guerra* (Madrid, 1985)

Morgan, Kenneth O. *Labour in Power 1945–1951* (Oxford, 1984)

Muñoz, Juan; Roldán, Santiago, & Serrano, Angel. *La internacionalización del capital en España 1959–1977* (Madrid, 1978)

Muñoz Tinoco, Concha. *Diego Hidalgo, un notario republicano* (Badajoz, 1986)

Muro Zegri, D. *La epopeya del Alcázar* (Valladolid, 1937)

Murray, Williamson. *The Change in the European Balance of Power, 1938–1939* (Princeton, 1984)

Murray, Williamson. *German Military Effectiveness* (Baltimore, 1992)

Nourry, Philippe. *Francisco Franco: la conquête du Pouvoir* (Paris, 1975)

Oliveros, Antonio L. *Asturias en el resurgimiento español (apuntes históricos y biográficos* (Madrid, 1935)

Oneto, José. *Arias entre dos crisis 1973–1975* (Madrid, 1975)

Oneto, José. *100 días en la muerte de Francisco Franco* (Madrid, 1975)

Ossorio y Gallardo, Angel. *Vida y sacrificio de Companys* 2ª edición (Barcelona, 1976)

Oven, W. von. *Hitler und der Spanische Bürgerkrieg. Mission und Schicksal der Legion Condor* (Tubingen, 1978)

Payne, Stanley G. *Falange: A History of Spanish Fascism* (Stanford, 1961)

Payne, Stanley G. *Franco: el perfil de la historia* (Madrid, 1992)

Payne, Stanley G. *Franco's Spain* (London, 1968)

Payne, Stanley G. *Politics and the Military in Modern Spain* (Stanford, 1967)

Payne, Stanley G. *The Franco Regime 1936–1975* (Madison, 1987)

Paxton, Robert O. *Vichy France: Old Guard and New Order 1940–1944* (London, 1972)

Peers, E. Allison. *Spain in Eclipse 1937–1943* (London, 1943)

Peñafiel, Jaime. *El General y su tropa. Mis recuerdos de la familia Franco* (Madrid, 1992)

Pike, David Wingeate. *Jours de gloire, jours de honte: le Parti Comuniste d'Espagne en France depuis son arrivée en 1939 jusque'à son départ en 1950* (Paris, 1984)

Pike, David Wingeate. *Les français et la guerre d'Espagne 1936–1939* (Paris, 1975)

Plenn, Abel. *Wind in the Olive Trees: Spain from the Inside* (New York, 1946)

Pons Prades, Eduardo. *Guerrillas españolas 1936–1960* (Barcelona, 1977)

Portero, Florentino. *Franco aislado: la cuestión española (1945–1950)* (Madrid, 1989)

Preston, Paul. *The Coming of the Spanish Civil War: Reform, Reaction and Revolution in the Second Republic 1931–1936* (London, 1978)

Preston, Paul. *Las derechas españolas en el siglo veinte: autoritarismo, fascismo, golpismo* (Madrid, 1986)

Preston, Paul. *The Politics of Revenge: Fascism and the Military in the 20th Century Spain* (London, 1990)

Preston, Paul, editor, *Revolution and War in Spain 1931–1939* (London, 1984)

Preston, Paul. *Salvador de Madariaga and the Quest for Liberty in Spain* (Oxford, 1987)

Preston, Paul, editor, *Spain in Crisis: Evolution and Decline of the Franco Regime* (Hassocks, 1976)

Preston, Paul. *The Spanish Civil War* (London, 1986)

Preston, Paul. *The Triumph of Democracy in Spain* (London, 1986)

Puzzo, Dante A. *Spain and the Great Powers, 1936–1941* (New York, 1962)

Qasim Bin Ahmad, *Britain, Franco Spain, and the Cold War, 1945–1950* (New York, 1992)

Quintanilla, Luis. *Los rehenes del Alcázar de Toledo* (Paris, 1967)

Ramírez, Luis [pseudonym of Luciano Rincón]. *Franco: la obsesión de ser, la obsesión de poder* (Paris, 1976)

Ramírez, Luis [pseudonym of Luciano Rincón]. *Nuestros primeros veinticinco años* (Paris, 1964)

Ramírez, Pedro J. *El año que murió Franco* (Barcelona, 1985)

Ramón-Laca, Julio de. *Bajo la férula de Queipo: como fue gobernada Andalucía* (Seville, 1939)

Reig Tapia, Alberto. *Ideología e historia: sobre la represión franquista y la guerra civil* (Madrid, 1984)

Rich, Norman. *Hitler's War Aims: Ideology, the Nazi State, and the Course of Expansion* 2 vols (London, 1973–4)

Río Cisneros, Agustín del. *Política internacional de España: El caso español en la ONU y en el mundo* (Madrid, 1946)

Risco, Alberto S.J., *La epopeya del Alcázar de Toledo* (San Sebastián, 1941)

Rivas Cherif, Cipriano. *Retrato de un desconocido: vida de Manuel Azaña* (Barcelona, 1980)

Rodríguez Aisa, María Luisa. *El Cardenal Gomá y la guerra de España: aspectos de la gestión pública del Primado 1936–1939* (Madrid, 1981)

Rodríguez Armada, Amandino, & Novais, José Antonio. *¿Quién mató a Julián Grimau?* (Madrid, 1976)

Rodríguez Moñino Soriano, Rafael. *La misión diplomática del XVII Duque de Alba en la Embajada de España en Londres (1937–1945)* (Valencia, 1971)

Romero, Luis. *Tres días de julio (18, 19 y 20 de 1936)* 2 edición (Barcelona, 1968)

Romero Maura, Joaquín. *The Spanish Army and Catalonia: The "Cu-Cut! Incident" and the Law of Jurisdictions, 1905–1906* (London, 1976)

Romero Maura, Joaquín. *"La rosa de fuego" El obrerismo barcelonés de 1899 a 1909* (Barcelona, 1975)

Ros, Samuel, & Bouthelier, Antonio. *A hombros de la Falange: Historia del traslado de los restos de José Antonio,* (Barcelona, 1940)

Rotvand, Georges. *Franco Means Business* (London, n.d. [1937])

Rubottom, R. Richard, & Murphy, J. Carter. *Spain and the United States Since World War II* (New York, 1984)

Ruhl, Klaus-Jörg. *Franco, Falange y III Reich* (Madrid, 1986)

Ruiz Ocaña, Carlos. *Los ejércitos españoles: las fuerzas armadas en la defensa nacional* (Madrid, 1980)

Saborit, Andrés. *Julián Besteiro* (Buenos Aires, 1967)

Sáez Alba, A. (pseudonym Alfonso Colodrón) *La otra "cosa nostra": la Asociación Católica Nacional de Propagandistas y el caso de El Correo de Andalucía* (Paris, 1974)

Salas Larrazábal, Jesús. *La guerra de España desde el aire* 2ª edición (Barcelona, 1972)

Salas Larrazábal, Ramón. *Historia del Ejército popular de la República* 4 vols (Madrid, 1973)

Salgado, Enrique. *Radiografía de Franco* (Barcelona, 1985)

Salva Miquel, Francisco, & Vicente, Juan, *Francisco Franco (historia de un español* (Barcelona, 1959)

Sánchez, José M. *The Spanish Civil War as Religious Tragedy* (Notre Dame, Indiana, 1987)

Sánchez del Arco, Manuel. *El sur de España en la reconquista de Madrid* (Seville, 1937)

Sánchez Blanco, Jaime. *La importancia de llamarse Franco: el negocio inmobiliario de doña Pilar* (Madrid, 1978)

Sánchez Silva, José María, & Saenz de Heredia, José Luis. *Franco . . . ese hombre* (Madrid, 1975)

Sánchez Soler, Mariano. *Villaverde: fortuna y caída de la casa Franco* (Barcelona, 1990)

Sánchez y García Sauco, Juan Antonio. *La revolución de 1934 en Asturias* (Madrid, 1974)

Sangróniz, José Antonio. *Marruecos. Sus condiciones físicas, sus habitantes y las instituciones indígenas* (Madrid, Sucesores de Rivadeneyra, 1921)

Saz Campos, Ismael. *Mussolini contra la II República: hostilidad, conspiraciones, intervención (1931–1936)* (Valencia, 1986)

Seco Serrano, Carlos. *Militarismo y civilismo en la España contemporánea* (Madrid, 1984)

Séguéla, Matthieu. *Pétain-Franco: les secrets d'une alliance* (Paris, 1992)

Senra, Alfonso. *Del 10 de agosto a la sala sexta del Supremo* (Madrid, 1933)

Servicio Histórico Militar (Coronel José Manuel Martínez Bande), *El final del frente norte* (Madrid, 1972)

Servicio Histórico Militar (Coronel José Manuel Martínez Bande), *La batalla de Pozoblanco y el cierre de la bolsa de Mérida* (Madrid, 1981)

Servicio Histórico Militar (Coronel José Manuel Martínez Bande), *La batalla del Ebro* 2 edición (Madrid, 1988)

Servicio Histórico Militar (Coronel José Manuel Martínez Bande), *El final de la guerra civil* (Madrid, 1985)

Servicio Histórico Militar (Coronel José Manuel Martínez Bande), *La batalla de Teruel* 2 edición (Madrid, 1990)

Servicio Histórico Militar, (Coronel José Manuel Martínez Bande) *La campaña de Andalucía* 2 edición (Madrid, 1986)

Servicio Histórico Militar (Coronel José Manuel Martínez Bande), *La campaña de Cataluña* (Madrid, 1979)

Servicio Histórico Militar (Coronel José Manuel Martínez Bande), *La gran ofensiva sobre Zaragoza* (Madrid, 1973)

Servicio Histórico Militar (Coronel José Manuel Martínez Bande), *La invasión de Aragón y el desembarco en Mallorca* (Madrid, 1989)

Servicio Histórico Militar (Coronel José Manuel Martínez Bande), *La llegada al mar* (Madrid, 1975)

Servicio Histórico Militar (Coronel José Manuel Martínez Bande), *La lucha en torno a Madrid* (Madrid, 1968)

Servicio Histórico Militar (Coronel José Manuel Martínez Bande), *La marcha sobre Madrid* (Madrid, 1968)

Servicio Histórico Militar (Coronel José Manuel Martínez Bande), *La ofensiva sobre Valencia* (Madrid, 1977)

Servicio Histórico Militar (Coronel José Manuel Martínez Bande), *Nueve meses de guerra en el norte* (Madrid, 1980)

Servicio Histórico Militar (Coronel José Manuel Martínez Bande), *Vizcaya* (Madrid, 1971)

Servicio Histórico Militar, *Historia de la guerra de liberación* (Madrid, 1945)

Shannon, Richard. *The Crisis of Imperialism 1865–1915* (London, 1974)

Shaw, Duncan. *Fútbol y franquismo* (Madrid, 1987)

Shirer, William L. *The Rise and Fall of the Third Reich* (London, 1959)

Shubert, Adrian. *The Road to Revolution in Spain: The Coal Miners of Asturias 1860–1934* (Urbana, 1987)

Silva, Carlos de. *General Millán Astray, el legionario* (Barcelona, 1956)

Silva, José Antonio. *Cómo asesinar con un avión* (Barcelona, 1981)

Simonnot, Philippe. *Le secret de l'armistice* (Paris, 1990)

Smyth, Denis. *Diplomacy and Strategy of Survival: British Policy and Franco's Spain, 1940–41* (Cambridge, 1986)

Somoza Silva, Lázaro. *El general Miaja: biografía de un heroe* (México D.F., 1944)

Sorel, Andrés. *Búsqueda, reconstrucción e historia de la guerrilla española del siglo XX, a través de sus documentos, relatos y protagonistas* (Paris, 1970)

Southworth, Herbert Rutledge. *Antifalange; estudio crítico de "Falange en la guerra de España" de Maximiano García Venero* (Paris, 1967)

Southworth, Herbert Rutledge. *El mito de la cruzada de Franco* (Paris, 1963)

Southworth, Herbert Rutledge. *Guernica! Guernica!: A Study of Journalism, Propaganda and History* (Berkeley, 1977)

Steer, G.L. *The Tree of Gernika: A Field Study of Modern War* (London, 1938)

Suárez Fernández, Luis. *Francisco Franco y su tiempo* 8 vols (Madrid, 1984)

Suárez Fernández, Luis. *Franco: la historia y sus documentos* 20 vols (Madrid, 1986)

Sueiro, Daniel. *El Valle de los Caídos: los secretos de la cripta franquista* 2 edición (Barcelona, 1983)

Sueiro, Daniel, & Díaz Nosty, Bernardo. *Historia del franquismo*, 2 vols, 2nd edition (Barcelona, 1985)

Taibo, Pablo Ignacio II. *Octubre 1934: el ascenso* vol 7 of *Historia general de Asturias* (Gijón, 1978)

Taibo, Pablo Ignacio II. *Octubre 1934: la caída* vol 8 of *Historia general de Asturias* (Gijón, 1978)

Talón, Vicente. *Arde Guernica* (Madrid, 1970)

Tamames, Ramón. *La República, la era de Franco* (Madrid, 1973)

Téllez, Antonio. *La guerrilla urbana. 1: Facerias* (Paris, 1974)

Téllez, Antonio. *La guerrilla urbana en España: Sabaté* (Paris, 1973)

Tello, José Angel. *Ideología y política: la Iglesia católica española (1936–1959)* (Zaragoza, 1984)

Terrón Montero, Javier. *La prensa de España durante el régimen de Franco* (Madrid, 1981)

Thomas, Hugh. *The Spanish Civil War* 3rd ed. (London, 1977)

Toquero, José María. *Franco y Don Juan: La oposición monárquica al franquismo* (Barcelona, 1989)

Townson, Nigel. 'The Collapse of the Centre: The Radical Republican Party during the Second Spanish Republic' Unpublished doctoral thesis, University of London, 1991,

Trevor-Roper, Hugh. *The Philby Affair* (London, 1968)

Trythall, J.W.D. *Franco* (London, 1970)

Tusell, Javier. *Las elecciones del Frente Popular* 2 vols (Madrid, 1971)

Tusell, Javier. *Franco y los católicos: la política interior española entre 1945 y 1957* (Madrid, 1984)

Tusell, Xavier. *La oposición democrática al franquismo 1939–1962* (Barcelona, 1977)

Tusell, Javier. *Radiografía de un golpe de Estado: el ascenso al Poder del General Primo de Rivera* (Madrid, 1987)

Tusell, Javier, & Calvo, José. *Giménez Fernández: precursor de la democracia española* (Madrid, 1990)

Tusell, Xavier, & García Queipo de Llano, Genoveva. *Franco y Mussolini: la política española durante la segunda guerra mundial* (Barcelona, 1985)

Ucelay da Cal, Enric. *La Catalunya populista. Imatge, cultura i política en la etapa republicana (1931–1939)* (Barcelona, 1982)

Ullman, Joan Connelly. *The Tragic Week: A Study of Anti-Clericalism in Spain 1875–1912* (Harvard, 1968)

Umbral, Francisco. *Leyenda del César visionario* (Barcelona, 1991)

Valdesoto, Fernando de. *Francisco Franco* (Madrid, 1945)

Valle, José María del. *Las instituciones de la República en exilio* (Paris, 1976)

Valls, Fernando. *La enseñanza de la literatura en el franquismo 1936–1951* (Barcelona, 1983)

Valls Montes, Rafael. *La interpretación de la Historia de España, y sus orígenes ideológicos, en el bachillerato franquista (1938–1953)* (Valencia, 1984)

Vázquez Montalbán, Manuel. *Autobiografía del general Franco* (Barcelona, 1992)

Vázquez Montalbán, Manuel. *Barcelonas* (London, 1992)

Vázquez Montalbán, Manuel. *Los demonios familiares de Franco* (Barcelona, 1978)

[Vázquez Montalbán, Manuel.] *El pequeño libro pardo del general* (Paris, 1972)

Venegas, José. *Las elecciones del Frente Popular* (Buenos Aires, 1942)

Vigón, General Jorge. *General Mola (el conspirador)* (Barcelona, 1957)

Vilanova, Antonio. *La defensa del Alcázar de Toledo (epopeya o mito)* (Mexico, 1963)

Vilar, Sergio. *La naturaleza del franquismo* (Barcelona, 1977)

Viñas, Angel. *La alemania nazi y el 18 de julio* 2ª edición (Madrid, 1977)

Viñas, Angel. *Guerra, dinero, dictadura: ayuda fascista y autarquía en la España de Franco* (Barcelona, 1984)

Viñas, Angel. *Los pactos secretos de Franco con Estados Unidos: bases, ayuda económica, recortes de soberanía* (Barcelona, 1981)

Viñas, Angel, Viñuela, Julio, Eguidazu, Fernando, Fernández Pulgar, Carlos & Florensa, Senen. *Política comercial exterior en España (1931–1975)* 2 vols (Madrid, 1979)

Watson, Bert Allan. *United States–Spanish Relations, 1939–1946* Unpublished Ph.D. dissertation, George Washington University, 1971

Weinberg, Gerhard L. *The Foreign Policy of Hitler's Germany: Diplomatic Revolution in Europe, 1933–1936* (Chicago, 1970)

Weinberg, Gerhard L. *World in the Balance: Behind the Scenes of World War II* (Hanover, New Hampshire, 1981)

Welles, Benjamin. *Spain: The Gentle Anarchy* (London, 1965)

Whealey, Robert H. *Hitler and Spain: the Nazi Role in the Spanish Civil War* (Lexington, Kentucky, 1989)

Whitaker, Arthur P. *Spain and the Defence of the West: Ally and Liability* (New York, 1961)

Woodward, Sir Llewellyn. *British Foreign Policy in the Second World War* 5 vols (London, 1970–1976)

Woolman, David S. *Rebels in the Rif: Abd el Krim and the Rif Rebellion* Stanford, 1969)

Wyden, Peter. *The Passionate War: The Narrative History of the Spanish Civil War, 1936–1939* (New York, 1983)

Yale, *Los últimos cien días: crónica de una agonía* (Madrid, 1975)

Ynfante, Jesús. *La prodigiosa aventura del Opus Dei: Génesis y desarrollo de la Santa Mafia* (Paris, 1970)

(ii) Articles

Allen, Jay. 'Blood flows in Badajoz', in Acier, Marcel, editor. *From Spanish Trenches: Recent Letters from Spain* (London, 1937)

Anon. 'Almogávares en el aire', *Revista de Aeronáutica* August 1942

Asensio Cabanillas, Carlos. 'El avance sobre Madrid y operaciones en el frente del centro', *La guerra de liberación nacional* (Zaragoza, 1961)

Avilés Farré, Juan. 'L'Ambassade de Lequerica et les rélations hispano–françaises 1939–1944', *Guerres Mondiales et Conflits Contemporains* No.158, April 1990

Avilés Farré, Juan. 'Lequerica, embajador franquista en París', *Historia 16*, No.160, August 1989

Avilés Farré, Juan. 'Vichy y Madrid. Las relaciones hispano–frances de junio de 1940 a noviembre de 1942', *Espacio, Tiempo y Forma: Revista de la Facultad de Geografía e Historia de la Universidad Nacional de Educación a distancia*, Serie V, tomo 2, 1989

Ben Jellon, Abd el Hajid. 'La participación de los mercenarios marroquíes en la guerra civil española (1936–1939)', *Revista Internacional de Sociología* Vol. 46, no.4, October–December 1988

Boyd, Carolyn P. ' "Responsibilities" and the Second Republic, 1931–1936' in Blinkhorn, Martin, ed., *Spain in Conflict 1931–1939: Democracy and its Enemies* (London, 1986)

Burdick, Charles B. ' "Moro": The Resupply of German Submarines in Spain, 1939–1942', *Central European History* Vol. 3, No.3, 1970

Busquets, Julio. 'La Unión Militar Española, 1933–1936', Historia 16, *La guerra civil* 24 vols (Madrid, 1986) III, pp. 86–99

Detwiler, Donald S. 'Spain and the Axis during World War II', *Review of Politics*, Vol. 33, No. 1, January, 1971.

Díaz, Elías. 'Miguel de Unamuno y la guerra civil' in *La voluntad del humanismo: homenaje a Juan Marichal*, ed. Ciplijauskaité, B. & Maurer, C. (Barcelona, 1990)

Duff, Katherine. 'Spain between the Allies and the Axis', in Arnold & Veronica Toynbee, editors, *Survey of International Affairs 1939–1946; The War and the Neutrals* (London, 1956)

Egido León, Angeles. 'Franco y las potencias del Eje. La tentación intervencionista de España en la segunda guerra mundial', *Espacio, Tiempo y Forma: Revista de la Facultad de Geografía e Historia de la Universidad Nacional de Educación a distancia*, Serie V, tomo 2, 1989

Eguidazu Palacios, Fernando. 'Factores monetarios y de balanza de pagos en la neutralidad española', *Revista de Estudios Internacionales*, no. 2, abril–junio 1984)

Ellwood, Sheelagh. 'La crisis de Salamanca: la Unificación', *Historia 16*, no. 132, April 1987

Ferrer Benimeli, José Antonio. 'Franco contra la masonería', *Historia 16*, año II, no. 15, julio de 1977

Gárate Córdoba, José María. '<<Raza>>, un guión de cine', *Revista de Historia Militar*, no. 40, 1976

Garmendia, José María. 'El Pacto de Santoña', in Garitaonandía, Carmelo, & Granja, José Luis de la. editors, *La guerra civil en el país vasco* (Bilbao, 1987)

González-Arnao Conde-Luque, Mariano. '¡Capturad al duque de Windsor!', *Historia 16*, no. 161, September 1989

Grau, Federico. 'Psicopatología de un dictador: entrevista a Carlos Castilla del Pino', *El Viejo Topo*, Extra no. 1, 1977

Halstead, Charles R. 'Un "Africain" méconnu: le colonel Juan Beigbeder', *Revue d'Histoire de la deuxième guerre mondiale*, Vol. 21, no. 83, 1971

Halstead, Charles R. 'Consistent and total peril from every side: Portugal and its 1940 Protocol with Spain', *Iberian Studies*, Vol. III, no. 1, Spring 1974

Halstead, Charles R. 'Diligent Diplomat: Alexander W. Weddell as American Ambassador to Spain, 1939–1942', *The Virginia Magazine of History and Biography*, Vol. 82, no. 1, January 1974

Halstead, Charles R. 'The dispute between Ramón Serrano Suñer and Alexander Weddell', *Rivista di Studi Politici Internazionali*, no. 3, 1974

Halstead, Charles R. 'Peninsular Purpose: Portugal and its 1939 Treaty of Friendship and Non-Aggression with Spain', *Il Politico: Rivista Italiana di Scienze Politiche* (Pavia) XLV, no. 2, 1980

Halstead, Charles R. 'A "Somewhat Machiavellian" Face: Colonel Juan Beigbeder As High Commissioner in Spanish Morocco, 1937–1939', *The Historian*, Vol. 37, 1974

Halstead, Charles R. and Halstead, Carolyn J. 'Aborted Imperialism: Spain's Occupation of Tangier 1940–1945', *Iberian Studies*, Vol. VII, no. 2, Autumn, 1978

Harrison, Joseph. 'Towards the liberalization of the Spanish economy, 1951–1959' in Holmes, Colin, & Booth, Alan. editors, *Economy and Society: European Industrialization and Its Social Consequences* (Leicester, 1990)

Juliá, Santos. 'Fracaso de una insurrección y derrota de una huelga: los hechos de octubre en Madrid', *Estudios de Historia social*, no. 31, October–December 1984, pp. 37–47

Kindelán, Alfredo. 'La aviación en nuestra guerra' in *La guerra de liberación nacional* (Zaragoza, 1961)

Lowi, Theodore J. 'Bases in Spain' in Stein, Harold, editor. *American Civil-Military Decisions: A Book of Case Studies* (Birmingham, Alabama, 1963)

Madariaga, María Rosa de. 'Imagen del moro en la memoria colectiva del pueblo español y retorno del moro en la guerra civil de 1936', *Revista Internacional de Sociología* Vol. 46 no. 4, October–December 1988, pp. 590–6

Marquina Barrio, Antonio. 'El atentado de Begoña', *Historia 16*, no. 76, August 1982

Marquina Barrio, Antonio. 'Conspiración contra Franco: el ejército y la injerencia extranjera en España: el papel de Aranda, 1939–1945', *Historia 16*, no. 72, April 1982,

Marquina Barrio, Antonio. 'La etapa de Ramón Serrano Suñer en el Ministerio de Asuntos Exteriores', *Espacio, Tiempo y Forma: Revista de la Facultad de Geografía e Historia de la Universidad Nacional de Educación a distancia*, Serie V, tomo 2, 1989

Marquina Barrio, Antonio. 'La Península ibérica y la planificación militar aliada', *Revista de Occidente*, no. 41, octubre 1984

Marquina Barrio, Antonio. 'Operación Torch: España al borde de la II guerra mundial', *Historia 16*, no. 79, noviembre 1982

Martínez Bande, José Manuel. 'Del alzamiento a la guerra civil verano de 1936: correspondencia Franco/Mola', *Historia y Vida*, no. 93, 1975

Monteath, Peter. 'Guernica Reconsidered: Fifty Years of Evidence', *War & Society*, Vol. 5, no. 1, May 1987

Morales Lezcano, Victor. 'Las causas de la no beligerancia española reconsideradas', *Revista de Estudios Internacionales*, vol. 5, no. 3, julio–septiembre 1984

Navarro Rubio, Mariano. 'La batalla de la estabilización' in *Anales de la Real Academia de Ciencias Morales y Políticas*, no. 53, 1976,

Navarro Rubio, Mariano. 'La batalla del desarrollo' in *Anales de la Real Academia de Ciencias Morales y Políticas*, no. 54, 1977,

Paz, Manuel de. 'Masonería y militarismo en el norte de Africa', Universidad Complutense, Cursos de Verano, El Escorial, 1988, *La masonería y su impacto internacional* (Madrid, 1989)

Pike, David Wingeate. 'Franco and the Axis Stigma', *Journal of Contemporary History*, vol. 17, no. 3, 1982,

Pike, David Wingeate. 'Franco et l'admission aux Nations Unies', *Guerres mondiales et conflits contemporains*, no. 162, Avril 1991

Preston, Paul. 'El discreto encanto del general Franco' in *Journal of the Association of Contemporary Iberian Studies*, Vol. 4, no. 1, Spring 1991

Preston, Paul. 'Franco and Hitler: The Myth of Hendaye 1940' in *Contemporary European History*, Vol. 1, Part 1, March 1992

Preston, Paul. 'Franco and the Hand of Providence' in *For Want of a Horse: Choice and Chance in History* edited by John M. Merriman. (Lexington, Massachusetts, Stephen Greene Press, 1985)

Preston, Paul. 'Franco et ses généraux (1939–1945)', *Guerres mondiales et conflits contemporains*, no. 162, Avril 1991

Preston, Paul. 'From Counter Revolution to Historical Accommodation: The Spanish Civil War and Historical Memory' in *Harvard University Center for European Studies Working Papers Series* no. 13, 1989.

Preston, Paul. 'General Franco Reassessed: Inertia and Risk, World War and Cold War, 1939–1953' in *Journal of the Association for Contemporary Iberian Studies*, Vol. 1, no. 1, Spring 1988.

Preston, Paul. 'Guerrilleros contra Franco' in *Historia del Franquismo*; edited by Justino Sinova. 2 volumes, (Madrid, Información y Prensa, 1985) Vol. 1

Preston, Paul. 'The PCE's Long Road to Democracy 1954–1977' in Kindersley, Richard, editor. *In Search of Eurocommunism* (London, 1981)

Preston, Paul. 'Spain' in Andrew Graham & Anthony Seldon, editors, *Government and economies in the postwar world: Economic policies and comparative performance, 1945–85* (London, Routledge, 1990)

Raguer, Hilari. 'Franco alargó deliberadamente la guerra', *Historia 16*, no. 170, June 1990

Raguer, Hilari. 'Magaz y los nacionalistas vascos (1936–1937)', *Letras de Deusto*, Vol. 16, no. 5, Mayo–Agosto 1986

Ramírez, Luis [pseudonym of Luciano Rincón]. 'Morir en el bunker' in *Horizonte español 1972* 3 vols (Paris, 1972) I

Reig Tapia, Alberto. 'El Caudillismo franquista', Casas Sánchez, José Luis. (editor), *La postguerra española y la segunda guerra mundial* (Córdoba, 1990)

Reig Tapia, Alberto. 'Guernica como símbolo' in Garitaonandía, Carmelo, & Granja, José Luis de la. editors, *La guerra civil en el país vasco* (Bilbao, 1987)

Ruhl, K.J. 'L'alliance à distance: les relations économiques germano–espagnoles de 1939 à 1945', *Revue d'histoire de la deuxième guerre mondiale*, no. 118, avril 1980

Salas Larrazabal, Jesús. 'El puente aereo del Estrecho', *Revista de Aeronáutica y Astronáutica*, September 1961, pp. 747–50

Sevilla Guzmán, Eduardo, & Giner, Salvador. 'Absolutismo despótico y dominación de clase: el caso de España' in *Cuadernos de Ruedo Ibérico*, (Paris) Nos 43–45, enero–junio 1975

Smyth, Denis. 'Les Chevaliers de Saint-George: la Grande-Bretagne et la corruption des généraux espagnols (1940–1942)', *Guerres mondiales et conflits contemporains*, no. 162, Avril 1991

Smyth, Denis. 'The Moor and the Money-lender: Politics and Profits in Anglo-German Relations with Francoist Spain 1936–1940' in Recker, Marie-Luise, editor, *Von der Konkurrenz zur Rivalität: Das britische-deutsche Verhältnis in den Ländern der europäischen Peripherie 1919–1939* (Stuttgart, 1986)

Smyth, Denis. 'Screening "Torch": Allied Counter-Intelligence and the Spanish Threat to the Secrecy of the Allied Invasion of French North Africa in November 1942' in *Intelligence and National Security*, Vol. 4, no. 2, April 1989

Tusell, Javier. 'La etapa Jordana (1942–1944)', *Espacio, Tiempo y Forma: Revista de la Facultad de Geografía e Historia de la Universidad Nacional de Educación a distancia*, Serie V, tomo 2, 1989.

Tusell, Javier. 'Franco no fue neutral', *Historia 16*, no. 141, 1988

Tusell, Javier. 'Un giro fundamental en la política española durante la segunda guerra mundial: la llegada de Jordana al Ministerio de Asuntos Exteriores', García Delgado, José Luis, editor, *El primer franquismo: España durante la segunda guerra mundial* (Madrid, 1989)

Viñas, Angel. 'Factores económicos externos en la neutralidad española', *Revista de Occidente*, no. 41, octubre 1984

Viñas, Angel. 'Las relaciones hispano–francesas, el Gobierno Daladier y la crisis de Munich' in *Españoles y franceses en la primera mitad del siglo XX* (Madrid 1986)

Wight, Martin. 'Spain and Portugal' in Toynbee, Arnold, & Ashton-Gwatkin, Frank T. *Survey of International Affairs 1939–1946: The World in March 1939* (London, 1952)

Whitaker, John. 'Prelude to World War: A Witness from Spain', *Foreign Affairs*, Vol 21, no. 1, October 1942

INDEX

968 INDEX